CARPENTRY

FLOYD VOGT / MICHAEL NAUTH

CARPENTRY

Third Canadian Edition

NELSON
EDUCATION

NELSON
EDUCATION

Carpentry, Third Canadian Edition
by Floyd Vogt and Michael Nauth

VP, Product and Partnership Solutions:
Anne Williams

Publisher, Digital and Print Content:
Jackie Wood

Senior Marketing Manager:
Alexis Hood

Technical Reviewer:
Kevin LeMay

Content Development Manager:
Toula Di Leo

Photo and Permissions Researcher:
Carrie McGregor

Production Project Manager:
Christine Gilbert

Production Service:
MPS Limited

Copy Editor:
Michael Kelly

Proofreader:
MPS Limited

Indexer:
Beverlee Day

Design Director:
Ken Phipps

Managing Designer:
Franca Amore

Interior Design:
Sharon Kish

Cover Design:
Sharon Kish

Cover Image:
Courtesy of Stiletto Tools

Art Coordinator:
Suzanne Peden

Illustrator(s):
Dave McKay; CrowleArt Group

Compositor:
MPS Limited

Library and Archives Canada Cataloguing in Publication Data

Vogt, Floyd, author
    Carpentry / Floyd Vogt, Michael Nauth. – Third Canadian edition.

Includes index.
ISBN 978-0-17-657042-2 (bound)

    1. Carpentry – Textbooks.
I. Nauth, Michael, 1953–, author
II. Title.

TH5604.V63 2016   694
C2015-903188-5

ISBN 13: 978-0-17-657042-2
ISBN 10: 0-17-657042-X

Contents

Photo by M. Nauth. Used with the permission of Brad Current.

Section 1

Tools and Materials 2

Photo by M. Nauth

Section 2

Rough Carpentry 218

Photo used with permission of Tieg Martin

Section 3

Exterior Finish   630

Photo by M. Nauth

Section 4

Preface

Welcome to *Carpentry*, Third Canadian Edition. This text is designed for students who are studying residential construction and who are pursuing careers in the trade either in post-secondary college programs, in apprenticeship programs, or at the senior high school level across Canada. The 2015 National Building Code of Canada is used as the point of reference. All measurements are given in both metric and imperial systems.

APPROACH

The core of the carpentry trade as practised in Canada is covered in four comprehensive sections: **Tools and Materials, Rough Carpentry, Exterior Finish,** and **Interior Finish.** Each section features step-by-step procedures for all stages of construction, important safety precautions, tips of the trade, and a look at anticipated future trends in the construction industry.

Section 1: Tools and Materials

This section describes how building materials from wood to engineered wood products to composites are shaped, produced, and applied in their various forms. The Canadian Plywood Association (CANPLY) is referenced. Wood I-joists are included in the discussion of engineered wood products. Information about cross-laminated timber (CLT) and its application in the construction industry has been added to the unit on engineered wood products. The latest fasteners and cordless tools and their batteries are described in their current use in the trade. The final unit on plans and codes has been revised to Canadian standards to provide the student with a basic understanding of how the building components come together to form a structure. The house plans used in Chapter 20 have been provided by **Cardel Homes.** The Ontario Building Code, 2012, is referenced, as well as the minimum requirements for occupancy.

Section 2: Rough Carpentry

The on-site work starts with locating the building, with the use of surveying instruments (automatic levels, theodolites, and total stations), in its exact position on the building lot. The site is then excavated. Foundation work in concrete and other materials, waterproofing, drainage tile placement, backfilling, and other associated work then gets the structure above grade. The "clean" work is next—floor framing, wall framing, and roof framing, using both new and old materials and methods. This unit has been expanded to emphasize the importance and the intricacies of roof work from both a mathematical and a geometric approach, including equal and unequal pitched hip roofs, dormer roofs, hexagonal and octagonal roofs, and gazebo roofs. Timber-frame construction is expanded with detailed drawings for timber joinery from Mohawk College in Hamilton, Ontario. Energy-efficient construction details (R-2000 Program, LEED® Canada for Homes rating system, and ENERGY STAR® for New Homes initiative) are emphasized with respect to energy and resource conservation and healthy indoor air quality.

Section 3: Exterior Finish

Exterior finish as it pertains to the surroundings— "fitting in" instead of sticking out—is discussed in terms of the availability and durability of various materials. The installation of roof coverings, sidings, doors and windows, cornice work, decks, and fences is also covered in this section. Special attention is paid to first and second planes of protection, end dams, air leakage control, ventilation, and the insulation values of materials.

Section 4: Interior Finish

The section on interior finishes begins by discussing how drywall is applied to walls and ceilings. It then describes how other finishes are applied to these surfaces, as well as to floors. The hanging of interior doors is discussed, followed by the installation of various trim mouldings to windows, doors, walls, and ceilings. The section then covers how to construct a finished set of stairs with a balustrade. The book ends by explaining how to install manufactured cabinets and countertops and how to construct countertops, unit boxes, and drawers from various materials.

Key Features

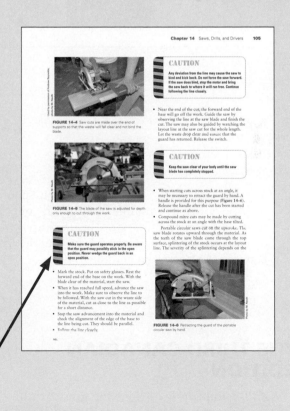

A **Success Story** opens each of the sections in the book, providing a look at the day-to-day job of a carpenter and the successes accomplished through dedication and education.

Safety Reminders and Cautions prominently feature the latest safety considerations so that readers can avoid the dangers of certain procedures, tools, and equipment in order to stay safe on the job.

Step by Step Procedures walk readers through the key tasks associated with specific residential building tasks, including On the Job tips of the trade.

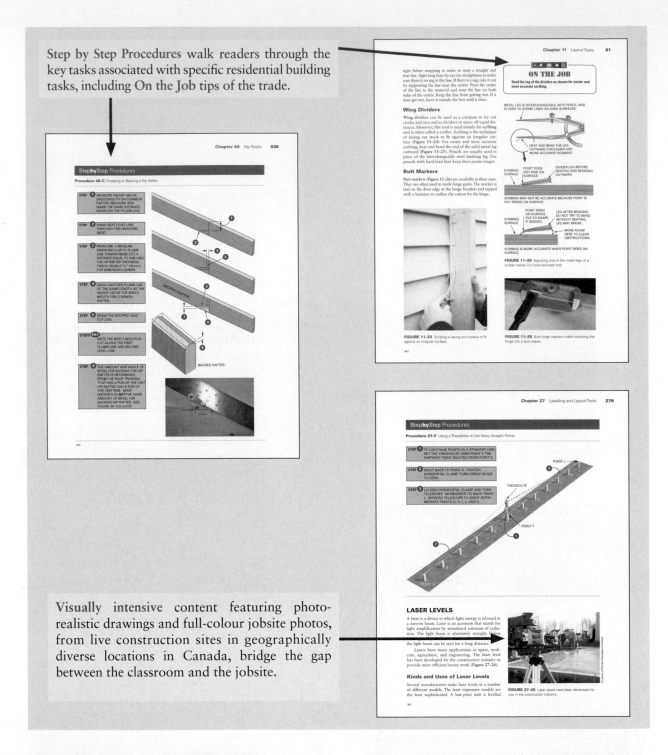

Visually intensive content featuring photo-realistic drawings and full-colour jobsite photos, from live construction sites in geographically diverse locations in Canada, bridge the gap between the classroom and the jobsite.

THE THIRD CANADIAN EDITION

- A *full-colour design* brings the jobsite to life—integrating photo-realistic drawings drawn to scale and on-the-job photos from construction projects across Canada and the United States.

- **Measurements** in the text are recorded in both metric and imperial. Some conversions are approximate, such as 4 feet (1.2 m), whereas others are exact, especially in building code references, such as $7\frac{7}{8}$ inches (200 mm) for the maximum rise for a set of stairs to a dwelling unit.

- *Safety* information retains its prominence in the text, including a section titled "Safety at the Worksite" in the Introduction, unit-opening *Safety Reminders* to alert students to potentially dangerous situations on the job, and *Cautions* to help prevent accidents when working with various tools of the trade.

- A section in the Introduction covers the critical *soft skills* required of successful and proficient carpenters. Coverage of important *organizations,* including the Carpenters' Union and Skills Canada, helps students keep up to date on industry expectations.

- A *Success Story* opens each of the four sections, highlighting a successful account of an individual working on carpentry jobs in different regions of the country, and providing insight into the industry.

- Structural and health and safety issues are aligned to the National Building Code of Canada (NBCC), 2015 edition.

New to this third Canadian edition are the following:

- New and revised drawings and photos on heavy concrete construction, including piles, precast, pre- and post-tensioning, fly-forms, slip forms, shoring, and more.

- A **What's Wrong with This Picture?** feature at the end of some units that includes an incorrect application or scenario as well as a corrected one.

- The **Deconstruct This** feature illustrates situations where a problem may or may not exist in the construction process, giving students the opportunity to consider what is right and wrong and to develop their critical thinking skills.

- An up-to-date Canadian path to obtaining training in preparation for taking the Inter-Provincial Red Seal Certificate of Qualification Exam.

- Improved coverage of stationary and portable power tools, the anti-kickback splitter and riving knife for the table saw, and setup for the jointer.

- Expanded coverage of insulated concrete forms (ICF), monopour ICF, rebar vibrator, and power screed.

- Added information on stair building, stairwell opening, landings, and circular stairs.

- Correct installation of steel studs and new techniques and construction aids for the application of drywall.

- Expanded CAD drawings to illustrate the determination of lengths and angles of rafters and sheathing for hip roofs, including setting the top of ridge and the drop and backing bevel for the hip rafter.

- The use of construction geometry to start with the plan view of roofs and to "unfold" them to illustrate actual roof surface.

- Revisions to the chapter on unequal pitched roofs and intersecting roofs (including dormer roofs), as well as hexagonal and octagonal roof rafter lengths, side cut angles, and backing angles.

- A complete set of construction drawings for timber framing (from Mohawk College) and structural insulated panels (SIPs) installation methods.

- Updates to energy-efficient housing, which highlights building science and its important impact on current practices and products such as increased air sealing, reduced thermal bridging, acoustical sealants, and heat recovery ventilators.

- A revised chart on Canada's climate zones for windows, including degree days.

- Methods to reduce sound transmission within the building envelope complete with Building Code references.

- Revised treatment of composite decking, air-gap at ledgers, Fast-Tube concrete piers, guardrails, and code references.

- Added discussion of drywall butt joints, ceiling drywall, 54-inch panels, and paper-faced corner bead.

- Added discussion of ceramic tile un-coupling membrane and shower bases.

- New photos for mitre saw compound angles and jigsaw templates for cutting crown moulding.

- Added discussion of cork, bamboo, and engineered flooring and OSB tiles for basement floors.

- Up-to-date information on frameless kitchen cabinets, European hardware, and composite, granite, and wood-edged countertops.

INSTRUCTOR RESOURCES

The **Nelson Education Teaching Advantage (NETA)** program delivers research-based instructor resources that promote student engagement and higher-order thinking to enable the success of Canadian students and educators. Visit Nelson Education's **Inspired**

Instruction website at *www.nelson.com/inspired* to find out more about NETA.

The following instructor resources have been created for *Carpentry*, Third Canadian Edition. Access these ultimate tools for customizing lectures and presentations at *www.nelson.com/instructor*.

NETA Test Bank

This resource was written by Tim Dorn, Okanagan College. It includes over 1500 multiple-choice questions written according to NETA guidelines for effective construction and the development of higher-order questions.

The NETA Test Bank is available in a new, cloud-based platform. **Nelson Testing Powered by Cognero®** is a secure online testing system that allows instructors to author, edit, and manage test bank content from anywhere Internet access is available. No special installations or downloads are needed, and the desktop-inspired interface, with its drop-down menus and familiar, intuitive tools, allows instructors to create and manage tests with ease. Multiple test versions can be created in an instant, and content can be imported or exported into other systems. Tests can be delivered from a learning management system, the classroom, or wherever an instructor chooses. Nelson Testing Powered by Cognero for *Carpentry*, Third Canadian Edition, can also be accessed through *www.nelson.com/instructor*.

NETA PowerPoint

Microsoft® PowerPoint® lecture slides for every chapter have been created by Brad MacDonald, Mohawk College. There is an average of 25 slides per chapter, featuring an outline of the chapter, key figures, drawings, photos, tables, and procedures from *Carpentry*, Third Canadian Edition. NETA principles of clear design and engaging content have been incorporated throughout, making it simple for instructors to customize the deck for their courses.

Image Library

This resource consists of digital copies of figures, drawings, photographs, and procedures used in the book. Instructors may use these jpegs to customize the NETA PowerPoint or create their own PowerPoint presentations.

Videos

Instructors can enhance the classroom experience with the exciting and relevant videos provided directly to students through MindTap. These videos have been selected to accompany *Carpentry*, Third Canadian Edition.

NETA Instructor Guide

The Instructor's Guide to accompany *Carpentry*, Third Canadian Edition, has been prepared by Brad MacDonald, Mohawk College. This guide contains many helpful tools to facilitate effective classroom presentations of material—comprehensive *Lesson Plans*, integrated PowerPoint® references and teaching tips, a review of important math, blueprint reading to help evaluate student knowledge at the start of the course, and *Answers to Review Questions*. A separate *Note to the Instructor* includes valuable information on effectively teaching construction students—everything from learning styles to safety on the worksite is covered in this introduction to the course.

DayOne Slides

Day One—Prof InClass is a PowerPoint presentation that instructors can customize to orient students to the class and their text at the beginning of the course.

MindTap

MindTap®

Offering personalized paths of dynamic assignments and applications, **MindTap** is a digital learning solution that turns cookie cutter into cutting edge, apathy into engagement, and memorizers into higher-level thinkers. MindTap enables students to analyze and apply chapter concepts within relevant assignments, and allows instructors to measure skills and promote better outcomes with ease. A fully online learning solution, MindTap combines all student learning tools—readings, multimedia, activities, and assessments—into a single Learning Path that guides the student through the curriculum. Instructors personalize the experience by customizing the presentation of these learning tools to their students, even seamlessly introducing their own content into the Learning Path.

STUDENT ANCILLARIES

MindTap

MindTap®

Stay organized and efficient with **MindTap**—a single destination with all the course material and study aids you need to succeed. Built-in apps leverage social media and the latest learning technology, including the following:

- ReadSpeaker will read the text to you.
- Flashcards are pre-populated to provide you with a jump start for review—or you can create your own.
- You can highlight text and make notes in your MindTap Reader. Your notes will flow into Evernote, the electronic notebook app that you can access anywhere when it's time to study for the exam.

- Videos to apply the concepts you learn in class.
- Self-quizzing allows you to assess your understanding.

Visit *www.nelson.com/student* to start using **MindTap**. Enter the Online Access Code from the card included with your text. If a code card is *not* provided, you can purchase instant access at *NELSONbrain.com*.

Workbook

A *workbook* that provides a wide range of practice problems to reinforce concepts learned in each chapter, as well as to prepare you for exams, is available for students. Question types include multiple choice, completion, and identification as well as critical math problems and soft skill activities. Instructors may assign these questions as homework to ensure full comprehension of the material.

Acknowledgments

The publisher and author of *Carpentry,* Third Canadian Edition, wish to sincerely thank those who contributed to this book. Our gratitude is extended to those reviewers who contributed to the revision: Hank Bangma, Thompson Rivers University; Barry Botham, St. Lawrence College; Donnie Brown, Holland College; Don Fishley, Durham College; Julie Lewis, George Brown College; Brad MacDonald, Mohawk College; and Ted Zak, University of the Fraser Valley.

We continue to benefit from the insightful comments of reviewers of previous editions: Hank Bangma, Thompson Rivers University; Chuck Barsony, Loyalist College; Jeff Chow, SIAST; Rick Dohl, BCIT; Tim Dorn, Okanagan College; Bruce Fergstad, Winnipeg Technical College; Murray Fleece, SIAST; Rob Gilchrist, Conestoga College; Steve Laing, Fanshawe College; Doug Laporte, George Brown College; Tom Newton, Camosun College; Rob Passmore, Canadore College; Dana Rushton, NSCC; Al Scott, Niagara College; and Kim Woodman, Georgian College. Their insights and recommendations were invaluable.

Special thanks to Kevin LeMay of Humber College, to Andris Balodis of Conestoga College, to Rick Dohl of the British Columbia Institute of Technology (BCIT), and to Greg Kenny of Algonquin College, for their significant input on necessary additions and changes to the Canadian author's submissions.

Thanks also to Jeff Acciaccaferro from Mohawk College for his help with the chapter on timber framing, both with the content and with the photographs.

The supplemental package could not have been completed without the help of four talented instructors—to Timothy Dorn for the Test Bank, to Brad MacDonald for the PowerPoint and the Instructor's Guide, to Kevin LeMay for MindTap, and to Kim Woodman for the Student Workbook, our many thanks.

And to the Nelson Education team, whose dedication to the project produced quality learning materials for aspiring carpenters everywhere—Jackie Wood, publisher; Toula Di Leo, content development manager; Carrie McGregor, permissions coordinator; Christine Gilbert, production project manager; and Michael Kelly, copy editor—thanks to you all!

A special thanks to all the many tradespeople who were willing to share their techniques and to pause in their tasks for pictures. Last and most, thanks to my wife, Barb, for her gracious support.

Michael Nauth

About the Authors

Floyd Vogt is a sixth-generation carpenter/builder. He was raised in a family with a business devoted to all phases of home construction, and began working in the family business at age fifteen.

After completing a B.A. in chemistry from the State University of New York at Oneonta, Mr. Vogt returned to the field as a self-employed remodeller. In 1985, he began teaching in the Carpentry program at the State University of New York at Delhi in Delhi, New York (*www.delhi.edu*). He has taught many courses, including Light Framing, Advanced Framing, Math, Energy Efficient Construction, Finish Carpentry, Finish Masonry, and Estimating. Currently, Mr. Vogt is a professor in construction design-build management, bachelor's degree program at Delhi. Course responsibilities include Residential Construction, AutoCAD, Construction Seminar, and Physical Science Applications. He has served as a carpentry regional coordinator for Skills-USA and postsecondary Skills-USA student advisor. He is currently a member of a local town planning board. He is available by e-mail at *vogtfh@delhi.edu*.

Michael Nauth was born in Guyana. His family immigrated to Canada in 1967. He finished high school in Scarborough, Ontario, and began working in construction in 1972, in both the residential and commercial sectors. He obtained his carpenter's licence in 1978 in Fort St. John, British Columbia. After moving to Ottawa, he started a residential construction company. He obtained his Ontario licence in the early 1990s. A few years later he passed the Inter-Provincial Certificate of Qualification Red Seal Exam.

Michael took training to teach in carpentry, receiving his diploma from McGill University in Montreal. He would later complete a B.Sc. in mathematics at Carleton University in Ottawa. He began teaching night school in carpentry at Algonquin College (Ottawa) in 1984; he was hired as a full-time professor of the Carpentry Apprenticeship programs. He is currently enrolled in the Master of Education in Educational Administration and Policy program at St. Francis Xavier University in Antigonish, Nova Scotia.

Michael has also served as a building supervisor, foreman, and board member with Habitat for Humanity—National Capital. He has worked with Global Village in Honduras, Nicaragua, and Yukon assisting in the building of homes, the establishment of septic systems, and the delivery of potable water. He has coordinated and judged carpentry competitions for Skills Canada–Ontario and for the Carpenters' Union, Local 93. Besides several other construction textbooks, he has reviewed Canada Mortgage and Housing Corporation's *Canadian Wood Frame House Construction*, 2005 edition. He has worked as a carpenter and a trainer in Russia and is currently engaged by Canada Wood to assist with training needs and quality assurance of the residential construction industry in China.

Michael is the proud father of four sons: Aaron, Anthony, Jason, and Jonathan. They have all helped Dad in the summers on the construction site and have experienced, firsthand, the adrenaline rush, the addictive buzz, and the sweet sleep that follows an honest, hard day's work.

Introduction

The history of carpentry goes back to 8000 B.C., when primitive people used stone axes to shape wood to build shelters. Stone Age Europeans built rectangular timber houses more than 100 feet (30 m) long—proving the existence of carpentry even at this early date. The Egyptians used copper woodworking tools as early as 4000 B.C. By 2000 B.C. they had developed bronze tools and were proficient in the drilling, dovetailing, mitring, and mortising of wood.

In the Roman Empire, two-wheeled chariots, called *carpentum* in Latin, were made of wood. A person who built such chariots was called a *carpentarius*, from which the English word *carpenter* is derived. Roman carpenters handled iron adzes, saws, rasps, awls, gouges, and planes.

During the Middle Ages, most carpenters were found in larger towns where work was plentiful. They would also travel with their tools to outlying villages or wherever there was a major construction project in progress. By this time, they had many efficient, steel-edged hand tools. During this period, skillful carpentry was required for the building of timber churches and castles.

In the twelfth century, carpenters banded together to form **guilds**. The members of the guild were divided into **masters, journeymen,** and **apprentices**. The master was a carpenter with much experience and knowledge who trained apprentices. The apprentice lived with the master and was given food, clothing, and shelter and worked without pay. After five to nine years, the apprentice became a journeyman who worked and travelled for wages. Eventually, a journeyman could become a master. Guilds were the forerunners of the modern labour unions and associations.

Starting in the fifteenth century, carpenters used great skill in constructing the splendid buildings of the Renaissance period and afterward. With the introduction of the balloon frame in the early nineteenth century, more modern construction began to replace the slower mortise-and-tenon frame. In 1873, electric power was used for the first time to drive machine tools. The first electric hand drill was developed in 1917, and by 1925, electric portable saws were being used.

At present, many power tools are available to the carpenter to speed up the work. Although the scope of the carpenter's work has been reduced by the use of manufactured parts, some of the same skills carpenters used in years past are still needed for the intricate interior finish work in buildings.

Carpenters construct and repair structures and their parts using wood, plywood, and other building materials. They lay out, cut, fit, and fasten the materials to erect the framework and apply the finish. They build houses, factories, banks, schools, hospitals, churches, bridges, dams, and other structures. In addition to new construction, a large part of the industry is engaged in the remodelling and repair of existing buildings.

The majority of workers in the construction industry are carpenters (**Figure I–1**). They are the first trade workers on the job, laying out excavation and building lines. They take part in every phase of the construction, working below the ground, at ground level, or at great heights. They are the last to leave the job when they put the key in the lock.

FIGURE I–1 Carpenters make up the majority of workers in the construction industry.

SPECIALIZATION

In large cities, where there is a great volume of construction, carpenters tend to specialize in one area of the trade. They may be specialists in rough carpentry, who are called rough carpenters or framers (**Figure I–2**).

FIGURE I–2 Some carpenters specialize in framing.

FIGURE I–4 Erecting concrete formwork may be a specialty.

Rough carpentry does not mean that the workmanship is crude. They typically measure to within $\frac{1}{16}$ inch (2 mm). Just as much care is taken in the rough work as in any other work. Rough carpentry will be covered eventually by the finish work or will be dismantled, as in the case of concrete form construction.

Finish carpenters specialize in applying exterior and interior finish, sometimes called trim (**Figure I–3**). Many materials require the finish carpenter to measure to tolerances of $\frac{1}{32}$ or $\frac{1}{64}$ of an inch (1 or 0.5 mm). Other specialties are constructing concrete forms (**Figure I–4**), laying finish flooring, building stairs, applying roofing (**Figure I–5**), insulating, and installing suspended ceilings.

In smaller communities, where the volume of construction is lighter, carpenters often perform tasks in all areas of the trade from the rough to the finish. Such communities may have workers who perform

FIGURE I–5 Some carpenters choose to do only roofing work.

all aspects of construction, including plumbing, electrical work, and excavating. The general carpenter needs a more complete knowledge of the trade than the specialist does.

REQUIREMENTS

Every occupation sets specific requirements for its workers. These include the skills necessary to perform each task and an attitude or mind-set the workers should have while on the job. Construction skills vary to fit the type of work being performed.

FIGURE I–3 Carpenters may be specialists in applying interior trim.

They tend to be clearly defined, and it is easy to determine whether a worker possesses these skills. The construction attitude is less tangible, but no less important. While work may continue if the workers have only the required skills, minor or severe slowdowns are inevitable if workers do not have the proper attitude.

Skill

Carpenters need to know how to use and maintain hand and power tools. They need to know the kinds, grades, and characteristics of the materials with which they work—how each can be cut, shaped, and most satisfactorily joined. Carpenters must be familiar with the many different fasteners available and choose the most appropriate ones for each task.

Carpenters should know how to lay out and frame floors, walls, stairs, and roofs. They must know how to install windows and doors and how to apply numerous kinds of exterior and interior finish. They must use good judgment to decide on proper procedures to do the job at hand in the most efficient and safest manner.

Carpenters need to be in good physical condition, because much of the work is done by hand and sometimes requires great exertion. They must be able to lift large sheets of plywood, heavy wood timbers, and bundles of roof shingles; they also have to climb ladders and work on scaffolds.

Carpenters need reading and math skills. Much of construction begins as an idea put on paper. A carpenter must be able to interpret these ideas from the written form to create the desired structure. This is done by reading prints and using a ruler. The quantity of material needed must be estimated using math and geometry. Accurate measurements and calculations speed the construction process and reduce waste of materials.

Communication skills are also very important. Carpenters must communicate with many people during the construction process. Work to be done as determined by the owner or architect must be accurately understood, otherwise costly delays and expenses may result. Efficiency of work relies heavily on workers' understanding of what others are doing and what work must be done next. Communication is vital for a jobsite to be safe.

Attitude

The proper construction attitude is not as clearly defined as job skills. Regional variations and requirements will affect the expected jobsite attitude. For example, in regions where heat is a concern, workers develop a steady rhythm to their work that survives the heat. In regions where winters are harsh, workers develop a faster style of work when the weather is warm, knowing they may have downtime in the winter. Other attitudes, such as having a good work ethic, are universal to all jobsites.

A good ethic is not easily defined in one sentence. It involves the person as a whole, the way he or she approaches life and work. A person with a good work ethic has respect and lets it show. Respect for the jobsite, co-workers, tools and materials, and themselves reveals care and concern for the construction process. Workers demonstrating this form of respect are safe and more pleasant to work alongside.

Workers with a good work ethic show up 15 minutes early, not a few minutes late. When finished with a task, they look for something else to do that promotes the job, even if it's using a broom. They perform their tasks as well as expected, up to the standard required for that application. They finish the task, not leaving something undone for someone else to fix. They are interested in learning, looking around at the work of others for ways to improve their own skills. They cooperate with other workers and tradespeople to make the jobsite a pleasant workplace. They are also honest with the material and their time, never cheating on the accepted method. They feel that nothing less than a first-class job is acceptable.

Jobsite humour makes work easier to do and more fun. Some tasks, by their very nature, are boring and unpleasant. Humour can make difficult jobs seem to get done faster. Unfortunately, humour has a bad side because humour for one person can be pain for another. Jobsite humour can be tasteless. It can single out a person, making him or her feel alienated. This type of humour can have severe effects on jobsite safety. Someone who feels like a victim will feel defensive and will not concentrate on what he or she is doing. During these times, unintended things can happen and someone may get hurt. Keep jobsite humour suitable for everyone present.

There is no replacement for teamwork on the jobsite. Someone once said, "Two people working together can outperform three people working alone." This is easily seen when one person holds material to be fastened or cut for another. But the difference between the team and individuals is more dramatic when each member of the team anticipates the next move of the other. For example, while holding a board that is being fastened, that person looks to see what else can be done. Is more material needed? Does the horse or ladder have to be moved into a better position? Will there be another tool needed? How can the tool best be handed to the co-worker? It can be easy to miss this type of teamwork when it is done without words and without being asked.

Another way a team works better is in the mental energy used on the job. Two pairs of eyes working on one task can ward off errors and mistakes. Two minds can find the better, faster, safer method. Teamwork is a major reason why people stay in construction for a lifetime.

BUILDING YOUR CAREER AS A CARPENTER

Across Canada, education and skills training fall under the jurisdiction of the provinces. This means that while there are various paths for apprentices to follow, depending on the province, all must serve time on the job under the tutelage of a journeyperson and pass the same Inter-Provincial Certificate of Qualification Red Seal Exam with a minimum score of 70 percent.

There are several paths toward the goal of journeyperson, but all of them require that the traveller achieve competence in both skill and knowledge. The first and most important path is that of **apprenticeship** (**Figure I–6**). The educational prerequisite for apprenticeship varies from grade 9 to grade 12 (minimum age 16). However, the Canadian Council of Directors of Apprenticeship (CCDA) advises that most employers require prospective apprentices to have grade 12 or the equivalent. Also, there is a provincially specified ratio of journeypersons to apprentices, in most cases 1:1.

To become an apprentice, the young worker must find a licensed employer who is willing to indenture him or her and then sign an agreement with that

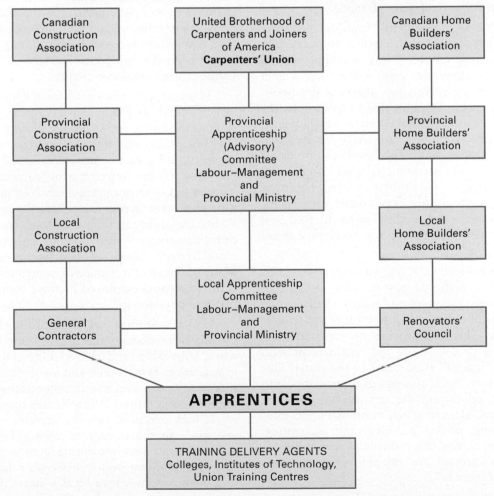

APPRENTICESHIP AND TRAINING SYSTEM OF THE CARPENTRY TRADE

FIGURE I–6 Employers in commercial and residential construction industry associations work together with labour and government organizations in apprenticeship programs.

employer and with the provincial ministry responsible for skills training. In Ontario, Carpentry Apprenticeship falls under the purview of the Ontario College of Trades. The apprentice's name then goes on a list, and he or she is called to go to school for the in-school portion of apprenticeship training. This type of training is called *block release* and can be in three or four blocks of six to eight weeks at a time. (Some of the Maritime provinces and Quebec have individualized learning packages rather than block release.) Usually, the apprentice takes a level of in-school training, works for a year, and then takes the next level. Human Resources and Skills Development Canada (HRSDC) is currently working with the provinces and territories to standardize the delivery of the in-school portion of apprenticeship training.

After the in-school training or the modularized training is complete, and all of the required on-the-job hours have been served, the apprentice is entitled to write the Certificate of Qualification Exam (C of Q) and obtain journeyperson status. Passing the Red Seal exam allows the carpenter to receive recognition and top dollar right across Canada.

Apprenticeship in-school training generally has a fee attached to it, though some provinces offer it as a free service. While attending school, the qualifying apprentice will receive Employment Insurance and in some cases travel and cost-of-living allowances. To help offset the cost to the apprentice, the federal government has launched the **Apprentice Incentive Grant (AIG)** for Red Seal trades, which offers $1000 per level to apprentices who have successfully completed level 1 and level 2 of the in-school portion of their training. There are also "loans for tools" and tax credits for apprentices. Upon successful completion of the Red Seal Exam, the "new" journeyperson receives a grant of $2000.

Another route to achieving journeyperson status is to complete pertinent postsecondary technician and technologist programs and receive credit for one or two levels of the in-school portion of apprenticeship training. This path provides novices with marketable skills to help them find employment with a company that will train them in the trade. This path costs more, but it also equips participants with "soft" skills relating to the construction industry, such as AutoCAD, surveying, business ethics, technical reporting, project planning, and accounting. Many employers find that individuals with this well-rounded background are valuable assets to their operations (**Figure I–7**).

A third way to get a carpenter's licence is referred to as the "Trade Qualifiers" way. This is intended for

FIGURE I–7 Many opportunities lie ahead for the apprentice carpenter.

older carpenters who have worked for many years on the job but who lack the financial or educational resources to go to school to get training. Having shown proof of the hours they have spent on the job, they are given the opportunity to take the C of Q exam. With sufficient preparation in trade calculations, safety training, and building code familiarization, many have succeeded. The C of Q is designed to test what the apprentice has learned on the job. Clearly, then, experience counts.

Many opportunities exist for the journeyperson carpenter. Advancement depends on dependability, skill, productivity, and ingenuity, among other characteristics. Carpentry foremen, construction superintendents, and general contractors usually rise from the ranks of the journeyperson carpenters. Many who start as apprentice carpenters eventually operate their own construction firms. A survey revealed that 90 percent of the top officials (presidents, vice presidents, owners, and partners) of construction companies who replied began their careers as apprentices. Many of the project managers, superintendents, and craft supervisors employed by these companies also began as apprentices.

National student organizations such as Skills Canada offer students training in leadership, teamwork, citizenship, and character development. The organization helps build and reinforce work attitudes, self-confidence, and communications skills for the future workforce. It emphasizes total quality at work, high ethical standards, superior work skills, and pride in the dignity of work. Skills Canada also promotes involvement in community service activities. Yearlong student activities culminate with skill competitions where local winners move up to regional, provincial, and then national and international levels.

SAFETY AT THE WORKSITE

Much of the work performed in construction carries risk with it. Although the risk will vary with the type of work, all construction workers, at some point, may be at risk of being maimed or killed on the job.

Tools are, by their very nature, dangerous to use. Each must be operated in a fashion suited to the design of the tool. Some tools can cause cuts, while others can kill. Safety cannot be taken for granted and must be built into the methods and processes being used for any particular task. Unusual situations leading to accidents can happen at any time. For example, a roofing carpenter can be thrown to the ground by a gust of wind, cleanup workers can be injured by falling objects from a scaffold, and trim carpenters can fall through a stairway opening that is under construction. Safety is like the air we breathe: It must be everywhere all of the time in order for workers to survive.

Jobsite safety is like team sportsmanship. Everyone on the team must work together and play by the same rules or someone is at risk of being injured. Safety is like silence: Just as everyone in a room must agree to be silent for silence to exist, so too everyone on the job must agree to be safe for safety to exist. One person can create a situation where dozens of workers will be at risk. All safety programs are successful only because workers join the team.

Safe Work Practices

Safe methods and practices learned correctly today will develop into habits that will become second nature for a lifetime. Always approach new tasks thoughtfully and carefully. If a new tool is being used, become familiar with its requirements. Adapt to the tool, because it cannot adapt to you. Always read and follow the directions associated with the tool or task. Read the manufacturer's recommendations for use and installation. Manufacturers clearly define the risks and recommended uses. Failure to follow their instructions is foolish and short-sighted. Always use the safety devices designed into a tool. They are there to protect the operator, not to make it harder to use the tool. Always remember where you are and the possible dangers of the worksite. Horseplay and practical jokes are distracting to everyone around. When horseplay happens, the work environment is less safe. Workers have suffered permanent back injuries while lifting an object too fast because someone touched them inappropriately. Respect is the key word to safety. Always listen and communicate fully. Think of possible ways for misunderstanding and clear them up. The only stupid question is the one not asked.

Safe Work Conditions

Safe work conditions are not hard to create, yet they can be easily neglected. Always maintain clear areas for working and walking. Store materials in an orderly fashion and dispose of waste materials as they are produced. If you stumble and struggle for a place to stand, then safety risks have increased. Always work to keep the area as dry as possible. Sweep water puddles away or make small trenches to divert the water. Keep electrical cords and tools dry. Always maintain tools and equipment according to manufacturer's recommendations. Equipment needs may include lubrication, cleaning, drying, or inspection. Always understand the risks associated with building materials. All workers are to have access to information on the hazards of materials they are working with through **Material Safety Data Sheets** (**MSDSs**). Read them. Always look for ways to keep fires from starting. Any possible heat source should be carefully studied and isolated from starting fires. Always be aware of the effect of air temperature on safety. Cold or frozen areas can be slippery to work on. Hot areas can have soft material that is easily marred. Always wear personal protective equipment. Eyes, ears, and airways should be protected when injury is possible. It is easier to understand why eyes are important to protect. Most of our perception of the world comes through our eyes. But ear, lung, and sinus damage often occurs slowly. Most times they do not show any effects for years. Personal protection is a personal responsibility.

Construction efficiency and jobsite safety require good listening skills. Listening to what is asked and performing that task efficiently is one form of good listening. It is just as important to listen to the sounds of the job. Most sounds are normal, some are not. A saw cutting wood, a mixer mixing mortar, a generator providing power, and the bangs, taps, and thuds of construction are examples of normal noises. As workers become acquainted and accustomed to these sounds, it is important for them to be able to tell when these sounds change. A change in sound can be an early warning signal that something is wrong. It could be that something is about to break or that someone is in trouble and needs assistance.

OCCUPATIONAL HEALTH AND SAFETY

The Canadian federal government states that all Canadians have the right to work in a safe and healthy environment. However, it is the provincial ministries and departments that have jurisdiction over jobsite safety. Each province has set regulations for health and safety practices on construction projects. Those who contravene these regulations can be punished by fines, job shutdowns, and even jail time. Every construction worker is protected by the *Workers' Compensation Act* and the Workplace Safety and Insurance Board (WSIB). Contact your provincial government offices or go online to your provincial government's website for more information.

Provincial construction safety associations have been established to provide research information, education, and training in the field of occupational health and safety. It is the employer's responsibility to ensure a safe and healthy workplace and to provide training for employees. It is the employee's responsibility to follow the safety rules, especially when it comes to wearing properly fitted personal protective equipment.

SUMMARY

Carpentry is a trade in which a great deal of self-satisfaction, pride, and dignity is associated with the work. It is an ancient trade and the largest of all trades in the building industry.

Skilled carpenters who have laboured to the best of their ability can take pride in their workmanship, whether the job was a rough concrete form or the finest finish in an elaborate staircase (**Figure I–8**). At the end of each working day, carpenters can stand back and actually see the results of their labour. As the years roll by, the buildings that carpenters' hands had a part in creating still can be viewed with pride in the community.

Courtesy of L. J. Smith Stair Systems

FIGURE I–8 After completing a complicated piece of work, such as this intricate staircase, carpenters can take pride in and view their accomplishment.

Section 1

Tools and Materials

Brad Current

Title: Red Seal journeyman carpenter and lead hand
Company: United Brotherhood of Carpenters and Joiners
of America, Local 93, Ottawa, Ontario

EDUCATION

Brad completed high school in 1996 in Lakefield, Ontario, and continued his education at Bishop's University in Lennoxville, Quebec. He majored in marketing and finance and graduated with a four-year B.B.A. degree.

He took his in-school apprenticeship training at Algonquin College, completing the basic level in 2005, the intermediate level in 2006, and the advanced level in 2007.

HISTORY

After graduating from Bishop's, Brad found employment in Montreal as a production manager in a clothing company. After two years he decided that working in an office was not what he wanted. He moved back to Ottawa and started working for a residential framing company. After a year in the residential sector, he moved to commercial work. At that point he was signed on by the Carpenters Union Local 93 and registered as an apprentice.

ON THE JOB

Brad's work experience has included residential framing, concrete formwork, scaffolding, rigging, and interior fit-up of light commercial buildings. He has framed floors, walls, and roofs for homes in Ottawa. His commercial experience has ranged from erecting scaffolding for large towers and office buildings and bridges, to constructing concrete forms for industrial and commercial buildings, to applying finishing trim and installing fixtures in office buildings. As a result of his varied background, he was chosen to represent his Local at the 14th Annual Ontario Provincial Carpentry Contest in Port Hope.

He completed the mandatory hours of on-the-job training and wrote and passed his Inter-Provincial Red Seal Certificate of Qualification Exam in August 2007.

BEST ASPECTS

Brad likes the physical, hands-on part of the job and the adrenaline rush that comes with working at heights. He stays fit and increases his strength by working out at the gym and playing rugby. He understands the team aspect of working in the trades— that is, working as a cohesive unit to complete specified tasks.

On the jobsite, he appreciates the fact that there is a new task every day and a constant change of scenery. Brad recently experienced another enjoyable part of his trade when he was able to contribute as a volunteer crew leader on the Habitat for Humanity "Women Build" in Stittsville (near Ottawa).

CHALLENGES

Brad finds that his biggest challenge is to learn and retain all the knowledge and skills that a good carpenter requires. The work skills expected of a carpenter are wide-ranging, and there are many different ways to approach the same task. Another challenge on the job is communication: among workers, and between worker and supervisor. Instructions must be clearly given and clearly heard.

FUTURE

Brad continues to grow in a leadership role. With the Carpenters Union, he works in northern Canada on reserves, hydro dams, and gold mines (1.5 m thick concrete head frame) where he trains apprentices, labourers, and even truck drivers. He is also teaching Concrete Formwork at Algonquin College where he integrates field trips to commercial projects with preparing fourth-year apprentices for the Red Seal Certification exam. Brad's ultimate goal is to be his own boss and operate his own business. He already has the business skills that will make him an asset for managing a construction company.

WORDS OF ADVICE

Success in the trades requires a commitment to the "long haul," a strong and fit worker, and mastery of arithmetic. "Brains and brawn," says Brad.

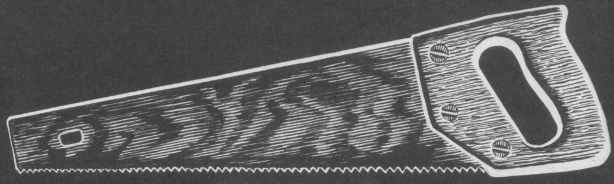

Unit 1

Wood and Lumber

Chapter 1 Wood

Chapter 2 Lumber

The construction material most often associated with a carpenter is wood. Wood has properties that make it the first choice in many applications in home construction. Wood is easy to tool and work with, is pleasing to look at and smell, and has strength that will last a long time.

Lumber is manufactured from the renewable resources of the forest. Trees are harvested and sawn into lumber in many shapes and sizes, having a variety of characteristics. It is necessary to understand the nature of wood to get the best results from the use of it. With this knowledge, the carpenter is able to protect lumber from decay, select it for appropriate use, work it with proper tools, and join and fasten it to the best advantage.

Wood comes from many tree species having many different characteristics. A good carpenter knows the different applications for specific wood species.

OBJECTIVES

After completing this unit, the student should be able to achieve the following:

- Name the parts of a tree trunk and state each part's function.
- Describe methods of cutting the log into lumber.
- Define hardwood and softwood, give examples of some common kinds, and list their characteristics.
- Explain moisture content at various stages of seasoning, tell how wood shrinks, and describe some common lumber defects.
- State the grades and sizes of lumber and compute board measure.

Chapter 1

Wood

The carpenter works with wood more than any other material and must understand its characteristics in order to use it intelligently. Wood is a remarkable substance. It can be more easily cut, shaped, or bent into almost any form than just about any other structural material. It is an efficient insulating material. Wood resists the flow of heat energy six times better than brick and fourteen times better than concrete of equal thickness.

There are many kinds of wood that vary in strength, workability, elasticity, colour, grain, texture, and smell. It is important to keep these qualities in mind when selecting wood. For instance, baseball bats, diving boards, and tool handles are made from hickory and ash because of their greater ability to bend without breaking (elasticity). Oak and maple are used for floors because of their beauty, hardness, and durability. Redwood, cedar, cypress, and teak are used in exterior situations because of their resistance to decay (**Figure 1–1**). Cherry, mahogany, and walnut are typically chosen for their beauty.

With proper care, wood will last indefinitely. It is a material with beauty and warmth that has thousands of uses. Wood is one of our greatest natural resources. With wise conservation practices, wood will always be in abundant supply. It is fortunate that we have perpetually producing forests that supply this major building material to construct homes and other structures that last for hundreds of years. When those structures have served their purpose and are torn down, the wood used in their construction can be salvaged and used again (recycled) in new building, remodelling, or repair. Wood is biodegradable and when it is considered not feasible for reuse, it is readily absorbed back into the earth with no environmental harm.

STRUCTURE AND GROWTH

Wood is a material cut from complex living organisms called trees. Trees are made up of many different kinds of cells and growth areas, which are visible in the cross-sectional view (**Figure 1–2**). Wood is made up of many hollow cells held together by a natural substance called lignin. The size, shape, and arrangement of these cells determine the strength, weight, and other properties of wood. Tree growth takes place in the cambium layer, which is just inside the protective shield of the tree, called the *bark*.

Courtesy of California Redwood Association

FIGURE 1–1 Redwood is often used for exterior trim and siding. *(Courtesy of California Redwood Association)*

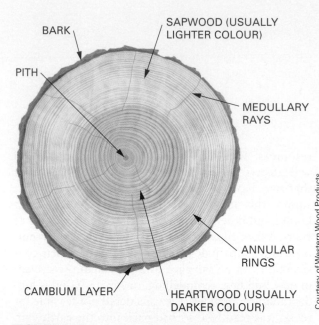

FIGURE 1–2 A cross-section of a tree showing its structure. *(Courtesy of Western Wood Products Association)*

Courtesy of Western Wood Products

The tree's roots absorb water that passes upward through the **sapwood** to the leaves, where it is combined with carbon dioxide from the air. Sunlight causes these materials to change into food, which is then carried down and distributed toward the centre of the trunk through the **medullary rays**.

As the tree grows outward from the **pith** (centre), the inner cells become inactive and turn into **heartwood**. This older section of the tree is the central part of the tree and usually is darker in colour and more durable than sapwood. The heartwood of cedar, cypress, and redwood, for instance, is extremely resistant to decay and is used extensively for outdoor furniture, patios, and exterior siding. Used for the same purposes, sapwood decays more quickly.

Each spring and summer, a tree adds new layers to its trunk. Wood grows rapidly in the spring; it is rather porous and light in colour. In summer, tree growth is slower; the wood is denser and darker, forming distinct rings. Because these rings are formed each year, they are called **annular rings**. By counting the dark rings, the age of a tree can be determined. By studying the width of the rings, periods of abundant rainfall and sunshine or periods of slow growth can be discerned. Some trees, such as the Douglas fir, grow rapidly to great heights and have very wide and pronounced annular rings. Mahogany, which grows in a tropical climate where the weather is more constant, has annular rings that do not contrast as much and sometimes are hardly visible.

HARDWOODS AND SOFTWOODS

Woods are classified as either **hardwood** or **softwood**. There are different methods of classifying these woods. The most common method of classifying wood is by its source. Hardwood comes from **deciduous** trees that shed their leaves each year. Softwood is cut from **coniferous**, or cone-bearing, trees, commonly known as evergreens (**Figure 1–3**). In this method of classifying wood, some of the softwoods may actually be harder than the hardwoods. For instance, fir, a softwood, is harder and stronger than basswood, a hardwood. There are other methods of classifying hardwoods and softwoods, but this method is the one most widely used.

FIGURE 1–3 Softwood is from cone-bearing trees, hardwood from broad-leaf trees.

Some common hardwoods are ash, birch, cherry, hickory, maple, mahogany, oak, and walnut. Some common softwoods are pine, fir, hemlock, spruce, cedar (**Figure 1–4**), cypress, and redwood.

Wood can also be divided into two groups according to cell structure. **Open-grained** wood has large cells that show tiny openings or pores in the surface. To obtain a smooth finish, these pores must be filled with a specially prepared paste wood filler. Examples of open-grained wood are oak, mahogany, and walnut. Some **close-grained** hardwoods are birch, cherry, maple (**Figure 1–5**),

FIGURE 1–4 Western red cedar.

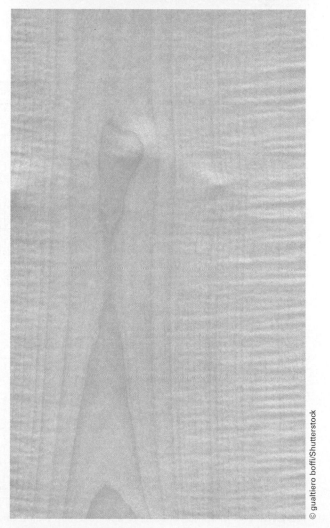

FIGURE 1–5 Maple.

and poplar. All softwoods are close grained. (See **Figure 1–6** for common types of softwoods and their characteristics and **Figure 1–7** for common hardwoods.)

Identification of Wood

Identifying different kinds of wood can be very difficult, because some closely resemble each other. For instance, ash and white oak are hard to distinguish from each other, as are some pine, hemlock, and spruce. Not only are they the same colour, but the grain pattern and weight are about the same. Only the most experienced workers are able to tell the difference.

It is possible to get some clues to identifying wood by studying the literature, but the best way to learn the different kinds of wood is by working with them. Each time you handle a piece of wood, examine it. Look at the colour and the grain; feel if it is heavy or light, if it is soft or hard; and smell it for a characteristic odour. Aromatic cedar, for instance, can always be identified by its pleasing, moth-repelling odour, if by no other means. After studying the characteristics of the wood, ask or otherwise find out the kind of wood you are holding, and remember it. In this manner, after a period of time, you will be able to identify easily those kinds of wood that are used regularly on the job. Identification of kinds of wood that are seldom worked with can be accomplished in the same manner, but, of course, the process will take a little longer. For more about Canadian trees and lumber, download *Tree Book: Learning to Recognize Trees of British Columbia*, by Roberta Parish and Sandra Thomson, published by the Ministry of Forests, Lands and Natural Resource Operations and available at *www.for.gov.bc.ca/hfd/library/documents/treebook/treebook.pdf*.

Softwoods									
Kind	**Colour**	**Grain**	**Hardness**	**Strength**	**Workability**	**Elasticity**	**Decay Resistance**	**Uses**	**Other**
Red Cedar	Dark Reddish Brown	Close Medium	Soft	Low	Easy	Poor	Very High	Exterior	Cedar Odour
Fir	Yellow to Orange Brown	Close Coarse	Medium to Hard	High	Hard	Medium	Medium	Framing Millwork Plywood	
Ponderosa Pine	White with Brown Grain	Close Coarse	Medium	Medium	Medium	Poor	Low	Millwork Trim	Pine Odour
Sugar Pine	Creamy White	Close Fine	Soft	Low	Easy	Poor	Low	Pattern-making Millwork	Large Clear Pieces
Western White Pine	Brownish White	Close Medium	Soft to Medium	Low	Medium	Poor	Low	Millwork Trim	
Southern Yellow Pine	Yellow Brown	Close Coarse	Soft to Hard	High	Hard	Medium	Medium	Framing Plywood	Much Pitch
Redwood	Reddish Brown	Close Medium	Soft	Low	Easy	Poor	Very High	Exterior	Light Sapwood
Spruce	Cream to Tan	Close Medium	Medium	Medium	Medium	Poor	Low	Siding Subflooring	Spruce Odour
Tamarack	Yellow Brown	Close Medium	Soft to Hard	High	Hard	Medium	High	Finishing Flooring	Fence Posts

FIGURE 1–6 Common species of softwood and their characteristics.

Hardwoods

Kind	Colour	Grain	Hardness	Strength	Workability	Elasticity	Decay Resistance	Uses	Other
Ash	Light Tan	Open Coarse	Hard	High	Hard	Very High	Low	Tool Handles Oars Baseball Bats	
Basswood	Creamy White	Close Fine	Soft	Low	Easy	Low	Low	Drawing Bds Veneer Core	Imparts No Taste or Odour
Beech	Light Brown	Close Medium	Hard	High	Medium	Medium	Low	Food Containers Furniture	
Birch	Light Brown	Close Fine	Hard	High	Medium	Medium	Low	Furniture Veneers	
Cherry	Lt. Reddish Brown	Close Fine	Medium	High	Medium	High	Medium	Furniture	
Hickory	Light Tan	Open Medium	Hard	High	Hard	Very High	Low	Tool Handles	
Lauan	Lt. Reddish Brown	Open Medium	Soft	Low	Easy	Low	Low	Veneers Underlayment Panelling	
Mahogany	Russet Brown	Open Fine	Medium	Medium	Excellent	Medium	High	Quality Furniture	
Maple	Light Tan	Close Medium	Hard	High	Hard	Medium	Low	Furniture Flooring	
Oak	Light Brown	Open Coarse	Hard	High	Hard	Very High	Medium	Flooring Boats	
Poplar	Greenish Yellow	Close Fine	Medium Soft	Medium Low	Easy	Low	Low	Furniture Underlayment Veneer Core	
Teak	Honey	Open Medium	Medium	High	Excellent	High	Very High	Furniture Boat Trim	Heavy Oily
Walnut	Dark Brown	Open Fine	Medium	High	Excellent	High	High	High-Quality Furniture	

FIGURE 1–7 Common species of hardwood and their characteristics.

Chapter 2

Lumber

MANUFACTURE OF LUMBER

Trees are selected in the forest, cut on site, and transported to the sawmill. When logs arrive at the sawmill, the bark is removed first. Then a huge bandsaw slices the log into large planks, which are passed through a series of saws. The saws slice, edge, and trim them into various dimensions, and the pieces become **lumber**.

Once trimmed of all uneven edges, the lumber is stacked according to size and grade and taken outdoors, where "stickering" takes place. Stickering is the process of restacking the lumber on small cross-sticks, which allows air to circulate between the pieces. This air-seasoning process may take six months to two years, due to the large amount of water found in lumber.

After air drying, the lumber is dried in huge ovens, or kilns. Once dry, the rough lumber is surfaced to standard sizes and shipped (**Figures 2–1 to 2–5**).

The long, narrow surface of a piece of lumber is called its edge; the long, wide surface is termed its side; and its extremities are called ends. The distance across the edge is called its thickness, across its side is called its width, and from end to end is called its length (**Figure 2–6**). The best-appearing side is called the face side, and the best-appearing edge is called the face edge.

In certain cases, the surfaces of lumber may acquire different names. For instance, the distance from the top to the bottom of a beam is called its depth, and the distance across its top or bottom may be called its width or its thickness. The length

FIGURE 2–3 The bark is stripped off the logs.

FIGURE 2–1 Trees are cut by individuals with chain saws.

FIGURE 2–4 The logs are sliced at the sawmill.

FIGURE 2–2 Trees are cut and de-limbed by machines.

FIGURE 2–5 The cut-lumber is stickered and stacked at the sawmill.

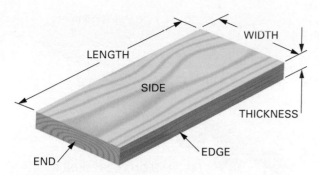

FIGURE 2–6 Lumber and board surfaces are distinguished by specific names.

of posts or columns, when installed, may be called their height.

Plain-Sawn Lumber

A common way of cutting lumber is called the **plain-sawn** method, in which the log is cut tangent to the annular rings. This method produces a distinctive grain pattern on the wide surface (**Figure 2–7**).

This method of sawing is the least expensive and produces greater widths. However, plain-sawn lumber shrinks more during drying and warps easily. Plain-sawn lumber is sometimes called slash-sawn lumber.

Quarter-Sawn Lumber

Another method of cutting the log, called quarter-sawing, produces pieces in which the annular rings are at or almost at right angles to the wide surface. Quarter-sawn lumber is less likely to warp and shrinks less and more evenly when dried. This type of lumber is durable because the wear is on the edge of the annular rings. Quarter-sawn lumber is frequently used for flooring.

A distinctive and desirable grain pattern can be produced in some wood, such as oak, by sawing the lumber along the length of the medullary rays. Quarter-sawn lumber is sometimes called vertical-grain or edge-grain (**Figure 2–8**).

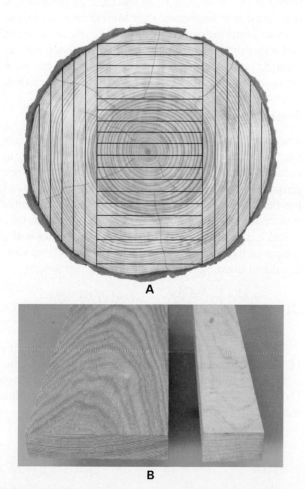

A

B

FIGURE 2–7 (A) Typical sawing approach for plain-sawn lumber. (B) Surface of plain-sawn lumber.

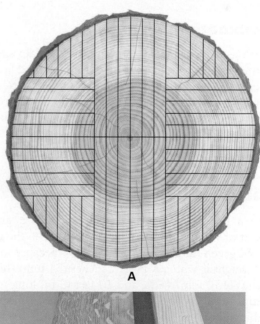

A

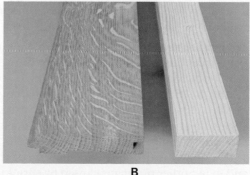

B

FIGURE 2–8 (A) Typical sawing approach for quarter-sawn lumber. (B) Surface of quarter-sawn lumber.

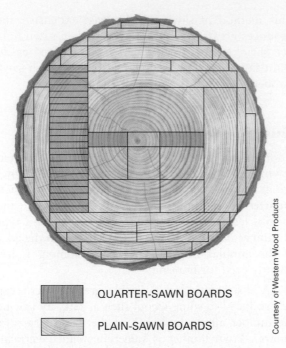

QUARTER-SAWN BOARDS

PLAIN-SAWN BOARDS

Courtesy of Western Wood Products

FIGURE 2–9 Typical sawing approach for combination-sawn lumber. *(Courtesy of Western Wood Products Association)*

Combination Sawing

Most logs are cut into a combination of plain-sawn and quarter-sawn lumber. With computers and laser-guided equipment, the sawyer determines how to cut the log with as little waste as possible in the shortest amount of time to get the desired amount and kinds of lumber (**Figure 2–9**).

MOISTURE CONTENT AND SHRINKAGE

When a tree is first cut down, it contains a great amount of water. Lumber, when first cut from the log, is called **green lumber** and is very heavy because most of its weight is water. A piece 2 inches (51 mm) thick, 6 inches (150 mm) wide, and 10 feet (3 m) long may contain as much as 4¼ gallons (19 litres) of water weighing about 35 pounds (16 kg) (**Figure 2–10**).

Green lumber should not be used in construction because it shrinks as it dries to the same moisture content as the surrounding air. This shrinking is considerable and unequal because of the large amount of water that leaves it. When it shrinks, it usually warps, depending on the way it was cut from the log. The use of green lumber in construction results in cracked ceilings and walls, squeaking floors, sticking doors, and many other problems caused by shrinking and warping of the lumber as it dries. Therefore, lumber must be dried to a suitable degree before it can be surfaced and used.

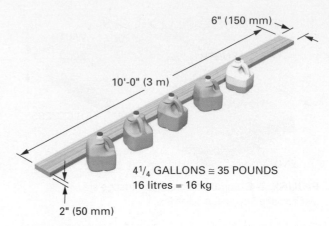

FIGURE 2–10 Green lumber contains a large amount of water.

Green lumber is also subject to decay. Decay is caused by *fungi,* low forms of plant life that feed on wood. This decay is commonly known as dry rot because it usually is not discovered until the lumber has dried. Decay will not occur unless wood moisture content is in excess of 19 percent. Wood construction maintained at moisture content of less than 20 percent will not decay. It is important that lumber with an excess amount of moisture be exposed to an environment that will allow the moisture to evaporate. Seasoned lumber must be protected to prevent the entrance of moisture that allows the growth of fungi and decay of the wood.

Moisture Content

The amount of liquid water in wood is significant because of the nature of wood cells. They grow in columns of tubes. Water fills these tubes during the growing process. The **moisture content (MC)** of lumber is expressed as a percentage and indicates how much of the weight of a wood sample is actually water. It is derived by determining the difference in the weight of the sample before and after it has been kiln-dried and dividing that number by the dry weight:

$$MC = \frac{\text{wood wet weight} - \text{wood dry weight}}{\text{wood dry weight}} \times 100$$

For example, if a wood sample weighs 16 ounces (454 g) prior to drying and 13 ounces (369 g) after drying, we assume that there were 3 ounces (85 g) (16 − 13 = 3) of water in the wood. To determine the moisture content of the sample before drying,

$$MC = \frac{16 - 13}{13} \times 100 = 23.0769\%$$

Thus, the moisture content of the sample before drying was roughly 23 percent. Lumber used for framing and exterior finish preferably should have an

FIGURE 2–11 Wood cells are long, hollow tubes that fill with water during tree growth.

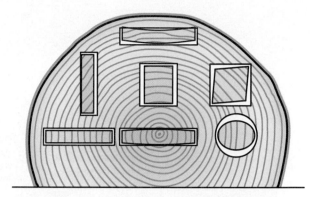

FIGURE 2–13 As lumber dries, the annular rings become shorter, sometimes causing wood to deform.

MC of 15 percent, not to exceed 19 percent. For interior finish, an MC of 10 to 12 percent is recommended.

Green lumber has water in the hollow part of the wood cells as well as in the cell walls (**Figure 2–11**). When wood starts to dry, the water in the cell cavities, called *free water*, is first removed. When all of the free water is gone, the wood has reached the **fibre-saturation point**—approximately 30 percent MC. No noticeable shrinkage of wood takes place up to this point.

As wood continues to dry, the water in the walls of the cells is removed and the wood starts to shrink. It shrinks considerably from its size at the fibre-saturation point to its size at the desired MC of less than 19 percent. The actual shrinkage of the lumber will vary with regional climate conditions and the MC of the lumber when delivered (**Figure 2–12**). Lumber at this stage is called *dry* or **seasoned** and now must be protected from getting wet.

It is important to understand not only that wood shrinks as it dries but also how it shrinks. Little shrinkage occurs along the length of lumber; therefore, shrinkage in that direction is not considered. Most shrinkage occurs along the length of each annular ring, with the longer rings shrinking more than the shorter ones. When viewing plain-sawn lumber in cross-section, it can be seen that the piece warps as it shrinks because of the unequal length of the annular rings. A cross-section of quarter-sawn lumber shows annular rings of equal length. Although the piece shrinks, it shrinks evenly with little warp (**Figure 2–13**). Wood warps as it dries according to the way it was cut from the tree. Cross-sectional views of the annular rings are different along the length of a piece of lumber; therefore, various kinds of warp result as lumber dries and shrinks.

When the moisture content of lumber reaches that of the surrounding air (about 10 to 12 percent MC), it is at **equilibrium moisture content**. At this point, lumber shrinks or swells only slightly with changes in the moisture content of the air. Realizing that lumber undergoes certain changes when moisture is absorbed or lost, the experienced carpenter uses techniques to deal with this characteristic of wood (**Figure 2–14**).

Drying Lumber

Lumber is either **air-dried**, **kiln-dried**, or a combination of both. In air drying, the lumber is stacked in piles with spacers, or stickers, placed between each layer to permit air to circulate through the pile. Kiln-dried lumber is stacked in the same manner but is dried in buildings, called kilns, that are like huge ovens (**Figure 2–15**).

Kilns provide carefully controlled temperatures, humidity, and air circulation to remove moisture.

Lumber Size	Actual Width	Width @ 19% MC	Width @ 11% MC	Width @ 8% MC
2 × 4	3½″ (89 mm)	3½″ (89 mm)	3⁷/₁₆″ (87 mm)	3³/₈″ (86 mm)
2 × 6	5½″ (140 mm)	5½″ (140 mm)	5³/₈″ (137 mm)	5⁵/₁₆″ (135 mm)
2 × 8	7¼″ (184 mm)	7¼″ (184 mm)	7¹/₈″ (181 mm)	7¹/₁₆″ (179 mm)
2 × 10	9¼″ (235 mm)	9¼″ (235 mm)	9¹/₁₆″ (230 mm)	9″ (228 mm)

Source: *U.S. Span Book for Major Wood Species. (Courtesy of Canadian Wood Council)*

FIGURE 2–12 Lumber dimensions change with the moisture content of the wood.

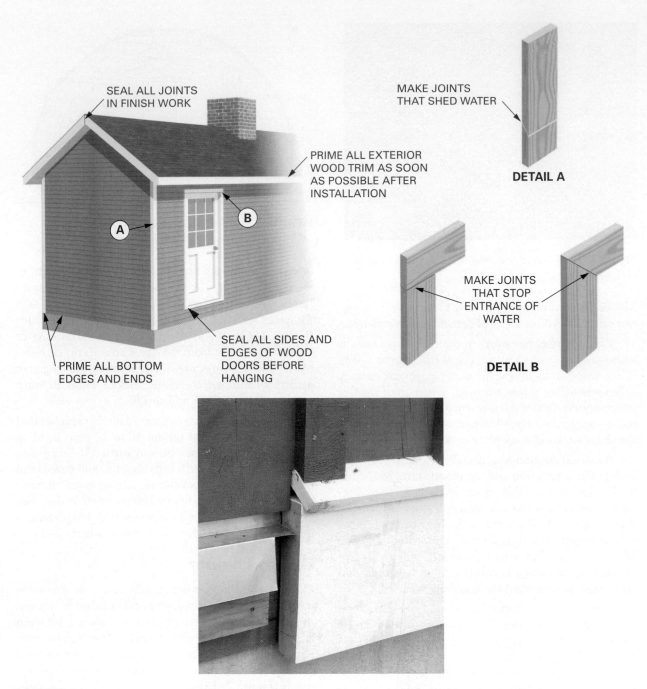

SEAL ALL JOINTS
IN FINISH WORK

MAKE JOINTS
THAT SHED WATER

DETAIL A

PRIME ALL EXTERIOR
WOOD TRIM AS SOON
AS POSSIBLE AFTER
INSTALLATION

A

B

MAKE JOINTS
THAT STOP
ENTRANCE OF
WATER

DETAIL B

SEAL ALL SIDES AND
EDGES OF WOOD
DOORS BEFORE
HANGING

PRIME ALL BOTTOM
EDGES AND ENDS

FIGURE 2–14 Techniques to prevent water from getting in behind the wood surface.

ON THE JOB

Techniques to prevent moisture from entrance into wood.

First, the humidity level is raised and the temperature is increased; the humidity is then gradually decreased. Kiln drying has the advantage of drying lumber in a shorter period of time, but is more expensive than air drying.

The recommended moisture content for lumber to be used for exterior finish at the time of installation is 12 percent, except in very dry climates, where 9 percent is recommended. Lumber with low moisture content (8 to 10 percent) is necessary for interior trim and cabinet work.

The moisture content of lumber is determined by the use of a **moisture meter** (**Figure 2–16**). Points on the ends of the wires of the meter are driven into the wood, and the moisture content is read off the meter.

FIGURE 2–15 Lumber is placed in a kiln to reduce moisture content.

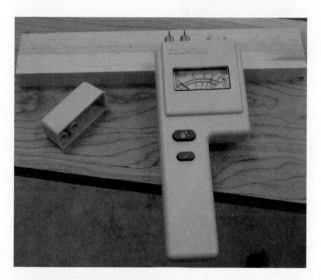

FIGURE 2–16 A moisture meter is used to determine moisture content in wood.

Experienced workers know when lumber is green (because it is much heavier than dry lumber), and they can estimate fairly accurately the moisture content of lumber simply by lifting it.

Lumber is brought to the planer mill, where it is dressed straight, smoothed, and uniformly sized. This process can be carried out whether the lumber is dry or green. Most construction lumber is surfaced on four sides (S4S) to standard thicknesses and widths. Some may be surfaced on only two sides (S2S) to required thicknesses.

Lumber Storage

Lumber should be delivered to the jobsite so that materials are accessible in the proper sequence; that is, those that are to be used first are on the top and those to be used last are on the bottom.

Lumber stored at the jobsite should be adequately protected from moisture and other hazards.

A common practice that must be avoided is placing unprotected lumber directly on the ground. Use short lengths of lumber (called *dunnage*) running at right angles to the length of the pile and spaced close enough to keep the pile from sagging and coming into contact with the ground. The pieces of dunnage need to be about 3 inches (75 mm) thick to permit the forks of a forklift to fit easily under the pile. The base on which the lumber is to be placed should be fairly level to keep the pile from falling over.

Protect the lumber with a tarp or other type of cover. Leave enough room at the bottom and top of the pile for circulation of air. A cover that reaches to the ground will act like a greenhouse, trapping ground moisture within the stack.

Keep the piles in good order. Lumber spread out in a disorderly fashion can cause accidents, as well as subject the lumber to stresses that may cause warping.

LUMBER DEFECTS

A defect in lumber is any fault that detracts from its appearance, function, or strength. One type of defect is called a **warp**. Warps are caused by, among other things, drying lumber too fast, careless handling and storage, or surfacing the lumber before it is thoroughly dry. A warp may be classified as a **crook** (also called a *crown*), a **bow**, a **cup**, or a **twist** (**Figure 2–17**).

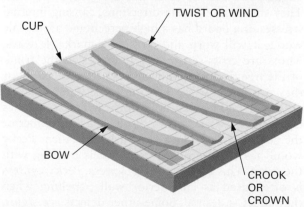

CUP

TWIST OR WIND

BOW

CROOK OR CROWN

FIGURE 2–17 Board deformations have names that depend on the type of warp.

TWIST

CUP

HEARTWOOD STAIN

PITH AND CHECK

WANE

ROT

CHECK

FIGURE 2–18 Examples of common wood defects.

Splits in the end of lumber running lengthwise and across the annular rings are called **checks** (**Figure 2–18**). Checks are caused by faster drying of the end than of the rest of the stock. Checks can be prevented to a degree by sealing the ends of lumber with paint, wax, or some other material during the drying period. A crack that runs parallel to and between the annular rings is called a **shake** and may be caused by weather or other damage to the tree.

The *pith* is the spongy centre of the tree. It contains the youngest portion of the lumber, called **juvenile wood**. Juvenile wood is the portion of wood that contains the first seven to fifteen growth rings. The wood cells in this region are not well-aligned and are therefore unstable when they dry. They shrink in different directions, causing internal stresses. If a board has a high percentage of juvenile wood, it will warp and twist in remarkable ways. **Knots** are cross-sections of branches in the trunk of the tree. Knots are not necessarily defects unless they are loose or weaken the piece. **Pitch pockets** are small cavities that hold pitch, which sometimes oozes out. A **wane** is bark on the edge of lumber or the surface from which the bark has fallen. **Pecky** wood has small grooves or channels running with the grain. This is common in cypress. Pecky cypress is often used as an interior wall panelling when that effect is desired. Some other defects are *stains*, *decay*, and *wormholes*.

LUMBER GRADES AND SIZES

Lumber grades and sizes are established by wood products associations, of which many wood mills are members. Member wood mills are closely supervised by the associations to ensure that standards are maintained. The **grade** stamp of the association is assurance that lumber grade standards have been met.

Member mills use the association grade stamp to indicate strict quality control. A typical grade stamp is shown in **Figure 2–19** and shows the association trademark (MLB, for Maritime Lumber Bureau), the mill number and association (99 NLGA, for National Lumber Grades Authority), the lumber grade (No. 1), the species of wood (S-P-F, for spruce-pine-fir), and

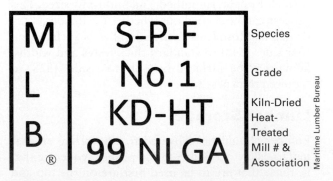

FIGURE 2–19 Typical softwood lumber grade stamp. *(Courtesy of Maritime Lumber Bureau)*

whether the wood was green or dry when it was planed (KD-HT, for kiln-dried heat-treated).

Softwood Grades

The largest softwood association of lumber manufacturers is the Western Wood Products Association (WWPA), which grades lumber in three categories: **boards** (under 2 inches [51 mm] thick), **dimension** (2 to 4 inches [51 to 102 mm] thick), and **timbers** (5 inches [127 mm] and thicker). The board group is divided into boards, sheathing, and form lumber. The dimension group is divided into light framing, studs, structural light framing, and structural joists and planks. Timbers are divided into beams and stringers. The three main categories are further classified according to strength and appearance, as shown in **Figure 2–20**.

Hardwood Grades

Hardwood grades are established by the National Hardwood Lumber Association. **Firsts and seconds (FAS)** is the best grade of hardwood and must yield about 85 percent clear cutting. Each piece must be at least 6 inches (150 mm) wide and 8 feet (2440 mm) long. The next best grade is called *select*. For this, the minimum width is 4 inches (102 mm) and the minimum length is 6 feet (1830 mm). **No. 1 common** allows even narrower widths and shorter lengths, with about 65 percent clear cutting.

Note that the price and quality of lumber are often linked. Some large home centres sell lumber at lower prices because it is of a lower quality. Understanding the information on lumber grade stamps will help the builder purchase material wisely.

Lumber Sizes

Rough lumber that comes directly from the sawmill is close in size to what it is called **nominal size**. There are slight variations to nominal size because of the heavy machinery used to cut the log into lumber. When rough lumber is planed, it is reduced in thickness and width to standard and uniform sizes. Its nominal size does not change even though the actual size does. Therefore, when *dressed* (surfaced), although a piece may be called a 2 × 4, its actual size is 1½ inches (38 mm) by 3½ inches (89 mm). The same applies to all surfaced lumber; the nominal size (what it is called) and the actual size are not the same. Hardwood lumber is usually purchased in the rough and then dressed straight, smoothed, and sized as needed.

BOARD MEASURE

Softwood lumber is usually purchased by specifying the number of pieces—thickness($''$) × width($''$) × length($'$) (i.e., 35 pieces−$2'' × 6'' × 16'$)—in addition to the grade. This is referred to as material list form. Often, when no particular lengths are required, the thickness, width, and total number of linear feet (length in feet) are ordered. The lengths of the pieces then may vary and are called *random* lengths. Another method of purchasing softwood lumber is by specifying the thickness, width, and total number of *board feet*. Lumber purchased in this manner may also contain random lengths.

Hardwood lumber is purchased by specifying the grade, thickness, and total number of board feet. Large quantities of both softwood and hardwood lumber are priced and sold by the board foot.

A **board foot** is a measure of lumber volume. It is defined as the volume of wood equivalent to a piece of wood that measures 1-inch thick by 1-foot wide by 1-foot long. It allows for a comparison of different-sized pieces. For example, the volume of wood in a 2 × 6 that is 1 foot long is also one board foot (see **Figures 2–21** and **2–22**).

To calculate board feet, use the following formula:

$$\text{bd ft} = \frac{\text{\# of Pieces} × \text{Thickness } ('') × \text{Width } ('') × \text{Length } (')}{12}$$

This formula uses the dimensions used to identify the board, for example, 2 × 6 × 12 feet. Always use the nominal dimensions, not the actual dimensions; for example, 16 pieces of 2 × 4 × 8 feet is 85⅓ board feet (bd ft).

$$\frac{\overset{4}{\cancel{16}} × 2 × 4 × 8}{\underset{3}{\cancel{12}}} = \frac{256}{3} = 85⅓ \text{ bd ft}$$

Note that this formula seems to ignore the basic rules of arithmetic by multiplying feet and inches together. This is because it calculates board feet, not cubic feet or cubic inches. In the metric system (Système International, SI), lumber is quantified by the cubic metre (m^3).

A GREEN INDUSTRY

GreenHouse Gas (GHG) emissions refer to the release of carbon dioxide (CO_2), methane (CH_4), nitrous oxide (N_2O), sulphur hexafluoride (SF_6),

Standard Lumber Sizes/Nominal, Dressed, Based on WWPA Rules

Product Description		Nominal Size		Dressed Dimensions		
				Thicknesses and Widths In. Surfaced		
		Thickness In.	Width In.	Surfaced Dry	Surfaced Unseasoned	Lengths Ft.
DIMENSION	S4S................... Other surface combinations are available. See "Abbreviations" below.	2 3 4	2 3 4 5 6 8 10 12 Over 12	1½ 2½ 3½ 4½ 5½ 7½ 9½ 11½ ¾ off normal	1⁹⁄₁₆ 2⁹⁄₁₆ 3⁹⁄₁₆ 4⅝ 5⅝ 7½ 9½ 11½ Off ½	Standard lengths are 6' and longer in multiples of 1'.
SCAFFOLD PLANK	Rough Full Sawn or S4S.... (Usually shipped unseasoned)	1¼ and thicker	8 and wider	If Dressed, refer to "DIMENSION" sizes		None specified
TIMBERS	Rough or S4S............. (Shipped unseasoned)	5 and larger		½ off nominal (S4S) See 3.20 of WWPA Grading Rules for Rough		Standard lengths are 6' and longer in multiples of 1'.

		Nominal Size		Dressed Dimensions		
		Thickness In.	Width In.	Thickness In.	Width In.	Lengths Ft.
DECKING	2" Single T&G.............	2	4 5 6 8 10 12	1½	3 4 5 6¾ 8¾ 10¾	Standard lengths are 6' and longer in multiples of 1'.
	3" and 4" Double T&G......	3 4	6	2½ 3½	5¼	
FLOORING	(D & M), (S2S & CM).......	⅜ ½ ⅝ 1 1¼ 1½	2 3 4 5 6	⁵⁄₁₆ ⁷⁄₁₆ ⁹⁄₁₆ ¾ 1 1¼	1⅛ 2⅛ 3⅛ 4⅛ 5⅛	3' or 6' and longer, generally shipped in multiples of 1'
CEILING AND PARTITION	(S2S & CM)...............	⅜ ½ ⅝ ¾	3 4 5 6	⁵⁄₁₆ ⁷⁄₁₆ ⁹⁄₁₆ ¹¹⁄₁₆	2⅛ 3⅛ 4⅛ 5⅛	4' depending on grade and longer, generally shipped in multiples of 1'
FACTORY AND SHOP LUMBER	S2S.....................	1 (4/4) 1¼ (5/4) 1½ (6/4) 1¾ (7/4) 2 (8/4) 2½ (10/4) 3 (12/4) 4 (16/4)	5" and wider (except 4" and wider in 4/4 No. 1 Shop and 4/4 No. 2 Shop, and 2" and wider in 5/4 and thicker No. 3 Shop)	¾ (4/4) 1⁵⁄₃₂ (5/4) 1¹³⁄₃₂ (6/4) 1¹⁹⁄₃₂ (7/4) 1¹³⁄₁₆ (8/4) 2⅜ (10/4) 2¾ (12/4) 3¾ (16/4)	Usually sold random width	Varies depending on grade.

Abbreviations

Abbreviated descriptions appearing in the size table are explained below.
S1S—Surfaced one side.
S2S—Surfaced two sides.

S4S—Surfaced four sides.
S1S1E—Surfaced one side, one edge.
S1S2E—Surfaced one side, two edges.
CM—Centre matched.

D & M—Dressed and matched.
T & G—Tongue and grooved.
Rough Full Sawn—Unsurfaced green lumber cut to full specified size.

Product Classification

	Thickness In.	Width In.		Thickness In.	Width In.
Board lumber	1"	2" or more	Beams & stringers	5" and thicker	more than 2" greater than thickness
Light framing	2" to 4"	2" to 4"	Posts & timbers	5" × 5" and larger	not more than 2" greater than thickness
Studs	2" to 4"	2" and wider			
Structural light framing	2" to 4"	2" to 4"	Decking	2" to 4"	4" to 12" wide
Structural joists & planks	2" to 4"	5" and wider	Siding	thickness expressed by dimension of butt edge	
			Mouldings	size at thickest and widest points	

Nailing Diagram

BOARD ON BOARD — 8" and wider, use 2 nails 3–4" apart

TONGUE AND GROOVE — 8" and wider, use 2 nails 3–4" apart

BOARD AND BATTEN — 8" and wider use 2 nails 3–4" apart in centre

CHANNEL RUSTIC — 8" and wider use 2 nails 3–4" apart at exposed edge

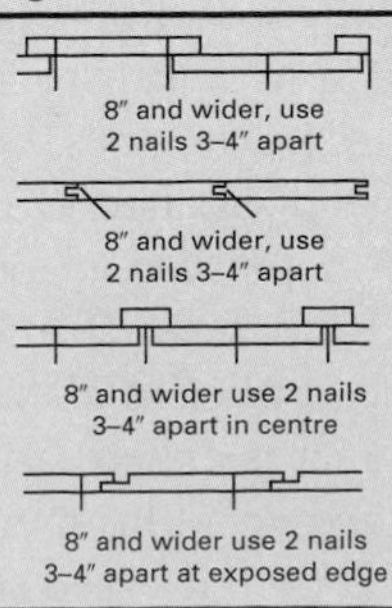

Standard lengths of lumber generally are 6 feet and longer in multiples of 1 foot.

FIGURE 2–20 Softwood lumber sizes. *(Courtesy of Western Wood Products Association)*

Courtesy of Western Wood Products Association

BOARD FEET = NUMBER OF PIECES X THICKNESS" X WIDTH" X LENGTH' ÷ 12

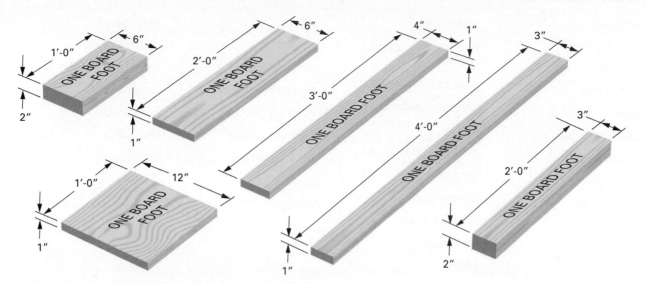

FIGURE 2–21 One board foot of lumber is a volume of wood; it can have many shapes.

FIGURE 2–22 Board foot measure allows lumber to be referred to in a volume of varying width and length boards.

per-fluorocarbons (PFCs) and hydro-fluorocarbons (HFCs) released by human activity (reported in megatonnes of CO_2 equivalent)into the atmosphere. GHG emissions are a global problem. Some countries are holding back on reductions until others agree to participate. In the meantime, the global levels are going up and the entire world is feeling the effects of climate change. The forest industry in Canada is setting the example for sustainability. For every tree that is harvested, more are planted. The clear cutting sins of the past are in the past and the industry is moving forward with definitive plans for the future. Forests are treated as a garden with the proper spacing of plants, thinning or weeding to reduce overgrowth, and management of

parasites. Climate change and warmer winters, and mature pine trees have led to the mountain pine beetle infestation that is killing pine trees by the millions. But the forest industry, armed with the knowledge that the dead trees still supply structurally strong lumber, though somewhat discoloured, and that the standing trees have a 'shelf life' of 8 to 12 years, is aggressively making use of what was previously considered useless and a targeted approach to reforestation has been initiated.

Operational Standards have been set by three organizations within Canada: Canadian Standards Association (CSA) and Sustainable Forestry Initiative (SFI), both under the umbrella of the Programme for the Endorsement of Forest Certification (PEFC) schemes, and the Forest Stewardship Council (FSC).

Wood structures are strong, durable, and resilient enough to withstand earthquakes and high winds. Wood materials are renewable, recyclable, biodegradable, and they store carbon. Trees take in CO_2 and release oxygen (O_2) into the atmosphere. Building with wood is good for the occupants, for the environment, and for Canada and the world.

http://www.nrcan.gc.ca/forests/video/16213

http://www.for.gov.bc.ca/hfp/mountain_pine _beetle/video/E_HarvestLevelsAndOpportunities .wmv

http://www.nrcan.gc.ca/forests/video/13552

http://mg-architecture.ca/wp-content/uploads/ 2012/09/Pages-from-Tall-Wood-Report.pdf

KEY TERMS

air-dried

annular rings

boards

board foot

bow

cambium layer

checks

close-grained

coniferous

crook

cup

deciduous

dimension

equilibrium moisture content

fibre-saturation point

firsts and seconds (FAS)

grade

green lumber

hardwood

heartwood

juvenile wood

kiln-dried

knots

lignin

lumber

lumber grades

medullary rays

moisture content (MC)

moisture meter

No. 1 common

nominal size

open-grained

pecky

pitch pockets

pith

plain-sawn

quarter-sawn

sapwood

sawyer

seasoned

shake

softwood

timbers

twist

wane

warp

REVIEW QUESTIONS

Select the most appropriate answer.

1. What is the centre of a tree in cross-section called?

    a. heartwood
    b. lignin
    c. pith
    d. sapwood

2. Where are new wood cells of a tree formed?

    a. heartwood
    b. bark
    c. medullary rays
    d. cambium layer

3. In which season is tree growth faster?

    a. spring
    b. summer
    c. fall
    d. winter

4. What is quarter-sawn lumber sometimes called?

    a. tangent-grained
    b. slash-sawn
    c. vertical-grained
    d. plain-sawn

5. When is lumber called "green?"

    a. when it is stained by fungi
    b. when the tree is still standing
    c. when it is first cut from the log
    d. when it has reached equilibrium moisture content

6. What percentage of moisture content must wood have before it will start decaying?

    a. over 15 percent
    b. over 19 percent
    c. over 25 percent
    d. over 30 percent

7. Which of the following terms refers to the state of wood when all of the free water in the cell **cavities** is removed, but before water is removed from the cell **walls**?

    a. fibre-saturation point
    b. 30 percent moisture content
    c. equilibrium moisture content
    d. shrinkage commencement point

8. To what level of dryness can air-dried lumber reach?

    a. not less than 8 to 10 percent moisture content
    b. not less than equilibrium moisture content
    c. not less than 25 to 30 percent moisture content
    d. not less than its fibre-saturation point

9. Which of the following is a commonly used and abundant softwood?

    a. ash
    b. fir
    c. basswood
    d. birch

10. Which of the following wood species is most naturally resistant to decay?

    a. pine
    b. spruce
    c. red cedar
    d. hemlock

WHAT'S WRONG WITH THIS PICTURE?

Carefully study **Figure 2–23** and think about what is wrong. Consider all possibilities. See **Figure 2–24.**

FIGURE 2–23 The bevel angle on the corner boards is oriented the wrong way, thereby allowing water to enter the joint and get behind the siding.

FIGURE 2–24 The bevel angle on the corner boards is oriented the correct way, thereby allowing water to run down the board and off the building.

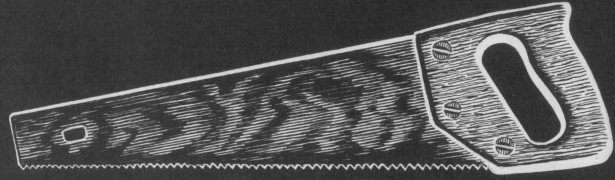

Unit 2
Engineered Panels

Chapter 3 Structural (Rated) Panels

Chapter 4 Non-Structural Panels

Growing concern over efficient use of forest resources led to the development of reconstituted wood. The resulting wood products are referred to as engineered lumber. One type of engineered lumber is manufactured from peeled logs and reconstituted wood, which is formed into large sheets referred to as engineered panels, commonly called plywood. The tradesperson must know the different kinds, sizes, and recommended applications in order to use these materials to the best advantage.

Plywood is a general term used to cover a variety of materials. A carpenter must understand the uses for the various materials to make optimum use of engineered panels.

OBJECTIVES

After completing this unit, the student should be able to describe the composition, kinds, sizes, grades, and several uses of the following:

- plywood
- oriented strand board
- composite panels
- particleboard
- hardboard (high-density fibreboard)
- medium-density fibreboard
- softboard (low-density fibreboard)

Chapter 3

Structural (Rated) Panels

The term **engineered panels** refers to manufactured products in the form of large reconstituted wood sheets, sometimes called **panels**. In some cases, the tree has been taken apart and its contents have been redistributed into sheet or panel form. The panels are widely used in the construction industry. They are also used in the aircraft, automobile, and boat-building industries, as well as in the making of road signs, furniture, and cabinets.

With the use of engineered panels, construction progresses at a faster rate because a greater area is covered in a shorter period of time (**Figure 3–1**). These panels, in certain cases, present a more attractive appearance and give more protection to a surface than does solid lumber. It is important to know the kinds and uses of various engineered panels in order to use them to the best advantage.

RATED PANELS

Many mills belong to associations that inspect, test, and allow mills to stamp the product to certify that it conforms to government and industrial standards. The grade stamp assures the consumer that the product has met the rigid quality and performance requirements of the association.

The largest association of this type, APA–The Engineered Wood Association (formerly called the American Plywood Association and still known by the acronym APA), is concerned not only with quality supervision and testing of **plywood** (cross-laminated wood veneer) but also of **composites** (veneer faces bonded to reconstituted wood cores) and non-veneered panels, commonly called **oriented strand board** (**OSB**) (**Figure 3–2**).

In Canada, the CertiWood™ Technical Centre (formerly known as CANPLY, the Canadian Plywood Association) represents manufacturers of engineered wood products. The trademarks of CANPLY appear only on products manufactured by member mills (**Figure 3–3**). Take a look at *http://www.canply .org/english/products/about.htm* and view the Certi-Wood video at *http://www.canply.org/english/ literature_media/video.htm*.

Plywood

One of the most extensively used engineered panels is plywood. Plywood is a sandwich of wood. Most plywood panels are made up of sheets of veneer (thin pieces), each called a **ply**. These plies, arranged in layers, are bonded under pressure with glue to form

FIGURE 3–1 Sheet material covers a greater area in a shorter period of time than does solid lumber.

FIGURE 3–2 APA performance-rated panels. *(Courtesy of APA–The Engineered Wood Association)*

FIGURE 3–3 The grade stamp is assurance of a high-quality, performance-rated panel. *(Courtesy of CERTIWOOD™–CANPLY)*

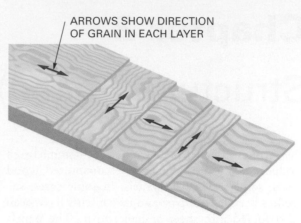

ARROWS SHOW DIRECTION OF GRAIN IN EACH LAYER

FIGURE 3–4 In the construction of plywood, the grains of veneer plies are placed at right angles to each other.

FIGURE 3–5 The veneer is peeled from the log like paper unwinding from a roll. *(Courtesy of APA–The Engineered Wood Association)*

FIGURE 3–6 Gluing and assembling plywood veneers into panels. *(Courtesy of APA–The Engineered Wood Association)*

a very strong panel. The plies are glued together so that the grain of each layer is at right angles to the next one. This cross-graining results in a sheet that is as strong as or stronger than the wood it is made from. Plywood usually contains an odd number of layers so that the face grain on both sides of the sheet runs in the direction of the long dimension of the panel (**Figure 3–4**). Softwood plywood is commonly made with three, five, or seven layers. Because of its construction, plywood is more stable with changes of humidity and is more resistant to shrinking and swelling than wood boards.

Manufacture of Veneer Core Plywood. Specially selected "peeler logs" are mounted on a huge lathe in which the log is rotated against a sharp knife. As the log turns, a thin layer is peeled off like paper unwinding from a roll (**Figure 3–5**). The entire log is used. The small remaining spindles are utilized for making other wood products.

The long ribbon of veneer is then cut into desired widths, sorted, and dried to a moisture content of 5 percent. After drying, the veneers are fed through glue spreaders that coat them with a uniform thickness. The veneers are then assembled to make panels (**Figure 3–6**). Large presses bond the assembly under controlled heat and pressure. From the presses, the panels are either left unsanded,

Veneer Grades

A	Smooth, paintable. Not more than 18 neatly made repairs, boat, sled, or router type, and parallel to grain, permitted. Wood or synthetic repairs permitted. May be used for natural finish in less demanding applications.
B	Solid surface. Shims, sled or router repairs, and tight knots to 1 inch (25 mm) across grain permitted. Wood or synthetic repairs permitted. Some minor splits permitted.
C Plugged	Improved C veneer with splits limited to $^1/_8$-inch (3 mm) width and knotholes or other open defects limited to ¼ × ½ inch (6 × 12.5 mm). Wood or synthetic repairs permitted. Admits some broken grain.
C	Tight knots to 1½ inch (38 mm). Knotholes to 1 inch (25 mm) across grain and some to 1½ inch (38 mm) if total width of knots and knotholes is within specified limits. Synthetic or wood repairs, discoloration and sanding defects that do not impair strength permitted. Limited splits allowed. Stitching permitted.
D	Knots and knotholes to 2½ inch (63 mm) width across grain and ½ inch (12.5 mm) larger within specified limits. Limited splits are permitted. Stitching permitted. Limited to Exposure 1 or interior panels.

APA - The Engineered Wood Association

FIGURE 3–7 Veneer letter grades define veneer appearance. *(Courtesy of APA–The Engineered Wood Association)*

touch-sanded, or smooth-sanded, and then cut to size, inspected, and stamped.

Veneer Grades. In declining order, the letters *A*, *B*, *C plugged*, *C*, and *D* are used to indicate the appearance quality of panel veneers. Two letters are found in the grade stamp of veneered panels. One letter indicates the quality of one face, while the other letter indicates the quality of the opposite face. The exact description of these letter grades is shown in **Figure 3–7**. Panels with B-grade or better veneer faces are always sanded smooth. Some panels, such as APA-Rated Sheathing, are unsanded because their intended use does not require sanding. Other panels used for such purposes as subflooring and underlayment require only a touch sanding to make the panel thickness more uniform.

Strength Grades. Softwood veneers are made of many different kinds of wood. These woods are classified in groups according to their strength and their usage within the structure of the plywood (**Figure 3–8**).

COMMON NAME	CSA 0121 DFP		CSA 0151				CSA 0153 POPLAR	
			CSP		ASPEN			
	Faces & Backs	Inner Plies	Faces & Backs	Inner Plies	Faces & Backs	Inner Plies	Faces & Backs	Inner Plies
Douglas fir	•	•		•		•		•
True fir*		•	•	•		•		•
Western white spruce*		•	•	•		•		•
Sitka spruce*		•	•	•		•		•
Lodgepole pine*		•	•	•		•		•
Western hemlock*		•	•	•		•		•
Western larch*		•	•	•		•		•
Trembling aspen		•		•	•	•	•	•
White birch		•	•	•		•		•
Balsam fir		•	•	•		•		•
Eastern spruce		•	•	•		•		•
Eastern white pine		•	•	•		•		•
Red pine		•	•	•		•		•
Jack pine		•	•	•		•		•
Ponderosa pine		•	•	•		•		•
Western white pine		•	•	•		•		•
Eastern hemlock		•	•	•		•		•
Tamarack		•	•	•		•		•
Yellow cedar		•	•	•		•		•
Western red cedar			•	•		•		•
Balsam poplar		•		•		•	•	•
Black cottonwood		•		•		•	•**	•**

Courtesy of CERTIWOOD™-CANPLY

* Permitted on the backs of 6, 8, 11 and 14 mm Good One Side DFP
** Not permitted in sheathing grades

FIGURE 3–8 Plywood is classified according to species permitted in CANPLY exterior plywood. *(Courtesy of CERTIWOOD™–CANPLY)*

Most of the construction-grade softwood plywood in Canada is made from spruce. For more about plywood, visit the Canadian Wood Council website, *www.cwc.ca*, and search for "plywood."

Oriented Strand Board

Oriented strand board (OSB) is a non-veneered performance-rated structural panel composed of small oriented (lined-up) strand-like wood pieces arranged in three to five layers, with each layer at right angles to the other (**Figure 3–9**). The cross-lamination of the layers achieves the same advantages of strength and stability as in plywood.

Manufacture of OSB.   Logs from specially selected species are debarked and sliced into strands that are between $^{25}/_{1000}$ and $^{30}/_{1000}$ of an inch (0.635 and 0.762 mm) thick, ¾ to 1 inch (19 to 25 mm) wide, and between 2½ and 4½ inches (63 to 114 mm) long. The strands are dried, loaded into a blender, coated with liquid resins, formed into a mat consisting of three or more layers of systematically oriented wood fibres, and fed into a press where, under high temperatures and pressure, they form a

FIGURE 3–9 Oriented strand board is often used for sheathing.

dense panel. As a safety measure, the panels have one side textured to help prevent slippage during installation.

Non-veneered panels have previously been sold with such names as waferboard, structural particleboard, and others. At present almost all panels manufactured with oriented strands or wafers are called oriented strand board. Various manufacturers have their own particular brand names for OSB, such as Oxboard, Aspenite, and many others. For more on the OSB, visit the Canadian Wood Council website, *www.cwc.ca*, and search for "OSB."

Composite Panels

Composite panels are manufactured by bonding veneers of wood to both sides of reconstituted wood panels. This product allows a more efficient use of wood while retaining the wood grain appearance on both faces of the panel.

Composite panels rated by APA are called *COMPLY* and are manufactured in three or five layers. A three-layer panel has a reconstituted wood core with wood veneers on both sides. A five-layer panel has a wood veneer in the centre as well as on both sides. OSB is glued to rigid foam insulation panels to form a "sandwich" called structural insulated panels, or SIPs. See Chapter 40 for more about this type of panel.

PERFORMANCE RATINGS

A performance-rated panel meets the requirements of the panel's end use. The three end uses for which panels are rated are floors, roofs, and concrete form work. Names given to designate end uses are *EASY T&G ROOF, EASY T&G FLOOR,* and *COFI FORM* (**Figure 3–10**). Panels are tested to meet standards in areas of resistance to moisture, strength, and stability.

Exposure Durability

CertiWood™ rates its CANPLY plywood in a chart according to its end use or typical application (**Figure 3–11**). The exterior panels are made from Douglas fir (DFP), Canadian softwood (CSP), aspen, or poplar. Some of the panels are either medium-density overlaid (MDO) or high-density overlaid (HDO).

Span Ratings

The span rating in the grade stamp on APA-Rated Sheathing appears as two numbers separated by a

Product*	Product Standard**	Grades**	Characteristics	Typical Applications
EASY T&G ROOF	DFP CSP	SHG or SEL	Milled with patented edge profile for easy installation and edge support without H-clips.	Roof sheathing and decking for residential, commercial, and industrial construction.
EASY T&G FLOOR	DFP CSP Aspen Poplar	SHG SEL SEL TF	Milled with a patented edge profile for fast, easy installation.	Floor and heavy roof sheathing for residential, commercial, and industrial construction.
COFI FORM PLUS and COFI FORM	DFP (limits on thickness and species of face and inner plies)	SEL G1S G2S SPECIALTY HDO MDO	Special construction Douglas Fir panels with greater stiffness and strength providing improved properties particularly in wet service conditions. Available in regular sanded and unsanded grades and speciality grades with resin-fibre overlays. Also available with factory-applied release agent.	Concrete forms and other uses where wet service conditions or superior strength requirements are encountered.

Grade*	Product**	Veneer Grades**			Characteristics	Typical Applications
		Face	Inner Plies	Back		
Good Two Sides (G2S) Sanded	DFP Poplar	A	C	A	Sanded. Best appearance both faces. May contain neat wood patches, inlays, or synthetic patching material.	Furniture, cabinet doors, partitions, shelving, concrete forms, and opaque paint finishes.
Good One Side (G1S)	DFP	A	C	C	Sanded. Best appearance one side only. May contain neat wood patches, inlays, or synthetic patching material.	Where appearance or smooth sanded surface of one face is important. Cabinets, shelving, concrete forms.
Select - Tight Face (SEL TF)	DFP	B***	C	C	Surface openings shall be filled and may be lightly sanded.	Underlayment and combined subfloor and underlayment. Hoarding. Construction use where sanded material is not required.
Select (SEL)	DFP Aspen Poplar CSP	B	C	C	Surface openings may be filled and may be lightly sanded.	
Sheathing (SHG)	DFP Aspen Poplar CSP	C	C	C	Unsanded. Face may contain limited size knots, knotholes, and other minor defects.	Roof, wall, and floor sheathing. Hoarding. Packaging. Construction use where sanded material is not required.
High-Density Overlaid (HDO)	DFP Aspen Poplar CSP	B***	C	B***	Smooth, resin-fibre overlaid surface. Further finishing not required.	Bins, tanks, boats, furniture, signs, displays, forms for architectural concrete.
Medium-Density Overlaid (MDO) MDO 1 Side	DFP Aspen Poplar CSP	C***	C	C	Smooth, resin-fibre overlaid surface. Best paint base.	Siding, soffits, panelling, built-in fitments, signs, any use requiring a superior paint surface.
MDO 2 Side	DFP Aspen Poplar CSP	C***	C	C***		

* All grades and products including overlays bonded with waterproof resin glue.
** For complete veneer and panel grade descriptions see CSA 0121 (DFP), CSA 0151 (CSP) and CSA 0153 (Poplar).
*** Indicates all openings are filled.

FIGURE 3–10 CANPLY Exterior Standard Plywood Grades. *(Courtesy of CERTIWOOD™–CANPLY)*

Courtesy of CERTIWOOD™–CANPLY CANPLY Exterior Standard Plywood Grades. http://www.canply.org/pdf/main/plywood_designfund.pdf, Table 4, p. 5.

slash, such as $^{32}/_{16}$ or $^{48}/_{24}$. The left number denotes the maximum recommended spacing of supports when the panel is used for roof or wall sheathing. The right number indicates the maximum recom- mended spacing of supports when the panel is used for subflooring (Figure 3–11). In both cases, the long dimension of the panel must be placed across three or more supports. A panel marked $^{800}/_{400}$, for

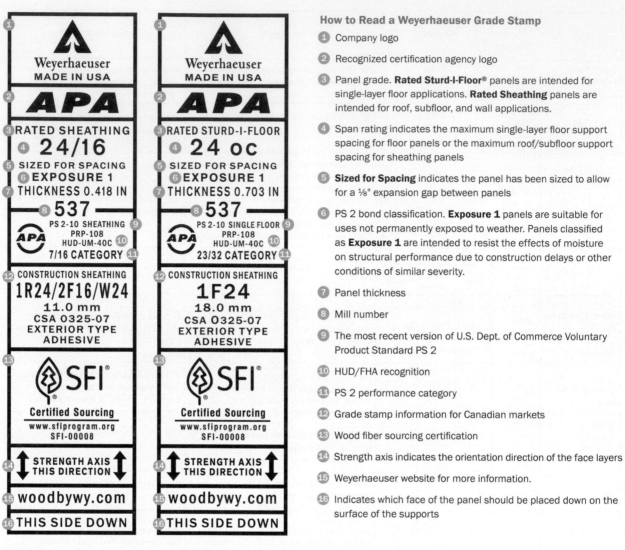

How to Read a Weyerhaeuser Grade Stamp

1 Company logo

2 Recognized certification agency logo

3 Panel grade. **Rated Sturd-I-Floor®** panels are intended for single-layer floor applications. **Rated Sheathing** panels are intended for roof, subfloor, and wall applications.

4 Span rating indicates the maximum single-layer floor support spacing for floor panels or the maximum roof/subfloor support spacing for sheathing panels

5 **Sized for Spacing** indicates the panel has been sized to allow for a ⅛" expansion gap between panels

6 PS 2 bond classification. **Exposure 1** panels are suitable for uses not permanently exposed to weather. Panels classified as **Exposure 1** are intended to resist the effects of moisture on structural performance due to construction delays or other conditions of similar severity.

7 Panel thickness

8 Mill number

9 The most recent version of U.S. Dept. of Commerce Voluntary Product Standard PS 2

10 HUD/FHA recognition

11 PS 2 performance category

12 Grade stamp information for Canadian markets

13 Wood fiber sourcing certification

14 Strength axis indicates the orientation direction of the face layers

15 Weyerhaeuser website for more information.

16 Indicates which face of the panel should be placed down on the surface of the supports

Courtesy of Weyerhaeuser

FIGURE 3–11 The grade stamp is assurance of a high-quality, performance-rated panel.

example, may be used for roof sheathing over rafters not more than 32 inches (800 mm) on centre, or for subflooring over joists not more than 16 inches (400 mm) on centre. See **Figure 3–12** for imperial-to-metric conversions.

CANPLY's Load-Span Information for DFP and CSP is based on load limits that are determined by deflection, bending, and shear. They rate the different thicknesses of plywood according to the face grain orientation and the spacing between the framing members.

Imperial	¼"	⁵/₁₆"	¹¹/₃₂"	³/₈"	⁷/₁₆"	¹⁵/₃₂"	½"	¹⁹/₃₂"	⅝"	²³/₃₂"	¾"	1"	1⅛"
Metric	6 mm	8 mm	9 mm	9.5 mm	11 mm	12 mm	12.5 mm	15 mm	16 mm	18 mm	19 mm	25.4 mm	28.5 mm
Imperial	²⁴/₁₆ OC		³²/₁₆ OC		20 OC		⁴⁸/₂₄ OC		16"	19.2"	24"	48"	
Metric	⁶⁰⁰/₄₀₀		⁸⁰⁰/₄₀₀		500 mm OC		¹²⁰⁰/₆₀₀		406 mm	488 mm	610 mm	1219 mm	

FIGURE 3–12 Imperial-to-metric conversion chart.

Chapter 4

Non-Structural Panels

PLYWOOD

All the rated products discussed in the previous chapter may also be used for non-structural applications. In addition, other plywood products, grade-stamped by the APA–The Engineered Wood Association, are available for non-structural use. They include sanded and touch-sanded plywood panels (**Figure 4–1**) and various other types of non-structural panels (**Figure 4–2**).

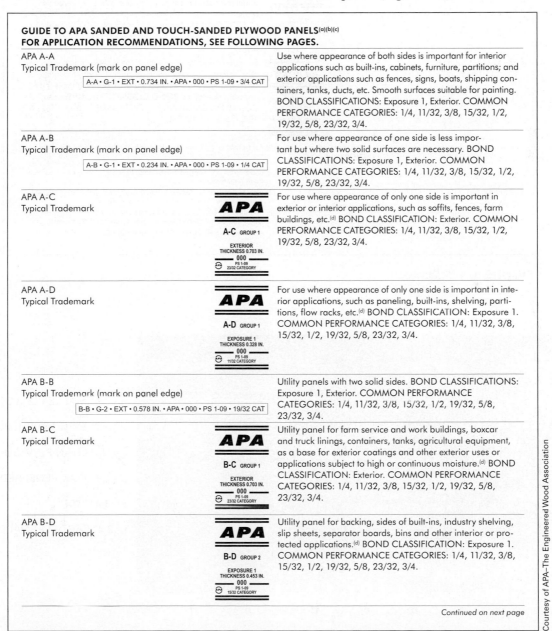

GUIDE TO APA SANDED AND TOUCH-SANDED PLYWOOD PANELS[a][b][c]
FOR APPLICATION RECOMMENDATIONS, SEE FOLLOWING PAGES.

APA A-A
Typical Trademark (mark on panel edge)

A-A • G-1 • EXT • 0.734 IN. • APA • 000 • PS 1-09 • 3/4 CAT

Use where appearance of both sides is important for interior applications such as built-ins, cabinets, furniture, partitions; and exterior applications such as fences, signs, boats, shipping containers, tanks, ducts, etc. Smooth surfaces suitable for painting. BOND CLASSIFICATIONS: Exposure 1, Exterior. COMMON PERFORMANCE CATEGORIES: 1/4, 11/32, 3/8, 15/32, 1/2, 19/32, 5/8, 23/32, 3/4.

APA A-B
Typical Trademark (mark on panel edge)

A-B • G-1 • EXT • 0.234 IN. • APA • 000 • PS 1-09 • 1/4 CAT

For use where appearance of one side is less important but where two solid surfaces are necessary. BOND CLASSIFICATIONS: Exposure 1, Exterior. COMMON PERFORMANCE CATEGORIES: 1/4, 11/32, 3/8, 15/32, 1/2, 19/32, 5/8, 23/32, 3/4.

APA A-C
Typical Trademark

APA
A-C GROUP 1
EXTERIOR
THICKNESS 0.703 IN.
000
PS 1-09
23/32 CATEGORY

For use where appearance of only one side is important in exterior or interior applications, such as soffits, fences, farm buildings, etc.[d] BOND CLASSIFICATION: Exterior. COMMON PERFORMANCE CATEGORIES: 1/4, 11/32, 3/8, 15/32, 1/2, 19/32, 5/8, 23/32, 3/4.

APA A-D
Typical Trademark

APA
A-D GROUP 1
EXPOSURE 1
THICKNESS 0.328 IN.
000
PS 1-09
11/32 CATEGORY

For use where appearance of only one side is important in interior applications, such as paneling, built-ins, shelving, partitions, flow racks, etc.[d] BOND CLASSIFICATION: Exposure 1. COMMON PERFORMANCE CATEGORIES: 1/4, 11/32, 3/8, 15/32, 1/2, 19/32, 5/8, 23/32, 3/4.

APA B-B
Typical Trademark (mark on panel edge)

B-B • G-2 • EXT • 0.578 IN. • APA • 000 • PS 1-09 • 19/32 CAT

Utility panels with two solid sides. BOND CLASSIFICATIONS: Exposure 1, Exterior. COMMON PERFORMANCE CATEGORIES: 1/4, 11/32, 3/8, 15/32, 1/2, 19/32, 5/8, 23/32, 3/4.

APA B-C
Typical Trademark

APA
B-C GROUP 1
EXTERIOR
THICKNESS 0.703 IN.
000
PS 1-09
23/32 CATEGORY

Utility panel for farm service and work buildings, boxcar and truck linings, containers, tanks, agricultural equipment, as a base for exterior coatings and other exterior uses or applications subject to high or continuous moisture.[d] BOND CLASSIFICATION: Exterior. COMMON PERFORMANCE CATEGORIES: 1/4, 11/32, 3/8, 15/32, 1/2, 19/32, 5/8, 23/32, 3/4.

APA B-D
Typical Trademark

APA
B-D GROUP 2
EXPOSURE 1
THICKNESS 0.453 IN.
000
PS 1-09
15/32 CATEGORY

Utility panel for backing, sides of built-ins, industry shelving, slip sheets, separator boards, bins and other interior or protected applications.[d] BOND CLASSIFICATION: Exposure 1. COMMON PERFORMANCE CATEGORIES: 1/4, 11/32, 3/8, 15/32, 1/2, 19/32, 5/8, 23/32, 3/4.

Continued on next page

Courtesy of APA–The Engineered Wood Association

FIGURE 4–1 Guide to APA sanded and touched-sanded plywood panels. (*Courtesy of APA–The Engineered Wood Association*)

continued

GUIDE TO APA SANDED AND TOUCH-SANDED PLYWOOD PANELS[a][b][c]
FOR APPLICATION RECOMMENDATIONS, SEE FOLLOWING PAGES.

APA UNDERLAYMENT Typical Trademark	 **APA** UNDERLAYMENT GROUP 1 EXPOSURE 1 THICKNESS 0.322 IN. 000 PS 1-09 11/32 CATEGORY	For application over structural subfloor. Provides smooth surface for application of carpet and pad and possesses high concentrated and impact load resistance. For areas to be covered with resilient flooring, specify panels with "sanded face."[e] BOND CLASSIFICATION: Exposure 1. COMMON PERFORMANCE CATEGORIES[f]: 1/4, 11/32, 3/8, 15/32, 1/2, 19/32, 5/8, 23/32, 3/4.
APA C-C PLUGGED[g] Typical Trademark	 **APA** C-C PLUGGED GROUP 2 EXTERIOR THICKNESS 0.451 IN. 000 PS 1-09 15/32 CATEGORY	For use as an underlayment over structural subfloor, refrigerated or controlled atmosphere storage rooms, pallet fruit bins, tanks, boxcar and truck floors and linings, open soffits, and other similar applications where continuous or severe moisture may be present. Provides smooth surface for application of carpet and pad and possesses high concentrated and impact load resistance. For areas to be covered with resilient flooring, specify panels with "sanded face."[e] BOND CLASSIFICATION: Exterior. COMMON PERFORMANCE CATEGORIES[f]: 11/32, 3/8, 15/32, 1/2, 19/32, 5/8, 23/32, 3/4.
APA C-D PLUGGED Typical Trademark	**APA** C-D PLUGGED GROUP 2 EXPOSURE 1 THICKNESS 0.451 IN. 000 PS 1-09 15/32 CATEGORY	For open soffits, built-ins, cable reels, separator boards and other interior or protected applications. Not a substitute for Underlayment or APA Rated Sturd-I-Floor as it lacks their puncture resistance. BOND CLASSIFICATION: Exposure 1. COMMON PERFORMANCE CATEGORIES: 3/8, 15/32, 1/2, 19/32, 5/8, 23/32, 3/4.

(a) Specific plywood grades, Performance Categories and bond classifications may be in limited supply in some areas. Check with your supplier before specifying.

(b) Sanded Exterior plywood panels, C-C Plugged, C-D Plugged and Underlayment grades can also be manufactured in Structural I (all plies limited to Group 1 species).

(c) Some manufacturers also produce plywood panels with premium N-grade veneer on one or both faces. Available only by special order. Check with the manufacturer. For a description of N-grade veneer, refer to the APA publication *Sanded Plywood*, Form K435.

(d) For nonstructural floor underlayment, or other applications requiring improved inner ply construction, specify panels marked either "plugged inner plies" (may also be designated "plugged crossbands under face" or "plugged crossbands" or "core"); or "meets underlayment requirements."

(e) Also available in Underlayment A-C or Underlayment B-C grades, marked either "touch sanded" or "sanded face."

(f) Some panels with Performance Categories of 1/2 and larger are Span Rated and do not contain species group number in trademark.

(g) Also may be designated APA Underlayment C-C Plugged.

FIGURE 4–1 *(Continued)*

a - 3/4" MAPLE PLYWOOD

b - 1/2" BIRCH APPLE PLY

c - 1/4" LUAN PLYWOOD

d - 3/4" OAK VENEERED MDF

e - 3/4" MDF

f - 3/4" PARTICLEBOARD

a   b   c   d   e   f

FIGURE 4–2 Various types of non-structural panels.

FIGURE 4–3 Particleboard is made from wood flakes, shavings, resins, and waxes.

Hardwood Plywood

Plywood is available with hardwood face veneers, of which the most popular are birch, oak, and lauan ("Philippine mahogany"). Beautifully grained hardwoods are sometimes matched in a number of ways to produce interesting face designs. Hardwood plywood is used in the interior of buildings for such things as wall panelling, built-in cabinets, and fixtures.

Particleboard

Particleboard is a reconstituted wood panel made of wood flakes, chips, sawdust, and planer shavings (**Figure 4–3**). These wood particles are mixed with an adhesive, formed into a mat, and pressed into sheet form. The kind, size, and arrangement of the wood particles determine the quality of the board.

The highest-quality particleboard is made of large wood flakes in the centre. The flakes become gradually smaller toward the surfaces, where finer particles are found. This type of construction results in an extremely hard board with a very smooth surface. Softer and lower-quality boards contain the same size of particles throughout. These boards usually have a rougher surface texture. In addition to the size, kind, and arrangement of the particles, the quality of the board is determined by the method of manufacture.

Particleboard Grades. The quality of particleboard is indicated by its density (hardness), which ranges from 28 to 55 pounds per cubic foot. Nonstructural particleboard is used in the construction industry for the construction of kitchen cabinets and countertops, and for the core of veneer doors and similar panels.

Fibreboards

Fibreboards are manufactured as *high-density,* *medium-density,* and *low-density* boards.

Hardboards

High-density fibreboard is called **hardboard** and is commonly known by the trademark *Masonite* regardless of the manufacturer. The hardboard industry makes almost complete use of the great natural resource of wood by utilizing the wood chips and board trimmings that were once considered waste.

Wood chips are reduced to fibres, and water is added to make a soupy pulp. The pulp flows onto a travelling mesh screen, where water is drawn off to form a mat. The mat is then pressed under heat to weld the wood fibres back together by utilizing lignin, the natural adhesive in wood.

Some panels are **tempered** (coated with oil and baked to increase hardness, strength, and water resistance). Saws with carbide-tipped blades trim the panels to standard sizes.

Sizes of Hardboard. The most popular thicknesses of hardboard range from $^1/_8$ to $^3/_8$ inch (3 to 9.5 mm). The most popular sheet size is 4 feet by 8 feet (1.2 by 2.4 m), although sheets can be ordered in practically any size.

Classes and Kinds of Hardboard. Hardboard is available in three different classes: tempered, standard, and service tempered (**Figure 4–4**). It may be obtained smooth-one-side (S1S) or smooth-two-sides (S2S). Hardboard is available in many forms, such as perforated, grooved, and striated.

Class	Surface	Nominal Thickness(in.)
1 Tempered	S1S and S2S	1/8 1/4
2 Standard	S1S and S2S	1/8 1/4 3/8
3 Service tempered	S1S and S2S	1/8 1/4 3/8

FIGURE 4–4 Kinds and thicknesses of hardboard.

Uses of Hardboard. Hardboard can be used inside or outside. It is used for exterior siding and interior wall panelling. It is also used extensively for cabinet backs and drawer bottoms. It can be used wherever a dense, hard panel is required. Because of the composition of hardboard, it is important to seal all sides and edges in exterior applications.

Because it is a wood-based product, hardboard can be sawed, routed, shaped, and drilled with standard woodworking tools. It can be securely fastened with glue, screws, staples, or nails.

Medium-Density Fibreboard

Medium-density fibreboard (MDF) is manufactured in a manner similar to that used to make hardboard, except that the fibres are not pressed as tightly together. The refined fibre (pioneered by Medite) produces a fine-textured, homogeneous board with an exceptionally smooth surface. Densities range from 28 to 65 pounds per cubic foot (450 to 1050 kg/m³). It is available in thicknesses ranging from 3/16 to 1½ inches (4.8 to 38 mm) and comes in widths of 4 and 5 feet (1.2 and 1.5 m). Lengths run from 6 to 18 feet (1.8 to 5.5 m).

MDF can be used for case goods, drawer parts, kitchen cabinets, cabinet doors, signs, and some interior wall finish. In response to health concerns for workers and off-gassing for end-users, the industry has manufactured MDF panels with no added formaldehyde (see medite-europe.com).

Softboard

Low-density fibreboard is called **softboard**. Softboard is very light and contains many tiny air spaces because the particles are not compressed tightly.

Courtesy of Armstrong World Industries

FIGURE 4–5 Softboards are used extensively for decorative ceiling tiles. *(Courtesy of Armstrong World Industries)*

The most common thicknesses range from ½ to 1 inch (12.5 to 25 mm). The most common sheet size is 4 feet by 8 feet (1.2 by 2.4 m), although many sizes are available.

Uses of Softboard. Because of their lightness, softboard panels are used primarily for insulating or sound-control purposes. They are used extensively as decorative panels in suspended ceilings and as ceiling tiles (**Figure 4–5**). Softboard can be used for exterior wall sheathing. This type may be coated or impregnated with asphalt to protect it from moisture during construction (called **tentest**, or donnaconna). For technical specifications about fibreboard, go to fiberboard.org and click Products and then Sheathing, or mslfibre.com/en and click Insulation and then ISOLEXT.

Softboard panels can easily be cut with a knife, hand saw, or power saw. Wide-headed nails, staples, or adhesives are used to fasten softboards in place, depending on the type and use.

OTHER

These chapters have been limited to engineered wood panels and boards. Many others are used in the construction industry besides those already mentioned. It is recommended that the student study *Sweet's Architectural File* to become better acquainted with the thousands of building material products on the market. This reference is well-known by architects, contractors, and builders, and is revised and published annually. *Sweet's* can be found online at *www.sweets.com.* Two publications, *Products for General Building and Renovations* and *Products for Home Building and Remodeling,* may be available at a school or city library.

DECONSTRUCT THIS

Carefully study **Figure 4–6** and think about what is wrong and/or what is right. Consider all possibilities. What construction practice or method is different in your area of the country?

FIGURE 4–6 Framing in the springtime.

KEY TERMS

composites

engineered panels

fibreboards

hardboard

high-density overlaid (HDO)

medium-density overlaid (MDO)

oriented strand board (OSB)

panels

particleboard

ply

plywood

softboard

tempered

tentest

REVIEW QUESTIONS

Select the most appropriate answer.

1. How many layers are there in 5/8-inch (16 mm) construction grade spruce plywood?

   a. three layers
   b. four layers
   c. five layers
   d. six layers

2. What is most construction-grade softwood plywood made of?

   a. cedar
   b. fir
   c. western pine
   d. spruce

3. What letter is used to indicate the best-appearing face veneer of a softwood plywood panel?

   a. A
   b. B
   c. E
   d. Z

4. Which of the following plywoods is a good choice for exterior wall sheathing?

   a. APA Structural Rated Sheathing, Exposure 1
   b. APA A-C, Exterior
   c. APA-Rated Sheathing, Exposure 2
   d. CD, Plugged, Exterior

5. What does the fraction 32/16 (800/400) on an APA grade stamp refer to?

   a. panel thickness
   b. maximum rafter spacing
   c. maximum wall stud spacing
   d. maximum wall stud and floor joist spacing

6. What may particleboard NOT rated as structural be used for?

   a. countertops
   b. subflooring
   c. wall sheathing
   d. roof sheathing

7. Where may hardboard be used?

   a. interior applications only
   b. exterior and interior applications
   c. applications protected from moisture
   d. cabinet and furniture work only

8. Where is most of the softboard used by the construction industry?

   a. underlayment for wall-to-wall rugs
   b. roof covering
   c. decorative ceiling panels
   d. interior wall finish

9. Which plywood should be used for an underlayment over structural subflooring where continuous or severe moisture may be present?

   a. A-C, Group 1, exterior plywood
   b. B-D, Group 2, Exposure 1 plywood
   c. Underlayment, Group 1, Exposure 1 plywood
   d. C-C Plugged, Group 2, exterior plywood

10. What is the recommended plywood for interior applications where the appearance of one side is important, such as for built-ins and cabinets?

   a. APA A-A
   b. APA A-B
   c. APA A-C
   d. APA A-D

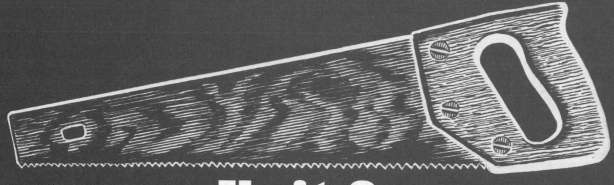

Unit 3

Engineered Wood Products

An increased demand for lumber and a diminishing supply of large old-growth trees has forced the wood industry to come up with solutions to ensure the survival of our natural resources. Structural lumber previously sawn from large logs is now produced from reconstituted wood in many shapes and sizes. These products are collectively referred to as engineered lumber.

OBJECTIVES

After completing this unit, the student should be able to describe the manufacture, composition, uses, and sizes of:

- laminated veneer lumber and cross-laminated timbers
- parallel strand lumber and laminated strand lumber
- engineered joists
- glue-laminated beams

 SAFETY REMINDER

Engineered wood is stronger than wood sawn from logs. Yet it must be handled, cut, and fastened properly to achieve the safe and desirable characteristics.

Chapter 5

Laminated Veneer Lumber and Cross-Laminated Timbers

Old-growth trees are large, tall, tight-grain trees that take more than 200 years to mature. The lumber produced from these trees is of the highest quality. Large-sized lumber can be efficiently cut from these trees. Due to centuries of logging, the number of old-growth trees is decreasing. Second- and third-growth trees, as well as trees planted during reforestation efforts, are more abundant. These are smaller and produce fewer large-sized pieces. Lumber from these trees sometimes has undesirable wood characteristics, such as a tendency to warp.

An inevitable result of the decreasing supply of large old-growth trees and the abundance of smaller trees is the development and use of **engineered wood products** (EWPs).

Engineered wood products are reconstituted wood products and assemblies designed to replace traditional structural lumber. Engineered wood products consume less wood and can be made from smaller trees than can traditional lumber. Traditional lumber processes typically convert 40 percent of a log to structural solid lumber. Engineered lumber processes convert up to 75 percent of a log into structural lumber. In addition, the manufacturing processes for engineered lumber consume less energy than those for solid lumber. Also, some EWPs make use of abundant, fast-growing species not currently harvested for solid lumber.

The final engineered wood product has greater strength and consequently can span greater distances. It is predicted that engineered lumber will be used more than solid lumber in the near future. It is important that present and future builders be thoroughly informed about engineered wood products.

This unit describes how engineered wood products are made, where they are used, and what sizes are available. Construction details for EWPs are shown in later units on floor, wall, and roof framing.

LAMINATED VENEER LUMBER

Laminated veneer lumber, commonly called LVL, is one of several types of engineered lumber products (**Figure 5–1**). It was first used to make airplane

FIGURE 5–1 Types of engineered lumber: wood I-joist, laminated strand lumber (LSL) rim board, laminated veneer lumber (LVL), LSL lintel, parallel strand lumber (PSL) beam and columns.

propellers during World War II. The world's first commercially produced LVL for building construction was patented as MICRO-LAM laminated veneer lumber in 1970. LVL is now widely used in wood frame construction.

Manufacture of LVL

Like plywood, LVL is a wood veneer product. The grain in each layer of veneer in LVL runs in the same direction, parallel to its length (**Figure 5–2**). This is

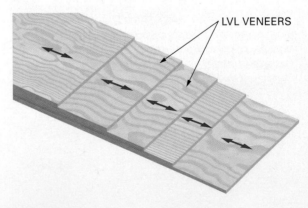

LVL VENEERS

FIGURE 5–2 In LVL, the grain in each layer of veneer runs in the same direction.

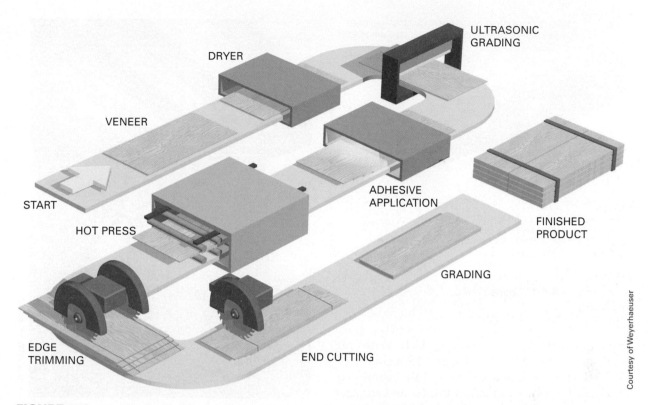

Courtesy of Weyerhaeuser

FIGURE 5–3 LVL manufacturing process, simplified. *(Courtesy of Weyerhaeuser)*

unlike plywood, in which each layer of veneer is laid with the grain at right angles to its neighbours.

Laminated veneer lumber is made from sheets of veneer peeled from logs, similar to the first step in the manufacture of plywood (see Figure 3–5). Douglas fir or southern pine is used because of its strength and stiffness. The veneer is peeled in widths of 27 or 54 inches (686 or 1372 mm) and from $1/10$ to $3/16$ inches (2.5 to 4.75 mm) in thickness. It is then dried, cut into sheets, ultrasonically graded for strength, and sorted (**Figure 5–3**).

The veneer sheets are laid in a staggered pattern so that the ends overlap. They are then permanently bonded together with an exterior-type adhesive in a continuous press under precisely controlled heat and pressure. Unlike plywood, the LVL veneers are *densified*; that is, the thickness is compressed and made more compact. Fifteen to twenty layers of veneer make up a typical 1¾-inch (44.5 mm) thick beam (**Figure 5–4**). The edges of the bonded veneers are then trimmed to specified widths and end-cut to specified lengths.

LVL Sizes and Uses

Laminated veneer lumber is manufactured up to 3½ inches (89 mm) thick, 18 inches (457 mm) wide, and 80 feet (24 m) long. The usual thicknesses are

FIGURE 5–4 Close-up view of LVL.

1½ inches (38 mm) and 1¾ inches (44.5 mm). The 1½-inch (38 mm) thickness is the same as nominal 2-inch (38 mm) framing lumber, while doubling the 1¾-inch (44.5 mm) thickness equals the width of

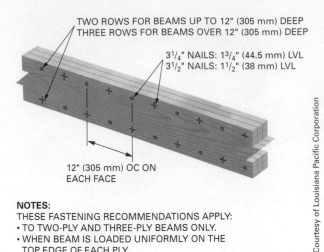

TWO ROWS FOR BEAMS UP TO 12" (305 mm) DEEP
THREE ROWS FOR BEAMS OVER 12" (305 mm) DEEP

3¼" NAILS: 1¾" (44.5 mm) LVL
3½" NAILS: 1½" (38 mm) LVL

12" (305 mm) OC ON
EACH FACE

NOTES:
THESE FASTENING RECOMMENDATIONS APPLY:
• TO TWO-PLY AND THREE-PLY BEAMS ONLY.
• WHEN BEAM IS LOADED UNIFORMLY ON THE
  TOP EDGE OF EACH PLY.

Courtesy of Louisiana Pacific Corporation

FIGURE 5–5 Recommended nailing pattern for fastening LVL beams together. *(Courtesy of Louisiana Pacific Corporation)*

nominal 4-inch (89 mm) framing. LVL widths are usually 9¼, 11¼, 11⅞, 14, 16, and 18 inches (235, 286, 302, 356, 406, and 457 mm). LVL beams may be fastened together to make a thicker and stronger beam (**Figure 5–5**).

Laminated veneer lumber is intended for use as high-strength, load-carrying beams to support the weight of construction over window and door openings, and in floor and roof systems (**Figure 5–6**). Although LVLs are heavier than dimension lumber, they are still easy to handle and suitable for use in concrete forming in manufactured housing and in many other specialties where a lightweight, strong beam is required. It can be cut with regular tools and requires no special fasteners. For technical specifications, see *http://www.woodbywy.com/trus-joist/microllam-lvl/*.

CROSS-LAMINATED TIMBER

Cross-laminated timber (CLT), developed in Switzerland in the early 1990s, improves stability and increases strength in framing systems. The added dimension of CLT is its multiple layers of wood, each layer oriented crosswise to the next, hence the name "cross-laminated." CLT can be used to support long spans in floors, walls, or roofs. CLT can arrive on-site prefinished, reducing on-site labour, and can either be applied to new construction or added to existing buildings. Visit naturallywood.com and search for "CLT" to find PDFs of technical data and videos of the product in action (**Figures 5–7** and **5–8**).

Courtesy of Weyerhaeuser

FIGURE 5–6 LVL is designed to be used for load-carrying beams and lintels.

www.naturallywood.com

FIGURE 5–7 Cross-laminated timbers.

FIGURE 5–8 Earth Systems Science Building, UBC.

FIGURE 5–9 CLT rig mat at 50 percent less weight and 25 percent more product per truckload than traditional rig mats.

According to Structurlam™, CLTs have the following benefits over other building materials (go to structurlam.com and select Products and then Cross Laminated Timber):

- Up to six times lighter than concrete
- Cost competitive against steel and concrete
- Reduces overall construction time
- One-third thinner than concrete
- Primarily requires carpentry skills and power tools

CLTs are also used as rig mats and access mats for large vehicles on job sites (**Figure 5–9**). See the video at rigmatsystems.com.

Chapter 6

Parallel Strand Lumber and Laminated Strand Lumber

PARALLEL STRAND LUMBER

Parallel strand lumber (PSL), commonly known by its brand name Parallam (**Figure 6–1**), was developed, like all engineered lumber products, to meet the needs of the building industry. PSL provides large-dimension lumber (beams, planks, and posts). PSL also utilizes small-diameter, second-growth trees, thus protecting the diminishing supply of old-growth trees.

FIGURE 6–1 Parallam® PSL, parallel strand lumber, is used as beams and posts to carry heavy loads. *(North Kamloops Library, photo by M. Nauth)*

North Kamloops Library, photo by M. Nauth

FIGURE 6–3 Close-up view of PSL.

Manufacture of PSL

Parallam PSL is manufactured by peeling the veneer of Douglas fir and southern pine from logs in much the same manner as for plywood and LVL. The veneer is then dried and clipped into *strands* (narrow strips) up to 8 feet (2.4 m) in length and ⅛ or ¹⁄₁₀ inch (3.2 or 2.5 mm) in thickness (**Figure 6–2**). Small defects are removed

and the strands are then coated with a waterproof adhesive. The oriented strands are fed into a rotary belt press and bonded using a patented microwave pressing process. The result is a continuous timber, up to 11 inches (279 mm) thick by 17 inches (432 mm) wide, which can then be factory-ripped into widths and thicknesses to fit builders' needs (**Figure 6–3**). The four surfaces are sanded smooth before the product is shipped.

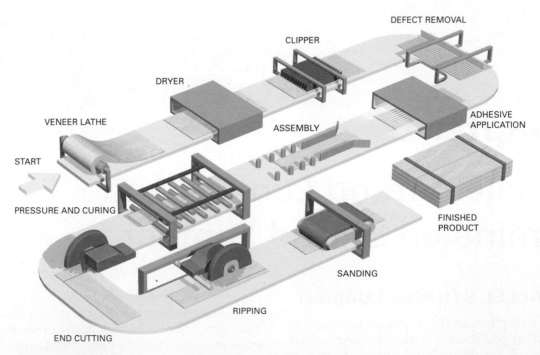

Courtesy of Weyerhaeuser

FIGURE 6–2 The manufacture of parallel strand lumber, simplified. *(Courtesy of Weyerhaeuser)*

PSL Sizes and Uses

PSL comes in many thicknesses and widths and is manufactured up to 66 feet (20.1 m) long. PSL is available in square and rectangular shapes for use as posts and beams. Beams are sold in a convenient 1¾-inch (44.5 mm) thickness for installation of single and multiple laminations. Solid 3½-inch (89 mm) thicknesses are compatible with 2 × 4 wall framing. A list of beam and post sizes is shown in **Figure 6–4.** Also available is Parallam 269, which measures 2¹¹/₁₆ inches (68 mm) thick.

Parallel strand lumber can be used wherever there is a need for a large beam or post. The differences between PSL and solid lumber are many. Solid lumber beams may have defects, such as knots, checks, and shakes, that weaken them, whereas PSL is consistent in strength throughout its length. PSL is readily available in longer lengths and its surfaces are sanded smooth, eliminating the need to cover them by boxing the beams. For technical specifications, visit *http://www.woodbywy.com/trus-joist/parallam-psl/TJ-9505.*

LAMINATED STRAND LUMBER

A registered brand name of laminated strand lumber (LSL) is *TimberStrand* (**Figure 6–5**). While LVL and PSL are made from Douglas fir and southern pine, LSL can be made from very small logs of practically any species of wood; its strands are much shorter than those of parallel strand lumber. At present, LSL is being manufactured from surplus, overmature aspen trees that usually are not large, strong, or straight enough to produce ordinary wood products.

Manufacture of LSL

The TimberStrand LSL manufacturing process begins by cleaning and debarking 8-foot (2.4 m) aspen logs (**Figure 6–6**). The wood is then cut into strands up to 12 inches (305 mm) long, dried, and treated with a resin. The treated strands are aligned parallel to one another to take advantage of the wood's natural strength. The strands are pressed into solid *billets* (large blocks) up to 5½ inches (140 mm) thick, 8 feet (2.4 m) wide, and 35 feet (10.67 m) long (**Figure 6–7**). Scraps from the process fuel the furnace that provides heat and steam for the plant.

PSL Column & Post Sizes	
Thickness	**Width**
3½″ (89 mm)	3½″ (89 mm)
3½″ (89 mm)	5¼″ (133 mm)
3½″ (89 mm)	7″ (178 mm)
5¼″ (133 mm)	5¼″ (133 mm)
5¼″ (133 mm)	7″ (178 mm)
7″ (178 mm)	7″ (178 mm)

PSL Beam Sizes	
Thickness	**Depths**
1¾″ (44.5 mm)	7¼″ (184 mm)
3½″ (89 mm)	9¼″ (235 mm)
5¼″ (133 mm)	11¼″ (286 mm)
7″ (178 mm)	11½″ (292 mm)
	11⅞″ (302 mm)
	12″ (305 mm)
	12½″ (318 mm)
	14″ (356 mm)
	16″ (406 mm)
	18″ (457 mm)

FIGURE 6–4 Available sizes of PSL used for posts and beams.

Trus Joist® TimberStrand® LSL Rim Board.
Courtesy of Weyerhaeuser

Light-Duty Structural Lintels

Stair Systems

Rim Board

Deck Ledger Attachment

FIGURE 6–5 The variety of uses of laminated strand lumber.

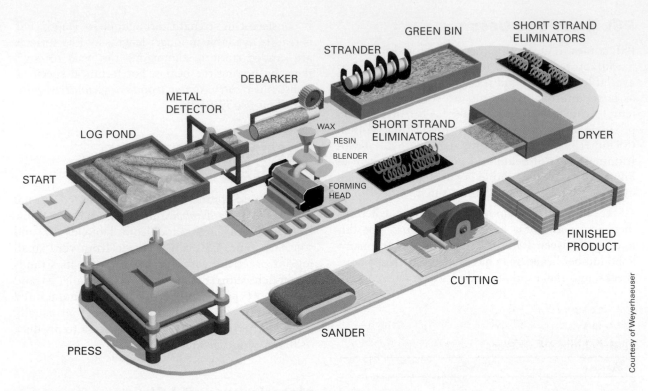

FIGURE 6–6 Laminated strand lumber can be manufactured from practically any species of wood. *(Courtesy of Weyerhaeuser)*

FIGURE 6–7 Close-up view of LSL.

LSL Sizes and Uses

The long billet is resawn and sanded to sizes as required by customers. It is used for a wide range of **millwork**, such as doors, windows, and virtually any product that requires high-grade lumber. It is also used for truck decks, manufactured housing, and some structural lumber, such as window and door headers. For technical specifications, see *http://www.woodbywy.com/trus-joist/timberstrand-lsl/*.

LSL is made from wood that is less strong and stiff than the wood used to make PSL. For this reason, LSL is only designed to serve as a product for light-duty structural lintels to carry lighter loads over shorter spans. It is often used as a **rim joist** attached to engineered floor joists. For more on engineered wood products, visit the Canadian Wood Council website at *www.cwc.ca* and search for "structural composite."

Chapter 7

Engineered Joists: Open Joist TRIFORCE®

OPEN JOIST TRIFORCE®

The Open Joist Triforce® floor **joist** was invented in the late 1980s as a substitute for open-metal web floor joists (**Figure 7–1**). **Open Joist** Triforce® is an engineered wood assembly that utilizes an efficient "I" shape for its diagonal conception, common in steel beams. This gives it tremendous strength in relation to its size and weight. Consequently, it is able to carry heavy loads over long spans while using considerably less wood than solid lumber used for the same load and span purposes.

The flanges of the joist are made of specially selected solid lumber—in some cases, **finger-jointed** lumber—for longer lengths of top and bottom flanges (**Figure 7–2**).

Courtesy of Open Joist TRIFORCE®

FIGURE 7–1 Open Joist Triforce® is available in many depths and lengths.

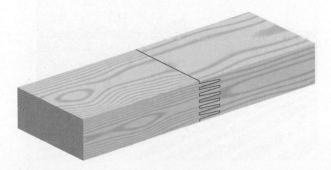

FIGURE 7–2 Finger joints are used to join the ends of short pieces of lumber to make a longer piece.

The Manufacturing Process

The manufacturing process of the Open Joist Triforce® floor joist consists of gluing top and bottom flanges to a series of vertical and diagonal webs using a finger-jointed system that is exclusive to Open Joist Triforce®. First, the vertical and diagonal webs are finger-jointed. One side of each of the top and bottom flanges is grooved to receive the vertical and diagonal webs. **Flanges** and **webs** are then assembled with a waterproof glue by pressure-fitting the webs into the flanges.

Cutting and grooving of joist components is automated to ensure precision crafting. A scanner then isolates any defective components so that they can be removed at the cutting stage prior to assembly. The final assembly produces a near-perfect product with very little adhesive residue and joist distortion (**Figure 7–3**). As with most other engineered wood products, Open Joist Triforce® is produced to an approximate equilibrium moisture content. After the required time for adequate curing, each Open Joist Triforce® is tested to 2.1 times its published design load (**Figure 7–4**).

Open Joist Triforce® — Sizes

Open Joist Triforce® webs are of various sizes (2 × 2, 3 × 2, 4 × 2, etc.) depending on their length or depth. Open Joist TRIFORCE® floor joists are available in depths from 9⅜ up to 16 inches (238 to 406 mm) (**Figure 7–5**). They are also manufactured from 3 to 30 feet (900 to 9150 mm) long, depending on their depths.

Open Joist Triforce® is intended for use as floor joists in the residential and commercial sectors. For span tables, see *http://www.hillsidelumber.net/Files/files/ojtech-77.pdf*.

Open Joist 2000® — Uses

The Open Joist 2000® floor joist has great flexibility for field installation because its light weight and open web system give the installer optimum facility when running electrical wires, plumbing pipes, and heating ducts through the floor joists (**Figure 7–6**).

FIGURE 7–3 The manufacturing process for the Open Joist TRIFORCE® is entirely automated, producing a high-precision product.

FIGURE 7–4 Each Open Joist Triforce® is tested to 2.1 times its published design load. *(Courtesy of Open Joist 2000 Inc.)*

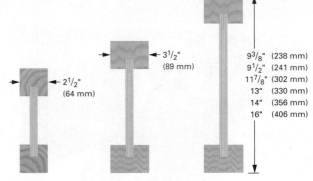

$2^1/_2$" (64 mm)

$3^1/_2$" (89 mm)

$9^3/_8$" (238 mm)
$9^1/_2$" (241 mm)
$11^7/_8$" (302 mm)
13" (330 mm)
14" (356 mm)
16" (406 mm)

FIGURE 7–5 Open Joist Triforce® is available in many depths and lengths.

FIGURE 7–6 Field installation of Open Joist TRIFORCE®. *(Courtesy of Open Joist 2000 Inc.)*

FIGURE 7–7 Open Joist TRIFORCE® is trimmable. *(Courtesy of Open Joist 2000 Inc.)*

The Open Joist Triforce® is easily cut to exact length, providing for a clean and easy field installation (**Figure 7–7**). For more information, search for "Open Joist 2000 from Toiture Mauricienne" at youtube.com.

WOOD-WEBBED JOISTS

Parallel chord trusses, similar in construction to triangular roof trusses, have the top and the bottom chord parallel to each other and wood webs on an angle in between, with metal or plywood gussets tying the members together at the connecting points. They can be used as floor trusses or as roof-trussed rafters. They are engineered at the truss plant to meet the building's specifications.

The Open Joist Triforce® is the only all-wood open-webbed joist. It consists of top and bottom chords that are parallel to each other and wood webs that are finger-jointed into the chords. The bearing ends have vertical 2 × 8 (38 × 184) blocks that can be trimmed down to a minimum of 2 inches (51 mm). The openings between the webs provide space for the subtrades to run heating, plumbing, electrical, and so on (**Figures 7–8** and **7–9**). For technical information, see *http://www.openjoisttriforce.com/open-joist-triforce/technical-information/*.

METAL-WEBBED JOISTS

The Space Joist TE® combines a wood I-joist end with metal webs that are pressed onto the sides of the top and bottom chords. The top and bottom

FIGURE 7–8 The Open Joist.

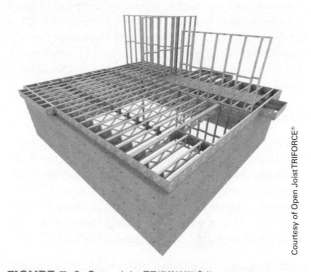

FIGURE 7–9 Open Joist TRIFORCE® floor system.

chords can be positioned either vertically or on the flat, depending on the load ratings. As with other engineered joists, these allow for greater spans, greater design flexibility, and the passage of mechanical systems within the joist space. View the Space Joist at *http://www.spacejoist.com/*.

Chapter 8

Glue-Laminated Lumber and Wood I-Joists

GLUE-LAMINATED LUMBER

Glue-laminated lumber, commonly called *glulam,* is constructed of solid lumber glued together, side against side, to make beams and joists of large dimensions that are stronger than natural wood of the same size (**Figure 8–1**). Even if it were possible, it would not be practical to make solid wood beams as large as most glulams. Glulams are used for structural purposes, but architectural appearance grade glulams are decorative as well, and in most cases, their surfaces are left exposed to show the natural wood grain.

Manufacture of Glulams

Glulam beams are made by gluing stacks of *lam stock into large beams,* **lintels,** and columns (**Figure 8–2**). Lams are the individual pieces in the glued-up stack. The lams, which are at approximate equilibrium moisture content, are glued together with exterior adhesives. Thus glulams are rated in terms of weather resistance (see *www.cwc.ca/products/glulam*).

Lam Layup. The lams are arranged in a certain way for maximum strength and minimum shrinkage. Different grade lams are placed where they will do the most good under load conditions.

High-grade *tension* lams are used in tension faces, and high-grade *compression* lams are used in compression faces.

Tension is a force applied to a member that tends to increase its length, while compression is a force tending to decrease its length (**Figure 8–3**). When a load is imposed on a glulam beam that is supported on both ends, the topmost lams are in compression while those at the bottom are in tension.

More economical lams are used in the lower stressed middle sections of the glulam. The sequence of lam grades, from bottom to top of a glulam, is referred to as *lam layup,* and it is a vitally important factor in glulam performance. Because of this lam layup, glulam beams come with one edge stamped "TOP." *Always remember to install glulam beams with the "TOP" stamp facing the sky* (**Figure 8–4**).

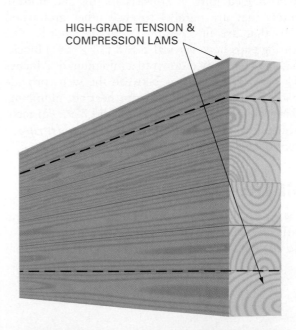

HIGH-GRADE TENSION & COMPRESSION LAMS

THE SEQUENCE OF LAM GRADES, FROM BOTTOM TO TOP OF A GLULAM, IS REFERRED TO AS A LAM LAYUP AND IS A VITALLY IMPORTANT FACTOR IN GLULAM PERFORMANCE.

FIGURE 8–2 Glulams are made by gluing stacks of solid lumber.

APA – The Engineered Wood Association

FIGURE 8–1 Glue-laminated lumber is commonly called glulam.

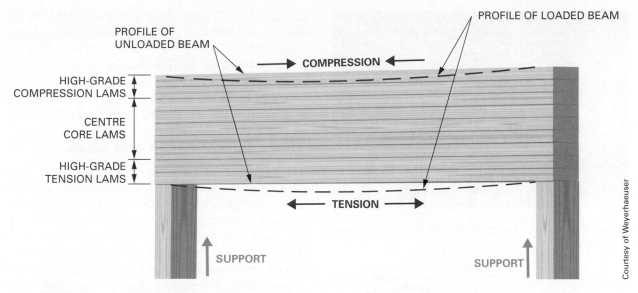

FIGURE 8–3 The load on a beam places lams in tension and compression.

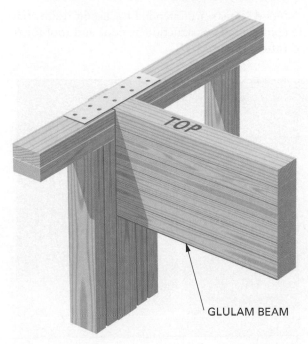

FIGURE 8–4 Glulam beams must be installed with the edge stamped "TOP" pointed toward the sky.

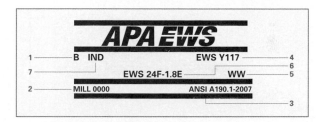

1) INDICATES STRUCTURAL USE:
   B - SIMPLE SPAN BENDING MEMBER.
   C - COMPRESSION MEMBER.
   T - TENSION MEMBER.
   CB- CONTINUOUS OR CANTILEVERED SPAN BENDING MEMBER.
2) MILL NUMBER.
3) IDENTIFICATION OF ANSI STANDARD A190.1, STRUCTURAL GLUED LAMINATED TIMBER.
4) CODE RECOGNITION OF QUALITY ASSURANCE AGENCY FOR GLUED STRUCTURAL MEMBERS.
5) APPLICABLE LAMINATING SPECIFICATION.
6) APPLICABLE COMBINATION NUMBER.
7) SPECIES OF LUMBER USED.
8) DESIGNATES APPEARANCE GRADE (INDUSTRIAL ARCHITECRAL, PREMIUM).

FIGURE 8–5 The grade stamp assures that the beam has met all the necessary requirements.

Glulam Grades

Canadian manufacturers of glulam are required to be qualified under CSA Standard 0177. This standard sets mandatory guidelines for equipment selection, manufacturing, testing, and record-keeping procedures (**Figure 8–5**).

Glulams are manufactured in three grades for appearance. The *industrial grade* is used in warehouses, garages, and other structures in which appearance is not of primary importance or when the beams are not exposed. *Commercial grade* is for projects where appearance is important, and the *quality grade* is used when appearance is critical (**Figure 8–6**). There is no difference in strength among the different appearance grades. All glulams with the same design values are rated for the same loadings, regardless of appearance grade.

Glulam Sizes and Uses

The dimensions of glulam beams are indicated by width, depth, and length (**Figure 8–7**). Widths range from 2½ to 8¾ inches (64 to 222 mm), depths from

FIGURE 8–6 Some glulam beams are also manufactured for appearance.

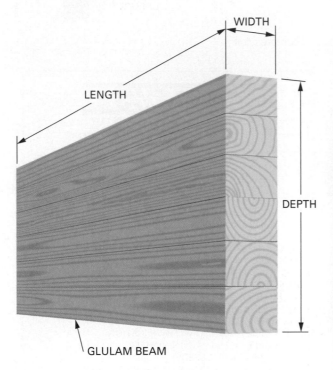

FIGURE 8–7 The dimensions of glulam beams are indicated by width, depth, and length.

6 to 28½ inches (150 to 724 mm), and lengths are generally available from 10 to 40 feet (3 to 12 m) in 2-foot (0.6 m) increments.

Various wood species are used to produce straight, curved, arched, and special shapes for all structures—from elegant homes and churches to large malls, warehouses, and civic centres.

WOOD I-JOISTS

The wood I-beam joist was invented in 1969 as a substitute for solid lumber joists. **Wood I-joists** are engineered wood assemblies that mirror the "I" shape of steel beams, which gives them tremendous strength in relation to their size and weight. So for less wood, they can carry the same load over the same span.

The top and bottom *flanges* of the joist are made of laminated veneer lumber, or specially rated finger-jointed solid wood lumber. The webs are most commonly made of oriented strand board (OSB).

Wood I-Joist Sizes and Uses

The webs of wood I-joists range in thickness from ³⁄₈ to ⁷⁄₁₆ inches (10 to 11 mm) and the flanges from 1½ to 3½ inches (38 to 89 mm). Joists are manufactured from depths of 9½ to 24 inches (240 to 600 mm). The joists with thicker webs and wider flanges are designed to carry heavier loads. Wood I-joists are available up to 80 feet (24.4 m) in length (**Figure 8–8**). For technical data, see *http://www.woodbywy.com/trus-joist/tji-joists/*.

Wood I-joists are intended for use in residential and commercial construction as floor and roof joists and rafters (**Figure 8–9**).

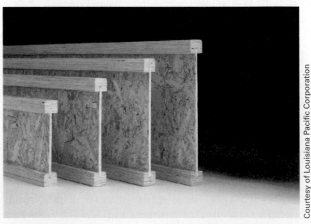

FIGURE 8–8 Wood I-joists are available in many sizes.

A

WARNING:
TEMPORARY CONSTRUCTION BRACING
REQUIRED FOR LATERAL SUPPORT BEFORE
DECKING IS COMPLETED. FAILURE TO USE
BRACING COULD RESULT IN SERIOUS INJURY
OR DEATH. SEE INSTALLATION GUIDE FOR
SPECIFICS.

B

Courtesy of Louisiana Pacific Corporation

FIGURE 8–9 Wood I-joists are used for rafters (A) and for floor joists (B).

FIRE SAFETY

Firefighter associations across Ontario have raised concerns about the performance of pre-engineered floor joists in a fire. The heat from the fire causes failure in sheet metal surface fasteners (metal gussets) and melts the glues in OSB webs, making the joists unstable and creating toxic fumes. Lumber floor joists burn in about 15 minutes whereas pre-engineered joists burn in 4 to 8 minutes. As a result, some fire departments have stopped entering front doors and are choosing to rescue people through windows, using ladders. Firefighters are calling for mandatory use of sprinklers in residential buildings.

The National Building Code of Canada requires a one-hour fire rating in floor-ceiling assemblies in multi-family structures that are four storeys or higher.

FIGURE 8–10 Trus Joist Flak Jacket Protection.

This has generally been achieved with a double-layer drywall system. The industry has responded with a coated I-joist that requires only a single layer of drywall (**Figure 8–10**). Some firefighter associations would like to see this requirement extended to all residential buildings with basements.

KEY TERMS

cross-laminated timber (CLT)

engineered wood products (EWPs)

finger-jointed

flanges

glue-laminated lumber

joist

laminated strand lumber (LSL)

laminated veneer lumber (LVL)

lintels

millwork

Open Joist

parallel strand lumber (PSL)

rim joist

webs

wood I-Joist

REVIEW QUESTIONS

Select the most appropriate answer.

1. Why is engineered wood produced today?
   a. because of the decreasing supply of old-growth trees
   b. because of the increasing supply of younger, smaller trees
   c. because of the need to make efficient use of natural resources
   d. all of the above.

2. How do LVL veneers differ from plywood?
   a. LVL veneers are made from deciduous wood.
   b. LVL veneers are densified.
   c. LVL veneers are diversified.
   d. LVL veneers are double-faced.

3. What is laminated veneer lumber often used for?
   a. high-strength load-carrying beams
   b. posts and columns
   c. low-strength rim joists
   d. floor joists

4. What is parallel strand lumber made from?
   a. Western red cedar
   b. Douglas fir
   c. Eastern hemlock
   d. Lodgepole pine

5. What is parallel strand lumber often used for?
   a. high-strength load-carrying beams
   b. posts and columns
   c. low-strength rim joists
   d. floor joists

6. What is laminated strand lumber often used for?
   a. high-strength load-carrying beams
   b. posts and columns
   c. low-strength rim joists
   d. floor joists

7. What are the webs of Open Joist TRIFORCE® made of?
   a. hardboard
   b. particleboard
   c. solid lumber
   d. strand board

8. What are the flanges of Open Joist TRIFORCE® made from?
   a. finger-jointed lumber
   b. laminated veneer lumber
   c. parallel strand lumber
   d. laminated strand lumber

9. What is the approximate moisture content of the lams used in glue-laminated lumber?

   a. the fibre-saturation point
   b. 19 percent moisture content
   c. equilibrium moisture content
   d. 30 percent moisture content

10. In glulam beams, where are specially selected tension lams placed?

    a. at the top of the beam
    b. at the bottom of the beam
    c. in the centre of the beam
    d. throughout the beam

WHAT'S WRONG WITH THIS PICTURE?

Carefully study **Figure 8–11** and think about what is wrong. Consider all possibilities.

FIGURE 8–11 The flange of a wood I-joist must never be cut.

FIGURE 8–12 Wood I-joists have perforated knockouts for plumbing pipes or electrical wiring, or another option is to attach the pipes to the bottom flange in situations where the ceiling will not be finished.

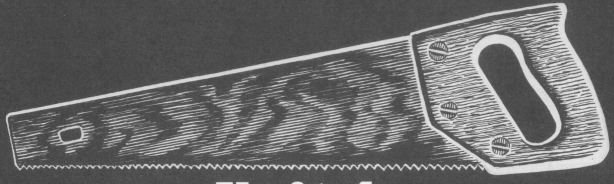

Unit 4
Fasteners

| Chapter 9 Nails, Screws, and Bolts | Chapter 10 Anchors and Adhesives |

The simplicity of fasteners can be misleading to students of construction. It is easy to believe only that nails are driven, screws are turned, and sticky stuff is used to glue. Although this tends to be true, joining material together so that it will last a long time is more challenging. Many times, a fastener is used for just one type of material. Some fasteners should never be used with certain other materials. The fastener selected often separates a quality job from a shoddy one.

Fasteners have been evolving for centuries. Today they come in many styles, shapes, and sizes, requiring different fastening techniques. It is important for the carpenter to know what types of fasteners are available, which securing technique should be employed, and how to wisely select the most appropriate fastener for various materials under different conditions.

There are many kinds and styles of fasteners made of various materials. Selecting the proper fastener is important for strength and durability.

OBJECTIVES

After completing this unit, the student should be able to name and describe the following commonly used fasteners and select them for appropriate use:
- nails
- screws and lag screws
- bolts
- solid and hollow wall anchors
- adhesives

Chapter 9
....
Nails, Screws, and Bolts

Nails, screws, and bolts are the most widely used of all fasteners. They come in many styles and sizes. The carpenter must know what is available and wisely select those most appropriate for fastening various materials under different conditions.

NAILS

There are hundreds of kinds of nails manufactured for just about any kind of fastening job. They differ according to purpose, shape, material, and coating, as well as in other ways. Nails are made of aluminum, brass, copper, steel, and other metals. Some nails are hardened so that they can be driven into masonry without bending. Only the most commonly used nails are described in this chapter (**Figure 9–1**).

Uncoated steel nails are called **bright** nails. Various coatings may be applied to reduce corrosion, increase holding power, and enhance appearance. To prevent rusting, steel nails are coated with *zinc*. These nails are called **galvanized** nails. They may be coated by being dipped in molten zinc (*hot-dipped galvanized nails*), or they may be electroplated with corrosion-resistant metal (*plated* nails). Hot-dipped nails have a heavier coating than plated nails and are more resistant to rusting. Many manufacturers specify that their products be fastened with hot-dipped nails because of the heavier coating.

When fastening metal that is going to be exposed to the weather, use nails of the same material. For example, when fastening aluminum, copper, or galvanized iron, use nails made of the same metal. Otherwise, a reaction with moisture and the two different metals, called **electrolysis**, will cause one of the metals to disintegrate over time.

When fastening some woods—such as cedar, redwood, and oak—that will be exposed to the weather, use stainless steel nails. Otherwise, a reaction between the acid in the wood and bright nails will cause dark stains to appear around the fasteners. For more information on the materials from which nails are made, visit the Canadian Wood Council website at *www.cwc.ca*, and search for "nails."

Kinds of Nails

Most nails, cut from long rolls of metal wire, are called **wire nails**. *Cut nails*, used only occasionally,

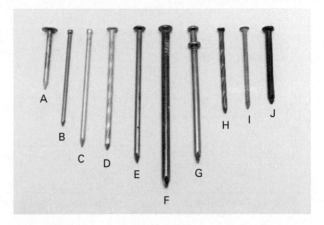

FIGURE 9–1 Kinds of commonly used nails: (A) roofing, (B) finish, (C) galvanized, (D) galvanized spiral, (E) box, (F) common, (G) duplex, (H) spiral, (I) coated box, and (J) masonry.

are wedge-shaped pieces stamped from thin sheets of metal. The most widely used wire nails are the common, box, and finish nails (**Figures 9–2 and 9–7**). **Figure 9–3** shows the metric and imperial sizes of nails.

Common Nails. Common nails are made of wire, are of heavy gauge, and have a medium-size head. They have a pointed end and a smooth shank. A barbed section just under the head increases the holding power of common nails.

Box Nails. Box nails are similar to common nails, except they are thinner. Because of their small gauge, they can be used close to edges and ends with less danger of splitting the wood. Many box nails are coated with resin cement to increase their holding power. **Ardox nails** or **spiral nails** will hold wood products together more securely than smooth nails and are permitted to be ¼ inch (6 mm) shorter than the minimums set by the National Building Code of Canada. They are commonly used in areas of high impact, such as subfloors.

Finish Nails. Finish nails are of light gauge with a very small head. They are used mostly to fasten interior trim. The small head is sunk into the wood with a nail set and covered with a filler. The small head of the finish nail does not detract from the appearance of a job as much as would a nail with a larger head.

Type of Nail	Head	Shank	Point	Material	Finishes and Coatings	Common Lengths	
						mm	in.
Common (spike) 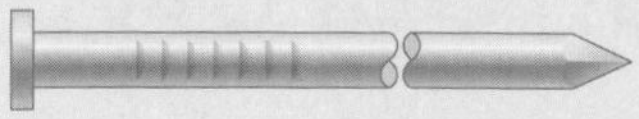	F	C, S	D	S, E	B	100–350	4–14
Eavestrough (spike) 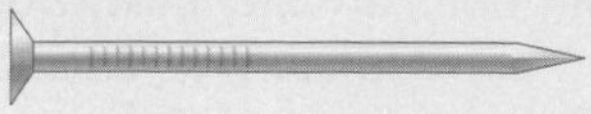	Cs, F	C, S	D, N	S	B, Ghd	125–250	5–10
Standard or Common 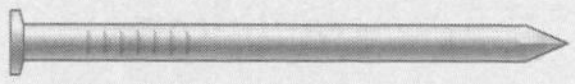	F	C, R, S	D	A, S, E	B, Ge	25–150	1–6
Box 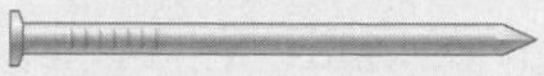	F, Lf	C, R, S	D	S	B, Pt, Ghd	19–125	3/4–5
Finishing 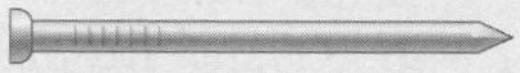	Bd	C, S	D	S	B, Bl	25–100	1–4
Flooring and Casing 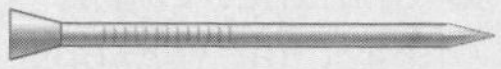	Cs	C, S	Bt, D	S	B, Bl, Ht	28–80	1-1/8–3-1/4
Concrete 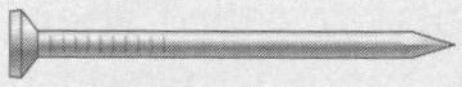	Cs	S	Con, Bt, D	Sc	Ht	13–75	1/2–3
Cladding and Decking 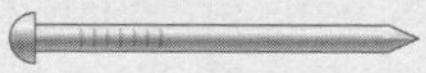	F, O	C, S	D	A, S	B, Ghd	50–63	2–2-1/2
Clinch 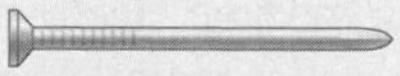	F, Lf	C, S	Db	S	B	19–63	3/4–2-1/2
Hardwood Flooring 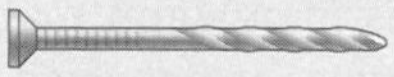	Cs	S	Bt	S	B, Ht	14–63	1-1/2–2-1/2
Gypsum Wallboard 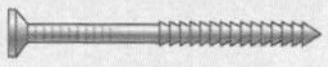	Dw, F	C, R, S	D, N	S	B, Bl, Ge	28–50	1-1/8–2
Underlay and Underlay Subfloor 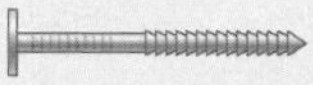	F, Cs	C, R	D	S	B, Ht	19–50	3/4–2
Roundwire Sash Pins 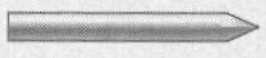	-	C	D	S	B	19–50	3/4–2
Roofing 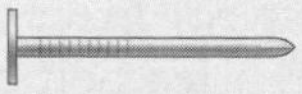	Lf, F	C	D	A, S	B, Ghd	19–50	3/4–2
Wood Shingle 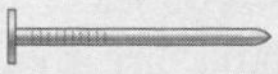	F	C, R, S	D	A, S	B, Ghd	31–44	1-1/4–1-3/4
Gypsum Lath 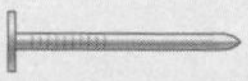	F	C, S	D, N	S	B, Bl, Ge	31	1–1/4
Wood Lath	F	C, S	D	S	Bl	25–28	1–1-1/8

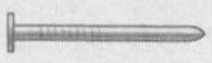

FIGURE 9–2 Nail types and sizes.

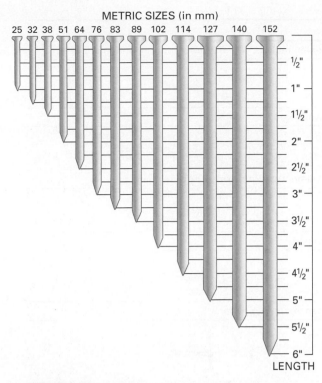

FIGURE 9–3 Metric and imperial sizes of nails.

FIGURE 9–5 (A) Finishing nails and (B) brad nails.

Casing Nails. Casing nails are similar to finish nails. Many carpenters prefer them to fasten exterior finish. The head is cone-shaped and slightly larger than that of the finish nail, but smaller than that of the common nail. The shank is the same gauge as that of the common nail.

Duplex Nails. On temporary structures, such as wood scaffolding and concrete forms, the **duplex nails** are often used. The lower head ensures that the piece is fastened tightly. The projecting upper head makes it easy to pry the nail out when the structure is dismantled (**Figure 9–4**).

Brads. Brads are small finishing nails (**Figure 9–5**). They are sized according to length in inches and gauge. Usual lengths are from ½ to 1½ inches (12.5 to 38 mm), and gauges run from #14 to #20 (1.6 mm to 0.8 mm in diameter). The higher the gauge number, the thinner the brad. Brads are used for fastening thin material, such as small moulding.

Roofing Nails. Roofing nails are short nails of fairly heavy gauge with wide, round heads. They are used for such purposes as fastening roofing material and softboard wall sheathing. The large head holds thin or soft material more securely.

Some roofing nails are coated to prevent rusting. Others are made from non-corrosive metals such as aluminum and copper. The shank is usually barbed to increase holding power. Usual sizes run from ¾ to 2 inches (19 to 51 mm). The gauge is not specified when ordering.

Masonry Nails. Masonry nails can be cut nails or wire nails (**Figure 9–6**). These nails are made from hardened steel to prevent them from bending when

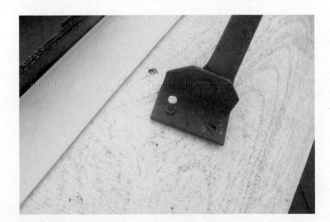

FIGURE 9–4 Duplex nails are used for temporary fastening. (Note: second nail was left out for clarity.)

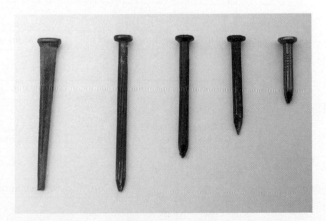

FIGURE 9–6 Masonry nails are made of hardened steel and may chip apart when driven.

Parts	Types	Abbr.	Remarks
Heads	Flat Counter-Sink	Cs	For nail concealment: light construction, flooring, and interior trim. 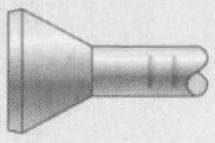
	Gypsum Wallboard	Dw	For gypsum wallboard. 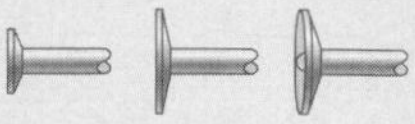
	Finishing	Bd	For nail concealment; cabinetwork, furniture. 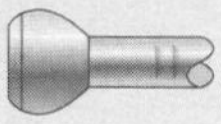
	Flat	F	For general construction. 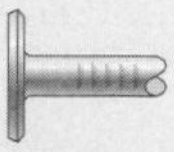
	Large Flat	Lf	For tear resistance; roofing paper. 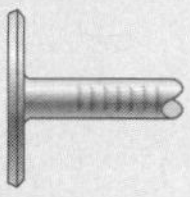
	Oval	O	For special effects; cladding and decking. 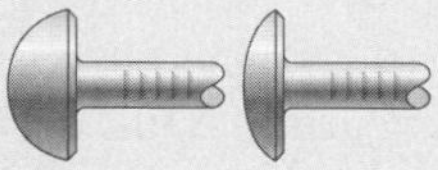
Shanks	Smooth	C	For normal holding power; temporary fastening. 
	Spiral or Helical	S	For greater holding power; permanent fastening. 
	Ringed	R	For highest holding power; permanent fastening. 
Points	Diamond	D	For general use, 35° angle; length about 1.5 x diameter. 
	Blunt Diamond	Bt	For harder wood species to reduce splitting, 45° angle. 
	Long Diamond	N	For fast driving, 25° angle; may tend to split harder species. 
	Duckbill	Db	For ease of clinching. 
	Conical	Con	For use in masonry; penetrates better than diamond. 

Courtesy of The Canadian Wood Council

FIGURE 9–7 Nail heads, shanks, and points.

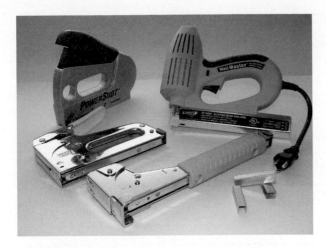

FIGURE 9–8 Staples can be driven by a variety of guns.

being driven into concrete or other masonry. The cut nail has a blunt point that tends to prevent splitting when it is driven into hardwood. Some masonry and flooring nails have round shanks of various designs for better holding power.

Staples. Staples are U-shaped fasteners used to secure a variety of materials. They come in a number of sizes and are designated by their length. Sold in boxes of 1,000 or more, they are driven by several types of tools, including the squeeze stapler, the hammer tacker, and electric staplers (**Figure 9–8**). They are used to fasten thin material.

Heavy-duty staples are similar in design but larger in size and gauge. These types of staples are often used to install some roofing materials or during cabinet construction (**Figure 9–9**).

FIGURE 9–9 Using an air stapler to fasten wood. WARNING! Be sure to keep other hand out of the line of fire.

SCREWS

The first wood screw (slotted) was invented in Germany by a clock maker who required fasteners that had greater holding power and that at the same time could be easily removed. Today, screws are still used for exactly the same reasons. Screws are defined by length, gauge, head, coating, kind of metal, and driver slot.

Kinds of Screws

The three most common screw heads are *flat*, used for counter-sinking flush with or below the surface; *oval*, used for partial countersinking; and *pan* or *round*, used when the head sits flat on the material it is fastening (**Figure 9–10**). The Robertson Screw company (established in Canada in 1908) has developed a flat-head screw that has four counter-sinking lugs under it that score out a recess for the head. The screw also has a Lo-Root® thread and the Aster® thread feature (**Figure 9–11**) that incorporates a serrated tooth thread design, which facilitates quicker starting, reduces splitting, and requires less drive torque. For more about the Robertson® screw, visit *http://www.robertsonscrew.com/*.

Screw Slots. Besides the straight slot, there are many other types of recesses for screws. In the building construction industry, the most commonly used are the *Phillips* for drywall and hardware and the *Robertson* (square drive) for all applications that require superior holding power. There are also the Recex and the Phillips Square Driv, both of which can be driven by either a Phillips or a Robertson® screwdriver. The Pozidriv is used mainly in Europe; the Torx is found on machine screws on power tools (see **Figures 9–12** to **9–15**).

More recently, Brad Wagner has developed the *LOX* screw, which comes collated and is driven by a specially designed screw gun. For high-torque applications such as metal frame houses, and for repetitive applications that require bending, such as subfloor installation, the LOX fastening system is proving to be the best so far. Installers are reporting less fatigue and better fastening, and driver bits are lasting eight to twelve times longer (**Figure 9–16**).

Screw Sizes

Screws usually range in length from ¼ to 4 inches (6 to 100 mm). The shanks have different diameters or gauges. Gauges run from 0 to 24, with the smaller numbers denoting thinner and usually shorter screws and the higher gauge numbers denoting longer, fatter

Part	Type	Use
Head Shapes	Flat	For countersinking flush with or below the surface. 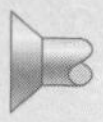
	Oval	For partial countersinking. 
	Pan	Recommended to replace round-headed screws; for use with washers or thin side pieces. 
Head Drive Shapes	Slot Recess	Common use.
		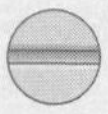 Slot
	Cross Recess	To minimize screwdriver slipout.
		 Phillips  Pozidriv
	Square Recess	To minimize screwdriver slipout.
		 Socket (Robertson)
Shanks	Double Lead	For faster turning; requires greater torque. 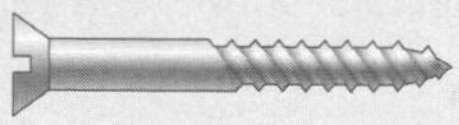
	Single Lead	For shorter screws (less than 25mm (1")). 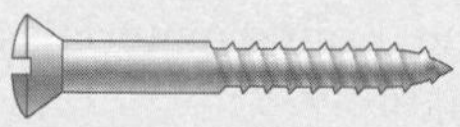
	Tapping	For better penetration; higher strength; designed for sheet metal but can be used with wood. 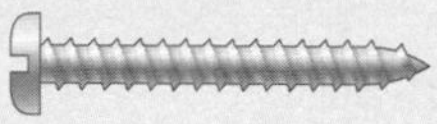
Points	Gimlet	For wood and some tapping screws. 
	Blunt	For some tapping screws.

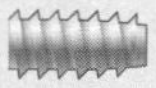

FIGURE 9–10 Types of screws.

Courtesy of The Canadian Wood Council

screws. For example, a kitchen cabinet hinge will use a $5/8$-inch (16 mm) No. 5 screw, whereas 2 × 6 (38 × 140 mm) decking lumber will require a 3½-inch (89 mm) No. 10 screw.

Sheet Metal Screws. Wood screws are threaded, in most cases, only partway to the head. The threads in sheet metal screws extend the full length of the screw and are much deeper. Sheet metal screws are used for fastening thin metal.

Another type of screw, used with power drivers, is the **self-tapping screw**, which is used extensively to fasten metal framing. This screw has a cutting edge

FIGURE 9–11 The Aster® thread.

Many special-purpose screws are available. Like nails, screws come in a variety of metals and coatings. Steel screws with no coating are called bright screws.

Lag Screws

Lag screws (**Figure 9–18**) are similar to wood screws except that they are larger and have a square or hex head designed to be turned with a wrench instead of a screwdriver. This fastener is used when great holding power is needed to join heavy parts and where a bolt cannot be used.

Lag screws are sized by diameter and length. Diameters range from ¼ to 1 inch (6 to 25 mm), with lengths from 1 to 12 inches (25 to 305 mm) and up. Shank and pilot holes are drilled to receive lag screws in the same manner as for wood screws (see Chapter 13). Place a flat washer under the head to prevent the head from digging into the wood as the lag screw is tightened down. Apply a little wax to the threads to allow the screw to turn more easily and to prevent the head from twisting off.

on its point, which eliminates the need to pre-drill a hole (**Figure 9–17**). It is important that drilling be completed before threading begins. Drill points are available in various lengths and must be equal to the thickness of the metal being fastened.

FIGURE 9–12 Pozidriv®.

FIGURE 9–13 Robertson® deck screws.

Courtesy of Robertson Screw Inc.

FIGURE 9–14 Recex Maxx Lo-root® screws with Aster® thread.

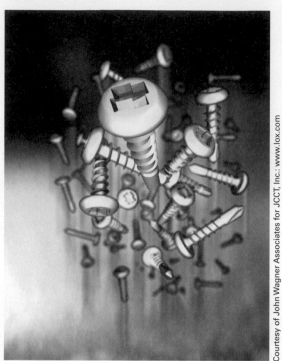

Courtesy of John Wagner Associates for JCCT, Inc.: www.lox.com

FIGURE 9–16 LOX® Screw.

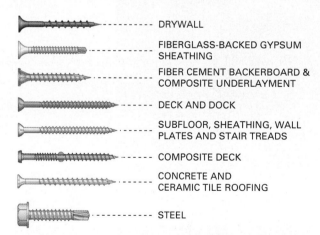

— DRYWALL

— FIBERGLASS-BACKED GYPSUM SHEATHING

— FIBER CEMENT BACKERBOARD & COMPOSITE UNDERLAYMENT

— DECK AND DOCK

— SUBFLOOR, SHEATHING, WALL PLATES AND STAIR TREADS

— COMPOSITE DECK

— CONCRETE AND CERAMIC TILE ROOFING

— STEEL

FIGURE 9–15 Screws come in a variety of styles for different applications.

FIGURE 9–17 The self-tapping screws have piercepoint or drill-point tips.

BOLTS

Most bolts are made of steel. To retard rusting, galvanized or stainless steel bolts are used. As with nails and screws, they are available in different kinds of metals and coatings. Many kinds are used for special

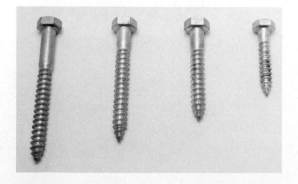

FIGURE 9–18 Lag screws are large screws with a square or hex head.

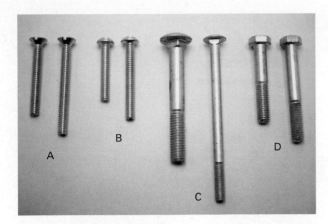

FIGURE 9–19 Commonly used bolts include (A) flat-head stove, (B) round-head stove, (C) carriage, and (D) machine.

purposes, but only a few are generally used. Commonly used bolts are the carriage, machine, and stove bolts (**Figure 9–19**).

Carriage Bolts

The **carriage bolt** has a square section under its oval head. The square section is embedded in wood and prevents the bolt from turning as the nut is tightened.

Machine Bolts

The **machine bolt** has a square or hex head. This is held with a wrench to keep the bolt from turning as the nut is tightened.

Stove Bolts

Stove bolts have either round or flat heads with a screwdriver slot. They are usually threaded all the way up to the head. *Machine screws* are very similar to stove bolts.

Bolt Sizes

Bolt sizes are specified by diameter and length. Carriage and machine bolts range from ¾ to 20 inches (19 to 500 mm) in length and from $^{13}/_{16}$ to ¾ inch (5 to 19 mm) in diameter. Stove bolts are small in comparison to other bolts. They commonly come in lengths from $^3/_8$ to 6 inches (9.5 to 152 mm) and from $^1/_8$ to $^3/_8$ inch (3 to 9.5 mm) in diameter.

Drill holes for bolts the same diameter as the bolt. Use flat washers under the head (except for carriage bolts) and under the nut to prevent the nut from cutting into the wood and to distribute the pressure over a wider area. Be careful not to overtighten carriage bolts. The head needs only be drawn snug, not pulled below the surface.

Chapter 10
Anchors and Adhesives

ANCHORS

Special kinds of fasteners used to attach parts to solid masonry and hollow walls and ceilings are called **anchors**. There are hundreds of types available. Those most commonly used are described in this chapter.

Solid Wall Anchors

Solid wall anchors can be classified as heavy, medium, or light duty. Heavy-duty anchors are used to install such things as machinery, hand rails, dock bumpers, and storage racks. Medium-duty anchors can be used for hanging pipe and ductwork, securing window and door frames, and installing cabinets. Light-duty anchors are used for fastening such things as junction boxes,

bathroom fixtures, closet organizers, small appliances, smoke detectors, and other lightweight objects.

Heavy-Duty Anchors. The **wedge anchor** (**Figure 10–1**) is used when high resistance to pullout is required. The anchor and hole diameter are the same, simplifying installation. The hole depth is not critical as long as the minimum is drilled. Proper installation requires cleaning out the hole.

The **sleeve anchor** (**Figure 10–2**) and its hole size are the same and the hole depth need not be exact. After the anchor is inserted in the hole, it is expanded by tightening the nut. This anchor can be used in material such as brick that may have voids or pockets.

The **drop-in anchor** (**Figure 10–3**) consists of an expansion shield and a cone-shaped, internal

INSERT – CLEAN HOLE, THEN DRIVE THE ANCHOR FAR ENOUGH INTO THE HOLE SO THAT AT LEAST SIX THREADS ARE BELOW THE TOP OF THE SURFACE OF THE FIXTURE.

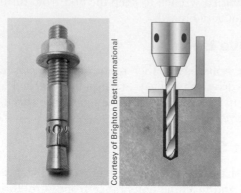
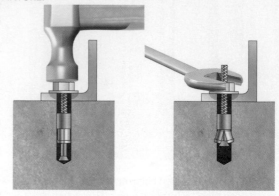

Courtesy of Brighton Best International

DRILL – SIMPLY DRILL A HOLE THE SAME DIAMETER AS THE ANCHOR. DO NOT WORRY ABOUT DRILLING TOO DEEP BECAUSE THE ANCHOR WORKS IN A "BOTTOMLESS HOLE." YOU CAN DRILL INTO THE CONCRETE WITH THE LOAD POSITIONED IN PLACE; SIMPLY DRILL THROUGH THE PREDRILLED MOUNTING HOLES.

ANCHOR – MERELY TIGHTEN THE NUT. RESISTANCE WILL INCREASE RAPIDLY AFTER THE THIRD OR FOURTH COMPLETE TURN.

FIGURE 10–1 The wedge anchor has high resistance to pullout.

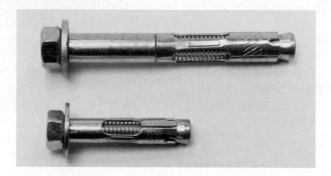

FIGURE 10–2 Sleeve anchors eliminate the problem of exact hole depth requirements.

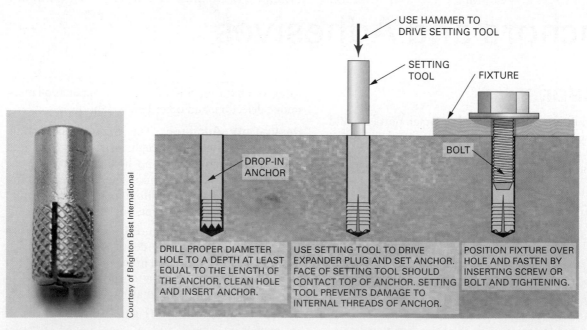

USE HAMMER TO DRIVE SETTING TOOL

SETTING TOOL

FIXTURE

DROP-IN ANCHOR

BOLT

Courtesy of Brighton Best International

DRILL PROPER DIAMETER HOLE TO A DEPTH AT LEAST EQUAL TO THE LENGTH OF THE ANCHOR. CLEAN HOLE AND INSERT ANCHOR.

USE SETTING TOOL TO DRIVE EXPANDER PLUG AND SET ANCHOR. FACE OF SETTING TOOL SHOULD CONTACT TOP OF ANCHOR. SETTING TOOL PREVENTS DAMAGE TO INTERNAL THREADS OF ANCHOR.

POSITION FIXTURE OVER HOLE AND FASTEN BY INSERTING SCREW OR BOLT AND TIGHTENING.

FIGURE 10–3 The drop-in anchor is expanded with a setting tool.

DRILL HOLE AND DRIVE ANCHOR WITH HAMMER
THROUGH FIXTURE AND INTO HOLE UNTIL FLUSH.

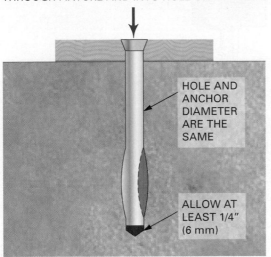

HOLE AND
ANCHOR
DIAMETER
ARE THE
SAME

ALLOW AT
LEAST 1/4"
(6 mm)

FIGURE 10–4 The split fast is a one-piece, all-steel
anchor for hard masonry.

FIGURE 10–6 Lag shields are designed for light- to
medium-duty fastening in masonry.

expander plug. The hole must be drilled at least equal
to the length of the anchor. A setting tool, supplied
with the anchors, must be used to drive and expand
the anchor. This anchor takes a machine screw or bolt.

Medium-Duty Anchors. Split fast anchors (**Figure 10–4**) are one-piece steel with two sheared
expanded halves at the base. When driven, these
halves are compressed and exert immense outward
force on the inner walls of the hole as they try to
regain their original shape. They come in both flat-
and round-head styles.

Single and *double expansion anchors* (**Figure 10–5**) are used with machine screws or bolts.

Drill a hole of recommended diameter to a depth
equal to the length of the anchor. Place the anchor
into the hole, flush with or slightly below the surface. Position the object to be fastened and bolt into
place. Once fastened, the object may be unbolted,
removed, and refastened, if desired.

The **lag shield** (**Figure 10–6**) is used with a lag
screw. The shield is a split sleeve of soft metal, usually
a zinc alloy. It is inserted into a hole of recommended
diameter and a depth equal to the length of the shield
plus ½ inch (12.5 mm) or more. The lag screw length
is determined by adding the length of the shield, the
thickness of the material to be fastened, plus ¼ inch
(6 mm). The tip of the lag screw must protrude from
the bottom of the anchor to ensure proper expansion. As the fastener is threaded in, the shield expands
tightly and securely in the drilled hole.

The *concrete screw* (**Figure 10–7**) utilizes specially fashioned high and low threads that cut into a
properly sized hole in concrete. Screws come in $^{13}/_{16}$

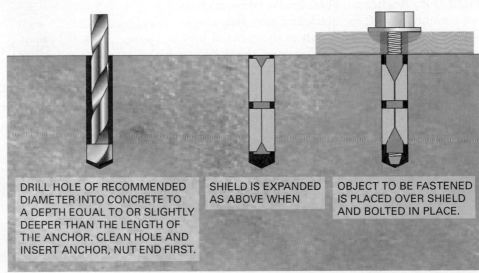

Courtesy of Brighton Best International

DRILL HOLE OF RECOMMENDED
DIAMETER INTO CONCRETE TO
A DEPTH EQUAL TO OR SLIGHTLY
DEEPER THAN THE LENGTH OF
THE ANCHOR. CLEAN HOLE AND
INSERT ANCHOR, NUT END FIRST.

SHIELD IS EXPANDED
AS ABOVE WHEN

OBJECT TO BE FASTENED
IS PLACED OVER SHIELD
AND BOLTED IN PLACE.

FIGURE 10–5 Two opposing wedges of the double expansion anchor pull toward each other, expanding the full length of
the anchor body.

 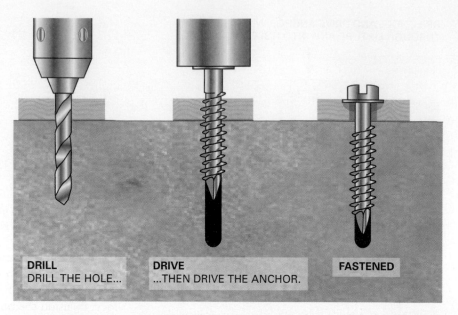

DRILL	DRIVE	FASTENED
DRILL THE HOLE...	...THEN DRIVE THE ANCHOR.	

FIGURE 10–7 Concrete screws eliminate the need for an anchor when fastening into concrete.

and ¼ inch (5 and 6 mm) diameters and up to 6 inches (152 mm) in length. The hole diameter is important to the performance of the screw. It is recommended that a minimum of 1-inch (25 mm) and a maximum of 1¾-inch (44 mm) embedment be used to determine the fastener length. The concrete screw system eliminates the need for plastic or lead anchors.

The *machine screw anchor* (**Figure 10–8**) consists of two parts. A lead sleeve slides over a threaded, cone-shaped piece. Using a special setting punch that comes with the anchors, the lead sleeve is driven over the cone-shaped piece to expand the sleeve and hold it securely in the hole. A hole of recommended diameter is drilled to a depth equal to the length of the anchor.

Light-Duty Anchors. Three kinds of **drive anchors** are commonly used for quick and easy fastening in solid masonry. They differ only in the material from which they are made. The *hammer drive anchor* (**Figure 10–9**) has a body of zinc alloy containing a steel expander pin. In the *aluminum drive anchor*, both the body and the pin are aluminum to avoid

the corroding action of electrolysis. The *nylon nail anchor* utilizes a nylon body and a threaded steel expander pin. All are installed in a similar manner.

Lead and *plastic anchors*, also called *inserts* (**Figure 10–10**), are commonly used for fastening

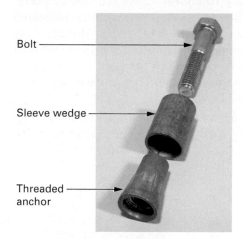

Bolt

Sleeve wedge

Threaded anchor

FIGURE 10–8 Machine screw anchors have a threaded bottom held by a sleeve wedged over the top.

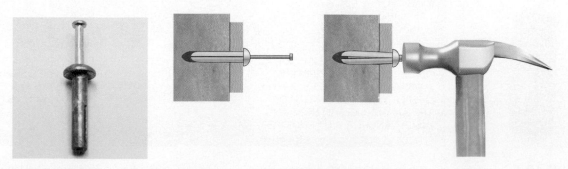

FIGURE 10–9 Hammer drive anchors come assembled for quick and easy fastening.

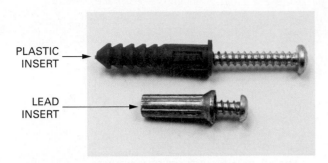

FIGURE 10–10 Lead and plastic anchors or inserts are used for light-duty fastening.

lightweight fixtures to masonry walls. These anchors have an unthreaded hole into which a screw is driven. The anchor is placed into a hole of recommended diameter and ¼ inch (6.5 mm) or more deeper than the length of the anchor. As the screw is turned, the threads of the screw cut into the soft material of the insert. This causes the insert to expand and tighten in the drilled hole. Ribs on the sides of the anchors prevent them from turning as the screw is driven.

Chemical Anchoring Systems. Threaded studs, bolts, and rebar (concrete reinforcing rod) may be anchored in solid masonry with a chemical bond using an *epoxy resin compound*. Two types of systems commonly used are the *epoxy injection* (**Figure 10–11**) and *chemical capsule* (**Figure 10–12**).

In the injection system, a dual cartridge is inserted into a tool similar to a caulking gun. The chemical is automatically mixed as it is dispensed. Small cartridges are available to accurately dispense epoxy from an ordinary caulking gun.

EPOXY INJECTION

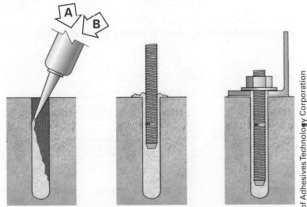

MIX THE TWO-COMPONENT ADHESIVE SYSTEM AND PLACE IN HOLE. PUSH THE ANCHOR ROD INTO THE HOLE AND ROTATE SLIGHTLY TO COAT WITH ADHESIVE. ALLOW TO CURE.

FIGURE 10–11 The epoxy injection system is designed for high-strength anchoring.

CHEMICAL CAPSULE ANCHOR

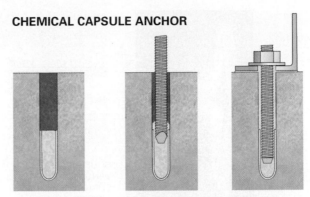

ATTACH THE ANCHOR ROD TO A ROTARY HAMMER ADAPTER. THE COMBINATION OF THE ROTATION AND HAMMERING ACTION MIXES THE CAPSULE CONTENTS TOGETHER. ALLOW TO CURE.

FIGURE 10–12 The chemical capsule anchoring system provides for easy, pre-measured application.

Chemical capsules contain the exact amount of all chemicals needed for one installation. Each capsule is marked with the appropriate hole size to be used. Drill holes according to diameter and depth indicated on each capsule. It is important to thoroughly clean and clear the hole of all concrete dust before inserting the capsule.

With all solid masonry anchors, follow the specifications in regard to hole diameter and depth, minimum embedment, maximum fixture thickness, and allowable load on anchor.

Hollow Wall Fasteners

Toggle Bolts. Toggle bolts (**Figure 10–13**) may have a wing or a tumble toggle. The wing toggle is fitted with springs, which cause it to open. The tumble toggle falls into its open position when passed through a drilled hole in the wall. The hole must be drilled large enough for the toggle of the bolt to slip

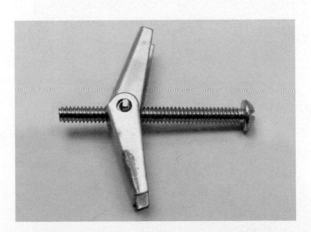

FIGURE 10–13 Toggle bolts are used for fastening in hollow walls.

Courtesy of Adhesives Technology Corporation

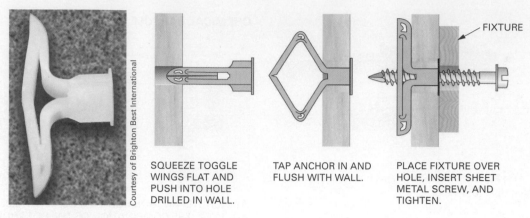

SQUEEZE TOGGLE WINGS FLAT AND PUSH INTO HOLE DRILLED IN WALL.

TAP ANCHOR IN AND FLUSH WITH WALL.

PLACE FIXTURE OVER HOLE, INSERT SHEET METAL SCREW, AND TIGHTEN.

FIXTURE

FIGURE 10–14 The plastic toggle hollow wall anchor.

through. A disadvantage of using toggle bolts is that, if removed, the toggle falls off inside the wall.

Plastic Toggles. The plastic toggle (Figure 10–14) consists of four legs attached to a body that has a hole through the centre and fins on its side to prevent turning during installation. The legs collapse to allow insertion into the hole. As sheet metal screws are turned through the body, they draw in and expand the legs against the inner surface of the wall.

Expansion Anchors. Hollow wall expansion anchors (Figure 10–15) consist of an expandable sleeve, a machine screw, and a fibre washer. The collar on the outer end of the sleeve has two sharp prongs that grip into the surface of the wall material. This prevents the sleeve from turning when the screw is tightened to expand the anchor. After the sleeve is expanded, the screw is removed, inserted through the part to be attached, and then screwed back into the anchor. Some types require that a hole be drilled, while other types have pointed ends that may be driven through the wall material.

Installed fixtures may be removed and refastened or replaced by removing the anchor screw without disturbing the anchor expansion. Anchors are manufactured for various wall board thicknesses. Make

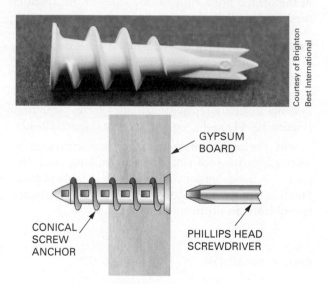

GYPSUM BOARD

CONICAL SCREW ANCHOR

PHILLIPS HEAD SCREWDRIVER

DRIVE ANCHOR IN WALL BY TURNING WITH SCREWDRIVER UNTIL HEAD IS FLUSH WITH SURFACE.

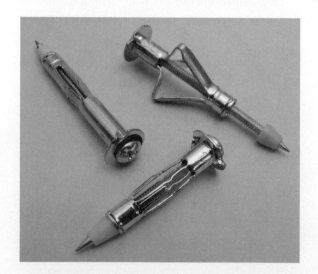

FIGURE 10–15 Hollow wall expansion anchors.

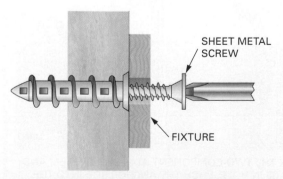

SHEET METAL SCREW

FIXTURE

PLACE FIXTURE OVER HOLE IN ANCHOR AND FASTEN WITH PROPER SIZE SHEET METAL SCREW.

FIGURE 10–16 The conical screw anchor is a self-drilling, hollow wall anchor for lightweight fastenings.

sure to use the right size of anchor for the wall thickness in which the anchor is being installed.

Conical Screws. The deep threads of the conical screw anchor (**Figure 10–16**) resist stripping out when screwed into gypsum board, strand board, and similar material. After the plug is seated flush with the wall, the fixture is placed over the hole and fastened by driving a screw through the centre of the plug.

Nylon Plugs. The nylon plug (**Figure 10–17**) is used for a number of hollow wall and some solid wall applications. A hole of proper diameter is drilled. The plug is inserted, and the screw is driven to draw or expand the plug.

Connectors

Widely used in the construction industry are devices called **connectors**. Connectors are metal pieces formed into various shapes to join wood to wood, or wood to concrete or other masonry. Connectors are available in hundreds of shapes and styles and have specific names depending on their function. Only a few are discussed here.

Wood-to-Wood. *Framing anchors* and *seismic* and *hurricane ties* (**Figure 10–18**) are used to join parts of a wood frame. *Post* and *column caps* and *bases* are used at the top and bottom of those members (**Figure 10–19**). *Joist hangers* and *beam hangers*

FIGURE 10–18 Framing ties and anchors are manufactured in many unique shapes.

Courtesy of Simpson Strong-Tie Company

Courtesy of Brighton Best International

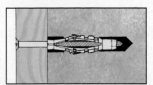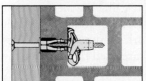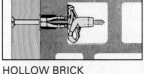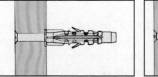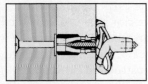

HOLLOW BRICK   PLASTER BOARD

HIGH VALUES IN PLASTER   AERATED CONCRETE

FIGURE 10–17 The nylon plug is used for many types of hollow wall fastening.

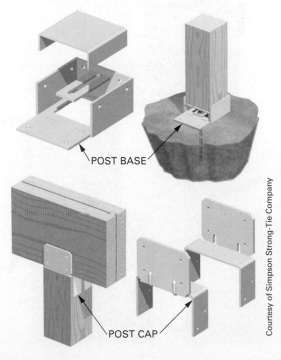

POST BASE

POST CAP

Courtesy of Simpson Strong-Tie Company

FIGURE 10–19 Caps and bases help fasten tops and bottoms of posts and columns.

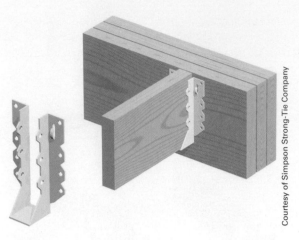

FIGURE 10–20 Hangers are used to support joists and beams.

Courtesy of Simpson Strong-Tie Company

are available in many sizes and styles (**Figure 10–20**). It is important to use the proper style, size, and quantity of nails in each hanger.

Wood-to-Concrete. Some wood-to-concrete connectors are *sill anchors*, *anchor bolts*, and *holddowns* (**Figure 10–21**). A *girder hanger* and a *beam*

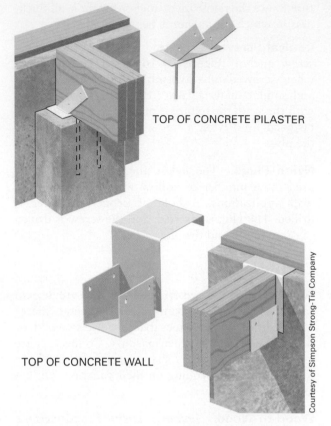

TOP OF CONCRETE PILASTER

TOP OF CONCRETE WALL

Courtesy of Simpson Strong-Tie Company

FIGURE 10–22 Girder and beam seats provide support from concrete walls.

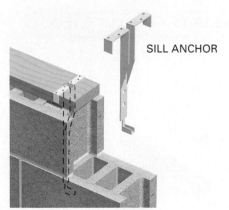

SILL ANCHOR

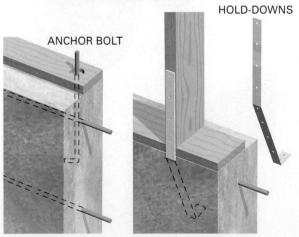

HOLD-DOWNS

ANCHOR BOLT

Courtesy of Simpson Strong-Tie Company

FIGURE 10–21 Sill anchors, anchor bolts, and holddowns connect frame members to concrete.

seat (**Figure 10–22**) make beam-to-foundation wall connections. *Post bases* come in various styles. They are used to anchor posts to concrete floors or footings.

Many other specialized connectors are used in frame construction. Some are described in the framing sections of this book.

ADHESIVES

The carpenter seldom uses any glue in the frame or exterior finish. Glue is used on some joints and other parts of the interior finish work. A number of mastics (heavy, paste-like adhesives) are used throughout the construction process.

Glue

White and Yellow Glue. Most of the glue used by the carpenter is the so-called white glue or yellow glue. The white glue is *polyvinyl acetate*; the yellow glue is *aliphatic resin*. Neither type is resistant to moisture. Both are fast setting, so joints should be made quickly after applying the glue. They are available under a number of trade names and are excellent for joining wood parts not subjected to moisture.

In 1994, Franklin International developed the first one-part PVA (polyvinyl acetate) to qualify as a Type II water-resistant glue. It is ideal for outdoor furniture, birdhouses, mailboxes, and so on. In 2004, the same company developed a Type I waterproof glue that is water solvent (before it cures) and nontoxic. It is safer to use than traditional waterproof glues. Also, it sands easily. For a more detailed discussion, see *https://www.canadianwoodworking .com/get-more/glues*.

Urethane Glue. Urethane glue is a fine all-purpose glue available for bonding wood, stone, metal, ceramics, and plastics. Its strong, waterproof bond cures with exposure to moisture in material and air. It can be used for interior or exterior work and does not become brittle over time. It tends to expand while curing, filling gaps and spaces in material.

Because urethane glue sticks to just about anything, care should be taken when working with it. It cannot be dissolved by common solvents and cleanup can be difficult. It often requires days to scrape and rub it from skin.

Hot Melt Glue. Hot melt glue, heated and applied with a glue gun, has been an important tool for the upholsterer, the carpet installer, and the hobbyist. For woodworkers, it has been used as a "third hand," but with new glue sticks on the market, it is being used as a permanent fastener. There are hot melt urethane glue sticks that will form secure bonds to wood, metal, glass, ceramic, and many other materials (**Figure 10–23**).

Contact Cement. Contact cement is so named because pieces coated with it bond on contact and need not be clamped under pressure. It is extremely important that pieces be positioned accurately before contact is made. Contact cement is widely

FIGURE 10–23 Hot melt glue used as a permanent fastener.

used to apply plastic laminates for kitchen countertops. It is also used to bond other thin or flexible material that otherwise might require elaborate clamping devices.

Mastics

Several types of mastics are used throughout the construction trades. They come in cans or cartridges used in hand or air guns and in large quantities that are trowelled into place. With these adhesives, the bond is made stronger and fewer fasteners are needed.

Construction Adhesive. One type of mastic is called *construction adhesive*. It is used in a glued floor system, described in a following unit on floor framing. It can be used in cold weather, even on wet or frozen wood. For technical specifications, go to lepageproducts.com and search for "construction adhesive." It is also used on stairs to increase stiffness and eliminate squeaks (**Figure 10–24**).

Panel Adhesive. *Panel adhesive* is used to apply such things as wall panelling, foam insulation, gypsum board, and hardboard to wood, metal, and masonry. It is usually dispensed with a caulking gun. It is important to use the adhesives matched for the material being installed.

Trowelled Mastics. Other types of mastics may be applied by hand for such purposes as installing vinyl base, vinyl floor tile, or ceramic wall tile. A notched trowel is usually used to spread the adhesive. The depth and spacing of the notches along the edges of the trowel determine the amount of adhesive left on the surface.

It is important to use a trowel with the correct notch depth and spacing. Failure to follow recommendations will result in serious consequences. Too much adhesive causes the excess to squeeze out onto the finished surface. This leaves no alternative but to remove the applied pieces, clean up, and start over. Too little adhesive may result in loose pieces.

FIGURE 10–24 Applying panel adhesive to joists with a caulking gun for plywood subfloor.

KEY TERMS

anchors	contact cement	lag shield	spiral nails
ardox nails	drive anchors	machine bolt	split fast anchors
box nails	drop-in anchor	masonry nails	staples
brads	duplex nails	mastics	stove bolts
bright	electrolysis	nylon plug	toggle bolts
carriage bolt	expansion anchors	plastic toggle	urethane glue
casing nails	finish nails	roofing nails	wedge anchor
common nails	galvanized	self-tapping screw	wire nails
conical screw	hot melt glue	sleeve anchor	
connectors	lag screws	solid wall anchors	

REVIEW QUESTIONS

Select the most appropriate answer.

1. What is the metric equivalent to a 2½-inch nail?

   a. 32 mm
   b. 51 mm
   c. 63 mm
   d. 76 mm

2. What are fasteners coated with zinc to retard rusting called?

   a. coated
   b. dipped
   c. electroplated
   d. galvanized

3. What are brads?

   a. types of screws
   b. small box nails
   c. small finishing nails
   d. types of stove bolts

4. When a moisture-resistant exterior glue is required, what should be used?

   a. white glue
   b. urethane glue
   c. yellow glue
   d. rubber cement

5. What do many carpenters prefer to use casing nails to fasten?

   a. interior finish
   b. exterior finish
   c. door casings
   d. roof shingles

6. What does the blunt point on the end of a cut nail help with?

   a. driving the nail straight
   b. preventing splitting of the wood
   c. holding the fastened material more securely
   d. starting the nail in the material

7. What type of fastener should be used to hold together temporary structures such as wood scaffolding and concrete forms?

   a. common nails
   b. duplex nails
   c. galvanized nails
   d. spikes

8. What is the common name for a fastener with tapered threads?

   a. bolt
   b. screw
   c. lag
   d. all of the above

9. Which of the following is an example of a solid wall anchor?

   a. wedge anchor
   b. conical screw
   c. toggle bolt
   d. all of the above

10. Which of the following does the term *connector* refer to?

    a. a metal device used to fasten wood to masonry
    b. wire used to make nails and screws
    c. a worker using a nail or screw gun
    d. the male end of an extension cord

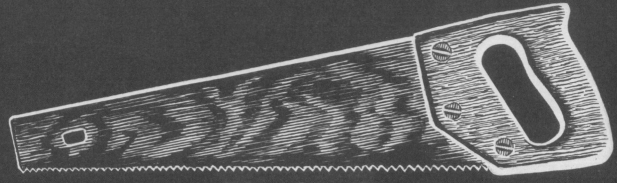

Unit 5

Hand Tools

One of the many benefits to working in the field of construction is the variety and diversity of tools available. Tools are the means by which construction happens.

Knowing how to choose the proper tool and how to keep it in good working condition is essential. A tradesperson should never underestimate the importance of tools and never neglect their proper use and care. Tools should be kept clean and in good condition. If they get wet on the job, dry them as soon as possible, and coat them with light oil to prevent them from rusting.

Carpenters are expected to have their own hand tools and to keep them in good working condition. Tools vary in quality, which is related to cost. Generally, expensive tools are of better quality than inexpensive tools. For example, inferior tools cannot be brought to a sharp, keen edge and will dull rapidly. They will bend or break under normal use. Quality tools are worth the expense. The condition of a tool reveals the attitude of the owner toward his or her profession.

OBJECTIVES

After completing this unit, the student should be able to:

- identify and describe the hand tools that are commonly used by the carpenter.
- use each of the hand tools in a safe and appropriate manner.
- sharpen and maintain hand tools in suitable working condition.

SAFETY REMINDER

Carpentry as a trade was created using hand tools, some having a long history. Each tool has a specific purpose and associated risk of use.

The use of tools requires the operator to be knowledgeable about how to safely manipulate the tools. This applies to hand tools as well as power tools. Safety is an attitude—an attitude of acceptance of a tool and all of its operational requirements. Safety is a blend of ability, skill, and knowledge—a blend that should always be present when working with tools.

Chapter 11

Layout Tools

LAYOUT TOOLS

Much of the work a carpenter does must first be laid out, measured, and marked. Layout tools are used to measure distances, mark lines and angles, test for depths, and align various materials into the proper positions.

The layout tools highlighted in this chapter will be calibrated in the imperial system of measure.

Measuring Tools

The ability to take measurements quickly and accurately must be mastered early in the carpenter's training. Practise reading the rule or tape to gain skill in fast and precise measuring.

Most industrialized countries use the metric system of measure (Système Internationale, SI). Linear metric measure centres on the metre, which is slightly longer than a yard. Smaller parts of a metre are denoted by the prefix *deci-* ($^1\!/\!_{10}$), which is used instead of *feet. Centi-* ($^1\!/\!_{100}$) and *milli-* ($^1\!/\!_{1000}$) are used instead of inches and fractions. The prefix *kilo-* represents 1,000 times larger, and *kilometres* is used instead of *miles.* For example, in metric measure a 2 × 4 is 38 mm (millimetres) × 89 mm. The metric system is easier to use than the imperial system because all measurements are in decimal form and there are no fractions.

Pocket Tapes. Most measuring done by tradespeople is done with a **pocket tape** (**Figure 11–1**). Notice in Figure 11–1 that the chrome tip of the tape slides on the three rivets. This displacement is equal to the thickness of the tip and provides exact readings of measurements when the tip is butted against a surface as well as when it is hooked onto a surface. These are painted steel ribbons wound around a spool with a spring inside. The spring returns the tape after it is extended. They are made as small as 3 feet but typical professional models are 16, 25, and 33 feet long.

They are divided into feet, inches, and sixteenths of an inch. They have clearly marked increments of 12 and 16 inches, the spacing for standard framing members, to speed up the layout. Markings are usually black for each 12 inches and red for every 16 inches. Some tapes also have small black dots at increments of 19.2 inches (**Figure 11–2**). This spacing is typically used only for layout of some engineered floor members. **Figure 11–3** shows a pocket tape measure with

Courtesy of Stanley Black & Decker

Photo courtesy of TEKTON

FIGURE 11–1 Pocket tape.

both imperial and metric measurements. Here we see that the metric measurements are in centimetres (cm) and that 1 inch is 2.5 cm or 25 mm.

Each inch on a tape is divided into fractions of an inch. Each fraction line has a name that must be memorized (see Figure 11–2). Most measuring done by a carpenter is to the nearest 16th, while a cabinet maker will work to a 32nd. A carpenter should be able to read a ruler quickly and accurately.

Steel tapes in 50- and 100-foot (15 and 30 m) lengths are commonly used to lay out longer measurements. They are not spring loaded, so they must be rewound by hand. The end of the tape has a steel ring with a folding hook attached. The hook can be unfolded to go over the edge of an object. It can also be left in the folded position and the ring placed over a nail when extending the tape. Remember to place the nail so that the *outside* of the ring, which is the actual end of the tape, reaches to the desired mark (**Figure 11–4**). Rewind the tape when not using it. If the tape is kinked, it will snap. Keep it out of water. If it gets wet, dry it thoroughly while rewinding.

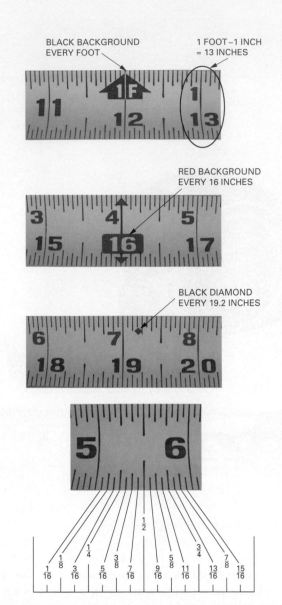

BLACK BACKGROUND
EVERY FOOT

1 FOOT–1 INCH
= 13 INCHES

RED BACKGROUND
EVERY 16 INCHES

BLACK DIAMOND
EVERY 19.2 INCHES

FIGURE 11–2 Tapes have colour-coded markings at 12-, 16- and 19.2-inch intervals. Each inch is typically broken into sixteenth of an inch increments.

FIGURE 11–3 Imperial/metric tape measure.

Courtesy of The Stanley Works

FIGURE 11–4 Steel tapes have ends with folding hooks. The carpenter must be aware of how to hold the tapes to maintain accuracy.

Squares

The carpenter has the use of a number of different kinds of squares to measure and lay out for square and other angle cuts.

Combination Squares. The combination square (**Figure 11–5**) consists of a movable blade, 1 inch (25 mm) wide and 12 inches (305 mm) long, that slides along the body of the square. It is used to lay out or test 90- and 45-degree angles. Hold the body of the square against the edge of the stock and mark along the blade (**Figure 11–6**). It can function as a depth gauge to lay out or test the depth of **rabbets**,

FIGURE 11–5 The body and blade of a combination square are adjustable.

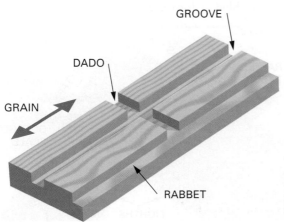

FIGURE 11–6 The combination square is useful for squaring and as a marking gauge. A pencil held in one hand is a quick way to draw a parallel line. Check the wood first to reduce the potential for splinters.

grooves, and **dadoes**. It can also be used with a pencil as a marking gauge to draw lines parallel to the edge of a board. Drawing lines in this manner is called *gauging* lines. Lines can also be gauged by holding the pencil and riding the finger along the edge of the board. Finger-gauging takes practice, but once mastered saves a lot of time. Be sure to check the edge of the wood for slivers first.

Speed Squares. Some carpenters prefer to use a triangular-shaped square known by the brand name **Speed Square** (**Figure 11–7**). Speed Squares are made of one-piece plastic or aluminum alloy and are available in two sizes. They can be used to lay out 90- and 45-degree angles and as guides for portable power saws. A degree scale allows angles to be laid out; other scales can be used to lay out rafters.

Try Squares. The **try square** (**Figure 11–8**) consists of a steel blade, graduated in inches or millimetres, that is fitted into a centre slot at the end of a handle, which is fastened rigidly at 90 degrees to the blade. It is used to measure, lay out, and verify the squareness of members and joints.

FIGURE 11–7 Speed Squares are used for layout of rafters and other angles.

Framing Squares. The **framing square**, often called the *steel* or *rafter square* (**Figure 11–9**), is an L-shaped tool made of thin steel or aluminum. The longer of the two legs is called the *blade* or *body* and is 2 inches (50 mm) wide and 24 inches (610 mm)

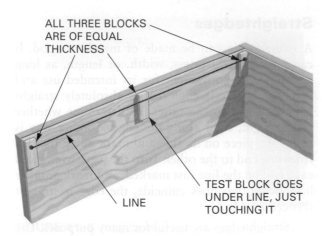

FIGURE 11–13 Use of a line and gauge blocks is an effective method to determine a straight line.

FIGURE 11–14 Trammel points are used to lay out arcs of large diameter.

Photo by M. Nauth

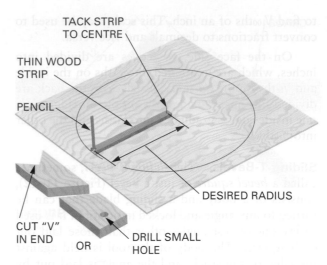

FIGURE 11–15 A thin strip of wood can be used to lay out circles or arcs.

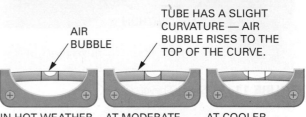

REGARDLESS OF CONDITIONS, THE AIR BUBBLE MUST BE CENTRED BETWEEN THE TWO LINES ON THE TUBE.

FIGURE 11–16 The bubble size of a carpenter's level can be affected by temperature.

other, which may have a pencil attached, is swung to lay out the circle or arc.

In place of trammel points, the same kinds of layouts can be made by using a thin strip of wood with a brad or small finish nail through it for a centre point. Measure from the end of the strip a distance equal to the desired radius. Drive the brad through the strip until the point comes through. Set the point of the brad on the centre, and hold a pencil against the end while swinging the strip to form the circle or arc (**Figure 11–15**). To keep the pencil from slipping, a small V can be cut on the end of the strip or a hole may be drilled near the end to insert the pencil. Make sure that measurements are taken from the bottom of the V or the centre of the hole.

Levels

In construction, the term **level** is used to indicate that which is *horizontal*, and the term **plumb** is used to mean the same as *vertical*. The term *level* also refers to a tool that is used to achieve both level and plumb.

Carpenter's Levels. The **carpenter's level** (**Figure 11–16**) is used to test both level and plumb

surfaces. Accurate use of the level depends on accurate reading. The air bubble in the slightly crowned glass tube of the level must be exactly centred between the lines marked on the tube. The tubes of a level are oriented in two directions for testing level and plumb. The number of tubes in a level depends on the level length and manufacturer.

Levels are made of wood or metal, usually aluminum. They come in various lengths from 12 to

ON THE JOB

Care must be taken not to drop the level because this could break the glass or disturb the accuracy of the level.

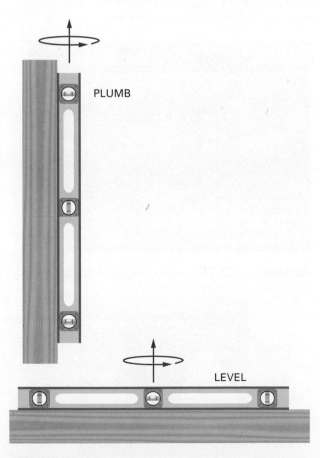

OUT OF PLUMB
BOWED EDGE
PLUMB
STRAIGHT EDGE
OUT OF PLUMB
LEVEL
STRAIGHT EDGE
BOWED EDGE
OUT OF LEVEL
OUT OF LEVEL

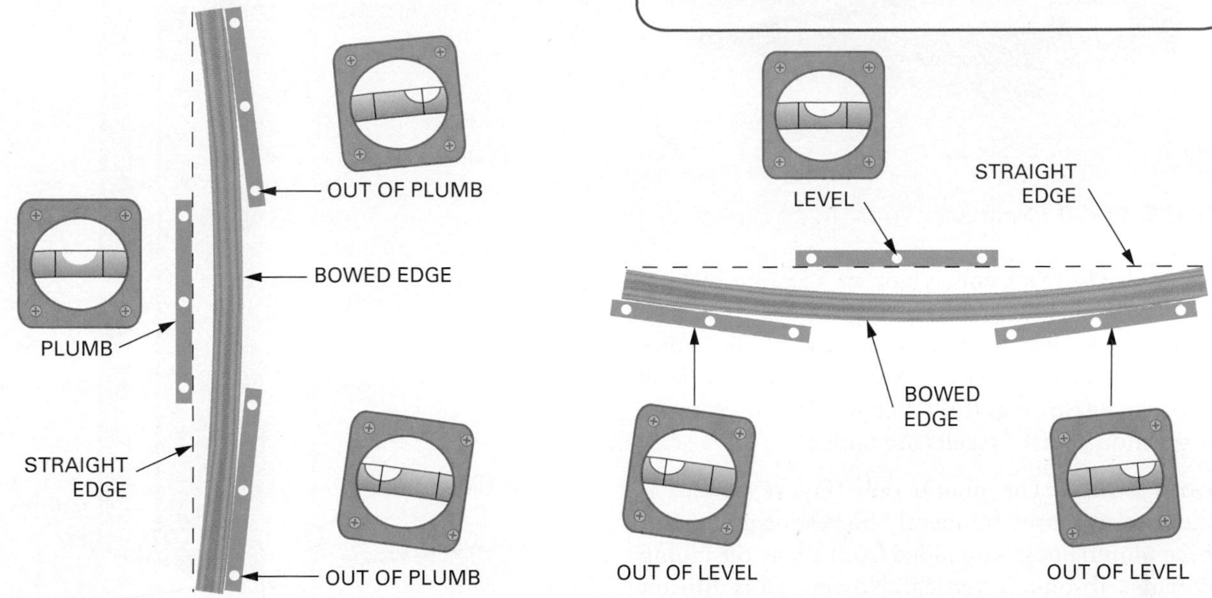

FIGURE 11–17 To be level or plumb for their entire length, pieces must be straight from end to end.

PLUMB
LEVEL

FIGURE 11–18 To check a level for accuracy the bubble should read exactly the same before and after rotating it.

78 inches (305 to 1981 mm). It is wise to use the longest level practical to improve accuracy.

An important point to remember is that level and plumb lines, or objects, must also be straight throughout their length or height. Parts of a structure may have their end points level or plumb with each other. If they are not straight in-between, however, they are not level or plumb for their entire length (**Figure 11–17**).

To check a level for accuracy, place it on a nearly level or plumb object that is firm. Note the exact position of the level on the object. Read the level carefully and remember where the bubble is located within the lines on the bubble tube. Rotate the level along its vertical axis and reposition it in the same place on the object (**Figure 11–18**). If the bubble reads the same as the previous measurement, then the level is accurate.

Line Levels. The line level (**Figure 11–19**) consists of one glass tube encased in a metal sleeve with hooks on each end. The hooks are attached to a stretched line, which is then moved up or down until the bubble

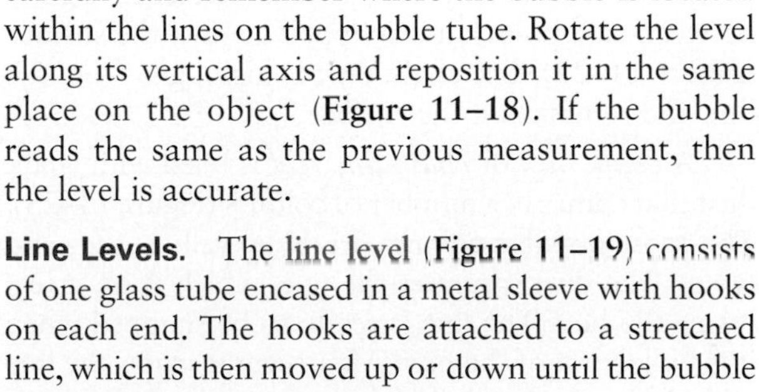

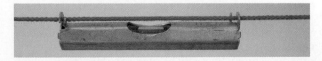

FIGURE 11–19 Line level.

FIGURE 11–20 Plumb bob.

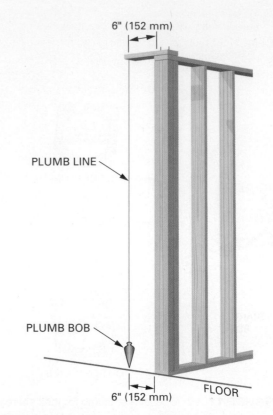

FIGURE 11–21 The post is plumb when the distance between it and the plumb line is the same.

is centred. However, this is not an accurate method and gives only approximate levelness. Care must be taken that the level be attached close to the centre of the suspended line, because the weight of the level causes the line to sag. If the line level is off centre to any great degree, the results are faulty.

Plumb Bobs. The **plumb bob** (**Figure 11–20**) is accurate and is used frequently for testing and establishing plumb lines. Suspended from a line, the plumb bob hangs absolutely vertical. However, it is difficult to use outside when the wind is blowing because it will move with the wind. Plumb bobs come in several different weights. Heavy plumb bobs stop swinging more quickly than lighter ones. Some have hollow centres that are filled with heavy metal to increase the weight without enlarging the size.

The plumb bob is useful for quick and accurate plumbing of vertical members of a structure (**Figure 11–21**). It can be suspended from a great height to establish a point that is plumb with another. Its only limitation is the length of the line.

Chalk Lines

Long straight lines are laid out by using a **chalk line**. A line coated with chalk dust is stretched tightly between two points and snapped against the surface (**Figure 11–22**). The chalk dust is dislodged from the line and remains on the surface.

A *chalk box* or *chalk line reel* is filled with chalk dust that comes in a number of colours (**Figure 11–23**). The most popular colours are blue, yellow, red, and white. The dust saturates the line, which is on a reel inside the box. The line is ready to be snapped when it is pulled out of the box. After several snaps, the line will need more chalk and will have to be reeled in to be recoated with chalk. Shaking or tapping the box helps recoat the line.

Chalk Line Techniques. When unwinding and chalking the line, keep it off the surface until snapped, otherwise many lines will be made on the surface, which could be confusing. Make sure that lines are stretched

FIGURE 11–22 Snapping a chalk line on a roof deck.

FIGURE 11–23 Chalk line reel.

tight before snapping in order to snap a straight and true line. Sight long lines by eye for straightness to make sure there is no sag in the line. If there is a sag, take it out by supporting the line near the centre. Press the centre of the line to the material and snap the line on both sides of the centre. Keep the line from getting wet. If it does get wet, leave it outside the box until it dries.

Wing Dividers

Wing dividers can be used as a compass to lay out circles and arcs and as dividers to space off equal distances. However, this tool is used mainly for **scribing** and is often called a *scriber*. Scribing is the technique of laying out stock to fit against an irregular surface (**Figure 11–24**). For easier and more accurate scribing, heat and bend the end of the solid metal leg outward (**Figure 11–25**). Pencils are usually used in place of the interchangeable steel marking leg. Use pencils with hard lead that keep their points longer.

Butt Markers

Butt markers (**Figure 11–26**) are available in three sizes. They are often used to mark hinge gains. The marker is laid on the door edge at the hinge location and tapped with a hammer to outline the cutout for the hinge.

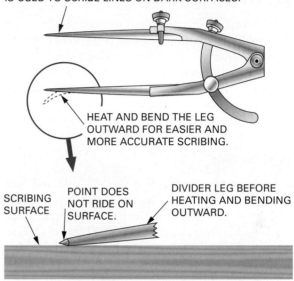

ON THE JOB

Bend the leg of the dividers as shown for easier and more accurate scribing.

METAL LEG IS INTERCHANGEABLE WITH PENCIL AND IS USED TO SCRIBE LINES ON DARK SURFACES.

HEAT AND BEND THE LEG OUTWARD FOR EASIER AND MORE ACCURATE SCRIBING.

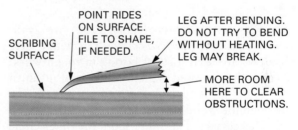

SCRIBING SURFACE

POINT DOES NOT RIDE ON SURFACE.

DIVIDER LEG BEFORE HEATING AND BENDING OUTWARD.

SCRIBING MAY NOT BE ACCURATE BECAUSE POINT IS NOT RIDING ON SURFACE.

POINT RIDES ON SURFACE. FILE TO SHAPE, IF NEEDED.

SCRIBING SURFACE

LEG AFTER BENDING. DO NOT TRY TO BEND WITHOUT HEATING. LEG MAY BREAK.

MORE ROOM HERE TO CLEAR OBSTRUCTIONS.

SCRIBING IS MORE ACCURATE WHEN POINT RIDES ON SURFACE.

FIGURE 11–25 Adjusting one of the metal legs of a scriber makes it a more accurate tool.

FIGURE 11–24 Scribing is laying out a piece to fit against an irregular surface.

FIGURE 11–26 Butt hinge markers make mortising the hinge into a door easier.

Chapter 12
Boring and Cutting Tools

BORING TOOLS

The carpenter is often required to cut holes in wood and metal. **Boring** denotes cutting large holes in wood. **Drilling** is often thought of as making holes in metal or small holes in wood. Boring tools include those that actually do the cutting and those used to turn the cutting tool. The hole size and the bit size are measured according to their diameters (see also Chapter 14).

Bit Braces

The **bit brace** (**Figure 12–1**) is used to hold and turn auger bits to bore holes in wood. Its size is determined by its *sweep* (the diameter of the circle made by its handle). Sizes range from 8 to 12 inches (203 to 305 mm). Most bit braces come with a ratchet that can be used when there is not enough room to make a complete turn of the handle.

Bits

Auger Bits. **Auger bits** (**Figure 12–2**) are available with coarse or fine *feed screws*. Bits with coarse feed screws are used for fast boring in rough work. Fine feed bits are used for slower boring in finish work. As the bit is turned, the feed screw pulls the bit through the wood so that little or no pressure on the bit is necessary. The *spurs* score the outer circle of the hole in advance of the *cutting lips*. The cutting lips lift the chip up and through the twist of the bit.

A full set of auger bits ranges in sizes from ¼ to 1 inch (6 to 25 mm), graduated in $^{1}/_{16}$-inch (2 mm) increments. The bit size is designated by the number of $^{1}/_{16}$-inch (2 mm) increments in its diameter. For instance, a #12 bit has 12 sixteenths. Therefore, it will bore a ¾-inch (19 mm) diameter hole.

Drills and Bits

Push Drills. Push drills are still being used by many cabinet makers and finish carpenters (**Figure 12–3**). Specially made bits are stored in the handle and they spin and cut into the wood as the drill is pushed repeatedly.

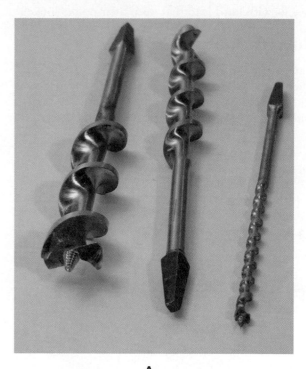

A

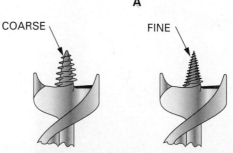

COARSE AND FINE AUGER BIT FEED SCREWS

B

FIGURE 12–2 (A) Auger bits and (B) coarse and fine feed screws.

FIGURE 12–1 Bit brace.

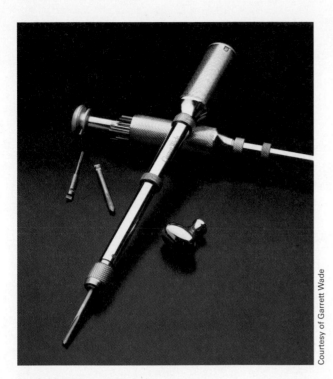

FIGURE 12–3 A push drill.

FIGURE 12–5 A brad point drill bit. *(Courtesy of Lee Valley Tools Ltd.; www.leevalley.com. © Copyright Lee Valley Tools Ltd.)*

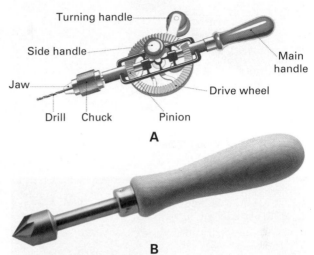

FIGURE 12–6 (A) Hand drill and countersink tool. (B) Countersink boring bit. *(Top: Reproduced with the permission of QA International, www.ikonet.com from the book "The Visual Dictionary." © QA International, 2003. All rights reserved. Bottom: Courtesy of Robert Larson Company)*

Twist Drills Bits. **Twist drills** range in size from $1/16$ to $1/2$ inch (1 to 13 mm) in increments of $1/64$ inch (1 mm) (**Figure 12–4**). High-speed twist bits are made of hardened steel that allows the bit to drill holes in various materials, including mild steel. The general rule is to drill in wood at high speed and lower pressure, but to drill in mild steel at slow speeds and high pressure. Cutting oil can be used to cool the bit, keeping it sharper longer. A brad point drill is used for greater precision when drilling holes in wood and wood products (**Figure 12–5**).

Hand Drills. **Hand drills** are operated manually and they will take a standard twist bit and a countersink bit. They are still being used for fine cabinet work (Figure 12-6).

Combination Drills. *Combination drills* and *countersinks* (**Figure 12–6**) are used to drill shank and pilot holes for screws and countersink in one operation (see Chapter 13, Figure 13–21).

CUTTING TOOLS

The carpenter uses many kinds of cutting tools and must know which to select for each job, as well as how to use, sharpen, and maintain them.

Edge-Cutting Tools

Wood Chisels. The **wood chisel** (**Figure 12–7**) is used to cut recesses in wood for such things as door hinges and locksets, as well as to make joints.

Chisels are sized according to the width of the blade and are available in widths of $1/8$ inch to 2 inches (3 to 51 mm). Most carpenters can do their work with a set consisting of chisels that are $1/4$, $1/2$, $3/4$, 1, and $1 1/2$ inches (6, 12, 19, 25, and 38 mm) in size.

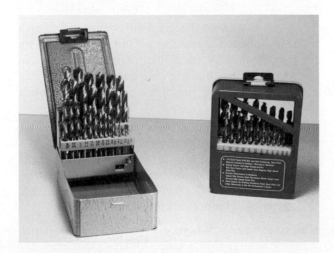

FIGURE 12–4 Twist bit sets have bits that may range in size from $1/16$ to $1/2$ inch (1 to 13 mm) in increments of $1/64$ inch (1 mm).

FIGURE 12–7 Wood chisel.

Used by permission of Andrew Reynolds; photo by M. Nauth

FIGURE 12–8 Secure the material and keep both hands in back of the chisel's cutting edge.

Firmer chisels have long, thick blades and are used on heavy framing. *Butt chisels* are short, with a thinner blade. They are preferred for finish work.

CAUTION

Improper use of chisels has caused many accidents. When using chisels, keep both hands behind the cutting edge at all times (Figure 12–8). When not in use, the cutting edge should be shielded. Never put or carry chisels or other sharp or pointed tools in pockets.

Bench Planes. Bench planes (Figure 12–9) come in several sizes. They are used for smoothing rough surfaces and bringing work down to the desired size. Large planes are used, for instance, on door edges to produce a straight surface over

Courtesy of Stanley Tools

FIGURE 12–9 The jack plane is a general-purpose bench plane.

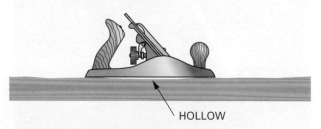

HOLLOW

FIGURE 12–10 Longer planes bridge hollows to allow for planing of long, straight edges.

a long distance. Long planes will bridge hollows in a surface and cut high spots better than a short plane (**Figure 12–10**). Small planes are more easily used for shorter work.

Bench planes are given names according to their length. The longest is called the *jointer*. In declining order are the *fore*, *jack*, and *smooth* planes. It is not necessary to have all the planes. The jack plane is 14 inches (356 mm) long and of all the bench planes is considered the best for all-around work.

Block Planes. Block planes are small planes designed to be held in one hand. They are often used to smooth the edges of short pieces and for trimming end grain to make fine joints (**Figure 12–11**).

Block planes are designed differently than bench planes. On bench planes, the cutting edge bevel is on the bottom side. On block planes, it is on the top. In addition, the bench plane iron has a plane iron cap attached to it, while the block plane iron has none (Figure 12–12).

Unlike bench planes, block planes are available with their blades set at a high angle or at a low angle. Most carpenters prefer the low-angle block plane because this type of plane cuts end grain more effectively. They also have a smoother cutting action and fit into the hand more comfortably.

Using Planes. When planing, have the stock securely held against a stop. Always plane with the grain. When starting, push forward while applying pressure downward on the toe (front). When the heel (back) clears the end, apply pressure downward

FIGURE 12–11 A block plane is small and often has a low blade angle.

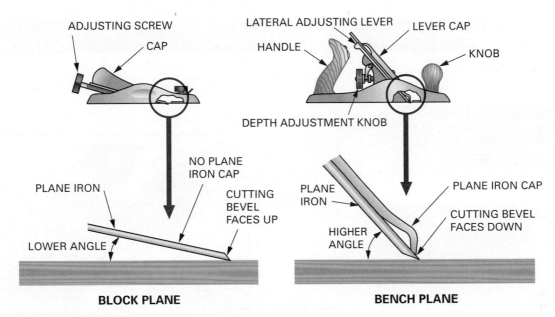

ADJUSTING SCREW

CAP

LATERAL ADJUSTING LEVER   LEVER CAP

HANDLE

KNOB

DEPTH ADJUSTMENT KNOB

PLANE IRON

NO PLANE IRON CAP

CUTTING BEVEL FACES UP

PLANE IRON

PLANE IRON CAP

CUTTING BEVEL FACES DOWN

LOWER ANGLE

HIGHER ANGLE

BLOCK PLANE

BENCH PLANE

FIGURE 12–12 Difference between a block plane and a bench plane.

on both ends while pushing forward. When the opposite end is approached, relax pressure on the toe and continue pressure on the heel until the cut is complete (**Figure 12–13**). This method prevents tilting the plane over the ends of the stock and helps ensure a straight, smooth edge.

Sharpening Chisels and Plane Irons. To produce a keen edge, the tool must be **whetted** (sharpened) using an oilstone or waterstone. Hold the tool on a well-oiled stone so that the bevel rests flat on it. Move the tool back and forth across the stone for a few strokes. Then make a few strokes

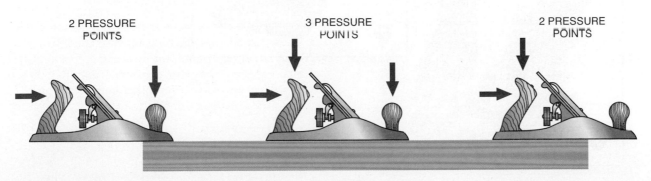

2 PRESSURE POINTS

3 PRESSURE POINTS

2 PRESSURE POINTS

FIGURE 12–13 Correct method for using a plane.

with the flat side of the chisel or plane iron held absolutely flat on the stone. Continue whetting in this manner until as keen an edge as possible is obtained. To obtain a keener edge, repeat the procedure on a finer stone or on a piece of leather. The edge is sharp when, after having whetted the bevel and before turning it over, no wire edge can be felt on the flat side (**Procedure 12–A**).

Chisels and plane irons do not have to be ground each time they need sharpening. Grinding is necessary only when the bevel has lost its concave shape by repeated whettings, the edge is badly nicked, or the bevel has become too short and blunt. The edge of a blade may be whetted many times before it needs grinding. After many whettings the bevel of the wood chisel or plane iron may need to be shaped by a grinding wheel. The bevel should have a concave surface, which is called a *hollow grind*. To obtain a hollow grind, a grinding attachment may be used.

CAUTION

Use safety goggles or otherwise protect your eyes when grinding.

StepbyStep Procedures

Procedure 12–A Sharpening a Plane Iron

STEP 1 ONE OR TWO PASSES WILL REMOVE ANY BURRS FROM THE BACK SIDE.

STEP 2 MANY PASSES ON BEVELLED SIDE WILL REMOVE MATERIAL TO SHARPEN BLADE. REPEAT STEP ONE AND THEN STEP TWO UNTIL DESIRED SHARPNESS IS ACHIEVED.

STEP 3 PULL BLADE OVER LEATHER STROP TO HONE THE EDGE. BOTH SURFACES SHOULD BE DONE IN A BACK-AND-FORTH MOTION.

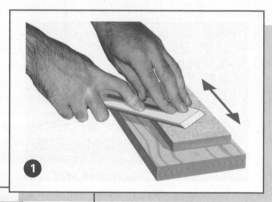

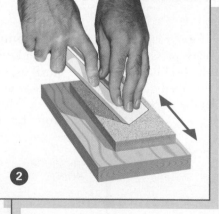

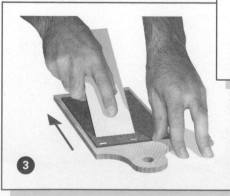

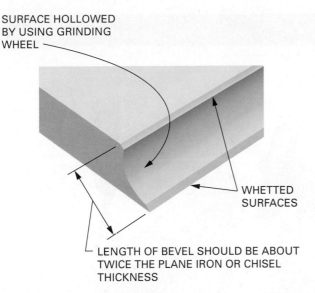

SURFACE HOLLOWED BY USING GRINDING WHEEL

WHETTED SURFACES

LENGTH OF BEVEL SHOULD BE ABOUT TWICE THE PLANE IRON OR CHISEL THICKNESS

FIGURE 12–14 Details of the cutting edge of chisels and planes.

If a grinding attachment is not available, hold the chisel or plane iron by hand on the tool rest at the proper angle. A general rule is that the width of the bevel is approximately twice the thickness of the blade (**Figure 12–14**). Move the blade up on the grinding wheel for a longer bevel and down for a shorter bevel.

Let the index finger of the hand holding the chisel ride against the outside edge of the tool rest as the chisel is moved back and forth across the revolving wheel. Dip the blade in water frequently to prevent overheating. Do not move the position of your index finger, to ensure that the tool can be replaced on the wheel at exactly the same angle to obtain a smooth hollow to the bevel. Grind the chisel or plane iron until an edge is formed (**Figure 12–15**). A burr or

wire edge will be formed on the edge on the flat side. This can be felt by lightly rubbing your thumb along the flat side toward the edge.

CAUTION

When using hatchets for driving fasteners, make sure there are no workers in the path of the backswing.

Snips. Straight tin snips are generally used to cut straight lines on thin metal, such as roof flashing and metal roof edging. Three styles of **aviation snips** are available for straight metal cutting and for left- and right-curved cuts (**Figure 12–16**). The colour of the handle denotes the differences in the design of the snips. Yellow handles are for straight cuts, green are for cutting curves to the right, and red are for cutting to the left.

Knives. A carpenter usually has a jackknife of good quality. The jackknife is used mostly for sharpening pencils and for laying out recessed cuts for some types of finish hardware, such as door hinges. The jackknife is used for laying out this type of work because a finer line can be obtained with it than with a pencil. Besides marking, it scores the layout

A

B

FIGURE 12–16 (A) Metal shears and (B) left-, straight-, and right-cutting aviation snips.

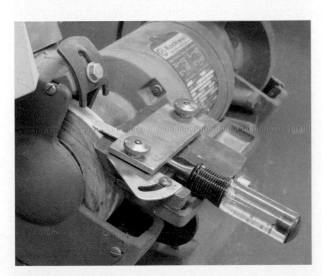

FIGURE 12–15 Grinding a wood chisel.

StepbyStep Procedures

Procedure 12–B Chiselling Square Edges

STEP 1 SCORE LAYOUT LINE WITH KNIFE.

STEP 2 LIFT CHIP WITH CHISEL.

STEP 3 CHIP BREAKS OFF AT SCORED LINE.

STEP 4 A SHOULDER IS PROVIDED TO REST THE CHISEL AGAINST WHEN DEEPENING THE CUT.

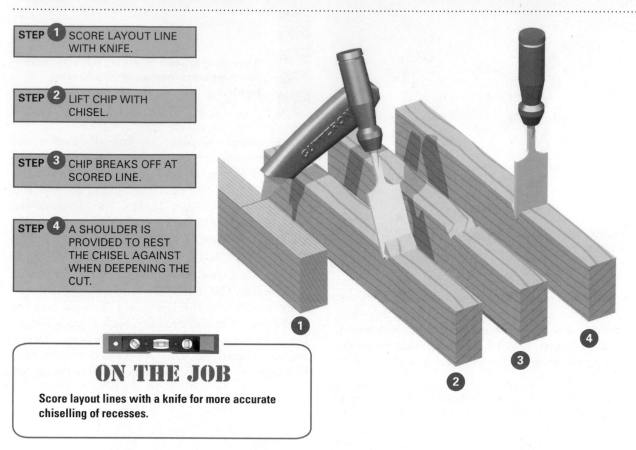

ON THE JOB

Score layout lines with a knife for more accurate chiselling of recesses.

line, which is helpful when chiselling the recess (**Procedure 12–B**).

The **utility knife** (**Figure 12–17**) is frequently used instead of a jackknife for such things as cutting gypsum board and softboards. Replacement blades are carried inside the handle.

Scrapers. The hand scraper (**Figure 12–18**) is very useful for removing old paint, dried glue, pencil, crayon, and other marks from wood surfaces. The scraper blades are reversible, removable, and replaceable. They dull quickly but can be easily sharpened by filing on the bevel and against the cutting edge.

Courtesy of OLFA North America

FIGURE 12–17 Utility knives.

Photo by M. Nauth

FIGURE 12–18 Hand scraper.

CAUTION

Be careful when filing so that the filing hand does not come in contact with the cutting edge. Also, care should be taken to evenly file the entire cutting edge and not to hollow out the centre of the blade or the outside corners.

FIGURE 12–19 Handsaws are still useful on the jobsite. Some handsaws are made with deeper teeth, which are designed to cut in both directions.

Tooth-Cutting Tools

The carpenter uses several kinds of saws to cut wood, metal, and other material. Each one is designed for a particular purpose.

Handsaws

Handsaws (**Figure 12–19**) used to cut across the grain of lumber are called **crosscut saws**. To cut with the grain, **ripsaws** are sometimes used. The difference in the cutting action is in the shape of the teeth. The crosscut saw has teeth shaped like knives. These teeth cut through the wood fibres to give a smoother action and surface when cutting across the grain. The ripsaw has teeth shaped like rows of tiny chisels that cut the wood ahead of them (**Figure 12–20**). Another design to handsaw teeth, called a **shark tooth saw**, makes the teeth longer and able to cut in both directions of blade travel.

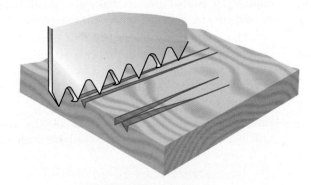

HOW A CROSSCUT SAW CUTS

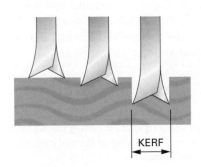

CROSS-SECTION OF CROSSCUT TEETH

KERF

CROSSCUT SAW CUTS

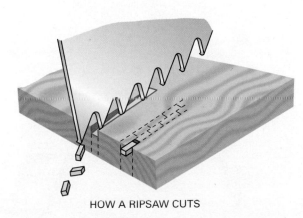

HOW A RIPSAW CUTS

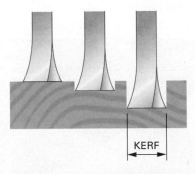

CROSS-SECTION OF RIP TEETH

KERF

Courtesy of Disston Precision, Inc.

RIPSAW CUTS

FIGURE 12–20 Cutting action of ripsaws and crosscut saws.

To keep the saw from binding, the teeth are *set*, that is, alternately bent, to make the saw cut or *kerf* wide enough to give clearance for the blade.

A seven-point or eight-point (number of tooth points to the inch) saw is designed to cut across the grain of framing and other rough lumber. A 10-point or 12-point saw is designed for fine crosscuts on finish work. The number of points is usually stamped on the saw blade at the heel.

Using Handsaws.   Stock is hand sawn with the face side up because the back side is splintered, along the cut, by the action of the saw going through the stock. This is not important on rough work. However, on finish work, it is essential to identify the face side of a piece and to make all layout lines and saw cuts with the face side up.

The saw cut is made on the waste side of the layout line by cutting away part of the line and leaving the rest. This takes some practice, especially when it is important to make thin layout lines rather than broad, heavy ones. Press the blade of the saw against the thumb when starting a cut. Make sure the thumb is above the teeth; steady it with the index finger, with the rest of the hand on the work. Move the thumb until the saw is aligned as desired and start the cut on the upstroke (**Figure 12–21**). Move the hand away when the cut is deep enough. When hand sawing, hold crosscut saws at about a 45-degree angle.

CAUTION

Do not use a ripsaw for cutting across the grain. It can jump at the start of the cut, possibly causing injury.

Most carpenters prefer to have their saws set and filed by sharpening shops. Sharpening handsaws requires special tools, much skill, and experience to do a professional job.

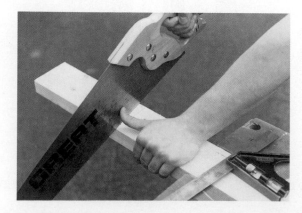

FIGURE 12–21 Starting a cut with a handsaw.

FIGURE 12–22 Compass saw.

Special-Purpose Saws

Compass and Keyhole Saws.   The compass saw is used to make circular cuts in wood (**Figure 12–22**). The keyhole saw is similar to the compass saw except that its blade is narrower, for making curved cuts of smaller diameter. To start the saw cut, a hole needs to be bored (except in soft material, when the point of the saw blade can be pushed through). Keyhole saws were once used to cut keyholes into doors for skeleton keys.

Coping Saws.   The coping saw (**Figure 12–23**) is used primarily to cut moulding to make coped joints. A *coped joint* is made by cutting and fitting the end of a moulding against the face of a similar piece. (Coping is explained in detail in Unit 26.) The coping saw is also used to make any small, irregular curved cuts in wood or other soft material.

The coping saw blade has fine teeth that may be installed with the teeth pointing either toward or away from the handle. Which is best depends on the operator and the situation. The blade cuts only in the direction the teeth point.

Hacksaws.   Hacksaws (**Figure 12–24**) are used to saw metal. Hacksaw blades are available with 18, 24,

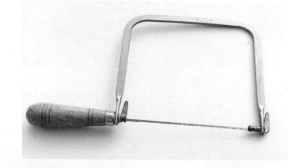

FIGURE 12–23 Coping saw.

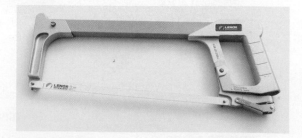

FIGURE 12–24 Hacksaw.

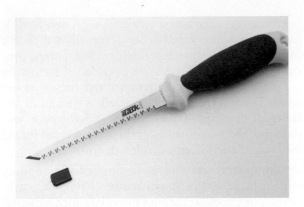

FIGURE 12–25 Wallboard saw.

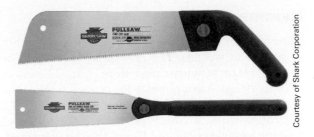

Courtesy of Shark Corporation

FIGURE 12–26 Pullsaws cut when they are pulled, which is the reverse of handsaws.

and 32 points to the inch. Coarse-toothed blades are used for fast cutting of thick metal. Fine-toothed blades are used for smooth cutting of thin metal. At least three teeth of the blade should be in contact with thin metal or cutting will be difficult. Make sure that blades are installed with the teeth pointing away from the handle.

Wallboard Saws. The wallboard saw (**Figure 12–25**) is similar to the compass saw but is designed especially for gypsum board. The point is sharpened to make self-starting cuts for electric outlets, pipes, and other projections. Another type with a handsaw handle is also used frequently.

Pullsaws. The Japanese-style **pullsaws** are gaining popularity. They have a unique design in that they cut on the up (pull) stroke instead of the down stroke (**Figure 12–26**). Some models cut in both directions. The pulling action of the pull saw is actually easier to use. It takes less effort and gives more control. It cuts fast and smooth with thin kerfs. Many styles are available for cutting rough and fine work.

Mitre Boxes. The mitre box (**Figure 12–27**) is used to cut angles of various degrees on finish lumber by swinging the saw to the desired angle and

FIGURE 12–27 A handsaw mitre box.

locking it in place. These cuts are called *mitres*. The joint between the pieces cut at these angles is called a mitred joint.

A mitred joint is made by cutting each piece at half the angle at which it is to be joined to another piece. For instance, if two pieces are to be joined with a mitred joint at 90 degrees to each other, each piece is cut at a 45-degree angle.

The mitre box has built-in stops to locate the saw to cut 90-, 67½-, 60-, and 45-degree angles, which are commonly used mitre angles. The back saw in a mitre box should be used only in the mitre box and for no other purpose.

Chapter 13
Fastening and Dismantling Tools

Discussed in this chapter are those tools used to drive nails and turn screws and other fasteners. Tools used to clamp, hold, pry, and dismantle workpieces are also included.

FASTENING TOOLS

The carpenter must decide which fastening tool to select and must be able to use it competently and safely for the job at hand.

Hammers

The carpenter's **claw hammer** is available in a number of styles and weights. The claws may be straight or curved. Head weights range from 7 to 32 ounces (198 to 907 g). Most popular for general work is the 16-ounce (454 g) curved claw hammer (**Figure 13–1**). For rough work, a 20- or 22-ounce (567 or 624 g) **framing hammer** (similar to **Figure 13–2**) is often used. This has a longer handle and may have a straight or curved claw. In some areas, a 28- or 32-ounce (794 or 907 g) framing hammer is preferred for extra driving power.

Hammer handles vary in styles and are made of wood, steel, or fibreglass. They also come in different lengths. The longer handles allow the carpenter to drive the nail harder and faster.

The hammerheads are smooth or serrated into a waffled surface. The waffle surface keeps the head from slipping off the nail, making nailing more effective. The direction of the driven nail can even be changed slightly by twisting the wrist. These hammers should be used exclusively for framing, because the wood surface is damaged by the waffle imprint that is left when the nail is seated.

Framing hammers with long, non-flexible handles and heavy heads have contributed to the incidence of repetitive strain injury (RSI) among production framers. Muscles are overloaded further by a tight grip, and inflammation often occurs at the elbow (epicondylitis or "tennis elbow") and at the shoulder (bursitis). Other RSIs that involve the tendons and the nerves occur at the wrist (carpal tunnel syndrome) and the elbow (ulna neuritis).

Manufacturers have responded by adding curved handles (axe-type) made of flexible, cushioning, or shock-absorbing materials such as wood and fibreglass. *Estwing®* has even redesigned the shape of the hammer. The inventors at *Stiletto®* found that they could reduce the weight of the hammerhead while maintaining the strength by using *titanium* steel. Their 14-ounce (400 g) hammer has the driving power of a regular 24-ounce (680 g) hammer (refer back to Figure 13–2). The laws of physics say that what drives the nail is the momentum of the head, which is a combination of its weight and its speed. A lighter hammer can be swung faster so that the same momentum is achieved. Add to that the shock absorption qualities of both titanium and the long wood handle, plus the lighter weight on the upswing, and you have less fatigue and less injury.

Courtesy of Cooper Industries. (Plumb® is a registered trademark of Cooper Industries, LLC.)

FIGURE 13–1 Curved claw hammer.

CAUTION

Do not use a waffle-headed hammer when driving anything other than nails. The striking surface of tools such as chisels or nail sets and steel stakes will become serrated if a waffle-headed hammer is used on them. These pieces of metal will eventually fly off, becoming dangerous shrapnel.

Courtesy of Stiletto Tools Inc.

Courtesy of Stiletto Tools Inc.

FIGURE 13–2 Straight claw framing hammer with a milled face.

Courtesy of Lee Valley Tools Ltd.; www.leevalley.com.
© Copyright Lee Valley Tools Ltd.

Courtesy of Stanley Black & Decker

FIGURE 13–3 (A) Double end nail set, and (B) Stanley nail set.

Nail Sets. Nail sets (**Figure 13–3**) are used to set nail heads below the surface. The most common sizes are $\frac{1}{32}$, $\frac{2}{32}$, and $\frac{3}{32}$ inch (0.8, 1.6, and 2.4 mm). The $\frac{1}{4}$-inch (6.4 mm) nail set is used to drive the large-headed nails typically used for exterior finish work. The size refers to the diameter of the tip. The surface of the tip is concave to prevent it from slipping off the nail head. If the tip becomes flattened, the nail set is more difficult to keep on the nail being driven.

CAUTION

Wear eye protection when using a hammer (Figure 13–4). Do not set hardened nails or hit the tip of the nail set with a hammer. This will cause the tip to flatten and may produce chips of flying metal. Do not hit the side of the nail set when setting nails in tongue-and-groove flooring.

Nailing Techniques. Hold the hammer firmly, close to the end of the handle, and hit the nail squarely. If the hammer frequently glances off the nail head, try cleaning the hammer face (**Figure 13–5**). As a general rule, use nails that are three times longer than the thickness of the material being fastened. To swing a hammer, use the entire arm and shoulder. It is important to use the wrist as well. During the latter part of the swing, as the hammer nears the nail, rotate the wrist quickly, to give more speed to the hammerhead. This increased speed generates more nail-driving force, all with less arm effort.

Toenailing is the technique of driving nails at an angle to fasten the end of one piece to another (**Figure 13–6**). It is used when nails cannot be driven

ON THE JOB

To help prevent glancing off the nail head when driving nails, clean the hammer face by rubbing it back and forth on a rough surface.

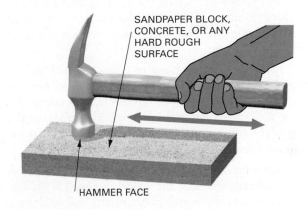

SANDPAPER BLOCK, CONCRETE, OR ANY HARD ROUGH SURFACE

HAMMER FACE

FIGURE 13–5 Roughing up the hammerhead face helps keep the hammerhead from glancing off the nail.

Used by permission of Tyler Forget; photo by M. Nauth

FIGURE 13–4 Wear eye protection when driving nails.

Used by permission of Andrew Reynolds; photo by M. Nauth

FIGURE 13–6 Toenailing is the technique of driving nails at an angle.

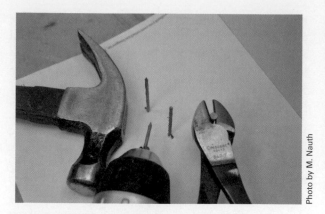

FIGURE 13–7 Methods to avoid splitting wood.

Photo by M. Nauth

FIGURE 13–8 Stagger and angle nails for greater strength and to avoid splitting the stock.

Used by permission of Andrew Reynolds; photo by M. Nauth

into the end, called **end-nailing**. Toenailing generally uses smaller nails than end nailing and offers greater withdrawal resistance of the pieces joined. Start the nail about ¾ to 1 inch (19 to 25 mm) from the end and at an angle of about 30 degrees from the surface.

Drive finish nails almost all the way. Then set the nail below the surface with a nail set to avoid making hammer marks on the surface. Set finish nails at least ⅛ inch (3 mm) deep so that the filler will not fall out.

In hardwood, or close to edges or ends, drill a hole slightly smaller than the nail shank to prevent the wood from splitting or the nail from bending. If a twist drill of the desired size is not available, cut the head off a finish nail of the same gauge and use it for making the hole.

Blunting or cutting off the point of the nail also helps prevent splitting the wood (**Figure 13–7**). The point spreads the wood fibres as the nail is driven, while the blunt end pushes the fibres ahead of it and reduces the possibility of splitting.

Holding the nail tightly with the thumb and as many fingers as possible while driving the nail in hardwood helps prevent bending the nail. Of course, hold the nail in this manner only as long as possible. Be careful not to glance the hammer off the nail and hit the fingers.

When nailing along the length of a piece, stagger the nails from edge to edge, rather than in a straight line. This avoids splitting and provides greater strength (**Figure 13–8**). Drive nails at an angle into end grain for greater holding power. This is called dovetail nailing. When fastening pieces side to side, nails are driven at an angle for greater strength (**Figure 13–9**). In addition, this may keep the nail points from protruding if using 3¼-inch (82 mm) nails to fasten 2-inch (38 mm) nominal stock.

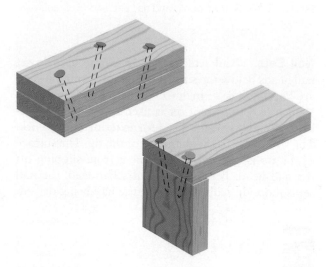

FIGURE 13–9 Driving nails at an angle increases holding power.

When it is necessary to start a nail higher than you can hold it, use the nail starter located in the head of many framing hammers.

Hatchets

Hatchets have a straight blade with a nail-pulling slot on one end and a nailing face on the other. Carpenters use three types of hatchets: the half-hatchet for sharpening stakes, constructing concrete form work, and rough framing; the wallboard hatchet, which has a convex, ringed face for nailing and dimpling drywall; and a **shingling hatchet** for splitting and nailing wood shingles and shakes (**Figure 13–10**).

Staplers

The heavy-duty staple gun is used for fastening building paper and housewrap on the exterior of the building and polyethylene air-vapour barrier on the

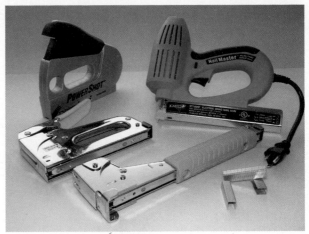

FIGURE 13–11 Staplers.

ening without "cam-out." The drivers are coded by number and colour. They are matched to screws that are number-coded according to the diameter of the shank (**Figure 13–12**).

The *Phillips*® is the next most common type of screwdriver used in the building construction

Top and bottom: Courtesy of Vaughan & Bushnell Manufacturing Company. Centre: Courtesy of Estwing

FIGURE 13–10 Half-hatchet, wallboard hatchet, and shingling hatchet.

interior. The hammer tacker speeds up this process. The strike tacker is used to fasten floor underlayment to the subfloor using divergent staples (**Figure 13–11**).

Screwdrivers

Screwdrivers are manufactured to fit specific types and sizes of the recesses found in screw heads. The length of the blade and the handle together ranges from 3 to 12 inches (76 to 305 mm). Today's handles have ergonomic shapes and a cushioned feel that helps reduce fatigue.

The most common type of screwdriver in Canada is the *Robertson*®, which was invented in Canada in 1908 by P.L. Robertson. This screwdriver fits snugly into the square socket in the head of the screw, which allows for one-handed operations and greater tight-

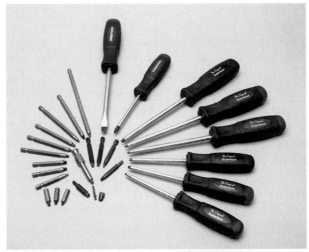

Colour	Screwdriver #	Screw Number	Screw Lengths
Orange	00	1, 2	¼"–⅞" (6–22 mm)
Yellow	0	3, 4	¼"–1½" (6–38 mm)
Green	1	5, 6, 7	⅜"–3" (9.5–76 mm)
Red	2	8, 9, 10	⅜"–4" (9.5–102 mm)
Black	3	12, 14	½"–6" (12–152 mm)
Black	4	16, 18, 20, 24	¾"–6" (19–152 mm)

Courtesy of Robertson Screw Inc.

FIGURE 13–12 Robertson® screwdriver and colour-coded screw table.

FIGURE 13–13 Phillips® screw.

FIGURE 13–15 Torx® drive system.

CORRECT THICKNESS   CORRECT WIDTH

FIGURE 13–14 The correct size screwdriver for the screw being driven is best.

FIGURE 13–16 Pozidriv® screw.

industry. It drives a screw with a cruciform recess head. The driver sizes range from #0 to #4. *Phillips®* screws are used mainly for fastening drywall and door and cabinet hardware. The *Phillips®* screw is very common in the United States, where it was patented in 1936 by Henry Phillips (**Figure 13–13**).

The slotted screwdriver is seldom used by carpenters, though slotted machine screws are still common in the mechanical subtrades. Screwdrivers should fit snugly, without play, into the slot of the screw being driven (**Figure 13–14**).

The *Torx®* drive system allows for greater torque when tightening. *Torx®* screws are found on many power tools and in motor vehicles (**Figure 13–15**).

In 1966 the Phillips Screw Company developed a variation of the cruciform recess called *Pozidriv®*. This has become an international standard and is widely used outside North America, especially in Europe (**Figure 13–16**).

To bridge the gap between the United States and Canada, the Phillips Screw Company developed the *Phillips Square-Driv®*, and the Robertson Screw Company developed the *Recex®*, both of which accommodate Phillips and Robertson screwdrivers (**Figure 13–17**).

In 2001, B. Wagner invented the LOX recess and drive system, which features eight points of contact. This provides positive engagement and extraordinary torque. This is a collated system available only for use with a power driver called Super Drive (**Figure 13–18**).

Screwdriver Bits. Screwdriver bits (**Figure 13–19**) are available in all the different shapes and sizes required to fit any screw on the market. They are designed to drive a screw using a drill or screw gun.

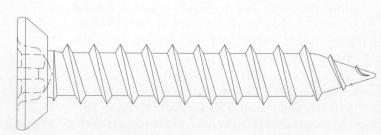

FIGURE 13–17 Phillips Square-Driv®.

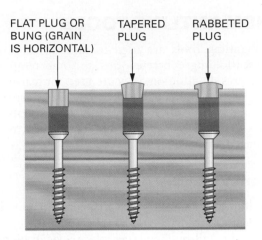

Courtesy of John Wagner Associates for JCCT, Inc.: www.lox.com

FIGURE 13–20 Counterbored holes may be plugged with various shapes of plugs.

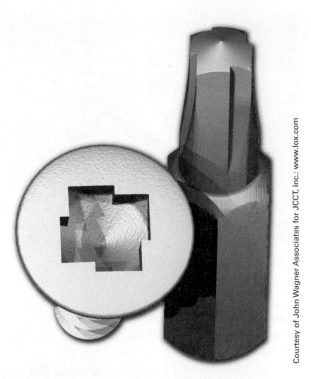

FIGURE 13–18 LOX recess and drive system.

Screwdriving Techniques. A screw should be selected so that two-thirds of its length penetrates into the piece it is gripping. For example, if you are fastening a piece of ½-inch (12.5 mm) plywood into a 2 × 4 (38 × 89), use a 1½-inch (38 mm) screw. Also, it is important to pre-drill the plywood to precisely fit the shank of the screw and to pre-drill a pilot hole into the lumber for the screw threads. The pilot hole should match the solid centre portion of the screw's threaded section. For flat-headed wood screws, the shank hole can be *countersunk* so that the head will be flush with the surface, or it can be **counterbored** so that the head will be below the surface. The hole can then be plugged with a flat wood plug (or bung), a tapered plug, or a rabbeted plug (**Figure 13–20**).

Today, the pilot hole, the shank hole, and the countersink are drilled using an all-in-one countersink drill bit, which has a collar to set the depth of the countersink or counterbore (**Figure 13–21**). The bit size must be matched to the screw size.

FIGURE 13–19 Screw gun drive bits for various screw head styles.

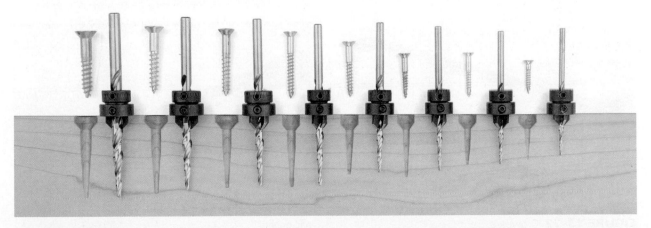

FIGURE 13–21 All-in-one countersink drill bits. *(Courtesy of Lee Valley Tools Ltd.; www.leevalley.com. © Copyright Lee Valley Tools Ltd.)*

DISMANTLING TOOLS

Dismantling tools are used to take down staging and scaffolding, concrete forms, and other temporary structures. In addition, they are used for tearing out sections of a building when remodelling. Carpenters must be skilled in the use of dismantling tools, and in the work, so that the dismantled members are not damaged any more than necessary.

Hammers

In addition to fastening, hammers are often used for pulling nails to dismantle parts. To increase leverage and make nail-pulling easier, place a small block of wood under the hammer head (**Figure 13–22**).

Bars and Pullers

The **crowbar** (**Figure 13–23**) is used to withdraw spikes and to pry when dismantling parts of a structure (**Figure 13–24**). They are available in lengths

FIGURE 13–24 Using a crowbar to pry stock loose.

from 12 to 36 inches (305 to 915 mm), with the 30-inch (750 mm) size often preferred for construction work.

Carpenters need a small **flat bar** or **prybar**, similar to those shown in **Figure 13–25**, to pry small work and pull small nails. To extract nails that have been driven home, a **nail claw**, commonly called a *cat's paw*, is used (**Figure 13–26**).

FIGURE 13–22 Pull a nail more easily by placing a block of wood under the hammer.

FIGURE 13–23 Crowbars.

FIGURE 13–25 A flat bar and a prybar. Both are used as nail pullers.

FIGURE 13–26 Nail claw or nail puller.

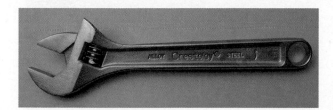

FIGURE 13–27 Adjustable wrench.

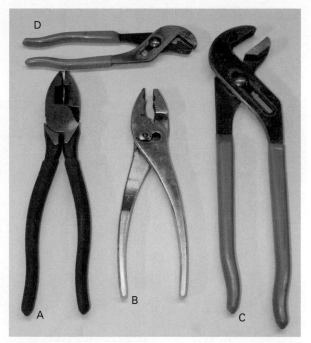

FIGURE 13–28 (A) Lineman's pliers, (B) slip joint pliers, and (C and D) tongue-and-groove pliers (straight jaw).

Holding Tools

To turn nuts, lag screws, bolts, and other objects, an **adjustable wrench** is often used (**Figure 13–27**). The wrench is sized by its overall length. The 10-inch (250 mm) adjustable wrench is the one most widely used.

For extracting, turning, and holding objects, a pair of pliers is often used. Many kinds of pliers are manufactured, including the three examples presented in **Figure 13–28**. **Lineman's pliers** are for use with cable and wire, including cutting and twisting. **Slip joint pliers** (also called combination pliers) has a sliding mechanism that can be adjusted for use in one of two positions. **Tongue-and-groove pliers** have an offset design with several positions to choose from, depending on the size of the object being plied.

Clamps come in a variety of styles and sizes (**Figure 13–29**). They are useful for holding objects together while they are being fastened, glued, and used as temporary guides. *Spring clamps* are quick and easy to set. They are spring-loaded for ease of closing their jaws. Simply squeezing the handles opens the jaws; releasing sets them. *C-clamps* are named for their shape. The size is designated by the throat opening. *Quick clamps* are named for the speed at which they can be adjusted and set. One side of the jaws is stationary and the other slides on the bar. After the material is placed in the jaws, the handles are squeezed to tighten and set the clamp.

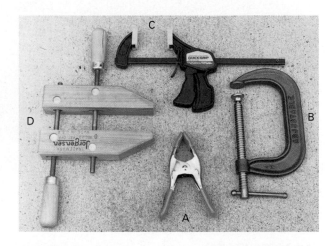

FIGURE 13–29 (A) Spring clamp, (B) C-clamp, (C) quick clamp, and (D) wood screw.

The small trigger is pulled to release the clamp. The *wood screw*, also called *parallel clamps*, is made of wood blocks and large screws. These clamps are used primarily for holding wood pieces while they are glued. It takes some practice to set this clamp quickly. The centre screw is turned one way to tighten and the other screw is turned the opposite way.

DECONSTRUCT THIS

Carefully study **Figure 13–30** and think about what is wrong and/or what is right. Consider all possibilities. What construction practice or method is different in your area of the country?

FIGURE 13–30 This photo shows a worker using a wood chisel to clean out a dado.

KEY TERMS

adjustable wrench	boring	coping saw	framing hammer
auger bits	butt markers	counterbored	framing square
aviation snips	carpenter's level	crosscut saws	grooves
bench plane iron	chalk line	crowbar	hacksaws
bench planes	clamps	dadoes	hand drills
bevel	claw hammer	drilling	hand scraper
bit brace	combination square	end-nailing	heel
block planes	compass saw	flat bar	keyhole saw

level	pocket tape	shingling hatchet	pliers
line level	prybar	slip joint pliers	trammel points
lineman's pliers	pullsaws	Speed Square	try square
mitre box	push drills	steel tapes	twist drills
mitred joint	rabbets	straightedge	utility knife
nail claw	ripsaws	straight tin snips	wallboard saw
nail sets	screwdrivers	toe	whetted
plumb	scribing	toenailing	wing dividers
plumb bob	shark tooth saw	tongue-and-groove	wood chisel

REVIEW QUESTIONS

Select the most appropriate answer.

1. Where does a safe worker attitude that promotes a safe jobsite come from?

   a. ability
   b. skill
   c. knowledge
   d. experienced mentors

2. When stretching a steel tape to lay out a measurement, where should you place the ring?

   a. with the 1-inch (25-mm) mark on the starting line
   b. with the end of the steel tape on the starting line
   c. with the inside of the ring on the starting line
   d. with the outside of the ring on the starting line

3. In construction, what does the term *plumb* mean?

   a. perfectly horizontal
   b. perfectly level
   c. perfectly straight
   d. perfectly vertical

4. What is the name of the large, L-shaped squaring tool that has tables stamped on it?

   a. framing square
   b. Speed Square
   c. bevel square
   d. combination square

5. What is the name of the layout tool that may be adjusted to serve as a marking gauge?

   a. framing square
   b. Speed Square
   c. bevel square
   d. combination square

6. What should be raised to adjust a carpenter's level into a level position when the bubble is found to be touching the right line?

   a. The right side should be raised.
   b. The left side should be raised.
   c. The left side should be lowered.
   d. The entire level should be raised.

7. When snapping a long chalk line, what should you make sure to do?

   a. dampen the string
   b. keep the string from sagging
   c. hold the string loosely
   d. let the string touch the surface as it unwinds

8. What tool is used to mark material to conform to an irregular surface?

   a. pen
   b. chisel
   c. scriber
   d. chalk line

9. What is the name of the one-handed plane with a low blade angle?

   a. block plane
   b. bench plane
   c. chisel
   d. plane iron

10. What does the colour of the handle on aviation snips indicate?

   a. which hand to use
   b. the direction in which curves can easily be cut
   c. the manufacturer
   d. what material may be easily cut

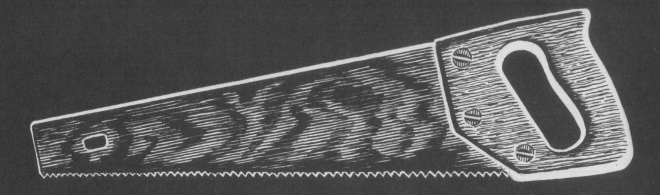

Unit 6

Portable Power Tools

Chapter 14   Saws, Drills, and Drivers
Chapter 15   Planes, Routers, Sanders,
                      and Plate Joiners

Chapter 16   Fastening Tools

The sound of construction has changed over the years. The rhythmic whoosh of a handsaw has virtually been replaced by the whir and ring of a circular saw. Power tools have been created to increase the productivity of most jobsite tasks.

The number and styles of power tools available today for the carpenter are vast and the list continues to grow. Power tools enable the carpenter to do more work in less time with less effort.

OBJECTIVES

After completing this unit, the student should be able to:

- state general safety rules for operating portable power tools.
- identify, describe, and safely use the following portable power tools: circular saws, jigsaws, reciprocating saws, drills, impact drivers, hammer-drills, screwdrivers, planes, routers, plate joiners, sanders, staplers, nailers, and powder-actuated drivers.

 SAFETY REMINDER

With increased speed and production comes an increase in personal risk. This danger can come from a spectrum of human shortcomings that range from a lack of knowledge and skill to overconfidence and carelessness. Safe operation of power tools requires knowledge and discipline.

Learn the safe operating techniques from the manufacturer's recommended instructions before operating any tool. Once you understand these procedures, follow them every time the tool is used. Don't take chances; life is too short as it is.

Being aware of the dangers of operating power tools is the first step in avoiding accidents. This begins with eye and ear protection.

Portable power tools are everywhere on the construction site today. All tools must be carefully used and properly maintained to keep workers using the tools and those nearby safe.

CAUTION

Safety is as important as breathing. Have a complete understanding of a tool before attempting to operate it. Read the manufacturer's operating instructions.

Maintain a proper attitude of safety at all times by following these guidelines:

- Wear eye and ear protection. Eyes and ears do not grow back.

- Do not become distracted by others or be distracting to others when tools are being operated.

- Do not wear loose-fitting clothes or jewellery that might become caught in the tool.

- Make sure the material being tooled is securely held and supported.

- Remember, using sharp bits and cutters is actually safer than using dull ones.

- Stay alert and develop an attitude of care and respect for yourself and others.

Use the proper power source:

- Electricity used to power a tool can be fatal to humans. Use ground fault circuit interrupters (GFCIs) at all times. These will trip before any electricity can leak out of a tool and cause a shock.

- Do not use frayed or badly worn power cords.

- Use properly sized extension cords that are rated for the power requirements of the tool.

- Avoid using a cord longer in length than necessary. Voltage to the tool drops as the cord gets longer.

- Keep extension cords out of the path of all construction traffic.

- Unplug the tool whenever touching the cutting surface of the tool.

Safety is a team sport. It requires all workers to take part.

Chapter 14

Saws, Drills, and Drivers

SAWS

The carpenter uses several kinds of portable saws for **crosscutting** and **ripping**, for making circular cuts, and for cutting openings in floors, walls, and ceilings.

Electric Circular Saws

Commonly called the circsaw, the portable electric **circular saw** (**Figure 14–1**) is used often by the carpenter. The circular saw blade is driven by an electric motor. The saw has a base that rests on the work to be cut. A handle with a trigger switch is provided for the operator to control the tool. The saw is adjustable for depth of cut. A retractable safety guard is provided over the blade, extending under the base.

The base may be tilted for making *bevel* cuts. Bevel or mitre cuts are those where the edge, end, or face of a board is cut at an angle. Edge bevels run along the length of the board or with the grain.

End mitres run along the width of the board or across the grain. Face mitres are angle cuts made on the face (**Figure 14–2**). Compound mitres are cuts with two angles, usually a combination of face and end mitres.

FIGURE 14–1 Using a portable electric circular saw to cut compound angles.

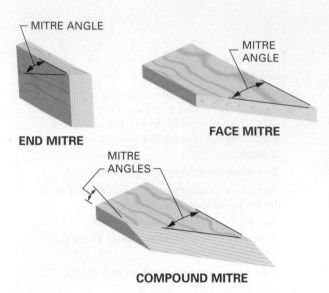

FIGURE 14–2 Edge, flat, and compound mitre cuts.

Saws are manufactured in many styles and sizes. The size is determined by the diameter of the blade, which ranges from 4½ to 16 inches (114 to 406 mm). The most common size of circular saw uses a 7¼-inch (184 mm) blade. The handle and switch may be located on the top or in back. The blade may be driven directly by the motor (sidewinder) or through a worm gear (**Figure 14–3**).

FIGURE 14–3 Direct-drive (sidewinder) and worm gear–drive portable electric circular saws.

A direct-drive saw can have the saw blade to the left or to the right of the electric motor. The forward end of the base is notched in two places to serve as a guide for following layout lines. One notch is used to follow layout lines when the base is tilted to 45 degrees and the other when the base is not tilted.

> # CAUTION
>
> **Make sure the saw blade is installed with the teeth pointing in the correct direction. The teeth of the saw blade projecting below the base should point toward the base as the blade rotates.**

To loosen the bolt that holds the blade in place, first unplug the saw and lock the arbor. This is done by pushing the arbor-locking slide or button of the saw, usually found between the blade shield and the handle. While pushing the slide, rotate the blade by hand until the slide locks the arbor. With the proper wrench, turn the bolt in the same direction as the rotation of the blade. To tighten, turn the bolt in the direction opposite the rotation of the blade.

On most models, a ripping bar that fits into the base is used for ripping narrow pieces parallel with the edge. The saw may also be guided by tacking or clamping a straightedge to the material and running the edge of the saw base against it. Allowance must be made for the distance from the saw blade to the edge of the saw base when positioning the straightedge.

Circular Saw Blades. Circular saw blades are available in a number of styles. The shape and number of teeth around the circumference of the blade determine their cutting action. Carbide-tipped blades are used more than high-speed steel blades. They stay sharper longer. More complete information on saw blades can be found in Unit 7.

Using the Portable Circular Saw. Safe and efficient cutting follows an established method:

- Make sure the work is securely held and that the waste will fall clear and not bind the saw blade (**Figure 14–4**).
- Adjust the depth of cut so that the blade just cuts through the work. Never expose the blade any more than is necessary (**Figure 14–5**). This will reduce the sawdust spray into the operator's face when cutting thinner material.

FIGURE 14-4 Saw cuts are made over the end of supports so that the waste will fall clear and not bind the blade.

FIGURE 14-5 The blade of the saw is adjusted for depth only enough to cut through the work.

Used by permission of Andrew Reynolds; photo by M. Nauth

Photo by M. Nauth

> ## CAUTION
> Make sure the guard operates properly. Be aware that the guard may possibly stick in the open position. Never wedge the guard back in an open position.

- Mark the stock. Put on safety glasses. Rest the forward end of the base on the work. With the blade clear of the material, start the saw.
- When it has reached full speed, advance the saw into the work. Make sure to observe the line to be followed. With the saw cut in the waste side of the material, cut as close to the line as possible for a short distance.
- Stop the saw advancement into the material and check the alignment of the edge of the base to the line being cut. They should be parallel.
- Follow the line closely.

> ## CAUTION
> Any deviation from the line may cause the saw to bind and kick back. Do not force the saw forward. If the saw does bind, stop the motor and bring the saw back to where it will run free. Continue following the line closely.

- Near the end of the cut, the forward end of the base will go off the work. Guide the saw by observing the line at the saw blade and finish the cut. The saw may also be guided by watching the layout line at the saw cut for the whole length. Let the waste drop clear and ensure that the guard has returned. Release the switch.

> ## CAUTION
> Keep the saw clear of your body until the saw blade has completely stopped.

- When starting cuts across stock at an angle, it may be necessary to retract the guard by hand. A handle is provided for this purpose (**Figure 14-6**). Release the handle after the cut has been started and continue as above.
- Compound mitre cuts may be made by cutting across the stock at an angle with the base tilted.

Portable circular saws cut on the upstroke. The saw blade rotates upward through the material. As the teeth of the saw blade come through the top surface, splintering of the stock occurs at the layout line. The severity of the splintering depends on the

FIGURE 14-6 Retracting the guard of the portable circular saw by hand.

Photo by M. Nauth

kind of blade used, the kind and thickness of the material being cut, and other factors. More splintering occurs when cutting across the grain than with the grain.

On *finish work*, that is, work that will ultimately be exposed to view, any splintering along the cut is unacceptable. One way to prevent this is to mark and cut from the back side. If it is not possible to cut from the back side, or if both sides may be exposed to view, mark the layout lines on the face side. Then score along the layout lines with a sharp knife. Make the cuts just outside the scored lines. Another way to avoid splintering is to place a strip of masking tape over the cut before marking it. The tape helps hold the wood fibres in place while being cut.

Making Plunge Cuts.   Many times it is necessary to make internal cuts in the material such as for sinks in countertops or openings in floors and walls. To make these cuts with a portable electric circular saw, the saw must be plunged into the material:

- Accurately lay out the cut to be made. Adjust the saw for depth of cut.
- Wearing eye protection, hold the guard open and tilt the saw up with the front edge of the base resting on the work. Have the saw blade over, and in line with, the cut to be made (**Figure 14–7**).
- Make sure the teeth of the blade are clear of the work and start the saw. Lower the blade slowly into the work, following the line carefully, until the entire base rests squarely on the material.
- Advance the saw into the corner. Release the switch and wait until the saw stops before removing it from the cut.

> **CAUTION**
>
> **Make sure the tool comes to a complete stop before withdrawing it from the material being cut.**

- Reverse the direction and cut into the corner. Again, wait until the saw stops before removing it from the cut.

> **CAUTION**
>
> **Never move the saw backward while cutting. The direction of the turning blade will make the saw want to rise up out of the cut, jumping backward across everything in its path.**

- Proceed in like manner to cut the other sides. Finish the cut into the corners with a handsaw or jigsaw.

Jigsaws

The jigsaw (sometimes called a sabre saw) is widely used to make curved cuts (**Figure 14–8**). The teeth of the blade generally point upward when installed, so the saw cuts on the upstroke. To produce a splinter-free cut on the face side, it is best to use a blade that cuts on the down stroke, typically used for sink cutouts in laminate countertops.

There are many styles and varieties of jigsaws. The length of the stroke along with the amperage of the motor determine its size and quality. Strokes range from ½ to 1 inch (12 to 25 mm). The longest stroke is the best for faster and easier cutting.

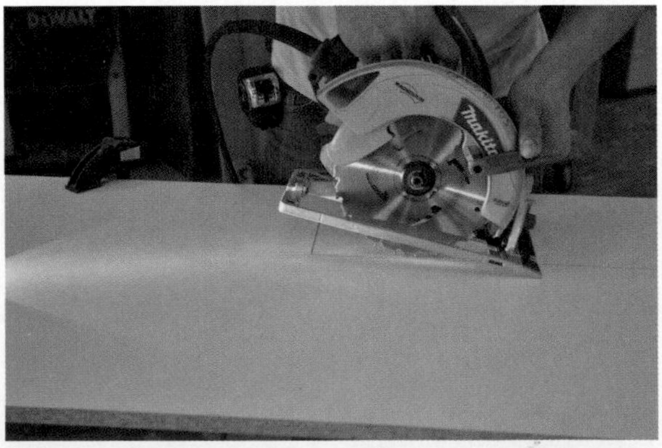

FIGURE 14–7 Making a plunge cut with a portable circular saw. First retract the guard, place the front edge of the saw base on the material, and then pivot, running the saw slowly into the material.

FIGURE 14–8 The jigsaw can make either straight or orbital cutting actions.

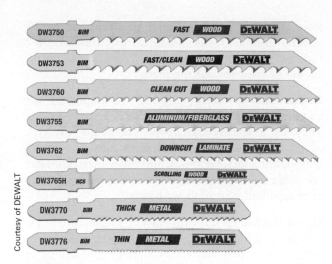

FIGURE 14–9 Jigsaw blade selection.

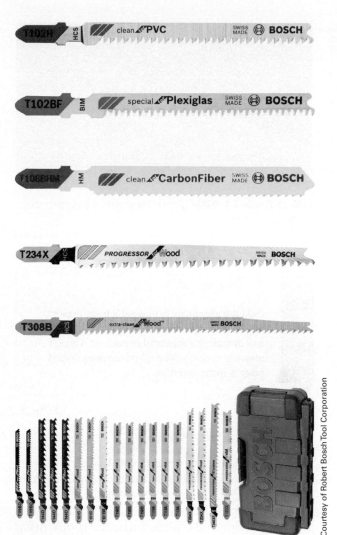

FIGURE 14–10 Jigsaw blade selection.

Some saws can be switched from straight up-and-down strokes to several orbital (circular) motions to provide more effective cutting action for various materials. The base of the saw may be tilted to make bevel cuts.

Blade Selection. Many blades are available for fine or coarse cutting in wood or metal (**Figure 14–9** and **14–10**). Wood-cutting blades have teeth that are from 6 to 12 points to the inch (25 mm). Blades with coarse teeth (fewer points to the inch) cut faster but rougher. Blades with more teeth to the inch may cut slower but produce a smoother cut surface. They do not splinter the work as much. Proper blade selection and use will produce the best cuts.

Using the Jigsaw. Follow a safe and established procedure:

- Outline the cut to be made. Secure the work either by hand, by tacking, by clamping, or by some other method.

- Using eye protection, hold the base of the saw firmly on the work. With the blade clear, squeeze the trigger.

- Push the saw into the work, following the line closely. Make the saw cut into the waste side, and cut as close to the line as possible without completely removing it.

- Keep the saw moving forward, holding the base down firmly on the work. This will allow the saw to cut faster and more efficiently by keeping saw vibration to a minimum. Turn the saw as necessary in order to follow the line to be cut. Feeding the saw into the work as fast as it will cut, but not forcing it, finish the cut. Keep the saw clear of your body until it has stopped.

Making Plunge Cuts. Plunge cuts are best made by first pre-drilling a hole that is larger than the blade of the jigsaw. Plunge cuts may be made with the jigsaw in a manner similar to that used with the circular saw:

- Tilt the saw up on the forward end of its base with the blade in line and clear of the work (**Figure 14–11**).

- Start the motor, holding the base steady. Very gradually and slowly, lower the saw until the blade penetrates the work and the base rests firmly on it.

- Cut along the line into the corner. Back up for about an inch, turn the corner, and cut along the other side and into the corner.

- Continue in this manner until all the sides of the opening are cut.

- Turn the saw around and cut in the opposite direction to cut out the corners.

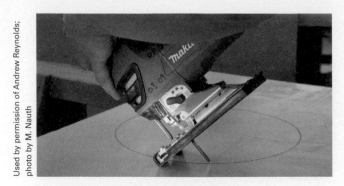

FIGURE 14–11 Making an internal cut by plunging the jigsaw.

> ## CAUTION
>
> Hold the saw firmly to prevent it from jumping when the blade makes contact with the material and to make a successful plunge cut. Thicker material may require a pilot hole to be drilled prior to blade insertion.

Reciprocating Saws

The reciprocating saw (**Figure 14–12**), also called a sabre saw by some manufacturers, is used primarily for *roughing in* work or remodelling. This work consists of cutting holes and openings for such things as pipes, heating and cooling ducts, and roof vents or cutting nails and removing studs or windows. It can be likened to a powered compass saw.

Most models have a variable speed of from 0 to 2,400 strokes per minute. Like jigsaws, some models may be switched to several *orbital* cutting strokes

FIGURE 14–12 Using the reciprocating saw to cut an opening in the subfloor.

from a straight back-and-forth to an orbital cutting action. Ordinarily, the orbital cutting mode is used for fast cutting in wood. The reciprocating stroke should be used for cutting metal.

Reciprocating Saw Blades. Common blade lengths run from 4 to 12 inches (101 to 305 mm). They are available for cutting practically any type of material. Blades are available to cut wood, metal, plaster, fibreglass, ceramics, and other material. They are made of a hardened steel that allows the blades to occasionally cut nails with ease.

Using the Reciprocating Saw. The reciprocating saw is used in a manner similar to the jigsaw. The difference is that the reciprocating saw is heavier and more powerful. It can be used more efficiently to cut through rough, thick material, such as walls when remodelling. With a long, flexible blade, it can be used to cut flush with a floor or along the side of a stud.

To use the saw, lines must be laid out and followed. The base or shoe of the saw is held firmly against the work whenever possible. Like the jigsaw, this reduces saw vibration and allows the saw to cut faster and more efficiently. To make cutouts, first drill a hole in the material. Then insert the blade, start the motor, and follow the layout lines. The blades can be reversed for cutting in confined areas.

DRILLS AND DRIVERS

Portable power drills, manufactured in a great number of styles and sizes, are widely used to drill holes and drive fasteners in all kinds of construction materials.

Drills

The drills used in the construction industry are classified as light duty or heavy duty. Light-duty drills usually have a *pistol-grip* handle. Heavy-duty drills have a *spade-shaped* or *D-shaped* handle (**Figure 14–13**).

Drill Sizes. The size of a drill is determined by the capacity of the *chuck*, its maximum opening. The chuck is that part of the drill that holds the cutting tool. The most popular sizes for light-duty models are ¼ and ⅜ inch (6 and 9.5 mm). Heavy-duty drills have a ½-inch (12 mm) chuck or larger.

Drill Speed and Rotation. Most drills have *variable speed* and *reversible* controls. Speed of rotation can be controlled from 0 to maximum rpm (revolutions per minute) by varying the pressure on the

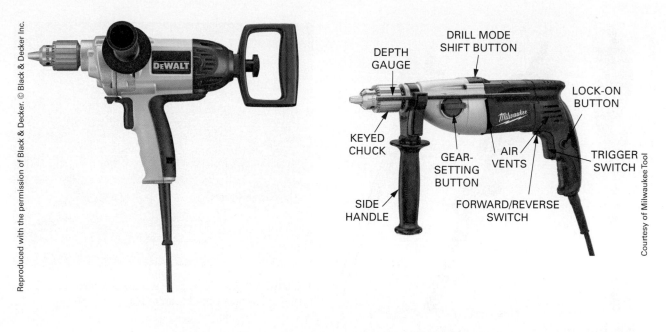

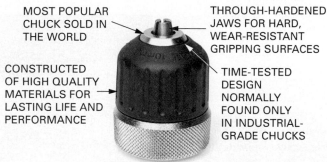

FIGURE 14–13 Portable electric drills, keyed and keyless chucks.

trigger switch. Slow speeds are desirable for drilling larger holes or holes in metal. Faster speeds are used for drilling smaller holes in softer material. A reversing switch changes direction of the rotation for removing screws or withdrawing bits and drills from clogged holes.

Bits and Twist Drills. Twist drills are used in electric drills to make small holes in wood or metal (**Figure 14–14**). For larger holes in wood, plastics, and composition materials, a variety of wood-cutting bits are used.

For boring holes in rough work, the *spade bit* is commonly used. For a hole with a cleaner edge in finish work, the *power bore* bit may be used (**Figure 14–15**). These bits make fine clean holes in wood with diameters from ¼ to 2⅛ inches (6 to 54 mm).

Forstner Bits. For holes that require a clean edge and a flat bottom, such as for a European-style cabinet hinge, the Forstner bit is used. The diameters of the bits range from ¼ to 4½ inches (6 to 114 mm).

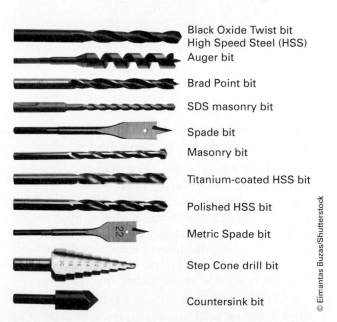

FIGURE 14–14 Different types of drill bits for various materials.

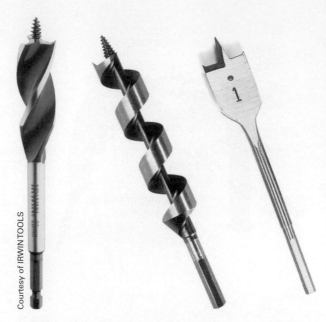

Courtesy of IRWIN TOOLS

FIGURE 14–15 Spade, auger, and self-feed wood bits are used to drill larger holes in wood and composites. *Courtesy of Irwin Industrial Tools.)*

FIGURE 14–16 Forstner-style wood bits.

FIGURE 14–17 Hole saws actually saw holes in material, leaving a circular centre plug.

FIGURE 14–18 Masonry bits have a carbide tip.

Using Portable Electric Drills.   Select the proper size and type of bit and insert it. Tighten the chuck with the chuck key or by holding the chuck of a *keyless* chuck. Holes in metal must be centre-punched because the drill will wander off centre.

More pressure is required for the larger bits and usually a drill press is used (**Figure 14–16**).

Other Drill Accessories.   Occasionally, carpenters may use **hole saws** (**Figure 14–17**). These saws cut holes from ⅝ inch to 6 inches (16 to 152 mm) in diameter. The difference with the holesaw is that the cut is "sawn" on the circumference of the hole; a centre hole is drilled slightly deeper, penetrating the material before the saw teeth. The rest of the cut is then made from the other side, thus avoiding splintering.

Masonry drill bits (**Figure 14–18**) have carbide tips for drilling holes in concrete, brick, tile, and other masonry. They are frequently used in portable power drills. They are more efficiently used in hammer-drills.

> ## CAUTION
>
> Hold small pieces securely by clamping or other means. When drilling through metal especially, the drill has a tendency to hang up when it penetrates the underside. If the piece is not held securely, the hangup will cause the piece to rotate with the drill. It could then hit anything in its path and possibly cause serious injury to a person before power to the drill can be shut off.

Place the bit on the centre of the hole to be drilled. Hold the drill at the desired angle and start the motor. Apply pressure as required, but do not force the bit. Drill into the stock, being careful not to wobble the drill. Failure to hold the drill steady may result in breakage of small twist drills.

CAUTION

While drilling a hole, withdraw the turning bit periodically to clear the shavings. This will keep the bit cooler and it will last longer. More important, it will help prevent the bit from jamming in the hole. Jamming causes the bit to stop suddenly. Personal injury may occur if the drill is powerful and the jammed bit is large. Therefore, be ready to release the trigger at any time.

Hammer-Drills

Hammer-drills (**Figure 14–19**) are similar to other drills. However, they can be changed to a hammering action as they drill, quickly making holes in concrete or other masonry. Some models deliver as many as 50,000 hammer blows per minute. Most popular are the ⅜- and ½-inch (9.5 and 12.5 mm) sizes.

A depth stop is usually attached to the side of the hammer-drill. It can be converted to a conventional drill by a quick-change mechanism. Most models have a variable speed of from 0 up to 2,600 rpm. The hammer-drill has the same type of chuck and is used in the same manner as conventional portable power drills.

Screw Guns

Screw guns or *drywall drivers* (**Figure 14–20**) are used extensively for fastening gypsum board to walls and ceilings with screws. They are similar in appearance to the light-duty drills, except for the chuck. They have a pistol-type grip for one-hand operation and controls for varying the speed and reversing the rotation.

The chuck is made to receive special screwdriver bits of various shapes and sizes. A screw gun has an adjustable nosepiece that surrounds the bit. When the forward end of the nosepiece touches the surface, the clutch is separated and the bit stops turning with a vibrating noise. Adjusting the nosepiece makes variations in the screw depth. Some screw guns use collated screws; others have extensions that facilitate the fastening of subfloor panels to floor joists (**Figure 14–21**).

Accessories are available for driving screws with different heads. Hex head nutsetters are available in magnetic or nonmagnetic styles, in sizes ranging from ³⁄₁₆ to ⅜ inch (5 to 9.5 mm).

Courtesy of Milwaukee Electric Tool

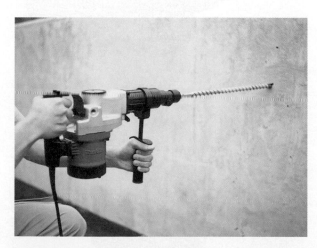

FIGURE 14–19 The hammer-drill is used to make holes in masonry.

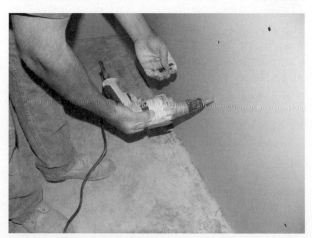

FIGURE 14–20 The drywall driver is used to fasten wallboard with screws.

Courtesy of John Wagner Associates for Makita USA, Inc.

FIGURE 14–21 Screw gun 05 series.

CORDLESS TOOLS

Cordless power tools are widely used due to their convenience, strength, and durability. The tools' power source is a removable, rechargeable battery that ranges in voltage from 4 to 36 volts. In general, the higher the voltage, the stronger the tool.

There are presently three types of rechargeable batteries: nickel-cadmium (Ni-Cad), nickel–metal hydride (Ni-MH), and lithium ion (Li-Ion). Manufacturers today are producing the same batteries in varying amperage-hours (amp-hours), a measure that indicates endurance of charge. Two other factors to consider are the torque of the tool and the durability of the components. Lithium ion batteries are the best available on the market.

Before investing in cordless tools, it is best to make a trip to the hardware store (or go online) and see what types of tools are available in each name brand. It makes sense to choose a brand and stick with it for the sake of common batteries, but this might not be always possible. It is important to check the compatibility of the newer versions of batteries with the older versions of tools. With the advances in tool and battery technology, a 12V drill with a 4 amp-hour (Ah) battery and brushless technology outperforms the 18V Ni-Cad drills of yesteryear. Several of the manufacturers now have a 5Ah 18V lithium ion battery that weighs the same as the 3Ah but has double the runtime.

Downtime is practically eliminated if chargers and spare batteries are kept on hand. Most chargers will charge various voltages in one hour and will stop charging when the battery is fully charged. One of the main disadvantages to cordless tools has been their performance in cold weather. Recently, Milwaukee Electric Tool has met this challenge with its Red Lithium batteries. Their claim is 40 percent more runtime, 20 percent more speed, 20 percent more torque (depending on the tool), and 50 percent more charges, but the claim of most interest to Canadians is that they operate down to –18°C (0°F). Cordless tools that work well in the winter are a real game changer. There is now the Red Lithium extra capacity (XC 4.0) battery, and the competition is pushing Milwaukee to produce an XC 5.0.

Cordless tools tend to make the jobsite safer by eliminating extension cords. Take care to set the tool down in a safe place after each use, especially when working on scaffolds. Have a look at the video on battery care at youtube.com (search for "How to look after Makita 18V Li-ion Batteries").

> ### CAUTION
>
> It is easy to think that because these tools are battery powered, they are safer to use than higher voltage tools with cords. While they are safe to use, the operator should never forget the proper techniques and requirements for using the tools. Always wear personal protective equipment.

Cordless drills come with variable-speed reversing motors and a positive clutch, which can be adjusted to drive screws to a desired torque and depth (**Figure 14–22**). Some models have a hammer-drilling

Courtesy of Milwaukee Tool

FIGURE 14–22 Cordless drill.

mode for drilling concrete (**Figure 14–23**). Cordless drills that are used for driving screws are being replaced by cordless impact drivers, which can drive many more screws per charge (**Figure 14–24**).

Cordless circular saws use 4½-, 6½-, and 7¼-inch blades (114, 165, and 184 mm) and perform in much the same way as corded tools (**Figure 14–25**). Newer models use two 18V batteries or one 36V battery for more torque and runtime.

Other cordless tools include the jigsaw (**Figure 14–26**), the reciprocating saw, the screwdriver, the finish nailer (**Figure 14–27**), the power plane, grinders, band saws (**Figure 14–28**), chain saws, plate joiners, and many others yet to come. Several combination kits are also available (**Figure 14–29**).

FIGURE 14–25 Milwaukee cordless circular saw.

FIGURE 14–23 Cordless hammer-drill with SDS bit.

FIGURE 14–26 Cordless jigsaw.

FIGURE 14–24 Cordless impact driver with brushless motor.

FIGURE 14–27 Cordless 16-gauge nailer.

FIGURE 14-28 Cordless band saw.

FIGURE 14-29 Cordless combination kit.

Chapter 15

Planes, Routers, Sanders, and Plate Joiners

PORTABLE POWER PLANES

Portable power planes make planing jobs much easier for the carpenter. Planing the ends and edges of hardwood doors and stair treads, for example, takes considerable effort with hand planes, even with razor-sharp cutting edges.

Jointer Planes

The jointer plane is used primarily to smooth and straighten long edges, such as when fitting doors in openings (**Figure 15–1**). It is manufactured in lengths up to 18 inches (456 mm). The electric motor powers a cutter head that may measure up to 3¾ inches (95 mm) wide. The planing depth, or the amount that can be taken off with one pass, can be set for 0 up to ⅛ inch (3 mm).

An adjustable fence allows planing of squares, bevelled edges to 45 degrees, or a chamfer (**Figure 15–2**). A rabbeting guide is used to cut rabbets up to ⅞ inch (22 mm) deep.

Operating Power Planes

The operation and feel of the power plane is similar to that of the bench plane. The major differences are the vibration and the ease of cutting with the power plane.

FIGURE 15–1 A portable electric jointer plane.

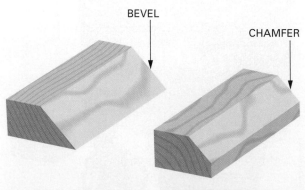

BEVEL

CHAMFER

A BEVEL IS A CUT AT AN ANGLE THROUGH THE TOTAL THICKNESS

A CHAMFER IS AN ANGLED CUT PARTWAY THROUGH THE THICKNESS

FIGURE 15–2 A bevel and chamfer.

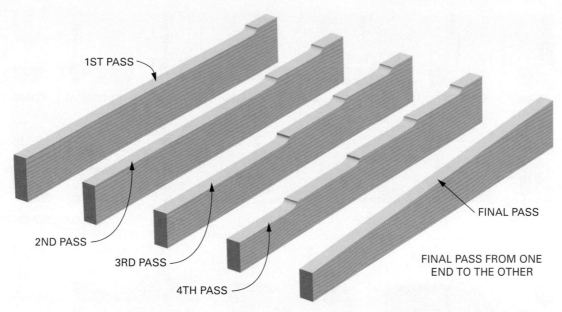

1ST PASS

2ND PASS

3RD PASS

4TH PASS

FINAL PASS

FINAL PASS FROM ONE END TO THE OTHER

DEPTH OF CUT EXAGGERATED FOR ILLUSTRATIVE PURPOSES

FIGURE 15–3 Technique for planing a taper.

CAUTION

Extreme care must be taken when operating power planes. There is no retractable guard, and the high-speed cutterhead is exposed on the bottom of the plane. Keep the tool clear of your body until it has completely stopped. Keep extension cords clear of the tool.

- Set the side guide to the desired angle, and adjust the depth of cut.
- Hold the toe (front) firmly on the work, with the plane cutterhead clear of the work.
- Start the motor. With steady, even pressure make the cut through the work for the entire length. Guide the angle of the cut by holding the guide against the side of the stock. Apply pressure to the toe of the plane at the beginning of the cut. Apply pressure to the heel (back) at the end of the cut to prevent tipping the plane over the ends of the work.
- To plane a **taper**, that is, to take more stock off one end than the other, make a number of passes. Each pass should be shorter than the preceding one. Lift the plane clear of the stock at the end of the pass. Make the last pass completely from one end to the other (**Figure 15–3**).

PORTABLE ELECTRIC ROUTERS

One of the most versatile portable tools used in the construction industry is the **router** (**Figure 15–4**). It is available in many models, ranging from ¼ hp to over 3 hp with speeds of 18,000 to 30,000 rpm. These tools have high-speed motors that enable the operator to make clean, smooth-cut edges.

The motor powers a ¼- or ½-inch (6 or 12.5 mm) chuck in which cutting bits of various sizes and shapes are held (**Figure 15–5**). An adjustable base is provided to control the depth of cut. A trigger or toggle switch controls the motor.

The router is used to make many different cuts, including grooves, dadoes, rabbets, and **dovetails**

FIGURE 15–4 Using a portable electric router.

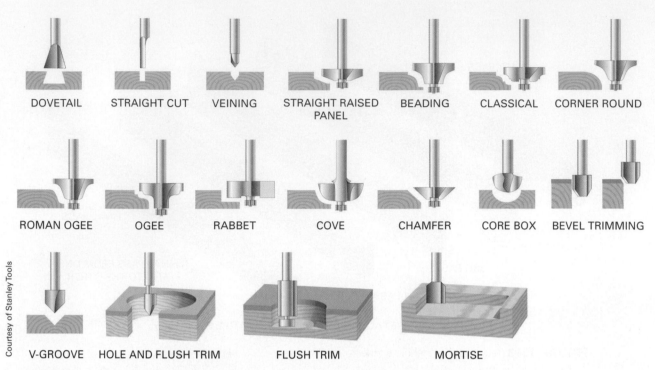

Courtesy of Stanley Tools

DOVETAIL STRAIGHT CUT VEINING STRAIGHT RAISED PANEL BEADING CLASSICAL CORNER ROUND

ROMAN OGEE OGEE RABBET COVE CHAMFER CORE BOX BEVEL TRIMMING

V-GROOVE HOLE AND FLUSH TRIM FLUSH TRIM MORTISE

FIGURE 15–5 Router bit selection guide.

(**Figure 15–6**). It is also used to shape edges and make cutouts, such as for sinks in countertops. It is extensively used with accessories, called templates, to cut recesses for hinges in door edges and door frames. When operating the router, it is important to be mindful of the bit at all times. Watch what you are cutting and keep the router moving. Stalling the movement of the router will cause the bit to burn or melt the material.

When routing small pieces, it is safer to use a router table to secure the router and, with the help of its accessories, to guide the piece through the bit. For plans on building a router table and fence, see *https://www.pinterest.com/explore/router-table/* and *http://www.woodworkerz.com/feature-filled-router-table-fence/*.

Laminate Trimmers

A light-duty specialized type of router is called a laminate trimmer. It is used almost exclusively for trimming the edges of *plastic laminates* (**Figure 15–7**). Plastic laminate is a thin, hard material used primarily as a decorative covering for kitchen and

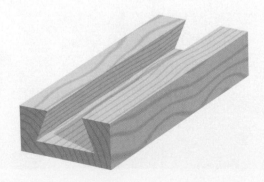

FIGURE 15–6 A dovetail cut is easily made with a router.

Photo by M. Nauth

FIGURE 15–7 The laminate trimmer is used to trim the edges of plastic laminates.

bathroom cabinets and countertops. (The installation of this material is described in Unit 29.)

Guiding the Router

Controlling the sideways motion of the router is accomplished by the following methods:

- By using a router bit with a pilot (guide) (**Figure 15–8**). The router bit can have a solid guide or it can have a ball-bearing guide that runs along the uncut portion of the material being routed.

- By guiding the edge of the router base against a straightedge (**Figure 15–9**). Be sure to keep the router tight to the straightedge, and do not rotate

the router during the cut, because its base may not be centred.

- By using an adjustable guide attached to the base of the router (**Figure 15–10**). The guide rides along the edge of the stock. Make sure the edge is in good condition. The router shown in Figure 15–10 is a plunge router.

- By using a template (pattern) with template guides attached to the base of the router (**Figure 15–11**). This is the method widely used for cutting recesses for door hinges. (This process is explained in detail in Unit 19.)

- By freehand routing, in which the sideways motion of the router is controlled by the operator only. Care should be taken during this operation.

FIGURE 15–8 Router bits may have a pilot bearing to guide the cut.

STRAIGHTEDGE IS ON OPERATOR'S RIGHT. AS ROUTER IS PULLED, ROTATION OF ROUTER TENDS TO KEEP IT AGAINST STRAIGHTEDGE. IF STRAIGHTEDGE WERE ON LEFT SIDE, ROUTER WOULD HAVE A TENDENCY TO PULL AWAY FROM THE STRAIGHTEDGE.

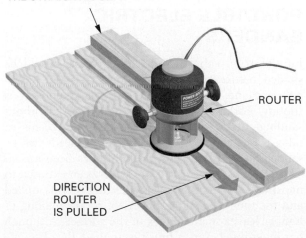

ROUTER

DIRECTION ROUTER IS PULLED

FIGURE 15–9 Using a straightedge to guide the router.

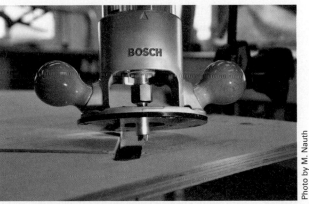

FIGURE 15–10 A guide attached to the base of the plunge router rides along the edge of the stock and controls the sideways motion of the router.

FIGURE 15–11 Guiding the router by means of a template and template guide.

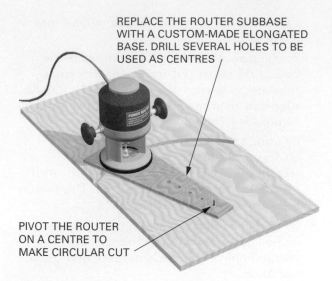

REPLACE THE ROUTER SUBBASE WITH A CUSTOM-MADE ELONGATED BASE. DRILL SEVERAL HOLES TO BE USED AS CENTRES

PIVOT THE ROUTER ON A CENTRE TO MAKE CIRCULAR CUT

FIGURE 15–12 Technique for making arcs using a router.

Courtesy of DELTA/PORTER-CABLE

FIGURE 15–13 Plate joiner and "biscuits."

- To make *circular* cuts, remove the subbase. Replace it, using the same screw holes, with a custom-made one in which one side extends to any desired length. Along a centreline make a series of holes to fasten the newly made subbase to the centre of the desired arc (**Figure 15–12**).

Using the Router. Before using the router, make sure that the power is disconnected. Follow the method outlined:

- Select the correct bit for the type of cut to be made.
- Insert the bit into the chuck. Make sure the chuck grabs at least ½ inch (12 mm) of the bit. Adjust the depth of cut.
- Control the sideways motion of the router by one of the methods previously described.
- Clamp the work securely in position. Plug in the cord.
- Lay the base of the router on the work with the router bit clear of the work. Start the motor.
- Advance the bit into the cut, pulling the router in a direction that is against the rotation of the bit. On outside edges and ends, the router is moved counterclockwise around the piece. When making internal cuts, the router is moved in a clockwise direction.

CAUTION

Finish the cut, keeping the router clear of your body until it has stopped. Be aware that the router bit is unguarded.

PLATE (OR BISCUIT) JOINERS

First developed by Lamello, this tool was called a biscuit joiner after the biscuit-like wafers that were used as mini-splines to join two pieces of wood together (**Figure 15–13**). The groove is cut by a circular blade and leaves a half-moon-shaped recess in each opposing piece. The recesses are filled with glue, a "biscuit" is inserted, and the two pieces are clamped together. Many different types of joints can be made, including butt joints and mitre joints. The advantage of this type of joinery, besides strength, is that there are no visible fasteners. Biscuit joints are often used for exposed gable ends of cabinets.

PORTABLE ELECTRIC SANDERS

Interior trim, cabinets, and other finish should be sanded before any finishing coats of paint, stain, polyurethane, or other material are applied. It is shoddy workmanship to coat finish work without sanding. In too many cases, expediency seems to take precedence over quality. Trim needs to be sanded because the grain probably has been *raised*. This happens because the stock has been exposed to moisture in the air between the time it was milled and the time of installation. Also, rotary planing of lumber leaves small ripples in the surface. Although hardly visible before a finish is applied, they become very noticeable later.

FIGURE 15–14 Using a portable electric belt sander.

Some finishing coats require more sanding than others. If a penetrating stain is to be applied, extreme care must be taken to provide a surface that is evenly sanded with no cross-grained scratches. If paints or transparent coatings are to be applied, surfaces need not be sanded as thoroughly. Portable sanders make sanding jobs less tedious.

Portable Electric Belt Sanders

The **belt sander** is used frequently for sanding cabinetwork and interior finish (**Figure 15–14**). The size of the belt determines the size of the sander. Belt widths range from 2½ to 4 inches (64 to 100 mm). Belt lengths vary from 14 to 24 inches (356 to 610 mm). The 3 × 21-inch (76 × 533 mm) belt sander is a popular, lightweight model. Some sanders have a bag to collect the sanding dust. Remember to wear eye protection. The small 2½ × 14 inch (64 to 356 mm) belt sander is suitable for one-handed use, and it also has an auxiliary handle at the front for two-handed operations.

Installing Sanding Belts. Sanding belts are usually installed by retracting the forward roller of the belt sander. An arrow is stamped on the inside of some sanding belts to indicate the direction in which the belt should run. The sanding belt should run *with*, not against, the lap of the joint (**Figure 15–15**). Sanding belts joined with butt joints may be installed in either direction. Install the belt over the rollers. Then release the forward roller to its operating position.

To keep the sanding belt centred as it is rotating, the forward roller can be tilted. Stand the sander on its back end. Hold it securely and start it. Turn the adjusting screw one way or the other to track the belt and centre it on the roller (**Figure 15–16**).

Using the Belt Sander. More work has probably been ruined by improper use of the portable belt sander than any other tool. It is wise to practise on scrap stock until enough experience in its use is gained to ensure an acceptable sanded surface. Care must be taken to sand squarely on the sander's pad.

FIGURE 15–15 Sanding belts should be installed in the proper direction, as indicated by the arrow on the inside of the belt.

FIGURE 15–16 The belt should be centred on its rollers by using the tracking screw.

Allowing the sander to tilt sideways or to ride on either roller results in a gouged surface.

> **CAUTION**
>
> Make sure the switch of the belt sander is off before plugging the cord into a power outlet. Some trigger switches can be locked in the "ON" position. If the tool is plugged in when the switch is locked in this position, the sander will travel at high speed across the surface. This could cause damage to the work and/or injury to anyone in its path.

- Secure the work to be sanded. Make sure the belt is centred on the rollers and is tracking properly.
- Hold the tool with both hands so that the edge of the sanding belt can be clearly seen.
- Start the machine. Place the pad of the sander flat on the work. Pull the sander back and lift it off just clear of the work at the end of the stroke.

> ## CAUTION
> Be careful to keep the electrical cord clear of the tool. Because of the constant movement of the sander, the cord may easily get tangled in the sander if the operator is not alert.

- Place the sander back on the material and bring the sander forward. Continue sanding using a skimming motion that lifts the sander just clear of the work at the end of every stroke. Sanding in this manner prevents overheating of the sander and helps to remove material evenly. It also allows debris to be cleared from the work, and the operator can see what has been done.
- Do not sand in one spot too long. Be careful not to tilt the sander in any direction. Always sand with the pad flat on the work. Do not exert excessive pressure. The weight of the sander is enough. Always sand with the grain to produce a smooth finish.
- Make sure the sander has stopped before setting it down. It is a good idea to lay it on its side to prevent accidental travelling.

Finishing Sanders

There are three basic types of finishing sanders; the ¼-sheet palm sander (**Figure 15–17**), the ⅓- to

Courtesy of Robert Bosch Inc.

FIGURE 15–17 Palm sander.

Courtesy of Metabo Canada Inc.

FIGURE 15–18 Finishing sander.

Courtesy of Milwaukee Electric Tool Corporation

FIGURE 15–19 Random orbital palm sander.

½-sheet finishing sander (**Figure 15–18**), and the random orbital palm sander (**Figure 15–19**). Some sheet sanders have an orbital motion that removes material quickly but leaves scratches. Others have an oscillating motion (straight back and forth) that leaves a clean finish. The random orbital sander randomly moves the centre of the rotating paper at high speeds; this allows the paper to sand in all directions at once. Besides removing material quickly and leaving a scratch-free finish, this type of sander has an easy place-and-remove hook-and-loop system for its paper, which comes with holes to direct the sanded dust into a bag.

To use the finishing sander, proceed as follows:

- Select the desired grit sandpaper. Attach it to the pad, making sure it is tight. A loose sheet will tear easily.
- Start the motor and sand the surface evenly, *slowly* pushing and pulling the sander with the grain. Let the action of the sander do the work. Do not use excessive pressure because this may overload the machine and burn out the motor. Always hold the sander flat on its pad.

Abrasives. The quality of the abrasives on the sandpaper is determined by the length of time it is

Description	Grit No.	Description	Grit No.
Very Fine	400	Coarse	80
	360		60
	320		50
	280		40
Fine	240	Very Coarse	36
	220		30
	180		24
	150		20
Medium	120		16
	100		12

FIGURE 15–20 Grits of coated abrasives.

able to retain its sharp cutting edges. Garnet is a reddish mineral often used as a natural abrasive on sandpaper. Synthetic abrasives include *aluminum oxide* and *silicon carbide*. Sandpaper coated with aluminum oxide is probably the most widely used for wood.

Grits. Sandpaper grit refers to the size of the abrasive particles. Sandpaper with large abrasive particles is considered coarse. Small abrasive particles are used to make fine sandpaper.

Sandpaper grits are designated by a grit number. The grit numbers range from No. 12 (coarse) to No. 400 (fine) (**Figure 15–20**). Commonly used grits are 60 or 80 for rough sanding, and 120 or 180 for finish sanding. Grit number is the number of grains of abrasive per square inch.

Sand with a coarser grit until a surface is uniformly sanded. Do not switch to a finer grit too soon. Do not use worn or clogged abrasives. Their use causes the surface to become glazed or burned.

Chapter 16
Fastening Tools

Portable fastening tools included in this chapter are **pneumatic** (powered by compressed air) and powder-actuated (drive fasteners with explosive powder cartridges). They are widely used throughout the construction industry for practically every fastening job from foundation to roof.

PNEUMATIC STAPLERS AND NAILERS

Pneumatic staplers and nailers are commonly called *guns* (**Figure 16–1**). They are used widely for quick fastening of framing, subfloors, wall and

roof sheathing, roof shingles, exterior finish, and interior trim. A number of manufacturers make a variety of models in several sizes for special fastening jobs. For instance, a *framing gun*, although used for many fastening jobs, is not used to apply interior trim.

Older models required frequent oiling, either by hand or by an oiler installed in the air line. With improvements in design, some newer models require only a few drops a day.

Remember to wear eye protection. It is also recommended to wear ear protection. Prolonged exposure to loud noises will damage the ear.

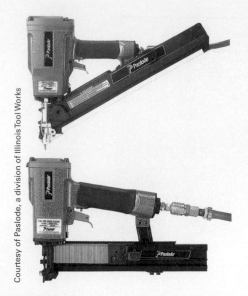

Courtesy of Paslode, a division of Illinois Tool Works

FIGURE 16–1 Pneumatic nailers and staplers are widely used to fasten building materials.

Courtesy of Hitachi Koki Canada

FIGURE 16–2 Strip nailer (gas).

Nailing Guns

Framers generally use a gun that will handle both a 2½-inch (64 mm) ardox nail and a 3¼-inch (83 mm) ardox nail. These will fulfill the requirements of the National Building Code of Canada for both sheathing and framing. These guns can be either a strip nailer (**Figure 16–2**) or a coil nailer (**Figure 16–3**). Strip nailers carry either a round-head nail or a clipped-head nail. There are also light-duty nailers, which are used for toe-nailing studs and fastening subfloor panels (**Figure 16–4**). Both headed and finish nails used in nailing guns come glued together in strips (**Figure 16–5**).

The *finish nailer* (15- and 16-gauge) (**Figure 16–6**) drives finish nails from ¾ to 2½ inches (19 to 63 mm) long. The 15- and 16-gauge nails have diameters of 1.83 mm and 1.63 mm respectively, the higher gauge

Courtesy of Makita Canada

FIGURE 16–3 Coil nailer.

having a smaller diameter. It can be used for the application of practically all kinds of exterior and interior finish work. It sets or flush-drives nails as desired. A nail set is not required, and the possibility of marring the wood is avoided. For a graphical discussion of trim nailers, go to *www.finehomebuilding.com* and

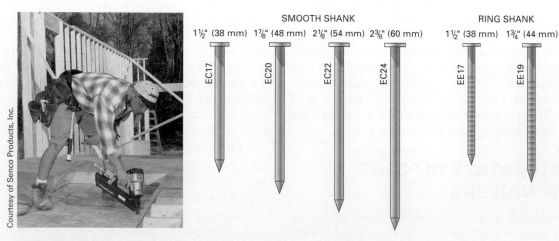

Courtesy of Senco Products, Inc.

SMOOTH SHANK

1½" (38 mm) 1⅞" (48 mm) 2⅛" (54 mm) 2⅜" (60 mm)

EC17 EC20 EC22 EC24

RING SHANK

1½" (38 mm) 1¾" (44 mm)

EE17 EE19

FIGURE 16–4 A light-duty nailer is used to fasten light framing, subfloors, and sheathing.

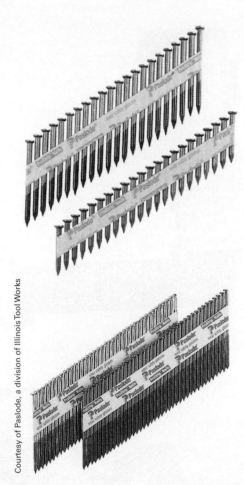

FIGURE 16–5 Both headed and finish nails used in nailing guns come glued together in strips.

SLIGHT-HEADED BRADS   MEDIUM-HEADED BRADS

⅝"   ¾"   1"   1"   1¼"   1½"   1⅝"

FIGURE 16–6 A light-duty brad nailer is used to fasten thin moulding and trim.

search for the video, "Choosing Trim Nailers: Which Gauge Finish Nailers Should You Own?"

The *brad nailer* (18-gauge, 1.02 mm in diameter) (**Figure 16–7**) drives T-headed brads ranging in length from ⅝ to 2⅛ inches (16 to 54 mm). It is used to fasten small mouldings and trim, cabinet door frames and panels, and other miscellaneous finish carpentry. It is important to remember to align the "T" head with the grain to avoid crushing the wood.

The *pin nailer* (23-gauge, 0.635 mm in diameter) fires headless or near-headless pins that result in very little splitting of the material and in tiny holes that need little filling (**Figure 16–8**). The pins range in length from ⅜ to 1⅜ inches (10 to 35 mm). The pin nailer is used for returns on trim work and for securing thin strips of mouldings. It is good practice to use glue to adhere the pieces. For a detailed discussion of pin nailers, go to *www.jlconline.com* and search the site's articles for "pin nailers" or to *www.woodworkersjournal.com* and search for "pin nailers."

The *coil roofing nailer* (**Figure 16–9**) is designed for fastening asphalt and fibreglass roof shingles.

It drives five different sizes of wide, round-headed roofing nails from ⅞ to 1¾ inches (22 to 44 mm). The nails come in coils of 120, which are easily loaded in a nail canister.

FIGURE 16–7 Brad nailer.

FIGURE 16–8 Pin nailer.

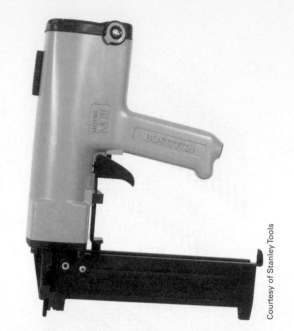

FIGURE 16–10 A pneumatic concrete nailer.

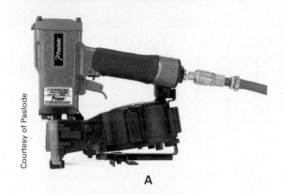

A

COIL ROOFING NAILS

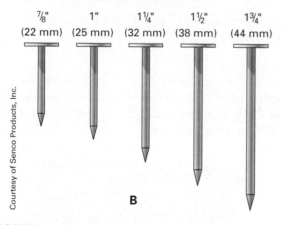

| ⅞" | 1" | 1¼" | 1½" | 1¾" |
| (22 mm) | (25 mm) | (32 mm) | (38 mm) | (44 mm) |

B

FIGURE 16–9 A coil roofing nailer is used to fasten asphalt roof shingles.

There are also pneumatic nailers that drive pins into concrete or steel. These nailers generally fasten thin material such as drywall metal track (**Figure 16–10**).

Staplers

Like nailing guns, *staplers* are manufactured in a number of models and sizes and are classified as light-, medium-, and heavy-duty.

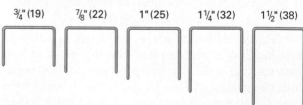

| ¾" (19) | ⅞" (22) | 1" (25) | 1¼" (32) | 1½" (38) |

FIGURE 16–11 The wide-crown stapler, being used in this photograph to fasten roof shingles, can be used to fasten a variety of materials.

A popular tool is the *roofing stapler* (**Figure 16–11**), which is used to fasten roofing shingles. It comes in several models and drives 1-inch (25 mm) wide-crown staples in lengths from ¾ to 1½ inches (19 to 38 mm).

FIGURE 16–12 Staples, like nails, come glued together in strips for use in stapling guns.

FIGURE 16–14 Each cordless gun comes in its own case with battery, battery charger, safety glasses, instructions, and storage for fuel cells.

It can also be used for fastening other materials, such as lath wire, insulation, furniture, and cabinets. The staples, like nails, come glued together in strips (**Figure 16–12**) for quick and easy reloading. Most stapling guns can hold up to 150 staples.

No single model stapler drives all widths and lengths of staples. Models are made to drive narrow-crown, intermediate-crown, and wide-crown staples. A popular model drives ⅜-inch (9.5 mm) intermediate-crown staples from ¾ inch to 2 inches (19 to 51 mm) in length.

Cordless Guns

Conventional pneumatic staplers and nailers are powered by compressed air. The air is supplied by an air compressor (**Figure 16–13**) through long lengths of air hose stretched over the construction site. The development of cordless nailing guns and cordless stapling guns (**Figure 16–14**) has eliminated the need for air compressors and hoses. The cordless gun utilizes a disposable fuel cell. A battery and spark plug power an internal combustion engine that forces a piston down to drive the fastener. Another advantage is the time saved in setting up the compressor, draining it at the end of the day, coiling the hoses, and storing the equipment.

The *cordless framing nailer* drives nails from 2 to 3¼ inches (51 to 83 mm) in length. Each fuel cell will deliver energy to drive about 1,200 nails. The battery will last long enough to drive about 4,000 nails before recharging is required.

The *cordless finish nailer* drives finish nails from ¾ to 2½ inches (19 to 64 mm) in length. It will drive about 2,500 nails before a new fuel cell is needed and about 8,000 nails before the battery has to be recharged.

The *cordless stapler* drives intermediate-crown staples from ¾ inch to 2 inches (19 to 51 mm) in length. It will drive about 2,500 staples with each fuel cell and about 8,000 staples with each charge of the battery. There are also gas-operated nailers, which are used to drive pins through steel-stud track into concrete or steel (**Figure 16–15**).

There are cordless nailers available that do not use fuel cells but instead use a more powerful battery (**Figure 16–16**).

FIGURE 16–13 Compressors supply air pressure to operate nailers and other pneumatic tools.

Courtesy of Hilti, Inc.

FIGURE 16–15 Gas-operated nailer.

Courtesy of RIDGID Tools

FIGURE 16–16 Cordless mechanized (not pneumatic) nailer.

Using Nailers and Staplers

Because of the many designs and sizes of staplers and nailers, you should study the manufacturer's directions and follow them carefully. Use the right nailer or stapler for the job at hand. Make sure all safety devices are working properly, and always wear eye protection. A work contact element allows the tool to operate only when this device is firmly depressed on a work surface and the trigger is pulled, promoting safe tool operation.

- Load the magazine with the desired size of staples or nails.
- Connect the air supply to the tool. For those guns that require it, make sure there is an oiler in the air supply line, adequate oil to keep the gun lubricated during operation, and an air filter to keep moisture from damaging the gun. Use the recommended air pressure. Larger nails require more air pressure than smaller ones.

CAUTION

Exceeding the recommended air pressure may cause damage to the gun or burst air hoses, possibly causing injury to workers.

- Press the trigger and tap the nose of the gun to the work. When the trigger is depressed, a fastener is driven each time the nose of the gun is tapped to the work. The fastener may also be safely driven by first pressing the nose of the gun to the surface and then pulling the trigger.
- Upon completion of fastening, disconnect the air supply.

CAUTION

Never leave an unattended gun with the air supply connected. Always keep the gun pointed toward the work. Never point it at other workers or fire a fastener except into the work. Serious injury can result from horseplay with the tool.

POWER-ACTUATED DRIVERS

Powder-actuated drivers (**Figure 16–17**) are used to drive specially designed pins into masonry or steel. They are used in a manner similar to firing a gun. Powder charges of various strengths drive the pin when detonated. Some powder-actuated drivers carry "clips" that allow for semi-automatic and fully automatic action (**Figure 16–18**).

FIGURE 16–17 Powder-actuated drivers are used for fastening into masonry or steel.

FIGURE 16–18 Powder-actuated driver with clip.

Courtesy of Hilti, Inc.

Drive Pins

Drive pins are available in a variety of sizes. Three styles are commonly used. The *headed* type is used for fastening material. The *threaded* type is used to bolt an object after the pin is driven. The *eyelet* type is used when attachments are to be made with wire (**Figure 16–19**).

Powder Charges

Powder charges are colour-coded according to strength. Learn the colour codes for immediate recognition of the strength of the charge. A stronger

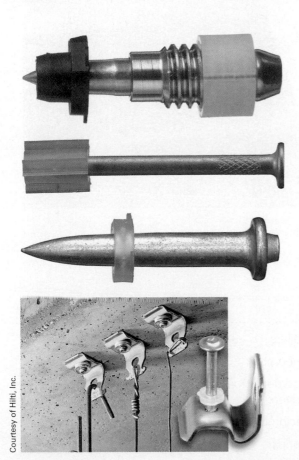

Courtesy of Hilti, Inc.

FIGURE 16–19 Drive pins.

charge is needed for deeper penetration or when driving into hard material. The strength of the charge must be selected with great care.

Because of the danger in operating these guns, some jurisdictions require the operator to be certified. Certificates may be obtained from the manufacturer's representative after a brief training course.

Using Powder-Actuated Drivers

- Study the manufacturer's directions for safe and proper use of the gun. Use eye and ear protection.

> **CAUTION**
>
> Treat this tool as if it were a gun. Never use it if the end that ejects the fastener is facing toward other workers.

- Make sure the drive pin will not penetrate completely through the material into which it is driven. This has been the cause of fatal accidents.
- To prevent ricochet hazard, make sure the recommended shield is in place on the nose of the gun. A number of different shields are available for special fastening jobs.
- Select the proper fastener for the job. Consult the manufacturer's drive pin selection chart to determine the correct fastener size and style.
- Select a powder charge of necessary strength. Always use the weakest charge that will do the job and gradually increase strength as needed. Load the driver with the pin first and the cartridge second.
- Keep the tool pointed at the work. Wear safety goggles. Press hard against the work surface, and pull the trigger. The resulting explosion drives the pin. Eject the spent cartridge.

> **CAUTION**
>
> If the gun does not fire, hold it against the work surface for at least 30 seconds. Then remove the cartridge according to the manufacturer's directions. Do not attempt to pry out the cartridge with a knife or screwdriver; most cartridges are rim-fired and could explode.

DECONSTRUCT THIS

Carefully study **Figure 16–20** and think about what is wrong and/or what is right. Consider all possibilities. What construction practice or method is different in your area of the country?

FIGURE 16–20 This photo shows a carpenter cutting a 2 × 6.

KEY TERMS

belt sander	grit	nailers	screw guns
chamfer	hammer-drills	pneumatic	staplers
circular saw	hole saws	powder-actuated drivers	stapling guns
cordless nailing guns	jigsaw		taper
cordless stapling guns	jointer plane	reciprocating saw	templates
crosscutting	laminate trimmer	ripping	
dovetails	masonry drill bits	router	

REVIEW QUESTIONS

Select the most appropriate answer.

1. To use a power tool properly, what should the operator NOT do?

   a. wear eye protection
   b. wear ear protection
   c. ignore the manufacturer's recommended instructions
   d. wear foot protection

2. What should NOT be done to the guard of a portable electric saw?

   a. It should not be lubricated.
   b. It should not be adjusted.
   c. It should not be retracted by hand.
   d. It should not be wedged open.

3. Which practice should be followed when selecting an extension cord for a power tool?

   a. use a longer cord to keep the cord from heating up
   b. keep the cord evenly spread out around the work area
   c. use one with a GFCI or plugged into a GFCI outlet
   d. use a cord that is rated for the amperage of the tool

4. When using a power tool for cutting, what should the operator wear?

   a. safety contact lenses and steel-toed work boots
   b. ear and eye protection
   c. stereo headphones
   d. all of the above

5. Which statement about sharp tools is correct?

   a. They put less stress on the operator than dull tools.
   b. They cut slower than dull tools.
   c. They are more dangerous to use than dull tools.
   d. They reduce the lifespan of the tools.

6. What is the jigsaw primarily used for making?

   a. curved cuts
   b. compound mitres
   c. cuts in drywall
   d. long straight cuts

7. What saw is primarily used for rough-in work?

   a. reciprocating saw
   b. circular saw
   c. sabre saw
   d. jig saw

8. What tool is best suited for drilling metal as well as wood?

   a. auger bit
   b. high-speed twist drill
   c. expansive bit
   d. speed bit

9. What should be done to produce a neat and clean hole in wood?

   a. use a fast-spinning, sharp bit
   b. use a slower travel speed
   c. finish the hole by drilling from the back side
   d. use a spade bit

10. When using the router to shape four outside edges and ends of a piece of stock, in what direction should the router be guided?

    a. direction with the grain
    b. clockwise direction
    c. counterclockwise direction
    d. direction against the grain

11. Because of the potential danger, which tool do some codes require the operator to be certified to use?

    a. powder-actuated driver
    b. hammer-drill
    c. cordless pneumatic nailer
    d. screw gun

12. Which direction should the saw arbor nuts that hold circular saw blades in position be rotated to loosen them?

    a. clockwise
    b. with the rotation of the blade
    c. counterclockwise
    d. against the rotation of the blade

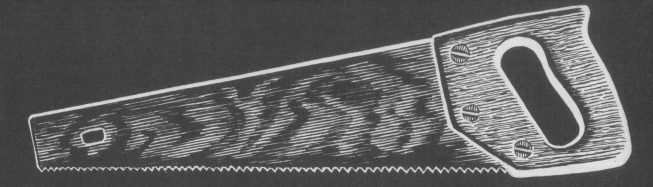

Unit 7
Stationary Power Tools

Many kinds of stationary power woodworking tools are used in wood mills and cabinet shops for specialized work. On the building site, usually only table and mitre saws are available for use by carpenters. These are not heavy-duty machines. They must be light enough to be transported from one job to another. These saws are ordinarily furnished by the contractor.

 SAFETY REMINDER

Stationary power tools are strong and durable, designed for heavy-duty tasks. While using power tools, the worker must keep a constant awareness of the risks of using the tool.

OBJECTIVES

After completing this unit, the student should be able to:

- describe different types of circular saw blades and select the proper blade for the job at hand.
- describe, adjust, and operate the power mitre saw to crosscut to length, making square and mitre cuts safely and accurately.
- describe, adjust, and operate the table saw safely to crosscut lumber to length, rip to width, and make mitres, compound mitres, dadoes, grooves, and rabbets.
- operate the table saw in a safe manner to taper, to rip, and to make cove cuts.

Chapter 17

Circular Saw Blades

All of the stationary power tools described in this unit use circular saw blades. To ensure safe and efficient saw operation, it is important to know which type to select for a particular purpose (**Figure 17–1**).

CIRCULAR SAW BLADES

The more teeth a saw blade has, the smoother the cut. However, a fine-toothed blade does not cut as fast. This means that the stock must be fed more slowly. A coarse-toothed saw blade leaves a rough surface but cuts more rapidly. Thus, the stock can be fed faster by the operator.

If the feed is too slow, the blade may overheat. This will cause it to lose its shape and wobble at high speed. This is a dangerous condition that must be avoided. An overheated saw will probably start to bind in the cut, possibly causing kickback and serious operator injury.

The same results occur when trying to cut material with a dull blade. Always use a sharp blade. Use fine-toothed blades for cutting thin, dry material, and coarse-toothed blades for cutting heavy, rough lumber.

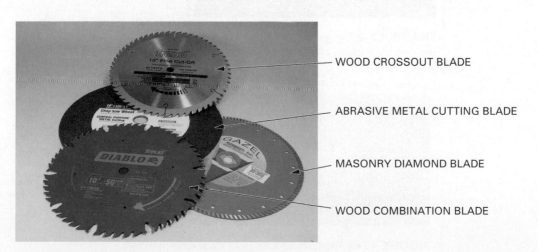

WOOD CROSSOUT BLADE

ABRASIVE METAL CUTTING BLADE

MASONRY DIAMOND BLADE

WOOD COMBINATION BLADE

FIGURE 17–1 Circular saw blades are made to cut different materials.

Types of Circular Saw Blades

Many different types of circular saw blades are used for various cutting applications. Blades are made to cut wood and wood products, plastic laminates, masonry, and steel. Each material has a different requirement for cutting, and the blades are designed accordingly (Figure 17–1).

Wood-cutting blades are diverse in design styles. They are loosely classified into two groups: high-speed steel blades and **tungsten carbide-tipped blades**. High-speed steel blades (HSS) are the original generation of circular blades. Teeth may be sharpened with a file and tend not to stay sharp as long as carbide-tipped blades. The carbide-tipped blades are superior and well-suited for cutting material that contains adhesives and other foreign material that rapidly dull high-speed steel blades. Tungsten carbide is a hard metal that can be sharpened only with diamond-impregnated grinding wheels. Most saw blades being used today are carbide-tipped. Resurfacing of the tips should be done by a professional blade sharpener.

In both groups of saw blades, the number and shape of the teeth vary to give different cutting actions according to the kind, size, and condition of the material to be cut. Masonry and steel cutting blades are made in general types as well as abrasive and steel types. Abrasive blades are made of composite materials that are consumed as the blade cuts. The blade diameter gets smaller with each cut, and it is easy to tell when the blade should be replaced. Steel blades designed to cut steel and masonry are carbide-tipped or diamond-coated. They tend to last longer and are more expensive than abrasive blades.

Classification of Circular Saw Blades

Circular Saw blades are classified as *rip*, *crosscut*, or *combination* blades. They may be given other names, such as plywood, panel, or flooring blades, but they are still in one of the three classifications.

Ripsaws. The ripsaw blade (**Figure 17–2**) usually has fewer teeth than crosscut or combination blades. As in the hand ripsaw, every tooth of the circular ripsaw blade is filed or ground at right angles to the face of the blade. This produces teeth with a cutting edge all the way across the tip of the tooth. These teeth act like a series of small chisels that cut and clear the stock ahead of them.

Use a ripsaw blade for cutting solid lumber with the grain when a smooth edge is not necessary. Also, use a ripsaw blade when cutting with the grain of unseasoned or green lumber and lumber of heavy dimension.

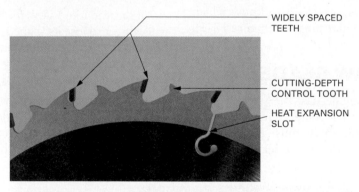

WIDELY SPACED TEETH

CUTTING-DEPTH CONTROL TOOTH

HEAT EXPANSION SLOT

FIGURE 17–2 The teeth of a rip blade have square-edged cutting tips with wide spaces between them. A carbide-tipped blade is shown.

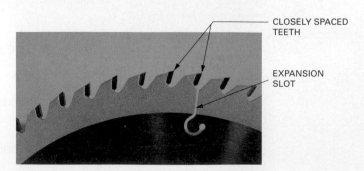

CLOSELY SPACED TEETH

EXPANSION SLOT

FIGURE 17–3 The teeth of a crosscut blade have bevelled sides and pointed tips.

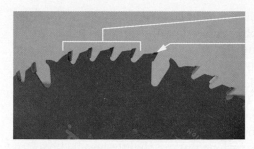

FIGURE 17-4 The teeth of a combination blade include both crosscut and rip teeth.

Crosscut Saws. The teeth of a crosscut circular saw blade are shaped like those of the crosscut handsaw (**Figure 17-3**). The sides of the teeth are alternately filed or ground on a bevel. This produces teeth that come up to points instead of edges.

Crosscut teeth slice through the wood fibres smoothly. A crosscut blade is an ideal blade for cutting across the grain of solid lumber. It also cuts plywood satisfactorily with little splintering of the cut edge.

Combination Saws. The combination blade is used when a variety of ripping and crosscutting is to be done. It eliminates the need to change blades for different operations.

There are several types of combination blades. One type has groups of teeth around its circumference (**Figure 17-4**). The leading tooth in each group is a rip tooth. The ones following are crosscut teeth.

Carbide-Tipped Blades

Carbide Blade Component Parts. The component parts of a carbide blade have functions that make the blade perform well (**Figure 17-5**). The teeth consist of bits of carbide and are silver-soldered to the metal blade. They should be the only part to actually touch the material. Carbide is made in different hardnesses for various applications. Drilling stone and masonry often requires hammering, thus softer carbide is used. The hardest carbide is used for cutting metal. The harder the carbide material, the more brittle the blade, which should be handled gently. A carbide blade can break like glass if it is dropped.

The *gullet* is a gap created to provide clearance for the material being removed, usually sawdust. The anti-vibration slots are cut in the blade with a laser to help keep the blade true while it is being run under heavy load. The anti-kickback teeth are designed to keep the tooth that follows them from biting too deeply into the material. Otherwise, the blade might kick back at the operator. The anti-kickback teeth usually are located before any large space in the blade. The hook angle is the angle at which the tooth is pitched forward. Larger hook angles cut more aggressively, making a faster but less smooth cut.

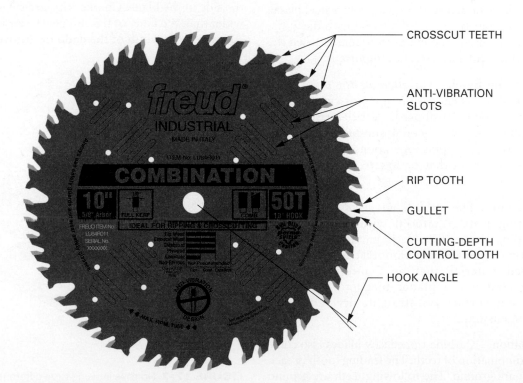

FIGURE 17-5 Parts of a typical carbide combination blade.

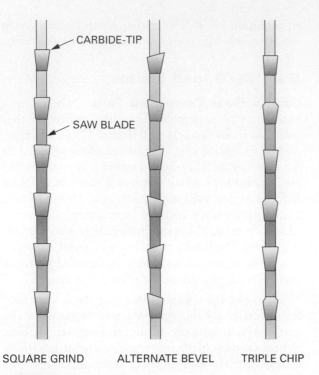

FIGURE 17–6 Three main tooth styles of carbide-tipped saw blades.

Carbide Tooth Style. The teeth of a carbide blade are ground to different shapes, giving the blade different cutting abilities (**Figure 17–6**).

Square Grind. The *square grind* (also known as raker teeth) is similar to the rip teeth in a steel blade. It is used primarily to cut solid wood with the grain. It can also be used on composition boards when the quality of the cut surface is not important.

Alternate Top Bevel. The *alternate top bevel* grind is used with excellent results for crosscutting solid lumber, plywood, hardboard, particleboard, fibreboard, and other wood composite products. It can also be used for ripping operations. However, the feed is slower than that of square or combination grinds.

Triple Chip. The *triple chip grind* is designed for cutting brittle material without splintering or chipping the surface. It is particularly useful for cutting an extremely smooth edge on plastic laminated material, such as countertops. It can also be used like a planer blade to produce a smooth cut surface on straight, dry lumber of small dimension.

Combination. Carbide-tipped saw blades also come with a combination of teeth. The leading tooth in each set is square ground. The following teeth are ground at alternate bevels. These saw blades are ideally suited for ripping and crosscutting solid lumber and all kinds of engineered wood products. Because of their versatility, these saw blades are probably the most widely used by carpenters.

Carbide-tipped teeth are not set (bent slightly outward). The carbide tips are slightly thicker than the saw blade itself. Therefore, they provide clearance for the blade in the saw cut. In addition, the sides of the carbide tips are slightly bevelled back to provide clearance for the tip.

Dado Blades

Dadoes, *grooves*, and *rabbets* can be cut using a single saw blade (**Figure 17–7**), but this would require making several passes of the material with the single blade. A *dado set*, which consists of more than one blade, is commonly used to make these cuts faster because only one pass needs to be made through the stock. One type, called the *stack dado head*, consists of two outside circular saw blades with several *chippers* of different thicknesses placed between (**Figure 17–8**).

Most dado sets make cuts from ⅛ to ¹³⁄₁₆ inch (3 to 21 mm) wide. This is done by using one or more of the blades and chippers together. Wide cuts are obtained by adding shims between chippers or by making more passes through the material. When installing this type of dado set, make sure the tips of the chippers are opposite the gullets and not against the side of the blade. Chipper tips are *swaged* (made wider than the body of the chipper) to ensure a clean cut across the width of the dado or groove.

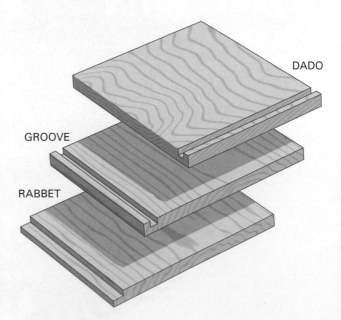

FIGURE 17–7 Notches in wood have different names depending on where they are located.

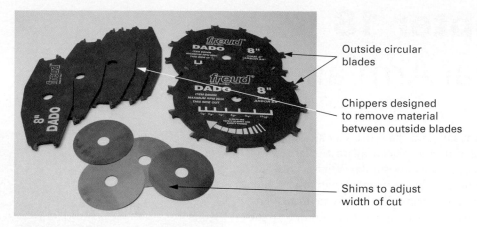

FIGURE 17–8 A stacked dado head set.

Outside circular blades

Chippers designed to remove material between outside blades

Shims to adjust width of cut

Another type of dado blade is the one-unit *adjustable dado head* (**Figure 17–9**), commonly called a *wobbler head*. This type can be adjusted for width of dado by rotating the sections of the head. The head can be adjusted by loosening the arbor nut and need not be removed from the saw arbor. This type of dado head does not make perfectly square-bottomed slots.

Removing and Replacing Circular Saw Blades

Carbide circular saw blades will at some time need to be sharpened or replaced. Several clues will help the operator realize that the blade is dull:

- Broken or chipped teeth are the most obvious sign of a dull blade. Remember that the teeth are brittle. Protect them from being struck by or dropped on hard objects.

- The feeding pressure of the material into the saw will seem to be increasing. The ability to notice this takes some experience with the saw.

- Burn marks are left on the material after the cut. Note that some material, such as cherry, burns easily even with a sharp blade. Keep the feed rate constant to reduce burning.

- Pitch and sawdust will build up on the blade. This buildup will cause blade drag, resulting in more heating up than normal.

Saw arbor shafts may have a right- or left-hand thread for the nut, depending on which side of the blade it is located. No matter what direction the arbor shaft is threaded, the arbor nut is loosened in the same direction in which the saw blade rotates (**Figure 17–10**). The arbor nut is always tightened against the rotation of the saw blade. This design prevents the arbor nut from loosening during operation.

FIGURE 17–9 A wobbler head dado blade.

FIGURE 17–10 Always unplug the saw before touching blade or cutters. The saw arbor nut is loosened by turning it in the same direction in which the blade rotates.

Chapter 18
Radial Arm and Mitre Saws

The radial arm saw and the mitre saw are similar in purpose. Materials are placed against a fence, and crosscutting operations are similar. While many cutting operations can be performed with the radial arm saw, it is used primarily for crosscutting. The motorized mitre saw is designed to crosscut at right angles as well as at compound mitre angles. The sliding compound mitre saw has the added feature of performing a dado cut.

RADIAL ARM SAWS

The major function of a **radial arm saw** (**Figure 18–1**) is crosscutting. The stock remains stationary while the saw moves across it. The operator does not have to push or pull the stock through the blade; instead, the saw is moved. The ends of long lengths are easily cut squared or mitred. The cut is made above the work and all layout lines are clearly visible. The radial arm saw can also be used for ripping operations.

The size of the radial arm saw is determined by the diameter of the largest blade that can be used. The arm of an industrial saw moves horizontally in a complete circle (**Figure 18–2**). The motor unit tilts

FIGURE 18–2 The arm of the radial saw moves horizontally in a complete circle.

to any desired angle and also rotates in a complete circle. The depth of cut is controlled by raising or lowering the arm. This flexibility allows practically any kind of cut to be made. Extreme care should be employed to follow all setup procedures exactly. Remember to wear eye and ear protection.

Crosscutting

For straight crosscutting, make sure the arm is at right angles to the fence. Adjust the depth of cut so that the teeth of the saw blade are about 1/16 inch (1.5 mm) below the surface of the table. With the saw all the way back and all guards in place, hold the stock against the fence. Make the cut by bringing the saw forward and cutting to the layout line (**Figure 18–3**).

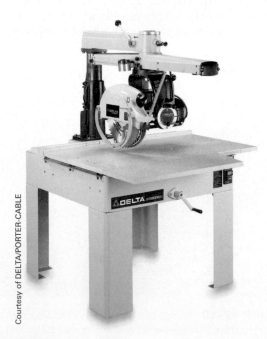

Courtesy of DELTA/PORTER-CABLE

FIGURE 18–1 The radial arm saw.

CAUTION

Do not cross your hands and arms. If the stock is on the left side of the blade, hold it with the left hand, using the right to pull the saw. If the stock is on the right, hold it with the right hand, using the left to pull the saw.

FIGURE 18-3 Straight crosscutting using the radial arm saw.

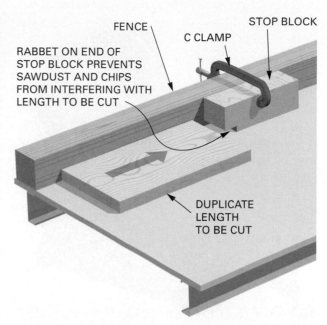

FENCE

C CLAMP

STOP BLOCK

RABBET ON END OF STOP BLOCK PREVENTS SAWDUST AND CHIPS FROM INTERFERING WITH LENGTH TO BE CUT

DUPLICATE LENGTH TO BE CUT

FIGURE 18-4 Technique for making a fence stop where sawdust will not interfere.

When crosscutting stock thicker than the capacity of the saw, cut through half the thickness. Then turn the stock over and make another cut.

> ## CAUTION
>
> - **Wear eye and ear protection. Also remember that loose clothing, such as gloves, is a hazard to the operator.**
> - **Watch the line of travel of the saw. Only the material being cut should be in the cutting line.**
> - **Pull saw gently into material. It will tend to feed itself into the stock, causing it to jam into the stock, so be prepared to resist this force.**
> - **Sharp blades tend to self-feed less than dull blades.**
> - **Bowed stock should be cut with the crown of the board down on the table. Otherwise the bow will tend to bind the blade as it is cut.**
> - **Replace the fence when the cuts in it are many or large. Material must be held securely by the fence.**

When the cut is complete, return the saw to the starting position, behind the fence. Turn off the power. As a safety precaution, all saws should automatically return to the retracted rest position if the operator should let go of the saw.

The radial arm saw can be set up to make repetitive cuts of precisely the same lengths (**Figure 18-4**). It can also be set up to rip narrow strips of stock. Because of the potential for the saw blade to jump forward into the cut, and because of the cumbersome size and weight of the machine, the radial arm saw has lost its place on the jobsite to the more portable motorized mitre saw. There is still a place for it in some furniture shops.

POWER MITRE SAW

The term **power mitre saw**, also called a *power mitre box*, refers to a tool that has a circular saw blade mounted above the base. It comes in a variety of styles and sizes. Operation of the saw is to push it down into the material using a chopping action. These tools are used in most phases of construction. A retractable blade guard provides operator protection and should be in place when the saw is being used.

The simplest of these designs is often referred to as a *chop box* (**Figure 18-5**). It comes in blades with 10-, 12-, 14-, and 16-inch (254, 304, 356, and 406 mm) diameters. The saw cannot be adjusted to cut angles; it can cut only in the up-and-down direction. To cut angles, the material itself must be clamped in place at the desired angle. Chop boxes are designed to cut metal, typically metal studs and pipe. An abrasive blade or a special carbide blade is used. Carbide blades cut cooler and do not spray sparks, as do abrasive blades.

FIGURE 18–6 Power mitre box. *(Reproduced with the permission of Black & Decker. © Black & Decker Inc.)*

In the more sophisticated models, the saw slides out on rails to increase the width of the cut. These models are referred to as *sliding compound mitre saws* (**Figure 18–7**). The 12-inch (305 mm) models can cut a 4½ × 12-inch (114 × 305 mm) block at 90 degrees, a 4½ × 8½-inch (114 × 216 mm) block at 45 degrees, and a 3½ × 8½-inch (89 × 216 mm) block to a compound 45-degree mitre. They can also be adjusted to cut compound angles that exceed 45 degrees. Some will cut up to 60 degrees in either the left or right direction. Some tilt for a bevel cut in both directions.

The sliding compound mitre saw (Figure 18–7) allows for cutting large crown mouldings with the stock lying flat on the base of the saw. The blade is adjusted to the compound angles using the preset stops on the saw. Check the manufacturer's

FIGURE 18–5 A chop box is designed to cut with a chopping action. *(Reproduced with the permission of Black & Decker. © Black & Decker Inc.)*

The power mitre saw designed for wood has more possible adjustments (**Figure 18–6**). The saw blade may be adjusted 45 degrees to the right or left. Some models may even tilt to the side: left, right, or both directions. It is possible with these saws to cut compound mitres. Positive stops are located in the saw-angle and the tilt-angle adjustments to help the operator cut common angles. Other angles are cut by reading the degree scale and locking in the desired position. Typical sizes include 8½-, 10-, and 12-inch (216, 254, and 304 mm) diameter blades.

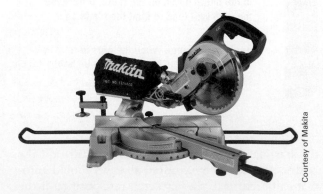

Courtesy of Makita

FIGURE 18–7 A cordless sliding compound mitre saw.

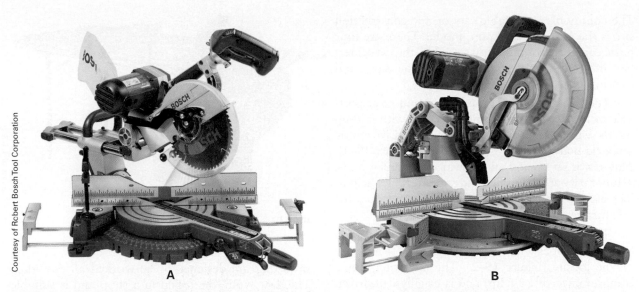

Courtesy of Robert Bosch Tool Corporation

FIGURE 18–8 (A) Sliding compound mitre saw with laser attachment; (B) dual-bevel glide mitre saw.

instructions for the saw to determine crown moulding cutting procedures. Some models have a laser attachment that lays a red line on the material before the cut is made, showing where the blade will enter the material (**Figure 18–8**). Bosch's new dual-bevel glide mitre saw incorporates an Axial-Glide™ system that provides consistent precision over the tool's life. The hinge system also allows for a compact workspace as it can be positioned close to a wall, saving up to 12 inches (305 mm) of workspace.

The operation of all of these types of mitre saws is primarily the same. Adjustments are made to the blade travel and then made secure with locking clamps. Material should be held firmly against the base and fence of the saw. Some models have a material hold-down clamp to ensure that the piece does not move during the cut. The saw is started and the blade eased into the material. With sliding models, the saw may be pulled toward the operator and then pushed down and back to cut the material. Rate of feed depends on the material being cut. Listen to the saw to determine the best feed rate. The saw should keep nearly its full rpm running sound when cutting; that is, it should not slow down much.

Chapter 19

Table Saws and Other Stationary Power Tools

The **table saw** is one of the most frequently used woodworking power tools. In many cases it is brought to the jobsite when the interior finish work begins. It is a useful tool because so many kinds of work can be performed with it. Common table saw operations, with different jigs to aid the process, are discussed in this chapter.

The other stationary power tools discussed in this chapter are more often found at the contractor's shop, where the materials for fixtures are milled, pre-built, and then brought to the site.

TABLE SAWS

The size of the table saw is determined by the diameter of the saw blade. It may measure from 8 inches (203 mm) up to 16 inches (406 mm).

The commonly used table saw on the construction site is the 10-inch (254 mm) model. There are four basic categories of table saws used by the carpenter: benchtop saws, contractor's saws, cabinet saws, and hybrid saws.

The **benchtop saw** is the lightest and most portable of the lot, and many models come with fold-up stands with wheels. It usually has a direct-drive motor, a plastic body, and an aluminum top (**Figure 19–1**). This saw is set up and taken away on a regular basis. Temporary out-feed tables and support tables help to increase the scope of the work that it can do and the safe manner in which it can be done.

The **contractor's saw** is a larger, heavier version but it also comes with wheels that allow for mobility on the jobsite (**Figure 19–2**). The induction motor is more powerful (1.5 hp) and is usually belt-driven and easily removable for transportation. It is wired for 110 V but can be adapted to run on 220 V.

The **cabinet saw** is a very heavy stationary saw that is meant to be fixed in a single location. The top is usually cast iron, and its massive weight and 3 to 5 hp motor (220 V) makes it more stable and more powerful (**Figure 19–3**). These are very accurate

FIGURE 19–3 Cabinet saw.

saws that are designed to be worked all day long. This saw would be found in a shop and is suitable for production cabinet work.

The **hybrid saw** sits between the contractor's saw and the cabinet saw (**Figure 19–4**). The motor is likely to be 1.75 hp, the top larger and made of steel, and the fence more substantial. It can also be fitted with wheels for portability and can be very prominent on commercial jobsites. All of the table saws mentioned will come with similar accessories and components (**Figure 19–5**).

The blade is adjusted for depth of cut and tilted up to 45 degrees by means of handwheels. A **rip fence** guides the work during ripping operations. A **mitre gauge** is used to guide the work when cutting square ends and mitres. The mitre gauge slides in grooves in the table surface. It may be turned and locked in any position up to 45 degrees.

A guard should always be placed on the blade to protect the operator. Most guards will come with a splitter and toothed anti-kickback pawls, and they will move with the blade as it goes up and down and tilts (**Figure 19–6**). Some guards are the overhead type

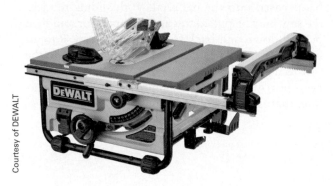

FIGURE 19–1 Benchtop (portable) table saw.

FIGURE 19–2 Contractor's saw.

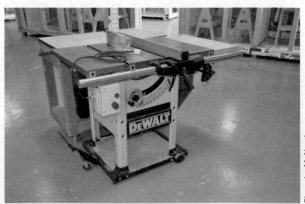

FIGURE 19–4 Hybrid table saw.

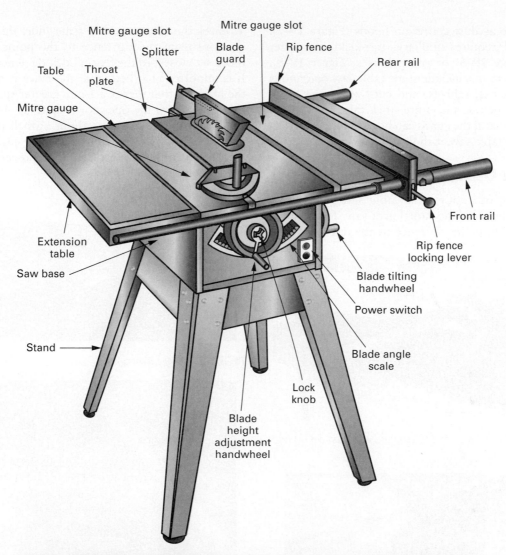

FIGURE 19–5 Diagram of a table saw (external). *(From Macdonald, Woodworking, 2E. © Delmar Learning, a part of Cengage Learning, Inc. Reproduced with permission.)*

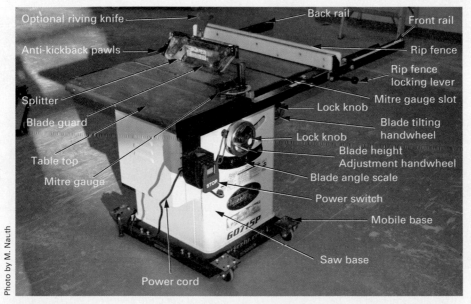

FIGURE 19–6 A table saw with guard and integral riving knife and anti-kickback pawls.

that double as dust collection hoods (**Figure 19–7**). These will require auxiliary anti-kickback/splitter units (**Figure 19–8**) or riving knives (**Figure 19–9**). Exceptions to this include some table saw operations such as dadoes, rabbets, and cuts where the blade does not penetrate the entire stock thickness. A general rule is that if the guard can be used for any operation on the table saw, it should be used.

Ripping Operations

For ripping operations, the table saw is easier to use and safer than the radial arm saw. To rip stock to width, adjust the rip fence to the desired width.

FIGURE 19–7 Overhead-type guard and dust collection.

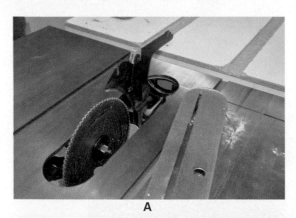

A

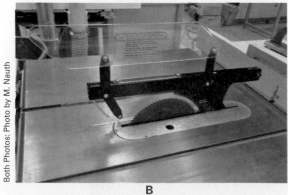

B

FIGURE 19–8 (A) Splitter, anti-kickback pawls, and (B) blade guard on a Powermatic table saw.

To check the accuracy of the scale under the *rip fence*, measure from the rip fence to the point of a saw tooth set closest to the fence. Lock the fence in place. It is important to check that the fence is parallel to the blade or that there's a 1 mm greater space on the side away from the operator (**Figure 19–10**). Adjust the height of the blade to about ¼ inch (6 mm) to ½ inch (12 mm) above the stock to be cut. With the stock clear of the blade, turn on the power.

FIGURE 19–9 Riving knife.

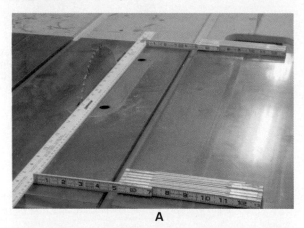

A

B

FIGURE 19–10 (A) Checking for parallelism between the blade and the mitre gauge slot. (B) Checking for parallelism between the mitre gauge slot and the fence. *(Both Photos: From Macdonald,* Woodworking, *2E. © Delmar Learning, a part of Cengage Learning, Inc. Reproduced with permission.)*

running on top of the fence, as in Figure 19–11. The riving knife and the anti-kickback teeth will keep the stock from drifting away from the fence and from lifting up off the tabletop. If the stock is not wide enough, use a *push stick* (**Figure 19–12**).

> ## CAUTION
>
> - **Make sure the stock is pushed all the way through the saw blade. Leaving the cut stock between the fence and a running saw blade may cause a kickback, injuring anyone in its path.**
> - **Use a push stick, especially when ripping narrow pieces.**
> - **Keep small cutoff pieces away from the running blade. Do not use your fingers to remove cutoff pieces; use the push stick or wait until the saw blade has stopped to do so.**
> - **Always use the rip fence for ripping operations. Never make freehand cuts.**
> - **Never reach over a running saw blade.**

Bevel Ripping

Bevel **ripping** is done in the same manner as straight ripping, except that the blade is tilted (**Figure 19–13**). The blade is adjustable from 0 to 45 degrees. Take care not to let the blade touch the rip fence. Also, keep the stock firmly in contact with the table. If the stock is allowed to lift off from the table during the cut, the width of the stock will vary.

Taper Ripping

Tapered pieces (one end narrower than the other) can be made with the table saw by using a **taper ripping jig.** The jig consists of a wide board with

> ## CAUTION
>
> - **Stand to either side of the blade's cutting line. Avoid standing directly behind the saw blade. Make sure no one else is in line with the saw blade in case of kickback.**
> - **Be prepared to turn the saw blade off at any moment should the blade bind.**
> - **A splitter and anti-kickback devices should be used when the cutting operation allows it.**

Hold the stock against the fence with the left hand. Push the stock forward with the right hand, holding the end of the stock (**Figure 19–11**). As the end approaches the saw blade, let it slip through the left hand. Remove the left hand from the work. Push the end all the way through the saw blade with the right hand, if the stock is of sufficient width (at least 5 inches [125 mm] wide). Keep your right hand

FIGURE 19–11 Using the table saw to rip lumber (guard has been removed for clarity).

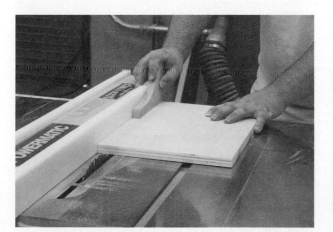

FIGURE 19–12 Use a push stick to rip narrow pieces (guard has been removed for clarity).

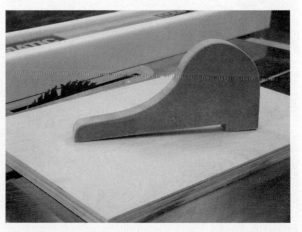

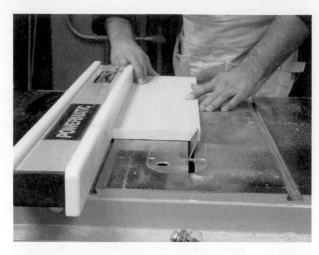

FIGURE 19–13 When bevel ripping, the blade is tilted (guard has been removed for clarity).

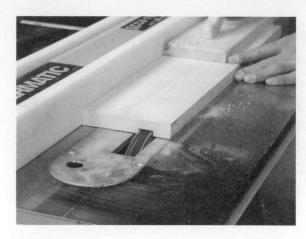

FIGURE 19–15 Cutting a groove with a dado head installed on the table saw.

the length and amount of taper cut out of one edge. The other edge is held against the rip fence. The stock to be tapered is held in the cutout of the jig. The taper is cut by holding the stock in the jig as both are passed through the blade (**Figure 19–14**).

By using taper ripping jigs, tapered pieces can be cut according to the design of the jig. A handle on the jig makes it safer to use. Also, if the jig is the same thickness as the stock to be cut, then the cutout section of the jig can be covered with a thin strip of wood to prevent the stock from flying out of the jig and back toward the operator.

Rabbeting and Grooving

Making *rabbets* is usually done using a dado head, which is composed of multiple blades and cutters. This tool makes fast, accurate grooves with one pass (**Figure 19–15**). Make sure that firm, even pressure

is applied to the table and fence. This will provide consistent and accurate depth cuts.

Rabbets and narrow *grooves* can also be made with a single blade when a dado head is unavailable or when only a few are needed. The operation takes two settings of the rip fence and two passes through the saw blade are required (**Figure 19–16**).

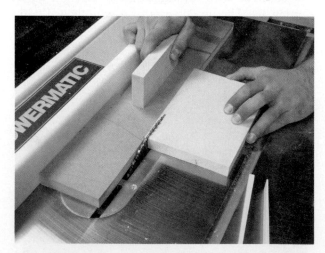

FIGURE 19–14 Using a taper ripping jig to cut identical wedges (guard has been removed for clarity).

FIGURE 19–16 Rabbeting an edge using a single saw blade by making two passes.

For narrow grooves, one or more passes are required. Move the rip fence slightly with each pass until the desired width of groove is obtained.

Crosscutting Operations

For most crosscutting operations, the *mitre gauge* is used. To cut stock to length with both ends squared, first check the mitre gauge for accuracy. Hold a framing square against it and the side of the saw blade. Usually the mitre gauge is operated in the left-hand groove. The right-hand groove is used only when it is more convenient.

Square one end of the stock by holding the work firmly against the mitre gauge with one hand while pushing the mitre gauge forward with the other hand. Measure the desired distance from the squared end. Mark on the front edge of the stock. Repeat the procedure, cutting to the layout line (**Figure 19–17**).

> ## CAUTION
>
> Do not use the mitre gauge and the rip fence together on opposite sides of the blade unless a stop block is used. Pieces cut off between the blade and the fence can easily bind and be hurled across the room or at the operator.

Cutting Identical Lengths

When a number of identical lengths need to be cut, first square one end of each piece. Clamp a

stop to an *auxiliary wood fence* installed on the mitre gauge. Place the square end of the stock against the stop block and then make the cut. Slide the remaining stock across the table until its end comes in contact with the stop block and then make another cut. Continue in this manner until the desired number of pieces are cut (**Figure 19–18**).

Another method for cutting identical lengths uses a block clamped to the rip fence in a location such that once the cut is made, there is clearance between the cut piece and the rip fence. The fence is adjusted so that the desired distance is between the face of the stop block and saw blade.

Square one end of the stock, slide the squared end against the stop block, and then make a cut. Continue making cuts in this manner until the desired number of pieces are obtained (**Figure 19–19**).

FIGURE 19–18 Cutting identical lengths using a stop block on an auxiliary fence of the mitre gauge (guard has been removed for clarity).

FIGURE 19–17 Using the mitre gauge as a guide to crosscut (guard has been removed for clarity).

FIGURE 19–19 Using the rip fence and block as a stop to cut identical lengths (guard has been removed for clarity).

StepbyStep Procedures

Procedure 19–A Making a Table Saw Jig to Cut 45-Degree Mitres

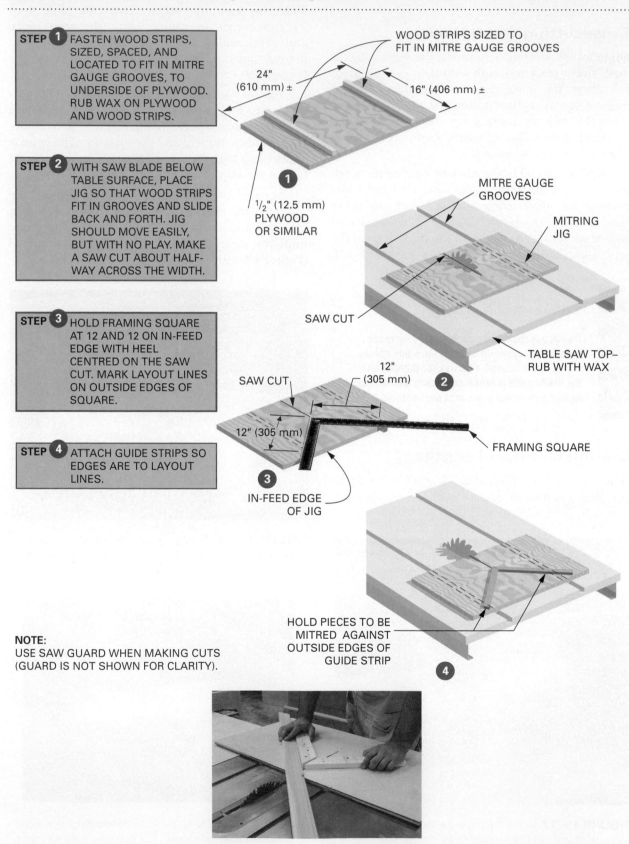

STEP 1 FASTEN WOOD STRIPS, SIZED, SPACED, AND LOCATED TO FIT IN MITRE GAUGE GROOVES, TO UNDERSIDE OF PLYWOOD. RUB WAX ON PLYWOOD AND WOOD STRIPS.

STEP 2 WITH SAW BLADE BELOW TABLE SURFACE, PLACE JIG SO THAT WOOD STRIPS FIT IN GROOVES AND SLIDE BACK AND FORTH. JIG SHOULD MOVE EASILY, BUT WITH NO PLAY. MAKE A SAW CUT ABOUT HALF-WAY ACROSS THE WIDTH.

STEP 3 HOLD FRAMING SQUARE AT 12 AND 12 ON IN-FEED EDGE WITH HEEL CENTRED ON THE SAW CUT. MARK LAYOUT LINES ON OUTSIDE EDGES OF SQUARE.

STEP 4 ATTACH GUIDE STRIPS SO EDGES ARE TO LAYOUT LINES.

WOOD STRIPS SIZED TO FIT IN MITRE GAUGE GROOVES

24" (610 mm) ±

16" (406 mm) ±

1/2" (12.5 mm) PLYWOOD OR SIMILAR

MITRE GAUGE GROOVES

MITRING JIG

SAW CUT

TABLE SAW TOP– RUB WITH WAX

SAW CUT

12" (305 mm)

12" (305 mm)

FRAMING SQUARE

IN-FEED EDGE OF JIG

HOLD PIECES TO BE MITRED AGAINST OUTSIDE EDGES OF GUIDE STRIP

NOTE:
USE SAW GUARD WHEN MAKING CUTS (GUARD IS NOT SHOWN FOR CLARITY).

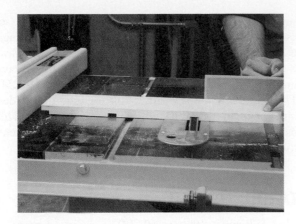

FIGURE 19–20 Cutting dadoes using a dado head.

Mitring

Flat mitres are cut in the same manner as square ends, except the mitre gauge is turned to the desired angle. *End mitres* are made by adjusting the mitre gauge to a square position and making the cut with the blade tilted to the desired angle.

Compound mitres are cut with the mitre gauge turned and the blade tilted to the desired angles.

A *mitring jig* can be used with the table saw. **Procedure 19–A** shows the construction and use of the jig. Use of such a jig eliminates turning the mitre gauge each time for left- and right-hand mitres.

Dadoing

Dadoing is done in a similar manner as crosscutting except a dado set is used (**Figure 19–20**). The dado set is used to cut only *partway* through the stock thickness.

TABLE SAW AIDS

A very useful aid is an *auxiliary fence*. A straight piece of ¾-inch (19 mm) plywood about 2 inches (50 mm) wide and as long as the ripping fence is screwed or bolted to the metal fence. When cuts must be made close to the fence, the use of an auxiliary wood fence prevents the saw blade from cutting into the metal fence. Also, the additional height provided by the fence gives a broader surface to steady wide work when its edge is being cut. The auxiliary wood fence also provides a surface on which to clamp **feather boards**. An auxiliary fence is also useful when attached to the mitre gauge.

Feather boards are useful aids to hold work against the fence as well as down on the table surface during ripping operations (**Figure 19–21**). Feather boards may be made of 1 × 6-inch (19 × 140 mm) nominal lumber, with one end cut at a 45-degree angle. Saw cuts are made in this end about ¼ inch (6 mm) apart. This gives the end some spring and allows pressure to be applied to the piece being ripped while also allowing the ripped piece to move.

A

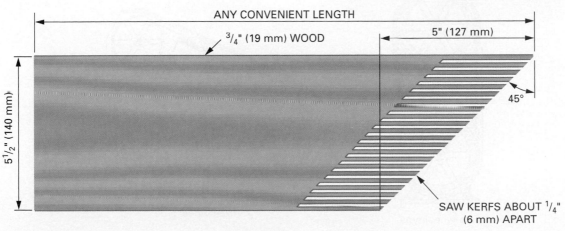

B

FIGURE 19–21 (A) Feather boards are useful aids to hold work during table saw operations (guard has been removed for clarity). Note that the locations of the feather boards do not cause the saw blade to bind. (B) Typical feather board design.

BAND SAWS

The band saw (**Figure 19–22**) consists of a narrow, flexible saw blade that is welded into a "band" or loop and is rotated around two or three wheels of equal diameter (**Figure 19–23**). The blade is kept on track by tensioning and aligning adjustments and by supporting guides that are positioned in place so as to just allow the passage of the rotating blade (**Figure 19–24**). Because of the narrow blade, the band saw is used to make curved, irregular cuts and to resaw lumber into boards or bevelled siding (**Figure 19–25**). The band saw comes with a mitre gauge that slides into a slot in the table top, which can be tilted. The band saw blades come in various widths with different tooth configurations (**Figure 19–26**).

DRILL PRESS

The drill press (**Figure 19–27**) is designed to produce holes in precise locations on the work piece. The wood is secured to the work table and the drill bit is slowly lowered into the exact desired location. The table can be lowered or raised, tilted, or rotated out of the way. A stop can be placed on the drill press arm in order to produce holes of exact depth. Many accessories are available to do other functions, such as drum sanding, cutting dowels, and making rosettes. (**Figures 19–28, 19–29**, and **19–30**). See the chart in **Figure 19–31** for recommended operating speeds.

JOINTERS

The jointer (**Figure 19–32**) is used primarily to produce "dressed" boards from rough lumber. It is also used to "joint" the edges of narrow boards prior to gluing them together to make one wider board.

It consists of an in-feed table, a rotating cutter head, and an out-feed table (**Figure 19–33**). Some rotating cutters are helical, with many four-sided

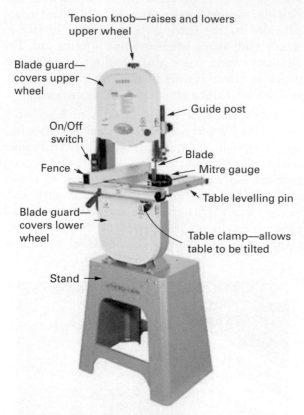

FIGURE 19–22 Diagram of a band saw (external). *(From Macdonald,* Woodworking, *2E. © Delmar Learning, a part of Cengage Learning, Inc. Reproduced with permission.)*

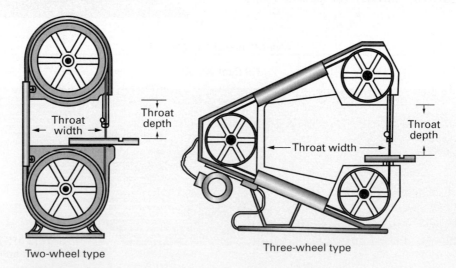

FIGURE 19–23 Two- and three-wheeled band saws. *(From Macdonald,* Woodworking, *2E. © Delmar Learning, a part of Cengage Learning, Inc. Reproduced with permission.)*

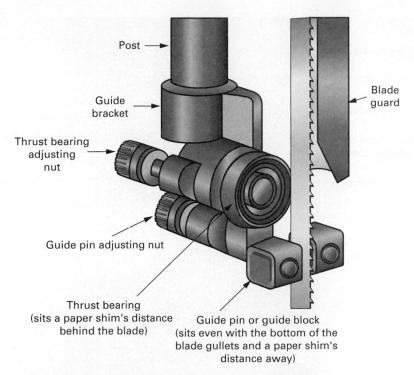

Post

Guide bracket

Thrust bearing adjusting nut

Guide pin adjusting nut

Blade guard

Thrust bearing (sits a paper shim's distance behind the blade)

Guide pin or guide block (sits even with the bottom of the blade gullets and a paper shim's distance away)

FIGURE 19–24 Thrust bearings and guide blocks support the band saw blade above and below the table. *(From Macdonald,* Woodworking, *2E. © Delmar Learning, a part of Cengage Learning, Inc. Reproduced with permission.)*

FIGURE 19–25 The band saw may be used to resaw thick boards into thinner pieces.

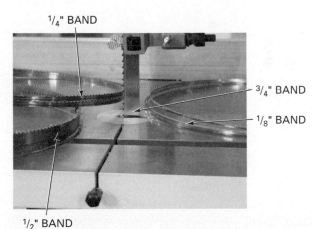

¼" BAND

¾" BAND

⅛" BAND

½" BAND

FIGURE 19–26 Bands come in various widths

cutters arranged in a spiral around the head (**Figure 19–34**). The two tables must be secured exactly parallel to each other, with the in-feed table slightly lower than the out-feed table ($\frac{1}{16}$ to $\frac{1}{8}$ of an inch [1.5 to 3 mm], depending on the hardness of the wood) (**Figure 19–35**). The work piece is held flat on the in-feed table with the aid of push plates; it is then fed into the rotating cutter head at a steady pace.

After the first pass, the wood should be inspected for chipping; if it has chipped, it is rotated end-for-end and fed into the jointer a second time. The rough lumber is passed several times consecutively until the entire surface of the face is flat and smooth (**Figure 19–36**).

Once the face of the lumber is smooth (**Figure 19–37**), the dressed face is held flat against the

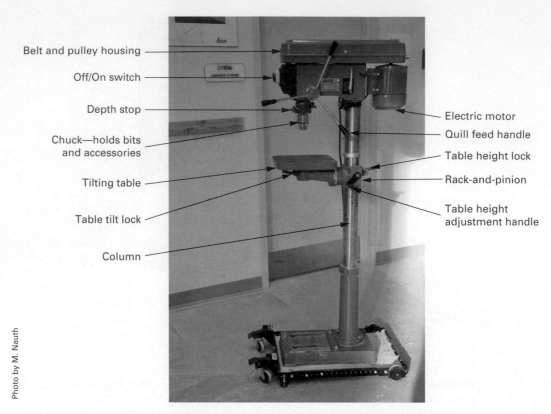

Belt and pulley housing

Off/On switch

Depth stop

Chuck—holds bits and accessories

Tilting table

Table tilt lock

Column

Electric motor

Quill feed handle

Table height lock

Rack-and-pinion

Table height adjustment handle

Photo by M. Nauth

FIGURE 19–27 Diagram of a drill press.

Circle cutter

Brad point bit

Countersink bit

Plug cutter

Forstner bit

FIGURE 19–28 Drill bits used on a drill press. *(From Macdonald, Woodworking, 2E. © Delmar Learning, a part of Cengage Learning, Inc. Reproduced with permission.)*

FIGURE 19–29 Drum sander on a drill press. *(From Macdonald, Woodworking, 2E. © Delmar Learning, a part of Cengage Learning, Inc. Reproduced with permission.)*

FIGURE 19–30 Rosette cutter. *(From Macdonald, Woodworking, 2E. © Delmar Learning, a part of Cengage Learning, Inc. Reproduced with permission.)*

fence, which is secured at 90 degrees. Repeated passes will produce a finished edge that is at right angles to the face. The piece is then ripped in the table saw about ⅛ of an inch (3 mm) wider than the desired measure. Passes are then made in the jointer to "true" the second edge. The final step is to place the board into the thickness planer (see the next section) and dress the opposite face parallel, smooth, and to the desired thickness.

For instructions on how to set up the jointer correctly, visit *http://www.ts-aligner.com/jointer.htm*.

Recommended operating speeds (RPM)

Accessory	Softwood (pine)	Hardwood (hard maple)
Twist drill bits		
$1/16" – 3/16"$	3000	3000
$1/4" – 3/8"$	3000	1500
$7/16" – 5/8"$	1500	750
$11/16" – 1"$	750	500
Brad-point bits		
$1/8"$	1800	1200
$1/4"$	1800	1000
$3/8"$	1800	750
$1/2"$	1800	750
$5/8"$	1800	500
$3/4"$	1400	250
$3/8"$	1200	250
$1"$	1000	250
Forstner bits		
$1/4" – 3/8"$	2400	700
$1/2" – 5/8"$	2400	500
$3/4" – 1"$	1500	500
$1 1/8" – 1 1/4"$	1000	250
$1 3/8" – 2"$	500	250
Hole saws		
$1" – 1 1/2"$	500	350
$1 5/8" – 2"$	500	250
$2 1/8" – 2 1/2"$	250–500	NR

Accessory	Softwood (pine)	Hardwood (hard maple)
Spade bits		
$1/4" – 1/2"$	2000	1500
$5/8" – 1"$	1750	1500
$1 1/8" – 1 1/2"$	1500	1000
Circle cutters		
$1 1/2" – 3"$	500	250
$3 1/4" – 6"$	250	250
Shear-cutting countersinks		
$1/4" – 3/8"$	1000	1000
	750	700
Countersinks		
2-flute	1400	1400
5-flute	1000	750
Countersink screw pilot bits		
All sizes	1500	1000
Taper drill bits with countersinks		
All sizes	500	250
Plug cutters		
All sizes	1000	500
Drum sanders		
Hard rubber	750	1500
Soft sleeveless	500	750
3" pneumatic	1750	1750

FIGURE 19–31 Drill press speed chart.

From Macdonald, Woodworking, 2E. © Delmar Learning, a part of Cengage Learning, Inc. Reproduced with permission.

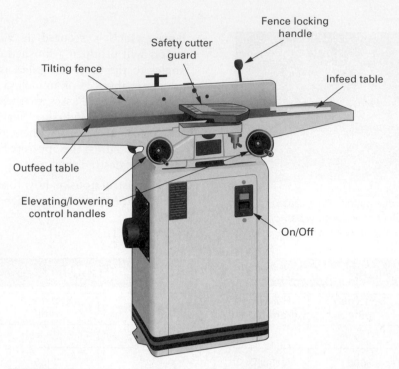

FIGURE 19–32 Diagram of a jointer. *(From Macdonald, Woodworking, 2E. © Delmar Learning, a part of Cengage Learning, Inc. Reproduced with permission.)*

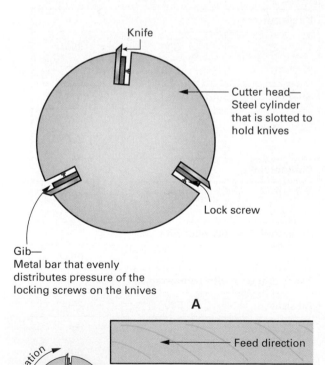

A

B

FIGURE 19–33 (A) Jointer knives are arranged radially in a cylindrical cutter head. (B) Jointer knives cut into the material opposite the feed direction. *(Both Figures: From Macdonald, Woodworking, 2E. © Delmar Learning, a part of Cengage Learning, Inc. Reproduced with permission.)*

FIGURE 19–34 A helical cutter head has many square or rectangular shaped cutters arrayed around the cutter head in a spiral.

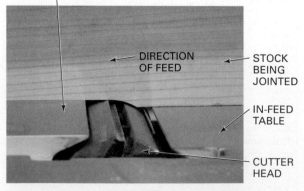

NOTE: STOCK TOUCHES BOTH TABLES DURING THE CUTTER OPERATION

FIGURE 19–35 The out-feed table is fixed to the height of the cutters and the in-feed adjusts for the depth of cut.

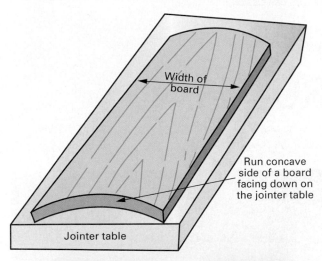

FIGURE 19–36 A professional 8-inch (200 mm) jointer.

Width of board

Run concave side of a board facing down on the jointer table

Jointer table

THICKNESS PLANERS

The thickness planer (Figure 19–38) is true to its name—it planes wood to a specified thickness, leaving a smooth surface that is parallel to the first face. It consists of three main components: a cutter head (with straight knives or a helical arrangement of four-sided cutters), in-feed and out-feed rollers, and a table that adjusts the gap between the cutters and the tabletop (Figure 19–39). The planer removes material from the top of the board so that the stock is fed in with the smooth face down. It is recommended that, for the first pass, the table be set at a thickness that is just less ($\frac{1}{32}$ of an inch, or 1 mm) than the thickest part of the stock to be dressed. Motorized in-feed rollers pull the wood into the cutters at the correct feed rate for the type of wood. See the operator's manual for proper setup, adjustments, and speeds (one example can be found at *http:// www.deltamachinery.com/downloads/manuals/ planers/22-450/Manual-En1349960.pdf.*)

After the first pass, the piece of stock should be checked for chipping; if it has chipped, it should be fed into the planer with the opposite end first. Successive passes are made until the board is at the desired thickness. Each pass should not remove more wood than allowed by the manufacturer's specifications.

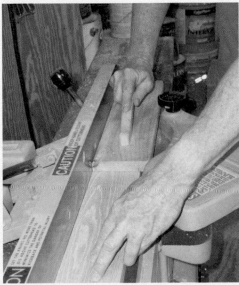

FIGURE 19–37 Joint first with the concave face of the board down on the in-feed table with downward pressure on the out-feed table. *(Figure and Photo: From Macdonald, Woodworking, 2E. © Delmar Learning, a part of Cengage Learning, Inc. Reproduced with permission.)*

FIGURE 19–38 A 20-inch (500 mm) thickness planer.

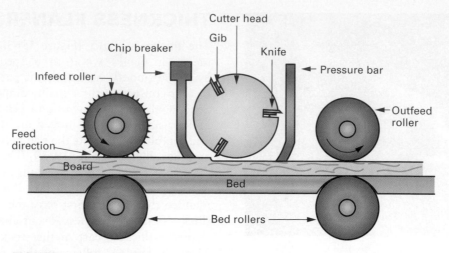

FIGURE 19–39 Cutting action of a planer. *(From Macdonald,* Woodworking, *2E. © Delmar Learning, a part of Cengage Learning, Inc. Reproduced with permission.)*

DECONSTRUCT THIS

Carefully study **Figure 19–40** and think about what is wrong and/or what is right. Consider all possibilities. What construction practice or method is different in your area of the country?

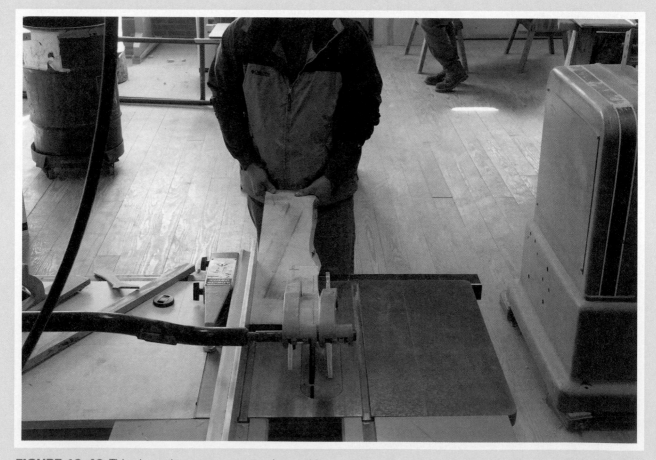

FIGURE 19–40 This photo shows a carpenter using a table saw.

KEY TERMS

band saw	contractor's saw	jointer	table saw
benchtop saw	crosscut circular saw blade	mitre gauge	taper ripping jig
bevel ripping	drill press	power mitre saw	thickness planer
cabinet saw	feather boards	radial arm saw	tungsten carbide-tipped blades
carbide blade	high-speed steel blades	rip fence	
combination blade	hybrid saw	ripsaw blade	

REVIEW QUESTIONS

Select the most appropriate answer.

1. To use a power tool properly, what should the operator NOT do?

   a. wear eye protection
   b. wear ear protection
   c. ignore the manufacturer's recommended instructions
   d. wear foot protection

2. What is the most frequently used blade in general carpentry?

   a. combination planer blade
   b. combination carbide-tipped blade
   c. square-grind carbide-tipped blade
   d. ripsaw blade

3. What is the alternate top bevel grind, carbide-tipped circular saw blade designed to do?

   a. rip solid lumber
   b. crosscut solid lumber
   c. cut in either direction of the grain
   d. cut green lumber of heavy dimension

4. What device is used with stationary saws to make repeated cuts of the same length?

   a. rabbet
   b. feather board
   c. stop block
   d. a pushstick

5. What saw is best suited for crosscutting and ripping material?

   a. power mitre box
   b. radial arm saw
   c. table saw
   d. chop saw

6. Which of the following is NOT essential to safe table saw use?

   a. Stand away from the line of the blade.
   b. Know the location of the shut-off switch.
   c. Keep wood scraps away from the blade.
   d. Wear work gloves at all times.

7. What is the name of the table saw guide used for cutting material with the grain?

   a. rip fence
   b. tilting arbor
   c. mitre gauge
   d. ripping jig

8. What tool is designed to hold a piece of stock safely while being cut with a table saw?

   a. mitre gauge
   b. push stick
   c. feather board
   d. all of the above

9. How should a dado be cut?

   a. across the grain
   b. with the grain
   c. in either direction of the grain
   d. close to the edge of the material

10. What is the table saw accessory that should not be used at the same time as a rip fence?

    a. push stick
    b. mitre gauge
    c. dado head
    d. feather board

11. Which tool is used to cut a circular design into the face of a board?

    a. table saw
    b. band saw
    c. drill press
    d. radial arm saw

12. Which tool is used to "dress" the edge of a board?

    a. table saw
    b. band saw
    c. jointer
    d. thickness planer

13. Which tool is used to cut the edge of a board parallel to its other edge?

    a. table saw
    b. band saw
    c. jointer
    d. thickness planer

14. Which tool is used to resaw 4 × 4 (89 × 89 mm) lumber into 1 × 4 (19 × 89 mm) boards?

    a. table saw
    b. band saw
    c. jointer
    d. thickness planer

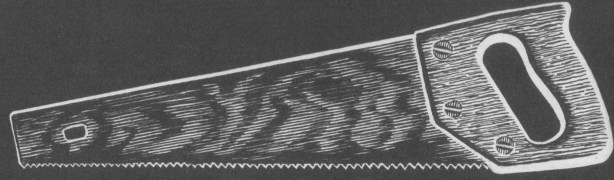

Unit 8

Architectural Plans and Building Codes

The ability to interpret architectural plans and to understand building codes, zoning ordinances, building permits, and inspection procedures is necessary for promotion on the job. With this ability, the carpenter has the competence needed to handle jobs of a supervisory nature, including those of foreman, superintendent, and contractor. All of the measurements on the sets of plans in Unit 8 will be in the imperial system.

OBJECTIVES

After completing this unit, the student should be able to:

- describe and explain the function of the various kinds of drawings contained in a set of architectural plans.
- demonstrate how specifications are used.
- read and use an architect's scale.
- identify various types of lines and read dimensions.
- identify and explain the meaning of symbols and abbreviations used on architectural plans.
- read and interpret plot, foundation, floor, and framing plans.
- locate and explain information found in exterior and interior elevations.
- identify and utilize information from sections and details.
- use schedules to identify and determine location of windows, doors, and interior finish.
- define and explain the purpose of building codes and zoning laws.
- explain the requirements for obtaining a building permit and the duties of a building inspector.

 SAFETY REMINDER

A set of building prints is the vehicle used to communicate to the tradesperson the desired structure. Failing to read and understand prints properly can have costly and potentially disastrous results.

Chapter 20

Understanding Architectural Plans

ARCHITECTURAL PLANS

Blueprinting was an early method of creating drawings and making copies. It used a water wash to produce prints with white lines against a dark blue background. Although these were true blueprints, the process is seldom used today. The word *blueprint* remains in use, however, to mean any copy of the original drawing.

A more recent method of making prints uses the original black line drawings of graphite or ink on translucent paper, cloth, or polyester film. Copies are usually made with a *diazo* printer. The copying process utilizes an ultraviolet light that passes through the paper but not the lines. The light causes a chemical reaction on the copy paper, except where the lines are drawn. The copy paper is then exposed to ammonia vapour. This vapour develops the remaining chemical into blue, black, or brown lines against a white background. The diazo printing process is faster and less expensive than the early blueprinting methods.

Today, most architectural plans, also called architectural drawings or construction drawings, are done using a computer-aided drafting (CAD) program. CAD drawings are produced more quickly and more easily than drawings done by hand. Changes to the plan can be made faster, allowing them to be easily customized to the desires of the customer. Once the plans are finished, they are sent to a plotter, which prints as many copies as needed.

Prints are drawn using symbols and a standardized language. To adequately read and interpret these prints, the builder must be able to understand the language of architectural plans. While this is not necessarily difficult, it can be confusing.

TYPES OF VIEWS

The architect can choose from any of several types of views to describe a building with clarity, and often multiple views are prepared. Because many people and many different trades are needed to build a house, different views are required that contain different information.

Pictorial View

Pictorial drawings are usually three-dimensional (3D) *perspective* or isometric views (**Figure 20–1**). The lines in a perspective view diminish in size as they converge toward vanishing points located on a line called the *horizon*. In an isometric drawing, the

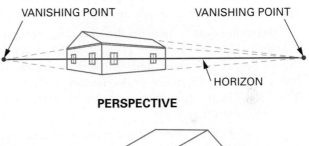

PERSPECTIVE

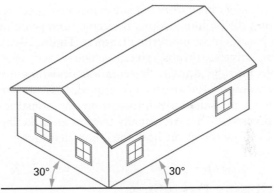

ISOMETRIC

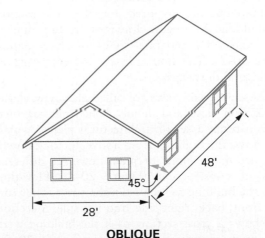

OBLIQUE

FIGURE 20–1 Pictorial views used in architectural drawings.

FIGURE 20–2 A presentation drawing is usually a perspective view.

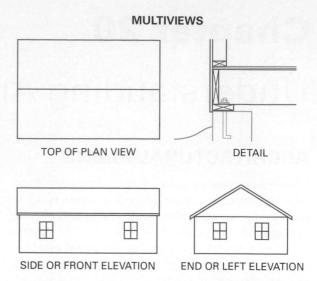

FIGURE 20–3 The two-dimensional views used in architectural drawings.

horizontal lines are drawn at 30-degree angles. All lines are drawn to actual scale. They do not diminish or converge as in perspective drawings.

The other type of pictorial drawing is the oblique drawing, cabinet or cavalier. The face of the object is drawn flat at full scale, and its depth is drawn at half-scale at 30 degrees or at 45 degrees to the face. This view distorts the image and thus is rarely used as a presentation drawing.

A *presentation drawing* is usually a perspective. It shows the building from a desirable vantage point to display its most interesting features (**Figure 20–2**). Walks, streets, shrubs, trees, vehicles, and even people may be drawn. Presentation drawings are often coloured for greater eye appeal. Presentation drawings provide little information for construction purposes. They are usually used as a marketing tool to show the client how the completed building will look.

Multiview Drawings

Different kinds of *multiview* drawings, also called **orthographics**, are required for the construction of a building. Multiview drawings are two dimensional and offer separate views of the building (**Figure 20–3**). The bottom view is never used in architectural drawings.

These different views are called plan view, elevation, section view, and details. When put together, they constitute a set of architectural plans. A **plan view** shows the building from above, looking down. There are many kinds of plan views for different stages of construction (**Figure 20–4**). **Elevation** shows the building as seen from the street. They can show front, back, right side, and left side. A **section view** shows a cross-section, as if the building were sliced open to reveal its skeleton (**Figure 20–5**). **Detail** views are blown up, zoomed-in views of various items to show a closer view.

Plan Views

Plot Plan. The *plot plan* shows information about the lot, such as property lines and directions, and measurements for the location of the building, walks, and driveways (**Figure 20–6**). It shows elevation heights and the direction of the sloping ground. The drawing simulates a view looking down from a considerable height. It is made at a small scale because of the relatively large area it represents. This view helps the builder locate the building on the site and helps local officials to estimate the impact of the project on the lot.

Foundation Plan. The *foundation plan* (**Figure 20–7**) shows a horizontal cut through the foundation walls. It shows the shape and dimensions of the foundation walls and footings. Windows, doors, and stairs located in this level are included. First-floor framing material size, spacing, and direction are sometimes included.

Floor Plan. The *floor plan* is a view of the horizontal cut made about 4 to 5 feet (1.2 to 1.5 m) above the floor. It shows the locations of walls and partitions, windows, and doors and fixtures appropriate for each room (**Figure 20–8**). Dimensions are included.

Framing Plan. *Framing plans* are not always found in a set of architectural drawings. When used, they may be of the floor or roof framing. They show the support beams and girders as well as the size, direction, and spacing of the framing members (**Figure 20–9**).

Elevations

Elevations are a group of drawings (**Figure 20–10**) that show the shape and finishes all sides of the exterior of a building. *Interior elevations* are

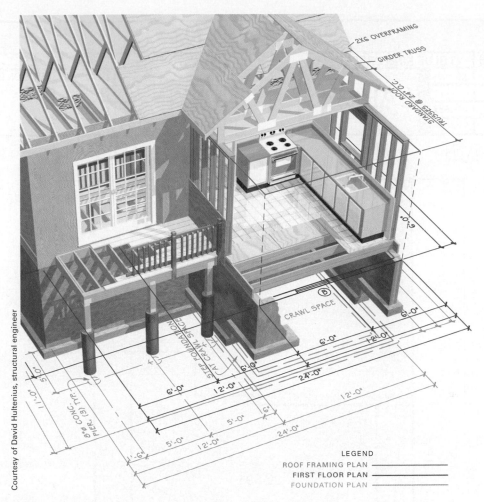

2X6 OVERFRAMING

GIRDER TRUSS

STANDARD ROOF TRUSSES @ 24" O.C.

9'-0"

CRAWL SPACE

Ⓑ

6'-0"

6'-0"

12'-0"

24'-0"

6'-0"

12'-0"

5'-0"

6"

STEP FOUNDATION AT CRAWL SPACE

6"x8" CONC. PIER. (3) TYP.

11'-0"

5'-0"

5'-0"

12'-0"

24'-0"

1'-6"

Courtesy of David Hultenius, structural engineer

LEGEND
ROOF FRAMING PLAN ─────────
FIRST FLOOR PLAN ─────────
FOUNDATION PLAN ─────────

FIGURE 20–4 Plan views are horizontal cut views through a building.

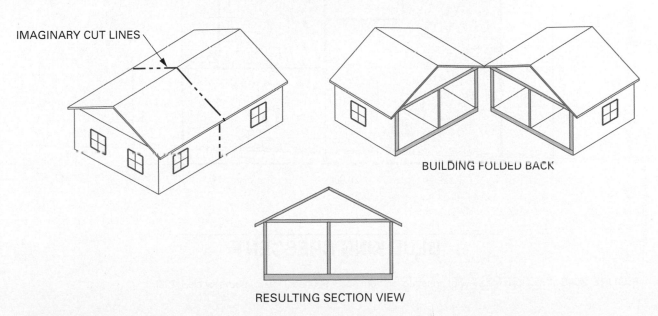

IMAGINARY CUT LINES

BUILDING FOLDED BACK

RESULTING SECTION VIEW

FIGURE 20–5 Section views are cutaway views.

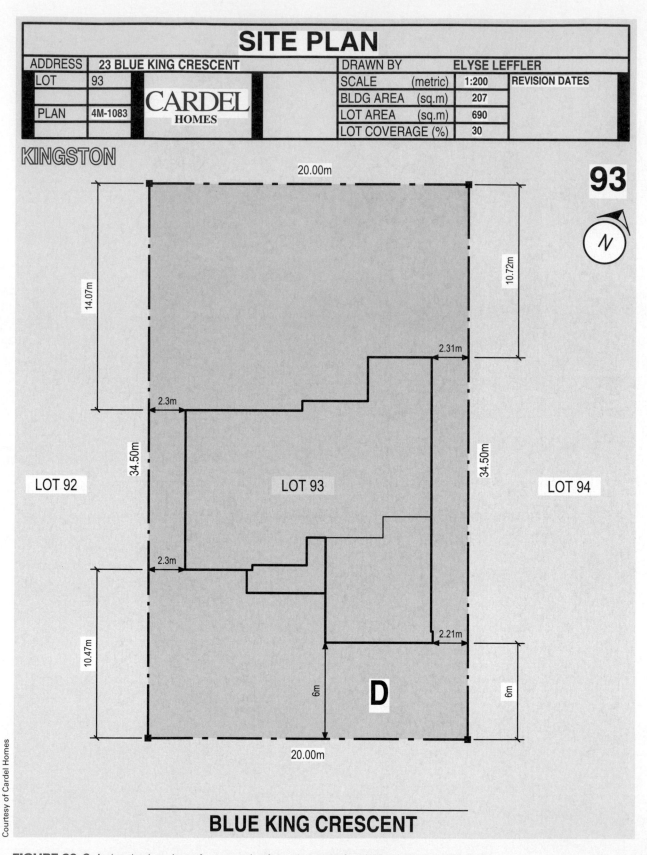

SITE PLAN

ADDRESS	23 BLUE KING CRESCENT		DRAWN BY	ELYSE LEFFLER	
LOT	93	**CARDEL** HOMES	SCALE (metric)	1:200	REVISION DATES
			BLDG AREA (sq.m)	207	
PLAN	4M-1083		LOT AREA (sq.m)	690	
			LOT COVERAGE (%)	30	

KINGSTON

93

20.00m

14.07m

10.72m

2.31m

2.3m

LOT 92   34.50m   LOT 93   34.50m   LOT 94

2.3m

10.47m

2.21m

6m   D   6m

20.00m

N

BLUE KING CRESCENT

FIGURE 20–6 A plot plan is a view of construction from about 500 feet (150 m) above (scale 1:200).

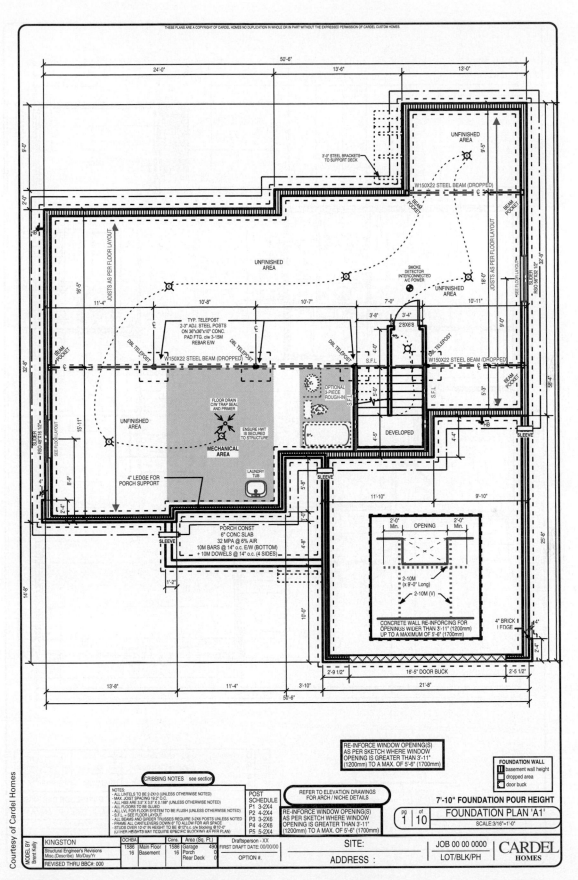

FIGURE 20–7 Foundation plan (scale 3/16″ = 1′ 0″).

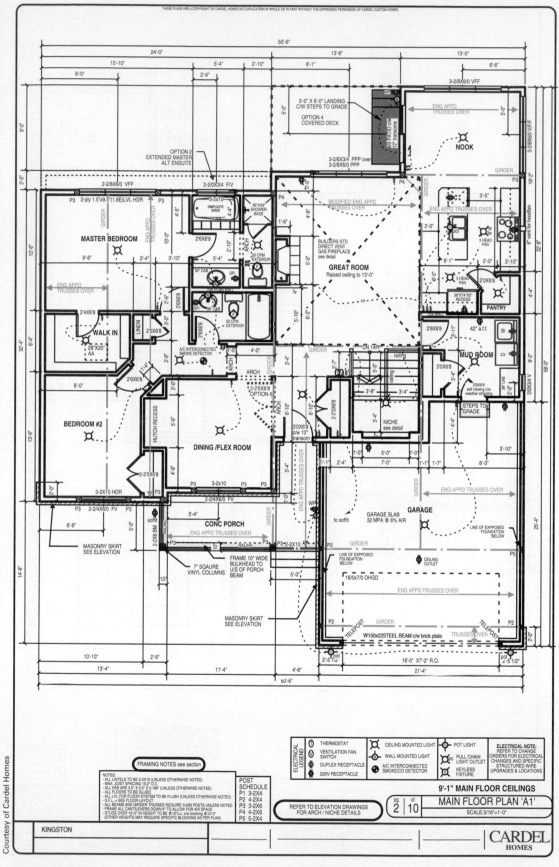

FIGURE 20–8 Main floor plan (scale 3/16″ = 1′ 0″).

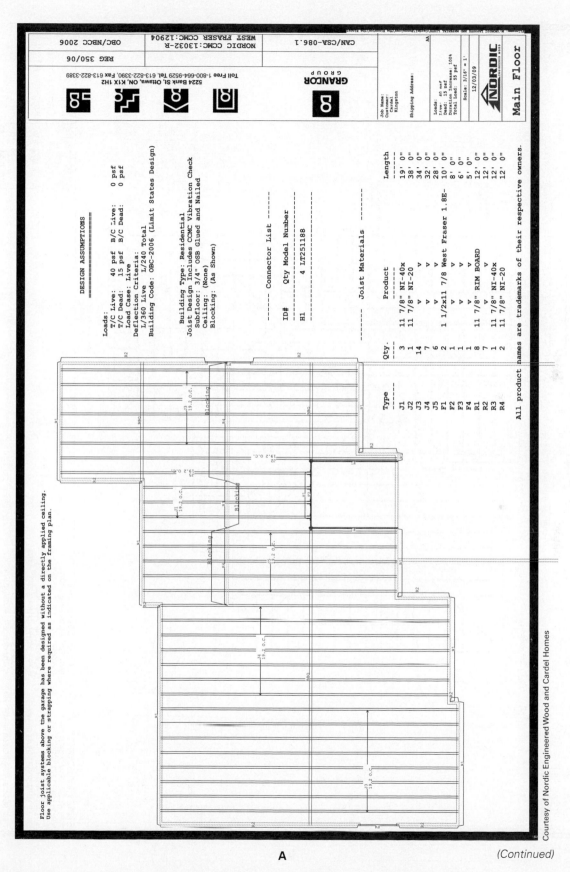

DESIGN ASSUMPTIONS

Loads:
T/C Live: 40 psf   B/C Live: 0 psf
T/C Dead: 15 psf   B/C Dead: 0 psf
Load Case: Live
Deflection Criteria:
L/360 Live   L/240 Total
Building Code: OBC-2006 (Limit States Design)

Building Type: Residential
Joist Design Includes CCMC Vibration Check
Subfloor: 3/4" OSB Glued and Nailed
Ceiling: (None)
Blocking: (As Shown)

------ Connector List ------
ID#   Qty Model Number

H1   4   LT251188

------ Joist Materials ------

Type	Qty.	Product	Length
J1	3	11 7/8" NI-40x	19' 0"
J2	1	11 7/8" NI-20	38' 0"
J3	14	"	34' 0"
J4	7	"	32' 0"
J5	6	"	28' 0"
F1	2	1 1/2x11 7/8 West Fraser 1.8E-	10' 0"
F2	1	"	8' 0"
F3	1	"	6' 0"
F4	1	"	5' 0"
R1	8	11 7/8" RIM BOARD	12' 0"
R2	7	"	12' 0"
R3	1	11 7/8" NI-40x	12' 0"
R4	2	11 7/8" NI-20	12' 0"

All product names are trademarks of their respective owners.

Floor joist systems above the garage has been designed without a directly applied ceiling.
Use applicable blocking or strapping where required as indicated on the framing plan.

GRANCOR
GROUP
5224 Bank St, Ottawa, ON, K1X 1H2
Toll Free 1-800-664-9529 Tel. 613-822-3390, Fax 613-822-3389

CAN/CSA-086.1	NORDIC CCMC:13032-R WEST FRASER CCMC:12904	OBC/NBCC 2006
		REG 350/06

Job Name:
Customer: Cardel
Kingston

Shipping Address:

Loads:
Live: 40 psf
Dead: 15 psf
Duration Increase: 100%
Total Load: 55 psf

Scale: 3/16" = 1'   12/03/09

NORDIC
ENGINEERED WOOD

Main Floor

Courtesy of Nordic Engineered Wood and Cardel Homes

A

(Continued)

FIGURE 20–9 (A) Main floor framing plan (scale 3/16″ = 1′ 0″). (B) Roof framing plan.

Courtesy of Grandor Truss and Cardel Homes

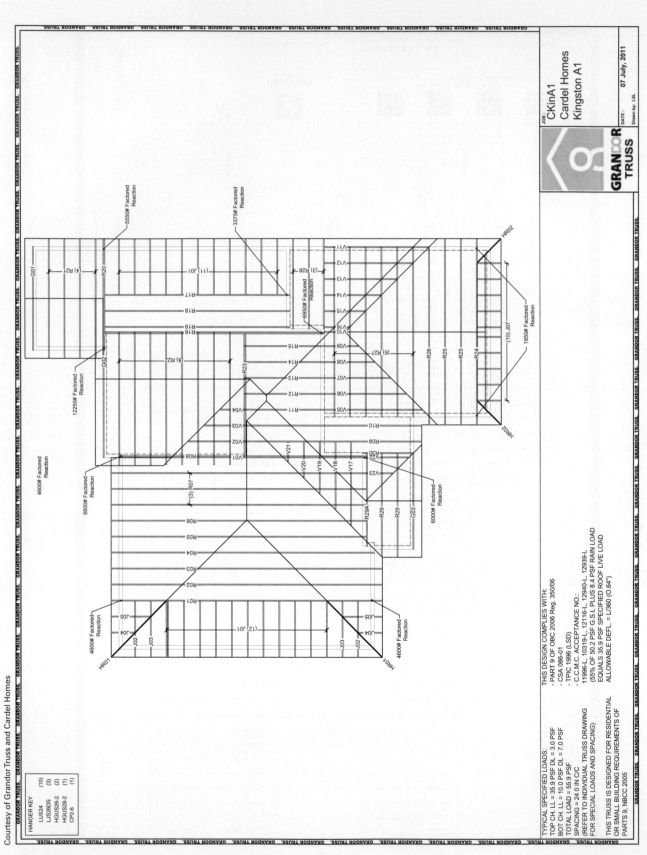

B

FIGURE 20–9 *(Continued)*

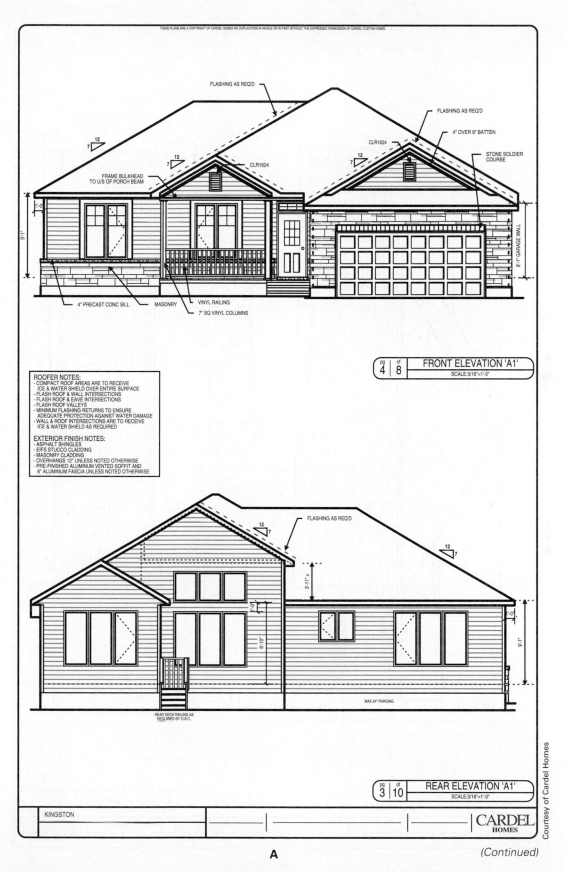

FIGURE 20–10 (A) Front and rear elevations (scale 3/16″ = 1′ 0″). (B) Side elevations (scale 3/16″ = 1′ 0″)

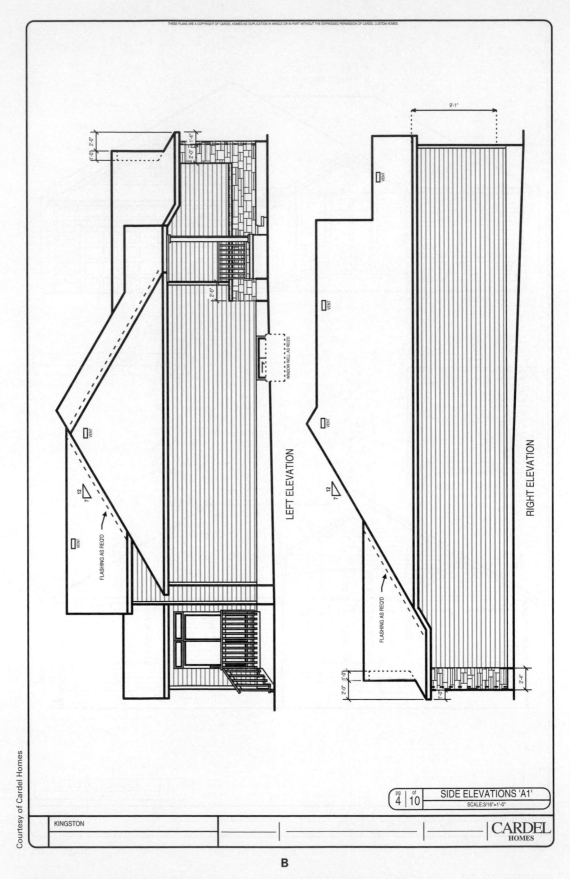

Courtesy of Cardel Homes

FIGURE 20–10 *(Continued)*

drawings of certain interior walls. The most common are kitchen and bathroom wall elevations. They show the design and size of cabinets built on the wall (**Figure 20–11**). Other walls that have special features, such as a fireplace, may require an elevation drawing. Occasionally found in some sets of plans are *framing elevations*. Similar to framing plans, they show the spacing, location, and sizes of wall framing members. No further description of framing drawings is required to be able to interpret them.

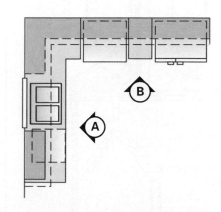

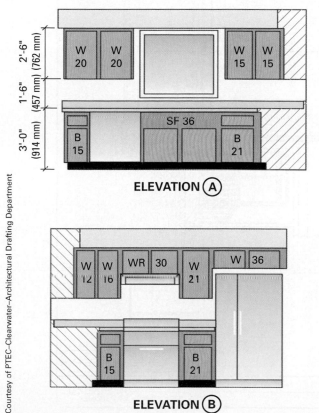

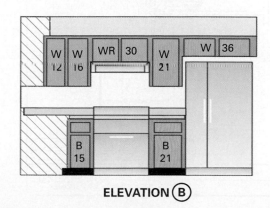

FIGURE 20–11 Interior wall elevations.

Courtesy of PTEC-Clearwater–Architectural Drafting Department

Section Views

A set of architectural plans may have many *section* views. Each is designed to reveal the structure or skeleton view of a particular part of the building (**Figure 20–12**). A section reference line is found on the plans or elevations to identify the section being viewed.

Details

To make parts of the construction clearer, it is usually necessary to draw details. Details are small parts drawn at a very large scale, even full size if necessary. Their existence is revealed on the plan and elevation views using a symbol (**Figure 20–13**). The location of the symbol shows the location of the vertical cut through the building. This symbol may have different shapes, yet all have numbers or letters that refer to the page where the detail is shown.

Other Drawings

Drawings relating to electrical work, plumbing, heating, and ventilating may be on separate sheets in a set of architectural plans. For smaller projects, separate plans are not always needed. All necessary information can usually be found on the floor plan. The carpenter is responsible for building to accommodate wiring, pipes, and ducts. He or she must be able to interpret these plans proficiently to understand the work involved.

SCHEDULES

Besides drawings, printed instructions are included in a set of drawings. **Window schedules** and **door schedules** (**Figure 20–14**) give information about the location, size, and kind of windows and doors to be installed in the building. Each of the different units is given a number or letter. A corresponding number or letter is found on the floor plan to show the location of the unit. Windows may be identified by letters and doors by numbers. The letters and numbers may be framed with various geometric figures, such as circles and triangles.

A **finish schedule** (**Figure 20–15**) may also be included in a set of plans. This schedule gives information on the kind of finish material to be used on the floors, walls, and ceilings of the individual rooms.

Window, door, and finish schedules are used for easy understanding and to conserve space on floor plans.

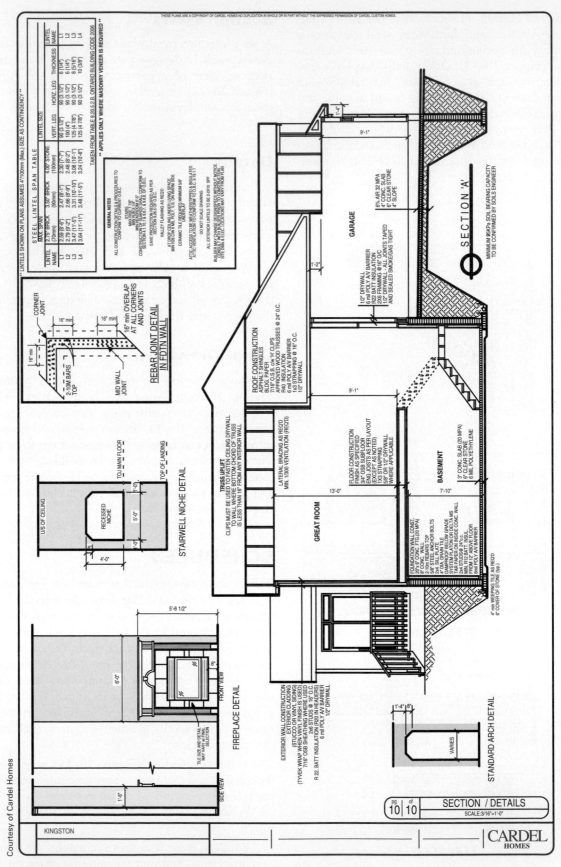

FIGURE 20–12 A section view of a vertical cut through part of the construction (scale 3/16″ = 1′ 0″).

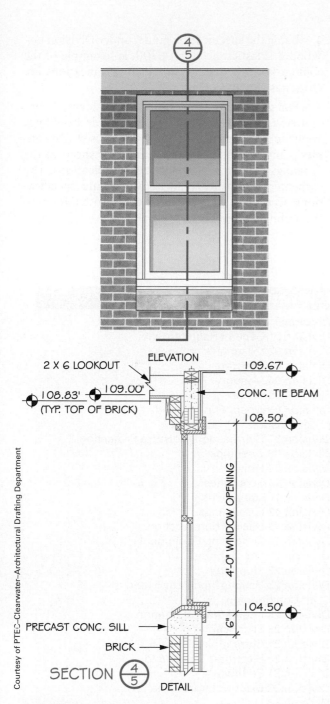

2 X 6 LOOKOUT

ELEVATION

109.67'

108.83'   109.00'
(TYP. TOP OF BRICK)

CONC. TIE BEAM

108.50'

4'-0" WINDOW OPENING

104.50'

6"

PRECAST CONC. SILL

BRICK

SECTION ④/⑤

DETAIL

Courtesy of PTEC–Clearwater–Architectural Drafting Department

FIGURE 20–13 Detail of a window.

SPECIFICATIONS

Specifications, commonly called *specs*, are written to give information that cannot be completely provided in the drawings or schedules. They supplement the working drawings with more complete descriptions of the methods, materials, and quality of construction. If there is a conflict, the specifications take precedence over the drawings. Any conflict should be pointed out to the architect so that corrections can be made.

Window Schedule				
Sym	Size	Model	Rough Open	Quantity
A	1' × 5'	Job Built	Verify	2
B	8' × 5'	W 4 N 5 CSM.	8'-0¾ × 5'-0⅞	1
C	4' × 5'	W 2 N 5 CSM.	4'-0¾ × 5'-0⅞	2
D	4' × 3⁶'	W 2 N 3 CSM.	4'-0¾ × 3'-6½	2
E	3⁸ × 3⁶	2 N 3 CSM.	3'-6½ × 3'-6½	2
F	6' × 4'	G 64 Sldg.	6'-0½ × 4'-0½	1
G	5' × 3⁶	G 536 Sldg.	5'-0½ × 3'-6½	4
H	4' × 3⁶	G 436 Sldg.	4'-0½ × 3'-6½	1
J	4' × 2'	A 41 Awn.	4'-0½ × 2'-0⅞	3

Door Schedule			
Sym	Size	Type	Quantity
1	3' × 6⁸	S.C. R.P. Metal Insulated	1
2	3' × 6⁸	S.C. Flush Metal Insulated	2
3	2⁸ × 6⁸	S.C. Self Closing	2
4	2⁸ × 6⁸	Hollow Core	5
5	2⁶ × 6⁸	Hollow Core	5
6	2⁶ × 6⁸	Pocket Sldg.	2

FIGURE 20–14 A typical window and door schedule.

Room	Floor					Walls				Ceil.		
	Vinyl	Carpet	Tile	Hardwood	Concrete	Paint	Paper	Texture	Spray	Smooth	Brocade	Paint
Entry					●							
Foyer		●				●			●		●	●
Kitchen		●					●			●		●
Dining				●		●			●		●	●
Family		●				●			●		●	●
Living		●				●		●			●	●
Mstr. Bath		●					●			●		●
Bath #2		●				●				●	●	●
Mstr. Bed		●				●		●			●	●
Bed #2		●				●				●	●	●
Bed #3		●				●				●	●	●
Utility	●					●				●	●	●

FIGURE 20–15 A typical finish schedule.

The amount of detail contained in the specs will vary, depending on the size of the project. On small jobs, they may be written by the architect. On larger jobs, a **specifications writer**, trained in the construction process, may be required. For complex commercial projects, a **specifications guide**, used by spec writers, has been developed by the Construction Specifications Institute (CSI). The MasterFormat 2012 structure has 49 major divisions, each containing a number of subdivisions (**Figure 20–16**). Several of the major divisions are reserved for future expansion. For more details, see *http://www.masterformat.com/revisions/* and *http://csinet.org/numbersandtitles.*

Using the specification guide, under Division 6—Wood & Plastics, Section 06200, an example of the content and the manner in which specifications are written is shown in **Figure 20–17**.

For some light commercial and residential construction, many sections of the spec guide would not apply and a shortened version is then used. On simpler plans, notations made on the same sheets as the drawings may take the place of specifications. The carpenter's ability to read notations and specifications accurately is essential in order to conform to the architect's design.

Pre-2004 CSI Version	2004 Version of CSI Divisions
DIVISION 1 GENERAL REQUIREMENTS	Division 0
DIVISION 2 SITE CONSTRUCTION	Division 01 General Requirements
DIVISION 3 CONCRETE	Division 02 Existing Conditions
DIVISION 4 MASONRY	Division 03 Concrete
DIVISION 5 METALS	Division 04 Masonry
DIVISION 6 WOOD AND PLASTICS	Division 05 Metals
DIVISION 7 THERMAL AND MOISTURE PROTECTION	Division 06 Wood, Plastics, and Composites
DIVISION 8 DOORS AND WINDOWS	Division 07 Thermal and Moisture Protection
DIVISION 9 FINISHES	Division 08 Openings
DIVISION 10 SPECIALTIES	Division 09 Finishes
DIVISION 11 EQUIPMENT	Division 10 Specialties
DIVISION 12 FURNISHINGS	Division 11 Equipment
DIVISION 13 SPECIAL CONSTRUCTION	Division 12 Furnishings
DIVISION 14 CONVEYING SYSTEMS	Division 13 Special Construction
DIVISION 15 MECHANICAL	Division 14 Conveying Equipment
DIVISION 16 ELECTRICAL	Division 21 Fire Suppression
	Division 22 Plumbing
	Division 23 Heating, Ventilating, and Air Conditioning
	Division 25 Integrated Automation
	Division 26 Electrical
	Division 27 Communications
	Division 28 Electronic Safety and
	Division 31 Earthwork
	Division 32 Exterior Improvements
	Division 33 Utilities
	Division 34 Transportation
	Division 40 Process Integration
	Division 41 Material Processing and Handling Equipment
	Division 42 Process Heating, Cooling, and Drying Equipment
	Division 43 Process Gas & Liquid Handling, Purification, & Storage Equipment
	Division 44 Pollution Control Equipment
	Division 45 Industry-Specific Manufacturing Equipment
	Division 48 Electrical Power Generation

FIGURE 20–16 The 2012 version of CSI formats has more details and divisions reserved for future use.

Division 6-Wood & Plastics

Section 06200—Finish Carpentry

General: This section covers all finish woodwork and related items not covered elsewhere in these specifications. The contractor shall furnish all materials, labor, and equipment necessary to complete the work, including rough hardware, finish hardware, and specialty items.

Protection of Materials: All millwork (finish woodwork*) and trim is to be delivered in a clean and dry condition and shall be stored to insure proper ventilation and protection from dampness. Do not install finish woodwork until concrete, masonry, plaster, and related work is dry.

Materials: All materials are to be the best of their respective kind. Lumber shall bear the mark and grade of the association under whose rules it is produced. All millwork shall be kiln dried to a maximum moisture content of 12%.

1. Exterior trim shall be select grade white pine, S4S.
2. Interior trim and millwork shall be select grade white pine, thoroughly sanded at the time of installation.

Installation: All millwork and trim shall be installed with tight-fitting joints and formed to conceal future shrinkage due to drying. Interior woodwork shall be mitered or coped at corners (cut in a special way to form neat joints*). All nails are to be set below the surface of the wood and concealed with an approved putty or filler.

*(Explanations in parentheses have been added to aid the student.)

FIGURE 20–17 Sample specifications following the CSI format.

ARCHITECTURAL PLAN LANGUAGE

Carpenters must be able to read and understand the combination of lines, dimensions, symbols, and notations on the architectural drawings. Only then can they build exactly as the architect has designed the construction. No deviation from the plans may be made without the approval of the architect.

Scales

It would be inconvenient and impractical to make full-sized drawings of a building. Therefore, they are drawn to scale. This means that each line in the drawing is reduced proportionally to a size that clearly shows the information and that can be handled conveniently. Not all drawings in a set of plans, or even on the same page, are drawn at the same scale. The scale of a drawing is stated in the title block of the page. It can also be listed directly below the drawing.

Architect's Scale Ruler. The architect's scale ruler is commonly used to scale lines when making drawings (**Figure 20–18**). It has a triangular cross-section giving space for two scales on each face.

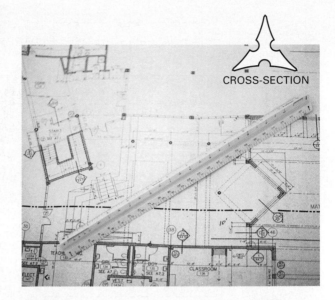

FIGURE 20–18 The architect's scale ruler is used to draw plans.

Six faces are produced, allowing room for a potential of six scales. But it doesn't stop there. The main scale is simply a ruler divided into 1-inch increments and fractions in $\frac{1}{16}$-inch increments. The other five scales actually show two scales in each. Each scale is doubled up with another scale. The desired scale used depends on which end of the scale is read.

Each of the five scales is divided into fractional increments depending on the number labelled at the end scale. For example, the ¼″ scale is read from right to left. At the other end, the ⅛″ scale is read from left to right (**Figure 20–19**). One scale is twice as big (or half as big, depending) as its counterpart. When drawing a line in ¼″ scale, each foot of actual size is drawn as ¼ inch. Read the scale in one direction from the starting end, being careful not to confuse it with the scale running in the opposite direction.

The scales are paired as follows:

1½″ = 1′-0″	and	3″ = 1′-0″
1″ = 1′-0″	and	½″ = 1′-0″
⅜″ = 1′-0″	and	¾″ = 1′ 0″
⅛″ = 1′-0″	and	¼″ = 1′-0″
³/₃₂″ = 1′-0″	and	³/₃₆″ = 1′-0″

On the end of each scale, a space representing one foot at that scale is divided up. This space is used when scaling off dimensions that include fractions of a foot. What these lines represent depends on the scale. In some cases, it is divided into fractions of an inch; in others, it is divided into whole inches. For example, in Figure 20–19, the 1½″ scale's smallest line represents an actual

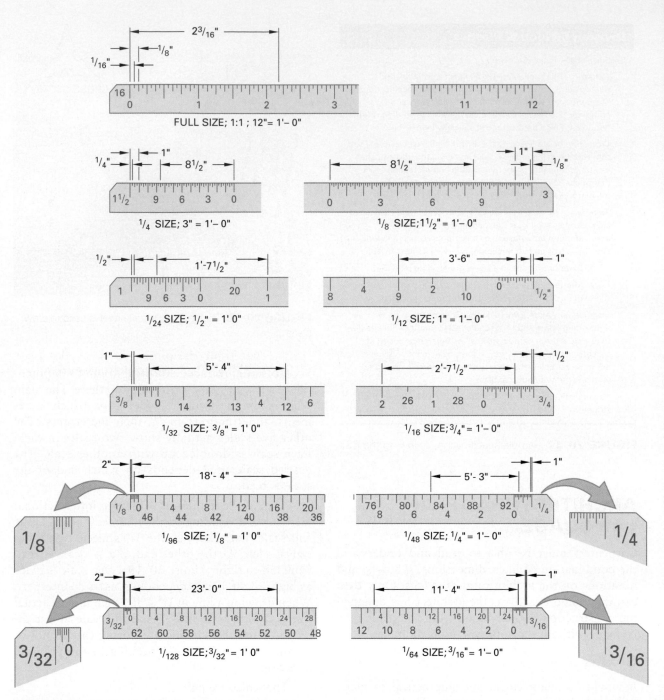

FIGURE 20–19 The 11 scales found on an architect's scale.

¼ inch. In the ⅜″ scale, the smallest line represents 1 actual inch.

The Metric Engineering Scale is the standard in most of the world. Metric plot plans are often drafted in ratios of 1:100. This scale is very close to the scale ⅛″ = 1′-0″ (1:96). Similarly, metric floor plans are drawn in a ratio of 1:50, close to a scale of ¼″ = 1′-0″ (1:48). Construction details are drawn to metric scales of 1:20, 1:10, or 1:5. It should be

noted that all dimensions in metric drawings are in millimetres. Therefore, it is not necessary to use the "mm" symbol.

Commonly Used Scales. Probably the most commonly used scale found on architectural plans is ¼ inch equals 1 foot. This is indicated as ¼″ = 1′-0″. It is often referred to as a *quarter-inch scale*. This means that every ¼ inch on the drawing will equal 1 foot in the building. Floor plans and exterior elevations for most residential buildings are drawn at this scale.

To show the location of a building on a lot and other details of the site, the architect may use a scale of $\frac{1}{16}'' = 1'\text{-}0''$. This reduces the size of the drawing to fit it on the paper. To show certain details more clearly, larger scales of $1\frac{1}{2}'' = 1'\text{-}0''$ or $3'' = 1'\text{-}0''$ are used. Complicated details may be drawn full size or half size. Other scales are used when appropriate. Section views showing the elevation of interior walls are often drawn at $\frac{1}{2}'' = 1'\text{-}0''$ or $\frac{3}{4}'' = 1'\text{-}0''$.

Drawing plans to scale is important. The building and its parts are shown in true proportion, making it easier for the builder to visualize the construction. However, the use of a scale rule to determine a dimension should be a last resort. Dimensions on plans should be determined either by reading the dimension or by adding and subtracting other dimensions to determine it. The use of a scale rule to determine a dimension will result in inaccuracies.

Types of Lines

Some lines in an architectural drawing look darker than others. They are broader so that they stand out clearly from other lines. This variation in width is called *line contrast*. This technique, like all architectural drafting standards, is used to make the drawing easier to read and understand (**Figure 20–20**).

Object Line. Lines that outline the object being viewed are broad, solid lines called *object lines*. These lines represent the portion of the building visible in this view.

Hidden Line. To indicate an important object not visible in the view, a *hidden line* consisting of short, fine, uniform dashes is used. Hidden lines are used only when necessary. Otherwise the drawing becomes confusing to read.

Centreline. *Centrelines* are indicated by a fine, long dash, then a short dash, then a long dash, and so on. They show the centres of doors, windows, partitions, and similar parts of the construction.

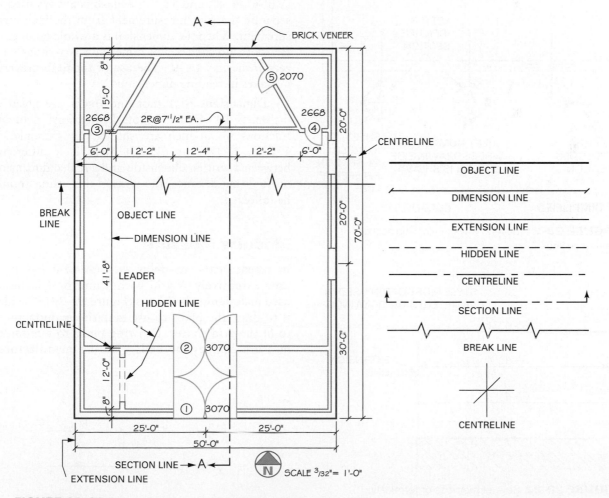

FIGURE 20–20 Types of lines on architectural drawings (scale 1:125).

Section Reference Line. A *section reference* or *cutting-plane line* is, sometimes, a broad line consisting of a long dash followed by two short dashes. At its ends are arrows. The arrows show the direction in which the cross-section is viewed. Letters identify the cross-sectional view of that specific part of the building. More elaborate methods of labelling section reference lines are used in larger, more complicated sets of plans (**Figure 20–21**). The sectional drawings may be on the same page as the reference line or on other pages.

Break Line. A *break line* is used in a drawing to terminate part of an object that, in actuality, continues. It can be used only when there is no change in the drawing at the break. Its purpose is to shorten the drawing to utilize space.

Dimension Line. A *dimension line* is a fine, solid line used to indicate the location, length, width, or thickness of an object. It is terminated with arrowheads, dots, or slashes (**Figure 20–22**).

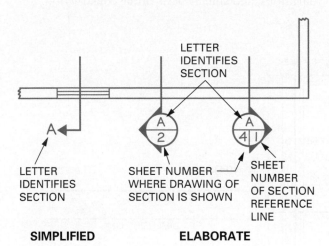

FIGURE 20–21 Several ways of labelling section reference lines.

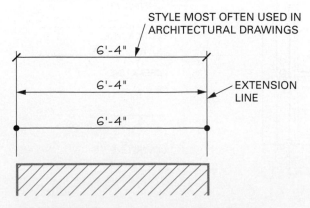

FIGURE 20–22 Several methods of terminating dimension lines.

Extension Line. *Extension lines* are fine, solid lines projecting from an object to show the extent of a dimension.

Leader Line. A *leader line* is a fine solid line. It terminates with an arrowhead and points to an object from a notation.

Dimensions

Dimension lines on a blueprint are generally drawn as continuous lines. The dimension appears above and near the centre of the line. All dimensions on vertical lines should appear above the line when the print is rotated ¼ turn clockwise. Extension lines are drawn from the object so that the end point of the dimension is clearly defined. When the space is too small to permit dimensions to be shown clearly, they may be drawn as shown in **Figure 20–23**.

Kinds of Dimensions. Dimensions on architectural blueprints are given in feet and inches, such as 3′-6″, 4′-8″, and 13′-7″. A dash is always used to separate the foot measurement from the inch measurement. When the dimension is a whole number of feet with no inches, the dimension is written with zero inches, as 14′-0″. The use of the dash prevents mistakes in reading dimensions.

Dimensions of 1 foot and under are given in inches, as 10″, 8″, and so on. Dimensions involving fractions of an inch are shown, for example, as 1′-0½″, 2′-3¾″, or 6½″. If there is a difference between a written dimension and a scaled dimension of the same distance, the written dimension should be followed.

Modular Measure

In recent years, **modular measurement** has been used extensively. A grid with a unit of 4 inches is used in designing buildings (**Figure 20–24**). The idea is to draw the plans to use material manufactured to fit the grid spaces. Drawing plans to a modular measure enables the builder to use manufactured

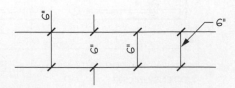

FIGURE 20–23 Methods of dimensioning small spaces.

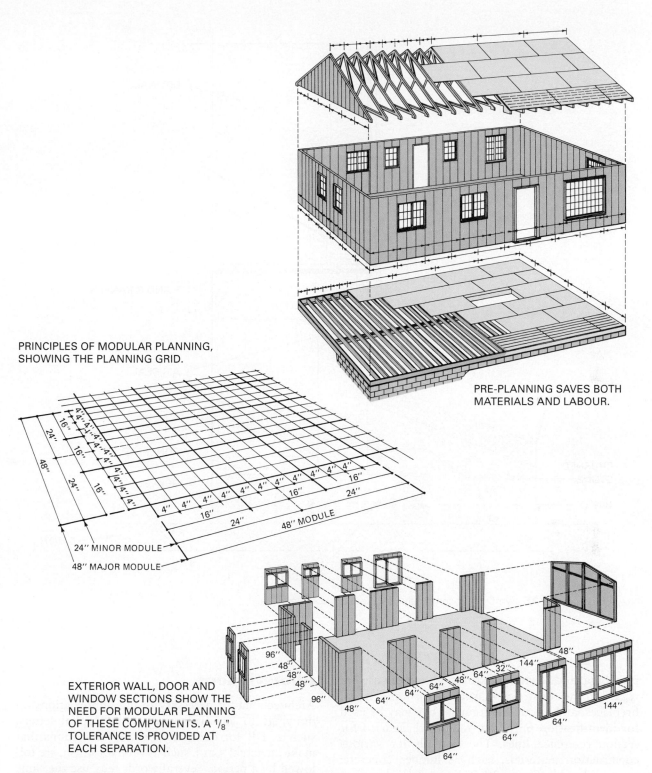

PRINCIPLES OF MODULAR PLANNING, SHOWING THE PLANNING GRID.

PRE-PLANNING SAVES BOTH MATERIALS AND LABOUR.

24″ MINOR MODULE

48″ MAJOR MODULE

48″ MODULE

EXTERIOR WALL, DOOR AND WINDOW SECTIONS SHOW THE NEED FOR MODULAR PLANNING OF THESE COMPONENTS. A ⅛″ TOLERANCE IS PROVIDED AT EACH SEPARATION.

FIGURE 20–24 Modular measurement uses a grid of 4 inches.

component parts with less waste, such as 4 × 8 sheet materials and manufactured wall, floor, and roof sections that fit together with greater precision.

When framing members are spaced according to modular measurement and window and door sizes

adhere to it, costs are reduced and materials are conserved.

Symbols

Symbols are used on drawings to represent objects in the building, such as doors, windows, cabinets,

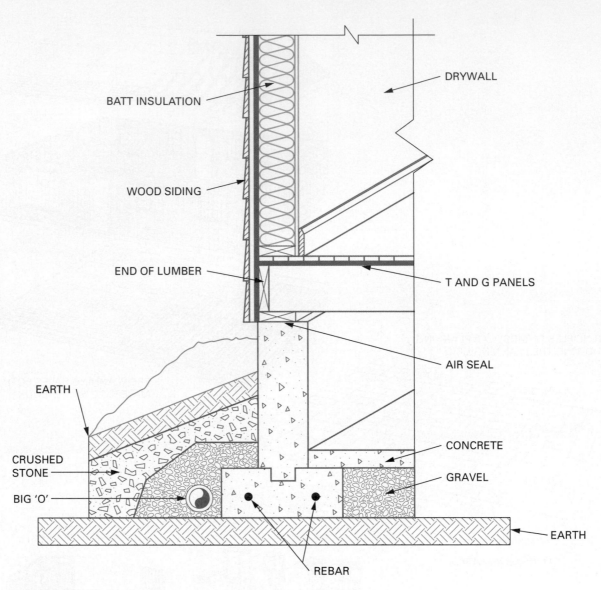

FIGURE 20–25 Symbols for commonly used construction materials.

plumbing, and electrical fixtures. Others are used in regard to the construction, such as for walls, stairs, fireplaces, and electrical circuits. They may be used for identification purposes, such as those used for section reference lines. The symbols for various construction materials, such as lumber, concrete, sand, and earth (**Figure 20–25**), are used when they make the drawing easier to read. (More detailed illustrations, descriptions, and uses of architectural symbols are presented in following units where appropriate.)

Abbreviations

Architects find it necessary to use abbreviations on drawings to conserve space. Only capital letters, such as DR for door, are used. Abbreviations that make an actual word, such as FIN. for finish, are followed by a period. Several words may use the same abbreviation, such as W for west, width, or with. The location of these abbreviations is the key to their meaning. A list of commonly used abbreviations is provided in **Figure 20–26**.

Access Panel	AP	Dressed and Matched	D & M	Plate	PL
Acoustic	ACST	Dryer	D	Plate Glass	PL GL
Acoustical Tile	AT	Electric Panel	EP	Platform	PLAT
Aggregate	AGGR	End to End	E to E	Plumbing	PLBG
Air Conditioning	AIR COND	Excavate	EXC	Plywood	PLY
Aluminum	AL	Expansion Joint	EXP JT	Porch	P
Anchor Bolt	AB	Exterior	EXT	Precast	PRCST
Angle		Finish	FIN	Prefabricated	PREFAB
Apartment	APT	Finished Floor	FIN FL	Pull Switch	PS
Approximate	APPROX	Firebrick	FBRK	Quarry Tile Floor	QTF
Architectural	ARCH	Fireplace	FP	Radiator	RAD
Area	A	Fireproof	FPRF	Random	RDM
Area Drain	AD	Fixture	FIX	Range	R
Asbestos	ASB	Flashing	FL	Recessed	REC
Asbestos Board	AB	Floor	FL	Refrigerator	REF
Asphalt	ASPH	Floor Drain	FD	Register	REG
Asphalt Tile	AT	Flooring	FLG	Reinforce or Reinforcing	REINF
Basement	BSMT	Fluorescent	FLUOR	Revision	REV
Bathroom	B	Flush	FL	Riser	R
Bathtub	BT	Footing	FTG	Roof	RF
Beam	BM	Foundation	FND	Roof Drain	RD
Bearing Plate	BRG PL	Frame	FR	Room	RM or R
Bedroom	BR	Full Size	FS	Rough	RGH
Blocking	BLKG	Furring	FUR	Rough Opening	RO
Blueprint	BP	Galvanized Iron	GI	Rubber Tile	R TILE
Boiler	BLR	Garage	GAR	Scale	SC
Book Shelves	BK SH	Gas	G	Schedule	SCH
Brass	BRS	Glass	GL	Screen	SCR
Brick	BRK	Glass Block	GL BL	Scuttle	S
Bronze	BRZ	Grille	G	Section	SECT
Broom Closet	BC	Gypsum	GYP	Select	SEL
Building	BLDG	Hardware	HDW	Service	SERV
Building Line	BL	Hollow Metal Door	HMD	Sewer	SEW
Cabinet	CAB	Hose Bibb	HB	Sheathing	SHTHG
Calking	CLKG	Hot Air	HA	Sheet	SH
Casing	CSG	Hot Water	HW	Shelf and Rod	SH & RD
Cast Iron	CI	Hot Water Heater	HWH	Shelving	SHELV
Cast Stone	CS	I Beam	I	Shower	SH
Catch Basin	CB	Inside Diameter	ID	Sill Cock	SC
Cellar	CEL	Insulation	INS	Single Strength Glass	SSG
Cement	CEM	Interior	INT	Sink	SK or S
Cement Asbestos Board	CEM AB	Iron	I	Soil Pipe	SP
Cement Floor	CEM FL	Jamb	JB	Specification	SPEC
Cement Mortar	CEM MORT	Kitchen	K	Square Feet	SQ FT
Center	CTR	Landing	LDG	Stained	STN
Center to Center	C TO C	Lath	LTH	Stairs	ST
Center Line	or CL	Laundry	LAU	Stairway	STWY
Center Matched	CM	Laundry Tray	LT	Standard	STD
Ceramic	CER	Lavatory	LAV	Steel	ST or STL
Channel	CHAN	Leader	L	Steel Sash	SS
Cinder Block	CIN BL	Length	L, LG or LNG	Storage	STG
Circuit Breaker	CIR BKR	Library	LIB	Switch	SW or S
Cleanout	CO	Light	LT	Telephone	TEL
Cleanout Door	COD	Limestone	LS	Terra Cotta	TC
Clear Glass	CL GL	Linen Closet	L CL	Terrazzo	TER
Closet	C, CL or CLO	Lining	LN	Thermostat	THERMO
Cold Air	CA	Living Room	LR	Threshold	TH
Cold Water	CW	Louvre	LV	Toilet	T
Collar Beam	COL B	Main	MN	Tongue and Groove	T & G
Concrete	CONC	Marble	MR	Tread	TR or T
Concrete Block	CONC B	Masonry Opening	MO	Typical	TYP
Concrete Floor	CONC FL	Material	MATL	Unfinished	UNF
Conduit	CND	Maximum	MAX	Unexcavated	UNEXC
Construction	CONST	Medicine Cabinet	MC	Utility Room	URM
Contract	CONT	Minimum	MIN	Vent	V
Copper	COP	Miscellaneous	MISC	Vent Stack	VS
Counter	CTR	Mixture	MIX	Vinyl Tile	V TILE
Cubic Feet	CU FT	Modular	MOD	Warm Air	WA
Cutout	CO	Mortar	MOR	Washing Machine	WM
Detail	DET	Moulding	MLDG	Water	W
Diagram	DIAG	Nosing	NOS	Water Closet	WC
Dimension	DIM	Obscure Glass	OBSC sL	Water Heater	WH
Dining Room	DR	On Center	OC	Waterproof	WP
Dishwasher	DW	Opening	OPNG	Weather Stripping	WS
Ditto	DO	Outlet	OUT	Weephole	WH
Double-Acting	DA	Overall	OA	White Pine	WP
Double Strength Glass	DSG	Overhead	OVHD	Wide Flange	WF
Down	DN	Pantry	PAN	Wood	WD
Downspout	DS	Partition	PTN	Wood Frame	WF
Drain	D or DR	Plaster	PL or PLAS	Yellow Pine	YP
Drawing	DWG	Plastered Opening	PO		

FIGURE 20–26 Commonly used abbreviations found in construction drawings.

Chapter 21
Floor Plans

A house starts out as an idea drawn on paper. The three-dimensional (3D) vision of the structure is converted into many two-dimensional (2D) views (**Figure 21–1**). These views must then be read and interpreted by the builder, who makes the idea come alive.

FLOOR PLANS

Floor plans (**Figure 21–2**) contain a substantial amount of information. They are used more than any other kind of drawing. After consideration of many factors that determine the size and shape of the building, floor plans are drawn first. Others, such as the foundation plan and elevations, are derived from it. They are generally drawn at a scale of ¼"= 1'-0" (1:50 mm). A separate plan is made for each floor of buildings with more than one storey.

Floor Plan Symbols

To make the plan as uncluttered as possible, numerous *symbols* are used. Recognition of commonly used

FIGURE 21–1 The 3D vision of a building is described on 2D paper.

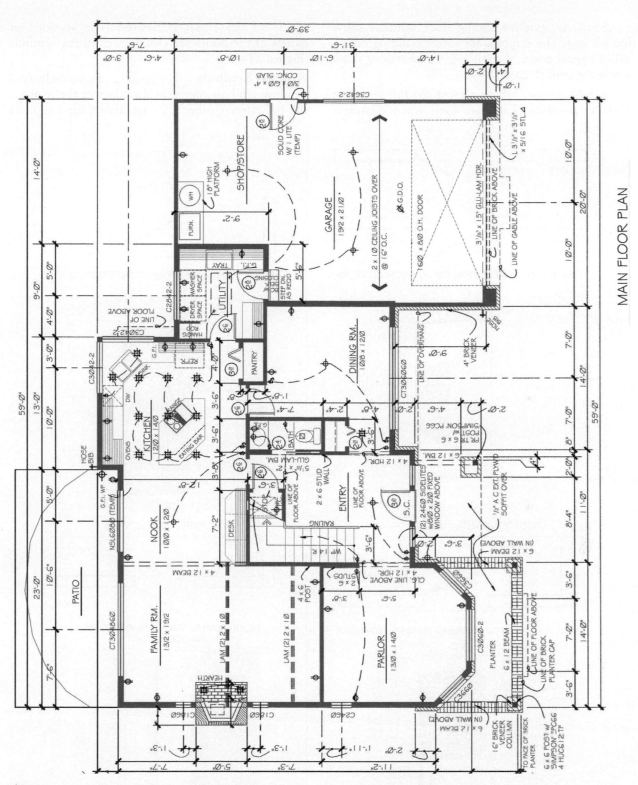

FIGURE 21–2 Floor plans contain a substantial amount of information (scale 1:50).

symbols makes it easier to read the floor plan, as well as other plans that use the same symbols. Symbols used in elevation and section drawings are different from plan symbols. They are described in later chapters.

Door Symbols. Symbols for exterior doors are drawn with a line representing the outside edge of the sill. Interior door symbols show no sill line. The symbols in **Figure 21–3** identify the **swing** and show on which side of the opening to hang the door.

Similarly, exterior sliding door symbols show the sill line. The symbols for interior sliding doors, called **bypass doors**, show none. A pocket door slides inside the wall (**Figure 21–4**).

Bifold doors open to almost the full width of a closet opening. They are used when complete

access to the closet is desired. The sections or panels of the doors are clearly seen in the symbols (**Figure 21–5**).

Window Symbols. The inside and outside lines of window symbols represent the edges of the window sill. In between, other lines are drawn for the panes

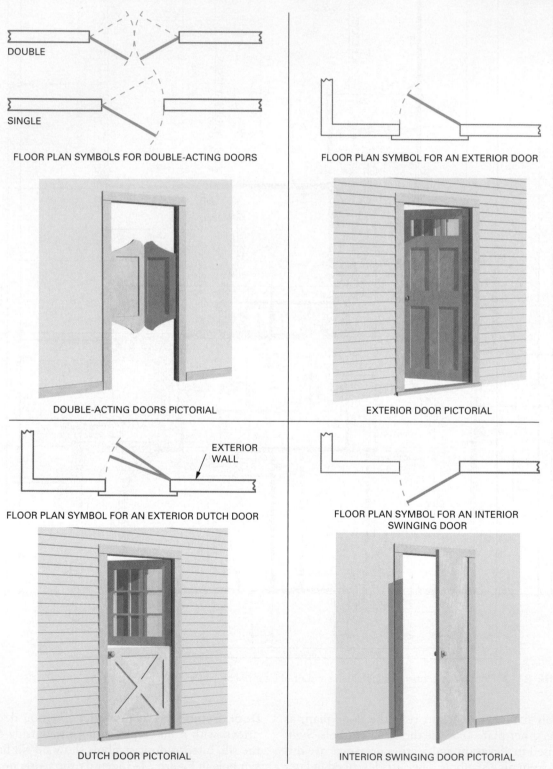

DOUBLE

SINGLE

FLOOR PLAN SYMBOLS FOR DOUBLE-ACTING DOORS

DOUBLE-ACTING DOORS PICTORIAL

FLOOR PLAN SYMBOL FOR AN EXTERIOR DOOR

EXTERIOR DOOR PICTORIAL

EXTERIOR WALL

FLOOR PLAN SYMBOL FOR AN EXTERIOR DUTCH DOOR

DUTCH DOOR PICTORIAL

FLOOR PLAN SYMBOL FOR AN INTERIOR SWINGING DOOR

INTERIOR SWINGING DOOR PICTORIAL

FIGURE 21–3 Floor plan symbols for exterior and interior swinging doors.

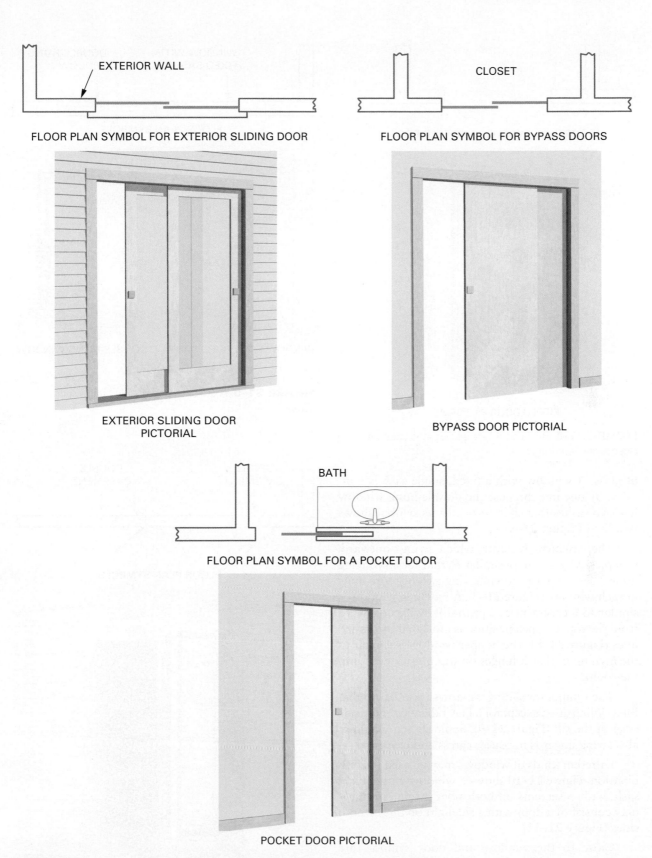

EXTERIOR WALL

FLOOR PLAN SYMBOL FOR EXTERIOR SLIDING DOOR

CLOSET

FLOOR PLAN SYMBOL FOR BYPASS DOORS

EXTERIOR SLIDING DOOR
PICTORIAL

BYPASS DOOR PICTORIAL

BATH

FLOOR PLAN SYMBOL FOR A POCKET DOOR

POCKET DOOR PICTORIAL

FIGURE 21–4 Symbols for exterior and interior sliding doors.

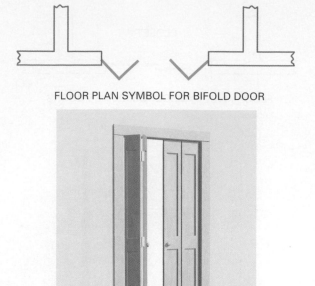

FLOOR PLAN SYMBOL FOR BIFOLD DOOR

BIFOLD DOOR PICTORIAL

FIGURE 21–5 Bifold doors are sometimes used on closets and wardrobes.

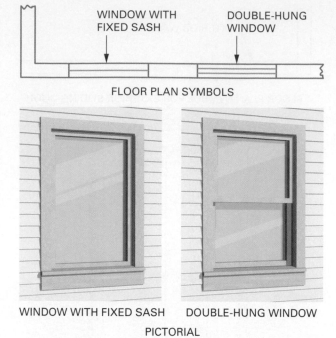

FLOOR PLAN SYMBOLS

WINDOW WITH FIXED SASH DOUBLE-HUNG WINDOW

PICTORIAL

FIGURE 21–6 Fixed sash and double-hung window symbols.

of glass. A window with a fixed, single sash is indicated by one line. Because the **double-hung window** has two sashes that slide vertically, its symbol shows two lines (**Figure 21–6**).

The **casement window**, which swings outward, is depicted by a symbol similar to that for a swinging door. These may be shown having two or more units in each window (**Figure 21–7**). An **awning window** is similar to a casement except that it swings outward from the top. Its open position is indicated by dashed lines (**Figure 21–8**). The **hopper window** is similar to the awning in that it hinges on top but it opens into the room.

The symbol for **sliding windows** is similar to that for a sliding door, except for a line indicating the inside edge of the sill (**Figure 21–9**). Some sliding windows also swing inwards for easier egress and cleaning.

Different kinds of windows may be used in combination. **Figure 21–10** shows a window with a fixed sash, with casements on both sides. Main entrances may consist of a door with a **sidelight** on one or both sides (**Figure 21–11**).

Close to the window and door symbols are letters and numbers that identify the units in the window and door schedules.

Structural Members. Openings without doors in interior walls for passage from one area to another

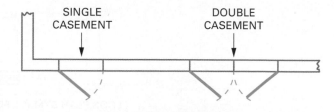

SINGLE CASEMENT DOUBLE CASEMENT

FLOOR PLAN SYMBOLS

DOUBLE CASEMENT PICTORIAL

FIGURE 21–7 Symbols for casement windows.

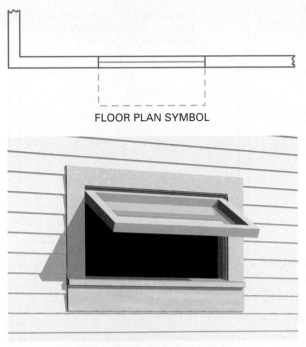

FLOOR PLAN SYMBOL

AWNING WINDOW PICTORIAL

FIGURE 21–8 The open position of an awning window is shown with dashed lines.

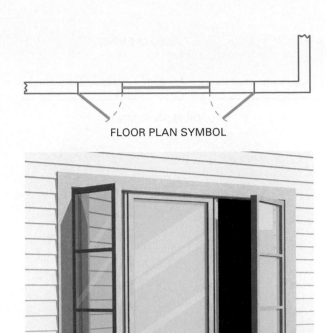

FLOOR PLAN SYMBOL

PICTORIAL

FIGURE 21–10 A window consisting of a fixed sash and casement units.

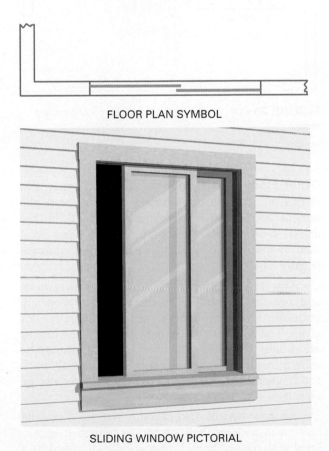

FLOOR PLAN SYMBOL

SLIDING WINDOW PICTORIAL

FIGURE 21–9 Sliding window floor plan symbol.

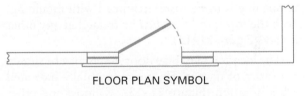

FLOOR PLAN SYMBOL

ENTRANCE DOOR FLANKED BY SIDELIGHTS
PICTORIAL

FIGURE 21–11 An entrance door flanked by sidelights.

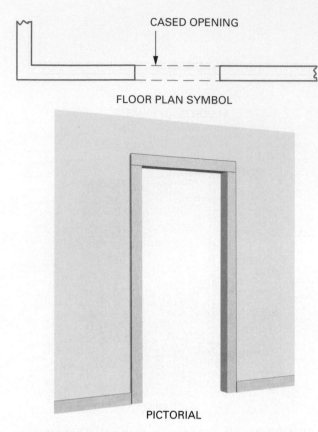

CASED OPENING

FLOOR PLAN SYMBOL

PICTORIAL

FIGURE 21–12 Dashed lines indicate an interior wall opening without a door.

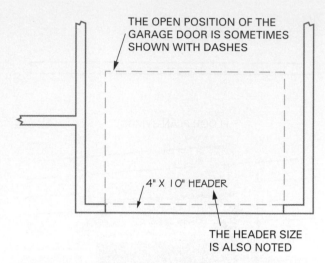

THE OPEN POSITION OF THE GARAGE DOOR IS SOMETIMES SHOWN WITH DASHES

4" X 10" HEADER

THE HEADER SIZE IS ALSO NOTED

FIGURE 21–13 The symbol for a garage door header is a dashed line.

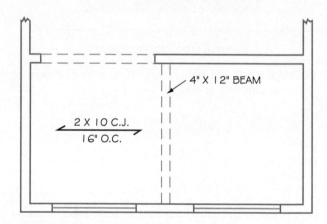

4" X 12" BEAM

2 X 10 C.J. 16" O.C.

FIGURE 21–14 Symbols for ceiling beams and ceiling joists.

are indicated by dashed lines. Notations are given if the opening is to be *cased* (trimmed with moulding) and if the top is to be *arched* or treated in any other manner (**Figure 21–12**).

The location of garage door *lintels* may be shown by a series of dashes. Their size is usually indicated with a notation (**Figure 21–13**). Window and other headers are shown in the same manner.

Ceiling beams above the cutting plane of the floor, which support the ceiling joists, are also represented by a series of dashes. *Ceiling joists* or **trusses** are identified in the floor plan with a double-ended arrow showing the direction in which they run. Their size and spacing are noted alongside the arrow (**Figure 21–14**).

Kitchen, Bath, and Utility Room. The locations of *bathroom* and *kitchen fixtures*, such as sinks, tubs, refrigerators, stoves, washers, and dryers, are shown by obvious symbols, abbreviations, and notations. The extent of the *base cabinets* is indicated by a line indicating the edge of the countertop. Objects such as dishwashers, trash compactors, and lazy susans are shown by a dashed line and notations or abbreviations. The *upper cabinets* are symbolized by dashed lines (**Figure 21–15**).

Other Floor Plan Symbols. The floor plan also shows the location of *stairways*. Lines indicate the outside edges of the **treads**. Also shown are the direction of travel and the number of **risers** (vertical distance from tread to tread) in the staircase (**Figure 21–16**).

The location and style of *chimneys*, *fireplaces*, and *hearths* are shown by the use of appropriate symbols. Fireplace dimensions are generally not given. The sizes of the chimney flue and the hearth are usually indicated. The fireplace material may be shown by symbols, according to the material specified (**Figure 21–17**). The kind of material may also be identified with a notation.

An *attic access*, also called a **scuttle**, is usually located in a closet, hall, or garage ceiling. It is outlined with dashed lines. It may also be identified by a notation (**Figure 21–18**).

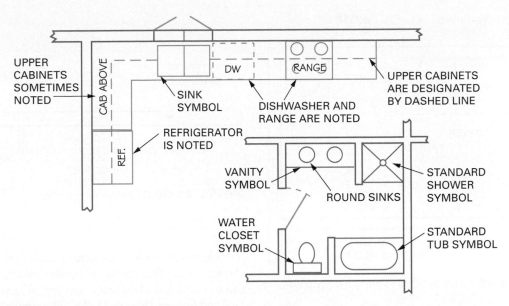

FIGURE 21–15 Floor plan symbols for kitchen and bath cabinets.

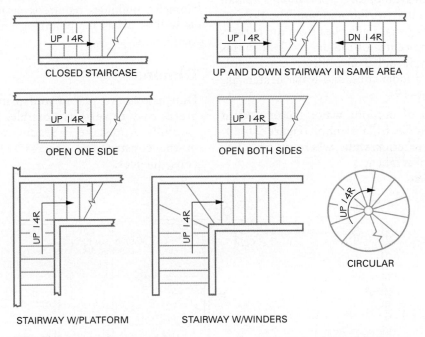

FIGURE 21–16 Stair symbols vary according to the style of the staircase.

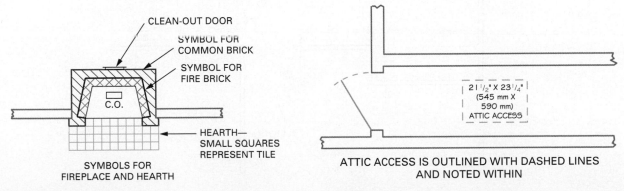

FIGURE 21–17 Symbols for fireplace and hearth.

FIGURE 21–18 Attic access is outlined with dashed lines.

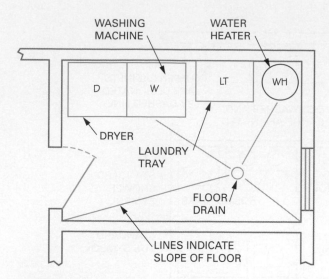

FIGURE 21–19 Some symbols found in a utility room.

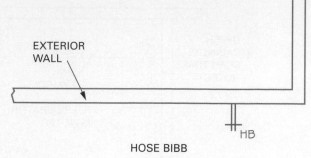

FIGURE 21–20 An exterior hose bibb.

The floor plan symbol for a *floor drain* is a small circle, square, or circle within a square. The slope of the floor is shown by straight lines from the corners of the floor to the centre of the drain. Floor drains are appropriately installed in utility rooms, where washers, dryers, laundry tubs, and water heaters are located (**Figure 21–19**).

The location of outdoor water faucets, called **hose bibbs**, is shown by a symbol (**Figure 21–20**) projecting from exterior walls where desired. For clarity, the symbol is labelled.

Electrical outlets, switches, and lights may be shown on the floor plan of simpler structures by the use of curved, dashed lines running to *switches*, *outlets*, and *fixtures* (**Figure 21–21**). The symbols for these electrical components are shown in **Figure 21–22**. Complex buildings require separate electrical plans, as well as prints for plumbing and for heating and ventilation.

Dimensions

Dimensions are placed and printed so that they are as easy to read as possible. Read dimensions carefully. A mistake in reading a dimension early in the construction process could have serious consequences.

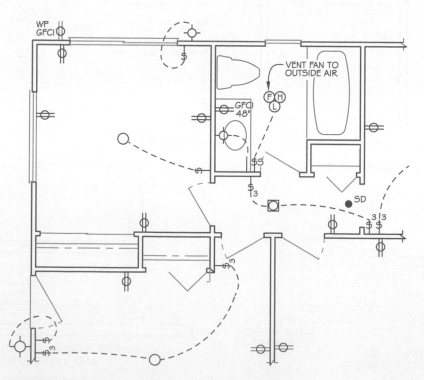

FIGURE 21–21 Part of a typical electrical floor plan.

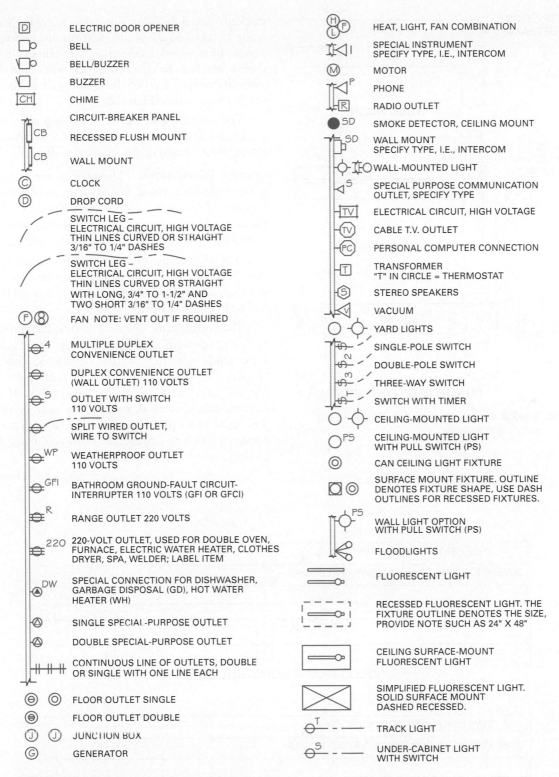

FIGURE 21–22 Electrical symbols used on plans.

Exterior Dimensions. On floor plans, the *overall* dimensions of the building are found on the outer dimension lines. In a *wood frame*, the dimensions are to the outside face of the frame. *Concrete block* walls are dimensioned to their outside face. *Brick* veneer walls are dimensioned to the outside face of the wood frame, with an added dimension and notation for the veneer (**Figure 21–23**).

Multiple dimensions lines are sometimes needed from the same corner. When this happens, they are

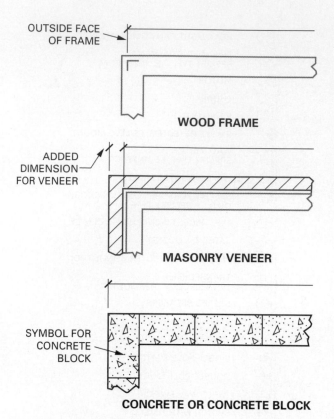

FIGURE 21–23 Overall dimensions are made to the structural portion of a building.

stacked. The dimension lines closest to the exterior walls are used to show the location of *windows* and *doors*. In a wood frame, they are dimensioned to their centreline. In concrete block walls, the dimensions are to the edges of the openings and also show the opening width (**Figure 21–24**).

The second dimension line from the exterior wall is used to locate the centreline of *interior partitions*, which intersect the exterior wall.

Interior Dimensions. Dimensions are given from the outside of a wood frame or from the inside of concrete block walls, to the centrelines or edges of **partitions**. Interior doors and other openings are dimensioned to their centreline, similar to exterior walls.

Not all interior dimensions are required. Some may be assumed. For instance, it can be clearly seen if a door is centred between two walls of a closet or hallway (**Figure 21–25**).

The minimum distance from the corner is typically determined by framing the jack and king studs starting at the corner of the room (**Figure 21–26**). This allows sufficient room for the door and **casing** to be applied. If wider custom casing is used, more room may be needed. The goal is to place the door in the corner with room to finish the corner, but not so close that the casing must be scribed to the wall.

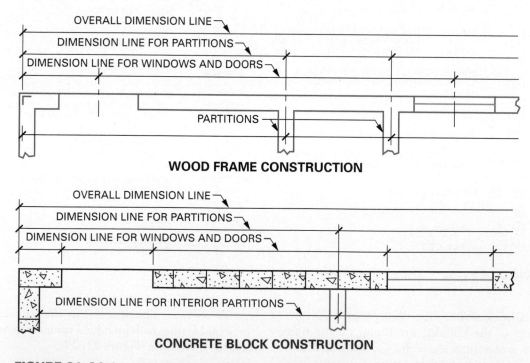

FIGURE 21–24 Standard practice for dimensioning windows, doors, partitions, and then the overall dimension.

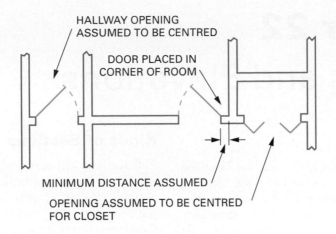

HALLWAY OPENING
ASSUMED TO BE CENTRED

DOOR PLACED IN
CORNER OF ROOM

MINIMUM DISTANCE ASSUMED

OPENING ASSUMED TO BE CENTRED
FOR CLOSET

FLOOR PLAN

FIGURE 21–25 Some dimensions are assumed to be centred or as close to the corner as possible.

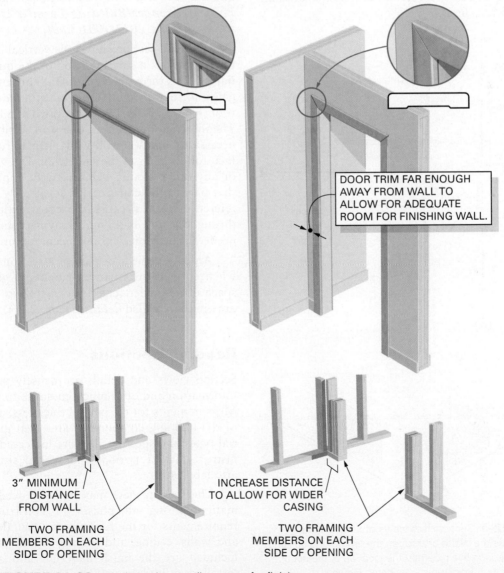

DOOR TRIM FAR ENOUGH
AWAY FROM WALL TO
ALLOW FOR ADEQUATE
ROOM FOR FINISHING WALL.

3" MINIMUM
DISTANCE
FROM WALL

TWO FRAMING
MEMBERS ON EACH
SIDE OF OPENING

INCREASE DISTANCE
TO ALLOW FOR WIDER
CASING

TWO FRAMING
MEMBERS ON EACH
SIDE OF OPENING

FIGURE 21–26 Locating a door to allow room for finish.

Chapter 22

Sections and Elevations

SECTIONS

Floor plans are views of a horizontal cut. Sections show *vertical* cuts called for on the floor, framing, and foundation plans (**Figure 22–1**). Sections provide information not shown on other drawings. The number and type of section drawings in a set of prints depend on what is required for a complete understanding of the construction. They are usually drawn at a scale of ⅜″= 1′-0″ or 1:25. Excellent section drawings and accompanying videos can be found at *http://web.ornl.gov/sci/buildingsfoundations/handbook/toc.shtml*.

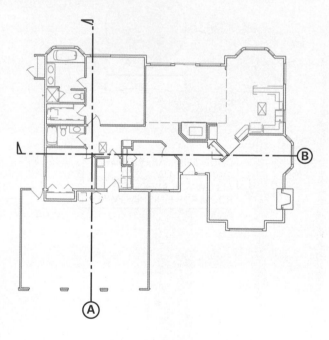

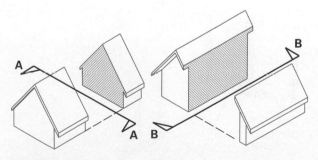

FIGURE 22–1 Sections are views of vertical cutting planes across the width or through the length of a building. Section reference lines identify the location of the section and the direction from which it is being viewed.

Kinds of Sections

Full sections cut across the width or through the length of the entire building (**Figure 22–2**). For a small residence, only one full section may be required to fully understand the construction. Commercial structures may require several full and many partial sections for complete understanding. A full set of plans, specifications, and pertinent information has been produced by TACBOC, the Toronto Area Chief Building Officials Committee, and can be found at *http://www.london.ca/business/Permit-Licences/Building-Permits/Documents/tacboc_details_2012_r001.3.pdf*.

A **partial section** shows the vertical relationship of the parts of a small portion of the building. Partial sections through exterior walls are often used to give information about materials from foundation to roof (**Figure 22–3**). They are drawn at a larger scale. The information given in one wall section does not necessarily apply for all walls. It may not apply, in fact, for all parts of the same wall. The wall section, or any other section being viewed, applies only to that part of the construction located by the section reference lines. Because the construction changes throughout the building, many section views are needed to provide clear and accurate information.

An enlargement of part of a section is required when enough information cannot be given in the space of a smaller scale drawing. These large-scale drawings are called *details* (**Figure 22–4**).

Reading Sections

Section views and details are densely packed with information and offer much guidance to the builder. Measurements for the height, thickness, and spacing of the building components are often given. Material types and specifications are indicated. Fastening instructions for normal and special situations are also labelled.

The section view may be referenced for information during all phases of construction. The requirements for the footings and foundation, floor and walls, ceiling, and roofing are all drawn. Also included are the finish materials and any special installation instructions. The location and size of all

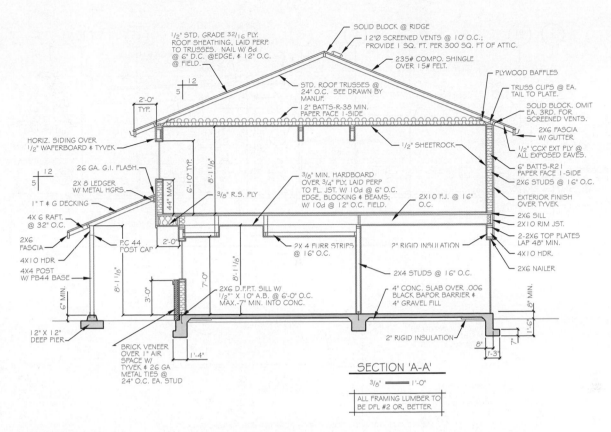

FIGURE 22–2 A typical full-section view of a residence.

building materials, such as steel, wood, masonry, and any other material used in the building, are shown.

ELEVATIONS

Elevations are orthographic drawings. They are usually drawn at the same scale as the floor plan. They show each side of the building as viewed from outside at a distance of about 100 feet (30 m). Generally four elevations, one for each side, are included in a set of drawings. They are titled *Front*, *Rear*, *Left Side*, and *Right Side*. They may also be titled according to the compass direction they face, that is, *North*, *South*, *East*, and *West*. From the exterior elevations, the general shape and design of the building can be determined (**Figure 22–5**).

Symbols

Elevation symbols are different from floor plan symbols for the same object. In elevation drawings, the symbols represent, as closely as possible, the actual object as it would appear to the eye. To make the drawing clearer, the symbols are usually identified with a notation.

The location of any steps, porches, dormers, skylights, and chimneys, although not dimensioned, can be seen in elevations. Foundation footings and walls

below the grade level may be shown with hidden lines. The kind and size of exterior siding, railings, entrances, and special treatments around doors and windows are shown (**Figure 22–6**).

The elevations show the windows and doors in their exact locations. Other openings, such as **louvres**, are shown in place. Their styles and sizes are identified by appropriate symbols and notations (**Figure 22–7**).

The type of roofing material, the roof slope, and the cornice style may also be determined from the exterior elevations (**Figure 22–8**).

Dimensions

In relation to other drawings, elevations have few dimensions. Some dimensions usually given are floor to floor heights, distance from grade level to finished floor, height of window openings from the finished floor, and distance from the ridge to the top of the chimney.

A number of other things may be shown on exterior elevations, depending on the complexity of the structure. Little information is given in elevations that cannot be seen in more detail in plans and sections. However, elevations serve an important purpose in making the total construction easier to visualize.

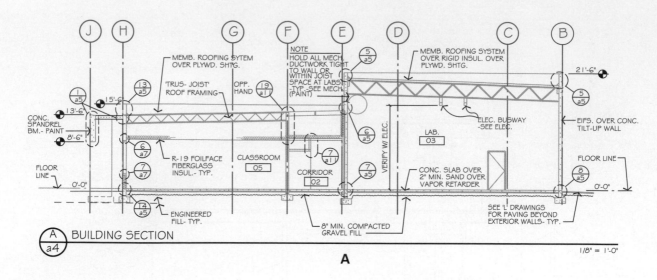

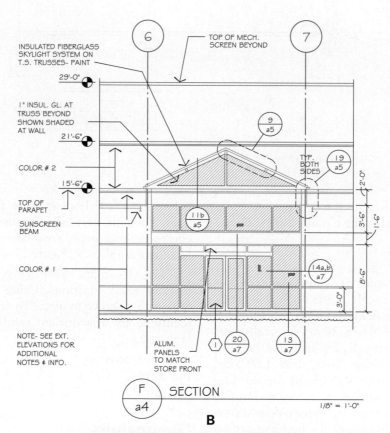

FIGURE 22–3 (A) A typical full-section and (B) partial section view of a commercial building.

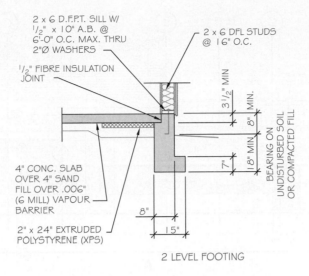

2 x 6 D.F.P.T. SILL W/
1/2" x 10" A.B. @
6'-0" O.C. MAX. THRU
2"Ø WASHERS

2 x 6 DFL STUDS
@ 16" O.C.

1/2" FIBRE INSULATION
JOINT

3 1/2" MIN

8" MIN.

18" MIN

7"

BEARING ON
UNDISTURBED SOIL
OR COMPACTED FILL

4" CONC. SLAB
OVER 4" SAND
FILL OVER .006"
(6 MILL) VAPOUR
BARRIER

8"

15"

2" x 24" EXTRUDED
POLYSTYRENE (XPS)

2 LEVEL FOOTING

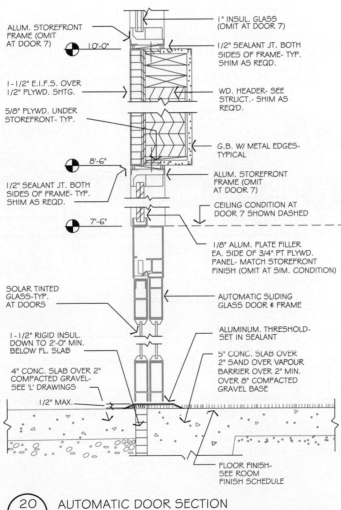

ALUM. STOREFRONT
FRAME (OMIT
AT DOOR 7)

10'-0"

1-1/2" E.I.F.S. OVER
1/2" PLYWD. SHTG.

5/8" PLYWD. UNDER
STOREFRONT- TYP.

8'-6"

1/2" SEALANT JT. BOTH
SIDES OF FRAME- TYP.
SHIM AS REQD.

7'-6"

SOLAR TINTED
GLASS-TYP.
AT DOORS

1-1/2" RIGID INSUL.
DOWN TO 2'-0" MIN.
BELOW FL. SLAB

4" CONC. SLAB OVER 2"
COMPACTED GRAVEL-
SEE 'L' DRAWINGS

1/2" MAX.

1" INSUL. GLASS
(OMIT AT DOOR 7)

1/2" SEALANT JT. BOTH
SIDES OF FRAME- TYP.
SHIM AS REQD.

WD. HEADER- SEE
STRUCT.- SHIM AS
REQ'D.

G.B. W/ METAL EDGES-
TYPICAL

ALUM. STOREFRONT
FRAME (OMIT
AT DOOR 7)

CEILING CONDITION AT
DOOR 7 SHOWN DASHED

1/8" ALUM. PLATE FILLER
EA. SIDE OF 3/4" PT PLYWD.
PANEL- MATCH STOREFRONT
FINISH (OMIT AT SIM. CONDITION)

AUTOMATIC SLIDING
GLASS DOOR & FRAME

ALUMINUM. THRESHOLD-
SET IN SEALANT

5" CONC. SLAB OVER
2" SAND OVER VAPOUR
BARRIER OVER 2" MIN.
OVER 8" COMPACTED
GRAVEL BASE

FLOOR FINISH-
SEE ROOM
FINISH SCHEDULE

20
a7

AUTOMATIC DOOR SECTION

8712\7-ADR1A   1 1/2" = 1'-0"

FIGURE 22–4 Details are a small part of a section drawn at a
large scale.

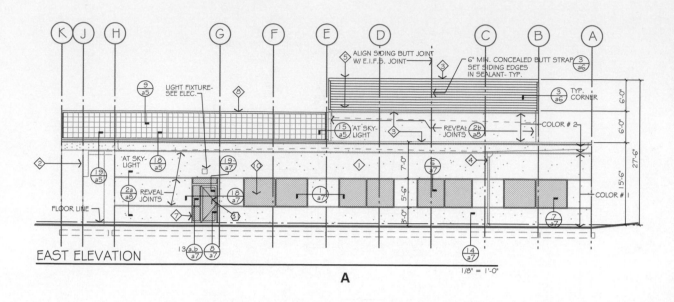

EAST ELEVATION

1/8" = 1'-0"

A

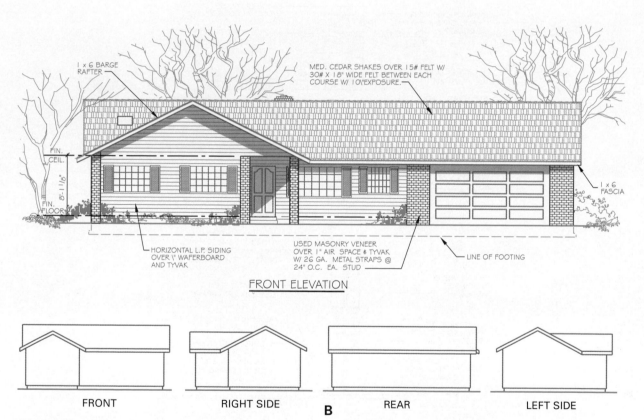

1 x 6 BARGE RAFTER

MED. CEDAR SHAKES OVER 15# FELT W/
30# X 18" WIDE FELT BETWEEN EACH
COURSE W/ 10" EXPOSURE.

FIN. CEIL.

8'-11⅝"

FIN. FLOOR

1 x 6 FASCIA

HORIZONTAL L.P. SIDING
OVER ½" WAFERBOARD
AND TYVAK

USED MASONRY VENEER
OVER 1" AIR SPACE & TYVAK
W/ 26 GA. METAL STRAPS @
24" O.C. EA. STUD

LINE OF FOOTING

FRONT ELEVATION

FRONT   RIGHT SIDE   **B**   REAR   LEFT SIDE

FIGURE 22–5 Elevations show the exterior of a building in (A) commercial and (B) residential construction.

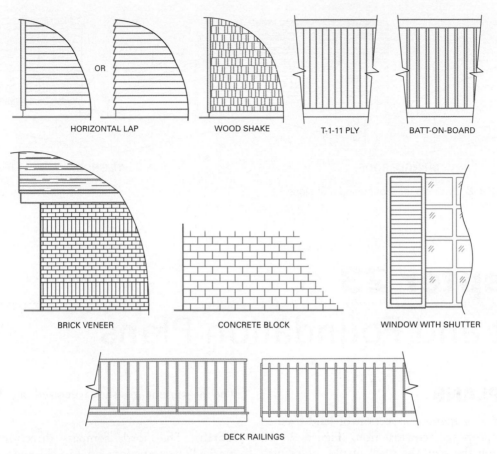

FIGURE 22–6 Symbols for siding, railings, and shutters.

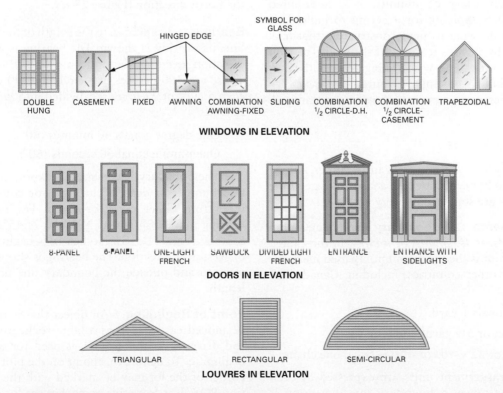

FIGURE 22–7 Symbols for windows, doors, and louvres.

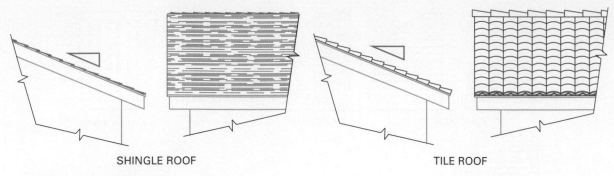

SHINGLE ROOF   TILE ROOF

FIGURE 22–8 Symbols for roofing and roof slope.

Chapter 23

Plot and Foundation Plans

PLOT PLANS

A plot plan is a map of a section of land used to show the proposed construction (**Figure 23–1**). Depending on the size, the scale of plot plans may vary from 1″= 10′ to 1″ = 200′ or 1:100 to 1:250. In practically every community, it is a required drawing when applying for a permit to build. It is a necessary drawing to plan construction that may be affected by various features of the land. The plan must show compliance with zoning and health regulations. Although plot plan requirements may vary with localities, certain items in the plan are standard.

Property Lines

The property line *measurements* and *bearings*, known as metes and bounds, show the shape and size of the parcel. They are standard in every plot plan.

Measurements.   The boundary lines are measured in *feet*, *yards*, *rods*, *chains*, or *metres*. In the United States, the foot is the most commonly used measurement; most other countries, including Canada, use metres.

   3 feet equals 1 yard.

   16½ feet or 5½ yards equal one rod.

   66 feet or 22 yards or 4 rods equal one chain.

Parts of measurement units are expressed as decimals. For instance, a boundary line dimension is expressed as 100.50 feet, not 100 feet, 6 inches.

The measurement is shown centred on, close to, and inside the line.

North.   The north compass direction is clearly marked on every plot plan. In a clear space, an arrow of any style labelled with the letter "N" is pointed in the north direction (**Figure 23–2**).

Bearings.   In addition to the length of the boundary line, its bearing is shown. The bearing is a compass direction given in relation to a *quadrant* of a circle. There are 360 degrees in a circle and 90 degrees in each quadrant. Degrees are divided into *minutes* and *seconds*.

   One degree equals 60 minutes (60′).

   One minute equal 60 seconds (60″).

   The boundary line bearing is expressed as a certain number of degrees clockwise or counterclockwise from either north or south. For instance, a bearing may be shown as N 30°W, N 45°E, S 60°E, S 30°W (**Figure 23–3**). No bearings begin with east or west as a direction. The bearing is shown centred close to and outside the boundary line opposite its length.

Point of Beginning.   An object that is unlikely to be moved easily, such as a large rock, tree, or iron rod driven into the ground, is used for a point of beginning. To denote this point on the plot plan, one corner of the lot may be marked with the abbreviation POB. It is from this point that the lot is laid out and drawn (**Figure 23–4**).

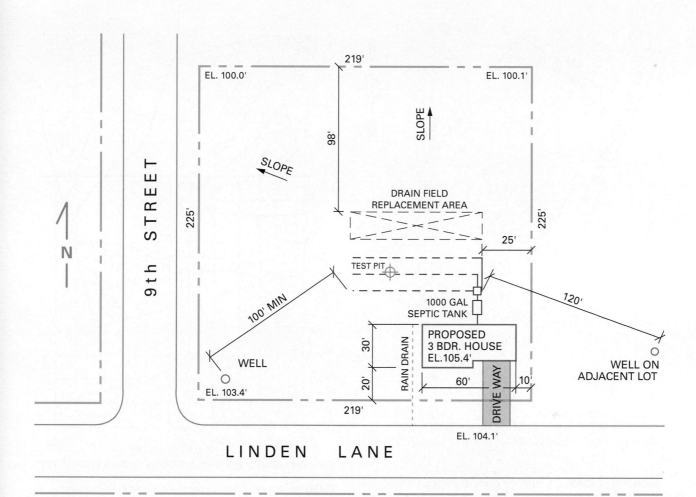

FIGURE 23–1 A typical plot plan (scale 1:750).

FIGURE 23–2 Typical north direction symbols.

Topography

Topography is the detailed description of the land surface. It includes any outstanding physical features and differences in *elevation* of the building site. Elevation is the height of a surface above sea level. It is expressed in decimal feet and $\frac{1}{100}$ of a foot or in metres and decimals.

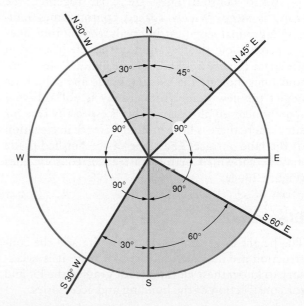

FIGURE 23–3 Method of indicating bearings for property lines.

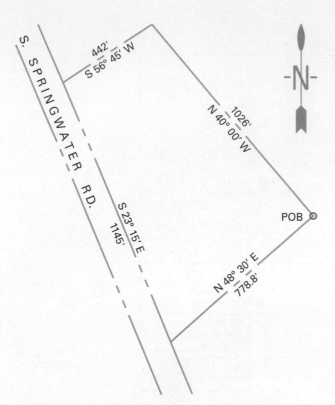

FIGURE 23–4 Measurements, bearings, and legal description of a parcel of land.

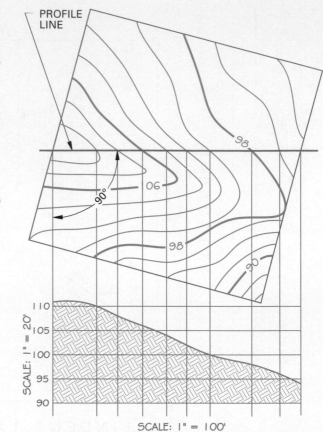

SECTION VIEW OF ELEVATION DRAWN FROM A PROFILE LINE

FIGURE 23–5 Contour lines show the elevation and slope of the land (scale 1:1000).

Contour Lines.

Contour lines are irregular, curved lines connecting points of the same elevation of the land. The vertical distance between contour lines is called the **contour interval**. It may vary depending on how specifically the contour of the land needs to be shown on the plot plan.

When contour lines are close together, the slope is steep. Widely spaced contour lines indicate a gradual slope. At intervals, the contour lines are broken and the elevation of the line inserted in the space (**Figure 23–5**). On some plans, dashed contour lines indicate the existing grade and solid lines depict the new grade. Topography is not always a requirement on plot plans. This is especially true for sites where there is little or no difference in elevation of the land surface. The slope of the finished grade may be shown by arrows instead of contour lines (**Figure 23–6**).

Elevations

The height of several parts of the site and the construction are indicated on the plot plan. It is necessary to know these elevations for grading the lot and for construction of the building and accessories.

Benchmark. Before construction begins, a reference point, called a **benchmark**, is established

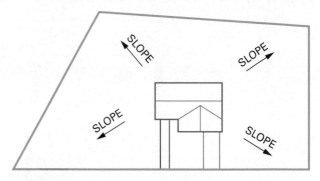

FIGURE 23–6 Arrows are sometimes used in place of contour lines to show the slope of the land.

on or close to the site. It is used for conveniently determining differences in elevation of various land and building surfaces.

The benchmark is established on some permanent object that will not be moved or destroyed, at least until the construction is complete. It may be the actual elevation in relation to sea level. It may also be given an arbitrary elevation of 100.00 feet (30 m). All points on the lot, therefore, would either be above,

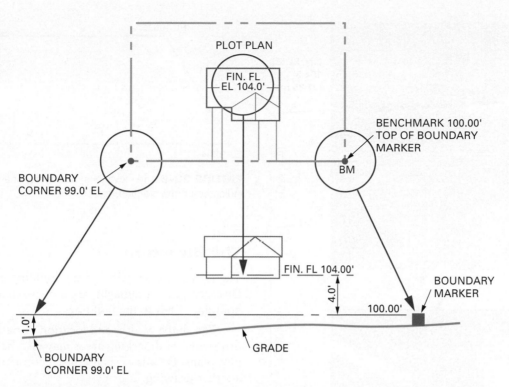

PLOT PLAN

FIN. FL
EL 104.0'

BENCHMARK 100.00'
TOP OF BOUNDARY
MARKER

BM

BOUNDARY
CORNER 99.0' EL

FIN. FL 104.00'

4.0'

BOUNDARY
MARKER

100.00'

1.0'

GRADE

BOUNDARY
CORNER 99.0' EL

FIGURE 23–7 A benchmark is a reference point used for determining differences in elevation.

level with, or below the benchmark (**Figure 23–7**). The location of the benchmark is clearly shown on the plot plan with the abbreviation BM.

Finish Floor. The elevation of the finished floor or the top of the foundation levels may be shown and noted on the plot plan. This elevation helps the contractor determine the bottom of the excavation. The bottom of excavation elevation must be calculated. To do this, subtract the building component heights from the finish floor height given. For example, what is the bottom of excavation elevation for the building shown in **Figure 23–8**?

To solve, add up the components:

$$10'' + 8'\text{-}0'' + 1'\text{-}2\tfrac{3}{4}'' = 10'\text{-}0\tfrac{3}{4}''$$

Convert ¾" to decimal feet.

$$\tfrac{3}{4} \div 12 = 0.0625$$

Add 0.0625 to 10 feet.

$$0.0625 + 10 = 10.0625 \text{ (rounded off to 10.06 feet)}$$

Next, subtract

$$104.50' - 10.06' = 94.44'$$

The bottom of excavation elevation is 94.44 feet.

Converting Decimals to Fractions. Calculations, especially those dealing with elevations and roof framing, require the carpenter to convert decimals of a foot to feet, inches, and 16ths of an inch

as found on the rule or tape. To convert, use the following method:

- Multiply a decimal of a foot by 12 (the number of inches in a foot) to get inches.
- Multiply any remainder decimal of an inch by 16 (the number of 16ths in an inch) to get 16ths of an inch. Round off any remainder to the nearest 16th of an inch.
- Combine whole feet, whole inches, and 16ths of an inch to make the conversion.

For example, convert the finished floor elevation of 104.65 feet to feet, inches, and 16ths of an inch.

1. Multiply 0.65 ft × 12 = 7.80 inches.
2. Multiply 0.80 inches × 16 = 12.8/16ths of an inch.
3. Round off 12.8/16ths of an inch to $\tfrac{13}{16}$ of an inch.
4. Combine feet, inches, and sixteenths = $104'\text{-}7\tfrac{13}{16}''$.

It is more desirable to remember the method of conversion rather than use conversion tables. Reliance on conversion tables requires access to the tables and knowledge of their use, encourages dependence on them, and results in helplessness without them.

The calculation in metric units for the excavation elevation is simply 31.85 m (or 31 850 mm) − 375 mm − 2440 mm − 254 mm = 28.781 m (or 28 781 mm).

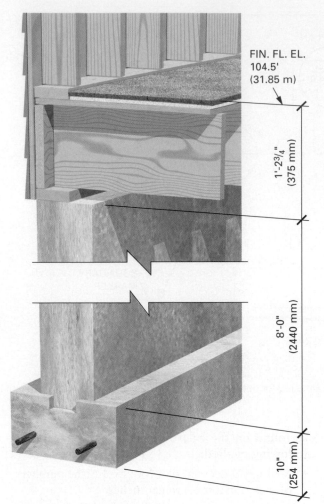

ADD COMPONENTS: 10" + 8'-0" + 1'-2³/₄" = 10'-0³/₄".

CONVERT TO DECIMAL FEET: ³/₄" ÷ 12 = 0.0625'.

ADD TO 10' = 10.0625 OR ROUNDED TO 10.06'.

SUBTRACT 104.50' – 10.06' = 94.44' IS THE EXCAVATION ELEVATION .

FIGURE 23–8 The distance from the finished floor elevation to the bottom of the footing may need to be calculated.

Other Elevations. In addition to contour lines, the elevation of each corner of the property is noted on the plot plan (**Figure 23–9**). The top of one of the boundary corner markers makes an excellent benchmark.

Existing and proposed roads adjacent to the property are shown, as well as any easements. An easement is a right-of-way strip running through the property. Easements are granted for various purposes, such as access to other property, storm drains, or utilities. Elevations of a street at a driveway and at its centreline are usually required. One of these is sometimes used as the benchmark.

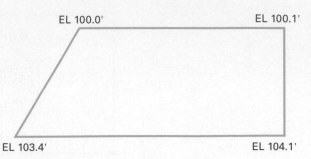

FIGURE 23–9 The elevations of property line corners are indicated on the plot line.

The Structure

The shape and location of the building are shown. Distances, called setbacks, are dimensioned from the boundary lines to the building.

The shape, width, and location of patios, walks, driveways, and parking areas may also be shown on plot plans. Details of their construction are found in another drawing.

A plot plan may also show any *retaining walls*. These walls are used to hold back earth to make more level surfaces instead of steep slopes.

Utilities. The water supply and public sewer connections are shown by noted lines from the structure to the appropriate boundary line. If a private sewer disposal system is planned, it is shown on the plot plan (**Figure 23–10**). Although there are many kinds of sewer disposal systems, those most commonly used consist of a septic tank and leach or drain field, which are usually subject to strict regulations with regard to location and construction.

The location of gas lines is shown, if applicable. Sometimes the location of the nearest utility pole is given. It is shown using a small solid circle as a symbol and is noted. Foundation drain lines leading to a storm drain, drywell, or other drainage may be shown and labelled.

Landscaping. The location and kind of existing and proposed trees are shown on the plot plan. Existing trees are noted, whether they are to be saved or removed. Those that are to be saved are protected by barriers during the construction process. Various symbols are used to show different kinds of trees (**Figure 23–11**).

Identification. Included on the plot plan are the name and address of the property owner, the title and scale of the drawing, and a legal description of the property (**Figure 23–12**).

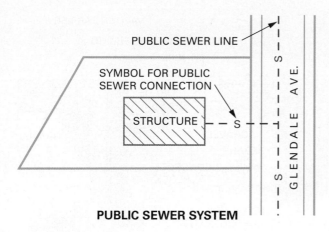

PUBLIC SEWER SYSTEM

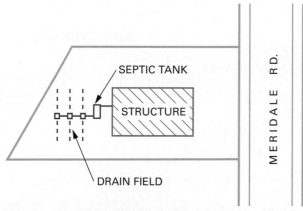

PRIVATE SEWER SYSTEM

FIGURE 23–10 The style of sewage disposal system is shown on the plot plan.

Foundation Plans

The foundation plan is drawn at the same scale as the floor plan. It is a view from above of a horizontal cut through the foundation. Great care

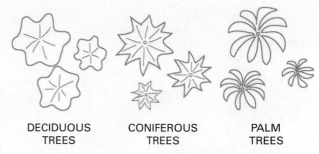

FIGURE 23–11 Various symbols are used to indicate different kinds of trees.

must be taken when reading the foundation plan so that no mistakes are made. A mistake in the foundation affects the whole structure, and generally requires adjustments throughout the construction process.

Two commonly used types of foundations are those having a crawl space, or basement below grade, and those with a concrete slab floor at grade level (**Figure 23–13**).

Crawl Space and Basement Foundations

The **crawl space** is the area enclosed by the foundation between the ground and the floor above. A minimum distance of 18 inches (457 mm) from the ground to the floor and 12 inches (305 mm) from the ground to the bottom of any beam is required. The ground is covered with a plastic sheet, called a **vapour barrier**, to prevent moisture rising from the ground from penetrating into the floor frame above.

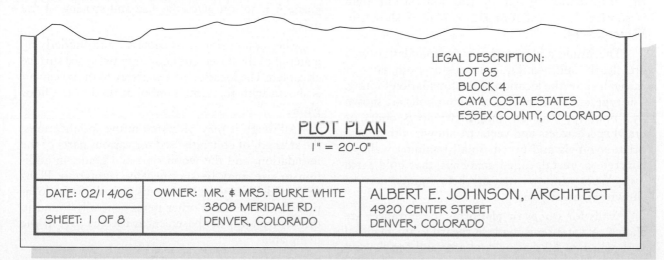

LEGAL DESCRIPTION:
LOT 85
BLOCK 4
CAYA COSTA ESTATES
ESSEX COUNTY, COLORADO

PLOT PLAN
1" = 20'-0"

DATE: 02/14/06

SHEET: 1 OF 8

OWNER: MR. & MRS. BURKE WHITE
3808 MERIDALE RD.
DENVER, COLORADO

ALBERT E. JOHNSON, ARCHITECT
4920 CENTER STREET
DENVER, COLORADO

FIGURE 23–12 Certain identification items are needed on the plot plan (scale 1:250).

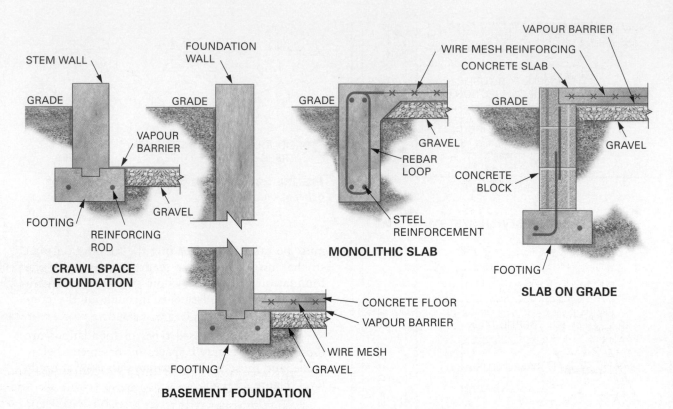

FIGURE 23–13 Commonly used foundation styles.

A foundation enclosing a basement is similar to that of a crawl space except the walls are higher, windows and doors may be installed, and a concrete floor is provided below grade. The basement may be used for additional living area, a garage, a utility room, or workshop.

Reading Plans. Whether the foundation supports a floor using closely spaced floor joists or more widely spaced post-and-beam construction, the information given in the foundation plan is similar. A typical foundation plan is shown in **Figure 23–14**.

The inside and outside of the foundation wall are clearly outlined. Dashed lines on both sides of the wall show the location of the foundation footing. The type, size, and spacing of anchor bolts are shown by a notation. Wall openings for windows, doors, or crawl space access and vents are shown with appropriate symbols and noted. Small retaining walls of concrete or metal, called **areaways**, that hold earth away from windows that are below grade may be shown (**Figure 23–15**).

Walls for *stoops* or platforms for entrances are shown. A notation is made in regard to the material with which to fill the enclosed area and cap the surface. Other footings shown by dashed lines include those for chimneys, fireplaces, and columns or posts.

Columns or posts support girders shown by a series of long and short dashes directly over the centre of the columns. A recess in the foundation wall, called a **beam pocket**, is used to support the ends of the girder is shown; the recess may also be called a girder pocket. Notations are made to identify all of these items (**Figure 23–16**).

Floor joist direction installed above the plan view is shown by a line with arrows on both ends, similar to those shown in the floor plan for ceiling joists. A notation gives the size and spacing of the joists.

The composition, thickness, and underlying material of the basement floor or crawl space surface are noted. The location of a stairway to the basement is shown with the same symbol as used in the floor plan.

Although it may be stated in the specifications, the strength of concrete used for various parts of the foundation, and the wood type and grade, in addition to size, may be specified by a notation. Plans for foundations with basements may show the location of furnaces and other items generally found in the floor plan if the basement is used as part of the living area.

Dimensions. It is important to understand how parts of the foundation are dimensioned. Foundation

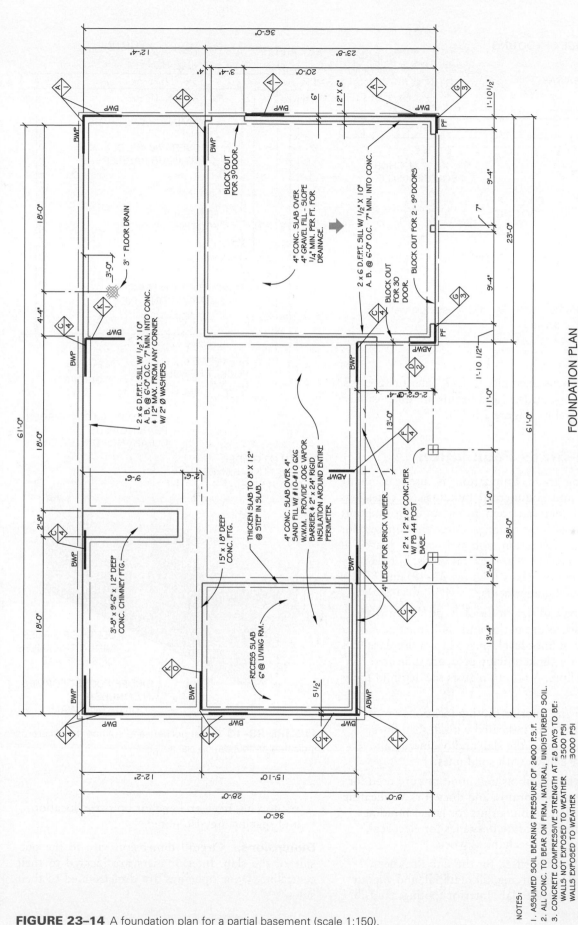

FOUNDATION PLAN
3/32" = 1'-0"

NOTES:
1. ASSUMED SOIL BEARING PRESSURE OF 2000 P.S.F.
2. ALL CONC. TO BEAR ON FIRM, NATURAL, UNDISTURBED SOIL.
3. CONCRETE COMPRESSIVE STRENGTH AT 28 DAYS TO BE:
   WALLS NOT EXPOSED TO WEATHER   2500 PSI
   WALLS EXPOSED TO WEATHER   3000 PSI
   PORCHES, STEPS AND GARAGE SLAB   3500 PSI
4. EXTEND FOOTINGS BELOW FROST LINE, (18" MIN. INTO NATURAL
   SOIL FOR 1 STORY AND 2 STORY CONSTRUCTION),
   FOOTINGS TO BE 6" THICK FOR 1 STORY, AND 7" THICK
   FOR 2 STORY CONSTRUCTION. ALL FOUNDATION WALLS
   TO BE 8" WIDE, UNLESS STEEL IS PROVIDED WITHIN 2"
   BUT NOT CLOSER THAN 1" FROM THE FACE OF THE WALL
   AWAY FROM THE SOIL. STEEL TO BE 2 - #3'S HORIZONTAL.
5. THE GRADE AWAY FROM THE FOUNDATION WALLS TO FALL
   A MIN. OF 6" WITHIN THE FIRST 10'.
6. SEE SCHEDULE FOR BRACING REQUIREMENTS.

FIGURE 23–14 A foundation plan for a partial basement (scale 1:150).

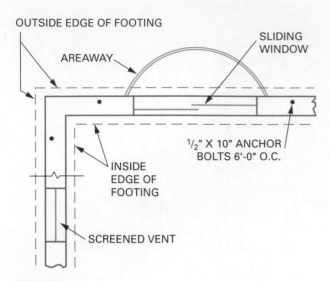

FIGURE 23–15 Partial plan of a foundation wall with various items indicated.

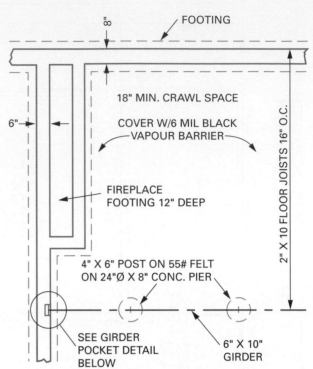

walls are dimensioned face to face. Interior footings, columns, posts, girders, and beams are dimensioned to their centreline (**Figure 23–17**).

Slab-on-Grade Foundation

The **slab-on-grade foundation** is used in many residential and commercial buildings. It takes less labour and material than foundations to support beam and joist floor framing. Often the concrete for the footing, foundation, and slab can be placed at the same time. There are several kinds, but slab-on-grade foundation plans (**Figures 23–18** and **23–19**) show common components:

- The shape and size of the slab are shown with solid lines, as are patios and similar areas. Changes in floor level, such as for a fireplace hearth or a sunken living area, are indicated by solid lines. A notation gives the depth of the recess.

- Exterior and interior footing locations ordinarily below grade are indicated with dashed lines. Footings outside the slab and ordinarily above grade are shown with solid lines.

- Appropriate symbols and notations are used for fireplaces, floor drains, and ductwork for heating and ventilation. Blueprints for larger buildings usually have separate drawings for electrical, plumbing, and mechanical work.

- Notations are written for the slab thickness, wire mesh reinforcing, fill material, and vapour barrier under the slab. Interior footing, mudsill,

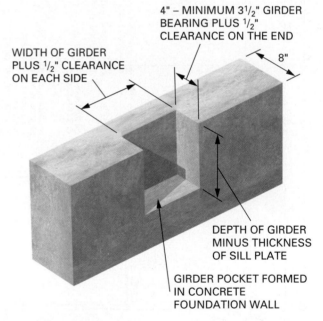

FIGURE 23–16 Girder pockets and column details are indicated on foundation plans.

reinforcing steel, and anchor bolt size, location, and spacing are also noted.

Dimensions. Overall dimensions are to the outside of the slab. Interior piers are located to their centreline. Door openings are dimensioned to their sides.

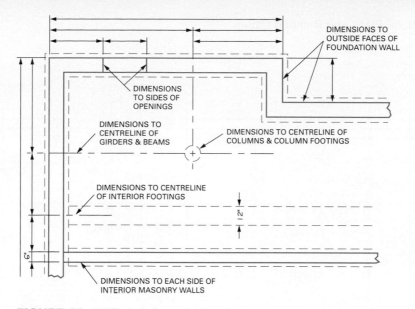

FIGURE 23-17 Typical dimensioning of crawl space and basement foundations.

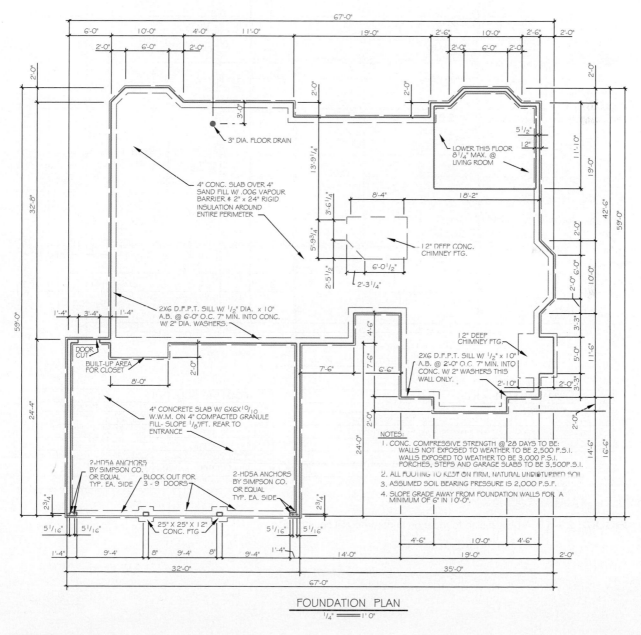

FIGURE 23-18 Slab-on-grade foundation plan (scale 1:50).

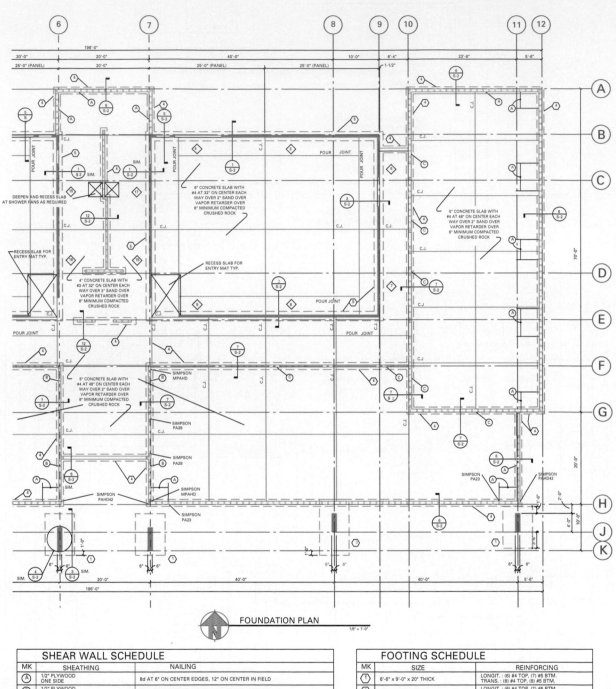

FOUNDATION PLAN
1/6" = 1'-0"

SHEAR WALL SCHEDULE		
MK	SHEATHING	NAILING
A	1/2" PLYWOOD ONE SIDE	8d AT 6" ON CENTER EDGES, 12" ON CENTER IN FIELD
B	1/2" PLYWOOD ONE SIDE	8d AT 4" ON CENTER EDGES, 12" ON CENTER IN FIELD
C	5/8" GYP. WALLBOARD ONE SIDE	6d COOLER OR WALLBOARD NAILS AT 7" ON CENTER TO ALL SUPPORTS (UNBLOCKED)

NOTES   1. USE COMMON NAILS U.O.N.
2. PROVIDE BLOCKING AT ALL UNSUPPORTED PLYWOOD EDGES.
3. WALLS NOTED ON PLAN ARE TYPE NOTED FULL LENGTH OF WALL (OR LENGTH SHOWN BY DIM. LINES)

FOOTING SCHEDULE		
MK	SIZE	REINFORCING
1	6'-6" x 9'-0" x 20" THICK	LONGIT. : (6) #4 TOP, (7) #5 BTM. TRANS. : (8) #4 TOP, (8) #5 BTM.
2	6'-6" x 10'-0" x 20" THICK	LONGIT. : (6) #4 TOP, (7) #5 BTM. TRANS. : (9) #4 TOP, (9) #5 BTM.
3	6'-6" x 12'-0" x 20" THICK	LONGIT. : (6) #4 TOP, (7) #7 BTM. TRANS. : (11) #4 TOP, (11) #5 BTM.
4	18" WIDE x 12" THICK x CONTINUOUS	(2) #4 CONTINUOUS BOTTOM
5	24" WIDE x 16" THICK	(2) #5 CONTINUOUS TOP AND BOTTOM
6	2'-0" X 2'-0" X 18" THICKENED SLAB	UNREINFORCED

FIGURE 23–19 A typical slab foundation in commercial construction.

Chapter 24

Building Codes and Zoning Regulations

Cities and towns have laws governing many aspects of new construction and remodelling. These laws protect the consumer and the community. Codes and regulations provide for safe, properly designed buildings in a planned environment. Contractors and carpenters should have knowledge of local zoning regulations and building codes.

ZONING REGULATIONS

Zoning regulations, generally speaking, deal with keeping buildings of similar size and purpose in areas for which they have been planned. They can also regulate the space in each of the areas. The community is divided into areas called zones, shown on *zoning maps*.

Zones

The names given to different zones vary from community to community. The zones are usually abbreviated with letters or a combination of letters and numbers. A large city may have 30 or more zoning districts.

There may be several *single-family residential zones*. Some zones have less-strict requirements than others. Other areas may be zoned as *multifamily residential*. They may be further subdivided into areas according to the number of apartments. Other residential zones may be set aside for *mobile home parks*, and those that allow a combination of *residences*, *retail stores*, *and offices*.

Other zones may be designated for the *central business district*, or various kinds of *commercial districts* and different *industrial* zones.

Lots

Zoning laws regulate buildings and building sites. Most cities specify a *minimum lot size* for each zone and a *maximum ground coverage* by the structure. The *maximum height* of the building for each zoning district is stipulated. A *minimum lot width* is usually specified as well as *minimum yards*.

Minimum yard refers to the distance that buildings must be kept from property lines. These distances are called setbacks. They are usually different for front, rear, and side.

Some communities require a certain amount of landscaped area, called green space, to enhance the site. In some residential zones, as much as half the lot must be reserved for green space. In a central business area, only 5 to 10 percent may be required.

In most zones, off-street parking is required. For instance, in single-family residential zones, room for two parking spaces on the lot is required.

Non-Conforming Buildings

Because some cities were in existence before the advent of zoning laws, many buildings and businesses may not be in their proper zone. They are called non-conforming. It would be unfair to require that buildings be torn down, or to stop businesses, in order to meet the requirements of zoning regulations.

Non-conforming businesses or buildings are allowed to remain. However, restrictions are placed on rebuilding. If partially destroyed, they may be allowed to rebuild, depending on the amount of destruction. If 75 percent or more is destroyed, they are not usually allowed to rebuild in the same manner or for the same purpose in the same zone.

Any hardships imposed by zoning regulations may be relieved by a variance. Variances are granted by a Zoning Board of Appeals within each community. A public hearing is held after a certain period of time. The general public and, in particular, those abutting the property are notified. The petitioner must prove certain types of hardship specified in the zoning laws before the zoning variance can be granted.

BUILDING CODES

Building codes regulate the design and construction of buildings by establishing minimum health and safety standards, accessibility, and fire and structural protection of buildings. They prevent such things as roofs being ripped off by high winds, floors collapsing from inadequate support, buildings settling because of a poor foundation, and tragic deaths from fire due to lack of sufficient exits from buildings. In addition to building codes, other codes govern the mechanical, electrical, and plumbing trades.

Some provinces have adopted the National Building Code of Canada 2015. Some use the national code supplemented with their own. Some have building codes that supersede the national code.

It is important to have a general knowledge of the building code used by a particular community. Construction superintendents and contractors must have extensive knowledge of the codes.

National Building Code of Canada 2015

The National Building Code of Canada (NBCC) 2015 is in an objective-based code format and it addresses the design and construction of new buildings and substantial renovations. It is revised and updated every five years and it sets minimum standards to which buildings must be constructed. Some provinces augment this code with more stringent requirements and publish the combination as a *Provincial Building Code*. A few cities have charters that allow them to publish their own building codes. For instance, the city of Vancouver requires sprinklers in residential settings, whereas the NBCC does not.

The NBCC is divided into two volumes:

Volume 1: Divisions A and C, which covers compliance, objectives, and functional statements as well as administrative provisions.

Volume 2: Division B, which covers acceptable solutions.

Division B has nine parts:

Part 1: General – Climatic Data, reference to Standards

Part 2: Reserved

Part 3: Fire Protection, Occupant Safety, and Accessibility

Part 4: Structural Design – live/dead loads; wind, rain, snow, earthquake, etc.

Part 5: Environmental Separation – protection from heat, cold, air, vapour, noise, precipitation, surface water, etc.

Part 6: Heating, Ventilation, and Air-Conditioning – ducting, piping, chimneys, carbon monoxide, etc.

Part 7: Plumbing Services – facilities, traps, venting, cleanouts, drainage, etc.

Part 8: Safety Measures at Construction and Demolition Sites

Part 9: Housing and Small Buildings ($<$ 600 m², 6460 ft²) – all related matters including concrete, wood, steel, interior/exterior finishes, mechanical, etc.

There are also four appendices: A (Explanatory Material), B (Fire Safety in High Buildings, C (Climatic Information), and D (Fire Performance Ratings).

The province of Ontario adds in Part 10 (Change of Use), Part 11 (Renovations), and Part 12 (Resource Conservation, energy and water efficiency). New home buyers are protected by the Tarion Warranty Corporation, and builders must enroll with the corporation before building a home or condominium. Offenders will be charged under the *Ontario New Home Warranties Plan Act*.

As part of the *Illustrated Code Series*, Anthony Boyko and Steven Penna have compiled *Housing and Small Buildings*. This is an illustrated version of Part 9 of the Ontario Building Code, 2012, containing detailed diagrams with commentary. This guide will help you interpret and apply Part 9 when you design, construct, or inspect homes and small buildings (**Figure 24–1**). The *Housing and Small Buildings* illustrated version has been endorsed by the Association of Architectural Technologists of Ontario (AATO) (Figure 24–1). Under the start-up section, it lists the typical steps in the plans, permits, and inspection process (**Figure 24–2**). There is provision in the code for the use of materials and systems that are not specifically addressed. The Building Materials Evaluation Commission (BMEC) will make an evaluation on a fee-for-service basis. The Canadian Construction Materials Centre (CCMC) of the National Research Council of Canada (NRC) is a designated body to evaluate materials.

Use of Residential Codes

In addition to structural requirements, major topics covered by residential codes include the following:

- Exit facilities, such as doors, halls, stairs, and windows as emergency exits, and smoke detectors
- Room dimensions, such as ceiling height and minimum area
- Light, ventilation, and sanitation, such as window size and placement, maximum limits of glass area, fans vented to the outside, and requirements for baths, kitchens, and hot and cold water

Use of Commercial Codes

Codes for commercial work are much more complicated than those for residential work. The structure must first be defined for code purposes. To define the structure, six classifications must be used.

1. The *occupancy group* classifies the structure according to how and by whom it will be used. The classification is designated by a letter, such

FIGURE 24–1 Illustrated Code Series, Housing and Small Buildings, OBC 2012, A Boyko & S. Penna

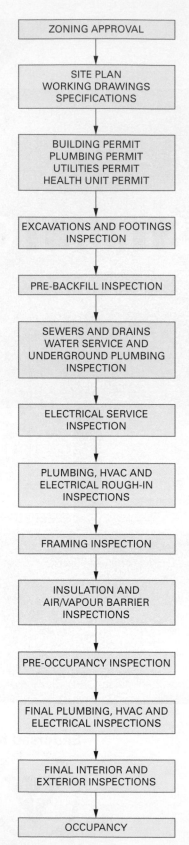

```
ZONING APPROVAL
        ↓
SITE PLAN
WORKING DRAWINGS
SPECIFICATIONS
        ↓
BUILDING PERMIT
PLUMBING PERMIT
UTILITIES PERMIT
HEALTH UNIT PERMIT
        ↓
EXCAVATIONS AND FOOTINGS
INSPECTION
        ↓
PRE-BACKFILL INSPECTION
        ↓
SEWERS AND DRAINS
WATER SERVICE AND
UNDERGROUND PLUMBING
INSPECTION
        ↓
ELECTRICAL SERVICE
INSPECTION
        ↓
PLUMBING, HVAC AND
ELECTRICAL ROUGH-IN
INSPECTIONS
        ↓
FRAMING INSPECTION
        ↓
INSULATION AND
AIR/VAPOUR BARRIER
INSPECTIONS
        ↓
PRE-OCCUPANCY INSPECTION
        ↓
FINAL PLUMBING, HVAC AND
ELECTRICAL INSPECTIONS
        ↓
FINAL INTERIOR AND
EXTERIOR INSPECTIONS
        ↓
OCCUPANCY
```

FIGURE 24–2 Plans, permits, and inspections.
(© *Queen's Printer for Ontario.* 2010 Code and Construction Guide for Housing, *published by the Ministry of Municipal Affairs and Housing. Reproduced with permission*)

as R, which includes not only single-family homes but also apartments and hotels.

2. The size and location of the building

3. The type of construction: Five general types are given numbers 1 through 5. Types 1 and 2 require that all structural parts be non-combustible. Construction in types 3, 4, or 5 can be made of masonry, steel, or wood.

4. The floor area of the building

5. The height of the building: Zoning regulations may also affect the height.

6. The number of people who will use the building, called the *occupant load*, which determines such things as the number and location of exits

Once the structure is defined, the code requirements may be studied.

BUILDING PERMITS

A **building permit** is needed before construction can begin. Application is made to the office of the local building official. The building permit application form (**Figure 24–3**) requires a general description of the construction, legal description and location of the property, estimated cost of construction, and information about the applicant. The process for getting building permits in Ontario is well described on the first five pages of ***http://www.london.ca/business/ Permit-Licences/Building-Permits/Documents/ tacboc_details_2012_r001.3.pdf***. In other provinces, contact the local building permit office for specific instructions.

Drawings of the proposed construction are submitted with the application. The type and kind of drawings required depend on the complexity of the building. For commercial work, usually five sets of plot plans and two sets of other drawings are required. The drawings are reviewed by the building inspection department. If all is in order, a permit (**Figure 24–4**) is granted upon payment of a fee. The fee is usually based on the estimated cost of the construction. Electrical, mechanical, plumbing, water, and sewer permits are usually obtained by subcontractors. The permit card must be displayed on the site in a conspicuous place until the construction is completed.

Why have requirements for building permits? The short answer is that permits are followed by **building inspections**, which ensure that structural, fire, and mechanical standards are followed to safeguard the health and safety of the workers and the occupants of the building, both present and future.

Application for a Permit to Construct or Demolish

This form is authorized under the Building Code Sentence 2.4.1.1A.(2).

For use by Principal Authority	
Application number:	Permit number (if different):
Date received:	Roll number:

Application submitted to: **City of Toronto**

District Offices:
☐ North York 416-395-7000 ☐ Toronto and East York 416-392-7539 ☐ Scarborough 416-396-7526 ☐ Etobicoke York 416-394-8002

A. Project information

Building number, street name		Unit number	Lot/con.
Municipality	Postal code	Plan number/other description	
Project value est. $		Area of work (m^2)	

B. Applicant

Applicant is: ☐ Owner or ☐ Authorized agent of owner

Last name	First name	Corporation or partnership		
Street address			Unit number	Lot/con.
Municipality	Postal code	Province	E-mail	
Telephone number ( )	Fax ( )		Cell number ( )	

C. Owner (if different from applicant)

Last name	First name	Corporation or partnership		
Street address			Unit number	Lot/con.
Municipality	Postal code	Province	E-mail	
Telephone number ( )	Fax ( )		Cell number ( )	

D. Builder (optional)

Last name	First name	Corporation or partnership (if applicable)		
Street address			Unit number	Lot/con.
Municipality	Postal code	Province	E-mail	
Telephone number ( )	Fax ( )		Cell number ( )	

E. Purpose of application

☐ New construction ☐ Addition to an existing building ☐ Alteration/repair ☐ Demolition ☐ Conditional Permit

Proposed use of building	Current use of building

Description of proposed work

F. Tarion Warranty Corporation (Ontario New Home Warranty Program)

i. Is proposed construction for a new home as defined in the *Ontario New Home Warranties Plan Act*? If no, go to section G. ☐ Yes ☐ No

ii. Is registration required under the *Ontario New Home Warranties Plan Act*? ☐ Yes ☐ No

iii. If yes to (ii) provide registration number(s): _____

G. Attachments

i. Attach documents establishing compliance with applicable law as set out in Article 1.1.3.3.

ii. Attach Schedule 1 for each individual who reviews and takes responsibility for design activities.

iii. Attach Schedule 2 where application is to construct on-site, install or repair a sewage system.

iv. Attach types and quantities of plans and specifications for the proposed construction or demolition that are prescribed by the by-law, resolution, or regulation of the municipality, upper-tier municipality, board of health or conservation authority to which this application is made.

H. Declaration of applicant

I _____ certify that:

(print name)

1. The information contained in this application, attached schedules, attached plans and specifications, and other attached documentation is true to the best of my knowledge.

2. I have authority to bind the corporation or partnership (if applicable).

_____   _____
Date   Signature of applicant

Personal information contained in this form and schedules is collected under the authority of subsection 8(1.1) of the *Building Code Act, 1992*, and will be used in the administration and enforcement of the *Building Code Act, 1992*. Questions about the collection of personal information may be addressed to: a) the Chief Building Official of the municipality or upper-tier municipality to which this application is being made, or, b) the inspector having the powers and duties of a chief building official in relation to sewage systems or plumbing for an upper-tier municipality, board of health or conservation authority to whom this application is made, or, c) Director, Building and Development Branch, Ministry of Municipal Affairs and Housing 777 Bay St., 2nd Floor. Toronto, M5G 2E5 (416) 585-6666.

06/07/05

(Continued)

FIGURE 24–3 A typical form used to apply for a building permit. (© *Queen's Printer for Ontario. 2010 Code and Construction Guide for Housing, published by the Ministry of Municipal Affairs and Housing. Reproduced with permission.*)

Schedule 1: Designer Information

Use one form for each individual who reviews and takes responsibility for design activities with respect to the project.

A. Project Information

Building number, street name		Unit no.	Lot/con.

Municipality	Postal code	Plan number/ other description

B. Individual who reviews and takes responsibility for design activities

Name	Firm

Street address		Unit no.	Lot/con.

Municipality	Postal code	Province	E-mail

Telephone number ( )	Fax number ( )	Cell number ( )

C. Design activities undertaken by individual identified in Section B. [Building Code Table 2.20.2.1]

☐ House ☐ HVAC – House ☐ Building Structural
☐ Small Buildings ☐ Building Services ☐ Plumbing – House
☐ Large Buildings ☐ Detection, Lighting and Power ☐ Plumbing – All Buildings
☐ Complex Buildings ☐ Fire Protection ☐ On-site Sewage Systems

Description of designer's work

D. Declaration of Designer

I _____ declare that (choose one as appropriate):
(print name)

☐ I review and take responsibility for the design work on behalf of a firm registered under subsection 2.17.4. of the Building Code. I am qualified, and the firm is registered, in the appropriate classes/categories.

Individual BCIN: _____

Firm BCIN: _____

☐ I review and take responsibility for the design work and am qualified in the appropriate category as an "other designer" under subsection 2.17.5. of the Building Code.

Individual BCIN: _____

Basis for exemption from registration: _____

☐ The design work is exempt from the registration and qualification requirements of the Building Code.

Basis for exemption from registration and qualification: _____

I certify that:
1. The information contained in this schedule is true to the best of my knowledge.
2. I have authority to bind the corporation or partnership (if applicable).

_____ _____
Date   Signature of Designer

* For the purposes of this form, "individual" means the "person" referred to in Clause 2.17.4.7.(1)(d), Article 2.17.5.1. and all other persons who are exempt from qualification under Subsections 2.17.4. and 2.17.5.

NOTE:

1. Firm and Individual BCIN numbers are not required for building permit applications submitted prior to January 1, 2006

2. Schedule 1 does not need to be completed by architects, or holders of a Certificate of Practice or a Temporary License under the *Architects Act.*

Application for a Permit to Construct or Demolish Schedule 1 06/07/05

FIGURE 24–3 *(Continued)*

BUILDING PERMIT
PERMIS DE CONSTRUCTION

Date of Issuance: Date de délivrance: **01-Jan-2012**	Application No / Demande n° : **A12-00001** Permit No / Permis n° : **120001** **2006 Building Code Review as amended**

Permission is Hereby Given To / Le présent permis est délivré à

Property Owner(s) / Propriétaire(s) : **John Doe**

Location / Lieu : **100   Main Street   Gloucester**

Lot Number / Numéro du lot : **10**

Permit Type / Type de permis : **Construction**

Project Description /
Description du projet : **Construct a 2 storey detached dwelling with attached 2 car garage**

Please contact the Inspector noted below prior to commencing construction /
Veuillez communiquer avec l'inspecteur mentionné ci-dessous avant de commencer les travaux

Building Inspector/ Inspecteur en bâtiment	**Building Inspector**	**613-580-2424**	**Ext./Poste**
Mechanical Inspector/ Inspecteur en mécanique	**Mechanical Inspector**	**613-580-2424**	**Ext./Poste**
Plumbing Inspector/ Inspecteur en plomberie	**Plumbing Inspector**	**613-580-2424**	**Ext./Poste**

Issued under the authority of /
Délivrance autorisée par

Chief Building Official /
Chef du service du bâtiment

The owner hereby, covenants and agrees with the Corporation of the City of Ottawa, that the owner will abide by and conform to the conditions and stipulations, in consideration of the above Permit. The owner hereby agrees to indemnify and save harmless the said Corporation of the City of Ottawa, and all the Officers, Servants and Agents thereof, from all claims, demands and damages, arising out of or incurred by reason of the execution of the work above referred to, or by reason of Permit above granted.

Le propriétaire soussigné, arrête et conviens avec la Ville d'Ottawa de se conformer aux conditions et aux clauses du permis ci-dessus, en contrepartie de sa délivrance. Le propriétaire conviens également d'indemniser la Ville d'Ottawa et ses dirigeants, employés et mandataires des réclamations, exigences et poursuites en dommages-intérêts liés à l'exécution des travaux mentionnés ci-dessus ou à la délivrance dudit permis.

Witness my hand this date:
Ce dont atteste ma signature en date du:

Issued To:
Délivré à: _____   _____

     (Please Print / En caractères d'imprimerie)   *Signature of owner or authorized agent*
     Signature du propriétaire ou de l'agent autorisé

POST THIS PERMIT IN A CONSPICUOUS PLACE
PRIÈRE D'AFFICHER EN UN ENDROIT BIEN EN VUE
For illustration purposes only

Courtesy of the City of Ottawa

FIGURE 24–4 A typical building permit.

Always verify with the local governing bodies, such as the Buildings Branch, the Conservation Authority, and the Heritage Branch, before planning any type of building project. The builder must adhere to the provincial building code and municipal bylaws, especially zoning bylaws. In most areas of the country, local municipalities will have a website, or questions regarding necessary permits will be answered over the phone. The city of Ajax has produced an excellent summary of the process of getting a permit that includes drawings and details. This can be found at *http://www.ajax.ca/en/doingbusinessinajax/resources/PD_BLD_DECKS_614.pdf.*

Besides local requirements, there are mandatory building code requirements. Building permits are required for projects such as the following:

- buildings with a base greater than 108 ft^2 (10 m^2)
- a porch, a garage, a carport, a sunroom
- decks greater than 108 ft^2 (10 m^2) or more than 24 inches (600 mm) above the surrounding ground
- finishing a basement
- any demolition
- any type of structural work
- plumbing and electrical alterations (except changing fixtures)

In most regions, building permits are not required for:

- buildings with a base less than 108 ft^2 (10 m^2)— for example, 3.16 × 3.16 m, 9 × 12 feet, 2.5 × 4 m, 10 × 10 feet
- roofing
- siding
- flooring
- painting
- fencing (other than fencing to enclose a pool)
- door or window replacement (not wider than the existing)

Again, always check with the local government when in doubt.

OTHER PERMITS

Besides building permits, the builder will be required to obtain other permits. These include sewer and septic, plumbing, electrical, oil and gas, and any others required by the municipality. Also, most places will have a central number to *call before you dig.* This is very important, as the excavator can inadvertently hit underground utilities and significantly damage them and cause injury to those nearby. If no central number is available, the builder will have to contact the hydro, gas, phone, and cable companies, and the municipality must be contacted about sewer and septic work.

FEES AND INSPECTIONS

Fees

Obtaining permits will mean paying fees. The fees cover the cost of the plans evaluation, the on-site inspections, and other related costs of doing business. Before plans are submitted for approval, it is important to check with the governing authorities to find out what they require. For an exterior residential building project, two sets of the following plans are generally required:

- site plan (usually recently surveyed)
- elevation plans (from four sides)
- foundation plan (including footings)
- floor plan(s)
- roof plan (for intersecting roofs)
- cross-section(s)
- structural framing details
- final grading and drainage plan

In Ontario, the homeowner is permitted to submit self-produced plans (drawn to scale with conventional notations and lines); otherwise, plans have to be drawn and submitted by a registered code agent (RCA). Contact your provincial Ministry of Municipal Affairs for more details or go to its website.

Inspections

Once the permit is obtained, the builder is given a laminated copy that is to be posted on site, in plain view, at all times. At various stages, an inspector must be called and the work verified before going on to the next stage. Usually 48 hours notice is required. Across Canada, some or all of the following inspections are required:

- excavation (required compaction to be signed off by the engineer)
- underground plumbing (prior to backfill, testing required)
- foundation (prior to backfill, waterproofing/damp-proofing, drain tile, lateral bracing, footing dimensions, etc.)
- plumbing—rough-in (venting, cleanouts, testing required)
- framing and mechanical (furnace ductwork, electrical, structural details; prior to insulation)

- insulation/air-vapour barrier (all cavities filled, air sealing of penetrations and junctures; prior to wall board)
- plumbing—final (testing required)
- occupancy—if move-in is prior to completion
- final—after all is complete—official Final Occupancy Permit issued

These are just general guidelines meant to give the reader a good idea about the building process. For a thorough explanation of the processes and procedures, contact the local governing authorities.

Some communities require many more inspections. These are designed to verify that the building meets the code of that area. For example, in the city of Vancouver, residential dwellings are inspected for sprinkler systems that are designed to be activated in case of fire. Energy Star houses are inspected for window glazing (low emissivity [low-e] and argon) and for energy ratings on all appliances.

It is the responsibility of the contractor to notify the building official when the construction is ready for a scheduled inspection. If all is in order, the inspector signs the permit card in the appropriate space and construction continues. If the inspector finds a code violation, it is brought to the attention of the contractor or architect for compliance.

These inspections ensure that construction is proceeding according to approved plans. They also make sure construction is meeting code requirements. This protects the future occupants of the building, as well as the general public. In most cases, a good rapport exists between inspectors and builders, enabling construction to proceed smoothly and on schedule (**Figure 24–5**).

A person may occupy an incomplete building provided the building falls within the scope of the Code and the following are complete and, operational, inspected, and tested where applicable:
- required exits
- handrails and guards
- fire alarms and detection systems
- fire separations
- water supply
- sewage disposal
- lighting and heating systems
- required fume barriers
- self-closing devices on doors between an attached or built-in garage and a dwelling unit
- water systems
- building drains
- building sewers
- draining systems
- venting systems
- and, the dwelling must conform to the Radon requirements of 9.1.1.7. in Division B of the Code*

Source: *2010 Code and Construction Guide for Housing*, 1-4.

FIGURE 24–5 Mandatory occupancy requirements. *(Queen's Printer for Ontario. 2010 Code and Construction Guide for Housing, published by the Ministry of Municipal Affairs and Housing. Reproduced with permission.)*

KEY TERMS

architect's scale ruler	bearing	building codes	casement window
areaways	benchmark	building inspections	casing
awning window	bifold doors	building permit	contour interval
beam pocket	blueprinting	bypass doors	contour lines

crawl space	isometric view	pocket door	specifications writer
detail	louvres	point of beginning	swing
door schedules	metes and bounds	risers	topography
double-hung window	modular measurement	scale	treads
easement	National Building Code of Canada (NBCC)	scuttle	trusses
elevation		section view	vapour barrier
finish schedule	non-conforming	setbacks	variance
floor plans	orthographics	sidelight	window schedules
full sections	partial section	slab-on-grade foundation	zones
green space	partitions	sliding windows	zoning regulations
hopper window	plan view	specifications	
hose bibbs	plot plan	specifications guide	

REVIEW QUESTIONS

Select the most appropriate answer.

1. What is a drawing view looking from the top down-ward called?

   a. elevation
   b. perspective
   c. plan
   d. section

2. What is the name for a drawing view showing a vertical cut through the construction?

   a. elevation
   b. perspective
   c. plan
   d. section

3. What scale is most commonly used for floor plans?

   a. ¼" = 1'-0"
   b. ¾" = 1'-0"
   c. 1½" = 1'-0"
   d. 3" = 1'-0"

4. What is the length of a line on a ¼" = 1'-0" scaled print that represents an actual distance of 14'-0"?

   a. 3½"
   b. 4½"
   c. 14"
   d. 56"

5. What does the symbol "FL" written on a set of prints mean?

   a. floor
   b. flush
   c. flashing
   d. all of the above

6. What is the best way to determine a dimension that is not written on a set of plans?

   a. use an architect's scale to measure
   b. calculate it
   c. read the specifications
   d. use the plot plan

7. How are centrelines indicated?

   a. by a series of short, uniform dashes
   b. by a series of long and then short dashes
   c. by a long dash followed by two short dashes
   d. by a solid, broad, dark line

8. Which of the styles of dimensioning below would most likely be found on a set of prints?

   a. 3'
   b. 3 ft
   c. 3'-0"
   d. 3 ft 0 ins

9. Where would the setback of a building from the property lines be found?

   a. floor plan
   b. plot plan
   c. elevation drawing
   d. foundation plan

10. Where do you look to find out which edge of a door is to be hinged?

    a. elevations
    b. specifications
    c. floor plan
    d. wall section

11. Where are the direction, size, and spacing of the first-floor joists found?

    a. foundation plan
    b. first-floor plan
    c. second-floor plan
    d. framing plan

12. Where is the finished floor height usually found?

    a. plot plan
    b. floor plan
    c. foundation plan
    d. framing plan

13. From where can an exterior wall stud height best be determined?

    a. floor plan
    b. framing elevation

    c. wall section
    d. specifications

14. Which set of prints is most helpful in determining the material installed behind a brick veneer?

    a. section view
    b. elevation
    c. foundation plan
    d. framing plan

15. What are the laws that guide what type of building may be built in a particular area?

    a. zoning regulations
    b. national building codes
    c. residential codes
    d. building permits

Section 2

Rough Carpentry

Scott Blair

Title: Red Seal journeyman carpenter and lead hand
Company: Blair Construction, Ottawa, Ontario

EDUCATION

Scott graduated from Rideau District High School in the town of Elgin, just south of Ottawa. After graduation, he started his carpentry apprenticeship with Kinch's Carpentry in Rideau Ferry, and he completed his apprenticeship with the same company. For the in-school portion of his training, he attended Algonquin College in Ottawa, where he excelled above his peers in all three levels of training. He is currently expanding the scope of his knowledge, as he is enrolled in business courses, including bookkeeping, in order to assist him with running his own company.

HISTORY

While Scott was in high school, he was "bitten by the carpentry bug." He recognized his innate ability to see the finished product, to de-construct it mentally into manageable pieces, and to bring it step-by-step to its completion. When he was in grade 11, he represented his school at a run-off competition at the college for a chance to compete in the Skills Ontario competition. Scott won, and he went on to win the High School Skills competition. In 2008, as an apprentice, he represented Algonquin College at the Post-Secondary Skills Ontario competition. He won that and went on to get a bronze medal at the Skills Canada competition in the same year. In 2009, he represented Canada at the World Skills Competition in Calgary, where he finished just one point shy of a medal of distinction.

Upon returning home from Skills Canada in 2008, Scott successfully completed the Red Seal Inter-Provincial Certificate of Qualification Exam. He spent much of the following year training for the world competition, with some time and expertise donated to Habitat for Humanity, putting the roof trusses and sheathing on its 2009 Ottawa house. After Calgary in 2009, he returned to start his own company, Scott Blair Construction. He currently has three crews working for him building custom homes in the Rideau Lakes/Smith's Falls area between Ottawa and Kingston, Ontario.

ON THE JOB

As an apprentice, Scott worked in the residential sector in a small town where every aspect of construction had to be learned and mastered. He was exposed to concrete foundations formed with insulated concrete forms (ICF), the open joist system, wall framing, roof trusses, exterior finishes, and the fine interior finishes. He thoroughly enjoyed the outdoor work, the camaraderie, the daily challenges, the planning, the organizing—putting the pieces together. Now, as the man in charge, he leads his crew by example, always willing to handle the tough jobs and taking the time to explain the wherefores and the whys to his guys.

BEST ASPECTS

"A job well done," says Scott. To have a job where you never get bored, where you are not sitting around inside an office, where you get to experience the "blood, sweat, and tears" of working in the elements— "What more could a guy ask for?" The best part of the job is the feeling you get when you stand back and admire the finished product, when the client is pleased, and, of course, when you get paid. There is the fun stuff—the toys, the truck and trailer, the Kubota, the backhoe. And then there's the paperwork, the Workers' Safety and Insurance Board (WSIB), liability insurance, employment insurance, HST, and so on. "It's a great day when you have a few laughs, listen to some good tunes, have a good lunch, do a good job, and everyone is safe and sound."

IMPORTANCE OF EDUCATION

"There's always something to learn." Carpentry involves a lot of on-site reading of plans and specifications as well as "homework," preparing and planning the next day's work. The obvious trade-related calculations are very important; so are the not-so-obvious construction geometry and communication. And when it comes to running a business, you can't forget bookkeeping, payroll, taxes, customer relations, and dealing with all the other subtrades and building officials.

FUTURE

Scott has his eye on the bigger world of commercial construction. More challenging work, more pieces to put together, more fun toys— heavy equipment and excavating—are all part of the allure.

WORDS OF ADVICE

"Don't be afraid of hard work and putting in extra hours; it will pay off in the long run. When working with more experienced carpenters, pay attention to every detail of their construction work—how they build something— and learn from them, because they may not explain why they are doing things a certain way. And ask lots of questions; if you don't ask, you won't know."

Unit 9

Jobsite Safety and Scaffolds

Chapter 25 Jobsite Safety and Construction Aids

Chapter 26 Scaffolds

The essential tools for jobsite safety involve attitude, knowledge, and skill. Workers must have a good attitude, be ready to learn and use their knowledge, and sharpen their skills every day in order for a jobsite to be safe. This affects all people on a jobsite.

Scaffolding and staging are terms that describe temporary working platforms. They are constructed at convenient heights above the floor or ground. They help workers to perform their jobs quickly and safely. This unit includes information on the safe erection of various kinds of working platforms and the building of several construction aids.

 SAFETY REMINDER

Scaffolds are involved with most jobsite accidents. Know how to safely erect, use, and dismantle scaffolding. Always remember where you are when working elevated levels.

OBJECTIVES

After completing this unit, the student should be able to:

- describe the characteristics of jobsite safety.
- identify personal protective equipment (PPE) used in construction.
- describe the function of various tools and equipment commonly found on a jobsite.
- describe the safe use of ladders, ladder jacks, and sawhorses.
- identify safety concerns with jobsite electricity.
- build a wooden ladder and sawhorse.
- erect and dismantle metal scaffolding in accordance with recommended safe procedures.
- safely set up, use, and dismantle pump jack scaffolding.
- name the parts of wood single-pole and double-pole scaffolding.
- build safe staging using roof brackets.

Chapter 25

Jobsite Safety and Construction Aids

Proper jobsite safety hinges on attitude, skill, and knowledge (A.S.K.). Never be afraid to ask questions when on the construction site. Your knowledge of safety rules and your safety skills as a tradesperson are required for your health and safety as well as that of your co-workers. Everyone has the right to go home at the end of a shift in the same shape as when they arrived.

JOBSITE ATTITUDE

Your attitude affects your work and the people around you. Jobsite attitude is created by every person on the job. Each person brings with them their viewpoint and approach to solving problems as they come to the job. All of these perspectives are blended into a "vibe" that pulses throughout the jobsite. But what does this mean really?

Attitude includes something as simple as using the correct tool for the job. It often seems quicker to use a tool because it is nearby, even if it is the wrong tool. The proper tool for a task is always safer and easier to use, even if it requires a few minutes to retrieve. For example, using a screwdriver from your pocket to pry apart materials may save a step, but the pry bar in your tool box works better, more easily, and more safely—and the material often comes apart with less effort and with less damage to the material.

Many times when a piece of material needs a little persuasion to go into place, the desire is to force it. Forcing is rarely a good idea. Re-cutting the material is good practice for doing a finer job. Also when tapping material into position, use a hammer. Not only do hammers work better than your hand, they do not get arthritis after a lifetime of hitting things.

Sometimes things do not go as planned. Often the same mistake is repeated two or three times in a row. Wasted material may start to accumulate. Do not give in to the alluring temptation of losing your temper. Throwing things will not help and is dangerous. Take the time to stop, breathe, and reflect on

a different path. If you are able to do this, you are a jobsite leader.

Cellphones are an excellent way to communicate on the job. Information is easy to transmit to other workers. Smartphones offer a fast solution to get technical information on a new material, for example. Unfortunately, they can also be a jobsite hazard. Across Canada, the use of handheld cellphones while driving has been banned because it can distract the driver from the road. Jobsites are a worse place to be distracted. Do not let anything keep you from paying attention to safe behaviours and actions.

The safety of all people on the job depends on the attitude of all people on the job. Safety is like silence. Silence will not happen unless everybody involved is quiet. A safe jobsite requires everyone to pay attention and contribute to proper jobsite protocol. Every person on a jobsite has responsibilities.

Know where you are at all times. Visualize your surrounding; be aware of the tools and people nearby. For example, never work so fast that you forget you are on a scaffold (**Figure 25–1**). Think about where you are placing your tools. It is often best to attach them to your tool belt.

Noise on the jobsite is normal (**Figure 25–2**). Jobsite sounds have a pattern and a rhythm. Almost every tool used in construction today makes noise. A safe worker listens to the noise tools make. Tools will

FIGURE 25–1 It is important to remember where you are at all times.

FIGURE 25–2 Jobsite noise is to be expected. It is important to recognize changes from normal.

often make a different sound when they are failing to perform as designed. Being alert to the tool's sounds is a first step in being safe.

Safety Responsibilities

The Canadian Centre for Occupational Health and Safety (CCOHS) has a vision: the elimination of work-related illnesses and injuries. Established in 1978, CCOHS promotes the physical, psychosocial, and mental health well-being of working Canadians with information, training, education, management systems, and solutions that support health, safety, and wellness programs. (For more detail, see *http://www.ccohs.ca/ ccohs.html*.) This federal corporation works together with provincial and international safety organizations to prevent illnesses, injuries, and fatalities in the Canadian workplace. A list of provincial and territorial organizations can be found on the Alberta Occupational Health & Safety (OH&S) site at *http://work .alberta.ca/occupational-health-safety/346.html*.

Provincial occupational health and safety (OH&S) organizations see safety as the responsibility of all personnel on the jobsite, employer and employees alike. Some of the responsibilities of the employers are as follows:

- Provide a workplace free from serious recognized hazards, and comply with standards, rules, and regulations issued under their various occupational health and safety acts (OH&SA).
- Examine workplace conditions to make sure they conform to applicable OH&SA standards.
- Make sure employees have and use safe tools and equipment and properly maintain this equipment.

- Establish or update operating procedures and communicate them so that employees follow safety and health requirements.
- Not to discriminate against employees who exercise their rights under the OH&SA act to seek a safer workplace.
- Correct cited violations by the deadline set in the OH&SA citation, and submit required abatement verification documentation.

Some of the responsibilities of the employees are as follows:

- Comply with the standards, rules, regulations, and orders issued by his/her employer.
- Use safety equipment, personal protective equipment, and other devices and procedures provided or directed by the employer.
- Report unsafe and unhealthful working conditions to the appropriate officials.

Falls are the number one cause of injuries on a jobsite. It should come as no surprise then that scaffolding and fall protection are most often cited by OH&S organizations for problems.

Teamwork

There is no replacement for teamwork on the jobsite. People working together can outperform individuals working alone (**Figure 25–3**). An example of teamwork is when one person holds material to be fastened or cut for another. The difference between a team and a group of individuals is more dramatic when each member of the team anticipates the next move of the other. For example, while holding a board for another, a good team member watches out for safety and asks questions like the following: Can this be done more safely? Does a sawhorse or ladder have to be moved into a better position? Will the tool being used cut only the desired material? Are

FIGURE 25–3 Teamwork makes big jobs easier.

FIGURE 25–4 Changes happen fast on the jobsite.

the actions of the team putting others at risk? Safety must be continuous and ongoing.

Ever-Changing Worksite

Jobsites are dangerous places to work. One of the reasons for this is that they change. From week to week, one can see how one phase or trade will finish and another trade will begin (**Figure 25–4**). The foundation crews finish up and move out allowing the framers to move in, or finish stair carpenters begin work on stairs, which creates a fall hazard that did not exist before. Constant change is part of the reality of working on a construction jobsite.

Changes also happen from moment to moment. A sawhorse or ladder may be removed from the room, or an unsafe situation is created while a worker's back is turned. Workers must always be aware of what is going on around them. Communication between workers is essential on the job. Let others know when something changes, no matter how small, no matter who caused the change.

PERSONAL PROTECTIVE EQUIPMENT (PPE)

Everyone on a jobsite must work and perform their tasks in a safe manner. This means operating tools and equipment appropriately and according to manufacturer's instructions. But safety begins with personal protective equipment (PPE). For a guide to complete PPE for home building, see *http://www.ihsa.ca/Free -Products/Downloads/M063-Homebuilding-Health -and-Safety-Manual.aspx*. See *http://www.wsps.ca/ WSPS/media/Site/Resources/Downloads/Personal _Protective_Equipment2_FINAL_1.pdf?ext=.pdf* for a general overview.

FIGURE 25–5 Safety glasses should be worn at all times.

Eye Protection

Wearing eye protection prevents injuries from flying particles and chemical splashes. They must be worn at all times, especially when there is the possibility of flying foreign objects, such as cutting, grinding, and nailing (**Figure 25–5**).

Safety glasses and goggles are made of impact-resistant plastic, side shields, and reinforced frames. In some instances, a face shield will cover the entire face. Workers who wear prescription eyeglasses must wear a set of prescription safety glasses or goggles that cover their existing glasses.

Laser light from laser levelling devices can cause permanent eye damage. Special safety glasses designed for the particular laser light frequency should be worn, as normal jobsite safety glasses will not protect your eyes from the laser light. Never look directly into a laser beam.

Welders and metal workers must shield themselves with special face and eye protection. These eye shields are very dark to protect from extreme bright radiation. When on a jobsite where welders are working, do not look directly at their work. It does not matter how far away you are, unprotected viewing of welding light will cause eye injury (**Figure 25–6**). See *http://www.ccohs.ca/oshanswers/ prevention/ppe/glasses.html* for more details.

Ear Protection

Jobsite workers are exposed to varying levels of noise. If you have to raise your voice for someone 3 feet (1 m) away to hear you, then the site is noisy, and ear protection should be worn. Excessive levels and/or extended periods of noise will result in permanent hearing loss. Hearing protection must be worn when noise levels or durations cannot be reduced.

FIGURE 25–6 Never look at welding arcs, even from a distance.

Ear plugs made of mouldable rubber, foam, or plastic are inserted into the ear. Ear muffs surround the ear and have soft cups that surround the ears. Some ear muffs are electronic, designed to cancel out impulse noises while allowing normal sounds to be heard.

OH&SA organizations have set limits of sound levels and the permissible exposure time limit. The perceived loudness doubles with an increase in 10 dB (decibels). Circular saws, for example, seem twice as loud as a backhoe. The louder the sound, the less time a worker should be exposed to the sound without wearing ear protection (**Figure 25–7**). See *http://www.ccohs.ca/oshanswers/prevention/ppe/ear_prot.html* for more details.

Head Protection

Workers must wear head protection at all times on a construction site. If there is no danger of objects falling from above, such as in the case of a flooring installer working in an enclosed area, the hard hat (and safety footwear) may be temporarily removed, but it must be placed at the door and worn as the worker leaves the building. Workers (and all visitors to the jobsite) must also wear head protection since they might bump their heads against fixed objects, such as exposed pipes or beams, and there is a possibility of accidental head contact with electrical hazards (**Figure 25–8**).

The following regulations are from Ontario's Infrastructure Health & Safety Association (IHSA)'s website, found at (*http://www.ihsa.ca/rtf/health_safety_manual/pdfs/equipment/Head_Protection.pdf*)

Hard hats are usually made of a lightweight plastic outer shell with shock-absorbing interior

Hours per day	Sound level (dB)	Examples
8.0	90	Backhoes
6.0	92	Diesel trucks
4.0	95	Hand Power tools
3.0	97	Bulldozers
2.0	100	Circular saws
1.5	102	Concrete saws
1.0	105	Jack hammers,
0.25	115	Sandblasting, Chainsaws, Rock concerts

FIGURE 25–7 Ear protection requirements depend on the amount of time exposed to various jobsite noises.

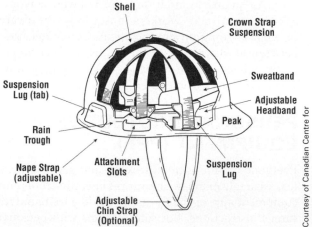

Courtesy of Canadian Centre for Occupational Health and Safety (CCOHS).

FIGURE 25–8 Hard hats protect from head injuries and electrical contact.

head straps. They must be properly maintained in order to make sure that they perform as expected. They need to be adjusted to fit properly. The hat should not bind, slip, fall off, or irritate the skin.

FIGURE 25–9 Hard hats are available with accessories.

FIGURE 25–10 Respirators are designed for different airborne hazards.

Although hard hats should never be painted, they may be purchased with a decorative overlay to the helmet. Ear protection may also be attached via slots in the rim of the hat (**Figure 25–9**).

Inspect your hard hat daily, looking for cracks, holes, or defects. Do not drill holes, paint, or place labels on the hat. Do not store the hat in direct sunlight since heat and sunlight can damage them. If it sustains an impact, the hard hat should be replaced.

Respiratory Protection

Air quality on a jobsite varies according to the work being done. A respirator is required to protect the workers when airborne hazards exist (**Figure 25–10**). Dry cutting of dust from masonry work produces silica dust that may cause lung diseases years later. Some solvents can produce vapours that are known to cause cancer.

Employers are expected to provide respirators that are applicable and suitable for the work situation. The best type of respirator depends on the conditions. Read the packaging to determine if the respirator will protect under the present work conditions. See *http://www.ccohs.ca/oshanswers/prevention/ppe/respslct.html* for more details.

Protective Clothing

There are many varieties of protective clothing to suit the work conditions (**Figure 25–11**). In general, jeans or denim are well suited for construction work.

FIGURE 25–11 Jobsite clothing should fit well and be made of durable materials.

FIGURE 25–12 Clothes protect the skin from the sun.

FIGURE 25–13 Gloves offer protection from jobsite hazards to the hands.

They offer protection from dust, abrasions, changing temperatures, and rough surfaces. Leather works well against heat and flames such as welding and metal cutting.

Clothes should be snug fitting to the body, not loose or hanging off, particularly at the waist. If they are loose, they will catch on material and equipment, potentially causing injury. Jewellery of any kind, including rings and watches, should not be worn while working. They can catch and direct sharp objects into the body and they often conduct electricity.

When working in hot regions it is important to protect one's self from the sun. Covering the skin with long sleeves and pants is better than working with bare skin (**Figure 25–12**). Use sunblock, dark safety glasses, and wide-brimmed hard hats. In extreme situations, avoid working during the times from 10 a.m. to 4 p.m., which is the most intense time of the day. During very hot days, steps must be followed to prevent dehydration. For more about working in extreme heat conditions, go to *http://www.ihsa.ca/ PDFs/Products/Id/W110.pdf*.

Hand and Foot Protection

A wide assortment of gloves is available today to protect hands from jobsite hazards. The type of work being done will affect the best choice of gloves to use (**Figure 25–13**). Leather gloves, for example, are good for general rough work and moderate heat. When working with diesel, nitrile gloves are recommended, not rubber gloves. It is important to check the glove manufacturer's recommendations against any particular jobsite hazard.

For gloves to provide maximum protection, they should be in good condition. Gloves that are stiff or discoloured may indicate they need to be replaced because chemicals have been absorbed into the glove material. Small pinhole leaks can allow electrical current to pass through the glove.

Work boots must be worn at all times on a construction site. They must meet minimum CSA standards for compression and impact and have a green triangle (see IHSA's regulations at *http://www .ihsa.ca/rtf/health_safety_manual/pdfs/equipment/ Foot_Protection.pdf*). Soles should be flexible yet stiff enough to resist penetrations from nails and sharp objects. Steel toes offer good impact protection.

Work boots are not made for all situations. Some offer good traction in dirty environments while poor in oily situations (**Figure 25–14**). Some insulate from electrical hazards and some offer a grounding to reduce the risk of static charge buildup. It is important to check the manufacturer's recommendations before purchasing. See *http://www .ccohs.ca/oshanswers/prevention/ppe/footwear .html* for more details.

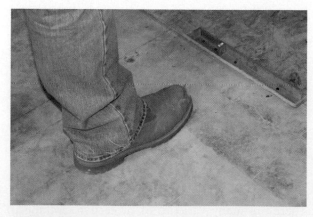

FIGURE 25–14 Work boots should have stiff soles and tread appropriate for the conditions.

FIGURE 25–15 Jobsites are often very busy with many activities.

FIGURE 25–17 Excavators provide superior deep digging.

JOBSITE ACTIVITY

Activity around a jobsite is usually busy and often hectic. Each site requires different types of equipment to make construction possible. Jobsite hazards to watch out for depend on the type of jobsite activity (**Figure 25–15**). To be safe, everyone must watch where they are and where they are going.

Excavation Equipment

Some sites require deep excavations, others only shallow trenches. Many types of equipment can be found on the site at some time. These machines make a lot of noise and are operated by a person. These operators often back up with little or no warning. While there is a beeping sound emitted when equipment backing up occurs, it can be quick. When working or walking near these operators, it is important to make eye contact with them (**Figure 25–16**). This way you know that they know you are nearby. Trust is one thing, but being safely sure is another.

Bulldozers provide the bulk of excavation work on residential sites. Their rolling tracks give them superior traction to push material out of the way. They carve layers of soil and rock, pushing them into piles or thicker layers. Topsoil is separated to be put back after construction is completed. Subsoil is loosened and removed as needed. They are limited to the depth of excavation by how large of an area they can operate. Excavations are much larger than the building.

When the project requires a deep hole, excavators are often used. They have a long arm with a bucket that curls under to loosen and remove material (**Figure 25–17**). The digging cycle involves pivoting in an axis. They can remove material from one direction and place it into a truck in the other direction. Excavators are also very handy for moving new material back into position. A good operator can make work on a jobsite much easier by reducing the amount of hand work necessary.

Backhoe-loaders are designed to be a universal tool. They can move loose material around the job and dig trenches (**Figure 25–18**). Often when pipe

FIGURE 25–16 Equipment operators need room to manoeuvre and nearby workers must be wary.

FIGURE 25–18 Backhoes can transport material and dig trenches.

FIGURE 25–19 Skid steers are highly manoeuvrable machines performing many labour-saving tasks.

FIGURE 25–20 Telehandlers can move material across varying terrain and to great heights.

or wire is buried in trenches, it must first be covered with a sandy material to protect it. The loader can bring new material from a remote location to the trench before it is filled in.

Backhoes and excavators operate on similar principles. They have many hydraulic pistons to move the arm as it digs. The smoothness and efficiency of operation can be seen when operators activate three or four valves at once. The arm behaves as though it were an extension of a human arm.

Skid steers are highly manoeuvrable machines. They perform many labour-saving tasks on the job (**Figure 25–19**). The machine can spin a full circle inside the space it takes up on the ground. The lift arms can accept many attachments to alter the skid steers' function. They include buckets or forks to transport material, posthole diggers, pneumatic hammers, and road sweepers.

A jobsite safety risk is the limited visibility of the operator. This machine allows for excellent view of the work being done in the front, but the view to the sides and back is poor. Whenever working around this machine, keep good eye contact with the operator. Do not get too close behind the machine. This problem is made worse by the fact that the machine can instantly change direction.

Forklifts or telehandlers can move pallets of material from delivery trucks to where it will be installed by workers. They provide excellent mobility in off-road, rough-terrain situations (**Figure 25–20**). They are four-wheel drive and can vary the way they steer from normal two-wheel steer to four-wheel, side-to-side steer. They can lift tonnes of materials to heights of 40 to 50 feet (12 to 15 m).

Self-propelled boom lifts are often used instead of scaffolding. They allow faster and more efficient work to heights from 10 to 150 feet (3 to 46 m)

(**Figure 25–21**). The operators can move and adjust the boom height from the end of the lift. Boom lifts come in many styles and sizes and are either electrically or engine propelled.

Fall Protection

Current IHSA (Infrastructure Health & Safety Association in Ontario) standards require fall protection

© Kondrachov Vladimir/Shutterstock

FIGURE 25–21 Boom lifts offer workers fast, safe access to varying working heights.

when workers are working at heights above 5 feet (1.5 m). These regulations allow the employer the option of a guardrail system or a personal fall protection system.

A guardrail system has a top rail at 40 inches (1 m or 39⅜″) high, a toe board at 4 inches (100 mm), and a mid-rail installed midway between the top rail and the platform. These requirements are for all open sides of the scaffold, except for those sides of the scaffold that are within 14 inches (356 mm) of the face of the building or open to a ladder. After a height of 8 feet (2.4 m), all platforms must be fully decked. The minimum scaffold plank is a 2 × 10 inch (50 × 254 mm), rough, and the minimum working platform is 20 inches (508 mm) wide. See *http://www.ccohs.ca/oshanswers/safety_haz/platforms/decks.html* and *http://www.ccohs.ca/oshanswers/safety_haz/platforms/components.html* for more details.

A typical personal fall protection system consists of five related parts: the harness, lanyard, lifeline, rope grab, and anchor (**Figure 25–22**). This system is designed to stop a worker after a fall has begun. The lanyard absorbs the shock of the fall. Failure of any one part means failure of the system. Therefore, constant monitoring of a lifeline system is a critical responsibility. It is easy for a system to lose its integrity, even on first use. Do not use the scaffold components as an anchor point of

FIGURE 25–23 Expected ladder capacities and uses can be found on the ladder rails.

the fall protection harness. See *http://www.ccohs.ca/oshanswers/prevention/ppe/belts.html* for more details.

Falls are the leading construction work site injury, not only from heights but also from holes/excavations, ladders, stairways, walkways, and wall openings. Do not let water accumulate on floors. Ladders should be inspected for structural defects, and safety feet spreaders and rungs must be free of oil, dirt, or grease. Note the load capacity of the ladder before using it (**Figure 25–23**).

Excavations are not only an area of fall risk but also a risk of cave-ins. Excavations where the depth is greater than the width are called trenches, and they will require adequate cave-in protection if they are more than 4 feet (1.2 m) deep. This protection is to be determined by a competent person. A competent person is one who has the authority to supervise safety conditions and make corrections if necessary. The soil type affects the type of cave-in protection. Loose soil, such as sand, will require excavation sidewall shoring or sloped excavation walls. See *http://www.ihsa.ca/PDFs/Products/Id/M026.pdf* for more details.

JOBSITE HOUSEKEEPING

Jobsite conditions vary from site to site. Some sites are small and congested with material where the project seems too large for the site. Other sites are spacious and easy to get around. Most sites are busy

Image courtesy of Capital Safety

FIGURE 25–22 Components of a personal fall protection system.

FIGURE 25–24 Jobsites are often very busy and congested.

FIGURE 25–25 An organized building site makes construction more efficient.

with workers and materials (**Figure 25–24**). On all jobsites, the condition of the site affects potential risks to personal safety.

Cleanliness

Cleanliness refers to general conditions that allow for a safe working environment. A general rule for jobsite safety is to keep the site clean and orderly.

- Gather up and remove debris continually and daily to keep the work site orderly.
- Plan for the adequate disposal of scrap, waste, and surplus materials.
- Keep emergency exits, stairways, passageways, ladders, and scaffolding free of material and supplies.
- Put away tools and equipment when they are not being used.
- Remove or bend over nails protruding from lumber.
- Keep hoses, power cords, and other tripping hazards from lying in heavily travelled walkways or areas.
- Ensure that structural openings are covered or adequately protected.
- Keep oily rags and debris in closed metal containers.
- Keep materials at least 5 feet (1.5 m) from openings, roof edges, excavations, or trenches.
- Secure material that is stored on roofs or on open floors.
- Do not throw tools or other materials.
- Do not raise or lower any tool or equipment by its own cable or supply hose.
- Maintain good air quality by not using compressed air to clean.

- Store toxic, flammable, and hazardous materials in a clearly marked, secure location.
- Stack and store material so that it will not fall over or be damaged.
- Report any and all potential safety hazards to superiors.

A clean jobsite allows potential hazards to be more easily seen before they can cause injury. Scrap material located in a central location can be reviewed occasionally to see if it could be reused. Lowering the amount of waste is a green and more sustainable approach to construction (**Figure 25–25**).

Material Handling

Materials are continually delivered to the jobsite. They often are unloaded and stored nearby until they are needed (**Figure 25–26**). When they arrive, care should be taken to keep them in good condition, out of harm's way. Even sturdy and durable

FIGURE 25–26 Materials are often delivered as they are needed.

FIGURE 25–27 Packing covers serve to protect material until it is needed.

materials like concrete blocks (or concrete masonry units, CMUs) should be handled appropriately. They are much easier to install if they are dry.

Materials should be kept in their shipping containers until needed and protected from moisture (**Figure 25–27**). Dispose of shipping materials immediately. Place materials on stickers or scraps to protect them from ground moisture.

Monitoring the jobsite inventory is easier when the materials are organized. A neat stack of boards clearly reveals the various lengths in the pile allowing the carpenter to select the most appropriate length. This helps to reduce waste.

Properly stacked and stored materials reduce risk of injury of workers and protect the materials from damage. Materials that fall over can injure everything and everyone nearby.

Lifting. When lifting materials and equipment, it is important not to exceed your limits. Back injuries often take a long time to heal. Most back injuries are caused by improper lifting techniques. It is easy to get into the correct habit for lifting. The general rule is to use your legs, not your back. The back should remain in a fixed, straight-line position, and your legs should raise the object. Bend your knees, grasp the item to be lifted securely, and then straighten the legs, keeping your back as straight as possible (**Figure 25–28**). Carry smaller loads in two hands to stay balanced. This reduces twisting stress on the back. Keep the load close to your body as you stand up straight.

Often materials and equipment are raised into position by a forklift or crane. Proper lifting of these objects is important to maintain a safe site. Loads should be strongly secured with ropes, cables, and slings (**Figure 25–29**). They should be properly

FIGURE 25–28 Keep the back as straight as possible when lifting.

FIGURE 25–29 Never work or walk under hoisted material.

balanced. Do not walk or work under loads being lifted. Accidents do happen, so take steps to avoid disastrous consequences.

JOBSITE ELECTRICITY

Electricity is a universal source of power. It must be handled properly to protect everyone on the job. Shocks from a power tool can be a minor inconvenience or could cause death. It is difficult to determine ahead of time which effect will result, so precautions should always be taken. Power tools should always be operated according to manufacturer's recommendations. If the cord is frayed where wires are exposed, the tool should be replaced.

Cords

Extension cords are used to provide greater reach of the tool from the source of power. They must be three-wire type and designed for hard or extra-hard usage. If they become damaged or frayed, they should be replaced. Taping damaged cords is not allowed.

Keep cords away from high traffic areas to protect them. This also allows the work area to be safe from tripping hazards. Never use cords to move a tool from one work level to another.

GFCI

Ground fault circuit interrupters (GFCIs) are designed to trip the power off if the electricity leaks out of a cord or tool. These devices are required outdoors and in work areas where moisture is present (**Figure 25–30**). They trip fast enough that shocks are

virtually eliminated. It is a good idea to use them for all situations.

Generators

Portable generators are internal combustion engines used to generate electricity. They are useful on jobsites for temporary power when local power is not available. Generators bring a variety of hazards to the site with them. They should be grounded to ensure the GFCI works properly. Refer to the manufacturer's recommendations.

Portable generators should be used only in well-ventilated areas. They produce carbon monoxide (CO), a deadly gas. A carbon monoxide detector should be installed if there is any doubt about air quality. Symptoms of CO poisoning include dizziness, headaches, nausea, and tiredness. If you experience any of these symptoms, seek fresh air and medical attention. Keep them away and downwind from the building (**Figure 25–31**).

The fuel used to run the generator is also a fire hazard. Generators have hot engine parts that

FIGURE 25–31 Portable gas-powered generators should be set up away from working areas.

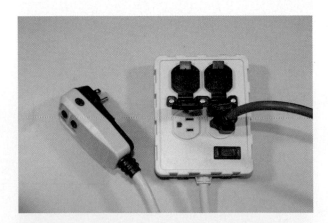

FIGURE 25–30 GFCI devices should be used on the jobsite to protect from potentially fatal electrical shocks.

could ignite the fuel during refuelling. Fuel should be stored in approved containers and handled as a hazardous material.

HAZARDOUS MATERIALS

Hazardous materials and wastes come in many forms and varieties. Understanding what risks you are potentially being exposed to depends largely on the type of work you are doing. Hazardous materials for framing crews may be few while there are many remodelling and demolition hazards. Learning the problems associated with all construction-material concerns is a lifetime endeavour.

MSDS

The material safety and data sheet (MSDS) is designed to keep workers informed of potential risks. Employers are required to have these sheets available for all employees. Construction workers should be well educated in the hazards they face. The MSDS provides a variety of information on all materials used (**Figure 25–32**). Product websites are an excellent place to find MSDS information for any building material.

Jobsite Waste

Jobsite waste is becoming more and more of a concern. Improving the way we handle jobsite waste is seen as a way of making the building process green. Cutting and fitting material usually produces waste (**Figure 25–33**). The issue is how can we reduce waste and make construction more efficient and sustainable.

What to do with the waste depends on its nature. Some waste is hazardous and must be discarded in

FIGURE 25–33 Cutting and fitting usually produces jobsite waste.

a particular way to protect the environment. Follow the MSDS requirements for each material or concern.

Other waste materials are reusable. Whole pieces should be sent back to the supplier, sold to a third party, or given away. If there is a significant amount of leftover material, the estimating math should be adjusted to ensure that the correct amount of each material is delivered to the site.

Scraps from normal cutting and fitting are sometimes useable. Reuse these items on-site whenever possible. Also protect them from damage and donate or sell them. Advertise reusable items in the newspaper, or conduct a yard sale at the jobsite. Take recyclable scrap to dealers for recycling (**Figure 25–34**). Simply sending usable scraps to the landfill is not a sustainable solution.

Packing materials used in shipping have a short lifespan. They quickly become waste. Request that

1. Product and manufacturer information	9. Physical and chemical properties
2. List of any hazardous ingredients in the product	10. Reactivity and stability of product
3. Physical characteristics of the product	11. Health hazards
4. First aid measures	12. Ecological and environmental information
5. Fire and explosive nature of product	13. Disposal concerns
6. Spill cleanup procedures	14. Transport information
7. Safe handling and storing information	15. Regulatory information
8. Personal exposure concerns	16. Special precautions

FIGURE 25–32 Material safety data sheets contain safety information on any particular material.

FIGURE 25–34 Some materials such as metal scraps are easily recycled.

vendors deliver materials in returnable containers. Wood pallets, for example, may be heavily constructed and reused by the material supplier. Choose materials that are delivered with minimal or no packaging. It is not hard to be green and sustainable. It only takes some thought.

FIRE PREVENTION

Construction by its very nature is a hazardous place where fires can happen easily. Discarded oily solvent rags left in an open pile may look neat and safe, but they can heat up over a few hours to the point at which they self-ignite. Always store these materials in closed metal containers.

Two words are used to describe materials' ability to burn—flammable and inflammable. These words both mean the same thing—the material will burn easily. Any material labelled with these words should be used thoughtfully. Protect them from heat sources that will cause them to ignite.

When utilizing equipment that produces heat, be sure the surrounding area is free of all potential fire hazards. Do not use heating devices like a salamander or other open-flamed heaters in an enclosed area. Make sure they vent to the outside air and are an appropriate distance from walls, ceilings, and floor. Fire extinguishers should be available whenever working with heat-producing equipment.

Steel fabrication often involves cutting and fastening with lots of heat and sparks (**Figure 25–35**). Flammable materials should be protected. Do not allow combustible materials and rubbish to accumulate on a jobsite. Compressed gas cylinders, full or empty, should be stored and kept away from excessive heat.

Flammable liquids such as gas, oil, paint and paint thinner, solvents, and grease should be kept in plugged or capped containers and stored in noncombustible areas. Never smoke near flammable or volatile materials.

LADDERS

Carpenters must often use ladders to work from or to reach working platforms above the ground. The most commonly used ladders are the **stepladder** and the **extension ladder**. They are usually made of wood, aluminum, or fibreglass. Make sure that all ladders are in good condition before using them.

Extension Ladders

To raise an extension ladder, the bottom of the ladder must be secured. This is done either with the help of another person or by placing the bottom against a solid object, such as the base of the wall. Pick up the other end. Walk forward under the ladder, pushing upward on each rung until the ladder is upright (**Figure 25–36**).

FIGURE 25–36 Raising an extension ladder using the base of the building to secure the bottom of the ladder.

FIGURE 25–35 Some cutting and fitting involves heat. Care should be taken to prevent fires.

With the ladder vertical and leaning toward the wall, extend the ladder by pulling on the rope with one hand while holding the ladder upright with the other. Raise the ladder to the desired height. Make sure the spring-loaded hooks are over the rungs on both sides. See *http://www.ccohs.ca/oshanswers/safety_haz/ladders/extension.html* for more details.

Lean the top of the ladder against the wall. Move the base out until the distance from the wall is about one-quarter of the vertical height. This will give the proper angle to the ladder. The proper angle for climbing the ladder can also be determined, as shown in **Figure 25–37**.

If the ladder is used to reach a roof or working platform, it must extend above the top support by at least 3 feet (900 mm).

> **CAUTION**
>
> **Be careful of overhead power lines when using ladders. Metal ladders conduct electricity. Contact with power lines could result in electrocution.**

When the ladder is in position, shim one leg (if necessary) to prevent wobbling, and secure the top of the ladder to the building. Check that the feet of the ladder are secure and will not slip. Then face the ladder to climb, grasping the side rails or rungs with both hands (**Figure 25–38**).

Stepladders

When using a stepladder, open the legs fully so that the brackets are straight and locked. Make sure the ladder does not wobble. If necessary, place a shim under the leg to steady the ladder. Never work above the recommended top step indicated by the manufacturer. This is usually the second step down from the top, not including the top as a step. Do not use the ledge in back of the ladder as a step. The ledge is used to hold tools and materials only. Move the ladder as necessary to avoid overreaching. Make sure that all materials and tools are removed from the ladder before moving it. See *http://www.ccohs.ca/oshanswers/safety_haz/ladders/step.html* for more details.

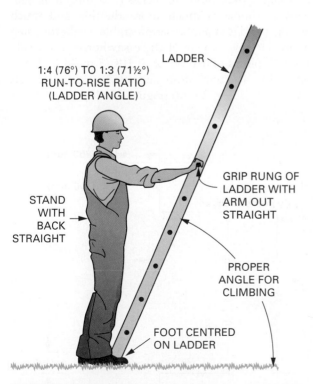

FIGURE 25–37 Techniques for finding the proper ladder angle before climbing.

LADDER

1:4 (76°) TO 1:3 (71½°) RUN-TO-RISE RATIO (LADDER ANGLE)

GRIP RUNG OF LADDER WITH ARM OUT STRAIGHT

STAND WITH BACK STRAIGHT

PROPER ANGLE FOR CLIMBING

FOOT CENTRED ON LADDER

FIGURE 25–38 Face the ladder when climbing. Hold on with both hands.

FIGURE 25–39 Ladder jacks are used to support scaffold plank for short-term, light repair work.

Ladder Jacks

Ladder jacks are metal brackets installed on ladders to hold scaffold planks (**Figure 25–39**). At least two ladders and two jacks are necessary for a section. Ladders should be heavy-duty, free from defects, and placed no more than 8 feet (2.4 m) apart. They should have devices to keep them from slipping.

The ladder jack should bear on the side rails in addition to the ladder rungs. If bearing on the rungs only, the bearing area should be at least 10 inches (254 mm) on each rung. No more than two persons should occupy any 8 feet (2.4 m) of ladder jack scaffold at any one time. The platform's width must not be less than 18 inches (457 mm). Planks must overlap the bearing surface by at least 10 inches (254 mm).

CONSTRUCTION AIDS

Sawhorses, work stools, ladders, and other construction aids are sometimes custom built by the carpenter on the job or in the shop.

Sawhorses

Sawhorses are used on practically every construction job. They support material that is being laid out or cut to size. They may be built of various materials, depending on the desired strength of the sawhorse. Light-duty horses may be made with a 2 × 4 (38 × 89 mm) top and 1 × 4 (19 × 89 mm) legs, while heavy-duty horses are made with a 2 × 6 (38 × 140 mm) top and 2 × 4 (38 × 89 mm) legs. Both use plywood for leg braces.

Sawhorses are constructed in a number of ways, according to the preference of the individual (**Procedure 25–A**). However, they should be of sufficient width, a comfortable working height, and light enough to be moved easily from one place to another. A typical sawhorse is 36 inches (914 mm) wide with 24-inch (610 mm) legs. A tall person may wish to make the leg 30 inches (762 mm) long.

An alternative to the traditional sawhorse is a collapsible version made of ¾-inch (19 mm) plywood and that stands about 30 inches (762 mm) in height. This sawhorse is much more durable and much stronger and is at a more comfortable height for using a circular saw. A pair of these sawhorses can easily hold 20 pieces of 16-foot 2 × 10s (5 m 38 × 235s). Also, they are much more suitable for cutting panels of 4′ × 8′ (1.2 × 2.4 m) (**Figure 25–40**).

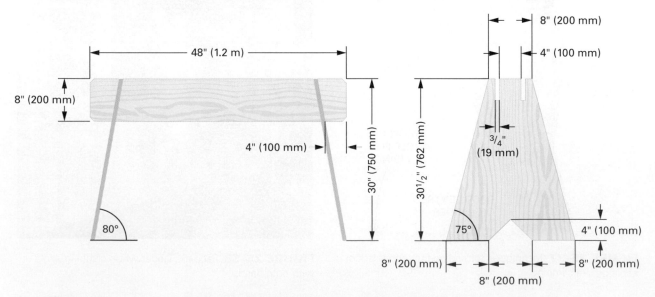

FIGURE 25–40 Interlocking collapsible sawhorses made with ¾-inch (19 mm) plywood.

StepbyStep Procedures

Procedure 25–A Constructing a Typical Sawhorse

STEP 1 CUT 2 X 6 (38 X 140) SAWHORSE TOP 36" (1 mm) LONG AND BEVEL BOTH EDGES OF EACH END AS SHOWN.

STEP 2 CUT FOUR LEGS TO 24-" (600-mm) LENGTH OR AS DESIRED, WITH BEVEL ON EACH END AT SAME ANGLE AS TOP.

STEP 3 FASTEN ALL FOUR LEGS TO SAWHORSE TOP.

STEP 4 HOLD PLYWOOD BRACE AS SHOWN AND MARK ITS LENGTH AT THE TOP EDGE.

FROM EACH MARK, LAY OUT SAME ANGLE AS TOP AND LEGS. CUT, MAKE DUPLICATE, AND FASTEN ONE ON EACH END OF HORSE FLUSH WITH OUTSIDE FACE OF LEGS. MOVE LEGS TO FIT THE PIECE.

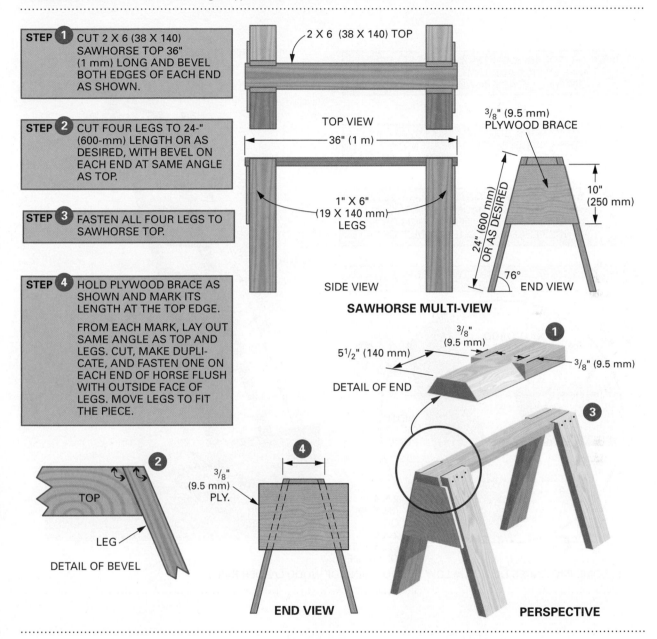

2 X 6 (38 X 140) TOP

TOP VIEW

36" (1 m)

1" X 6" (19 X 140 mm) LEGS

SIDE VIEW

SAWHORSE MULTI-VIEW

3/8" (9.5 mm) PLYWOOD BRACE

10" (250 mm)

24" (600 mm) OR AS DESIRED

76°

END VIEW

5 1/2" (140 mm)

3/8" (9.5 mm)

3/8" (9.5 mm)

DETAIL OF END

TOP

LEG

DETAIL OF BEVEL

3/8" (9.5 mm) PLY.

END VIEW

PERSPECTIVE

Job-Made Ladders

At times it is necessary to build a ladder on the job (**Procedure 25–B**). These are usually straight ladders that are no more than 24 feet (7.3 m) in length. The side rails are made of clear, straight-grained 2 × 4 (38 × 89 mm) or 2 × 6 (38 × 140 mm) stock spaced 16 to 24 inches (406 to 610 mm) apart. Rungs are cut from 2 × 4 (38 × 89 mm) stock and inset into the edges of the side rails not more than ¾ inch (19 mm).

Filler blocks are sometimes used on the rails between the rungs. Rungs are uniformly spaced at 12 inches (305 mm) top to top.

The Ontario Occupational Health and Safety Act provides construction regulations for ladders (**Figure 25–41**). For information on the construction safety regulations in your area, contact your provincial occupational health and safety association.

StepbyStep Procedures

Procedure 25–B Constructing a Job-Built Ladder

STEP 1 TEMPORARILY FASTEN RAILS OF LADDER SIDE BY SIDE AND LAY OUT DADOES 12" (300 mm) OC AS SHOWN BELOW.

STEP 2 CUT DADOES, CUTTING ONLY 3/4" (19 mm) DEEP. ANY DEEPER WILL WEAKEN THE RAIL.

STEP 3 CUT RUNGS 16–24" (400 TO 610 mm) LONG. FASTEN RUNGS IN DADOES WITH 3 1/2" (89 mm) NAILS, KEEPING ENDS OF RUNGS FLUSH WITH OUTSIDE FACE OF RAILS.

STEP 4 CUT CLEATS TO FIT BETWEEN RUNGS.

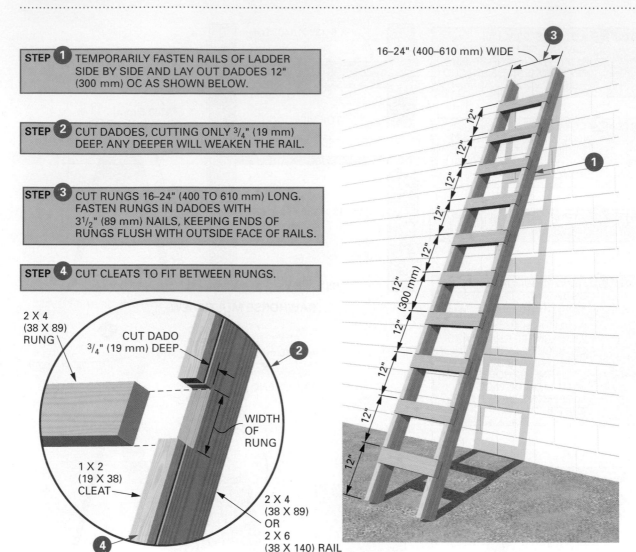

2 X 4 (38 X 89) RUNG

CUT DADO 3/4" (19 mm) DEEP

WIDTH OF RUNG

1 X 2 (19 X 38) CLEAT

2 X 4 (38 X 89) OR 2 X 6 (38 X 140) RAIL OF LADDER

16–24" (400–610 mm) WIDE

12" (300 mm)

NOTE: SOME PROVINCES DO NOT ALLOW THE NOTCHING OF WOOD LADDER RAILS.

SAFETY SUMMARY

When you arrive at the jobsite, run a pre-work safety check. As you put on your hard hat and personal safety equipment, look for anything on the job that has changed from the previous day. Check to see that your tools and equipment are in good working order. Keep your work site orderly and clean. Watch for anything that interferes with your safe work environment. Safety is an activity that never stops.

Means of Access

80. A ladder used as a regular means of access between levels of a structure,

    (a) shall extend at the upper level at least 900 millimetres (3′) above the landing or floor;

    (b) shall have a clear space of at least 150 millimetres (6″) behind every rung;

    (c) shall be located so that an **<u>adequate</u>** landing surface that is clear of obstructions is available at the top and bottom of the ladder; and

    (d) shall be secured at the top and bottom to prevent movement. O.Reg. 213/91, s.80.

Wooden Ladder

81. (1) A wooden ladder,

    (a) shall be made of wood that is straight-grained and free of loose knots, sharp edges, splinters and shakes; and

    (b) shall not be painted or coated with an opaque material. O.Reg. 213/91, s.81(1).

  (2) The side rails of a wooden ladder of the cleat type,

    (a) shall be not less than 400 millimetres (16″) and not more than 610 millimetres (24″) apart; and

    (b) shall measure not less than,

        (i) 38 millimetres (2″) by 89 millimetres (4″) if the ladder is 5.8 metres (19′) or less long, or

        (ii) 38 millimetres (2″) by 140 millimetres (6″) if the ladder is more than 5.8 metres (19′) long. O.Reg. 213/91, s.81(2).

  (3) The rungs of a wooden ladder of the cleat type,

    (a) shall measure not less than,

        (i) 19 millimetres (1″) by 64 millimetres (3″) if the side rails are 400 millimetres (16″) apart, or

        (ii) 19 millimetres (1″) by 89 millimetres (4″) if the side rails are more than 400 millimetres (16″) and not more than 610 millimetres (24″) apart; and

    (b) shall be braced by filler blocks that are 19 millimetres (3⁄4″) thick and are located between the rungs. O.Reg. 213/91, s.81(3).

Ontario Occupational Health and Safety Act, Construction Regulations

FIGURE 25–41 Regulations for ladder construction.

Chapter 26

Scaffolds

Scaffolds are an essential component of construction because they allow work to be performed at various elevations. However, they also can create one of the most dangerous working environments. According to Workplace & Safety Insurance Board statistics, one in six lost-time injuries results from falls (see ***http://www.wsps.ca/Information-Resources/Topics/Working-at-Heights.aspx***). The United States Occupational

Safety and Health Administration (OSHA) reports that in construction, falls are the number-one killer, and 40 percent of those injured in falls had been on the job less than one year. From 2004 to 2013, WorkSafeBC collated statistics on workplace injuries. Many construction workers fell from heights, including from ladders, scaffolds, stairs, roofs, and stationary vehicles. This represented 22 percent of all claims but 50 percent of all claim costs, with 1509 workers falling from a ladder and 1465 falling from a roof or from scaffolding (**Figure 26–1**). The flipbook with the complete 2013 statistics can be found at *http://www.worksafebc.com/publications/reports/statistics_reports/assets/flipbook/2013/index.html#18/z*. The point of these studies is that accidents do not just happen; they are caused.

These findings are similar to those reported by the federal Canadian Centre for Occupational Health and Safety, which administers Canadian occupational health and safety regulations under the Canada Labour Code. For information on the construction safety regulations in your area, contact your provincial occupational health and safety association.

Scaffolds must be strong enough to support workers, tools, and materials. They must also provide an extra safety margin. The standard safety margin requirement is that all scaffolds must be capable of supporting at least four times the maximum intended load.

Those who erect scaffolding must be familiar with the different types and construction methods of scaffolding in order to provide a safe working platform for all workers. The type of scaffolding depends on its location, the kind of work being performed, the distance above the ground, and the load it is required to support. No job is so important as to justify risking one's safety and life. All workers deserve to be able to return to their families without injury.

Employers are responsible for ensuring that workers are trained to erect and use scaffolding. One level of training is required for workers, such as painters, to work from the scaffold. A higher level of training is required for workers involved in erecting, disassembling, moving, operating, repairing, maintaining, or inspecting scaffolds.

The employer is required to have a **competent person** to supervise and direct the scaffold erection. This individual must be able to identify existing and predictable hazards in the surroundings as well as working conditions that are unsanitary, hazardous, or dangerous to employees. This person also has authorization to take prompt corrective measures to eliminate such hazards. A competent person has the authority to take corrective measures and stop

		Fall from a Ladder		Fall from Scaffolding	
Sub-sectors	General construction		40%	General construction	65%
	Retail		13%	Other services	5%
	Other services		11%	Metal and non-metallic mineral products	4%
	Accommodation, food, and leisure services		6%	Wood and paper products	3%
	Metal and non-metallic mineral products		4%	Retail	3%
Injury groups	Sprains and strains of joints and adjacent muscles		40%	Sprains and strains of joints and adjacent muscles	46%
	Contusion with intact skin surface		14%	Fracture of lower limb	14%
	Fracture of upper limb		13%	Contusion with intact skin surface	10%
	Fracture of lower limb		10%	Fracture of spine and trunk	9%
	Fracture of spine and trunk		7%	Fracture of upper limb	8%
Age	15 to 24		13%	15 to 24	16%
	25 to 34		20%	25 to 34	27%
	35 to 44		21%	35 to 44	23%
	45 to 54		27%	45 to 54	23%
	55 to 64		17%	55 to 64	10%
	65+		2%	65+	2%
Severity of injuries	Total time-loss claims		4259	Total time-loss claims	1902
	Average claim cost		$48 000	Average claim cost	$71 000
	Average work days lost		104	Average work days lost	126
Serious injury claims	Percentage of injury claims		57%	Percentage of injury claims	58%
	Average cost per serious injury claim		$82 000	Average cost per serious injury claim	$119 000
	Average work days lost		169	Average work days lost	207

FIGURE 26–1 Workplace injury data: Falls from ladders and scaffolding.

work if need be to ensure that scaffolding is safe to use.

WOOD SCAFFOLDS

Wood scaffolds are single pole or double pole. They are used when working on walls. The single-pole scaffold is used when it can be attached to the wall and does not interfere with the work (**Figure 26–2**). The double-pole scaffold is used when the scaffolding must be kept clear of the wall for the application of materials or for other reasons (**Figure 26–3**). Wood scaffolds are designated as light-, medium-, or heavy-duty scaffolds, according to the loads they are required to support.

Scaffolding Terms

Poles.   The vertical members of a scaffold are called poles. All poles should be set plumb. They should bear on a footing of sufficient size and strength to spread the load. This prevents the poles from settling (**Figure 26–4**). If wood poles need to be spliced for additional height, the ends are squared so that the upper pole rests squarely on the lower pole. The joint is scabbed on at least two adjacent sides. Scabs should be at least 4 feet (1.2 m) long. The scabs are fastened to the poles so that they overlap the butted ends equally (**Figure 26–5**).

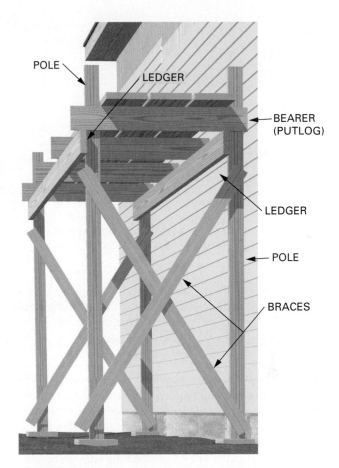

FIGURE 26–3 A double-pole wood scaffold.

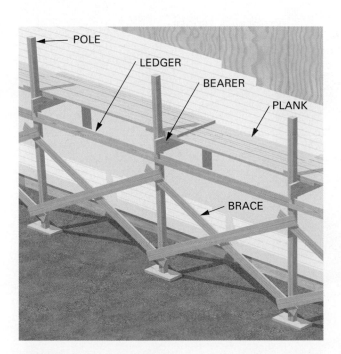

FIGURE 26–2 A light-duty single-pole scaffold. Guardrails are required when the scaffolding is over 10 feet (3 m) in height.

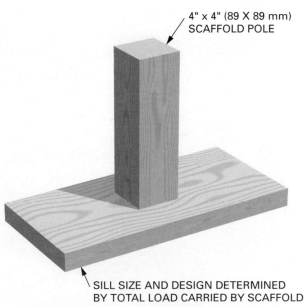

FIGURE 26–4 The bottom ends of scaffold poles are set on footings or pads to prevent them from sinking into the ground.

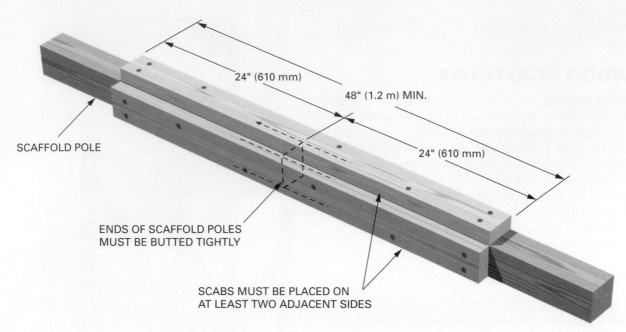

SCAFFOLD POLE

24" (610 mm)

48" (1.2 m) MIN.

24" (610 mm)

ENDS OF SCAFFOLD POLES
MUST BE BUTTED TIGHTLY

SCABS MUST BE PLACED ON
AT LEAST TWO ADJACENT SIDES

FIGURE 26–5 Splicing a wooden scaffold pole for additional height.

Bearers. **Bearers** or **putlogs** are horizontal load-carrying members. They run from building to pole in a single-pole staging. In double-pole scaffolds, bearers run from pole to pole at right angles to the wall of the building. They are set with their width oriented vertically. They must be long enough to project a few inches outside the staging pole.

When placed against the side of a building, bearers must be fastened to a notched *wall ledger*. At each end of the wall, bearers are fastened to the corners of the building (**Figure 26–6**).

Ledgers. A **ledger** runs horizontally from pole to pole. They are parallel with the building and support

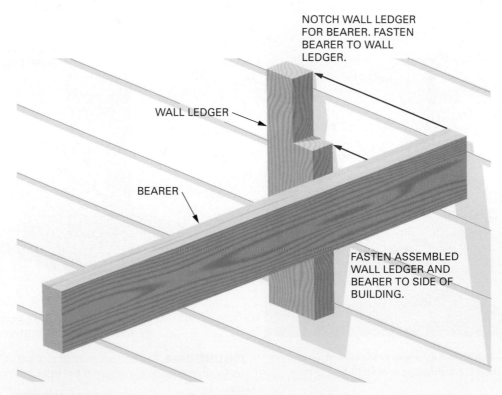

NOTCH WALL LEDGER
FOR BEARER. FASTEN
BEARER TO WALL
LEDGER.

WALL LEDGER

BEARER

FASTEN ASSEMBLED
WALL LEDGER AND
BEARER TO SIDE OF
BUILDING.

FIGURE 26–6 Bearers must be fastened in a notch of a wall ledge for placement against the side of a building.

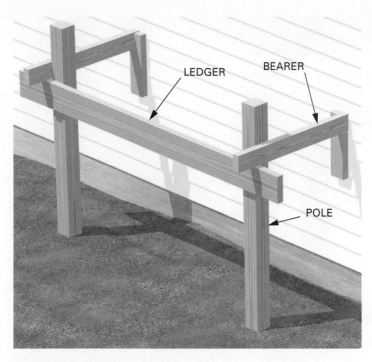

FIGURE 26–7 Ledgers run horizontally from pole to pole and support the bearers.

the bearers. Ledgers must be long enough to extend over two pole spaces. They must be overlapped at the pole and not spliced between them (**Figure 26–7**).

Braces. **Braces** are diagonal members. They stiffen the scaffolding and prevent the poles from moving or buckling. Full diagonal face bracing is applied across the entire face of the scaffold in both directions. On medium- and heavy-duty double-pole scaffolds, the inner row of poles is braced in the same manner. Cross-bracing is also provided between the inner and outer sets of poles on all double-pole scaffolds. All braces are spliced on the poles (**Figure 26–8**).

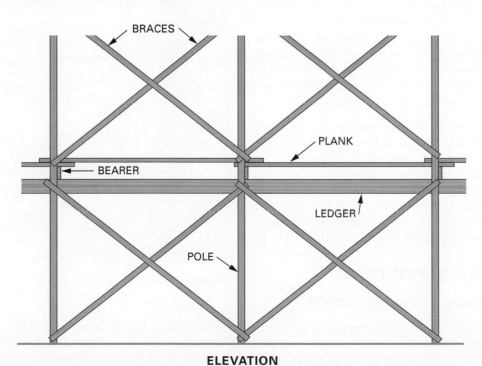

ELEVATION

FIGURE 26–8 Diagonal bracing is applied across the entire face of the scaffold.

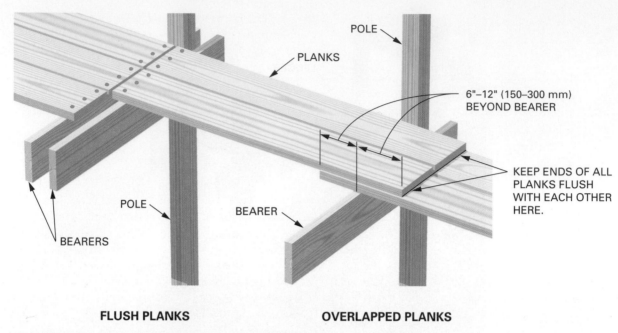

FLUSH PLANKS   **OVERLAPPED PLANKS**

FIGURE 26–9 Recommended placement of scaffold planks.

Plank. Staging planks rest on the bearers. They are laid with the edges close together so that the platform is tight. There should be no spaces through which tools or materials can fall. All planking should be scaffold grade or its equivalent. Planking should have the ends banded with steel to prevent excessive checking.

Overlapped planks should extend at least 6 inches (150 mm) beyond the bearer. Where the end of planks butt each other to form a flush floor, the butt joint is placed at the centre line of the pole. Each end rests on separate bearers. The planks are secured to prevent movement. End planks should not overhang the bearer by more than 12 inches (300 mm) (**Figure 26–9**).

Guardrails. Guardrails are installed on all open sides and ends of scaffolds that are more than 5 feet (1.5 m) in height. The top rail is usually 2 × 4 (38 × 89 mm) lumber and is fastened to the poles 40 to 44 inches (1 to 1.12 m) above the working platform. A middle rail of 1 × 6 (19 × 140 mm) and a toe board with a minimum height of 3½ inches (89 mm) are also installed (**Figure 26–10**).

METAL SCAFFOLDING

Metal Tubular Frame Scaffold

Metal tubular frame scaffolding consists of manufactured *end frames* with folding *cross braces*, *adjustable screw legs*, *baseplates*, *platforms*, and *guardrail hardware* (**Figure 26–11**). They are erected in sections that consist of two end frames and two cross braces and

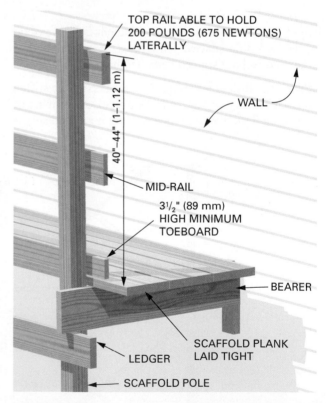

FIGURE 26–10 Typical wood guardrail specifications.

typically come in 5 × 7-foot (1.5 × 2.1 m) modules. Frame scaffolds are easy to assemble, which can lead to carelessness. Because untrained erectors may think scaffolds are just stacked up, serious injury and death can result from a lack of training.

FIGURE 26–11 A typical metal tubular frame scaffold.

End frames consist of posts, horizontal bearers, and intermediate members. End frames come in a number of styles, depending on the manufacturer. Frames can be wide or narrow. Some are designed for rolling tower scaffolds. Other frames have an access ladder built into the end frame (**Figure 26–12**).

Cross braces rigidly connect one scaffold member to another member. Cross braces connect the bottoms and tops of frames. This diagonal bracing keeps the end frames plumb and provides the rigidity that allows them to attain their designed strength. The braces are connected to the end frames using a variety of locking devices (**Figure 26–13**).

Regulations require the use of baseplates on all supported or ground-based scaffolds (**Figure 26–14**) in order to transfer the load of scaffolding, material, and workers to the supporting surface. It is extremely important to distribute this load over an area large enough to reduce the point load on the ground. If the scaffold sinks into the ground when it is being used, accidents could occur. Therefore, baseplates should sit on and be nailed to a

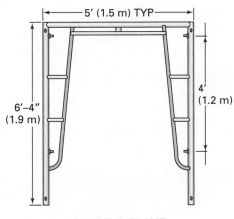

OPEN END FRAME

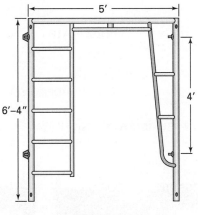

WALK-THROUGH FRAME
WITH BUILT-IN LADDER

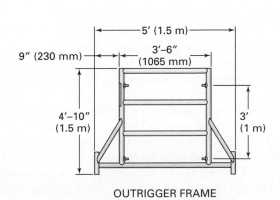

OUTRIGGER FRAME

SIDEWALK CANOPY FRAME

FIGURE 26–12 Examples of typical metal tubular end frames.

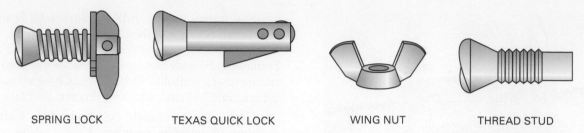

SPRING LOCK   TEXAS QUICK LOCK   WING NUT   THREAD STUD

FIGURE 26–13 Typical locking devices used to connect cross braces to end frames.

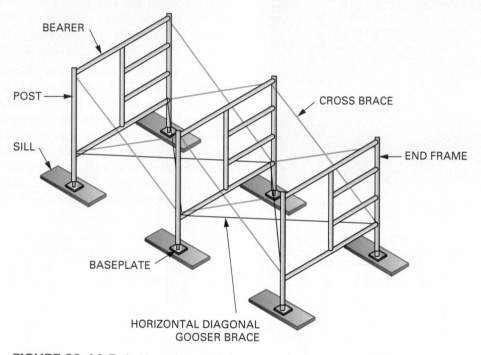

BEARER

POST

SILL

CROSS BRACE

END FRAME

BASEPLATE

HORIZONTAL DIAGONAL
GOOSER BRACE

FIGURE 26–14 Typical baseplate setup for a metal tubular frame scaffold.

FIGURE 26–15 Baseplates and jackscrews should rest on mud sills.

mud sill (**Figure 26–15**). A mud sill is typically a 2 × 10 (38 × 235 mm) board approximately 18 to 24 inches (457 to 610 mm) long. On soft soil it may need to be longer and/or thicker.

To level an end frame while erecting a frame scaffold, screw jacks may be used. At least one-third of the screw jack must be inserted in the scaffold leg. Lumber may be used to crib up the legs of the scaffold (**Figure 26–16**). Cribbing height is restricted to equal the length of the mud sill. Therefore, using a 20-inch (508 mm) 2 × 10 (38 × 235 mm) mud sill, the crib height is limited to 20 inches (508 mm).

A guardrail system is a vertical fall-protection barrier consisting of, but not limited to, top rails, mid-rails, toeboards, and posts (**Figure 26–17**). It prevents employees from falling off a scaffold platform or walkway. A guardrail system is required when the working height is 5 feet (1.5 m) or more. Guardrail systems must have a top-rail capacity of 200 pounds (90 kg) applied downward or horizontally. The top rail must be between 40 and 44 inches (1 and 1.12 m) above the work deck, with the mid-rail installed midway between the upper guardrail and the platform surface.

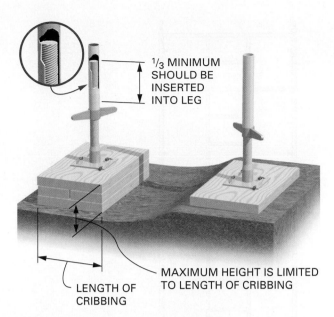

FIGURE 26–16 Cribbing, interlocked blocking may be used to level the ground under the scaffold.

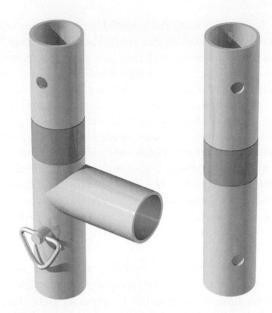

FIGURE 26–18 Coupling pins to join end frames.

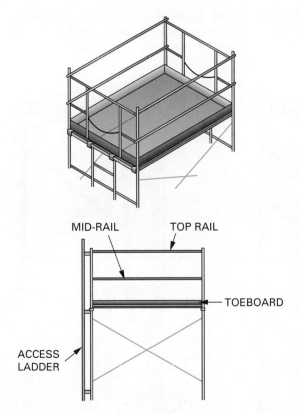

FIGURE 26–17 Typical guardrail system for a metal tubular frame scaffold.

FIGURE 26–19 Coupling locking devices to prevent scaffold uplift.

If workers are on different levels of the scaffold, toeboards must be installed as an overhead protection for lower-level workers. Toeboards are typically 1 × 4-inch (19 × 89 mm) boards installed touching the platform. If materials or tools are stacked up higher than the toeboards, screening must be installed. Moreover, all workers on the scaffold must wear hard hats.

Coupling pins are used to stack the end frames on top of each other (**Figure 26–18**). They have holes in them that match the holes in the end-frame legs; these holes allow locking devices to be installed. Workers must ensure that the coupling pins are designed for the scaffold frames in use.

The scaffold end frames and platforms must have uplift protection installed when a potential for uplift exists. Installing locking devices through the legs of the scaffold and the coupling pins provides this protection (**Figure 26–19**). If the platforms are not equipped with uplift protection devices, they can be tied down to the frames with number nine steel-tie wire.

Safe access onto the scaffold is required for both erectors and users of the scaffolds. Frames may be used as a ladder only if they are designed as such. Frames meeting such design guidelines must have level horizontal members that are parallel and are not more than 16¾ inches (425 mm) apart vertically (**Figure 26–20**). Scaffold erectors may climb end-frame rungs that are spaced up to 22 inches (560 mm). Platform planks should not extend over the end frames where end-frame rungs are used as a ladder access point. The cross braces should never be used as a means of access or egress. Attached ladders and stair units may be used (**Figure 26–21**). A rest platform is required for every 35 feet (10.7 m) of ladder.

Manually propelled mobile scaffolds use wheels or casters in the place of baseplates (**Figure 26–22**). Casters have a designed load capacity that should never be exceeded. Mobile scaffolds made from tubular end frames must use diagonal horizontal braces, or gooser braces (see Figure 26–14), to keep the mobile tower frame square.

Side brackets are light-duty extension pieces used to increase the working platform (**Figure 26–23**).

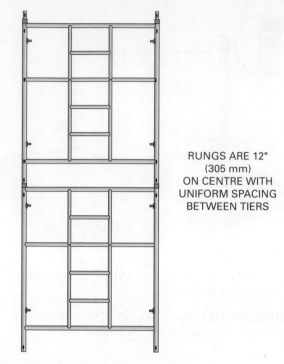

RUNGS ARE 12"
(305 mm)
ON CENTRE WITH
UNIFORM SPACING
BETWEEN TIERS

FIGURE 26–20 The rungs of an end frame must be spaced no more than 16¾ inches (425 mm) apart if users are going to use the frame as an access ladder.

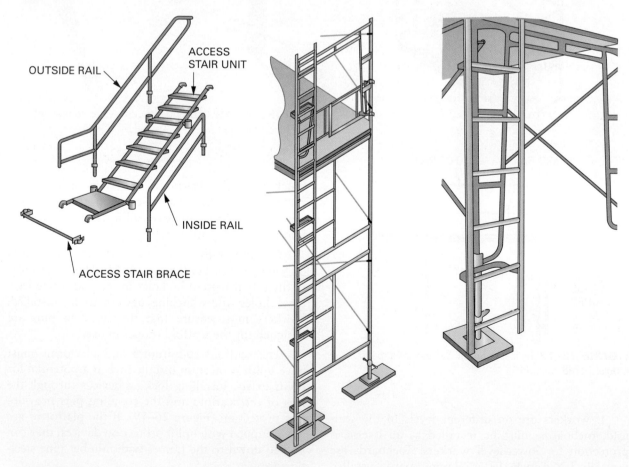

OUTSIDE RAIL

ACCESS STAIR UNIT

INSIDE RAIL

ACCESS STAIR BRACE

FIGURE 26–21 Typical access ladder and stairway.

FIGURE 26–22 Casters replace a baseplate to transform a metal tubular frame scaffold into a mobile scaffold.

They are designed to hold personnel only and are not to be used for material storage. When side brackets are used, the scaffold must have tie-ins, braces, or outriggers to prevent the scaffold from tipping.

Hoist arms and wheel wells are sometimes attached to the top of the scaffold to hoist scaffold parts to the erector or material to the user of the scaffold (**Figure 26–24**). The load rating of these hoist arms and wheel wells is typically no more than 100 pounds (45 kg). The scaffold must be secured from overturning at the level of the hoist arm, and workers should never stand directly under the hoist arm when hoisting a load. They should stand a slight distance away, but not too far to the side, because this will increase the lateral or side-loading force on the scaffold.

Scaffold Inspection

A competent person is required to inspect all scaffolds at the beginning of every work shift.

Visual inspection of scaffold parts should take place at least five times: before erection, during erection, during scaffold use, during dismantling, and before scaffold parts are put back in storage. All damaged parts should be red-tagged and removed from service and then repaired or destroyed as required. Things to look for during the inspection process include the following:

Broken and excessively rusted welds

Split, bent, or crushed tubes

Cracks in the tube circumference

Distorted members

Excessive rust

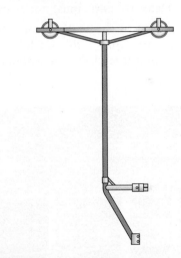

FIGURE 26–24 A hoist that attaches to the top of a scaffold is used to raise material and equipment.

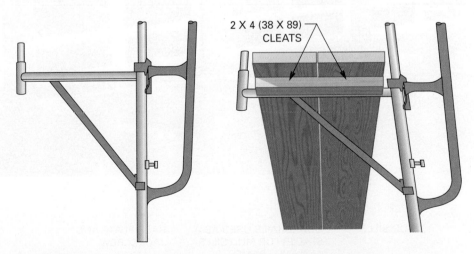

FIGURE 26–23 Side brackets are used to extend a scaffold work platform. These brackets should be used only for workers and never for material storage.

Damaged brace locks

Lack of straightness

Excessively worn rivets or bolts on braces

Split ends on cross braces

Bent or broken clamp parts

Damaged threads on screw jacks

Damaged caster brakes

Damaged swivels on casters

Corrosion of parts

Metal fatigue caused by temperature extremes

Leg ends filled with dirt or concrete

Scaffold Erection Procedure

The first thing to be done during the erection procedure is to inspect all scaffold components delivered to the jobsite. Defective parts must not be used.

The foundation of the scaffold must be stable and sound, able to support the scaffold and four times the maximum intended load without settling or displacement.

Always start erecting the scaffold at the highest elevation, which will allow the scaffold to be levelled without any excavating by installing cribbing, screw jacks, or shorter frames under the regular frames. The scaffold must always be level and plumb. Lay out the baseplates and screw jacks on mud sills so that the guardrails and end frames with cross braces can be properly installed (**Figure 26–25**).

Stand one of the end frames up and attach the cross braces to each side, making sure the correct length of cross braces has been selected for the job. Connect the other end of the braces to the second end frame. All scaffold legs must be braced to at least one other leg (**Figure 26–26**). Make sure that all brace connections are secure. If any of these mechanisms is not in good working order, replace the frame with one that has properly functioning locks.

Use a level to plumb and level each frame (**Figure 26–27**). Remember that the construction regulations require that all tubular welded frame scaffolds be plumb and level. Adjust screw jacks or cribbing to level the scaffold. As each frame is added, keep the scaffold bays square with each

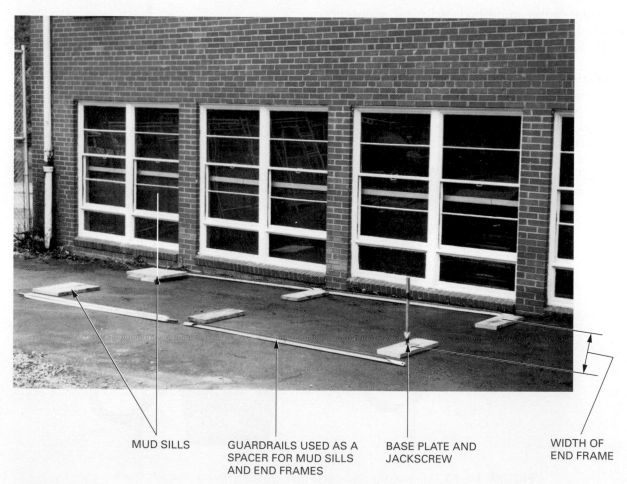

MUD SILLS   GUARDRAILS USED AS A SPACER FOR MUD SILLS AND END FRAMES   BASE PLATE AND JACKSCREW   WIDTH OF END FRAME

FIGURE 26–25 During scaffold erection, the baseplates are spaced according to guardrail braces.

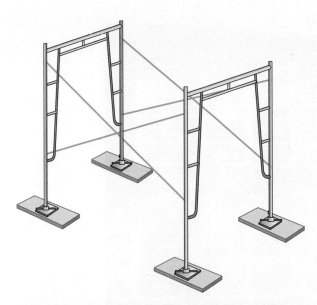

FIGURE 26–26 Cross braces connect end frames, keeping them rigid and plumb.

FIGURE 26–27 End frames must be level. Level and plumb begins on the first row of scaffolding.

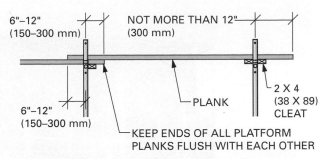

FIGURE 26–28 Recommended placement for scaffold planks.

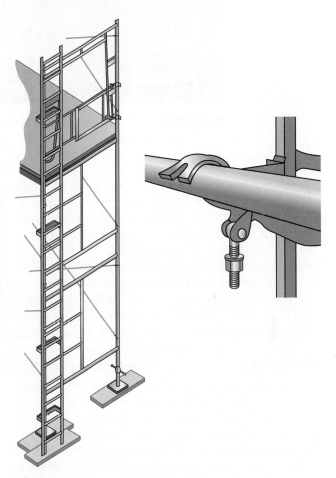

FIGURE 26–29 Attachable ladders are connected so that the bolt attaches from the bottom.

other. Repeat this procedure until the scaffold run is erected. Remember, if the first level of scaffolding is plumb and level, the remaining levels will be more easily assembled.

The next step is to place the planks on top of the end frames. All planking must meet the construction regulations and be in good condition. If planks that do not have hooks are used, they must extend over their end supports by at least 6″ (150 mm) and not more than 12″ (300 mm). A cleat should be nailed to both ends of wood planks to prevent plank movement (**Figure 26–28**). Platform laps must be at least 12″ (300 mm), and all platforms must be secured from movement. Hooks on planks also have uplift protection installed on the ends.

It is a good practice to plank each layer fully as the scaffold is erected. If the deck is to be used only for erecting, then a minimum of two planks can be used. However, full decking is preferred, because it is a safer method for the erector.

Before the second lift is erected, the erector must provide an access ladder. Access may be on the end frame, if it is so designed, or an attached ladder. If the ladder is bolted to a horizontal member, the bolt must face downward (**Figure 26–29**). Next, the second level of frames may be hung temporarily over the ends of the first frames and then installed onto the coupling pins of the first-level frames (**Figure 26–30**).

FIGURE 26–30 The next lift of end frames may be preloaded by hanging them on the previous end frame.

Special care must be taken to ensure proper footing and balance when lifting and placing frames. Construction safety regulations require erector fall protection—a full body harness attached to a proper anchor point on the structure—when it is feasible and not a greater hazard to do so. You can attach fall protection to the scaffold during erection.

After the end frames have been set in place and braced, they should have uplift protection pins installed through the legs and coupling pins. Wind, side brackets, and wheel wells can cause uplift, so it is a good practice to pin all scaffold legs together.

The remaining scaffolding is erected in the same manner as the first. Remember, all work platforms must be fully decked and have a guardrail system or personal fall-arrest system installed before they can be turned over to the scaffold users. If the scaffold is three times its minimum base dimension, it must be restrained from tipping by guying, tying, bracing, or equivalent means (**Figure 26–31**). The scaffold is not allowed to tip into or away from the structure.

After the scaffold is complete, it is inspected again to make sure that it is plumb, level, and square before turning it over for workers to use. The inspection should also include checking that all legs are on baseplates or screw jacks and mud sills (if required), ensuring the scaffolding is properly braced with all brace connections secured, and making sure all tie-ins are properly placed and secured, both at the scaffold and at the structure. All platforms must be fully planked with proper decking and in good shape.

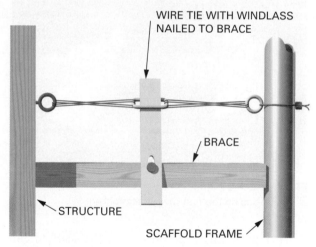

FIGURE 26–31 A wire tie and a windlass may be used to secure the scaffold tightly to the building.

Toeboards and/or screening should be installed as needed. Check that end and/or side brackets are fully secured and that any overturning forces are compensated for. All access units are inspected to ensure that they are correctly installed and ladders and stairs are secured. Again, workers on the scaffold must wear hard hats.

After the scaffolding passes all inspections, it is ready to be turned over to the workers. Remember that this scaffolding must be inspected by a competent person at the beginning of each work shift and after any occurrence, such as a high wind or a rainstorm, that could affect its structural integrity.

Scaffold Capacity

All scaffolds and their components must be capable of supporting, without failure, their own weight and at least four times the maximum intended load applied or transmitted to them. Erectors and users of scaffolding must never exceed this safety factor.

Erectors and users of the scaffold must know the maximum intended load and the load-carrying capacities of the scaffold they are using. The erector must also know the design criteria and the intended use of the scaffold.

When erecting a frame scaffold, the erector should know the load-carrying capacities of its components. The rated leg capacity of a frame may never be exceeded on any leg of the scaffold. Also, the capacity of the top horizontal member of the end frame, called the bearer, may never be exceeded. Remember, it is possible to overload the bottom legs of the scaffold without overloading the bearer or top horizontal member of any frame. It is also possible to overload the bearer or top horizontal member of the frame scaffold and not overload the leg of that same scaffold. Erectors must pay careful attention to the load capacities of all scaffold components.

If the scaffold is covered with weatherproofing plastic or tarps, the lateral pressure applied to the scaffold will dramatically increase. Consequently, the number of tie-ins attached to prevent overturning must be increased. Additionally, any guy wires added for support will increase the downward pressure and weight of the scaffold.

Construction safety regulations state that supported scaffolds with a ratio larger than three-to-one (3:1) of the height to narrow base width must be restrained from tipping by guying, tying, bracing, or equivalent means. Guys, ties, and braces must be installed at locations where horizontal members support both inner and outer legs. Guys, ties, and braces must be installed according to the scaffold manufacturer's recommendations or at the closest horizontal member to the 3:1 height. For scaffolds greater than 3 feet (0.91 m) wide, the vertical locations of horizontal members are repeated every 26 feet (7.9 m). The top guy, tie, or brace of completed scaffolds must be placed no further than the 3:1 height from the top. Such guys, ties, and braces must be installed at each end of the scaffold and at horizontal intervals not to exceed 30 feet (9.1 m). The tie or standoff should be able to take pushing and pulling forces so that the scaffold does not fall into or away from the structure.

The supported scaffold poles, legs, post, frames, and uprights should bear on baseplates, mud sills, or other adequate, firm foundation. Because the mud sills have more surface area than baseplates, sills distribute loads over a larger area of the foundation. Sills are typically wood and come in many sizes. Erectors should choose a size according to the load and the foundation strength required. Mud sills made of 2×10-inch (51×254 mm) full-thickness or nominal lumber should be 18 to 24 inches (457 to 610 mm) long and centred under each leg (**Figure 26–32**).

The loads exerted onto the legs of a scaffold are not equal. Consider a scaffold with two loads on two adjacent platforms (**Figure 26–33**). Half of load A is carried by end frame #1 and the other half is carried by #2. Half of load B is carried by end frames #2 and #3. End frame #2 carries two half loads or one full load, which is twice the load of end frames #1 and #3. At no time should the manufacturer's load rating for its scaffolding be exceeded.

Scaffold Platforms

The scaffolding's work area must be fully planked between the front uprights and the guardrail supports in order for the user to work from the scaffold. The planks should not have more than a 1-inch (25 mm) gap between them unless it is necessary to fit around uprights such as a scaffold leg. If the platform is planked as fully as possible, the remaining gap between the last plank and the uprights of the guardrail system must not exceed 9½ inches (240 mm).

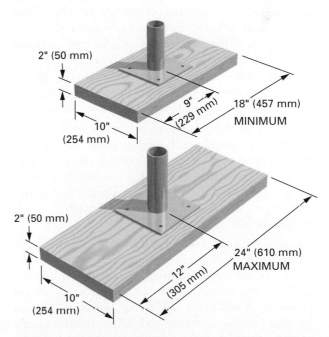

FIGURE 26–32 Baseplates should be centred on the mud sills.

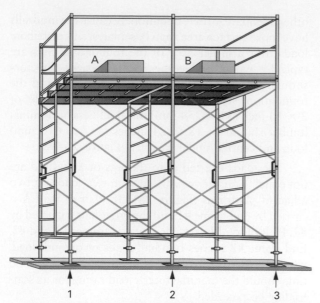

THE CENTRE END FRAME LABELLED #2 CARRIES TWICE
THE LOAD OF EACH OF THE END FRAMES
LABELLED #1 AND #3

FIGURE 26–33 The inner frames, such as #2 shown here, often carry twice the load of the frames located at the end of the scaffold.

Scaffold platforms must be at least 18 inches (460 mm) wide with a guardrail system in place. In areas where they cannot be 18 inches (460 mm) wide, they will be as wide as is feasible.

Planking for the platforms, unless cleated or otherwise restrained by hooks or equivalent means, should extend over the centreline of their support at least 6″ (150 mm) and not more than 12″ (300 mm). If the platform is overlapped to create a long platform, the overlap shall occur only over supports and should not be less than 12 inches (300 mm) unless the platforms are nailed together or otherwise restrained to prevent movement.

When fully loaded with personnel, tools, and/or material, the wood plank used to make the platform must never deflect more than one-sixtieth (1/60) of its span. In other words, a 2 × 10 (48 × 248 mm) rough plank that spans 7 feet (2.1 m) should not deflect more than 1⅜-inch (84″ ÷ 60) or 35 mm (2100 ÷ 60) or one-sixtieth of the span.

Any solid sawn wood planks should be scaffold-grade lumber as set out by the grading rules for the species of lumber being used. A recognized lumber-grading association, such as the Western Wood Products Association (WWPA) or the National Lumber Grades Authority (NLGA), establishes these grading rules. A grade should be stamped on the scaffold-grade plank, indicating that it meets safety and industry requirements for scaffold planks. Two of the most common wood species used for scaffold planks are southern yellow pine and Douglas fir.

Scaffold Access

A means of access must be provided to any scaffold platform that is 2 feet (610 mm) above or below a point of access. Such means include a hook-on or attachable ladder, a ramp, or a stair tower and are determined by the competent person on the job.

If a ladder is used, it should extend 3 feet (900 mm) above the platform and be secured both at the top and bottom. Hook-on and attachable ladders should be specifically designed for use with the type of scaffold used, have a minimum rung length of 11½ inches (292 mm), and have uniformly spaced rungs with a maximum spacing between rung length of 16¾ inches (425 mm). Sometimes a stair tower can be used for access to the work platform, usually on larger jobs (**Figure 26–34**). A ramp can also be used as access to the scaffold or the work platform. When using a ramp as a path for a wheelbarrow, it is important to remember that a guardrail system or fall protection is required at 4 feet (1.2 m) above a lower level.

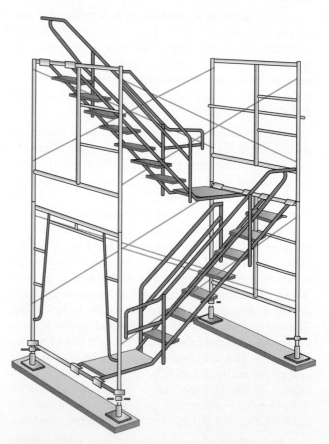

FIGURE 26–34 Scaffold access may be provided by a stair tower.

The worker using the scaffold can sometimes access the work platform using the end frames of the scaffold itself. According to regulations, the end frame must be specifically designed and constructed for use as ladder rungs. The rungs can run up the centre or to one side of the end frame; some have the rungs all the way across the end frame. Scaffold users should never climb any end frame unless the manufacturer of that frame has designated it to be used for access.

Scaffold Use

Scaffolds must not be loaded in excess of their maximum intended load or rated capacities, whichever is less. Workers must know the capacity of scaffolds they are erecting and/or using. Before the beginning of each work shift, or after any occurrence that could affect a scaffold's structural integrity, the competent person must inspect all scaffolds on the job.

Employees must not work on scaffolds covered with snow or ice except to remove the snow or ice. Generally, work on or from scaffolds is prohibited during storms or high winds. Debris must not be allowed to accumulate on the platforms. Makeshift scaffold devices, such as boxes or barrels, must not be used on the scaffold to increase workers' working height. Stepladders should not be used on the scaffold platform.

Fall Protection

Current construction safety regulations for scaffolding require **fall protection** when workers are working at heights above 10 feet (3 m). This regulation applies to both the user of the scaffold and the erector or dismantler of the scaffold. These regulations allow the employer the option of a guardrail system or a personal fall protection system. The fall protection system most often used is a complete guardrail system. A guardrail system has a top rail 40 to 44 inches (1 to 1.12 m) above the work deck, with a mid-rail installed midway between the top rail and the platform. The work deck should also be equipped with a toeboard.

A typical personal fall protection system consists of five related parts: the harness, lanyard, lifeline, rope grab, and anchor (**Figure 26–35**). The failure of

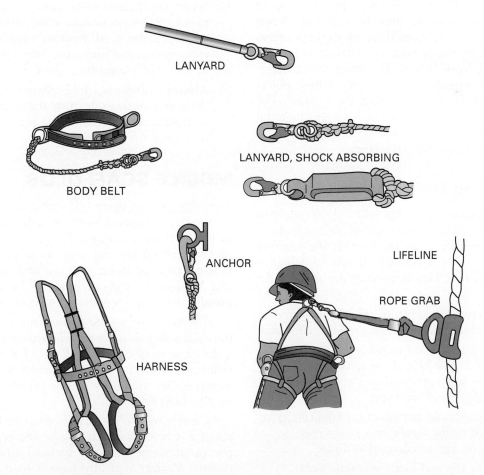

LANYARD

BODY BELT

LANYARD, SHOCK ABSORBING

ANCHOR

LIFELINE

ROPE GRAB

HARNESS

FIGURE 26–35 Components of a personal fall protection system.

any one part means failure of the system. Therefore, constant monitoring of a lifeline system is a critical responsibility. It is easy for a system to lose its integrity almost immediately, even on first use.

The construction regulations recognize that sometimes fall protection may not be possible for erectors. As the scaffold increases in length, the personal fall-arrest system may not be feasible because of its fixed anchorage and the need for employees to traverse the entire length of the scaffold. Additionally, fall protection may not be feasible due to the potential for lifelines to become entangled or to create a tripping hazard for erectors or dismantlers as they traverse the scaffold. Do not use the scaffold components as an anchor point of the fall-protection harness. Construction safety regulations put the responsibility of when to use fall protection, both for the user of the scaffold and the erector, on the competent person.

Falling Object Protection

According to industry standards and safety requirements, workers must wear hard hats when erecting a scaffold. In addition to hard hats, protection from potential falling objects may be required. When material on the scaffold could fall on workers below, some type of barricade must be installed to prevent that material from falling. The construction regulations list toeboards as part of the falling object protection for the workers below the scaffold. The toeboard can serve two functions: It keeps material on the scaffold and keeps the workers on the scaffold platform if they happen to slip.

Dismantling Scaffolds

Many guidelines and rules for erection also apply to scaffold dismantling. However, dismantling requires additional precautions to ensure that the scaffold will come down in a controlled, safe, and logical manner. Important factors to consider include the following:

1. Check every scaffold before dismantling. Any loose or missing ties or bracing must be corrected.

2. If a hoist is to be used to lower the material, the scaffold must be tied to the structure at the level of the hoist arm to dispel any overturning effect of the wheel and rope.

3. The erector should be tied off for fall protection, as required by the regulations, unless it is infeasible or a greater hazard to do so.

4. Start at the top and work in reverse order, following the step-by-step procedures for

erection. Leave the work platforms in place as long as possible.

5. Do not throw planks or material from the scaffold. This practice will damage the material and presents overhead hazards for workers below.

6. Building tie-ins and bracing can be removed only when the dismantling process has reached that level or location on the scaffold. An improperly removed tie can cause the entire scaffold to overturn.

7. Remove the ladders or the stairs only as the dismantling process reaches that level. Never climb or access the scaffold by using the cross braces.

8. As the scaffold parts come off the scaffold, they should be inspected for any wear or damage. If a defective part is found, it should be tagged for repair and not used again until inspected by the competent person.

9. Dismantled parts and materials should be organized, stacked, and placed in bins or racks out of the weather.

10. Secure the disassembled scaffold equipment to ensure that no unauthorized, untrained employees use it. All erectors must be trained, experienced, and under the supervision and direction of a competent person.

11. Always treat the scaffold components as if a life depended on them, because the next time the scaffold is erected, someone's life will indeed be depending on it being sound.

MOBILE SCAFFOLDS

The rolling tower, or mobile scaffold, is widely used for small jobs, generally not more than 20 feet (6 m) in height (**Figure 26–36**). The components of the mobile scaffold are the same as those for the stationary frame scaffold, with the addition of casters and horizontal diagonal bracing. There are additional restrictions on rolling towers as well.

The height of a rolling tower must never exceed three times the minimum base dimension. For example, if the frame sections are 5×7 feet (1.5×2.1 m), the rolling tower can be only 15 feet (4.5 m) high. When outriggers are used on a mobile tower, they must be used on both sides.

Casters on mobile towers must be locked with positive wheel swivel locks or the equivalent to prevent movement of the scaffold while it is stationary. Casters typically have a load capacity of 600 pounds (272 kg) each, and the legs of a frame

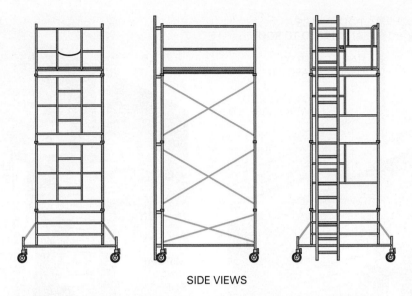

SIDE VIEWS

FIGURE 26–36 Typical setup for a mobile scaffold.

scaffold can hold 2000 to 3000 pounds (900 to 1360 kg) each. Care must be taken not to overload the casters.

Never put a cantilevered work platform, side bracket, or well wheels on the side or end of a mobile tower. Mobile towers can tip over if used incorrectly. Mobile towers must have horizontal, diagonal, or gooser braces at the base to prevent racking of the tower during movement (**Figure 26–37**). Metal hook planks also help prevent racking if they are secured to the frames.

The force to move the scaffold should be applied as close to the base as practicable, but not more than 5 feet (1.5 m) above the supporting surface. The casters must be locked after each movement before

beginning work again. Employees are not allowed to ride on rolling tower scaffolds during movement unless the height-to-base width ratio is two-to-one or less. Before the scaffold is moved, each employee on the scaffold must be made aware of the move. Caster and wheel stems shall be pinned or otherwise secured in scaffold legs or adjustment screws. The surface that the mobile tower rolls on must be free of holes, pits, and obstructions and must be within 3 degrees of level. Use a mobile scaffold only on firm floors.

PUMP JACK SCAFFOLDS

Pump jack scaffolds consist of 4 × 4 (89 × 89 mm) poles, a pump jack mechanism, and metal braces for each pole (**Figure 26–38**). The braces are attached to the pole at intervals and near the top. The arms of the bracket extend from both sides of the pole at 45-degree angles. The arms are attached to the sidewall or roof to hold the pole steady.

The scaffold is raised by pressing on the foot pedal of the pump jack (**Figure 26–39**). The mechanism has brackets on which to place the scaffold plank. Other brackets hold a guardrail or platform. Reversing a lever allows the staging to be pumped downward.

Pump jack scaffolds are used widely for siding, where staging must be kept away from the walls, and when a steady working height is desired. However, pump jack scaffolds have their limitations. They should not be used when the working load exceeds 500 pounds (225 kg). No more than two persons are permitted at one time between any two supports.

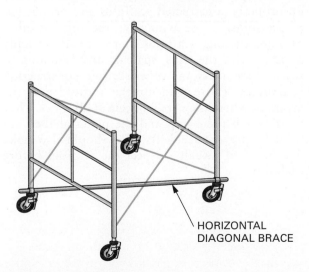

HORIZONTAL
DIAGONAL BRACE

FIGURE 26–37 The horizontal diagonal brace (or gooser) is used to keep the tower square when it is rolled.

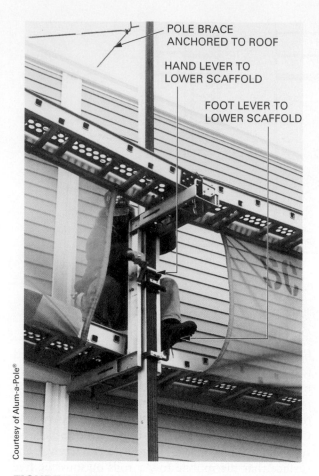

POLE BRACE
ANCHORED TO ROOF

HAND LEVER TO
LOWER SCAFFOLD

FOOT LEVER TO
LOWER SCAFFOLD

Courtesy of Alum-a-Pole®

FIGURE 26–38 Components of a pump jack system.

POLE

GUARDRAIL
WORKBENCH

PUMP JACK

END GUARDRAIL
SYSTEM

MID-RAIL
AND
TOEBOARD
SYSTEM

FOOT LEVER FOR
RAISING SCAFFOLD

WORK
PLATFORM

HAND
CRANK FOR
LOWERING

FOOT LEVER
FOR LOWERING
SCAFFOLD

Courtesy of Alum-a-Pole®

FIGURE 26–39 Pump jacks are raised by pressing the foot lever.

Wood poles must not exceed 30 feet (9 m) in height. Braces must be installed at a maximum vertical spacing of not more than 10 feet (3 m).

To pump the scaffold past a brace location, temporary braces are used. The temporary bracing is installed about 4 feet (1.2 m) above the original bracing. Once the scaffold is past the location of the original brace, it can be reinstalled. The temporary brace is then removed.

Wood pump jack poles are constructed of two 2 × 4s (38 × 89s) nailed together. The nails should be 3 inches (75 mm), and no less than 12 inches (305 mm) apart, staggered uniformly from opposite outside edges.

SCAFFOLD SAFETY

The safety of those working at a height depends on properly constructed scaffolds and proper use of scaffolds. Those who have the responsibility of constructing scaffolds must be thoroughly familiar with the sizes, spacing, and fastening of scaffold members and other scaffold construction techniques. Those who use the scaffold must be aware of where they are working at all times. They must watch their step and the material they are using to prevent accidents. They should also make it a habit to visually inspect the scaffold before each use.

DECONSTRUCT THIS

Carefully study **Figure 26–40** and think about what is wrong and/or what is right. Consider all possibilities. What construction practice or method is different in your area of the country?

Image Courtesy of Capital Safety

FIGURE 26–40 This photo shows workers using personal safety harnesses.

KEY TERMS

bearers	fall protection	mud sill	scaffolds
braces	guardrails	poles	single pole
competent person	ladder jacks	pump jack	staging planks
double pole	ledger	putlogs	stepladder
extension ladder	mobile scaffold	sawhorses	

REVIEW QUESTIONS

Select the most appropriate answer.

1. What are the vertical members of a scaffold called?

   a. columns
   b. piers
   c. poles
   d. uprights

2. What do bearers support?

   a. ledgers
   b. planks
   c. rails
   d. braces

3. What is the minimum size for scaffold planks?

   a. 2 × 6 rough (48 × 146 mm)
   b. 2 × 8 rough (48 × 197 mm)
   c. 2 × 10 rough (48 × 248 mm)
   d. 2 × 12 rough (48 × 298 mm)

4. What are the minimum and maximum distances that scaffold planks can extend beyond the bearers?

   a. 3 to 6 inches (75 to 150 mm)
   b. 3 to 8 inches (75 to 200 mm)
   c. 6 to 8 inches (150 to 200 mm)
   d. 6 to 12 inches (150 to 300 mm)

5. What is used to hold metal tubular frame scaffolding rigidly plumb?

   a. end frames
   b. putlogs
   c. cross braces
   d. cribbing

6. What part of a scaffold platform protects workers below from falling objects?

   a. toeboard
   b. mid-rail
   c. top rail
   d. posts

7. Which workers are allowed to climb an access ladder for a metal tubular scaffold that has its rungs spaced 18 inches (460 mm) apart?

   a. scaffold users only
   b. scaffold erectors only
   c. scaffold erectors and dismantlers only
   d. competent persons only

8. What is the maximum deflection that a loaded wooden scaffold plank can have?

   a. $\frac{1}{6}$ of the span
   b. $\frac{1}{16}$ of the span
   c. $\frac{1}{20}$ of the span
   d. $\frac{1}{60}$ of the span

9. By what factor can the height of a mobile scaffold exceed its minimum base dimension?

   a. three times
   b. four times
   c. five times
   d. six times

10. At what height of scaffold platform are guardrails required?

    a. 10 feet (3 m) in height
    b. 16 feet (5 m) in height
    c. 20 feet (6 m) in height
    d. 24 feet (7.3 m) in height

11. On top of what are tubular scaffold end frames installed level and plumb?

    a. baseplates
    b. mud sills
    c. cribbing
    d. putlogs

12. What is the minimum number of times a scaffold should be visually inspected?

    a. two times
    b. three times
    c. four times
    d. five times

13. Besides an approved ladder, what else can workers climb to access the work platform of a scaffold?

    a. the ladder built into the end frame
    b. cross braces
    c. the horizontal bearing points of the scaffold
    d. the mid- and top rails of the guardrail

14. How many times the intended load should a scaffold be able to support?

    a. four times
    b. five times
    c. ten times
    d. twenty times

15. Who is responsible for safety on a jobsite?

    a. the local safety inspector
    b. the general contractor
    c. scaffold erectors
    d. every worker on the job

16. How does metal tubular scaffolding differ from pump jacks?

    a. Metal tubular scaffolding often requires toeboards.
    b. Metal tubular scaffolding is usually installed on a mud sill.
    c. Metal tubular scaffolding has braces.
    d. Metal tubular scaffolding has working platforms at fixed heights, whereas pump jacks can be continuously adjusted.

Unit 10

Building Layout

Chapter 27 Levelling and Layout Tools   **Chapter 28** Laying Out Foundation Lines

Before construction begins, lines must be laid out showing the location and elevation of the building foundation. Accuracy in laying out these lines is essential in order to comply with local zoning ordinances. In addition, accurate layout lines provide for a foundation that is level, square, and set to specified dimensions. Accuracy in the beginning makes the work of the carpenter and other construction workers easier later. Layout for the location of a building and its component parts must be done properly. Failure to do so can be costly and time-consuming.

OBJECTIVES

After completing this unit, the student should be able to:

- establish level points across a building area using a water level and using a carpenter's hand spirit level in combination with a straightedge.
- accurately set up and use an automatic level and a laser level for levelling, determining, and establishing elevations.
- accurately set up and use a theodolite to establish a straight line, turn angles, and establish the corners of a foundation.
- lay out building lines by using the Pythagorean theorem method for calculating diagonals, the tangent function for calculating diagonal angles, and the theodolite for establishing corners and checking the layout for accuracy.
- complete a closed traverse using a theodolite and checking with trigonometry.
- build batter boards.

Chapter 27

Levelling and Layout Tools

Building layout requires levelling lines as well as laying out various angles over the length and width of the structure. It is interesting to realize that no matter where you go, plumb points toward the centre of the Earth. Level is always perpendicular (at right angles) to plumb (**Figure 27–1**). The construction industry uses many tools to achieve level and plumb. The carpenter must be able to set up, adjust, and use a variety of levelling and layout tools.

LEVELLING TOOLS

Several tools, ranging from simple to state-of-the-art tools, are used to level the layout. More sophisticated levelling and layout tools, although preferred, are not always available.

Levels and Straightedges

If sophisticated levelling tools are not available, simple levelling tools may be used to level a building area. Such tools include a *carpenter's hand level* and a long straightedge. This levelling process begins at some point or stake that is at the desired elevation. Stakes are then placed across the building area to the desired distance from the starting point. This method can be an accurate, although time-consuming, method of levelling over a long distance. It can also be done by one person. Care should be taken to be sure each step is performed properly because slight errors, multiplied by each succeeding step, can grow into large ones.

Begin by selecting a length of lumber for the straightedge, sighting it carefully to make sure that it is straight. It should be wide enough that it would not bend under the weight of the level when placed on its edge and supported only on its ends. Place the straightedge on edge with one end on the first stake or a surface at the desired elevation. Drive a second stake at the other end slightly higher than level. Reposition the straightedge to the top of each stake. Place the level on top and carefully drive the second stake until level is achieved. Remember it is easier to drive the

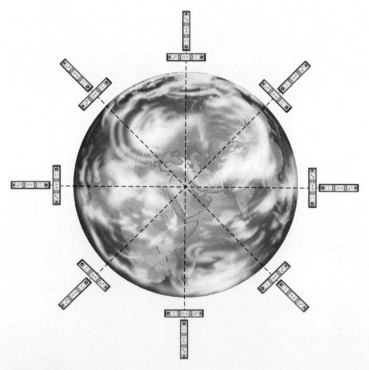

FIGURE 27–1 Plumb always points to the centre of the Earth, and level is a local reference perpendicular to plumb.

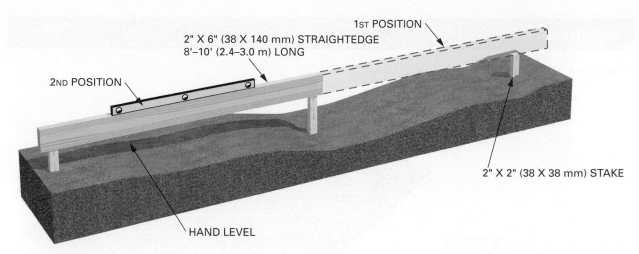

2ND POSITION

2" X 6" (38 X 140 mm) STRAIGHTEDGE
8'–10' (2.4–3.0 m) LONG

1ST POSITION

2" X 2" (38 X 38 mm) STAKE

HAND LEVEL

FIGURE 27–2 Levelling with a straightedge from stake to stake.

stake farther than it is to raise it, so don't go too far. Recheck the levelness of the two stakes.

Continue across the building area to the desired distance by moving the straightedge one stake at a time. Use the last stake driven as the new starting stake (**Figure 27–2**). Place the other end on another driven stake and drive until the straightedge is again level.

If the ground is so hard that stakes are difficult to drive, a crow or shale bar may be used (**Figure 27–3**). This will also help to keep the stakes straight with relative ease. Continue moving the straightedge from stake to stake until the desired distance is levelled. String a taut line to check that the tops of the stakes are in the same plane.

If you want to level to the corners of building layouts, start by driving a stake near the centre so that its top is to the desired height. Level from the centre stake to each corner in the manner described in **Procedure 27–A**.

Water Levels

A *water level* is a very accurate tool, dating back centuries. It is used for levelling from one point to another. Its accuracy, within a pencil point, is based on the principle that water seeks its own level (**Figure 27–4**).

One commercial model consists of 50 feet (15 m) of small-diameter, clear vinyl tubing and a small tube storage container. A built-in reservoir holds the

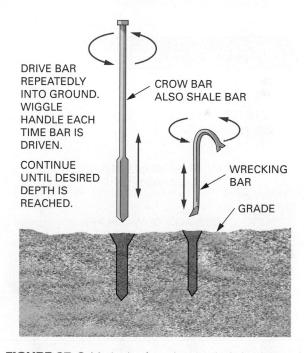

DRIVE BAR REPEATEDLY INTO GROUND. WIGGLE HANDLE EACH TIME BAR IS DRIVEN.

CROW BAR ALSO SHALE BAR

CONTINUE UNTIL DESIRED DEPTH IS REACHED.

WRECKING BAR

GRADE

FIGURE 27–3 Methods of starting a stake in hard ground.

coloured water that fills the tube. One end is held to the starting point. The other end is moved down until the water level is seen and marked on the surface to be levelled (**Figure 27–5**).

Although highly accurate, the water level is somewhat limited by the length of the plastic tube. However, extension tubings are available. Also, though slightly inconvenient, the water level may be moved from point to point.

A water level is accurate only if both ends of the tube are open to the air and there are no air bubbles in the length of the tube. Because both ends must be open, there may occasionally be some loss of liquid. However, this is replenished by the

Procedure 27–A Levelling Corners Using a Level

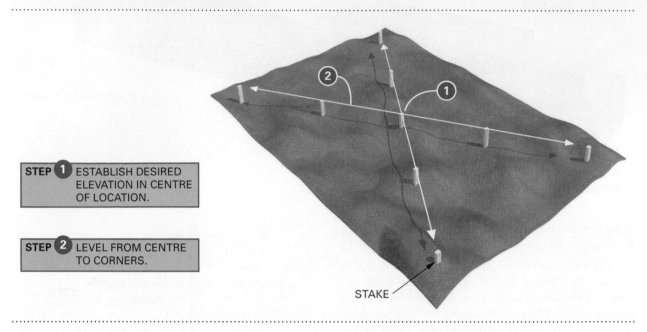

STEP 1 ESTABLISH DESIRED ELEVATION IN CENTRE OF LOCATION.

STEP 2 LEVEL FROM CENTRE TO CORNERS.

STAKE

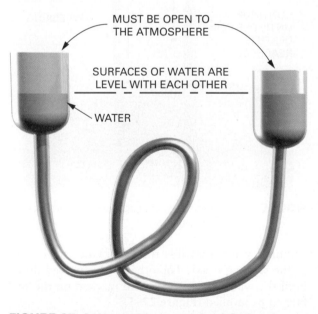

MUST BE OPEN TO THE ATMOSPHERE

SURFACES OF WATER ARE LEVEL WITH EACH OTHER

WATER

FIGURE 27–4 Water seeks its own level. Both ends of the water level must be open to the atmosphere.

reservoir. Any air bubbles can be easily seen with the use of coloured water.

In spite of these drawbacks, the water level is an extremely useful, inexpensive, simple tool for levelling from room to room, where walls obstruct views; down in a hole; or around obstructions.

Another advantage is that levelling can be done by one person.

OPTICAL LEVELS

Commonly used instruments for levelling, plumbing, and angle layout are optical levels, which include the builder's level and transit-level.

Builder's Levels and Transit-Levels

The builder's level and the transit-level with four levelling screws and a separate split bubble are "old" technology. Because three distinct points establish a two-dimensional plane, all instruments in Canada have three levelling screws. The builder's level has a *telescope* to which a *spirit level* is mounted. The telescope is fixed in a horizontal position. It can rotate 360 degrees for measuring horizontal angles but cannot be tilted up or down. The transit-level is different because its telescope can be moved up and down 45 degrees in each direction.

Automatic Levels

Automatic levels and *automatic transit-levels* (**Figure 27–6**) are similar to those previously described except that they have an internal *compensator*. This compensator uses gravity to maintain

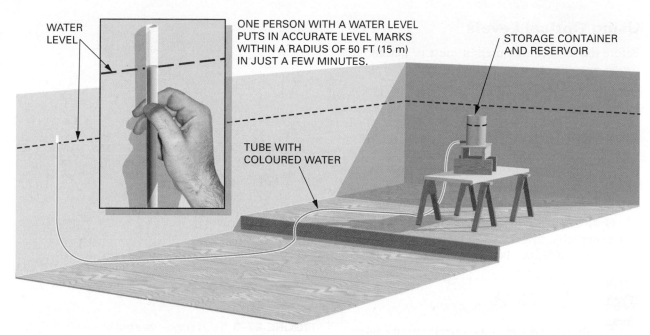

WATER LEVEL

ONE PERSON WITH A WATER LEVEL PUTS IN ACCURATE LEVEL MARKS WITHIN A RADIUS OF 50 FT (15 m) IN JUST A FEW MINUTES.

STORAGE CONTAINER AND RESERVOIR

TUBE WITH COLOURED WATER

FIGURE 27–5 The water level is a simple yet effective levelling tool.

a true level line of sight. Even if the instrument is jarred, the line of sight stays true because gravity does not change. Automatic levels are the levels mainly used in Canada.

Many models of levelling instruments are available. To become familiar with more sophisticated levels, study the manufacturers' literature. No matter what type of level is used, the basic procedures are the same.

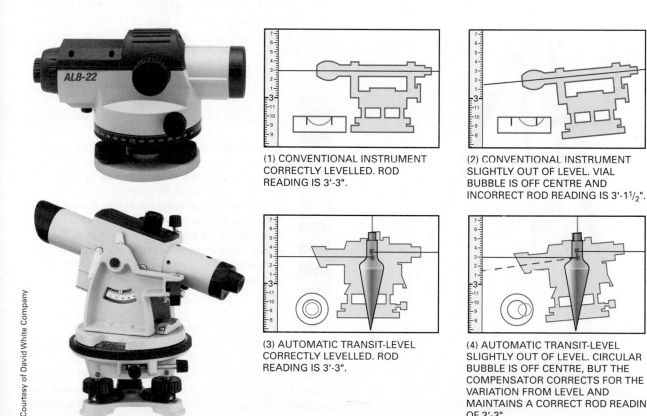

Courtesy of David White Company

AL8-22

(1) CONVENTIONAL INSTRUMENT CORRECTLY LEVELLED. ROD READING IS 3'-3".

(2) CONVENTIONAL INSTRUMENT SLIGHTLY OUT OF LEVEL. VIAL BUBBLE IS OFF CENTRE AND INCORRECT ROD READING IS 3'-1½".

(3) AUTOMATIC TRANSIT-LEVEL CORRECTLY LEVELLED. ROD READING IS 3'-3".

(4) AUTOMATIC TRANSIT-LEVEL SLIGHTLY OUT OF LEVEL. CIRCULAR BUBBLE IS OFF CENTRE, BUT THE COMPENSATOR CORRECTS FOR THE VARIATION FROM LEVEL AND MAINTAINS A CORRECT ROD READING OF 3'-3".

FIGURE 27–6 Automatic levels and automatic transit-levels level themselves when set up nearly level.

Using Optical Levels

Before the level can be used, it must be placed on a *tripod* or some other solid support and levelled.

Setting Up and Adjusting the Level. The telescope is adjusted to a level position by means of three *levelling screws* that rest on a *base levelling plate*. In higher quality levels, the base plate is part of the instrument. In less expensive models, the base plate is part of the tripod. For detailed photos of the parts and the setup of the automatic level, visit *http://nikon.com/about/feelnikon/recollections/r27_e/index.htm*.

Open and adjust the legs of the tripod to a convenient height (chin height). Spread the legs of the tripod well apart, and firmly place its feet into the ground.

> ## CAUTION
>
> On a smooth surface it is essential that the points on the feet hold without slipping (Figure 27–7). Make an equilateral triangle (60°) out of ¾-inch (19 mm) plywood and place small holes or depressions for the tripod points to fit into in each corner. The plywood can be shaped to reduce its mass (Figure 27–8). Or attach wire or light chain to the lower ends of each leg.

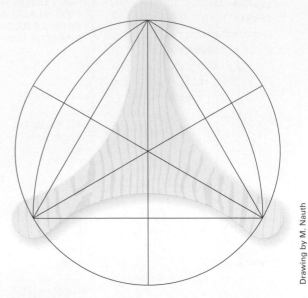

FIGURE 27–8 Tripod stand made out of ¾-inch (19 mm) plywood.

Drawing by M. Nauth

When set up, the top of the tripod should be close to level. Sight by eye and tighten the tripod wing nuts. With the top of the tripod close to level, adjustment of the instrument is made easier.

Lift the instrument from its case by the frame. Note how it is stored so that it can be replaced in the case in the same position. Make sure the horizontal clamp screw is loose so that the telescope revolves freely. While holding onto the frame, secure the instrument to the tripod.

> ## CAUTION
>
> Care must be taken not to damage the instrument. Never use force on any parts of the instrument. All moving parts turn freely and easily by hand. Excessive pressure on the levelling screws may damage the threads of the base plate. Unequal tension on the screws will cause the instrument to wobble on the base plate, resulting in levelling errors. Periodically use a toothbrush dipped in light instrument oil to clean and lubricate the threads of the adjusting screws.

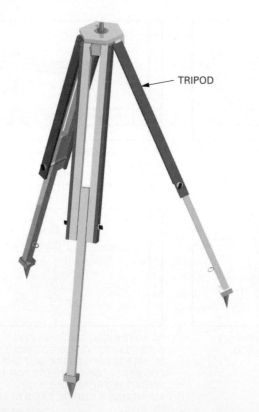

TRIPOD

FIGURE 27–7 Make sure the feet of the tripod do not slip on smooth or hard surfaces.

Initial setup of the instrument is important. Line up the telescope directly over two of the levelling screws. Then turn the screws in opposite directions with forefingers and thumbs. Move the thumbs toward or away from each other, as the case may be, to move the air bubble in the circular vial so that it is centred left to right (**Figure 27–9**). The bubble will always move in the same direction as your left

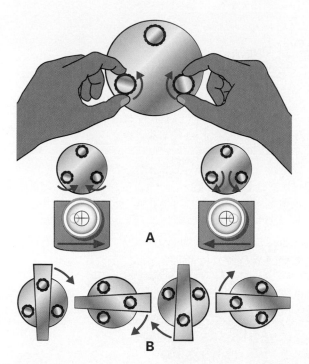

FIGURE 27–9 (A) Level the instrument by moving thumbs toward or away from each other. (B) The instrument is level when the bubble remains centred as the telescope is revolved in a complete circle.

thumb is moving. Turn the remaining levelling screw to move the air bubble into the centre of the target circle on the vial. The internal compensator will keep it level as you rotate the telescope.

CAUTION

Do not leave unattended an instrument that has been set up near moving equipment.

Sighting the Level. To sight an object, rotate the telescope and sight over its top, aiming it at the object. Look through the telescope. Focus it by turning the focusing knob one way or the other until the object becomes clear. Keep both eyes open. This eliminates squinting, does not tire the eyes, and gives the best view through the telescope.

CAUTION

If the lenses need cleaning, dust them with a soft brush or rag. Do *not* rub the dirt off. Rubbing may scratch the lens coating.

When looking into the telescope, vertical and horizontal *crosshairs* are seen. They enable the target

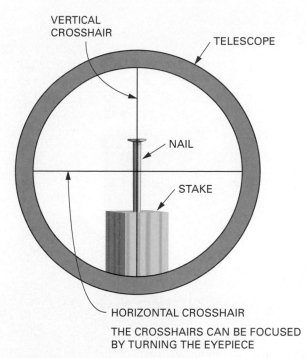

FIGURE 27–10 When looking into the telescope, vertical and horizontal crosshairs are seen.

to be centred properly (**Figure 27–10**). The crosshairs themselves can be brought into focus by turning the eyepiece one way or the other. Centre the crosshairs on the object by moving the telescope left or right. A fine adjustment can be made by tightening the horizontal clamp screw and turning the horizontal tangent screw one way or the other. The horizontal crosshair is used for reading elevations. The vertical crosshair is used when laying out angles and aligning vertical objects.

Levelling

When the instrument is levelled, a given point on the line of sight is exactly level with any other point. Any line whose points are the same distance below or above the line of sight is also level (**Figure 27–11**). To level one point with another, a helper must hold a *target* on the point to be levelled. A reading is taken. The target is then moved to selected points that are brought to the same elevation by moving those points up or down to get the same reading.

Targets. A tape is often used as a target. The end of the tape is placed on the point to be levelled. The tape is then moved up or down until the same mark is read on the tape as was read at the starting point.

Because of its flexibility, the tape may need to be backed up by a strip of wood to hold it rigid (**Figure 27–12**).

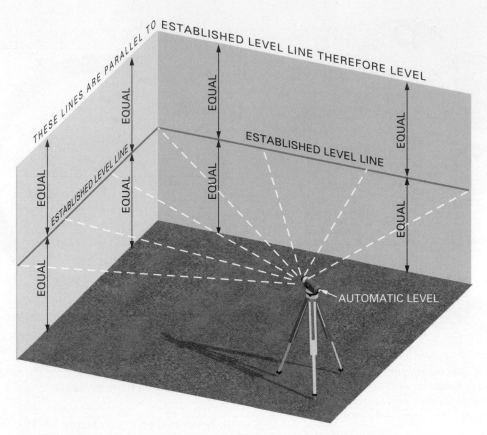

FIGURE 27–11 Any line parallel to the established level line is also level.

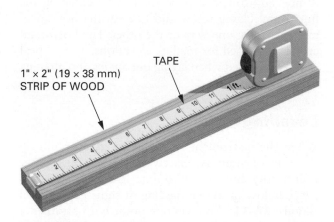

FIGURE 27–12 A tape can be backed up by a strip of wood to make it a stiff and steady target.

The simplest target is a plain 1 × 2-inch (19 × 38 mm) strip of wood. The end of the stick is held on the starting point of desired elevation. The line of sight is marked on the stick. The end of the stick is then placed on top of various points. They are moved up or down to bring the mark to the same height as the line of sight (**Procedure 27–B**). A stick of practically any length can be used.

Cut the stick to a length so that the mark to be sighted is a noticeable distance off from the centre of its length. It is then immediately noticeable if the stick is inadvertently turned upside down (**Figure 27–13**). If, for some reason, it is not desirable to cut the stick, clearly mark the top and bottom ends.

Levelling Rods. For longer sightings, the *levelling rod* is used because of its clearer graduations. A variety of rods are manufactured of wood or fibreglass for several levelling purposes. They are made with two or more sections that extend easily and lock into place. Rods vary in length—from two-section rods extending 9'-0" (3 m) up to seven-section rods extending 25'-0" (8 m).

The builder's rod (Architectural) has feet, inches, and eighths of an inch. The graduations are $\frac{1}{8}$ inch wide and $\frac{1}{8}$ inch apart. The engineer's rod is very similar but the scale is slightly different. It is in feet, tenths, and hundredths of a foot (see Figure 27–28). Instead of inches, the number markings represent a tenth of a foot. The smaller graduations are $\frac{1}{100}$ of a foot wide and $\frac{1}{100}$ of a foot apart, which is slightly smaller than $\frac{1}{8}$ inch. They are both designed for easy reading. An oval-shaped, red-and-white, movable target is available to fit

StepbyStep Procedures

Procedure 27–B Establishing Level Points

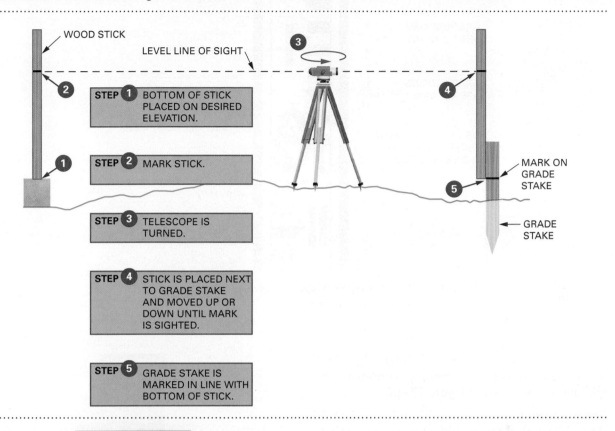

WOOD STICK

LEVEL LINE OF SIGHT

STEP 1 BOTTOM OF STICK PLACED ON DESIRED ELEVATION.

STEP 2 MARK STICK.

STEP 3 TELESCOPE IS TURNED.

STEP 4 STICK IS PLACED NEXT TO GRADE STAKE AND MOVED UP OR DOWN UNTIL MARK IS SIGHTED.

STEP 5 GRADE STAKE IS MARKED IN LINE WITH BOTTOM OF STICK.

MARK ON GRADE STAKE

GRADE STAKE

ON THE JOB

Keep the stick to an acceptable length so that the mark is not close to being centred on its length.

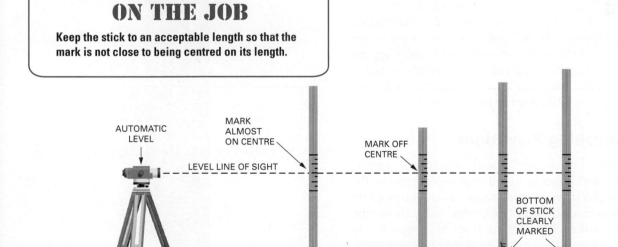

AUTOMATIC LEVEL

MARK ALMOST ON CENTRE

MARK OFF CENTRE

LEVEL LINE OF SIGHT

BOTTOM OF STICK CLEARLY MARKED

BOT

V

A MARK APPROXIMATELY CENTRED ON A STICK USED FOR A LEVELLING ROD CAN CAUSE ERRORS BY MISTAKENLY TURNING THE STICK UPSIDE DOWN

TO PREVENT MISTAKES, CUT THE STICK TO A LENGTH SO THE MARK WILL BE FAR OFF CENTRE

OR MARK THE BOTTOM OF THE STICK WITH MARKS THAT CAN BE CLEARLY UNDERSTOOD

OR BOTH

FIGURE 27–13 Techniques for creating an easy-to-use marking stick.

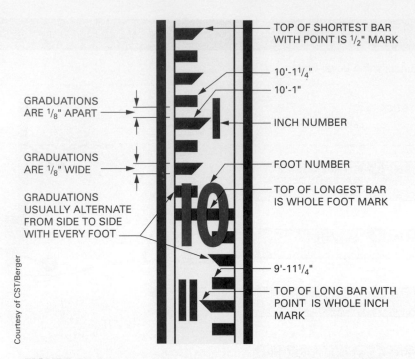

Courtesy of CST/Berger

TOP OF SHORTEST BAR
WITH POINT IS 1/2" MARK

10'-11/4"

10'-1"

INCH NUMBER

FOOT NUMBER

TOP OF LONGEST BAR
IS WHOLE FOOT MARK

9'-111/4"

TOP OF LONG BAR WITH
POINT IS WHOLE INCH
MARK

GRADUATIONS
ARE 1/8" APART

GRADUATIONS
ARE 1/8" WIDE

GRADUATIONS
USUALLY ALTERNATE
FROM SIDE TO SIDE
WITH EVERY FOOT

FIGURE 27–14 The builder's levelling rod (Archtectural) is marked in feet, inches, and eighths of an inch.

on any rod for easy reading (**Figure 27–14**). A metric rod, graduated in metres, centimetres, and millimetres, is shown in **Figure 27–15.**

Communication. A responsible rod operator holds the rod vertical and faces the instrument so that it can be read with ease and accuracy. Sighting distances are not usually over 100 to 150 (30 to 45 m), yet sometimes voice commands cannot be used. Hand signals are then given to the rod operator to move the target as desired by the instrument operator. Usually, appropriate hand signals are given even when distances are not great. Shouting on the jobsite is unnecessary and unprofessional, and it creates confusion.

Establishing Elevations

Many points on the jobsite, such as the depth of excavations, the heights of foundation footing and walls, and the elevation of finish floors, are required to be set at specified elevations or grades. These elevations are established by starting from the **benchmark**. The benchmark is a point of designated elevation. The instrument operator records elevations and rod readings in a field book to make calculations.

Height of the Instrument (HI). When it is necessary to set a point at some definite elevation, first determine the **height of the instrument (HI)**. To find HI, place the rod on the benchmark and add the reading to the elevation of the benchmark

Courtesy of Butler Survey Supplies Ltd.

FIGURE 27–15 A metric rod.

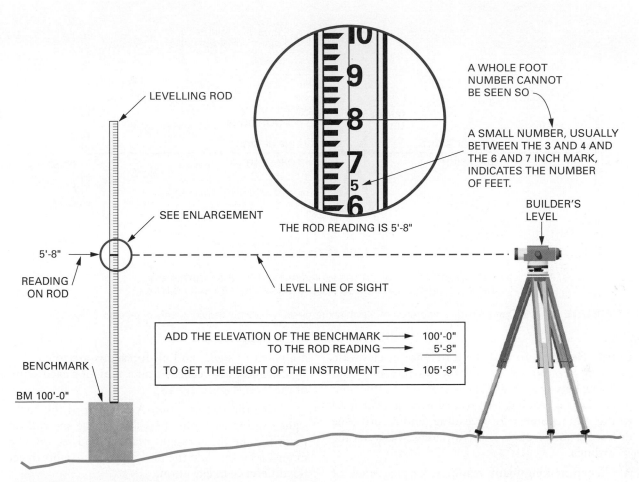

LEVELLING ROD

A WHOLE FOOT NUMBER CANNOT BE SEEN SO

A SMALL NUMBER, USUALLY BETWEEN THE 3 AND 4 AND THE 6 AND 7 INCH MARK, INDICATES THE NUMBER OF FEET.

THE ROD READING IS 5'-8"

SEE ENLARGEMENT

BUILDER'S LEVEL

5'-8"

READING ON ROD

LEVEL LINE OF SIGHT

ADD THE ELEVATION OF THE BENCHMARK →	100'-0"
TO THE ROD READING →	5'-8"
TO GET THE HEIGHT OF THE INSTRUMENT →	105'-8"

BENCHMARK

BM 100'-0"

FIGURE 27–16 Determining the height of the instrument.

(**Figure 27–16**). For instance, if the benchmark has an elevation of 100.00 feet and the rod reads 5'-8", then the HI is 105'-8".

Grade Rod. What must be read on the rod when its base is at the desired elevation is called the **grade rod.** This is found by subtracting the desired elevation from the height of the instrument. For instance, if the elevation to be established is 102'-0", subtract it from 105'-8" (HI) to get 3'-8" (the grade rod). The rod operator places the rod at the desired point. He or she then moves it up or down, at the direction of the instrument operator, until the grade rod of 3'-8" is read on the rod. The base of the builder's rod is then at the desired elevation (**Figure 27–17**). A mark, drawn at the base of the rod on a stake or other object, establishes the elevation.

Determining Differences in Elevation

Differences in elevation need to be determined for such tasks as grading driveways, sidewalks, and parking areas; laying out drainage ditches; plotting contour lines; and estimating cut and fill

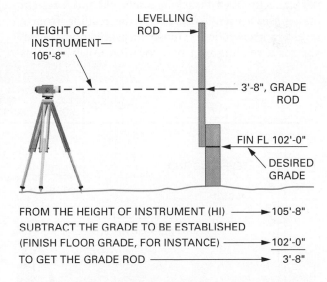

HEIGHT OF INSTRUMENT— 105'-8"

LEVELLING ROD

3'-8", GRADE ROD

FIN FL 102'-0"

DESIRED GRADE

FROM THE HEIGHT OF INSTRUMENT (HI) →	105'-8"
SUBTRACT THE GRADE TO BE ESTABLISHED (FINISH FLOOR GRADE, FOR INSTANCE) →	102'-0"
TO GET THE GRADE ROD →	3'-8"

THE "GRADE ROD" IS WHAT THE ROD MUST READ WHEN ITS BASE IS AT THE DESIRED GRADE OR HEIGHT.

FIGURE 27–17 Calculating the grade rod and establishing a desired elevation.

requirements. The difference in elevation of two or more points is easily determined with the use of the automatic level.

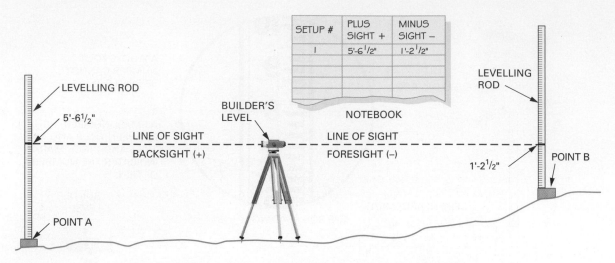

THE DIFFERENCE BETWEEN THE PLUS SIGHT AND MINUS SIGHT IS 4'-4".
THE PLUS SIGHT IS LARGER, SO POINT B IS 4'-4" HIGHER THAN POINT A.

FIGURE 27–18 Determining a difference of elevation between two points requiring only one setup.

Single Setup. To find the difference in elevation between two points, set up the instrument about midway between them. Place the rod on the first point. Take a reading, and record it. Swing the level to the other point, take a reading, and record. The difference in elevation is the difference between the recordings.

When making many readings, keeping track of the readings becomes more difficult. Using the surveying technique of tracking *backsight* and *foresight* makes this easier. Backsight is the reading from a level to a known or previously measured point. Foresight is a reading to a new location. All backsights are *plus* (+) *sights* and all foresights are *minus* (−) *sights*. These are recorded on a table for each setup of the level (**Figure 27–18**).

Place the rod on point A and record the backsight reading as a plus (+) sight. Place the rod on point B, take the foresight reading, and record as a minus (−) sight. Add the plus and minus sights to get the difference in elevation.

Sometimes a reading is made above the level line of the level. To keep this straight a rule is applied: For any backsight or foresight reading where the rod is flipped upside down, the plus or minus sign is reversed (**Figure 27–19**).

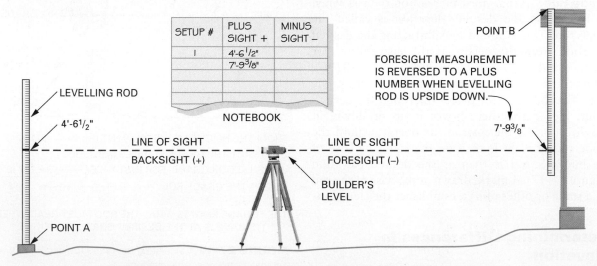

A MINUS SIGHT BECOMES A PLUS SIGHT IF THE ROD IS READ UPSIDE DOWN.
POINT B IS 12'-3⁷/₈" HIGHER THAN POINT A.

FIGURE 27–19 For any readings taken with the rod upside down, the plus and minus signs of the sighting measurements are reversed.

Multiple Setup. Sometimes the difference in elevation of two points is too great or the distance is too far apart. Then it is necessary to make more than one instrument setup to determine the difference. The procedure is similar to a series of one-setup operations until the final point is reached.

> ## CAUTION
>
> **When carrying a tripod-mounted instrument, handle with care. Carry it in an upright position. Do not carry it over the shoulder or in a horizontal position. Be careful not to bump the instrument when going through buildings or close quarters.**

Record all backsights as plus sights and all foresights as minus sights, unless the rod is upside down when read (**Figure 27–20**). Find the sum of all minus sights and all plus sights. The difference between them is the difference in elevation of the beginning and ending points. If the sum of the plus sights is larger, then the end point is higher than the starting point. If the sum of the minus sights is larger, then the end point is lower than the starting point. The surveyor's method of recording notes in a field book is shown in **Figure 27–21**.

Measuring and Laying Out Angles

To measure or lay out angles, the instrument that is most commonly used is the **theodolite**. Most theodolites have a built-in *optical plummet* that allows the operator to sight to a point below, exactly plumb with the centre of the instrument. This enables quick and accurate setups over a point (**Figure 27–22**).

The older instruments use a plumb bob. A hook, centred below the instrument, is provided for suspending a plumb bob. The plumb bob is used to place the level directly over this point.

Setting Up Over a Point. Suspend the plumb bob from the instrument. Secure it with a slip knot. Move the tripod and instrument so that the plumb bob appears to be over the point.

Press the legs of the tripod into the ground. Lower the plumb bob by moving the slip knot until it is about ¼ inch (6 mm) above the point on the ground. The final centring of the instrument can be made by loosening any two adjacent levelling screws and slowly shifting the instrument until the plumb bob is directly over the point (**Figure 27–23**). Retighten the same two levelling screws that were previously loosened, and level the instrument. Shift the instrument on the base plate until the plumb bob is directly over the point. Check the levelness of the instrument. Adjust, if necessary. This procedure is easier with an optical plummet.

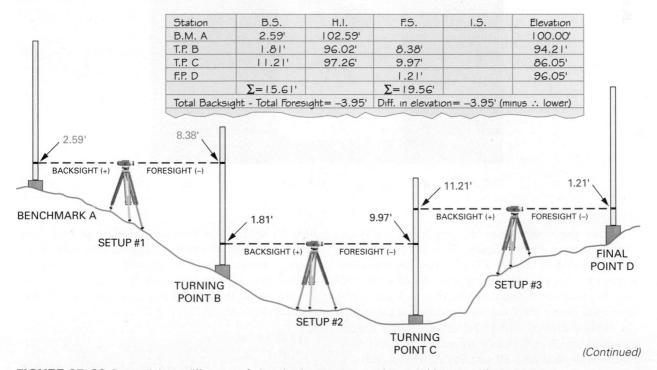

Station	B.S.	H.I.	F.S.	I.S.	Elevation
B.M. A	2.59'	102.59'			100.00'
T.P. B	1.81'	96.02'	8.38'		94.21'
T.P. C	11.21'	97.26'	9.97'		86.05'
F.P. D			1.21'		96.05'
	Σ=15.61'		Σ=19.56'		
Total Backsight − Total Foresight= −3.95'			Diff. in elevation= −3.95' (minus ∴ lower)		

FIGURE 27–20 Determining a difference of elevation between two points requiring more than one setup.

(Continued)

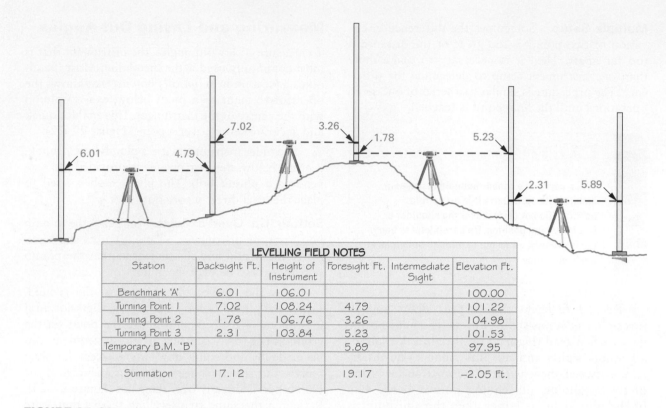

LEVELLING FIELD NOTES					
Station	Backsight Ft.	Height of Instrument	Foresight Ft.	Intermediate Sight	Elevation Ft.
Benchmark 'A'	6.01	106.01			100.00
Turning Point 1	7.02	108.24	4.79		101.22
Turning Point 2	1.78	106.76	3.26		104.98
Turning Point 3	2.31	103.84	5.23		101.53
Temporary B.M. 'B'			5.89		97.95
Summation	17.12		19.17		−2.05 Ft.

FIGURE 27–20 *(Continued)*

Photo by M. Nauth

FIGURE 27–21 Organized notations are placed in a field book.

Most theodolites that are purchased today in Canada have a digital display. They are called *electronic theodolites* (**Figure 27–24**). The improvement in accuracy, the ease of reading, the ease of setup, the short learning curve, and the reduction in price makes them ideal instruments for the construction company. The circle is divided into 360 degrees (°), each degree is divided into 60 minutes or **arc minutes** ('), and each minute is divided into 60 seconds or **arc seconds** ("). Most theodolites used on the construction site will have an accuracy of between 5 and 30 seconds (**Figure 27–25**). An example of adding and subtracting degrees, minutes, and seconds, and conversion to decimal degrees, is given below. Most scientific calculators are able to do this as well as conversions to and from decimal degrees.

$$135° \ 25' \ 36''$$
$$+ \ 29° \ 13' \ 42''$$
$$\overline{164° \ 38' \ 78''} = 164° \ 38' \ (1' \ 18'') = \mathbf{164° \ 39' \ 18''}$$

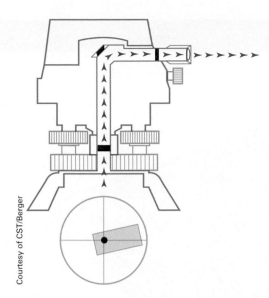

FIGURE 27–22 Some instruments have a device called an optical plummet for setting the instrument directly over a point.

HANGING THE PLUMB BOB USING A SLIP KNOT

TO HANG THE PLUMB BOB, ATTACH CORD TO THE PLUMB BOB HOOK ON THE TRIPOD AND KNOT THE CORD AS ILLUSTRATED

FIGURE 27–23 To locate the instrument directly over a point, a plumb bob is suspended from the level.

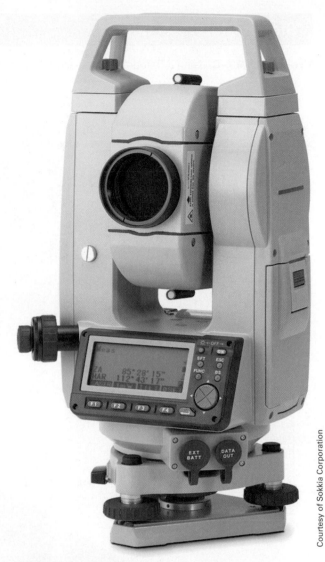

FIGURE 27–24 An electronic theodolite.

$$
\begin{array}{r}
135°\ 25'\ 36'' \\
-\ 29°\ 13'\ 42'' \\
\end{array}
\qquad
\begin{array}{r}
135°\ 24'\ 96'' \\
-\ 29°\ 13'\ 42'' \\
\hline
106°\ 11'\ 54'' \\
\end{array}
$$

$$115°\ 26'\ 42'' = 115°\ 26'_{\ 42/60'} = 115°\ 26.7'$$
$$= 115°_{\ 26.7/60°} = \mathbf{115.445°}$$

$$115.445° = 115°\ (0.445 \times 60)' = 115°\ 26.7'$$
$$= 115°\ 26'\ (0.7 \times 60)'' = 115°\ 26'\ 42''$$

Measuring a Horizontal Angle. After levelling the instrument over the point of an angle, called its *vertex*, loosen the horizontal clamp screw. Rotate the instrument until the vertical crosshair is nearly in line with a distant point on one side of the angle. Tighten the clamp screw. Then turn the tangent screw to line up the vertical crosshair exactly with the point. Set the horizontal angle to zero (0SET).

FIGURE 27–25 An electronic theodolite, 5-second accuracy.

StepbyStep Procedures

Procedure 27–C Using a Theodolite to Measure Horizontal Angles

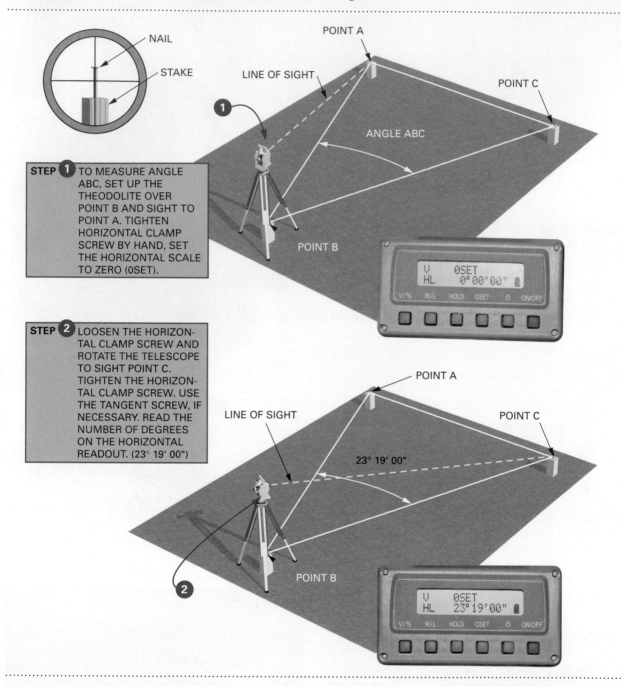

NAIL

STAKE

POINT A

LINE OF SIGHT

POINT C

ANGLE ABC

POINT B

1

STEP 1 TO MEASURE ANGLE ABC, SET UP THE THEODOLITE OVER POINT B AND SIGHT TO POINT A. TIGHTEN HORIZONTAL CLAMP SCREW BY HAND, SET THE HORIZONTAL SCALE TO ZERO (0SET).

V   0SET
HL   0°00'00"

V/% R/L HOLD OSET ON/OFF

STEP 2 LOOSEN THE HORIZONTAL CLAMP SCREW AND ROTATE THE TELESCOPE TO SIGHT POINT C. TIGHTEN THE HORIZONTAL CLAMP SCREW. USE THE TANGENT SCREW, IF NECESSARY. READ THE NUMBER OF DEGREES ON THE HORIZONTAL READOUT. (23° 19' 00")

POINT A

LINE OF SIGHT

POINT C

23° 19' 00"

POINT B

2

V   0SET
HL   23°19'00"

V/% R/L HOLD OSET ON/OFF

If the point is above or below the line of sight and a theodolite is not available, sight a straightedge held plumb from the point. Or, sight the line of a plumb bob line suspended over the point. If using a transit-level, release the locking lever and tilt the telescope to sight the point.

Next, loosen the clamp screw. Swing the telescope until the vertical crosshair lines up with a point on the other side of the angle. Tighten the horizontal clamp screw. Then turn the tangent screw for a fine adjustment, if necessary. Read the degrees on the digital readout (**Procedure 27–C**).

StepbyStep Procedures

Procedure 27–D Using a Theodolite to Lay Out a 90-Degree Angle

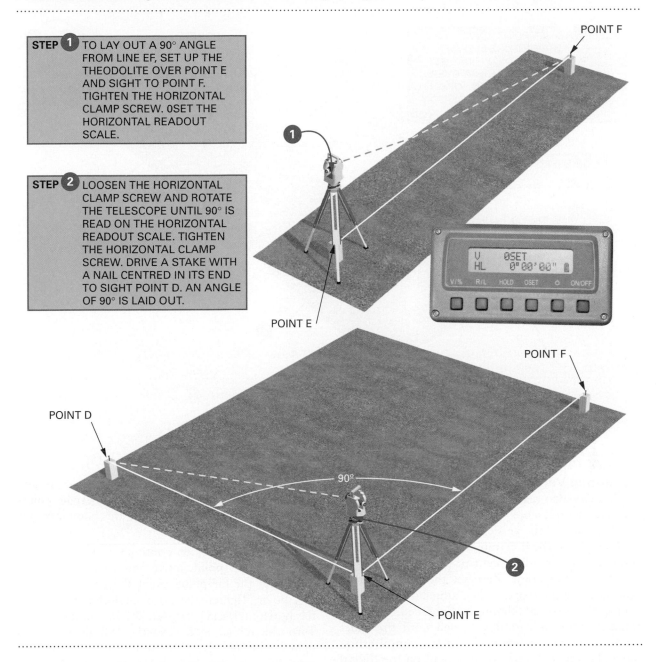

STEP 1 TO LAY OUT A 90° ANGLE FROM LINE EF, SET UP THE THEODOLITE OVER POINT E AND SIGHT TO POINT F. TIGHTEN THE HORIZONTAL CLAMP SCREW. 0SET THE HORIZONTAL READOUT SCALE.

STEP 2 LOOSEN THE HORIZONTAL CLAMP SCREW AND ROTATE THE TELESCOPE UNTIL 90° IS READ ON THE HORIZONTAL READOUT SCALE. TIGHTEN THE HORIZONTAL CLAMP SCREW. DRIVE A STAKE WITH A NAIL CENTRED IN ITS END TO SIGHT POINT D. AN ANGLE OF 90° IS LAID OUT.

POINT F

POINT E

POINT F

POINT D

POINT E

If the theodolite is turned to the left (counter-clockwise), the angle is read as 23° 19′ 00″, and if it is turned to the right (clockwise), the angle is read as 336° 41′ 00″.

Laying Out a Horizontal Angle.   Centre and level the instrument over the vertex of the angle to be laid out. Sight the telescope on a distant point on one side

of the angle. Tighten the horizontal clamp, and zero-set the digital readout (0SET). Loosen the clamp. Then turn the telescope until close to the desired number of degrees. Tighten the clamp. Use the tangent screw and make a fine adjustment until the index reads exactly the desired number of degrees, as in 90° in the example. Sight the vertical crosshair to lay out the other side of the angle (**Procedure 27–D**).

StepbyStep Procedures

Procedure 27–E Using a Theodolite to Set Points in a Straight Line

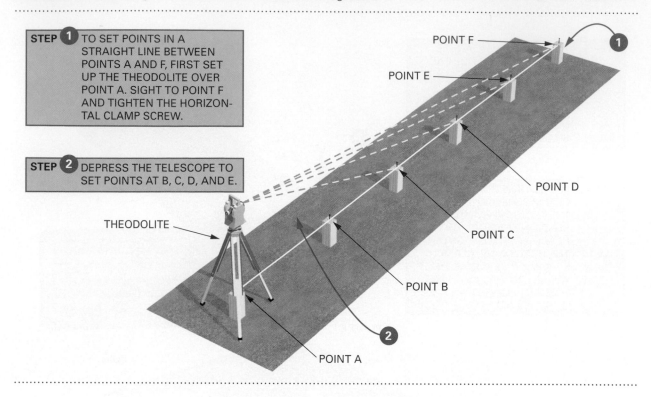

STEP 1 TO SET POINTS IN A STRAIGHT LINE BETWEEN POINTS A AND F, FIRST SET UP THE THEODOLITE OVER POINT A. SIGHT TO POINT F AND TIGHTEN THE HORIZONTAL CLAMP SCREW.

STEP 2 DEPRESS THE TELESCOPE TO SET POINTS AT B, C, D, AND E.

THEODOLITE

POINT F

POINT E

POINT D

POINT C

POINT B

POINT A

Measuring Vertical Angles. The electronic theodolite can also read vertical angles. By tilting and rotating the telescope, set the horizontal crosshair on the point of the vertical angle being measured. Tighten the vertical clamp. Then turn the tangent screw for a fine adjustment to place the crosshair exactly on the point. Zero-set the vertical angle readout. The telescope is then aimed at the second point, the vertical clamp is tightened, and the tangent screw is carefully turned until the exact point is sighted in line with the horizontal crosshair. The vertical angle is then read on the digital readout screen, similar to the reading of horizontal angles.

Setting Points in a Straight Line. It may be necessary to set points in a straight line, such as on a property boundary line. Set up the instrument over the one point. Rotate and tilt the telescope to sight the other point fairly closely. Tighten the horizontal clamp. Then turn the tangent screw until the vertical crosshair is exactly on the far point. With the

horizontal clamp tight, release the lock levers and depress the telescope, as required, to sight points between the corners and along the boundary line (**Procedure 27–E**).

If it is necessary to continue in a straight line beyond the far point, move and set up the instrument over the far point, point F. Sight back to the first point. Tighten the horizontal clamp. Zero-set the horizontal scale. Loosen the horizontal clamp. Turn the telescope to exactly 180 degrees, and tighten the horizontal clamp again. Depress or raise the telescope as needed to sight additional points in a straight line (**Procedure 27–F**). Instruments with shorter telescopes are able to "transit" over from one side to the other without a horizontal rotation.

Storing the Instrument. If the instrument gets wet, dry it before returning it to its case. Keep it in its carrying case when it is not being used or when being transported in a vehicle over long distances.

StepbyStep Procedures

Procedure 27–F Using a Theodolite to Set Many Straight Points

STEP 1 TO CONTINUE POINTS IN A STRAIGHT LINE: SET THE THEODOLITE OVER POINT F, THE FARTHEST POINT SIGHTED FROM POINT A.

STEP 2 SIGHT BACK TO POINT A. TIGHTEN HORIZONTAL CLAMP. TURN CIRCLE SCALE TO ZERO.

STEP 3 LOOSEN HORIZONTAL CLAMP AND TURN TELESCOPE 180 DEGREES TO SIGHT POINT L. DEPRESS TELESCOPE TO SIGHT INTERMEDIATE POINTS G, H, I, J, AND K.

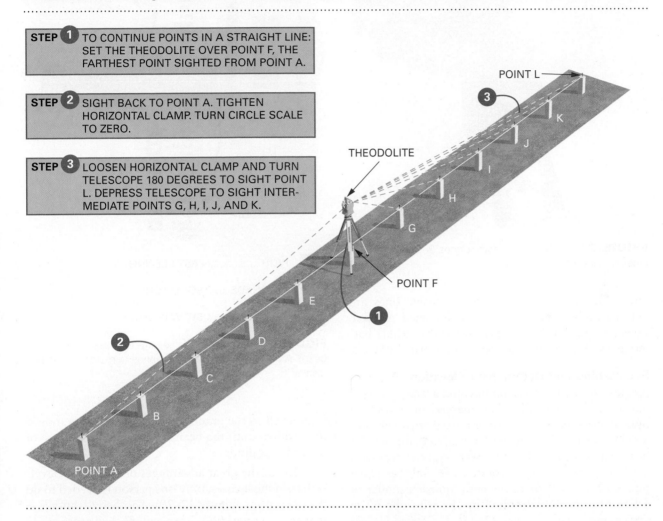

LASER LEVELS

A laser is a device in which light energy is released in a narrow beam. Laser is an acronym that stands for light amplification by stimulated emission of radiation. The light beam is absolutely straight. Unless interrupted by an obstruction or otherwise disturbed, the light beam can be seen for a long distance.

Lasers have many applications in space, medicine, agriculture, and engineering. The laser level has been developed for the construction industry to provide more efficient layout work (**Figure 27–26**).

Kinds and Uses of Laser Levels

Several manufacturers make laser levels in a number of different models. The least expensive models are the least sophisticated. A low-price unit is levelled

FIGURE 27–26 Laser levels have been developed for use in the construction industry.

FIGURE 27–27 The laser beam rotates 360 degrees, creating a level plane of light.

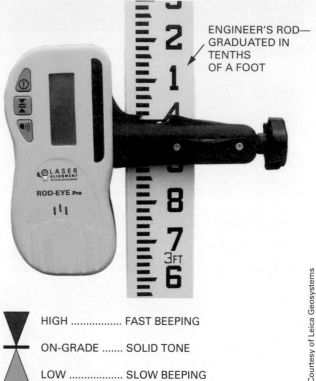

ENGINEER'S ROD— GRADUATED IN TENTHS OF A FOOT

HIGH .... FAST BEEPING

ON-GRADE .... SOLID TONE

LOW .... SLOW BEEPING

Courtesy of Leica Geosystems

FIGURE 27–28 An electronic target senses the laser beam. An audio feature provides tones to match the visual display.

and adjusted manually. More expensive ones are automatically adjusted to and maintained in level. Power sources include batteries, a rechargeable battery pack, or an AC/DC converter for 110 or 220 volts.

Establishing and Determining Elevations. A simple, easy-to-use model is mounted on a tripod or solid flat surface and levelled like manual or automatic optical instruments. The laser is turned on. It will emit a red beam, usually ⅜ inch (9.5 mm) in diameter. The beam rotates through a full 360 degrees, creating a level *plane* of light. As it rotates, it establishes equal points of elevation over the entire jobsite, similar to a line of sight being rotated by the telescope of an optical instrument (**Figure 27–27**).

Depending on the quality of the instrument, the laser head may rotate at various revolutions per second (rps), up to 40 rps. Its quality also determines its working range. This may vary from a 75 to 1000-foot (23 to 305 m) radius.

Laser beams are difficult to see outdoors in bright sunlight. To detect the beam, a battery-powered electronic *sensor* target, also called a *receiver* or *detector*, is attached to the levelling rod or stick. Most sensors have a visual display with a selectable audio tone to indicate when it is close to or on the beam (**Figure 27–28**). In addition to electronic sensor targets, specially designed targets are used for interior work, such as installing ceiling grids and levelling floors.

The procedures for establishing and determining elevations with laser levels are similar to those with optical instruments. To establish elevations, the sensor

is attached to the grade rod. The grade rod is moved up or down until the beam indicates that the base of the rod is at grade.

One of the great advantages of using laser levels is that, in most cases, only one person is needed to do the operations (**Figure 27–29**). Another advantage is that certain operations are accomplished more easily in less time.

Courtesy of Leica Geosystems

FIGURE 27–29 When using laser levels, only one person is required for levelling operations.

FIGURE 27–30 Levelling suspended ceiling grids. The beam is viewed through the target.

FIGURE 27–31 Placed on its side, the laser level is used to lay out and align partitions and walls.

FIGURE 27–32 Attaching a laser detector to a bulldozer allows the operator to level the ground quickly.

Special Horizontal Operations. For levelling *suspended ceiling grids*, the laser level is mounted to an adjustable grid mount bracket. Once the first strip of angle trim has been installed, the laser and bracket can be attached easily and clamped into place.

The unit is levelled. The height of the laser beam is then adjusted for use with a special type of magnetic or clip-on target. Magnetic targets are used on steel grids. Clip-on targets are used on aluminum or plastic. Once the laser is set up, the ceiling grid is quickly and easily levelled using the rotating beam as a reference and viewing the beam through the target (**Figure 27–30**). Sprinkler heads, ceiling outlets, and similar objects can be set in a similar manner.

Special Vertical Operations. Most laser levels are designed to work lying on their side using special brackets or feet. The unit is placed on the floor and levelled. To lay out a partition, rotate the head of the laser downward. Position the laser beam over one end of the partition. Align the beam toward the far point by turning the head toward and adjusting to the second point. Recheck the alignment and turn on to rotate the beam. The beam will be displayed as continuous straight and plumb lines on the floor, ceiling, and walls to both align and plumb the partition at the same time (**Figure 27–31**). Mounted on excavation equipment, lasers simplify and speed up construction work (**Figure 27–32**).

Layout Operations. Some laser levels emit a plumb line of light projecting upward from the top at a right angle (90 degrees) to the plane of the rotating beam. The plumb reference beam allows one person to lay out 90-degree cross-walls or building lines. Set the unit on its side as for laying out a partition.

The reference beam establishes a 90-degree corner with the rotating laser beam.

Plumbing Operations. The vertical beam provides a ready reference for plumbing posts, columns, elevator shafts, slip forming, and wherever a plumb reference beam is required.

Mount the laser unit on a tripod over a point of known offset from the work to be plumbed. Suspend

AVOID EXPOSURE

Laser Light Is Emitted from This Aperture

APERTURE LABEL

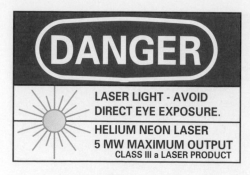

DANGER

LASER LIGHT - AVOID
DIRECT EYE EXPOSURE.

HELIUM NEON LASER
5 MW MAXIMUM OUTPUT
CLASS III a LASER PRODUCT

WARNING LABEL

FIGURE 27–33 Warning labels must be attached to every laser instrument.

a plumb bob from the centre of the tripod directly over the point. Apply power to the unit. The vertical beam that is projected is ready for use as a reference. Move the top of the object until it is offset from the beam the same distance as the bottom point.

The heads on some laser units can be tilted so that the rotating beam produces a plane of light at an angle to the horizontal. Such units are used for laying out slopes. Other units are manufactured for special purposes such as pipe-laying and tunnel guidance. Marine laser units, with ranges up to 10 miles (16 km), are used in port and pier construction and offshore work.

Laser Safety

With a little common sense, the laser can be used safely. All laser instruments are required to have warning labels attached (**Figure 27–33**). The following are safety precautions for laser use:

- Only trained persons should set up and operate laser instruments.

- Never stare directly into the laser beam or view it with optical instruments.

- When possible, set the laser up so that it is above or below eye level.

- Turn the laser off when not in use.

- Do not point the laser at others.

TOTAL STATIONS

Total stations are instruments that measure angles and distances in an instant. They are a combination of laser transit and measuring tape. Angles are measured internally and displayed in an easy-to-read digital form (**Figures 27–34** and **27–35**). Distances are determined by a system called electronic distance measuring (EDM). The instrument emits an electromagnetic pulse, laser, or infrared

Courtesy of Sokkia Canada

FIGURE 27–34 A total station.

FIGURE 27–35 A total station is a surveying tool designed to quickly and accurately measure distance and angles.

FIGURE 27–37 Prism is held plumb by using a bull's eye bubble.

FIGURE 27–38 Survey data points are measured by pressing a button.

FIGURE 27–36 A prism is used to reflect the light beam back to the total station.

beam. The beam is reflected back to the instrument. Some units use a **prism** as a target (**Figure 27–36**). The prism is held plumb over a desired point on the ground as the measurement is taken. Plumb is determined by a **bull's eye bubble** located on the prism rod (**Figure 27–37**).

An on-board computer then turns this information into distance and angles. This information makes a 3D measurement of the prism's location from the instrument. As more points are measured, data is collected (**Figure 27–38**). Computerized data collectors enable the measurements to be downloaded to the CAD file used to create a building.

The power and awe of the total station is in the blending of the survey and the structural drawing of the project. The entire site can be uploaded into the total station. It is then possible to locate the 3D position of any point of the jobsite. The information may also be converted to a topographical survey map.

Models vary in price and sophistication. Better models are able to measure distances of over 3000 feet (1 km) with an accuracy of ⅛ inch (2 to 3 mm). They can measure angles to within one

arc-second (1/3600 of a degree). This equates to a pencil-lead thickness at about 320 feet (100 m). They can cost $2,500 to $30,000 for the entire package.

Higher-end models are robotic in nature. Only one person is needed to collect survey data points. The instrument is set in a strategic place, and the operator then walks the prism around the site. When the operator is ready to take a reading, a button is pushed emitting an infrared signal to the instrument. It then rotates to find the signal, positioning the laser on the prism toward the prism. When ready, the operator pushes another button, and the data point distance and angles are recorded. As the operator moves to a new point, the instrument is then able to follow the prism as it moves and be instantly ready for a new reading.

It is important for the operator to read and follow the manufacturer's instructions. Because these tools are very expensive, using them is best done under the guidance of an experienced person. Small mistakes can be very costly. Understanding what to do with the data and using the available software require learning surveying methods and techniques as well as computer programs.

Chapter 28

Laying Out Foundation Lines

Before any layout can be made, the builder must determine the dimensions of the building and its location on the site from the plot plan. This task often falls to a professional surveyor, but in some areas may be done by any qualified person. Care should be taken to position the building properly to avoid costly changes. It is usually the carpenter's responsibility to lay out building lines.

STAKING THE BUILDING

Proper layout begins with locating the property corners, and then placing stakes at each corner of the building. Some lots are large enough that the building is measured from only one property line, while other lots are small enough to make it necessary to check all property lines with the building (**Figure 28–1**).

Begin by finding the survey rods that mark the corners of the property. Do not guess where the property lines are. Sometimes it is a good idea to stretch and secure lines between each corner, laying out all the property boundary lines.

Locate the front building line by measuring in from the front property line the specified front setback. Measure from both ends of the building line and drive a stake at each end. Stretch a line between

these stakes to better show the front edge of the building (**Figure 28–2**).

Along the front building line, measure in from the side property line the specified side setback (**Figure 28–3**). Drive a stake, Stake A, firmly into the ground. Place a nail in the top of the stake to aid in making precise measurements. From this nail,

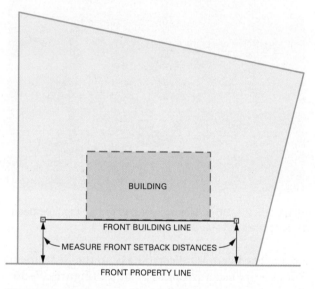

FIGURE 28–2 Locating the front building line.

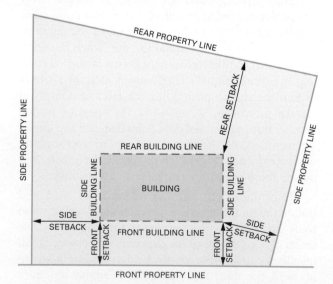

FIGURE 28–1 Locating the building on a lot from the dimensions found on the plot plan.

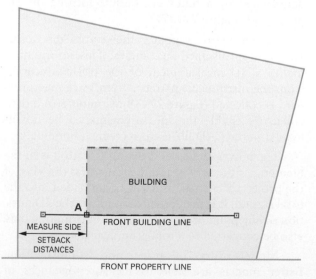

FIGURE 28–3 Measure from the side property line to locate the first building corner stake, Stake A.

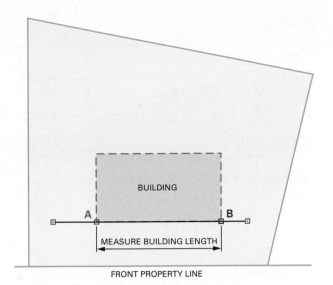

FIGURE 28–4 Measure along the building line to locate the second building corner stake, Stake B.

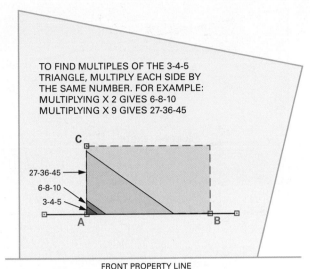

FIGURE 28–5 Use multiples of the 3-4-5 right triangle to place the third building corner stake, Stake C.

measure the front dimension of the building along the front building line. Drive Stake B directly under the front building line string. Drive a nail in the top of the stake marking the exact length of the building (**Figure 28–4**).

The third stake, Stake C, is placed to locate the back corner of the building. It must be square or at a right angle to the front building line. This stake on the rear building line may be located using one of at least three methods. First, you could use an optical instrument such as a transit-level. It is set up directly over Stake A and sighted to Stake B, aligning the crosshairs. Then the telescope is rotated exactly 90 degrees. Stake C is located by using a tape measure and the crosshairs at the same time. (See Procedure 27–D.) The second method is to use the 3-4-5 method. This process uses multiples of 3-4-5 to create larger right triangles. Multiplying each side of a 3-4-5 triangle by the same number creates a larger triangle that also has a right angle. For example, 3-4-5 multiplied by 9 gives a 27-36-45 right triangle (**Figure 28–5**).

The third method is faster and more accurate. Two tapes are used to measure the building width from Stake A and the **diagonal** of the building from Stake B at the same time. To determine the diagonal of the building, the **Pythagorean theorem** ($a^2 + b^2 = c^2$) is used. For example, consider a building whose length is $a = 40'$ and width is $b = 32'$. The diagonal equals c. Using the Pythagorean theorem, $c^2 = 40^2 + 32^2 = 2{,}624$. Taking the square root of the diagonal c gives us 51.2249939.

To convert to feet-inches to the nearest sixteenth, subtract 51 feet to leave the decimal. Convert the

decimal 0.2249939 to inches by multiplying by 12: $0.2249939' \times 12 = 2.6999268''$. Subtract 2 inches and write it down with 51 feet to make 51'-2". Convert 0.6999268" to a fraction by multiplying by 16, the desired denominator: $0.6999268'' \times 16 = 11.1988/16$th, which rounds off to $^{11}/_{16}''$. Thus, the diagonal of a $32' \times 40'$ rectangle is $51'$-$2^{11}/_{16}''$. Using two tapes, position Stake C, which is located where the two tapes cross in **Figure 28–6**.

PYTHAGOREAN THEOREM
$c^2 = a^2 + b^2$

FOR EXAMPLE, IF LENGTH = 40' AND WIDTH = 32', THEN DIAGONAL EQUALS c.

$c^2 = 40^2 + 32^2 = 2624$
$c = 51.2249939' = 51' + 0.2249939'$
$0.2249939' \times 12 = 2.6999268'' = 2'' + 0.6999268''$
$0.6999268'' \times 16 = 11.1988/16^{th}$ WHICH ROUNDS OFF TO $^{11}/_{16}''$
THUS THE DIAGONAL OF A 32' X 40' RECTANGLE EQUALS $51' - 2^{11}/_{16}''$.

FRONT PROPERTY LINE

THE DIAGONAL OF A 10 X 12 m RECTANGLE IS THE SQUARE ROOT OF ($10^2 + 12^2$), WHICH IS $\sqrt{244}$ OR 15.621 m OR 15621 mm.

FIGURE 28–6 Use the Pythagorean theorem to place the third building corner stake, Stake C. Two tapes are used to measure the width and diagonal at the same time.

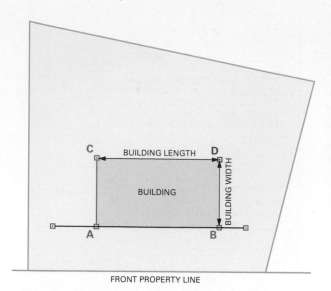

FIGURE 28–7 Measure from established front and rear corners (Stakes B and C) to locate the last building corner stake, Stake D.

When the locating of Stake C is completed, drive a nail in the top of the stake marking the rear corner exactly. Using two tapes, locate Stake D by measuring the building length from Stake C and the width from Stake B (**Figure 28–7**). Secure the stake and drive a nail in its top to mark exactly the other rear corner. Check the accuracy of the work by measuring widths, lengths, and diagonals. The diagonal measurements should be the same (**Figure 28–8**).

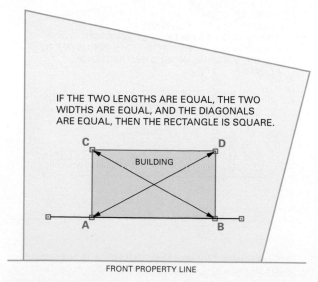

FIGURE 28–8 If the length and width measurements are accurate and the diagonal measurements are equal, then the corners are square.

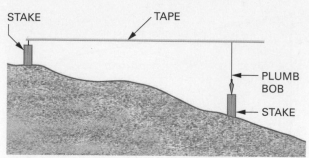

FIGURE 28–9 When layouts are done on sloping land, measurements must be taken on the level.

All measurements must be made on the level. If the land slopes, the tape is held level with a plumb bob suspended from it (**Figure 28–9**).

Layout of irregularly shaped buildings may seem complicated at first, but they are laid out using the same fundamental principles just outlined. The irregularly shaped building may be staked out from a large rectangle (**Procedure 28–A**). The large rectangle corners should align with as many of the building corners as possible. Stake the outermost corners for distance and square, then pull a string to make the large rectangle. The intermediate corner stakes are then located by measuring from the four corners along the strings. Then the final (inside-corner) stakes are measured and located using two tapes.

Placing stakes accurately often requires the builder to move them slightly while installing them. This can sometimes be a nuisance. To speed the process, large nails may be driven into the ground through small squares of thin cardboard (**Figure 28–10**). The cardboard serves to make the nail head more visible. This process allows for faster erecting of batter boards later.

Laying out the foundation lines is also done with a theodolite. The instrument is set up and levelled directly over the left front corner of the property line (point **P**). Knowing the side and front setback distances, the distance to **A** is calculated using the Pythagorean theorem, and the angle from the front property line to **A** is calculated using the trigonometric function called tangent (TAN) (**Figure 28–11**). Similarly, the angles and distances to **B**, **C**, and **D** are calculated relative to the front property line. The theodolite is sighted using the **P** to **R** lot line, and the horizontal angle is set to zero (0SET). A counter-clockwise angle and distance from **P** will establish all the points required. The four corners of the foundation are

StepbyStep Procedures

Procedure 28–A Establishing Multiple Corners for an Irregularly Shaped Building

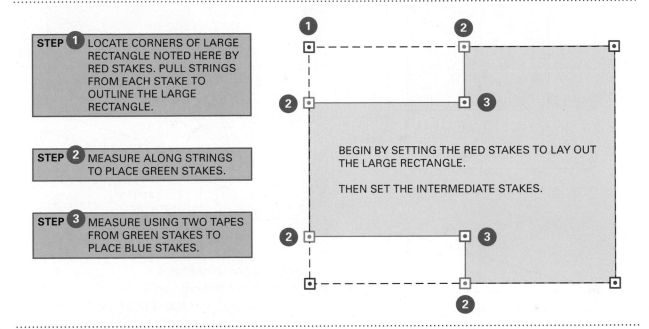

STEP 1 LOCATE CORNERS OF LARGE RECTANGLE NOTED HERE BY RED STAKES. PULL STRINGS FROM EACH STAKE TO OUTLINE THE LARGE RECTANGLE.

STEP 2 MEASURE ALONG STRINGS TO PLACE GREEN STAKES.

STEP 3 MEASURE USING TWO TAPES FROM GREEN STAKES TO PLACE BLUE STAKES.

BEGIN BY SETTING THE RED STAKES TO LAY OUT THE LARGE RECTANGLE.

THEN SET THE INTERMEDIATE STAKES.

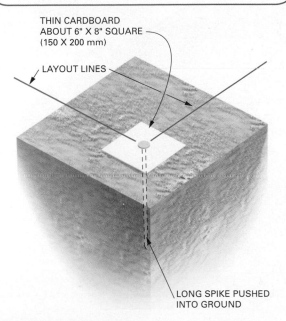

ON THE JOB

Use a long spike driven in the ground through a small square of cardboard to mark corners of a building layout.

THIN CARDBOARD ABOUT 6" X 8" SQUARE (150 X 200 mm)

LAYOUT LINES

LONG SPIKE PUSHED INTO GROUND

FIGURE 28–10 Technique for marking the ground location of stakes.

established by turning their angles and **chaining** (also called **steel-taping**) their distances from the instrument. With a total station, the coordinates of each point are uploaded from the CAD drawings to the data collector and then converted automatically to the angles and distances that are used to lay out the said points, thereby reducing the chances of error.

Erecting Batter Boards

Before excavating is done, batter boards are installed to allow the layout stakes to be reinstalled quickly after the excavation process. **Batter boards** are wood frames built behind the stakes to which building layout lines are secured. Batter boards consist of horizontal members, called **ledgers**. These are attached to stakes driven into the ground. The ledgers are fastened in a level position to the stakes, usually at the same height as the **foundation wall** (**Figure 28–12**). Batter boards are built in the same way for both residential and commercial construction.

Batter boards are erected in such a manner that they will not be disturbed during excavation. Drive batter board stakes into the ground a minimum of 4 feet (1.2 m) outside the building lines at each corner. When setting batter boards for large construction,

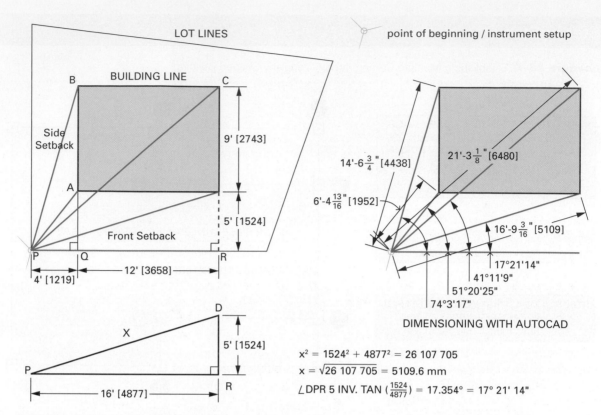

LOT LINES

point of beginning / instrument setup

BUILDING LINE

B   C

Side
Setback

9' [2743]

A

5' [1524]

Front Setback

P   Q   R

4' [1219]

12' [3658]

DIMENSIONING WITH AUTOCAD

14'-6$\frac{3}{4}$" [4438]   21'-3$\frac{1}{8}$" [6480]

6'-4$\frac{13}{16}$" [1952]

16'-9$\frac{3}{16}$" [5109]

17°21'14"

41°11'9"

51°20'25"

74°3'17"

D

X

5' [1524]

P

16' [4877]   R

$x^2 = 1524^2 + 4877^2 = 26\ 107\ 705$

$x = \sqrt{26\ 107\ 705} = 5109.6$ mm

$\angle$ DPR 5 INV. TAN $\left(\frac{1524}{4877}\right) = 17.354° = 17°\ 21'\ 14"$

FIGURE 28–11 Foundation layout with a theodolite.

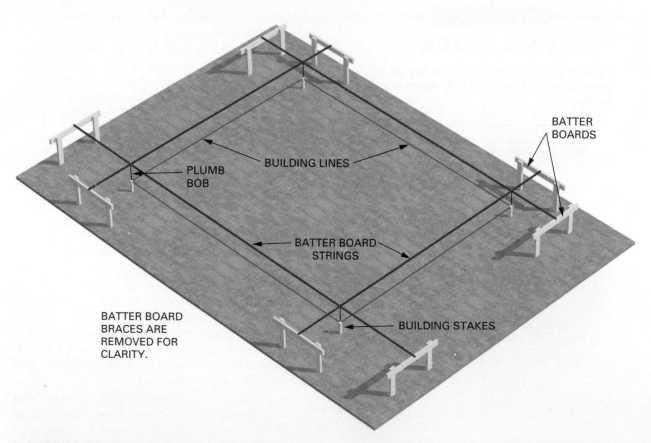

BATTER
BOARDS

BUILDING LINES

PLUMB
BOB

BATTER BOARD
STRINGS

BATTER BOARD
BRACES ARE
REMOVED FOR
CLARITY.

BUILDING STAKES

FIGURE 28–12 Batter boards are installed behind the building stakes and nearly level with the top of the foundation.

increase this distance. This will allow room for heavy excavating equipment to operate without disturbing the batter boards. In loose soil or when stakes are higher than 3 feet (1 m), they must be braced (**Figure 28–13**).

Set up the builder's level about centre on the building location. Sight to the benchmark and record the sighting. Determine the difference between the benchmark sighting and the height of the ledgers. Sight and mark each corner stake at the specified elevation. Attach ledgers to the stakes so that the top edge of each ledger is on the mark. Brace the batter boards for strength, if necessary.

Stretch lines between batter boards directly over the nail heads in the original corner stakes. Locate the position of the lines by suspending a plumb bob directly over the nail heads. When the lines are accurately located, make a saw cut on the outside corner of the top edge of the ledger. This prevents the layout lines from moving when stretched and secured. Be careful not to make the saw cut below the top edge (**Figure 28–14**). Saw cuts are also often made on batter boards to mark the location of the foundation footing. The footing

width usually extends outside and inside the foundation wall.

Check the accuracy of the layout by again measuring the diagonals to see if they are equal. If not, make the necessary adjustment until they are equal (**Figure 28–15**). Once the excavation is completed, the building stakes may be relocated. Reattach the batter board strings in the appropriate saw kerfs. Use a plumb bob to determine the building stake location (**Figure 28–16**).

For practice and training with the theodolite, a five-sided closed traverse is set out in **Figure 28–17**. This exercise will allow the apprentice to practise and improve his or her technique of instrument setup, which involves setting up over a specific point using the optical plummet, levelling the instrument, turning angles, and chaining (steel-taping) of distances.

In an open area, the theodolite is set up over a set point **A**. A distance of 9'-10" (2996.73 mm) is chained from **A** in a convenient direction and the point **B** is established. The theodolite is then moved to point **B**, backsighted to point **A**, set to horizontal zero (0SET), and then turned left (counter-clockwise) through 125° 2′ 56″ and

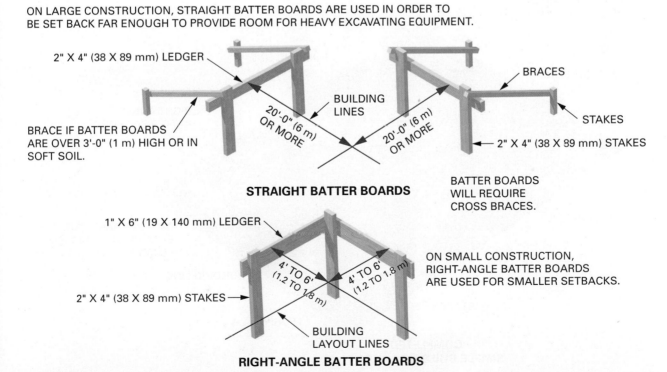

ON LARGE CONSTRUCTION, STRAIGHT BATTER BOARDS ARE USED IN ORDER TO BE SET BACK FAR ENOUGH TO PROVIDE ROOM FOR HEAVY EXCAVATING EQUIPMENT.

2" X 4" (38 X 89 mm) LEDGER

BRACES

BUILDING LINES

BRACE IF BATTER BOARDS ARE OVER 3'-0" (1 m) HIGH OR IN SOFT SOIL.

20'-0" (6 m) OR MORE

20'-0" (6 m) OR MORE

STAKES

2" X 4" (38 X 89 mm) STAKES

STRAIGHT BATTER BOARDS

BATTER BOARDS WILL REQUIRE CROSS BRACES.

1" X 6" (19 X 140 mm) LEDGER

4' TO 6' (1.2 TO 1.8 m)

4' TO 6' (1.2 TO 1.8 m)

2" X 4" (38 X 89 mm) STAKES

BUILDING LAYOUT LINES

RIGHT-ANGLE BATTER BOARDS

ON SMALL CONSTRUCTION, RIGHT-ANGLE BATTER BOARDS ARE USED FOR SMALLER SETBACKS.

FIGURE 28–13 Batter boards are placed back far enough so they will not be disturbed during excavation operations.

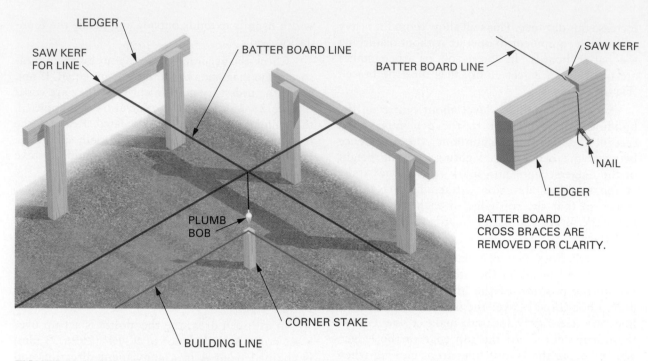

FIGURE 28–14 Saw kerfs are usually made in the ledger board to keep the strings from moving out of line.

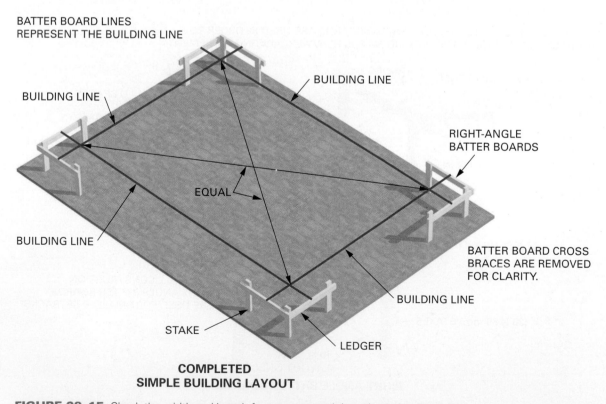

COMPLETED SIMPLE BUILDING LAYOUT

FIGURE 28–15 Check the width and length for accuracy and then check the diagonals.

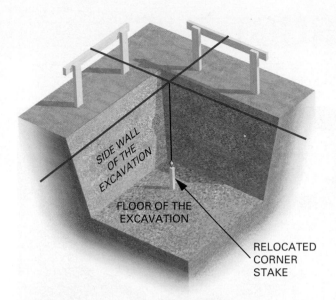

locked horizontally. The distance of 15'-15/16" (4595.26 mm) is chained from point **B**, sighted through the telescope (vertically), and the point **C** is set at the intersection of the line of sight and the steel tape. Continue around the perimeter in the same fashion in order to establish the points **D** and **E**. The final setup over the point **E** will measure the angle **AED**, and the distance from **E** to **A** is measured with the steel tape. The exact calculations for these measurements can also be done with AutoCAD and trigonometry (Figure 28–17). The results of the exercise can be matched against the calculated values and percentage accuracy can be calculated.

FIGURE 28–16 Building corner stakes are easily repositioned later in the excavation using the batter board strings and a plumb bob.

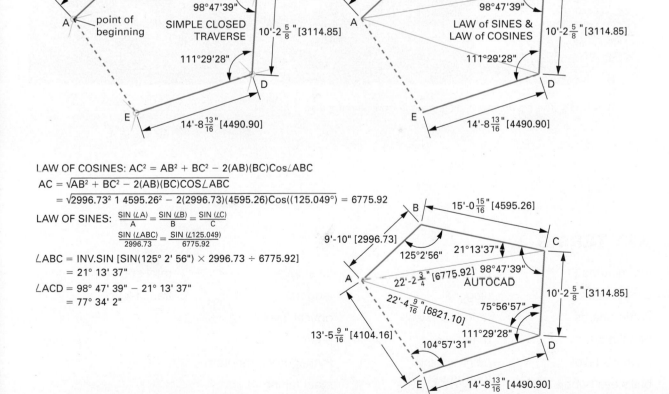

LAW OF COSINES: AC² = AB² + BC² − 2(AB)(BC)Cos∠ABC

AC = √AB² + BC² − 2(AB)(BC)COS∠ABC

    = √2996.73² 1 4595.26² − 2(2996.73)(4595.26)Cos((125.049°) = 6775.92

LAW OF SINES: SIN (∠A)/A = SIN (∠B)/B = SIN (∠C)/C

    SIN (∠ABC)/2996.73 = SIN (∠125.049)/6775.92

∠ABC = INV.SIN [SIN(125° 2' 56") × 2996.73 ÷ 6775.92]

    = 21° 13' 37"

∠ACD = 98° 47' 39" − 21° 13' 37"

    = 77° 34' 2"

FIGURE 28–17 Theodolite exercise to lay out a closed traverse and checking obtained distances and angles with AutoCAD and trigonometry.

DECONSTRUCT THIS

Carefully study **Figure 28–18** and think about what is wrong and/or what is right. Consider all possibilities. What construction practice or method is different in your area of the country?

FIGURE 28–18 This photo shows a construction jobsite dumpster.

KEY TERMS

arc minutes (')	closed traverse	laser level	total stations
arc seconds (")	diagonal	ledgers	transit-level
batter boards	foundation wall	optical levels	
benchmark	grade rod	prism	
builder's level	height of the instrument (HI)	Pythagorean theorem	
bull's eye bubble		steel taping	
chaining	laser	theodolite	

REVIEW QUESTIONS

Select the most appropriate answer.

1. How is the location of the building on the lot determined?

   a. from the foundation plan
   b. from the floor plan
   c. from the architect
   d. from the plot plan

2. What is the reference for establishing elevations on a construction site called?

   a. starting point
   b. reference point
   c. benchmark
   d. sight mark

3. What is the builder's level ordinarily used for?

   a. laying out straight lines
   b. reading elevations
   c. reading vertical angles
   d. determining datum

4. When levelling an optical level instrument with three levelling screws, in which direction are the opposite levelling screws turned?

   a. opposite directions, noting the motion of your thumbs
   b. same direction, noting the motion of your thumbs
   c. opposite directions, noting the motion of your forefinger
   d. same direction, noting the motion of your fingers

5. What is rotated to focus the crosshairs of an optical levelling instrument for individual users?

   a. focus knob on top of the telescope
   b. eyepiece ring
   c. tangent screw
   d. base plate

6. Which of the following is NOT used with an automatic level to check the level of a foundation wall?

   a. pocket tape backed up by a strip of wood
   b. grade rod
   c. piece of wood with graduations marked on it
   d. plumb bob

7. After a transit-level is moved and set up at a second location, toward what point is a backsight reading?

   a. previous point and is a plus measurement
   b. unknown point and is a plus measurement
   c. previous point and is a minus measurement
   d. unknown point and is a minus measurement

8. If the first reading of a rod at point A is 6' (1.8 m) and the second reading at point B is 2' (0.6 m), then which point is higher, and by how much?

   a. Point A is higher than point B by 4' (1.2 m).
   b. Point A is higher than point B by 8' (2.4 m).
   c. Point B is higher than point A by 4' (1.2 m).
   d. Point B is higher than point A by 8' (2.4 m).

9. What is the elevation difference between point A and point D where the backsight rod readings are 36", 48", and 59", and the foresight rod readings are 28", 42", and 60"?

   a. Point A is 13" higher than point D.
   b. Point A is 13" lower than point D.
   c. Point A is 23" higher than point D.
   d. Point A is 23" lower than point D.

10. What is the elevation difference between point A and point D where the backsight rod readings are 0.9 m, 1.2 m, and 1.5 m, and the foresight readings are 0.8 m, 1.1 m, and 1.6 m?

    a. Point A is 0.1 m higher than point D.
    b. Point A is 0.1 m lower than point D.
    c. Point A is 1.1 m higher than point D.
    d. Point A is 1.1 m lower than point D.

11. Which of the following do NOT use laser levelling tools?

    a. carpenters
    b. heavy equipment operators
    c. plumbers
    d. welders

12. What is the diagonal of a rectangle whose dimensions are 30 feet × 40 feet?

    a. 45 feet
    b. 50 feet
    c. 60 feet
    d. 70 feet

13. What is the diagonal of a rectangle whose dimensions are 32 feet × 48 feet?

    a. 57'-$^{11}/_{16}$"
    b. 57'-8¼"
    c. 57'-11"
    d. 60'-0"

14. What is the diagonal of a rectangle whose dimensions are 18 × 24 m?

    a. 27 m
    b. 28 m
    c. 30 m
    d. 34 m

15. What is the diagonal of a rectangle whose dimensions are 19.8 × 26.4 m?

    a. 27.8 m
    b. 29.8 m
    c. 32.6 m
    d. 33 m

Unit 11

Concrete Form Construction

Concrete formwork is usually the responsibility of the carpenter, whether constructing wood forms or erecting a forming system. The forms must meet specified dimensions and be strong enough to withstand tremendous pressure. The carpenter must know that the forms will be strong enough before the concrete is placed. If a form fails to hold during concrete placement, workers may be killed or injured, and costly labour and materials are lost. Formwork is unique in construction in that it is built for a specific purpose and then must be dismantled after the concrete cures. The carpenter must keep this in mind when designing the joints and seams in the formwork so that the material can be taken apart easily and salvaged.

OBJECTIVES

After completing this unit, the student should be able to:

- describe the composition of concrete and factors affecting its strength, durability, and workability.

- describe techniques used for the proper placement and consolidation of concrete.

- job-mix a batch of concrete and explain the method of and reasons for making a slump test.

- estimate required quantities of concrete.

- explain the reasons for reinforcing concrete and describe the materials used.

- detail the stages of site work in heavy concrete construction.

- describe the various types of foundations in heavy construction.

- describe various types of forming used in heavy construction.

- lay out and construct forms for footings, slabs, walks, and driveways.

- lay out and construct concrete forms for walls.

- lay out and construct concrete forms for pilasters and columns.

- lay out and construct concrete forms for beams.

- lay out and build concrete forms for stairs.

- lay out and construct a preserved wood foundation (PWF).

- distinguish between precast and cast-in-place concrete.

- distinguish between pre-tensioning and post-tensioning.

Chapter 29

Characteristics of Concrete

Understanding the characteristics of concrete is essential for the construction of reliable concrete forms, the correct handling of freshly mixed material, and the final quality of hardened concrete.

PRECAUTIONS ABOUT USING CONCRETE

Avoid prolonged contact with fresh concrete or wet cement because of possible skin irritation. Wear protective clothing when working with newly mixed concrete. Wash skin areas that have been exposed to wet concrete as soon as possible. If any cementitious material gets into the eyes, flush immediately with water and get medical attention.

CONCRETE

Concrete is an internationally used building material. It can be formed into practically any shape for construction of buildings, bridges, dams, and roads. Improvements over the years have created a product that is strong, durable, and versatile (**Figure 29–1**).

Composition of Concrete

Concrete is a mixture of **Portland cement (PC)**, fine and coarse *aggregates*, water, air, and various *admixtures*. Aggregates are fillers, usually sand, gravel, or stone (**Figure 29–2**). Admixtures are materials or chemicals added to the mix to achieve certain desired qualities.

When these concrete ingredients are mixed, the Portland cement and water form a paste. A chemical reaction, called **hydration**, begins within this paste.

The hydration process produces heat, which may be either a benefit or a problem during the curing process. This reaction causes the cement to set or harden, referred to as **curing**. As the paste hardens it binds the sand, which in turn binds the small and large aggregates. Together they form a strong, durable, and airtight mass called concrete.

Hydration (hardening) of the concrete is a relatively slow process. Initial setting, to the point of being able to walk on it, may take hours. It often takes a month or more for the concrete to achieve 90 percent of its full strength. Hydration continues for many years.

This process is easily disrupted, so it is important to provide and maintain the proper conditions on site. If rapid evaporation or freezing occurs during

FIGURE 29–1 Concrete is widely used in the construction industry.

FIGURE 29–2 Concrete is made up of varying sizes of aggregates with cement binding them together.

the curing of concrete, the cement hydration process will stop. If this happens, the concrete will never achieve full design strength.

Portland Cement. In 1824, Joseph Aspdin, an Englishman, developed an improved type of cement. He named it Portland cement because it produced a concrete that resembled stone found on the Isle of Portland, England. Portland cement is a fine grey powder.

Portland cement is usually sold in plastic-lined paper bags that hold one cubic foot (0.03 m³). Each bag weighs 88 pounds or 40 kilograms. The cement can still be used even after a long period of time, as long as it remains dry. Eventually, however, it will absorb moisture from the air. It should not be used if it contains lumps that cannot be broken up easily. This condition is known as pre-hydration.

Types of Portland Cement. In 2004, the Canadian Standards Association (CSA) published CSA A3001, "Cementitious Materials for Use in Concrete," which changed the nomenclature and designations for types of cement, as summarized in the table below. The National Building Code of Canada (NBCC) 2010, references CSA A3000-08 and CSA A23.1-09, includes Portland-limestone cements (PLC) that have up to 15 percent of limestone as opposed to the 5 percent used in Portland cement. The production of PLC reduces greenhouse gas emissions by 10 percent. PLC should not be used where the concrete will be exposed to sulphate. Conclusive tests have shown that concrete with PLC has the same strength and the same durability as concrete with PC. PLC has been used successfully in Europe for 25 years.

Canadian Types CSA A3001-3			Descriptions of the Types of Hydraulic Cement	Previous Type Designations		U.S. Type Designations	
Portland Cement				Portland Cement	Blended Cement	ASTM C150	ASTM C1157
PC	Blended	PLC					
GU	GUb	GUL	General Use	10	10E-x	I	Gu
MS	MSb	-	Moderate Sulphate-Resistant	20	20E-x	II	MS
MH	MHb	MHL	Moderate Heat-Resistant	20	20E-x	II	MH
HE	HEb	HEL	High Early-Strength	30	30E-x	III	HE
LH	LHb	LHL	Low Heat of Hydration	40	40E-x	IV	LH
HS	HSb	-	High Sulphate-Resistant	50	50E-x	V	HS

TABLE 29–1 Types of Portland Cement

Blended hydraulic cements include the following supplementary cementitious materials: blast furnace slag (S), silica fume (SF), natural pozzolans (N), and fly ash.

White Portland cement, used for decorative purposes, differs only in colour from grey cement. There are many other types of cement, each with its specific purpose, ranging from underwater use, to sealing oil wells, to air-entraining (Type IA).

Visit *http://www.cement.org/tech/index.asp* and *http://www.toolbase.org/index.aspx* for a more complete exposition of cement, concrete, admixtures, aggregates, and so on, as well as a virtual tour of a cement plant.

Water. Water is the part of the concrete mixture that starts the hydration process. It also allows the concrete mixture to flow easily into place. Water shall be clean and free of injurious amounts of oil, organic matter, sediment, or any other deleterious material [NBCC 9.3.1.5.(1)]. The chemical reaction of hydration can be negatively affected by water impurities. A good general rule is that if it is safe to drink, it is safe to use in concrete.

The amount of water used can affect the quality and strength of the concrete. The amount of water in relation to the amount of Portland cement is called the water-to-cement ratio. This ratio should be kept as small as possible while allowing for workability. Adding excessive water to a mix simply to increase the flow can weaken the concrete and must be avoided. Refer to CAN/CSA A23.1 "Concrete Materials and Methods of Concrete Construction" and NBCC 9.3.1.2–9.3.1.9 for site-batched concrete.

Aggregates. Aggregates have no cementing value of their own. They serve only as filler but are important ingredients. They constitute from 60 to 80 percent of the concrete volume. They must be clean, well-graded, and free of injurious amounts of organic and other deleterious material [NBCC 9.3.1.4.(1) (b)].

Fine aggregate consists of particles ¼ inch (6 mm) or less in diameter. Sand is the most commonly used fine aggregate. *Coarse* aggregate, usually gravel or crushed stone, usually comes in sizes of ½, ¾, 1, 1½, and 2 inches (12.5, 16, 19, 25, 38, and 50 mm). When strength is the only consideration, ¾ inch (19 mm) is the optimum size for most aggregates. Large-size aggregate uses less water and cement. Therefore, it is more economical. However, the maximum aggregate size must not exceed the following:

- 1/5 of the smallest dimension of the unreinforced concrete
- 1/3 of the depth of unreinforced slabs on the ground (**Figure 29–3**)

MAXIMUM AGGREGATE SIZE

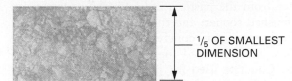

1/5 OF SMALLEST DIMENSION

SECTION THROUGH UNREINFORCED FOOTING

SLAB

1/3 OF DEPTH OF SLAB

SECTION THROUGH SLAB ON GROUND

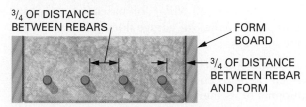

3/4 OF DISTANCE BETWEEN REBARS

FORM BOARD

3/4 OF DISTANCE BETWEEN REBAR AND FORM

SECTION THROUGH REINFORCED FOOTING

FIGURE 29–3 Aggregate size depends on the final dimensions of the concrete.

- ¾ of the clear spacing between reinforcing bars or between the bars and the form

Fine and coarse particles must be in a proportion that allows the finer particles to fill the spaces between the larger particles. The aggregate should be clean and free of dust, loam, clay, or organic matter.

Admixtures. Admixtures, chemicals that are added to the concrete mixture, perform six main functions: water reduction, set-retarding, early strength development, slump improvement or workability, air-entrainment, and corrosion inhibition. As well, they serve to add colour. These chemicals include detergents (air-entrainment), polymers (super-plasticizer), sugar (retarder), calcium chloride (accelerator), fly ash (workability and strength), and metal oxides (colour).

Air. An important advance was made with the development of **air-entrained concrete**. It is produced by using air-entraining Portland cement or an admixture. The intentionally made air bubbles are very small. Billions of them are contained in a cubic yard (cubic metre) of concrete.

The introduction of air into concrete was designed to increase its resistance to freezing and thawing. It also has other benefits: Less water and sand are required; the workability is improved;

shrinkage cracks are minimized; separation of the water from the paste is reduced; the concrete can be finished sooner; and the concrete is more water-tight than ordinary concrete. Air-entrained concrete is now recommended for almost all concrete projects. Concrete used for garage and carport floors and exterior steps shall have air entrainment of 5 to 8 percent [NBCC 9.3.1.6 (2)].

Concrete Strength. Concrete strength is measured in pounds per square inch (psi) or megapascals (MPa). Test cylinders are compressed in large machines to determine how much stress they will withstand before breaking. Typical concrete strengths are 2200 psi (15 MPa) for footings and foundation walls, 2900 psi (20 MPa) for basement slabs, and 4650 psi (32 MPa) for garage floors, driveways, and exterior steps.

The overall strength of the concrete is determined by many things. It is affected by the relative amounts of its ingredients. Water is required for the reaction to begin, but too much water weakens the mix. Also, increasing the amount of Portland cement in the mix increases its strength. Aggregate sizes can also affect the strength.

Strength of concrete begins with the mix, but it is also affected by how the concrete is placed and cured. Generally, a slow cure is best because the hydration process is slow.

Mixing Concrete

Concrete mixtures are designed to achieve the desired qualities of strength, durability, and workability in the most economical manner. The mix will vary according to the strength and other desired qualities, plus other factors such as the method and time of curing.

For very large jobs, such as bridges and dams, concrete mixing plants may be built on the jobsite. This is because freshly mixed concrete is needed on an around-the-clock basis. The construction engineer calculates the proportions of the mix and supervises tests of the concrete.

Ready-Mixed Concrete. Most concrete used in construction is *ready-mixed* concrete. It is sold by the cubic yard or cubic metre (1 m³ = 1.31 yds³). There are 27 cubic feet to the cubic yard. The purchaser usually specifies the amount and the desired strength of the concrete. The concrete supplier is then responsible for mixing the ingredients in the correct proportion to yield the desired strength.

Sometimes, the purchasers specify the proportion of the ingredients. They then assume responsibility for the design of the mixture.

The ingredients are accurately measured at the plant with computerized equipment. The mixture is delivered to the jobsite in *transit-mix trucks* (**Figure 29–4**). The truck contains a large revolving

FIGURE 29–4 Ready-mixed concrete is delivered in trucks.

drum, capable of holding from 1 to 10 cubic yards (0.76 to 7.6 m^3). There is a separate water tank with a water-measuring device. The drum rotates to mix the concrete as the truck is driven to the construction site.

Specifications require that each batch of concrete be delivered within 1½ hours after water has been added to the mix. If the jobsite is a short distance away, water is added to the cement and aggregates at the plant. If the jobsite is farther away, the proper amount of water is added to the dry mix from the water tank on the truck as it approaches or arrives at the job.

Job-Mixed Concrete. A small job may require that the concrete be mixed on the site either by hand or with a powered concrete mixer. All materials must be measured accurately. For measuring purposes, remember that a cubic foot of water weighs about 62½ pounds and contains approximately 7½ gallons. One litre (1L) or one cubic decimetre (1 dm^3, 1000 cm^3) of water weighs one kilogram (1 kg).

All ingredients should be thoroughly mixed according to the proportions shown in **Figure 29–5**. A little water should be put in the mixer before the dry materials are added. The water is then added uniformly while the ingredients are mixed from one to three minutes.

Concrete Reinforcement

Concrete has high **compressive strength**. This means that it resists being crushed. It has, however, low **tensile strength**. It is not as resistant to bending or pulling apart. Steel bars, called **rebars**, are used in concrete to increase its tensile strength. The concrete is then called **reinforced concrete**. Fibre-reinforced concrete (FRC) uses short thin fibres (usually glass [GFRC], steel, nylon, or spun-basalt fibres) that are added to the concrete during the final mixing stage to add tensile strength. The added strength depends on the type of fibre, the percent volume of the fibres, the aspect ratio of the fibres (length-to-diameter), and the orientation of the fibres in the mix. Steel fibres with an

aspect ratio of 100 are used in precast beams and columns, where they can be aligned using magnets. Other fibres are generally oriented randomly. Fibre-reinforcement is mainly used in *Shotcrete*, where the concrete is sprayed using compressed air. Also called *gunite*, this type of concrete has only lightweight aggregate and is useful in curved wall and thin wall applications, as well as for repairs.

Rebars. Steel rebars work well in concrete because steel expands and contracts at a similar rate as concrete, thus eliminating internal stresses when the ambient temperature changes. The steel used in rebar is usually *deformed*, which means that the surface has ridges that increase the bond between the concrete and the steel, which in turn allows stresses to be transmitted from one material to the other. Also, the alkali property of the concrete provided by calcium carbonate (lime) causes a film to form on the surface of the steel, making it more resistant to corrosion. Rebar comes in standard sizes, identified by numbers that indicate the diameter in eighths. For instance, a #6 rebar has a diameter of 6/8 or ¾ inch (**Figure 29–6**).

The size, location, and spacing of rebars are determined by engineers and shown on the plans. The rebars are positioned inside the form before the concrete is placed (**Figure 29–7**). Rebars in bridges and roads that experience a salty environment, such as

Bar #	Bar Ø (inches)	"Soft" Metric Sizes (mm)	Weight (lb./ft.)	Weight (kg/m)	Nominal Cross-Sectional Area (mm²)
2	¼	6	0.17	0.023	32
3	⅜	10	0.38	0.052	129
4	½	13	0.67	0.092	129
5	⅝	16	1.04	0.143	208
6	¾	19	1.50	0.207	284
7	⅞	22	2.04	0.282	387
8	1	25	2.67	0.368	509

Bar Size Can. Metric	Bar Ø (inches)	Bar Ø (mm)	Weight (lb./ft.)	Weight (kg/m)	Nominal Cross-Sectional Area (mm²)
10 M	.445	11.3	0.530	0.785	100
15 M	.630	16.0	1.05	1.570	200
20 M	.768	19.5	1.58	2.355	300
25 M	.992	25.2	2.64	3.925	500
30 M	1.177	29.9	3.69	5.495	700
35 M	1.406	35.7	5.27	7.850	1000

Possible Strength (MPa)	Water : Cement Ratio - Water (Litres - L)	Portland Cement (40 kg bags - L)	Fine Aggregates : Sand (L)	Coarse Aggregate (20 mm)(L)
32	0.45 - 12.6	1 - 28	1.75 - 49	2.5 - 70
20	0.65 - 18.2	1 - 28	1.75 - 49	2.5 - 70
15	0.70 - 19.6	1 - 28	1.75 - 49	2.5 - 70

FIGURE 29–5 Formulas for several concrete mixtures—see Table 9.3.1.7 (NBCC).

FIGURE 29–6 Numbers and sizes of commonly used reinforcing steel bars.

Photo by M. Nauth

FIGURE 29–7 Rebar is placed after the forms are created.

seawater or salting during the winter months, have an epoxy coating. This protects the rebar from rusting, and thus helps prevent premature concrete failure. The tasks of cutting, bending, placing, and tying rebars require workers who have been trained in that trade.

FRP rebar can be as strong as steel but it has the advantage of being corrosion-proof and thus extends the life of the structure. It requires less of a protective concrete covering, and so FRP-reinforced concrete can be lighter. FRP rebar is used in structures that need to allow radio waves to enter unaltered and in buildings with MRI (magnetic resonance imaging) machines. Because of its limited fire resistance, there is a need for protective cladding.

Wire Mesh. Welded wire mesh (or welded wire fabric, WWF) is used to reinforce concrete floor slabs resting on the ground, driveways, and walks. It is identified by the gauge or cross-sectional area and spacing of the wire. Common gauges are #6, #8, and #10. The wire is usually spaced to make 6-inch squares (152 × 152 mm) (**Figure 29–8**).

Welded wire mesh is laid in the slab above the vapour barrier, if used, before the concrete is placed. It is spliced by lapping one full square plus 2 inches (50 mm).

Placing Concrete

Prior to placing the concrete, sawdust, nails, and other debris should be removed from inside the forms. The inside surfaces are brushed or sprayed with oil to make form removal easier. No oil should be allowed to get on the steel reinforcement. This will reduce the steel/cement bond, thereby reducing the tensile strength provided by the steel. Form oil should also be kept off of the footings. Also, before concrete is placed, the forms and ground inside the forms are moistened with water. This is done to prevent rapid absorbing of water from the concrete.

Concrete is *placed*, not poured. Water should never be added so that concrete flows into forms without working it. Adding water alters the water-to-cement ratio on which the quality of the concrete depends.

Slump Test. A slump test is made by supervisors on the job to determine the consistency of the concrete. The concrete sample for a test should be taken just before the concrete is placed. The *slump cone* is first dampened. It is placed on a flat surface and filled to about one-third of its capacity with the fresh concrete. The concrete is then *rodded* by moving a #5 (16 mm) metal rod up and down 25 times over the entire surface. Two more approximately equal layers are added to the cone. Each layer is rodded in a similar manner, with the rod penetrating the layer below. The excess is screeded from the top. The cone is then turned over onto a flat surface and removed by carefully lifting it vertically in three to seven seconds. The cone is gently placed beside the concrete, and the amount of slump is measured between the top of the cone and the concrete (**Figure 29–9**). The test should be completed within two to three minutes.

New Mesh Size (ins)	Cross-Sectional Area M (.01 ins²)	Old Mesh Gauge	Metric Size	Metric Area M (mm²)	Diameter of wire (mm)
2 × 2	W4.0 × 4.0	4 × 4	50 × 50	MW25.8 × 25.8	5.73
4 × 4	W2.9 × 2.9	6 × 6	102 × 102	MW18.7 × 18.7	4.88
4 × 4	W1.4 × 1.4	10 × 10	102 × 102	MW9.1 × 9.1	3.40
6 × 6	W3.4 × 3.4	5 × 5	152 × 152	MW21.7 × 21.7	5.26
6 × 6	W5.4 × 5.4	2 × 2	152 × 152	MW34.9 × 34.9	6.67
6 × 6	W6.3 × 6.3	1 × 1	152 × 152	MW40.6 × 40.6	7.19
6 × 6	W7.4 × 7.4	0 × 0	152 × 152	MW47.6 × 47.6	15.57
12 × 12	W5.4 × 5.4	2 × 2	305 × 305	MW34.9 × 34.9	6.67

FIGURE 29–8 Size, gauge, and diameter of examples of welded wire mesh.

FIGURE 29–9 A slump test shows the wetness of a concrete mix.

Types of Construction	Slump in Inches (mm)	
	Maximum	Minimum
Reinforced foundation walls & footings	3 (75)	1 (25)
Plain footings, caissons & substructure walls	3 (75)	1 (25)
Beams & reinforced walls	4 (100)	1 (25)
Building columns	4 (100)	1 (25)
Pavement & slabs	3 (75)	1 (25)
Heavy mass concrete	2 (50)	1 (25)

FIGURE 29–10 Recommended slumps for different kinds of construction.

Changes in slump should be corrected immediately. The table shown in **Figure 29–10** shows recommended slumps for various types of construction. Concrete with a slump greater than 6 inches (152 mm) should not be used unless a slump-increasing admixture has been added. *Note:* The slump test has been replaced by verifying the ratio of water to cementing materials ratio as noted on-site by the mixed batch card provided by the concrete plant or by a third-party evaluation test.

Placing of Concrete in Forms. The concrete truck should get as close as possible. Concrete is placed by chutes where needed. It should not be pushed or dragged any more than necessary. It should not be dropped more than 4 to 6 feet (1.2 to 1.8 m), and should be dropped vertically and not angled. Drop chutes should be used in high forms to prevent the buildup of dry concrete on the side of the form or reinforcing bars above the level of the placement. Drop chutes also prevent separation caused by concrete striking and bouncing off the side of the form. Pumps are used on large jobs that need concrete continuously over long distances or to heights up to 500 feet (150 m).

Concrete must not be placed at a rapid rate, especially in high forms. The amount of pressure at any point on the form is determined by the height and weight of the concrete above it. Pressure is not affected by the thickness of the form (**Figure 29–11**).

A slow rate of placement allows the concrete nearer the bottom to begin to harden. Once concrete hardens it cannot exert more pressure on the forms, even though liquid concrete continues to be placed above it (**Figure 29–12**). The use of stiff concrete with a low slump, which acts less like a liquid, will transmit less pressure. Rapid placing leaves the concrete in the bottom still in a fluid state. It will exert great lateral pressure on the forms at the bottom. This may cause the form ties to fail or the form to deflect excessively.

Concrete should be placed in forms in layers of not more than 12 to 18 inches (305 to 457 mm) thick. Each layer should be placed before setting occurs in the previous layer. The layers should be thoroughly consolidated (**Figure 29–13**). The rate of concrete placement in high forms should be carefully controlled.

Consolidation of Concrete. To eliminate voids or honeycombs in the concrete, it should be thoroughly worked by hand-spading, or vibrated after it goes into the form. Vibrators make it possible to use a stiff mixture that would be difficult to consolidate by hand.

An immersion type vibrator, called a **spud vibrator**, has a metal tube on its end. This tube vibrates at a rapid rate. It is commonly used in construction to vibrate and consolidate concrete. Vibration makes the concrete more fluid and able to move, allowing trapped air to escape. This will prevent the formation of air pockets, honeycombs, and cold joints. The operator should be skilled in the use of the vibrator, keeping it moving vertically up and down, uniformly vibrating the entire pour. Over-vibrating should not be done, because vibrating increases the lateral pressure on the form (**Figure 29–14**). For tips and suggestions on

RATE (FT/HR)	PRESSURES OF VIBRATED CONCRETE (PSF)			
	50°F (10°C)		70°F (21°C)	
	COLUMNS	WALLS	COLUMNS	WALLS
1	330	330	280	280
2	510	510	410	410
3	690	690	540	540
4	870	870	660	660
5	1050	1050	790	790
6	1230	1230	920	920
7	1410	1410	1050	1050
8	1590	1470	1180	1090
9	1770	1520	1310	1130
10	1950	1580	1440	1170

PRESSURE INCREASES AS PLACEMENT RATE INCREASES.
PRESSURE INCREASES AT LOWER TEMPERATURES.

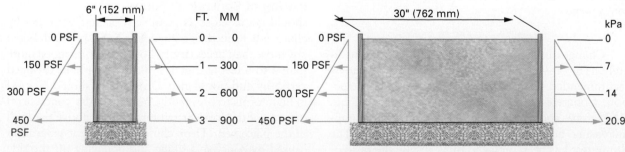

PSF = POUNDS PER SQUARE FOOT; kPa = KILOPASCALS (1 kPa = 20.9 PSF)

FIGURE 29–11 The height of concrete being poured affects the amount of pressure against the forms. Pressure is not affected by the thickness of the wall.

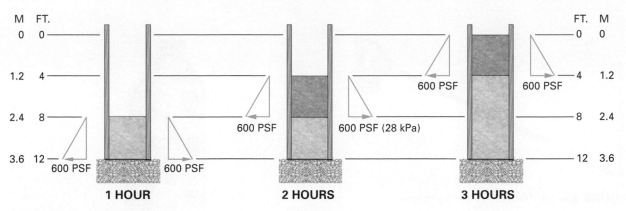

FIGURE 29–12 Once concrete sets, it does not exert more pressure on the form, even though liquid concrete continues to be placed above it.

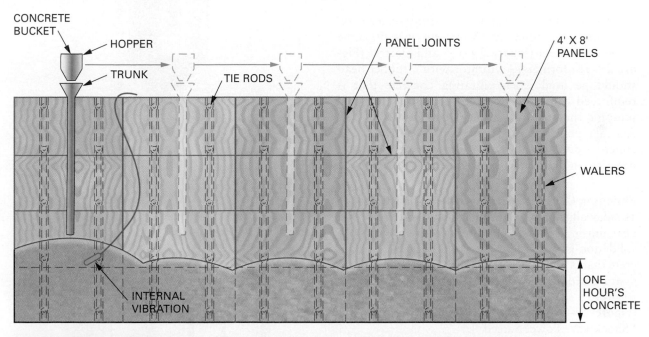

FIGURE 29–13 A safe, consistent pour rate is accomplished using drop chutes to place concrete in internally vibrated layers.

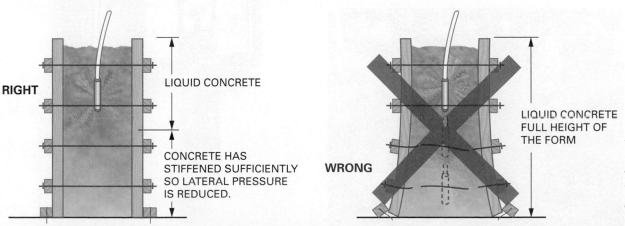

VIBRATE ONLY TO THE DEPTH OF THE FRESHLY PLACED CONCRETE. INSERTING THE VIBRATOR TOO FAR WILL CAUSE THE CONCRETE AT THE BOTTOM OF THE FORM TO REMAIN IN A LIQUID STATE LONGER THAN EXPECTED. THIS WILL RESULT IN HIGHER THAN EXPECTED LATERAL FORM PRESSURE AND MAY CAUSE THE FORM TO FAIL. THE DEPTH OF VIBRATION SHOULD JUST PENETRATE THE PREVIOUS LAYER OF CONCRETE BY A FEW INCHES (~50 mm).

FIGURE 29–14 Avoid excessive vibration of concrete.

FIGURE 29–15 Cordless concrete vibrator.

FIGURE 29–17 A rubber head protects epoxy-coated rebar from damage.

the proper use of vibrators, visit *http://oztec.com/ Tips_and_suggestions.htm.* Concrete vibrators come in both cordless (**Figure 29–15**) and electric (**Figure 29–16**) formats. A vibrator with a rubber head should be used when vibrating concrete that is reinforced with epoxy-coated rebar so as to avoid scraping the coating off the rebar (**Figure 29–17**). Another power tool that is used to consolidate a concrete pour is the **rebar vibrator**, which is very practical in tight quarters (**Figure 29–18**).

Finishing Concrete. Most concrete flatwork, such as sidewalks, driveways, and floor slabs, requires a neat finish. The first step after placement and consolidation is to screed, or strike off, the excess concrete that is above the forms. This is done by moving a straightedge back and forth across the top of the forms in a sawing action or by using a power screed (**Figure 29–19**) (go to *youtube.com* and search for "Shockwave Power Screed").

FIGURE 29–16 An electric concrete vibrator.

FIGURE 29–18 A rebar vibrator.

Courtesy of Crown Construction Equipment

FIGURE 29–19 A concrete power screed.

Bullfloating is next, and this is similar to a farmer drawing a harrow over a freshly ploughed field in an effort to break up the clods and level the ground. The bullfloat is a long-handled straight-edge that is pulled across the concrete in order to level and smooth the surface. Where space is too limited for a bullfloat, a handle-less tool called a Darby is used. The concrete is then floated with a wooden or metal float or with a circular power finisher. This will set small aggregate below the surface and bring the "cream" or mortar to the top. The final stage is steel-trowelling for a smooth finish, or "brooming" for a slip-resistant surface. Colouration and texturing impressions can be added as per manufacturer's directions.

Curing Concrete

Concrete hardens and gains strength because of a chemical reaction called hydration. All the desirable properties of concrete are improved, the longer this process takes place. A rapid loss of water from fresh concrete can stop the hydration process too soon, weakening the concrete. Curing prevents loss of moisture, allowing the process to continue so that the concrete can gain strength.

Concrete is *cured* either by keeping it moist or by preventing loss of its moisture for a period of time. For instance, if moist-cured for seven days, its strength is up to about 60 percent of full strength. A month later, the strength is 95 percent; it is up to full strength in about three months. Air-cured concrete will reach only about 55 percent after three months and will never attain design strength. In addition, a rapid loss of moisture causes the concrete to shrink, resulting in cracks. Curing should

be started as soon as the surface is hard enough to resist marring.

Methods of Curing. Flooding, misting, or constant sprinkling of the hardened surface with water is the most effective method of curing concrete. Curing can also be accomplished by keeping the forms in place; covering the concrete with burlap, straw, sand, or other material that retains water; and wetting it continuously.

In hot weather, the main concern is to prevent rapid evaporation of moisture. Sunshades or wind-breaks may need to be erected. The formwork may be allowed to stay in place or, the concrete surface may be covered with plastic film or other water-proof sheets. The edges of the sheets are overlapped and sealed with tape or covered with planks. Liquid curing chemicals may be sprayed or mopped on to seal in moisture and prevent evaporation. However, manufacturers' directions for their use should be carefully followed.

Curing Time. Concrete should be cured for as long as practical, which means that the moisture should be prevented from evaporating or freezing. The curing time depends on the temperature. At or above, curing should take place for at least three days. At or above 10°C (50°F), concrete is cured for at least five days. Near freezing, there is practically no hydration and concrete takes considerably longer to gain strength. There is no strength gain while concrete is frozen. When thawed, hydration resumes with appropriate curing. If concrete is frozen within the first 24 hours after being placed, permanent damage to the concrete is almost certain. In cold weather, *accelerators* that shorten the setting time are sometimes used. Protect concrete from freezing for at least four days after being placed by providing insulation or artificial heat, if necessary (**Figure 29–20**). When the air temperature is below 5°C (41°F), concrete should be kept at a temperature of not less than 10°C (50°F) or not more than 25°C (77°F) while being mixed and placed, and maintained at a temperature of not less than 10°C (50°F) for 72 hours after placing (Boyko, A., & Penna, S., *House Construction, Illustrated Code Series,* Ont. Building Code, 2012).

Forms may be removed after the concrete has set and hardened enough to maintain its shape. This time will vary depending on the mix, temperature, humidity, and other factors.

A

B

C

D

FIGURE 29–20 Methods of protecting concrete during curing: (A) misting; (B) burlap; (C) polyethylene sheets; (D) straw.

Chapter 30

Forms for Footings, Slabs, Walks, and Driveways

Concrete for footings, slabs, walks, and driveways is placed directly on the ground. The supporting soil must be suitable for the type of concrete work placed on it. It must be well drained. The weather affects the placement of concrete, so it is important to check the local environmental conditions.

FOOTING FORMS

The **footing** for a foundation provides a base on which to spread the load of a structure over a wider area of the soil. For foundation walls, the most typical type is a *continuous* or *spread* footing (**Figure 30–1**).

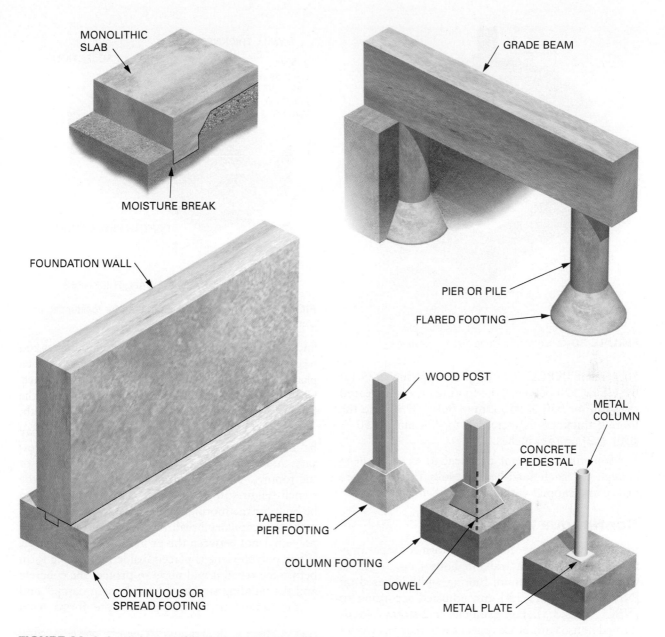

FIGURE 30–1 Several types of footings are constructed to support foundations.

The National Building Code of Canada (NBCC) requires that the soil be stable and have an allowable bearing pressure of 75 kPa for foundations of concrete, unit masonry, and wood frame, and of 100 kPa for foundations with flat insulating concrete forms with a maximum of 2 storeys of 3 metres maximum height each [NBCC 9.15.1.1]. Footings shall rest on undisturbed *soil*, *rock*, or compacted granular *fill* [NBCC 9.15.3.2].

To provide support for columns and posts, pier footings of square, rectangular, circular, or tapered shape are used. The base area of a column footing with columns spaced 3m (9'-10") or less for a 1, 2, and 3-storey house is 0.40 m² (4.31 ft²), 0.75 m² (8.07 ft²), and 1 m² (10.76 ft²), respectively [NBCC Table 9.15.3.4]. If the footings are square, then taking the square root of the area would give the pad sizes of 632 × 632 mm (25" × 25"), 866 × 866 mm (34" × 34") and 1 m × 1 m (39³/₈" × 39³/₈"). The thickness of the footing will depend on the base area of the post or plate. The post is centred and the thickness of the footing must be at least the dimension of the projection of the footing beyond the post; for example, a 150 mm (6") post on a 866 mm (34") footing would leave 358 mm (14") of footing on each side and would therefore require at least a 358-mm-thick (14") footing (**Figure 30–2**).

When column spacing exceeds 3 m (9'-10"), the area of the column footing is increased in proportion to the increase in the spacing; for example, 3.6 m (11'-9¾") is 20 percent greater than 3 m (9'-10") and therefore the footing area is increased by

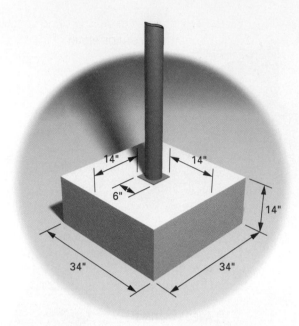

FIGURE 30–2 Column footing size calculations.

20 percent [NBCC 9.15.3.7]. Multiplying 0.75 m²
by 1.2 (or 120 percent), the footing area is increased
to 0.9 m² or 950 × 950 mm (37⅜″ × 37⅜″) and the
footing thickness is increased to 400 mm [(950 −
150) ÷ 2] or 15¾ inches.

Sometimes it is not practical to excavate deep
enough to reach load-bearing soil. Then **piles** are
driven and capped with a *grade beam*.

Continuous Wall Footings

In most cases, the footing is formed separately
from the foundation wall. In houses up to 9.8 m
wide (32′), the minimum footing width is based on
[NBCC Table 9.15.3.4] and adjusted according to
[NBCC 9.15.3.5]. For example, a 2-storey wood-
framed house with brick veneer on both levels would
require a base footing of 350 mm (13¾″) plus an
adjustment of 2 × 65 mm or 130 mm (5⅛″) for a
total of 480 mm (19″). The footing depth is often
equal to the wall thickness (**Figure 30–3**).

For larger buildings, architects or engineers
design the footings to carry the load. Usually these
footings are strengthened by reinforcing rods of specified size and spacing. For specific soil conditions other
than bedrock, a competent soils engineer must specify
the size of footings to suit the building. It is important
that the builder have the building site checked by an
engineer before going ahead with construction.

Frost Line. In areas where frost occurs, footings
must be located below the **frost line**. The frost line is
the point below the surface to which the ground typically freezes in winter. Because water expands when
frozen, foundations whose footings are above the frost

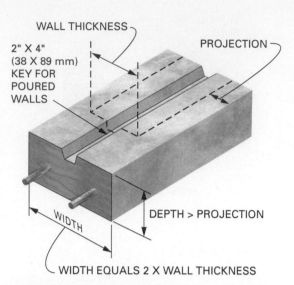

FIGURE 30–3 Typical spread footing for residential
construction.

line will heave and buckle when the ground freezes and
heaves. In extreme northern climates, footings must be
placed as much as 6 feet (1.8 m) below the surface.
Check with the local municipality to ensure that the
footings are placed below the frost-penetration depth.

In areas where the soil is stable, footings may
be *trench poured* where no formwork is necessary. A trench is dug to the width and depth of
the footing. The concrete is carefully placed in the
trench (**Figure 30–4**). In other cases, forms need to
be built for the footing. In all cases it is important to
provide a capillary break (such as 6 mil (0.15 mm)
polyethylene) between the ground and the concrete.
When polyethylene is placed inside the footing form
before the rebar, it will serve to protect the concrete
and the building structure from "rising damp," and
at the same time, it will protect the forms from

CAPILLARY BREAK NEEDS TO BE
INSTALLED BEFORE THE REBAR

FIGURE 30–4 In stable soils, the soil acts as the
footing form.

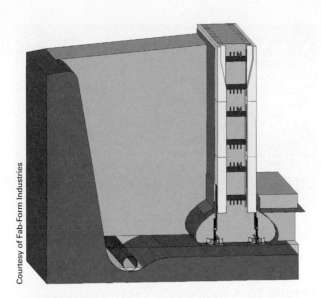

FIGURE 30–5 Fastfoot fabric footings.

Courtesy of Fab-Form Industries

getting covered with wet concrete. Go to *youtube .com* and search for "Grab-R Footing System" and think about placement of the polyethylene and the benefits that would be derived.

Fastfoot®, a fabric membrane, was invented in Canada, and it doubles as the form and as the capillary break, as well as using and wasting significantly less wood (**Figure 30–5**). For more information, go to *youtube.com* and search for "FastFoot Monopour Testimonial-Econ Group" and then go to *http:// www.fab-form.com/fastfoot/fastfootRisingDamp .php* for a clear explanation.

Locating Footings

To locate the footing forms, stretch lines on the batter boards in line with the outside of the footing. This is done by noting where the line should be on the batter board for the corner of the building. Then move the strings far enough away from the building to allow for the extra footing width (**Figure 30–6**). Suspend a plumb bob from the batter board lines at each corner. Drive stakes and attach lines to represent the outside surface of the footing.

Steel stakes, **spreaders**, and braces are manufactured for use in building footing forms and other edge formwork. They come with pre-punched nail holes for easy fastening (**Figure 30–7**).

Building Wall Footing Forms

Stakes are used to hold the sides in position. Fasten the sides by driving nails through the stakes.

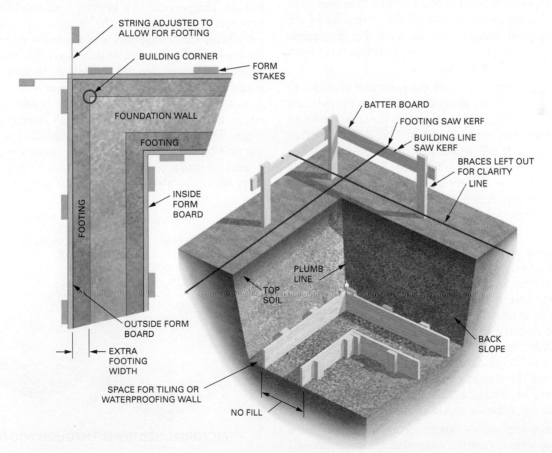

FIGURE 30–6 The footing is located by suspending a plumb bob from the batter board lines.

STEEL STAKES

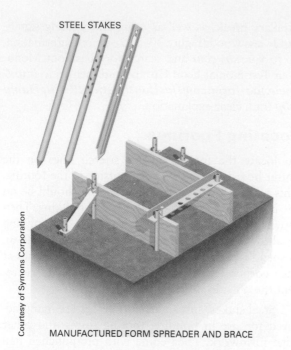

Courtesy of Symons Corporation

MANUFACTURED FORM SPREADER AND BRACE

FIGURE 30–7 Steel stakes, spreaders, and braces are manufactured for use in footing form construction.

Use duplex nails for easy removal. Various methods are used to build wall footing forms. One way is to erect the outside form around the perimeter of the building, and then return to assemble the inside form. This process will provide a strong, level form system (**Procedure 30–A**).

Set form stakes around the perimeter spaced 4 to 6 feet (1.2 to 1.8 m) apart, depending on the firmness of the soil. Use a gauge block that is the same thickness as the form board to set stakes. This will ensure that the string does not touch any previous stakes while the current one is being adjusted. Snap a chalk line on the stakes at slightly above the proper elevation. Fasten the form board to the stakes and tap the stakes down until the form is level. Erect the form boards for the outside of the footing in this manner, all around.

Before erecting the inside forms, cut a number of spreaders. Spreaders serve to tie the two sides together and keep them the correct distance apart. Nail one end to the top edges of the outside form. Erect the inside forms in a manner similar to that used in erecting the outside forms. Place stakes for the inside forms opposite those holding the outside form. Level across from the outside form to determine the height of the inside form. Fasten the spreaders as needed across the form at intervals to hold the form the correct distance apart.

Brace the stakes where necessary to hold the forms straight. In many cases, no bracing is necessary. Footing forms are sometimes braced by shovelling

CAPILLARY BREAK
TO BE ADDED

FIGURE 30–8 Forms must be braced as necessary.

earth or placing large stones against the outside of the forms. **Figure 30–8** shows a typical setup for braced footing forms.

Keyways. A keyway is formed in the footing by pressing 2 × 4 (38 × 89 mm) lumber into the fresh concrete (**Figure 30–9**). The keyway form is bevelled on both edges for easy removal after the concrete has set. The purpose of a keyway is to provide a lock between the footing and the foundation wall. This joint helps the foundation wall resist the pressure of the back-filled earth against it. It also helps retard the seepage of water into the basement. In some cases, where the design of the keyway is not so important, 2 × 4 (38 × 89 mm) pieces are not bevelled on the edges, but are pressed into the fresh concrete at an angle. This cold joint is also accomplished with the use of 49 feet (15 m) vertical rebar dowels spaced at 4 feet (1.2 m) on centre [NBCC 9.15.4.4.1(c)]. A capillary break can also be introduced at this juncture.

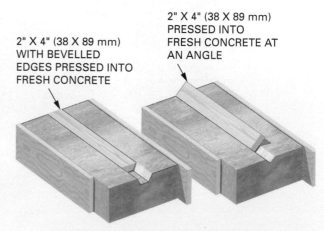

2" X 4" (38 X 89 mm) WITH BEVELLED EDGES PRESSED INTO FRESH CONCRETE

2" X 4" (38 X 89 mm) PRESSED INTO FRESH CONCRETE AT AN ANGLE

PICTORIAL SECTIONS THROUGH FOOTING

FIGURE 30–9 Methods of forming keyways in the footing.

StepbyStep Procedures

Procedure 30–A Technique for Setting Footing Forms After Excavation

STEP 1 MOVE BATTER BOARD STRINGS TOWARD THE OUTSIDE TO ALLOW FOR EXTRA WIDTH OF FOOTING.

STEP 2 SUSPEND PLUMB BOB TO LOCATE OUTSIDE EDGES OF FOOTING. DRIVE TWO STAKES AT EACH CORNER AND STRETCH LINES FROM CORNER TO CORNER.

STEP 3 DRIVE INTERMEDIATE STAKES OUTSIDE OF THE STRETCH LINE USING A GAUGE BLOCK OF EQUAL THICKNESS TO THE FORM BOARD.

STEP 4 SNAP A LINE ON THE INSIDE FACE OF THE STAKES AT THE HEIGHT OF THE TOP OF THE FOOTING.

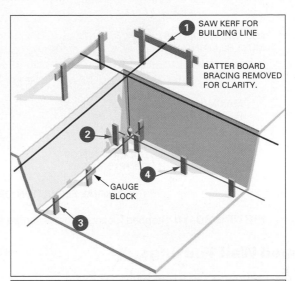

SAW KERF FOR BUILDING LINE

BATTER BOARD BRACING REMOVED FOR CLARITY.

GAUGE BLOCK

STEP 5 NAIL THE FORM BOARD TO THE STAKE SLIGHTLY HIGHER THAN THE CHALK LINE.

STEP 6 LEVEL THE FORM BOARDS WITH A LASER OR OPTICAL TRANSIT BY TAPPING EACH STAKE DOWN AS NEEDED.

STEP 7 CUT AND ATTACH SPREADERS TO OUTSIDE FORM BOARD.

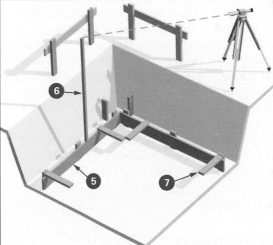

STEP 8 ATTACH THE INSIDE FORM BOARD TO THE SPREADERS AS INSIDE STAKES ARE PLACED.

STEP 9 USE A LEVEL TO POSITION THE HEIGHT OF THE INSIDE FORM.

STEP 10 BRACE AS NEEDED WITH STAKES DRIVEN AT AN ANGLE AND NAIL TO THE FORMS.

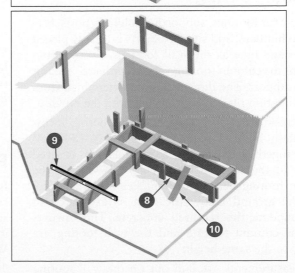

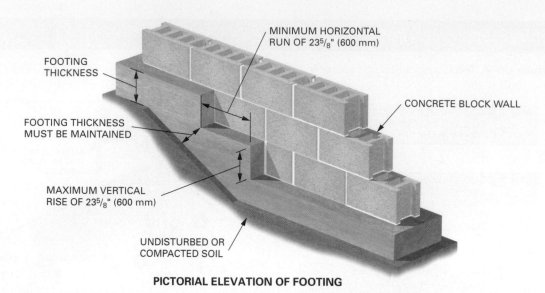

FOOTING THICKNESS

FOOTING THICKNESS MUST BE MAINTAINED

MAXIMUM VERTICAL RISE OF 23⅝" (600 mm)

MINIMUM HORIZONTAL RUN OF 23⅝" (600 mm)

CONCRETE BLOCK WALL

UNDISTURBED OR COMPACTED SOIL

PICTORIAL ELEVATION OF FOOTING

FIGURE 30–10 Stepped footings must be properly dimensioned.

Stepped Wall Footings

When the foundation is to be built on sloped land, it is sometimes necessary to step the footing. The footing is formed at different levels, to save material. In building stepped footing forms, the thickness of the footing must be maintained. The vertical and horizontal footing distances are adjusted so that a whole number of blocks or concrete forms can easily be placed into that section of the footing without cutting. The vertical part of each step should not exceed $23\frac{5}{8}$ inches (600 mm), and the horizontal part of the step must be a minimum of $23\frac{5}{8}$ inches (600 mm) [NBCC 9.15.3.9.(1)(b)] (**Figure 30–10**). Vertical boards are placed between the forms to retain the concrete at each step.

Column Footings

Concrete for footings, supporting columns, posts, fireplaces, chimneys, and similar objects is usually placed at the same time as the wall footings. The size and shape of the column footing vary according to what it has to support. The dimensions are determined from the foundation plan in concurrence with the NBCC.

In residential construction, these footing forms are usually built by nailing 2 × 8 (38 × 184 mm) pieces together in square, rectangular, or tapered shapes to the specified size (**Figure 30–11**). If the column footing needs to be thicker than the form, then the ground below the form is excavated to accommodate the required concrete. This ensures that the column footing and the wall footing are poured to the same height.

Measurements are laid out on the wall footing forms to locate the column footings. Lines are

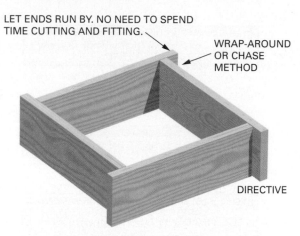

LET ENDS RUN BY. NO NEED TO SPEND TIME CUTTING AND FITTING.

WRAP-AROUND OR CHASE METHOD

DIRECTIVE

A RECTANGULAR FORM FOR A COLUMN FOOTING

FIGURE 30–11 Construction of column footings.

stretched from opposite sides of the wall footing forms to locate the position of the forms. They are laid in position corresponding to the stretched lines (**Figure 30–12**). Stakes are driven. Forms are usually fastened in a position so that the top edges are level with the wall footing forms.

FORMS FOR SLABS

Building forms for slabs, walks, and driveways is similar to building continuous footing forms. The sides of the form are held in place by stakes driven into the ground. Forms for floor slabs are built level. Walks and driveways are formed to shed water. Usually 2 × 4 (38 × 89 mm) or 2 × 6 (38 × 140 mm) lumber is used for the sides of the form.

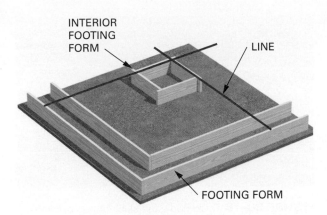

INTERIOR
FOOTING
FORM

LINE

FOOTING FORM

FIGURE 30–12 Locating interior footings using a string.

Slab-on-Grade

In warm climates, where frost penetration into the ground is not very deep, little excavation is necessary. The first floor may be a concrete slab placed directly on the ground. This is commonly called a **slab-on-grade foundation** (**Figure 30–13**). With improvements in the methods of construction, the need for lower construction costs, and the desire to give the structure a lower profile, slabs-on-grade are being used more often in all climates.

Basic Requirements

The construction of concrete floor slabs should meet certain basic requirements:

1. The finished floor level must be high enough so that the finish grade around the slab can be sloped away for good drainage. The top of the slab should be no less than 6 inches (152 mm) above the finish grade [NBCC 9.15.4.6(1)].

2. All topsoil in the area in which the slab is to be placed must be removed. A base for the slab consisting of 4 to 6 inches (100 to 152 mm) of gravel, crushed stone, or other approved material must be well-compacted in place.

3. The soil under the slab may be treated with chemicals for control of termites, but caution is advised. Such treatment should be done only by those thoroughly trained in the use of these chemicals.

4. All mechanicals (water and sewer lines, heating ducts, and other utilities) that are to run under the slab must be installed.

5. A moisture barrier must be placed under the concrete slab to prevent soil moisture from rising through the slab. The moisture barrier should be a heavy plastic film, such as 6 mil (0.15 mm) polyethylene, preferably black because it would hardly deteriorate over time, or other material having equal or superior resistance to the passage of moisture. It should be strong enough to resist puncturing during the placing of the concrete. Joints in the moisture barrier must be lapped at least 4 inches (100 mm) and sealed. A layer of sand may be applied to protect the membrane during concrete placement.

6. Where necessary to prevent heat loss through the floor and foundation walls, waterproof, rigid insulation (extruded polystyrene, or XPS) is installed around the perimeter of the slab.

7. The slab should be reinforced with 6×6 inch (152×152 mm), W1.4 $\times$ 1.4 (MW9.1 $\times$ 9.1) welded wire mesh, or by other means to provide

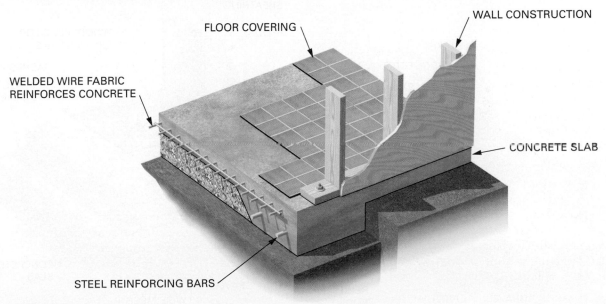

FLOOR COVERING

WALL CONSTRUCTION

WELDED WIRE FABRIC
REINFORCES CONCRETE

CONCRETE SLAB

STEEL REINFORCING BARS

FIGURE 30–13 Slab-on-grade foundation.

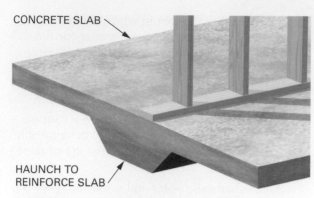

CONCRETE SLAB

HAUNCH TO REINFORCE SLAB

FIGURE 30–14 The slab is haunched under load-bearing walls.

FIGURE 30–16 Thickened edge slabs of two buildings.

equal or superior reinforcing. The concrete slab must be at least 4 inches (100 mm) thick and *haunched* (made thicker) under load-bearing walls (**Figure 30–14**).

Monolithic Slabs

A combined slab and foundation is called a **monolithic slab** (**Figure 30–15**). This type of slab is also referred to as a *thickened edge slab*. It consists of a shallow footing around the perimeter. The perimeter is placed at the same time as the slab. The slab and footing make up a one-piece integral unit. The bottom of the footing must be placed at least 12 inches (305 mm) below the finish grade, or according to the approved construction drawings.

Forms for monolithic slabs are constructed using stakes and edge form boards, plank, or steel manufactured especially for forming slabs and similar objects. The construction procedure is similar

to that for wall footing forms (see Procedure 30–A, page 311) except that the inside form board is omitted (**Figure 30–16**).

- From the batter board lines, plumb down to locate the building corner stakes. Stretch lines from these corner stakes and mark the soil to outline the perimeter of the building. Also mark where mechanical trenches must be dug. Excavate the trench as needed.

- Re-establish the batter board lines and locate corner stakes. Stretch lines on these corner stakes at the desired elevation.

Drive intermediate stakes using a gauge block to allow for the form thickness. Snap lines and fasten form boards to the stakes. Level the form with a transit-level or builder's level. Brace the form as required, remembering that most of the side pressure will be from the concrete. Install reinforcement and mechanicals as required.

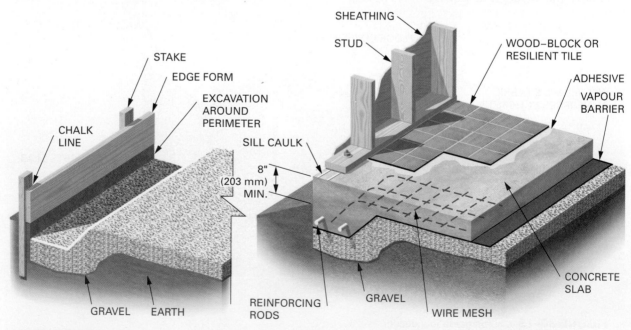

STAKE
EDGE FORM
EXCAVATION AROUND PERIMETER
CHALK LINE
SILL CAULK
8" (203 mm) MIN.
GRAVEL   EARTH
REINFORCING RODS

SHEATHING
STUD
WOOD–BLOCK OR RESILIENT TILE
ADHESIVE
VAPOUR BARRIER
GRAVEL   WIRE MESH
CONCRETE SLAB

FIGURE 30–15 Typical monolithic slab-on-grade construction.

Independent Slabs

In areas where the ground freezes to any appreciable depth during winter, the footing for the walls of the structure must extend below the frost line. If slab-on-grade construction is desired in these areas, the concrete slab and foundation wall may be separate. This type of slab-on-grade is called an **independent slab**. It may be constructed in a number of ways, according to conditions (**Figure 30–17**).

If the frost line is not too deep, the footing and wall may be an integral unit and placed at the same time. In colder climates, the foundation wall and footing are formed and placed separately. The wall may be set on piles or on a continuous footing, the forming of which has been described previously.

Slab Insulation

In colder climates, insulation is required under the slab. The thickness, amount, and location are governed by local energy codes. The only insulation board suited for underground applications is extruded polystyrene. It is produced by various manufacturers in pink, green, blue, and grey colours. It has closed foam cells that keep water from getting into the insulation.

Thin strips of extruded polystyrene may be used between slabs. These strips serve as isolation joints that allow the slabs to expand and contract with temperature and not interfere with each other.

FORMS FOR WALKS AND DRIVEWAYS

Forms for walks and driveways are usually built so that water will drain from the surface of the concrete. In these cases, grade stakes must be established and grade lines carefully followed (**Figure 30–18**).

Establish the grade of the walk or driveway on stakes at both ends. Stretch lines tightly between the end stakes. Drive intermediate stakes. Fasten the edge pieces to the stakes, following the line in a manner similar to making continuous footing forms.

Forms for Curved Walks and Driveways

In many instances, walks and driveways are curved. Special metal forms can be purchased to easily form curves, or wood forms can be constructed from ¼-inch (6 mm) plywood or hardboard. They may be used for small-radius curves. If using plywood, install it with the grain vertical for easier bending without breaking. Wetting the stock sometimes helps the bending process.

For curves of a long radius, 1 × 4 (19 × 89 mm) lumber can be used and satisfactorily bent if the curve is not too tight. Lumber can also be curved by making saw kerfs spaced closely together. The lumber is then bent until the saw kerfs close (**Figure 30–19**).

The spacing of the kerfs affects the radius of the curve. Closer kerfs yield a tighter bend. The number of kerfs affects the length of the arc—more kerfs equal more curve length.

To determine the kerf spacing, measure the radius distance from the end of the form (**Procedure 30–B**). Bend the form to close the kerf. The height of the board when the end is raised is the spacing of the kerfs.

The length of the form that has kerfs cut into it can be determined from the circumference of the curve.

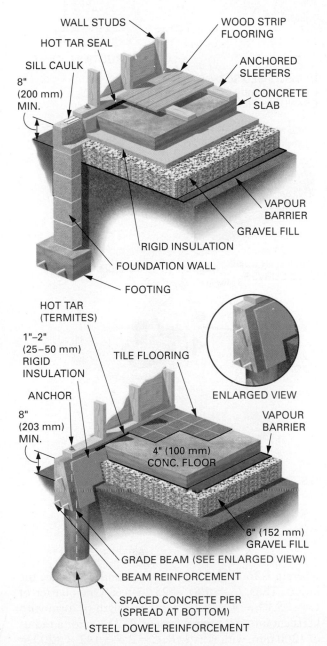

WALL STUDS
HOT TAR SEAL
SILL CAULK
8" (200 mm) MIN.
WOOD STRIP FLOORING
ANCHORED SLEEPERS
CONCRETE SLAB
VAPOUR BARRIER
GRAVEL FILL
RIGID INSULATION
FOUNDATION WALL
FOOTING

HOT TAR (TERMITES)
1"–2" (25–50 mm) RIGID INSULATION
ANCHOR
8" (203 mm) MIN.
TILE FLOORING
4" (100 mm) CONC. FLOOR
ENLARGED VIEW
VAPOUR BARRIER
6" (152 mm) GRAVEL FILL
GRADE BEAM (SEE ENLARGED VIEW)
BEAM REINFORCEMENT
SPACED CONCRETE PIER (SPREAD AT BOTTOM)
STEEL DOWEL REINFORCEMENT

FIGURE 30–17 Independent slabs are constructed in a number of ways.

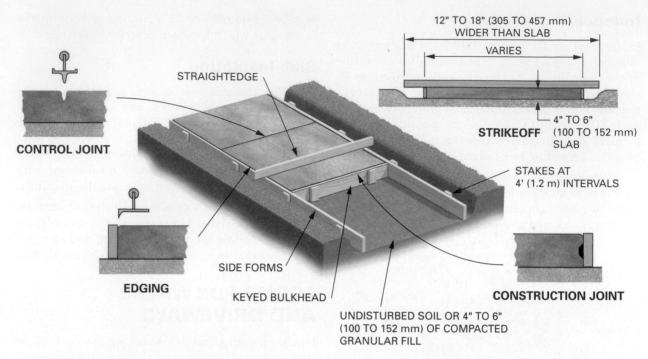

FIGURE 30–18 Typical way of forming of slabs for walkways.

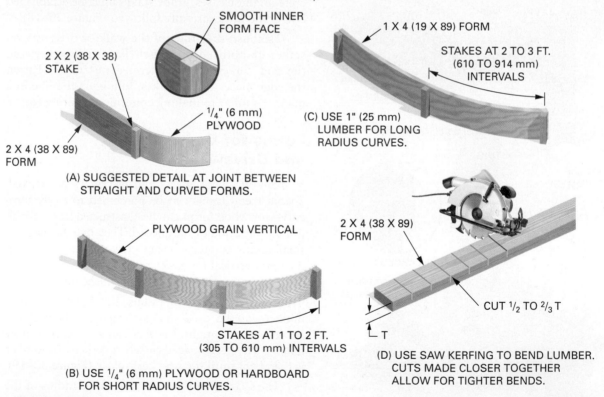

FIGURE 30–19 Forming curved edge slabs.

First calculate the circumference of the whole circle that the curve would make if it were stretched out to a full circle. Then determine what portion of a whole circle the curve occupies; that is, one-quarter, one-eighth, etc. For example, what is the length of the form if it is a one-quarter bend with a radius of 4 feet (1200 mm)? The formula to determine circumference is $C = \pi d$, where π is found with a calculator and $d =$ twice the radius. Thus, $C = \pi(8) = 25.13$ feet. One-quarter of $C = 25.13 \div 4 = 6.28$ feet. The length of form with kerfs cut into it is approximately 6'-3". Or, for a radius of 1200 mm, with $C = 2\pi R$, $C = 2 \times 3.142 \times 1200 = 7540$ mm. One-quarter of that is 1885 mm, and the form with kerfs is approximately 2 m long.

StepbyStep Procedures

Procedure 30–B Making Radius Forms

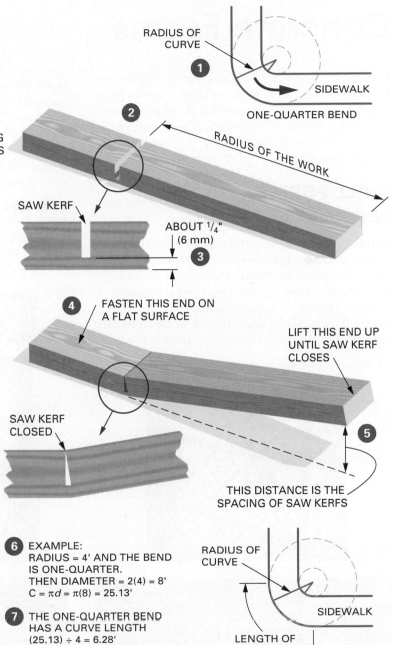

STEP 1 DETERMINE THE RADIUS OF THE CURVE AND THE AMOUNT OF CURVE IN THE BEND, THAT IS, 4" (100 mm) RADIUS AND ONE-QUARTER BEND.

NOTE: SELECT GOOD-QUALITY OR GRADE OF LUMBER FOR BENDING FORMS. KNOTS AND OTHER DEFECTS MIGHT CAUSE THE BEND TO BREAK.

STEP 2 FROM ONE END OF THE FORM, MEASURE A DISTANCE EQUAL TO THE RADIUS.

STEP 3 MAKE A SAW CUT TO WITHIN 1/4" (6 mm) OF THE BOTTOM OF THE FORM.

STEP 4 FASTEN OR HOLD THE STOCK AND BEND THE END UP TO CLOSE THE KERF.

STEP 5 MEASURE THE DISTANCE THE END HAS RISEN. THIS IS THE KERF SPACING.

STEP 6 CALCULATE THE CIRCUMFERENCE OF A WHOLE CIRCLE WITH THE DESIRED RADIUS.

STEP 7 DIVIDE THE WHOLE BY THE PORTION OF THE CURVE DESIRED.

1 RADIUS OF CURVE
SIDEWALK
ONE-QUARTER BEND
RADIUS OF THE WORK

2

SAW KERF
3 ABOUT 1/4" (6 mm)

4 FASTEN THIS END ON A FLAT SURFACE
LIFT THIS END UP UNTIL SAW KERF CLOSES

SAW KERF CLOSED

5 THIS DISTANCE IS THE SPACING OF SAW KERFS

6 EXAMPLE:
RADIUS = 4' AND THE BEND IS ONE-QUARTER.
THEN DIAMETER = 2(4) = 8'
$C = \pi d = \pi(8) = 25.13'$

7 THE ONE-QUARTER BEND HAS A CURVE LENGTH
$(25.13) \div 4 = 6.28'$
APPROXIMATELY 6'-3"

RADIUS OF CURVE
SIDEWALK
LENGTH OF CURVE

Chapter 31

Wall, Column, and Insulated Concrete Forms

Foundation walls and columns are usually formed by using **panels** rather than building forms in place, piece by piece. Panel construction simplifies the erection and stripping of formwork. Panels also reduce the cost of forming by allowing the material to be re-used several times.

WALL FORMS

Wall Form Components

Various kinds of panels and panel systems are used. Some concrete panel systems are manufactured of steel, aluminum, or wood. They are designed to be used many times. Specially designed hardware is used for joining, spacing, aligning, and bracing the panels (**Figure 31–1**). Care should be taken to keep the form system clean after the concrete is placed.

Form Panels. Panels are placed side by side to form the inside and outside of the foundation walls. Panel sizes vary with the manufacturer. Standard sizes for manufactured panels are 2 × 8 feet (0.6 × 2.4 m). Wood panels are often 4 × 8 feet (1.2 × 2.4 m). Narrower panels of several widths are available to be used as fillers when the space is too narrow for a standard size panel.

Snap Ties. Snap ties hold the wall forms together at the desired distance apart. They support both sides against the lateral pressure of the concrete (see formwork in Figure 31–15). They allow the side pressure

Courtesy of Duraform, Ltd.

FIGURE 31–1 Styles of formwork for concrete walls.

on the outer panel to be carried or cancelled out by the pressure on the inner panel. These ties reduce the need for external bracing and greatly simplify the erection of wall forms. The design of a form for a particular job, including the spacing of the ties and studs, is decided by a structural engineer.

These ties are called snap ties. After removal of the form, the projecting ends are snapped off slightly inside the concrete surface. A special snap tie wrench is used to break back the ties (**Procedure 31–A**). The small remaining holes are easily filled.

The Strip Easy Ties use easy-to-handle $^{11}/_{16}$-inch (18 mm) plywood forms (**Figure 31–2**). When forming residential or light commercial, Strip Easy Ties enable you to set up and strip quickly. This system requires that slots be cut in the form ply at the specified intervals, and then the ties are inserted and flat steel bars with rounded edges hold the ties in place. The tabs in the ties keep the form ply the set distance apart.

Because of the great variation in the size and shape of concrete forms, a large number of snap tie styles are used (**Figure 31–3**). For instance, *flat ties* of various styles are used with some manufactured panels. For heavier formwork, *coil ties* and reusable *coil bolts* are used (**Figure 31–4**). For each kind of tie, there are also several sizes and styles. There are hundreds of kinds of form hardware. To become better acquainted with form hardware, study manufacturers' catalogues.

Walers. The snap ties run through and are wedged against form members called **walers**. Walers are doubled 2 × 4 (38 × 89 mm) pieces with space between them of about ½ inch (12 mm). They may be horizontal or vertical. Walers are spaced at right angles to the panel frame members. The number and spacing depend on the style of the form and the pressure exerted on the form (**Figure 31–5**).

The vertical spacing of the snap ties and walers depends on the height of the concrete wall. The vertical spacing is closer together near the bottom. This is because there is more lateral pressure from the concrete there than at the top (**Figure 31–6**).

For low wall forms less than 4 feet (1.2 m) in height, the panel may be laid horizontally with vertical walers spaced as required (**Figure 31–7**).

Care of Forms. After the concrete placed in the footing has hardened sufficiently (sometimes three days), the forms are removed and cleaned. Any concrete clinging to the form should be scraped off. They are easier to clean at this stage rather than later, when the concrete has reached full strength. Forms may be oiled to protect them and stored, ready for reuse.

The forms last longer if they are handled properly. Place them gently and do not drop them. Damaged corners from dropping the panel will affect

A

Photo by M. Nauth

B

FIGURE 31–2 A strip-ease tie holds inner and outer forms from spreading when the concrete is placed.

the surface appearance of the concrete. Forms are also easier to use if they are not twisted or broken.

Preparing for Wall Form Assembly

Locating the Forms. Lines are stretched on the batter boards in line with the outside of the foundation wall. A plumb bob is suspended from the layout lines to the footing. Marks that are plumb with the layout lines are placed on the footing at each corner. A chalk line is snapped on the top of the footing between the corner marks outlining the outside of the foundation wall.

StepbyStep Procedures

Procedure 31–A Breaking Snap Ties

STEP 1 SLIDE THE SNAP TIE WRENCH UP AGAINST THE TIE SO THAT THE FRONT OF THE WRENCH IS TOUCHING THE CONCRETE.

STEP 2 KEEPING THE FRONT OF THE WRENCH TIGHT AGAINST THE CONCRETE, PUSH THE HANDLE END TOWARD THE CONCRETE WALL SO THAT THE TIE IS BENT OVER AT APPROXIMATELY A 90º ANGLE.

STEP 3 ROTATE THE WRENCH AND TIE END 1/4 TO 1/2 TURN, BREAKING OFF THE TIE END.

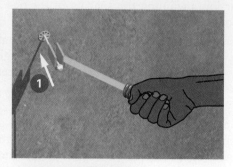

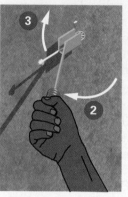

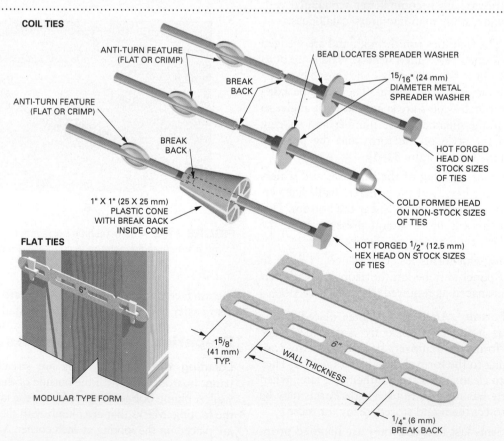

FIGURE 31–3 A large variety of snap ties are manufactured.

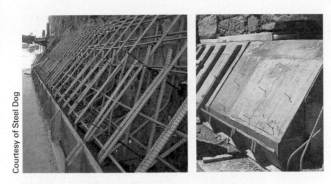

FIGURE 31–4 Steel Dog® Snap-Coil Ties, used for a one-sided battered wall.

FIGURE 31–5 Walers are easily installed on forming systems when special hardware is used.

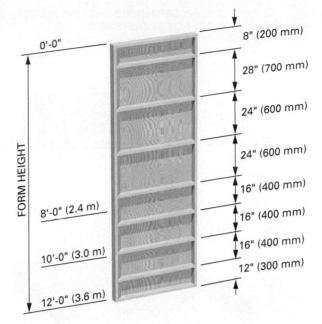

FIGURE 31–6 Horizontal panel stiffeners are placed closer together near the bottom than they are at the top. Dimensions are for purposes of illustration only.

Installing Plates. Sometimes panels are set on 2 × 4 (38 × 89 mm) or 2 × 6 (38 × 140 mm) lumber plates. Plates provide a positive online wall pattern. They can also level out rough areas on the footing. Plates function to locate the position and size of pilasters and changes in wall thickness, corners, and other variations in the wall (**Figure 31–8**). The outer plate is fastened to the footing using masonry nails or pins

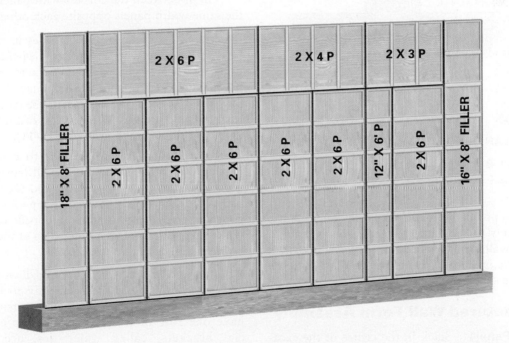

FIGURE 31–7 Forms can be laid horizontally or vertically as needed.

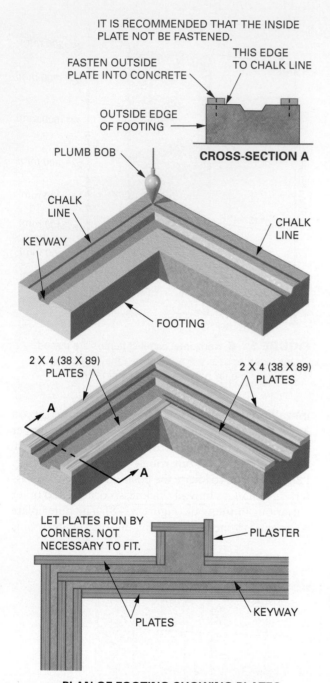

IT IS RECOMMENDED THAT THE INSIDE
PLATE NOT BE FASTENED.

FASTEN OUTSIDE
PLATE INTO CONCRETE

THIS EDGE
TO CHALK LINE

OUTSIDE EDGE
OF FOOTING

CROSS-SECTION A

PLUMB BOB

CHALK
LINE

CHALK
LINE

KEYWAY

FOOTING

2 X 4 (38 X 89)
PLATES

2 X 4 (38 X 89)
PLATES

A

A

LET PLATES RUN BY
CORNERS. NOT
NECESSARY TO FIT.

PILASTER

KEYWAY

PLATES

PLAN OF FOOTING SHOWING PLATES

FIGURE 31–8 Attaching plates to the footing before forming walls.

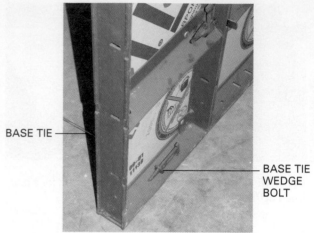

BASE TIE

BASE TIE
WEDGE
BOLT

FIGURE 31–9 Base ties hold the panel bottoms from spreading.

inside of the wall. Lay the panels needed for the outside of the wall around the walls of the excavation. The faces of all panels should be oiled or treated with a chemical releasing agent. This provides a smooth face to the hardened concrete and makes stripping of the forms easy.

Panels may be assembled in inside/outside pairs. But if rebar is required in the wall, then erect the outside wall forms first. This makes installing and tying rebar easier. Set panels in place at all corners first. Set base snap ties into slots while placing the panel and attach the outside corners (**Figure 31–9**). Nail the panel into the plate with duplex nails. Make sure the corners are plumb by testing with a hand level and brace.

Fill in between the corners with panels, keeping the same width panels opposite each other.

Placing Snap Ties. Place snap ties in the dadoes between panels as work progresses. Tie panels together using the wedge bolts or the connection designed by the manufacturer (**Figure 31–10**). Snap ties must be positioned as each panel is placed. Be careful not to leave out any snap ties. Wedge bolts are not hammered tight, only drawn up snug (**Figure 31–11**).

Use filler panels as necessary to complete the outside wall section. Brace the wall temporarily as needed. Install rebar as needed, and then erect the panels for the inside of the wall. Keep joints between panels opposite to those for the outside of the wall (**Figure 31–12**). Insert the other ends of the snap ties between panels as they are erected.

Installing Walers. In a typical 8-foot-high (2.4 m) wall, walers are used only to help keep the forms straight. Typically the outside panels have one row of walers near the top edge. Special brackets, called waler ties, are used to attach walers to two rows of 2 × 4s (38 × 89s)

driven by a powder-actuated tool. The inner plate is fastened only to the concrete form. This is done to allow the wall form thickness to swell slightly when the concrete is placed. This movement is due to the slack in the snap ties being taken up.

Manufactured Wall Form Assembly

Erecting Panels. Stack in the centre of the excavation the number of panels necessary to form the

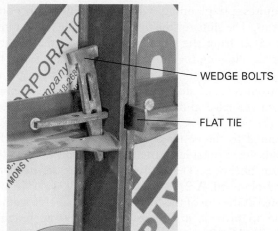

FIGURE 31-10 Form panels are connected by the flat ties, which are secured by special bolts and connectors.

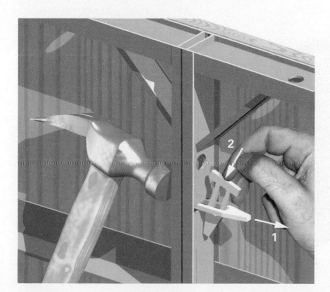

FIGURE 31-11 Tapping the side as the wedge is pushed down adequately tightens the form, yet leaves it loose enough to be easily removed later.

FIGURE 31-12 Panels should be assembled with opposing joints so that snap ties can be easily installed.

(Figure 31-13). Let the bracket come between them and wedge into place.

Forming Pilasters. The wall may be formed at intervals for the construction of **pilasters**. These are thickened portions of the wall that serve to give the wall more lateral (side-to-side) strength. They may also provide support for beams. They are formed with inside and outside corners in the usual manner. If the pilaster is large, longer snap ties are necessary (**Figure 31-14**).

Erecting Wood Wall Forms

Erecting Panels. Stack in the centre of the excavation the number of panels necessary to form the inside of the wall. Lay the panels needed for the outside of the wall around the walls of the excavation. The faces of all panels should be oiled or treated with a chemical releasing agent. This provides a smooth face to the hardened concrete and makes stripping of the forms easy.

Panels may be assembled in inside/outside pairs. But if rebar is required in the wall, erect the outside

FIGURE 31-13 Special brackets used to attach walers.

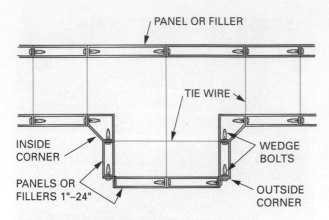

PANEL OR FILLER

TIE WIRE

INSIDE CORNER

WEDGE BOLTS

PANELS OR FILLERS 1"–24"

OUTSIDE CORNER

FIGURE 31–14 Pilasters may be formed like any other intersection. Note that inside the pilaster, tie wire is used to hold panel seams.

wall forms around the perimeter first. This makes installing and tying rebar easier. Set panels in place at all corners first. Nail the panel into the plate with duplex nails (**Figure 31–15**). Make sure the corners are plumb by testing with a hand level and brace.

Fill in between the corners with panels, keeping panels of the same width opposite each other. Place snap ties in the dadoes between panels as work progresses. Tie panels together by driving U-shaped clamps over the edge 2 × 4s (38 × 89s) or by nailing them together with duplex nails. Use filler panels as necessary to complete each wall section. Brace the wall temporarily as needed.

Placing Snap Ties. After the panels for the outside of the wall have been erected, place snap ties in the inter-

mediate holes. Be careful not to leave out any snap ties. Erect the panels for the inside of the wall. Keep joints between panels opposite to those for the outside of the wall. Insert the other end of the snap ties between panels and in intermediate holes as panels are erected (**Figure 31–16**). If the concrete is to be reinforced, the rebars are tied in place before the inside panels are erected.

Installing Walers. When all panels are in place, install the walers. Let the snap ties come through them and wedge into place. Care must be taken when installing and driving snap tie wedges (**Figure 31–17**). Let the ends of the walers extend by the corners of the formwork. Reinforce the corners with vertical 2 × 4s (38 × 89s) (**Figure 31–18**). This is called yoking the corners.

Strip-Ease. Wall forming with Strip-Ease ties requires pre-planning when using new plywood forms. The slots for the ties on the inside forms must be offset from the slots for the ties on the outside forms, allowing for the concrete wall thickness. The plywood sheets are placed vertically, and the top and bottom rows of slots are placed 8 inches (200 mm) from the edge and 20 inches (510 mm) on centre in the field. A 2 × 4 (38 × 89 mm) bottom plate is secured to the concrete footing, set back from the chalk-lined outside of the concrete wall by the plyform thickness. The outside forms are set in place and plumbed. A 2 × 4 (38 × 89 mm) waler is fastened at the top of the form to tie the panels together and to provide a nailing surface for the bracing (**Figure 31–19**).

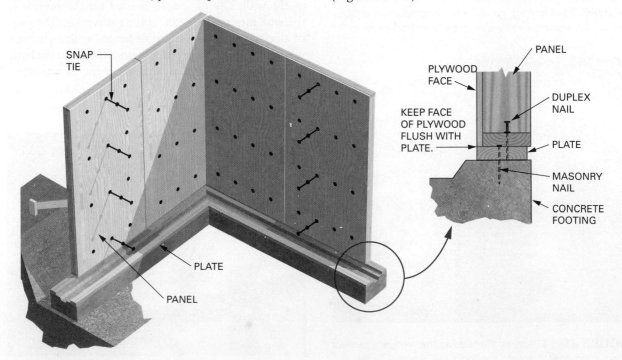

SNAP TIE

PANEL

PLYWOOD FACE

DUPLEX NAIL

KEEP FACE OF PLYWOOD FLUSH WITH PLATE.

PLATE

MASONRY NAIL

CONCRETE FOOTING

PLATE

PANEL

FIGURE 31–15 Forming begins at the corners.

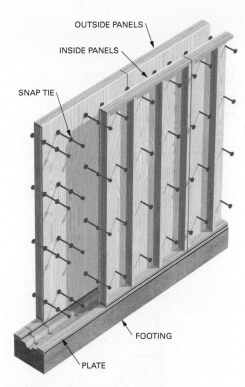

OUTSIDE PANELS

INSIDE PANELS

SNAP TIE

FOOTING

PLATE

FIGURE 31–16 Inside panels of the form are assembled later with snap ties.

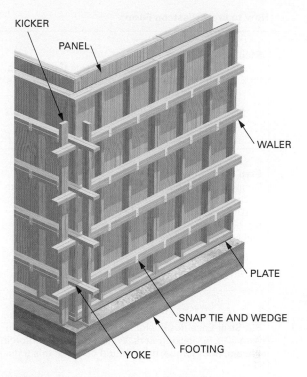

KICKER

PANEL

WALER

PLATE

SNAP TIE AND WEDGE

FOOTING

YOKE

FIGURE 31–18 Yokes are vertical members that hold outside form corners together.

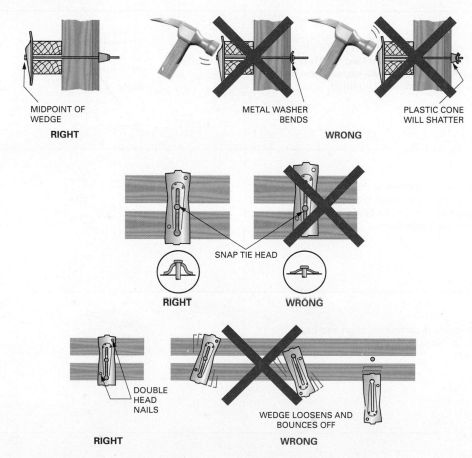

MIDPOINT OF WEDGE

RIGHT

METAL WASHER BENDS

WRONG

PLASTIC CONE WILL SHATTER

SNAP TIE HEAD

RIGHT

WRONG

DOUBLE HEAD NAILS

WEDGE LOOSENS AND BOUNCES OFF

RIGHT

WRONG

FIGURE 31–17 Care must be taken when installing snap tie wedges.

How to Erect Westcon Forms

Step 1 Nail a 2 x 4 plate to footing ¾" outside the building line. Set and plumb outside corners of form panels.	
Step 2 Connect top corners with 2 x 4 walers (same length as the bottom plate), and completely wrap the outside forms of the building. Set panels and nail top and bottom with 2" common nails. The result will be a square foundation.	2" x 4" WALER
Step 3 Put Strip-Ease ties in slots in the panels and secure with ¼" x ¾" locking bars. Window bucks, reinforcing offsets, door jambs, etc., should be set at this time.	
Step 4 Locking bars must overlap the adjacent edges of the plywood forms and enter the next tie. This, and offset spacing of the bars, will increase the form stability when the concrete is poured.	Right / Wrong
Step 5 Set inside panels, starting with the corners. Feed the ties through the slots. Insert the bars by pulling them through the ties. No further bracing or walering is required on the inside forms.	
Step 6 When the inside is complete, align and brace the outside form. Note the small amount of bracing and walering.	

Courtesy of Northland Construction

(Continued)

FIGURE 31–19 Wall forming using Strip-Ease ties.

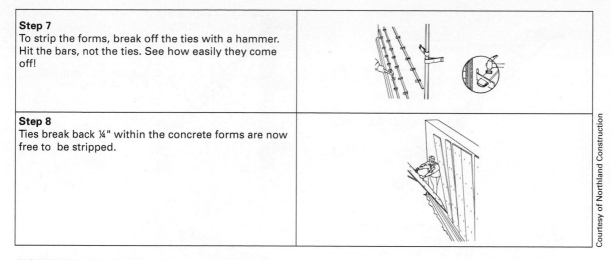

Step 7
To strip the forms, break off the ties with a hammer. Hit the bars, not the ties. See how easily they come off!

Step 8
Ties break back ¼" within the concrete forms are now free to be stripped.

Courtesy of Northland Construction

FIGURE 31–19 *(Continued)*

Forming Pilasters. The wall may be formed at intervals for the construction of pilasters. These are thickened portions of the wall that serve to give the wall more lateral (side-to-side) strength. They may strengthen the wall or provide support for beams. They may be constructed on the inside or outside of the wall. In the pilaster area, longer snap ties are necessary (**Figure 31–20**).

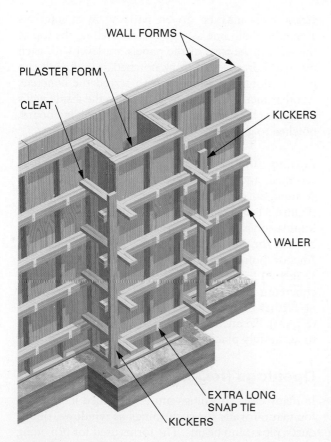

FIGURE 31–20 Formwork for a pilaster.

Completing the Wall Forms

Straightening and Bracing. Brace the walls inside and outside as necessary to straighten them. Wall forms are easily straightened by sighting by eye along the top edge from corner to corner. Another method of straightening is to use line and gauge blocks, stretching a line from corner to corner at the top of the form over two blocks of the same thickness. Move the forms until a test block of equal thickness passes just under the line (**Figure 31–21**).

A special adjustable form brace and aligner are used to position and hold wall forms (**Figure 31–22**). This allows for easy prying when straightening the formwork simply by rotating the turnbuckle. Braces are cut with square ends and nailed into place against **strongbacks**. Strongbacks are placed across walers at right angles wherever braces are

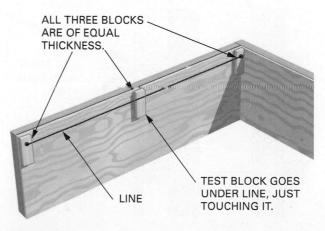

FIGURE 31–21 Straightening the wall form with a line and test block.

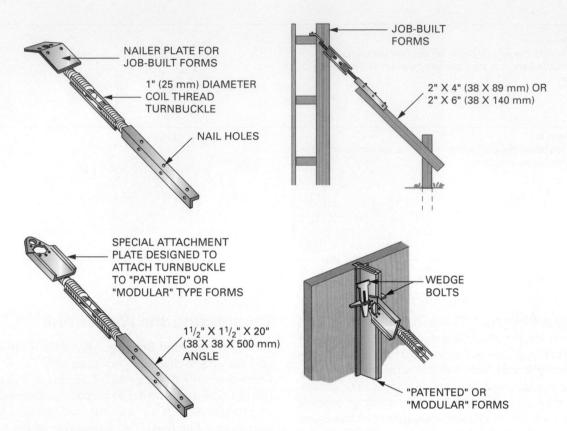

FIGURE 31–22 Adjustable braces are available with forming systems.

needed. The sharp corner of the square ends helps hold the braces in place. It also allows easy prying with a bar to tighten the braces and move the forms (**Figure 31–23**). There is no need to make a bevel cut on the ends of braces.

Levelling. After the wall forms have been straightened and braced, chalk lines are snapped on the inside of the form for the height of the foundation wall.

FIGURE 31–23 Cut braces with square ends. This allows for easy prying when straightening the formwork.

Grade nails may be driven partway in at intervals along the chalk line as a guide for levelling the top of the wall. If the tops of the panels are level with each other, a short piece of stock notched at both ends can be run along the panel tops to screed the concrete. Another method is to fasten strips on the inside walls along the chalk line and use a similar screeding board, notched to go over the strips (**Figure 31–24**).

Setting Anchor Bolts. As soon as the wall is screeded, anchor bolts are set in the fresh concrete. A number of styles and sizes are manufactured (**Figure 31–25**). The type is usually specified on the foundation plan. Care must be taken to set the anchor bolts at the correct height and at specified locations. Anchor bolts must have a minimum diameter of ½ inch (12.7 mm) and must be embedded into the concrete a minimum of 4 inches (100 mm) and spaced no farther than 8 feet (2.4 m) apart [NBCC 9.23.6.1 (2), (3)]. An anchor bolt *template* is sometimes used to accurately place the bolts (**Figure 31–26**).

Openings in Concrete Walls

In many cases, openings must be formed in concrete foundation walls for such things as windows, doors, ducts, pipes, and beams. The forms used for providing the larger openings are called **blockouts** or **bucks**.

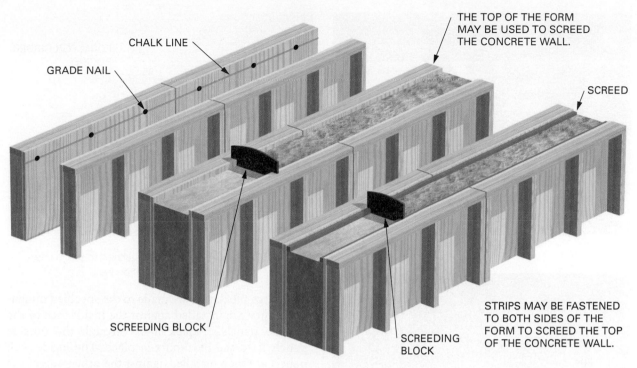

FIGURE 31–24 Methods to establish the top surface to the concrete in a wall form.

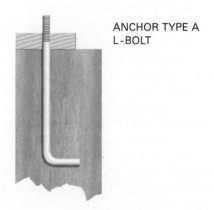

FIGURE 31–25 Typical anchor bolts.

Constructing Bucks. The buck is usually made of 2-inch (38 mm) dimension lumber. Its width is the same as the thickness of the foundation wall. Nailing blocks or strips are often fastened to the outside of the bucks. These are bevelled on both edges to lock them into the concrete when the form is stripped. They provide for the fastening of window and door frames in the openings. Intermediate pieces may be necessary in bucks for large openings to withstand the pressure of the concrete against them (**Figure 31–27**). When PVC, fibreglass, or metal windows are used, the sashes are removed; the window frames are furred out to the concrete wall thickness, braced, and

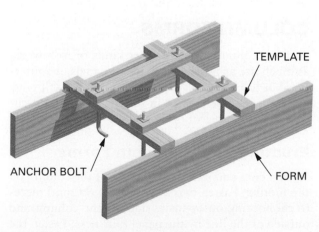

FIGURE 31–26 Templates are sometimes used to accurately place anchor bolts in fresh concrete.

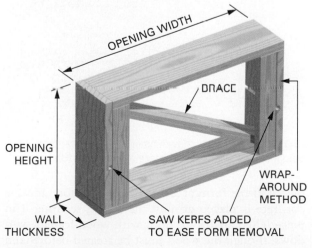

FIGURE 31–27 Construction of a typical window buck.

FIGURE 31–28 Windows installed inside the form before concrete is placed, before (A) and after (B).

placed in the form at the specified location; and the concrete is placed around the frame, securing it in place. This can also be done with wood-framed windows, though the wood must be treated beforehand (**Figure 31–28**).

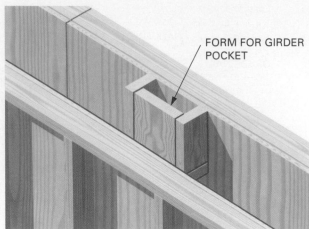

FIGURE 31–29 A small box is attached to the inside form for a girder pocket in a foundation wall.

Large blockouts are made to the specified dimension. They are installed against the inside face of the outside panels. Duplex nails through the outside panels hold the blockouts in place. The inside wall panels are then installed against the other side of the blockouts. Nails are driven through the inside wall panels to secure the bucks on the inside.

Girder Pockets. Girder pockets are recesses in the top of the foundation wall. They are sometimes required to receive the ends of girders (beams). A box of the size needed is made and fastened to the inside at the specified location (**Figure 31–29**). Because it is near the top of the form, not much pressure from the liquid concrete is exerted against it. For a wood beam, care must be taken to construct the box to allow for beam bearing plus ½ inch (12.7 mm) deep, beam width plus 1 inch (25.4 mm) wide, and beam height minus 1½ inches (38 mm) high. This will accommodate an air space on the end and sides and the thickness of the sill plate. The pocket for a steel beam is constructed so that the beam will sit flush with the top of the concrete wall.

COLUMN FORMS

To form columns, like all other kinds of formwork, as much use as possible is made of panels, manufactured or job built, to simplify erection and stripping. Columns may be formed in square, rectangular, circular, or a number of other shapes.

Erecting Wood Column Forms

Stretch lines and mark the location of the column on the footing. Fasten two 2 × 4 (38 × 89 mm) pieces to the footing on opposite sides of the column and outside of the line by the panel thickness. Fasten the other two to the overlapping ends of the first pair in a similar location (**Figure 31–30**).

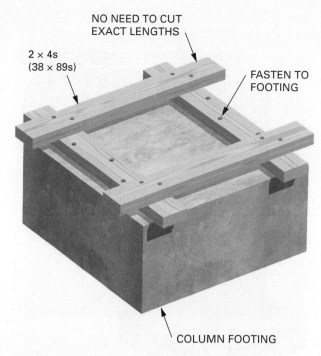

NO NEED TO CUT
EXACT LENGTHS

2 × 4s
(38 × 89s)

FASTEN TO
FOOTING

COLUMN FOOTING

FIGURE 31–30 A yoke is constructed on the column footing to start the forming of a column.

Build and erect two panels to form the thickness and height of the column. A **cove moulding** of desired size may be fastened on the edges of this panel to form a radius to the corners of the concrete column. A **quarter-round** moulding may be used for a cove shape. Triangular-shaped strips of wood can be used to form a chamfer on the corners of the column (**Figure 31–31**).

The face of the column may be decorated with **flutes** by fastening vertical strips of **half-round** moulding spaced on the panel faces. In addition, *form liners* are often used. They provide various wood, brick, stone, and many other textures in the face of the concrete wall or column.

Build panels for the opposite sides of the column to overlap the previously built panels. Plumb and nail the corners together with duplex nails. Install 2 × 4 (38 × 89 mm) yokes around the column forms, letting their ends extend beyond the corners. Nail them together where they overlap. Yoke the column closer together at the bottom. The number and spacing of yokes depend on the height of the column. These details are specified by the form designer. Install vertical 2 × 4s (38 × 89s) between the overlapping ends of the yokes (**Figure 31–32**). Brace the formwork securely to hold it plumb.

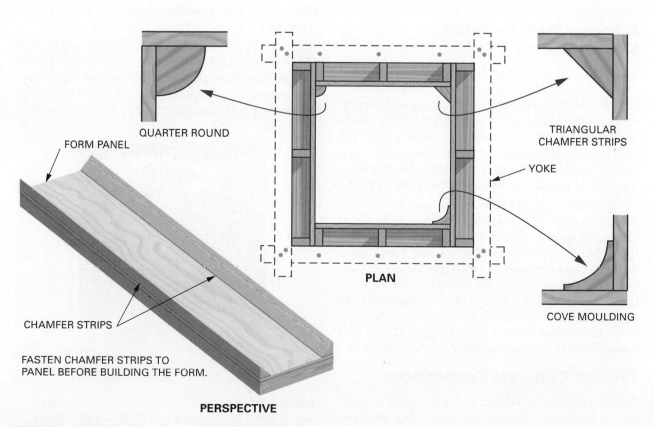

QUARTER ROUND

FORM PANEL

TRIANGULAR
CHAMFER STRIPS

YOKE

CHAMFER STRIPS

COVE MOULDING

FASTEN CHAMFER STRIPS TO
PANEL BEFORE BUILDING THE FORM.

PLAN

PERSPECTIVE

FIGURE 31–31 The corners of a concrete column can be formed in several ways.

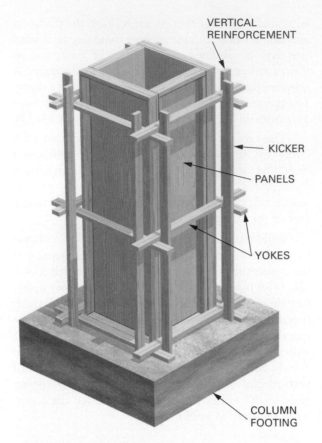

FIGURE 31–32 A completed concrete column form.

VERTICAL REINFORCEMENT

KICKER

PANELS

YOKES

COLUMN FOOTING

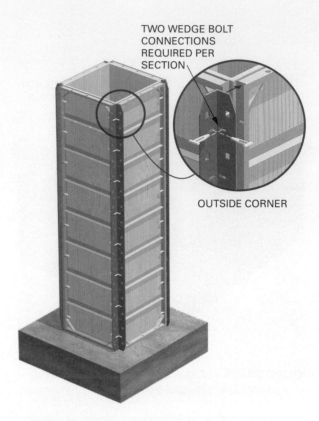

TWO WEDGE BOLT CONNECTIONS REQUIRED PER SECTION

OUTSIDE CORNER

FIGURE 31–33 Yokes are not necessary to form columns when forming systems are used.

Erecting Manufactured Column Forms

Manufactured forms can be used instead of wood forms. Various width forms can be assembled with outside corners (**Figure 31–33**). These are erected in the same fashion as for wall forms. Wedge bolts secure the panels to the outside corners. Nail the panels to the plate fastened to the footing. No snap ties are needed. Concrete corners can be shaped in the manner previously discussed. (See **Figure 31–34**.)

Steel forms are available to form circular columns (**Figure 31–35**). They bolt together and need only to be braced. Another system uses heavy-duty cardboard tubes. These can be assembled on base forms for a monolithic column and footing. After concrete placement, the forms are stripped by peeling off the cardboard. This makes this form a one-time-use-only form (**Figure 31–36**). Fast-Tube™ is a fabric column form that is inexpensive and easy to erect (**Figure 31–37**). Visit *www.fab-form.com/fast-tube/fast-tubeOverview.php*.

Precast Concrete Foundations

Precast concrete foundations provide a fast alternative to traditional foundation types. They are manufactured off-site and delivered on truck. They are

then craned into place by a specially trained crew (**Figure 31–38**). Setup time is typically one day.

Sections are constructed with insulation, 1-inch (25-mm) extruded polystyrene, and have a ribbed construction that allows for additional insulation. This makes for an energy-efficient foundation. The sections are bolted together, and each seam is caulked

Courtesy of Peri Framework Systems Inc.

FIGURE 31–34 VARIO QUATTRO column formwork; VARIO QUATTRO–equipped with VARIO GB 80 brackets for the creation of on-site concreting platforms

COLUMN CAPITAL FORM

CIRCULAR COLUMN FORM

FIGURE 31–35 Manufactured circular column and column capital forms simplify forming.

FIGURE 31–36 Some manufactured forms are designed to be used only once.

Courtesy of Fab-Form Industries

FIGURE 31–37 Fast-Tube fabric concrete column form.

FIGURE 31–38 Precast concrete foundations are delivered on trucks and craned into place.

Insulated Concrete Forms

Insulated concrete forms (ICF) are made of varying thicknesses of expanded (EPS) or extruded (XPS) polystyrene, usually from 2 to 4 inches (51 to 102 mm). Units can be shaped into *blocks* with a metal or plastic snap tie system (**Figure 31–40**). The block sizes vary depending on manufacturer. Other systems are bundled as *planks* with high-density polyethylene ties, and still others are packaged as *panels* that are transported flat and that fold out on-site (**Figure 31–41**). Most have fastening strips included on their surfaces

to produce a watertight seal (**Figure 31–39**). A concrete slab is later placed as a floor and this keeps the bottom of the walls from moving toward the inside during the backfilling of the foundation.

TOP SEAMS ARE BOLTED

ALL JOINTS ARE CAULKED

FIGURE 31–39 Precast concrete panels are caulked and bolted together to make a watertight seal.

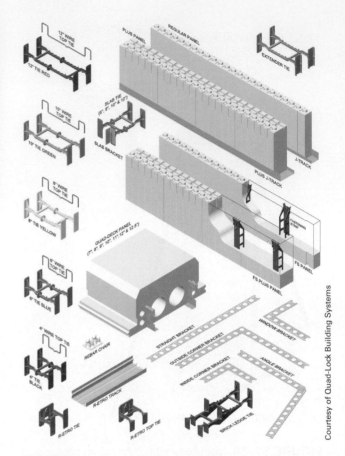

Courtesy of Quad-Lock Building Systems

FIGURE 31–41 Components of a plank ICF system.

Courtesy of IntegraSpec

FIGURE 31–40 An ICF system with scaffold bracing.

to make it easy to screw on a wall finish later. The benefit of these systems is that the finished product is a heavily insulated wall that reduces heating and cooling requirements. Also, no disassembly of forms is necessary. The "forms" are lightweight. and an entire foundation can easily be erected in a day.

Assembly begins one course at a time with corner and straight blocks. The next course is set so that the vertical seams are broken. This helps to interlock the components for strength. Rebar is installed as needed after each layer is installed. Special reinforced tape or twine is sometimes used to help hold the courses together. Window and door bucks are inserted at specified locations, as well as beam pockets.

The success of this type of system depends on the method of placing the concrete. Placement of the concrete is done in levels or lifts with a pump truck. Only one course is filled at a time, giving the concrete time to hydrate or harden slightly. Typically, by the time placement is done for the entire perimeter, the first section is ready to support another lift.

Fastfoot® Monopour now has a system that provides ICFs with a footing base that allows for a monolithic pour. See *www.fab-form.com/fastfootMp/ fastfootMpOverview.php* (**Figure 31–42**).

For more information on ICFs (**Figure 31–43**), visit these websites: *www.nudura.com*; *www.integraspec. com*; *www.arxx.com*; *www.quadlock.com/products*; and *www.cement.org/basics/concreteproducts_icf.asp*.

Courtesy of Fab-Form Industries

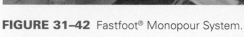

FIGURE 31–42 Fastfoot® Monopour System.

All Photos: Photos by M. Nauth; courtesy of Habitat for Humanity–NCR, Canada

FIGURE 31–43 ICF used for a foundation wall.

Chapter 32

Concrete Stair Forms and Preserved Wood Foundations

STAIR FORMS

It may be necessary to refer to Chapters 37 and 38 on stair framing for definitions of stair terms, types of stairs, stair layout, and methods of stair construction.

Concrete stairs may be suspended or supported by earth (**Figure 32–1**). Each type may be constructed between walls or have open ends.

Forms for Earth-Supported Stairs

Before placing concrete, stone, gravel, or other suitable fill is graded to provide proper thickness to the stairs. It should not be overly thick, or concrete will be wasted. It may be necessary to lay out the stairs before the supporting material is placed.

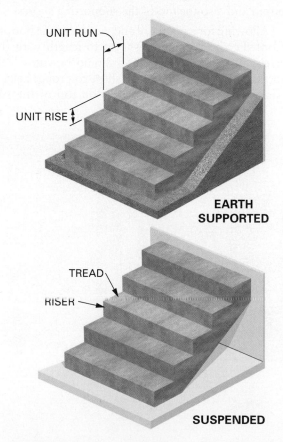

FIGURE 32–1 Concrete stairs may be formed on supporting soil or suspended.

Forming between Existing Walls. When earth-supported stairs are formed between two existing walls, the rise and run of each step are laid out on the inside of the existing walls. *Rise* is the height of each step above the one below. The vertical portion of each concrete step is called the *riser*. *Run* is the horizontal distance of a step. It is roughly the width of each step or tread. Boards are ripped to width to correspond to the height of each riser. The board is wedged and secured in place with its inside face aligned to the riser layout line. The top and bottom edges of the board are aligned with tread layout lines. After the riser boards are secured in position, they are braced from top to bottom and at mid-span. This keeps them from bowing outward due to the pressure of the concrete (**Figure 32–2**).

The riser forms are bevelled on the bottom edge, leaving about ⅛ to ¼ inch (3 to 6 mm) untouched. Bevelling the bottom edge of the plank permits the mason to trowel the entire surface of the tread. Otherwise, the bottom edge of the riser form will leave its impression in the concrete tread.

Forming Stairs with Open Ends.   In cases where the ends of the stairs are to be open, panels are erected on each end. It does no harm if the panels are larger than needed; the panels only need to be plumb. The distance between them is the desired width of the stairs. The end panels are then firmly fastened and braced in position.

The risers and treads are laid out on the inside surfaces of the panels. Cleats (short strips of wood) are fastened at each riser location. An alternative to individual cleats is an inverted cut stringer. Allowances must be made for the thickness of the riser form board. Screws or duplex nails should be used to make stripping the form easier. The space should then be filled and compacted with the supporting material to the proper level. Any necessary reinforcing should be installed (**Figure 32–3**).

The boards used to form the risers are ripped to width and cut to length. They are then fastened with duplex nails or screws through the side panels in a position against the cleats, with the top and bottom

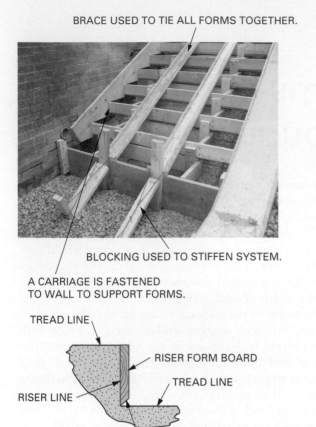

BRACE USED TO TIE ALL FORMS TOGETHER.

BLOCKING USED TO STIFFEN SYSTEM.

A CARRIAGE IS FASTENED
TO WALL TO SUPPORT FORMS.

TREAD LINE

RISER FORM BOARD

TREAD LINE

RISER LINE

BOTTOM OF RISER FORM
BOARD IS BEVELLED TO PERMIT
FINISHING OF TREAD.

FIGURE 32–2 Concrete stairs may be formed between existing walls. Note that the bottom edge of the riser form is bevelled.

edges aligned to the tread layout lines. The riser forms are then braced from top to bottom at intervals between the two ends (**Figure 32–4**).

Forms for Suspended Stairs

Forms for suspended stairs are more difficult to build. Instead of earth support, a form needs to be built on the bottom to support the stair slab. With proper design and reinforcement, the stairs are strong enough to support themselves in addition to the weight of the traffic. As with earth-supported stairs, suspended stairs may be formed between existing walls, with open ends, or both.

Forming between Existing Concrete Walls. A cross-section of the formwork for suspended stairs between existing walls, with the form members identified, is shown in **Figure 32–5**. First, lay out the treads, risers, and stair slab bottom on both of the walls between which the stairs run. This can be done by using a hand level, ruler, and chalk line (**Figure 32–6**).

The form for the bottom of the stair slab is then laid out. Allow for the thickness of the plywood deck, the width of the supporting joists, and the depth of the *horses*. Snap a line on both walls to indicate their bottom and also the top of the supporting *shores*.

Allowing for a plank sill and wedges at the bottom end of the shores, cut the shores to length with the correct angle at the top. They should be cut a little short to allow for wedging the shore to proper height. Install them at specified intervals on top of the sill.

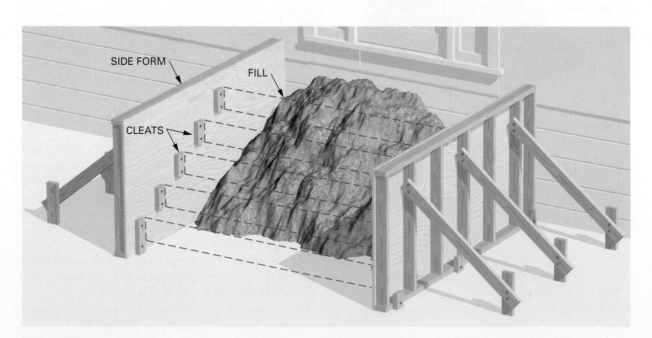

SIDE FORM

FILL

CLEATS

FIGURE 32–3 End panels are braced firmly in position. The stairs are laid out, cleats fastened to them, and the space filled with supporting material. Note: Cleats may be installed 1 inch (25 mm) out of plumb to emulate stair nosing.

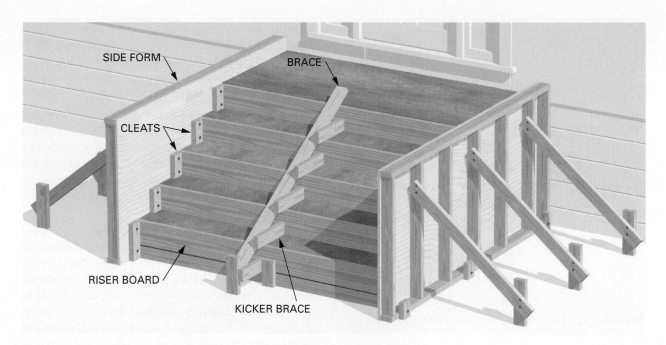

FIGURE 32–4 Typical form construction for concrete stairs having open ends.

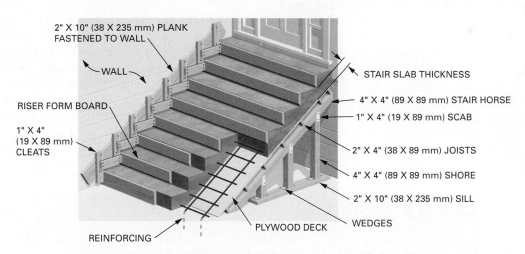

FIGURE 32–5 Cross-section of formwork for suspended stairs between walls.

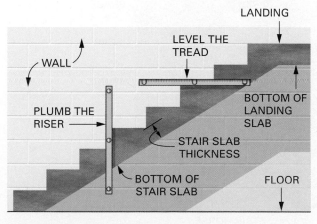

FIGURE 32–6 Make the layout for suspended stairs on the inside of the walls.

Brace in position, then place and fasten the horses on top with **scabs** or **gusset** plates. Scabs are short lengths of narrow boards fastened across a joint to strengthen it. Install joists at right angles across the horses at the specified spacing (Figure 32–7). Fasten the plywood form in position on top of the joists. Use only as many fasteners as needed to hold the plywood in place for easier stripping later. Wedge the shoring as necessary to bring the surface of the plywood to the layout line. Fasten the wedges in place so that they will not move. The plywood should now be oiled to facilitate stripping. Install the reinforcing rods.

Rip the riser boards to width. Cut to length, allowing for wedges. Bevel the outside bottom edge of the treads to allow for easy levelling of the

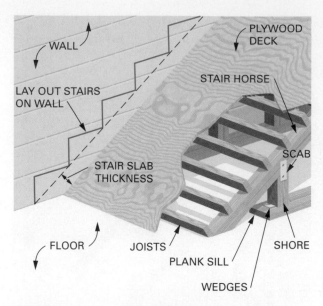

FIGURE 32-7 The framework for the bottom form of the suspended stairs.

concrete at each tread. Install the riser form boards to the layout lines. Wedge in position as shown previously in Figure 32-2.

Another method, shown previously in Figure 32-5, secures the riser form with cleats fastened to a 2 × 10 (38 × 235 mm) attached to the wall. The 2 × 10 (38 × 235 mm) is attached about an inch (25 mm) above the stair layout. Project the riser layout lines upward on the planks. Fasten cleats to the plank. Then brace the cleats and fasten the riser board to them.

Forming with Open Ends. A completed form for suspended stairs with open ends is shown in **Figure 32-8**. This style of form sets the side panels for the stairs on the stair horse. The riser layout is made on the side panels and then set in place on top of the stair horse.

Laying Out the Side Forms. First, measure and snap a line on the side form up from the bottom edge a distance equal to the thickness of the slab. The inside corner intersection of the treads and risers lies along this line. Lay out the risers and treads above this line. The stair slab thickness is on the bottom edge of the form panels.

A pitch board speeds up the layout process (**Procedure 32-A**). It is a triangular piece of plywood or hardboard for which the tread width and the riser height are the legs of a right triangle. To mark the tread and riser locations, use the pitch board held to the previously snapped slab line. To maintain accuracy, mark off the snapped line at intervals equal to the calculated hypotenuse of the pitch board, that is, $\sqrt{(\text{Unit Run}^2 + \text{Unit Rise}^2)}$.

Laying Out the Stair Horses. The length of the stair horse and shores can be determined from the side form panels. Lay the side panel down on a floor surface. Snap a line that is parallel to the bottom edge of the side form. The distance away from the form bottom is equal to the thickness of the stair plywood deck and supporting joists (**Figure 32-9**).

From the side panel, extend the bottom floor line and the top plumb lines, making a large triangle on

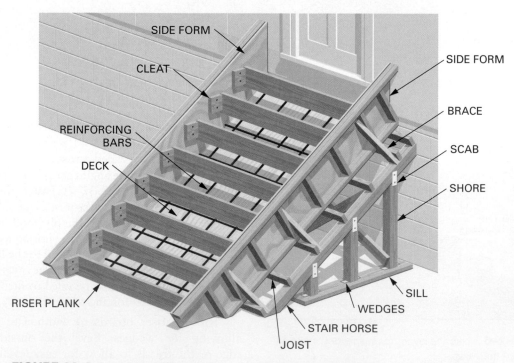

FIGURE 32-8 A completed form for suspended stairs with open ends.

StepbyStep Procedures

Procedure 32–A Laying Out a Set of Stairs Using a Pitch Board

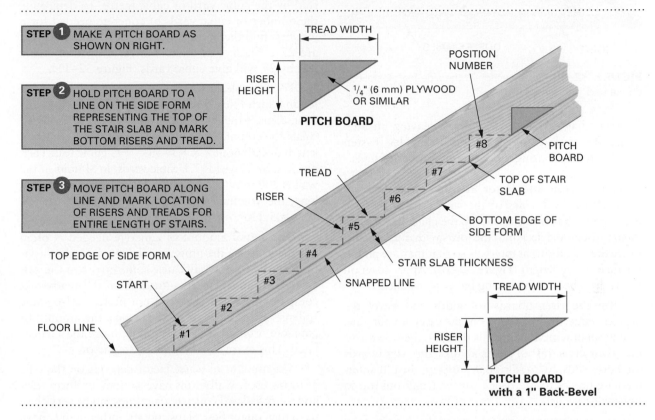

STEP 1 MAKE A PITCH BOARD AS SHOWN ON RIGHT.

STEP 2 HOLD PITCH BOARD TO A LINE ON THE SIDE FORM REPRESENTING THE TOP OF THE STAIR SLAB AND MARK BOTTOM RISERS AND TREAD.

STEP 3 MOVE PITCH BOARD ALONG LINE AND MARK LOCATION OF RISERS AND TREADS FOR ENTIRE LENGTH OF STAIRS.

TREAD WIDTH

RISER HEIGHT

¹/₄" (6 mm) PLYWOOD OR SIMILAR

PITCH BOARD

POSITION NUMBER

PITCH BOARD

#8
#7
#6
#5
#4
#3
#2
#1

TOP OF STAIR SLAB

BOTTOM EDGE OF SIDE FORM

STAIR SLAB THICKNESS

SNAPPED LINE

TREAD
RISER

TOP EDGE OF SIDE FORM
START
FLOOR LINE

TREAD WIDTH

RISER HEIGHT

PITCH BOARD
with a 1" Back-Bevel

the floor. Lay out the thickness of the stair horse and shores, including the sill. Measure and cut the length of the stair horse and shore. Cut one stair horse for each side to length at the angles indicated.

Installing Horses and Joists. Temporarily support and brace the horses in position. Install shores in a manner similar to that of closed stairs. Brace shores and horses firmly in position. Fasten joists at

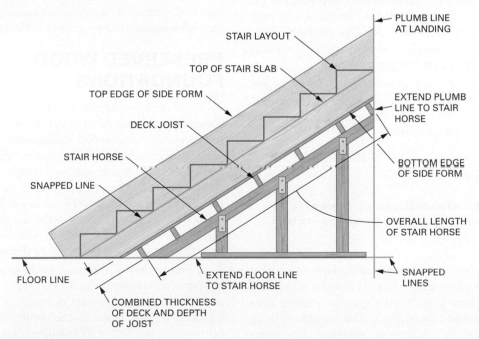

PLUMB LINE AT LANDING
STAIR LAYOUT
TOP OF STAIR SLAB
TOP EDGE OF SIDE FORM
DECK JOIST
STAIR HORSE
SNAPPED LINE
EXTEND PLUMB LINE TO STAIR HORSE
BOTTOM EDGE OF SIDE FORM
OVERALL LENGTH OF STAIR HORSE
FLOOR LINE
EXTEND FLOOR LINE TO STAIR HORSE
SNAPPED LINES
COMBINED THICKNESS OF DECK AND DEPTH OF JOIST

FIGURE 32–9 Method for finding the length and cuts of the stair horses.

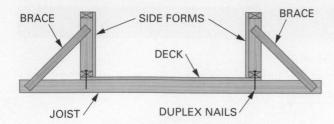

FIGURE 32–10 Cross-section of the side form installed on the deck and braced.

designated intervals to the horses, leaving an adequate and uniform overhang on both sides. Fasten decking to joists using a minimum number of nails.

Installing Side and Riser Forms. Snap lines on the deck for both sides of the stairs. Stand the side forms up with their inside face to the chalk line. Fasten them through the deck into the joists with duplex nails. Brace the side forms at intervals so that they are plumb for their entire length (**Figure 32–10**). Apply form oil to the deck before reinforcing bars are installed.

Rip the riser boards to width and bevel the bottom edge. Install them on the riser layout line. Fasten them with duplex nails through the side forms into their ends. Install cleats against the riser boards on both ends for additional support. Install intermediate braces to the riser form boards from top to bottom, if needed.

Economy and Conservation in Form Building

Economical concrete construction depends on the reuse of forms. Forms should be designed and built to facilitate stripping and reuse. Use panels to build forms whenever possible. Use only as many nails as necessary, to make stripping forms easier.

Care must be taken when stripping forms to prevent damage to the panels so that they can be reused. Stripped forms should be cleaned of all adhering concrete and stacked neatly.

CAUTION

Remove all protruding nails to eliminate the danger of stepping on or brushing against them.

Long lengths of lumber can often be used without trimming. Random length boards can extend beyond the forms. There is no need to spend a lot of time cutting lumber to exact length. The important thing is to form for the concrete to specified dimensions without spending too much time in unnecessary fitting.

Estimating Concrete Quantities

Sometimes it is the carpenter's responsibility to order concrete for the job. Ready-mix concrete is sold by the cubic yard or cubic metre. To determine the number of cubic yards of concrete needed for a job, first find the number of cubic feet. Because there are 27 cubic feet in one cubic yard, dividing cubic feet by 27 will give cubic yards (**Figure 32–11**).

For example, how many cubic yards will be needed for an 8-inch-thick, 8-foot-high, and 36-foot-long wall? First, convert the thickness to feet $(8 \div 12)'$. Then calculate the volume by multiplying thickness × width × length or $0.6666667 \times 8 \times 36 = 192$ cubic feet. Then dividing by 27 yields 7.1 cubic yards. In SI units, if the wall is 200 mm (0.2 m) thick, 2.4 m high, and 10.9 m long, the volume of concrete is $V = 0.2 \times 2.4 \times 10.9 = 5.2$ m^3.

The actual amount of concrete needed is often not the same as the amount calculated. Slight variations in the forms may cause some errors in the calculated volume. The quantities from the ready-mix company are often close but not perfect. Also some spillage will occur. For these reasons, the amount of concrete ordered should be a little more than calculated. This extra is called the **waste factor**.

The amount of waste factor depends on the type of forms used. Wall forms have smooth, uniform sides and the calculated concrete quantities can be very close to actual quantities. Slabs, on the other hand, have a bottom surface that is irregular, making the actual thickness measurement difficult to determine. Therefore, take several thickness measurements and average them. Also, add a higher waste factor than you would for wall forms. In general, concrete quantities are merely estimates, and experience is the best teacher.

PRESERVED WOOD FOUNDATIONS

All wood decays naturally when not completely dry or completely submersed in water. To prevent or retard this process, some wood species are injected with preservatives while under pressure (**Figure 32–12**). Sometimes the lumber is incised with a series of knife cuts to allow for deeper penetration of the preservative. When **preserved wood foundation (PWF)** lumber is cut on the end, this exposes the untreated inside of the wood, which must then be treated with a copper naphthanate field-cut preservative (**Figure 32–13**).

Lumber that is approved for in-ground use must be stamped accordingly and must state the name of the preservative used. Currently in use are CCA and ACA (chromated and ammoniacal copper arsenates) for below-grade use and ACQ (alkaline copper quaternary) and CA (copper azole) for above-ground use. CCA,

27 CUBIC FEET

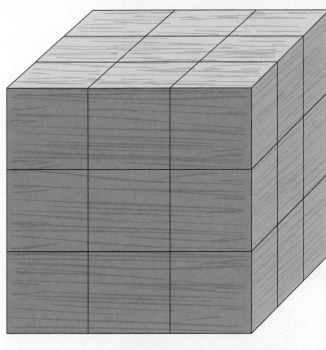

1728 CUBIC INCHES

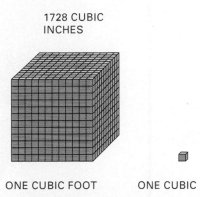

ONE CUBIC YARD

ONE CUBIC FOOT   ONE CUBIC INCH

FIGURE 32–11 Relative sizes of cubic measurements.

Courtesy of iLevel by Weyerhaeuser

FIGURE 32–12 Preservatives are forced into lumber under pressure in large cylindrical tanks.

Photo by M. Nauth

FIGURE 32–13 Treating the end cut of PWF lumber.

ACA, and ACQ have a green finish; CA is brownish. Pentachlorophenol is still used for utility poles as it softens the wood and facilitates climbing. Black, oily creosote is the preservative of choice for railway ties.

A full and complete guide to designing, specifying the components, and constructing a PWF is found in *Permanent Wood Foundations*, published by the Canadian Wood Council (**Figure 32–14**).

Foundation System Materials

PWF systems resemble the familiar wood-frame construction of top and bottom plates, studs, and sheathing, with a few strategic differences, the most obvious being the "green" finish.

- All wood footings, studs, plates, sills, cripples, etc., located lower than 8 inches (203 mm) above grade must be designated "PWF" (**Figure 32–15**).
- All plywood sheathing that is fully or partially less than 8 inches (203 mm) below grade must be treated and approved for in-ground use.
- All nails and metal connectors in contact with PWF material must be hot-dipped galvanized or stainless steel and must meet the size and spacing requirements of the Canadian Standards Association's specifications for PWFs.
- Caulking that is used between panel joints, between wall and footing, and at the top of the plywood cover plate must be either a

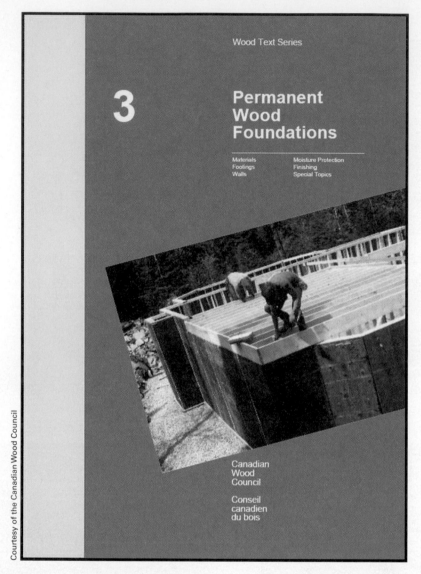

FIGURE 32–14 *Permanent Wood Foundations* by the Canadian Wood Council.

STANDARD CSA 0322 UNDER WHICH MATERIAL IS QUALIFIED

CCA (or ACA) – TYPE OF PRESERVATIVE USED

FIRST TWO DIGITS IDENTIFY THE TREATING PLANT. THE LAST TWO DIGITS IDENTIFY THE YEAR OF TREATMENT.

USE APPROVED FOR THE MATERIAL

COMPANY NAME

FIGURE 32–15 Typical grade stamp for lumber approved for in-ground use.

chemical-curing elastomeric or a solvent-curing butyl-polyisobutylene polymer base.

- Exterior moisture barriers can be 6 mil (0.15 mm) polyethylene (preferably black for longevity), a bituminous damp-proofing coating, a sprayed-on waterproofing membrane, a dimpled waterproof drainage mat, or equivalent.

- The granular drainage layer beneath the footings is generally ¾-inch (19 mm) clear gravel or pea gravel.

- Backfill must be a free-draining material (like sand) that is free of debris and rocks larger than 6 inches (152 mm).

Basic Construction

The site for the PWF is chosen for good drainage. The base of the excavation is sloped toward the location of the proposed sump pit (usually under the basement stairs). The wood footings (2 × 10 or 38 × 235) are usually placed on a levelled granular drainage layer that is a minimum of 5 inches (127 mm) deep. If the gravel bed exceeds 8 inches (203 mm), it must be compacted. Concrete footings can also be used, but they must either be placed on a granular drainage layer or they must include transverse water passages to relieve the buildup of hydrostatic pressure under the basement floor.

The wall framing proceeds in the usual way, with a sole plate, studs, and two top plates. Any openings in the foundation wall will require special framing techniques, since the main forces that act on the PWF come from the backfill, and the wall must be designed to withstand lateral force more so than vertical force. Interior corners are especially susceptible to the force of surrounding earth. Therefore, they must be braced and secured significantly. The sheathing is applied to the wall frame with construction adhesive, nailed with corrosion-resistant nails, and caulked between the joints. The walls are stood up, fastened together, and waterproofed with a suitable membrane. The top of the membrane is sealed

according to the manufacturer's specifications, or a 12-inch (305 mm) PWF plywood cover plate is applied and caulked at the top.

Basement floors can be wood floors that rest on 2 × 4 (38 × 89 mm) sleepers on the gravel bed, or wood floors that are suspended above the gravel bed, or a concrete slab floor. The wood floors must have the floor joists aligned with the wall studs to provide for maximum resistance to the lateral force of the backfill. Column footings must be located and installed before the floor is placed. The column footings can consist of either wood pads or concrete pads. An alternative to columns and beams is to place an interior footing and erect an interior load-bearing wall with appropriate headers located at the openings.

To resist the lateral force of the backfill, it is necessary to install framing straps from the PWF to the rim joist of the first floor (**Figure 32–16**). The straps

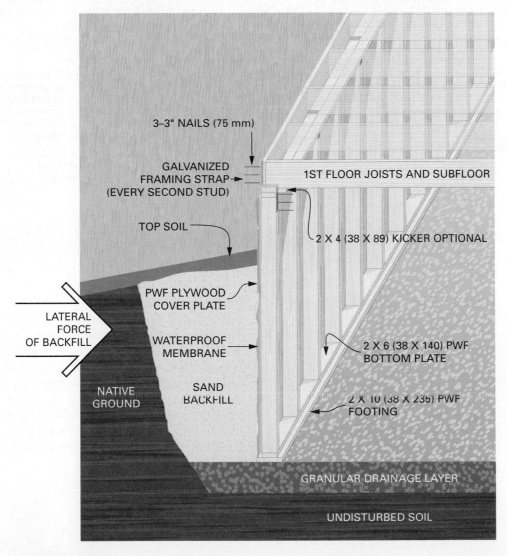

FIGURE 32–16 Lateral force resistance of a preserved wood foundation.

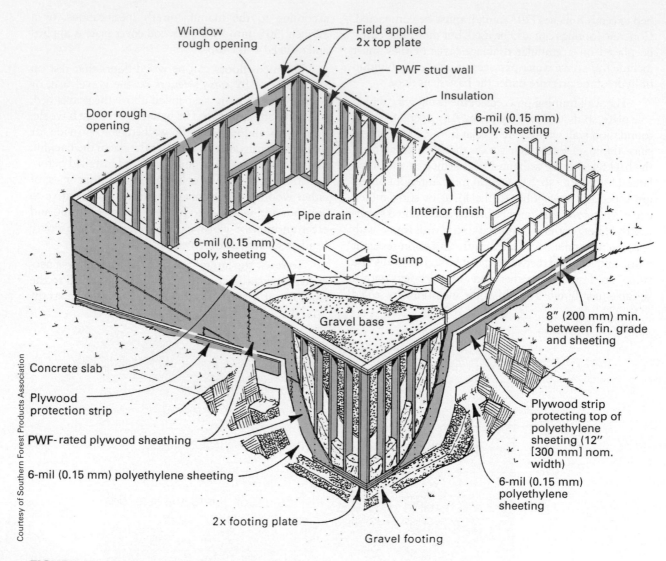

Courtesy of Southern Forest Products Association

FIGURE 32–17 Details of a preserved wood foundation.

are placed on every second stud and wrap over the top plate and then back up the outside of the rim joist. Specially designed headers and stairwell beams are placed around the stairwell to prevent deflection. The subfloor must be glued and securely nailed to the joists and around the stairwell opening before backfilling can take place.

A PWF (**Figure 32–17**) offers many advantages over more conventional foundations (**Figure 32–18**). The same crew frames the foundation and the rest of

the house, thereby eliminating wait times. The walls can be insulated, wired, and finished in much the same way as the rest of the house, thus producing a warm living space and effectively doubling the area of the home. Larger windows can be designed in to provide bedrooms built to code. A walk-out basement door can also be added, as long as the framing requirements are met. A wood floor adds to the comfort and warmth of the basement. PWFs are very popular in the north and in the western provinces of Canada.

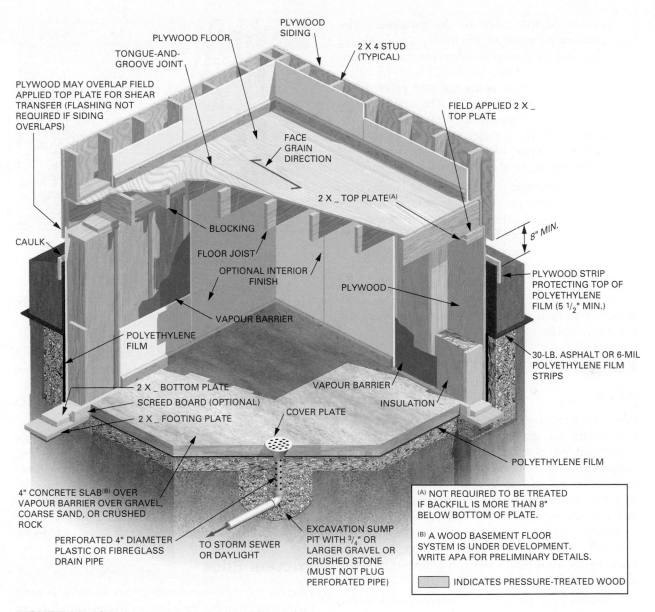

PLYWOOD SIDING

PLYWOOD FLOOR

TONGUE-AND-GROOVE JOINT

2 X 4 STUD (TYPICAL)

PLYWOOD MAY OVERLAP FIELD APPLIED TOP PLATE FOR SHEAR TRANSFER (FLASHING NOT REQUIRED IF SIDING OVERLAPS)

FIELD APPLIED 2 X _ TOP PLATE

FACE GRAIN DIRECTION

2 X _ TOP PLATE(A)

CAULK

8" MIN.

BLOCKING

FLOOR JOIST

OPTIONAL INTERIOR FINISH

PLYWOOD

PLYWOOD STRIP PROTECTING TOP OF POLYETHYLENE FILM (5 1/2" MIN.)

VAPOUR BARRIER

POLYETHYLENE FILM

30-LB. ASPHALT OR 6-MIL POLYETHYLENE FILM STRIPS

2 X _ BOTTOM PLATE

SCREED BOARD (OPTIONAL)

2 X _ FOOTING PLATE

COVER PLATE

VAPOUR BARRIER

INSULATION

POLYETHYLENE FILM

4" CONCRETE SLAB(B) OVER VAPOUR BARRIER OVER GRAVEL, COARSE SAND, OR CRUSHED ROCK

PERFORATED 4" DIAMETER PLASTIC OR FIBREGLASS DRAIN PIPE

TO STORM SEWER OR DAYLIGHT

EXCAVATION SUMP PIT WITH 3/4" OR LARGER GRAVEL OR CRUSHED STONE (MUST NOT PLUG PERFORATED PIPE)

(A) NOT REQUIRED TO BE TREATED IF BACKFILL IS MORE THAN 8" BELOW BOTTOM OF PLATE.

(B) A WOOD BASEMENT FLOOR SYSTEM IS UNDER DEVELOPMENT. WRITE APA FOR PRELIMINARY DETAILS.

INDICATES PRESSURE-TREATED WOOD

FIGURE 32–18 Typical details of an all-wood foundation.

Chapter 33

Concrete Forming for the ICI Sector (Industrial, Commercial, and Institutional)

Building construction in the industrial, commercial, and institutional (ICI) sector is on a grander scale. No longer are the buildings covered by Part 9 of the NBCC (which covers buildings with a maximum floor area of 600 m² [6458 ft²] and a limit of 3 storeys), but for new buildings they must comply with the *Acceptable Solutions* of Division B, Parts 1, 3, 4, 5, 6, 7, and 12. Farm buildings fall under the National Farm Building Code of Canada.

Because of the scope of the projects, architects and engineers of all disciplines are employed to design and take responsibility for the structure. Every component of the building has to be calculated and tested before being put into place. Every truckload of concrete is tested. The scaffolding is diagrammed and engineered. The building plans consist of hundreds of pages and are accompanied by a detailed "book" of specifications. The project can extend from several months to several years.

Heavy equipment such as cranes and excavators and pile-drivers are present at some point during the construction. Many sub-trades are on the site at the same time, performing different tasks. On-site safety is paramount, and there is a site safety officer on the larger projects. There are regular site meetings in the site office, and all workers must have the required safety certifications for the job. Apprentices work closely with their mentors, and everyone looks out for everyone else.

FOUNDATION WORK

Piles

Piles are structural foundation members that are driven into the ground to solid bearing, and are engineered to provide vertical and/or horizontal support for super-imposed heavy construction and to resist lateral and uplift loads. They can be hammered in, pushed in, or vibrated in depending on the soil conditions, noise requirements, and proximity to existing structures. They are classified as bearing piles, friction piles, cohesion piles, and sheet piles. There are several videos of pile work at *www .hammersteel.com/videos.html.*

Bearing piles are made of concrete, steel, wood, or a combination, and they are driven down to load-bearing soil, or as specified by the engineer.

Concrete piles are either cast in place or precast. Cast-in-place piles are either *shelled* or *shell-less*; that is, the concrete is cast into a corrugated steel casing and the casing remains, or the casing is removed as the concrete is poured into firm and cohesive soil (**Figure 33–1**). The casing either has a solid core called a **mandrel** that is removed after driving in order to fill it with concrete, or an auger-like device that removes the earth as it is being driven. In the auger-cast system, the drilling machine is fitted with a hollow-stemmed auger that allows it to push concrete through the auger as the bit is withdrawn.

Pre-cast piles have a **pile head**, a **pile foot**, and a body that has a rebar cage on the inside for tensile strength. The head is protected from the pile hammer with a **pile cap** and the foot is protected with a **pile shoe.**

FIGURE 33–1 Shell-less cast-in-place concrete pile (auger-cast).

The body is tapered to facilitate penetration into the ground.

Steel piles are either H-shaped or pipe-shaped columns. Pipe piles have a steel boot that keeps the earth out as they are being driven and they can be filled with concrete after being set into place. H-shaped piles, called *soldier piles*, are used in conjunction with wood timbers as retaining walls (**Figure 33–2**).

Wood piles last hundreds of years in the right environment. They will not decay under water and are used in tandem with other materials that will not rot in soil or air. The wood is usually Douglas fir-larch (D.Fir-L) or hemlock-fir (Hem-Fir) and is pressure-treated with creosote or chromate copper arsenate (CCA). South American green heart is used for docks.

Friction Piles

Friction piles do not reach down to solid bearing but instead transfer their loads through skin friction. They are also known as *floating pile foundations.*

FIGURE 33–2 Timbers used as "lagging" in H-shaped piles.

Cohesion Piles

These piles also transmit most of their load by friction, but they are placed closely together and this process compresses the adjacent soil. They are thus also referred to as *compaction piles* or *grouped piles*, and are sometimes capped together with concrete to support super-imposed point loads.

Sheet Piles

Steel sheet piling is Z-shaped and interlocking, and serves as retaining walls and to contain materials that are injurious to the environment (**Figure 33–3**). Sheet piling is also used for a *cofferdam*, a water-tight compartment constructed in water, below and above the surface, in which workers perform bridge pier construction and repairs. It is available in wood, concrete, vinyl, aluminum, and fibreglass.

Caissons

When it is impracticable to reach bearing soil, *caisson concrete columns*, 20 inches (508 mm) and more in diameter, are created by pouring concrete around a rebar cage that is suspended in holes that are bored to a stable level of earth. Caissons are much larger than piles, and often they are drilled with steel casings that are lowered in sections as the depth is increased. For greater support, the bottom of the caisson is *belled* to create a pedestal-like concrete column.

Mat Foundation

A mat foundation (or *raft* foundation) is a slab foundation that is placed below grade to "float" on non-load-bearing soil in cases where stable soil cannot be economically reached. The excavation removes soil that is equal in weight to the proposed building, and then a thick, reinforced concrete slab is poured to provide support for a series of columns that must not be allowed to settle differentially. The mat foundation is usually 3 to 10 feet (1 to 3 m) thick.

Spread Foundation

Similar to a footing in residential construction, the commercial spread foundation uses a much wider spread footing and therefore is much thicker. The design must be engineered.

Flatwork

Some commercial concrete slabs are produced in three stages (**Figure 33–4**). First, the concrete piers or caissons are placed deep into the ground, as per plan. Then a concrete *grade beam* is cast in place on top of the piers. And after the mechanicals and steel reinforcement are positioned, then the concrete slab is placed. All components are connected together with rebar and isolation joints and expansion joints are set and/or cut with a concrete saw.

Courtesy of Hammer & Steel

FIGURE 33–3 Sheet piles being pushed into place by a Z-Pile Pusher.

Courtesy of Fab-Form Industries

FIGURE 33–4 Heavy construction concrete slab showing rebar grid and Fastfoot fabric form.

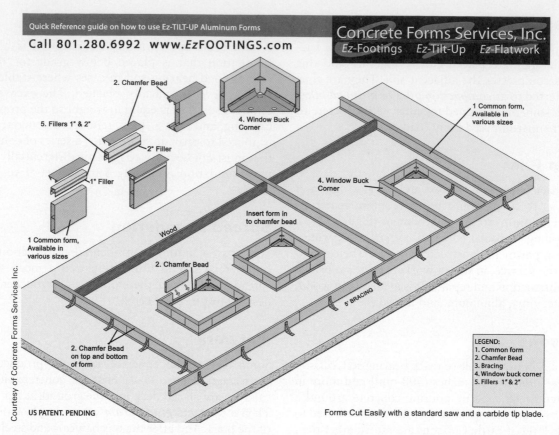

FIGURE 33–5 Forming for tilt-up concrete walls.

For a monolithic slab-on-grade construction in light commercial construction, it is common to thicken the slab at the perimeter to distribute the imposed loads of the walls and roof. The "footing" is incorporated with the floor slab to allow for one placement of the concrete. The engineer will set the size and number of rebar stirrups, as well as the floor grid or the welded wire fabric, and the specifications for the concrete mixture will be clearly defined. In colder climates, high-density extruded polystyrene (XPS) is placed under and around the slab, and sometimes an XPS "apron" will be sloped away from the sides of the slab perimeter. A general rule is that for the dimension the XPS projects from the foundation, the ground under the perimeter is protected from freezing to that depth.

Tilt-Up Concrete Construction

Tilt-up concrete construction follows the same rules as other concrete construction—geological evaluation, structural engineering, solid base, etc.—and it is similar in concept to platform framing in wood-frame construction. The floor slab is cast in place on the foundation and the walls are formed and cast on the floor and then "tilted up" into place (**Figure 33–5**).

The concrete for the walls must be reinforced to withstand not only building and environmental stresses but also lifting stresses (**Figure 33–6**). Also, before the concrete is placed, inserts and embeds are installed. *Embeds* are pre-fabricated steel plates with lugs that are cast into the panel to attach it to the footing, other panels, or the roof system, or for attachment of building accessories after the shell is completed. They can be attached to the side forms

FIGURE 33–6 Tilt-up concrete walls.

Courtesy of Bob Moore Construction Inc.

FIGURE 33–7 Connecting a brace to an *insert*.

if they are on the panel edges, or they can be wired to the reinforcing. *Inserts* provide the attachment points for lifting hardware and braces (**Figure 33–7**). They usually are sized by the supplier, who also should furnish engineering drawings showing insert locations (*www.tilt-up.org/basics/inserts.php*).

The surfaces of the concrete can be textured in the casting bed and the walls insulated in a "sandwich" application. The concrete is placed and consolidated using conventional methods. After the set time for the concrete, the walls are craned into position and braced plumb by attaching adjustable braces to the inserts. The braces remain in position until the roof and roof deck are securely installed. The lifting operation can potentially be dangerous, so it is important that *all* workers on-site be included in the "Site Talk" on the set procedure and on each one's responsibility.

Lift-Slab

Lift-slab concrete construction is used in square or rectangular buildings that have successive floors that are separated and supported by columns. Construction starts with the foundation and ground slab being

placed in location as per plans and specifications. The next step is to position in place all of the perimeter and internal columns for the building. These can be steel or pre-cast concrete columns. Then the floor slabs and the roof slab are cast on top of the ground floor slab, beginning with the second floor and ending with the roof slab. Each slab is kept separate from the other with a resin material called a *bond-break*. The slabs are also kept separate from the columns with isolation joint material. Hydraulic jacks are then placed on top of the columns and, in unison, they gently lift each slab into position. The slabs are held in position with shear rods and, in the case where the columns are steel, steel collars are welded in place directly under the slab to serve as permanent supports.

Lift-slab operations are also potentially dangerous. In 1987, 28 workers were killed at Bridgeport, Connecticut, as a result of a collapse during a lift-slab operation. There is now an OSHA (Occupational Safety & Health Association) standard in the United States for lift-slab construction.

PRE-CAST CONCRETE

Pre-cast concrete, as opposed to cast-in-place, refers to concrete that is formed and cast under controlled conditions, either in a factory or on a level area close to the site where it will be positioned. First used in the United States on a bridge in Philadelphia in 1950, pre-cast has become popular in today's heavy construction. Many structures contain components of the same dimensions and these can be mass-produced and then brought to a site, thereby avoiding the unpredictability of weather. In hotel construction, entire rooms are precast and then stacked to form the building (**Figure 33–8**).

Pre-casting also benefits the industry and the environment. Forms last much longer, as the work is done in a controlled and strictly supervised setting. Architectural concrete is used and the resulting smooth finish eliminates the need for additional finishing.

Pre-Stressed Concrete

The bridge to Prince Edward Island is pre-cast and *pre-stressed* concrete (**Figure 33–9**). In normal concrete work, the tensile strength of the steel rebar adds to the compressive strength of the concrete to produce a material that is strong in both compression and tension when placed under load. *Pre-tensioning* refers to tensioning the steel before the concrete is placed, which creates compressive stresses in the cured concrete member that will balance the tensile stresses that it will experience when placed under load. This allows designers

Courtesy of Peri® Forming System

FIGURE 33–8 Pre-cast concrete, Golden Ears Bridge, Vancouver, BC.

Photo by M. Nauth

FIGURE 33–9 Confederation Bridge to Prince Edward Island, 13 km long, pre-cast and pre-stressed.

Courtesy of Portland Cement Association

FIGURE 33–10 Banded tendons follow column lines with transverse tendons spaced uniformly.

to specify lighter and thinner concrete structures without sacrificing strength (For more information about pre-stressed concrete, refer to *www.cement.org/basics/concreteproducts_prestressed.asp*.)

Post-Tensioning

Pre-stressed concrete can also be tensioned after the concrete hardens. In *post-tensioning*, the concrete is cast around plastic ducts or sheaths and it does not come into contact with the steel tendons (**Figure 33–10**). Once the concrete is sufficiently cured, the steel tendons are stretched against the ends of the unit and anchored off externally, thus placing the concrete into compression and providing

greater strength under load. Post-tensioning is used for cast-in-place concrete and for large girders, floor slabs, shells, roof, and pavements.

HEAVY CONSTRUCTION FORMING

Forming for heavy construction is mainly done with propriety forming systems, such as ALUMA and PERI. The largest in the world is PERI Formwork Systems Inc, with the parent company PERI GmbH based in Weissenhorn, Germany. PERI is located in six cities across Canada.

FIGURE 33–11 Water tank reservoir, Barrie, ON.

FIGURE 33–12 Scaffolding at Lester B. Pearson International Airport, Toronto ON.

An important part of site work is scaffolding. Safety, stability, security, ease of installation, and access are of the essence. Examples of large scaffolding installations are shown in **Figures 33–11, 33–12,** and **33–13.**

Fly Forms

When casting identical floors in succession, the floor slab form is supported on a system of joists and girders, which are supported on adjustable "props" or *shores*. After the concrete is placed and then sufficiently set, the adjustable props are dropped and "flown" by crane and cables to the next level, hence the name *fly forms* (**Figure 33–14**). The freshly placed concrete floor is supported by *reshores* until such time as it reaches the required strength.

Slip Forms

Slip forms, used extensively for bridge piers and towers, are forms that climb upwards to provide for continuous placement of architectural concrete.

FIGURE 33–13 Harry Rosen expansion project, Toronto, ON. MULTIPROPs and GT 24 girders were used to maintain a clean storefront during construction.

Courtesy of Peri Formwork Systems Inc.

FIGURE 33–14 Fly form: RCMP headquarters, Surrey, BC.

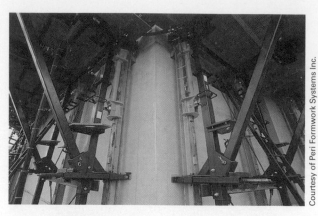

Courtesy of Peri Formwork Systems Inc.

FIGURE 33–16 A view of the ACS hoisting gear that moves the climbing unit without the use of an intermediate anchor usually required for the next pour. Viaduc de Millau Motorway Bridge, France.

As the design dictates, the forms can be adjusted to curve in or out as they climb. They are surrounded by a secure working platform that moves with the forms (**Figures 33–15** and **33–16**).

The Viaduc de Millau Motorway Bridge in France provided a site for the various aspects of heavy concrete construction (**Figure 33–17**). See the bridge under construction in animation at *www.peri.ca/en/projects.cfm*.

Courtesy of Peri Formwork Systems Inc.

FIGURE 33–15 The PERI climbing scaffold systems CB 240 together with VARIO GT24 wall formwork provide the solution for construction of piers. PERI UP Stairs provides access from ground to the pour level. 16 Mile Creek Bridge, Oakville, ON.

Courtesy of Peri Formwork Systems Inc.

FIGURE 33–17 A view of the Viaduc de Millau in France under construction.

DECONSTRUCT THIS

Carefully study **Figure 33–18** and think about what is wrong and/or what is right. Consider all possibilities. What construction practice or method is different in your area of the country?

FIGURE 33–18 This photo shows a back-filled foundation and footing.

KEY TERMS

admixtures	concrete	half-round	pilasters
aggregates	cove moulding	hydration	pile cap
air-entrained concrete	curing	independent slab	pile foot
bearing piles	flutes	lift-slab	pile head
blockouts	footing	mandrel	pile shoe
bucks	frost line	mat foundation	piles
cleats	girders	monolithic slab	pitch board
compressive strength	gusset	panels	Portland cement (PC)

preserved wood foundation (PWF)	rise	slump test	tensile strength
	run	snap ties	tilt-up
quarter-round	scabs	spreaders	walers
rebars	screeded	spud vibrator	waste factor
rebar vibrator	slab-on-grade foundation	strongbacks	yoking
reinforced concrete			

REVIEW QUESTIONS

Select the most appropriate answer.

1. What is concrete made of?

   a. Portland cement, sand, and water
   b. Portland cement, sand, gravel, and water
   c. mortar mix, sand, and water
   d. mortar mix, sand, gravel, and water

2. What are the steel rods that are placed in concrete to increase its tensile strength?

   a. reinforcing bars
   b. aggregates
   c. reinforcing nails
   d. duplex nails

3. Why are the inside surfaces of forms oiled?

   a. to protect the forms from moisture
   b. to prevent the loss of moisture from concrete
   c. so the forms can be stripped more easily
   d. to prevent honeycombs in the concrete

4. Why are keyways often put in spread footings?

   a. for unlocking the forms for easy removal
   b. to increase the compressive strength of concrete
   c. to keep the form boards from spreading
   d. to provide a stronger joint between footing and foundation

5. Which statement about rapid placing of concrete is correct?

   a. It omits the need for vibrating.
   b. It will burst the forms.
   c. It keeps the aggregate from separating.
   d. It reduces voids and honeycombs.

6. What could happen if footings were NOT placed below the frost line?

   a. The foundation could settle.
   b. The foundation might heave and crack.
   c. Excavation would be difficult in winter.
   d. There would be problems with form construction.

7. Why are spreaders used for footing forms?

   a. to allow easy placement of the concrete
   b. to keep the forms straight
   c. because they are easier to fasten
   d. because they maintain the proper footing width

8. Which of the following is NOT a consideration when preparing to size and dimension for a step in a footing?

   a. the concrete block used in the foundation
   b. the form used to pour the foundation wall
   c. the building code requirements
   d. the strength of the concrete mix

9. What length must the horizontal surface of a stepped footing be?

   a. at least 4 feet (1.2 m)
   b. at least twice the vertical distance
   c. at least the vertical distance
   d. at least the thickness of the footing

10. What is the typical order of installation of a manufactured forming system?

    a. inside forms, outside forms, reinforcing bars, then snap ties
    b. inside forms, reinforcing bars, snap ties, then outside forms
    c. outside forms, snap ties, inside forms, then reinforcing bars
    d. outside forms, snap ties, reinforcing bars, then inside forms

11. Which of the following is NOT one of the reasons walers are used on concrete wall forms?

    a. to stiffen the forms
    b. to straighten the forms
    c. to strengthen the forms
    d. to plumb the forms

12. Which of the following is NOT applicable to the curing of a concrete slab?

    a. It must be protected from freezing before it cures.
    b. It must be kept moist so that it does not cure too quickly.
    c. It must be protected on the underside by a sub-slab vapour retarder.
    d. It must be reinforced with welded wire fabric.

13. What volume of concrete should be ordered for a 6-inch (152 mm) slab that measures 24 × 36 feet (7.32 × 10.98 m)?

    a. 16 cubic yards (12.22 m³)
    b. 36 cubic yards (27.54 m³)
    c. 192 cubic yards (146.88 m³).
    d. 432 cubic yards (330.48 m³)

14. When erecting footing forms, what procedure follows locating the corner stakes with a plumb bob and batter board strings?

    a. installation of the spreaders
    b. installation of the outside form boards
    c. installation of the inside form boards
    d. installation of the reinforcing bars

15. What is caused by overvibrating concrete while placing it in wall forms?

    a. voids and honeycombs
    b. the aggregate rises to the top
    c. extra side pressure on the forms that could cause form failure
    d. separation of the components of the concrete mix

16. Which of the following is placed on the outside of the concrete forms?

    a. blockouts
    b. bucks
    c. girder pocket forms
    d. walers

17. How do you establish a level line along which to screed the top of a foundation?

    a. space nails along a chalk line
    b. vibrate the top surface so it will flow level
    c. use a chalk line only
    d. add enough water to the concrete so that it will flow level

18. What volume of concrete should be ordered for a 150 mm slab that measures 11.5 × 15.2 m?

    a. 26.22 m³
    b. 262.2 m³
    c. 2622 m³
    d. 26220 m³

19. How far must bearing piles be driven down?

    a. to bedrock
    b. to undisturbed soil, as pre-determined by the municipality
    c. to load-bearing soil, as specified by the engineer
    d. to twice the frost depth

20. A cofferdam, used in bridge construction, is constructed below and above the surface of water using which type of piles?

    a. cohesion
    b. friction
    c. grouped
    d. sheet

21. When are fly forms used?

    a. when casting tilt-up walls on the ground
    b. when casting identical floors in succession
    c. when casting cantilevered beams on tall buildings
    d. when casting water tank reservoirs

22. Which of the following types of architectural concrete placement require the use of slip forms?

    a. continuous
    b. intermittent
    c. structural
    d. smooth

23. In which country are PERI Formwork Systems developed?

    a. U.S.A.
    b. Germany
    c. Japan
    d. Canada

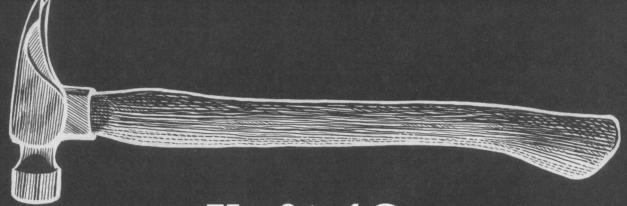

Unit 12

Floor Framing

Wood frame construction is used for residential and light commercial construction for important reasons of economy, durability, and variety. The cost for wood frame construction is generally less than for other types of construction. Fuel and air-conditioning expenses are reduced because wood frame construction provides better insulation.

Wood frame homes are very durable. If properly maintained, a wood frame building will last indefinitely. Many existing wood frame structures are hundreds of years old.

Because of the ease with which wood can be cut, shaped, fitted, and fastened, many different architectural styles are possible. In addition to single-family homes, wood frame construction is used for all kinds of low-rise buildings, such as apartments, condominiums, offices, motels, warehouses, and manufacturing plants.

OBJECTIVES

After completing this unit, the student should be able to:

- describe platform, balloon, and post-and-beam framing, and identify framing members of each.
- describe several energy and material conservation framing methods.
- build and install beams, erect columns, and lay out sills.
- lay out and install floor joists.
- frame openings in floors.
- lay out, cut, and install bridging.
- apply subflooring.
- describe several stairway designs.
- define terms used in stair framing.
- determine the unit rise and unit run of a stairway.
- determine the length of and frame a stairwell.
- lay out a stair carriage and frame a straight stairway.
- lay out and frame a stairway with a landing.
- lay out and frame a stairway with winders.
- lay out and frame service stairs.

 SAFETY REMINDER

Floor framing creates an elevated horizontal plane. Construction of the floor must proceed in a logical and thoughtful manner to reduce the risk of falling.

Chapter 34

Types of Frame Construction

There are several methods of framing a building. Some types are used less often today but still exist, so knowledge of them is necessary when remodelling. Other types are relatively new and knowledge about them is not widespread. Some wood frames are built using a combination of types. New designs utilizing engineered lumber are increasing the height and width to which wood frame structures can be built.

PLATFORM FRAME CONSTRUCTION

The **platform frame**, sometimes called the *western* frame, is most commonly used in residential construction (**Figure 34–1**). In this type of construction, the floor is built and the walls are erected on top of it. When more than one storey is built, the second-floor platform is erected on top of the walls of the first storey.

At each floor level, a flat surface is provided on which to work. A common practice is to assemble wall framing units on the floor and then tilt the units up into place (**Figure 34–2**).

Effects of Shrinkage

Lumber shrinks mostly across width and thickness. A disadvantage of the platform frame is the relatively large amount of settling caused by the shrinkage of the large number of horizontal load-bearing frame members. However, because of the equal amount of horizontal lumber, the shrinkage is more or less equal throughout the building. To reduce shrinkage, only framing lumber with the proper moisture content should be used (MC ≤ 19 percent).

BALLOON FRAME CONSTRUCTION

In **balloon frame** construction, the wall *studs* and first-floor *joists* rest on the sill plate. The second-floor joists rest on a 1 × 4 (19 × 89 mm) **ribbon**

that is cut in flush with the inside edges of the studs (**Figure 34–3**). This type of construction is used less often today, but a substantial number of structures built with this type of frame are still in use.

Effects of Shrinkage

Shrinkage of lumber along the length of a board is insignificant compared to shrinkage that can occur across the width of the board. Therefore, in the balloon frame, settling caused by shrinkage of lumber is held to a minimum in the exterior walls. This is because the studs are continuous from sill to top **plate**. To prevent unequal settling of the frame due to shrinkage, the studs of **bearing partitions** rest directly on the *girder or beam*.

FIRESTOPS

Firestop blocking is material installed to slow the movement of fire and smoke within smaller cavities of the building frame during a fire. It is sometimes called **draftstop blocking** or *fireblocking*. This allows the occupants more time to get out of a burning building and can reduce the overall damage. Firestops must be installed in many places.

In a wood frame, a firestop in a wall might consist of dimension lumber blocking between studs. Fire blocks are required when the studs exceed 9'-10" (3 m) and the wall cavity is *not* filled with insulation. In a standard platform frame, the wall plates act as firestops.

Firestops must be installed in the following locations:

- In all stud walls, partitions, and furred spaces at ceiling and floor levels.
- Between stair carriages at the top and bottom. (*Carriages* are stair framing members. They are sometimes called *stair horses*.)
- Around chimneys, fireplaces, vents, pipes, and at ceiling and floor levels with non-combustible material.

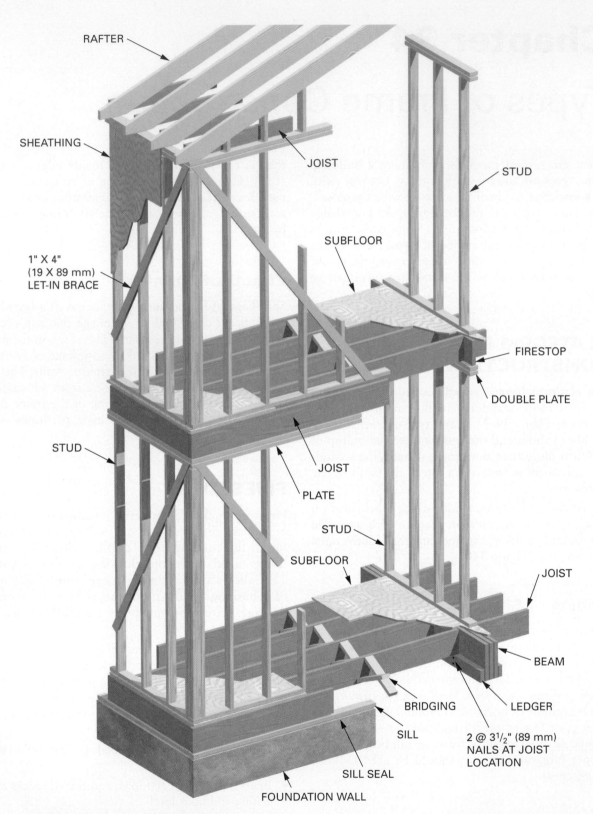

RAFTER

SHEATHING

JOIST

STUD

1" X 4"
(19 X 89 mm)
LET-IN BRACE

SUBFLOOR

FIRESTOP

DOUBLE PLATE

STUD

JOIST

PLATE

STUD

SUBFLOOR

JOIST

BEAM

LEDGER

BRIDGING

2 @ 3¹/₂" (89 mm)
NAILS AT JOIST
LOCATION

SILL

SILL SEAL

FOUNDATION WALL

FIGURE 34–1 Platform frame construction.

FIGURE 34–2 Platform frame buildings are built one floor at a time.

- The space between floor joists at the sill and beam.
- All other locations as required by building codes (**Figure 34–4**).

POST-AND-BEAM FRAME CONSTRUCTION

The *post-and-beam* frame uses fewer but larger pieces. Large timbers, widely spaced, are used for joists, posts, and roof beams. **Matched boards** (tongue and grooved) are often used for floors and roof sheathing (**Figure 34–5**).

Floors

APA-rated Sturd-I-Floor 48 **on centre** (OC), which is 1³/₃₂ inches (28 mm) thick, may be used on floor joists that are spaced 4 feet (1.2 m) OC instead of matched boards (**Figure 34–6**). In addition to being nailed, the plywood panels are glued to the floor beams with construction adhesive applied with caulking guns. The use of matched planks allows the floor beams to be more widely spaced.

Walls

Exterior walls of a post-and-beam frame may be constructed with widely spaced posts. This allows wide expanses of glass to be used from floor to ceiling. Usually some sections between posts in the wall are studded at close intervals, as in platform framing. This provides for door openings, fastening for finish, and wall **sheathing**. In addition, close spacing of the studs permits the wall to be adequately braced (**Figure 34–7**).

Roofs

The post-and-beam frame roof is widely used. The exposed roof beams and sheathing on the underside are attractive. Usually the bottom surface of the roof planks is left exposed to serve as the finished ceiling. Roof planks come in 2-, 3-, and 4-inch (50, 75, and 100 mm) nominal thicknesses. Some are **end** matched as well as **edge** matched. Some buildings may have a post-and-beam roof while the walls and floors are conventionally framed.

The post-and-beam roof may be constructed with a *longitudinal* frame. The beams run parallel

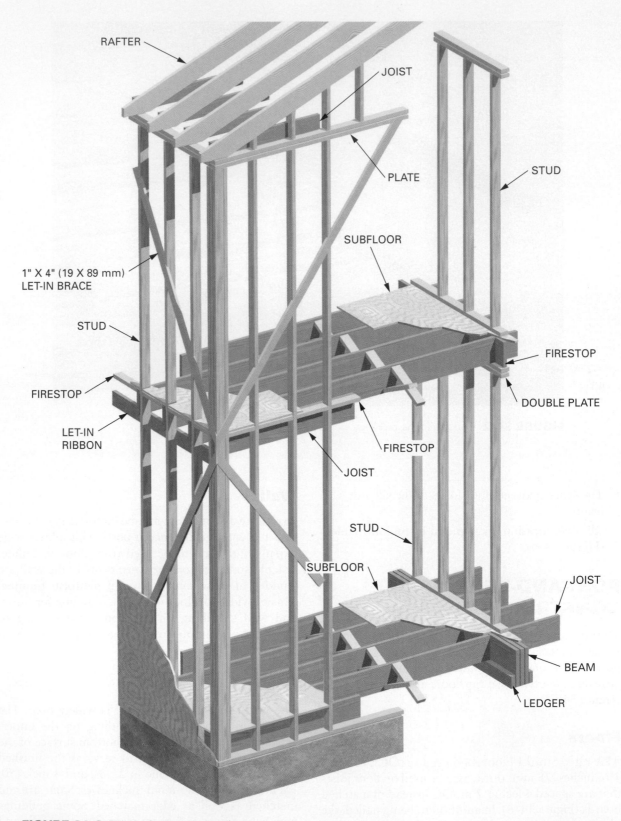

FIGURE 34–3 Balloon frame construction.

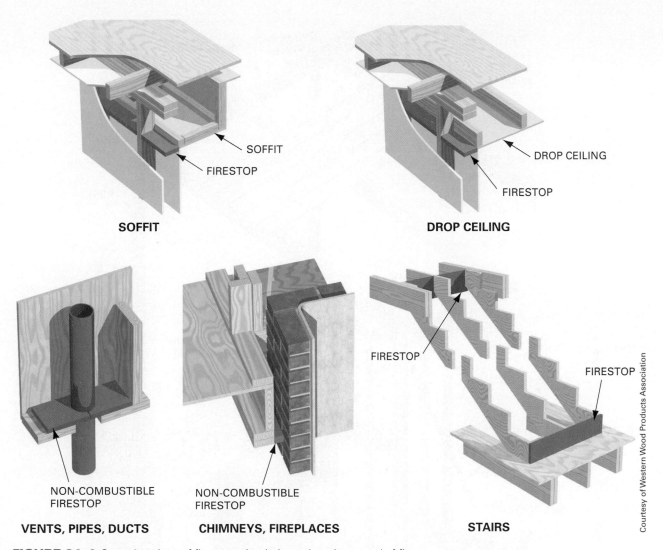

SOFFIT

SOFFIT
FIRESTOP

DROP CEILING

DROP CEILING
FIRESTOP

VENTS, PIPES, DUCTS

NON-COMBUSTIBLE
FIRESTOP

CHIMNEYS, FIREPLACES

NON-COMBUSTIBLE
FIRESTOP

STAIRS

FIRESTOP
FIRESTOP

Courtesy of Western Wood Products Association

FIGURE 34–4 Some locations of firestops that help to slow the spread of fire.

to the ridge beam. Or they may have a *transverse* frame. The beams run at right angles to the ridge beam, similar to roof rafters.

The ridge beam and longitudinal beams, if used, are supported at each end by posts in the end walls. They must also be supported at intervals along their length. This prevents the side walls from spreading and the roof from sagging (**Figure 34–8**). One of the disadvantages of a post-and-beam roof is that interior partitions and other interior features must be planned around the supporting roof beam posts.

Because of the fewer number of pieces used, a well-planned post-and-beam frame saves material and labour costs. Care must be taken when erecting

the frame to protect the surfaces and make well-fitting joints on exposed posts and beams. Glulam beams are well-suited for this and are frequently used in post-and-beam construction. A number of metal connectors are used to join members of the frame (**Figure 34–9**).

ENERGY AND MATERIAL CONSERVATION FRAMING METHODS

There has been much concern and thought about conserving energy and materials in building construction. Several systems have been devised that differ from conventional framing methods.

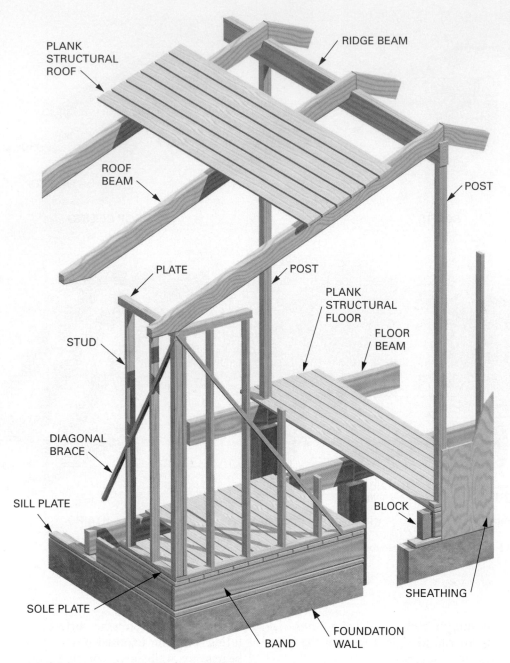

FIGURE 34–5 The post-and-beam frame.

They conserve energy and use less material and labour. Check provincial and local building codes for limitations.

Floors

For maximum savings, a single layer of ¾-inch (19 mm) tongue-and-grooved plywood is used over joists. In-line floor joists are used to make installation of the plywood floor easier (**Figure 34–10**). The use of adhesive when fastening the plywood floor is recommended. Gluing increases stiffness and prevents squeaky floors (**Figure 34–11**).

Walls

Walls are required to have two planes of protection against the weather. In **Figure 34–12**, they are provided

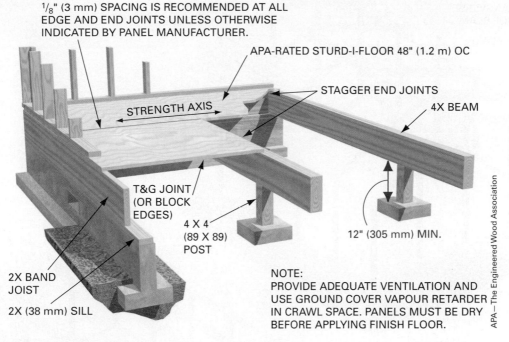

FIGURE 34–6 Floor beams may be spaced 4 feet (1.2 m) OC when 1³/₃₂-inch-thick (28 mm) panels are used for a floor.

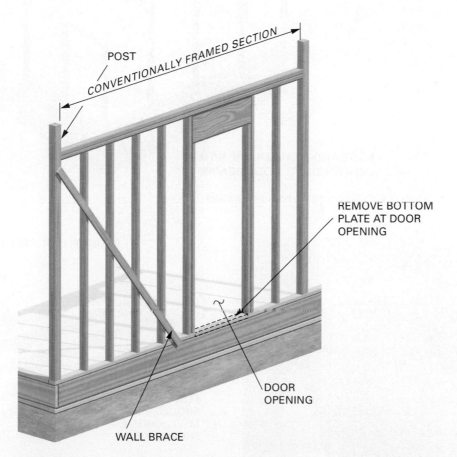

FIGURE 34–7 Sections of the exterior walls of a post-and-beam wall may need to be conventionally framed.

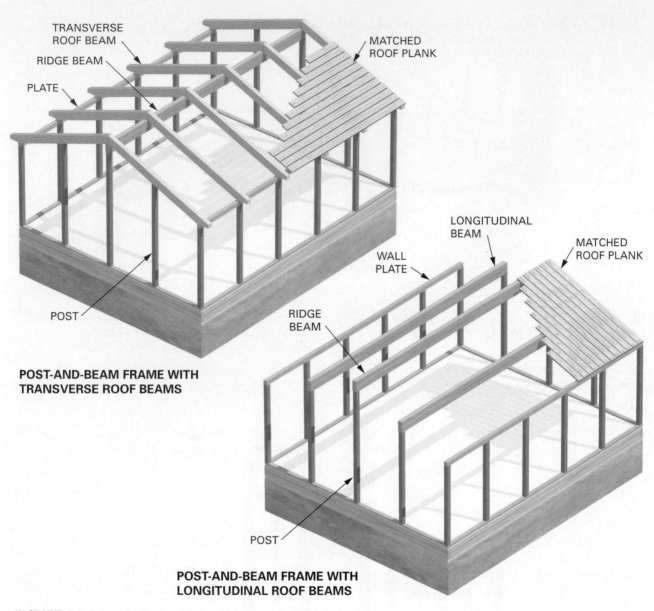

POST-AND-BEAM FRAME WITH
TRANSVERSE ROOF BEAMS

POST-AND-BEAM FRAME WITH
LONGITUDINAL ROOF BEAMS

FIGURE 34–8 Longitudinal and transverse post-and-beam roofs.

FIGURE 34–9 Metal connectors are specially made to join glulam beams.

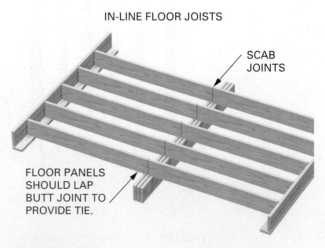

FIGURE 34–10 In-line floor joists make installation of plywood subflooring simpler.

FIGURE 34–11 Using adhesive when fastening subflooring makes the floor frame stiffer, stronger, and quieter.

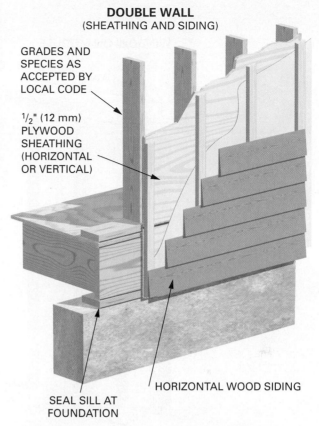

DOUBLE WALL
(SHEATHING AND SIDING)

GRADES AND SPECIES AS ACCEPTED BY LOCAL CODE

1/2" (12 mm) PLYWOOD SHEATHING (HORIZONTAL OR VERTICAL)

HORIZONTAL WOOD SIDING

SEAL SILL AT FOUNDATION

FIGURE 34–12 Siding and housewrap serve as two planes of protection (NBCC 9.27.2.2).

by the housewrap paper placed over the wall sheathing and by the siding.

Wall openings are planned to be located so that at least one side of the opening falls on an OC stud. Whenever possible, window and door sizes are selected so that the rough opening width is a multiple of the module (**Figure 34–13**). Also, locate partitions at OC wall stud positions if possible.

Roofs

Roof systems can be modified to improve the insulation over the exterior walls. The raised heel is built into the truss by the manufacturer (**Figure 34–14**). It raises the roof slightly to allow for full-thickness insulation at the eaves. Most areas in Canada also use 2 × 6 (38 × 140 mm) wall studs to increase the wall insulation.

House Depths

House depths that are not evenly divisible by four waste floor framing and sheathing. Lumber for floor joists is produced in increments of 2 feet (610 mm).

Assuming the beam remains in the centre of the building, a house 25 feet (7.6 m) wide would require 14-foot-long (4.2 m) floor joists. These joists could be used uncut to make a building 28 feet (8.5 m) wide. If the beam was installed offset from the centre, 12- and 14-foot (3.7 and 4.2 m) joists could be used to span 25 feet (7.6 m). Either way, material is wasted.

Full-width subfloor panels can be used without cutting on buildings that are 24, 28, and 32 feet (7.3, 8.5, and 9.8 m) wide. This decreases construction time and saves money.

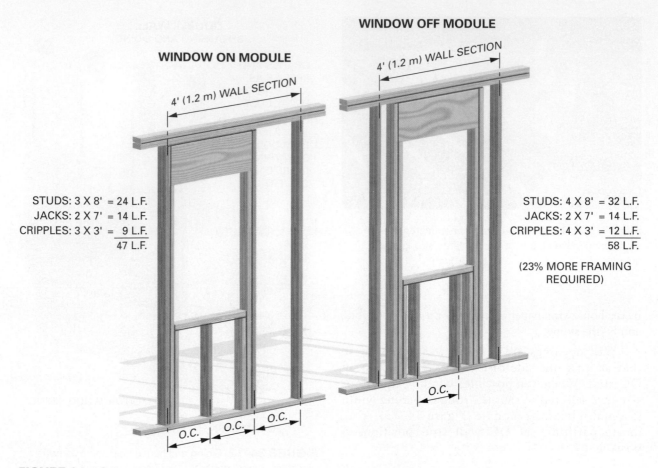

WINDOW ON MODULE

4' (1.2 m) WALL SECTION

STUDS: 3 X 8' = 24 L.F.
JACKS: 2 X 7' = 14 L.F.
CRIPPLES: 3 X 3' = 9 L.F.
_____
47 L.F.

O.C. O.C. O.C.

WINDOW OFF MODULE

4' (1.2 m) WALL SECTION

STUDS: 4 X 8' = 32 L.F.
JACKS: 2 X 7' = 14 L.F.
CRIPPLES: 4 X 3' = 12 L.F.
_____
58 L.F.

(23% MORE FRAMING
REQUIRED)

O.C.

FIGURE 34–13 To conserve materials, locate wall openings so that they fall on the OC studs.

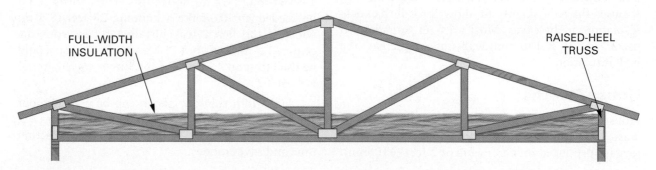

FULL-WIDTH
INSULATION

RAISED-HEEL
TRUSS

FIGURE 34–14 Modified truss design accommodates thick ceiling insulation without compressing at eaves.

Chapter 35

The Substructure: Columns, Beams, and Sill Plates

A floor frame consists of members fastened together to support the loads a floor is expected to bear. The floor frame is started after the foundation has been placed and has hardened. A straight and level floor frame makes it easier to frame and finish the rest of the building.

The next several chapters will focus on platform frame construction. It is important that the floor frame be constructed on a substructure that transfers building loads to the foundation walls and to the footings that bear on undisturbed soil.

DESCRIPTION AND INSTALLATION OF FLOOR FRAME MEMBERS

In the usual order of installation, the floor frame consists of *girders* or *beams*, *posts* or *columns*, *sill plates*, *joists*, **bridging**, and *subflooring* (**Figure 35–1**).

Description of Beams

Beams are horizontal load-bearing members that support the inner ends of floor joists. Girders are heavy beams that support other smaller beams. Several types of beams are commonly used.

Kinds of Beams. Beams may be made of solid wood or built up of three or more 2 × (38 ×) planks or laminated veneer lumber (LVL) or glulam (**Figure 35–2**). Sometimes, wide flange, I-shaped steel beams are used.

Determining the Size of Beams and Other Structural Members. The sizes and end bearings of engineered beams such as glulam and LVL beams are best determined by the manufacturer or supplier. Steel beams are sized by structural engineers. Common sizes of steel I-beams can be found in provincial or national building codes. Also, tables for built-up wood beams are found in the building code with specified end bearings

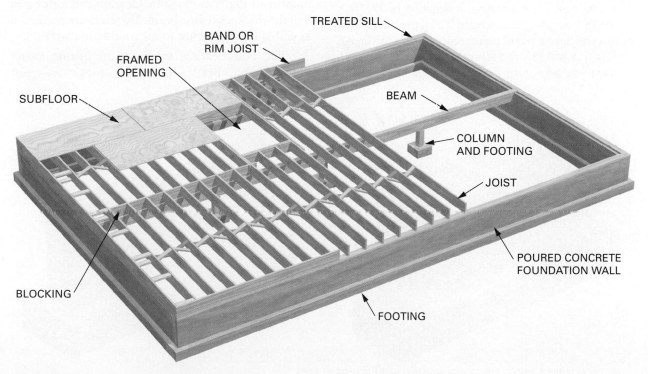

FIGURE 35–1 A floor frame of platform construction.

FIGURE 35–2 Large glulam beams are often used as beams and lintels.

(NBCC Tables A-8 to A-10). However, more complete tables are found in the *Span Book* published by the Canadian Wood Council. The *Span Book* also has tables for glulam beams.

Steel I-beams are sized according to flange width, beam depth, and weight per unit length. For example, a wide-flange steel beam that is 6 inches deep and weighs 15 pounds per foot is designated a W6 × 15. The same beam has a metric designation of W150 × 22, or 150 mm deep and weighs 22 kg per metre. Steel beams will have a wood plate secured to the top flange, usually a 2 × 6 (38 × 140 mm), to provide a surface for the toenailing of the floor joists (**Figure 35–3**).

Span tables for steel beams are found in the National Building Code of Canada (NBCC Tables A-20 to A-29) and on page 283, Table 18, of the PDF edition of *Canadian Wood-Frame House Construction* found at *http://www.cmhc-schl.gc.ca/odpub/pdf/61010.pdf?fr=1414012864483*.

Built-up beam sizes depend on the supported joist length, the post spacing, and the number of storeys they support. Supported joist length refers to one-half the sum of the joist span on either side of the beam. Also, beam strength depends on the wood species and grade; most common is spruce-pine-fir (SPF) #1 and #2. The strength of the beam will vary as the width times the depth squared (wd^2). Thus, a 3-ply 2 × 12 (38 × 286 mm) beam will generally be stronger than a 4-ply 2 × 10 (38 × 235 mm) beam because $3 × 2 × 12^2$ (864) is greater than $4 × 2 × 10^2$ (800).

Built-up Beams. Built-up beams consist of a minimum of three members fastened together with nails or bolts. The NBCC requires that a minimum of two 3½-inch (89 mm) nails be placed 4 to 6 inches (100 to 152 mm) in from the end of each member and that two rows of nails be fastened at 18 inches (457 mm) OC (**Figure 35–4**). Butt joints in each ply must be located within 6 inches (152 mm) of ¼ point of clear span (unsupported distance) or over a support. Butt joints must be cut square and true to maximize integrity of load transfer. No two adjacent butt joints can be at the same joint location, and the number of joints at the same location cannot exceed half the thickness of the beam. No joints are permitted at end ¼ points (closest to the foundation wall).

The end bearing for built-up beams varies with the number of plies, the supported joist

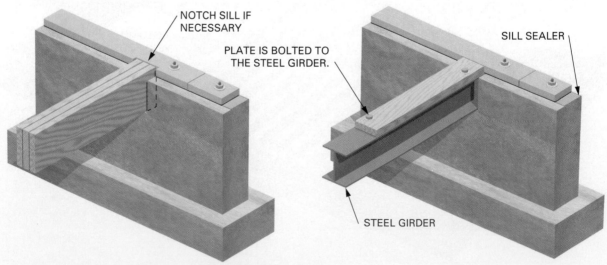

NOTCH SILL IF NECESSARY

PLATE IS BOLTED TO THE STEEL GIRDER.

SILL SEALER

STEEL GIRDER

NOTE: ALL SILLS ARE FLUSH WITH NAILING SURFACE OF BEAMS.

FIGURE 35–3 Variations in girder and sill installations.

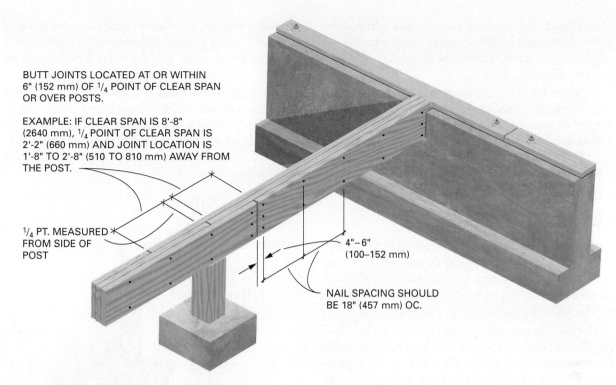

BUTT JOINTS LOCATED AT OR WITHIN 6" (152 mm) OF ¼ POINT OF CLEAR SPAN OR OVER POSTS.

EXAMPLE: IF CLEAR SPAN IS 8'-8" (2640 mm), ¼ POINT OF CLEAR SPAN IS 2'-2" (660 mm) AND JOINT LOCATION IS 1'-8" TO 2'-8" (510 TO 810 mm) AWAY FROM THE POST.

¼ PT. MEASURED FROM SIDE OF POST

4"–6" (100–152 mm)

NAIL SPACING SHOULD BE 18" (457 mm) OC.

FIGURE 35–4 Spacing of fasteners and seams of a built-up beam made with dimension lumber.

length, and the number of floors that are being supported.

- *Beams supporting one floor*: 3-ply beams with supported lengths greater than 13'-9" (4.2 m) require 4½-inch (114 mm) bearing; all other beams require 3-inch (75 mm) bearing.
- *Beams supporting two floors*: 3-ply beams require 4½-inch (114 mm) bearing; 4- and 5-ply beams with supported lengths greater than 9'-10" (3 m) require 4½-inch (114 mm) bearing; all other beams require 3-inch (75 mm) bearing.
- *Beams supporting three floors*: All beams require 4½-inch (114 mm) bearing.

Beam Location. The ends of the beam are usually supported by a *pocket* formed in the foundation wall. The pocket should provide at least a 3-inch (75 mm) bearing for the beam. It should be wide enough to provide ½-inch (12.5 mm) clearance on both sides and at the end. This allows any moisture to be evaporated by circulation of air. Thus, no moisture will get into the beam, which would cause decay of the wood.

The pocket is formed deep enough to provide for shimming the beam to its designated height. A steel bearing plate may be *grouted* in under the beam while it is supported temporarily (**Figure 35–5**). **Grouting** is the process of filling in a small space with a thick paste of cement. Wood **shims** are not usually suitable for use under beams. The weight imposed on them compresses the wood, causing the beam to sink below its designated level. The wood beam should

be separated from the concrete by a moisture break such as polyethylene or a steel shim.

Installing Beams

Steel girders usually come in one piece and are set in place with a crane (minimum bearing of 3½ inches [89 mm]). A solid wood beam is often installed in a similar manner but is pieced together with half-lap joints. Joints are made near the posts or columns.

Wood beams are usually built up from dimension lumber and erected in sections. This process begins by building one section at a time. One end is then set in the pocket in the foundation wall. The other

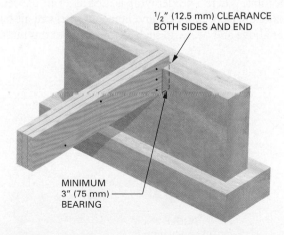

½" (12.5 mm) CLEARANCE BOTH SIDES AND END

MINIMUM 3" (75 mm) BEARING

FIGURE 35–5 A beam pocket of a foundation wall should be large enough to provide air space around the end and sides of the beam.

end is placed and fastened on a braced temporary support. Continue building and erecting sections on posts until the beam is completed into the opposite pocket (see Figure 35–5).

Sight the beam by eye from one end to the other or use a stringline. Place wedges under the temporary supports to straighten the beam. Permanent posts or columns are usually installed after the beam has some weight imposed on it by the *floor joists*. Temporary posts should be strong enough to support the weight imposed on them until permanent ones are installed.

SILL PLATES

Sill plates, also called *mudsills*, are horizontal members of a floor frame. They bear directly on the foundation wall and provide a bearing for *floor joists*. It is prudent that the sill be made with a decay-resistant material such as pressure-treated lumber.

The sill plate is secured to the foundation wall with anchor bolts or other proprietary anchors (which must be compatible with wood preservatives). Anchor bolts are usually L-shaped and must be embedded at least 4 inches (100 mm) into the concrete. Maximum spacing is 7'-10½" (2.4 m), and the anchor bolts must be located 6 to 12 inches (152 to 305 mm) in from the end of each piece of sill plate (**Figure 35–6**).

The National Building Code of Canada requires a full bed of mortar on an uneven foundation wall to level the sill plate. Or the sill can be placed directly on top of a level foundation. If the sill plate is less than 6 inches (150 mm) above grade, it must be pressure-treated or separated from the concrete by a continuous seal of caulking or an acceptable gasket. Some builders run a bead of caulking (zigzagged), a layer of roofing felt or sill gasket, another bead of caulking (or acoustical sealant), and then the sill plate. This meets the intent of the NBCC to provide a moisture break and an air seal at the joint. Where a header wrap is used as an air barrier, it shall be sealed, clamped, and lapped to the wall air barrier above and below in such a way as to be continuous.

Sills must be installed so that they are straight, level, and to the specified dimension of the building. The level of all other framing members depends on the care taken with the installation of the sill.

Sometimes the outside edge of the sill is flush with the outside of the foundation wall. Sometimes it is set in the thickness of the wall sheathing, depending on custom or design. In the case of brick-veneered exterior walls, the sill plate may be set back even farther (**Figure 35–7**).

Installing Sills

Remove washers and nuts from the anchor bolts. Snap a chalk line on the top of the foundation wall in line

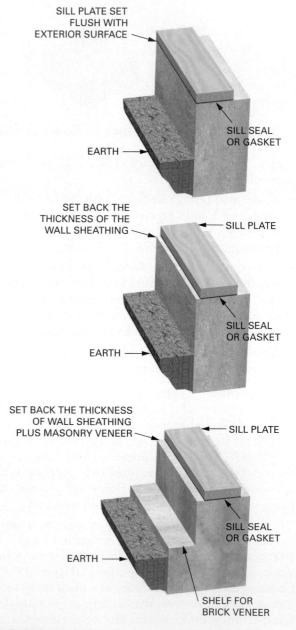

FIGURE 35–7 A sill plate may be located with different setbacks from the foundation edge.

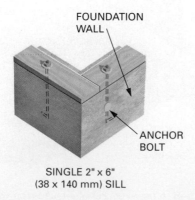

FIGURE 35–6 An anchor bolt should be located between 6 and 12 inches (152 and 305 mm) from the end of each plate.

StepbyStep Procedures

Procedure 35–A Installing a Sill Plate on Foundation with Anchor Bolts

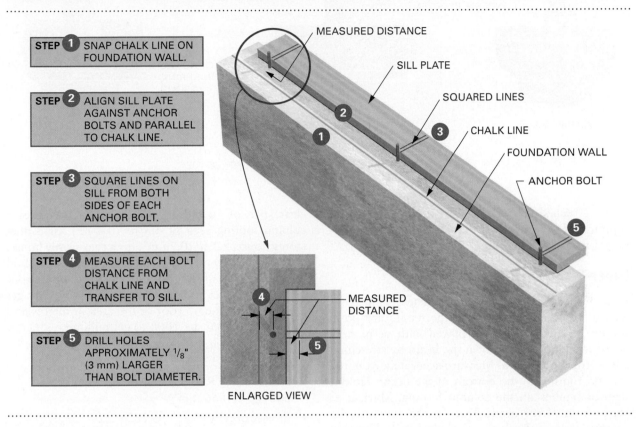

STEP 1 SNAP CHALK LINE ON FOUNDATION WALL.

STEP 2 ALIGN SILL PLATE AGAINST ANCHOR BOLTS AND PARALLEL TO CHALK LINE.

STEP 3 SQUARE LINES ON SILL FROM BOTH SIDES OF EACH ANCHOR BOLT.

STEP 4 MEASURE EACH BOLT DISTANCE FROM CHALK LINE AND TRANSFER TO SILL.

STEP 5 DRILL HOLES APPROXIMATELY 1/8" (3 mm) LARGER THAN BOLT DIAMETER.

MEASURED DISTANCE

SILL PLATE

SQUARED LINES

CHALK LINE

FOUNDATION WALL

ANCHOR BOLT

MEASURED DISTANCE

ENLARGED VIEW

with the inside edge of the sill. These lines must be of the correct dimension and must form a squared rectangle.

Cut the sill sections to length. Hold the sill in place against the anchor bolts. Square lines across the sill on each side of the bolts. Measure the distance from the centre of each bolt to the chalk line. Transfer this distance at each bolt location to the sill by measuring from the inside edge (**Procedure 35–A**).

Bore holes in the sill for each anchor bolt. Bore the holes at least 1/8 inch (3 mm) oversize to allow for adjustments. Place the sill sections in position over the anchor bolts and check that the inside edges of the sill sections are on the chalk line. Remove the sill plates, install the predetermined sill air seal, and then reposition the sill plates. Replace the nuts and washers. Be careful not to overtighten the nuts, especially if the concrete wall is still green. This may crack the wall.

COLUMNS

Beams may be supported by framed walls, wood posts, or steel columns (**Figure 35–8**). Metal plates are used at the top and bottom of the columns to distribute the load over a wider area. The plates have predrilled holes so that they may be fastened to the

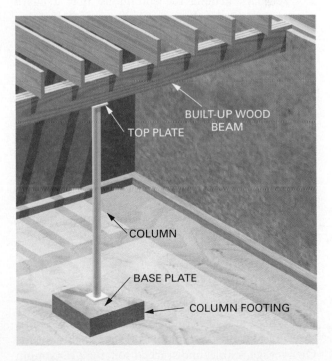

FIGURE 35–8 Typical column supporting a beam.

BUILT-UP WOOD BEAM

TOP PLATE

COLUMN

BASE PLATE

COLUMN FOOTING

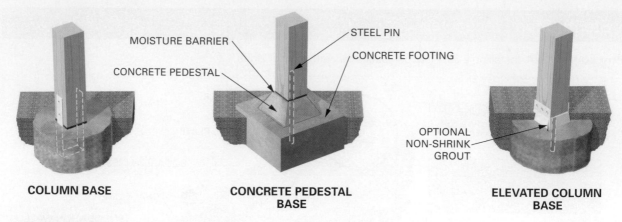

MOISTURE BARRIER

STEEL PIN

CONCRETE PEDESTAL

CONCRETE FOOTING

OPTIONAL NON-SHRINK GROUT

COLUMN BASE

CONCRETE PEDESTAL BASE

ELEVATED COLUMN BASE

FIGURE 35–9 The bottoms of wood posts sometimes rest on pedestal-type footings.

beam. Notched sections prevent the columns from slipping off the plates. Column size should be determined from the blueprints.

Installing Columns

After the floor joists are installed and before any more weight is placed on the floor, the temporary posts supporting the beam are replaced with permanent posts or columns. Straighten the beam by stretching a line from end to end. Measure accurately from the column footing to the bottom of the beam. Hold a strip of lumber on the column footing. Mark it at the bottom of the beam. Transfer this mark to the column. Deduct the thickness of the top and bottom column plate.

To mark around the column so that it has a square end, wrap a sheet of paper around it. Keeping the edges even, mark along the edge of the paper.

Cut through the metal along the line using a hacksaw, reciprocating saw, or circular saw with a metal cutting blade. Install the columns in a plumb position under the beam and centred on the footing. Fasten the top plates to the beam with lag screws. If the beam is steel, then holes must be drilled. The plates are then bolted to the beam, or they may be welded to the beam. The bottoms of the columns are held in place when the finish concrete basement floor is placed around them.

Wood posts are installed in a similar manner, except that their bottoms are placed on a pedestal footing (**Figure 35–9**).

COLUMN FOOTINGS

A column is supported on a column footing or pier pad, whose area depends on the load the column is carrying and on the type of soil on which it bears. On average, stable, undisturbed soil, for a maximum

clear span of 10 feet (3 m), the NBCC requires a column footing area of 4.3 ft² (0.4 m²) for a one-storey house, 8.1 ft² (0.75 m²) for a two-storey house, and 10.8 ft² (1.0 m²) for a three-storey house. To find the dimensions of the footings, take the square root of the area in feet and multiply the result by 12 to get inches, or the square root of the area in metres and multiply the result by 1000 to get millimetres.

- one-storey: $\sqrt{4.3}$ ft² = 2.07, × 12 = 24.88″ → 25″ × 25″ (632 mm)
- two-storey: $\sqrt{8.1}$, = 2.84, × 12 = 34.09″ → 34″ × 34″ (866 mm)
- three-storey: $\sqrt{10.8}$, = 3.28, × 12 = 39.37″ → 40″ × 40″ (1000 mm)

To determine the footing thickness, subtract the width of the base plate from the width of the footing and divide by two. For example, a 25 × 25-inch (632 × 632 mm) footing with a column base plate of 6 × 6 inches (152 × 152 mm) is required to be at least 9.5 inches (240 mm) thick, → 25″ − 6″ = 19″ ÷ 2 = 9.5″ (632 − 152 = 480 ÷ 2 = 240).

If the clear span between the posts exceeds 10′ (3 m), increase the area of the pad by the percentage increase of clear span. The increased area means a larger pad, which in turn means a thicker pad, unless it is reinforced. For example, if the span is increased from 10 feet (3 m) to 14 feet (4.2 m), that is an increase of 40 percent or 0.4. The area of the column footing is therefore increased by 40 percent or is multiplied by 1.4. Taking the square root of the area will give the length of the side of the pad. The new calculations are as follows:

- 4.3 ft² × 1.4 = 6.02 ft² (0.56 m²)
- $\sqrt{6.02}$, = 2.45′, × 12 = 29.44, or 30″ → the new pad is 30 × 30 inches (750 × 750 mm)
- the revised thickness: 30″ − 6″ = 24″, ÷ 2 = 12″ (300 mm) thick

Chapter 36

The Floor Frame: Joists, Openings, Bridging, and the Subfloor

FLOOR JOISTS

Floor joists are horizontal members of a frame. They rest on and transfer the load to sills and beams. In residential construction, dimension lumber placed on edge has traditionally been used. Wood I-joists, with lengths up to 80 feet (24.4 m), are being specified more often today (**Figure 36–1**). Steel framing in the form of joists and walls is sometimes used. In general, when the price of lumber increases significantly or when termites are a problem, steel framing is the solution.

Joists are generally spaced 16 inches (406 mm) OC in conventional framing. They may be spaced 12, 19.2, or 24 inches (305, 488, or 610 mm) OC, depending on the type of construction and the load. The joist spacing is designed so that an 8-foot (2.4 m) section of plywood is fully supported at its ends. Dividing 96 by the number of joist spaces in that section gives the OC (**Figure 36–2**). The size of floor joists should be determined from the construction drawings.

Joist Framing at the Sill

Joists should rest on at least 1½ inches (38 mm) of bearing. In platform construction, the ends of floor joists are capped with a *band joist*, also called a *rim*

joist, **box header**, or **joist header** (**Figure 36–3**). The use of wood I-joists requires sill construction as recommended by the manufacturer for satisfactory performance of the frame (**Figure 36–4**).

Joist Framing at the Beam

If joists are lapped over the beams, the minimum amount of lap is 3 inches (75 mm), or beam width, to allow for adequate nailing surface between lapped joists. The maximum overhang of joists at the beam is 12 inches (305 mm) to eliminate floor squeaking. These squeaks are caused when walking on the floor at joist midspan; the overhung joist end raises up, rubbing against the lapped joist. There is no need to lap wood I-joists. They come in lengths long enough to span the building. However, they may need to be supported by beams, depending on the span and size of the wood I-joists. No matter how the joists are framed over the beam, draftstop blocking is recommended. It should be installed using full-width framing lumber between joists on top of the beam (**Figure 36–5**).

Sometimes, to gain more headroom, joists may be framed into the side of the beam. There are a number of ways to do this. Joist hangers must be

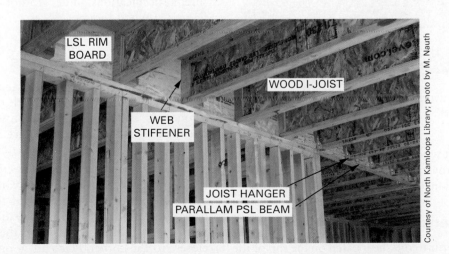

LSL RIM BOARD

WEB STIFFENER

WOOD I-JOIST

JOIST HANGER
PARALLAM PSL BEAM

Courtesy of North Kamloops Library; photo by M. Nauth

FIGURE 36–1 Engineered lumber makes a strong and flat floor system.

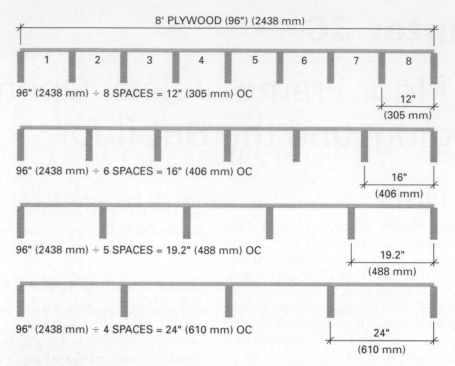

8' PLYWOOD (96") (2438 mm)

| 1 | 2 | 3 | 4 | 5 | 6 | 7 | 8 |

96" (2438 mm) ÷ 8 SPACES = 12" (305 mm) OC

12"
(305 mm)

96" (2438 mm) ÷ 6 SPACES = 16" (406 mm) OC

16"
(406 mm)

96" (2438 mm) ÷ 5 SPACES = 19.2" (488 mm) OC

19.2"
(488 mm)

96" (2438 mm) ÷ 4 SPACES = 24" (610 mm) OC

24"
(610 mm)

FIGURE 36–2 On-centre spacing is determined by dividing the length of a sheet of plywood by the number of joist spaces under it.

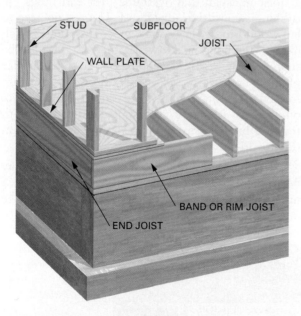

STUD   SUBFLOOR

JOIST

WALL PLATE

BAND OR RIM JOIST

END JOIST

FIGURE 36–3 Typical framing near the sill using dimension lumber.

used to support wood I-beams. Web stiffeners should be applied to the beam ends if the hanger does not reach the top flange of the beam (see Figure 36–1).

Notching and Boring of Joists

Notches are permitted to be located on the top of the joist if they are located within one-half of the joist

depth from the edge of bearing and are *not* deeper than one-third of the joist depth, unless the depth of the joist is increased by the depth of the notch. No notches are permitted on the bottom of joists.

Holes bored in joists for piping or wiring should not be larger than one-quarter of the joist depth. They should not be closer than 2 inches (50 mm) to the top or bottom of the joist (**Figure 36–6**).

Some wood I-joists are manufactured with perforated knockouts in the web along its length. This allows for easy installation of wiring and pipes. To cut other size holes in the web, consult the manufacturer's specifications guide. Do not cut or notch the flanges of wood I-joists.

Laying Out Floor Joists

The locations of floor joists are marked on the sill plate. A squared line marks the side of the joist. An *X* to one side of the line indicates on which side of the line the joist is to be placed (**Figure 36–7**).

Floor joists must be laid out so that the ends of *plywood subfloor* sheets fall directly on the centre of floor joists. Start the joist layout by measuring the joist spacing from the end of the sill. Measure back one-half the thickness of the joist. Square a line across the sill. This line indicates the side of the joist closest to the corner. Place an *X* on the side of the line on which the joist is to be placed.

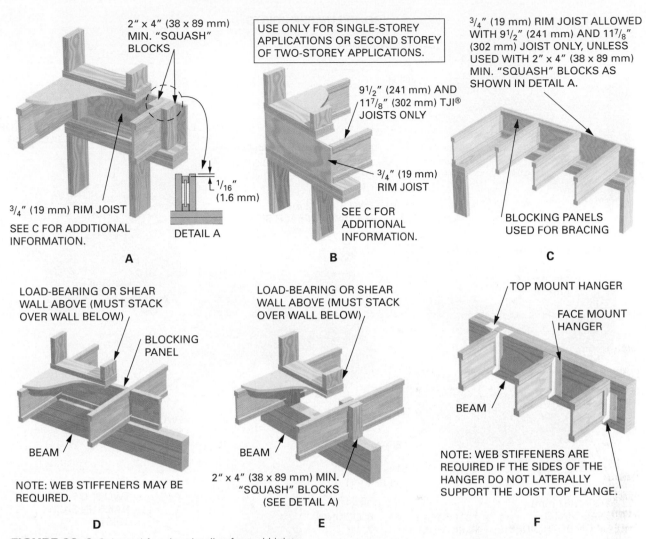

2" x 4" (38 x 89 mm)
MIN. "SQUASH"
BLOCKS

USE ONLY FOR SINGLE-STOREY
APPLICATIONS OR SECOND STOREY
OF TWO-STOREY APPLICATIONS.

3/4" (19 mm) RIM JOIST ALLOWED
WITH 9 1/2" (241 mm) AND 11 7/8"
(302 mm) JOIST ONLY, UNLESS
USED WITH 2" x 4" (38 x 89 mm)
MIN. "SQUASH" BLOCKS AS
SHOWN IN DETAIL A.

9 1/2" (241 mm) AND
11 7/8" (302 mm) TJI®
JOISTS ONLY

3/4" (19 mm)
RIM JOIST

1/16"
(1.6 mm)

DETAIL A

3/4" (19 mm) RIM JOIST

SEE C FOR ADDITIONAL
INFORMATION.

SEE C FOR
ADDITIONAL
INFORMATION.

BLOCKING PANELS
USED FOR BRACING

A

B

C

LOAD-BEARING OR SHEAR
WALL ABOVE (MUST STACK
OVER WALL BELOW)

BLOCKING
PANEL

LOAD-BEARING OR SHEAR
WALL ABOVE (MUST STACK
OVER WALL BELOW)

TOP MOUNT HANGER

FACE MOUNT
HANGER

BEAM

BEAM

BEAM

NOTE: WEB STIFFENERS MAY BE
REQUIRED.

2" x 4" (38 x 89 mm) MIN.
"SQUASH" BLOCKS
(SEE DETAIL A)

NOTE: WEB STIFFENERS ARE
REQUIRED IF THE SIDES OF THE
HANGER DO NOT LATERALLY
SUPPORT THE JOIST TOP FLANGE.

D

E

F

FIGURE 36–4 Selected framing details of wood I-joists.

From the squared line, measure and mark the spacing of the joists along the length of the building. Place an *X* on the same side of each line as for the first joist location (**Procedure 36–A**).

When measuring for the spacing of the joists, use a tape stretched along the length of the building. Most professional tapes have predominant markings for repetitive layout: black rectangles for 12- and 24-inch layouts, red rectangles for 16-inch, and small black diamonds for 19.2-inch layouts. Using a tape in this manner is more accurate. Measuring and marking each space individually with a rule or framing square generally causes a gain in the spacing. If the spacing is not laid out accurately, the plywood subfloor may not fall in the centre of some floor joists. Time will then be lost either cutting the plywood back or adding strips of lumber to the floor joists (**Figure 36–8**).

Laying Out Floor Openings. After marking floor joists for the whole length of the building,

study the plans for the location of floor openings. Mark the sill plate, where joists are to be doubled, on each side of large floor openings. Identify the layout marks that are not for full-length floor joists. Shortened floor joists at the ends of floor openings are called tail joists. They are usually identified by changing the *X* to a *T* or a *C* for cripple joist. Lay out for partition supports, or wherever doubled floor joists are required (**Figure 36–9**). Instead of doubling the floor joists under partition walls, the carpenter can instead install 2 × 4 (38 × 89 mm) blocking at 48 inches (1.2 m) OC between the adjacent joists. This will allow access to the wall cavity for wiring and other mechanical inserts. Check the mechanical drawings to make adjustments in the framing to allow for the installation of plumbing and heating runs.

Lay out the floor joists on the beam and on the opposite wall. If the joists are in-line, *X*s are made on the same side of the mark on both the beam and the sill plate on the opposite wall. If the joists are

SOLID LUMBER JOISTS

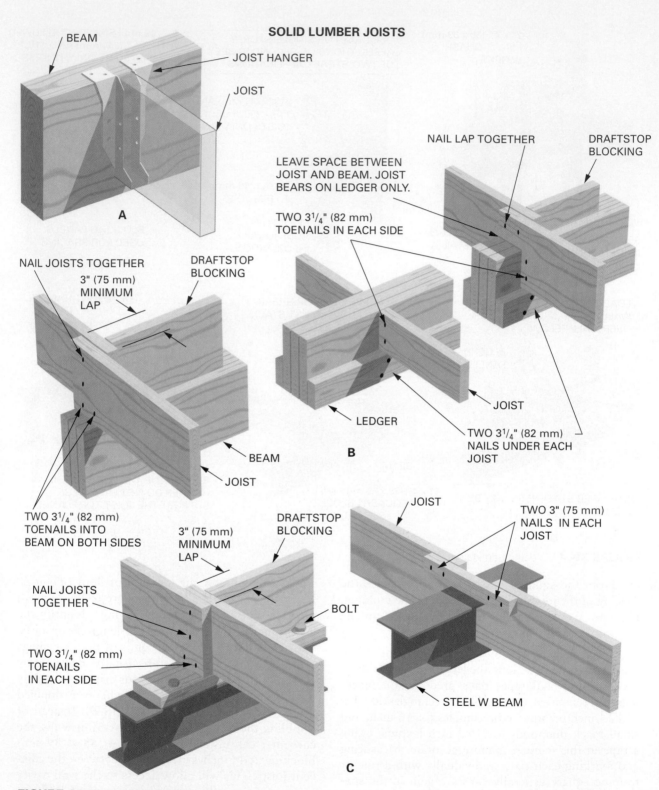

FIGURE 36–5 Various possible framing details at a beam.

lapped, a mark is placed on both sides of the line at the beam and on the opposite side of the mark on the other wall (**Figure 36–10**). These marks may be changed to make it easier to tell which mark is for the joist toward the front of the house and which is for the back.

Installing Floor Joists

Stack the necessary number of full-length floor joists at intervals along both walls. Each joist is carefully sighted along its length by eye. Any joist with a severe crook or other warp should not be

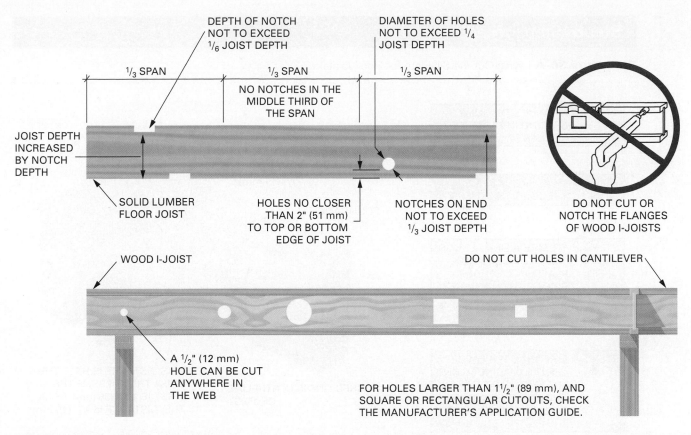

FIGURE 36-6 Allowable notches, holes, and cutouts in floor joists (see NBCC 9.23.5.1 and 5.2).

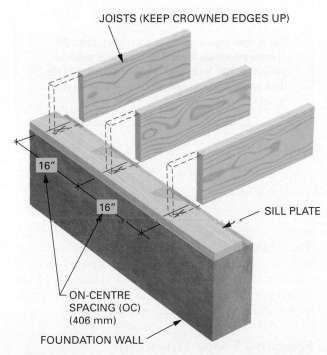

FIGURE 36-7 A line is drawn to mark the edge of a joist; an *X* is marked to indicate on which side of the line the joist is to be placed.

used. Joists are installed with the crowned (convex) edge up.

Keep the end of the floor joist in from the outside edge of the sill plate by the thickness of the rim joist. Toenail the joists to the sill and beam with 3¼-inch (82 mm) common nails. Nail the joists together if they lap at the beam (**Figure 36–11**). When all floor joists are in position, they are sighted by eye from end to end and straightened. They may be held straight by strips of 1 × 3s (19 × 64s) tacked to the top of the joists in about the middle of the joist span.

Wood I-joists are installed using standard tools. They can be easily cut to any required length at the jobsite. A minimum bearing of 1¾ inches (45 mm) is required at joist ends and 3½ inches (89 mm) over the beam. The wide, straight wood flanges on the joist make nailing easier, especially with pneumatic framing nailers (**Figure 36–12**). Nail joists at each bearing with one 3¼-inch (82 mm) nail on each side. Keep nails at least 1½ inches (38 mm) from the ends to avoid splitting. (Use Table 9.23.3.4 from the National Building Code of Canada (NBCC) for nailing requirements.)

StepbyStep Procedures

Procedure 36–A Laying Out the Sill Plate for Floor Joists

STEP ❶ MEASURE THE JOIST SPACING IN FROM THE CORNER.

STEP ❷ MEASURE BACK ½ THE JOIST THICKNESS.

STEP ❸ SQUARE A LINE ACROSS THE SILL PLATE AND PLACE AN *X* ON THE SIDE OF THE LINE WHERE THE JOIST IS TO BE PLACED.

STEP ❹ CONTINUE THE ON-CENTRE SPACING ALONG THE LENGTH OF THE BUILDING. USE A STEEL TAPE STRETCHED OVER THE ENTIRE LENGTH.

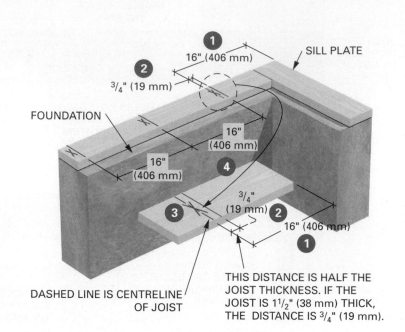

SILL PLATE

16" (406 mm)

¾" (19 mm)

FOUNDATION

16" (406 mm)

16" (406 mm)

¾" (19 mm)

16" (406 mm)

DASHED LINE IS CENTRELINE OF JOIST

THIS DISTANCE IS HALF THE JOIST THICKNESS. IF THE JOIST IS 1½" (38 mm) THICK, THE DISTANCE IS ¾" (19 mm).

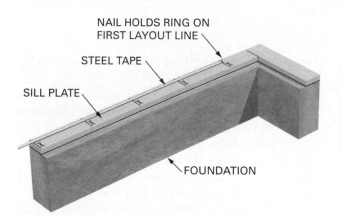

NAIL HOLDS RING ON FIRST LAYOUT LINE

STEEL TAPE

SILL PLATE

FOUNDATION

ON THE JOB

Use a steel tape when laying out floor joists. An inaccurate layout leads to cutting back subfloor panels so their ends will fall on the centres of the floor joists. A steel tape prevents gaining on the spacing.

FIGURE 36–8 Using a steel tape for layout reduces the possibility of step-off errors.

DOUBLING FLOOR JOISTS

For added strength, doubled floor joists must be securely fastened together. Their top edges must be even. In most cases, the top edges do not lie flush with each other. They must be brought even before they can be nailed together.

To bring them flush, toenail down through the top edge of the higher one, at about the centre of their length. At the same time squeeze both together tightly by hand. Use as many toenails as necessary,

spaced where needed, to bring the top edges flush (**Procedure 36–B**). Usually no more than two or three nails are needed. Then fasten the two pieces securely together. Drive nails from both sides, staggered from top to bottom, about 2 feet (610 mm) apart. Angle nails slightly so that they do not protrude.

Framing Floor Openings

Large openings in floors should be framed before floor joists are installed. This is because room

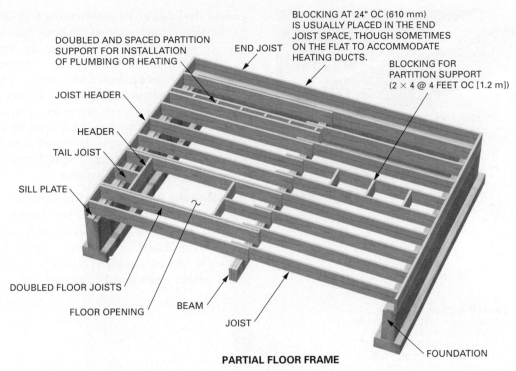

DOUBLED AND SPACED PARTITION SUPPORT FOR INSTALLATION OF PLUMBING OR HEATING

END JOIST

BLOCKING AT 24" OC (610 mm) IS USUALLY PLACED IN THE END JOIST SPACE, THOUGH SOMETIMES ON THE FLAT TO ACCOMMODATE HEATING DUCTS.

BLOCKING FOR PARTITION SUPPORT (2 × 4 @ 4 FEET OC [1.2 m])

JOIST HEADER

HEADER

TAIL JOIST

SILL PLATE

DOUBLED FLOOR JOISTS

FLOOR OPENING

BEAM

JOIST

FOUNDATION

PARTIAL FLOOR FRAME

FIGURE 36–9 Typical framing components of a floor system.

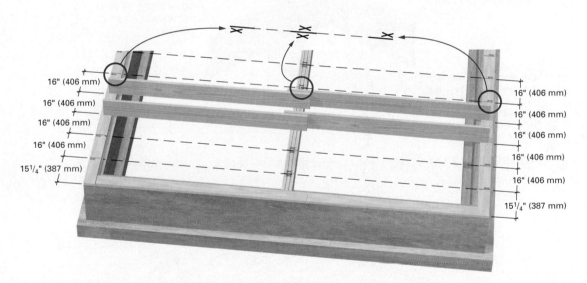

16" (406 mm)
16" (406 mm)
16" (406 mm)
16" (406 mm)
15¼" (387 mm)

16" (406 mm)
16" (406 mm)
16" (406 mm)
16" (406 mm)
15¼" (387 mm)

FIGURE 36–10 Floor joist layout lines span the entire width of the building. If the joists are lapped at the beam, then the *X*s are marked on different sides of the line.

is needed for end-nailing. To frame an opening in a floor, first fasten the trimmer joists in place. Trimmer joists are full-length joists that run along the inside of the opening. Mark the location of the *headers* on the trimmers. Headers are members of the opening that run at right angles to the floor joists. They must be doubled if they are more than 3'-11¼" (1.2 m) long.

Cut *headers* to length by taking the measurement at the sill between the trimmers. Taking the

measurement at the sill where the trimmers are fastened, rather than at the opening, is standard practice. A measurement between trimmers taken at the opening may not be accurate. There may be a bow in the trimmer joists (**Figure 36–13**).

Place two headers, one for each end of the opening, on the sill between the trimmers. Transfer the layout of the tail joists on the sill to the headers. Fasten the first header on each end of the opening in position by driving nails through the side of the trimmer into the

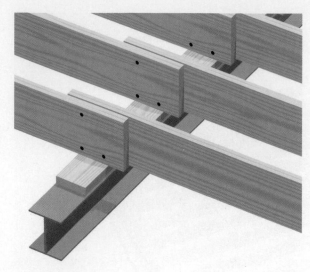

FIGURE 36–11 Floor joists may be lapped at the girder.

ends of the headers. Be sure that the first header is the header that is farthest from the floor opening. Fasten the tail joists in position. Double up the headers. Finally, double up the trimmer joists. **Procedure 36–C** shows the sequence of operations used to frame a floor opening. This particular sequence allows you to end-nail the members rather than toenailing them. Use joist hangers as required by local codes (see Figure 41–18, p. 453). Joist hangers must be installed with the correct amount, size, and type of nails. Roofing nails are **NOT** acceptable for joist hangers.

Installing the Rim Joist. After all the openings have been framed and all floor joists are fastened, install the rim joist. This closes in the ends of the floor joists. Rim joists may be lumber of the same size as the floor joists. They also may be a single or double layer of laminated stand lumber when wood I-joists are used.

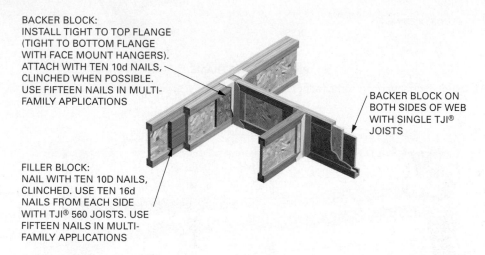

BACKER BLOCK:
INSTALL TIGHT TO TOP FLANGE
(TIGHT TO BOTTOM FLANGE
WITH FACE MOUNT HANGERS).
ATTACH WITH TEN 10d NAILS,
CLINCHED WHEN POSSIBLE.
USE FIFTEEN NAILS IN MULTI-
FAMILY APPLICATIONS

BACKER BLOCK ON
BOTH SIDES OF WEB
WITH SINGLE TJI®
JOISTS

FILLER BLOCK:
NAIL WITH TEN 10D NAILS,
CLINCHED. USE TEN 16d
NAILS FROM EACH SIDE
WITH TJI® 560 JOISTS. USE
FIFTEEN NAILS IN MULTI-
FAMILY APPLICATIONS

FIGURE 36–12 (A) Doubled wood I-joist requires blocking between the webs.

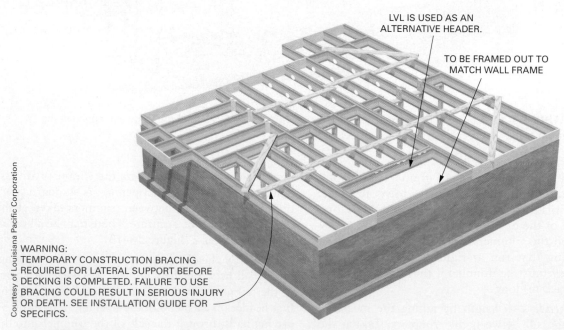

LVL IS USED AS AN
ALTERNATIVE HEADER.

TO BE FRAMED OUT TO
MATCH WALL FRAME

Courtesy of Louisiana Pacific Corporation

WARNING:
TEMPORARY CONSTRUCTION BRACING
REQUIRED FOR LATERAL SUPPORT BEFORE
DECKING IS COMPLETED. FAILURE TO USE
BRACING COULD RESULT IN SERIOUS INJURY
OR DEATH. SEE INSTALLATION GUIDE FOR
SPECIFICS.

FIGURE 36–12 (B) Floor framing using wood I-joists.

Step**by**Step Procedures

Procedure 36–B Aligning the Top Edges of Joists

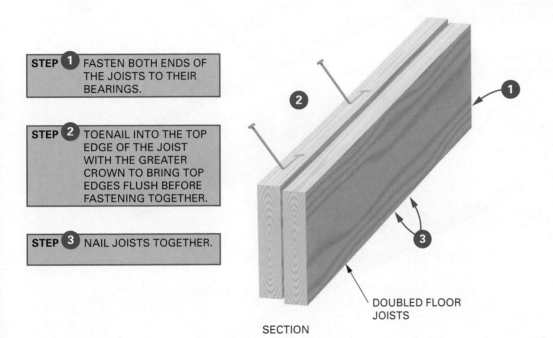

ON THE JOB

Bring the tops of double floor joists flush by first toenailing the edges before nailing them together.

STEP 1 FASTEN BOTH ENDS OF THE JOISTS TO THEIR BEARINGS.

STEP 2 TOENAIL INTO THE TOP EDGE OF THE JOIST WITH THE GREATER CROWN TO BRING TOP EDGES FLUSH BEFORE FASTENING TOGETHER.

STEP 3 NAIL JOISTS TOGETHER.

DOUBLED FLOOR JOISTS

SECTION

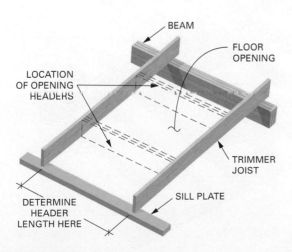

BEAM

FLOOR OPENING

LOCATION OF OPENING HEADERS

TRIMMER JOIST

DETERMINE HEADER LENGTH HERE

SILL PLATE

FIGURE 36–13 Length of the header for a floor opening should be measured at the sill, not at the midspan.

Fasten the rim joist into the end of each floor joist. If wood I-joists are used as floor joists, drive one nail into the top and bottom flange. The rim joist is also toenailed to the sill plate at about 6-inch (150 mm) intervals.

BRIDGING

Bridging is installed in rows between floor joists at intervals not exceeding 6′-11″ (2.1 m). For instance, floor joists with spans of 8 to 14 feet (2.4 to 4.2 m) need one row of bridging near the centre of the span. Its purpose is to distribute a concentrated load on the floor over a wider area. Floor joists that span more than 14 feet (4.2 m) require two rows of bridging.

Bridging may be solid wood, wood cross-bridging, metal cross-bridging, or 1 × 3 (19 × 64 mm)

Step**by**Step Procedures

Procedure 36–C Installing Framing Members around a Floor Opening

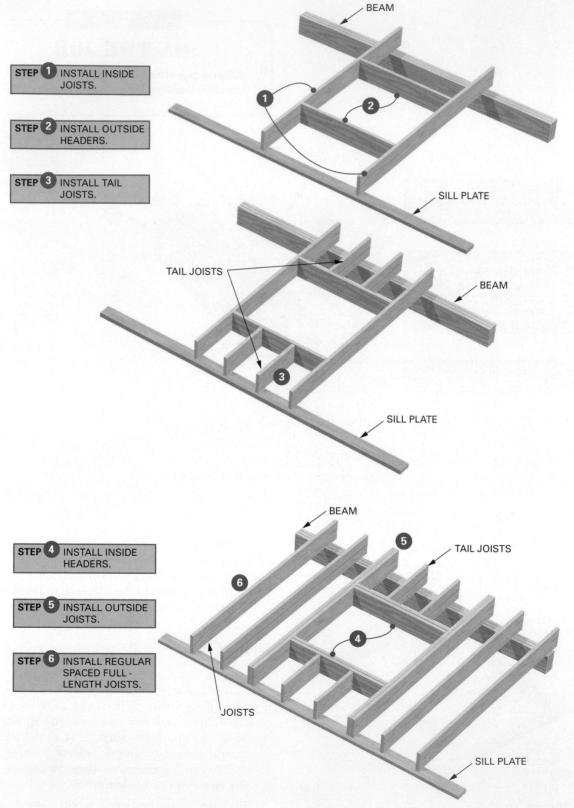

STEP 1 INSTALL INSIDE JOISTS.

STEP 2 INSTALL OUTSIDE HEADERS.

STEP 3 INSTALL TAIL JOISTS.

BEAM

SILL PLATE

TAIL JOISTS

BEAM

SILL PLATE

STEP 4 INSTALL INSIDE HEADERS.

STEP 5 INSTALL OUTSIDE JOISTS.

STEP 6 INSTALL REGULAR SPACED FULL - LENGTH JOISTS.

BEAM

TAIL JOISTS

JOISTS

SILL PLATE

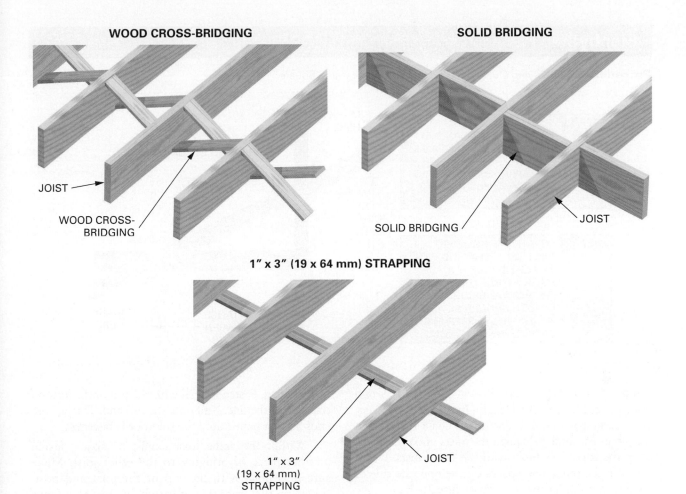

WOOD CROSS-BRIDGING

JOIST

WOOD CROSS-BRIDGING

SOLID BRIDGING

SOLID BRIDGING

JOIST

1″ x 3″ (19 x 64 mm) STRAPPING

1″ x 3″ (19 x 64 mm) STRAPPING

JOIST

FIGURE 36–14 Types of bridging.

nailed to the bottom of joists (**Figure 36–14**). Usually, solid wood bridging is the same size as the floor joists. It is installed in an offset fashion to permit end-nailing.

Wood cross-bridging should be at least nominal 1 × 3 (19 × 64 mm) lumber with two 2¼-inch (57 mm) nails at each end. It is placed in double rows that cross each other in the joist space.

Metal cross-bridging is available in different lengths for particular joist size and spacing. It is usually made of 18-gauge steel, and is ¾ inch (19 mm) wide. It comes in a variety of styles. It is applied in a way similar to that used for wood cross-bridging (see *http://www.homeadditionplus.com/framing-info/Cross_Bracing_Floor_Joists.htm*).

Laying Out and Cutting Wood Cross-Bridging

Wood cross-bridging may be laid out using a framing square. Determine the actual distance between floor joists and the actual depth of the joist. For example, 2 × 10 floor joists 24 inches OC measure 22½ inches

between them. The actual depth of the joist is 9¼ inches (235 mm). Or for a 38 × 235 mm floor joist at 610 mm OC, measure 572 mm between them.

Hold the framing square on the edge of a piece of bridging stock. Make sure the 9¼-inch (235 mm) mark of the tongue lines up with the upper edge of the stock. Also make sure the 22½-inch (572 mm) mark of the blade lines up with the lower edge of the stock. Mark lines along the tongue and blade across the stock.

Rotate the square, keeping the same face up. Realign the square to the previous marks and then mark along the tongue (**Procedure 36–D**). The bridging may then be cut using a power mitre box. Tilt the blade and use a stop set to cut duplicate lengths.

Installing Bridging

Determine the centreline of the bridging. Snap a chalk line across the tops of the floor joist from one end to the other. Square down from the chalk line to the bottom edge of the floor joists on both sides.

StepbyStep Procedures

Procedure 36–D Laying Out Cross-Bridging Using a Framing Square

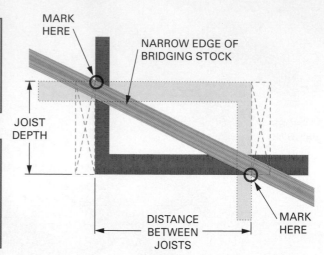

STEP 1 HOLD SQUARE IN FIRST POSITION AS INDICATED BY THE DARKER SQUARE. MARK STOCK ALONG BLADE AND TONGUE. THESE MARKS ARE MADE ON OPPOSITE EDGES OF BRIDGING EDGE.

STEP 2 ROTATE THE SQUARE TO THE POSITION AS INDICATED BY THE LIGHTER SQUARE. MARK ALONG THE TONGUE. THE ACTUAL BRIDGING LENGTH SHOULD BE ABOUT $\frac{1}{4}$" (6 mm) SHORTER FOR EASE OF INSTALLATION.

MARK HERE

NARROW EDGE OF BRIDGING STOCK

JOIST DEPTH

DISTANCE BETWEEN JOISTS

MARK HERE

Solid Wood Bridging. To install solid wood bridging, cut the pieces to length. Install pieces in every other joist space on one side of the chalk line. Fasten the pieces by nailing through the joists into their ends. Keep the top edges flush with the floor joists. Install pieces in the remaining spaces on the opposite side of the line in a similar manner (**Procedure 36–E**).

Wood Cross-Bridging. To install wood cross-bridging, start two 2¼-inch (57 mm) nails in one end of

the bridging. Fasten it flush with the top of the joist on one side of the line. Nail only the top end. The bottom ends are not nailed until the subfloor is fastened.

Within the same joist cavity or space, fasten another piece of bridging to the other joist. Make sure it is flush with the top of the joist and positioned on the other side of the chalk line. Also, leave a space between the bridging pieces where they form the *X* to minimize floor squeaks. Continue installing

StepbyStep Procedures

Procedure 36–E Installing Solid Bridging

STEP 1 INSTALL SOLID BRIDGING IN EVERY OTHER SPACE.

STEP 2 FILL IN THE REMAINING SPACES.

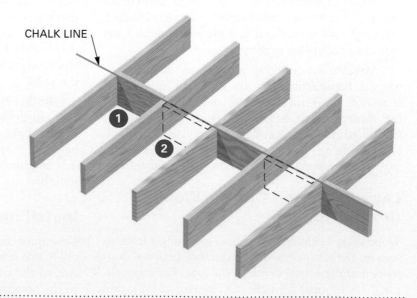

CHALK LINE

StepbyStep Procedures

Procedure 36–F Installing Cross-Bridging

STEP 1 SNAP CHALK LINE ACROSS TOPS OF FLOOR JOISTS IN CENTRE OF BRIDGING ROW.

STEP 2 SQUARED LINES MAY BE DRAWN DOWN FROM CHALK LINE ON BOTH SIDES OF FLOOR JOISTS.

STEP 3 FASTEN TOP ENDS OF BRIDGING SO THEY OPPOSE EACH OTHER ON THE SAME SIDE OF THE CHALK LINE.

STEP 4 LEAVE BOTTOM ENDS LOOSE UNTIL SUBFLOOR IS APPLIED. THEN FASTEN SO EDGE LINES UP WITH SQUARED LINE. ALSO, LEAVE A SPACE BETWEEN THE BRIDGING PIECES WHERE THEY CROSS.

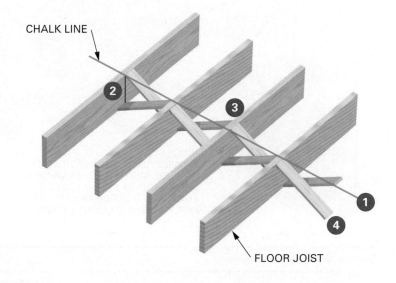

CHALK LINE

FLOOR JOIST

bridging in the other spaces, but alternate so that the top ends of the bridging pieces are opposite each other where they are fastened to the same joist (**Procedure 36–F**). Wood cross-bridging is usually nailed off at the bottom after the drywall is installed so that the house has had time to settle under load and the floor joist lumber has mostly dried out.

Metal Cross-Bridging. Metal cross-bridging is fastened in a manner similar to that used for wood cross-bridging. The method of fastening may differ according to the style of the bridging. Usually the bridging is fastened to the top of the joists through predrilled holes in the bridging. Because the metal is thin, nailing to the top of the joists does not interfere with the subfloor.

Some types of metal cross-bridging have steel prongs that are driven into the side of the floor joists. This bridging can be installed from below to layout lines made previous to the installation of the subfloor.

SUBFLOORING

Subflooring is used over joists to form a working platform. This is also a base for finish flooring, such as hardwood flooring, or underlayment for carpet

or resilient tiles. APA-Rated Sheathing Exposure 1 is generally used for subflooring in a two-layer floor system. APA-Rated Sturd-I-Floor panels are used when a single-layer subfloor and underlayment system is desired. Blocking is required under the joints of these panels unless tongue-and-groove edges are used.

APPLYING PANEL SUBFLOORING

Starting at the corner from which the floor joists were laid out, measure in 4 feet (1219 mm). Note that tongue-and-groove plywood subfloor is only 47½ inches (1207 mm) wide. Snap a line across the tops of the floor joists from one end to the other. Start with a full panel. Fasten the first row to the chalk line and align the joists to the correct spacing before nailing the panel.

Start the second row with a half-sheet to stagger the end joints (**Figure 36–15**). Continue with full panels to finish the row. Leave a ⅛-inch (3 mm) space between panel edges. All end joints are made over joists.

Continue laying and fastening plywood sheets in this manner until the entire floor is covered (**Procedure 36–G**). Leave out sheets where there are to be openings in the floor. Snap chalk lines across

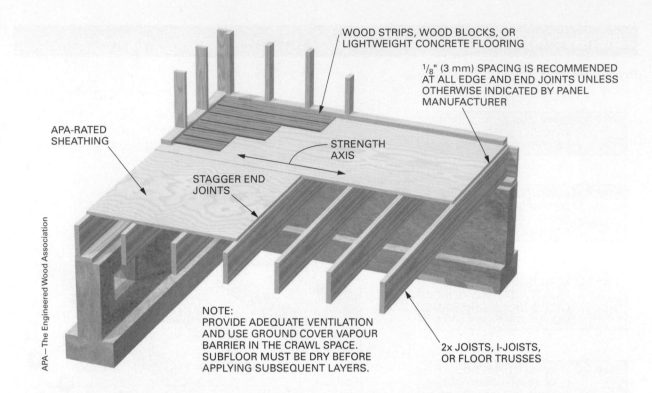

WOOD STRIPS, WOOD BLOCKS, OR
LIGHTWEIGHT CONCRETE FLOORING

$1/8"$ (3 mm) SPACING IS RECOMMENDED
AT ALL EDGE AND END JOINTS UNLESS
OTHERWISE INDICATED BY PANEL
MANUFACTURER

APA-RATED
SHEATHING

STRENGTH
AXIS

STAGGER END
JOINTS

APA—The Engineered Wood Association

NOTE:
PROVIDE ADEQUATE VENTILATION
AND USE GROUND COVER VAPOUR
BARRIER IN THE CRAWL SPACE.
SUBFLOOR MUST BE DRY BEFORE
APPLYING SUBSEQUENT LAYERS.

2x JOISTS, I-JOISTS,
OR FLOOR TRUSSES

Nailing Recommendations

Type of Panel	Recommended Nail and Type	Panel Edges[a]	Intermediate Supports
APA-Rated Sturd-I-Floor— *Glue-nailed installation*	*Ring- or screw-shank nails*		
STURD-I-FLOOR 16, 20, 24 oc, 3/4" thick or less	2" [b] (51 mm)	12"	12"
STURD-I-FLOOR 24 oc, 7/8" or 1" thick	2 1/2" [b] (63 mm)	6"	12"
STURD-I-FLOOR 32, 48 oc, 32" spans	2 1/2" [b] (63 mm)	6"	12"
STURD-I-FLOOR 48 oc, 48" spans	2 1/2" [c] (63 mm)	6"	6"
APA-Rated Sturd-I-Floor— *Nailed only installation*	*Ring- or screw-shank nails*		
STURD-I-FLOOR 16, 20, 24 oc. 3/4" thick or less	2" (51 mm)	6"	12"
STURD-I-FLOOR 24, 32 oc, 7/8" or 1" thick	2 1/2" (63 mm)	6"	12"
STURD-I-FLOOR 48 oc, 32" spans	2 1/2" [c] (63 mm)	6"	12"
STURD-I-FLOOR 48 oc, 48" spans	2 1/2" [c] (63 mm)	6"	6"
APA-Rated Sheathing— *Subflooring*	*Common, smooth, ring- or screw-shank*		
7/16" to 1/2" thick	2" (51 mm)	6"	12"
7/8" thick or less	2 1/2" (63 mm)	6"	12"
Thicker panels	3" (76 mm)	6"	6"
APA-Rated Sheathing— *Wall sheathing*			
1/2" thick or less	2" (51 mm)	6"	12"
Over 1/2" thick	21/2" (63 mm)	6"	12"
APA-Rated Sheathing— *Roof sheathing*	*Common, smooth, ring- or screw-shank[c]*		
5/16" to 1" thick	2 1/2" (63 mm)	6"	12"[d]
Thicker panels	2 1/2" ring- or screw-shank or 76-mm common smooth	6"	12"[d]
APA-Rated Siding— *Applied directly to studs or over nonstructural sheathing*	*Hot dipped galvanized box, siding or casing*		
1/2" thick or less	2" (51 mm)	6"	12"
Over 1/2" thick	2 1/2" (63 mm)	6"	12"

(a) Fasteners shall be located 3/8" (9 mm) from panel edges.
(b) 2 1/2" (63 mm) common nails may be substituted if ring- or screw-shank nails are not available.
(c) 3" (76 mm) common nails may be substituted if supports are well seasoned.
(d) For spans 48" (1.2 m) or greater, space nails 6" (150 mm) at all supports.

FIGURE 36–15 Nailing specifications for APA panels.

StepbyStep Procedures

Procedure 36–G Layout Procedure for Installing Plywood Subfloor

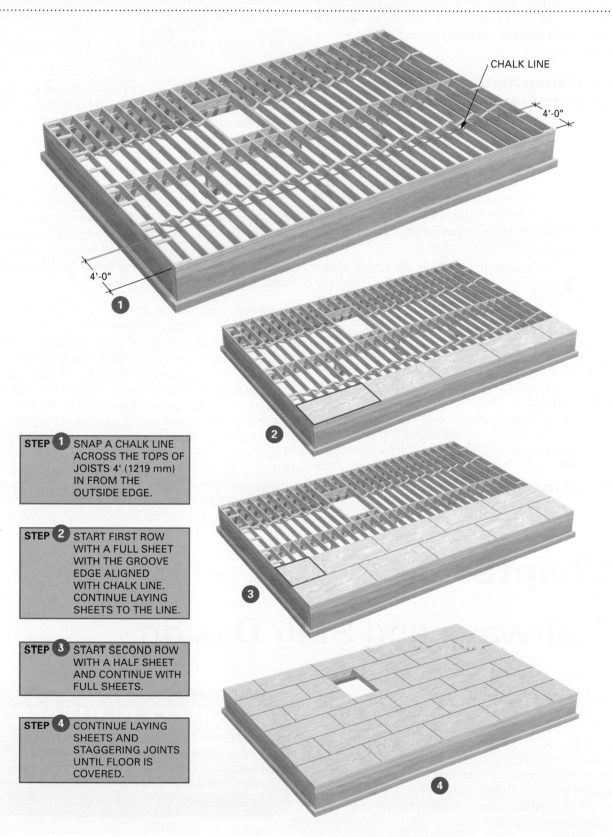

CHALK LINE

4'-0"

4'-0"

STEP 1 SNAP A CHALK LINE ACROSS THE TOPS OF JOISTS 4' (1219 mm) IN FROM THE OUTSIDE EDGE.

STEP 2 START FIRST ROW WITH A FULL SHEET WITH THE GROOVE EDGE ALIGNED WITH CHALK LINE. CONTINUE LAYING SHEETS TO THE LINE.

STEP 3 START SECOND ROW WITH A HALF SHEET AND CONTINUE WITH FULL SHEETS.

STEP 4 CONTINUE LAYING SHEETS AND STAGGERING JOINTS UNTIL FLOOR IS COVERED.

the edges and ends of the building. Trim overhanging plywood with a circular saw.

For a solid and squeak-free installation, it is recommended that subfloor adhesive be applied over the joists and in the groove of the tongue-and-groove panels and that the panels be fastened with low-root subfloor screws.

Estimating Material Quantities for Floor Framing

To estimate the material needed for the floor frame, the length and width of the building are needed.

Beam.   The amount of beam material required is determined by the beam length. If the beam is steel, then the quantity required is merely the actual length of the beam. If, however, the beam is built up, then a calculation must be performed to arrive at the amount of beam material required.

For a built-up beam, multiply the length of the beam by the number of plies in the beam. This will provide the quantity of material required for a cost estimate. Before ordering the pieces for a built-up beam, it will be necessary to determine the joint locations in the beam according to the ¼ points.

Sills.   The number of sill pieces depends on the building perimeter and the length of each sill piece. Take the building perimeter and divide by the sill piece length. For example, if a building is 28 × 48 feet and sill pieces are 12 feet long, then the number of pieces is calculated as follows: 2(28 + 48) = 152 feet (perimeter) ÷ 12 = 12.667 or 13 pieces.

Floor Joists.   To find the number of full-length joists, divide the length of the building by the spacing in terms of feet. For example, if the spacing is 16 inches, then it is 16 ÷ 12 = 1.333 feet. So a 40-foot-long building divided by 1.333 = 30 pieces. The rim joist material is the building perimeter divided by the length of material used. For example, if the perimeter is 160 feet and the material used is 12 feet long, then the band joist material is 160 ÷ 12 = 13.333 or 14 pieces. Add extra for headers as needed.

Bridging.   Bridging quantity depends on the style of bridging. The linear feet of solid bridging is simply the length of the building times the number of rows of bridging. Linear feet of cross-bridging is determined by taking the number of joists times 3 feet for 16-inch OC joists. This number is arrived at because 3 feet of bridging is needed for each joist cavity. Four feet is needed for 19.2-inch OC, and 5 feet is needed for 24-inch OC. Then multiply times the number of rows of bridging needed. For example, if two rows of bridging are needed for 30 full-length 16-inch OC joists, then 30 × 3 × 2 = 180 linear feet of bridging must be purchased.

Subfloor.   Subfloor is determined from the square footage of the building. Multiply building length times building width to determine the square footage of the building. The number of panel pieces needed is this area divided by the square feet per panel. For example, if the building is 30 × 50 feet and standard 4 × 8 panels are used, then the number of pieces is 30 × 50 = 1,500 divided by 32 (from 4 × 8 = 32 square feet per panel) = 46.875 or 47. Note that since tongue-and-groove panels are only 47½ inches wide instead of 48, they do not cover a full 32 square feet. Add 1 percent more to compensate. For example, 46.875 sheets + 1 percent of 46.875 = 47.34 or 48 sheets.

Chapter 37

Stairways and Stair Design

Staircases can be a showcase for carpenters to demonstrate their skill and talent. They are often intricate and ornate, requiring close cutting and fitting. Stairs must be carefully designed and laid out to ensure safe passage and ease of use. Also, many stair dimensions must comply with national and local codes. Building a set of stairs challenges the carpenter to work at his or her best and can be a source of great reward in pride of workmanship.

SAFETY REMINDER

Stairways must obviously be built strong enough to support all intended loads. They must also be built to exact standards to ensure they are safe to use by many different-sized people.

FIGURE 37–1 Staircases are often the showpiece of a building's interior.

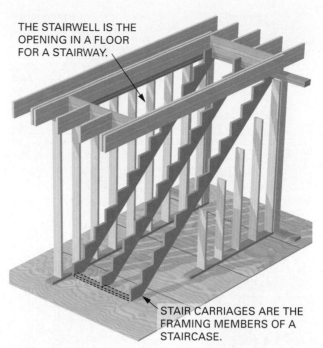

THE STAIRWELL IS THE OPENING IN A FLOOR FOR A STAIRWAY.

STAIR CARRIAGES ARE THE FRAMING MEMBERS OF A STAIRCASE.

FIGURE 37–2 Typical framing details for stairs.

A set of stairs or a **staircase** can be an outstanding feature of an entrance (**Figure 37–1**). A staircase can add beauty and grace to a room, generally affecting the character of the entire interior.

A set of stairs generally refers to one or more flights of steps leading from one level of a structure to another. *Staircase* is a term usually saved to refer to a finished set of stairs that has architectural appeal. Stairs are further defined as finish or service stairs. Finish stairs extend from one habitable level of a house to another. Service stairs extend from a habitable to a non-habitable level, typically a basement.

Stairs are also governed closely by national and local building codes. Codes set limits on total stair height, step width, step rise, step depth, and the acceptable amount of variation between steps. These codes are designed to ensure that a set of stairs will be safe for use by anyone.

Stairs, like rafters, are built by the most experienced carpenter on the job. Many design concepts and terms must be understood to successfully build a set of stairs. In addition to the fine carpentry of a staircase, consideration must be given to framing of the stairwell, or the opening in the floor through which a person must pass when climbing and going down the stairs (**Figure 37–2**). The stairwell is framed at the same time as the floor.

TYPES OF STAIRWAYS

A straight stairway is continuous from one floor to another. There are no turns or landings. Platform stairs have intermediate landings between floors. Platform-type stairs sometimes change direction at the landing. An L-type platform stairway changes direction 90 degrees. A U-type platform stairway changes direction 180 degrees.

Platform stairs are installed in buildings that have a high floor-to-floor level. They also provide a temporary resting place. They are a safety feature in case of falls. The landing is usually constructed at roughly the middle of the staircase.

A winding staircase gradually changes direction as it ascends from one floor to another. In many cases, only a part of the staircase winds. Winding stairs may solve the problem of a shorter straight horizontal run (**Figure 37–3**).

A stairway constructed between walls is called a **closed stairway**. Closed stairways are more economical to build, but they add little charm or beauty to a building. A stairway that has one or both sides open to the room is called an **open stairway**. Open stairways have more parts and pieces, adding to the charm and beauty of the stairs.

The terms used in stair framing have similarities with those used for rafters (**Figure 37–4**).

Total rise. The vertical distance between finish floors.

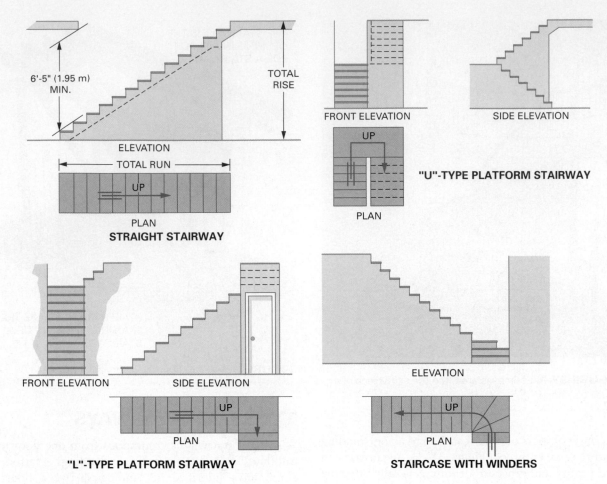

FIGURE 37–3 Various types of stairways.

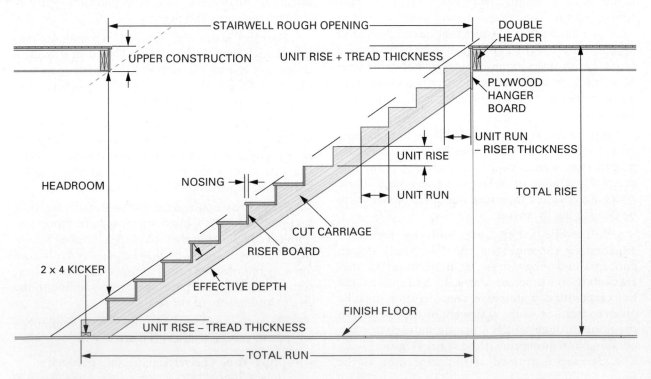

FIGURE 37–4 Terms used in stair framing.

Total run. The total horizontal distance that the stairway covers.

Riser. The finish material used to cover the unit rise distance.

Unit rise. The vertical distance from one step to another.

Tread. The horizontal finish material used to make up the step on which the feet are placed when ascending or descending the stairs.

Unit run. The horizontal distance between the faces of the risers.

Nosing. The nosing is that part of the tread that extends beyond the face of the riser. It is not part of the calculations for stairs, but rather an add-on to treads.

Stair carriage. A stair carriage, sometimes called a stair horse, provides the main strength for the stairs. It is usually a nominal 2 × 10 (38 × 235 mm) or 2 × 12 (38 × 286 mm) framing member cut to support the treads and risers.

Stair stringer. The stringer is the finish material applied to cover the stair carriage. It can also take the place of the stair carriage when side walls are used to support the stairs.

Stairwell. A stairwell is an opening in the floor for the stairway to pass through. It provides adequate headroom for persons using the stairs.

Headroom. The headroom of a set of stairs is the vertical distance above the stairs from a line drawn from nosing to nosing to the finished upper construction.

STAIR DESIGN

Stairs in residential construction are at least 34 inches (860 mm) wide (**Figure 37–5**). The maximum height of a single flight of stairs is 12 feet (3.7 m), unless a platform is built in to break up the continuous run. This platform must be as long as the stair is wide but need not be longer than 4 feet (1.2 m).

Staircases must be constructed at a proper angle for maximum ease in climbing and for safe descent. The relationship of the rise and run determines this angle (**Figure 37–6**). The preferred angle is between 30 and 38 degrees. Stairs with a slope of less than 20 degrees waste a lot of valuable space. Stairs with a slope that is excessively steep (50 degrees or over) are difficult to climb and dangerous to descend.

The NBCC limits residential stair construction to a minimum rise of $4^{7}/_{8}$ inches (125 mm), a maximum rise of $7^{7}/_{8}$ inches (200 mm), a minimum run of $8^{1}/_{4}$ inches (210 mm), a maximum run of 14 inches (355 mm), and a minimum tread width

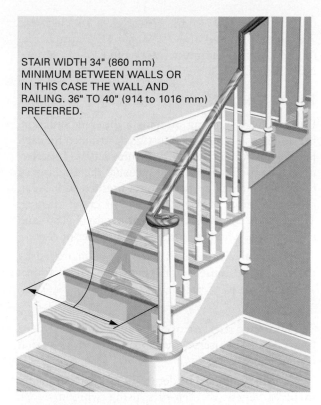

STAIR WIDTH 34" (860 mm) MINIMUM BETWEEN WALLS OR IN THIS CASE THE WALL AND RAILING. 36" TO 40" (914 to 1016 mm) PREFERRED.

FIGURE 37–5 Recommended stair widths.

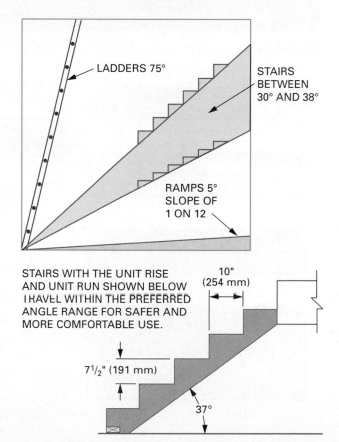

LADDERS 75°

STAIRS BETWEEN 30° AND 38°

RAMPS 5° SLOPE OF 1 ON 12

STAIRS WITH THE UNIT RISE AND UNIT RUN SHOWN BELOW TRAVEL WITHIN THE PREFERRED ANGLE RANGE FOR SAFER AND MORE COMFORTABLE USE.

10" (254 mm)

$7^{1}/_{2}$" (191 mm)

37°

FIGURE 37–6 Appropriate angles and dimensions for stairs.

of 9¼ inches (235 mm). However, a rise between 7 inches (180 mm) and 7½ inches (190 mm) is recommended for maximum safety and ease of use. (See NBCC 9.8.2 to 9.8.8.)

Riser Height

Because every building presents a different situation, the unit rise and unit run must be determined for each building. The unit rise is determined first. To determine the individual rise, first measure the total rise of the stairway from finished floor to finished floor. Next, find the number of risers that will fit in the opening. This is done by dividing the total rise by 7 (180) and choosing the whole number as the number of risers.

Alternatively, the unit rise may be increased by decreasing the number of risers by one. For example:

- 106.75″ ÷ 14 = 7.625 or 7⅝″; 2711 mm ÷ 14 = 193.6 mm

Both unit rises are acceptable within the NBCC limits. Using 13 risers would produce a unit rise that is unacceptable.

Another method of determining the riser height involves repetitive step-offs using wing dividers.

EXAMPLE   A total rise (from finish floor to finish floor) of 8′-10¾″ (2711 mm) is measured.

- Change 8′-10¾″ to decimal inches.
  → 106.75″ (2711 mm).
- Divide total rise by 7 (180)
  → 106.75 ÷ 7 = **15**.25 (2711 ÷ 180 = **15**.06).
- Choose 15 risers.
- Divide total rise by the number of risers.
  → 106.75 ÷ 15 = 7.12 or 7⅛; or in SI units
  → 2711 ÷ 15 = 180.7 mm as actual unit rise.

Stand a *storey pole* vertically. This could be a 1 × 2 (305 × 610 mm) strip of lumber. Mark the total rise, allowing for the finish floor to finish floor height. Set the dividers to 7⅞-inch (200 mm) rise spacing. Step off the rod over the entire total rise. If the last step is not even, adjust the dividers slightly smaller. Repeat this process as often as necessary (**Procedure 37–A**).

StepbyStep Procedures

Procedure 37–A Determining Stair Riser Height Using the Step-Off Method

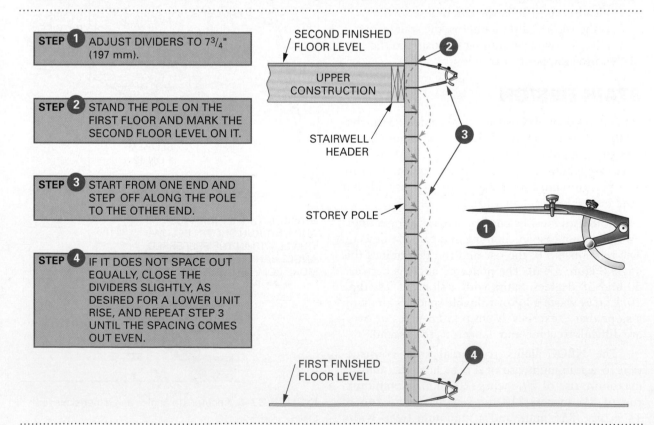

STEP **1** ADJUST DIVIDERS TO 7³/₄″ (197 mm).

STEP **2** STAND THE POLE ON THE FIRST FLOOR AND MARK THE SECOND FLOOR LEVEL ON IT.

STEP **3** START FROM ONE END AND STEP OFF ALONG THE POLE TO THE OTHER END.

STEP **4** IF IT DOES NOT SPACE OUT EQUALLY, CLOSE THE DIVIDERS SLIGHTLY, AS DESIRED FOR A LOWER UNIT RISE, AND REPEAT STEP 3 UNTIL THE SPACING COMES OUT EVEN.

SECOND FINISHED FLOOR LEVEL

UPPER CONSTRUCTION

STAIRWELL HEADER

STOREY POLE

FIRST FINISHED FLOOR LEVEL

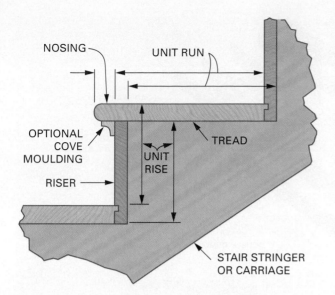

FIGURE 37–7 The tread unit run does not include the nosing.

Unit Run

The unit run is measured from the face of one riser to the next. It does not include the nosing (**Figure 37–7**). It needs to be adjusted to create a proper stair angle. There are several rules to follow to accomplish this.

17–18 Method. First, the sum of the unit rise and the unit run should be between 17 and 18 inches. For example, if the unit rise is 7⅜ inches, then the minimum unit run may be 17 inches minus 7⅜ inches.

This equals 9⅜ inches. The maximum tread width may be 18 inches minus 7⅜ inches. This equals 10⅝ inches. In SI units, this is referred to as the 430–460 rule. A 187 mm unit rise would have a unit run between 243 and 273 mm.

24–25 Method. The second rule for determining the unit run is found in many building codes, which state that the sum of two unit rises and one unit run shall not be less than 24 inches nor more than 25 inches (**Figure 37–8**). With this formula, a unit rise of 7⅜ inches calls for a minimum unit run of 9¼ inches and a maximum of 10¼ inches. In SI units, this is referred to as the 610–635 rule. A 187 mm unit rise would have a unit run between 236 and 261 mm.

A third method, which always maintains a stair angle of 34 degrees, is to multiply the unit rise by 1.5.

- 7″ (178 mm) unit rise × 1.5 → 10½″ (267 mm) unit run

Variations in Stair Steepness

Decreasing the rise increases the run of the stairs. This makes the stairs easier to climb but uses up more space. Increasing the rise height decreases the run. This makes the stairs steeper and more difficult to climb but uses less space. The carpenter must use good judgment when adapting the rise and run dimensions to fit the space in which the stairway is constructed. In general, a rise height of 7½ inches (191 mm) and a unit run of 10 inches (254 mm) create a safe, comfortable stairway.

TWO FORMULAS ARE USED TO DETERMINE THE UNIT RUN FOR STAIRS
AFTER UNIT RISE HAS BEEN DETERMINED.

17–18 METHOD

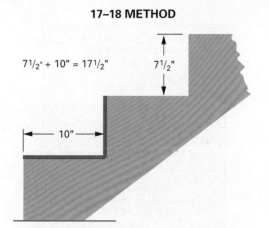

7½″ + 10″ = 17½″   7½″

10″

ONE UNIT RISE PLUS ONE UNIT RUN SHOULD
EQUAL BETWEEN 17 AND 18″.

24–25 METHOD

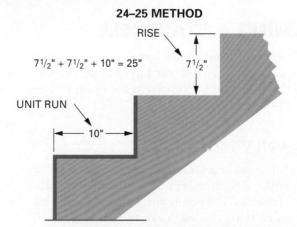

RISE

7½″ + 7½″ + 10″ = 25″   7½″

UNIT RUN

10″

THE SUM OF TWO UNIT RISES AND ONE UNIT RUN
SHOULD EQUAL BETWEEN 24 AND 25″.

NOTE: CHECK LOCAL CODES FOR MAXIMUM ALLOWABLE
RISE AND MINIMUM ALLOWABLE RUN (NBCC Table 9.8.4.1).

FIGURE 37–8 Two techniques for determining the unit run with the desired unit rise.

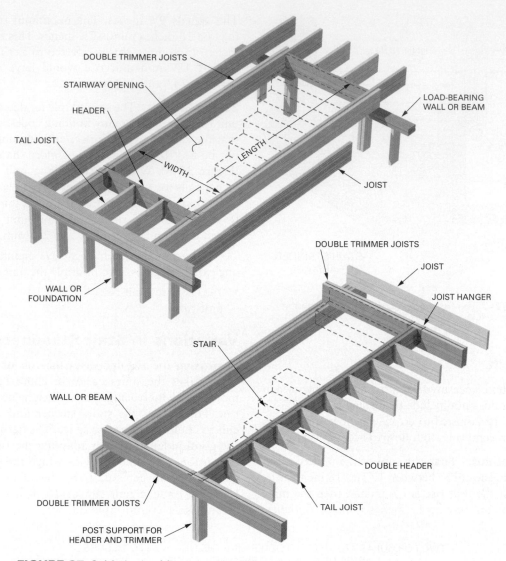

DOUBLE TRIMMER JOISTS

STAIRWAY OPENING

HEADER

TAIL JOIST

LOAD-BEARING WALL OR BEAM

WIDTH

LENGTH

JOIST

WALL OR FOUNDATION

DOUBLE TRIMMER JOISTS

JOIST

JOIST HANGER

STAIR

WALL OR BEAM

DOUBLE HEADER

DOUBLE TRIMMER JOISTS

TAIL JOIST

POST SUPPORT FOR HEADER AND TRIMMER

FIGURE 37–9 Methods of framing stairwells.

FRAMING A STAIRWELL

A stairwell is framed in the same manner as any large floor opening, as discussed in Chapter 36. Several methods of framing stairwells are illustrated in **Figure 37–9**.

Stairwell Width

The width of the stairwell depends on the width of the staircase. The drawings show the finish width of the staircase. However, the stairwell must be made wider than the staircase to allow for wall and stair finish (**Figure 37–10**). Extra width will be required for a handrail and other finish parts of an open staircase that makes a U-turn on the landing above. The carpenter must be able to determine the width of the stairwell by studying the prints for size, type, and placement of the stair finish before framing the stairs.

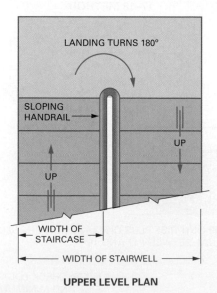

LANDING TURNS 180°

SLOPING HANDRAIL

UP

UP

WIDTH OF STAIRCASE

WIDTH OF STAIRWELL

UPPER LEVEL PLAN

FIGURE 37–10 The stairwell must be made wider than the staircase.

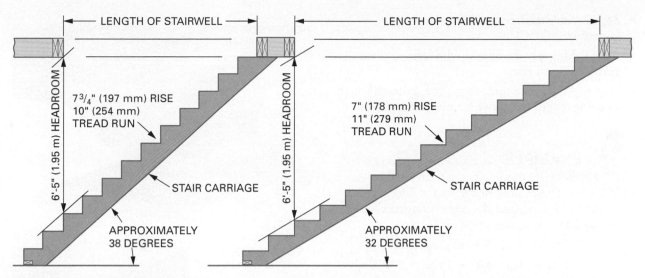

FIGURE 37–11 Stair angles affect the stairwell opening length.

Length of the Stairwell

The length of the stairwell depends on the slope of the staircase. Stairs with a low angle require a longer stairwell to provide adequate headroom (**Figure 37–11**). The NBCC requires a minimum of 6′-5″ (1.95 m) for headroom in residences; however, this is the minimum and more headroom is preferred.

To find the minimum length of the stairwell, add the desired headroom to the thickness of the upper floor construction (UC). This sum is sometimes referred to as the *total headroom*. The upper construction consists of the finish floor, the subfloor, the floor joist, and the ceiling finish (**Procedure 37–B**). Then apply the formula in the following steps:

- *Step 1.* Measure thickness of upper floor construction.

StepbyStep Procedures

Procedure 37–B Determining Stairwell Length

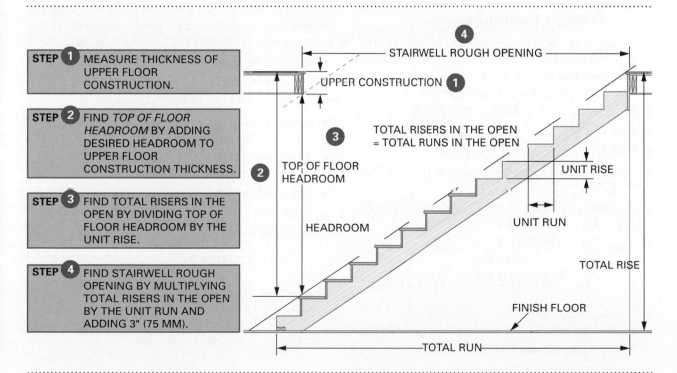

STEP ① MEASURE THICKNESS OF UPPER FLOOR CONSTRUCTION.

STEP ② FIND *TOP OF FLOOR HEADROOM* BY ADDING DESIRED HEADROOM TO UPPER FLOOR CONSTRUCTION THICKNESS.

STEP ③ FIND TOTAL RISERS IN THE OPEN BY DIVIDING TOP OF FLOOR HEADROOM BY THE UNIT RISE.

STEP ④ FIND STAIRWELL ROUGH OPENING BY MULTIPLYING TOTAL RISERS IN THE OPEN BY THE UNIT RUN AND ADDING 3″ (75 MM).

- *Step 2.* Add headroom (HR) to upper construction (UC).
- *Step 3.* Multiply the sum by the ratio of the unit run to the unit rise.
- *Step 4.* Add 3″ (75 mm) to get the stairwell rough opening length (to allow for finish materials).

EXAMPLE A stairway has a unit rise of 7½ inches (191 mm) and a unit run of 10 inches (254 mm). The desired headroom is 7 feet or 84 inches (2134 mm), and the upper construction is measured at 11¾ inches (298 mm).

- *Step 1.* UC 11¾″ (298 mm) measured.
- *Step 2.* HR + UC = 84″ + 11¾″ = 95.75″ (2432 mm).
- *Step 3.* 95.75″ × (10″ ÷ 7.5″) = 127.67″ (3234 mm).
- *Step 4.* 127.67″ (3234 mm) + 3″ (76 mm) = 130.67″ (3310 mm) = 10′-10¹¹/₁₆″ (3.31 m).

This process assumes the carriage is framed to the upper construction, where the header of the stairwell acts as the top riser. If the top tread is framed flush with the second floor, add another tread width to the length of the stairwell (**Figure 37–12**).

Additional floor space may be obtained by framing the subfloor past the header. This addition is framed at the same angle as the stairs (**Figure 37–13**).

Stair Design Example

Calculate a set of stairs given a total rise of 8′-6″ (102″), desired headroom of 6′-8″ (80″), and the upper construction of 11¼″, using 17½″ as the average sum of the unit rise and the unit run.

- Number of risers = 102″ ÷ 7 = **14**.57.
- Unit rise = 102″ ÷ 14 = 7.29″ → 7⁵/₁₆″.
- Unit run = 17½″ − 7⁵/₁₆″ = 10³/₁₆″.
- Tread width = Unit run + Nosing = 10³/₁₆″ + 1″ = 11³/₁₆″.
- Total run = Unit run × (# of risers − 1) = 10³/₁₆″ × (14 − 1) = 132⁷/₁₆″ = 11′-0⁷/₁₆″.
- Stairwell = $\dfrac{(\text{HR} + \text{UC}) \times \text{Unit run}}{\text{Unit rise}}$

  $= \dfrac{91.25 \times 10.1875}{7.29} = 127.52 \,(10′\text{-}7½″)$
- Stairwell R.O. = 10′7½″ + 3″ = 10′-10½″

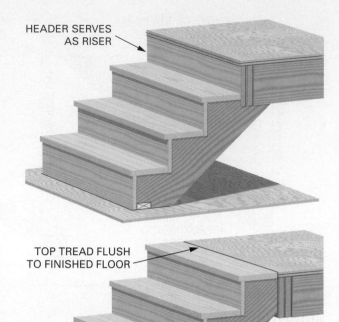

HEADER SERVES AS RISER

TOP TREAD FLUSH TO FINISHED FLOOR

FIGURE 37–12 Methods of framing stairs at the stairwell header.

SI. Calculate a set of stairs given a total rise of 2591 mm, desired headroom of 2032 mm, and the upper construction of 286 mm, using 445 as the average sum of the unit rise and the unit run.

- Number of risers = 2591 ÷ 180 = 14.39.
- Unit rise = 2591 ÷ 14 = 185 mm.
- Unit run = 445 − 185 = 260 mm.
- Tread width = Unit run + Nosing = 260 mm + 25 mm = 285 mm.
- Total run = Unit run × (# of risers − 1) = 260 mm × (14 − 1) = 3380 mm.
- HR + UC = 286 mm + 2032 mm = 2318 mm.
- Stairwell = $\dfrac{(\text{HR} + \text{UC}) \times \text{Unit run}}{\text{Unit rise}}$

  $= \dfrac{2318 \times 260}{185} = 3258 \text{ mm}$
- Stairwell R.O. = 3258 mm + 76 mm = 3334 mm.

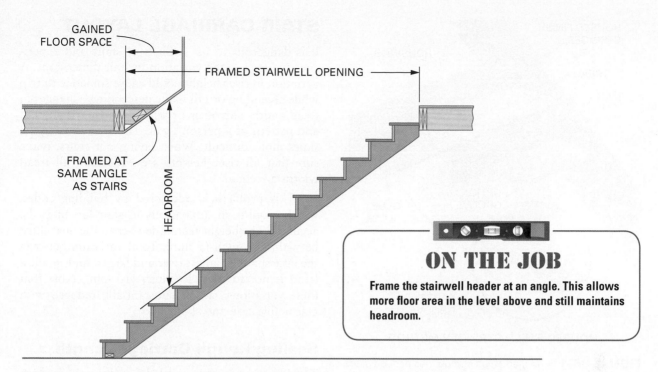

FIGURE 37–13 Technique for increasing the upper floor space while maintaining required headroom.

Labels in figure:
GAINED FLOOR SPACE
FRAMED STAIRWELL OPENING
FRAMED AT SAME ANGLE AS STAIRS
HEADROOM
ON THE JOB
Frame the stairwell header at an angle. This allows more floor area in the level above and still maintains headroom.

Chapter 38

Stair Layout and Construction

All stairs are laid out in approximately the same way. The use, location, and cost of stairs determine the way they are built. Regardless of the kind of stairs, where they are located, or how much they cost, care should be taken in their layout and construction.

METHODS OF STAIR CONSTRUCTION

There are two principal methods of stair construction. The **housed stringer** is laid out and cut to be a finished product. It is often fabricated off-site in a shop and installed near the conclusion of the construction process. The job-built staircase uses stair carriages that are usually built on-site. These are finished later as the construction process comes to an end.

Housed Stringer Staircase

For the housed stringer staircase, the framing crew frames only the stairwell. The staircase is installed when the house is ready for finishing. Dadoes are routed into the sides of the finish stringer. They *house* (make a place for) and support the risers and treads (**Figure 38–1**).

Occasionally, the finish carpenter builds a housed stringer staircase on the jobsite. A router and jig are used to dado the stringers. Then the treads, risers, and other stair parts are cut to size. The staircase is then assembled either in place or as a unit and then installed into place. Stair carriages are not required when the housed stringer method of construction is used. Building of housed stringer stairs is described in more detail in Chapter 82.

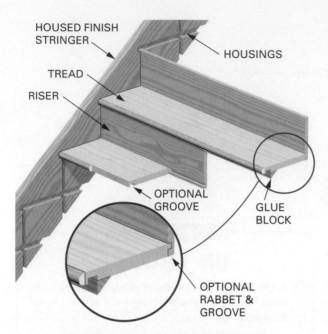

HOUSED FINISH STRINGER METHOD

FIGURE 38–1 A housed finish stringer has dadoed sides to accept treads and risers.

Job-Built Staircase

The *job-built staircase* uses stair carriages that are installed when the structure is being framed. The carriage is laid out and cut with risers and treads fastened to the cutouts (**Figure 38–2**). This style uses temporary rough treads installed for easy access to upper levels during construction. Later the carriage is fitted with finish treads and risers with other stair trim.

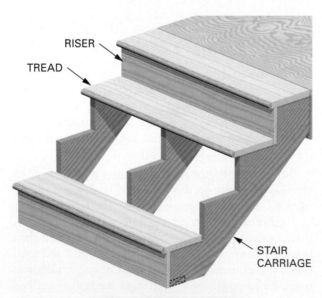

STAIR CARRIAGE METHOD

FIGURE 38–2 Stair carriages have notches cut to support treads and risers.

STAIR CARRIAGE LAYOUT

It is dangerous to use a flight of stairs and experience an unexpected variation in stair dimensions. A variation in riser height could cause someone to trip while ascending or fall while descending. Changes in tread width, narrower or wider, change the rhythm and pattern of a person's gait. This makes using the stairs more difficult. When laying out stairs, make sure that all riser heights are equal and all tread widths are equal.

This problem is addressed by building codes, which require all dimensions in stair layout to be accurate. The height from one riser to the next must be within $3/16$ inch (5 mm). Total variation between the largest and smallest riser and largest and smallest tread is not to exceed $3/8$ inch (10 mm). Note that these variations can be substantially reduced with reasonable care and skill.

Scaling Rough Carriage Length

The length of lumber needed for the stair carriage is often determined using the Pythagorean theorem. It also can be found by scaling across the framing square. Use the edge of the square that is graduated in 12ths of an inch. Mark the total rise on the tongue. Then mark the total run on the blade. Scale off in between the marks (**Figure 38–3**).

EXAMPLE   A stairway has a total rise of 8'-9" and a total run of 12'-3". What is the length of material needed to build the carriage?

Pythagorean Theorem
- Change dimensions to decimals.
  8'-9" = 8.75'; 12'-3" = 12.25'
- Substitute into $a^2 + b^2$ and c^2 and solve.
  $8.75^2 + 12.25^2 = c^2 = 226.625$
  $c^2 = \sqrt{226.62} = 15.05'$
- Round up to nearest even number.
  15.05 → 16 feet

Scaling
Locate 8–$9/12$ths and 12–$3/12$ths on framing square. Measuring across the square between these dimensions results in a reading of a little over 15. At a scale of 1 inch = 1 foot, board length ordered should be 16 feet.

SI. A stairway has a total rise of 2667 mm and a total run of 3734 mm. What is the length of material needed to build the carriage?

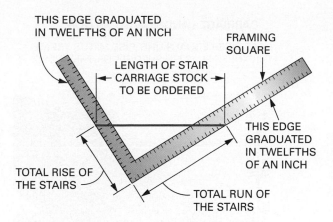

FIGURE 38–3 The framing square may be used to estimate the rough length of a stair carriage.

Pythagorean Theorem. Let the carriage length be x.

$x = \sqrt{2667^2 + 3734^2} = 4589$ mm or 5 m of stock.

Stepping Off the Stair Carriage

Place the stair carriage stock on a pair of sawhorses. Sight the stock for a crowned edge. Set gauges on the framing square with the rise on the tongue and the unit run on the blade (see Figure 46–7 for types of gauges). Lines laid out along the tongue will be plumb lines. Those laid out along the blade will be level lines when the stair carriage is in its final position.

Using the framing square gauges against the top edge of the carriage, step off on the carriage the necessary number of times. Mark both rise and run along the outside of the tongue and blade. Lay out enough level and plumb lines to include the top tread and the lower finished floor line (**Figure 38–4**). The layout lines represent the bottom of the tread and backside of the riser.

Equalizing the Bottom Riser

The bottom of the carriage must be adjusted to allow for the finished floor. This amount depends on how the finish material is applied. This will make the bottom riser equal in height to all of the other risers when the staircase is finished. This process is known as dropping the stair carriage.

If the carriage rests on the finish floor, then the first riser is cut shorter by the thickness of the tread stock (**Figure 38–5**). If the carriage rests on the sub-floor, the riser height must be adjusted using the tread and finished floor thicknesses. To achieve the first riser height dimension, take the riser height minus the tread thickness plus finished floor thickness.

Sometimes the finish floor and tread stock are the same thickness. In this special case, nothing is cut off the bottom end of the stair carriage.

Equalizing the Top Riser

The top of the carriage must be properly located against the upper construction to maintain the unit rise. This location depends on the style of the stairs. When the header of the upper construction acts as

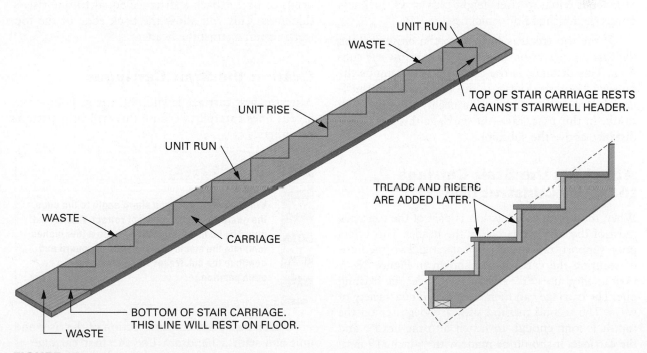

FIGURE 38–4 A completed stair carriage layout.

CARRIAGE LANDS ON FINISHED FLOOR

FIRST RISER EQUALS UNIT RISE MINUS TREAD THICKNESS.

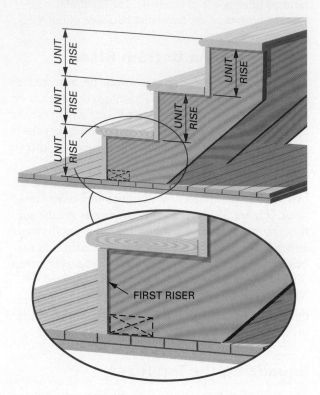

CARRIAGE LANDS ON SUBFLOOR

FIRST RISER EQUALS UNIT RISE MINUS TREAD THICKNESS PLUS FINISHED FLOOR THICKNESS.

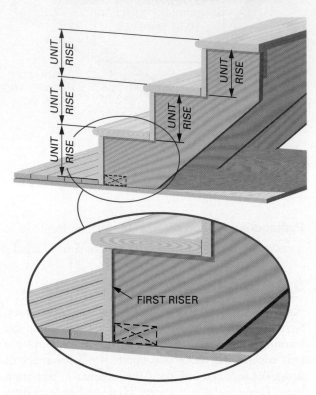

FIGURE 38–5 Adjusting the stair carriage bottom to equalize the first riser.

the top riser, the top of the carriage is roughly one riser height below the subfloor (**Figure 38–6**). This distance is equal to riser height plus tread thickness minus the finished floor thickness.

If the top tread is flush with the finished floor, the carriage elevation is only slightly below the subfloor. This distance is the tread thickness minus the finished floor thickness. If the finished floor thickness is larger than the tread, a negative number will result. In this case, raise the carriage this calculated distance above the subfloor.

Attaching the Stair Carriage to Upper Construction

When the upper construction is part of the top riser, most of the carriage is below the header. This offers poor support and room for fastening. Two methods of securing the carriage are shown in **Figure 38–7**. One method uses an extra header under the existing one. The carriage can then be attached in a variety of ways. The second method uses a wider riser on the top. It is long enough to fasten into the header and the carriage. It should be made with ¾-inch (19 mm) plywood for strength.

If no finished riser is desired, the carriage must be adjusted. With no riser board, the plumb line needs to be cut back a distance equal to one riser's thickness. This will allow the back edge of the top tread to rest against the header.

Cutting the Stair Carriages

After the first carriage is laid out, cut it. Follow the layout lines carefully, because this will be a pattern for others.

CAUTION

When making a cut at a sharp angle to the edge, the guard of the saw may not retract. Retract the guard by hand until the cut is made a few inches (cm) into the stock. Then release the guard and continue the cut. Never wedge the guard in an open position.

Finish the cuts at the intersection of the riser and unit run with a handsaw. Use the first carriage as a pattern. Lay out and cut as many other carriages

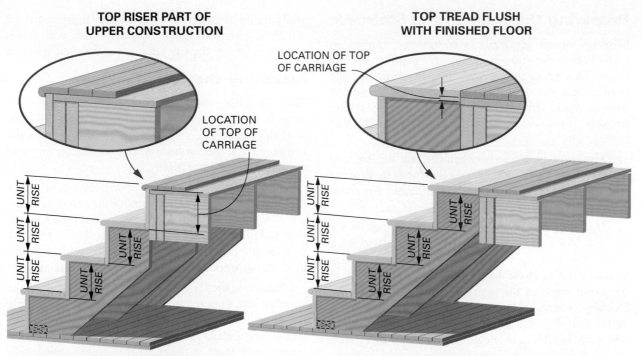

TOP RISER PART OF UPPER CONSTRUCTION

LOCATION OF TOP OF CARRIAGE

UNIT RISE

TOP OF CARRIAGE IS LOCATED BELOW SUBFLOOR. DISTANCE EQUALS UNIT RISE PLUS TREAD THICKNESS MINUS FINISHED FLOOR THICKNESS.

TOP TREAD FLUSH WITH FINISHED FLOOR

LOCATION OF TOP OF CARRIAGE

UNIT RISE

TOP OF CARRIAGE IS LOCATED BELOW SUBFLOOR. DISTANCE EQUALS TREAD THICKNESS MINUS FINISHED FLOOR THICKNESS.

FIGURE 38–6 Locating the top of the carriage on the header.

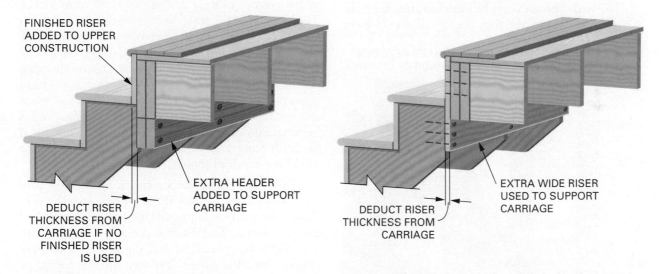

FINISHED RISER ADDED TO UPPER CONSTRUCTION

EXTRA HEADER ADDED TO SUPPORT CARRIAGE

DEDUCT RISER THICKNESS FROM CARRIAGE IF NO FINISHED RISER IS USED

EXTRA WIDE RISER USED TO SUPPORT CARRIAGE

DEDUCT RISER THICKNESS FROM CARRIAGE

FIGURE 38–7 Methods of framing the stair carriage to the stairwell header.

as needed. Three carriages are often used for residential staircases of average width. For wider stairs, the number of carriages depends on such factors as whether or not risers are used and the thickness of the tread stock. If the carriages are supported, their minimum thickness is 1 inch (25 mm); 1½ inches (38 mm) if unsupported. For single dwellings, the maximum space between carriages is 35¹³⁄₁₆ inches (900 mm) unless the risers support the treads and then it can be increased to 47¼ inches (1.2 m). In all other cases, the maximum space is 23⅝ inches (600 mm; see NBCC 9.8.9.4).

FRAMING A STRAIGHT STAIRWAY

If the stairway is either completely closed or closed on one side, the walls must be prepared before the stair carriages are fastened in position.

Preparing the Walls of the Staircase

Gypsum board (drywall) is sometimes applied to walls before the stair carriages are installed against them. This procedure saves time. It eliminates the need to cut the drywall around the cutouts of the stair carriage. This method also requires no blocking between the studs to fasten the ends of the drywall panels.

However, blocking between the studs in back of the stair carriage provides backing for fastening the stair trim. Lack of it may cause difficulty for those who apply the finish.

If the drywall is to be applied after the stairs are framed, blocking is required between studs in back of the stair carriage to fasten the ends of the gypsum board. Snap a chalk line along the wall sloped at the same angle as the stairs. Be sure the top of the blocking is sufficiently above the stair carriage to be useful. Install 2 × 6 (38 × 140 mm) or 2 × 8 (38 × 184 mm) blocking, on edge, between and flush with the edges of the studs. Their top edge should be to the chalk line (**Figure 38–8**). Another option is to secure a 2 × 4 (38 × 89 mm) to the studs so that it is flush with the bottom of the carriage. This provides a space for the drywall and room for a finish skirt board.

Housed staircases may be installed after the walls are finished. Sometimes they are installed before the

walls are finished. In such a case, they must be furred out away from the studs. This allows the wall finish to extend below the top edge of the finished stringer.

Installing the Stair Carriage

When installing the stair carriage, fasten the first carriage in position on one side of the stairway. Attach it at the top to the stairwell header. Make sure the distance from the subfloor above to the top unit run is correct (see Figure 38–6).

You can also draw a line along the underside of the carriage onto each stud, remove the carriage, fasten a 2 × 4 (38 × 89 mm) to the studs above the marked line, and then fasten the carriage to the 2 × 4. This will provide for the drywall to be installed with a simple angle cut as opposed to matching the cut carriage.

Fasten a kicker to the bottom front end of the carriage and then secure the carriage to the sole plate of the wall and with intermediate fastenings into the studs. Drive nails near the bottom edge of the carriage, away from cutouts. This prevents splitting of the triangular sections.

Fasten a second carriage on the other wall in the same manner as the first. If the stairway is to be open on one side, fasten the carriage at the top and at the bottom using a 2 × 4 (38 × 89 mm) kicker. The location of the stair carriage on the open end of a stairway is in relation to the position of the handrail. First, determine the location of the centre line of the handrail. Then position the stair carriage on the open side of a staircase. Make sure its outside face will be in a line plumb with the centre line of the handrail when it is installed (**Figure 38–9**). The height of handrail is between 865 mm (34″) and 965 mm (38″) measured vertically from a line touching the tread nosing to the top of the handrail (NBCC 9.8.7.4 & .5).

Fasten intermediate carriages at the top into the stairwell header and at the bottom into the subfloor. The intermediate carriage has to be notched over the kicker. Test the unit run and riser cuts with a straightedge placed across the outside carriages (**Figure 38–10**). About halfway up the flight, or where necessary, fasten a temporary riser board. This straightens and maintains the spacing of the carriages (**Figure 38–11**).

If a wall is to be framed under the stair carriage at the open side, fasten a bottom plate to the subfloor plumb with the outside face of the carriage. Lay out the studs on the plate. Cut and install studs under the carriage in a manner similar to that used to install gable studs. Be careful to keep the carriage straight. Do not crown it up in the centre (**Figure 38–12**). Install rough lumber treads on the carriages until the stairway is ready for the finish treads.

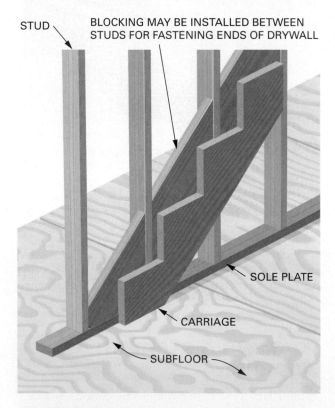

STUD

BLOCKING MAY BE INSTALLED BETWEEN STUDS FOR FASTENING ENDS OF DRYWALL

SOLE PLATE

CARRIAGE

SUBFLOOR

FIGURE 38–8 Preparation of the wall for application of drywall before stair carriages are installed.

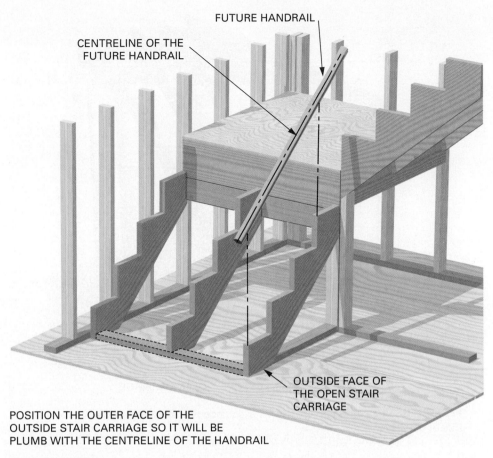

FUTURE HANDRAIL

CENTRELINE OF THE
FUTURE HANDRAIL

OUTSIDE FACE OF
THE OPEN STAIR
CARRIAGE

POSITION THE OUTER FACE OF THE
OUTSIDE STAIR CARRIAGE SO IT WILL BE
PLUMB WITH THE CENTRELINE OF THE HANDRAIL

FIGURE 38–9 The outside stair carriage is located plumb under the handrail.

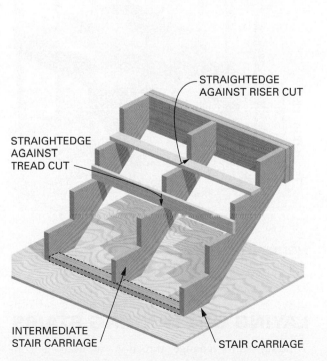

STRAIGHTEDGE
AGAINST RISER CUT

STRAIGHTEDGE
AGAINST
TREAD CUT

INTERMEDIATE
STAIR CARRIAGE

STAIR CARRIAGE

FIGURE 38–10 The alignment of tread and riser cuts on carriages must be checked with a straightedge.

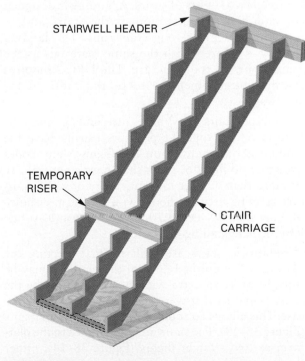

STAIRWELL HEADER

TEMPORARY
RISER

STAIR
CARRIAGE

FIGURE 38–11 A temporary riser about halfway up the flight straightens and maintains the carriage spacing.

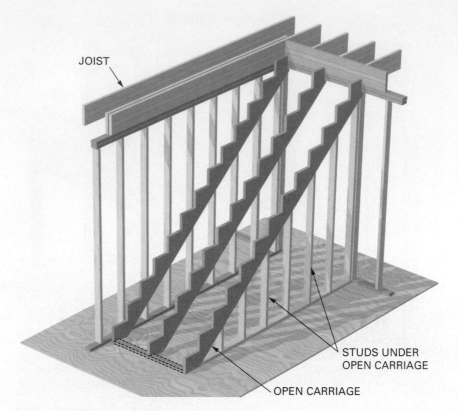

FIGURE 38–12 Studs are typically installed under the stair carriage on the open side of the stairway.

STAIRWAY LANDINGS

A stair landing is an intermediate platform between two flights of stairs. A landing is designed for changing the direction of the stairs and as a resting place for long stair runs. The landing usually is floored with the same materials as the main floors of the structure. The NBCC requires the minimum dimension to be the width of the stairway.

L-type stairs may have the landing near the bottom or the top. U-type stairs usually have the landing about midway on the flight. Many codes state that no flight of stairs shall have a vertical rise of more than 12 feet (3.7 m). Therefore, any staircase running between floors with a vertical distance of more than 12 feet (3.7 m) must have at least one intermediate landing or platform.

Platform stairs are built by first erecting the platform. The finished floor of the platform may be thought of as an extra wide tread. It should be the same height as if it were a finish tread in the staircase. This allows an equal riser height for both flights (**Figure 38–13**). The stairs are then framed to the platform as two straight flights (**Figure 38–14**). Either the stair carriage or the housed stringer method of construction may be used.

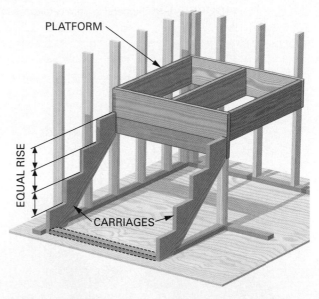

FIGURE 38–13 The top side of the stair platform is located as if it were a tread.

LAYING OUT WINDING STAIRS

Winding stairs change direction without a conventional landing. They often will allow for two extra risers in the space normally occupied by the

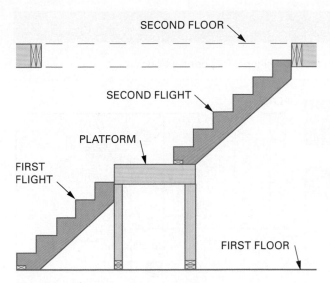

FIGURE 38–14 Stair carriages are framed to the platform as two straight flights of stairs.

landing. This is particularly useful for stairwells that are too small for normal stair construction. Winders, however, are not recommended, for safety reasons. Some codes state that they may be used in individual dwelling units, if the required tread width is provided along an arc. This is called the *line of travel*. It is a certain distance from the side of the stairway where the treads are narrower. The IRC (International Residential Code) states the line of travel is 12 inches (305 mm) from the narrow edge of the tread. The minimum tread width at the line of travel is 11 inches (279 mm) and the minimum width at the narrow edge is 6 inches (152 mm) (**Figure 38–15**). The National Building Code of Canada does not refer to the "line of

travel" for winder stairs. Stairs within dwelling units are permitted to contain winders that converge to a centre point with the following requirements:

- a maximum turn of 90 degrees.
- winders from 30 degrees to 45 degrees.
- adjacent winders must have the same angle.
- where there is more than one set of winders in a single stairway between adjacent floor levels, such winders shall be separated in plan view by at least 47¼ inches (1200 mm).
- tread nosing of 1″ (25 mm).

The IRC does not contradict the NBCC.

To lay out a winding turn of 90 degrees, draw a full-size winder layout on the floor directly below where the winders are to be installed (**Procedure 38–A**). On a closed staircase, the walls at the floor line represent its sides. For stairs open on one side, lay out lines on the floor representing the outside of the staircase. The wall represents the inside. Swing an arc, showing the line of travel. Use the outside corner of the wall under the outer carriage or layout lines as centre. The radius of the arc may be 12 to 18 inches (305 to 457 mm), as the codes permit.

From the same centre, lay out the width of the narrow end of the treads in both directions. Square lines from the end points to the opposite side of the staircase. Divide the arc into equal parts. Project lines from the narrow end of the tread, through the intersections at the arc, to the wide end at the wall. These lines represent the faces of the risers. Draw lines parallel to these to indicate the riser thickness. Plumb these lines up the wall to intersect with the unit run for each winder.

The cuts on the stair carriage for the winding steps are obtained from the full-size layout. Lay out and cut the carriage. Fasten it to the wall. Install rough treads until the stairs are ready for finishing.

If one side of the staircase is to be open, a newel post is installed. Then, the risers are mitred to or mortised into the post (**Figure 38–16**). A **mortise** is a rectangular cavity in which the riser is inserted. Newel posts are part of the stair finish. They are described in more detail in Chapters 81 and 83.

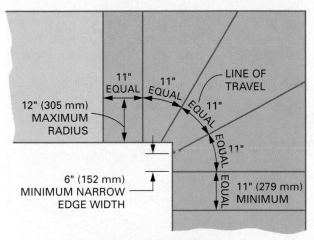

PLAN OF WINDERS

IRC (INTERNATIONAL RESIDENTIAL CODE) REQUIREMENTS FOR WINDERS

FIGURE 38–15 International Residential Code specifications for winding stairs.

CIRCULAR STAIRS

Circular stairs will either follow a complete circle, as in the case of *spiral stairs*, or they will turn through part of a circle, giving the appearance of curved stairs. Sometimes the curve will follow an ellipse instead of a circle. The calculations for unit rise will be the

StepbyStep Procedures

Procedure 38–A Laying Out Winding Stairs (according to the International Residential Code)

STEP ① 1 BEGIN WITH LAYOUT OF STAIR WIDTH AND THE LOWER TREADS. MARK THEM TOWARD THE PLATFORM.

STEP ② 2 FROM POINT A OF LAST TREAD, MEASURE 6" (152 mm) TOWARD PLATFORM AND MARK IT AS POINT B.

STEP ③ 3 MEASURE FROM POINT A ALONG TREAD LINE 12" (305 mm) TO POINT C.

STEP ④ 4 SWING AN 11" (279 mm) ARC FROM POINT C TOWARD WHERE POINT D WILL BE. SWING A 12" (305 mm) ARC FROM POINT B TO LOCATE POINT D.

STEP ⑤ 5 DRAW LINE TO WALL FROM POINT B THROUGH POINT D. THIS IS THE TREAD LINE FOR THE NEXT STEP.

STEP ⑥ 6 SWING A 6" (152 mm) ARC FROM POINT B TO LOCATE POINT E ON THE STAIR WIDTH LINE.

STEP ⑦ 7 SWING AN 11" (279 mm) ARC FROM POINT D TOWARD WHERE POINT F WILL BE. SWING A 12" (305 mm) ARC FROM POINT E TO LOCATE POINT F.

STEP ⑧ 8 DRAW LINE TO WALL FROM POINT E THROUGH POINT F.

STEP ⑨ 9 REPEAT THIS PROCEDURE FOR THE REMAINING WINDERS.

STEP ⑩ 10 FIRST NORMAL TREAD WILL BE 11" (279 mm) FROM LAST WINDER MEASURED 12" (305 mm) OUT AS SEEN BETWEEN G AND H.

STEP ⑪ 11 LAY OUT REMAINING NORMAL TREADS.

STEP ⑫ 12 DRAW PLUMB LINES UP THE WALL TO LOCATE RISER LINES FOR EACH STEP.

PLAN

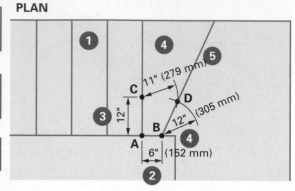

PLAN

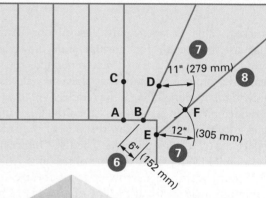

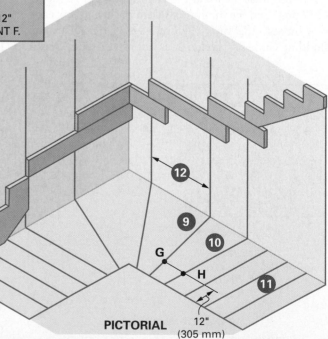

PICTORIAL

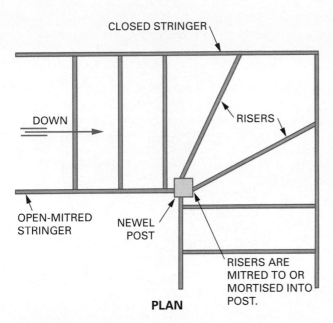

FIGURE 38–16 Risers of open-sided winders are mitred against or mortised into a newel post.

same as for a straight-run set of stairs. The total rise will be divided up into equal parts that measure between 7 and $7\frac{7}{8}$ inches (178 and 200 mm). The unit run, however, is a little different. Because of the curve, each tread will be wedge-shaped, with the narrow end following the inside radius of the stairs and the wide end following the outside radius. The general practice is to make sure that there is a wide-enough walking surface at the *line of travel*, which is located between 12 and 16 inches (305 and 406 mm) away from the inner handrail. The NBCC requires a minimum of 6 inches (152 mm) at the narrow end of the tread.

To determine the unit run for each carriage and the tread size, the carpenter must consult the plans. The angle through which the stairs will turn and the width of the stairs will guide the calculations. If the angle is a part of a complete circle, determine the ratio by dividing it into 360 degrees. This will give the divisor, which you will use to get the length of the inner and outer arcs of the stairs. The circumference of a full circle is calculated by the formula $C = 2BR$, where B is equal to 3.142 and R is equal to the radius. Divide the inner and outer circumference by the divisor determined above and that will give the arc lengths for the stairs.

Remember that there will be one less run than the number of rises. Divide the arcs by the number of runs and that will give the narrow and wide ends of the unit run. Nosing of 1 inch (25 mm) will be added to these numbers to get the tread size. The unit run at the line of travel is calculated in the same way

EXAMPLE

120° Circular Stairs – Imperial Example

- Total rise = $9'\text{-}9\frac{5}{8}'' = 117.625''$
  $117.625 \div 7 = 16.80$
  Number of risers = **16**; number of runs = **15**
- Unit rise = $117.625 \div 16 = 7.35 = 7\frac{3}{8}''$
- Inside radius = $5' = 60''$
- Stair tread width = $3'\text{-}6'' = 42''$
- Outside radius = $60'' + 42'' = 102''$
- Line of travel = $1'\text{-}3'' = 15''$
- Inside arc = $(2 \times B \times 60'') \div 3 = 125.66''$
- Outside arc = $(2 \times B \times 102'') \div 3 = 213.63''$
- Line of travel arc = $(2 \times B \times 75'') \div 3 = 157.08''$
- Unit run (inside) = $125.66'' \div 15 = 8\frac{3}{8}''$
- Unit run (line/travel) = $157.08'' \div 15 = \mathbf{10\frac{1}{2}''}$
- Unit run (outside) = $213.63'' \div 15 = \mathbf{14\frac{1}{4}''}$

120° Circular Stairs – Metric Example

- Total rise = 2988 mm
  $2988 \div 180 = 16.6$
  Number of risers = **16**; number of runs = **15**
- Unit rise = $2988 \div 16 = \mathbf{187\ mm}$
- Inside radius = 1524 mm
- Stair tread width = 1070 mm
- Outside radius = $1524 + 1070 = 2594$ mm
- Line of travel = 380 mm
- Inside arc = $(2 \times B \times 1524) \div 3 = 3192$ mm
- Outside arc = $(2 \times B \times 2594) \div 3 = 5433$ mm
- Line of travel arc = $(2 \times B \times 1904) \div 3 = 3988$ mm
- Unit run (inside) = $3192 \div 15 = \mathbf{213\ mm}$
- Unit run (line/travel) = $3988 \div 15 = \mathbf{266\ mm}$
- Unit run (outside) = $5433 \div 15 = \mathbf{362\ mm}$

(Figure 38–17). The worked example is for a set of circular stairs that turns through 120°.

SERVICE STAIRS

Service stairs, typically used for basement stairs, are built as a quick set of stairs. They are not considered finish stairs and are often built without risers. Two carriages are used with nominal 2 × 10 (38 × 235 mm) treads cut between them. The carriages are not always

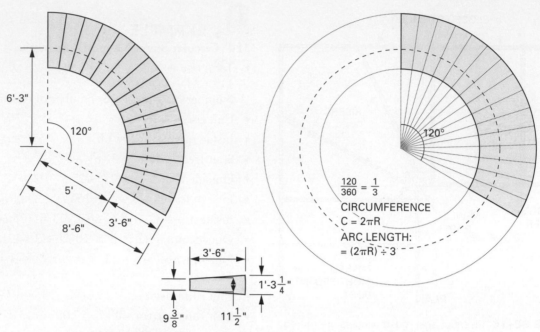

APPROXIMATE TREAD SIZE (INCLUDING 1" NOSING)

FIGURE 38–17 Circular stairs that turn through 120 degrees.

$$\frac{120}{360} = \frac{1}{3}$$

CIRCUMFERENCE
$C = 2\pi R$
ARC LENGTH:
$= (2\pi R) \div 3$

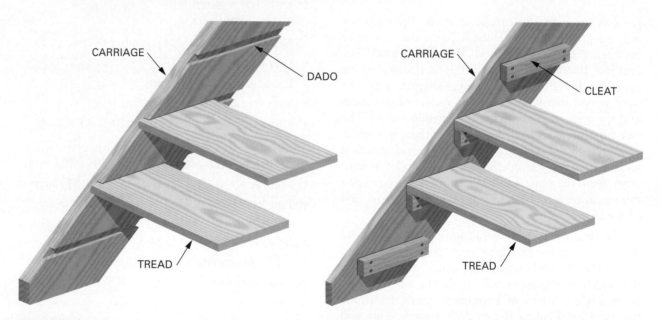

FIGURE 38–18 Service stair carriages may be dadoed or cleated to support the treads.

cut out, like those previously described. They may be dadoed to receive the treads. An alternative method is to fasten cleats to the carriages to support the treads (**Figure 38–18**).

Lay out the carriages in the usual manner. Cut the bottoms on a level line to fit the floor. Cut the tops along a plumb line to fit against the header of the stairwell. "Drop" the carriages as necessary to provide a starting riser with a height equal to the rest of the risers.

Dadoed Carriages

If the treads of rough stairs are to be dadoed into the carriages, lay out the thickness of the tread on the stringer below the layout line. The top of the tread is to the original layout line. Mark the depth of the dado (no greater than one-third of the thickness of the carriage) on both edges of the carriage. Set the circular saw to cut the depth of the dado. Make cuts along the layout lines for the top and bottom of each tread. Then make a series of saw cuts between those just made.

Chisel from both edges toward the centre, removing the excess to make the dado. Apply construction adhesive, position and clamp the treads into the dadoes, and nail the carriages into the treads. Assemble the staircase. Fasten the assembled staircase in position. Locate the top tread at a height to obtain an equal riser at the top. The lower end of a basement stairway is sometimes anchored by installing a kicker plate, which is fastened to the floor (**Figure 38–19**).

Cleated Carriages

If the treads are to be supported by cleats, measure down from the top of each tread a distance equal to its thickness. Draw another level line. Fasten 1 × 3 (19 × 64) cleats (or 2 × 4s) to the carriages with screws and construction adhesive. Make sure their top edges are to the bottom line. Fasten the carriages in position. Cut the treads to length. Install the treads between the carriages so that the treads rest on the cleats in a bed of construction adhesive. Fasten the treads by nailing the stair carriage into their ends.

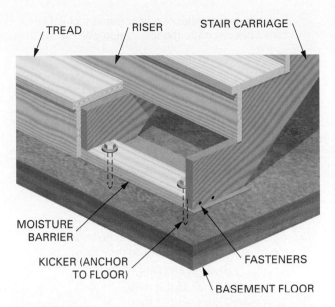

FIGURE 38–19 A kicker plate may be used to anchor a carriage to floor.

KEY TERMS

balloon frame	firestop blocking	nosing	skirt board
bearing partitions	floor joists	on centre (OC)	stair carriage
box header	grouting	open stairway	stair horse
bridging	handrail	plate	staircase
circular stairs	housed stringer	platform frame	stairwell
cleats	joist header	ribbon	stringer
closed stairway	kicker	ridge	tail joists
draftstop blocking	matched boards	sheathing	trimmer joists
edge	mortise	shims	web stiffeners
end	newel post	sill plates	winders

REVIEW QUESTIONS

Select the most appropriate answer.

1. Why is a platform frame easy to erect?
   a. Only one-storey buildings are constructed with this type of frame.
   b. Each platform may be constructed on the ground.
   c. At each level, a flat surface is provided on which to work.
   d. Fewer framing members are required.

2. What is a heavy beam that supports the inner ends of floor joists often called?
   a. pier
   b. stud
   c. girder
   d. sill

3. What is the member of a floor frame that is fastened directly on to the foundation wall called?

   a. pier
   b. beam
   c. stud
   d. sill

4. How can you quickly and accurately mark a square end on a round column?

   a. Use a square.
   b. Measure down from the other end several times.
   c. Use a pair of dividers.
   d. Wrap a piece of paper around it.

5. When the ends of floor joists rest on a supporting member, what bearing should they have?

   a. at least 4 inches (100 mm)
   b. at least 3½ inches (89 mm)
   c. at least 2½ inches (63 mm)
   d. at least 1½ inches (38 mm)

6. If floor joists lap over a girder, what is the minimum lap they should have?

   a. 2 inches (50 mm)
   b. 3 inches (75 mm)
   c. 6 inches (152 mm)
   d. 12 inches (305 mm)

7. What is it important to do when installing dimension lumber floor joists?

   a. Toenail them to the sill with at least two 2½-inch (63 mm) nails.
   b. Have the crowned edges up.
   c. End-nail them to a band joist with at least three 3¼-inch (82 mm) nails.
   d. Toenail them to the beam with at least two 2½-inch (63 mm) nails.

8. For best order of installation for the members of a floor opening, what would be installed next after the inside trimmer?

   a. outside trimmer
   b. tail joists
   c. inside header
   d. outside header

9. Which of the following is NOT a requirement for boring holes into dimension lumber floor joists?

   a. no closer than 2 inches (50 mm) to the edge of the joist
   b. no larger than ¼ of joist depth
   c. depth of member increased by size of hole
   d. no farther than ½ of joist depth away from edge of support

10. How long should the bearing points of a beam be?

    a. at least as long as the beam is deep (wide)
    b. at least 4 inches (100 mm)
    c. at least 5 inches (127 mm)
    d. at least 6 inches (152 mm)

11. What is the maximum nail spacing for an engineered panel subfloor on 16-inch OC (406 mm) floor joists?

    a. 6 inches (152 mm) on the edge and 6 inches (152 mm) on intermediate supports
    b. 6 inches (152 mm) on the edge and 8 inches (203 mm) on intermediate supports
    c. 6 inches (152 mm) on the edge and 12 inches (305 mm) on intermediate supports
    d. 8 inches (203 mm) on the edge and 8 inches (203 mm) on intermediate supports

12. When laying out dimension joists on 16-inch (406 mm) centres, what is the distance between the first and second joist?

    a. 13 inches (330 mm)
    b. 13¾ inches (349 mm)
    c. 15¼ inches (387 mm)
    d. 16 inches (406 mm)

13. Why is pressure treatment done on lumber?

    a. to improve decay resistance
    b. to improve pressure resistance
    c. to improve nail-holding strength
    d. to match the lumber to the colour of grass

14. What is the rounded outside edge of a tread that extends beyond the riser called?

    a. housing
    b. coving
    c. turnout
    d. nosing

15. What does the unit rise of a stair refer to?

    a. vertical part of a step
    b. riser
    c. calculation from total rise and number of risers
    d. the slope or angle of the stairs

16. What does the unit run of a stair refer to?

    a. tread width without nosing
    b. horizontal portion of each step plus the nosing
    c. stairwell opening
    d. horizontal portion of each step minus the nosing

Answer the following questions according to the NBCC.

17. What is the minimum width for stairways in residential construction?

    a. 36 inches (910 mm)
    b. 34 inches (860 mm)
    c. 32 inches (810 mm)
    d. 30 inches (760 mm)

18. What is the maximum riser height for residential stairs?

    a. 7 inches (178 mm)
    b. 7½ inches (190 mm)
    c. 7⅞ inches (200 mm)
    d. 8¼ inches (210 mm)

19. What is the minimum run for residential stairs?

    a. 7⅞ inches (200 mm)
    b. 8¼ inches (210 mm)
    c. 9¼ inches (235 mm)
    d. 10 inches (250 mm)

20. What is the maximum run for residential stairs?

    a. 10 inches (250 mm)
    b. 12 inches (300 mm)
    c. 13 inches (330 mm)
    d. 14 inches (356 mm)

21. What is the minimum riser height for residential stairs?

    a. 7 inches (178 mm)
    b. 6½ inches (165 mm)
    c. 6 inches (150 mm)
    d. 5 inches (125 mm)

22. What is the minimum tread width for residential stairs?

    a. 7¼ inches (184 mm)
    b. 9¼ inches (235 mm)
    c. 11¼ inches (286 mm)
    d. 12 inches (305 mm)

23. What is the minimum effective depth for the carriage in residential stairs?

    a. 3½ inches (89 mm)
    b. 3 inches (76 mm)
    c. 2½ inches (64 mm)
    d. 2 inches (51 mm)

24. What is the minimum headroom for residential stairs?

    a. 7' (2134 mm)
    b. 6'-10" (2083 mm)
    c. 6'-8" (2032 mm)
    d. 6'-5" (1950 mm)

25. What is the minimum number of risers for a set of stairs with a total rise of 9'-2½" (2800 mm)?

    a. 15
    b. 14
    c. 13
    d. 12

26. Given that the unit rise is 7½ inches (190 mm) and that the tread thickness is 1¼ inches (30 mm), what is the riser height of the first step?

    a. 7½ inches (190 mm)
    b. 7 inches (178 mm)
    c. 6½ inches (165 mm)
    d. 6¼ inches (160 mm)

WHAT'S WRONG WITH THIS PICTURE?

Carefully study **Figure 38–20** and think about what is wrong. Consider all possibilities.

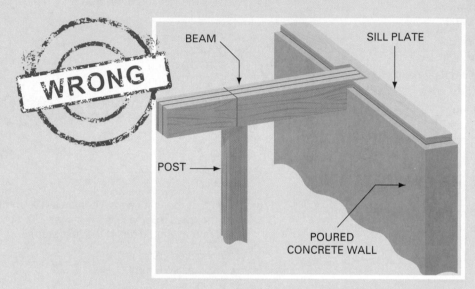

FIGURE 38–20 The diagram shows a built-up beam. Look closely at the beam and notice that all of the splices are located over the post. While this may intuitively make sense, The National Building Code of Canada requires that the joints be within 6 inches (150 mm) of the ¼ point of clear span, with a joint over the support also being acceptable.

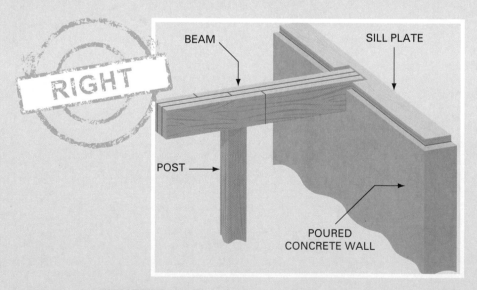

FIGURE 38–21 This diagram complies with the NBCC.

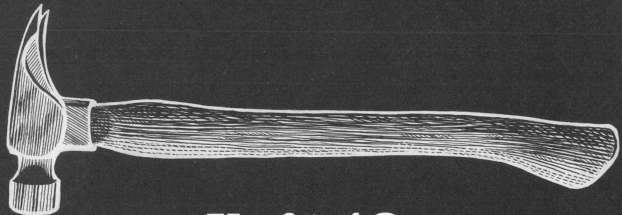

Unit 13
Exterior Wall Framing

Chapter 39 Exterior Wall: Components and Their Functions	**Chapter 40** Exterior Wall: Construction Sequence

Wall framing methods vary across the country and are affected by regional characteristics. These variations are easy to adjust for when the carpenter has an understanding of basic framing methods and practices. Exterior walls must be constructed to the correct height, corners braced plumb, walls straightened from corner to corner, and window and door openings framed to specified size. The techniques described in this unit will enable the apprentice carpenter to frame exterior walls with competence.

Setting the exterior wall frame is the first step in defining the outline of the house. Stack material close enough to the work area, yet not in the way of future work.

OBJECTIVES

After completing this unit, the student should be able to:

- identify and describe the function of each part of the wall frame.
- determine the length of exterior wall studs and the size of rough openings, and lay out a storey pole.
- build corner posts and partition intersections, and describe several methods of forming them.
- lay out the wall plates.
- construct and erect wall sections to form the exterior wall frame.
- plumb, brace, and straighten the exterior wall frame.
- apply wall sheathing.
- describe the construction of wood foundation walls.

Chapter 39

Exterior Wall: Components and Their Functions

The wall frame consists of a number of different parts. The student should know the name, function, location, and usual size of each member. Sometimes the names given to certain parts of a structure may differ according to the geographical area. For that reason, some members may be identified with more than one term.

PARTS OF AN EXTERIOR WALL FRAME

An exterior wall frame consists of plates, studs, lintels, sills, jack studs, trimmers, corner posts, partition intersections, ribbons, and braces (**Figure 39–1**). For nailing requirements, see National Building Code of Canada (NBCC) Table 9.23.3.4.

Plates

The top and bottom horizontal members of a wall frame are called *plates*. The bottom member is called a sole plate. It is also referred to as the *bottom plate*. The top member is called a **top plate**. Top plates usually consist of doubled 2-inch (38 mm) stock. In a balloon frame, the sole plate is not used. Instead, the studs rest directly on the sill plate.

Studs

Studs are vertical members of the wall frame. They run full-length between plates. *Jack studs* or *trimmers* are shortened studs that line the sides of an opening. They extend from the bottom plate up to the underside of the lintel. *Cripple* studs are shorter members above and below an opening, which extend from the top or the bottom plates to the opening.

Studs are usually 2 × 4s (38 × 89s), but 2 × 6s (38 × 140s) are used for extra insulation or when the supported load requires it. Studs are usually placed 16 inches (406 mm) OC. With energy conservation a key issue in construction, stud spacing is moving to 24 inches (610 mm) OC where structurally permitted.

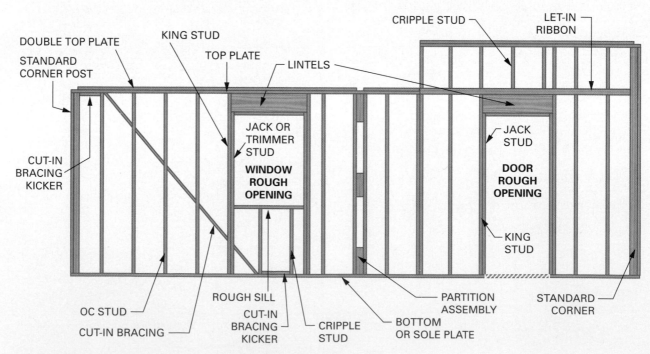

FIGURE 39–1 Typical component parts of an exterior wall frame.

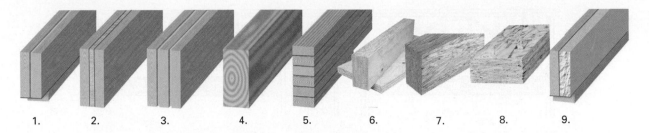

1. A BUILT-UP LINTEL WITH A 2 X 4 OR 2 X 6 LAID FLAT ON THE BOTTOM
2. A BUILT-UP LINTEL WITH A ½" SPACER SANDWICHED IN BETWEEN
3. A BUILT-UP LINTEL FOR A 6" WALL
4. A LINTEL OF SOLID SAWN LUMBER
5. GLULAM BEAMS ARE OFTEN USED FOR LINTELS.
6. A BUILT-UP LINTEL OF LAMINATED VENEER LUMBER
7. PARALLEL STRAND LUMBER MAKES EXCELLENT LINTELS.
8. LAMINATED STRAND LUMBER IS USED FOR LIGHT-DUTY LINTELS.
9. ENERGY-EFFICIENT LINTEL WITH RIGID FOAM INSULATION

FIGURE 39–2 Types and styles of lintels.

This allows the studs to align with the trusses and eliminates the requirement for a second top plate. Therefore, much less wood is used and less thermal bridging, because wood has a low R-value (RSI). To better accommodate the interior finishes, the walls are insulated and sealed with polyethylene and acoustical sealant, and then strapped with 2 × 3s (38 × 63s). This allows electrical boxes to be placed on the inside of the poly, thereby maintaining a complete air seal.

Lintels

Lintels run at right angles to studs. They form the top of window, door, and other wall openings, such as fireplaces. Lintels must be strong enough to support the load above the opening. The depth of the lintel depends on the width of the opening. As the width of the opening increases, so must the strength of the lintel. Check NBCC 9.23.12.3 and Tables A-12 to A-16 for sizing and framing requirements for lintels.

Kinds of Lintels. Solid or built-up lumber may be used for lintels. For 2 × 4 (38 × 89 mm) walls, two pieces of 2-inch (38 mm) lumber with ½-inch (12.5 mm) plywood or strand board sandwiched in between them gives the lintel the full 3½-inch (89 mm) thickness of the wall. In 2 × 6 (38 × 140 mm) walls, three pieces of 2-inch (38 mm) lumber with two pieces of ½-inch (12.5 mm) plywood or strand board in between makes up the 5½-inch (140 mm) wall thickness (**Figure 39–2**).

Much engineered lumber is now being used for window and door opening lintels. Figures 5–6, 6–1, and 8–1 (pages 38, 40, and 46, respectively) show the use of laminated veneer lumber, parallel strand lumber, and glulam beams as opening lintels (**Figure 39–3**).

In many buildings, when the opening must be supported without increasing the lintel size, the top

GLULAM

PSL (PARALLEL STRAND LUMBER)

LSL (LAMINATED STRAND LUMBER)

FIGURE 39–3 Lintels may be made of glulam, parallel strand lumber, and laminated strand lumber.

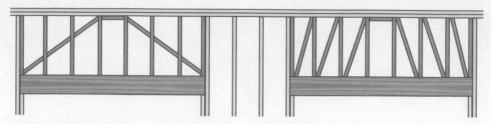

FIGURE 39–4 Two methods of trussing a large opening. Must be engineered.

of a wall opening may be trussed to provide support (**Figure 39–4**). Alternatively, the lintel is placed at the top of the framed wall and secured to the top plates, and the area below the lintel is framed in to the top of the rough opening. However, when the opening is fairly close to the top plate, the depth of the lintel is increased. This completely fills the space between the plate and the top of the opening. In this case, the same size of lintel is usually used for all wall openings, regardless of the width of the opening (**Figure 39–5**). This eliminates the need to install short cripple studs above the lintel.

Rough Sills

Forming the bottom of a window opening, at right angles to the studs, are members called **rough sills**. They usually consist of a single 2× (38×) (38 mm) thickness. They carry the weight of the windows. Many carpenters prefer to use a double thickness rough sill when the opening is greater than 4 feet (1.2 m). This provides more surface on which to fasten window trim in the later stages of construction. Rough sills are supported by cripple studs, which are sometimes cut at a slight angle at the top to slope the sill to facilitate drainage.

Jack Studs (Trimmers)

Jack studs (**trimmers**) are shortened studs that support the lintels. They are fastened to the king studs

on each side of the opening. In window openings, the trimmer should be installed full length from lintel to bottom plate (**Figure 39–6**). They should not be cut to allow the sill to fit to the king stud. Door jack studs are installed the same as window trimmers (**Figure 39–7**).

Corner Posts

Corner posts are the same length as studs. They are constructed in a manner that provides an outside and an inside corner on which to fasten the exterior and interior wall coverings. They may be constructed in several ways.

One method uses a three-stud corner where the two corner studs on the long wall are nailed together to form an "L" shape. The second method is designed to increase the amount of insulation in the corner. One stud is removed and replaced by a strip of ½" plywood or OSB that serves as drywall backing (**Figure 39–8**).

Partition Intersections

Wherever interior **partitions** meet an exterior wall, extra studs need to be put in the exterior wall. This provides wood for fastening the interior wall covering in the corner. In most cases, the *partition intersection* is made of two studs nailed to the edge of

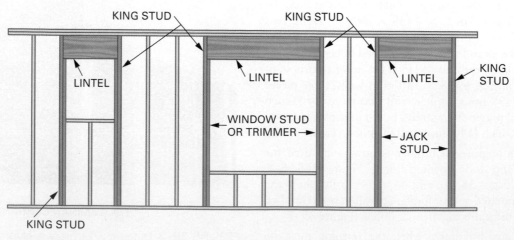

FIGURE 39–5 It is common practice to use the same lintel height for all wall openings.

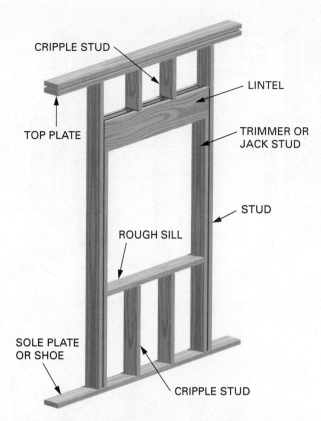

FIGURE 39–6 Typical framing for a window opening.

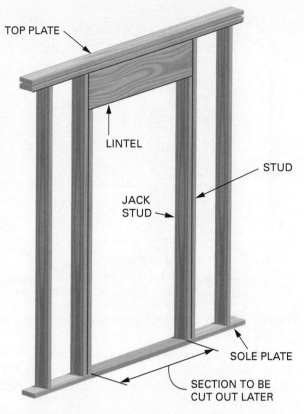

FIGURE 39–7 Typical framing for a door opening.

2×4 (38×89 mm) blocks about a foot long (300 mm). One block is placed at the bottom, one at the top, and one about centre on the studs (**Figure 39–9**).

Another method is to maintain the regular spacing of the studs. Blocking is then installed between them wherever partitions occur. The block is set back from the inside edge of the stud the thickness of a board. A 1×6 (19×140 mm) board is then fastened vertically on the inside of the wall so that it is centred on the partition.

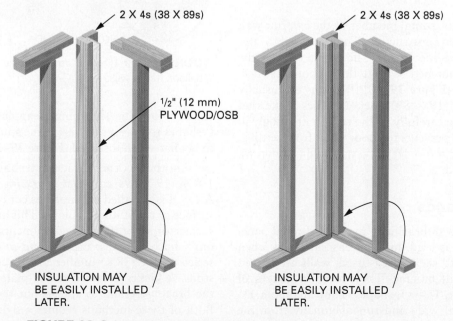

FIGURE 39–8 Methods of fabricating corner posts.

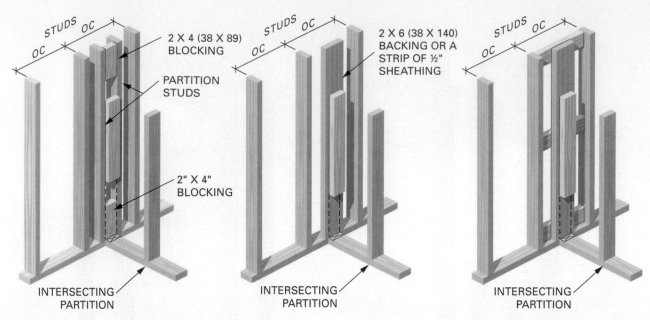

FIGURE 39–9 Methods of fabricating partition intersections.

Another method is to nail a continuous 2 × 6 (38 × 140 mm) backer to a full-length stud. The edges of the backer project an equal distance beyond the edges of the stud.

In all three methods it is important to place a 12-inch (305 mm) strip of polyethylene between the partition and the exterior wall, as well as between the two top plates of the interior walls of the top storey of the house. This strip will later be sealed to the wall and ceiling polyethylene air/vapour barrier and will provide continuity of the barrier.

Ribbons

A ribbon is a horizontal member of the exterior wall frame in balloon construction. They are used to support the second-floor joists. The inside edge of the wall studs is notched so that the ribbon lies flush with the edge (**Figure 39–10**). Ribbons are usually made of 1 × 4 (19 × 89) stock. Notches in the stud should be made carefully so the ribbon fits snugly in the notch. This prevents the floor joists from settling. If the notch is cut too deep, the stud will be unnecessarily weakened.

Corner Braces

Generally, no wall bracing is required if rated panel wall sheathing is used. In other cases, such as when insulating board sheathing is used, walls are braced with metal wall bracing. They come in gauges of 22 to 16 in flat, T-, or L-shapes. They are about 1½ inches (40 mm) wide and run diagonally from the top to the bottom plates. They are nailed to the stud

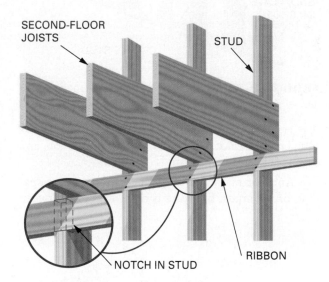

FIGURE 39–10 Ribbons are used to support floor joists in balloon frame walls.

edges before the sheathing is applied. The T- and L-shapes require a saw kerf in the stud to allow them to lay flat when installed (**Figure 39–11**).

Another corner bracing technique is to use 1 × 4s (19 × 89s) called *let-in bracing* (**Figure 39–12**). A 1 × 4 is installed into notches cut out of the inside surface of the studs and plates. This allows the inside surface of all of the wall components to be continuously flush. The last method, *cut-in bracing*, uses a series of 2× (38×) lumber blocks cut between the studs. A kicker is nailed to the plate at the ends of the bracing system to spread out the racking load. Both of these methods require a tight fit to achieve maximum stiffness.

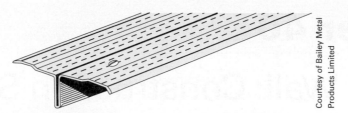

FIGURE 39–11 Bailey 'T' used for diagonal metal brace for wood-framed wall.

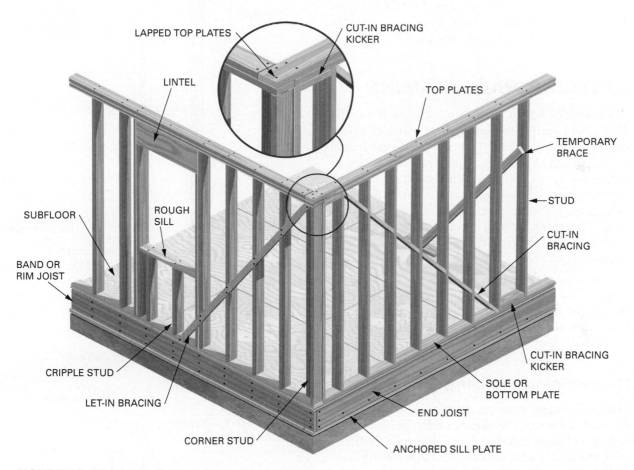

FIGURE 39–12 Wood wall bracing may be cut-in or let-in.

Chapter 40

Exterior Wall: Construction Sequence

Careful construction of the wall frame makes the application of the exterior and interior finishes easier. It also reduces problems for those who apply them later. Stud sizes are found in the NBCC Tables A-30–33, 9.23.10.1(2). Lintel sizes are found in Tables A-13–16, 9.23.12.3(1) & (2).

EXTERIOR WALL FRAMING

The standard height of a rough ceiling is usually 8'-1" (2464 mm). Subtracting three plates of a total thickness of 4½ inches (114 mm), the stud length is 92½ inches (2350 mm). Studs can be purchased precut to length, called *precut studs*, to save the carpenter time and wasted material. Note, however, that precut studs are usually 92⅝ inches (2353 mm) long. The extra ⅛ inch (3 mm) length provides insurance that the finished material will be easily installed later.

Sometimes the ceiling height is not the standard height. The section view of the house plans will specify the finished floor to finished ceiling height. From this number, the length of the stud must be calculated.

Determining the Length of Studs

The stud length must be calculated so that, after the wall is framed, the distance from finish floor to ceiling will be as specified in the drawings. To determine the stud length, the thickness of the finish floor and the finished ceiling thickness below the ceiling joist must be known.

Stud length for platform framing is found by adding these measurements (**Figure 40–1**):
- Finished floor to ceiling height
- Ceiling thickness (includes *furring strips* if used)
- Finished floor thickness

Then deduct the total plate thickness.

Stud length for balloon frame construction is found by adding these measurements (**Figure 40–2**):
- Finished floor to ceiling height of both storeys
- Ceiling thickness of both storeys (includes *furring strips* if used)

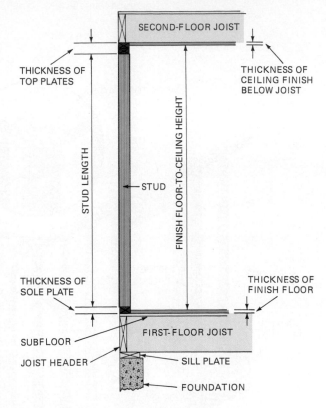

① ADD THICKNESS OF CEILING FINISH AND FINISH FLOOR TO CEILING HEIGHT

② SUBTRACT THE COMBINED THICKNESSES OF THE PLATES TO FIND THE STUD LENGTH

FIGURE 40–1 Determining stud length in platform construction.

- Finished floor thickness of both storeys
- Subfloor thickness of both storeys
- Width of floor joists of both storeys

Then deduct the top plate total thickness.

Ribbon height to support second-floor joists is found by adding the following:
- Finished floor to ceiling height of first floor
- First-floor ceiling thickness (includes *furring strips* if used)
- Finished floor thickness of first floor
- Subfloor thickness of first floor
- Width of floor joists of first floor

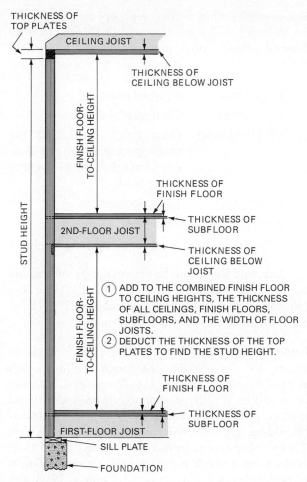

FIGURE 40–2 Determining stud length in balloon frame construction.

Determining Rough Opening Size (RO)

A **rough opening** is an opening framed in the wall in which to install doors and windows. The width and height of rough openings are usually indicated in the plans or in the door and window schedule provided by the manufacturer. It is the carpenter's responsibility to determine the rough opening size for the particular unit from the information given. The door and window schedule contains the kind, style, manufacturer's model number, size of each unit, and rough opening dimensions.

Rough Openings for Doors

The rough opening for a door must be large enough to accommodate the door, door frame, and space for shimming the frame to a level and plumb position. Usually, ½ inch (12 mm) is allowed for shimming insulating, and air sealing, on all sides, between the door frame and the rough opening. The amount allowed for the door frame itself depends on the thickness of the door frame beyond the door.

Care must be taken not to make the rough opening oversized. If the opening is made too large, the window or door finish may not cover it.

The sides and top of a door frame are called **jambs**. Jambs may vary in thickness. Sometimes *rabbeted* wood jambs are used. The rabbet is that part of the jamb that the door stops up against. For interior doors, nominal 1-inch (19 mm) lumber is used for the jamb. A separate **stop** is applied (**Figure 40–3**). Steel

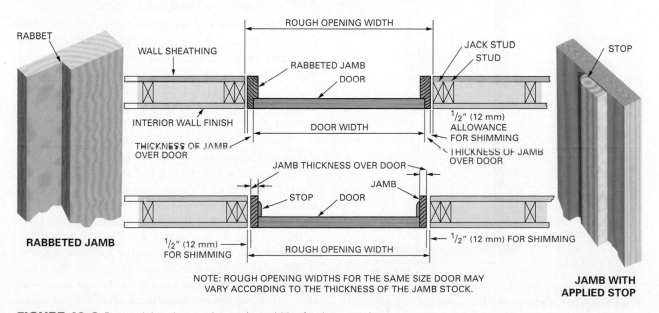

NOTE: ROUGH OPENING WIDTHS FOR THE SAME SIZE DOOR MAY VARY ACCORDING TO THE THICKNESS OF THE JAMB STOCK.

FIGURE 40–3 Determining the rough opening width of a door opening.

jambs have the door stop built-in similar to a rabbeted wood jamb.

The bottom member of the door frame is called a **threshold**. Thresholds may be hardwood, metal, or a combination of wood and metal. The type of threshold and its thickness must be known in order to figure the rough opening height.

Door Rough Opening Height. Rough opening height is determined by adding five dimensions (**Figure 40–4**). The rough opening heights for all openings in a house are usually the same, so only one rough opening height needs to be calculated. Because the wall rests on the subfloor, the subfloor is the starting point:

- Finished floor thickness
- Door threshold thickness (if none, add 1 inch [25 mm] for swing clearance under the door)
- Door height
- Head (top) jamb thickness
- Shim space (usually ½ inch [12 mm])

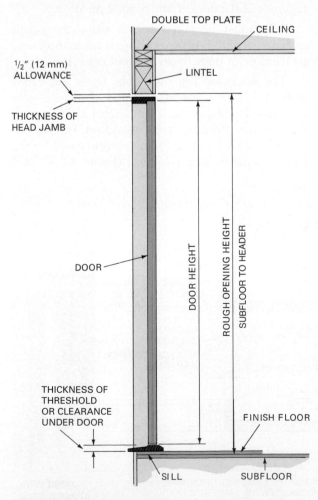

FIGURE 40–4 Determining the rough opening height of a door opening.

For example, what is the rough opening height for an 80-inch (2032 mm) door with a ¾-inch (19 mm) finished floor, no threshold, and a ¾-inch (19 mm) jamb?

¾″ (19 mm)	Finished floor
1″ (25 mm)	Clearance under door
80″ (2032 mm)	Door height (including the small space between the door and jambs that allows opening of the door)
¾″ (19 mm)	Jamb thickness
+ ½″ (12 mm)	Shim space
83″ (2107 mm)	Total rough opening height

Jack stud length can be determined by subtracting bottom plate thickness, which is typically 1½ inches (38 mm). In this case, the jack stud length is 83″ − 1½″ = 81½″ (2108 mm − 38 mm = 2070 mm).

Door Rough Opening Width. Refer back to Figure 40–3 to see that the rough opening width is also found by adding five dimensions. Because door widths vary from room to room, a shorthand method is used. The rough opening width is found by adding 2½ inches (63 mm) to the door width. This number comes from adding ½ inch (12.5 mm) of shim space and ¾ inch (19 mm) of jamb thickness to both sides of the door; or 12.5 mm + 19 mm = 31.5 mm times two = 63 mm. Note that if the jamb thickness is not ¾ inch (19 mm), then the 2½″ (63 mm) number must be adjusted accordingly. Thus, if the door is 30 inches (760 mm) wide, then the rough opening width for this door is 30″ + 2½″ = 32½″ (760 mm + 63 mm = 823 mm).

Today, doors and windows are manufactured as units that include the jambs and the thresholds. Unit sizes and outside dimensions are available from the manufacturer as well as recommended rough openings. It is recommended that a ½-inch (12 mm) space be provided around the units to allow for shims and insulation.

Standard insulated steel doors for residential installation have an RO height of 83¼ inches (2115 mm). If a 2 × 10 (38 × 235 mm) lintel with a 2× (38×) plate on its underside is used for the door opening, the resultant RO using precut studs will be a perfect fit. Because of this, most houses that are built with 8-foot (2.4 m) ceilings use 2 × 10 lintels for all doors and windows so that the tops are aligned. (Care must be taken to check the RO of the patio doors—sometimes less than an entrance door—so that the windows on the shared wall will be set at the same height.)

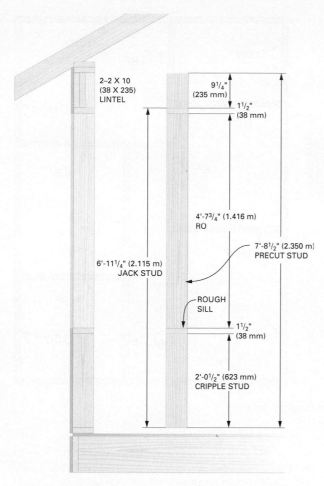

FIGURE 40–5 Marking a king stud to determine the lengths of the jack and cripple studs.

To determine the length of the jack stud, simply lay a piece of 2 × 10 (38 × 235 mm) right-angled across and flush to the top of the precut stud and mark the cut line. Once the jack or trimmer is cut, mark off the thickness of the 2× (38×) plate, measure down the RO height of a given window, again mark off 1½ inches (38 mm) for the rough sill, and measure the remaining dimension to get the length of the cripple studs (**Figure 40–5**). This simplified method is similar to the storey pole and avoids the possibility of mathematical errors.

WALL LAYOUT

Wall construction begins with careful layout of all wall components. This is usually done on the top and bottom plates. The layout of the walls usually begins at the same corner of the building where the floor joists began. This will help ensure that the load-bearing studs are directly over the joist below.

Before layout can begin, the carpenter must first determine if the wall to be laid out is a load-bearing or non-load-bearing wall. Note that exterior walls are referred to as *walls* and interior walls are referred to as *partitions*.

Determining Wall Type

The load-bearing walls (LBW) are usually built first. They support the ceiling joists and rafters and typically run the length of the building. Non-load-bearing walls (NLBW) are end walls and run parallel with the joists. Interior partitions are also load- and non-load-bearing. They are load-bearing partitions (LBP) if they run perpendicular to the joists and support the ends of the joists above. All other partitions are considered non-load-bearing partitions (NLBP) (**Figure 40–6**).

Each type of wall has a slightly different layout characteristic. To reduce confusion, remember that all centreline dimensions for openings are measured from the building line, which is on the outside edge of the exterior framing. Layout must take this fact into account (**Figure 40–7**). **Figure 40–8** notes the similarities and differences of laying out walls and partitions.

Laying Out the Plates

To lay out the plates, measure in on the subfloor, at the corners, the thickness of the wall. Snap lines on the subfloor between the marks. This is done so that the wall can be erected to a straight line later. *Tack* the plates in position so that their inside edges are to the chalk line. Do not drive nails home (they have to be pulled later). Use only as many as are needed to hold the pieces in place. Plan plate lengths so that joints between them fall in the centre of a full-length stud for the convenient erection of wall sections later (**Figure 40–9**).

Wall Openings. From the blueprints, determine the centreline dimension of all the openings in the wall. Lay these out on the plate. Then mark for the king and jack studs by measuring in each direction from the centreline one-half the width of the rough opening (**Figure 40–10**). Recheck the rough opening measurement to be sure it is correct. Square lines at these points across the plates. Mark an O on the side of each line away from the centreline. The O represents the jack stud, but a T for trimmer or a J for jack can also be used. It makes little difference what marks are used as long as the builder understands what they mean.

From the squared lines, measure away from the centreline of the jack stud thickness. Square lines across. Mark Xs on the side of the line away from

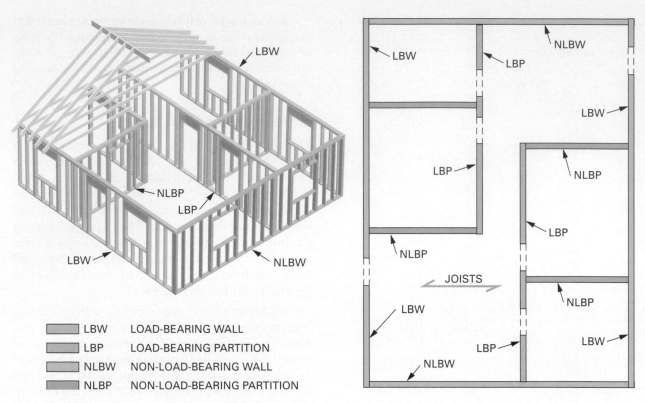

	LBW	LOAD-BEARING WALL
	LBP	LOAD-BEARING PARTITION
	NLBW	NON-LOAD-BEARING WALL
	NLBP	NON-LOAD-BEARING PARTITION

FIGURE 40–6 Walls in a building have different functions and characteristics.

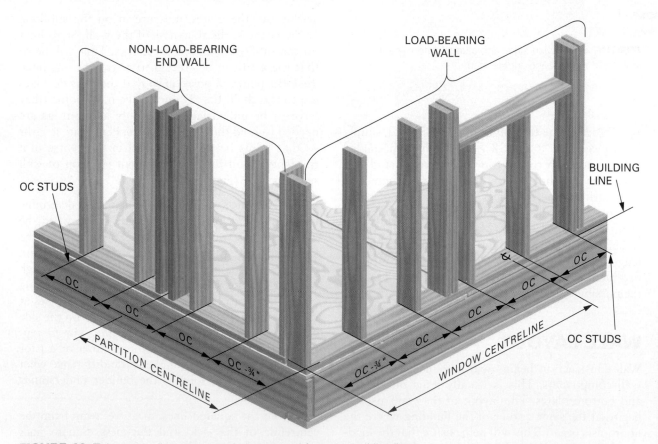

FIGURE 40–7 Layout wall components are measured from the building line.

LAYOUT VARIATIONS FOR WALLS AND PARTITIONS

	MEASURE TO OC STUDS	MEASURE TO CENTRELINES OF OPENINGS
LOAD-BEARING WALL (LBW)	FROM END OF PLATE	FROM END OF PLATE
NON-LOAD-BEARING WALL (NLBW)	INCLUDE WIDTH OF ABUTTING WALL AND SHEATHING THICKNESS	INCLUDE WIDTH OF ABUTTING WALL
LOAD-BEARING PARTITION (LBP)	INCLUDE WIDTH OF ABUTTING WALL	INCLUDE WIDTH OF ABUTTING WALL
NON-LOAD-BEARING PARTITION (NLBP)	FROM END OF PLATE	INCLUDE WIDTH OF ABUTTING WALL

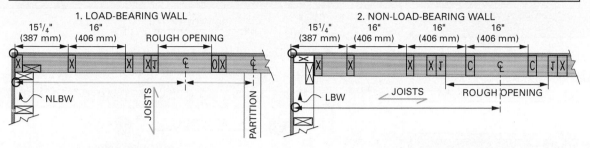

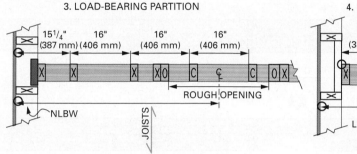

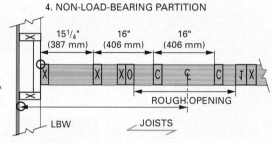

FIGURE 40–8 Layout details for four types of walls.

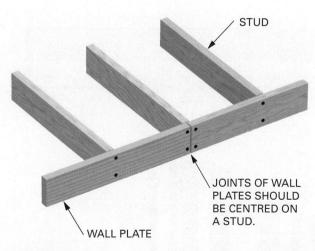

FIGURE 40–9 Joints in the plates should fall at the centre of a stud.

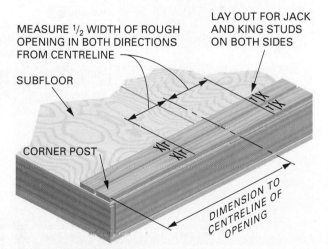

REMEASURE THE ROUGH OPENING WIDTH AS A CHECK

FIGURE 40–10 Laying out a rough opening width.

centre for the king studs on each side of the openings (Figure 40–10).

Partition Intersections

On architectural prints, interior partitions usually are dimensioned to their centreline. Mark on the plates the centreline of all partitions intersecting the wall (**Figure 40–11**). From the centrelines, measure in each direction one-half the partition stud thickness. Square lines across the plates. Mark Xs on the side of the lines away from centre for the location of partition intersection studs.

Step**by**Step Procedures

Procedure 40–A Laying Out a Typical Wall Section

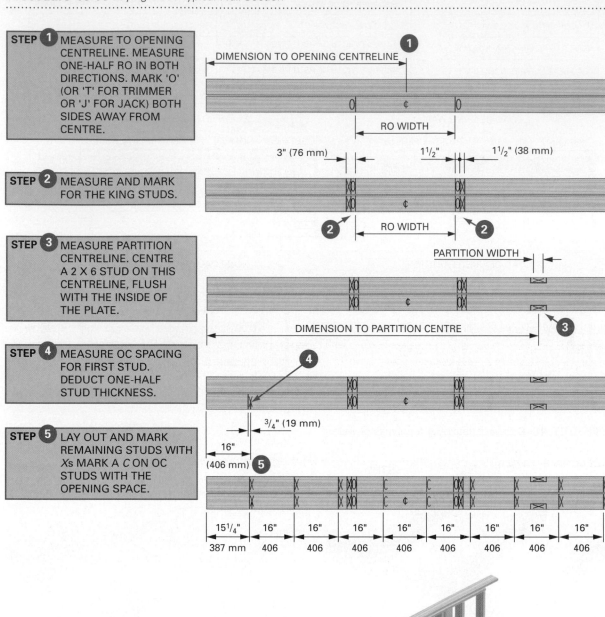

STEP 1 MEASURE TO OPENING CENTRELINE. MEASURE ONE-HALF RO IN BOTH DIRECTIONS. MARK 'O' (OR 'T' FOR TRIMMER OR 'J' FOR JACK) BOTH SIDES AWAY FROM CENTRE.

DIMENSION TO OPENING CENTRELINE

RO WIDTH

3" (76 mm)   1½"   1½" (38 mm)

STEP 2 MEASURE AND MARK FOR THE KING STUDS.

RO WIDTH

STEP 3 MEASURE PARTITION CENTRELINE. CENTRE A 2 X 6 STUD ON THIS CENTRELINE, FLUSH WITH THE INSIDE OF THE PLATE.

PARTITION WIDTH

DIMENSION TO PARTITION CENTRE

STEP 4 MEASURE OC SPACING FOR FIRST STUD. DEDUCT ONE-HALF STUD THICKNESS.

¾" (19 mm)

16" (406 mm)

STEP 5 LAY OUT AND MARK REMAINING STUDS WITH Xs MARK A C ON OC STUDS WITH THE OPENING SPACE.

15¼"   16"   16"   16"   16"   16"   16"   16"   16"
387 mm   406   406   406   406   406   406   406   406

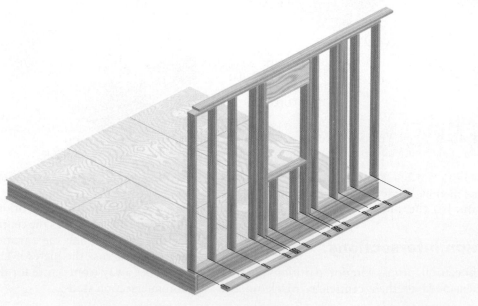

LAYOUT VARIATIONS FOR WALLS AND PARTITIONS

	MEASURE TO OC STUDS	MEASURE TO CENTRELINES OF OPENINGS
LOAD-BEARING WALL (LBW)	FROM END OF PLATE	FROM END OF PLATE
NON-LOAD-BEARING WALL (NLBW)	INCLUDE WIDTH OF ABUTTING WALL AND SHEATHING THICKNESS	INCLUDE WIDTH OF ABUTTING WALL
LOAD-BEARING PARTITION (LBP)	INCLUDE WIDTH OF ABUTTING WALL	INCLUDE WIDTH OF ABUTTING WALL
NON-LOAD-BEARING PARTITION (NLBP)	FROM END OF PLATE	INCLUDE WIDTH OF ABUTTING WALL

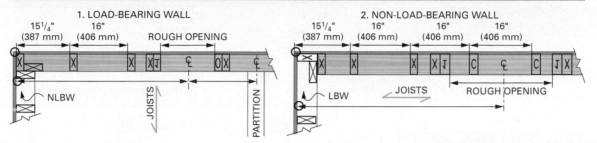

FIGURE 40–8 Layout details for four types of walls.

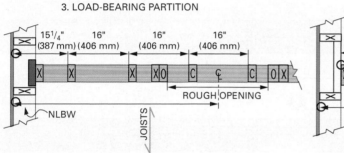

FIGURE 40–9 Joints in the plates should fall at the centre of a stud.

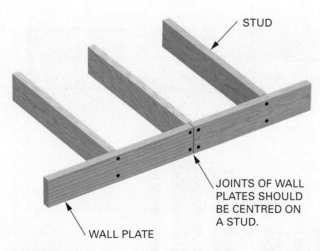

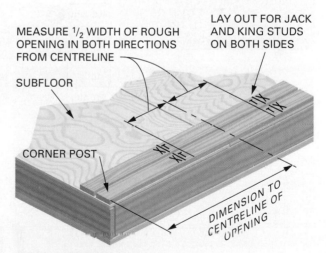

REMEASURE THE ROUGH OPENING WIDTH AS A CHECK

FIGURE 40–10 Laying out a rough opening width.

centre for the king studs on each side of the openings (Figure 40–10).

Partition Intersections

On architectural prints, interior partitions usually are dimensioned to their centreline. Mark on the plates the centreline of all partitions intersecting the wall (**Figure 40–11**). From the centrelines, measure in each direction one-half the partition stud thickness. Square lines across the plates. Mark Xs on the side of the lines away from centre for the location of partition intersection studs.

StepbyStep Procedures

Procedure 40–A Laying Out a Typical Wall Section

STEP 1 MEASURE TO OPENING CENTRELINE. MEASURE ONE-HALF RO IN BOTH DIRECTIONS. MARK 'O' (OR 'T' FOR TRIMMER OR 'J' FOR JACK) BOTH SIDES AWAY FROM CENTRE.

STEP 2 MEASURE AND MARK FOR THE KING STUDS.

STEP 3 MEASURE PARTITION CENTRELINE. CENTRE A 2 X 6 STUD ON THIS CENTRELINE, FLUSH WITH THE INSIDE OF THE PLATE.

STEP 4 MEASURE OC SPACING FOR FIRST STUD. DEDUCT ONE-HALF STUD THICKNESS.

STEP 5 LAY OUT AND MARK REMAINING STUDS WITH Xs MARK A C ON OC STUDS WITH THE OPENING SPACE.

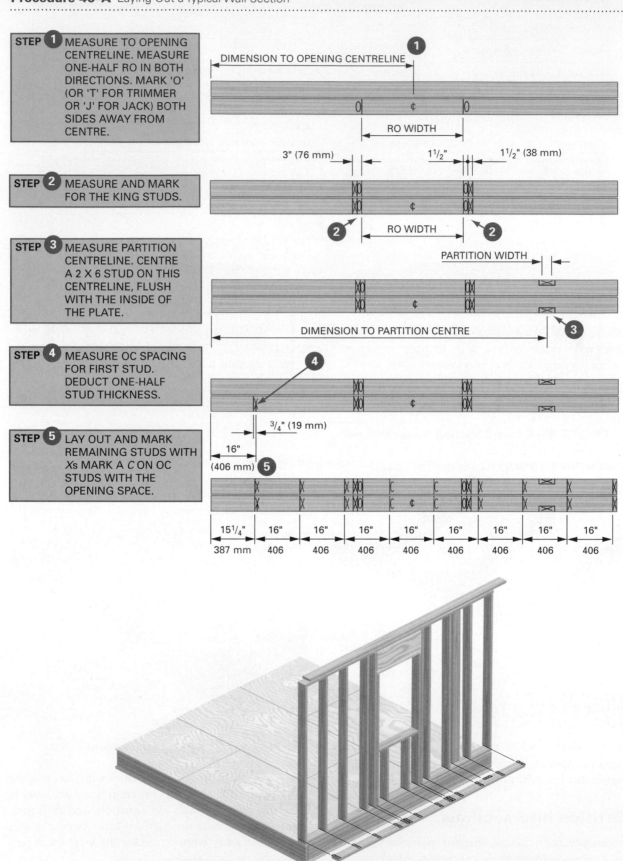

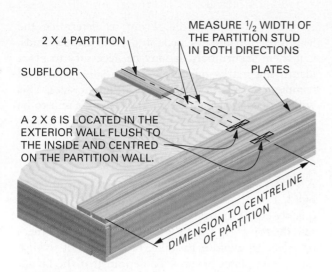

2 X 4 PARTITION

SUBFLOOR

MEASURE ½ WIDTH OF THE PARTITION STUD IN BOTH DIRECTIONS

PLATES

A 2 X 6 IS LOCATED IN THE EXTERIOR WALL FLUSH TO THE INSIDE AND CENTRED ON THE PARTITION WALL.

DIMENSION TO CENTRELINE OF PARTITION

FIGURE 40-11 Laying out a partition intersection.

Studs and Cripple Studs

After all openings and partitions have been laid out, start laying out all full-length studs and cripple studs. Proceed in the same manner and from the same end as laying out floor joists. This keeps studs directly in line with the joists below.

Measure in from the outside corner the regular stud spacing. From this mark, measure, in one direction or the other, one-half the stud thickness. Square a line across the plates. Place an *X* on the side of the line where the stud will be located.

Stretch a tape along the length of the plates from this first stud location. Square lines across the plates at each specified stud spacing. Place *X*s on the same side of the line as the first line.

Where openings occur, mark the OC studs with a *C* instead of an *X*. This will indicate the location of cripple studs (**Procedure 40-A**). All regular and cripple studs should line up with the floor joists below. When laying out the opposite wall, start from the same end as the first wall to keep all framing lined up with joists below.

ASSEMBLING AND ERECTING WALL SECTIONS

The usual method of framing the exterior wall is to precut the wall frame members, assemble the wall frame on the subfloor, and erect the frame. With a small crew and without special equipment, the walls are raised section by section. When the frame is erected, the corners are plumbed and braced. Then the walls are straightened between corners. They are also braced securely in position.

To prevent problems with the installation of the finish work later, it is important to keep the edges of the frame members flush wherever they join each other.

Precutting Wall Frame Members

Full-Length Studs. Studs are often purchased precut to length for a standard 8′-1″ (2464 mm) wall height. Some builders buy in such high volume that lumber suppliers will also precut lintels and jack and cripple studs. Usually, framing members, other than studs, are cut to length on the job. A power mitre saw is an effective tool for cutting studs and other framing members to length. Set a stop the desired distance from the saw blade to cut duplicate lengths. If this type of saw is not available, a jig can be made for a portable electric circular saw to cut duplicate lengths of framing (**Figure 40-12**). Reject any studs that are severely warped. They may be cut into shorter lengths for blocking.

Corner Posts and Partition Intersections. Corner posts and partition intersections are often made up ahead of time to speed the assembly. Corners may be made by nailing two full-length studs together where one stud is rotated at a right angle. This detail allows for more insulation in the corner, thereby making the building more energy-efficient (**Figure 40-13**).

Partition intersections may be made using ladder blocking, which again allows for more insulation in exterior walls. In this case, no extra framing layout is needed because the ladder blocking is installed between on-centre studs. Another method also allows for insulation to be easily installed later. A 2 × 6 (38 × 140 mm) is nailed at right

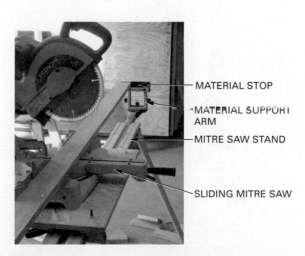

MATERIAL STOP

MATERIAL SUPPORT ARM

MITRE SAW STAND

SLIDING MITRE SAW

FIGURE 40-12 Techniques for making a cutoff jig for a mitre saw.

FIGURE 40–13 Construction of corner posts.

angles to a stud. A third method is similar to that used for corner posts except for the way the blocks are placed. They are placed in a similar location between two full-length studs, yet with blocks on their edge (**Figure 40–14**).

Lintels, Rough Sills, Jack Studs, and Trimmers. Cut all lintels and rough sills. Their length can be determined from the layout on the plates.

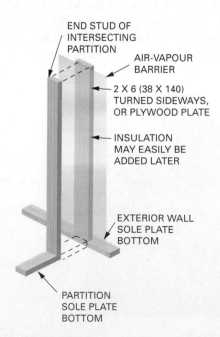

FIGURE 40–14 Construction of partition intersections.

Make a storey pole. From it, determine the length of all trimmers, jack studs, and cripple studs. Cut them accordingly. It may be necessary to place identifying marks on lintels, rough sills, jacks, and trimmers if rough openings are different sizes. This will assist in locating the window or door unit to be placed in each rough opening.

Assembling Wall Sections

A variety of wall assembly procedures are used to build a wall frame. The following technique will quickly create a strong wall (**Procedure 40–B**). Separate the top plate from the bottom plate. Stand them on edge. To avoid a mistake, be careful not to turn one of the plates around. Be certain that the layout lines on top and bottom plates line uvp. Place all full-length studs, corner posts, and partition intersections in between them.

Very few studs are absolutely straight from end to end, so each stud crown will be faced the same way. Sight each full-length stud. It will be difficult for those who apply the interior finish if no attention is paid to the manner in which studs are installed in the wall. A stud that is installed with its crowned edge out next to one with its crowned edge in will certainly present problems later. For this reason, some builders come back after the wall is built to flatten and adjust the crowns of the wall studs. Studs bowing inward are shaved with a power plane and studs bowing outward are shimmed with strips of heavy cardboard (**Figure 40–15**).

Assemble the window and door openings first to ensure easy nailing. Fasten the lintels, jacks, rough sills, and cripples in position, then fasten the king

PLANED STUD EDGE   SHIMMED STUD EDGE

FIGURE 40–15 Warped studs can be adjusted to make a flat, straight wall.

StepbyStep Procedures

Procedure 40–B Assembling a Wall Section

....

STEP 1 PULL TACK NAILS AND SEPARATE PLATES.

STEP 2 ASSEMBLE THE ROUGH OPENING FRAME. NAIL ALL MEMBERS TO EACH OTHER AND THE PLATES.

STEP 3 INSTALL REMAINING STUDS. NAIL ON THE DOUBLED TOP PLATE. LEAVE GAPS WHERE INTERSECTING PLATES WILL FIT.

STEP 4 ALIGN FRAME TO CHALK LINE AND ADJUST IT TO BE SQUARE. TACK IT TO THE SUBFLOOR AND INSTALL PERMANENT BRACING.

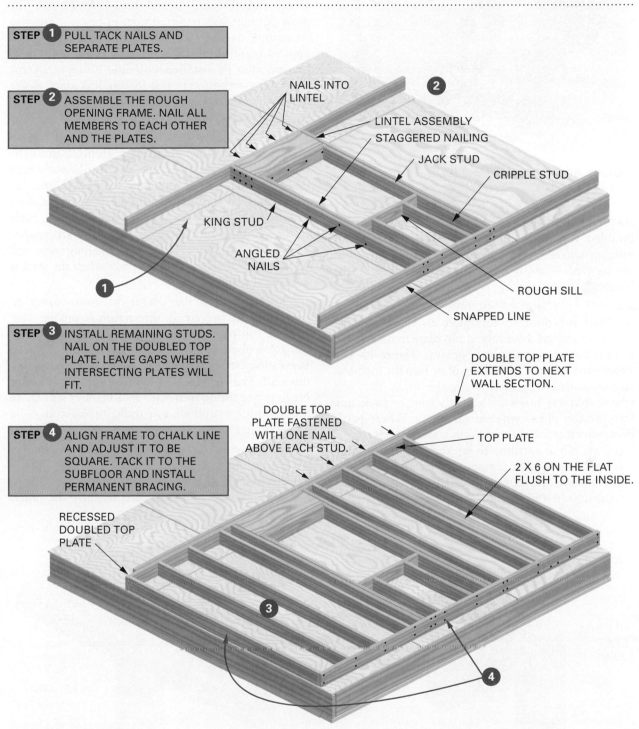

NAILS INTO LINTEL

LINTEL ASSEMBLY

STAGGERED NAILING

JACK STUD

CRIPPLE STUD

KING STUD

ANGLED NAILS

ROUGH SILL

SNAPPED LINE

DOUBLE TOP PLATE EXTENDS TO NEXT WALL SECTION.

DOUBLE TOP PLATE FASTENED WITH ONE NAIL ABOVE EACH STUD.

TOP PLATE

2 X 6 ON THE FLAT FLUSH TO THE INSIDE.

RECESSED DOUBLED TOP PLATE

....

studs. If lintels meet the top plate, nail through the plate into the lintel. Also fasten the jacks to the king studs by driving the nails at an angle. This is a stronger nailing technique and eliminates any pro-truding nails. A pneumatic framing nailer makes the work easier (**Figure 40–16**). Fasten each stud, corner post, and partition intersection in the proper posi-tion by driving two 3½-inch (89 mm) nails through

FIGURE 40–16 Pneumatic nailers are industry standard tools for framing.

the plates into the ends of each member. Nail the doubled top plate to the other top plate. Leave notches where the intersecting walls and partitions are located. Some builders toenail the studs to the bottom plate later, as the wall is erected. Toenails are typically 2½-inch (63 mm) nails.

Nail the doubled top plate to the top plate. Recess or extend the doubled top plate to make a lap joint at the corners and intersections (**Figure 40–17**). Be sure to nail the doubled top plate into the top plate so that nails are located above the studs. This will ensure that any holes drilled for wiring or plumbing into the top plates will not hit a nail. Where partitions intersect the exterior wall, a space is left for the top plate of the partition to lap the plate of the exterior wall. Lapping the plates of the interior partitions with those of the exterior walls ties them together. This results in a more rigid structure.

Bracing Walls

There are several methods of creating a strong wall section that will withstand **racking** loads. Plywood and oriented strand board (OSB) are the most popular, but let-in and cut-in bracing are also sometimes used. In any case, wall bracing may be applied before the wall is erected. Let-in and cut-in bracing must be installed with tight-fitting seams for maximum strength. For this reason, it is much easier to do when the wall is lying down.

Before installing permanent wall bracing, the wall section should be squared while it is lying on the deck. To do this, align the bottom plate to the previously snapped chalk line on the subfloor where the inside edge of the wall plate will rest. Adjust the ends of the bottom plate into their proper positions lengthwise (**Figure 40–18**). Toenail the sole plate to the subfloor with 2½-inch (63 mm) nails spaced about every 6 to 8 feet (1.8 to 2.4 m) along what will be the top side of the bottom plate when the wall is in its final position.

Measure both the diagonals from corner to corner. If they are equal, the section is square. Toenail the top plate to the subfloor from the top side using one or two nails. These are used simply to keep the wall square and will be removed before standing the wall. The section is now ready for sheathing or bracing. Sheathing is applied in a manner similar to that used for a subfloor, or parallel to the studs.

If let-in bracing is used, place the brace in position on top of the studs and plates at about a 45-degree angle. Make sure top and bottom plates are covered by the brace. Mark along the side of the brace at each stud and plate. Remove the brace.

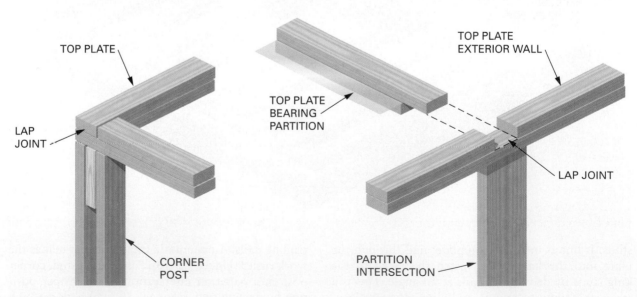

FIGURE 40–17 Doubled top plates are lapped at all intersections of wall sections.

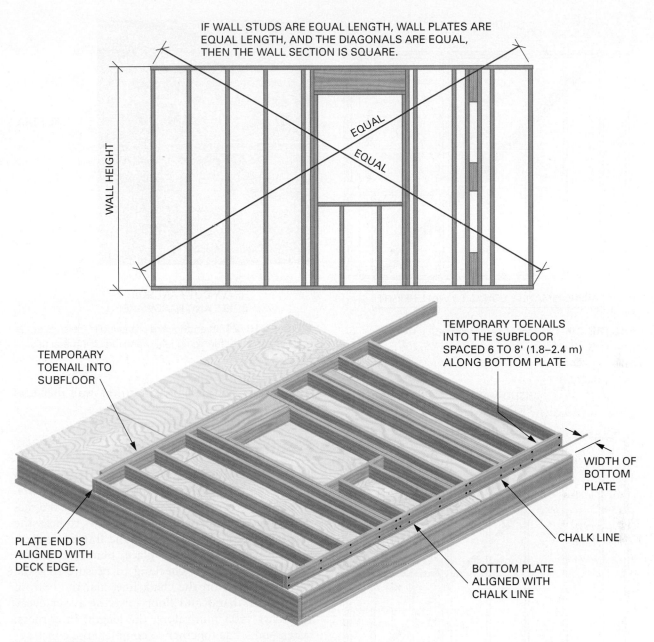

IF WALL STUDS ARE EQUAL LENGTH, WALL PLATES ARE EQUAL LENGTH, AND THE DIAGONALS ARE EQUAL, THEN THE WALL SECTION IS SQUARE.

WALL HEIGHT

EQUAL

EQUAL

TEMPORARY TOENAILS INTO THE SUBFLOOR SPACED 6 TO 8' (1.8–2.4 m) ALONG BOTTOM PLATE

TEMPORARY TOENAIL INTO SUBFLOOR

WIDTH OF BOTTOM PLATE

CHALK LINE

PLATE END IS ALIGNED WITH DECK EDGE.

BOTTOM PLATE ALIGNED WITH CHALK LINE

FIGURE 40–18 Squaring a wall section before erecting the frame.

Using a portable electric circular saw with the blade set for the depth of the notch, make multiple saw cuts between the layout lines. With the claw of a straight-claw hammer, knock out and trim the remaining waste from the notch. Fasten the brace in the notches using two 2½-inch (63 mm) common nails in each framing member (**Figure 40–19**).

Cut-in bracing has two blocks nailed to the plates, called *kickers*, at each end of the brace. These serve to transfer the racking load to the plates. First, snap a line at about a 45-degree angle on top of the stud edges. Use a speed square to determine the angle to cut the pieces. Fasten them into place with 3½-inch (89 mm) nails, two at each end. The kicker

nails should be angled so that they do not protrude (**Figure 40–20**).

In regions where seismic activity is severe, such as Vancouver, bracing takes on a new meaning. Earthquakes occur with such severity that engineers must design the buildings to protect the occupants of the house long enough for them to escape the shaking building. **Shear walls** are built into the building (**Figure 40–21**). These are framed of 2×6s (38×140s) anchored to the slab with OSB heavily nailed to them. Metal strapping is nailed to the building, anchoring the building to the shear wall. The large wood/metal anchors are bolted to long $^{15}/_{16}$-inch (25 mm) diameter bolts with $^{5}/_{8}$-inch (16 mm)

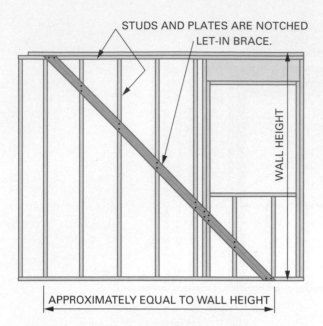

STUDS AND PLATES ARE NOTCHED
LET-IN BRACE.

WALL HEIGHT

APPROXIMATELY EQUAL TO WALL HEIGHT

FIGURE 40–19 A let-in corner brace.

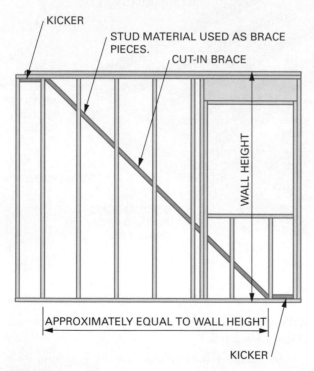

KICKER

STUD MATERIAL USED AS BRACE
PIECES.

CUT-IN BRACE

WALL HEIGHT

APPROXIMATELY EQUAL TO WALL HEIGHT

KICKER

FIGURE 40–20 A cut-in brace.

anchor bolts spaced closely together. The shear wall bottom plates are sometimes 3 inches (75 mm) thick (**Figure 40–22**).

To provide shear resistance in walls that have large openings, such as a garage door, metal frames are used. They are bolted to large 1-inch (25 mm) diameter bolts anchored 3 feet (914 mm) into concrete. These frames are made with 1/8-inch-thick (3 mm)

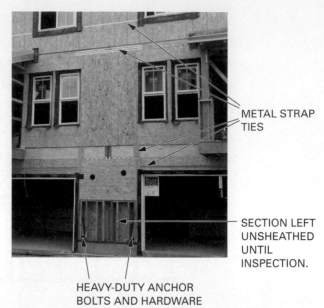

METAL STRAP
TIES

SECTION LEFT
UNSHEATHED
UNTIL
INSPECTION.

HEAVY-DUTY ANCHOR
BOLTS AND HARDWARE

FIGURE 40–21 In earthquake-prone areas, seismic shear walls are constructed using heavy anchors and metal ties.

steel that is bent and folded into the wall thickness (**Figure 40–23**).

Erecting Wall Sections

To erect the wall, remove the toenails from the top plate while leaving the toenails in the bottom plate. The bottom toenails will remain until after the section is erected; they will serve as hinges to keep the bottom plate in position while the frame is raised. Lift the wall section into place, plumb, and temporarily brace. After checking to be sure that the bottom plate is on the chalk line, nail the bottom plate to the band and floor joists or about every 16 inches (405 mm) along the length. In corners, fasten end studs together to complete the construction of the corner post. A completed corner post provides surfaces to fasten both exterior and interior wall finish.

Brace each section temporarily as erected. Fasten one end of a brace to a 2 × 4 (38 × 89 mm) block that has been nailed to the subfloor and the other end to the side of a stud (**Figure 40–24**).

Bracing Walls Temporarily

If the walls have not been previously braced while being framed on the subfloor, all corners must be plumbed and temporarily braced. Install braces for both sides of each corner. Fasten the top end of the brace to the top plate near the corner post on the inside of the wall.

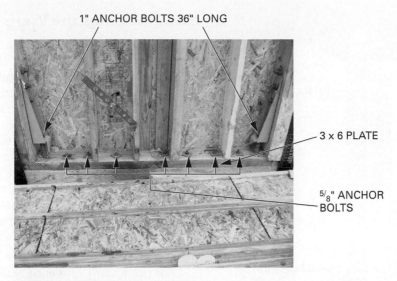

FIGURE 40-22 The wall frame can be anchored to the foundation for seismic resistance.

FIGURE 40-23 Shear strength for walls with large openings can be supplied by metal panels bolted to the foundation.

FIGURE 40-24 Temporary braces hold the frame erect during construction.

Temporary braces are fastened to the inside of the wall. They can remain in position until the exterior wall sheathing is applied. Sometimes they remain until it is absolutely necessary to remove them for the application of the interior wall finish. Care must be taken not to let the ends of the braces extend beyond the corner post or top plate. This might interfere with the application of wall sheathing or subsequent ceiling or roof framing.

Methods of Plumbing. Plumb the corner post and fasten the bottom end of the brace. For accurate plumbing of the corner posts, use a 6-foot (1.8 m) level with accessory aluminum blocks attached to each end. The blocks keep the level from resting against the entire surface of the corner post. This

prevents any bow or irregularity in the surface from affecting the accurate reading of the level.

An alternative method is to use the carpenter's hand level in combination with a long straightedge. On the ends of the straightedge, two blocks of equal thickness are fastened to keep the edge of the straightedge away from the surface of the corner post (**Figure 40-25**). A plumb bob, transit-level, or laser level may also be used to plumb corner posts.

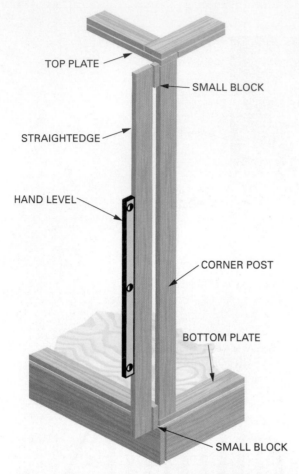

TOP PLATE

SMALL BLOCK

STRAIGHTEDGE

HAND LEVEL

CORNER POST

BOTTOM PLATE

SMALL BLOCK

FIGURE 40–25 Plumbing should be done from the plates and requires a special level or a straightedge with blocks.

Straightening the Walls

After the corner posts have been plumbed and the top plates doubled, the tops of the walls must be straightened and braced. This can be done with a line and gauge blocks (see Figure 31–21, p. 327). Nail 2 × 4s (38 × 89s) at about a 45-degree angle. This will require a length of at least 12 feet (3.6 m) for an 8-foot (2.4 m) wall. Nail the brace into each plate and twice more into the studs at midspan. Another method is to straighten the brace by eye. After a little practice, eyeing for straightness is fast and surprisingly accurate over distances of less than 40 feet (12 m). As one person sights the top plate for straightness by getting the eye as close as possible to the plate, another person nails the brace.

For particularly stubborn wall sections that are difficult to move into plumb, a *spring brace* can be used. Variations of this brace can be used to move the wall top inward or outward.

To create an outward thrust, set a 12- to 16-foot- (4 to 5 m) long 2 × 4 (38 × 89 mm) or 2 × 6 (38 × 140 mm) against the top plate over a stud (**Procedure 40–C**). The width of the board should be facing up. Nail a block to the subfloor behind the brace. Push down on the brace to make it arch. This will push the wall out. If more outward push is needed, nail the top of the brace to the stud while it is arched. Then lift the middle of the brace to straighten it. This will create a tremendous outward force. Care should be taken not to break anything.

To bring the wall inward, nail a 16-foot (5 m) 1 × 6 (19 × 140 mm) to the top plate and the subfloor. These nails should be set firmly in a floor joist. Lift the brace to an arch. Block the arch with a short piece when the wall is plumb. Care should be taken to make sure the brace does not break.

Wall Sheathing

Wall sheathing covers the exterior walls. It may consist of boards, rated panels, fibreboard, gypsum board, or rigid foam board.

Boards. Before plywood was invented, boards were used predominantly but are seldom used today. Many buildings in existence today have board sheathing. The boards were applied diagonally or horizontally. The diagonally sheathed walls require no other bracing. They made the frame stiffer and stronger than boards applied horizontally. If used today, nail with two 2½-inch (63 mm) common nails at each stud for 6- and 8-inch (140 and 184 mm) boards, or three nails at each stud for 10- and 12-inch (235 and 286 mm) boards. End joints must fall over the centre of studs. They must

StepbyStep Procedures

Procedure 40–C Plumbing Stubborn Wall Sections with Spring Braces

OUTWARD THRUST SPRING BRACE

STEP 1 SET A 2 X 4 (38 X 89 mm), 12 OR 16' (4 OR 5 m) LONG AGAINST WALL WITH TOP AGAINST UPPER TOP PLATE.

STEP 2 NAIL BLOCK TO THE SUBFLOOR BEHIND THE BRACE.

STEP 3 PUSH BRACE DOWN AT MIDSPAN SO THAT TOP END SLIDES DOWN AND DIGS INTO STUD.

STEP 4 IF MORE WALL MOVE-MENT IS NEEDED, NAIL TOP WHILE BRACE IS BENT, THEN PICK UP ON BRACE TO STRAIGHTEN IT.

NOTE: DO NOT EXCEED MATERIAL STRENGTH WHERE SOMETHING MIGHT BREAK, CAUSING PERSONAL INJURY.

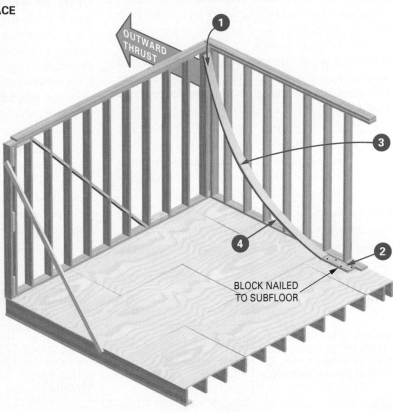

BLOCK NAILED TO SUBFLOOR

INWARD THRUST SPRING BRACE

STEP 1 NAIL THE UPPER END OF A 1 X 6 (19 X 140), 16-FOOT (5 m) BRACE TO THE UNDERSIDE OF THE TOP PLATE.

STEP 2 NAIL LOWER END OF BRACE TO FLOOR JOIST WITH SEVERAL 3$\frac{1}{2}$ " (89 mm) NAILS. CAUTION: THIS END MUST BE SECURELY FASTENED SO THAT IT WILL NOT COME LOOSE WHEN UNDER STRESS.

STEP 3 USE A SHORT 2 X 4 (38 X 89 mm) AS A POST, ARC THE BRACE AT MIDSPAN. SECURE THE POST WITH A NAIL.

NOTE: DO NOT EXCEED MATERIAL STRENGTH WHERE SOMETHING MIGHT BREAK, CAUSING PERSONAL INJURY.

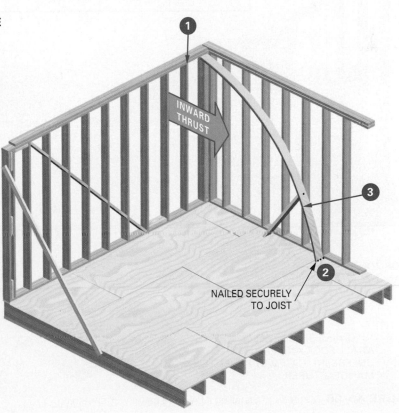

NAILED SECURELY TO JOIST

be staggered so that no two successive end joints fall on the same stud.

Rated Panels. APA-rated wall sheathing panels are used most often today. They may be applied horizontally or vertically. There is no need for corner braces. A minimum ³⁄₈-inch (9.5 mm) thickness is recommended when the sheathing is to be covered by exterior siding. Greater thicknesses are recommended when the sheathing also acts as the exterior finish siding.

Use 2-inch (50 mm) nails spaced 6 inches (152 mm) apart on the edges and 12 inches (305 mm) apart on intermediate studs for panels ½ inch (12 mm) thick or less. Use 2½-inch (63 mm) nails for thicker sheathing panels (**Figure 40–26**).

Rated sheathing panels are sometimes used in combination with rigid foam or gypsum board. When panels are applied vertically on both sides of the corner, no other corner bracing may be necessary.

Other Sheathing Panels. *Fibreboard* sheathing panels are available as wall sheathing (**Figure 40–27**). They are about ⅛-inch-thick (3 mm) panels with widths and lengths similar to those of plywood. They are made of specially treated, long wood fibres from recycled products. These fibres are pressed into plies that are pressure laminated with a water-resistant adhesive.

They come in three grades of strength. Green panels are for nonstructural applications, red is for permanent wall bracing in the corners, and blue is used where stronger bracing is required. Nails must be 1¼-inch (32 mm) roofing nails or 1-inch (25 mm) crown staples. Nail spacing for the green non-structural panels is the same as for APA-rated

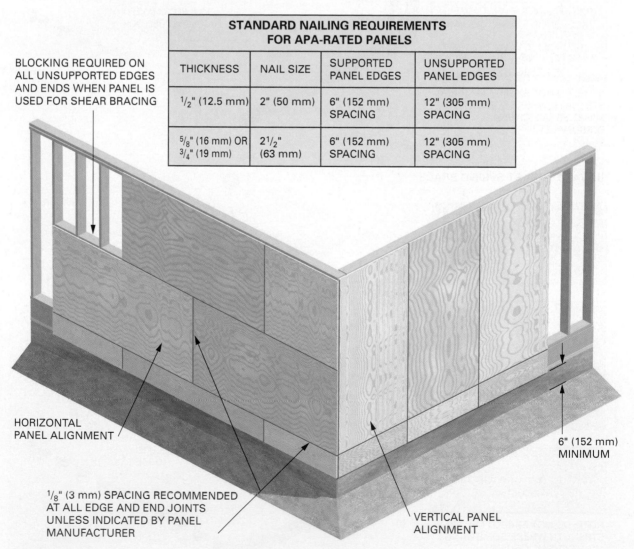

STANDARD NAILING REQUIREMENTS FOR APA-RATED PANELS			
THICKNESS	NAIL SIZE	SUPPORTED PANEL EDGES	UNSUPPORTED PANEL EDGES
¹⁄₂" (12.5 mm)	2" (50 mm)	6" (152 mm) SPACING	12" (305 mm) SPACING
⁵⁄₈" (16 mm) OR ³⁄₄" (19 mm)	2¹⁄₂" (63 mm)	6" (152 mm) SPACING	12" (305 mm) SPACING

BLOCKING REQUIRED ON ALL UNSUPPORTED EDGES AND ENDS WHEN PANEL IS USED FOR SHEAR BRACING

HORIZONTAL PANEL ALIGNMENT

¹⁄₈" (3 mm) SPACING RECOMMENDED AT ALL EDGE AND END JOINTS UNLESS INDICATED BY PANEL MANUFACTURER

VERTICAL PANEL ALIGNMENT

6" (152 mm) MINIMUM

FIGURE 40–26 Methods of installing APA-rated panel wall sheathing.

FIGURE 40–27 Thin sheets of ⅛-inch (3 mm) fibreboard may be used as wall sheathing.

FIGURE 40–28 Rigid foam insulation may be used for exterior wall sheathing.

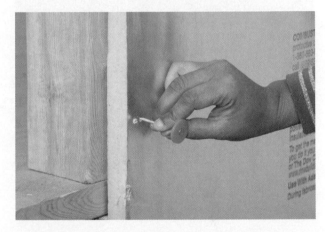

FIGURE 40–29 Rigid foam requires special fasteners with large plastic-capped heads.

panels, 6 inches (152 mm) on edges and 12 inches (305 mm) in the field. Structural panels are nailed 3 inches (75 mm) on edges and 6 inches (152 mm) in the field.

It is recommended that panels be nailed along one stud at a time. This serves to help keep ripples out of the sheathing. Nailing begins at one corner, nailing the first stud. Then nail along the plates to the next stud, which is in turn nailed off. Nailing four corners first should be avoided.

Gypsum sheathing consists of a treated gypsum filler between sheets of water-resistant paper. Usually ½ inch (12 mm) thick, 2 feet (610 mm) wide, and 8 feet (2.4 m) long, the sheets have matched edges to provide a tighter wall. Because of the soft material, galvanized wall board nails must be used to fasten gypsum sheathing.

Space the nails about 4 inches (100 mm) around the edges and 8 inches (203 mm) in the centre. Gypsum board sheathing is used when a more fire-resistant sheathing is required.

Rigid foam sheathing is used when greater wall insulation is desired (**Figure 40–28**). Rated panels are used in the corners to give the building adequate stiffness. It may be applied in thicknesses of ½ or 1 inch (12 or 25 mm). It may be foil-faced to increase the thermal barrier. One disadvantage of using these panels as sheathing is that they cannot be used as a nail base for siding. Any siding material applied must be fastened to the studs.

Application of Sheathing Panels. Sheathing panels are installed in a manner similar to that used for installing subfloors. If needed, snap lines to keep the plates and the edges of panels aligned. Panels should be installed with as few seams as possible. Seams should be gapped ⅛ inch (3 mm) on the edges of the panels and 1/16 inch (1.5 mm) on the ends,

or according to the manufacturer's specifications (**Figure 40–29**).

STRUCTURAL INSULATED PANELS

Another method of constructing a wall or a roof section is to use **structural insulated panels (SIPs)**. They consist of two outer skin panels and an inner core of insulating foam (**Figure 40–30**). This sandwich of materials forms a rigid unit. Most panels use **oriented strand board (OSB)**, for their facings. OSB makes panels in sheets up to 12 × 36 feet (3.7 × 11.0 m). The cores of SIPs are mostly made from expanded polystyrene (EPS), but sometimes extruded polystyrene (XPS) or urethane foam is used.

The core and the two skins are not structurally strong by themselves. But when adhered together under pressure, a stiff and strong structural unit is created. The result is that no frame is required. Panel

FIGURE 40–30 SIPs are made from thin panel skins and a foam core. *(Courtesy of the Structural Insulated Panel Association)*

FIGURE 40–31 SIPs are assembled without traditional wood framework. *(Courtesy of the Structural Insulated Panel Association)*

FIGURE 40–32 SIPs are generally assembled using heavy equipment. *(Courtesy of the Structural Insulated Panel Association)*

FIGURE 40–33 SIPs are sealed at each seam. *(Courtesy of the Structural Insulated Panel Association)*

manufacturers supply splines, adhesives, and fasteners to erect their systems (**Figure 40–31**).

SIPs are produced in thicknesses from 4½ to 12¼ inches (108 to 312 mm), with R-values of 16 to 53 (RSI 3.42 to 9.33). The panel sizes range from 4 × 8 feet (1.2 × 2.4 m) to 9 × 24 feet (2.7 × 7.3 m).

Blank panels are available, but SIPs are typically fabricated at the manufacturing plant with all door and window openings cut out. They are delivered to the site ready to install, with a shop drawing of how to assemble the panels by numbers stamped on the panels. This means faster assembly than wood-framed walls and roofs. SIPs have electrical chases precut in the core to accept electrical wiring. Large panels will require a crane, boom truck, or forklift to set them in place (**Figure 40–32**).

The fire-rated performance of SIPs is similar to other wood-frame structures. Residential structures will typically be covered with ½-inch (12 mm) drywall to provide a half-hour fire rating. Light commercial and multi-family units will require extra layers or fire-code drywall to meet the building code standard.

Building with SIPs offers several advantages and benefits over stick framing. It is easy to achieve an airtight seal with SIPs (**Figure 40–33**), which decreases the energy consumption of the building. Overall, SIPs use less wood than standard framing. The OSB panels are made from small, fast-growing trees, and so they use the natural resource of wood more efficiently than sawn lumber. They also reduce the amount of waste produced during construction.

ESTIMATING MATERIALS FOR EXTERIOR WALLS

To estimate the amount of material needed for exterior walls, first determine the total linear feet of exterior wall. Then figure one stud for every linear foot of wall, if spaced 16 inches (406 mm) on centre. This allows for the extras needed for corner posts, partition intersections, trimmers, door jacks, and blocking openings.

For plates, multiply the total linear feet of wall by three (one sole plate and two top plates). Add 5 percent for waste in cutting.

For lintels and rough sills, calculate the size for each opening. Add together the material needed for different sizes.

For wall sheathing, first find the total area to be covered. To find the total number of square feet of wall area, multiply the total linear feet of wall by the wall height. Deduct the area of any large openings. Disregard small openings.

To find the number of sheathing panels, divide the total wall area to be covered by the number of square feet in each sheet. For instance, if a panel measures 4 by 8 feet (1.2 by 2.4 m), it contains 32 feet^2 (3 m^2). Divide the wall area to be covered by 32 to find the number of panels required. Another way to determine the number of plywood wall panels is to divide the perimeter of the building (in feet) by 4, because the panels are each 4 feet (1.2 m) wide × 8 feet (2.4 m) tall.

In metric units, count one stud for every 300 mm of total wall length if the stud layout is 400 mm OC. Wall plates will be three times the exterior wall perimeter, and the number of sheathing panels will be the perimeter (in metres) divided by 1.2, since the panels are each 1.2 × 2.4 m.

DECONSTRUCT THIS

Carefully study **Figure 40–34** and think about what is wrong and/or what is right. Consider all possibilities. What construction practice or method is different in your area of the country?

FIGURE 40–34 This photo shows a wall being framed on a deck.

KEY TERMS

corner posts	lintels	ribbon	studs
cripple studs	oriented strand board (OSB)	shear walls	threshold
gypsum board		sole plate	top plate
jack studs	partitions	stop	trimmers
jambs	racking	storey pole	wall sheathing
jig	rough opening	structural insulated panels (SIPs)	
kicker	rough sills		

REVIEW QUESTIONS

Select the most appropriate answer.

1. What are the top and bottom horizontal members of a wall frame called?

   a. lintels
   b. plates
   c. trimmers
   d. sills

2. What is the horizontal wall member supporting the load over an opening called?

   a. lintel
   b. rough sill
   c. plate
   d. truss

3. What are shortened studs above and below an opening called?

   a. shorts
   b. lames
   c. cripples
   d. stubs

4. What should be installed with diagonal cut-in bracing?

   a. kickers
   b. backing
   c. blocking
   d. 1 × 4s (19 × 89s)

5. A door jamb is 19 mm thick. Allowing for 12 mm on each side for shimming the frame, what is the rough opening width for a door that is 810 mm wide?

   a. 810 mm
   b. 848 mm
   c. 867 mm
   d. 872 mm

6. A doorjamb is ¾" thick. Allowing ½" on each side for shimming the frame, what is the rough opening width for a door that is 2'-8" wide?

   a. 2'-9½"
   b. 2'-10½"
   c. 2'-11½"
   d. 3'-½"

7. Which measurement is typically shown by a storey pole?

   a. the length of lintels
   b. the length of rough sills
   c. the length of jack studs
   d. the width of the rough opening

8. When laying out plates for walls and partitions, where do measurements for centrelines of openings start from?

   a. end of the plate
   b. outside edge of the abutting wall
   c. building line
   d. nearest intersecting wall

9. When laying out plates for any OC wall or partition stud, where does the measurement begin from?

   a. end of the plate
   b. abutting wall or partition
   c. opening centrelines
   d. depends on the type of wall or partition

10. What is the rough opening height for an interior door that is 2032 mm wide if the shim space is 12 mm, the jamb thickness is 19 mm, and the clearance between the bottom of the door and the floor is 12 mm?

    a. 2032 mm
    b. 2063 mm
    c. 2075 mm
    d. 2094 mm

11. What should be used for a corner stud that allows ample room for insulation in the corner?

    a. three small blocks
    b. a stud that is rotated from the others in the wall
    c. three full studs nailed as a post
    d. a vertical strip of ½" plywood fastened to the side of the end stud of the non-load-bearing wall

12. How are exterior walls usually aligned before ceiling joists are installed?

    a. using only a carpenter's level
    b. using a line stretched between two blocks and testing with a gauge block
    c. using a plumb bob dropped to the bottom plate at intervals along the wall
    d. sighting along the length of the wall using a builder's level

13. What makes partition walls "bearing" partitions?

    a. They have a single top plate.
    b. They carry no load.
    c. They are constructed like bearing walls.
    d. They are erected after the roof sheathing is installed.

14. From where does the on-centre layout of studs for a bearing partition start?

    a. the outside of the exterior wall
    b. the inside of the exterior wall
    c. the end of the bottom plate
    d. the end of the top plate

15. What are spring braces?

    a. temporary wall braces
    b. permanent wall braces
    c. temporary ceiling braces
    d. permanent ceiling braces

16. What is the rough opening height of a door opening for a 6'-8" door if the finish floor is ¾" thick, ½" clearance is allowed between the door and the finish floor, and the jam thickness is ¾"?

    a. 6'-9"
    b. 6'-9½"
    c. 6'-10"
    d. 6'-10½"

17. What type of plywood is typically used for wall sheathing?

    a. CDX
    b. AC
    c. BC
    d. hardwood

18. What type of permanent wall bracing is used most often in construction today?

    a. APA-rated sheathing
    b. cut-in
    c. let-in
    d. metal T bracing

19. Approximately how many 16-inch OC exterior wall studs are needed for a rectangular house that measures 28 × 48 × 8 feet high?

    a. 76
    b. 152
    c. 1344
    d. 10 752

20. Approximately how many pieces of wall sheathing (1200 mm × 2400 mm) are required to sheath a building that is 36 m × 44 m × 2.4 m? Do not make allowances for openings or gable ends.

    a. 134
    b. 140
    c. 1320
    d. 1386

WHAT'S WRONG WITH THIS PICTURE?

Carefully study **Figure 40–35** and think about what is wrong. Consider all possibilities.

FIGURE 40–35 The joint in the first top plate falls beside the stud. This is poor building practice and should be avoided.

FIGURE 40–36 The joint in the first top plate falls correctly over the centre of the stud and the second top plate spans the splice. The roof truss sits directly over the stud.

Unit 14

Interior Rough Work

Chapter 41   Interior Partitions and Ceiling Joists
Chapter 42   Backing and Blocking

Chapter 43   Steel Framing

Interior rough work is constructed on the inside of a structure and later covered by some type of finish work. The interior rough work described in this unit includes the installation of *partitions*, *ceiling joists*, *furring strips*, and *backing* and *blocking*. The term *rough work* does not imply that the work is crude. It is a kind of work that will eventually be covered by other material. Careful construction of the rough frame makes application of the finish work easier and less complicated.

OBJECTIVES

After completing this unit, the student should be able to:

- assemble, erect, brace, and straighten bearing partitions.
- determine and make rough openings for doors.
- lay out, cut, and install ceiling joists.
- lay out and erect non-bearing partitions and install backing in walls for fixtures.
- describe various components of light-gauge steel framing.
- lay out and frame light-gauge steel interior partitions.

 SAFETY REMINDER

As tools and materials are gathered for interior wall framing, jobsite organization becomes very important for safety and efficiency. Keep waste material outside the work area in organized piles for removal.

Chapter 41

Interior Partitions and Ceiling Joists

Partitions and *ceiling joists* constitute some of the interior framing. Ceiling joists tie the exterior side walls together and support the ceiling finish.

Partitions supporting a load are called *load-bearing partitions*. Partitions that merely divide the area into rooms are called *non-load-bearing partitions* (**Figure 41–1**). See sections 5 through 7 of the National Building Code of Canada (NBCC) for design and areas of rooms in indoor spaces and requirements for ceiling heights, windows, glass, etc.

PARTITIONS

Load-bearing partitions (LBPs) support the inner ends of ceiling or floor joists. They are placed directly over the girder or the bearing partition in the lower level. If several bearing partitions are used on the same floor, supported girders or walls are placed directly under each or within $23^5/8$ inches (600 mm) of a supported floor or within $35\frac{1}{2}$ inches (900 mm) of a supported ceiling [NBCC 9.23.9.8 (4) and (5)].

Non-load-bearing partitions (NLBPs) are built to divide the space into rooms of varying size. They carry no structural load from the rest of the building, only the load of the partition material itself. They may be placed anywhere on the subfloor where joist reinforcement has been provided. They often run parallel to the joists and are nailed to blocking between joists (minimum 2×4 (38×89) at 4 feet (1.2 m) on centre [NBCC 9.23.9.8(2)]) (**Figure 41–2**).

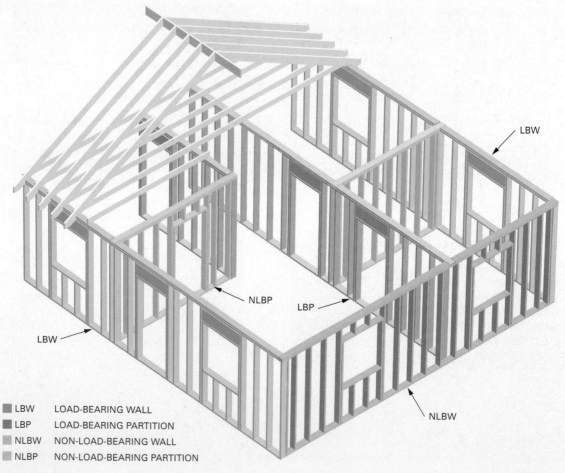

▪ LBW	LOAD-BEARING WALL
▪ LBP	LOAD-BEARING PARTITION
▪ NLBW	NON-LOAD-BEARING WALL
▪ NLBP	NON-LOAD-BEARING PARTITION

FIGURE 41–1 Walls are exterior and partitions are interior. Either may be load-bearing or non-load-bearing.

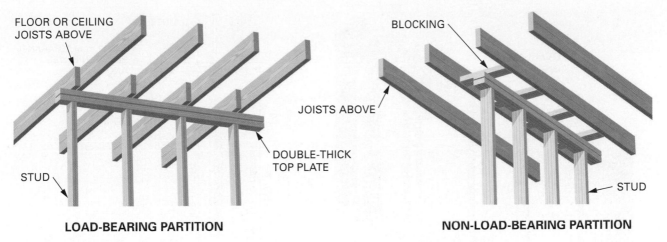

LOAD-BEARING PARTITION

FLOOR OR CEILING JOISTS ABOVE

STUD

DOUBLE-THICK TOP PLATE

NON-LOAD-BEARING PARTITION

BLOCKING

JOISTS ABOVE

STUD

FIGURE 41–2 Non-load-bearing partitions merely divide an area into rooms. Load-bearing partitions support the weight of the floor or ceiling above.

Partitions are erected in a manner similar to that used for exterior walls. A double top plate is used to tie the wall and partition intersections together. If *roof trusses* are used, partitions have doubled top plates that are thinner than the other walls (**Figure 41–3**). This is done so that the trusses touch only the bearing walls where the roof load is transferred to the foundation. All partitions are non-load bearing in a home with standard roof trusses. More on roof trusses are discussed in Chapter 45.

In all cases, when partitions meet an exterior wall, there needs to be a 12-inch (305 mm) strip of 6-mil (0.15 mm) polyethylene placed between the partition and the wall. When the partitions are just below the insulated ceiling, the same strip of "poly" is placed between the two top plates. This is done to provide continuity of the air-vapour barrier.

Lintels on LBPs, as in walls, must be strong enough to support the intended load. Lintels on NLBPs may be constructed with 2 × 4s (38 × 89s) on the flat with cripple studs above (**Figure 41–4**).

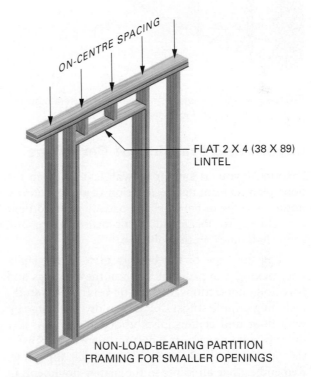

ON-CENTRE SPACING

FLAT 2 X 4 (38 X 89) LINTEL

NON-LOAD-BEARING PARTITION FRAMING FOR SMALLER OPENINGS

FIGURE 41–4 A method for framing a non-load-bearing lintel.

This saves material, because the only load on the partition is the wall finish material.

Bathroom and kitchen walls sometimes must be made thicker to accommodate plumbing. Sometimes 2 × 6 (38 × 140 mm) plates and studs are used, or a double 2 × 4 (38 × 89 mm) partition is erected. Still another wall variation is to use 2 × 6 (38 × 140 mm) or 2 × 8 (38 × 184 mm) plates with alternated 2 × 4 (38 × 89 mm) studs (**Figure 41–5**). This allows fibreglass insulation to be woven between the studs for increased soundproofing. Also, if the wall thickness needs to be increased only slightly, furring strips or resilient channel may be added to the edges of the studs and plates.

FIGURE 41–3 The double top plates of non-load-bearing walls and partitions may be thinner than those of load-bearing walls and partitions.

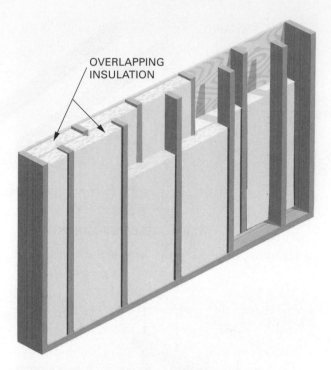

FIGURE 41–5 Sometimes a wide plate is used to build stagger-stud walls for increased soundproofing.

Layout and Framing of Partitions

Partition layout is similar to wall layout. From the floor plan, determine the location of the partitions, noting that the dimensions are usually given to their centrelines. To locate all partitions, measure and snap chalk lines on the subfloor.

Lay the edge of the bottom plate to the chalk line, tacking it in position. Lay out the openings and partition intersections, then all on-centre (OC) studs, including cripple studs. LBP OC studs should line up with floor and ceiling joists above and below. Lay the top plate next to the bottom plate and transfer the layout of the bottom plate to the top plate. Remember that all joints in the plates should end in the centre of a stud.

Rough Opening Door Sizes. The rough opening sizes are found in the same way they are found for exterior walls (**Figure 41–6**). The width for a door opening is found by adding to the door width, twice the thickness of the jamb stock, and twice the ½-inch (12 mm) shim space (one for each side).

The rough opening height is found by adding five measurements from the door cross-section. They include the following from the subfloor up:

- The thickness of the finish floor
- The ½-inch (12 mm) clearance between the finish floor and the bottom of the door
- The height of the door

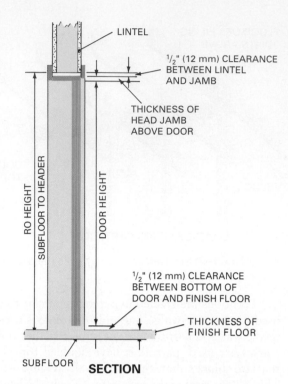

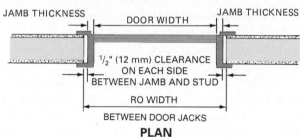

FIGURE 41–6 Figuring the rough opening size for an interior door.

- The thickness of the head jamb
- The ½-inch (12 mm) shim space between the head jamb and the rough lintel

Framing Partitions. For ease of erecting, construct the longer partitions first. Then construct shorter cross-partitions, such as for closets, later. There is no hard and fast rule for constructing partitions; experience will allow the best process to emerge.

Pull the tack nails and separate the top and bottom plates. Place all full-length studs, corner posts, and partition intersections on the floor. Make sure their crowned edges run in the same direction. Nail the framing members around an opening first, and then the OC studs, to allow fast and easy assembly. Raise the section into position, locking the lapping top plates into position. Plumb and brace the wall section and nail the bottom plate to the floor about 16 inches (406 mm) apart or into floor joists where possible.

In some regions of Canada, the partition bottom plates, with the door plates removed, are nailed securely in place and the top plates are only end-nailed to the studs. The second top plate is secured to the top plate with nails located over the studs so that the mechanical subtrades need not worry about hitting nails with drill bits. The frame is then raised with the studs dangling. It is secured at the end, and the studs are toenailed to the bottom plate with four 2¼-inch (57 mm) nails. The end stud that butts against another wall can be straightened by toenailing through its edge into the centre block of the partition intersection (**Figure 41–7**).

Other Openings. Besides door openings, the carpenter must frame openings for heating and air-conditioning ducts, medicine cabinets, electrical panels,

> ## ON THE JOB
>
> **Straighten the edge of end studs in interior partitions that intersect with another wall to frame straight inside corners.**

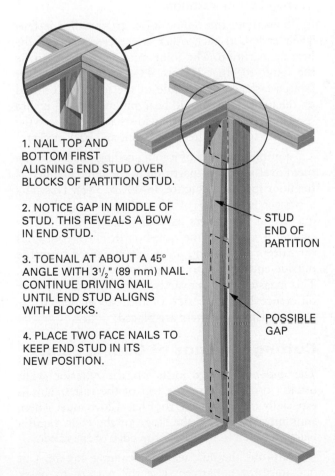

1. NAIL TOP AND BOTTOM FIRST ALIGNING END STUD OVER BLOCKS OF PARTITION STUD.

2. NOTICE GAP IN MIDDLE OF STUD. THIS REVEALS A BOW IN END STUD.

3. TOENAIL AT ABOUT A 45° ANGLE WITH 3½" (89 mm) NAIL. CONTINUE DRIVING NAIL UNTIL END STUD ALIGNS WITH BLOCKS.

4. PLACE TWO FACE NAILS TO KEEP END STUD IN ITS NEW POSITION.

STUD END OF PARTITION

POSSIBLE GAP

FIGURE 41–7 Technique for straightening a crowned stud.

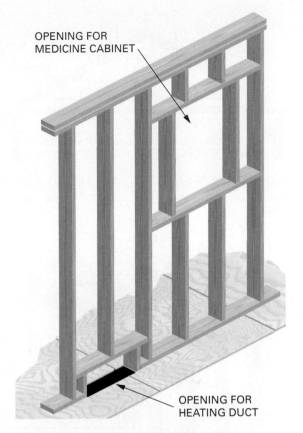

OPENING FOR MEDICINE CABINET

OPENING FOR HEATING DUCT

FIGURE 41–8 Miscellaneous openings in interior partitions are framed with non-load-bearing lintels.

and other similar items. If the items do not fit in a stud space, the stud must be cut and a lintel installed. When ducts run in a wall through the floor, the bottom plate and subfloor must be cut out (**Figure 41–8**). The reciprocating saw is a useful tool for making these cuts.

CEILING JOISTS

Ceiling joists generally run from the exterior walls to the bearing partition across the width of the building. Construction design varies according to geographic location, traditional practices, and the size and style of the building. The size of ceiling joists is based on the span, spacing, load, and the kind and grade of lumber used. Determine the size and spacing from the plans, or from the NBCC 9.23.4.2(1), Table A-3 (metric only), or from *The Span Book* (Canadian Wood Council).

Methods of Installing Joists

In a conventionally framed roof, the *rafters* and the ceiling joists form a triangle. Framing a triangle is a common method of creating a strong and rigid building. The weight of the roof and weather is transferred from the roof to the exterior walls (**Figure 41–9**). The rafters are located over the studs

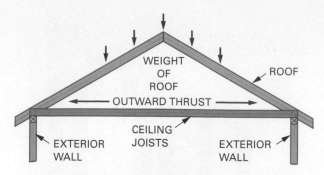

FIGURE 41–9 Ceiling joists tie the roof frame together into a triangle, which resists the outward thrust caused by the rafters.

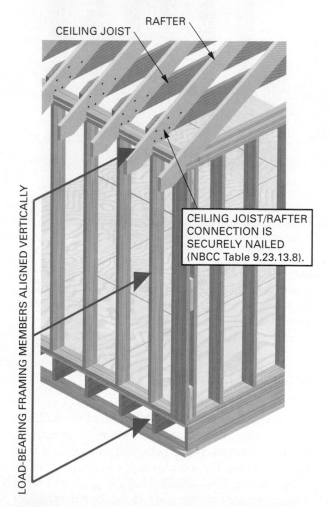

FIGURE 41–10 Ceiling joists are located so that they can be fastened to the side of rafters and over the vertically aligned structural frame.

and the ceiling joists are fastened to the sides of the rafters (**Figure 41–10**). This binds the rafters and ceiling joists together into a rigid triangle and keeps the walls from spreading outward due to the weight of the roof. The entire roof load is transferred to the foundation through vertically aligned framing members.

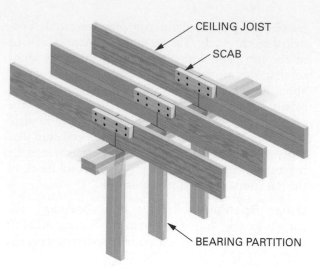

FIGURE 41–11 The joint of in-line ceiling joists must be scabbed at the bearing partition.

Ceiling joists may be made from engineered lumber and purchased in long lengths so that the rafter-ceiling triangle is easily formed. Typically, though, the ceiling joist lengths are half of the building width and therefore must be joined over a beam or bearing partition.

Sometimes the ceiling joists are installed in-line. Their ends butt each other at the centre line of the bearing partition. The joint must be *scabbed* to tie the joints together (**Figure 41–11**). **Scabs** are short boards fastened to the side of the joist and centred on the joint. They should be a minimum of 24 inches (610 mm) long. In-line ceiling joists are attached to the same side of each rafter pair (front and back rafter).

Another method of joining ceiling joists is to lap them over a bearing partition in the same manner as for floor joists (see Figure 36–11, p. 382). This puts a stagger in the line of the ceiling joist and consequently in the rafters as well (**Figure 41–12**). This stagger is visible at the ridgeboard. The layout lines for rafters and ceiling joists are measured from the outside end wall onto the top plate (**Figure 41–13**). This measurement is exactly the same, with the only difference being the side of the line on which the ceiling joists and rafters are placed.

Cutting the Ends of Ceiling Joists

The ends of ceiling joists on the exterior walls usually project above the top of the rafter. This is especially true when the roof has a low slope. These ends must be cut to the slope of the roof, slightly below (¼ inch [6 mm]) the top edge of the rafter.

Lay out the cut, using a framing square. Cut one joist for a pattern. Use the pattern to mark the rest. Make sure when laying out the joists that

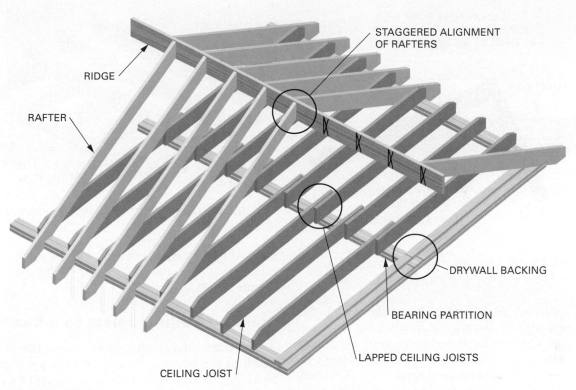

RIDGE

RAFTER

STAGGERED ALIGNMENT
OF RAFTERS

DRYWALL BACKING

BEARING PARTITION

LAPPED CEILING JOISTS

CEILING JOIST

FIGURE 41–12 When ceiling joists are lapped, it causes a stagger in the rafters, which is visible at the ridge.

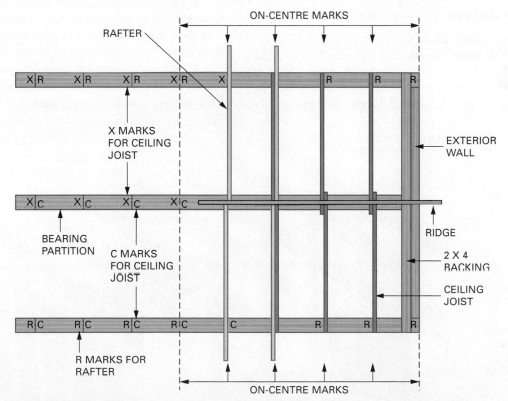

ON-CENTRE MARKS

RAFTER

X MARKS
FOR CEILING
JOIST

EXTERIOR
WALL

BEARING
PARTITION

C MARKS
FOR CEILING
JOIST

RIDGE

2 X 4
BACKING

CEILING
JOIST

R MARKS FOR
RAFTER

ON-CENTRE MARKS

NOTE: ON-CENTRE LINES FOR BOTH EXTERIOR WALLS AND BEARING PARTITIONS ARE ALL THE SAME.

FIGURE 41–13 Layout lines on all plates are the same measurements; only the positions of the rafters and ceiling joists vary.

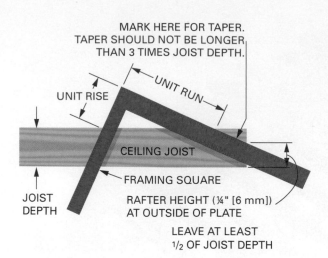

FIGURE 41–14 A framing square is used to mark the slope of a tapered cut on a ceiling joist.

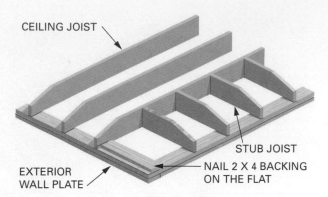

FIGURE 41–15 Stub joists are used for low-pitched hip roofs.

you sight each for a crown. Make the cut on the crowned edge so that edge is up when the joists are installed. Cut the taper on the ends of all ceiling joists before installation. Make sure the length of the taper cut does not exceed three times the depth of the member. Also make sure that the end of the joist remaining after cutting is at least half the member's width (**Figure 41–14**).

Stub Joists

Usually, ceiling joists run parallel to the end walls and in the same direction as the roof rafters. When low-pitched hip roofs are built, stub joists are used. These are joists that are shorter and run perpendicular to the normal joists. The use of stub

joists allows clearance for the bottom edge of the rafters that run at right angles to the end wall (**Figure 41–15**).

Framing Ceiling Joists to a Beam

In many cases, the bearing partition does not run the length of the building because of large room areas. Some type of beam is then needed to support the inner ends of the ceiling joists in place of the supporting wall. Similar in purpose and design to a girder, the beam may be of built-up, solid lumber or engineered lumber.

If the beam is to project below the ceiling, it is installed in the same manner as a lintel for an opening. The joists are then installed over the beam in the same manner as over the bearing partition (**Figure 41–16**).

If the ceiling is to be flush and continuous through the partition, then the ceiling joists are cut and fitted to a flush beam (**Figure 41–17**). The beam

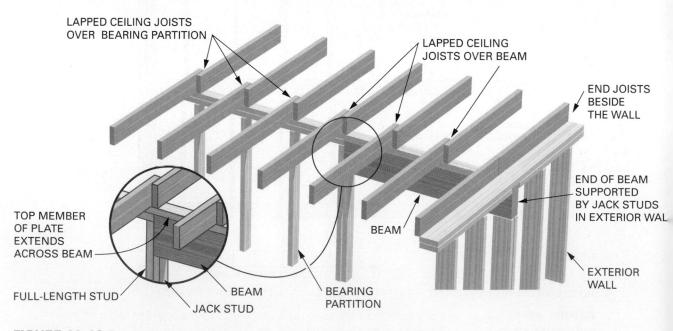

FIGURE 41–16 Support for ceiling joists may be placed as a lintel in the wall below.

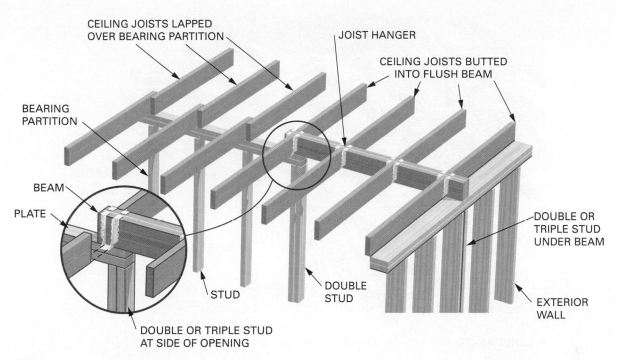

FIGURE 41–17 Support for ceiling joists may be a flush beam. This creates a flush ceiling through the partition below.

is supported by the bearing partition and exterior wall. The joists are usually set in **joist hangers**. Adhesive may be used in the joist hanger to eliminate any squeaks that might occur between the joist and ceiling joist (**Figure 41–18**).

Openings

Openings in ceiling joists may need to be made for such things as chimneys, attic access (scuttle), or disappearing stairs. Large openings are framed in the same manner as for floor joists. For small openings, there is no need to double the joists or headers (**Figure 41–19**). According to the National Building Code of Canada, the minimum finished opening for the attic access (scuttle) in a single family residential unit is

$21\frac{1}{2} \times 23\frac{1}{2}$ inches (546×597 mm), but the size that is in general use is 20×28 inches (508×711 mm).

Ribbands and Strongbacks

Ceilings are made stiffer by installing *ribbands* and *strongbacks*. Ribbands are 2×4s (38×89s) installed flat on the top of ceiling joists. They are placed at midspan to stiffen the joists as well as to keep the spacing uniform. They should be fastened with $3\frac{1}{2}$-inch (89 mm) nails and long enough to be attached to the end walls. With the addition of a 2×6 (38×140 mm) installed on edge, the ribband becomes a strongback. A strongback is used

FIGURE 41–18 Floor squeaks created by the joist hanger and joist intersection can be eliminated with caulk.

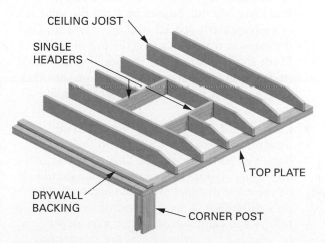

FIGURE 41–19 Joists and headers need not be doubled for small ceiling openings.

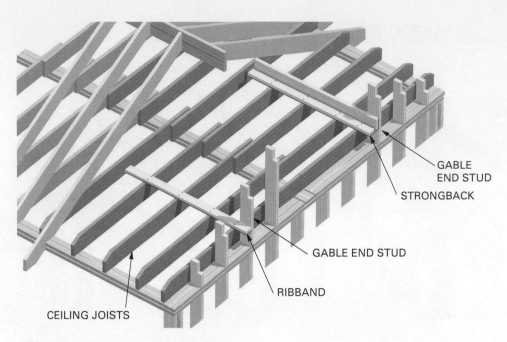

GABLE END STUD

STRONGBACK

GABLE END STUD

RIBBAND

CEILING JOISTS

FIGURE 41–20 Ribbands and strongbacks are sometimes used to stiffen and straighten ceiling joists.

when extra support and stiffness are required on the ceiling joists (**Figure 41–20**).

Layout and Spacing of Ceiling Joists

Roof rafters rest on the plate directly over the regularly spaced studs in the exterior wall. Ceiling joists are installed against the side of the rafters and fastened to them. Spacing of the ceiling joists and rafters should be the same so that they can be tied together at the plate line.

- Start the ceiling joist layout from the same corner of the building where the floor joists and wall stud were laid out. Square up from the same side of each regularly spaced stud or cripple stud in the exterior wall and across the top of the plate. Mark an *R* on the side of the line over top of the stud for the rafter and an *X* or a *C* on the other side of the line for the location of the ceiling joists.

- Layout lines on the bearing partition and on the opposite exterior wall are on the same layout as on the first wall. This is similar to floor joist layout. These layout lines should all be the same distance from the end wall. The only difference is the location of the marks for rafters and ceiling joists. They vary depending on whether the ceiling joists are continuous

or lapped at the load-bearing partition (**Figure 41–21**).

- Continuous and butted joists are placed on the same side of the layout line. The layout marks for both exterior walls are exactly the same. The load-bearing partition has only the joist layout line. This places the rafters on the same side of the joist.

- Lapped joist layout marks are reversed, similar to floor joists. Because the joists lap at the load-bearing partition, the layout on the opposite exterior wall is reversed from the first wall. Place on the opposite exterior wall an *R* and a *C* or an *X* on either side of the line but opposite from the first exterior wall. This allows for the lapped joist and creates staggered rafters. The layout for the bearing partition shows where front and back joists lap.

Installing Ceiling Joists

The ceiling joists on each end of the building are placed to allow for installation of **gable end** studs. The last ceiling joist is actually nailed to the gable studs that sit on the exterior wall. It is nailed from the inside of the building. These end joists are installed butted in-line regardless of how the other joists are laid out (**Figures 41–22** and **41–23**).

In addition to other functions, these end joists provide fastening for the ends of the ceiling finish.

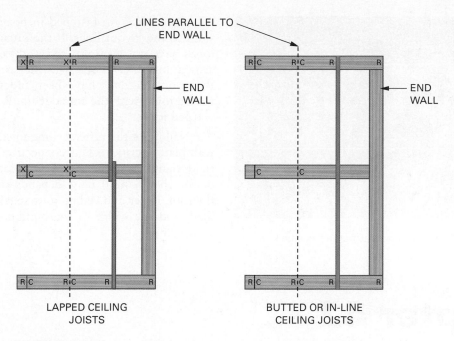

FIGURE 41–21 Layout lines for ceiling joists are the same as for lapped and in-line joists.

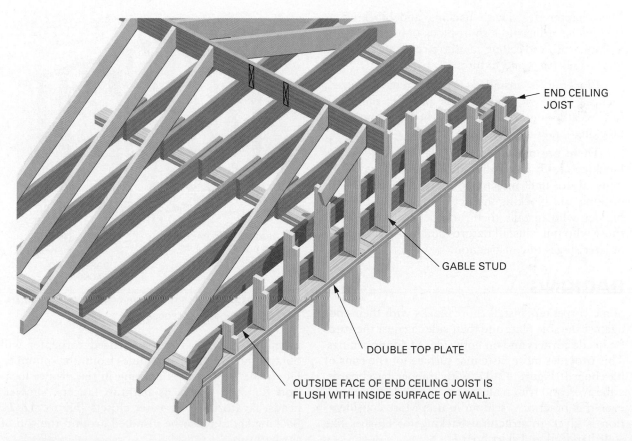

FIGURE 41–22 The end ceiling joist is attached to the gable end studs.

Photo by M. Nauth

FIGURE 41–23 A gable end wall framed as a unit.

All other joists are fastened in position with their sides to the layout lines. If the outside ends of the joists have not been tapered, sight each joist for a crown. Install each one with the crowned edge up. If the outside ends have been tapered, install the ceiling joists with the cut edge up. Reject any badly warped joists.

Nailing of the rafter, ceiling joist, collar tie, and wall plate is critical. This connection establishes the rigid triangle that supports the roof load. The number of 3-inch (75 mm) nails depends on the slope of the roof, rafter and ceiling joist spacing, roof snow-load, and the width of the building (NBCC Table 9.23.13.8).

Chapter 42

Backing and Blocking

This chapter deals with backing and blocking. A *backing* is, ordinarily, a short block of lumber placed in floor, wall, and ceiling cavities to provide fastening for various parts and fixtures.

Blocking is installed for different purposes, such as to provide support for parts of the structure, weather-tightness, and fire-stopping. Sometimes blocking serves as backing.

There are many places in the structure where blocking and backing should be placed. It would be unusual to find directions for the placement of backing and blocking in a set of plans. It is the wise builder who installs them, much to the delight of those who must install fixtures and finish of all types in later stages of construction.

BACKING

Some carpenters install short blocks with their ends against the sole plate and their sides against the studs on inside corners and on both sides of door openings. This provides more fastening surface for the ends of baseboard (**Figure 42–1**). With thinner baseboards, adhesives, and trim nailers now in popular use, however, this practice is seldom in use. More consideration is given to structural backing for finishes like stair handrail and skirt boards.

Much backing is needed in bathrooms. Plumbing rough-in work varies with the make and style of

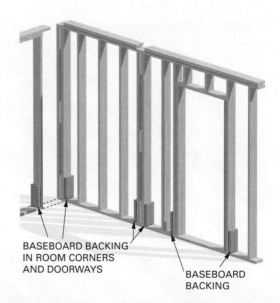

BASEBOARD BACKING IN ROOM CORNERS AND DOORWAYS

BASEBOARD BACKING

FIGURE 42–1 Backing is sometimes installed at corners and door openings for baseboard.

plumbing fixtures. The experienced carpenter will obtain the rough-in schedule from the plumber. He or she then installs backing in the proper location for such things as bathtub faucets, shower-heads, lavatories, and water closets (**Figure 42–2**). Backing should also be installed around the top of the bathtub.

In the kitchen, backing should be provided for the tops and bottoms of wall cabinets and for the

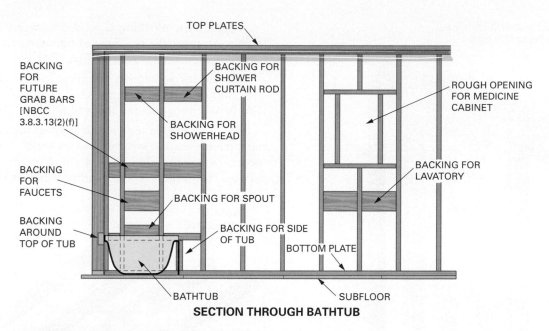

TOP PLATES

BACKING FOR FUTURE GRAB BARS [NBCC 3.8.3.13(2)(f)]

BACKING FOR SHOWER CURTAIN ROD

BACKING FOR SHOWERHEAD

ROUGH OPENING FOR MEDICINE CABINET

BACKING FOR FAUCETS

BACKING FOR SPOUT

BACKING FOR LAVATORY

BACKING AROUND TOP OF TUB

BACKING FOR SIDE OF TUB

BOTTOM PLATE

BATHTUB

SUBFLOOR

SECTION THROUGH BATHTUB

FIGURE 42–2 Typical backing needed in bathrooms.

tops of base cabinets. If the ceiling is to be built down to form a *soffit* at the tops of wall cabinets, backing should be installed to provide fastening for the soffit (**Figure 42–3**). Note that firestop blocking must be installed if the soffit is framed before the drywall is installed.

A homeowner will appreciate the thoughtfulness of the builder who provides backing in appropriate locations in all rooms for the fastening of curtain and drapery hardware (**Figure 42–4**).

BLOCKING

Some types of blocking have already been described in earlier units. See Chapter 36 for the installation of solid lumber blocking used for bridging.

When floor panels are used as a combination subfloor and **underlayment** (under carpet and pad), the panel edges must be tongue-and-grooved or supported on 2-inch (38 mm) lumber blocking installed between joists (**Figure 42–5**).

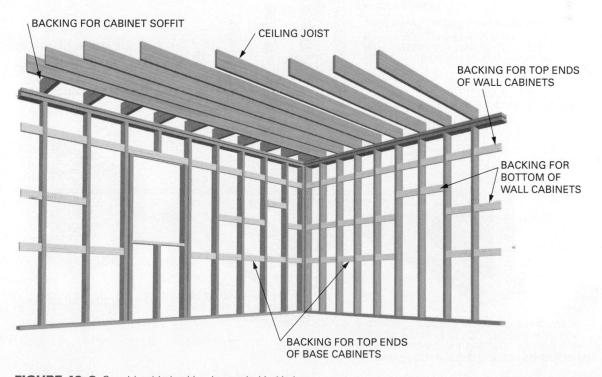

BACKING FOR CABINET SOFFIT

CEILING JOIST

BACKING FOR TOP ENDS OF WALL CABINETS

BACKING FOR BOTTOM OF WALL CABINETS

BACKING FOR TOP ENDS OF BASE CABINETS

FIGURE 42–3 Considerable backing is needed in kitchens.

Ladder-type blocking is needed between ceiling joists to support the top ends of partitions that run parallel to and between joists (**Figure 42–6**).

Blocks are sometimes installed to support the back edge of bathtubs (**Figure 42–7**). This stiffens the joint between the tub and the wall finish.

Wall blocking is required to support the edge of rated panels used in structural shear walls. It is also used in wood foundations and panels permanently exposed to weather, for weather-tightness. When

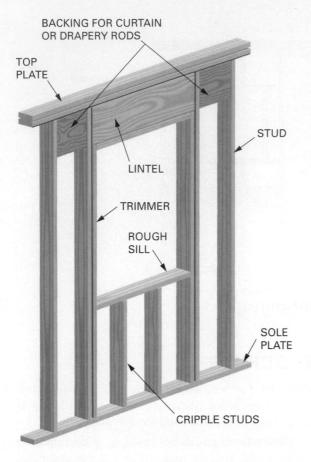

FIGURE 42–4 Installing backing around windows allows for easy installation of curtain and drapery hardware.

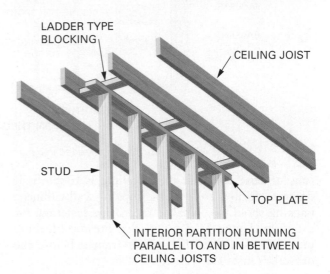

FIGURE 42–6 Ladder-type blocking provides support for the top plates of interior partitions.

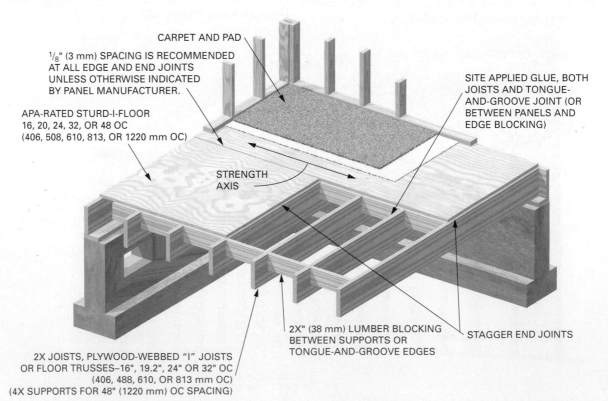

FIGURE 42–5 The edges of APA-Rated Sturd-I-Floor panels must be tongue-and-grooved or supported by blocking.

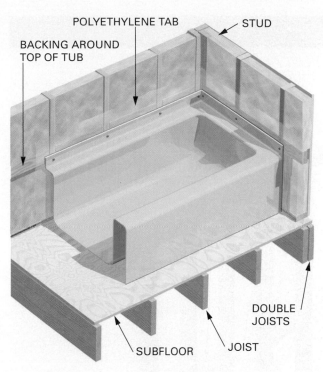

FIGURE 42–7 labels: POLYETHYLENE TAB, STUD, BACKING AROUND TOP OF TUB, DOUBLE JOISTS, JOIST, SUBFLOOR

SUPPORT OF BATHTUB AGAINST WALL FRAMING

FIGURE 42–7 Backing provides increased support between tubs and finish material.

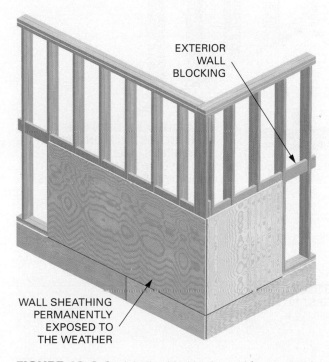

EXTERIOR WALL BLOCKING

WALL SHEATHING PERMANENTLY EXPOSED TO THE WEATHER

FIGURE 42–8 Straight-line blocking is used for structural or exterior wall sheathing.

used for these purposes, blocking must be installed in a straight line (**Figure 42–8**).

Fire blocking is required between studs that are greater than 9'-10" (3 m) in length when the wall

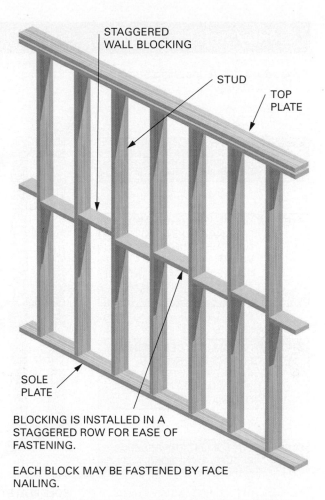

STAGGERED WALL BLOCKING, STUD, TOP PLATE, SOLE PLATE

BLOCKING IS INSTALLED IN A STAGGERED ROW FOR EASE OF FASTENING.

EACH BLOCK MAY BE FASTENED BY FACE NAILING.

INSTALL BLOCKING IN EVERY OTHER SPACE FIRST. THEN, INSTALL IN REMAINING SPACES.

FIGURE 42–9 Blocking used for fire-stopping may be installed in a staggered fashion for easier nailing.

cavity is not insulated. The fire blocks also serve to stiffen the studs and strengthen the structure. This type of blocking may be installed in a staggered fashion (**Figure 42–9**). Insulated wall cavities do not require fire blocking.

Blocking between studs is required in walls over 8'-1" (2464 mm) high. The purpose is to stiffen the studs and strengthen the structure. The required blocking also functions as a firestop in stud spaces. Blocking for these purposes may be installed in staggered fashion (Figure 42–9).

INSTALLING BACKING AND BLOCKING

Install blocking in a staggered row. Fasten by nailing through the studs into each end of each block in the same manner as staggered solid wood bridging, described previously. Installing blocking in a straight

StepbyStep Procedures

Procedure 42–A Installing Straight, In-Line Blocking

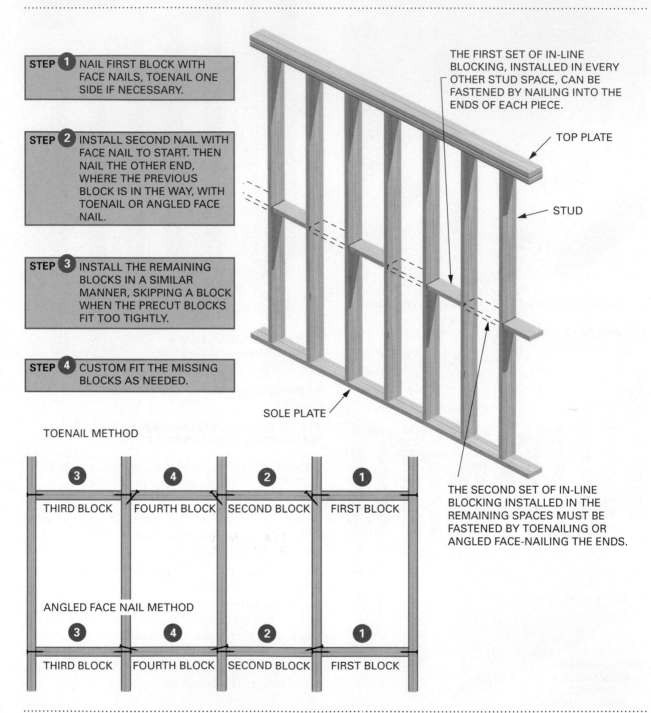

STEP ❶ NAIL FIRST BLOCK WITH FACE NAILS, TOENAIL ONE SIDE IF NECESSARY.

STEP ❷ INSTALL SECOND NAIL WITH FACE NAIL TO START. THEN NAIL THE OTHER END, WHERE THE PREVIOUS BLOCK IS IN THE WAY, WITH TOENAIL OR ANGLED FACE NAIL.

STEP ❸ INSTALL THE REMAINING BLOCKS IN A SIMILAR MANNER, SKIPPING A BLOCK WHEN THE PRECUT BLOCKS FIT TOO TIGHTLY.

STEP ❹ CUSTOM FIT THE MISSING BLOCKS AS NEEDED.

THE FIRST SET OF IN-LINE BLOCKING, INSTALLED IN EVERY OTHER STUD SPACE, CAN BE FASTENED BY NAILING INTO THE ENDS OF EACH PIECE.

TOP PLATE

STUD

SOLE PLATE

THE SECOND SET OF IN-LINE BLOCKING INSTALLED IN THE REMAINING SPACES MUST BE FASTENED BY TOENAILING OR ANGLED FACE-NAILING THE ENDS.

TOENAIL METHOD

❸ THIRD BLOCK ❹ FOURTH BLOCK ❷ SECOND BLOCK ❶ FIRST BLOCK

ANGLED FACE NAIL METHOD

❸ THIRD BLOCK ❹ FOURTH BLOCK ❷ SECOND BLOCK ❶ FIRST BLOCK

line is more difficult. The ends of some pieces may be toenailed or face-nailed at an angle (**Procedure 42–A**). Snap a line across the framing. Square lines in from the chalk line on the sides of the studs.

It may be helpful to use a short post on one side of each stud to support the blocking while the end is being toenailed. Start the toenails in the end of the block before positioning it (**Figure 42–10**).

Backing may also be installed in a continuous length by notching the studs and fastening into its edges (**Figure 42–11**).

ON THE JOB

When driving toenails, start the nails partway and then use something to hold the block in position while you drive the nails.

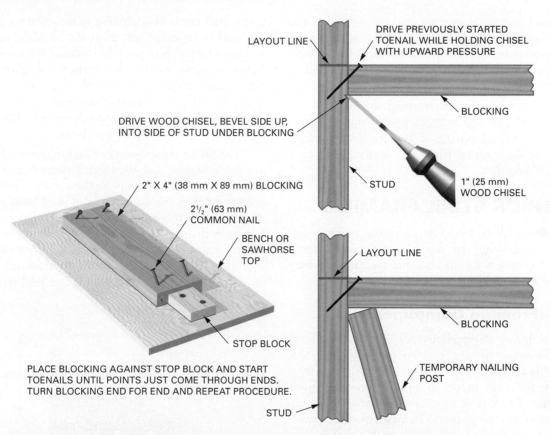

DRIVE PREVIOUSLY STARTED TOENAIL WHILE HOLDING CHISEL WITH UPWARD PRESSURE

LAYOUT LINE

BLOCKING

DRIVE WOOD CHISEL, BEVEL SIDE UP, INTO SIDE OF STUD UNDER BLOCKING

STUD

1" (25 mm) WOOD CHISEL

2" X 4" (38 mm X 89 mm) BLOCKING

2¹/₂" (63 mm) COMMON NAIL

BENCH OR SAWHORSE TOP

STOP BLOCK

PLACE BLOCKING AGAINST STOP BLOCK AND START TOENAILS UNTIL POINTS JUST COME THROUGH ENDS. TURN BLOCKING END FOR END AND REPEAT PROCEDURE.

LAYOUT LINE

BLOCKING

TEMPORARY NAILING POST

STUD

FIGURE 42–10 Techniques for toenailing blocking between studs.

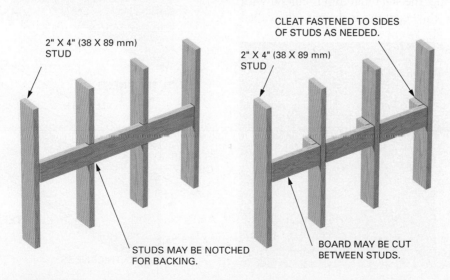

2" X 4" (38 X 89 mm) STUD

CLEAT FASTENED TO SIDES OF STUDS AS NEEDED.

2" X 4" (38 X 89 mm) STUD

STUDS MAY BE NOTCHED FOR BACKING.

BOARD MAY BE CUT BETWEEN STUDS.

FIGURE 42–11 Various ways to install continuous backing for plumbing and other fixtures.

Chapter 43

Steel Framing

Light steel framing is used for structural framing and interior non-load-bearing partitions (**Figure 43–1**). Carpenters often frame interior partitions and apply *furring channels* of steel. This chapter is limited to their installation.

The strength of steel framing members of the same design and size may vary with the manufacturer. The size and spacing of steel framing members should be determined from the drawings or by a structural engineer.

INTERIOR STEEL FRAMING

The framing of steel interior partitions is quite similar to the framing of wood partitions. Different kinds of fasteners are used. Some special tools may be helpful.

Steel Framing Components

All steel frame components are cold-formed steel, which is made by one of two methods. The first method is to *press-brake* a steel section into shape. Most shapes are made using this process unless otherwise noted. The other method is to continuously *roll-form* the shape from a coil of steel. This is known as *cold-rolled (CR) steel*.

Before forming, the steel material is coated with zinc. Steel with this coating is commonly called

galvanized steel. This coating is designed to protect the steel from corrosion until the building is made weather-tight. It is important to detail the air-vapour barrier exactly so that no moisture is allowed to accumulate in the exterior wall. The bottom track is ideally shaped to collect condensing water, and tracks and the bottom of studs have been known to rust away, especially in humid areas.

The major components of an interior steel frame are *studs*, *tracks*, and *channels*. Fasteners and accessories are needed to complete the system. The names used to refer to the dimensions of steel framing vary slightly from those used for wood (**Figure 43–2**). Length is the same for both wood and steel framing. The thickness of wood is typically 1½ inches (38 mm), whereas the thickness for steel is the thickness of the steel sheet used to make the stud. The steel term that

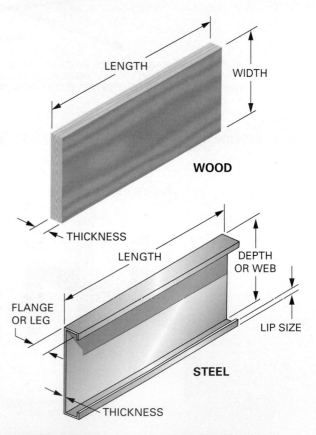

FIGURE 43–2 The names for steel stud material are somewhat different from those of wood.

FIGURE 43–1 Steel may be used for load-bearing and non-load-bearing building frames.

is similar to the *thickness* of a piece of wood is *leg* or *flange* size. A wood stud *width* is similar to steel stud *depth*, which is also referred to as the **web**.

Studs. The thicknesses of steel studs used for non-load-bearing partitions are typically 18, 27, and 33 mil, where a mil is 1/1000 of an inch. They are also referred to as 25, 22, and 20 gauge, where the larger number denotes the smaller thickness. These studs are used for non-load-bearing partitions. These sizes in metric are 0.46, 0.69, and 0.84 mm. The stud web has punchouts at intervals through which to run pipes and conduit.

Studs come in widths of 1⅝, 2½, 3⅝, 4, and 6 inches, with 1- and 1¼-inch leg thicknesses (41, 63, 92, 102, and 152 mm, with 25 and 32 mm leg thicknesses). Studs are available in stock lengths of 8, 9, 10, 12, and 16 feet (2.4, 2.7, 3.0, 3.7, and 5.0 m) (**Figure 43–3**). Custom lengths up to 28 feet (8.5 m) are also available.

Track. The top and bottom horizontal members of a steel-framed wall are called *track*s or *runners*. They are installed on floors and ceilings to receive the studs. They are manufactured by thickness, widths,

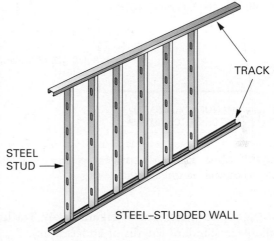

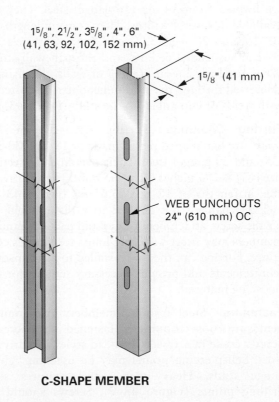

Thickness of Cold-Formed Steel Members			
Mils (1/1000ths inch)	Gauge Number	Metric (mm)	Intended Use
18	25	0.5	Nonstructural
27	22	0.7	Nonstructural
33	20	0.8	Nonstructural
43	18	1.1	Structural
54	16	1.4	Structural
68	14	1.7	Structural
97	12	2.5	Structural

FIGURE 43–3 Steel studs come in several widths, lengths, and thicknesses.

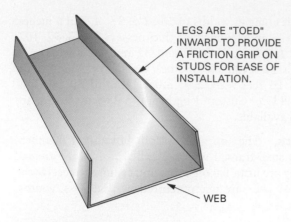

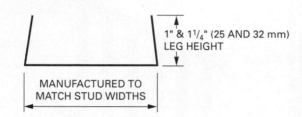

FIGURE 43–4 Cold-formed channels called *track* are the top and bottom plates of steel-stud walls.

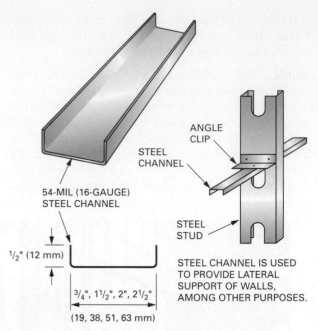

FIGURE 43–5 Cold-rolled channels are used to stiffen the framing members of walls and ceilings.

and leg size to match studs (**Figure 43–4**). Track is available in standard lengths of 10 feet (3 m).

Channels. Steel **cold-rolled channels (CRCs)** are formed from 54-mil (16-gauge) steel. They are available in several widths. They come in lengths of 10, 16, and 20 feet (3, 5, and 6 m). Channels are used in suspended ceilings and through wall studs. When used for lateral bracing of walls, the channel is inserted through the stud punchouts. It is fastened with welds or clip angles to the studs (**Figure 43–5**).

Furring Channels. **Furring channels** (or *hat track*) are hat-shaped pieces made of 18- and 33-mil (25- and 20-gauge) steel. Their overall cross-section size is $^7/_8 \times 2^9/_{16}$ inches (22 × 65 mm). They are available in lengths of 12 feet (3.65 m) (**Figure 43–6**). Furring channels are applied to walls and ceilings for the screw attachment of gypsum panels. Framing members may exceed spacing limits for various coverings. Furring can then be installed to meet spacing requirements and provide necessary support for the surfacing material.

Fasteners. Steel framing members and components are most commonly fastened with screws. Screws come in a variety of head styles and driving slots. Self-piercing points may be used on lighter gauge studs. Heavy-gauge steel requires self-drilling points (**Figure 43–7**). Screws should be about ½ inch (12 mm) longer than the materials

being fastened. A minimum of three threads should penetrate the steel.

Plywood may be attached using specially designed pneumatic nails. Powder-actuated fasteners are often used to attach the framed wall to other support material such as steel or masonry.

Layout and Framing of Steel Partitions

Lay out steel-framed partitions as you would wood-framed partitions. Snap chalk lines on the floor. Plumb up from partition ends. Snap lines on the ceiling. Make sure that partitions will be plumb. Using a laser level is an efficient way to lay out floor and ceiling lines for partitions.

Lay out the stud spacing and the wall opening on the bottom track. The top track is laid out after the first stud away from the wall is plumbed and fastened.

Installing Track. Fasten track to floor and ceiling so that one edge is to the chalk line (**Figure 43–8**). Make sure both floor and ceiling track are on the same side of the line. Leave openings in floor track for door frames. Allow for the width of the door and thickness of the door frame. Tracks are usually fastened into concrete with powder-driven fasteners. Stub concrete nails or masonry screws may also be used. Fasten into wood with 1¼-inch (32 mm) oval head screws.

Attach the track with two fasteners about 2 inches (50 mm) from each end and a maximum

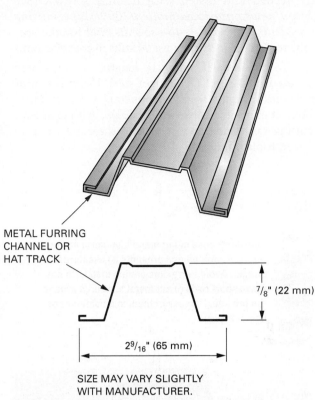

METAL FURRING
CHANNEL OR
HAT TRACK

$7/8$" (22 mm)

$2^9/_{16}$" (65 mm)

SIZE MAY VARY SLIGHTLY
WITH MANUFACTURER.

FIGURE 43–6 Furring channels (hat track) are used in both ceiling and wall installations.

SELF-DRILLING   SELF-PIERCING

FIGURE 43–7 Screws to fasten steel members are either self-drilling or self-piercing.

of 24 inches (610 mm) OC in between. At corners, extend one track to the end. Then butt or overlap the other track (**Figure 43–9**). It is not desirable or necessary to make mitred joints.

Install backing, if necessary, between joists or *trusses* where the top track will be attached. Plumb up from the bottom track to the ceiling backing to locate the top track. Snap lines

FIGURE 43–8 Bottom track is fastened to the floor next to chalk lines.

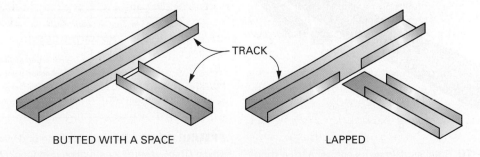

TRACK

BUTTED WITH A SPACE   LAPPED

FIGURE 43–9 Track may be butted or overlapped at intersections.

as needed and fasten with framing screws. (See *http://news.ontario.ca/mol/en/2009/12/permitting -residential-drywallers-to-use-stilts.html* for the specific requirements for the use of stilts in construction.)

To cut metal framing to length, tin snips may be used on 18-mil (25-gauge) steel. Using tin snips becomes difficult on thicker metal. A power mitre box, commonly called a *chop saw*, with a metal-cutting saw blade or a **steel stud shear** are the preferred tools (**Figure 43–10**).

CAUTION

The sharp ends of cut metal can cause serious injury. A pointed end presents an even greater danger. Avoid mitre cuts on thin metal. Do not leave short ends of cut metal scattered around the job site. Dispose of them in a container as you cut them.

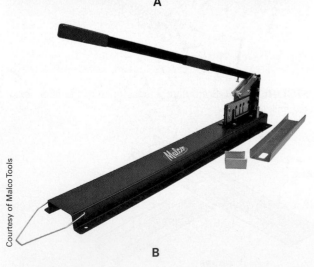

A

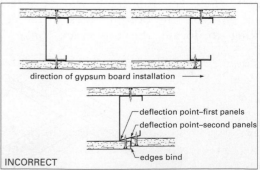

Courtesy of Malco Tools

B

FIGURE 43–10 Steel studs may be cut using (A) a chop saw or (B) a steel stud shear.

Installing Studs. Cut the necessary number of full-length studs needed. For ease of installation, cut them about ¼ inch (6 mm) short. Install studs at partition intersections and corners. Fasten to bottom and top track. Use ³⁄₈ -inch (9.5 mm) self-drilling pan head screws. If moisture may be present where a stud butts an exterior wall, place a strip of asphalt felt between the stud and the wall.

Place the first stud in from the corner between track. Fasten the bottom in position at the layout line. Plumb the stud. Using a magnetic level can be very helpful. Clamp the top end when plumb, and fasten to the top track. Lay out the stud spacing on the top track from this stud (**Procedure 43–A**).

Place all full-length studs in position between track with the open side facing the starting point of the layout (**Figure 43–11**). The web punchouts should be aligned vertically. This provides for lateral bracing of the wall and the running of plumbing and wiring (**Figure 43–12**). Fasten all studs except those on each side of door openings securely to top and bottom track.

Wall Openings

Several methods are used to frame around door and window openings. One method involves installing wood jack studs and sills (**Figure 43–13**). This allows for conventional installation of interior finishes. A second method requires wood door

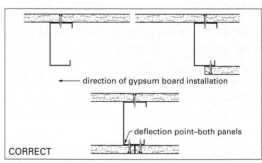

FIGURE 43–11 Correct application of drywall to steel studs. (The Gypsum Construction Handbook, *Sixth Edition. Reproduced with permission of John Wiley & Sons, Inc.)*

StepbyStep Procedures

Procedure 43–A Assembling a Steel Stud Wall

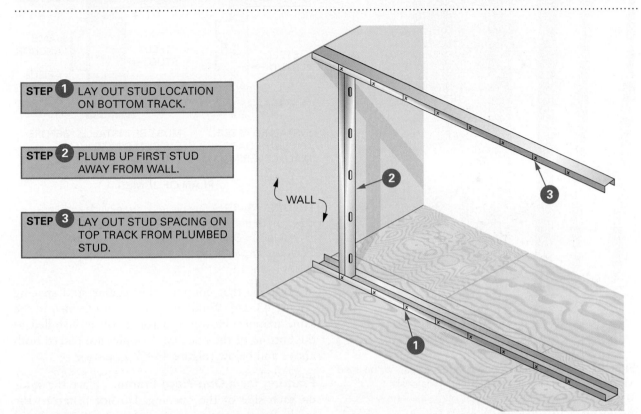

STEP 1 LAY OUT STUD LOCATION ON BOTTOM TRACK.

STEP 2 PLUMB UP FIRST STUD AWAY FROM WALL.

STEP 3 LAY OUT STUD SPACING ON TOP TRACK FROM PLUMBED STUD.

WALL

FIGURE 43–12 Punchouts should line up for bracing and mechanicals.

frames to be installed with screws to the steel frame (**Figure 43–14**). The third method uses only the steel studs with a metal door frame called a **steel buck**. There are two styles of bucks.

A one-piece metal door frame must be installed before the gypsum board is applied. A three-piece, knocked-down frame is set in place after the wall covering is applied (**Figure 43–15**).

Framing for a Three-Piece Frame. First, place full-length studs on each side of the opening in a plumb position. Fasten securely to the bottom and

FIGURE 43–13 Window and door openings may be lined with wood for easier fastening of finish materials.

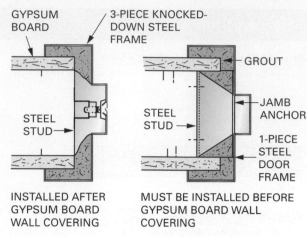

INSTALLED AFTER GYPSUM BOARD WALL COVERING

MUST BE INSTALLED BEFORE GYPSUM BOARD WALL COVERING

PLAN OF JAMBS

FIGURE 43–15 A one-piece or three-piece knocked-down metal door frame may be used in steel wall framing. (The Gypsum Construction Handbook, *Sixth Edition. Reproduced with permission of John Wiley & Sons, Inc.*)

top plates. Cut a length of track for use as a lintel. Cut up to 6 inches (152 mm) longer than the width of the opening. Cut the flanges in appropriate places. Bend the web to fit over the studs on each side of the opening. Fasten the fabricated lintel to the studs at the proper height. Install jack studs over the opening

in positions that continue the regular stud spacing (**Figure 43–16**). Window openings are framed in the same manner. However, a rough sill is installed at the bottom of the opening. Cripples are placed both above and below (**Figure 43–17**).

Framing for a One-Piece Frame.   Place the studs on each side of the opening. Do not fasten to the track. Set the one-piece door frame in place. Level the door frame lintel by shimming under a jamb, if necessary. Fasten the bottom ends of the door jambs to the floor in the proper location. Fasten the studs to the door jambs. Then, fasten the studs to the bottom track. Plumb the door frame. Clamp the stud to

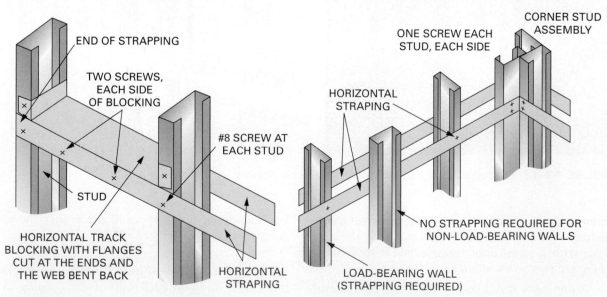

FIGURE 43–14 Wall strapping and bracing are installed with the ends well secured.

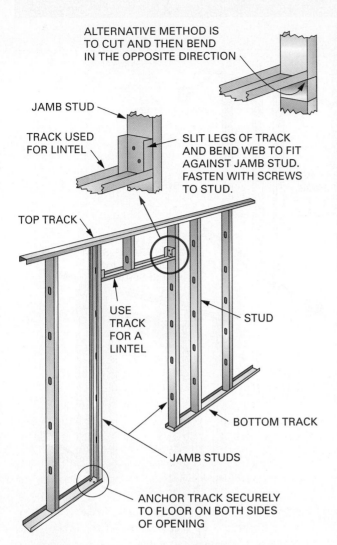

ALTERNATIVE METHOD IS TO CUT AND THEN BEND IN THE OPPOSITE DIRECTION

JAMB STUD

TRACK USED FOR LINTEL

SLIT LEGS OF TRACK AND BEND WEB TO FIT AGAINST JAMB STUD. FASTEN WITH SCREWS TO STUD.

TOP TRACK

USE TRACK FOR A LINTEL

STUD

BOTTOM TRACK

JAMB STUDS

ANCHOR TRACK SECURELY TO FLOOR ON BOTH SIDES OF OPENING

FIGURE 43–16 An opening framed for a three-piece, knocked-down, metal door frame.

FIGURE 43–17 Rough sills are installed similarly to non-load-bearing lintels.

the top track, and fasten with screws. Install lintel and jack studs in the same manner as described previously (**Procedure 43–B**).

The steel framing described above is suitable for average weight doors up to 2′-8″ (813 mm) wide. For wider and heavier doors, the framing should be strengthened by using 33-mil (20-gauge) steel framing. Also, double the studs on each side of the door opening (**Figure 43–18**).

Chase Walls

A **chase wall** is made by constructing two closely spaced, parallel walls for the running of plumbing, heating and cooling ducts, and similar items. They

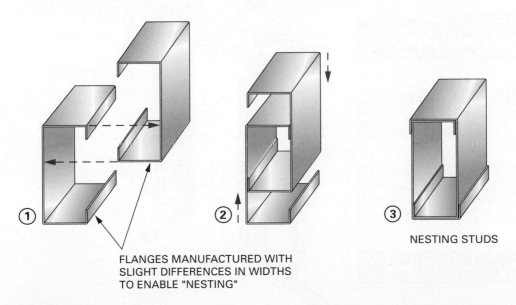

① ② ③ NESTING STUDS

FLANGES MANUFACTURED WITH SLIGHT DIFFERENCES IN WIDTHS TO ENABLE "NESTING"

FIGURE 43–18 Steel studs may be nested to create a stronger frame for larger, heavier doors.

StepbyStep Procedures

Procedure 43–B Assembling a One-Piece Door Frame

STEP 1 FASTEN TOP AND BOTTOM TRACKS IN POSITION. LEAVE SPACE IN BOTTOM TRACK FOR OPENING.

STEP 2 FASTEN ALL STUDS IN POSITION BUT LEAVE JAMB STUDS LOOSE AT THE BOTTOM.

STEP 3 MOVE JAMB STUDS OUT OF THE WAY.

STEP 4 SET DOOR FRAME IN OPENING. LEVEL LINTEL AND FASTEN BOTTOM OF FRAME TO FLOOR AT CORRECT WIDTH.

STEP 5 FASTEN JAMB STUDS ON BOTH SIDES TO SIDE JAMBS OF DOOR FRAME.

STEP 6 FASTEN JAMB STUDS ON BOTH SIDES TO BOTTOM TRACK.

STEP 7 PLUMB SIDE JAMB OF DOOR FRAME AND FASTEN JAMB STUDS TO TOP TRACK.

STEP 8 INSTALL LINTEL AND JACK STUDS.

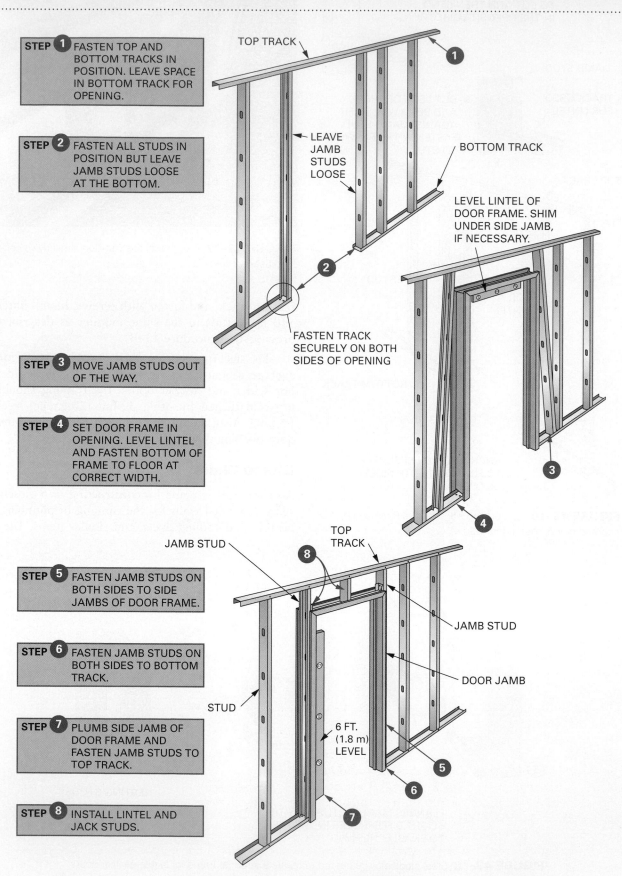

TOP TRACK

LEAVE JAMB STUDS LOOSE

BOTTOM TRACK

LEVEL LINTEL OF DOOR FRAME. SHIM UNDER SIDE JAMB, IF NECESSARY.

FASTEN TRACK SECURELY ON BOTH SIDES OF OPENING

JAMB STUD

TOP TRACK

JAMB STUD

DOOR JAMB

STUD

6 FT. (1.8 m) LEVEL

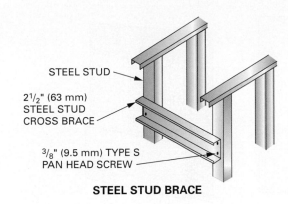

STEEL STUD

2¹/₂" (63 mm)
STEEL STUD
CROSS BRACE

³/₈" (9.5 mm) TYPE S
PAN HEAD SCREW

STEEL STUD BRACE

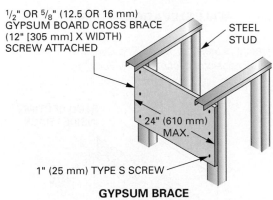

¹/₂" OR ⁵/₈" (12.5 OR 16 mm)
GYPSUM BOARD CROSS BRACE
(12" [305 mm] X WIDTH)
SCREW ATTACHED

STEEL
STUD

24" (610 mm)
MAX.

1" (25 mm) TYPE S SCREW

GYPSUM BRACE

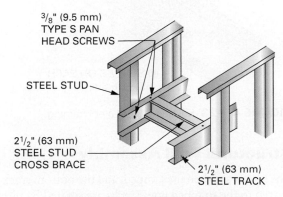

³/₈" (9.5 mm)
TYPE S PAN
HEAD SCREWS

STEEL STUD

2¹/₂" (63 mm)
STEEL STUD
CROSS BRACE

2¹/₂" (63 mm)
STEEL TRACK

STEEL STUD AND TRACK BRACE

FIGURE 43–19 Chase wall construction details.

are constructed in the same manner as described previously. However, the spacing between the outside edges of the wall frames must not exceed 24 inches (610 mm).

The studs in each wall should be installed with the flanges running in the same direction. They should be directly across from each other. The walls should be tied to each other either with pieces of 12-inch-wide (305 mm) gypsum board or short lengths of steel stud. If the wall studs are not opposite each other, install lengths of steel stud horizontally inside both walls. Tie together with shorter lengths of stud material spaced 24 inches (610 mm) on centre (**Figure 43–19**). Wall ties should be spaced 48 inches (1220 mm) on centre vertically.

Installing Metal Furring

Metal furring may be used on ceilings applied at right angles to joists. They may be applied vertically or horizontally to framed or masonry walls. Space metal furring channels a maximum of 24 inches (620 mm) on centre.

Ceiling Furring. Metal furring channels may be attached directly to structural ceiling members or suspended from them. For direct attachment, saddle tie with double-strand 43-mil (18-gauge) wire to each member (**Figure 43–20**). Leave a 1-inch (25 mm) clearance between ends of furring and walls. Metal furring channels may be spliced. Overlap the ends by at least 8 inches (203 mm). Tie each end with wire. Steel studs may be used, with their open side up for furring, when supporting framing is widely spaced. Several methods of utilizing metal furring channels or steel studs in suspended ceiling applications are shown in **Figure 43–21**.

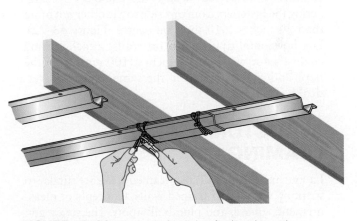

FIGURE 43–20 Method of splicing furring channels.

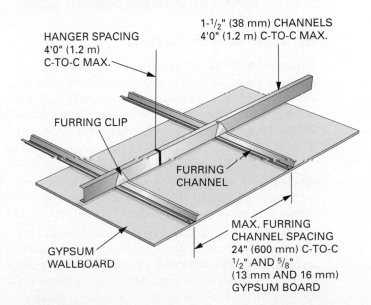

HANGER SPACING
4'0" (1.2 m)
C-TO-C MAX.

1-¹/₂" (38 mm) CHANNELS
4'0" (1.2 m) C-TO-C MAX.

FURRING CLIP

FURRING
CHANNEL

MAX. FURRING
CHANNEL SPACING
24" (600 mm) C-TO-C
¹/₂" AND ⁵/₈"
(13 mm AND 16 mm)
GYPSUM BOARD

GYPSUM
WALLBOARD

FIGURE 43–21 Metal channels are also used in suspended ceiling applications.

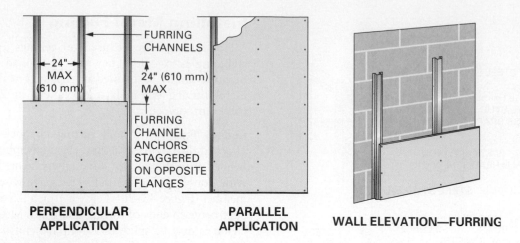

FIGURE 43–22 Furring channels may be attached directly to masonry walls. (The Gypsum Construction Handbook, *Sixth Edition. Reproduced with permission of John Wiley & Sons, Inc.*)

Wall Furring. Vertical application of steel furring channels is preferred. Secure the channels by staggering the fasteners from one side to the other not more than 24 inches (610 mm) on centre (**Figure 43–22**). For horizontal application on walls, attach furring channels not more than 4 inches (100 mm) from the floor and ceiling. Fasten in the same manner as vertical furring.

STRUCTURAL STEEL FRAMING

Layout and framing of steel-framed walls are similar to working with wood-framed walls. Concepts of measurement, square, and plumb all apply. The screw gun is used extensively. The assembly techniques between structural and non-structural steel walls vary slightly.

Structural Wall Layout

Structural walls are constructed in the same fashion as wood walls. Layout begins by snapping lines on the subfloor where walls are to be constructed. Next, the top and bottom tracks are cut for length. Walls should be no longer than what the work crew can lift into place.

Splices in the track are done between studs. The splicing block must be at least 6 inches (152 mm) long with eight #8 screws (**Figure 43–23**). Thus, splices cannot be located within 3 inches (75 mm) of a stud.

Tracks, clamped together back-to-back, are placed on the deck for layout. The location of studs and openings is marked with a permanent marker on the flanges. Studs must align with framing members above and below. This allows the load to be positively transferred to the foundation. The corner studs and other wall intersections are also marked.

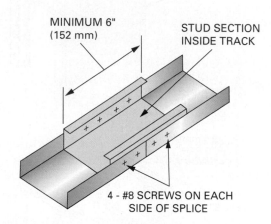

FIGURE 43–23 Splicing track in steel framing.

Structural Wall Assembly

The tracks are then unclamped, and the bottom track is moved to the snapped line. Tracks are rotated by tilting them up. The bottom track is secured to the subfloor with temporary screws through the flanges. Studs are then placed into the bottom and top tracks. They should all be facing the same direction and have the knockouts aligned. Placing the coloured ends of the studs into the bottom track will ensure that this happens.

The studs must touch the web of the track. This is done by tapping the tracks tight to the stud before fastening. The stud and track flanges must be temporarily clamped together to hold them tight as the screw is driven. Otherwise, an undesirable space may result between the flanges. Load-bearing studs require four screws, two at each end (**Figure 43–24**).

The wall is lying on its side at this point, so only half of the screws can be driven. After lifting the wall into position, fasten the second pair of screws in the stud. Also, fasten screws through the bottom track into the framing below at 12 inches (305 mm) on centre.

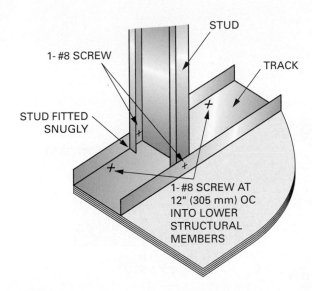

FIGURE 43-24 Load-bearing steel studs require four screws.

King studs are positioned with the web side facing the centre of the opening. Jack studs are reversed. Thus, the jack studs and king studs are placed back-to-back to allow for their webs to be screwed together with #8 screws at about 24 inches (610 mm) on centre vertically. An additional track may be installed to cover the C-shape of the jack. This simplifies attaching the jamb of the door to the jack stud. High wind or seismic areas may require two king studs on each side.

Lintels. *Lintels* are designed to carry loads from above to around the opening. This may be done by using methods called box, back-to-back, or L-lintel (**Figure 43-25**). Box and back-to-back use the same material; only the lintel pieces are reversed. L-lintels use one or two L-shaped pieces applied over the cripple studs over the opening.

Back-to-back lintels are fastened through the webs with two #8 screws at 24 inches (610 mm) on centre. The ends of the lintel are attached to the web of the king stud with a clip angle and four to eight screws, depending on lintel span. For example, if the lintel is less than 8 feet (2.4 m) then four screws may be used.

Box lintels are installed with tracks attached to the bottom and top. These are attached with two #8 screws at 24 inches (610 mm) on centre. Ends of the lintels are attached with clip angles with four to eight screws, depending on lintel span. For insulated walls, the box lintel must be installed as it is assembled.

Framing Intersections. Framing intersections and corners follows similar wood-framing techniques (**Figure 43-26**). Corners may be assembled in two fashions; both provide adequate support of interior finish. Style A in must have a 2-inch (50 mm) space, as illustrated in Figure 43-26. This allows insulation, if necessary, to be installed later. There is also sufficient room to run wire around the corner. Style B is better if the wall must be insulated, but this method makes running wires a little difficult. Style C shows the partition intersections made with a 2 × 6 (38 × 140) attached. This allows for excellent support of interior finish, room for insulation, and running wires.

Intersecting walls must be tied together. This may be done with gusset plates of the same thickness as the track, with four #8 screws into each wall track. Alternatively, the tracks may be overlapped and secured with four #8 screws.

Window Framing. Framing around a window rough opening may be done with the addition of track material wrapping the opening (**Figure 43-27**). Dimension lumber may also be installed to wrap the rough opening. A standard 2 × (38 ×) lumber is screwed into the track lining of the rough opening or directly to the jack stud. Both of these framing methods simplify later window installation and finish by allowing residential wood carpentry techniques.

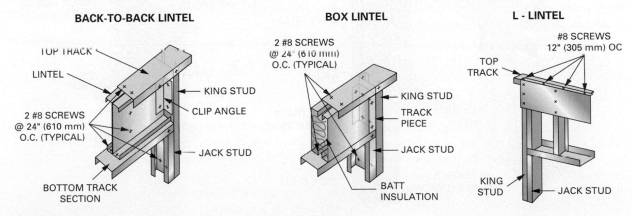

FIGURE 43-25 Supports over openings in structural steel framing.

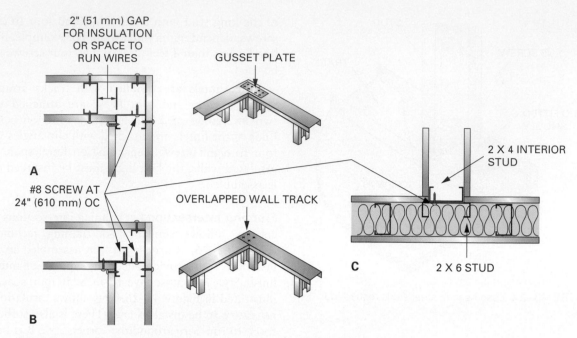

FIGURE 43–26 Details for wall framing at corners and intersections.

Horizontal Wall Bracing. Walls 8 feet (2.4 m) high or less require one row of horizontal bracing at mid-span to resist stud twist when under load. This is similar to floor bridging and may be accomplished by strapping. Walls between 8 and 12 feet (2.4 and 3.6 m) require two rows evenly spaced vertically. Strapping is installed on two sides of the stud and must be at least 1½ inches (38 mm) wide and 35 mil (1 mm) thick. It must be pulled taut and fastened with one #8 screw in each stud on each side.

Both ends of the wall bracing must be carefully secured to ensure that the twisting force is adequately resisted. One method is by two screws

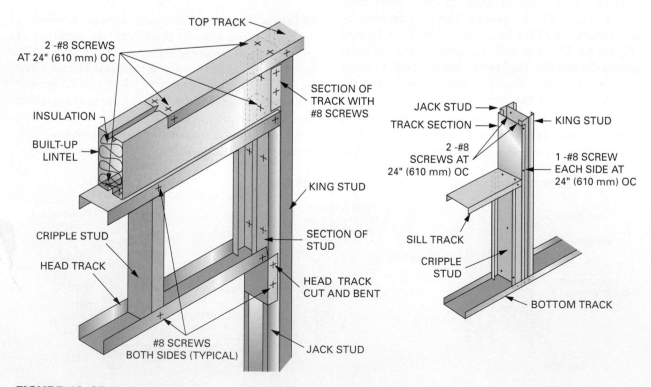

FIGURE 43–27 Track may be used to wrap a window rough opening.

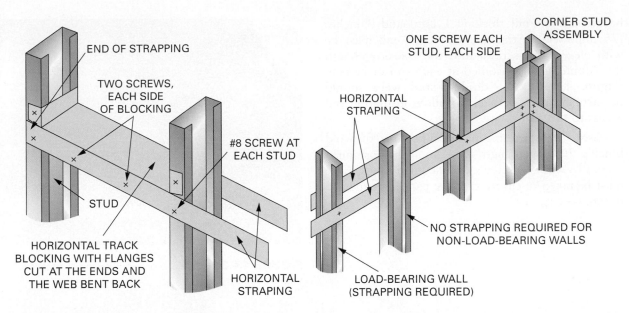

ONE SCREW EACH
STUD, EACH SIDE

CORNER STUD
ASSEMBLY

END OF STRAPPING

TWO SCREWS,
EACH SIDE
OF BLOCKING

HORIZONTAL
STRAPING

#8 SCREW AT
EACH STUD

STUD

HORIZONTAL TRACK
BLOCKING WITH FLANGES
CUT AT THE ENDS AND
THE WEB BENT BACK

HORIZONTAL
STRAPING

NO STRAPPING REQUIRED FOR
NON-LOAD-BEARING WALLS

LOAD-BEARING WALL
(STRAPPING REQUIRED)

FIGURE 43-28 Wall strapping and bracing is installed with the ends well secured.

into a corner stud assembly (**Figure 43–28**). Another method uses a block of track material: Track is cut, bent, and fastened between two studs. Strapping is then attached with #8 screws. If structural sheathing is used, the exterior side strap may be omitted, but interior side bracing is still needed.

Wind Bracing. Wall sections may be wind braced using metal straps (**Figure 43–29**). Bracing must

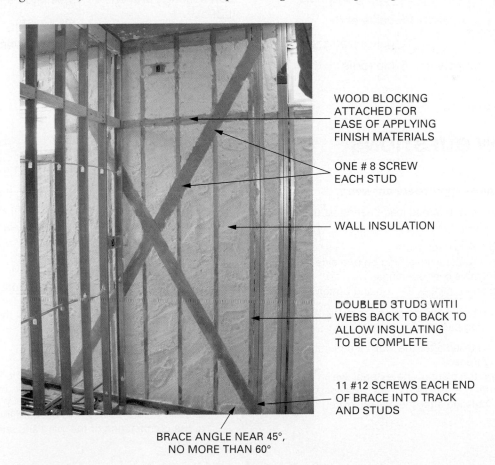

WOOD BLOCKING
ATTACHED FOR
EASE OF APPLYING
FINISH MATERIALS

ONE # 8 SCREW
EACH STUD

WALL INSULATION

DOUBLED STUDS WITH
WEBS BACK TO BACK TO
ALLOW INSULATING
TO BE COMPLETE

11 #12 SCREWS EACH END
OF BRACE INTO TRACK
AND STUDS

BRACE ANGLE NEAR 45°,
NO MORE THAN 60°

FIGURE 43–29 Wind bracing is installed from the top of the wall to the bottom.

be at least 45 mil thick (1.1 mm) and 3 inches (75 mm) wide. Attachment at each end must be with eleven #12 screws into the tracks and into the doubled studs installed at each end of bracing (**Figure 43–30**). The doubled-stud webs should be installed back to back to allow insulation full access to the wall cavity.

Structural sheathing may be substituted for wind bracing. Its longer-length axis should be positioned vertically with #8 screws or pneumatic pins. Care must be taken to ensure that the panels are tight to the studs.

FIGURE 43–30 Each end of wind bracing is attached with eleven screws.

KEY TERMS

backing

baseboard

blocking

chase wall

cold-rolled channels (CRCs)

flange size

furring channels

gable end

galvanized steel

hip roofs

joist hangers

load-bearing partitions (LBPs)

non-load-bearing partitions (NLBPs)

scabs

steel buck

steel stud shear

stub joists

underlayment

web

REVIEW QUESTIONS

Select the most appropriate answer.

1. How are wood-framed load-bearing partitions constructed?

   a. with a single top plate
   b. with structural sheathing on one side
   c. with lintels over openings
   d. with steel track at the top and bottom

2. How or when is the doubled top plate of a load-bearing partition installed?

   a. It is installed after the walls are plumbed and straightened.
   b. It laps the plate of the exterior wall.
   c. It butts the top plate of the exterior wall.
   d. It is installed after the ceiling joists.

3. What must be done when attaching the end stud of partitions that butt against another wall?

   a. The end stud must be straightened as it is nailed to the intersecting wall.
   b. The end stud must be fastened to the intersecting wall with screws.
   c. The end stud is usually left out until the drywall is secured.
   d. The end stud is secured ONLY at the top and bottom.

4. What supports a roof load?

   a. the triangle frame formed by the collar ties and rafters
   b. framing members aligned vertically
   c. ceiling joists resisting the outward thrust of rafters
   d. the combination of the wind and snow load

5. When are ceiling joists typically installed?

   a. after the walls are plumbed and aligned
   b. after the rafters are installed
   c. after the collar ties are installed
   d. after the roof decking is applied

6. Why are the ends of ceiling joists cut to the pitch of the roof?

   a. for easy application of the wall sheathing
   b. so they will not project above the rafters
   c. to mark the crowned edges up
   d. to resist the outward thrust of the rafters

7. Where are stub joists located?

   a. at the ends of the ceiling joist frame for a hip roof
   b. at the ends of the ceiling joist frame for a gable roof
   c. at the end of the header for a stairwell opening
   d. at the end of the header for an attic opening

8. Which of the following is acceptable as a finished attic access opening?

   a. 16 × 20 inches (406 × 508 mm)
   b. 16 × 24 inches (406 × 610 mm)
   c. 20 × 24 inches (508 × 610 mm)
   d. 21½ × 23½ inches (546 × 597 mm)

9. What are non-load-bearing partition lintels usually constructed from?

   a. 2 × 4s (38 × 89s) installed on the flat with cripples above
   b. a doubled 2 × 4 (38 × 89) with a plywood spacer
   c. a doubled 2 × 6 (38 × 140) with a plywood spacer
   d. a doubled 2 × 8 (38 × 184) with a plywood spacer

10. What are blocking and backing NOT used for?

    a. to prevent the spread of fire in the wall cavity
    b. to secure cabinets to stud walls
    c. to secure non-load-bearing partitions to ceiling joists
    d. to secure load-bearing partitions to ceiling joists

11. Which measurement system for sizing steel studs for strength and thickness uses numbers in a seemingly reverse order?

    a. the metric system
    b. mil or 1/1000th
    c. gauge
    d. the imperial system

12. What are the actual dimensions of a nominal 2 × 4 steel stud?

    a. 1½ × 3½ inches (38 × 89 mm)
    b. 1½ × 3⅝ inches (38 × 92 mm)
    c. 1⅝ × 3½ inches (41 × 89 mm)
    d. 1⅝ × 3⅝ inches (41 × 92 mm)

13. Which of the following is NOT done when working with steel framing?

    a. Special self-tapping screws are used to fasten track to studs.
    b. Top plates are doubled.
    c. Bottom track is secured to the floor.
    d. The open side of the stud faces the starting point of stud layout.

14. Which of the following tools is NOT used to position the top track of non-load-bearing steel partitions?

    a. a magnetic level
    b. a laser plumb
    c. a plumb bob
    d. a line level

Unit 15

Roof Framing

The ability to lay out rafters and frame all types of roofs is an indication of an experienced carpenter. On most jobs, the lead carpenter lays out the different rafters, while workers make duplicates of them. Those persons aspiring to supervisory positions in the construction field must know how to frame various kinds of roofs.

 SAFETY REMINDER

Roof framing involves sloped and often slippery surfaces, while also requiring greater mental energy to perform more complicated framing details. Maintain a constant awareness of your surroundings and what your feet are doing as you focus on roofing details.

OBJECTIVES

After completing this unit, the student should be able to:

- describe the joints and joinery techniques for timber frame construction.
- determine lengths and angles for hexagonal and octagonal gazebo roofs.
- describe several roof types and define roof framing terms.
- describe the members of gable, gambrel, hip, intersecting, and shed roofs.
- lay out a common rafter and erect gable, gambrel, and shed roofs.
- lay out and install gable studs.
- lay out the members of and frame equal- and unequal-pitch hip and intersecting roofs.
- erect a trussed roof.
- apply roof sheathing.

Chapter 44

Roof Types, Components, and Terminology

Roofs may be framed in a stick-built fashion using rafters and ridgeboards, or they may be framed with trusses. Trusses will be discussed later, in Chapter 45. Careful thought and patience are required for a carpenter who wants to become proficient at roof framing. Knowledge of diverse roof types and the associated terms are essential to framing roofs. Carpenters demonstrate their craftsmanship when constructing a roof frame.

ROOF TYPES

Several roof styles are in common use. These roofs are described in the following material and are illustrated in **Figure 44–1**.

Gable Roof. The gable roof is the most common roof style. Two sloping roof surfaces meet at the top. They form triangular shapes at each end of the building called *gables*.

Shed Roof. The shed roof slopes in one direction. It is sometimes referred to as a *lean-to*. It is com-

monly used on additions to existing structures. It is also used extensively on contemporary homes.

Hip Roofs. The hip roof slopes upward to the ridge from all walls of the building. This style is used when the same overhang is desired all around the building. The hip roof eliminates construction of gable ends.

Intersecting Roof. An intersecting roof is required on buildings that have wings. Where two roofs intersect, valleys are formed. This requires several types of rafters.

Gambrel Roof. The gambrel roof is a variation of the gable roof. It has two slopes on each side instead of one. The lower slope is much steeper than the upper slope. It is framed somewhat like two separate gable roofs.

Mansard Roof. The mansard roof is a variation of the hip roof. It has two slopes on each of the four sides. It is framed somewhat like two separate hip roofs.

Butterfly Roof. The butterfly roof is an inverted gable roof. It resembles two shed roofs with their low ends placed against each other.

Other Roofs. Other roof styles are a combination of the styles just mentioned. The shape of the roof can be one of the most distinctive features of a building.

ROOF FRAME MEMBERS

A roof frame may consist of a ridgeboard and common, hip, or valley rafters. It may also have hip jacks, valley jacks, cripple jack rafters, collar ties, and gable end studs (**Figure 44–2**). Each of these components may be laid out and cut using similar mathematical principles and theory. Rafters are the sloping members of the roof that support the roof covering.

Ridgeboard. The ridgeboard is the uppermost member of a roof. Although not absolutely necessary, the ridgeboard simplifies the erection of the roof. It provides a place for the upper ends of rafters to be secured before the sheathing is applied.

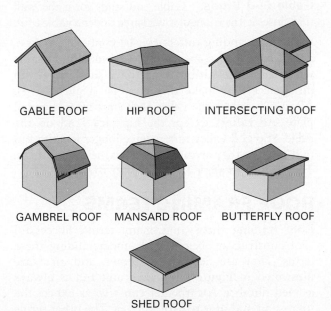

GABLE ROOF   HIP ROOF   INTERSECTING ROOF

GAMBREL ROOF   MANSARD ROOF   BUTTERFLY ROOF

SHED ROOF

FIGURE 44–1 Many roof styles can be used for residential buildings.

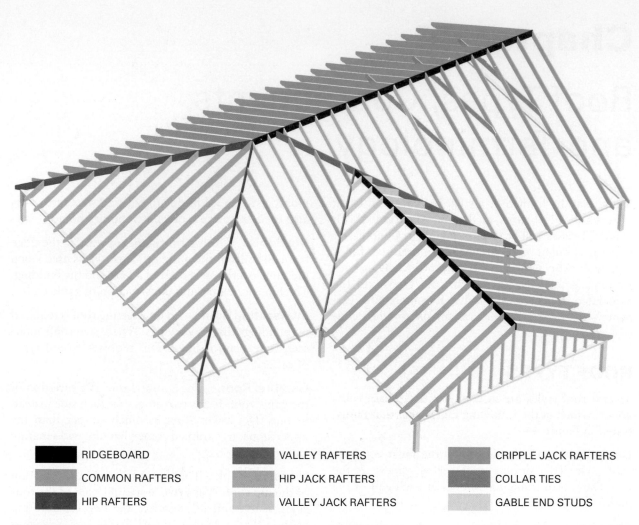

■	RIDGEBOARD	■	VALLEY RAFTERS	■	CRIPPLE JACK RAFTERS
■	COMMON RAFTERS	■	HIP JACK RAFTERS	■	COLLAR TIES
■	HIP RAFTERS	■	VALLEY JACK RAFTERS	■	GABLE END STUDS

FIGURE 44–2 There are many different components to a roof system.

Common Rafters. Common rafters are so named because they are the most common rafter. They make up the major portion of the roof, spanning from the ridgeboard to the wall. They extend at right angles from the plate to the ridge. They are used as a basis or starting point for all other rafters.

Hip Rafters. Hip rafters form an intersection of two roof sections. They project out from the roof plane forming an outside corner. They usually run at a 45-degree angle to the plates.

Valley Rafters. Valley rafters also create the intersection of two roof sections but project inward, forming an inside corner. Valley and hip rafters are similar in theory and layout.

Jack Rafters. Jack rafters come in three types: hip jacks, valley jacks, and cripple jacks. They are essentially common rafters that are cut shorter to land on a hip, valley, or both.

Collar Ties. Collar ties are horizontal members that add strength to the common rafter. They are raised and shortened ceiling joists.

Gable End Studs. Gable end studs form the wall that closes in the triangular wall area under a gable roof.

Understanding rafters can be confusing at first, and it may take some time to become comfortable with them. The various types have similarities and differences that make them unique (**Figure 44–3**). All rafters except the valley jack start from the wall plate. All rafters except the hip jack land on the ridge. Hip and valley rafters have longer lengths and tails because they are not parallel to the other rafters. They run at an angle from the plate to the ridge.

ROOF FRAMING TERMS

Roof framing theory has many terms. Successful roof construction begins with understanding these terms. They are defined as follows, and most are illustrated in **Figure 44–4**. The unit run is always a fixed number. A carpenter can always expect the total span and unit rise to be given. The other terms and measurements are marked with a square or calculated. These terms will also be discussed in more detail in later chapters.

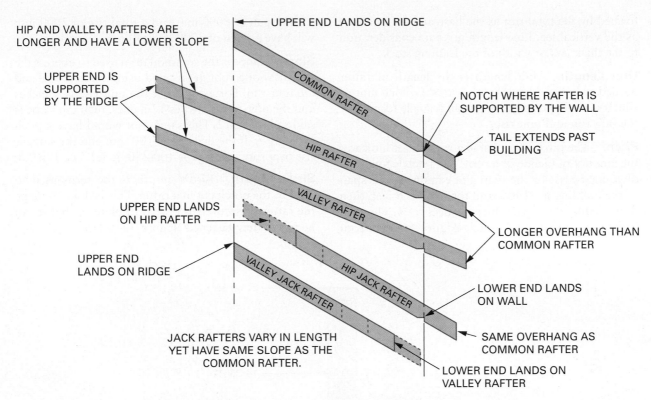

HIP AND VALLEY RAFTERS ARE LONGER AND HAVE A LOWER SLOPE

UPPER END IS SUPPORTED BY THE RIDGE

UPPER END LANDS ON RIDGE

COMMON RAFTER

NOTCH WHERE RAFTER IS SUPPORTED BY THE WALL

TAIL EXTENDS PAST BUILDING

HIP RAFTER

VALLEY RAFTER

UPPER END LANDS ON HIP RAFTER

LONGER OVERHANG THAN COMMON RAFTER

UPPER END LANDS ON RIDGE

VALLEY JACK RAFTER

HIP JACK RAFTER

LOWER END LANDS ON WALL

JACK RAFTERS VARY IN LENGTH YET HAVE SAME SLOPE AS THE COMMON RAFTER.

SAME OVERHANG AS COMMON RAFTER

LOWER END LANDS ON VALLEY RAFTER

FIGURE 44–3 The various types of rafters have many similarities and differences.

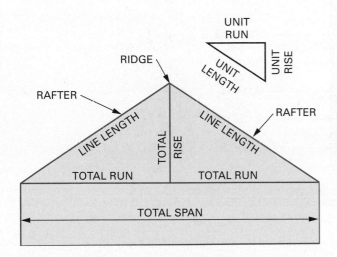

RIDGE

RAFTER

UNIT RUN

UNIT LENGTH

UNIT RISE

LINE LENGTH

RAFTER

LINE LENGTH

TOTAL RISE

TOTAL RUN

TOTAL RUN

TOTAL SPAN

FIGURE 44–4 The terms associated with roof theory.

Unit Run. The unit run is a standardized horizontal distance. It is the number that is used as a base for the roof angle, and it is a horizontal distance under a rafter. This distance is always 12 inches for a common rafter. It is 16.97 inches (or 17 inches) for hip and valley rafters. In metric roof framing, the standard unit run is 250 mm for the common rafter and 353.6 mm (or 354 mm) for the hip and valley rafter.

Total Span. The total span of a roof is the horizontal distance covered by the roof. This is usually the width of the building measured from the outer faces of the framing.

Unit Rise. The unit rise is the number of inches the roof will rise vertically for every unit of run. For example, if the unit rise is 6, then a common rafter will rise 6 inches for every 12 inches it covers horizontally. This number is typically shown on a triangular symbol that is found on elevations and section views. Similarly, in metric, the unit rise is the number of millimetres the roof will rise for every 250 mm of run. For example, if the unit rise is 125, the common rafter will rise 125 mm for every 250 mm it runs horizontally.

Total Run. The total run of a rafter is the total horizontal distance over which the rafter slopes. This is usually one-half the span of the roof.

Total Rise. The total rise is the vertical distance that the roof rises from plate to ridge. Total rise may be calculated by multiplying the unit rise by the number of units of run of the rafter. For example, if the unit rise is 6 and the run is 13 feet, the number of unit runs is 13, and then the total rise is $6 \times 13 = 78$ inches. This measurement is not to the top of the rafter, but rather to some point inside the rafter.

A metric example would be a roof with a run of 4000 mm ($4000 \div 250 = 16$ unit runs). A unit rise of 125 will have a total rise of 125 mm $\times$ 16, or 2000 mm.

Theoretical Line Length. The line length of a rafter is the length of the rafter from the plate to the ridge. It is the hypotenuse (longest side) of the right triangle formed by the total run as the base and the total rise as the vertical leg. Line length gives no consideration to the thickness or width of the framing stock.

Unit Length. Unit length is the length of rafter needed to cover a horizontal distance of one unit of run. It is the hypotenuse of a right triangle formed by the unit run and unit rise.

Pitch. The pitch of a roof is a fraction. It indicates the amount of incline of a roof. The pitch is found by dividing the rise by the span. For example, if the span of the building is 32 feet and total rise is 8 feet, then 8 divided by 32 is $8/32$, which reduces to ¼. The roof is then said to have a ¼ pitch (**Figure 44–5**). A roof

with a span of 8000 mm and a total rise of 2000 mm will have a pitch of 2000 ÷ 8000 or a ¼ pitch.

Slope. Slope is the common term used to express the steepness of a roof. It is stated using the unit rise and the unit run. For example, if the unit rise is 4 inches and the unit run is always 12 inches, then the slope is said to be 4 in 12. This same roof would have a pitch of $4/24$ or ⅙. If the unit rise is 100 mm and the slope is rise over run, the slope is 100/250 or 1:2.5 or 1 in 2.5.

Bird's Mouth. Bird's mouth is the term used to refer to the notch cut in a rafter. This is done to make the rafter sit securely on the wall plate so that it can be adequately fastened (**Figure 44–6**).

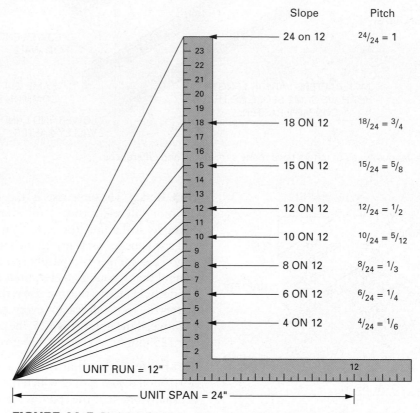

Slope	Pitch
24 on 12	$24/24 = 1$
18 ON 12	$18/24 = 3/4$
15 ON 12	$15/24 = 5/8$
12 ON 12	$12/24 = 1/2$
10 ON 12	$10/24 = 5/12$
8 ON 12	$8/24 = 1/3$
6 ON 12	$6/24 = 1/4$
4 ON 12	$4/24 = 1/6$

UNIT RUN = 12"

UNIT SPAN = 24"

FIGURE 44–5 Pitch is a fraction of rise over span in lowest terms, whereas slope is always stated as unit rise over 12.

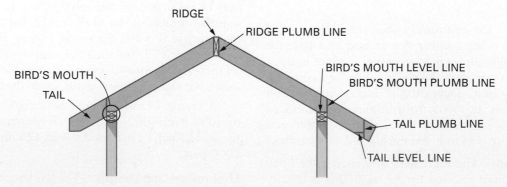

RIDGE
RIDGE PLUMB LINE
BIRD'S MOUTH
TAIL
BIRD'S MOUTH LEVEL LINE
BIRD'S MOUTH PLUMB LINE
TAIL PLUMB LINE
TAIL LEVEL LINE

FIGURE 44–6 A bird's mouth is a notch for the rafter to sit on the wall.

Rafter Stand. Rafter stand or *height above plate* (HAP) refers to the part of the rafter stock that remains after the bird's mouth has been notched out. This distance is measured vertically (plumb).

Plumb Line. A plumb line is any line on the rafter that is vertical when the rafter is in position. There is a plumb line at the ridge, at the wall plate, and usually at the end of the tail. They are marked using a square. A framing square is marked along the tongue. A speed square is marked along the edge of the square where the inch ruler is located.

Level Line. A level line is any line on the rafter that is horizontal when the rafter is in position. It is marked along the blade of the framing square when laying out level cuts on rafters. A speed square is marked along the long edge of the square where the degree scale is located after lining up the alignment guide with the plumb line.

DETERMINING RAFTER LENGTHS

Three methods are available for finding the length of a rafter: the estimated, step-off, and calculated methods. They are used for different reasons and have varying degrees of speed and accuracy.

Estimated Rafter Length

With the estimated method, the carpenter uses two framing squares to scale a distance that the rafter will cover. It serves only as an estimate for ordering the proper length of material (**Figure 44–7**).

The blade of the square represents the total run of the rafter. The tongue represents the total rise. Using a scale of one inch to one foot, locate the run and rise. For example, a run of 10 feet becomes 10 inches and a rise of 5 feet becomes 5 inches. Then measure the shortest distance between the blade and tongue. This distance in inches is easily converted to feet. Add extra if the rafter overhangs the wall of the building.

EXAMPLE If the span of a building is 28 feet and the unit rise is 6 inches, what is the estimated length? The run of the rafter is one-half the span, 28 ÷ 2 = 14 feet. Therefore, the number of unit runs is 14. The total rise of a rafter is 14 × 6 = 84 inches. Divide this by 12 to convert it to feet, 84 ÷ 12 = 7 feet. Locate 7 inches on the tongue of the square and 14 inches on the blade. Measure the distance between these points. This is approximately 15⅝ inches. Converting back to feet, the rough rafter length from plate to ridge is 15⅝ feet. A stock length of 16 feet would have to be used for the rafter, if there is no overhang.

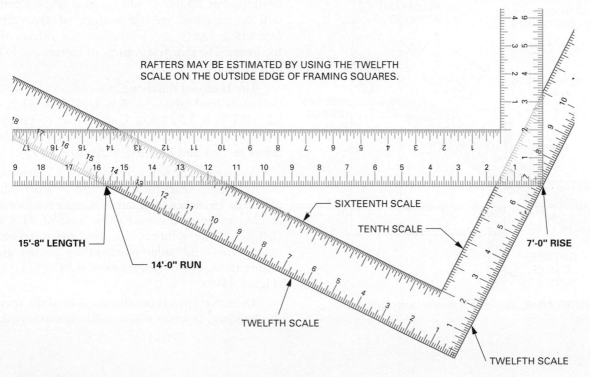

RAFTERS MAY BE ESTIMATED BY USING THE TWELFTH SCALE ON THE OUTSIDE EDGE OF FRAMING SQUARES.

15'-8" LENGTH

14'-0" RUN

TWELFTH SCALE

SIXTEENTH SCALE

TENTH SCALE

7'-0" RISE

TWELFTH SCALE

FIGURE 44–7 Rafter length may be estimated using two framing squares.

A metric example would be, "What is the approximate length of the rafter for a building that is 8000 mm wide with a unit rise of 125 mm?" The total run is 4000 mm (4000 ÷ 250 = 16 unit runs). Thus the total rise is 125 × 16 = 2000 mm. Divide these numbers by 10, and then measure diagonally across from 20 cm to 40 cm to get 44.7 cm, which is then multiplied by 10 to give 447 cm or 4.47 m. You will need rafter stock 5 m long.

Step-off Rafter Length Method

The step-off method uses a framing square to step off the length of a rafter. It can be used for most types of rafters. The step-off method is based on the unit of run (12 inches for the common rafter). The rafter stock is stepped off for each unit of run until the desired number of units or parts of units are stepped off (**Figure 44–8**). If the rafter has a total run of 16 feet, for example, the square is moved 16 times. This method works well in most situations but it can cause errors in the length because small incremental errors that add up occur during each step.

The metric unit run is 250 mm, so 250 is set on the body of the square and the unit rise of 125 mm is

set on the tongue, and the rafter stock is stepped off 16 times for a total run of 4000 mm.

SIZING RAFTER MEMBERS

To begin sizing rafters, information about the roof from the set of prints is consulted. The geographic location of the building is also considered with regard to snow. Charts are then read following a similar process as for joists.

Structural Terms

Structural terms for rafters are the same as for joists. These terms include on-centre spacing, material strength, intended load (i.e., live and dead load PSF [pounds per square foot] or kPa [kilopascals]; 1 kPa = 20.9 PSF), and deflection limits.

The major difference between joists and rafters is the span. Both spans are measured along a horizontal distance. For rafters, this distance is the horizontal projection of the rafter in the plan view, not the length of the rafter. The distance is the total run of the rafter, usually one-half the width of the building.

The intended load for rafters is more broadly considered than for joists. Dead load designs are the same. Dead load of 10 PSF (0.5 kPa) is used where only the rafters are to be installed, and dead load of 20 PSF (1 kPa) is used when a ceiling will be attached to the bottom of the rafters. Deflection limits are less strict for rafters than for joists. The stiffness ratings of rafters are L/180 and L/240.

The National Building Code of Canada (NBCC) uses snow load ranges of 20.9, 31.3, 41.8, 52.2, and 62.7 PSF (1.0, 1.5, 2.0, 2.5, and 3.0 kPa). The *Span Book* from the Canadian Wood Council extends its tables to 73.1 and 83.5 PSF (3.5 and 4.0 kPa).

As an exercise using S-P-F #1 & #2, determine the maximum span for 2 × 6 (38 × 140) rafters placed at 16 inches (406 mm) OC on a roof located in Ottawa given the roof snow load of 41.8 PSF (2.0 kPa). The solution is to find the intersection of the row and column according to the given information. The correct answer is 11′-7″ (3.53 m) (**Figure 44–9**).

The rafter span is usually one-half of the span of the building. In cases where collar ties are used, or

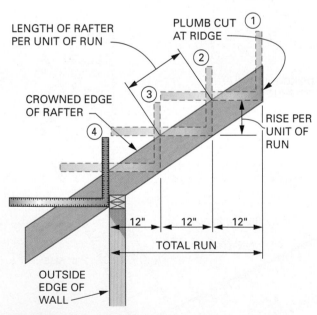

FIGURE 44–8 Step-off method of determining rafter length.

Spruce-Pine-Fir #1 & #2 – Roof Rafter Spans

Rafter Size	Maximum Span (m)								
	Specified Snow Load (kPa)								
	1.0			1.5			2.0		
	Rafter Spacing (mm)			Rafter Spacing (mm)			Rafter Spacing (mm)		
	305	406	610	305	406	610	305	406	610
38 × 89	3.11	2.83	2.47	2.72	2.47	2.16	2.47	2.24	1.96
38 × 140	4.90	4.45	3.89	4.28	3.89	3.40	3.89	3.53	3.08
38 × 184	6.44	5.85	5.11	5.62	5.11	4.41	5.11	4.64	3.89
38 × 235	8.22	7.47	6.38	7.18	6.52	5.39	6.52	5.82	4.75
38 × 286	10.00	9.06	7.40	8.74	7.66	6.25	7.80	6.76	5.52

Spruce-Pine-Fir #1 & #2 – Roof Rafter Spans

Rafter Size	Maximum Span (ft-in)								
	Specified Snow Load (PSF)								
	20.9			31.3			41.8		
	Rafter Spacing (inches)			Rafter Spacing (inches)			Rafter Spacing (inches)		
	12	16	24	12	16	24	12	16	24
2 × 4	10'-3"	9'-3"	8'-1"	8'-11"	8'-1"	7'-1"	8'-1"	7'-4"	6'-5"
2 × 6	16'-1"	14'-7"	12'-9"	14'-0"	12'-9"	11'-2"	12'-9"	11'-7"	10'-1"
2 × 8	21'-1"	19'-2"	16'-9"	18'-5"	16'-9"	14'-6"	16'-9"	15'-3"	12'-9"
2 × 10	27'-0"	24'-6"	20'-11"	23'-7"	21'-5"	17'-8"	21'-5"	19'-1"	15'-7"
2 × 12	32'-10"	29'-9"	24'-3"	28'-8"	25'-2"	20'-6"	25'-7"	22'-2"	18'-1"

Source: *The Span Book 2009* and NBCC Table A-6. Courtesy of Canadian Wood Council.

FIGURE 44–9 Sample roof rafter span tables for S-P-F #1 & #2 from the *Span Book* and NBCC Table A-6.

for struts or dwarf walls, the span can be cut in half (**Figure 44–10**).

Calculated Rafter Length

The calculated rafter length method uses the rafter tables located on a framing square and a calculator. It is the most accurate way of determining rafter length. The procedure can be confusing at first, but once it is understood, any and all rafter lengths can be determined. This includes hip, valley, and jack rafters. The following chapters of this book will focus mainly on the calculated method.

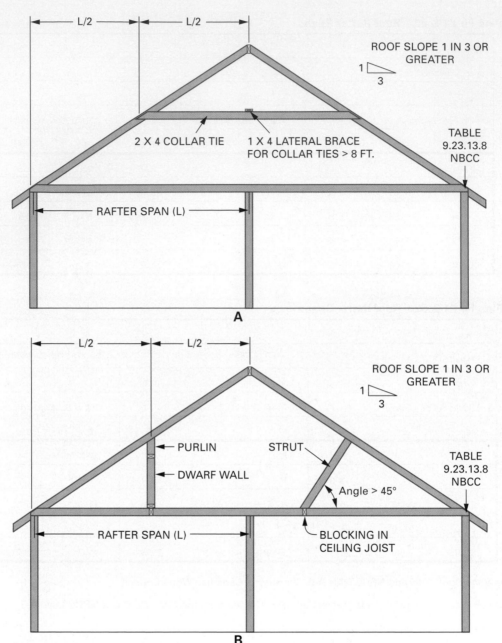

FIGURE 44–10 (A) Collar ties reduce rafter span by one-half. (B) Struts and dwarf walls reduce rafter span by one-half.

Chapter 45

Roof Trusses

Roof trusses are used extensively in residential and commercial construction (**Figure 45–1**). They are designed to support the roof over a wide span, up to 100 feet (30 m), and eliminate the need for load-bearing partitions. They require less wood to manufacture and less time to install, and can reduce overall construction costs. Some styles, however, allow for little or no usable attic space.

TRUSS CONSTRUCTION

A roof truss consists of upper and lower *chords* and diagonal members called *webs*. The upper chords act as rafters, the lower chords serve as ceiling joists, and the webs are braces and stiffeners that replace collar ties. All truss members are fastened and joined securely with metal gusset plates (**Figure 45–2**). Plywood gussets may be used on jobsite-built trusses.

> ### CAUTION
>
> Trusses are designed with smaller member sizes than are rafter-ceiling joist systems. This causes higher stresses in the roof system members. Never cut any webs or chords of a truss unless directed by an engineer. Also, installing trusses can be very dangerous. Lives have been lost while installing trusses improperly. For these reasons, care must be employed by engineers in their designs and carpenters in their installation practices.

FIGURE 45–1 Trusses are used extensively in residential construction.

FIGURE 45-2 The members of roof trusses are securely fastened with metal gussets.

Truss Design

Most trusses are made in fabricating plants. They are transported to the jobsite. Trusses are designed by engineers to support prescribed loads. Trusses may also be built on the job, but approved designs must be used. The carpenter should not attempt to design a truss. Approved designs and instructions for job-built trusses are available from the American Plywood Association, the Truss Plate Institute, and the Canada Mortgage and Housing Corporation.

The most common truss design for residential construction is the *Fink* truss (**Figure 45-3**). Other truss shapes are designed to meet special requirements (**Figure 45-4**). In the colder regions of Canada,

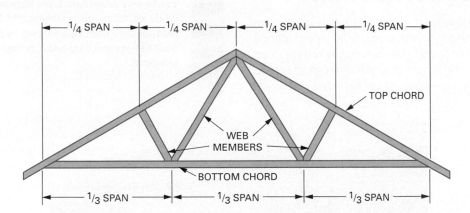

FIGURE 45-3 The Fink truss design is widely used in residential construction.

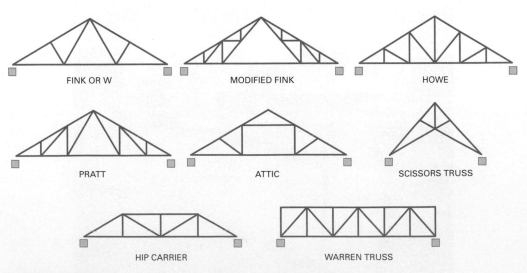

FIGURE 45-4 Various truss designs for special requirements.

ENERGY-HEEL TRUSS
(Ventilation baffles not shown for clarity.)

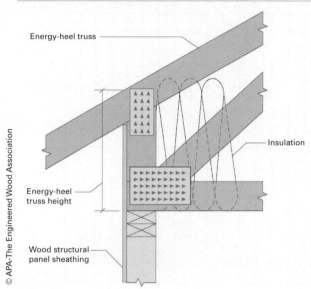

© APA-The Engineered Wood Association

FIGURE 45–5 Raised-heel truss.

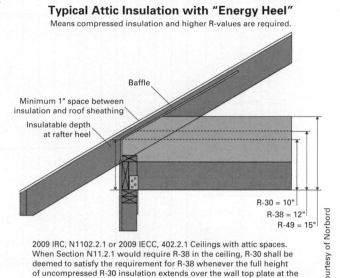

Typical Attic Insulation with "Energy Heel"
Means compressed insulation and higher R-values are required.

Baffle

Minimum 1" space between insulation and roof sheathing

Insulatable depth at rafter heel

R-30 = 10"
R-38 = 12"
R-49 = 15"

Courtesy of Norbord

2009 IRC, N1102.2.1 or 2009 IECC, 402.2.1 Ceilings with attic spaces. When Section N11.2.1 would require R-38 in the ceiling, R-30 shall be deemed to satisfy the requirement for R-38 whenever the full height of uncompressed R-30 insulation extends over the wall top plate at the eaves. Similarly, R-38 shall be deemed to satisfy the requirement for R-49...

raised-heel trusses (**Figure 45–5**) are being used to accommodate the greater levels of insulation required by the National Building Code of Canada (NBCC).

Erecting a Trussed Roof

Carpenters are more involved in the erection than the construction of trusses. Trusses are delivered to the jobsite using specially designed trucks. They should be unloaded and stored on a flat, dry surface. A print is provided showing the location of all trusses. A drawing of each truss is also provided that outlines important installation points (**Figure 45–6**).

The erection and bracing of a trussed roof is a critical stage of construction. Failure to observe recommendations for erection and bracing could cause a collapse of the structure. This could result in loss of life or serious injury, not to mention the loss of time and material.

The recommendations contained herein are technically sound. However, they are not the only method of bracing a roof system. They serve only as a guide. The builder must take necessary precautions during handling and erection to ensure that trusses are not damaged, which might reduce their strength.

Small trusses, which can be handled by hand, are placed upside down, hanging on the wall plates,

toward one end of the building. Trusses for wide spans require the use of a crane to lift them into position. One at a time, each truss is lifted, fastened, and braced in place. The end truss is installed first by swinging it up into place and bracing it securely in a plumb position (**Figure 45–7**).

First Truss Installation

The end truss is usually installed first, but not always. In either case, installation of the first truss requires great care. Trusses must be adequately braced in position, held secure, and plumb. They must be temporarily braced with enough strength to support the tip-over force of the trusses to follow. This may be achieved by bracing to securely anchored stakes driven into the ground or by bracing to the inside floor under the truss (**Figure 45–8**). These braces should be located directly in line with all rows of continuous top chord lateral bracing, which will be installed later. All bracing should be securely fastened with the appropriate size and quantity of nails, remembering that lives depend on the bracing doing its intended job.

Temporary Bracing of Trusses

As trusses are set in place, they are nailed to the plate. Metal framing connectors are usually applied

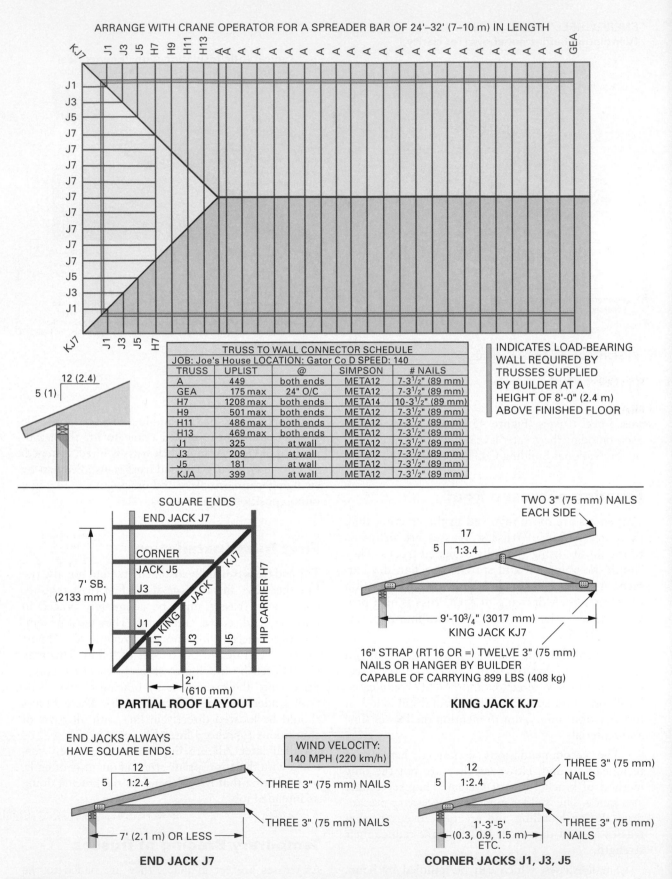

ARRANGE WITH CRANE OPERATOR FOR A SPREADER BAR OF 24'–32' (7–10 m) IN LENGTH

TRUSS TO WALL CONNECTOR SCHEDULE				
JOB: Joe's House LOCATION: Gator Co D SPEED: 140				
TRUSS	UPLIST	@	SIMPSON	# NAILS
A	449	both ends	META12	7-3½" (89 mm)
GEA	175 max	24" O/C	META12	7-3½" (89 mm)
H7	1208 max	both ends	META14	10-3½" (89 mm)
H9	501 max	both ends	META12	7-3½" (89 mm)
H11	486 max	both ends	META12	7-3½" (89 mm)
H13	469 max	both ends	META12	7-3½" (89 mm)
J1	325	at wall	META12	7-3½" (89 mm)
J3	209	at wall	META12	7-3½" (89 mm)
J5	181	at wall	META12	7-3½" (89 mm)
KJA	399	at wall	META12	7-3½" (89 mm)

INDICATES LOAD-BEARING WALL REQUIRED BY TRUSSES SUPPLIED BY BUILDER AT A HEIGHT OF 8'-0" (2.4 m) ABOVE FINISHED FLOOR

12 (2.4)
5 (1)

SQUARE ENDS
END JACK J7
CORNER JACK J5
KJ7
JACK
J3
KING JACK
HIP CARRIER H7
J1
7' SB.
(2133 mm)
J1
J3
J5
2'
(610 mm)

PARTIAL ROOF LAYOUT

TWO 3" (75 mm) NAILS EACH SIDE
17
5 | 1:3.4
9'-10¾" (3017 mm)
KING JACK KJ7
16" STRAP (RT16 OR =) TWELVE 3" (75 mm) NAILS OR HANGER BY BUILDER CAPABLE OF CARRYING 899 LBS (408 kg)

KING JACK KJ7

END JACKS ALWAYS HAVE SQUARE ENDS.
12
5 | 1:2.4
THREE 3" (75 mm) NAILS
THREE 3" (75 mm) NAILS
7' (2.1 m) OR LESS

WIND VELOCITY:
140 MPH (220 km/h)

END JACK J7

12
5 | 1:2.4
THREE 3" (75 mm) NAILS
THREE 3" (75 mm) NAILS
1'-3'-5'
(0.3, 0.9, 1.5 m)
ETC.

CORNER JACKS J1, J3, J5

FIGURE 45–6 Trusses are often delivered with an engineered set of prints showing truss labels and locations.

FIGURE 45–7 Trusses are often lifted into place with a crane for speed and safety.

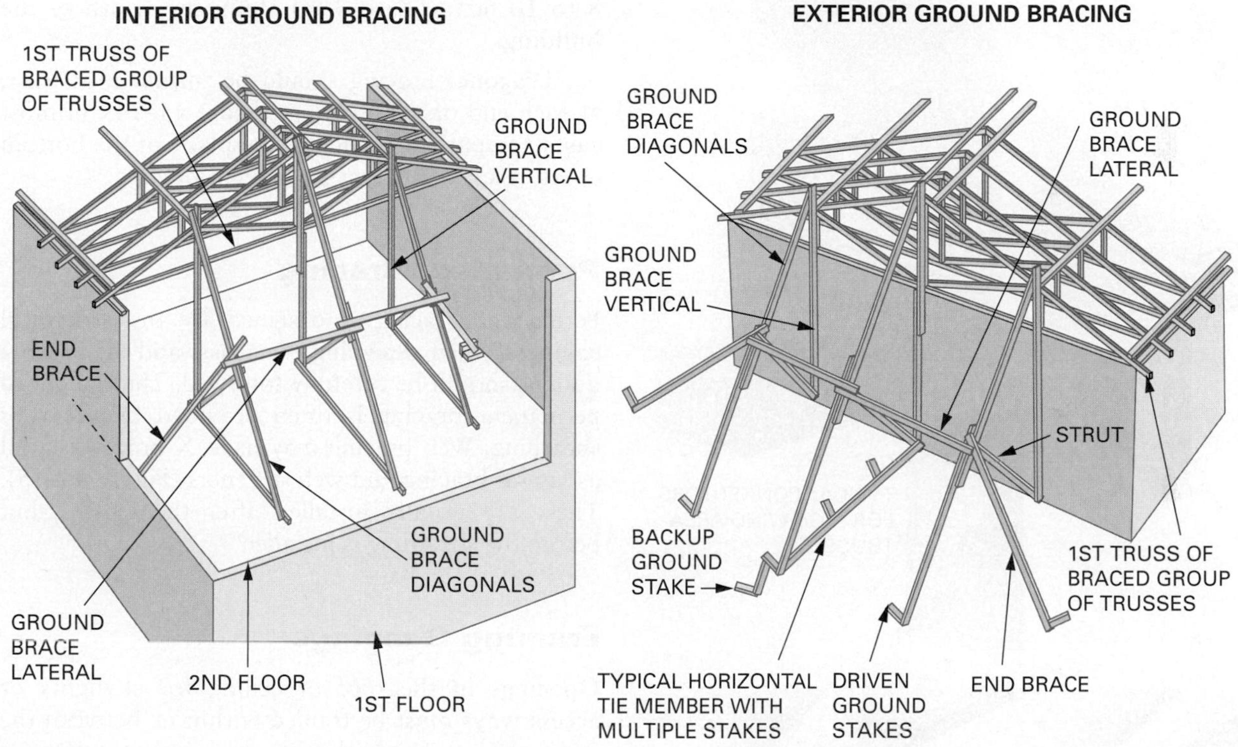

INTERIOR GROUND BRACING

1ST TRUSS OF
BRACED GROUP
OF TRUSSES

GROUND
BRACE
VERTICAL

END
BRACE

GROUND
BRACE
DIAGONALS

GROUND
BRACE
LATERAL

2ND FLOOR

1ST FLOOR

EXTERIOR GROUND BRACING

GROUND
BRACE
DIAGONALS

GROUND
BRACE
VERTICAL

GROUND
BRACE
LATERAL

STRUT

1ST TRUSS OF
BRACED GROUP
OF TRUSSES

BACKUP
GROUND
STAKE

TYPICAL HORIZONTAL
TIE MEMBER WITH
MULTIPLE STAKES

DRIVEN
GROUND
STAKES

END BRACE

FIGURE 45–8 The first set of trusses must be well-braced before the erection of other trusses.

(**Figure 45–9**). Information on these connectors may be found on the connector schedule of the set of prints. Sufficient temporary bracing must be applied to secure trusses until the finish material is applied and/or until permanent bracing is installed. Temporary bracing should be no less than 2 × 4 (38 × 89 mm) lumber as long as practical, with a minimum length of 8 feet (2.4 m). The 2 × 4s (38 × 89s) should be fastened with two 3½-inch (89 mm) nails at every intersection and should be overlapped by two trusses (**Figure 45–10**). Exact spacing of the

trusses should be maintained. Adjusting trusses later, while possible, is time-consuming and risky.

Temporary bracing must be applied to three planes of the truss assembly: the top chord or sheathing plane, the bottom chord or ceiling plane, and the vertical web plane at right angles to the bottom chord (**Figure 45–11**).

Top Chord Bracing. Continuous lateral bracing should be installed within 6 inches (152 mm) of the ridge and at about 8- to 10-foot (2.4 to 3 m)

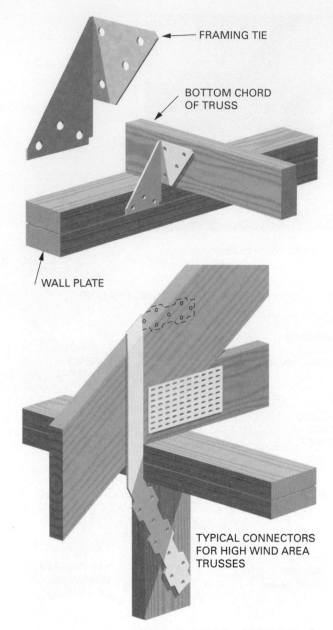

FIGURE 45–9 Metal framing ties are used to fasten trusses to the wall plate.

intervals between the ridge and wall plate. Diagonal bracing should be set at approximately 45-degree angles between the rows of lateral bracing. It forms triangles that provide stability to the plane of the top chord (**Figure 45–12**).

Web Plane Bracing. Temporary bracing in the plane of the web members is made up of diagonals placed at right angles to the trusses from top to bottom chords (**Figure 45–13**). They usually become permanent braces of the web member plane.

Bottom Chord Bracing. To maintain the proper spacing on the bottom chord, continuous lateral bracing for the full length of the building must be applied. The bracing should be nailed to the top of the bottom chord at intervals no greater than 8 to 10 feet (2.4 to 3 m) along the width of the building.

Diagonal bracing should be installed, at least, at each end of the building (**Figure 45–14**). In most cases, temporary bracing of the plane of the bottom chord is left in place as permanent bracing.

Permanent Bracing

Permanent bracing is designed by the structural engineer when designing the truss, and all bracing designs should be carefully followed. The top chord permanent bracing is often provided by the roof sheathing. Web bracing may have X braces as well as lateral bracing and web stiffeners (**Figure 45–15**). These are usually installed after the trusses, but before the sheathing is installed.

Framing Openings

Openings in the roof or ceiling for skylights or access ways must be framed within or between the trusses. The chords, braces, and webs of a truss system should never be cut or removed unless directed to do so by an engineer. Simply installing headers between trusses will create an opening. The sheathing or ceiling finish is then applied around the opening.

ROOF SHEATHING

Roof sheathing is applied after the roof frame is complete. Sheathing gives rigidity to the roof frame. It also provides a nailing base for the roof covering. Rated panels of plywood and strand board are commonly used to sheath roofs.

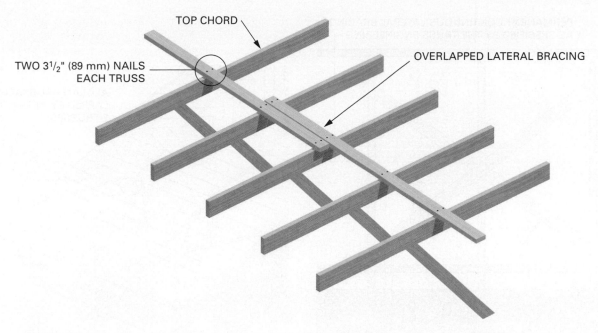

FIGURE 45–10 All bracing must be nailed with 3½-inch (89 mm) common nails.

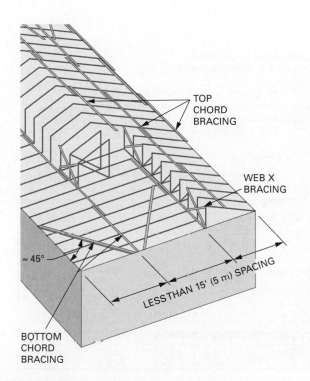

FIGURE 45–11 Temporary bracing secures three planes of trusses.

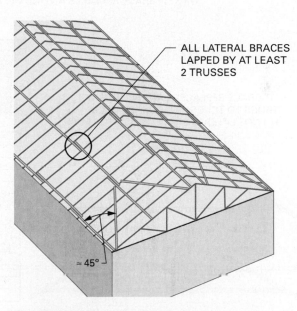

FIGURE 45–12 Typical temporary bracing of the top chord plane.

Panel Sheathing

Plywood and other rated panel roof sheathing is laid with the face grain running perpendicular to the rafter or top chord for greater strength (**Figure 45–16**). End joints are made on the framing member and staggered, similar to subfloor. Nails are 2 inches (50 mm) long and spaced 6 inches (152 mm) apart on the ends and 12 inches (305 mm) apart on intermediate supports (NBCC Table 9.23.3.5).

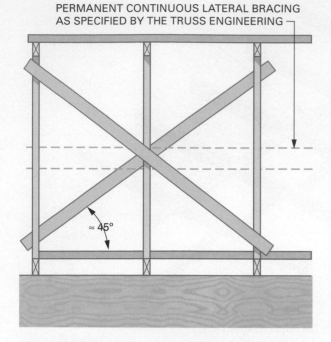

PERMANENT CONTINUOUS LATERAL BRACING
AS SPECIFIED BY THE TRUSS ENGINEERING

≈ 45°

FIGURE 45–13 Bracing of the web member plane prevents lateral movement of the trusses.

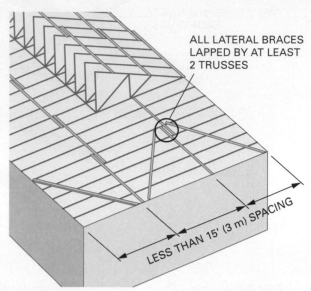

ALL LATERAL BRACES
LAPPED BY AT LEAST
2 TRUSSES

LESS THAN 15' (3 m) SPACING

FIGURE 45–14 Typical bracing of the top chord plane.

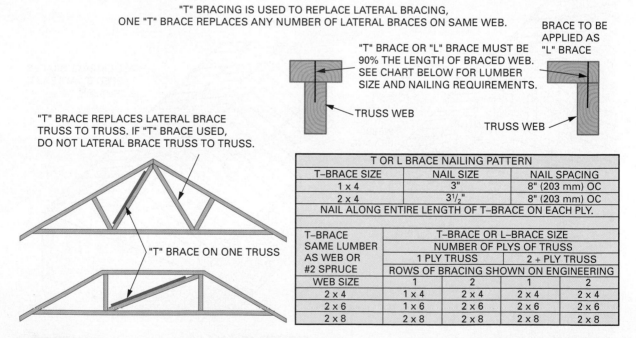

"T" BRACING IS USED TO REPLACE LATERAL BRACING,
ONE "T" BRACE REPLACES ANY NUMBER OF LATERAL BRACES ON SAME WEB.

"T" BRACE OR "L" BRACE MUST BE
90% THE LENGTH OF BRACED WEB.
SEE CHART BELOW FOR LUMBER
SIZE AND NAILING REQUIREMENTS.

BRACE TO BE
APPLIED AS
"L" BRACE

TRUSS WEB

TRUSS WEB

"T" BRACE REPLACES LATERAL BRACE
TRUSS TO TRUSS. IF "T" BRACE USED,
DO NOT LATERAL BRACE TRUSS TO TRUSS.

"T" BRACE ON ONE TRUSS

T OR L BRACE NAILING PATTERN		
T-BRACE SIZE	NAIL SIZE	NAIL SPACING
1 x 4	3"	8" (203 mm) OC
2 x 4	3½"	8" (203 mm) OC
NAIL ALONG ENTIRE LENGTH OF T–BRACE ON EACH PLY.		

T–BRACE SAME LUMBER AS WEB OR #2 SPRUCE	T-BRACE OR L-BRACE SIZE			
	NUMBER OF PLYS OF TRUSS			
	1 PLY TRUSS		2 + PLY TRUSS	
	ROWS OF BRACING SHOWN ON ENGINEERING			
WEB SIZE	1	2	1	2
2 x 4	1 x 4	2 x 4	2 x 4	2 x 4
2 x 6	1 x 6	2 x 6	2 x 6	2 x 6
2 x 8	2 x 8	2 x 8	2 x 8	2 x 8

FIGURE 45–15 Some webs require stiffeners to be added by the carpenter.

Additional nails are added according to local codes for increased strength. Some areas require extra nails in the sheathing to protect from uplift caused by high winds. These nailing zones put more nails along the perimeter of the roof and at the corners of the roof. Other areas require more nails on special trusses that are designed to improve shear resistance (**Figure 45–17**).

FIGURE 45–16 Sheathing a trussed roof with plywood.

RED LINES INDICATE AREAS OF INCREASED NAILING.

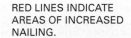

FIGURE 45–17 Sheathing in seismic-prone areas has extra nails.

Adequate blocking, tongue-and-groove edges, or other suitable edge support such as *panel clips* or *H-clips* must be used when spans exceed the indicated value of the plywood roof sheathing. Panel clips are small metal pieces shaped like a capital H. They are used between adjoining edges of the plywood between rafters (**Figure 45–18**). One clip is typically used for 24-inch (610 mm) spans and two panel clips are used for 32- and 48-inch (813 and 1220 mm) spans. If OSB (oriented strand board) is used as sheathing, two clips are required for 24-inch (610 mm) spans.

The ends of sheathing may be allowed to slightly and randomly overhang at the *rakes* until sheathing is completed. A chalk line is then snapped between the ridge and plate at each end of the roof. The sheathing ends are trimmed straight to the line with a circular saw.

CAUTION

Sawdust from a circular saw on roof sheathing can be very slippery. Make sure of your footing as you cut.

Plank Sheathing

Plank decking is used in post-and-beam construction where the roof supports are spaced farther apart. Plank roof sheathing may be of 2-inch (38 mm) nominal thickness or greater, depending on the span. Usually both edges and ends are matched. The plank roof often serves as the finish ceiling for the rooms below.

ESTIMATING

Common Rafters for Gable Roof. Divide the length of the building by the spacing of the rafters. Add one, as a starter, and then multiply the total by two.

EXAMPLE A building is 42 feet long. The rafter spacing is 16 inches on centre (OC). Divide 42 by $1\frac{1}{3}$ (16 inches divided by 12 equals $1\frac{1}{3}$ feet) to get $31\frac{1}{2}$ spaces. Change $31\frac{1}{3}$ to the next whole number to get 32. Add 1 to make 33. Multiplying by 2 equals 66 rafters. Or if the building is 13 000 mm long and the rafter spacing is 400 mm OC, divide 13 000 by 400, which gives 32.5 or 33 times 2 to get 66 rafters.

Common Rafters with Hip or Valley Rafters. Use the same procedure as for a gable roof, but add two rafters for each hip and valley rafter. This will allow material for the common and jack rafters.

Hip and Valley. Count the number of hips and valleys.

Ridgeboard for Gable Roof. Take the length of the building plus the rake overhang. Divide this sum by the length of the material to be used for the ridge. Round up to the next whole number.

Ridgeboard for Hip Roof. Subtract the width of the building from the length of the building.

APA PANEL ROOF SHEATHING

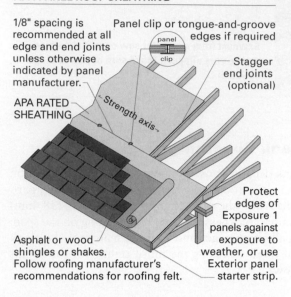

1/8" spacing is recommended at all edge and end joints unless otherwise indicated by panel manufacturer.

Panel clip or tongue-and-groove edges if required

Stagger end joints (optional)

APA RATED SHEATHING

Strength axis

Asphalt or wood shingles or shakes. Follow roofing manufacturer's recommendations for roofing felt.

Protect edges of Exposure 1 panels against exposure to weather, or use Exterior panel starter strip.

Notes:
1. Cover sheathing as soon as possible with roofing felt for extra protection against excessive moisture prior to roofing application.

2. For pitched roofs, place screened surface or side with skid-resistant coating up if OSB panels are used. Keep roof surface free of dirt, sawdust and debris, and wear skid-resistant shoes when installing roof sheathing.

3. For buildings with conventionally framed roofs (trusses or rafters), limit the length of continuous sections of roof area to 80 feet maximum during construction, to allow for accumulated expansion in wet weather conditions. Omit roof sheathing panels in each course of sheathing between sections, and install "fill in" panels later to complete roof deck installation prior to applying roofing.

RECOMMENDED MINIMUM FASTENING SCHEDULE FOR APA PANEL ROOF SHEATHING (Increased nail schedules may be required in high wind zones and where roof is engineered as a diaphragm.)

Panel Performance Category[c]	Size[d]	Nailing[a][b] Maximum Spacing (in.)	
		Supported Panel Edges[e]	Intermediate
3/8 – 1	8d	6	12[f]
1-1/8	8d or 10d	6	12[f]

(a) Use common smooth or deformed shank nails for panels with Performance Category 1 or smaller. For 1-1/8 Performance Category panels, use 8d ring- or screw-shank or 10d common smooth-shank nails.

(b) Other code-approved fasteners may be used.

(c) For stapling asphalt shingles to Performance Category 3/8 and thicker panels, use staples with a 15/16-inch minimum crown width and a 1-inch leg length. Space according to shingle manufacturer's recommendations.

(d) See Table 5, page 14, for nail dimensions.

(e) Supported panel joints shall occur approximately along the centerline of framing with a minimum bearing of 1/2". Fasteners shall be located 3/8 inch from panel edges.

(f) For spans 48 inches or greater, space nails 6 inches at all supports.

Courtesy of APA-The Engineered Wood Association

FIGURE 45–18 Recommendations for the application of APA panel roof sheathing.

The actual length could be as much as $3^5/_8$ inches (95 mm) longer, so add enough to compensate.

Gable End Studs. Divide the width of the building by the OC spacing. Add two extra studs to allow for two gable ends.

Trusses. Divide the building length by the OC spacing, usually 2 feet (610 mm). Subtract one. This is the number of common trusses. Two additional gable end trusses will be needed for ends. *Note:* The number of hip trusses is determined by the manufacturer.

Bracing Material. Divide the building width by four to get the number of rows of top, bottom, and web bracing. Round this number up to the nearest whole number. Divide by the bracing length, typically 16 feet (5 m). Round to the nearest whole

number. Add extra for ground bracing of the first truss.

Sheathing Gable Roof. Round up the rafter length to the nearest 2 feet (610 mm). Round up the ridgeboard length to the nearest even number of feet (i.e., 14.7 rounds up to 16). Multiply these numbers together and then double the result to account for the other side. Divide the total by 32 (the number of square feet in a sheathing panel). Round up to the nearest whole number. For metric measure, multiply the rafter length (in metres) times the length of the ridge (in metres) times 2 and then divide by 2.88 (m²) and add 5 percent to get the total number of sheets required.

Sheathing Hip Roof. Calculate the number of panels as if this roof were a gable roof. Then add another 5 percent for waste. There is some waste when cutting triangles.

Gable End Sheathing. Calculate the total rise of the roof by multiplying the unit rise times the rafter run (one-half the building width), then divide by 12 to change the answer to feet. Multiply this number by the building width and divide by 32 (panel square feet). In metric, multiply the total rise of the roof times the width of the building and divide by 2.88. Add 10 percent for waste. This will provide enough material for two gable ends.

Chapter 46

Gable and Gambrel Roofs

The equal-pitched gable roof is the most commonly used roof style. Gable roofs have an equal slope on both sides, intersecting the ridge in the centre of the span (**Figure 46–1**). This roof is the simplest to frame. Only one type of rafter, the common rafter, needs to be laid out.

Like the gable roof, the gambrel roof is symmetrical (see Figure 46–1). Each side of the building has rafters with two different slopes. They have different unit rises. Their runs may or may not be the same depending on the desired roof style.

The saltbox is a gable roof with rafters having different slopes on each side of the ridge (**Figure 46–2**). One rafter slopes faster to reach to the ridge from a lower floor.

COMPONENTS OF A COMMON RAFTER

The common rafter requires several cuts, which have to be laid out. The cuts and lines of a rafter have several names (**Figure 46–3**). The cut at the top is called the *plumb cut* or *ridge cut*. It fits against the ridgeboard. The notch cut at the plate is called the *bird's mouth*, *seat cut*, or *heel*. It fits against the top and outside edge of the wall plate. It consists of a plumb line and a level line layout. These lines are named according to the notch name used. For example, if the notch is referred to as a seat cut, then it is laid out with a seat plumb line and a seat level line.

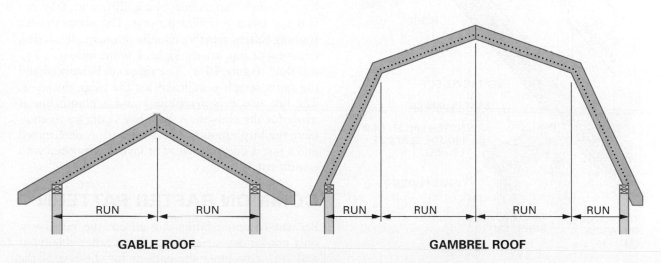

GABLE ROOF   **GAMBREL ROOF**

FIGURE 46–1 The gable and gambrel roofs have opposing rafters that are symmetrical.

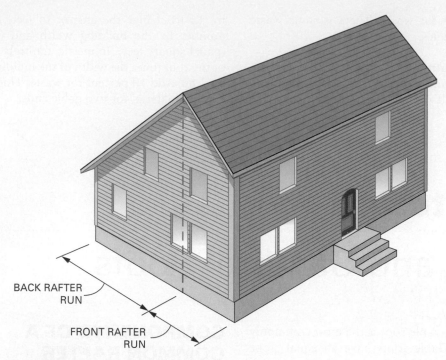

FIGURE 46–2 The ridge of the saltbox roof is off centre.

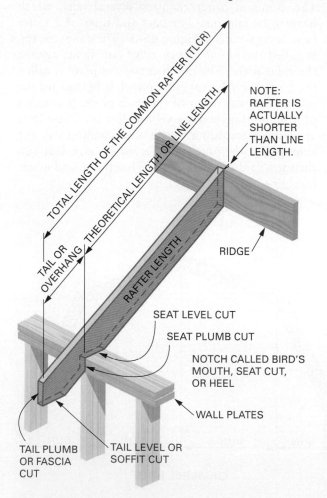

FIGURE 46–3 Names of the cuts and lines of a common rafter.

The distance between the ridge cut and the seat cut is referred to as the *rafter length*. It is also called the *line length of the common rafter (LLCR)* or theoretical line length of the rafter. This distance must be determined by the carpenter.

At the bottom end of the rafter is the tail or overhang, which extends beyond the building. It supports the **fascia** and **soffit**. The plumb line is called the *tail plumb line* or *fascia cut*. The *tail level line* is also referred to as the *soffit cut*.

COMMON RAFTER LAYOUT

Even though rafters may look different, they are laid out using a similar process. The principles for framing rafters, whether equally or unequally sloped, common or hip, are the same. Layout follows a logical flow (**Figure 46–4**). The ridge cut is marked and the rafter length is adjusted for the ridge thickness. The line length is determined and a plumb line is made for the seat cut. A level line is drawn to complete the bird's mouth. The tail length is determined and a fascia line is drawn. The layout is finished with a soffit cut.

COMMON RAFTER PATTERN

Because common rafters are all cut the same way, only one layout is required. Once a rafter is laid out and cut, it becomes the pattern for the rest of the rafters. It should be tested and verified to fit before cutting additional rafters.

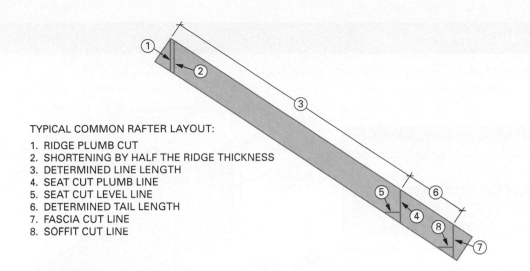

TYPICAL COMMON RAFTER LAYOUT:

1. RIDGE PLUMB CUT
2. SHORTENING BY HALF THE RIDGE THICKNESS
3. DETERMINED LINE LENGTH
4. SEAT CUT PLUMB LINE
5. SEAT CUT LEVEL LINE
6. DETERMINED TAIL LENGTH
7. FASCIA CUT LINE
8. SOFFIT CUT LINE

FIGURE 46–4 Rafters are laid out in a step-by-step sequence.

Select the straightest piece of stock for the pattern. Lay it across two sawhorses. Sight the stock along the edge for straightness and mark the crowned edge. This edge will become the top side, where the roof sheathing is attached. Further discussion of rafters will assume the carpenter is standing near the rafter's lower edge, opposite of the crowned edge.

Laying Out the Ridge Cut

Place the square down on the side of the stock at its left end. Hold the tongue of the square with the left hand and the blade with the right hand. Move the square until the outside edge of the tongue and the edge of the stock line up with the specified rise in inches (or mm). Make sure the blade of the square and the edge of the stock line up with the unit run or 12 inches (or 250 mm). Slide the square to the left until it reaches the end of the rafter stock. Recheck the alignment and mark along the outside edge of the tongue for the plumb cut at the ridge (**Figure 46–5**).

When using a speed square, place the pivot point of the square on the top edge of the rafter. Rotate the square with the pivot point touching the rafter. Looking in the rafter scale window, align the edge of the rafter with the number that corresponds to the rise per unit of run desired. Note that there are two scales, one for a common rafter and one for hips and valleys. Mark the plumb line along the edge of the square that has the inch ruler marked on it.

Rafter Shortening Due to Ridge

The length of the rafter must be shortened when a ridgeboard is inserted between abutting rafters (**Procedure 46–A**). The total shortening is equal to the thickness of the ridgeboard, thus each rafter will be shortened one-half the ridgeboard's thickness.

To shorten the rafter, measure at a right angle from the ridge plumb line a distance equal to one-half the thickness of the ridgeboard. Lay out another plumb line at this point. This will be the cut line. Note

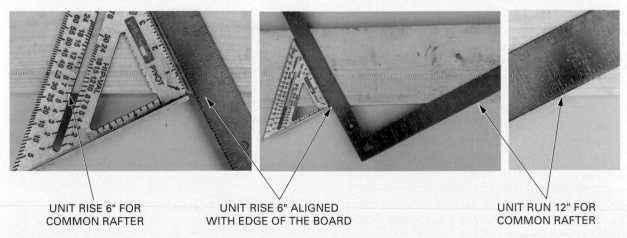

UNIT RISE 6" FOR
COMMON RAFTER

UNIT RISE 6" ALIGNED
WITH EDGE OF THE BOARD

UNIT RUN 12" FOR
COMMON RAFTER

FIGURE 46–5 Laying out the plumb cut of the common rafter at the ridge.

StepbyStep Procedures

Procedure 46–A Shortening a Rafter That Abuts a Ridgeboard

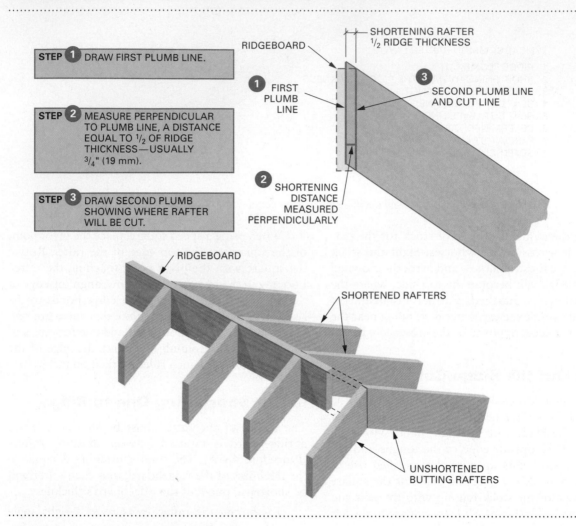

STEP 1 DRAW FIRST PLUMB LINE.

STEP 2 MEASURE PERPENDICULAR TO PLUMB LINE, A DISTANCE EQUAL TO ¹/₂ OF RIDGE THICKNESS—USUALLY ³/₄" (19 mm).

STEP 3 DRAW SECOND PLUMB SHOWING WHERE RAFTER WILL BE CUT.

SHORTENING RAFTER ¹/₂ RIDGE THICKNESS

RIDGEBOARD

1 FIRST PLUMB LINE

3 SECOND PLUMB LINE AND CUT LINE

2 SHORTENING DISTANCE MEASURED PERPENDICULARLY

RIDGEBOARD

SHORTENED RAFTERS

UNSHORTENED BUTTING RAFTERS

EXAMPLE Assume the slope of the roof is 6 on 12. Hold the square so that the 6-inch mark on the tongue and the 12-inch mark on the blade line up with the top edge of the rafter stock. If the slope is 1:2, set stair gauges at 125 mm on the tongue and 250 mm on the body. Mark along the outside edge of the tongue of the square. Make a second plumb line for practice, using a speed square. Place the speed square on the rafter stock and rotate the square, keeping the pivot point touching the edge of the rafter. Align the edge of the rafter with the number 6 on the common scale in the rafter scale window. Mark the plumb line along the edge of the square that has the ruler marked on it. Both layouts should look like that shown in Figure 46–4.

that shortening is always measured at right angles to the ridge cut, regardless of the slope of the roof.

COMMON RAFTER LENGTH

The length of a common rafter can be determined in two ways, via the step-off or the calculation method. The step-off method requires close and fine measuring as each increment is made. The calculation method uses a calculator and the rafter tables found on a rafter square. Both methods can be used to determine the length of any kind of a rafter, but the calculation method is faster and more accurate.

Step-Off Method

The step-off method is based on the unit of run (12 inches or 250 mm for the common rafter). The rafter stock is stepped off for each unit of run until the desired number of units or parts of units have been stepped off.

First, lay out the ridge cut. Keeping the square in place, mark where the blade intersects with the top edge of the rafter. Hold the square at the same angle and slide it to the right until the tongue lines up with this mark. Move the square in a similar manner until the total run of the rafter is laid out (**Procedure 46–B**). Mark a plumb line along the tongue of the square at the last step. This line will be parallel to the ridge cut.

Step**by**Step Procedures

Procedure 46–B Using the Step-Off Method for Rafter Layout

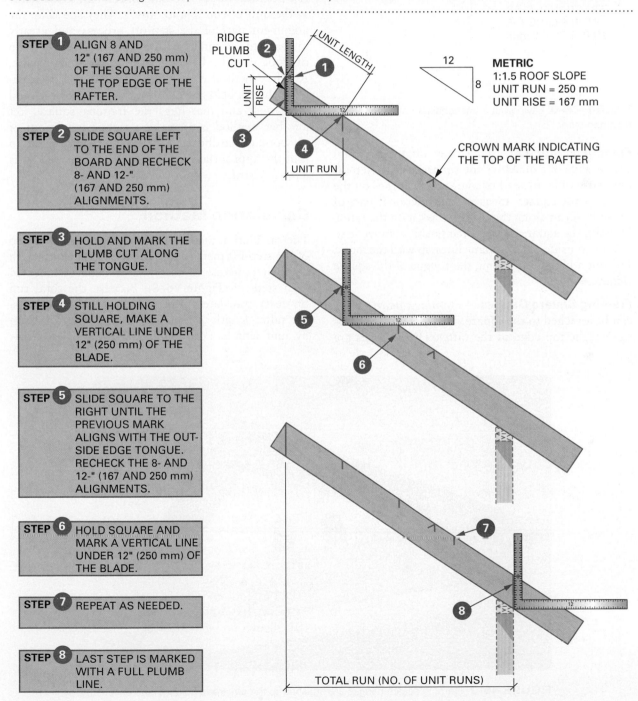

| STEP 1 | ALIGN 8 AND 12" (167 AND 250 mm) OF THE SQUARE ON THE TOP EDGE OF THE RAFTER. |

| STEP 2 | SLIDE SQUARE LEFT TO THE END OF THE BOARD AND RECHECK 8- AND 12-" (167 AND 250 mm) ALIGNMENTS. |

| STEP 3 | HOLD AND MARK THE PLUMB CUT ALONG THE TONGUE. |

| STEP 4 | STILL HOLDING SQUARE, MAKE A VERTICAL LINE UNDER 12" (250 mm) OF THE BLADE. |

| STEP 5 | SLIDE SQUARE TO THE RIGHT UNTIL THE PREVIOUS MARK ALIGNS WITH THE OUT-SIDE EDGE TONGUE. RECHECK THE 8- AND 12-" (167 AND 250 mm) ALIGNMENTS. |

| STEP 6 | HOLD SQUARE AND MARK A VERTICAL LINE UNDER 12" (250 mm) OF THE BLADE. |

| STEP 7 | REPEAT AS NEEDED. |

| STEP 8 | LAST STEP IS MARKED WITH A FULL PLUMB LINE. |

RIDGE PLUMB CUT

UNIT LENGTH

UNIT RISE

UNIT RUN

12

8

METRIC
1:1.5 ROOF SLOPE
UNIT RUN = 250 mm
UNIT RISE = 167 mm

CROWN MARK INDICATING THE TOP OF THE RAFTER

TOTAL RUN (NO. OF UNIT RUNS)

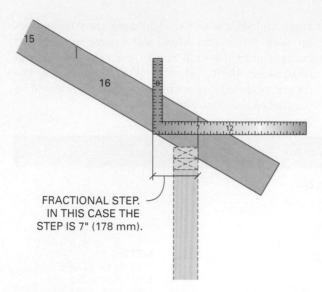

FIGURE 46–6 Laying out a fractional part of a run for a common rafter.

Fractional Step-Off. First, step off for the total whole units of run. Move the square as if to step off one more time. Instead of marking 12 inches on the blade of the square, measure the fractional part of the unit of run along the blade. Mark it on the rafter. Holding the square in the same position, move it so that the outside of the tongue lines up with the mark. Lay out a plumb line along the tongue of the square (**Figure 46–6**).

Framing Square Gauges. *Framing square* gauges can be attached to the square. They are used as stops against the top edge of the rafter. These gauges are

EXAMPLE If the rafter has a total run of 16′-7″, step off 16 times. Hold the square for the 17th step. Mark along the blade at the 7-inch mark. With the square held at the same angle, move it to the mark. Lay out the plumb cut of the bird's mouth by marking along the tongue.

attached to the tongue of the square for the desired rise per foot (or per 250 mm) of run and to the blade at the unit of run (**Figure 46–7**). They speed up the alignment part of the step-off process and greatly increase the accuracy of the overall layout. A strip of wood can also be clamped to the framing square. Another method is to make a double rail of plexiglass that is held together by three short bolts and wing nuts and that hugs the framing square. Yet another method is to make a right triangle out of plywood or hardboard and fasten a strip of wood to it at the appropriate markings. This is referred to as a *pitch board*.

Calculation Method

The calculation method uses the idea that each step of the step-off method has a unit length. Unit length is the hypotenuse of the right triangle included in each step-off (**Figure 46–8**). Because the total run (in feet) translates into the number of step-offs, the rafter length is found by multiplying total run by unit length. For example, if the unit length is

Photo by M. Nauth

FIGURE 46–7 Framing square gauges are attached to the square to hold it in the exact position for repetitive layout marks.

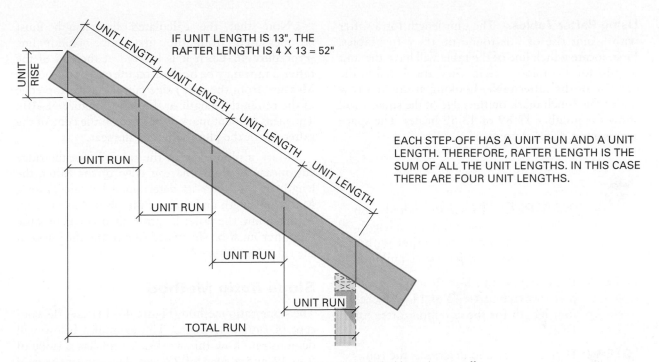

IF UNIT LENGTH IS 13", THE
RAFTER LENGTH IS 4 X 13 = 52"

UNIT RISE

UNIT LENGTH

UNIT LENGTH

UNIT LENGTH

UNIT LENGTH

UNIT RUN

UNIT RUN

UNIT RUN

UNIT RUN

TOTAL RUN

EACH STEP-OFF HAS A UNIT RUN AND A UNIT
LENGTH. THEREFORE, RAFTER LENGTH IS THE
SUM OF ALL THE UNIT LENGTHS. IN THIS CASE
THERE ARE FOUR UNIT LENGTHS.

FIGURE 46–8 Unit length is the hypotenuse of the right triangle included in each step-off.

13 inches and the number of step-offs is 4, then the rafter length is $13 \times 4 = 52$ inches.

It may seem strange to multiply inches times feet. Normally in arithmetic, this will give a result that has no meaning. But in this case, the run is 4, a number of units, not 4 feet. Total run must always look like feet, but it is always simply the number of unit runs (step-offs) under the rafter.

Rafter tables provide rafter information in six rows for roof slopes of 2 through 18 inches of unit rise. The table is stamped on the blade of a rafter (framing) square or sometimes printed in small booklets that come with the square (**Figure 46–9**). The inch-marks 2 through 18 represent the unit rises available with the information directly below each number.

The first row of the rafter table is the common rafter length per unit run or, simply stated, *unit length or unit common rafter (UCR)*. This number comes from using the Pythagorean theorem, $a^2 + b^2 = c^2$ on the unit triangle. The unit rise and unit run are the *a* and *b* with the unit length being *c*. All unit lengths are calculated this way. Knowing this, the unit lengths of unusual unit rises, such as $6\tfrac{3}{4}$ inches, can be calculated.

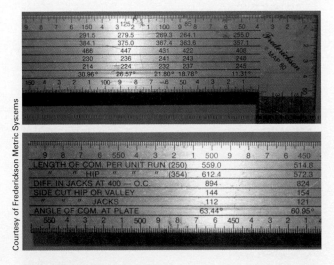

FIGURE 46–9 Rafter tables are found on the framing square.

EXAMPLE Suppose the unit rise of a rafter is 7 inches. What is the unit length? Unit run is always 12 inches for a common rafter.

$a^2 + b^2 = c^2$

$7^2 + 12^2 = c^2$

$49 + 144 = 193 = c^2$

$c = \sqrt{193} = 13.892443$ inches

Unit length 13.89 inches rounded to the nearest hundredths.

Similarly in metric, a unit rise of 100 mm and a unit run of 250 mm will give a unit rafter of 269.3 mm. This number is found on the face of the body of the Frederickson metric framing square, under 100 mm, line one.

Using Rafter Tables.   The unit length for a rafter with a unit rise of 7 is found on the rafter tables. First, locate which line of the table will have the unit lengths for a common rafter. They are placed in the top line of the rafter table. Looking at the first row under the 7-inch mark on the edge of the square will show the number 13 89 or 13.89 inches. The space represents a decimal point.

EXAMPLE   To use the calculation method to determine common rafter line length, find the line length of a common rafter with a unit rise of 8 inches for a 28-foot-wide building.

STEP 1: Read below the 8-inch mark in the first row to find 14 42 or 14.42 inches unit length for the common rafter (or UCR).

STEP 2: Divide 28′ by 2 to determine the run.

28′ ÷ 2 = 14′ or 14 units of run.

STEP 3: Multiply unit length (or UCR) times number of units of run.

14.42″ × 14 = 201.88″

That is 201 inches plus a fraction or 16′-9″ + a fraction of an inch.

STEP 4: Convert decimal to sixteenths.

0.88 × 16 = 14.08

STEP 5: Round off to nearest whole sixteenths.

14.08 ⇒ $^{14}/_{16}$, which reduces to $^7/_8$ inches.

STEP 6: Add to determine rafter length.

201 + $^7/_8$″ = 201$^7/_8$″ (16′-9$^7/_8$″) ⇒

line length of the common rafter (LLCR).

For a metric example, find the length of the common rafter with a unit rise of 150 mm for a 9000-mm wide building.

STEP 1: Read on the face of the framing square below 150, line 1, to find 291.5 mm for the unit length of the common rafter (UCR).

STEP 2: Divide 9000 mm by 2 to get a total run of 4500 mm. Divide 4500 mm by 250 mm to get 18 unit runs.

STEP 3: Multiply the unit length (UCR) by the number of unit runs.

291.5 × 18 = 5247 mm ⇒

line length of the common rafter (LLCR).

Note that the calculated line length must be measured from the first ridge plumb line (**Procedure 46–C**). If it is located at the end of the rafter, a tape may be hooked on the end of the board. Measure from the first ridge cut along the top edge of the rafter, the length of the rafter (**Figure 46–10**). The measurement mark should be on the edge of the rafter. Mark the plumb line for the seat.

Mark a plumb cut at the ridge. From the ridge cut, measure, along the top edge of the rafter, the length of the rafter as determined by calculations. Mark the length and make a plumb line at the seat. At this point, the rafter length is theoretical, because the rafter must be shortened due to the thickness of the ridge (see Figure 46–10).

Slope-Ratio Method

The slope-ratio method (**Figure 46–11**) uses the concept of similar triangles. The examples below will demonstrate how this works. A roof has a slope of 9 in 12 and a span of 28 feet. The run of the roof is 28 ÷ 2 = 14 feet. This will form a right triangle with a base (run) of 14 feet and a rise of x. This triangle is similar to the roof slope triangle, which has a base of 12 inches and a rise of 9 inches. The two

EXAMPLE

IMPERIAL: Span = 28′, slope = $^9/_{12}$

$\frac{x}{14} = \frac{9}{12}$

$14 \times \frac{x}{14} = \frac{9}{12} \times 14$

$x = \frac{126}{12} = 10.5$ *feet* (total rise)

To find the line length of the common rafter (LLCR):

$\sqrt{run^2 + rise^2}$

$\sqrt{14^2 + 10.5^2} = \sqrt{196 + 110.25}$

$\sqrt{306.25} = 17.5$ *feet* or 17′6″ (LLCR).

Or, using the length of the hypotenuse of the unit triangle

$\frac{llcr}{15} = \frac{14}{12}; llcr = \frac{14 \times 15}{12} = 17.5$ *feet* or 17′6″

METRIC: Span = 8500 mm, slope = 1:1.333

$\frac{x}{4250} = \frac{1}{1.333}$

$4250 \times \frac{x}{4250} = \frac{1}{1.333} \times 4250$

$x = \frac{4250}{1.333} = 3187.5$ *mm* (total rise)

To find the line length of the common rafter (LLCR):

$\sqrt{run^2 + rise^2}$

$\sqrt{4250^2 + 3187.5^2} = 5312.5$ mm (LLCR).

StepbyStep Procedures

Procedure 46–C Using the Calculation Method for Common Rafter Layout

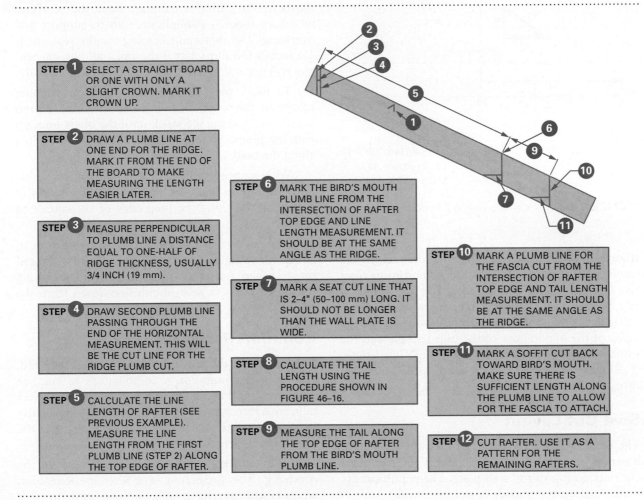

STEP 1 SELECT A STRAIGHT BOARD OR ONE WITH ONLY A SLIGHT CROWN. MARK IT CROWN UP.

STEP 2 DRAW A PLUMB LINE AT ONE END FOR THE RIDGE. MARK IT FROM THE END OF THE BOARD TO MAKE MEASURING THE LENGTH EASIER LATER.

STEP 3 MEASURE PERPENDICULAR TO PLUMB LINE A DISTANCE EQUAL TO ONE-HALF OF RIDGE THICKNESS, USUALLY 3/4 INCH (19 mm).

STEP 4 DRAW SECOND PLUMB LINE PASSING THROUGH THE END OF THE HORIZONTAL MEASUREMENT. THIS WILL BE THE CUT LINE FOR THE RIDGE PLUMB CUT.

STEP 5 CALCULATE THE LINE LENGTH OF RAFTER (SEE PREVIOUS EXAMPLE). MEASURE THE LINE LENGTH FROM THE FIRST PLUMB LINE (STEP 2) ALONG THE TOP EDGE OF RAFTER.

STEP 6 MARK THE BIRD'S MOUTH PLUMB LINE FROM THE INTERSECTION OF RAFTER TOP EDGE AND LINE LENGTH MEASUREMENT. IT SHOULD BE AT THE SAME ANGLE AS THE RIDGE.

STEP 7 MARK A SEAT CUT LINE THAT IS 2–4" (50–100 mm) LONG. IT SHOULD NOT BE LONGER THAN THE WALL PLATE IS WIDE.

STEP 8 CALCULATE THE TAIL LENGTH USING THE PROCEDURE SHOWN IN FIGURE 46–16.

STEP 9 MEASURE THE TAIL ALONG THE TOP EDGE OF RAFTER FROM THE BIRD'S MOUTH PLUMB LINE.

STEP 10 MARK A PLUMB LINE FOR THE FASCIA CUT FROM THE INTERSECTION OF RAFTER TOP EDGE AND TAIL LENGTH MEASUREMENT. IT SHOULD BE AT THE SAME ANGLE AS THE RIDGE.

STEP 11 MARK A SOFFIT CUT BACK TOWARD BIRD'S MOUTH. MAKE SURE THERE IS SUFFICIENT LENGTH ALONG THE PLUMB LINE TO ALLOW FOR THE FASCIA TO ATTACH.

STEP 12 CUT RAFTER. USE IT AS A PATTERN FOR THE REMAINING RAFTERS.

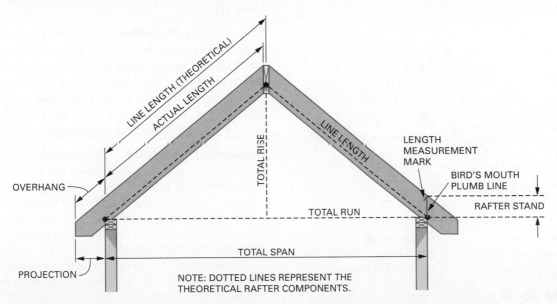

FIGURE 46–10 The line length is theoretical and measured from the first plumb line drawn.

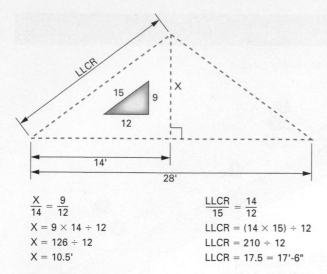

$$\frac{X}{14} = \frac{9}{12}$$

X = 9 × 14 ÷ 12

X = 126 ÷ 12

X = 10.5'

$$\frac{LLCR}{15} = \frac{14}{12}$$

LLCR = (14 × 15) ÷ 12

LLCR = 210 ÷ 12

LLCR = 17.5 = 17'-6"

FIGURE 46–11 Slope-ratio method: imperial.

triangles are set next to each other, oriented in the same fashion, and a ratio is set up to find *x*, the total rise of the rafter. Using the total rise and the total run, you can calculate the line length of the rafter using the Pythagorean theorem. Or you can use the value of the hypotenuse of the unit triangle and the similar triangle's ratio to find the rafter's line length. (See **Figure 46–12** for a slope-ratio method example using the metric system.)

Seat Cut Layout

The seat cut or *bird's mouth* of the rafter is a combination of a level cut and a plumb cut. The level cut rests on top of the wall plate. The plumb cut fits snugly against the outside edge of the wall.

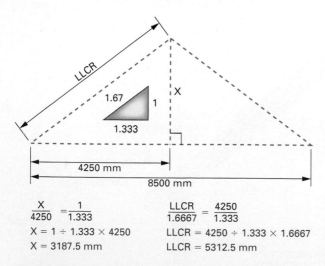

$$\frac{X}{4250} = \frac{1}{1.333}$$

X = 1 ÷ 1.333 × 4250

X = 3187.5 mm

$$\frac{LLCR}{1.6667} = \frac{4250}{1.333}$$

LLCR = 4250 ÷ 1.333 × 1.6667

LLCR = 5312.5 mm

FIGURE 46–12 Slope-ratio method: metric *(systéme internationale).*

The seat cut level line is not a precise measurement. The main concern is that it be perpendicular to the plumb line and not cut too deeply. Position the level so that there is sufficient stock, about two-thirds of the rafter remaining above the seat cut, to ensure there is enough strength to support any overhang. The minimum bearing for the seat cut is 1½ inches (38 mm), but if the stock allows, carpenters will generally match the width of the wall plate.

To mark the seat level line, hold the framing square in the same manner as for marking plumb lines. Slide it along the stock until the blade lines up with the plumb line. Mark the appropriate level line along the blade.

With a speed square, hold the alignment guide of the square in line with the plumb line previously drawn. Mark along the long edge of the square to achieve level lines for seat cuts (**Figure 46–13**).

Sometimes it is desirable to cut the seat cut to allow for wall sheathing (**Figure 46–14**). Extend the level line a distance equal to the wall sheathing thickness. Draw a new plumb line down from the end of the line.

Common Rafter Tails

The tail cut is the cut at the end of the rafter tail. It may be a plumb cut, a combination of cuts, or a square cut (**Figure 46–15**). Sometimes the rafter tails are left to run wild. This means they are slightly longer than needed. They are cut off later, after the roof frame is erected.

On most plans, the projection is given, not the overhang. The projection is a level measurement, whereas the overhang is a sloping measurement along the rafter.

Stepping-Off Projection. To mark the fascia cut, place a framing square on the rafter to mark a plumb cut. Align the square using the unit rise on the tongue and unit run on the blade. Larger projections will require the square to be moved to the lower edge of the rafter. **Figure 46–16** shows a fascia cut example when the unit rise is 8 inches and the projection is 18 inches.

Calculating Overhang. The overhang may also be calculated and measured using the same formula for determining rafter length: unit length times run. Convert the projection to a run by changing inches to feet. In this case, 18 inches is 1.5 units of run and the unit length is 14.42 inches (see Figure 46–16). Thus the overhang measurement is 1.5 × 14.42 = 21.63 inches or 21⅝ inches. In metric, a projection of 460 mm is 460/250 or 1.8 units of run. With the unit length of 291.5, this gives an overhang of 525 mm. This number is measured along the rafter edge.

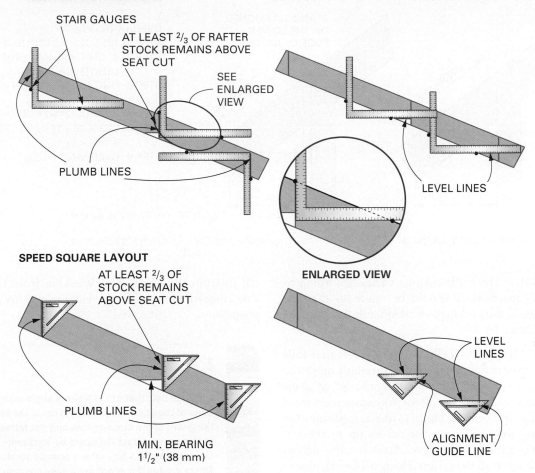

FRAMING SQUARE LAYOUT

STAIR GAUGES

AT LEAST ²/₃ OF RAFTER STOCK REMAINS ABOVE SEAT CUT

SEE ENLARGED VIEW

PLUMB LINES

LEVEL LINES

ENLARGED VIEW

SPEED SQUARE LAYOUT

AT LEAST ²/₃ OF STOCK REMAINS ABOVE SEAT CUT

PLUMB LINES

MIN. BEARING 1¹/₂" (38 mm)

LEVEL LINES

ALIGNMENT GUIDE LINE

FIGURE 46–13 Techniques for using a framing and speed square to lay out a rafter.

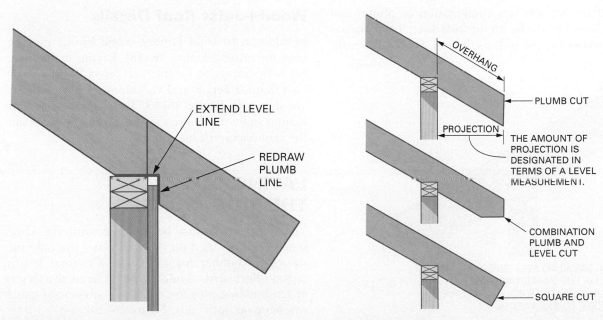

EXTEND LEVEL LINE

REDRAW PLUMB LINE

FIGURE 46–14 Sometimes seat cuts are made deeper to allow for wall sheathing.

OVERHANG

PLUMB CUT

PROJECTION

THE AMOUNT OF PROJECTION IS DESIGNATED IN TERMS OF A LEVEL MEASUREMENT.

COMBINATION PLUMB AND LEVEL CUT

SQUARE CUT

FIGURE 46–15 Various tail cut styles for the common rafter.

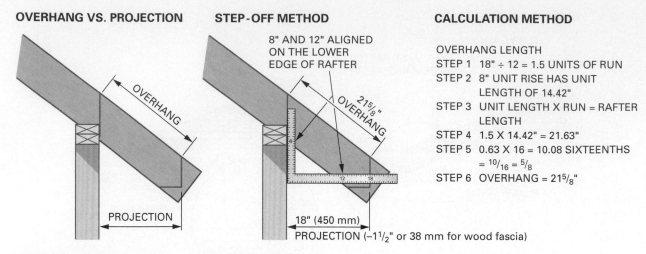

OVERHANG VS. PROJECTION

OVERHANG

PROJECTION

STEP-OFF METHOD

8" AND 12" ALIGNED
ON THE LOWER
EDGE OF RAFTER

$21^5/8$ "

OVERHANG

18" (450 mm)

PROJECTION (–$1^1/2$" or 38 mm for wood fascia)

CALCULATION METHOD

OVERHANG LENGTH
STEP 1   18" ÷ 12 = 1.5 UNITS OF RUN
STEP 2   8" UNIT RISE HAS UNIT
              LENGTH OF 14.42"
STEP 3   UNIT LENGTH X RUN = RAFTER
              LENGTH
STEP 4   1.5 X 14.42" = 21.63"
STEP 5   0.63 X 16 = 10.08 SIXTEENTHS
              = $^{10}/_{16}$ = $^5/_8$
STEP 6   OVERHANG = $21^5/_8$"

FIGURE 46–16 Laying out the tail of the common rafter using the projection or overhang calculation.

Soffit Cut. The soffit cut line varies according to the finish material. It should be made so that the plumb line is long enough to adequately support the fascia (**Figure 46–17**).

Cutting Wild Rafter Tails. To cut wild rafter tails after the rafters are installed, measure and mark the two rafters on either end for the amount of overhang. This is usually a level measurement from the outside of the wall studs to the tail plumb line (**Procedure 46–D**). Plumb the marks up to the top edge of the rafters with a level. Stretch a line across the top edges of all the rafters. Using a T-bevel, plumb down on the side of each rafter from the chalk line. Use a circular saw to cut each rafter. Cutting down along the line is the easiest way to make the cut.

If the tail cut is a combination of plumb and level cuts, make the plumb cuts first. Then snap a line across the cut ends as desired. Level each soffit cut in from the chalk line. Working from the outside toward the wall is the easiest way to accomplish this.

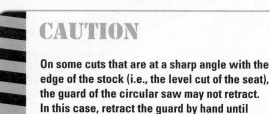

CAUTION

On some cuts that are at a sharp angle with the edge of the stock (i.e., the level cut of the seat), the guard of the circular saw may not retract. In this case, retract the guard by hand until the cut is made a few inches into the stock. Never wedge the guard in an open position to overcome this difficulty.

Wood I-Joist Roof Details

In addition to solid lumber, wood I-joists may be used for rafters (**Figure 46–18**). Layout is the same for both dimension lumber and wood I-joists. Some roof framing details and variations for wood I-joists are shown in **Figure 46–19**. To determine sizes for various spans and more specific information, consult the manufacturer's literature.

LAYING OUT THE RIDGEBOARD

The ridgeboard must be laid out with the same spacing as was used on the wall plate. Use only the straightest lumber for the ridge. The total length of the ridgeboard should be the same as the length of the building plus the necessary amount of gable overhang on both ends (**Figure 46–20**).

From the end of the ridgeboard, measure and mark the required amount of gable end overhang.

FASCIA

CUT IS MADE TO KEEP THE BOTTOM
EDGE OF THE RAFTER ABOVE THE
FASCIA, BUT LOW ENOUGH TO GIVE
SUPPORT TO THE FASCIA.

FIGURE 46–17 Soffit cuts are positioned to allow for adequate support of fascia.

StepbyStep Procedures

Procedure 46–D Optional Method for Cutting Rafter Tails

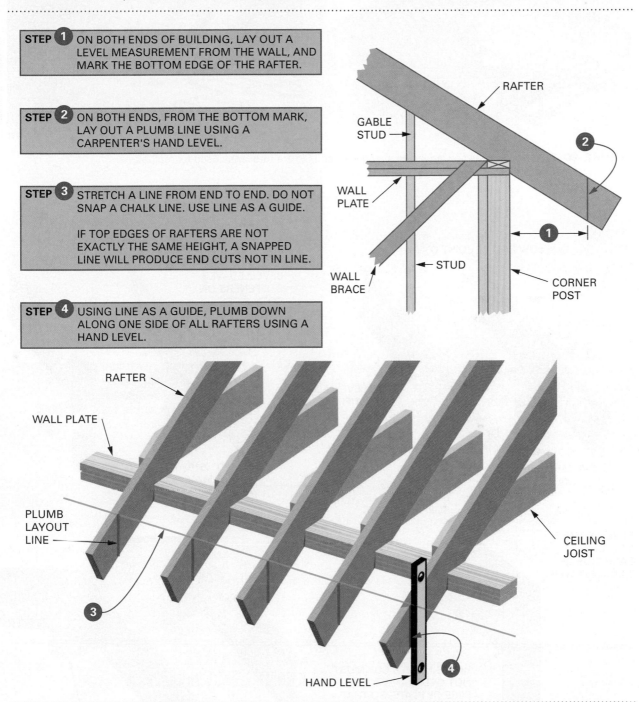

STEP ❶ ON BOTH ENDS OF BUILDING, LAY OUT A LEVEL MEASUREMENT FROM THE WALL, AND MARK THE BOTTOM EDGE OF THE RAFTER.

STEP ❷ ON BOTH ENDS, FROM THE BOTTOM MARK, LAY OUT A PLUMB LINE USING A CARPENTER'S HAND LEVEL.

STEP ❸ STRETCH A LINE FROM END TO END. DO NOT SNAP A CHALK LINE. USE LINE AS A GUIDE.

IF TOP EDGES OF RAFTERS ARE NOT EXACTLY THE SAME HEIGHT, A SNAPPED LINE WILL PRODUCE END CUTS NOT IN LINE.

STEP ❹ USING LINE AS A GUIDE, PLUMB DOWN ALONG ONE SIDE OF ALL RAFTERS USING A HAND LEVEL.

RAFTER

GABLE STUD

WALL PLATE

WALL BRACE

STUD

CORNER POST

RAFTER

WALL PLATE

PLUMB LAYOUT LINE

CEILING JOIST

HAND LEVEL

This distance will vary with the style of the house. The second rafter is nailed on this line. Continue the layout of remaining rafters, marking on both sides of the ridgeboard. If the top of the ridgeboard remains flat, then its height above the top of the wall plate is the rise of the roof plus the rafter stand minus the unit rise over 16 (**Figure 46–21**). Using the roof slope triangle (run = 12″, rise = 6″) and the small imaginary triangle on top of the ridge (run = ¾″, rise = *x*″) as similar triangles, calculate the *drop* of

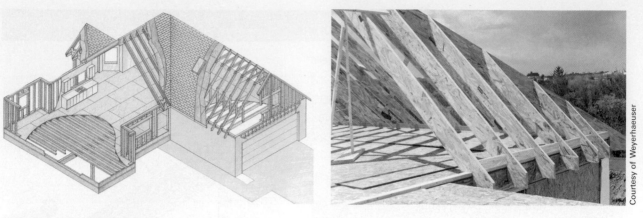

FIGURE 46–18 Wood I-joists and laminated strand lumber (LSL) are frequently used for roof rafters.

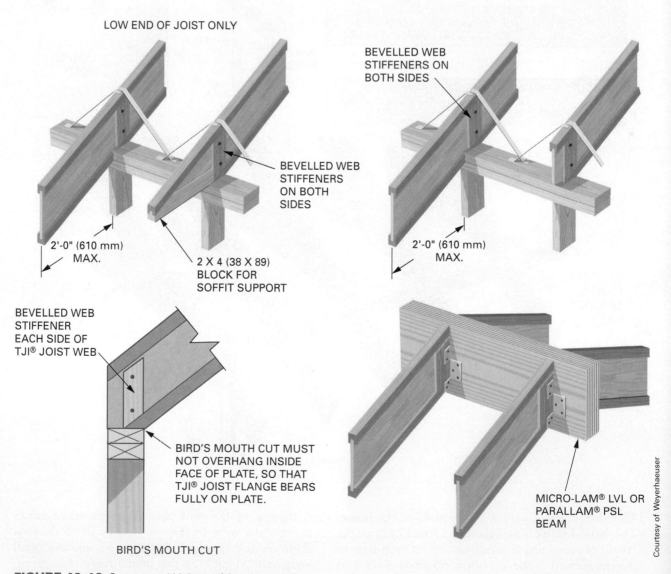

LOW END OF JOIST ONLY

BEVELLED WEB STIFFENERS ON BOTH SIDES

BEVELLED WEB STIFFENERS ON BOTH SIDES

2'-0" (610 mm) MAX.

2 X 4 (38 X 89) BLOCK FOR SOFFIT SUPPORT

2'-0" (610 mm) MAX.

BEVELLED WEB STIFFENER EACH SIDE OF TJI® JOIST WEB

BIRD'S MOUTH CUT MUST NOT OVERHANG INSIDE FACE OF PLATE, SO THAT TJI® JOIST FLANGE BEARS FULLY ON PLATE.

BIRD'S MOUTH CUT

MICRO-LAM® LVL OR PARALLAM® PSL BEAM

Courtesy of Weyerhaeuser

FIGURE 46–19 Some wood I-joist roof framing details.

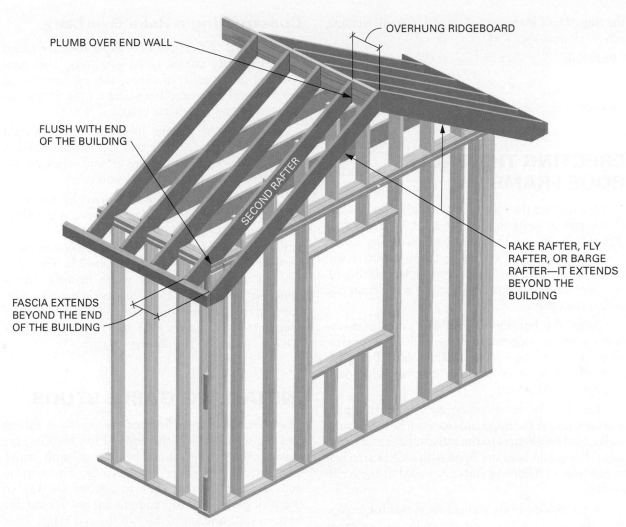

FIGURE 46–20 Ridgeboards may overhang the end of the building.

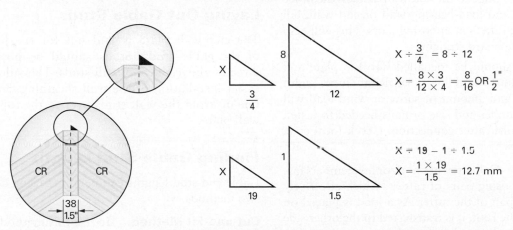

SETTING THE TOP OF THE RIDGEBOARD

$X \div \frac{3}{4} = 8 \div 12$

$X = \frac{8 \times 3}{12 \times 4} = \frac{8}{16}$ OR $\frac{1"}{2}$

$X \div 19 = 1 \div 1.5$

$X = \frac{1 \times 19}{1.5} = 12.7$ mm

TOP OF BEVELLED RIDGEBOARD = TOTAL RISE + RAFTER STAND

FOR A $1\frac{1}{2}"$ OR 38 mm THICK RIDGE

TOP OF 'FLAT' RIDGEBOARD = TOTAL RISE + RAFTER STAND − (UNIT RISE /16)"; or
− (19/Slope Factor) mm

FIGURE 46–21 Determining the top of the ridgeboard.

the ridgeboard (metric: roof slope 1:2, small triangle, run = 19 mm, rise = x).

Imperial: $\frac{x}{3/4} = \frac{6}{12}$; $x = \frac{6 \times 3/4}{12} = \frac{6 \times 3}{12 \times 4}$

$= \frac{6}{4 \times 4} = \frac{6}{16} = \frac{3}{8}''$

Metric: $\frac{x}{19} = \frac{1}{2}$; $x = \frac{19}{2} = 9.5$ mm

ERECTING THE GABLE ROOF FRAME

Prepare to erect the roof by placing plywood on top of the ceiling joists to serve as a safe work surface (**Procedure 46–E**). Tack them in place for safety. Place the ridgeboard on top in the direction it will be installed. Take care not to reverse the ridgeboard, because the layout is usually not the same from one end to the other.

Align the bottom edge of the rafter with the bottom of the ridgeboard. This will allow greater support of the entire rafter. This also allows for greater airflow over the top of the ridgeboard through the ridge vent.

Erection is most efficiently done by three workers, one at the ridge and one each at the bearing walls. Nail two rafters to the ridge at each end of the ridge. Raise and hold the ridge with rafters attached in position as the lower ends are nailed at the bird's mouth.

A temporary brace sometimes is helpful to support the ridge and two rafters while nailing takes place. Position and nail opposing pairs of rafters to complete the outline of the roof frame. Next, plumb and brace the ridge to the end wall. Attach the brace to the ridge and load-bearing wall or end wall. Fill in remaining rafters in opposing pairs. This will help keep the ridgeboard straight.

Rafters should be toenailed into the plate and face-nailed into the ceiling joist. Roof slope, width of building, and amount of snow or wind load will affect the number and size of nails needed to fasten the ceiling joist/rafter connection. Check local code requirements.

Collar Ties. Collar ties are horizontal members fastened to opposing pairs of rafters, which effectively reduce the span of the rafter. As a load is placed on one side of the roof, it is transferred to the other side through the collar tie. In effect, the load on one rafter supports the load on the other rafter and vice versa.

Install collar ties to every third rafter pair or as required by drawings or codes. The length of a collar tie varies, but they are usually about one-third to one-half the building span.

Constructing a Rake Overhang

The first and last rafters of a gable roof are called rake rafters, *fly rafters*, or *barge rafters*. They support the finish material. They may be constructed using several methods, depending on the amount of gable overhang and rafter length.

They should always be made using straight rafter pieces because they will form the roof edge. This is easily visible from the ground. They do not need to have the bird's mouth cut out.

Rake rafters are often supported by **lookouts**. Lookouts are framed in before the roof is sheathed. One method uses 2 × 4s (38 × 89s) on the flat notched into the second rafter. The notched-in members are called *outriggers* (**Figure 46–22**). The other lookout method uses rafter width material and is strongest (**Figure 46–23**). This method looks similar to a cantilevered joist system. Lookouts may be spaced 16 to 32 inches (406 to 813 mm) on centre (OC) according to the desired strength.

INSTALLING GABLE STUDS

The triangular areas formed by the rake rafters and the wall plate at the ends of the building are called gables. They must be framed with studs. These studs are called **gable studs**. The bottom ends are cut square. They fit against the top of the wall plate. The top ends fit snugly against the bottom edge and inside face of the end rafter. They are cut off flush with or below the top edge of the rafter.

Laying Out Gable Studs

The end wall plate is laid out for the location of the gable studs. Studs should be positioned directly above the end wall studs. This allows for easier installation of the wall sheathing. Square a line up from the wall studs over to the top of the wall plate.

Finding Gable Stud Length

Gable end stud lengths can be found using either of two methods.

Cut-and-Fit Method. To use this method, stand each stud plumb in place on the layout line and then mark it along the bottom and top edge of the rafter (**Figure 46–24**). Remove the stud. Use a scrap piece of rafter stock to mark the depth of cut on the stud. Mark and cut all studs in a similar manner.

StepbyStep Procedures

Procedure 46–E Assembling Gable Rafters

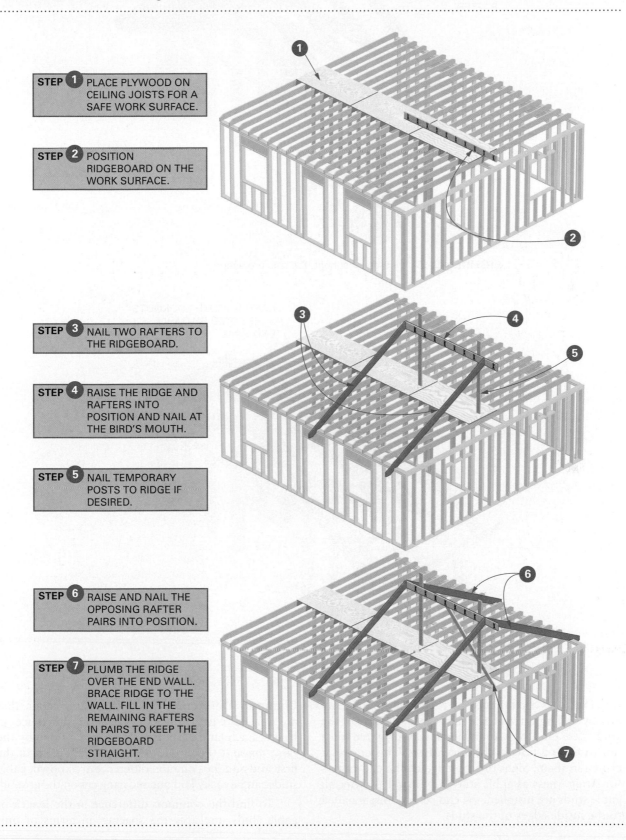

STEP 1 PLACE PLYWOOD ON CEILING JOISTS FOR A SAFE WORK SURFACE.

STEP 2 POSITION RIDGEBOARD ON THE WORK SURFACE.

STEP 3 NAIL TWO RAFTERS TO THE RIDGEBOARD.

STEP 4 RAISE THE RIDGE AND RAFTERS INTO POSITION AND NAIL AT THE BIRD'S MOUTH.

STEP 5 NAIL TEMPORARY POSTS TO RIDGE IF DESIRED.

STEP 6 RAISE AND NAIL THE OPPOSING RAFTER PAIRS INTO POSITION.

STEP 7 PLUMB THE RIDGE OVER THE END WALL. BRACE RIDGE TO THE WALL. FILL IN THE REMAINING RAFTERS IN PAIRS TO KEEP THE RIDGEBOARD STRAIGHT.

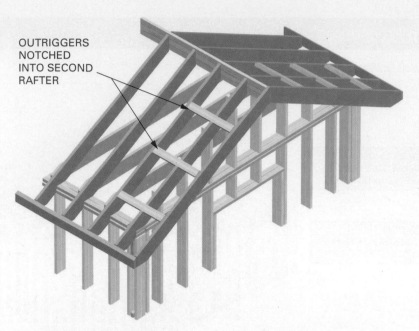

OUTRIGGERS
NOTCHED
INTO SECOND
RAFTER

FIGURE 46–22 Outriggers support the rake projection.

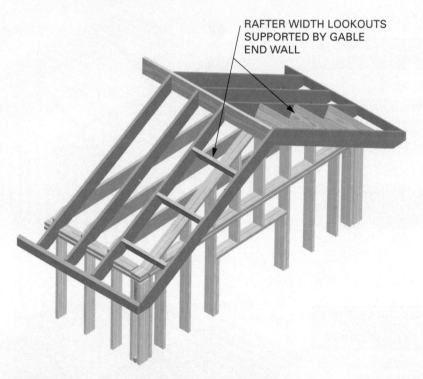

RAFTER WIDTH LOOKOUTS
SUPPORTED BY GABLE
END WALL

FIGURE 46–23 Lookouts support the rake projection.

The studs are fastened by toenailing to the plate and by nailing through the rafter into the edge of the stud. Care must be taken when installing gable studs not to force a bow in the end rafters. This creates a crown in them. Sight the top edge of the end rafters for straightness as gable studs are installed. After all gable studs are installed, the end ceiling joist is nailed to the inside edges of the studs.

Common Difference Method. Gable studs that are spaced equally have a common difference in length. Each stud is shorter than the next one by the same amount (**Figure 46–25**). Once the length of the first stud and the common difference are known, gable studs can be easily laid out and gang cut on the ground.

To find the common difference in the length of gable studs, multiply the spacing, in terms of unit

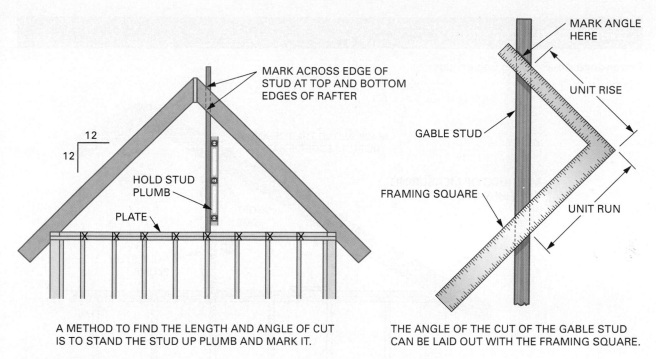

A METHOD TO FIND THE LENGTH AND ANGLE OF CUT
IS TO STAND THE STUD UP PLUMB AND MARK IT.

THE ANGLE OF THE CUT OF THE GABLE STUD
CAN BE LAID OUT WITH THE FRAMING SQUARE.

FIGURE 46–24 Cut-and-fit method of finding the length and cut of a gable stud.

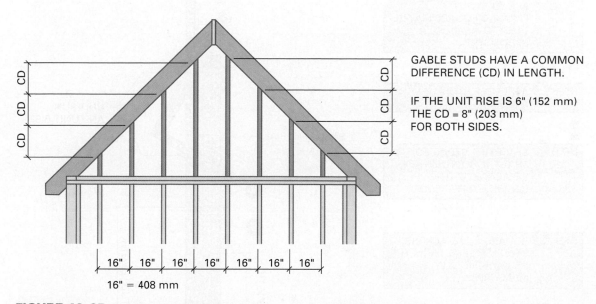

GABLE STUDS HAVE A COMMON
DIFFERENCE (CD) IN LENGTH.

IF THE UNIT RISE IS 6" (152 mm)
THE CD = 8" (203 mm)
FOR BOTH SIDES.

16" = 408 mm

FIGURE 46–25 Equally spaced gable studs have a common difference in length.

run, by the unit rise of the roof. For example, if the stud spacing is 16 inches OC and the unit rise is 6, what is the common difference in length?

STEP 1: Convert OC spacing to units of run.

$$16 \div 12 = 1\frac{1}{3} \text{ feet or units of run}$$

STEP 2: Multiply unit rise by units of run.

$$6 \times 1\frac{1}{3} = 8 \text{ inches common difference}$$

Next, determine the length of a stud. A framing square works well to measure the distance plumb and accurately. Mark it with the angle for the notch as described previously, using a framing square. Assemble the stud material necessary for one side of the gable. Lay them on horses next to each other where the bottom of each stud is separated from its neighbour by one common difference in length (**Procedure 46–F**).

StepbyStep Procedures

Procedure 46–F Cutting Gable Studs

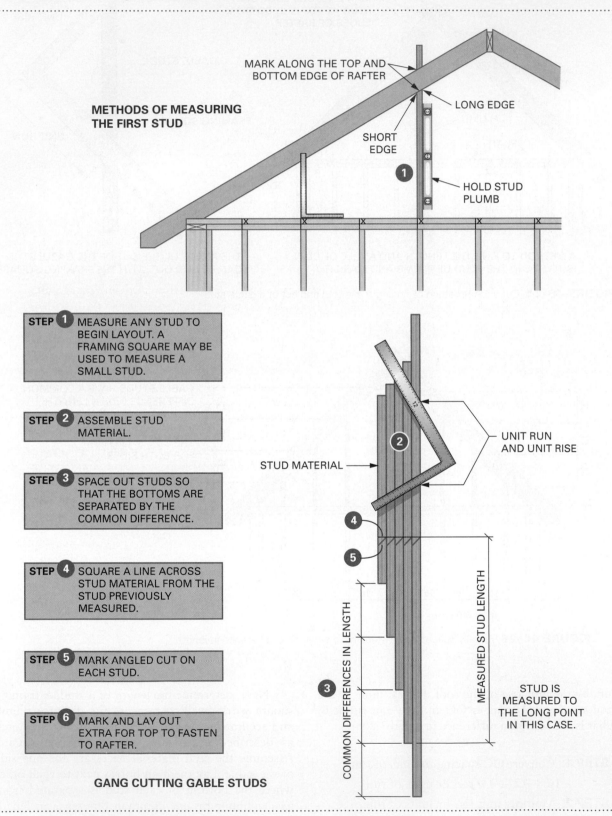

MARK ALONG THE TOP AND
BOTTOM EDGE OF RAFTER

**METHODS OF MEASURING
THE FIRST STUD**

LONG EDGE

SHORT
EDGE

HOLD STUD
PLUMB

1

STEP 1 MEASURE ANY STUD TO BEGIN LAYOUT. A FRAMING SQUARE MAY BE USED TO MEASURE A SMALL STUD.

STEP 2 ASSEMBLE STUD MATERIAL.

STEP 3 SPACE OUT STUDS SO THAT THE BOTTOMS ARE SEPARATED BY THE COMMON DIFFERENCE.

STEP 4 SQUARE A LINE ACROSS STUD MATERIAL FROM THE STUD PREVIOUSLY MEASURED.

STEP 5 MARK ANGLED CUT ON EACH STUD.

STEP 6 MARK AND LAY OUT EXTRA FOR TOP TO FASTEN TO RAFTER.

STUD MATERIAL

UNIT RUN
AND UNIT RISE

2

4

5

3

COMMON DIFFERENCES IN LENGTH

MEASURED STUD LENGTH

STUD IS
MEASURED TO
THE LONG POINT
IN THIS CASE.

GANG CUTTING GABLE STUDS

Using the first measured stud as a reference, square a line across the ganged studs. Mark the angle for each stud the same as the first. Be sure to measure and mark to the same point for each, that is, to the long point or short point of each notch.

Toenail each stud to the plate and face-nail into the rafter. Check to be sure the rafter does not bow upward as each stud is installed.

EXAMPLE Gable studs are spaced 16 inches OC. The roof rises 8 inches per foot of run. Change 16 inches to $1\frac{1}{3}$ feet. Multiply $1\frac{1}{3}$ by 8 to get 10.666. The common difference in length for gable end studs here is $10\frac{2}{3}$ inches or roughly $10\frac{11}{16}$ inches. Similarly, in metric, a roof has a slope of 1 in 2.5 and the gable studs are spaced at 400 mm OC. The common difference is therefore determined by dividing 400 by 2.5, giving an answer of 160 mm.

FRAMING A GAMBREL ROOF

A gambrel roof is one where each side of the ridge-board has two sloping rafters. The rafters of a true gambrel roof are chords of a semicircle whose diameter is the width of the building (**Figure 46–26**). To frame a true gambrel roof, the rafter lengths are calculated using the Pythagorean theorem or determined from a full-scale layout on a large, flat surface, such as the subfloor. From the layout, rafter lengths and angles can be determined.

The calculated method involving the Pythagorean theorem may at first appear intimidating, but it really is only repetitive steps of the same process. Before calculations can begin, some information

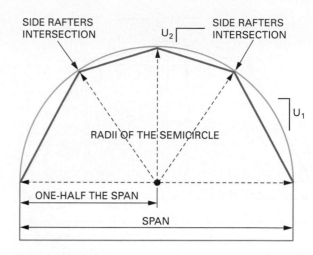

FIGURE 46–27 Rafter intersections are located a radius from the centre of the span.

must be obtained from the plans or the architect. The measurements needed are the building span and the horizontal distances that the side rafter intersections are from the building's centre and building line. Another method, which will not be described here, uses the ceiling height created at the rafter intersections. It is helpful to note that the total rise of the roof is equal to half the span. This happens because the height is also a radius of the semicircle. The distance from the side rafter intersection to the centre of the building is also a radius (**Figure 46–27**). Note that there are only two different rafters, because they come in pairs. This is because a gambrel roof is symmetrical. Bringing all of this information together shows that right triangles are formed and the Pythagorean theorem can be used to solve for rafter lengths (**Figures 46–28** and **46–29**).

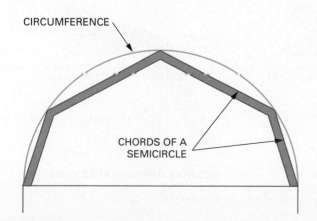

FIGURE 46–26 The slopes of a true gambrel roof form chords of a semicircle.

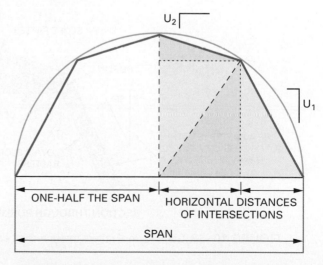

FIGURE 46–28 Various right triangles are formed by horizontal and vertical lines and radii of the semicircle.

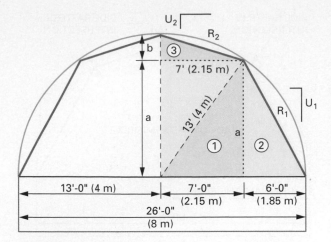

FIGURE 46–29 Starting information gathered for a gambrel calculation.

If the rafters are to be framed to a **purlin** or a knee wall, as shown in **Figure 46–30**, the unit rise for each rafter may be found by dividing the total rise of each rafter, in inches, by its total run. For example, in Figure 46–29, U_1 is found by first converting (a) to inches, 10.954 feet × 12 = 131.448 inches, and then dividing it by 6, the run, to get 21.908 or $21^{15}/_{16}$ inches unit rise. U_2 is found similarly, (b), in inches, divided by 7. That is, 2.046 feet × 12 = 24.552 inches, and then divide by 7 to get 3.507 or 3½ inches unit rise.

Sometimes the slope of gambrel rafters is given in the drawings. The slopes, then, may not actually be chords of a semicircle (**Figure 46–31**). In this case, the rafter lengths and cuts can also be found with a full-size layout. They are usually laid out with a

EXAMPLE Find the lengths of the rafters labelled R_1 and R_2 for a gambrel roof system, given that the span is 26 feet and the horizontal distances of the side intersections are 6 and 7 feet (Figure 46–29).

1. $a^2 + b^2 = c^2$; therefore $a^2 = c^2 - b^2$
2. From triangle 1
$$a^2 = 13^2 - 7^2 = 169 - 49 = 120$$
$$a = \sqrt{120} = 10.954'$$
3. From triangle 2
$$R_1 = \sqrt{a^2 + 6^2} = \sqrt{10.952^2 + 6^2} = \sqrt{120 + 36}$$
$$R_1 = \sqrt{156} = 12.490' = 12'\text{-}5^7/_8''$$
4. From triangle 3
$$b = \text{radius} - a = 13' - 10.954' = 2.046'$$
$$R_2 = \sqrt{b^2 + 7^2} = \sqrt{2.046^2 + 7^2} = \sqrt{4.186 + 49}$$
5. $R_2 = \sqrt{53.186} = 7.293' = 7'\text{-}3½''$

A similar gambrel roof in metric measure—given that the span is 8 m, the run is 4 m, and the horizontal distances are 2.15 and 1.85 m:

1. $a^2 = 4^2 - 2.15^2$; $a = 3.37$ m
2. $R_1^2 = a^2 + 1.85^2$; $R_1 = 3.85$ m
3. $b = \text{radius} - a = 4 - 3.37 = 0.63$ m
4. $R_2^2 = b^2 + 2.15^2$; $R_2 = 2.24$ m

The unit rise for R_1 can be found by dividing a by 1.85 and multiplying by 250 → 3.37 ÷ 1.85 × 250 = 465 mm. In the same way, the unit rise for R_2 is found by → 0.63 ÷ 2.15 × 250 = 73 mm.

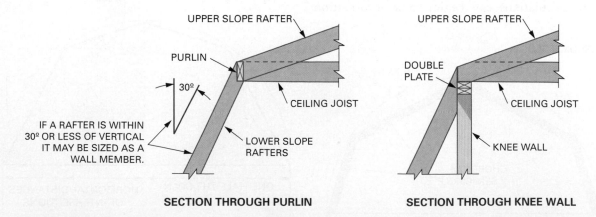

SECTION THROUGH PURLIN   **SECTION THROUGH KNEE WALL**

FIGURE 46–30 Gambrel roof rafters may intersect each other at a purlin or knee wall.

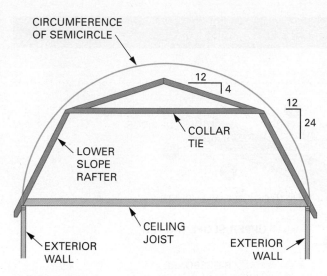

FIGURE 46–31 In some cases, the slope of the gambrel rafters is given in the drawings.

Gambrel Rafter Layout

Determine the construction at the intersection of the two slopes. Usually gambrel roof rafters meet at a continuous member, similar to the ridge, called a *purlin*. Any structural member that runs at right angles to and supports rafters is called a purlin. Sometimes, the rafters of a gambrel roof meet at the top of a *knee wall*. In the lower and steeper slope, the rafters may be sized as wall members rather than

framing square, similar to the layout considerations of gable roof rafters.

roof members if the rafter is within 30 degrees of vertical (Figure 46–30).

Determine the run and rise of the rafters for both slopes. Find their line length in the same way as for rafters in a gable roof. Lay out plumb lines at the top and bottom of rafters for each slope. Make the seat cut on the lower rafter. Notch all rafters where they intersect at the purlin (**Procedure 46–G**). Because of the steep slope of the lower roof, the level cut of the seat cannot be made the full width of the plate. At least ⅔ of the width of the rafter stock must remain.

Erecting the Gambrel Roof

If a purlin is used, fasten a rafter to each end of a section of purlin. Raise the assembly. Fasten the lower end of the rafters against ceiling joists and on the top plate of the wall. Brace the section under each end of the purlin. Continue framing sections until the other end of the building is reached. Place temporary and adequate bracing under the purlin where needed. If a knee wall is used, build, straighten, and brace the wall.

Frame the roof by installing both lower and upper slope rafters opposite each other. This maintains equal stress on the frame. Plumb and brace the roof after a few rafters have been installed. Sight along the ridge and purlin or knee wall for straightness as framing progresses. After all rafters have been erected, install ceiling joists.

The open ends formed by the gambrel roof are framed with studs in the same manner as installing studs in gable ends.

StepbyStep Procedures

Procedure 46–G Laying Out Gambrel Rafters

UPPER SLOPE RAFTER

STEP ❶ LAY OUT PLUMB LINE.

STEP ❷ SHORTEN ¹/₂ THICKNESS OF RIDGE.

STEP ❸ LAY OUT LINE LENGTH.

STEP ❹ MARK PLUMB LINE.

STEP ❺ DRAW LEVEL LINE THE WIDTH OF KNEE WALL PLATE.

LOWER SLOPE RAFTER

STEP ❶ LAY OUT PLUMB LINE— NO SHORTENING.

STEP ❷ LAY OUT LINE LENGTH.

STEP ❸ DRAW PLUMB LINE.

STEP ❹ LAY OUT LEVEL LINE OF SEAT CUT. LEAVE ON MINIMUM OF ²/₃ WIDTH OF RAFTER STOCK.

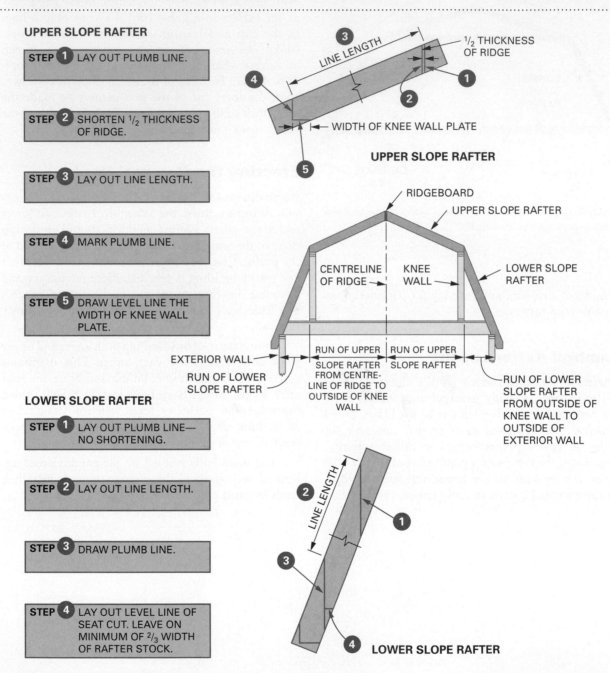

UPPER SLOPE RAFTER

LOWER SLOPE RAFTER

Chapter 47

Shed, Dormer, and Porch Roofs

The shed roof slopes in only one direction. It is relatively easy to frame. A shed roof may be free-standing or one edge may rest on an exterior wall while the other edge butts against an existing wall (**Figure 47–1**). Shed and other type roofs are also used on **dormers**. A dormer is a framed projection above the plane of the roof. It contains one or more windows for the purpose of providing light and

ventilation or for enhancement of the exterior design (**Figure 47–2**).

FRAMING A SHED ROOF

A shed roof is made of rafters that are common. A shed rafter run is at a right angle to the plate line, and the unit of run is 12 inches (250 mm). The total

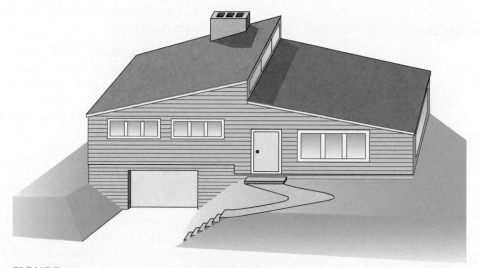

FIGURE 47–1 A building with a clerestory roof.

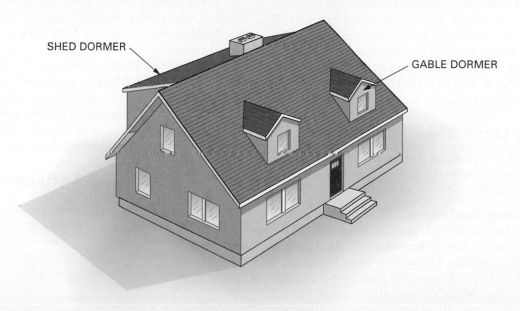

SHED DORMER

GABLE DORMER

FIGURE 47–2 A roof with gable and shed dormers.

FREESTANDING SHED ROOF

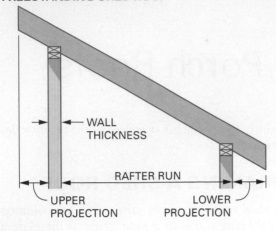

ATTACHED SHED ROOF

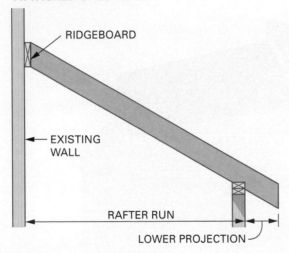

FIGURE 47–3 Styles of shed roofs.

run of the rafter is determined as for other common rafters—from where it bears on other members. If both ends rest on exterior walls, the run is the width of the building minus the thickness of one of the walls (**Figure 47–3**). If one end rests on another wall, the run is from wall to wall. These are laid out similar to common rafters.

Laying Out a Shed Roof Rafter

Layout for a shed rafter begins at one end, such as the upper one, and progresses to the other end. The process is similar to that for any rafter.

Freestanding Shed Rafter. To lay out a rafter for a freestanding shed roof, mark a plumb line at the upper end of the board (**Procedure 47–A**). This will be the fascia cut line. Deduct for fascia thickness, if appropriate. Step off the upper projection and mark the plumb line. Mark a reasonable seat cut line for the upper bird's mouth and note the height above the bird's mouth.

Calculate the line length and measure it along the rafter from the second plumb line. Draw the third plumb line. Measure down along this plumb line the height above the bird's mouth. Draw another seat cut. Step off the lower projection and draw the final plumb line. Level cuts may be added to each fascia cut as is done for common rafters.

The rafters may be installed with the tails left running "wild." The overhang would then be trimmed after the rafters have been installed. This procedure is described in Procedure 46–D.

Attached Shed Rafter. Attached shed rafters are laid out in the same way as for common rafters (**Procedure 47–B**). The only difference is the shortening for the ridge. A common rafter shortens by deducting one-half the ridge thickness. For an attached shed rafter, however, the shortening is a full ridge thickness.

Shed Rafter Erection. Shed rafters are toenailed to the plates or ridge at the designated spacing. It is important to keep the bird's mouth snug up against the walls. In addition to nailing the rafters, metal framing connectors may be required. This includes all types of framed roofs.

Dormers

Dormers typically have either a gable or a shed roof. A gable dormer roof is framed like an intersecting gable roof with valleys and jacks. A shed dormer is simpler in design and often framed to the ridge of the main roof. This is typically done to gain sufficient rafter slope (**Figure 47–4**).

When framing openings in the main roof for dormers, the rafters on both sides of the opening are doubled. Some dormers have their front walls directly over the exterior wall below. In this case, much of the load of the dormer is transferred into the load-bearing wall below. If dormers are placed such that the front wall is recessed or if the dormers are partway up the main roof, the main roof must be strengthened. Double headers are also added at the top and bottom of the opening.

OTHER ROOF FRAMING

A number of other types of roof framing exist, but they are all related in some way to the framing theory previously described. To solve these various roof framing situations, begin by returning to the basics of understanding rafter length as compared with run and rise.

Sometimes a gable roof or dormer is framed to an already existing building. Ridgeboards and valley

StepbyStep Procedures

Procedure 47–A Laying Out a Freestanding Shed Rafter (Double Overhang)

STEP 1 MARK A PLUMB LINE AT THE UPPER END OF THE BOARD.

STEP 2 STEP OFF THE WALL THICKNESS AND THE UPPER PROJECTION AND MARK THE SECOND PLUMB LINE.

STEP 3 MARK A REASONABLE SEAT CUT LINE FOR THE UPPER BIRD'S MOUTH AND NOTE THE HEIGHT ABOVE THE BIRD'S MOUTH.

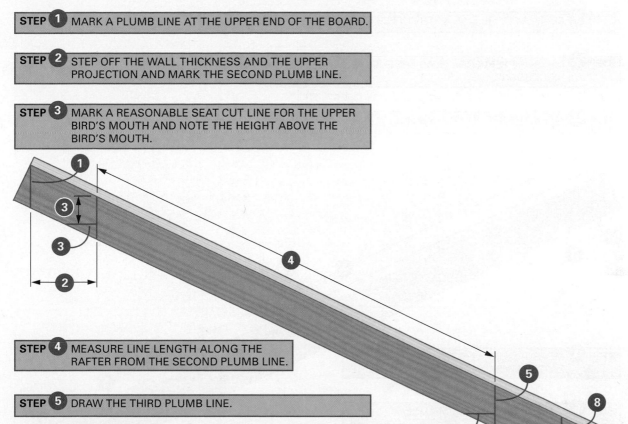

STEP 4 MEASURE LINE LENGTH ALONG THE RAFTER FROM THE SECOND PLUMB LINE.

STEP 5 DRAW THE THIRD PLUMB LINE.

STEP 6 DRAW SEAT CUT AFTER MEASURING DOWN ALONG PLUMB LINE THE HEIGHT ABOVE THE BIRD'S MOUTH.

STEP 7 STEP OFF THE LOWER PROJECTION.

STEP 8 DRAW THE FINAL PLUMB LINE.

jack rafters are cut to fit to the existing sheathing line and not to valley rafters. Shed rafters may also be cut to existing roofs as well.

Fitting a Gable Dormer Ridge to Roof Sheathing

The ridge must be fitted level into the existing roof. The simplest way to determine the ridge length is to first cut a slope on the end of the ridge, and then install it with common rafters and adjust it until it fits. The ridge is later cut for length after plumbing up from the end wall.

The ridge slope cut layout uses the unit rise for the existing roof (**Procedure 47–C**). Hold the unit rise and 12 on a framing square and adjust to the end of the board. Mark the ridgeboard on the blade side of the square.

Fitting Valley Jack Rafters to Roof Sheathing

To lay out a valley jack framed to a roof of the same slope (**Procedure 47–D**):

1. Cut as many pairs of common rafters as are required for the minor roof, complete with

StepbyStep Procedures

Procedure 47–B Laying Out an Attached Shed Rafter (Single Overhang)

STEP 1 MARK A PLUMB LINE AT THE UPPER END OF THE BOARD.

STEP 2 MEASURE PERPENDICULAR THE FULL THICKNESS OF RIDGEBOARD.

STEP 3 MARK THE SECOND PLUMB LINE.

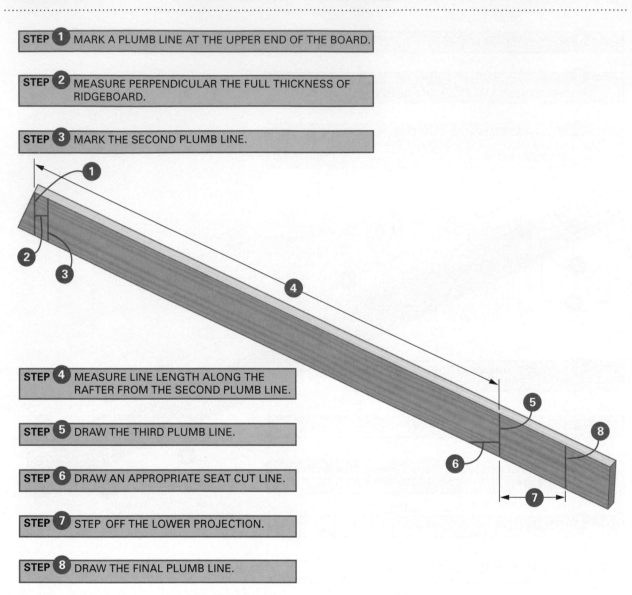

STEP 4 MEASURE LINE LENGTH ALONG THE RAFTER FROM THE SECOND PLUMB LINE.

STEP 5 DRAW THE THIRD PLUMB LINE.

STEP 6 DRAW AN APPROPRIATE SEAT CUT LINE.

STEP 7 STEP OFF THE LOWER PROJECTION.

STEP 8 DRAW THE FINAL PLUMB LINE.

ridge plumb cut, bird's mouth, and tail cut to match the eave of the main roof.

2. Nail a ledger across the top of a pair of the rafters to allow for the insertion of the ridgeboard.

3. Erect the pair in place over the end wall and brace plumb.

4. Place the ridge with the angled-cut end onto the main roof and the other end in the space between the pair of erected rafters.

5. Level the ridge and install blocking under the sheathing where the ridge meets the main roof. Secure the ridge to the pair of common rafters.

6. Mark off the end of the ridge for rake projection. From this mark, lay out for the remaining pairs of common rafters on the ridge, and then transfer the layout marks to the side wall plates.

7. Complete the installation of the rest of the common rafters. This will centre the ridgeboard on the main roof.

8. Locate the point on the eave of the main roof where the sheathing of the minor roof will meet the existing sheathing. Snap a chalk line from this point to the nearer edge of the ridgeboard. (This line is the location of the underside of the new sheathing.)

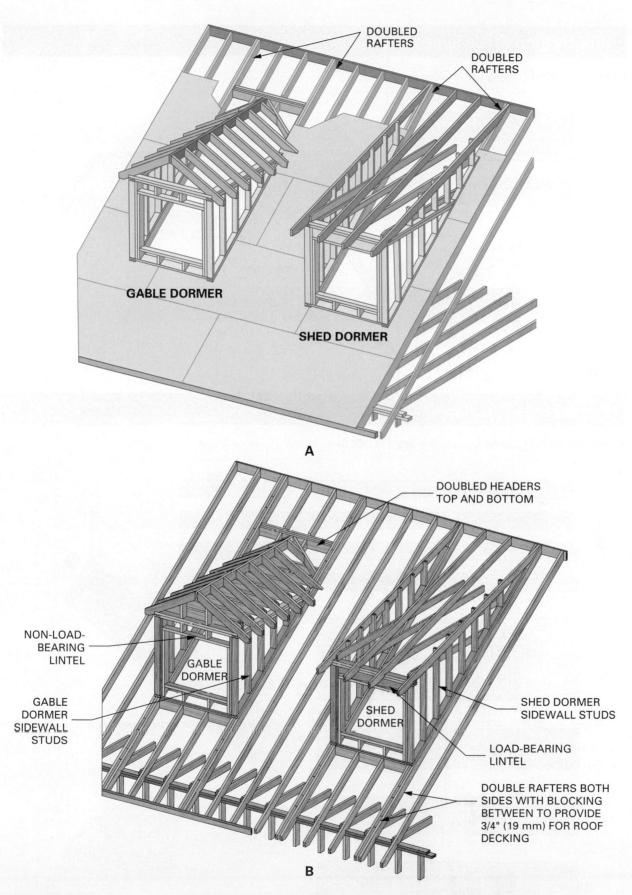

DOUBLED
RAFTERS

DOUBLED
RAFTERS

GABLE DORMER

SHED DORMER

A

DOUBLED HEADERS
TOP AND BOTTOM

NON-LOAD-
BEARING
LINTEL

GABLE
DORMER

GABLE
DORMER
SIDEWALL
STUDS

SHED
DORMER

SHED DORMER
SIDEWALL STUDS

LOAD-BEARING
LINTEL

DOUBLE RAFTERS BOTH
SIDES WITH BLOCKING
BETWEEN TO PROVIDE
3/4" (19 mm) FOR ROOF
DECKING

B

FIGURE 47–4 (A) Two styles of framed dormers, and (B) flat roofs may be framed in two styles.

StepbyStep Procedures

Procedure 47–C Laying Out a Gable Dormer Ridge

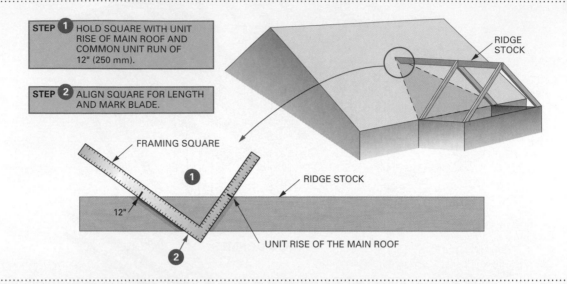

STEP 1 HOLD SQUARE WITH UNIT RISE OF MAIN ROOF AND COMMON UNIT RUN OF 12" (250 mm).

STEP 2 ALIGN SQUARE FOR LENGTH AND MARK BLADE.

RIDGE STOCK

FRAMING SQUARE

①

12"

②

RIDGE STOCK

UNIT RISE OF THE MAIN ROOF

StepbyStep Procedures

Procedure 47–D Laying Out a Valley Jack Framed to a Roof of the Same Slope

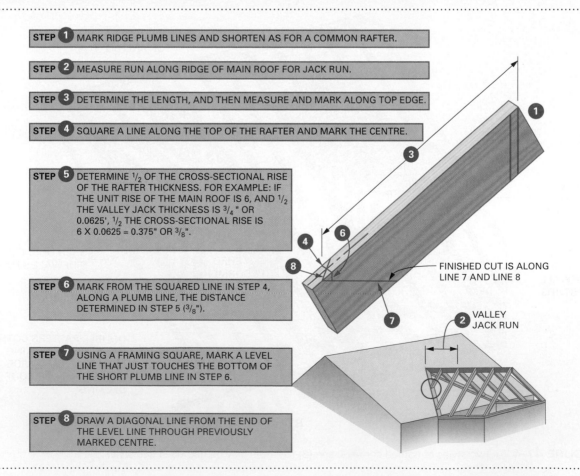

STEP 1 MARK RIDGE PLUMB LINES AND SHORTEN AS FOR A COMMON RAFTER.

STEP 2 MEASURE RUN ALONG RIDGE OF MAIN ROOF FOR JACK RUN.

STEP 3 DETERMINE THE LENGTH, AND THEN MEASURE AND MARK ALONG TOP EDGE.

STEP 4 SQUARE A LINE ALONG THE TOP OF THE RAFTER AND MARK THE CENTRE.

STEP 5 DETERMINE $1/2$ OF THE CROSS-SECTIONAL RISE OF THE RAFTER THICKNESS. FOR EXAMPLE: IF THE UNIT RISE OF THE MAIN ROOF IS 6, AND $1/2$ THE VALLEY JACK THICKNESS IS $3/4$" OR 0.0625', $1/2$ THE CROSS-SECTIONAL RISE IS 6 X 0.0625 = 0.375" OR $3/8$".

STEP 6 MARK FROM THE SQUARED LINE IN STEP 4, ALONG A PLUMB LINE, THE DISTANCE DETERMINED IN STEP 5 ($3/8$").

STEP 7 USING A FRAMING SQUARE, MARK A LEVEL LINE THAT JUST TOUCHES THE BOTTOM OF THE SHORT PLUMB LINE IN STEP 6.

STEP 8 DRAW A DIAGONAL LINE FROM THE END OF THE LEVEL LINE THROUGH PREVIOUSLY MARKED CENTRE.

①

③

④ ⑥

⑧

FINISHED CUT IS ALONG LINE 7 AND LINE 8

⑦

② VALLEY JACK RUN

9. Place a 2 × 8 (38 × 184 mm) along this line to provide a solid base for the valley jack rafters. The 2 × 8 can be bevelled to match the slope of the minor roof (e.g., 12/12 slope or 1:1 would require a 45-degree bevel), *or* it can be left as is but set back from the line by the required displacement.

To lay out a valley jack to a roof of a different slope:

1. Hold the blade of the framing square flush to the far side of the last common rafter and mark the 2 × 8 where the toe of the tongue of the square makes contact with its upper edge. This will give the 16 inches (406 mm) OC location for the first valley jack rafter. Reverse the square for 24 inches (610 mm) OC (**Figure 47–5**).

2. Measure from the layout mark on the ridgeboard to this mark on the 2 × 8. This is the long-to-long dimension of the first valley jack rafter.

3. On a piece of rafter stock, draw the ridge plumb line on the tongue side of the square, mark off the above dimension, and draw the horizontal cut line on the blade side of the square.

4. Make the ridge plumb cut with circular saw set at right angles. Then change the angle of the saw to match the slope of the main roof and cut

along the horizontal cut line (e.g., if the main roof is 6/12 or 1:2, set the saw angle to 26½°) (**Figure 47–6**).

5. Follow the same procedure to locate and cut the other valley jacks. Install collar ties to every second pair of valley jacks.

Calculating Dormers—Lengths and Angles

In order to frame a dormer roof onto an existing roof, you first need to know the roof slope for the main roof, the slope of the dormer roof, and the span of the dormer roof. In the example that follows (**Figures 47–7** and **47–8**), the slope of the main roof is 11 in 12 (1:1.091), the slope of the dormer roof is 8 in 12 (1:1.5), and the span of the dormer is 6 feet (1800 mm). The 11/12 plumb angle is 47.5°, so the plumb cut on the valley jack rafters is 47.5°, and the ridgeboard bevels to a point at 47.5°. Remember to set the mitre saw angle to 42.5° (90°–47.5°). The 11/12 level angle is 42.5° and the 8/12 roof angle is 33.7°, so the seat cuts of the valley jacks are made using those angles. (Set the mitre angle at 47.5°, tilt the saw to 33.7°, lay the valley jack flat on the table top of the saw, and then make the cut.)

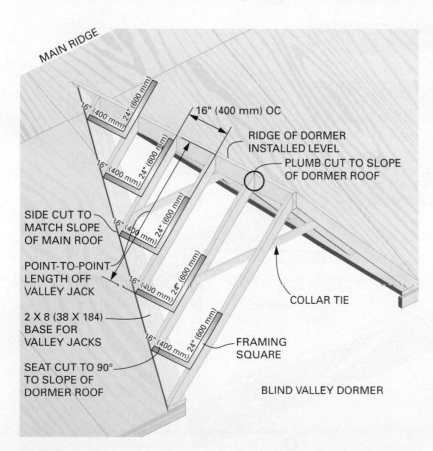

FIGURE 47–5 Framing valley jack rafters onto a roof with a different slope.

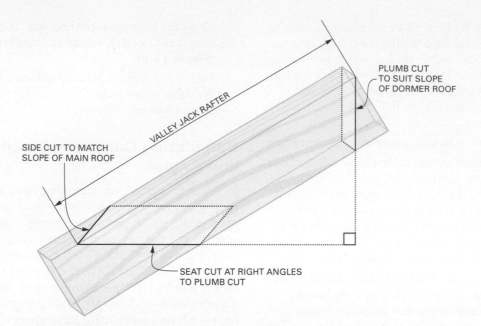

FIGURE 47–6 The seat cut of the valley jack rafter is cut with the circular saw blade set at the same angle as the slope of the main roof.

StepbyStep Procedures

Procedure 47–E Laying Out a Shed Dormer Rafter

STEP 1 DRAW A LEVEL LINE TO THE SLOPE OF THE SHED ROOF.

STEP 2 REVERSE THE SQUARE AND HOLD ON THE LEVEL LINE TO THE SLOPE OF THE MAIN ROOF.

STEP 3 DRAW CUTTING LINE ALONG THE BLADE OF THE SQUARE. EXTEND THE LINE ACROSS ENTIRE SIDE OF RAFTER.

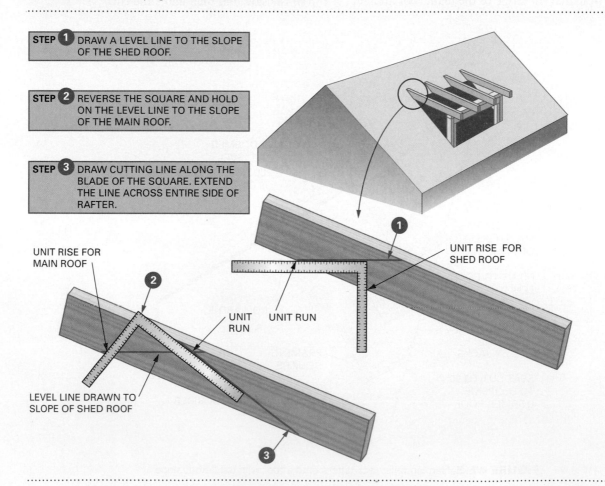

DORMER CALCULATIONS:

Dormer Span = 6', Run = 3' (36"), Rise = 11" x 3 = 33"

LLCR = 16.28" x 3 = 48.84" = 4'$\frac{7}{8}$"

Ridge = $\dfrac{(12 \times \text{Dormer Rise})}{\text{Roof Unit Rise}}$ = $\dfrac{12 \times 33}{8}$ = 49$\frac{1}{2}$"

$\mathcal{C}_L$ = $\sqrt{(\text{Ridge}^2 + \text{Rise}^2)}$ = $\sqrt{(49.5^2 + 33^2)}$ = 59.49" = 59$\frac{1}{2}$"

Valley = $\sqrt{(\mathcal{C}_L^2 + \text{Run}^2)}$ = $\sqrt{(59.49^2 + 36^2)}$ = 69.54" = 69$\frac{9}{16}$"

X = $\dfrac{(12 \times \text{Run})}{\text{Valley}}$ = $\dfrac{(12 \times 36)}{69.54}$ = 6.21" = 6$\frac{3}{16}$"

$\angle$Y = ARCTAN $\left(\dfrac{\text{Roof Unit Rise}}{X}\right)$ = INV TAN $\left(\dfrac{8}{6.21}\right)$ = 52.2°

Valley Board Bevel = 90° −52° = 38°

FIGURE 47–7 Dormer calculations.

Given that the span of the dormer is 6 feet (1800 mm), the run is 3 feet (900 mm) and the rise is 33 inches (825 mm). Using similar triangles and the roof slope of the main roof, the ridge is determined to be 49.5 inches (33 × 1.5) or 1237.5 mm

(825 × 1.5). Using the Pythagorean theorem, the centreline of the dormer roof is calculated to be 59.49 inches (1487.3 mm). Using this number and the run of the dormer, by Pythagoras, the length of the valley is calculated to be 69.54 inches (1738.4 mm). These values can now be used to lay out the isosceles triangle (the footprint) of the dormer onto the main roof. The valley board can be left with its edges square, or it can be cut at a bevel to match the dormer slope. Follow the use of similar triangles and the tangent function to determine the angle:

$$[\tan^{-1}\left(\dfrac{1738.4}{1.5 \times 900}\right) = 52.2°]$$

Fitting a Shed Dormer Rafter to Roof Sheathing

The top ends of shed dormer rafters may be fit against the main roof, which has a steeper slope. The cuts are laid out by using a framing square as outlined in **Procedure 47–E**.

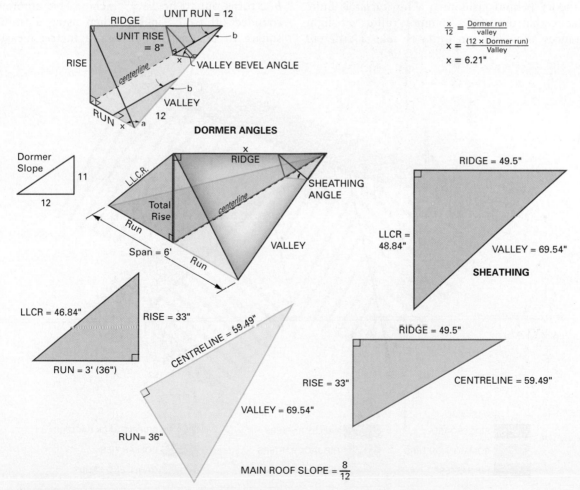

$$\dfrac{x}{12} = \dfrac{\text{Dormer run}}{\text{valley}}$$
$$x = \dfrac{(12 \times \text{Dormer run})}{\text{Valley}}$$
$$x = 6.21"$$

FIGURE 47–8 Determining the lengths and angles of the framing components of dormers.

Chapter 48

Hip Roofs

The hip roof is a little more complicated to frame than the gable roof. Two additional kinds of rafters need to be laid out.

To frame the hip roof it is necessary to lay out not only common rafters and a ridge but also hip rafters and **hip jack rafters** (**Figure 48–1**). Hip rafters form the outside corners where two sloping roof planes meet. They are longer than common rafters and their slope is at a lower angle. Hip jacks are common rafters that must be cut shorter to land on a hip rafter. They all have a common difference in length.

HIP RAFTER THEORY

The theory behind fashioning a hip rafter is quite similar to that used for a common rafter, yet slight differences allow the hip rafter to take a different shape in the roof. The length of the hip rafter can be estimated using a framing square.

Unit Run

Both hip and common rafters have a total run that always runs horizontal under the rafters. A common rafter run forms a 90-degree angle with the plates and the hip run forms a 45-degree angle. Therefore, a hip rafter run covers more distance than does a common run, but the number of units of run is the same. And the total rise is the same.

The hip unit run forms the diagonal of a square created by the common unit run (**Figure 48–2**). Using the Pythagorean theorem with 12 as a and b, c turns out to be 16.97 inches. This amount is rounded off to 17 inches when using a framing square to lay out the rafter. In metric measure,

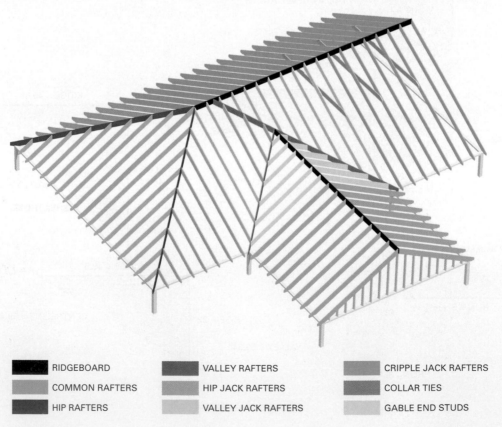

■ RIDGEBOARD		■ VALLEY RAFTERS		■ CRIPPLE JACK RAFTERS	
■ COMMON RAFTERS		■ HIP JACK RAFTERS		■ COLLAR TIES	
■ HIP RAFTERS		■ VALLEY JACK RAFTERS		■ GABLE END STUDS	

FIGURE 48–1 Members of a hip roof frame.

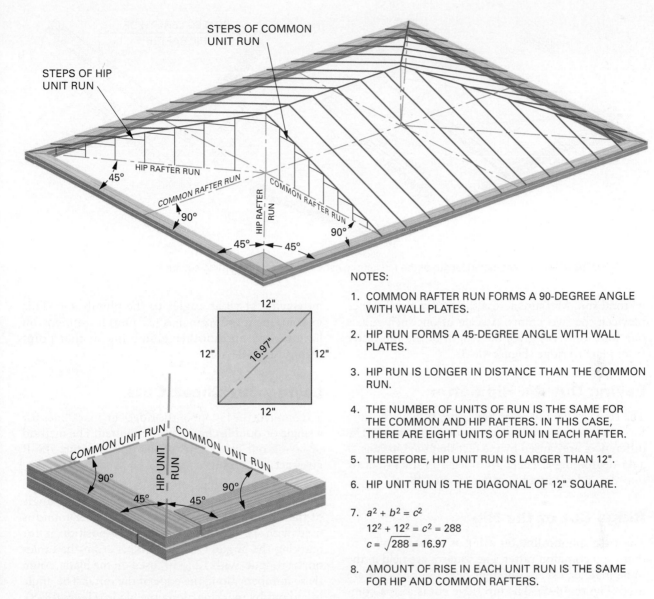

NOTES:

1. COMMON RAFTER RUN FORMS A 90-DEGREE ANGLE WITH WALL PLATES.

2. HIP RUN FORMS A 45-DEGREE ANGLE WITH WALL PLATES.

3. HIP RUN IS LONGER IN DISTANCE THAN THE COMMON RUN.

4. THE NUMBER OF UNITS OF RUN IS THE SAME FOR THE COMMON AND HIP RAFTERS. IN THIS CASE, THERE ARE EIGHT UNITS OF RUN IN EACH RAFTER.

5. THEREFORE, HIP UNIT RUN IS LARGER THAN 12".

6. HIP UNIT RUN IS THE DIAGONAL OF 12" SQUARE.

7. $a^2 + b^2 = c^2$
   $12^2 + 12^2 = c^2 = 288$
   $c = \sqrt{288} = 16.97$

8. AMOUNT OF RISE IN EACH UNIT RUN IS THE SAME FOR HIP AND COMMON RAFTERS.

FIGURE 48–2 The unit of run of the hip rafter is 16.97 inches, which is rounded to 17 inches for layout purposes.

the square has sides of 250 mm and a diagonal of 353.55, or 354 mm. Therefore, the unit hip rafter is 354 mm.

Unit Rise

Both hip and common rafters start at the wall plate and slope up to meet at the ridge, which is at a fixed height. Each rafter has the same number of unit runs under it. Therefore, the rise for each unit run must be the same.

A framing square can be used to lay out a hip rafter. The square is held the same as for a common rafter. The unit rise is held on the tongue and the unit run on the blade. The difference is the unit run is now 17 inches, not 12 (or 354, not 250 mm).

Estimating the Rough Length of the Hip Rafter

The rough length of the hip rafter can be found in the manner previously described for the common rafter. Let the tongue of the square represent the length of common rafter. Let the blade of the square represent

EXAMPLE The common rafter length is 15 feet. Its total run is 12 feet. A measurement across these points scales off to 19′-2½″. It will be necessary to order 20-foot lengths for the hip rafters, unless extra is needed for overhang.

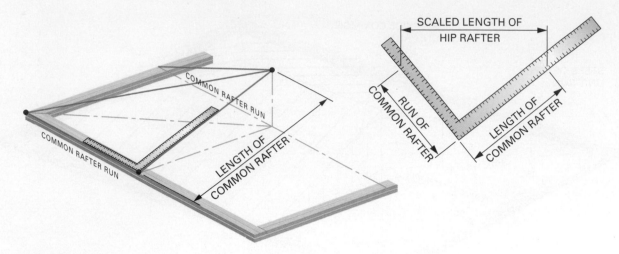

FIGURE 48–3 An estimated length of the hip rafter by scaling across the framing square.

its total run. Measure across the square the distance between the two points. A scale of one inch equals one foot (or 1:10) gives the length of the hip rafter from plate to ridge (**Figure 48–3**).

Laying Out the Hip Rafter

The steps to lay out a hip rafter are similar to those for a common rafter with some differences. The differences are caused by the fact that the hip runs at a 45-degree angle. There are at least 11 lines to draw in order to lay out a hip rafter (**Figure 48–4**).

Ridge Cut of the Hip

The ridge plumb line on a hip is drawn similar to that for a common rafter. The squares are held in the same manner, except different numbers or scales are used (**Figure 48–5**). The hip ridge cut is also a compound angle called a **cheek cut** or *side cut*. A *single cheek* cut or a *double cheek* cut may be made on the hip rafter according to the way it is framed at the ridge (**Figure 48–6**).

Shortening the Hip

The rafter must be shortened before the cheek cuts are laid out. The hip rafter is shortened by one-half the 45-degree thickness of the ridgeboard. The rafter must be shortened due to the ridge before the cheek cuts are laid out (**Figure 48–7**). This is due to the fact that the hip run is at a 45-degree angle.

To lay out the ridge cut, select a straight length of stock for a pattern. Lay it across two sawhorses. Mark a plumb line at the left end. Hold the tongue of the square at the rise and the blade of the square at 17 inches (or 354 mm), the unit of run for the hip rafter. Shorten the rafter by

measuring at right angles to the plumb line. This measurement is 1¹⁄₁₆ inches (27 mm) for dimension lumber. Lay out another plumb line at that point (**Procedure 48–A**).

Laying Out Cheek Cuts

To complete the layout of the ridge cut, mark lines for a single or double cheek cut as required. The method of laying out cheek cuts shown in **Procedure 48–B** gives accurate results.

The bottom line of the rafter tables on the framing square can be used to determine the angle of the side cuts of hip rafters. The number found is used with 12 (250) on a square to position it for marking the angle. Use the number from the tables on the tongue, with 12 (250) used on the blade. Align these numbers along the edge of the rafter. The angle is laid out by marking along the blade (**Figure 48–8**).

The same example in Figure 48–8 with a ⁴⁄₁₂ slope becomes a 1:3, or 183.3:250 in metric units. The side cut for the hip is given as 243 on the Frederickson metric square, and the square is held at 250 and 243, with the line being drawn on the 250 side. (Note that the angle is always less than 45 degrees.)

Another way to get the side cut is to draw the ridge plumb line at the end of the rafter stock (which must be at least 2 inches [51 mm] wider than the common rafter stock), shorten by ½ the 45-degree thickness of the common rafter stock +½ the thickness of the hip rafter stock (1¹³⁄₁₆ inches or 46 mm), and then follow the second plumb cut line with the circular saw set at a 45-degree bevel.

Or simpler still, draw a ridge plumb line about 1 inch (25 mm) from the end of the rafter, square the line over the top, and draw the same line on the other side of the rafter. With the bevel of the saw set

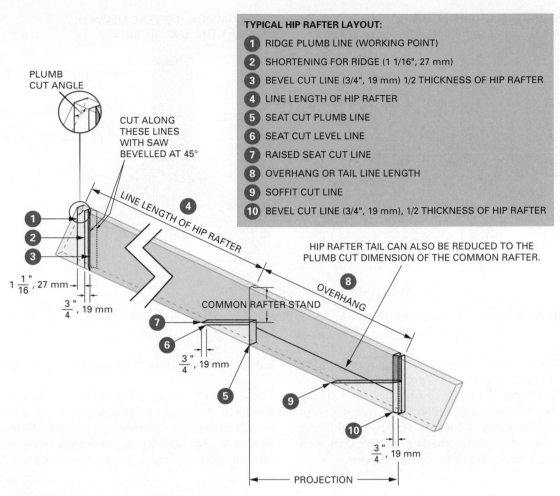

TYPICAL HIP RAFTER LAYOUT:

1 RIDGE PLUMB LINE (WORKING POINT)

2 SHORTENING FOR RIDGE (1 1/16", 27 mm)

3 BEVEL CUT LINE (3/4", 19 mm) 1/2 THICKNESS OF HIP RAFTER

4 LINE LENGTH OF HIP RAFTER

5 SEAT CUT PLUMB LINE

6 SEAT CUT LEVEL LINE

7 RAISED SEAT CUT LINE

8 OVERHANG OR TAIL LINE LENGTH

9 SOFFIT CUT LINE

10 BEVEL CUT LINE (3/4", 19 mm), 1/2 THICKNESS OF HIP RAFTER

PLUMB CUT ANGLE

CUT ALONG THESE LINES WITH SAW BEVELLED AT 45°

LINE LENGTH OF HIP RAFTER

HIP RAFTER TAIL CAN ALSO BE REDUCED TO THE PLUMB CUT DIMENSION OF THE COMMON RAFTER.

OVERHANG

COMMON RAFTER STAND

$1\frac{1}{16}$", 27 mm

$\frac{3}{4}$", 19 mm

$\frac{3}{4}$", 19 mm

$\frac{3}{4}$", 19 mm

PROJECTION

FIGURE 48–4 Layout for a hip rafter is done in steps.

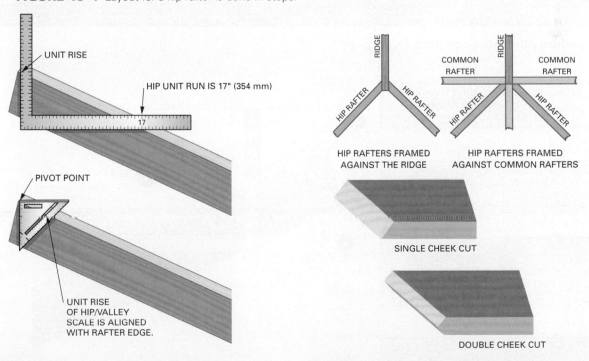

UNIT RISE

HIP UNIT RUN IS 17" (354 mm)

17

PIVOT POINT

UNIT RISE OF HIP/VALLEY SCALE IS ALIGNED WITH RAFTER EDGE.

FIGURE 48–5 Hip plumb lines are drawn in a fashion similar to that for common rafters.

RIDGE

RIDGE

COMMON RAFTER

COMMON RAFTER

HIP RAFTER

HIP RAFTER

HIP RAFTER

HIP RAFTER

HIP RAFTERS FRAMED AGAINST THE RIDGE

HIP RAFTERS FRAMED AGAINST COMMON RAFTERS

SINGLE CHEEK CUT

DOUBLE CHEEK CUT

FIGURE 48–6 Single or double cheek cuts are used depending on the method of framing at the ridge.

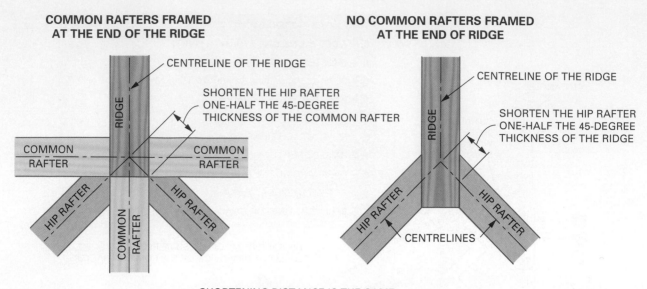

COMMON RAFTERS FRAMED AT THE END OF THE RIDGE

NO COMMON RAFTERS FRAMED AT THE END OF RIDGE

CENTRELINE OF THE RIDGE

SHORTEN THE HIP RAFTER ONE-HALF THE 45-DEGREE THICKNESS OF THE COMMON RAFTER

RIDGE

COMMON RAFTER

COMMON RAFTER

HIP RAFTER

HIP RAFTER

COMMON RAFTER

CENTRELINE OF THE RIDGE

SHORTEN THE HIP RAFTER ONE-HALF THE 45-DEGREE THICKNESS OF THE RIDGE

RIDGE

HIP RAFTER

HIP RAFTER

CENTRELINES

SHORTENING DISTANCE IS THE SAME.

FIGURE 48–7 Amount to shorten the hip rafter for either method of framing at the ridge.

at 45 degrees, cut along both lines to form the point of the hip rafter. From this point, the line length of the hip rafter is measured off; then the seat cut plumb line is drawn and then shortened back toward the top of the rafter before the seat cut level line is drawn. The side cuts for the tail can be taken with the sliding T-bevel from the top of the rafter.

Laying Out the Hip Rafter Length

The length of the hip rafter is laid out in a manner similar to that used for the common rafter. It may be found by the step-off method or calculated. Remember always to start any layout for length from the first ridge plumb line, the one before any shortening.

Step**by**Step Procedures

Procedure 48–A Marking a Rafter Plumb Line

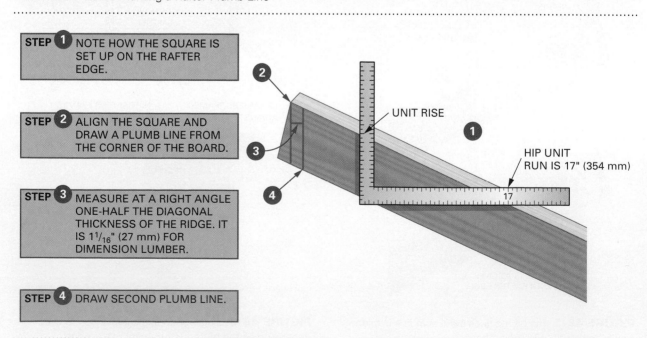

STEP ① NOTE HOW THE SQUARE IS SET UP ON THE RAFTER EDGE.

STEP ② ALIGN THE SQUARE AND DRAW A PLUMB LINE FROM THE CORNER OF THE BOARD.

STEP ③ MEASURE AT A RIGHT ANGLE ONE-HALF THE DIAGONAL THICKNESS OF THE RIDGE. IT IS 1 1/16" (27 mm) FOR DIMENSION LUMBER.

STEP ④ DRAW SECOND PLUMB LINE.

UNIT RISE

HIP UNIT RUN IS 17" (354 mm)

17

StepbyStep Procedures

Procedure 48–B Making Cheek Cuts Using the Measurement Method

SINGLE CHEEK LINE

STEP 1 SQUARE A LINE ACROSS THE TOP EDGE FROM THE SECOND PLUMB LINE. THE SECOND PLUMB LINE IS 1$\frac{1}{16}$" (27 mm) PERPENDICULARLY AWAY FROM THE FIRST PLUMB LINE—ONE-HALF THE 45-DEGREE THICKNESS OF THE HIP RAFTER.

STEP 2 MARK THE CENTRE OF THE TOP EDGE.

STEP 3 FROM THE SECOND LINE, MEASURE AT A RIGHT ANGLE ONE-HALF THICKNESS OF THE HIP RAFTER. IT IS $\frac{3}{4}$" (19 mm) FOR DIMENSION LUMBER.

STEP 4 DRAW THIRD PLUMB LINE.

STEP 5 DRAW A DIAGONAL LINE ACROSS THE TOP FROM THE THIRD LINE THROUGH THE CENTRELINE.

DOUBLE CHEEK LINES

STEP 1 SQUARE ACROSS FROM THE THIRD PLUMB LINE.

STEP 2 DRAW A DIAGONAL LINE FROM SQUARED LINE THROUGH THE CENTRELINE.

STEP 3 A FOURTH PLUMB LINE MAY BE DRAWN AS A CUT LINE.

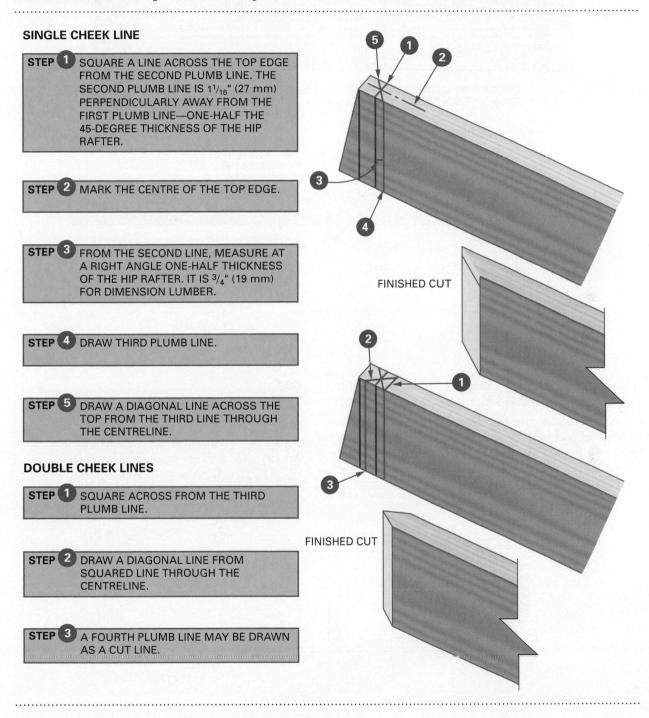

FINISHED CUT

FINISHED CUT

Stepping Off the Hip. In the step-off method, the number of steps for the hip rafter is the same as for the common rafter in the same roof. The same rise is used, but the unit of run for the hip is 17, not 12 (354, not 250).

For example, for a roof with a rise of 6 inches per foot of run, the square is held at 6 and 12 for the common rafter, and 6 and 17 for the hip rafter of the same roof. In metric, for a roof with a unit rise of 150 mm, the

CHEEK CUTS USING RAFTER TABLES

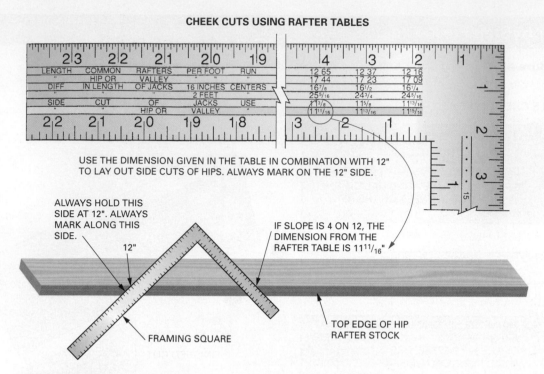

USE THE DIMENSION GIVEN IN THE TABLE IN COMBINATION WITH 12" TO LAY OUT SIDE CUTS OF HIPS. ALWAYS MARK ON THE 12" SIDE.

ALWAYS HOLD THIS SIDE AT 12". ALWAYS MARK ALONG THIS SIDE.

12"

IF SLOPE IS 4 ON 12, THE DIMENSION FROM THE RAFTER TABLE IS 11¹¹/₁₆"

FRAMING SQUARE

TOP EDGE OF HIP RAFTER STOCK

FIGURE 48–8 Cheek cuts using the rafter table method.

square would be held at 150 and 250 for the common rafter and at 150 and 354 for the hip rafter.

If the total run of the common rafter contains a fractional part of 12, then the hip run must contain the same fractional part of 17. In other words, if the common run is 12½ steps of 12 inches, then the hip run is 12½ steps of 17 inches.

EXAMPLE If the common run distance is 15′-9″, then the run is 15⁹/₁₂ or 15³/₄ steps. The hip run is the same number of steps. In this case, 15 steps are made. The last step is ³/₄ of 17 or 12³/₄ inches along the blade (**Figure 48–9**).

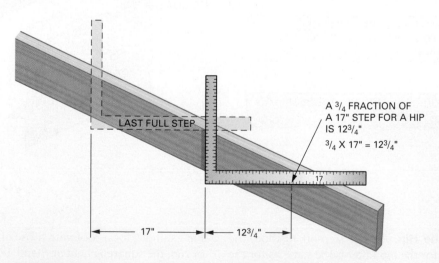

LAST FULL STEP

A ³/₄ FRACTION OF A 17" STEP FOR A HIP IS 12³/₄"

³/₄ X 17" = 12³/₄"

17"

17"

12³/₄"

FIGURE 48–9 Stepping off a fractional unit of run on the hip rafter.

Using Rafter Tables. Finding the length of the hip using the tables found on the framing square is similar to the process used for finding the length of the common rafter. However, the numbers from the second line are used instead of the first line. These numbers are the unit length of the hip rafter in inches or length per unit of run (or in millimetres per 250 mm of run).

As with common rafters, the numbers in the rafter tables can be calculated using the Pythagorean theorem. The unit rise and unit run are used as *a* and *b* to find *c*, unit length. A common rafter uses unit rise and 12 inches, where a hip uses unit rise and 16.97 inches. It is important not to use the rounded-off number of 17 in these calculations.

If the total run of the rafter contains a fractional part of a foot, multiply the figure in the tables by the whole number of feet plus the fractional part changed to a decimal. For instance, if the total run of

the rafter is 15′-6″, multiply the figure in the tables by 15½ changed to 15.5.

Lay out the length obtained along the top edge of the hip rafter stock. This must be done from the first plumb line before shortening and cheek cuts are made (**Figure 48–10**).

Laying Out the Seat Cut of the Hip Rafter

Like the common rafter, the seat cut of the hip rafter is a combination of plumb and level cuts. The height above the bird's mouth on a hip rafter is the same distance as the common rafter. To locate where the seat level line should be, first measure down along the plumb line from the top of the common rafter to its seat cut. This measurement is called the rafter stand. Then mark that same measurement on the hip. No consideration needs to be given to fitting it around the corner of the wall.

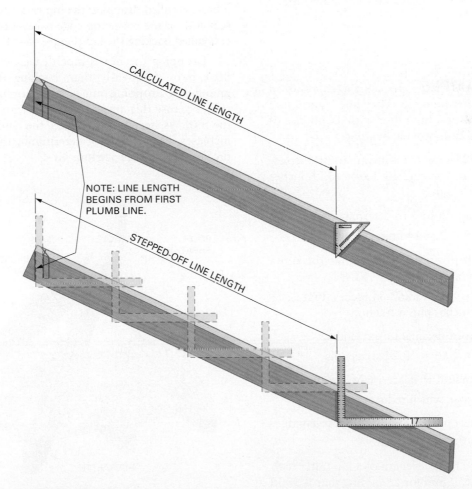

NOTE: LINE LENGTH BEGINS FROM FIRST PLUMB LINE.

CALCULATED LINE LENGTH

STEPPED-OFF LINE LENGTH

17

FIGURE 48–10 Line length measurement begins at the first plumb line.

EXAMPLE What is the unit length for a hip rafter with a unit rise of 7 inches?

$$a^2 + b^2 = c^2$$
$$7^2 + 16.97^2 = c^2$$
$$49 + 287.9809 = 336.9809 = c^2$$
$$c = \sqrt{336.9809} = 18.35704 \text{ inches}$$

Unit length = 18.36 inches rounded to the nearest hundredth. This number is also found in the rafter table: On the second line under the 7 is 18 36, which is 18.36.

As with the common rafter, the hip length is then found by multiplying the unit length by the total run of a common rafter. This is because the number of units of run under a common rafter is the same as for a hip.

Similarly in metric, the unit hip rafter for a roof with a slope of 100:250 (1:2.5) is found by adding 100^2 to 353.55^2 and taking the square root of the sum to get 367.4 mm. The framing square is held at 100 and 367.4 and stepped off the same number of times as the common rafter.

EXAMPLE To use the calculation method to determine hip rafter line length, find the line length of a hip rafter with a unit rise of 8 inches for a 28-foot-wide building.

STEP 1: Read below the 8-inch mark in the second row to find 18 76, or 18.76 inches unit length.

STEP 2: Divide 28 by 2 to determine the run.
$$28 \div 2 = 14 \text{ units of run}$$

STEP 3: Multiply unit length times the units of run.
$$18.76 \times 14 = 262.64 \text{ inches (that is,} $$
$$262 \text{ inches plus a fraction)}$$

STEP 4: Convert decimal to sixteenths.
$$0.64 \times 16 = 10.24 \text{ sixteenths}$$

STEP 5: Round off to nearest whole sixteenths.
$$10.24 \Rightarrow {}^{10}/_{16}, \text{ which reduces to } {}^5/_8 \text{ inches}$$

STEP 6: Collect information into rafter length.
$$262 + {}^5/_8'' = 262{}^5/_8''$$

METRIC: Find the line length of a hip rafter with a unit rise of 150 mm for a building that is 9000 mm wide.

Span = 9000 mm; run = 9000 ÷ 2 = 4500 mm; number of unit runs = 4500 ÷ 250 = 18

STEP 1: Read below 150 mm on the second line of the metric square to find 384.1 mm unit length for the hip rafter.

STEP 2: Multiply the unit length by the number of unit runs.
$$384.1 \times 18 = 6913.8 \text{ mm}$$

Therefore, the total line length of the hip rafter is 6913.8 mm.

Backing and Dropping the Hip. When making the level cut of the seat, special consideration must be given. The hip rafter is at the intersection of the two roof slopes. The centre of the top edge is in perfect alignment with the other rafters, but the top outside corners of the top edge surface stick above the sheathing. To remedy this situation, the seat level line is cut deeper, a process called **dropping the hip** rafter (or raising the seat cut), or the rafter top edge is bevel planed, a process called **backing the hip** (**Figure 48–11**).

Dropping the hip is much easier to do, so it is done more frequently than backing the hip. The amount of dropping must be determined for each roof, because this amount changes with the slope of the roof. As steepness increases, the amount of drop increases. The process for determining the amount of drop is shown in **Procedure 48–C**.

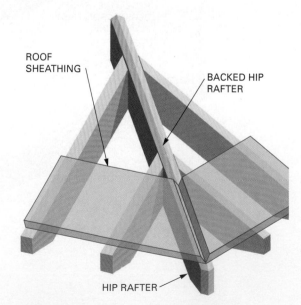

FIGURE 48–11 The hip rafter must be adjusted to allow for the roof sheathing.

Step**by**Step Procedures

Procedure 48–C Dropping or Backing a Hip Rafter

STEP ❶ MEASURE HEIGHT ABOVE BIRD'S MOUTH ON COMMON RAFTER. MEASURE AND MARK THE SAME DISTANCE DOWN ON THE PLUMB LINE.

STEP ❷ DRAW SEAT LEVEL LINE THROUGH THIS MEASUREMENT.

STEP ❸ FROM LINE 1, MEASURE PERPENDICULAR TO PLUMB LINE TOWARD RIDGE CUT, A DISTANCE EQUAL TO ONE-HALF THE HIP RAFTER THICKNESS. THIS IS USUALLY $^3/_4$" (19 mm) FOR DIMENSION LUMBER.

STEP ❹ DRAW ANOTHER PLUMB LINE OF THE SAME LENGTH AS THE HEIGHT ABOVE THE BIRD'S MOUTH FOR COMMON RAFTER.

STEP ❺ DRAW THE DROPPED SEAT CUT LINE.

STEPS ❻&❼ NOTE THE BIRD'S MOUTH IS CUT ALONG THE FIRST PLUMB LINE AND SECOND LEVEL LINE.

STEP ❽ THE AMOUNT AND ANGLE OF BEVEL FOR BACKING THE HIP RAFTER IS DETERMINED FROM THE ROOF TRIANGLE THAT HAS A RUN OF THE UNIT HIP RAFTER AND A RISE OF THE UNIT RISE. DROP DISTANCE IS **NOT** THE SAME AMOUNT OF BEVEL FOR BACKING HIP RAFTER. (SEE FIGURE 48–12A and B)

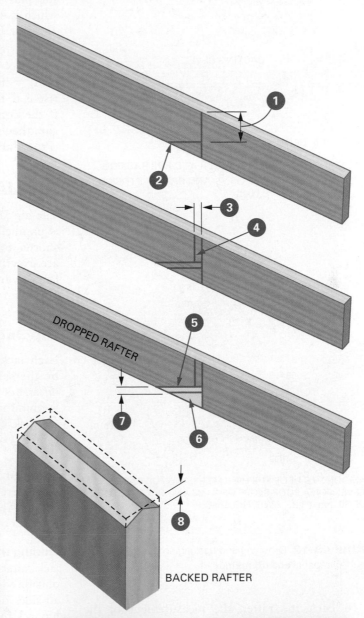

DROPPED RAFTER

BACKED RAFTER

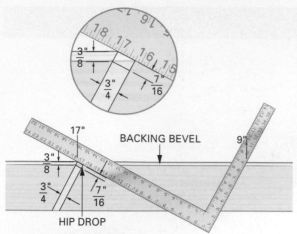

AMOUNT TO DROP THE HIP RAFTER AND FOR BACKING BEVEL FOR A ROOF SLOPE OF $\frac{9}{12}$ AND RAFTER STOCK OF $1\frac{1}{2}"$. (N.B. ...*NOT* THE SAME)

A

AMOUNT TO DROP THE HIP RAFTER AND FOR BACKING BEVEL FOR A ROOF SLOPE OF 1:1.33 AND RAFTER STOCK OF 38 mm. (N.B. ...*NOT* THE SAME)

B

FIGURE 48–12 Dropping and backing amounts for hip rafter: (A) imperial and (B) metric.

To back the rafter, the measurement of the drop is NOT used to determine the bevel. The difference between the two measurements is shown in **Figure 48–12A** and **B**. The angles used for cutting the hip roof are shown in **Figure 48–13A** and **B** and in the geometric layout in **Figure 48–14**.

LAYING OUT THE TAIL OF THE HIP RAFTER

The hip tail can be found in one of two ways: calculate the overhang or calculate the projection. Hip tail projection is determined with the fractional step-off

method, as shown earlier in Figure 48–9. For example, if the common rafter projection is 15 inches, then the run is $15 \div 12 = 1.25$, or 1¼ units of run. The hip tail has the same run. To step this off, mark one full step plus a quarter step or ¼ × 16.97 = 4.25 inches, which is 4¼ inches. The total hip projection here is 16.97 or 17 + 4¼ = 21¼ inches.

The overhang calculation uses the same formula as rafter length: unit length times run (tail) = line length (tail). For example, if the unit rise is 6, then the unit length of a hip is 18 inches. If the common rafter projection is 1.25 units of run, then the overhang is 18 × 1.25 = 22½ inches (**Procedure 48–D**).

HIP JACK RAFTERS

The hip jack rafter is a shortened common rafter. It is parallel to the common rafter, but is simply cut shorter to meet the hip. The seat cut and the tail are exactly like those of a common rafter, only the length varies, as shown earlier in Figure 48–1 (see also Figures 48–13 and 48–14).

Hip Jack Length

Each jack is shorter or longer than the next set by the same amount. This is called the *common difference* (**Figure 48–15**). The common difference is found in the rafter tables on the framing square for jacks 16 and 24 inches (406 and 610 mm) on centre (OC).

Once the length of one jack is determined, the length of all others can be found by making each set shorter or longer by the common difference. To find the length of any jack, its total run must be known.

Finding the Total of the Run. The total run of any jack rafter and its distance from the corner along the outside edge of the wall plate form a square. Because all sides of a square are equal, the total run of any hip jack rafter is equal to its distance from the corner of the building. To determine the jack run, measure the distance from the corner to the centre of where the jack will sit on the wall plate (**Figure 48–16**).

Find the length of the hip jack rafter by the step-off method, or use the rafter tables in the same way as for common rafters. The only difference is that the total run of the jack found above is used. Use the common rafter pattern to lay out the jack tail.

Hip Jack Shortening. Because the hip jack rafter meets the hip rafter at a 45-degree angle, it must be

StepbyStep Procedures

Procedure 48–D Laying Out a Hip Rafter Tail

STEP 1 UNIT RISE = 6, THEN UNIT LENGTH IS 18 FROM THE RAFTER TABLES.
UNIT RISE = 125 mm
UNIT HIP RAFTER = 375.0 mm

STEP 2 PROJECTION IS 15".
15 ÷ 12 = 1.25 UNITS OF RUN.
PROJECTION = 400 mm
400 ÷ 250 = 1.6 UNITS OF RUN.

STEP 3 OVERHANG LENGTH:
18 X 1.25 = 22$\frac{1}{2}$".
HIP RAFTER TAIL = 375.0 X 1.6 mm = 600 mm

STEP 4 MEASURE THE OVERHANG LENGTH.

STEP 5 DRAW FASCIA PLUMB LINE.

STEP 6 SQUARE A LINE ACROSS THE TOP OF THE RAFTER FROM THIS LINE AND MARK THE CENTRE.

STEP 7 FROM LINE 6, MEASURE PERPENDICULAR TOWARD THE RIDGE A DISTANCE EQUAL TO ONE-HALF THE HIP THICKNESS.

STEP 8 DRAW A SECOND PLUMB LINE AND SQUARE A LINE ACROSS THE TOP.

STEP 9 DRAW DIAGONALS ACROSS THE TOP FROM THE ENDS OF THE SQUARED LINE THROUGH THE CENTRE MARK.

PLUMB LINE OF HIP RAFTER SEAT CUT

FINISHED CUT

THE STOCK OF THE HIP RAFTER (VALLEY RAFTER) IS USUALLY ONE SIZE BIGGER THAN THAT OF THE COMMON RAFTER. THEREFORE, THE TAIL "HANGS" DOWN BELOW THE BOTTOM OF THE FASCIA BOARD. IT IS COMMON PRACTICE TO RIP THE TAIL OF THE HIP RAFTER (VALLEY RAFTER) TO THE DIMENSION OF THE PLUMB CUT OF THE COMMON RAFTER; ALTERNATIVELY, IT CAN BE SQUARED BACK TO THE BUILDING SO THAT IT DOES NOT PROJECT BELOW THE BOTTOM OF THE FASCIA BOARD.

shortened one-half the 45-degree angle or diagonal thickness of the hip rafter stock. Measure the distance at right angles to the original plumb line toward the tail. Draw another plumb line (**Procedure 48–E**).

Hip Jack Cheek Cuts. The hip jack rafter has a single cheek cut where it lands on the hip rafter. Square the shortened plumb line across the top of the rafter. Draw an intersecting centreline along the top edge. Measure back toward the tail at right angles from the second plumb line a distance equal to one-half the jack rafter stock. Draw another plumb line. On the top edge, draw a diagonal from the third plumb line through the intersection of the centreline. Take care when drawing the diagonal

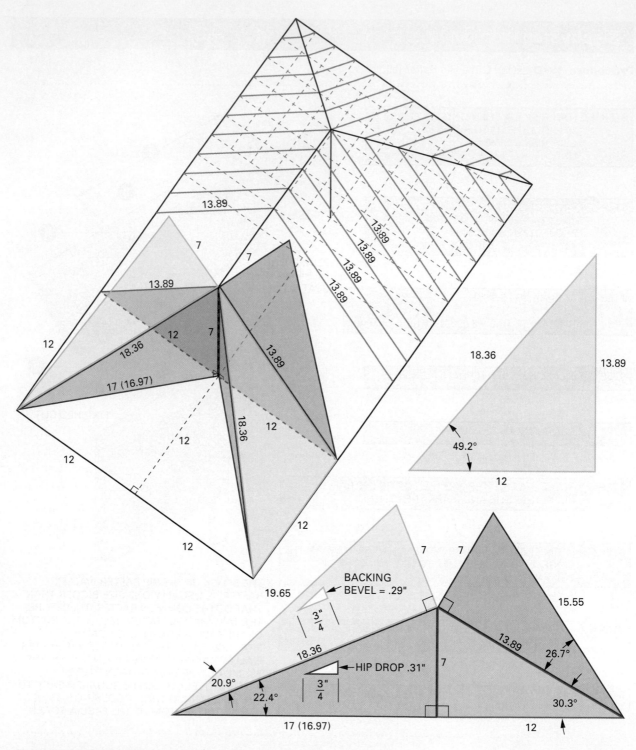

COMMON RAFTER PLUMB CUT ANGLE = 30.3° (59.7°)
HIP RAFTER PLUMB CUT ANGLE = 22.4° (67.6°)
HIP RAFTER DROP = .31" OR 7.8 mm
HIP RAFTER BACKING BEVEL AMOUNT = .29" OR 7.3 mm
HIP RAFTER BACKING BEVEL ANGLE = 20.9°
PLYWOOD FACE ANGLE AT HIP = 49.2°
PLYWOOD MITRE ANGLE AT HIP = 20.9°
N.B. FOR HIP AND JACK SIDE CUTS, SET SAW BASE AT 45°

FIGURE 48–13A Roof angles for a hip roof with a slope factor of 1:1.71 or 7 in 12.

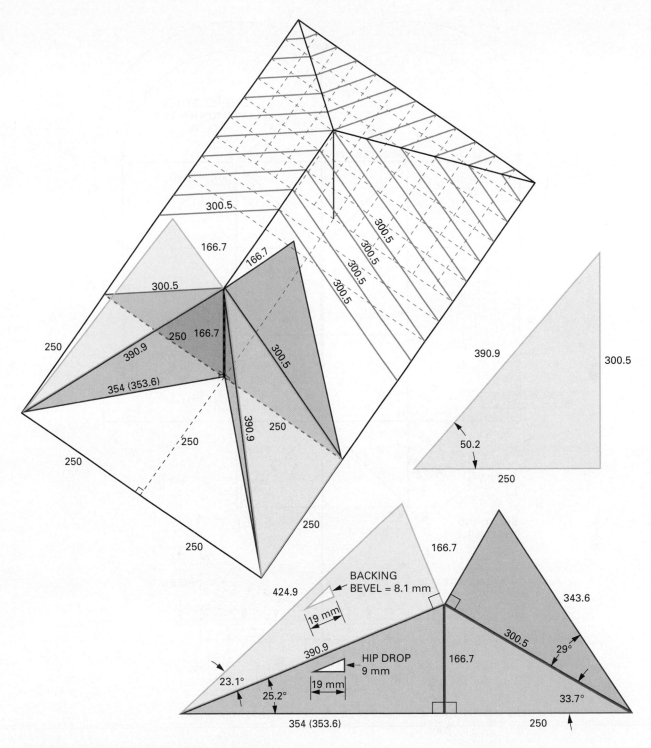

COMMON RAFTER PLUMB CUT ANGLE = 33.7° (56.3°)
HIP RAFTER PLUMB CUT ANGLE = 25.2° (64.8°)
HIP RAFTER DROP = 9.0 mm
HIP RAFTER BACKING BEVEL AMOUNT = 8.1 mm
HIP RAFTER BACKING BEVEL ANGLE = 23.1°
PLYWOOD FACE ANGLE AT HIP = 50.2°
PLYWOOD MITRE ANGLE AT HIP = 23.1°
N.B. FOR HIP AND JACK SIDE CUTS, SET SAW BASE to 45°

FIGURE 48–13B Roof angles for a hip roof with a slope factor of 1:1.5 or 8 in 12.

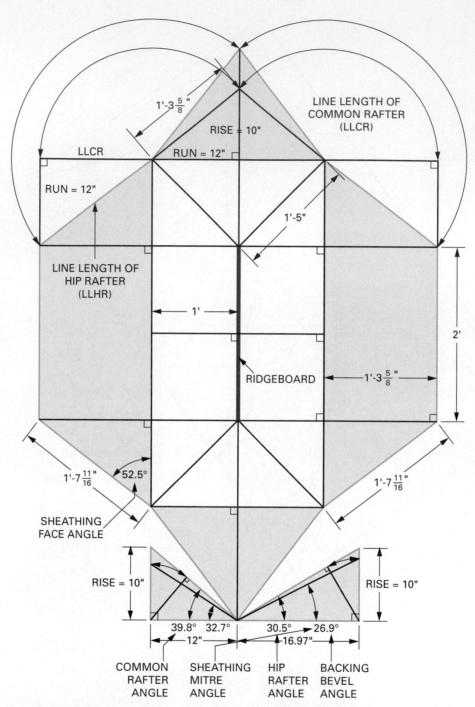

FIGURE 48–14 Geometric layout of the hip roof with associated angles.

line on the top edge. The direction of the diagonal depends on which side of the hip the jack rafter is framed.

Determine the common difference of hip jack rafters from the rafter tables under the inch mark that coincides with the slope of the roof.

The common difference may also be calculated. For rafters spaced 24 inches OC, the common difference is equal to the common rafter length for two units of run. For rafters spaced 16 inches OC, the common difference is equal to the common rafter length for $1\frac{1}{3}$ units of run (**Procedure 48–F**).

Similarly for metric roof framing, the common difference for jack rafters at 400 mm OC is 1.6 × the unit common rafter, at 500 mm OC. It is 2 × the unit CR, and at 600 mm OC it is 2.4 × the unit CR.

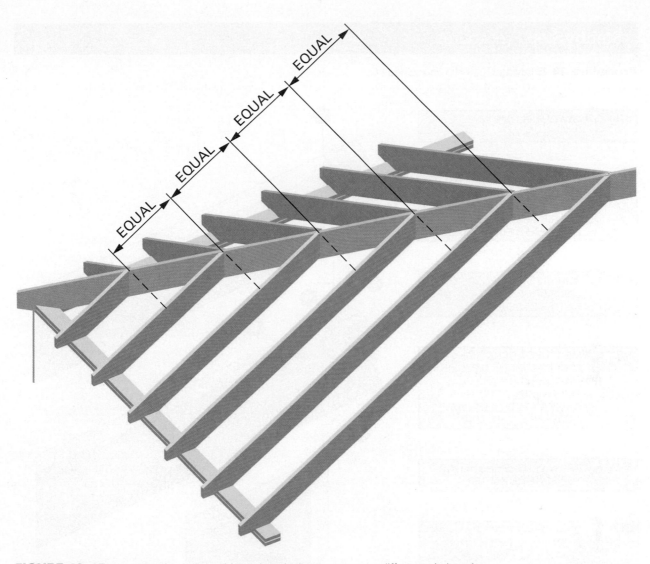

FIGURE 48–15 Hip jack rafters, like gable end studs, have a common difference in length.

Once the common difference is determined, measure the distance successively along the top edge of the pattern for the longest jack rafter. This pattern is then used to cut all the jack rafters necessary to frame that section of roof. This process is repeated for each section on jacks. Some jacks are framed in pairs, depending on the width and length dimensions of the house.

Hip Roof Ridge Length

The hip roof ridge length is shorter than for a gable roof on the same sized building. To find the

ridgeboard length, it is helpful to visualize the run of the major components of a roof (**Figure 48–17**). The length of the building under the ridgeboard is made up of two common runs and the ridgeboard length. Also, two common runs are equal to the width. To find the (theoretical) line length of the ridgeboard, simply subtract the width from the length.

The actual length is longer than the line length. The amount to add depends on two styles of framing. One style has a common rafter framed at the end of the ridgeboard. The other has no common rafter at the end (**Figure 48–18**).

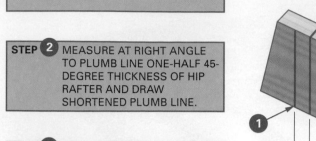

StepbyStep Procedures

Procedure 48–E Laying Out a Hip Jack Cheek Cut

STEP 1 LAY OUT PLUMB LINE.

STEP 2 MEASURE AT RIGHT ANGLE TO PLUMB LINE ONE-HALF 45-DEGREE THICKNESS OF HIP RAFTER AND DRAW SHORTENED PLUMB LINE.

STEP 3 SQUARE SHORTENED PLUMB LINE ACROSS TOP EDGE OF RAFTER STOCK.

STEP 4 MEASURE AT RIGHT ANGLE FROM SHORTENED PLUMB LINE ONE-HALF THE THICKNESS OF THE JACK RAFTER STOCK AND DRAW ANOTHER PLUMB LINE.

STEP 5 DRAW CENTRELINE ALONG TOP EDGE OF RAFTER.

STEP 6 DRAW DIAGONAL FROM LAST PLUMB LINE THROUGH CENTRELINE. THIS DIAGONAL LINE CAN ALSO BE PRODUCED BY USING THE NUMBER ON LINE 5 OF THE RAFTER TABLES FOUND ON THE FRAMING SQUARE.

STEP 7 THE AMOUNT AND ANGLE OF BEVEL FOR BACKING THE HIP RAFTER IS DETERMINED FROM THE ROOF TRIANGLE THAT HAS A RUN OF THE UNIT HIP RAFTER AND A RISE OF THE UNIT RISE.

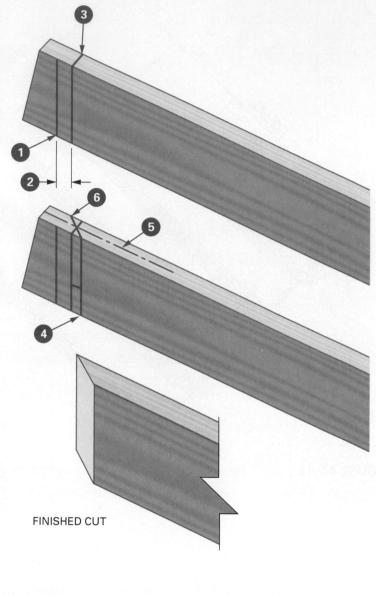

FINISHED CUT

The amount to add in the first case is one-half the thickness of the common rafter at each end. This is ¾ inch (19 mm) for dimension lumber. The amount to add when no common rafter is framed at the end is one-half the thickness of the common rafter plus one-half the diagonal thickness of the hip. For dimension lumber, this is ¾ + 1¹/₁₆ = 1¹³/₁₆ inches (46 mm) added to each end.

StepbyStep Procedures

Procedure 48–F Calculating Jack Rafter Common Difference

THE JACK RAFTER SPACING AND RUN OF ITS COMMON DIFFERENCE IN LENGTH
FORM THE SIDES OF A SQUARE AND ARE THEREFORE EQUAL TO EACH OTHER.

EXAMPLE: IF THE UNIT RISE IS 6" AND THE RAFTER SPACING IS 16", CALCULATE
THE COMMON DIFFERENCE IN LENGTH FOR A JACK.

STEP 1 UNIT RISE = 6" THEN THE UNIT LENGTH IS 13.42" FOR A COMMON RAFTER.

STEP 2 CONVERT 16" OC SPACING TO UNITS OF RUN: 16 ÷ 12 = 1/3 UNITS.

STEP 3 MULTIPLY UNIT LENGTH X RUN: 13.42 X 1 1/3 = 17.89".

STEP 4 COMMON DIFFERENCE IS 17 7/8".

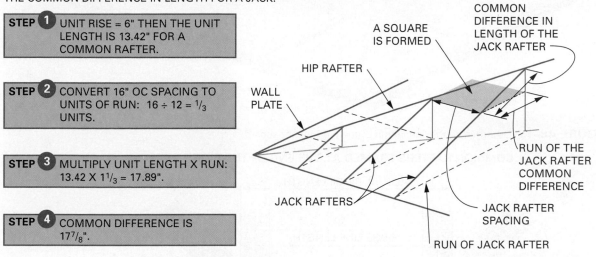

COMMON DIFFERENCE IN LENGTH OF THE JACK RAFTER

A SQUARE IS FORMED

HIP RAFTER

WALL PLATE

JACK RAFTERS

RUN OF THE JACK RAFTER COMMON DIFFERENCE

JACK RAFTER SPACING

RUN OF JACK RAFTER

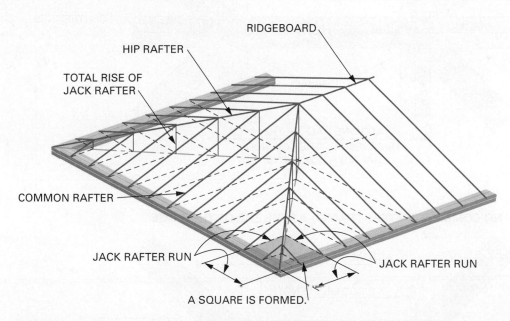

RIDGEBOARD

HIP RAFTER

TOTAL RISE OF JACK RAFTER

COMMON RAFTER

JACK RAFTER RUN

JACK RAFTER RUN

A SQUARE IS FORMED.

FIGURE 48–16 The total run of any hip jack is equal to its distance from the corner of the building.

Raising the Hip Roof

Before raising the hip roof, the ridgeboard must first be laid out for the location of the common rafters (**Procedure 48–G**). Measure from the outside corner the common run distance. Mark this length on the plate. Deduct the distance that the ridgeboard was lengthened on one end. This is ¾ or 1 13/16 inches (19 or 46 mm) for dimension lumber. Measure from

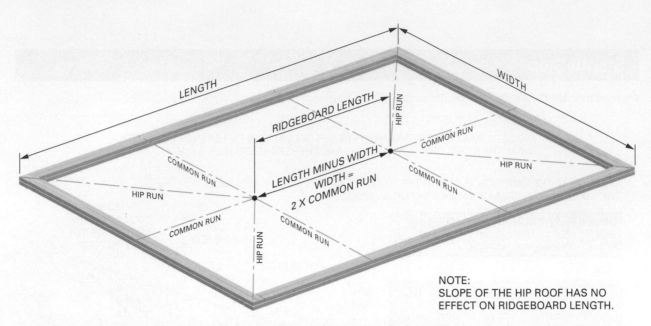

FIGURE 48–17 Determining the line length of the hip roof ridgeboard.

COMMON RAFTERS FRAMED AT THE END OF THE RIDGE

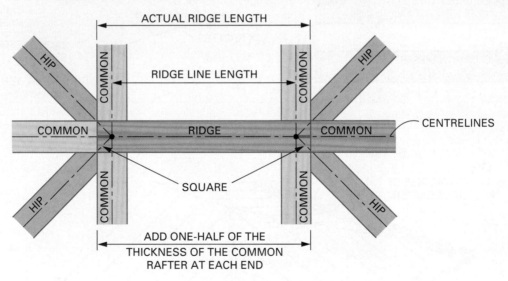

NO COMMON RAFTERS FRAMED AT THE END OF THE RIDGE

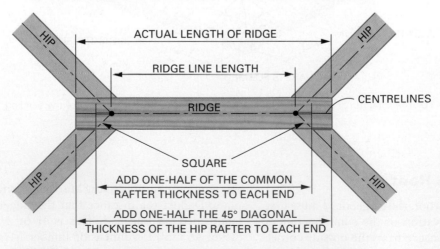

FIGURE 48–18 Determining the actual length of the hip roof ridge.

StepbyStep Procedures

Procedure 48–G Laying Out the Hip Ridgeboard

STEP 1 MEASURE ALONG WALL FROM THE CORNER THE DISTANCE EQUAL TO THE COMMON RUN. MARK ON THE WALL PLATE.

STEP 2 DEDUCT BACK TOWARD THE CORNER THE AMOUNT ADDED TO THE RIDGEBOARD (DETERMINED FROM FIGURE 48–18).

STEP 3 MEASURE FROM THIS MARK TO THE SIDE OF THE NEXT COMMON RAFTER.

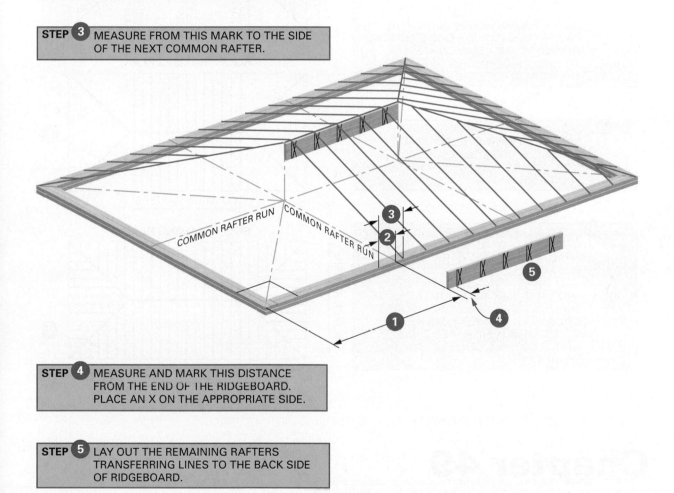

COMMON RAFTER RUN COMMON RAFTER RUN

STEP 4 MEASURE AND MARK THIS DISTANCE FROM THE END OF THE RIDGEBOARD. PLACE AN X ON THE APPROPRIATE SIDE.

STEP 5 LAY OUT THE REMAINING RAFTERS TRANSFERRING LINES TO THE BACK SIDE OF RIDGEBOARD.

this second mark to the next common rafter. This distance can then be transferred to the ridgeboard, measured from the end. Mark an X on the appropriate side.

Erect the common roof section in the same manner as described previously for a gable roof. If the hip rafters are framed against the ridge, install them next. If they are framed against the common

rafters, install the common rafters against the end of the ridge. Then install the hip rafters.

Fasten jack rafters to the plate and to hip rafters in pairs. As each pair of jacks is fastened, sight the hip rafter along its length. Keep it straight. Any *bow* in the hip is straightened by driving the jack a little tighter against the bowed-out side of the hip as the roof is framed (**Procedure 48–H**).

StepbyStep Procedures

Procedure 48–H Erecting the Hip Roof Frame

STEP 1 ERECT THE RIDGE WITH ONLY AS MANY COMMON RAFTERS AS NEEDED.

STEP 2 INSTALL THE HIP RAFTERS. THE HIP RAFTERS EFFECTIVELY BRACE THE ASSEMBLY.

NOTE: THIS PROCEDURE SHOWS THE HIPS FRAMED TO THE RIDGE. IF HIPS ARE FRAMED TO THE COMMON RAFTERS, INSTALL THE COMMONS TO THE END OF THE RIDGE BEFORE THE HIPS.

STEP 3 INSTALL THE REMAINING COMMON RAFTERS IN PAIRS OPPOSING EACH OTHER. SIGHT THE TOP EDGE OF THE RIDGE FOR STRAIGHTNESS AS FRAMING PROGRESSES.

STEP 4 INSTALL THE HIP JACK RAFTERS IN PAIRS OPPOSING EACH OTHER. SIGHT THE HIP FOR STRAIGHTNESS AS JACKS ARE INSTALLED.

NOTE: IT IS BEST TO INSTALL A PAIR OF JACKS ABOUT HALFWAY UP THE HIP TO STRAIGHTEN IT. THE HIP CAN BE STRAIGHTENED BY DRIVING THE JACK, ON THE CROWN SIDE OF THE BOW, DOWN TIGHTER.

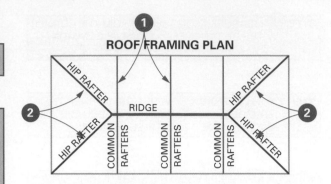

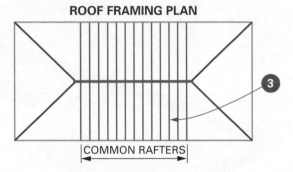

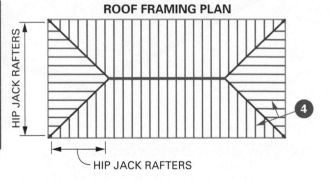

Chapter 49

Intersecting Roofs with Equal Pitch

Buildings of irregular shape, such as L-, H-, or T-shaped buildings, require a roof for each section. These roofs meet at an inside corner intersection, where **valleys** are formed. They are called *intersecting roofs*. The roof may be a gable, a hip, or a combination of types.

THE INTERSECTING ROOF

The intersecting roof requires the layout and installation of several kinds of rafters not previously described (**Figure 49–1**). Some buildings have sections of different widths referred to as

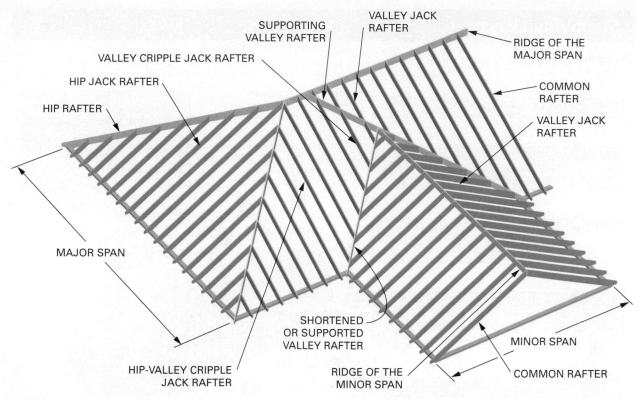

FIGURE 49–1 Members of the intersecting roof frame.

major spans and **minor spans**. A major span is the width of the main part of the building, whereas a minor span refers to the width of any extension to the building.

Valley rafters form the inside corner intersection of two roofs. If the heights of both roofs are different, two kinds of valley rafters are required.

Supporting valley rafters run from the plate to the ridge of the main roof. Their run is one-half of the major span.

Shortened or supported valley rafters run from the plate to the supporting valley rafter. Their run is one-half of the minor span.

Valley jack rafters run from the ridge to the valley rafter.

Valley cripple jack rafters run between the supporting and shortened or supported valley rafter.

Hip-valley cripple jack rafters run between a hip rafter and a valley rafter.

Confusion concerning the layout of so many different kinds of rafters can be eliminated by remembering the following:

- The length of any kind of rafter can be found using its run.
- The run of each rafter is found on the plan view of the roof framing diagram.

- Hip and valley rafters are similar in that they run at 45 degrees to the building line. Common, jack, and cripple jack rafters are similar in that they run at right angles to the building line (orthogonally).
- The method previously described for laying out cheek cuts works on all rafters for any slope.

The amount of shortening is always measured at right angles to the plumb cut.

Supporting Valley Rafter Layout

The layout of a supporting valley rafter is more similar than not to that of a hip rafter. The unit of run for both is 17 inches (354 mm). The total run for both is the same, one-half the major span. The valley rafter is shortened by half the 45 degree or diagonal thickness of the ridgeboard (**Procedure 49–A**). Cheek cuts are made at the ridge, depending on how the building is framed.

The line length of either valley is found by the step-off method or by the calculation method, using the formula unit length × run.

Laying Out a Valley Rafter Seat Cut

Valley rafters are not dropped or backed like hip rafters. Instead, the seat cut line is made with

StepbyStep Procedures

Procedure 49–A Laying Out a Valley Rafter

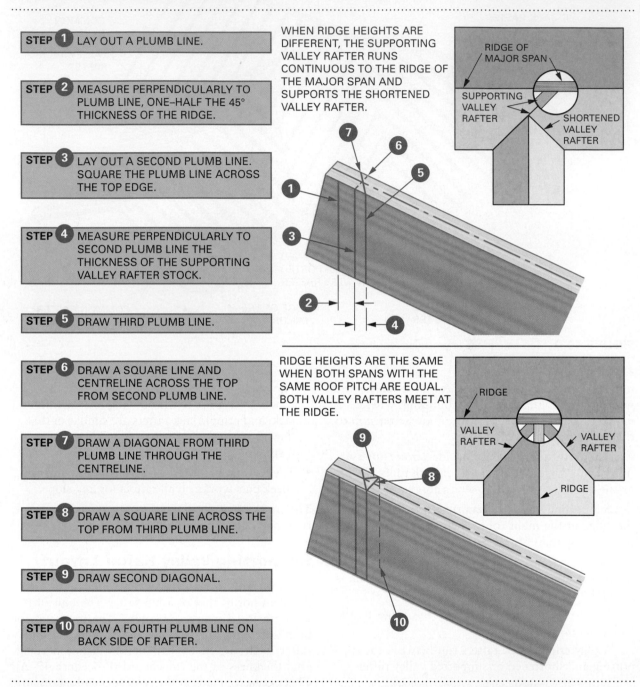

STEP 1 LAY OUT A PLUMB LINE.

STEP 2 MEASURE PERPENDICULARLY TO PLUMB LINE, ONE–HALF THE 45° THICKNESS OF THE RIDGE.

STEP 3 LAY OUT A SECOND PLUMB LINE. SQUARE THE PLUMB LINE ACROSS THE TOP EDGE.

STEP 4 MEASURE PERPENDICULARLY TO SECOND PLUMB LINE THE THICKNESS OF THE SUPPORTING VALLEY RAFTER STOCK.

STEP 5 DRAW THIRD PLUMB LINE.

STEP 6 DRAW A SQUARE LINE AND CENTRELINE ACROSS THE TOP FROM SECOND PLUMB LINE.

STEP 7 DRAW A DIAGONAL FROM THIRD PLUMB LINE THROUGH THE CENTRELINE.

STEP 8 DRAW A SQUARE LINE ACROSS THE TOP FROM THIRD PLUMB LINE.

STEP 9 DRAW SECOND DIAGONAL.

STEP 10 DRAW A FOURTH PLUMB LINE ON BACK SIDE OF RAFTER.

WHEN RIDGE HEIGHTS ARE DIFFERENT, THE SUPPORTING VALLEY RAFTER RUNS CONTINUOUS TO THE RIDGE OF THE MAJOR SPAN AND SUPPORTS THE SHORTENED VALLEY RAFTER.

RIDGE OF MAJOR SPAN

SUPPORTING VALLEY RAFTER

SHORTENED VALLEY RAFTER

RIDGE HEIGHTS ARE THE SAME WHEN BOTH SPANS WITH THE SAME ROOF PITCH ARE EQUAL. BOTH VALLEY RAFTERS MEET AT THE RIDGE.

RIDGE

VALLEY RAFTER

VALLEY RAFTER

RIDGE

side cuts or lengthened. This is done so that the valley will clear the inside corner of the wall plate (**Procedure 49–B**).

To lay out a valley bird's mouth, first measure the line length and draw the seat plumb line. Then measure and mark down along this line, from the top edge of the rafter, a distance equal to the height above the bird's mouth on the common rafter. This makes the height above the bird's mouth the same for all rafters. Draw a level line at this point to make a bird's mouth. From the plumb line, measure at a right angle toward the tail a distance equal to one-half the valley rafter thickness. Draw a new plumb line and extend to it to the previously drawn seat cut line. On the bottom

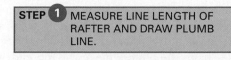

StepbyStep Procedures

Procedure 49–B Laying Out a Valley Seat Cut

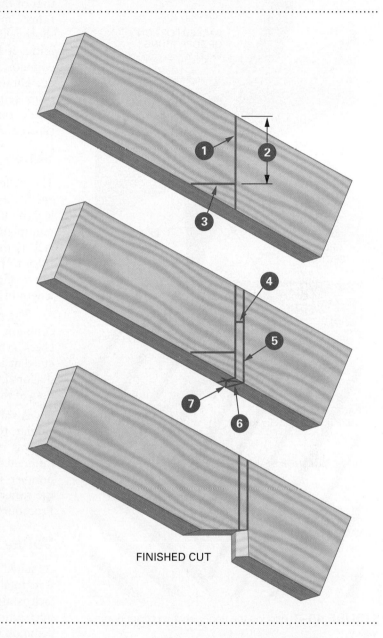

STEP 1 MEASURE LINE LENGTH OF RAFTER AND DRAW PLUMB LINE.

STEP 2 MEASURE DOWN THE SAME DISTANCE AS HEIGHT ABOVE BIRD'S MOUTH OF COMMON RAFTER.

STEP 3 DRAW LEVEL LINE FOR THE SEAT CUT.

STEP 4 MEASURE PERPENDICULAR TO PLUMB LINE TOWARD THE TAIL, A DISTANCE THAT IS ONE-HALF THE VALLEY THICKNESS. THIS IS 3/4" (19 mm) FOR DIMENSION LUMBER.

STEP 5 DRAW SECOND PLUMB LINE.

STEP 6 DRAW TWO SQUARED LINES ACROSS BOTTOM FROM THE TWO PLUMB LINES.

STEP 7 FIND THE CENTRE OF THE FIRST SQUARED LINE AND DRAW DIAGONALS FOR CHEEK CUTS.

NOTE: THE BIRD'S MOUTH MAY BE CUT SQUARE FROM THE SECOND PLUMB LINE.

FINISHED CUT

edge of the rafter, draw two squared lines from the two plumb lines. Find the centre of the first line and draw the appropriate diagonals.

There is no reason to drop or back a valley rafter as is done for hip rafters. The top corners of a valley are actually below the roof sheathing. The centreline of the top edge of the valley is aligned with the roof sheathing.

However, one edge of a supporting valley must be backed. This portion is located between the ridge-boards of the major and minor spans (**Figure 49–2**). Only the lower edge facing the wall plate is backed.

Laying Out a Valley Rafter Tail

The length of the valley rafter tail is found in the same manner as that used for the hip rafter. It may be stepped off, as shown in Figure 48–9 or calculated as shown in Procedure 48–D. Step off the same number of times, from the plumb line of the seat cut, as stepped off for the common rafter. However, use a unit run of 17 (354 mm) instead of 12 (250 mm).

The tail cut of the valley rafter is a double cheek cut. It is similar to that of the hip rafter, but angles outward instead. From the tail plumb line, measure

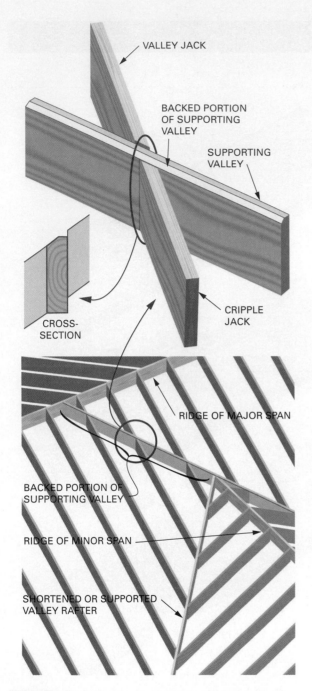

VALLEY JACK

BACKED PORTION
OF SUPPORTING
VALLEY

SUPPORTING
VALLEY

CROSS-
SECTION

CRIPPLE
JACK

RIDGE OF MAJOR SPAN

BACKED PORTION OF
SUPPORTING VALLEY

RIDGE OF MINOR SPAN

SHORTENED OR SUPPORTED
VALLEY RAFTER

FIGURE 49–2 Backing the upper section of supporting valley.

at right angles to the plumb line toward the tail, one-half the thickness of the rafter stock. Lay out another plumb line. Square both plumb lines across the top of the rafter. Draw diagonals from the centre (**Procedure 49–C**).

Shortening a Valley

The shortened or supported valley differs from the supporting valley only in length and ridge cut. The

length of the shortened valley is found by using the run of the common rafter of the minor span.

The plumb cut at the top end is different from that of the supporting valley. Because the two valley rafters meet at right angles, the cheek cut of the shortened valley is a squared mitre, not a compound mitre. It looks similar to the ridge plumb line of a common rafter. To shorten this rafter, measure back, at right angles from the plumb line at the top end, a distance that is half the thickness of the supporting valley rafter stock. Lay out another plumb line (**Procedure 49–D**).

Valley Jack Layout

The valley jack is a common rafter that is shortened at its bottom end where it meets the valley. The length of the valley jack is found, like any rafter, by multiplying unit length × run. The total run of any jack is the horizontal distance measured along the ridge to that rafter from the upper end of its valley rafter (**Figure 49–3**). Remember to use unit run and length of a common rafter for the valley jack.

The ridge cut of the valley jack is the same as a common rafter. It is shortened in the same way. The cheek cut against the valley rafter is a single cheek cut that is shortened, like other rafters meeting at a diagonal, by deducting one-half the 45-degree thickness of the valley rafter.

A valley jack has a common difference in length, similar to hip jacks. The layout for other valley jacks is made by adding or deducting the common difference from the jack length (**Figure 49–4**). This number is found on the third and fourth line of the rafter tables or may be calculated, as shown in Procedure 48–F.

Valley Cripple Jack Layout

As stated before, the length of any rafter can be found if its total run is known. The run of the valley cripple jack is always twice its horizontal distance from the intersection of the valley rafters (**Figure 49–5**). Use the common rafter tables or step off in a manner similar to that used for common rafters to find its length. Shorten each end by one-half the 45-degree thickness of the valley rafter stock (**Procedure 49–E**). Noting that they angle in opposite directions, make a single cheek cut.

Hip-Valley Cripple Jack Rafter Layout

All hip-valley cripple jacks cut between the same hip and valley rafters are the same. This is because the hip and valley rafters are parallel. To determine the length of the rafter, first find its total run. The run of a hip-valley cripple jack rafter is equal to the plate

StepbyStep Procedures

Procedure 49–C Laying Out a Valley Rafter Tail

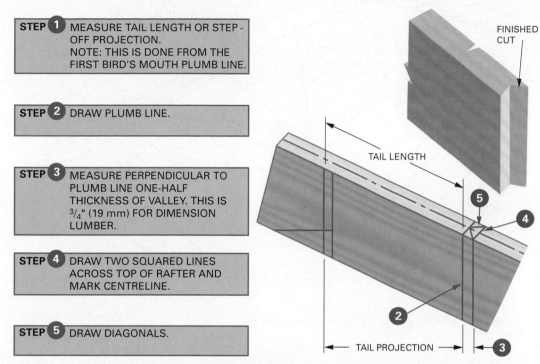

STEP 1 MEASURE TAIL LENGTH OR STEP-OFF PROJECTION.
NOTE: THIS IS DONE FROM THE FIRST BIRD'S MOUTH PLUMB LINE.

STEP 2 DRAW PLUMB LINE.

STEP 3 MEASURE PERPENDICULAR TO PLUMB LINE ONE-HALF THICKNESS OF VALLEY. THIS IS $3/4$" (19 mm) FOR DIMENSION LUMBER.

STEP 4 DRAW TWO SQUARED LINES ACROSS TOP OF RAFTER AND MARK CENTRELINE.

STEP 5 DRAW DIAGONALS.

THE STOCK OF THE HIP RAFTER (VALLEY RAFTER) IS USUALLY ONE SIZE BIGGER THAN THAT OF THE COMMON RAFTER. THEREFORE, THE TAIL "HANGS" DOWN BELOW THE BOTTOM OF THE FASCIA BOARD. IT IS COMMON PRACTICE TO RIP THE TAIL OF THE HIP RAFTER (VALLEY RAFTER) TO THE DIMENSION OF THE PLUMB CUT OF THE COMMON RAFTER. ALTERNATIVELY, IT IS SQUARED BACK TO THE BUILDING SO THAT IT DOES NOT PROJECT BELOW THE BOTTOM OF THE FASCIA BOARD.

StepbyStep Procedures

Procedure 49–D Laying Out Shortened or Supported Valley Rafter Ridge Plumb Cut

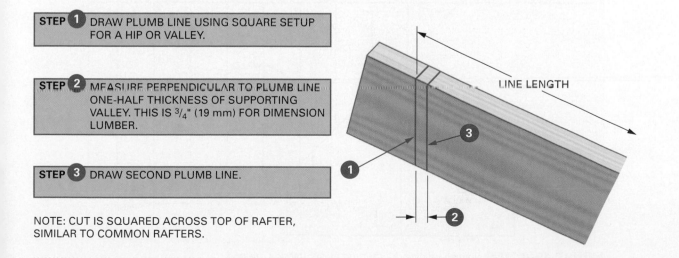

STEP 1 DRAW PLUMB LINE USING SQUARE SETUP FOR A HIP OR VALLEY.

STEP 2 MEASURE PERPENDICULAR TO PLUMB LINE ONE-HALF THICKNESS OF SUPPORTING VALLEY. THIS IS $3/4$" (19 mm) FOR DIMENSION LUMBER.

STEP 3 DRAW SECOND PLUMB LINE.

NOTE: CUT IS SQUARED ACROSS TOP OF RAFTER, SIMILAR TO COMMON RAFTERS.

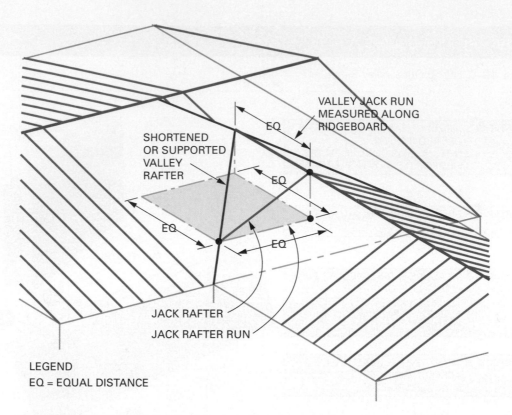

FIGURE 49–3 Determining the valley jack rafter run.

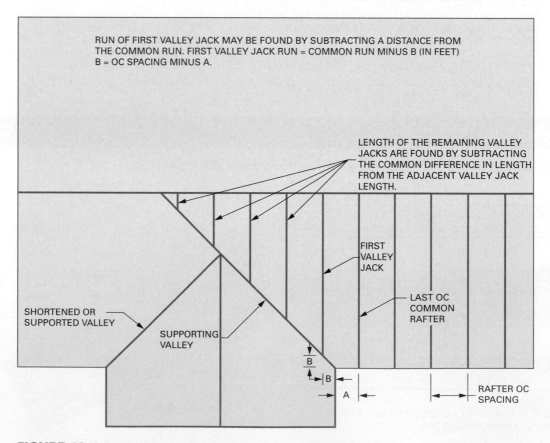

FIGURE 49–4 Determining a valley jack rafter run by subtraction method.

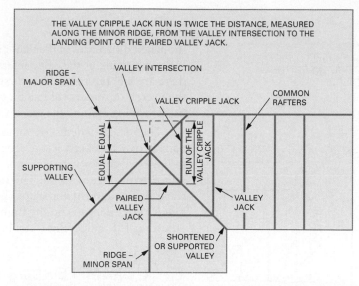

THE VALLEY CRIPPLE JACK RUN IS TWICE THE DISTANCE, MEASURED ALONG THE MINOR RIDGE, FROM THE VALLEY INTERSECTION TO THE LANDING POINT OF THE PAIRED VALLEY JACK.

VALLEY INTERSECTION

RIDGE – MAJOR SPAN

VALLEY CRIPPLE JACK

COMMON RAFTERS

SUPPORTING VALLEY

EQUAL EQUAL

RUN OF THE VALLEY CRIPPLE JACK

PAIRED VALLEY JACK

VALLEY JACK

RIDGE – MINOR SPAN

SHORTENED OR SUPPORTED VALLEY

FIGURE 49–5 Determining a valley cripple jack rafter run.

StepbyStep Procedures

Procedure 49–E Laying Out a Valley Cripple Jack Rafter

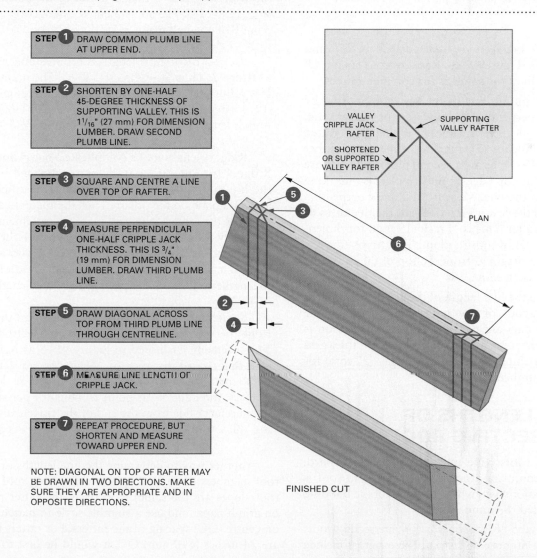

STEP ① DRAW COMMON PLUMB LINE AT UPPER END.

STEP ② SHORTEN BY ONE-HALF 45-DEGREE THICKNESS OF SUPPORTING VALLEY. THIS IS 1 1/16" (27 mm) FOR DIMENSION LUMBER. DRAW SECOND PLUMB LINE.

STEP ③ SQUARE AND CENTRE A LINE OVER TOP OF RAFTER.

STEP ④ MEASURE PERPENDICULAR ONE-HALF CRIPPLE JACK THICKNESS. THIS IS 3/4" (19 mm) FOR DIMENSION LUMBER. DRAW THIRD PLUMB LINE.

STEP ⑤ DRAW DIAGONAL ACROSS TOP FROM THIRD PLUMB LINE THROUGH CENTRELINE.

STEP ⑥ MEASURE LINE LENGTH OF CRIPPLE JACK.

STEP ⑦ REPEAT PROCEDURE, BUT SHORTEN AND MEASURE TOWARD UPPER END.

NOTE: DIAGONAL ON TOP OF RAFTER MAY BE DRAWN IN TWO DIRECTIONS. MAKE SURE THEY ARE APPROPRIATE AND IN OPPOSITE DIRECTIONS.

VALLEY CRIPPLE JACK RAFTER

SUPPORTING VALLEY RAFTER

SHORTENED OR SUPPORTED VALLEY RAFTER

PLAN

FINISHED CUT

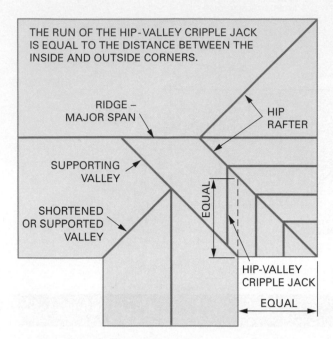

THE RUN OF THE HIP-VALLEY CRIPPLE JACK IS EQUAL TO THE DISTANCE BETWEEN THE INSIDE AND OUTSIDE CORNERS.

RIDGE – MAJOR SPAN

HIP RAFTER

SUPPORTING VALLEY

SHORTENED OR SUPPORTED VALLEY

EQUAL

HIP-VALLEY CRIPPLE JACK

EQUAL

FIGURE 49–6 Determining the run of the hip-valley cripple jack rafter.

line distance between the seat cuts of the hip and valley rafters (**Figure 49–6**). Remember to lay out all jack plumb lines the same as for common rafters.

Draw a plumb line at the top end. Shorten by measuring perpendicular toward the lower end, one-half 45-degree thickness of the hip rafter. This is $1^1/_{16}$ inches (27 mm) for dimension lumber (**Procedure 49–F**). Draw a second plumb line. Square a line over the top edge of the cripple, also marking the centre of the line. Measure and mark perpendicular toward the lower end, one-half the thickness of the cripple rafter. This is ¾ inch (19 mm) for dimension lumber. Draw a third plumb line. Across the top edge, draw a diagonal from the top of the third line through the centre line.

Measure the line length of the cripple and repeat the exact layout of the upper end. This is done because the cuts are parallel. The only exception is that the shortening is one-half 45-degree thickness of the valley rafter, which is $1^1/_{16}$ inches (27 mm) for dimension lumber.

RIDGE LENGTHS OF INTERSECTING ROOFS

Intersecting roofs have more than one ridgeboard. To reduce confusion, the width of the main roof is called the major span, while the width of the building wings is called the minor span.

The lengths of ridgeboards for intersecting roofs depend on a number of factors. These factors include

the height differences of the major and minor ridges, the style of framing at the ridge intersection, and whether there are any hip rafters. The first step is to determine the overall or theoretical length, and then modify it to fit, depending on how the intersection is framed.

Theoretical lengths of minor ridges come from adding two numbers. Ridge length equals the length of the extension or wing plus one-half the width of the minor span (**Figure 49–7**). The minor common rafter run is one-half the minor span. The actual length of the minor ridges depends on the framing style. The typical adjustments to ridge length can be seen in **Figure 49–8**.

EXAMPLE Using the information in Figure 49–8 and dimension lumber framing, the theoretical length of Ridge 1 is $49' - 24' = 24'$. Adjustments, including adding ¾" at both points A, give an actual Ridge 1 (R1) length of $24'$-$1½"$.

Ridge 2 theoretical length is found by adding the wing length to one-half minor span. This is $9' + 12' = 21'$. The adjustment is a deduction at point B of one-half major ridge thickness or 21-¾" $= 20'$-$11¼"$.

Ridge 3 theoretical length is found similarly to Ridge 2. Thus $5' + 6.5' = 11'$-$6"$. The necessary adjustment at point C is to deduct one-half the 45-degree thickness of the valley or $1^1/_{16}"$. Actual length of Ridge 3 is $11'$-$6"$ minus $1^1/_{16}" = 11'$-$4^{15}/_{16}"$.

Ridge 4 length looks complicated but is not. Ridge 4 is a side of a parallelogram formed by the hip and valley on two sides and the ridge and wall plate on the others. Thus the theoretical length is the same as the wall plate measurement, or simply 5 feet. Adjustments are made at either end. At point C it is shortened by $1^1/_{16}"$ and at point B it is lengthened by $1^{13}/_{16}"$. The net result is to lengthen Ridge 4 by ¾", giving an actual length of $5'$-¾".

Sometimes we may want to measure on the ridge the location of a valley/ridge intersection. Measurement 1 is noted in Figure 49–8 as M1. The theoretical distance of Measurement 1 is $6'$-$0"$ from the wall plate distance. Add ¾" at point A and add $5/_{16}"$ at point E. Thus the actual measurement line from the end of the ridgeboard is $6'$ plus ¾" plus $5/_{16}" = 6'$-$1^1/_{16}"$.

Another way to determine the line lengths of the roof members of an equal pitch intersecting roof (all roof slopes are the same) is to draw the rafter plan on graph paper and use a suitable scale to match the on-centre (OC) spacing of the rafters. For rafters that are 24 inches (610 mm) OC, it would be best to use

StepbyStep Procedures

Procedure 49–F Laying Out a Hip-Valley Cripple Jack Rafter

STEP 1 DRAW COMMON PLUMB LINE AT UPPER END.

STEP 2 SHORTEN ONE-HALF 45-DEGREE THICKNESS OF SUPPORTING VALLEY. THIS IS 1¹/₁₆" (27 mm) FOR DIMENSION LUMBER. DRAW SECOND PLUMB LINE SQUARE AND CENTRE A LINE OVER TOP OF RAFTER.

STEP 3 MEASURE PERPENDICULAR ONE-HALF CRIPPLE JACK THICKNESS. THIS IS ³/₄" (19 mm) FOR DIMENSION LUMBER. DRAW THIRD PLUMB LINE.

STEP 4 DRAW DIAGONAL ACROSS TOP FROM THIRD PLUMB LINE THROUGH CENTRELINE.

STEP 5 MEASURE LINE LENGTH OF HIP - VALLEY CRIPPLE JACK.

STEP 6 REPEAT PROCEDURE, BUT SHORTEN USING ONE-HALF 45-DEGREE THICKNESS OF VALLEY. THIS IS 1¹/₁₆" (27 mm) FOR DIMENSION LUMBER.

NOTE: DIAGONALS ON TOP OF RAFTER MAY BE DRAWN IN TWO DIRECTIONS. MAKE SURE THEY ARE APPROPRIATE AND IN SAME DIRECTION.

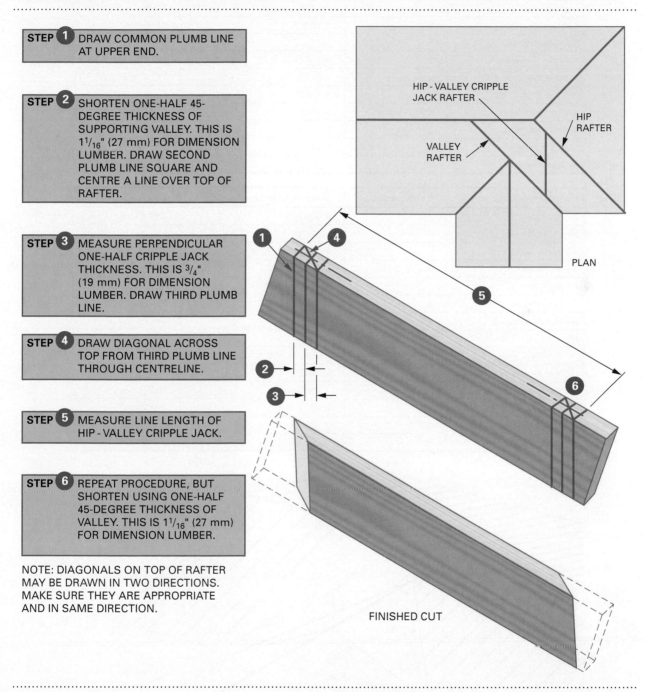

HIP - VALLEY CRIPPLE JACK RAFTER

HIP RAFTER

VALLEY RAFTER

PLAN

FINISHED CUT

¼-inch graph paper and let 1 square be equal to 1 foot. For rafters that are 500 mm OC, it would be best to let 1 square be equal to 250 mm (**Figure 49–9**).

For all rafters that run orthogonally (either vertically or horizontally), you would count the number of squares (or dots) that are covered (i.e., the run of the rafter) and multiply this number by the unit common rafter found on the first line of the rafter tables. For all rafters that run diagonally, you would count the number of squares (or dots) that are covered and multiply this number by the unit hip or valley rafter found on the second line of the rafter tables. For the ridgeboards, the number of squares (or dots) covered would be the actual line length dimension. This would give the following table:

Roof Component	# of Squares	Imperial System: Line 1 or Line 2	Line Length
a – Common Rafter	8	× 13.89″	111.12″ = 9′-3 ⅛″
b – 1st Hip Jack R.	6	× 13.89″	83.34″ = 6′-11 5/16″
c – 2nd Hip Jack R.	4	× 13.89″	55.56′ = 4′-7 9/16″
d – 3rd Hip Jack R.	2	× 13.89″	27.78″ = 2′-3 ¾″
e – Hip Rafter	8	× 18.36″	146.88″ = 12′-2 ⅞″
f – Supporting Valley R.	8	× 18.36″	146.88″ = 12′-2 ⅞″
g – Main Ridgeboard	8		8′
h – Valley Jack Rafter	2	× 13.89″	27.78″ = 2′-3 ¾″
i – Hip-Valley Cripple JR	2	× 13.89″	27.78″ = 2′-3 ¾″
j – Valley Cripple Jack R.	4	× 13.89″	55.56′ = 4′-7 9/16″
k – Shortened Valley R.	4	× 18.36″	73.44″ = 6′-1 7/16″
l – Valley Jack Rafter	2	× 13.89″	27.78″ = 2′-3 ¾″
m – Common Rafter	4	× 13.89″	55.56′ = 4′-7 9/16″
n – Minor Ridgeboard	10		10′

Roof Component	# of Squares	Metric System: Line 1 or Line 2	Line Length
a – Common Rafter	8	× 320.2 mm	2561.6 mm
b – 1st Hip Jack R.	6	× 320.2 mm	1921.2 mm
c – 2nd Hip Jack R.	4	× 320.2 mm	1280.8 mm
d – 3rd Hip Jack R.	2	× 320.2 mm	640.4 mm
e – Hip Rafter	8	× 406.2 mm	3249.6 mm
f – Supporting Valley R.	8	× 406.2 mm	3249.6 mm
g – Main Ridgeboard	8		2 m
h – Valley Jack Rafter	2	× 320.2 mm	640.4 mm
i – Hip-Valley Cripple JR	2	× 320.2 mm	640.4 mm
j – Valley Cripple Jack R.	4	× 320.2 mm	1280.8 mm
k – Shortened Valley R.	4	× 406.2 mm	1624.8 mm
l – Valley Jack Rafter	2	× 320.2 mm	640.4 mm
m – Common Rafter	4	× 320.2 mm	1280.8 mm
n – Minor Ridgeboard	10		2.5 m

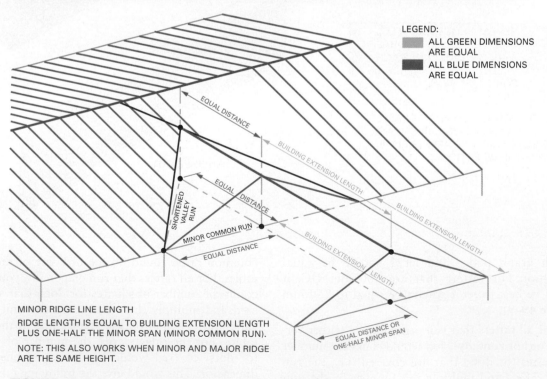

LEGEND:

ALL GREEN DIMENSIONS ARE EQUAL

ALL BLUE DIMENSIONS ARE EQUAL

MINOR RIDGE LINE LENGTH

RIDGE LENGTH IS EQUAL TO BUILDING EXTENSION LENGTH PLUS ONE-HALF THE MINOR SPAN (MINOR COMMON RUN).

NOTE: THIS ALSO WORKS WHEN MINOR AND MAJOR RIDGE ARE THE SAME HEIGHT.

FIGURE 49–7 Determining the line length of the minor ridge.

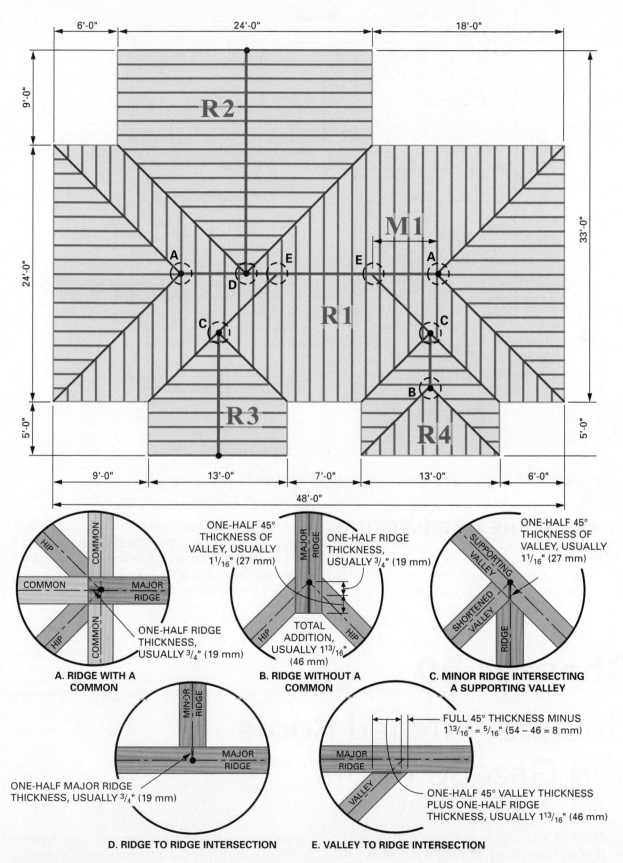

FIGURE 49–8 Ridgeboard line lengths are adjusted depending on the style of framing.

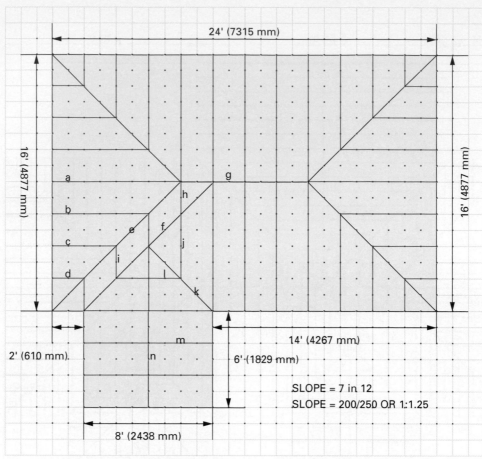

INTERSECTING ROOFS — EQUAL PITCH

FIGURE 49–9 Finding rafter lengths using graph paper and scaling.

ERECTING THE INTERSECTING ROOF

The intersecting roof is raised by framing opposing members of the main span first. Then, the valley rafters are installed. To prevent bowing the ridge of the main span, install rafters to oppose the valley rafters. Install common and jack rafters in sets opposing each other. Sight members of the roof as framing progresses, keeping all members in a straight line.

Chapter 50

Unequal Pitched Roofs and Gazebo Roofs

Equal pitched hip roofs have the same slope on all four roof surfaces. **Unequal pitched** hip roofs have the two longer roof surfaces (trapezoidal in shape) at the same slope and the two end roof surfaces (triangular in shape) at a different slope. Calculating the rafter lengths for this type of roof requires the carpenter to return to the concepts of Pythagoras' theorem and similar triangles that were used to produce the rafter tables.

The main difference in the calculation order that must be followed is in the layout. For equal-pitched roofs, the layout (on graph paper) does not need to include the projection of the roof overhang, because all line lengths are made to the building line, the bird's mouths are drawn in, and the tails are added on as indicated on the plans. However, for an unequal pitched roof, in order to maintain an even projection on all sides as well as level soffits, it is necessary to do all rafter layouts to the **fascia line** (**Figure 50–1**).

For this section, it will be necessary to separate imperial calculations from metric calculations, because the two systems have different rafter tables. First, we will look at the unequal pitched hip roof in the imperial system in plan view. The run of the rafters and the ridge have been labelled by letters. The total lengths (TL) of the rafters will be calculated using the rafter tables as well as similar triangles and Pythagoras' theorem.

The total span (from the fascia line) is 16 feet. The SIDE run is therefore 8 feet.

A. The run of the SIDE common rafter (CR) is therefore 8 feet or 8 unit runs.
   - Using the rafter tables, line 1, under 5″, we read 13″.
   - The line length of the SIDE common rafter = 13″ × 8 = 104″, or 8′-8″.
   - TLCRs = 13″ × 8 = 104″ = 8′-8″.

B. Ridge = 10 feet (given)
   - END run = (total fascia length – ridge) ÷ 2 = (20′ – 10′) ÷ 2 = 10′ ÷ 2 = 5′

C. The run of the END common rafter (CR) is therefore 5 feet, or 5 unit runs (rises).
   - The total rise (for the sides) = unit rise × number of rises (runs) = 5″ × 8 = 40″ (3′-4″)
   - END unit rise = total rise ÷ number of rises = 40″ ÷ 5 = 8″, or $^8/_{12}$ slope
   - Using the rafter tables, line 1, under 8″, we read 14.42″.

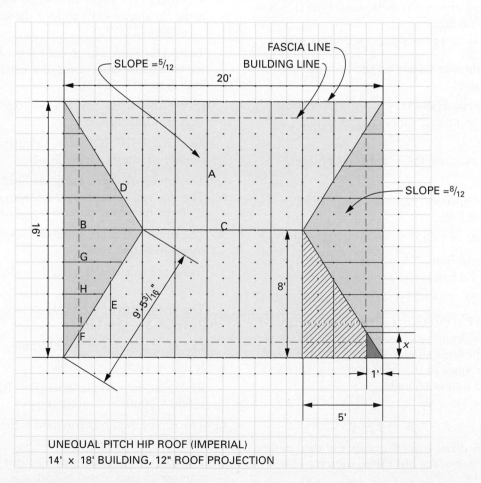

UNEQUAL PITCH HIP ROOF (IMPERIAL)
14' × 18' BUILDING, 12" ROOF PROJECTION

FIGURE 50–1 Unequal pitched hip roof—14′ × 18′ building—12″ roof projection.

- The common rafter (at the end) = 14.42″ × 5 = 72.1″ = 6′-0⅛″.
- TLCRs = 14.42″ × 5 = 72.1″ = 6′-0⅛″.

To find the length of the hip rafter, we must first calculate its run using Pythagoras. From the large right triangle that includes the run of the hip rafter as its hypotenuse, we get . . .
(Run of HR)² = 5² + 8² = 89

D. Run of HR = $\sqrt{89}$ = 9.43....′ = 9′-5³⁄₁₆″ = 9.434 feet

The rise of the hip rafter is 40″ (3.33′), the same as the common rafter's.

TLHR = $\sqrt{(3.33^2 \times 9.43^2)}$ = 10.007′ = 10′-0¹⁄₁₆″

Because the first hip jack rafter on the side is 2′ away from the common rafter, its run can be calculated using similar triangles:

$^{run}/_8 = {}^3/_5$; run(HJR#1) = 8 × 3 ÷ 5 = 4.8′

Using the unit CR on the tables,

E. The run of the 1st SIDE hip jack rafter = 4.8 feet

TLHJR$_s$ #1 = 13″ × 4.8 = 62.4″ = 5′-2³⁄₈″

F. In similar fashion, $^{run}/_8 = {}^1/_5$; run(HJR#1) = 8 × 1 ÷ 5 = 1.6 feet

LLHJR$_s$ #2 = 13″ × 1.6 = 20.8″ = 1′-8¹³⁄₁₆″

For the hip jack rafters on the ends, we use similar ratios.

$^{run}/_6 = {}^8/_5$; run(HJR#1) = 6 × 5 ÷ 8 = 3.75′

G. The run of the first END hip jack rafter = 3.75 feet

TLHJR$_e$ #1 = 14.42″ × 3.75 = 54.075″ = 4′-6¹⁄₁₆″

H. In similar fashion, $^{run}/_4 = {}^8/_5$; run(HJR#2) = 4 × 5 ÷ 8 = 2.5 feet

TLHJR$_e$ #2 = 14.42″ × 2.5 = 36.05″ = 3′-0¹⁄₁₆″

I. In similar fashion, $^{run}/_2 = {}^8/_5$; run(HJR#3) = 2 × 5 ÷ 8 = 1.25 feet

TLHJR$_e$ #3 = 14.42² × 1.25 = 18.025″ = 1′-6″

N.B. The hip rafters are offset from the corners of the building. To calculate this offset, we look at the large shaded triangle and the small shaded triangle in the right bottom corner and again use similar triangles.

x:1 = 8:5; $^x/_1 = {}^8/_5$

x = 8 × 1 ÷ 5 = 1.6′ = 19³⁄₁₆″

To get the offset from the corner of the building to the centre of the hip rafter, subtract 12″: 19³⁄₁₆″ – 12″ = 7³⁄₁₆″.

If both sets of common rafters (end and side) are cut from the same stock, and the vertical distance above the bird's mouth (sometimes referred to as the "stand" or "height above plate" [HAP]) is kept the same, you will see that the end common rafters are above the wall plate. Thus the wall plate has to be raised. To determine the amount that the plate has to be raised, we consider the two different slopes and the common projection:

The rafter tail on the ⁵⁄₁₂ side will drop 5″ over the 12″ projection.

The rafter tail on the ⁸⁄₁₂ side will drop 8″ over the 12″ projection.

The difference in drop is 3″, therefore the end wall plate will have to be raised by **3″** to support the bird's mouths of the end common and hip jack rafters. Alternatively, wider rafter stock can be used and the HAP increased by 3″.

To find side cut angles and shortening amount for the rafters, it is simplest to do a full-scale drawing on graph paper (**Figures 50–2** and **50–3**).

The plumb cut angle for the end hip jack rafters is the ⁸⁄₁₂ angle (~34°) and for the side it is the ⁵⁄₁₂ angle (~22½°).

To find the ridge plumb cut for the hip rafter, the triangle would consist of a run of 9.434′ and a rise of 3.333′, or:

INV. TAN (3.333 ÷ 9.434) = **70.5° or 19.5°** set angle for the mitre saw

For a METRIC ROOF, the calculations follow the same pattern (**Figure 50–4**).

The total span (from the fascia line) is 4000 mm. The side run is therefore 2000 mm or 2000 ÷ 250 = 8 unit runs. Slope = ¹⁵⁰⁄₂₅₀ or ¹⁄₁.₆₇

A. The run of the SIDE common rafter (CR) is therefore 8 unit runs.

Using the rafter tables, line 1, under 150″, we read 291.5.

The line length of the SIDE common rafter = 291.5 × 8 = 2332 mm.

LLCR$_s$ = 291.5 × 8 = **2332 mm**

B. Ridge = 3000 mm (given)

END run = (total fascia length – Ridge) ÷ 2 = (5000 – 3000) ÷ 2 = 1000 mm

C. The run of the END common rafter (CR) is 1000 mm, or 4 unit runs (rises).

The total rise (for the sides) = unit rise × number of rises (runs) = 150 × 8 = **1200 mm**

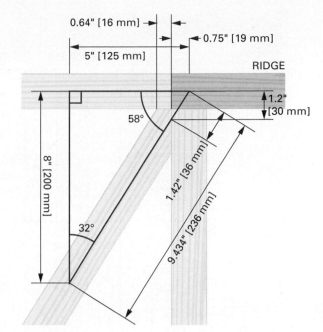

A, B – COMMON RAFTERS
D – HIP RAFTER
G, E – HIP JACK RAFTERS

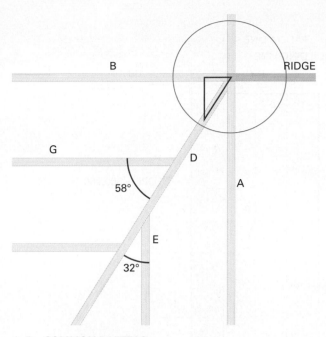

FIGURE 50–2 Using a full-scale drawing: Side cut angles/shortenings.

THE MEASUREMENT OF THE ANGLE
CAN BE FOUND USING THE
TAN BUTTON ON THE CALCULATOR:
$\angle a = 5, \div, 8, =, $ INV,TAN $= 32°$
$125, \div, 200, =, $ INV,TAN $= 32°$
FOR D.A.L. CALCULATORS, THE ORDER
WOULD BE 2nd TAN, $(, 5, \div, 8,), = 32°$

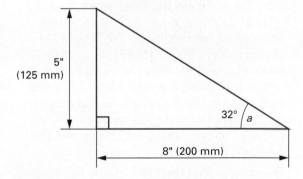

FIGURE 50–3 Side cut angles for the hip jack rafters.

END unit rise = total rise ÷ number of rises = $1200 \div 4 = 300$ mm

∴ slope = $^{300}/_{250}$ $(^1/_{.83})$

Using the rafter tables, line 1, under 300, we read 390.5 (mm).

The common rafter (at the end) = $390.5 \times 4 = 1562$ mm

$TLCR_e = 390.5 \times 4 = \textbf{1562 mm}$

To find the length of the hip rafter, we must first calculate its run using Pythagoras. From the large right triangle that includes the run of the hip rafter as its hypotenuse, we get

$(\text{Run of HR})^2 = 2000^2 + 1000^2$

D. Run of HR $= \sqrt{5\ 000\ 000} = 2236$ mm

The rise of the hip rafter is 1200 mm, the same as the common rafter.

$TLHR = \sqrt{(1200^2 + 2236^2)} = \sqrt{6439696} = \textbf{2537.5 mm}$

Because the first hip jack rafter on the side is 250 mm away from the common rafter, its run can be calculated using similar triangles:

$^{run}/_{500} = ^{2000}/_{1000}$; run(HJR#1) $= 2000 \times 500 \div 1000 = 1000$ mm

number of unit runs $= 1000 \div 250 = 4$ unit runs

E. The run of the first SIDE hip jack rafter = 4.8 feet

Using the unit CR on the tables,

$TLHJR_s \#1 = 291.5 \times 4 = \textbf{1166 mm}$

F. Similarly, $^{run}/_{500} = ^{1000}/_{1000}$; run(HJR#2) $= 1000 \times 500 \div 1000 = 500$ mm

$TLHJR_s \#2 = 291.5 \times 4 = \textbf{583 mm}$

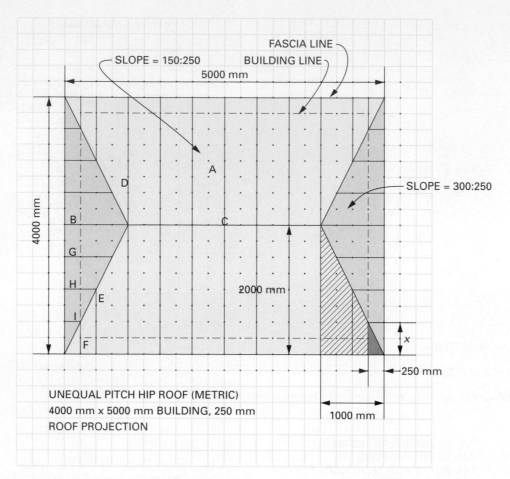

FIGURE 50–4 Unequal pitched hip roof: 4000 × 5000 mm building—250 mm roof projection.

For the hip jack rafters on the ends, we use similar ratios:

$^{run}/_{1500} = {}^{1000}/_{2000}$; run(HJR#1) = 1000 × 1500 ÷ 2000 = 750 mm, or 3 unit runs

G. The run of the first END hip jack rafter = 750 mm, or 3 unit runs

TLHJR$_e$ #1 = 390.5 × 3 = **1171.5 mm**

H. Similarly, $^{run}/_{1000} = {}^{1000}/_{2000}$; run(HJR#2) = 1000 × 1000 ÷ 2000 = 500 mm

TLHJR$_e$ #2 = 390.5 × 2 = **781 mm**

I. Similarly, $^{run}/_{500} = {}^{1000}/_{2000}$; run(HJR#3) = 500 × 1000 ÷ 2000 = 250 mm

TLHJR$_e$ #3 = 390.5 × 1 = **390.5 mm**

N.B. The hip rafters are offset from the corners of the building. To calculate this offset, we look at the large shaded triangle and the small shaded triangle in the right bottom corner and again use similar triangles.

$^x/_{250} = {}^{2000}/_{1000}$; x = 2000 × 250 ÷ 1000 = 500 mm

To get the offset from the corner of the building to the centre of the hip rafter, subtract 250 mm from the 500 mm to get **250 mm**.

If both sets of common rafters (end and side) are cut from the same stock, and the vertical distance above the bird's mouth (sometimes referred to as the "stand" or "height above plate" [HAP]) is kept the same, you will see that the end common rafters are above the wall plate. Thus the wall plate has to be raised. To determine the amount the plate has to be raised, we consider the two different slopes and the common projection:

The rafter tail on the 150/250 side will drop 150 mm over the 250 projection.

The rafter tail on the 300/250 side will drop 300 mm over the 250 projection.

The difference in drop is 150 mm. Therefore the end wall plate will have to be raised by 150 mm to support the bird's mouths of the end common and hip jack rafters. Alternatively, wider rafter stock can be used and the HAP increased by 150 mm.

To find side cut angles and shortening amount for the rafters, it is simplest to do a full-scale drawing on graph paper (Figure 50–2).

The plumb cut angle for the end hip jack rafters is the $\frac{300}{250}$ angle (~50°) and for the side it is the $\frac{150}{250}$ angle (~31°).

To find the ridge plumb cut for the hip rafter, the triangle would consist of a run of 2236 mm and a rise of 1200 mm or:

INV. TAN (1200 ÷ 2236) = **62°**, or **28°** set angle for the mitre saw

UNEQUAL PITCHED INTERSECTING ROOFS

For intersecting roofs with a major and minor roof having different spans but ridgeboards at the same height, this will also be a case of unequal pitch. The principles, however, are the same, whether these relate to the amount of offset, the amount to raise the walls, or the lengths and angles of the hip, valley, and jack rafters. It is always best to draw the roof, to scale, using graph paper, and to remember to calculate lengths from the fascia line. **Figures 50–5** and **50–6** illustrate the procedure.

For the imperial measure roof, the 12/12 side has a run of 8 feet and therefore a rise of 8 feet. So the other sides of the roof will have a rise of 8 feet. With a slope of 16/12 and this 8-foot rise, the ends of the roof will have to rise 6 times (8′ or 96″ ÷ 16), so it will run 6 times or it will have a run of 6 feet. This means that the ridge will be 12 feet long (24′ − 6′ − 6′). If the slope for the minor roof is unknown, divide the total rise by the run of the addition (8′ ÷ 6 = 16″), which gives the unit rise and hence the slope. All other calculations follow the pattern set for the unequal pitched hip roof.

Note that the projection is 2 feet. The rafter tail on the 12/12 side will drop by 24 inches, and the tail on the 16/12 end will drop 32 inches. The difference of 8 inches is the amount by which to raise the wall plates that support the 16/12 section of the roof.

For the metric roof, the 300/250 side has 8 runs of 250 mm and therefore 8 rises of 300 mm or 2400 mm. Therefore, the other parts of the roof will rise 2400 mm. With a slope of 400/250, the ends of the roof will have to rise 6 times (2400 ÷ 400), so it

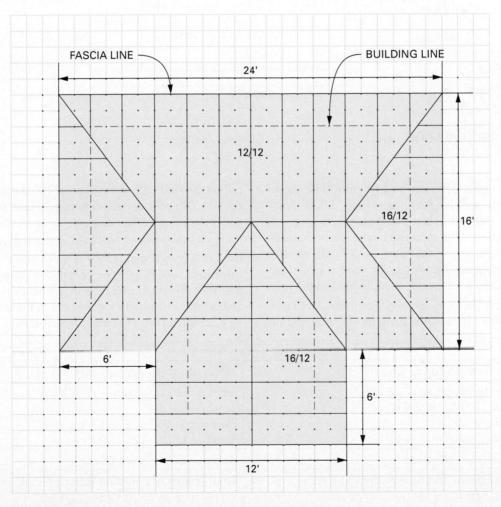

FIGURE 50–5 Intersecting roofs with unequal pitch—imperial.

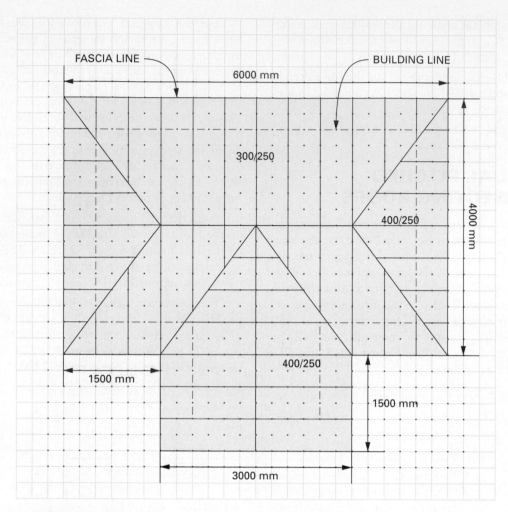

FIGURE 50–6 Intersecting roofs with unequal pitch—metric.

will run 6 times × 250 mm, or 1500 mm. This means that the ridge will be 3000 mm long (6000 mm − 1500 mm − 1500 mm). If the slope for the minor roof is unknown, divide the total rise by the run of the addition (2400 mm ÷ 6 = 400 mm), which will give the unit rise and hence the slope. All other calculations follow the pattern set for the unequal pitched hip roof.

Note that the projection is 500 mm. The rafter tail on the 300/250 side will drop by 600 mm and the

tail on the 400/250 end will drop 800 mm. The difference of 200 mm is the amount by which to raise the wall plates that support the steeper section of the roof.

In the case of an L-shaped intersecting roof, where the slope on the major roof is given and the slope on the minor roof is to be determined, it is again important that all layout and measurements be made to the fascia line. The rafters for the roof in **Figure 50–7** can be calculated (imperial and SI) as follows:

	Major Roof (to Fascia Line)		*Minor Roof (to Fascia Line)*
	Span = 16′; total run = 8′ (96″) Slope = 10/12 (1:1.2) Unit rise = 10″ Total rise = 10″ × 8 = 80″		Span = 12′; total run = 6′ (72″) Total rise = 80″ Unit rise = 80″ ÷ 6 = 13.33 (40/3) Slope = 13.33/12 (1:0.9)
A	Ridge = 32′ − 6′ = 26′	B	Ridge = 22′ − 8′ = 14′
C	Total length of common rafter (TLCR) = $\sqrt{(96^2 + 80^2)}$ = 124.96″ = $10′4\frac{15}{16}″$	H	Total length of common rafter (TLCR) = $\sqrt{(72^2 + 80^2)}$ = 107.63″ = $8′11\frac{5}{8}″$

D	Line length of valley jack rafter, LLVJR#1: Run: $\frac{D}{4} = \frac{8}{6}$; $D = \frac{8 \times 4}{6} = 5.33' = 64''$ LLVJR#1 $= \frac{124.96 \times 64}{96} = 83.31'' = 6'11\frac{5}{16}''$	I	Line length of valley jack rafter, LLVJR#1: Run: $\frac{I}{6} = \frac{6}{8}$; $I = \frac{6 \times 6}{8} = 4.5' = 54''$ LLVJR#1 $= \frac{107.63 \times 54}{72} = 80.72'' = 6'8\frac{3}{4}''$
E	Line length of valley jack rafter, LLVJR#2: Run: $\frac{E}{2} = \frac{8}{6}$; $E = \frac{8 \times 2}{6} = 2.66\ldots$ ft $= 32''$ LLVJR#2 $= \frac{124.96 \times 32}{96} = 41.65'' = 3'5\frac{5}{8}''$	J	Line length of valley jack rafter, LLVJR#2: Run: $\frac{J}{4} = \frac{6}{8}$; $J = \frac{6 \times 4}{8} = 3' = 36''$ LLVJR#2 $= \frac{107.63 \times 36}{72} = 53.82'' = 4'5\frac{13}{16}''$
F G	Hip jack rafter #1 = D Hip jack rafter #2 = E	K	Line length of valley jack rafter, LLVJR#3: Run: $\frac{K}{2} = \frac{6}{8}$; $K = \frac{6 \times 2}{8} = 1.5' = 18''$ LLVJR#3 $= \frac{107.63 \times 18}{72} = 26.91'' = 2'2\frac{7}{8}''$
L M	Total length of valley rafter = Total length of hip rafter = TLVR: Run of VR (HR): $\sqrt{(96^2 + 72^2)} = 120'' = 10'$ Rise of VR (HR) $= 80''$ TLVR $= \sqrt{(120^2 + 80^2)} = 144.22'' = 12'\frac{1}{4}'' =$ TLHR		
	Valley (hip) offset: $\frac{x}{12} = \frac{6}{8}$; $x = \frac{6 \times 12}{8} = 9''$ Offset $= 12'' - 9'' = 3''$		
	Rafter tail drop (12″ projection) = 10″		Rafter tail drop (12″ projection) = 13.33...″
	Amount to raise the walls of the steeper roof: $13.33'' - 10'' = 3.33'' = 3\frac{5}{16}''$		

	Major Roof (to Fascia Line)		*Minor Roof (to Fascia Line)*
	Span = 4800; total run = 2400 mm Slope = 1:1.2 Total rise = 2400/1.2 = 2000 mm		Span = 3600; total run = 1800 mm Total rise = 2000 Slope = 2000/1800 = 1:0.9
A	Ridge = 9600 − 1800 = 7800 mm	B	Ridge = 6600 − 2400 = 4200 mm
C	Total length of common rafter (TLCR) = $\sqrt{(2400^2 + 2000^2)} = 3124.10$ mm	H	Total length of common rafter (TLCR) = $\sqrt{(1800^2 + 2000^2)} = 2690.72$ mm
D	Line length of valley jack rafter, LLVJR#1: Run: $\frac{D}{1200} = \frac{2400}{1800}$; $D = \frac{2400 \times 1200}{1800} = 1600$ mm Rise $= \frac{1600}{1.2} = 1333.33\ldots$ mm LLVJR#1 $= \sqrt{(1600^2 + 1333.33^2)} = 2082.7$	I	Line length of valley jack rafter, LLVJR#1: Run: $\frac{I}{1800} = \frac{1800}{2400}$; $I = \frac{1800 \times 1800}{2400} = 1350$ mm Rise $= \frac{1350}{0.9} = 1500$ mm LLVJR#1 $= \sqrt{(1350^2 + 1500^2)} = 2018$ mm
E	Line length of valley jack rafter, LLVJR#2: Run: $\frac{E}{600} = \frac{2400}{1800}$; $E = \frac{2400 \times 600}{1800} = 800$ mm Rise $= \frac{800}{1.2} = 666.66\ldots$ mm LLVJR#2 $= \sqrt{(800^2 + 666.66^2)} = 1041.36$ mm	J	Line length of valley jack rafter, LLVJR#2: Run: $\frac{J}{1200} = \frac{1800}{2400}$; $J = \frac{1800 \times 1200}{2400} = 900$ mm Rise $= \frac{900}{0.9} = 1000$ mm LLVJR#2 $= \sqrt{(900^2 + 1000^2)} = 1345.36$ mm
F G	Hip jack rafter #1 = D Hip jack rafter #2 = E	K	Line length of valley jack rafter, LLVJR#3: LLVJR#3 $= \sqrt{(450^2 + 500^2)} = 672.68$ mm
L M	Total length of valley rafter = Total length of hip rafter = TLVR: Run of VR (HR): $\sqrt{(1800^2 + 2400^2)} = 3000$ mm Rise of VR (HR) = 2000 mm TLVR $= \sqrt{(3000^2 + 2000^2)} = 3605.55$ mm = TLHR		
	Valley (hip) offset: $\frac{x}{300} = \frac{1800}{2400}$; $x = \frac{1800 \times 300}{2400} = 225$ mm Offset = 300 − 225 = 75 mm		
	Rafter tail drop (300 projection) = 250 mm		Rafter tail drop (300 projection) = 333.3 mm
	Amount to raise the walls of the steeper roof: 333.33 mm − 250 mm = 83.33 mm		

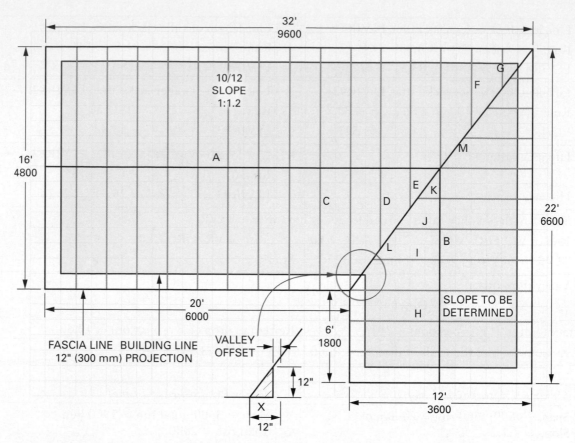

FIGURE 50–7 L-shaped intersecting roof with two different spans—plan view.

In the perspective drawing in **Figure 50–8**, with all lengths to the fascia lines:

AB = FG = ED, run of the minor roof

BC = GD = FE, run of the major roof

FA = GB = DC, total rise from the fascia

GE = run of the valley rafter

$AE = \sqrt{FE^2 + FA^2}$ = common rafter of the major roof

$CE = \sqrt{ED^2 + DC^2}$ = common rafter of the minor roof

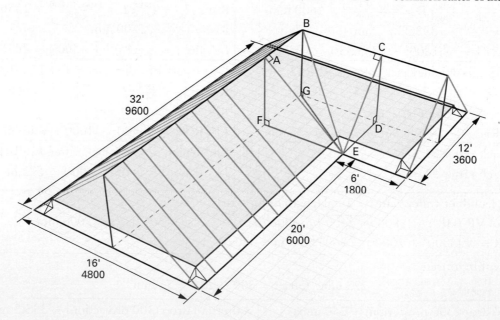

FIGURE 50–8 L-shaped intersecting roof with two different spans—perspective view.

$BE = \sqrt{GE^2 + GB^2}$ = total length of the valley rafter
$BE = \sqrt{AE^2 + AB^2}$ = total length of the valley rafter
$BE = \sqrt{CE^2 + BC^2}$ = total length of the valley rafter

GAZEBO ROOFS

There are two popular shapes for **gazebos**—the regular *hexagon* and the *octagon*. The hexagon has six equal sides and six equal angles, and it is made up of six equilateral triangles (60° angles). The first step before constructing the roof is to draw the plan view. The lines that meet at the centre represent the run of the rafters (**Figure 50–9**). The plan view remains the same regardless of the roof slope. From the run and the roof slope, the line lengths and the cut angles of the rafters can be determined.

The determination of the line lengths can be done mathematically or geometrically. The unit 30°–60°–90° triangle has sides of $1 - \sqrt{3} - 2$ respectively, and the run of the common and jack rafters can be calculated by similar triangles. Given the slope of the roof, the total rise can be determined, and then the line length of the common rafter can be calculated using the Pythagorean theorem. Or, knowing the slope of the roof (10/12), the unit common rafter is read on the rafter tables on the framing square under 10 inches (15.62″), and this value is multiplied by the number of unit runs of the

rafter. *Remember that the unit common rafter is the hypotenuse of the roof slope triangle.* The line length of the common rafter can be used to develop the actual roof sheathing triangles, which will produce the line length of the hip rafter. The line lengths of the hip jack rafters can now be determined by similar triangles, as in rafter **D** in Figure 50–9. The plumb cut and side cut angles of all of the rafters, as well as the sheathing face and bevel angles, can be determined from the plan view and the actual roof triangle (**Figure 50–10**). Deductions for rafter thickness are easily determined from full-scale drawings of the intersections.

The hip rafters (running from the outside corner to the peak) can either be dropped or backed. The unit hip rafter is calculated using similar triangles, and for the hexagon is 13.86 inches, or 13⅞ inches. A good way to remember this number is to associate it with the 7/12 roof triangle, which is very close to the 60° of the hexagon and has a hypotenuse of 13.89 inches. With the framing square set at the unit rise (8″) on the tongue, and at 13⅞ inches on the body, the values for the hip rafter drop and for the backing bevel can be determined (**Figure 50–11**). These are not the same values.

Similarly, for the metric **hexagonal roof**, the determination of the line lengths can be done mathematically or geometrically (**Figure 50–12**). The unit 30°–60°–90° triangle has sides of $1 - \sqrt{3} - 2$ respectively,

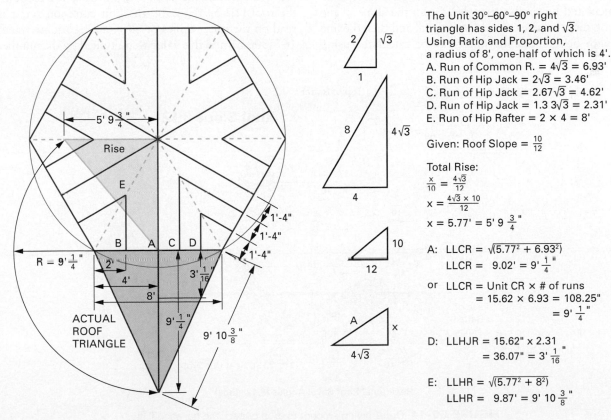

FIGURE 50–9 Plan view of the hexagonal roof (imperial).

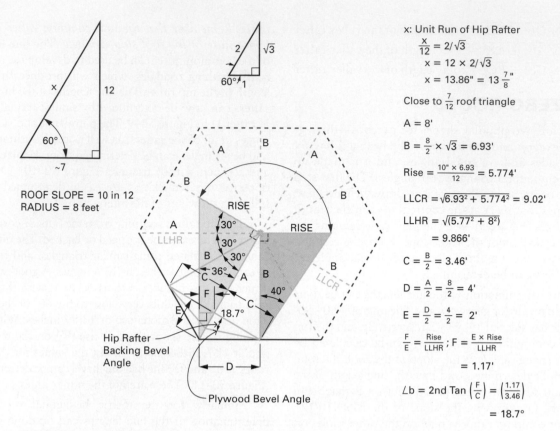

x: Unit Run of Hip Rafter

$$\frac{x}{12} = 2/\sqrt{3}$$

$$x = 12 \times 2/\sqrt{3}$$

$$x = 13.86" = 13\frac{7}{8}"$$

Close to $\frac{7}{12}$ roof triangle

$$A = 8'$$

$$B = \frac{8}{2} \times \sqrt{3} = 6.93'$$

$$Rise = \frac{10" \times 6.93}{12} = 5.774'$$

$$LLCR = \sqrt{6.93^2 + 5.774^2} = 9.02'$$

$$LLHR = \sqrt{(5.77^2 + 8^2)}$$

$$= 9.866'$$

$$C = \frac{B}{2} = 3.46'$$

$$D = \frac{A}{2} = \frac{8}{2} = 4'$$

$$E = \frac{D}{2} = \frac{4}{2} = 2'$$

$$\frac{F}{E} = \frac{Rise}{LLHR}; F = \frac{E \times Rise}{LLHR}$$

$$= 1.17'$$

$$\angle b = 2nd\ Tan\left(\frac{F}{C}\right) = \left(\frac{1.17}{3.46}\right)$$

$$= 18.7°$$

ROOF SLOPE = 10 in 12
RADIUS = 8 feet

FIGURE 50–10 Determining hexagonal roof angles for a 10/12 roof slope (imperial).

and the run of the common and jack rafters can be calculated by similar triangles. Given the slope of the roof, the total rise can be determined, and then the line length of the common rafter can be calculated using the Pythagorean theorem. Or, knowing the slope of the roof (1:1.5, 167:250), the unit common rafter is read on the rafter tables on the framing square under 167 (300.5), and this value is multiplied by the number

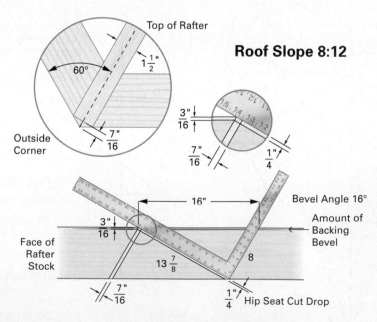

Hexagonal Roof Bevel Angles (8/12 slope)

FIGURE 50–11 Determining amount to drop or back the hexagonal hip rafter for a 10/12 roof slope (imperial).

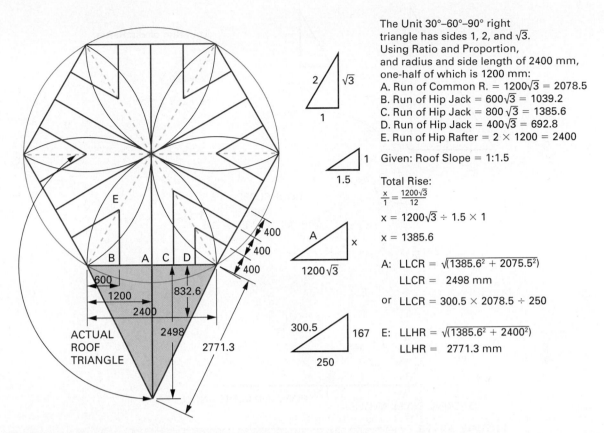

The Unit 30°–60°–90° right triangle has sides 1, 2, and $\sqrt{3}$. Using Ratio and Proportion, and radius and side length of 2400 mm, one-half of which is 1200 mm:

A. Run of Common R. = $1200\sqrt{3}$ = 2078.5
B. Run of Hip Jack = $600\sqrt{3}$ = 1039.2
C. Run of Hip Jack = $800\sqrt{3}$ = 1385.6
D. Run of Hip Jack = $400\sqrt{3}$ = 692.8
E. Run of Hip Rafter = 2 × 1200 = 2400

Given: Roof Slope = 1:1.5

Total Rise:
$$\frac{x}{1} = \frac{1200\sqrt{3}}{12}$$

$x = 1200\sqrt{3} \div 1.5 \times 1$

$x = 1385.6$

A: $LLCR = \sqrt{(1385.6^2 + 2075.5^2)}$
   $LLCR = 2498$ mm

or $LLCR = 300.5 \times 2078.5 \div 250$

E: $LLHR = \sqrt{(1385.6^2 + 2400^2)}$
   $LLHR = 2771.3$ mm

FIGURE 50–12 Plan view of the hexagonal roof (metric).

of unit runs of the rafter (total run divided by 250). *Remember that the unit common rafter is the hypotenuse of the roof slope triangle.*

The line length of the common rafter can be used to develop the actual roof sheathing triangles, which will produce the line length of the hip rafter. The line lengths of the hip jack rafters can now be determined by similar triangles, as in rafter **D** in Figure 50–9. The plumb cut and side cut angles of all of the rafters, as well as the sheathing face and bevel angles, can be determined from the plan view and the actual roof triangle (**Figure 50–13**). Deductions for rafter thickness are easily determined from full-scale drawings of the intersections.

The hip rafters (running from the outside corner to the peak) can either be dropped or backed. The unit hip rafter is calculated using similar triangles, and for the hexagon is 288.7 or 289 mm. With the framing square set at the unit rise (167) on the tongue, and at 289 on the body, the values for the hip rafter drop and for the backing bevel can be determined (**Figure 50–14**). These are not the same values.

Determining the rafter lengths and angles for the **octagonal roof** requires following the same procedure. The octagon has eight equal sides and eight equal angles, and it is made up of eight isos-

celes triangles (67½° angles). The first step before constructing the roof is to draw the plan view. The lines that meet at the centre represent the run of the rafters (**Figure 50–15**). The plan view remains the same regardless of the roof slope. From the run and the roof slope, the line lengths and the cut angles of the rafters can be determined.

The determination of the line lengths can be done mathematically or geometrically. The unit 22½°–67½°–90° triangle has sides of 1 – tan (67½°) – 1/cos(67½°), respectively, and the run of the common and jack rafters can be calculated by similar triangles. Given the slope of the roof, the total rise can be determined, and then the line length of the common rafter can be calculated using the Pythagorean theorem. Or, knowing the slope of the roof (10/12), the unit common rafter is read on the rafter tables on the framing square under 10 inches (15.62″), and this value is multiplied by the number of unit runs of the rafter. *Remember that the unit common rafter is the hypotenuse of the roof slope triangle.* The line length of the common rafter can be used to develop the actual roof sheathing triangles, which will produce the line length of the hip rafter. The line lengths of the hip jack rafters can now be determined by similar triangles, as in rafter **D** in Figure 50–15. The plumb cut and side

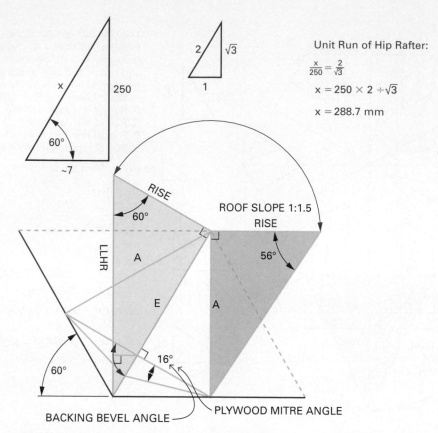

Unit Run of Hip Rafter:

$$\frac{x}{250} = \frac{2}{\sqrt{3}}$$

$$x = 250 \times 2 \div \sqrt{3}$$

$$x = 288.7 \text{ mm}$$

FIGURE 50–13 Determining hexagonal roof angles geometrically (metric).

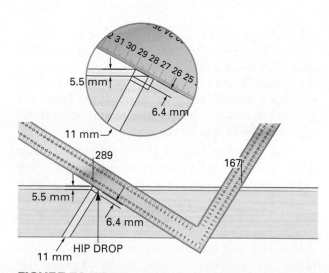

FIGURE 50–14 Determining amount to drop or back the hexagonal hip rafter (metric).

cut angles of all of the rafters, as well as the sheathing face and bevel angles, can be determined from the plan view and the actual roof triangle (**Figure 50–16**). Deductions for rafter thickness are easily determined from full-scale drawings of the intersections.

The hip rafters (running from the outside corner to the peak) can either be dropped or backed. The unit

hip rafter is calculated using similar triangles and for the octagon is 12.989 inches, or 13. A good way to remember this number is to associate it with the 5/12 roof triangle, which is very close to the 67½° of the octagon and has a hypotenuse of 13 inches. With the framing square set at the unit rise (10″) on the tongue, and at 13″ on the body, the values for the hip rafter drop and for the backing bevel can be determined (**Figure 50–17**). These are not the same values.

Similarly for the metric octagonal roof, the determination of the line lengths can be done mathematically or geometrically (**Figure 50–18**). The unit 22½°–67½°–90° triangle has sides of 1 – tan (67½°) – 1/cos(67½°), respectively, and the run of the common and jack rafters can be calculated by similar triangles. Given the slope of the roof, the total rise can be determined, and then the line length of the common rafter can be calculated using the Pythagorean theorem. *Remember that the unit common rafter is the hypotenuse of the roof slope triangle.*

The line length of the common rafter can be used to develop the actual roof sheathing triangles, which will produce the line length of the hip rafter. The line lengths of the hip jack rafters can now be determined by similar triangles, as in rafter **D** in Figure 50–18. The plumb cut and side cut angles of all of the rafters,

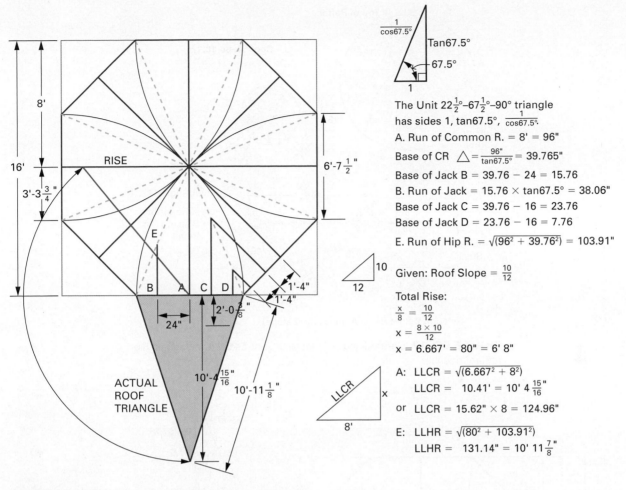

The Unit $22\frac{1}{2}°$–$67\frac{1}{2}°$–$90°$ triangle has sides 1, tan67.5°, $\frac{1}{\cos67.5°}$.

A. Run of Common R. = 8' = 96"

Base of CR △ = $\frac{96"}{\tan67.5°}$ = 39.765"

Base of Jack B = 39.76 − 24 = 15.76

B. Run of Jack = 15.76 × tan67.5° = 38.06"

Base of Jack C = 39.76 − 16 = 23.76

Base of Jack D = 23.76 − 16 = 7.76

E. Run of Hip R. = $\sqrt{(96^2 + 39.76^2)}$ = 103.91"

Given: Roof Slope = $\frac{10}{12}$

Total Rise:

$\frac{x}{8} = \frac{10}{12}$

$x = \frac{8 \times 10}{12}$

$x = 6.667' = 80" = 6' 8"$

A: LLCR = $\sqrt{(6.667^2 + 8^2)}$

LLCR = 10.41' = 10' $4\frac{15}{16}$"

or LLCR = 15.62" × 8 = 124.96"

E: LLHR = $\sqrt{(80^2 + 103.91^2)}$

LLHR = 131.14" = 10' $11\frac{7}{8}$"

FIGURE 50–15 Plan view of the octagonal roof (imperial).

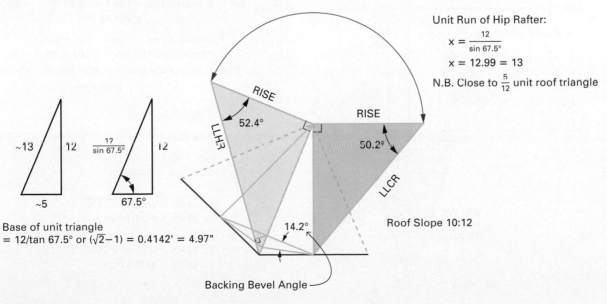

Unit Run of Hip Rafter:

$x = \frac{12}{\sin 67.5°}$

$x = 12.99 = 13$

N.B. Close to $\frac{5}{12}$ unit roof triangle

Base of unit triangle
= 12/tan 67.5° or ($\sqrt{2}$−1) = 0.4142' = 4.97"

Roof Slope 10:12

FIGURE 50–16 Determining octagonal roof angles (imperial), 10/12 slope.

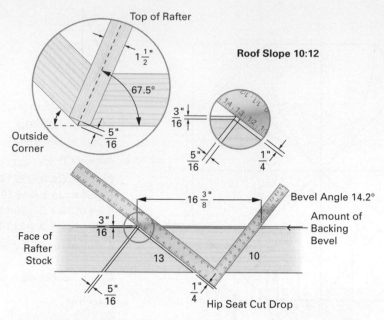

Roof Slope 10:12

Octagonal Roof Bevel Angles (10/12 slope)

FIGURE 50–17 Determining amount to drop or back the octagonal hip rafter (imperial).

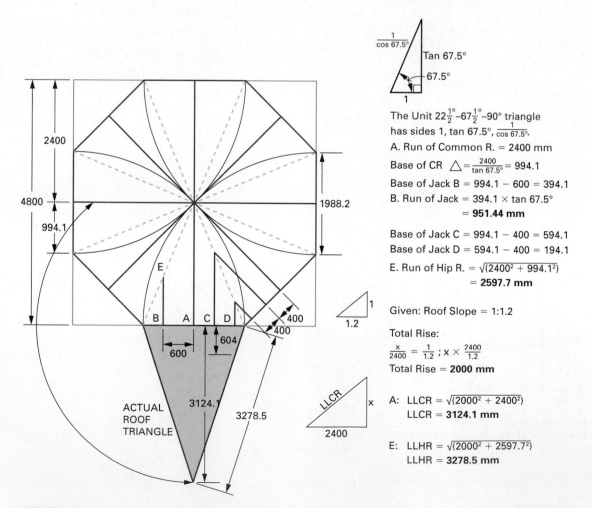

The Unit $22\frac{1}{2}°$–$67\frac{1}{2}°$–90° triangle has sides 1, tan 67.5°, $\frac{1}{\cos 67.5°}$.

A. Run of Common R. = 2400 mm

Base of CR $\triangle = \frac{2400}{\tan 67.5°} = 994.1$

Base of Jack B = 994.1 − 600 = 394.1

B. Run of Jack = 394.1 × tan 67.5°
= **951.44 mm**

Base of Jack C = 994.1 − 400 = 594.1

Base of Jack D = 594.1 − 400 = 194.1

E. Run of Hip R. = $\sqrt{(2400^2 + 994.1^2)}$
= **2597.7 mm**

Given: Roof Slope = 1:1.2

Total Rise:

$\frac{x}{2400} = \frac{1}{1.2}$; $x \times \frac{2400}{1.2}$

Total Rise = **2000 mm**

A: LLCR = $\sqrt{(2000^2 + 2400^2)}$
LLCR = **3124.1 mm**

E: LLHR = $\sqrt{(2000^2 + 2597.7^2)}$
LLHR = **3278.5 mm**

FIGURE 50–18 Plan view of the octagonal roof (metric).

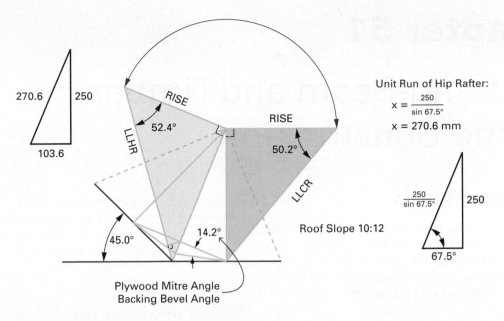

FIGURE 50–19 Determining octagonal roof angles (metric), 1:1.2 slope.

as well as the sheathing face and bevel angles, can be determined from the plan view and the actual roof triangle (**Figure 50–19**). Deductions for rafter thickness are easily determined from full-scale drawings of the intersections.

The hip rafters (running from the outside corner to the peak) can either be dropped or backed.

The unit hip rafter is calculated using similar triangles and for the octagon is 270.6 or 271 mm. With the framing square set at the unit rise (208) on the tongue, and at 271 on the body, the values for the hip rafter drop and for the backing bevel can be determined (**Figure 50–20**). These are not the same values.

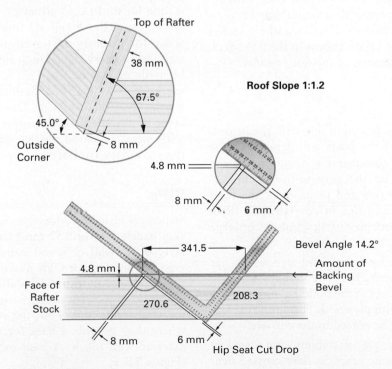

FIGURE 50–20 Amount to drop or back the octagonal hip rafter (metric).

Chapter 51

Post-and-Beam and Timber Frame Construction

POST-AND-BEAM

As mentioned in Chapter 34, post-and-beam frame construction uses larger but fewer members than conventional framing. The structure essentially consists of horizontal or sloping beams that rest on vertical posts, which in turn bear on the foundation.

Floors

The floors in a post-and-beam building are generally independent of the walls (Figure 34–5). They consist of wood beams or girders that are spaced on 4′, 5′-4″, 6′, or 8′ centres (1.2, 1.6, 1.8, or 2.4 m). The beams can be solid wood (Figure 34–6), built-up (Figure 35–4), LVL (Figure 5–4), PSL (Figure 6–3), or glulam (Figure 8–2). These beams can rest on the foundation, or they can bear on special metal connectors that have been engineered to carry the load. After the floor beams are in place and secured, they are covered with $1^3/_{32}$-inch (28 mm) tongue-and-groove (T&G) plywood or with thick T&G or double T&G planks that are 1½, 2½, and 3½ inches (38, 64, and 89 mm). The thickness of the floor deck will depend on the spacing of the floor beams.

Walls

The posts are usually placed in line with the beams, and again the size of the posts will depend on the spacing. There are also metal connectors for the post-to-foundation or the post-to-beam connection. Because the beams and posts carry the load of the entire structure, the spaces between them can accommodate floor-to-ceiling windows or large skylights.

To prevent racking from wind stresses, it has become common practice to conventionally frame in the spaces at the corners of the building and to cover the framed section with plywood. This limits the size of windows that can be placed in the corners.

Nonetheless, because of stability and speed of erection, platform-frame practices have crossed over into post-and-beam construction. This is most readily observed with posts that are tied together with top and bottom plates and with blocking being placed between the post and the foundation (Figure 34–5).

Roofs

The roof structure can have the roof beams running the length of the building or across the length of the building. A **longitudinal roof** will have parallel, horizontal roof beams that rest on posts of varying heights that are on the ends of the building. A **transverse roof** will have parallel, sloping roof beams that rest on posts of equal heights in the side walls of the building and on a **ridge beam** that is supported by taller posts in the end walls. A series of heavy metal connectors secure anchorage for the roof beams (Figure 34–9).

The roof beams are usually covered with T&G planks, which are finished on the underside as a ceiling finish. In cold climates, another roof frame is constructed on top of this roof to provide a space for insulation as well as for a vapour barrier and to carry the roofing finish. This frame can be made of 2 × 6, 2 × 8, 2 × 10, or even 2 × 12 stock (38 × 140, 184, 235, or 286 mm), depending on the level of insulation desired. Alternatively, a layer of thick extruded polystyrene held in place with metal or wood battens can be placed before the secondary roof deck.

SIPs

Structural insulated panels, or SIPs, are thick, stress-skin panels (~6″ or 152 mm) consisting of an inner layer of drywall or wood panel, an outer layer of wood or plywood or OSB, and a core of polystyrene, polyurethane, or other insulating materials. These interlocking panels can be placed on the outsides of the wall frame and/or on the roof of the building, where they provide a base for the interior and the exterior wall finish as well as for the roof finish (**Figure 51–1**).

FIGURE 51–1 Structural insulated panel system.

FIGURE 51–2 Trailhead outdoors store (Ottawa)— exterior "bent." *(Photo by M. Nauth; courtesy of Trailhead, Ottawa; www.trailhead.ca)*

A crane is used to position the SIPS, but it takes only a short time before the entire structure is enclosed and insulated. Panels are designed and cut to fit at the factory and then assembled on-site. The "look" of the post-and-beam is preserved, labour costs and time are greatly reduced, and the building is protected from the weather in short order. Care must be taken on-site to introduce chases for the mechanical trades while the panels are being installed. Openings are cut in later as per plan.

TIMBER FRAME CONSTRUCTION

Long before we had engineered metal connectors, power tools, and cranes, large barns were being raised all across North America. Wood joinery on a large scale, performed with specialized hand tools, was used to tie together posts and beams and braces. This work was done close to the ground on a horizontal plane; these "locked" arches or bents were then raised in succession and held together and apart by roof purlins and wall plates (**Figures 51–2** and **51–3**).

The basic hand tools of the timber framer include the 1½- to 2-inch (38 to 50 mm) socket or tang framing chisel about 12 inches (305 mm) long; a framing mallet, 32 to 48 ounces (907 to 1361 g) in weight, that looks like half a rolling pin; a framing square 16/24 inches (406/610 mm); a rip and a cross-cut handsaw; two block planes (low angle and standard); a rabbet plane (for shoulders); and an auger (**Figures 51–4, 51–5,** and **51–6**).

FIGURE 51–3 Lateral tie beam using joinery and brackets. *(Photo by M. Nauth; courtesy of Trailhead, Ottawa; www.trailhead.ca)*

FIGURE 51–4 A basic chisel.

Courtesy of Woodcraft Supply, LLC

FIGURE 51–5 Barr timber framing socket chisels.

FIGURE 51–6 Tang framing chisels. *(Courtesy of Lee Valley Tools Ltd.; www.leevalley.com.)*

Power tools that are useful include the ½-inch (12 mm) drill, the circular saw, and perhaps a chain mortiser to speed up the process.

The basic framing members of the bent are the principal posts, wall plates, tie beams, rafters, ridge beam, purlins, king post, collar ties, and knee braces (**Figure 51–7**).

The basic joint for connecting the members of the bent is the mortise and tenon. There are several variations of this joint—housed, through, shouldered, offset, half-dovetailed, and so on (Figure 51–8). To lock the bents together with purlins and summer beams, a dovetail joint is used.

A mastery of layout tools is a must for the timber framer. The tape, framing square, sliding T-bevel, marking gauge, try square, and so on are all staples on the site or in the shop. A rudimentary knowledge of trigonometry and geometry will assist with accuracy when laying out roof angles and arches.

Each member of the bent or frame must be checked and rechecked before cutting and chiselling. Once you are satisfied with the layout marks, cut the joints on enough pieces for one bent and dry-fit the members. Then re-measure to verify accuracy.

TIMBER FRAMING PROJECT (IN-SCHOOL)

Professor Jeff Acciaccaferro, from Mohawk College in Hamilton, Ontario, leads his apprentices in the construction of a timber frame building. They start with very detailed plans drawn by Robert Deeks, and they work together in groups to produce two *bents*, which are then connected together with purlins, the ridge beam, and braces (**Figures 51–8 to 51–19**). Each tenon has to fit perfectly into its respective mortise, so great care must be taken with the chisel work. Working together on various parts of a single construction masterpiece encourages teamwork and creates camaraderie (**Figures 51–20 and 51–21**).

After a timber frame structure is erected and covered, the owner strives to leave as much and as many of the timbers as possible exposed to view (Figure 51–22). Being housed in the presence of these great timbers lends a feeling of warmth and security (**Figures 51–22 to 51–24**). For a thorough treatment of timber framing, check out *A Timber Framer's Workshop* by Steve Chappell (Fox Maple Press, ISBN 1-889269-00-X).

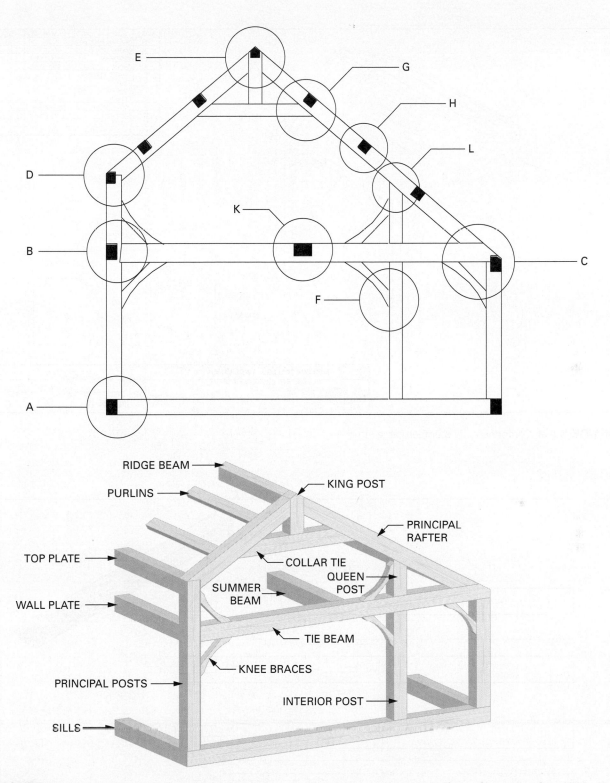

FIGURE 51–7 The basic framing members of the bent. (Timber Framer's Workshop *by Steve Chappell, a Corn Hill book, Courtesy of Fox Maple Press*)

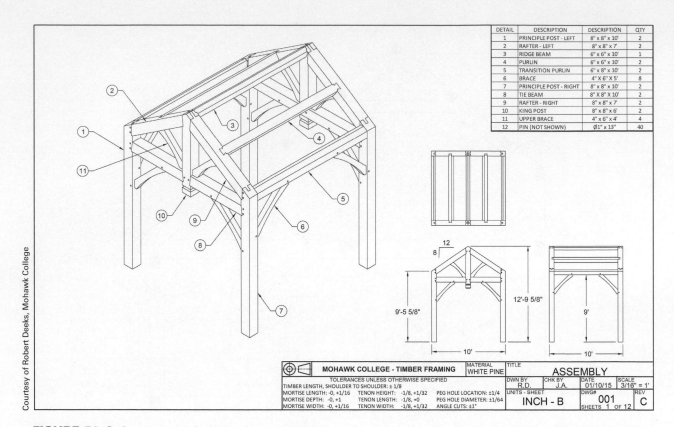

DETAIL	DESCRIPTION	DESCRIPTION	QTY
1	PRINCIPLE POST - LEFT	8" x 8" x 10'	2
2	RAFTER - LEFT	8" x 8" x 7'	2
3	RIDGE BEAM	6" x 6" x 10'	1
4	PURLIN	6" x 6" x 10'	2
5	TRANSITION PURLIN	6" x 8" x 10'	2
6	BRACE	4" X 6" X 5'	8
7	PRINCIPLE POST - RIGHT	8" x 8" x 10'	2
8	TIE BEAM	8" X 8" X 10'	2
9	RAFTER - RIGHT	8" x 8" x 7'	2
10	KING POST	8" x 8" x 6'	2
11	UPPER BRACE	4" x 6" x 4'	4
12	PIN (NOT SHOWN)	Ø1" x 13"	40

MOHAWK COLLEGE - TIMBER FRAMING
MATERIAL WHITE PINE
TITLE ASSEMBLY
TOLERANCES UNLESS OTHERWISE SPECIFIED
TIMBER LENGTH, SHOULDER TO SHOULDER: ± 1/8
MORTISE LENGTH: -0, +1/16   TENON HEIGHT: -1/8, +1/32   PEG HOLE LOCATION: ±1/4
MORTISE DEPTH: -0, +1   TENON LENGTH: -1/8, +0   PEG HOLE DIAMETER: ±1/64
MORTISE WIDTH: -0, +1/16   TENON WIDTH: -1/8, +1/32   ANGLE CUTS: ±1°
DWN BY R.D.   CHK BY J.A.   DATE 01/10/15   SCALE 3/16" = 1'
UNITS - SHEET INCH - B   DWG# 001   REV C
SHEETS 1 OF 12

Courtesy of Robert Deeks, Mohawk College

FIGURE 51–8 Components of a timber frame structure.

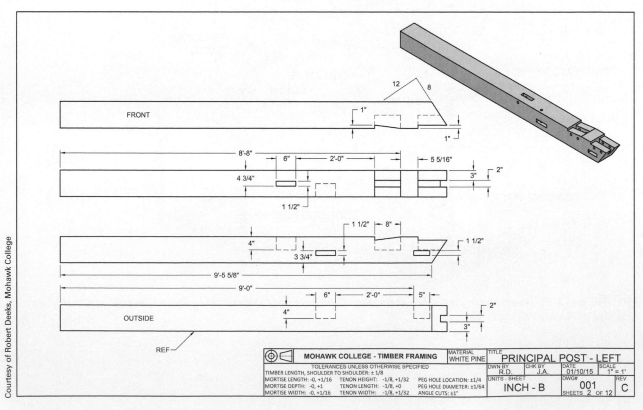

MOHAWK COLLEGE - TIMBER FRAMING
MATERIAL WHITE PINE
TITLE PRINCIPAL POST - LEFT
TOLERANCES UNLESS OTHERWISE SPECIFIED
TIMBER LENGTH, SHOULDER TO SHOULDER: ± 1/8
MORTISE LENGTH: -0, +1/16   TENON HEIGHT: -1/8, +1/32   PEG HOLE LOCATION: ±1/4
MORTISE DEPTH: -0, +1   TENON LENGTH: -1/8, +0   PEG HOLE DIAMETER: ±1/64
MORTISE WIDTH: -0, +1/16   TENON WIDTH: -1/8, +1/32   ANGLE CUTS: ±1°
DWN BY R.D.   CHK BY J.A.   DATE 01/10/15   SCALE 1" = 1'
UNITS - SHEET INCH - B   DWG# 001   REV C
SHEETS 2 OF 12

Courtesy of Robert Deeks, Mohawk College

FIGURE 51–9 Principal posts.

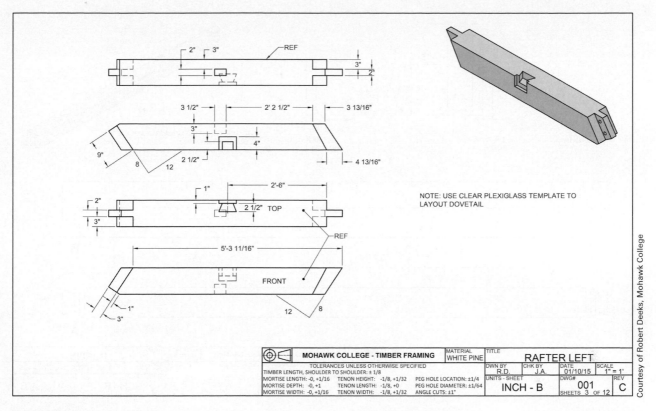

FIGURE 51-10 Rafter left.

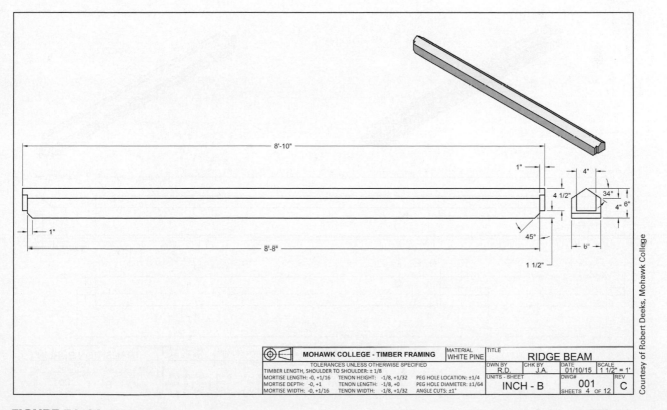

FIGURE 51-11 Ridge beam.

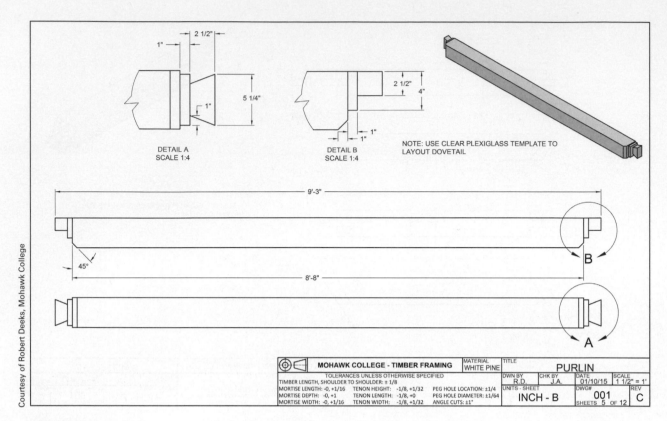

FIGURE 51–12 Purlin.

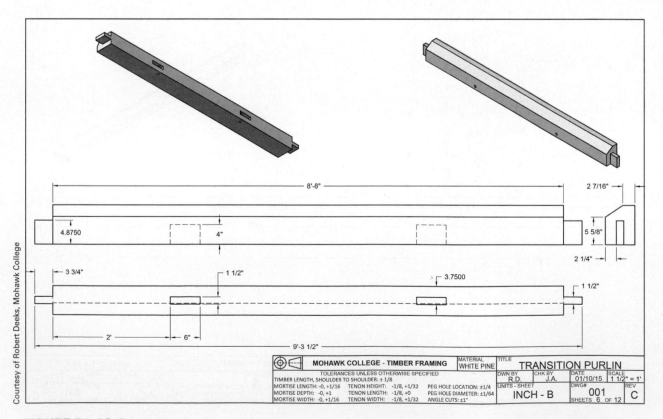

FIGURE 51–13 Transition purlins.

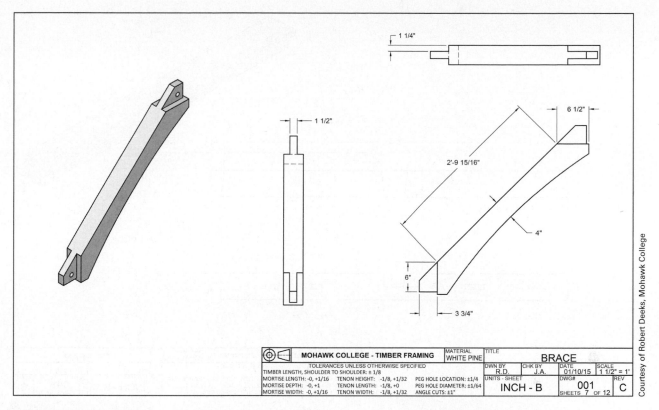

	MOHAWK COLLEGE - TIMBER FRAMING	MATERIAL WHITE PINE	TITLE BRACE			
	TOLERANCES UNLESS OTHERWISE SPECIFIED TIMBER LENGTH, SHOULDER TO SHOULDER: ± 1/8		DWN BY R.D.	CHK BY J.A.	DATE 01/10/15	SCALE 1 1/2" = 1'
	MORTISE LENGTH: -0, +1/16 TENON HEIGHT: -1/8, +1/32 PEG HOLE LOCATION: ±1/4 MORTISE DEPTH: -0, +1 TENON LENGTH: -1/8, +0 PEG HOLE DIAMETER: ±1/64 MORTISE WIDTH: -0, +1/16 TENON WIDTH: -1/8, +1/32 ANGLE CUTS: ±1"		UNITS - SHEET INCH - B		DWG# 001 SHEETS 7 OF 12	REV C

FIGURE 51–14 Brace.

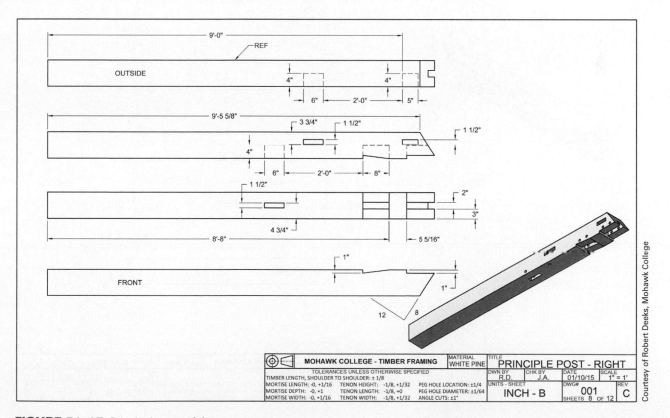

	MOHAWK COLLEGE - TIMBER FRAMING	MATERIAL WHITE PINE	TITLE PRINCIPLE POST - RIGHT			
	TOLERANCES UNLESS OTHERWISE SPECIFIED TIMBER LENGTH, SHOULDER TO SHOULDER: ± 1/8		DWN BY R.D.	CHK BY J.A.	DATE 01/10/15	SCALE 1" = 1'
	MORTISE LENGTH: -0, +1/16 TENON HEIGHT: -1/8, +1/32 PEG HOLE LOCATION: ±1/4 MORTISE DEPTH: -0, +1 TENON LENGTH: -1/8, +0 PEG HOLE DIAMETER: ±1/64 MORTISE WIDTH: -0, +1/16 TENON WIDTH: -1/8, +1/32 ANGLE CUTS: ±1"		UNITS - SHEET INCH - B		DWG# 001 SHEETS 8 OF 12	REV C

FIGURE 51–15 Principal post—right.

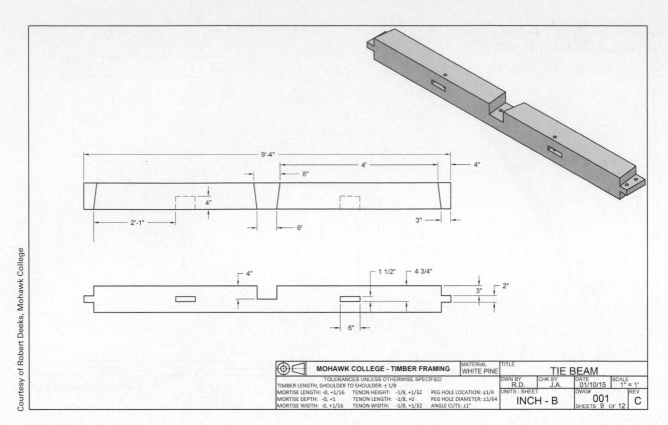

FIGURE 51–16 Tie beam.

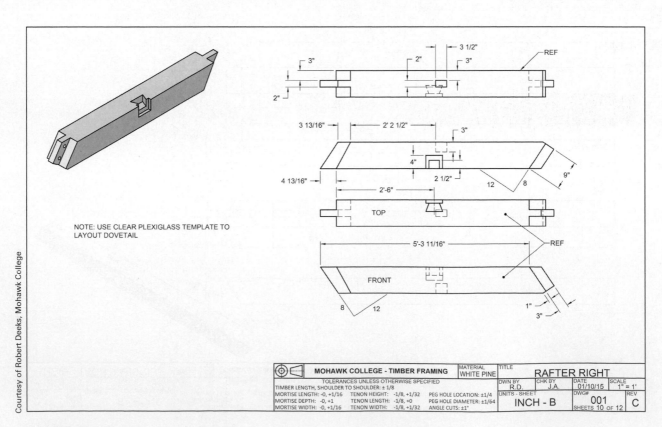

NOTE: USE CLEAR PLEXIGLASS TEMPLATE TO
LAYOUT DOVETAIL

FIGURE 51–17 Rafter right.

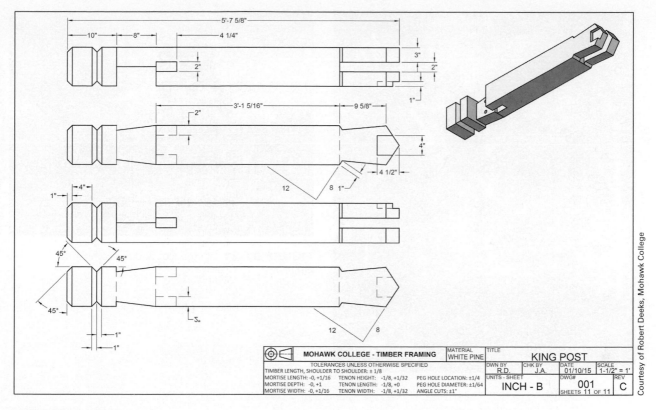

FIGURE 51–18 King post.

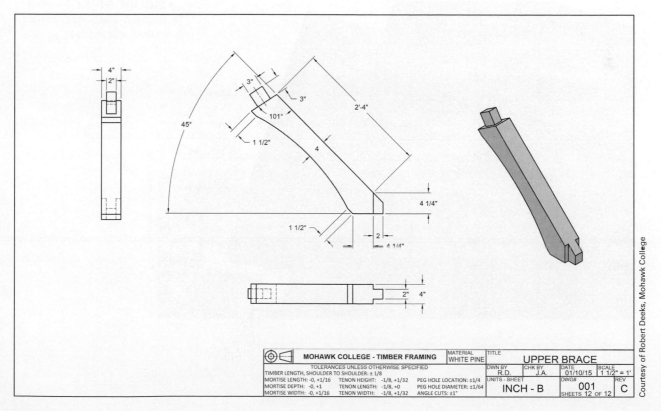

FIGURE 51–19 Upper brace.

Courtesy of Jeff Acciaccaferro, Mohawk College

FIGURE 51–20 The completed timber frame.

Courtesy of Jeff Acciaccaferro, Mohawk College

FIGURE 51–21 Apprentices and the completed structure.

FIGURE 51–22 Finished timber frame porch, detailing, ceiling, and roofing.

Photo by M. Nauth; courtesy of Dan Brigham

FIGURE 51–23 Working under the canopy.

FIGURE 51–24 Coming together—beams at 45 degrees.

KEY TERMS

backing the hip

bents

bird's mouth

butterfly roof

cheek cut

collar ties

common rafters

cripple jacks

dormers

dropping the hip

fascia

fascia line

gable roof

gable studs

gambrel roof

gazebos

hexagonal roof

hip jack rafters

hip jacks

hip rafters

hip roof

hip-valley cripple jack rafters

intersecting roof

jack rafters

knee wall

lateral bracing

level line

line length

lookouts

longitudinal roof

major spans

mansard roof

minor spans

octagonal roof

pitch

plumb line

purlin

rafter stand

rafter tables

rake rafters

ridge beam

ridgeboard

seat cut

shed roof

shortened or supported valley rafters

slope

soffit

supporting valley rafters

tail cut

total rise

total run

total span

transverse roof

unequal pitched

unit length

unit rise

unit run

valleys

valley cripple jack rafters

valley jack rafters

valley jacks

valley rafters

REVIEW QUESTIONS

Select the most appropriate answer.

1. What type of roof has the fewest number of different sized members?

   a. gable
   b. hip
   c. gambrel
   d. mansard

2. What is the term used to represent horizontal distance under a rafter?

   a. run
   b. span
   c. line length
   d. pitch

3. What type of rafter spans from a hip rafter to the ridge?

   a. hip jack
   b. valley jack
   c. valley
   d. cripple jack

4. What type of rafter spans from a wall plate to the ridge and runs perpendicular to the ridge?

   a. hip
   b. valley
   c. common
   d. hip jack

5. What is the recommended minimum "stand" of the common rafter relative to the heel plumb cut of the rafter?

   a. one-quarter
   b. one-half
   c. two-thirds
   d. three-quarters

6. What is the jack rafter most similar to?

   a. common rafter
   b. hip rafter
   c. valley rafter
   d. gable end stud

7. What is the total run of any hip jack rafter equal to?

   a. its distance along the plate from the outside corner
   b. its distance along the plate from the inside corner
   c. its line length
   d. its common difference in length

8. What is the total run of the shortened valley rafter equal to?

   a. the span of the minor roof
   b. the span of the major roof
   c. the total run of the common rafter of the minor roof
   d. the total run of the hip rafter of the minor roof

9. What is the total run of the hip-valley cripple jack rafter equal to?

   a. one-half the run of the hip jack rafter
   b. one-half the run of the longest valley jack rafter
   c. the distance from the outside corner to the inside corner of the building
   d. the difference in run between the supporting and shortened valley rafters

10. In order to determine the length of any rafter of a roof frame with a specified slope, what must be determined?

    a. total run
    b. unit rise
    c. unit run
    d. ridgeboard length

11. What roof framing member provides the most tipping resistance for roof trusses?

    a. diagonal brace
    b. lateral brace
    c. web
    d. gusset

12. Which member of the truss requires a lateral brace if its length exceeds 6 feet (1.83 m)?

    a. tension web
    b. compression web
    c. top chord
    d. king post

13. What is the line length of a common rafter on a 5 in 12 slope if the building is 28'-0" wide?

    a. 70 inches
    b. 140 inches
    c. 168 inches
    d. 182 inches

14. What is the common difference in the length of gable studs spaced 24 inches OC for a roof with a slope of 8 on 12?

    a. 8 inches
    b. 12 inches
    c. 16 inches
    d. 20 inches

15. What is the hip projection, in inches, if the common rafter projection is 6 inches?

    a. 6 inches
    b. 7 inches
    c. 8½ inches
    d. 11 inches

16. What is the line length of a hip rafter with a unit rise of 6 inches and a total run of 12 feet?

    a. 72 inches
    b. 144 inches
    c. 216 inches
    d. 224 inches

17. What is the typical amount that a dimension lumber hip rafter is shortened because of the ridge?

    a. ¾ inch
    b. $1^1/_{16}$ inches
    c. 1½ inches
    d. $1^{13}/_{16}$ inches

18. What is the length of a ridgeboard for a hip roof installed on a rectangular building measuring 28 × 48 feet if all of the rafters are 1½ inches thick?

    a. 14'-1½"
    b. 20 feet
    c. 20'-1½"
    d. 24 feet

19. Excluding the rake rafters, what is the estimated number of gable rafters (at 16" OC) for a rectangular building measuring 28 × 48 feet?

    a. 28 gable rafters
    b. 48 gable rafters
    c. 72 gable rafters
    d. 74 gable rafters

20. Excluding the roof overhangs, what is the estimated number of 4 × 8-feet sheathing panels needed for a hip roof installed on a rectangular building that measures 28 × 48 feet that has a unit rise of 6 inches?

    a. 24 panels
    b. 25 panels
    c. 48 panels
    d. 51 panels

21. What is the line length of a common rafter measured from the centreline of the ridge to the outside edge of the wall plate, given that the building is 9000 mm wide and the roof slope is 1:2.5?

    a. 1800 mm
    b. 4510 mm
    c. 4846.6 mm
    d. 9694.8 mm

22. What is the common difference in the length of gable studs spaced 600 mm OC for a roof with a slope of 200/250?

    a. 200 mm
    b. 480 mm
    c. 510 mm
    d. 751 mm

23. What is the hip projection, in millimetres, if the projection of the common rafter is 200 mm?

    a. 200 mm
    b. 240 mm
    c. 250.6 mm
    d. 282.2 mm

24. What is the line length of a hip rafter for a roof with a slope of 1:2 and a total building span of 7000 mm?

    a. 5377.4 mm
    b. 5250 mm
    c. 4081 mm
    d. 1239 mm

25. Given that the ridgeboard is 38 × 284, how much should the hip rafter be shortened after its line length is marked off?

    a. 19 mm
    b. 27 mm
    c. 38 mm
    d. 54 mm

26. Given that all of the material is 38 mm thick, what is the actual length of the ridgeboard for a hip roof installed on a 4800 mm × 8400 mm rectangular building?

    a. 3600 mm
    b. 3638 mm
    c. 4838 mm
    d. 8400 mm

27. Excluding the rake rafters, how many common rafters, placed at 400 mm OC, are required for the gable roof on a rectangular building that measures 4800 mm × 8400 mm?

    a. 21 common rafters
    b. 22 common rafters
    c. 42 common rafters
    d. 44 common rafters

28. What is the area of a hip roof with a slope of 200 in 250 that is to be installed on a rectangular building that measures 5000 mm × 8400 mm, without any overhang?

    a. 42.0 m²
    b. 26.9 m²
    c. 53.8 m²
    d. 84.0 m²

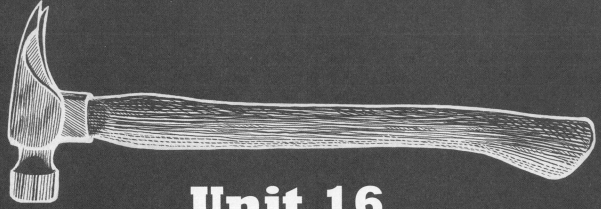

Unit 16

Energy-Efficient Housing

Thermal insulation prevents the loss of heat in buildings during cold seasons. It also resists the passage of heat into air-conditioned buildings in hot seasons. Moisture in the air may cause severe problems for a building. Vapour retarders are essential to a long-lived building. Adequate ventilation must be provided in the living environment and within the building materials. Ventilation encourages evaporation of harmful moisture formed in living spaces and within the insulation. Acoustical insulation reduces the passage of sound from one area to another.

Insulation and ventilation should be thought of together. If a house has insulation it must also have allowances for ventilation. This will ensure that the insulation performance is maximized while also making the house pleasant to live in.

Cost is also a factor. With the depletion of non-renewable fossil fuels, the cost of energy has soared, and controlling the consumption of precious resources has become central to the thinking of the Western world. In addition, greenhouse gas emissions and global warming have become driving forces in the building industry.

OBJECTIVES

After completing this unit, the student should be able to:

- describe how insulation works and define insulating terms and requirements.
- describe the commonly used insulating materials and state where insulation is placed.
- properly install various insulation materials.
- explain relative humidity and moisture migration.
- explain the need for ventilating a structure.
- explain the kinds and purpose of vapour retarders and how they are applied.
- describe various methods of construction to reduce the transmission of sound.

Chapter 52

Building Science: Principles and Practice

A BRIEF HISTORY

In the 1970s, the Western world was confronted with an energy crisis that startled governments into realizing that energy supplies were not endless. The world's population was exploding, and more people with more money wanting more creature comforts caused demand for fuel to begin to outstrip supply.

In 1980 the Canadian government, through the Department of Energy, Mines, and Resources (now Natural Resources Canada [NRCan]), set up the Super Energy Efficient Home program. In 1982, the Government of Canada officially launched the R-2000 Program. From its inception, it was more than a mere energy efficiency program. The strength of the R-2000 initiative has been that, while it is largely focused on energy efficiency, it helped to define and formulate the "house-as-a-system" concept. R-2000 has been the "best in class" energy-efficient label and the most comprehensive environmental standard in Canada for nearly 30 years. The R-2000 standard, effective July 2012, can be found at *http://www.nrcan.gc.ca/ energy/efficiency/housing/new-homes/5089*.

In 1993, in conjunction with Canada Mortgage and Housing Corporation, ten "advanced houses" were built across Canada that added to energy efficiency the standard of healthy indoor air quality and environmentally responsive selection of materials and construction practices. In 2005, Canada joined the United States and enacted the ENERGY STAR® for New Homes initiative. The measuring stick used to qualify homes is greenhouse gas (GHG) emissions. The ENERGY STAR® brand is administered and promoted in Canada by NRCan on behalf of the U.S. Environmental Protection Agency. The ENERGY STAR® for New Homes initiative promotes a cost-effective and flexible method for building energy-efficient new homes.

The **ENERGY STAR® for New Homes initiative** qualifies homes that are approximately 25 percent more energy efficient than a conventionally built home. The increased energy efficiency of these homes translates into reduced energy costs for homeowners. Typical energy efficiency measures for these new homes include insulation upgrades; higher-performance windows; better draft proofing; more efficient heating, hot water, and air conditioning systems; and ENERGY STAR® qualified appliances (if supplied by the builder). ENERGY STAR® qualified new homes are built to strict technical specifications by ENERGY STAR® for New Homes builders and undergo random quality assurance audits during construction. The ENERGY STAR technical specifications chart (2015) for doors, windows, and skylights (**Figure 52-1**) uses Heating Degree Day (HDD) from the Canadian Model National Building Code (2010). The map divides Canada into three zones but you can find the HDDs for your specific location by entering its Postal Code at *http://oee.nrcan.gc.ca/energy-zones-climatiques/*. The three zones are described below:

Zone 1 (HDDs < 3500): in light green; southwestern B.C. and Vancouver Island.

Zone 2 (3500 < HDDs < 6000): in dark green; southeastern, central and northern B.C. except the extreme northern part; southern and central Alberta; southern Saskatchewan and extreme southern Manitoba; the extreme southern part of northwestern Ontario; central and southern Ontario; south-western Quebec; all of the Maritimes and Newfoundland except for the northern part of the Northern Peninsula.

Zone 3 (HDDs > 6000): in light blue; the extreme northern part of B.C. and northern Alberta; central and northern Saskatchewan; all of Manitoba except the extreme southern part; northern Ontario except the extreme southern part of northwestern Ontario; all of Quebec except the south-west; the northern part of the Northern Peninsula; all of Labrador, Nunavut, N.W.T., and the Yukon.

The new Energy Star for New Homes Standard, effective April 2014, can be found at *http://www .nrcan.gc.ca/energy/efficiency/housing/new-homes/ energy-star/14178#a1*.

The **LEED® Canada for Homes** (Leadership in Energy and Environmental Design) has a rating system that measures the overall performance of a

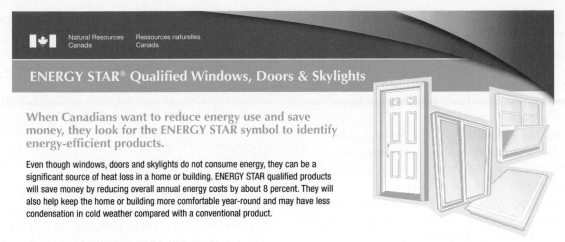

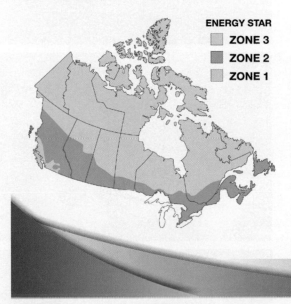

Natural Resources Canada
Ressources naturelles Canada

ENERGY STAR® Qualified Windows, Doors & Skylights

When Canadians want to reduce energy use and save money, they look for the ENERGY STAR symbol to identify energy-efficient products.

Even though windows, doors and skylights do not consume energy, they can be a significant source of heat loss in a home or building. ENERGY STAR qualified products will save money by reducing overall annual energy costs by about 8 percent. They will also help keep the home or building more comfortable year-round and may have less condensation in cold weather compared with a conventional product.

How do these products qualify for ENERGY STAR?

To be ENERGY STAR qualified, products must meet specific energy efficiency levels that have been set for three climate zones in Canada. In addition, all products must be certified for their energy efficiency by an accredited agency.

The three climate zones were developed by using heating degree-days, a measure of annual average temperature. The efficiency levels indicate how well a window, door or skylight insulates against the cold and how well it uses the sun's heat to supplement the heating system of a home or building. Because the climate becomes progressively colder from Zone 1 to Zone 3, the levels are more stringent for each successive zone. This means that models that qualify for Zone 3 also qualify for Zones 1 and 2.

Climate zones

ENERGY STAR
- ZONE 3
- ZONE 2
- ZONE 1

Criteria

Products are rated on either their U-factor or their Energy Rating (ER). The U-factor is a measure of the rate of heat loss. The lower the number, the slower the heat loss. The ER is a formula that includes the U-factor, air leakage and the benefit of potential solar gain. The higher the value, the higher the potential annual energy savings. Products must also have an air leakage rate of ≤1.5 litres per second per square metre of product area.

Windows and doors (effective February 1, 2015)				
Zone	Heating degree-day range	Minimum Energy Rating (unitless)	or	Maximum U-factor (W/m²·K)
1	<3 500	25	or	1.60
2	3 500 to <6 000	29	or	1.40
3	≥6 000	34	or	1.20

Skylights* (effective February 1, 2015)		
Zone	Heating degree-day range	Maximum U-factor W/m²·K
1	<3 500	2.60
2	3 500 to <6 000	2.40
3	≥6 000	2.20

*The requirement for tubular-type skylights for all three zones is 2.60.

Canadá

Climate zone map divided into four zones.

Zone A, in light green, in southwestern B.C. and Vancouver Island.

Zone B, in dark green, central and southeastern B.C., southwestern Alberta, extreme southern Saskatchewan, central and southern Ontario, southwestern Quebec, all of the Maritimes, and Newfoundland except for the Avalon Peninsula.

Zone C, in light blue, southern and central Yukon, extreme southwestern N.W.T., northern B.C., central and northern Alberta, all of Saskatchewan except for the extreme south, all of Manitoba, most of northern Ontario except for the far north, central Quebec, southern and central Labrador, and the Avalon Peninsula.

Zone D, in purple in the northern Yukon, all of N.W.T. except for the extreme southern portion, all of Nunavut, extreme northern Ontario, northern Quebec, and northern Labrador.

FIGURE 52–1 Canada's climate zones for window specifications. *("ENERGY STAR® Qualified Windows, Doors & Skylights." Natural Resources Canada, 2014. Reproduced with the permission of the Department of Natural Resources, 2015.)*

home in the following eight categories, according to a point system, to a maximum of 136 points:

1. **Innovation & Design Process (ID):** Special design methods, unique regional credits, measures not currently addressed in the rating system, and exemplary performance levels (maximum 11 points)

2. **Location & Linkages (LL):** The placement of homes in socially and environmentally responsible ways in relation to the larger community (max. 10 points)

3. **Sustainable Sites (SS):** The use of the entire property so as to minimize the project's impact on the site (max. 22 points)

4. **Water Efficiency (WE):** Water-efficient practices, both indoor and outdoor (max. 15 points)

5. **Energy & Atmosphere (EA):** Energy efficiency, particularly in the building envelope and heating and cooling design (max. 38 points)

6. **Materials & Resources (MR):** Efficient utilization of materials, selection of environmentally preferable materials, and minimization of waste during construction (max. 16 points)

7. **Indoor Environmental Quality (EQ):** Improvement of indoor air quality by reducing the creation of and exposure to pollutants (max. 21 points)

8. **Awareness & Education (AE):** The education of the homeowner, tenant, and/or building manager about the operation and maintenance of the green features of a LEED home (max. 3 points)

To be LEED-certified, a house needs to accumulate 45 points. LEED-silver needs 60 points, LEED-gold 75 points, and LEED-platinum 90 points. For more information, go to *cagbc.org* and click on LEED and then Green Homes.

Energy use and costs are reduced when buildings are insulated. Unfortunately, when a building is made more airtight and more energy efficient, a negative side effect occurs. This effect is caused by moisture.

Moisture is the enemy of a building. Roofing and siding are installed to protect the building from water. Water can also cause the interior to be musty and mouldy, and the insulation to degrade and perform poorly, and can potentially cause structural failures. Moisture in and around a building must be understood and dealt with for the building to function properly and last a long time.

Energy-efficient construction is desirable to reduce energy costs and make a building more comfortable. But the problems of excess moisture within a building must be addressed. Solutions include controlling air leakage, using vapour retarders (also called vapour barriers), and building drying potential into building systems.

BUILDING SCIENCE BASICS

Energy efficiency required a new, scientific approach, which led to the **house-as-a-system** approach to dealing with building components. All parts of the system interact together—the change of one part will affect all other parts. Now all parts, including the building envelope, building materials, mechanical systems, appliances, finishes and furnishings, and occupants, are considered not just separately, but also how they impact the system as a whole. How do heat, moisture, and air actually flow, and how can they be controlled (**Figure 52–2**)?

FIGURE 52–2 Various poppers use different types of heat transfer to pop corn.

Heat Flow

- **Conduction:** Heat energy transfer through solids by molecular agitation. Some materials, such as metals, are good conductors and act as heat "sinks." Other materials are poor conductors or good insulators. These resist the flow of heat and thus have a high resistance or R-value (RSI). Examples are high-density foam board and extruded polystyrene.

- **Convection:** Heat transfer in fluids such as water or air. When air is heated, it becomes less dense and rises, causing cool air to move in to take its place. This process occurs within wall cavities, so batt insulation is installed to interrupt the convection loop. The installation must be near perfect so as to minimize air pockets, which will promote convection loops.

- **Radiation:** Heat transfer by electromagnetic waves. Here heat is transferred from a source to an object without heating up the space in between. An example is the sun heating the Earth and leaving space cold. Radiant barriers such as foil-backed products and low-emissivity (low-e) coatings on window panes are now commonly found in Canadian homes.

Heat loss by conduction is determined by the equation $Q = \dfrac{A\Delta T}{RSI}$

Where Q = the rate of heat flow in watts;

A = the area of the surface under consideration in m²;

ΔT = the difference in temperature °C between inside and outside; and

RSI = the resistance to heat flow, in m² × °C/W

For example, the heat loss through 10 m² (107 ft²) of wall area with an RSI value of 2.1 (R12) when the indoor temperature is 20°C and the outdoor temperature is −20°C is

$Q = \dfrac{10*40}{2.1} = 190.5\text{W}$ or equivalent to **two 100-W light bulbs.**

For 10 m² of attic area with an RSI value of 7.0 (R40),

$Q = \dfrac{10*40}{7} = 57\text{W}$ or equivalent to **one 60-W light bulb.**

For 10 m² of low-e, argon-filled, double-glazed window with RSI 0.53 (R3),

$Q = \dfrac{10*40}{0.53} = 750\text{W}$ or equivalent to **eight 100-W light bulbs.**

Note that heat is a form of energy and radiates equally in all directions away from its source.

Therefore, *heat loss by conduction* is a function of the heat-resistance property (RSI- or R-value) of the materials that make up the envelope.

Heat transfer by electromagnetic waves or *heat loss by radiation* can be reduced by inserting radiant barriers into the building envelope to reflect the waves. Aluminum foil bubble packs, paper-backed foil sheets, steel doors, and low-e coatings on window glazings are examples of radiant barriers.

Heat loss by convection or air movement is the major area of concern. **Air leakage control (ALC)** has become priority #1 with custom home builders, because where the air goes, the heat goes with it. Today the emphasis in national and provincial building codes is on air barriers and air-sealing details, especially at the interfaces between dissimilar materials (e.g., the junction between the sill plate and the top of the concrete wall). For the air to move, there must be pressure difference as well as a hole. The pressure difference almost always exists between inside and outside the house, so it is up to the builder to plug the holes. The pressurized rainscreen method of siding application helps to reduce the pressure differential in front of and behind the cladding (see Figure 53–11, page 614).

Moisture Flow

Moisture flows or moves in four ways:

- *By gravity:* Water or rain flows downhill, seeking the path of least resistance. The builder must recognize this, and use this principle to his or her advantage. Roofs, sills, ledges, and drip caps must be adequately sloped (steeper is better); the grade around the building must be sloped away from the foundation (minimum 1 in 12); a top layer of impervious material must be placed over the backfill before the top soil and sod; eavestroughs, downspouts, and splash-guard drains should be installed; drainage tile should be covered with a filter cloth; and free-flowing backfill should be placed next to the concrete footings.

- *By capillary action:* **Capillary action** is the movement of water uphill as well as sideways, by "wicking," as can be observed by placing a sheet of paper towel vertically over a puddle. Knowing this, the builder inserts "capillary breaks" between the footing and the undisturbed soil, between the brick veneer and the concrete wall, between the concrete foundation and the backfill, between the basement concrete slab and the ground below, and any place where water will sit and then wick up an adjacent surface

- *By air flow:* Where the air goes, the moisture goes with it. This knowledge encourages air barrier detailing and sealing.

FIGURE 52–3 Water can exist in three forms or states. Moisture in warm air condenses when it comes in contact with a cold surface.

- *By vapour diffusion:* This is the least worrisome of the four and can easily be dealt with by applying polyethylene sheets to the inside surfaces of the house frame or by using vapour diffusion retarder (VDR) paint.

CONDENSATION

Water exists in solid, liquid, and gaseous forms. Most familiar are the solid and liquid forms, which we all know as ice and water. Gaseous water is called steam when it is very hot and **vapour** when it is cool (**Figure 52–3**). Vapour is normally invisible. **Condensation**

is when the vapour falls out of the air onto cooler surfaces. **Dew point** is a term used to identify the air temperature when condensation has occurred. Fog is an example of air that has reached its dew point.

Relative Humidity

The amount of moisture held in the air is referred to as **relative humidity**. This amount can vary with the temperature of the air. Warm air can hold more moisture than cool air. Consider four identical 1-cubic-foot (28.3 L) containers of air, each having about five drops (0.00047 pound, or 0.00021 kg) of water suspended in the air as vapour (**Figure 52–4**).

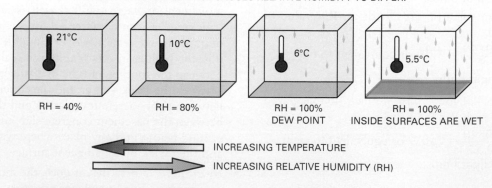

FIGURE 52–4 Relative humidity changes with air temperature.

The relative humidity (RH) in each container depends on the temperature of the air. At 21°C, the RH would be 40 percent, or put another way, the air would be holding 40 percent of the moisture it could hold at that temperature.

As the temperature drops, the air is less able to hold moisture. When the temperature reaches 6°C, the air can hold no more than the five drops; thus, the relative humidity is 100 percent. The air is completely saturated with moisture. A slight lowering of temperature will cause the air to be at the dew point and the container walls begin to feel moist.

Dew point is not always at 6°C; it can be any temperature where the RH reaches 100 percent. The dew point occurs when the air can no longer hold any more moisture.

Moisture in Buildings

Building materials perform best and last longest when the relative humidity averages below 50 percent, but the RH in the living environment is often much higher.

Moisture in the form of water vapour inside a building comes from many activities. It is produced by cooking, bathing, washing, drying, and cleaning, as well as by many other sources. Reducing the moisture production within the building is one step in solving the problem, but can only go so far.

Older, poorly insulated homes did not have the interior moisture problems that we have today. They were drafty enough to dry the house during the heating season. Often, humidifiers were operated to add moisture to the air. Today, tighter homes may require dehumidifiers during the heating seasons (**Figure 52–5**).

If water vapour moves into the thermal envelope (insulated walls, ceilings, and floors), it will condense on cooler surfaces (**Figure 52–6**). Inside the wall, this contact point may be on the inside surface of the exterior wall sheathing. In a crawl space, condensation can form on the floor frame and subfloor. In attics, the ceiling joists and roof frame can become saturated with moisture.

Controlling moisture in the building is essential. Condensation of water vapour inside walls, attics, roofs, and floors can lead to serious problems.

Reducing the production of moisture within the house is ultimately the responsibility of the homeowner. Homeowners must be educated about the problems of moisture. Proper maintenance of exhaust piping for clothes dryers and using bathroom and kitchen fans regularly are important. If such devices are defective, constricted, or unused, the moist air will not be removed.

Moisture Problems

- High relative humidity and a warm environment will allow mould and mildew to grow. This can be seen in bathrooms and basements.

- When insulation absorbs water, the dead air spaces in the insulation may become filled with water. Insulation may compress when it gets wet and will not return to its original shape. This significantly reduces the insulation R-value (RSI-value).

- Uncontrolled moisture can move through the wall. This may cause exterior paint to blister and peel.

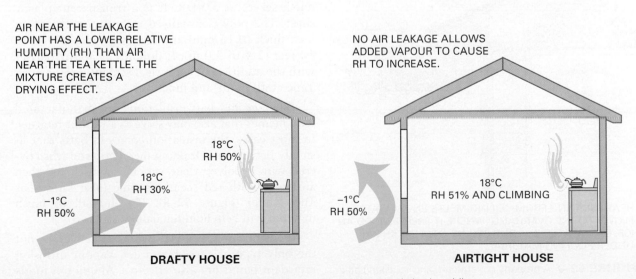

AIR NEAR THE LEAKAGE POINT HAS A LOWER RELATIVE HUMIDITY (RH) THAN AIR NEAR THE TEA KETTLE. THE MIXTURE CREATES A DRYING EFFECT.

18°C RH 50%

18°C RH 30%

−1°C RH 50%

DRAFTY HOUSE

NO AIR LEAKAGE ALLOWS ADDED VAPOUR TO CAUSE RH TO INCREASE.

18°C RH 51% AND CLIMBING

−1°C RH 50%

AIRTIGHT HOUSE

FIGURE 52–5 Comparison of drafty and airtight houses with respect to relative humidity.

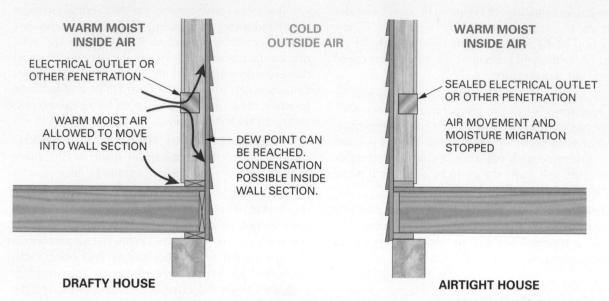

FIGURE 52–6 Moisture migration into building materials is caused mostly by air leakage.

- A warm attic will cause the formation of ice dams at the eaves. After a heavy snowfall, lost heat from the building causes the snow next to the roof to melt. Water then runs down the roof and freezes on the cold roof overhang, forming ice buildup. As this continues, an ice dam is formed. This causes water to back up on the roof, under the shingles, and into the walls and ceiling (**Figure 52–7**).

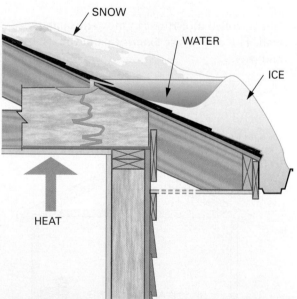

HEAT ESCAPING FROM CEILING MELTS SNOW. WATER FLOWS TO THE OVERHANG, WHERE IT FREEZES INTO AN ICE DAM. WATER BUILDUP BEHIND DAM BACKS UP UNDER ROOFING MATERIAL.

FIGURE 52–7 A properly constructed and ventilated attic will keep ice dams from forming.

PREVENTION OF MOISTURE PROBLEMS

The goal is not to remove all moisture, because this would be virtually impossible and undesirable from the standpoint of human comfort. The goal is to reduce moisture migration into the building components and remove excess moisture through ventilation.

Vapour Diffusion Retarder

A **vapour diffusion retarder** (VDR) is a material used to slow the flow of airborne moisture from passing through building materials. Polyethylene sheeting installed behind the drywall or interior finish serves as a VDR. It is a transparent plastic sheet. The poly is usually 6 mils or $6/1000$ths of an inch thick (0.15 mm) and comes in widths of 8 to 10 feet (2.4 to 3.0 m). It doubles as an air barrier with the addition of acoustical sealant and/or Tuck Tape at all joints and junctions.

Blanket and batt insulations are manufactured with facings that also may serve as a vapour retarder. In most cases, the insulation comes in batts that fit snugly between the framing members, and it is covered with a polyethylene vapour retarder and then air-sealed with red sheathing tape (also known as Tuck Tape) (**Figure 52–8**). The drywall or finish material will help hold the flanges tight.

Polyethylene sheeting, while effective, is not the only type of vapour retarder. Vapour diffusion retardant paints are also effective. Aluminum foil is the most effective type of vapour retarder.

© 2015 ROXUL Inc.

FIGURE 52–8 The vapour retarder material that covers the batt insulation must be overlapped and stapled to the stud and sealed with Tuck Tape to be an effective air/ vapour retarder.

Air Flow

For air to flow or move, there must be a pressure drop across the "barrier" as well as a hole. The amount of air that moves will depend on the size of the hole or opening and the difference in pressure. Four factors cause a pressure difference:

- **Wind effect:** The wind will create a positive pressure on the windward side of the house and a negative pressure on the leeward side of the house (blowing and sucking), similar to the forces experienced when a large truck passes a small car on the highway.

- **Stack effect:** This is caused by temperature differences. The air around a heat source warms up, becomes less dense, and rises. When cool air moves in to take its place, a convection current is created. This is why a basement floor is cold in winter and the top floor ceilings are warm. This effect can be addressed by adding ceiling fans to push the warm air down.

- **Flue effect:** Furnace chimneys, wood stoves, and fireplaces all punch holes in the building envelope. When these are in use, they actively draw air out of the house; when not in use, they leak warm air out of the house. This creates a negative pressure in the house that causes outdoor air to be drawn in. Airtight dampers reduce this effect.

- **Mechanical exhaust effect:** Appliances such as dryers, central vacuums, bathroom/kitchen fans, and other ventilation equipment all empty the house of warm indoor air. This causes cold outdoor air to be drawn in. In a "tight" house, they can cause back-drafting and the spillage of combustion products such as smoke and carbon monoxide (CO). All combustion equipment must now have a dedicated air supply and must provide

all required make-up air. It is very important to have properly balanced ventilation equipment so that air in equals air out.

An example of the impact of a simple bathroom fan and its effect on air movement follows. If you have a 1500-square-foot (140 m²) house, slab on grade, with 8-foot (2.4 m) walls, it will hold 12 000 cubic feet (336 m³) of air. If the bathroom fan is rated at 120 cfm (cubic feet per minute) or 56 litres per second, it will take 100 minutes to empty the house; that is, it will empty more than 14 times each day. Consider the cost of reheating the cold air that enters to take its place.

Air Leakage

Most excess moisture migration into building components occurs by air leakage. If air is allowed to move into a wall section, moisture will be included. Thus, if air movement into the wall section is minimized, so is moisture migration.

Many methods and techniques are available to reduce air leakage. Some involve only a little extra time and money:

- Apply sealant under exterior wall plates before they are stood up to seal the plates to the subfloor.

- Install wall sheathing panels with adhesive applied to studs and plates to seal the sheet perimeter.

- Apply sheathing tape to the seams of wall sheathing.

- Seal all penetrations in the framing members. These include those created by plumbing, electrical, heating, air-conditioning, and ventilating installations.

- Install foamed-in-place insulation.

- Install drywall panels with construction adhesive applied to studs and plates to seal the sheet perimeter.

It is not necessary to use every technique to reduce air leakage; each will make a difference. For example, an effective approach could be to seal the wall plates to the subfloor, seal holes and penetrations, and seal wall sheathing seams with sheathing tape.

Ventilation

Ventilation is the exchange of air to allow for drying and improved air quality. This often must take place both inside the living environment and within the building components.

Some believe that buildings can be built too airtight. This is not an accurate statement. More to the point, buildings must control unwanted air leakage and ventilate unwanted moisture. With airtight construction techniques, the energy costs of the building are reduced. With proper ventilation, the building is comfortable and will last a long time.

Ventilation can be achieved by either allowing the natural flow of air or by using a fan. The interior living areas are vented with fans installed in rooms where air quality needs to be improved. These areas include bathrooms, kitchens, and laundry rooms. Using them daily exhausts the moisture-laden air to the outside.

Drying Potential

Materials of a building are exposed to severe weather. The building should be constructed such that building materials can dry easily. This is referred to as **drying potential**. Drying will allow condensed moisture and wind-driven water to be removed by evaporation. It can be achieved by constructing natural ventilation in the building.

Proper ventilation depends on where the building is built. The relative humidity of the region must be considered. In cool climes where the building will require heat, the inside air usually has a higher RH than the outside air. In warm, moist climes, the outside air usually has the higher RH.

If ventilation takes place, the outside air brought in is adjusted to the same temperature as the inside air. This may be accomplished by heating or cooling. When air is warmed, its ability to hold moisture increases and its RH is reduced. This has a drying effect. If the air must be cooled, its ability to hold moisture goes down and its RH goes up. This will possibly cause dampness.

In crawl spaces under floors, where the air is usually cooler than the outside air, no ventilation should be installed. Here, warmer outside air entering the crawl space area would be cooled. This would raise the relative humidity, causing more moisture problems.

Drying happens more effectively when the outside air is cooler. The best solution for crawl spaces is to install the vapour retarder on the ground. A sheet of polyethylene will inhibit moisture from leaving the ground and getting into the floor system above. A layer of sand over the sheet will protect it from accidental perforations.

Attic Ventilation

With a well-insulated ceiling and adequate ventilation, attic temperatures are lower and excess moisture is removed.

On roofs where the ceiling finish is attached to the roof rafters, insulation is usually installed between the rafters. An adequate air space of at least 1 inch (25 mm) must be maintained between the insulation and the roof sheathing. The air space must be connected to air inlets in the soffit and outlets at the ridge (**Figure 52–9**). Failure to do so may result in reduced shingle life, formation of ice dams at the eaves, and possible decay of the roof frame.

Types of Ventilators

There are many types of ventilators. Their location and size are factors in providing adequate ventilation.

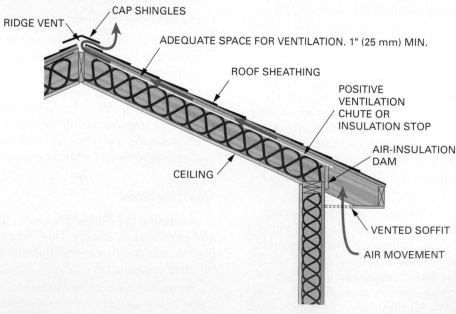

FIGURE 52–9 Method of providing ventilation when the entire rafter cavity is filled with insulation.

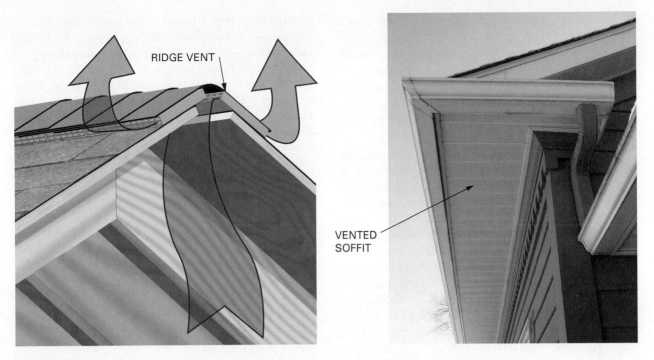

FIGURE 52-10 Ridge and soffit vents work together to provide adequate attic ventilation.

Ventilating Gable Roofs. The best way to vent an attic is with a combination of ridge and soffit vents (**Figure 52–10**). Each rafter cavity is vented from soffit to ridge. The roof sheathing is cut back about 1 inch (25 mm) from the ridge on each side, and the vent material is nailed over this slot.

Cap shingles then can be nailed directly to the vent installed over the vent space. Perforated material or screen vents are installed in the soffits to provide the entry point for the ventilation. Positive-ventilation chutes should be installed to prevent any air obstructions caused by the ceiling insulation near the eaves (**Figure 52–11**). This system

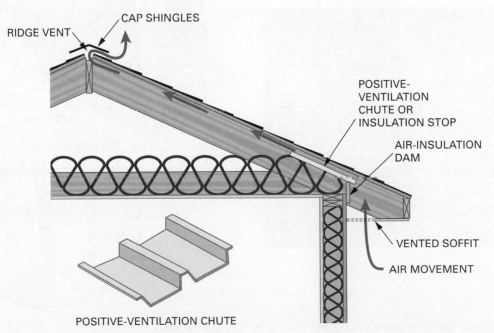

FIGURE 52-11 Positive-ventilation chutes maintain air space between compressed insulation and roof sheeting.

can adequately vent the attic space of a house that is up to 50 feet (15 m) wide.

Triangularly shaped louvre vents are sometimes installed in both end walls of a gable roof. They come in various shapes and sizes and are installed as close to the roof peak as possible. Their effectiveness depends on the prevailing wind direction.

The minimum free-air area for attic ventilators is based on the ceiling area of the rooms below (NBCC 9.19.1.2). For roofs that are constructed with roof joists (low slope or cathedral roofs), the minimum unobstructed vent area is 1/150th of the insulated ceiling area, with vent area equally distributed on opposite sides of the building, and with at least 25 percent of the vent area in the eaves and at least 25 percent near the top of the roof space. For roofs that have a slope of 1:6 or steeper (2/12) and that have an attic space, the vent area is 1/300th of the insulated ceiling area. For example, if the ceiling is 1500 square feet (140 m²), the vent area is 5 square feet (0.46 m²), with at least 1.25 square feet (0.12 m²) of the vent area located near the top or near the bottom of the attic space.

Ventilating Hip Roofs. Hip roofs should have additional continuous venting along each hip rafter. This allows each hip-jack rafter cavity to be vented. When cutting a 2½-inch-wide (63 mm) slot for the vents, it is recommended to leave a 1-foot (305 mm) section of sheathing uncut between every 2 feet (610 mm) of slot section (**Figure 52–12**). This allows for adequate ventilation while maintaining the integrity sheathing for the hip roof.

Reducing Air Leakage

Airtight sheathing is installed by adding construction adhesive. Apply a continuous bead on the studs and plates where the perimeter of the panel will fit. Nail the panel as required by nailing codes. Another technique uses sheathing tape, which is installed so that it covers every panel seam. Either technique is completed with a bead of adhesive applied under the bottom plate before the wall is raised.

Airtight drywall is installed in a similar fashion as airtight sheathing (**Figure 52–13**). Adhesive is applied continuously to the studs and plates along

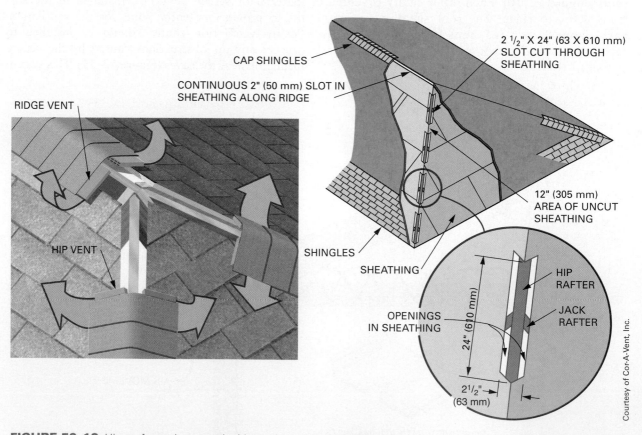

FIGURE 52–12 Hip roofs can be vented with continuous ridge and hip vents.

Courtesy of Cor-A-Vent, Inc.

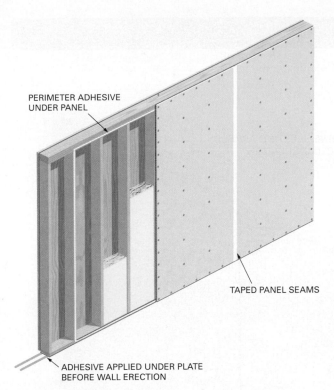

FIGURE 52–13 Airtight drywall may be achieved with adhesive placed on the perimeter of panel.

PERIMETER ADHESIVE UNDER PANEL

TAPED PANEL SEAMS

ADHESIVE APPLIED UNDER PLATE BEFORE WALL ERECTION

the sheet perimeter. The plate should also be sealed to the subfloor before the wall is erected.

Foamed-in-place insulation creates an airtight thermal envelope. It can be installed only by insulation contractors with special equipment (**Figure 52–14**).

All penetrations should be sealed before the interior finish is applied. No crack or hole is too small to be sealed (**Figure 52–15**). Interior and exterior walls are equally important. Any air movement in the structure can cause air leakage and moisture migration.

Polyethylene sheeting is installed by unrolling a length long enough to cover the wall. Add extra length and width to ensure proper coverage. All seams should be along a nailing surface such as a plate or stud.

Partially unfold the section to expose the long edge. Staple it along the top plate about 6 to 12 inches (152 to 305 mm) apart, letting the rest drape down to the floor. Next, begin along the stud in the middle of the sheeting. Smooth out wrinkles downward as it is stapled. The result is that the staples will form a large T shape in the sheeting. Have someone pull each unfastened corner snug to smooth wrinkles. Staple as needed (**Procedure 52–A**).

Carefully fit it around all openings. Lap all joints by several inches (~100 mm), keeping them on a surface nailing. Repair any tears with sheathing tape. Cut off the excess at the floor line.

Some carpenters cut the film out of openings after the drywall finish is applied. This assures a more positive seal. The retarder should be fitted tightly around outlet boxes. Add a ribbon of sealing compound around outlets and switch boxes.

Courtesy of Richard Harrington

FIGURE 52–14 Foamed-in-place insulation provides an excellent R-value and air-tightness.

FIGURE 52–15 All penetrations within a house should be sealed.

Procedure 52–A Installing a Polyethylene Film Vapour Retarder

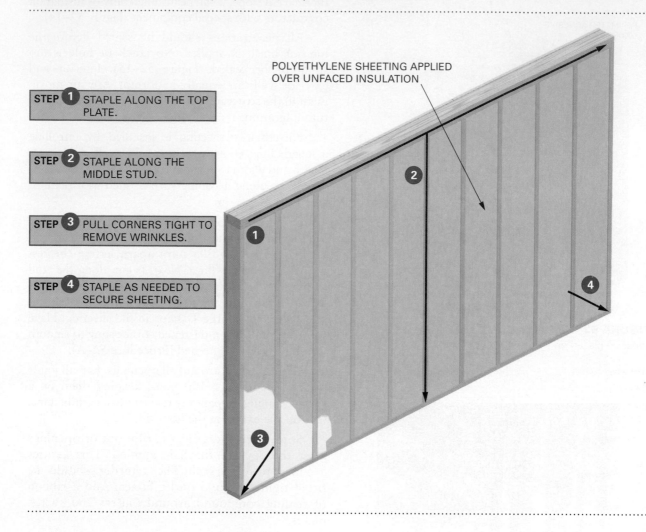

POLYETHYLENE SHEETING APPLIED OVER UNFACED INSULATION

STEP 1 STAPLE ALONG THE TOP PLATE.

STEP 2 STAPLE ALONG THE MIDDLE STUD.

STEP 3 PULL CORNERS TIGHT TO REMOVE WRINKLES.

STEP 4 STAPLE AS NEEDED TO SECURE SHEETING.

Chapter 53

Progressive Practices: Framing and Air Sealing

Knowledge of building science prompts the builder to translate theory into practice. Knowing the ways in which heat, air, and moisture move into and out of a building motivates tradespeople to rethink the ways that things have always been done and to work together to minimize energy consumption, reduce damage to components, and prolong the life of structures.

Starting from the bottom, a polyethylene moisture barrier wraps the concrete footing, where it serves as a capillary break. Similarly, a gravel bed and a layer of polyethylene are placed under the basement slab to form a capillary break. A bond break is placed between the concrete basement slab and the foundation walls. The joint is later sealed to prevent soil gases from entering the living space. Water passages are placed through the footing to prevent hydrostatic pressure from building up under the slab. The foundation wall is waterproofed by wrapping it with a dimple mat, by applying a bituminous roofing membrane, or by spraying on latex coatings.

To prevent heat loss and to add to the thermal mass of a building, the foundation walls are insulated from the outside with an appropriate product. Better still, the foundation can be poured using insulated concrete forms (ICFs). To keep water away from the foundation, the finish grade is sloped (min. 1:12). The backfill is a free-flowing soil (like sand) and is capped with a layer of clay before topsoil and sod are placed as a ground cover. At the footing level, drain tile surrounds the perimeter and is connected either to the storm sewer or to a sump pit that has a pump that pumps water away from the building. Below-grade insulation boards double as drainage boards.

The wood sill plate must be separated from the concrete. For this purpose, a layer of caulking and a layer of gasket material followed by another layer of caulking will stop air leakage and moisture migration. The sill plate is cut to precise dimensions and installed straight and square along a line that is 2 inches (50 mm) in from the floor frame line. It is anchored in the conventional way. It is very important that the anchor bolts be placed accordingly and that the top of the foundation wall be trowelled smooth and level (**Figure 53–1**).

The rim board is then placed flush with the edge of the sill plate on top of a bead of sealant. The subfloor is fastened to the floor frame using construction adhesive, especially on the perimeter joists. The bottom plate for a 2 × 6 (38 × 140 mm) framed wall can then be a 2 × 4 (38 × 89 mm) that is placed over a bead of acoustical sealant. This provides space for 2 inches (50 mm) of rigid insulation at an area that is usually cold and drafty. On the inside of the rim joists, double the thickness of the outside rigid insulation is placed, sealed, and covered with drywall.

To reduce thermal bridging, stud centres are maximized. Also, the second top plate is eliminated and metal ties are used where the top plates join.

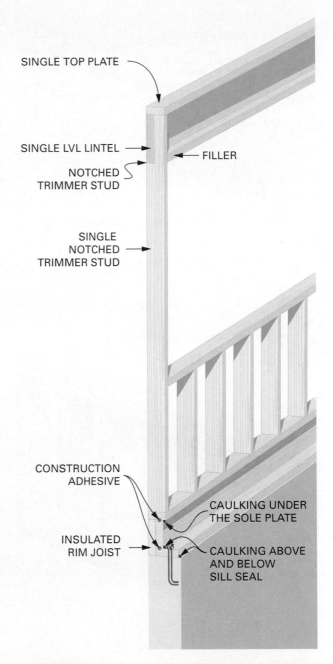

SINGLE TOP PLATE

SINGLE LVL LINTEL

FILLER

NOTCHED TRIMMER STUD

SINGLE NOTCHED TRIMMER STUD

CONSTRUCTION ADHESIVE

CAULKING UNDER THE SOLE PLATE

INSULATED RIM JOIST

CAULKING ABOVE AND BELOW SILL SEAL

FIGURE 53–1 Framing and air sealing techniques.

For multi-storey houses, lintels above openings are eliminated. Instead, structural strength is provided either by a double rim joist, a single laminated veneer lumber (LVL) rim joist, or by a single LVL lintel at the opening locations (**Figure 53–2**). The top-floor walls have studs at 24 inches (610 mm) OC, and the trusses are aligned with them. They also have single LVL lintels and notched trimmer studs, which support the lintels and remove the requirement for double studs at the openings. It is prudent to

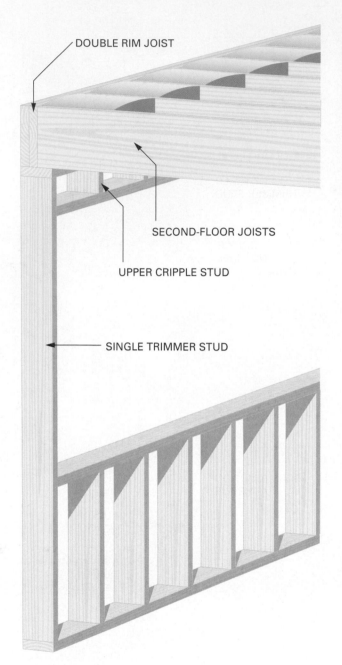

FIGURE 53–2 Double floor joists, or LVLs, in the second-floor rim location; no lintels over the window below.

DOUBLE RIM JOIST

SECOND-FLOOR JOISTS

UPPER CRIPPLE STUD

SINGLE TRIMMER STUD

FIGURE 53–3 Parallel strand lumber (PSL) columns and lintels are shallower than stud depth and allow for insulation cover and therefore less thermal bridging.

have double studs and blocking beside entry doors. Check codes and bylaws to ensure that requirements for the prevention of forced entry are met.

Support for beam trusses is provided by smaller engineered wood columns, which allow for insulation backing and reduce thermal bridging (**Figure 53–3**). Exterior and interior corners reduce the buildup of wood members. Drywall backing is provided by plywood strips or by drywall clips. Alternatively, interior partitions can be held back by ¾ inch

(19 mm) to allow for the continuous installation of the polyethylene, the acoustical sealant, and the drywall on the exterior wall.

It is important to place a bead of resilient sealant (or acoustical sealant) under wall plates and perimeter joists, because depressurization tests have shown that placing the sealant afterward, at the wall/floor junction, results in significantly more air leakage.

After the trusses or rafters have been installed, insulation baffles are placed between each successive pair before the roof sheathing is applied. This directs the air flow from the eaves up to the roof vents. It also prevents air from flowing through the insulation and "wind-washing" its R-value (RSI) down. There are other products on the market that will serve the same purpose.

The housewrap or air-barrier paper is installed next. The paper is cut in the shape of a capital letter **I**. The side tabs are folded into the building and stapled to the jack studs. At the top of the opening, the paper is slit 8 inches(203 mm) upward at a 45-degree angle at each corner. Then the flap that is created is folded up and stapled (**Figure 53–4**). A piece of bevelled siding is placed on the rough sill. Then a peel-and-stick membrane is applied to cover the siding and up each side by about 6 inches (152 mm), and folded over onto the housewrap (**Figure 53–5**). In some installations, a pressure-treated (PT) wood strip (called a *back dam*) is nailed to the rough sill and then covered with the self-sealing membrane. The width of the strip is the width of the wall frame minus the thickness of

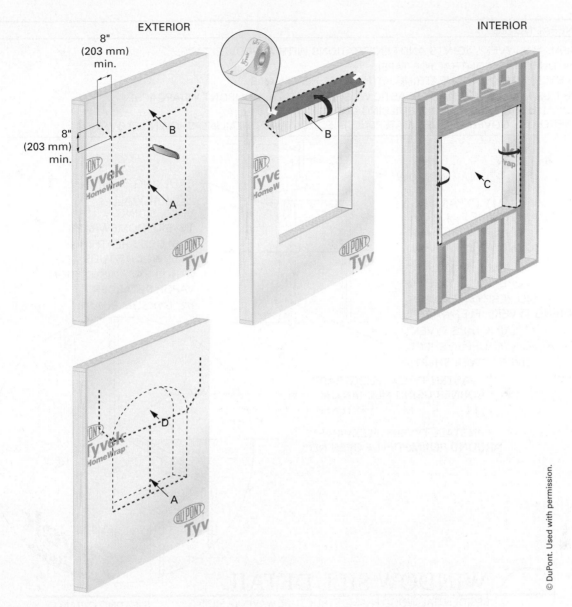

EXTERIOR

8"
(203 mm)
min.

8"
(203 mm)
min.

INTERIOR

© DuPont. Used with permission.

FIGURE 53–4 Installing weather barrier before window installation. (*Continued*)

the window unit (**Figure 53–6**). Sealant is applied to the edges of the top and sides of the opening, and the window is installed and secured as per manufacturer's instructions (**Figure 53–7**). The peel-and-stick membrane is then applied to seal the window to the housewrap on the sides and to the wall sheathing on the top (**Figure 53–8**). A drip cap is installed; the top flap of housewrap is then folded down and taped to the drip cap (**Figure 53–9**). The space between the window and the rough sill is insulated and sealed from the inside but is left open at the bottom so that water can drain freely to the outside. Doors are dealt with in a similar fashion (**Figure 53–10**).

For more detailed information on the installation of housewrap and rough openings, visit *http://www2.dupont.com/Tyvek_Weatherization/en_US/tech_info/install.html*.

The rest of the housewrap is installed, secured, lapped, and taped on the sides and on the top, layered over the bottom roll. At the foundation wall, a metal flashing is installed on the brick ledge and the housewrap is placed over it and taped to it. For conventional sidings, it is best to strap the walls to provide an air space that will act as a **rainscreen** by equalizing the pressure on both surfaces of the siding. At the top of the wall, the water-resistant barrier (WRB) is sealed to the tab of WRB that was

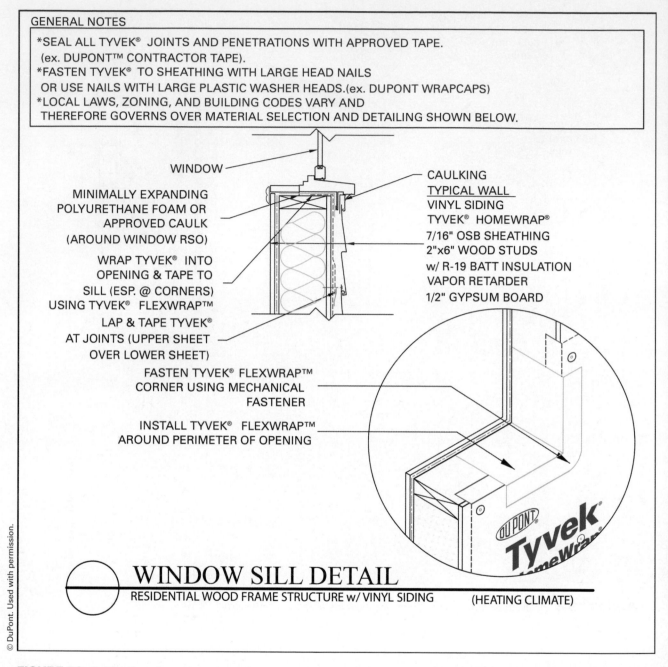

GENERAL NOTES

*SEAL ALL TYVEK® JOINTS AND PENETRATIONS WITH APPROVED TAPE.
 (ex. DUPONT™ CONTRACTOR TAPE).
*FASTEN TYVEK® TO SHEATHING WITH LARGE HEAD NAILS
 OR USE NAILS WITH LARGE PLASTIC WASHER HEADS.(ex. DUPONT WRAPCAPS)
*LOCAL LAWS, ZONING, AND BUILDING CODES VARY AND
 THEREFORE GOVERNS OVER MATERIAL SELECTION AND DETAILING SHOWN BELOW.

WINDOW

MINIMALLY EXPANDING
POLYURETHANE FOAM OR
APPROVED CAULK
(AROUND WINDOW RSO)

WRAP TYVEK® INTO
OPENING & TAPE TO
SILL (ESP. @ CORNERS)
USING TYVEK® FLEXWRAP™

LAP & TAPE TYVEK®
AT JOINTS (UPPER SHEET
OVER LOWER SHEET)

FASTEN TYVEK® FLEXWRAP™
CORNER USING MECHANICAL
FASTENER

INSTALL TYVEK® FLEXWRAP™
AROUND PERIMETER OF OPENING

CAULKING
TYPICAL WALL
VINYL SIDING
TYVEK® HOMEWRAP®
7/16" OSB SHEATHING
2"x6" WOOD STUDS
w/ R-19 BATT INSULATION
VAPOR RETARDER
1/2" GYPSUM BOARD

WINDOW SILL DETAIL
RESIDENTIAL WOOD FRAME STRUCTURE w/ VINYL SIDING   (HEATING CLIMATE)

FIGURE 53–4 (Continued)

previously installed between the two top plates at the framing stage.

The National Building Code of Canada (NBCC) recognizes that standard siding products provide only a single plane of protection from the elements, especially wind-driven rain. Together with resilient sealants at the intersections with other building components, the siding will provide reasonable protection but there is an increased requirement for ongoing maintenance. The rainscreen assemblies provide both a first and a second plane of protection (**Figure 53–11**). The cladding protects against almost all of the precipitation and the second plane forms an air barrier system that will drain away any water that penetrates the first plane. The second plane of protection generally consists of a breathable (NOT waterproof) sheathing membrane that forms a drainage plane

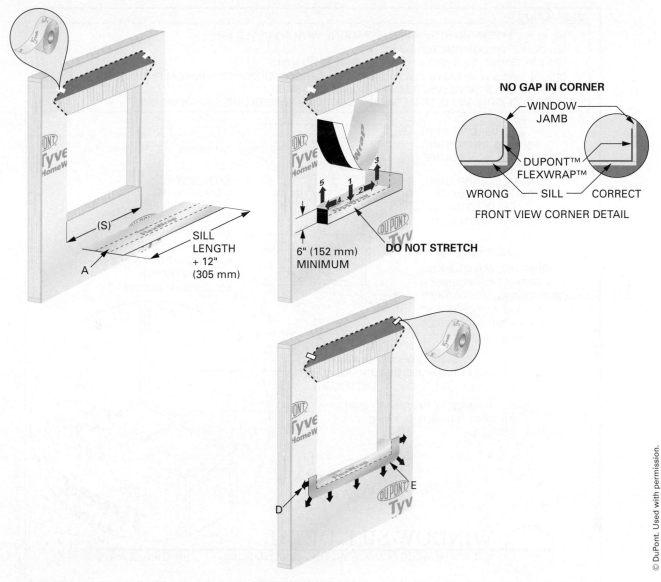

FIGURE 53–5 Sealing the rough sill to the housewrap. (*Continued*)

that, together with drip caps and base flashings, will direct the water off of the envelope and away from the building. The simple rainscreen includes ¾-inch (19 mm) vertical strapping (rot-resistant material with corrosion-resistant fasteners) placed over the membrane. The pressure-equalized rainscreen sections the walls into 20-foot-wide (6 m) areas and seals each area at the top and the sides (the corners) and provides vents in each section to equalize the pressure behind and in front of the cladding. The vents have screens to keep out insects.

The rest of the air sealing is done from the inside. The electrical boxes are sealed plastic boxes that hug the wire as it is pulled into the box. The box has a ring that seals the poly to the box, and the flange of the ring has a gasket material on it that provides an airtight seal to the drywall. The polyethylene air barrier is lapped by at least a stud space and is sealed and fastened to the stud, to the top plate, and to the bottom plate. At the openings, the spaces are filled with backer rod, batt insulation, low-expansion foam, or a combination of the above. The poly is then sealed and secured to the back of the jamb or jamb extension. This can also be accomplished by applying red contractors' tape (Tuck Tape) to bridge the gap between the jamb

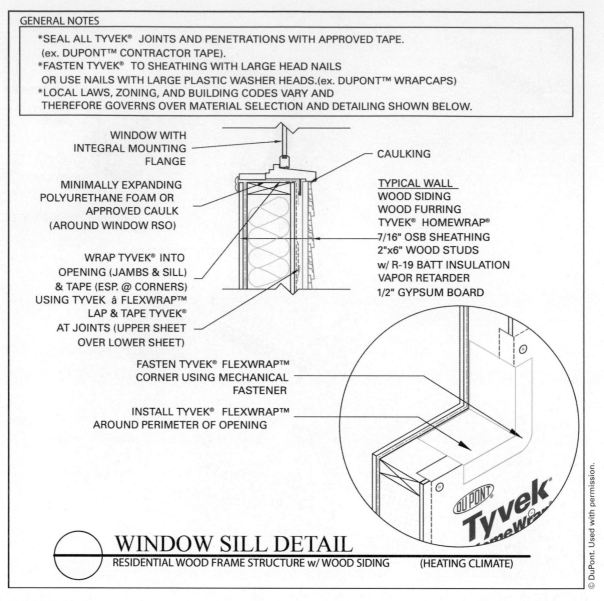

GENERAL NOTES

*SEAL ALL TYVEK® JOINTS AND PENETRATIONS WITH APPROVED TAPE.
(ex. DUPONT™ CONTRACTOR TAPE).
*FASTEN TYVEK® TO SHEATHING WITH LARGE HEAD NAILS
OR USE NAILS WITH LARGE PLASTIC WASHER HEADS.(ex. DUPONT™ WRAPCAPS)
*LOCAL LAWS, ZONING, AND BUILDING CODES VARY AND
THEREFORE GOVERNS OVER MATERIAL SELECTION AND DETAILING SHOWN BELOW.

WINDOW WITH INTEGRAL MOUNTING FLANGE

CAULKING

MINIMALLY EXPANDING POLYURETHANE FOAM OR APPROVED CAULK (AROUND WINDOW RSO)

TYPICAL WALL
WOOD SIDING
WOOD FURRING
TYVEK® HOMEWRAP®
7/16" OSB SHEATHING
2"x6" WOOD STUDS
w/ R-19 BATT INSULATION
VAPOR RETARDER
1/2" GYPSUM BOARD

WRAP TYVEK® INTO OPENING (JAMBS & SILL) & TAPE (ESP. @ CORNERS) USING TYVEK â FLEXWRAP™ LAP & TAPE TYVEK® AT JOINTS (UPPER SHEET OVER LOWER SHEET)

FASTEN TYVEK® FLEXWRAP™ CORNER USING MECHANICAL FASTENER

INSTALL TYVEK® FLEXWRAP™ AROUND PERIMETER OF OPENING

WINDOW SILL DETAIL
RESIDENTIAL WOOD FRAME STRUCTURE w/ WOOD SIDING   (HEATING CLIMATE)

FIGURE 53–5 (*Continued*)

FIGURE 53–6 Installation of a back dam and the self-sealing membrane over building paper: Kunming, China.

extension and the drywall, which in turn is covered by the casing. Ceiling poly is sealed and secured to the wall poly. Any protrusions into the attic area (such as electrical wires and plumbing pipes) are foamed and sealed as airtight as possible.

After or before the drywall is installed, a blower door test is performed to determine the degree of air infiltration at a preset depressurization level (**Figure 53–12**). Only after the air-tightness target has been achieved will the house be certified—either as ENERGY STAR or as R-2000. For example, an R-2000 house is permitted to have an air leakage rate of only 1.5 ACH (air changes per hour) at 50 pascals or less of depressurization.

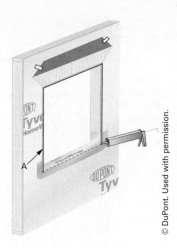

FIGURE 53–7 Applying sealant at the top and sides of the opening before window installation.

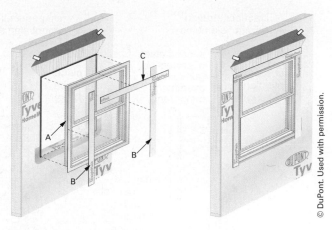

FIGURE 53–8 Sealing the window to the housewrap and sheathing.

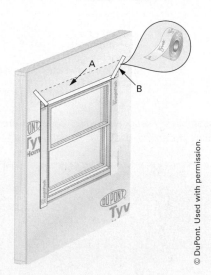

FIGURE 53–9 Sealing the housewrap to the drip cap.

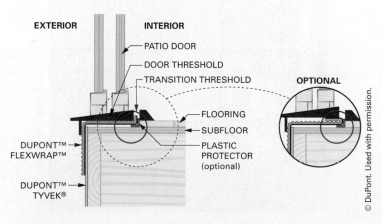

FIGURE 53–10 Sealing under the door threshold.

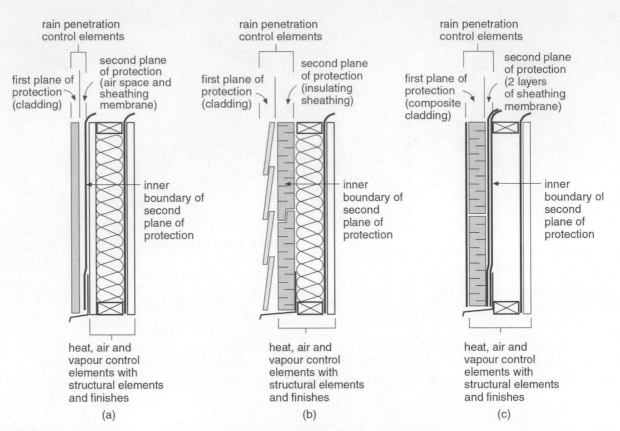

FIGURE 53–11 Generic rainscreen assemblies. *(National Building Code of Canada Figure A-9.27.2. Reproduced with the permission of the National Research Council of Canada, copyright holder.)*

FIGURE 53–12 A blow door tests the air-tightness of a building.

Chapter 54

Thermal and Acoustical Insulation

All materials used in construction have some insulating value. **Insulation** is a material that interrupts or slows the transfer of heat. Heat transfer is a complex process involving three mechanisms: conduction, convection, and radiation.

The best way to slow heat transfer within a house depends on where the house is located. In southern climes, protection from solar radiation is important, whereas in northern climes this importance is low. In both climes, protection from conduction and convection are important.

HOW INSULATION WORKS

Insulating materials create a space between two surfaces, thereby breaking contact and reducing conduction. They trap air and slow convection. They also can either reflect or absorb some radiation.

Air is an excellent insulator if confined to a small space and kept still. Insulation materials are designed to keep air motionless in small pockets. Insulation effectiveness increases as the air spaces become smaller in size and greater in number. Millions of tiny air cells, trapped in its unique cellular structure, make wood a better insulator than concrete, steel, or aluminum.

All insulating materials are manufactured from materials that are themselves poor insulators. For example, fibreglass insulation is made from glass, which is a good conductor of heat. The improved insulation value comes from air trapped within the insulating material. Insulation also provides resistance to sound travel.

THERMAL INSULATION

Among the materials used for insulating are glass fibres, mineral fibres (rock), organic fibres (paper), and plastic. Aluminum foil is also used. It works by reflecting heat instead of stopping air movement.

Resistance Value of Insulation

The **R-value (RSI)** of insulation is a number that indicates its measured resistance to the flow of heat. The higher the R-value, the more efficient the material is in retarding the passage of heat (**Figure 54–1**). R-values are clearly printed on insulation packages (**Figure 54–2**).

To get RSI value, multiply R-value by 0.176. For example, R-20 = 20 × 0.176 = RSI-3.5. To get RSI/mm, multiply R/inch × 0.0069.

Insulation Requirements

The rising costs of energy coupled with the ecological need to conserve have resulted in higher R-value recommendations for new construction than in previous years. The NBCC has minimum insulation levels listed by region by degree-days. A **degree-days rating** is the sum of the difference of the mean temperature and 18°C for all the days that have a daily average temperature below 18°C. This information is used to determine the R-value of insulation installed in walls, ceilings, and floors. Insulation requirements vary according to the average low temperature.

In warmer climates, less insulation is needed to conserve energy and provide comfort in the cold season. However, air-conditioned homes should also receive more insulation in walls, ceilings, and floors. This ensures economy in the operation of air-conditioning equipment in hot climates.

Comfort and operating economy are dual benefits. Insulating for maximum comfort automatically provides maximum economy of heating and cooling operations.

Where Heat Is Lost

The amount of heat lost from the average house varies with the type of construction. The principal areas and approximate amount of heat loss for a typical house with moderate insulation are shown in **Figure 54–3**.

Houses of different architectural styles vary in their heat loss characteristics. A single-storey house, for example, contains less exterior wall area than a two-storey house, but has a proportionately greater ceiling area. Greater heat loss through floors is experienced in homes erected on concrete slabs or unheated crawl spaces, unless these areas are well-insulated.

Foundation Materials	
8″ concrete block (2-hole core)	1.11
12″ concrete block (2-hole core)	1.28
8″ lightweight aggregate block	2.18
12″ lightweight aggregate block	2.48
Common brick	0.20/inch
Sand or stone	0.08/inch
Concrete	0.08/inch
Structural Materials	
Softwood	1.25/inch
Hardwood	0.91/inch
Steel	0.0032/inch
Aluminum	0.00070/inch
Sheathing Materials	
½″ plywood	0.63
⅝″ plywood	0.78
¾″ plywood	0.94
½″ aspenite, OSB	0.91
¾″ aspenite, OSB	1.37
Insulating Materials	
Batts and Blankets	
3½″ fibreglass	11
6″ fibreglass	19
8″ fibreglass	25
12″ fibreglass	38
3½″ high-density fibreglass	13
5½″ high-density fibreglass	21
8½″ high-density fibreglass	30
10″ high-density fibreglass	38
fibreglass	3.17/inch

Loose Fill	
Fibreglass	3.17/inch
Cellulose	3.70/inch
Reflective Foil	
Foil-faced bubble pack	1.0
1-layer foil	0.22
4-layer foil	11
Rigid Foam	
Expanded polystyrene foam (bead board)	4.0/inch
Extruded polystyrene foam	5.0/inch
Polyisocyanurate/urethane foam	7.2/inch
Spray Foams	
Low-density polyurethane	3.60/inch
High-density polyurethane	7.0/inch
Finish Materials	
Wood shingles	0.87
Vinyl siding	Negligible
Aluminum siding	Negligible
Wood siding (½″ × 8″)	0.81
½″ hardboard siding	0.36
Polyethylene film	Negligible
Builder's felt (15#)	0.06
²⁵⁄₃₂″ hardwood flooring	0.68
Vinyl tile (1/8″)	0.05
Carpet and pad	1.23
¼″ ceramic tile	0.05
⅜″ gypsum board	0.32
½″ gypsum board	0.45
⅝″ gypsum board	0.56
½″ plaster	0.09
Asphalt shingles	0.27

Courtesy of Richard Harrington

FIGURE 54–1 Insulating R-values of various building materials.

FIGURE 54–2 The R-value is broadly stamped on insulation packaging.

The transfer of heat through un-insulated ceilings, walls, or floors can be reduced almost any desired amount by installing insulation. Maximum quantities in these areas can cut heat losses by up to 90 percent. The use of 2 × 6 (38 × 140 mm) studs in exterior walls permits installation of 6-inch-thick (152 mm) insulation. This achieves an R-19 value (RSI 3.34), or, with improved insulation, an R-21 value (RSI 3.7).

Windows and doors are generally sources of great heat loss. **Weatherstripping** around windows and doors reduces heat loss. Heat loss through glass surfaces can be reduced 50 percent or more by installing double- or triple-glazed windows. This is referred to as **insulated glass**. Improved insulated glass design uses argon gas between two panes of glass. The trapped gas improves the R-value significantly. A **low-emissivity coating** on the inner glass surface also improves the window performance. This coating allows ultraviolet rays from the sun to pass but reflects the inside infrared (radiant) rays back into the building. Windows having both of these features are called low-e argon windows.

Adding a **storm window** or storm door is also effective. Adding additional layers of trapped air creates additional thermal resistance.

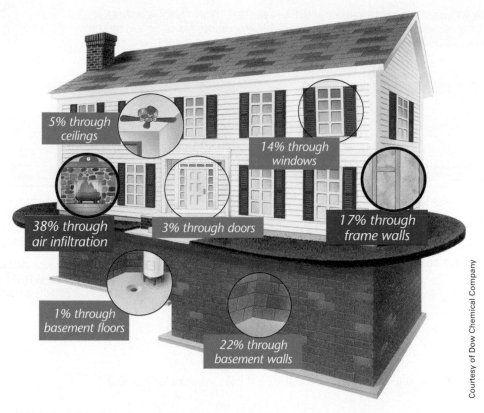

FIGURE 54–3 Typical heat loss for a house built with moderate insulation.

Constructing homes to be more airtight is an effective method of reducing energy needs. Because heat is lost through air leakage, anything that stops air movement will improve energy efficiency. An additional benefit of air-tightness includes reducing moisture migration into building components. Moisture migration is covered further in Chapter 52.

TYPES OF INSULATION

Insulation is manufactured in a variety of forms and types. These are commonly grouped as *flexible*, *loose-fill*, *rigid*, *reflective*, and *miscellaneous*.

Flexible Insulation

Flexible insulation is manufactured in *blanket* or *batt* form. Blanket insulation comes in rolls. Widths are suited to 16- and 24-inch (406 to 610 mm) stud and joist spacing. The usual thickness is from 3½ to 5½ inches (89 to 140 mm). The body of the blanket is made of fluffy rockwool or glass fibres. Blanket insulation is un-faced or faced with asphalt-laminated kraft paper or aluminum foil with flanges on both edges for fastening to studs or joists.

These facings may be considered a **vapour retarder**, also called a **vapour barrier**, but installation of the flanges must be airtight to function as a vapour retarder. These facings should always face the warm side of the wall.

Batt insulation (**Figure 54–4**) is made of the same material as blanket insulation. It comes in thicknesses up to 12 inches (305 mm). Widths are for standard

FIGURE 54–4 Batt insulation is made up to 12 inches thick.

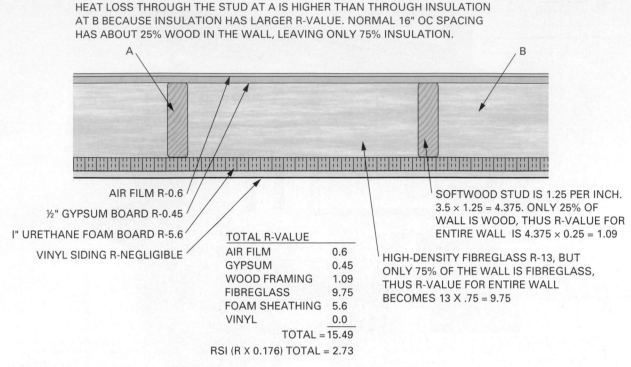

HEAT LOSS THROUGH THE STUD AT A IS HIGHER THAN THROUGH INSULATION AT B BECAUSE INSULATION HAS LARGER R-VALUE. NORMAL 16" OC SPACING HAS ABOUT 25% WOOD IN THE WALL, LEAVING ONLY 75% INSULATION.

A

B

AIR FILM R-0.6

½" GYPSUM BOARD R-0.45

I" URETHANE FOAM BOARD R-5.6

VINYL SIDING R-NEGLIGIBLE

TOTAL R-VALUE
AIR FILM	0.6
GYPSUM	0.45
WOOD FRAMING	1.09
FIBREGLASS	9.75
FOAM SHEATHING	5.6
VINYL	0.0
TOTAL =	15.49

RSI (R X 0.176) TOTAL = 2.73

SOFTWOOD STUD IS 1.25 PER INCH. 3.5 × 1.25 = 4.375. ONLY 25% OF WALL IS WOOD, THUS R-VALUE FOR ENTIRE WALL IS 4.375 × 0.25 = 1.09

HIGH-DENSITY FIBREGLASS R-13, BUT ONLY 75% OF THE WALL IS FIBREGLASS, THUS R-VALUE FOR ENTIRE WALL BECOMES 13 X .75 = 9.75

FIGURE 54–5 R-values of a wall section vary according to the actual path of heat loss.

stud and joist spacing. Lengths are either 47 or 93 inches (1.19 or 2.36 m). The total insulation or R-value (RSI) of a wall takes into consideration all the different materials that make up the finished wall assembly (**Figure 54–5**).

Loose-Fill Insulation

Loose-fill insulation is usually composed of materials in bulk form. It is supplied in bags or bales. It is placed by pouring, blowing, or packing by hand. Materials include rockwool, fibreglass, and cellulose.

Loose-fill insulation is suited for use between ceiling joists and hollow-core masonry walls (**Figure 54–6**). It is also used in the sidewalls of existing houses that were not insulated during construction.

Rigid Insulation

Rigid insulation is usually a foamed plastic material in sheet or board form (**Figure 54–7**). The material is available in a wide variety of sizes, with widths up to 4 feet (1.2 m) and lengths up to 12 feet (3.6 m). The most common types are made from polystyrene and polyurethane.

Expanded polystyrene (EPS), or white bead board, has good insulating qualities. It is sometimes

installed on the inside of dry basement walls. If expanded polystyrene comes in contact with water it will absorb moisture. The absorbed moisture will significantly reduce the R-value. Therefore, this foam should not be installed in damp areas.

Extruded polystyrene (XPS) comes in a variety of colours, depending on the manufacturer. It is often installed as insulation for footings and basement walls. Extruded polystyrene is a closed-cell foam that will not absorb moisture, keeping its R-value while wet. It may be installed in wet locations and is sometimes used for flotation in docks.

Courtesy of Pat Dundon, Dundon Insulation Inc.

FIGURE 54–6 Loose-fill cellulose insulation being blown into place.

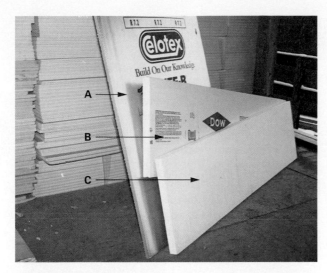

FIGURE 54–7 Types of rigid insulation boards: (A) foil-faced polyurethane; (B) extruded polystyrene; (C) expanded polystyrene.

Polyurethane or poly-isocyanurate foam is usually made with a facing of foil or building paper. It has a superior R-value of 7.2 per inch (RSI 0.050/mm). It is used under roof decks and as wall sheathing. Polyurethane foam should be used only on dry areas. It absorbs moisture when in contact with water. Polyurethane insulation is also found in insulated exterior entrance doors and garage doors.

Reflective Insulation

Foil conducts heat well but reflects radiant energy. It is typically used in warm climates to protect the building from the sun and reflected heat. Reflective insulation usually consists of foil bonded to a surface of some other material such as drywall. It is effective only when installed with a minimum of ¾ inch (19 mm) of space.

Miscellaneous Insulation

Foamed in place insulation is sometimes used. A urethane foam is produced by mixing two chemicals together. It is injected into place and expands on contact with the surface (**Figure 54–8**). This product requires special equipment, provided by an insulating contractor. It not only adds superior insulation value but also adds rigidity to the overall structure. It forms a structural bond between the sheathing and studs, giving racking strength. It also has superior soundproofing qualities that result from its air-sealing properties (**Figure 54–9**).

Photo by M. Nauth. Courtesy of Jeff and Alison Armstrong.

FIGURE 54–8 Foamed-in-place poly-isocyanurate insulation: (top) application and (bottom) trimmed with a handsaw.

Courtesy of Pat Dundon, Dundon Insulation Inc.

FIGURE 54–9 High-density urethane insulation being applied between ceiling joists.

Spray foams are available in aerosol cans. They are used to seal and insulate gaps between doors and window units and the wall frame. They are also used to seal mechanical penetrations such as those made by electricians. Cans are typically inverted and sprayed into gaps. As the foam reacts and expands, it fills the space, creating an airtight seal.

Other types of insulating materials include lightweight *vermiculite* and *perlite* aggregates. Vermiculite is made by superheating mica, which causes it to expand quickly, creating air spaces. It is able to withstand high temperatures and can be used around chimneys. Perlite is made from heating volcanic glass, causing it to expand. It is sometimes used in plaster and concrete to make it lighter and more thermal resistant.

WHERE AND HOW TO INSULATE

Any building surfaces that separate conditioned air, heated or cooled, from unconditioned air must be insulated. This will save energy and money over the life of the building. Insulation should be placed in all outside walls and top-floor ceilings. In houses with unheated crawl spaces, insulation should be placed between the floor joists. Collectively, these surfaces are called the **thermal envelope**.

Great care should be exercised when installing insulation. Insulation that is not properly installed can render the material useless and a waste of time and money. Pay attention to details around outlets, pipes, and any obstructions. Make the insulation conform to irregularities by cutting and piecing without bunching or squeezing. Keep the natural fluffiness of the insulation intact at all times. Voids in insulation of only 5 percent can create an overall efficiency reduction of 25 percent. This is caused by the fact that heat will move to the colder areas. Insulation should be installed neatly. Generally, if insulation looks neat, it will perform well.

In houses with flat or low-pitched roofs, insulation should be used in the ceiling area only if sufficient space is left above the insulation for ventilation. Insulation is used along the perimeter of houses built on slabs when required (**Figure 54–10**).

Installing Flexible Insulation

Cut the batts or blankets with a knife. Make lengths slightly oversize by about an inch at the top and bottom. Measure out from one wall a distance equal to the desired lengths of insulation. Draw a line on the floor. Roll out the material from the wall. Compress the insulation with a straightedge on the line. Cut it with a sharp knife on a cutting board to protect the floor (**Figure 54–11**). Cut the necessary number of lengths.

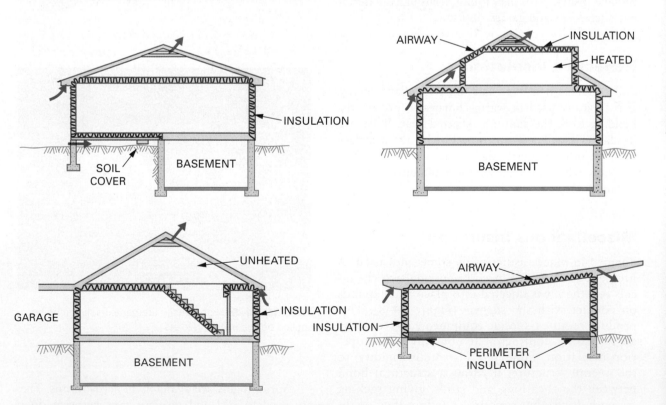

FIGURE 54–10 Various configurations of the thermal envelope.

FIGURE 54–11 Flexible insulation is compressed with a straightedge as it is cut.

FIGURE 54–13 Spaces around doors and windows are filled with spray-foam insulation to seal them to the rough opening.

> ### CAUTION
>
> **Always protect lungs, sinuses, eyes, and skin from insulation fibres. Long-term effects could be severe.**

Place the batts or blankets between the studs. Cut the batts to the required shape, a little oversize in order to friction-fit the insulation into place (**Figure 54–12**).

Fill any spaces around windows and doors with spray-can foam. Foam will fill the voids with an airtight seal and protect the house from air leakage (**Figure 54–13**). After the foam cures, flexible insulation may be added to fill the remaining space. Do not pack the insulation tightly. Squeezing or compressing it reduces its effectiveness. If the insulation has no covering or stapling tabs, it is friction-fitted between the framing members.

Ceiling Insulation

Ceiling insulation is installed by stapling it to the ceiling joists or by friction-fitting it between them. If furring strips have been applied to the ceiling joists, the insulation is simply laid on top of the furring strips. In most cases, the use of un-faced insulation is recommended. This makes it easier to determine proper fit of the insulation and also lowers the cost of materials. Extend the insulation across the top plate.

In most areas of North America, attics should be well-ventilated. This ensures that any moisture and heat that escapes the building will not be trapped in the attic space. As warm/moist air is vented out, cool air from the outside replaces it. This has a drying effect on the house. This venting usually begins at the eaves and ends with ridge or attic vents. Therefore, the ceiling insulation must not block this venting.

It may be necessary to compress the insulation against the top of the wall plate to permit the free flow of air. An air-insulation dam should also be installed to protect the insulation from air movement inside the fibres. Otherwise, the insulation's R-value and performance will be reduced (**Figure 54–14**).

FIGURE 54–12 Cutting the batt insulation to shape and friction-fitting it into place.

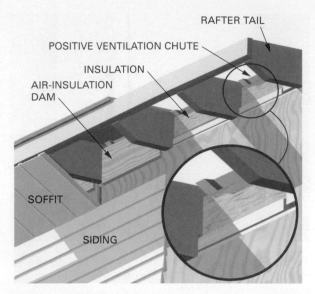

FIGURE 54–14 Air-insulation dams protect the insulation from air infiltration.

Insulating Floors over Crawl Spaces

Flexible insulation installed between floor joists over a crawl space should be un-faced because the facing could act as a vapour retarder and introduce the risk of trapped moisture. Un-faced batts or blankets may be secured in place with an air barrier material stapled to the joists (**Figure 54–15**). This supports the insulation from falling out and protects it from air movement.

A vapour barrier should be installed over the soil under crawl spaces. This will restrict moisture from leaving the soil and entering the crawl space cavity. The easiest material to install is a sheet of polyethylene.

Installing Loose-Fill Insulation

Loose-fill insulation is typically blown into place with special equipment. The surface of the material

between ceiling joists is levelled to the desired depth (**Figure 54–16**). Care should be taken to evenly distribute the required amount of insulation. This amount is determined from the instructions and proportions printed on the bales of insulation. Take care not to fill the soffit in the eave areas. This will restrict the attic ventilation and allow the moisture to accumulate.

Installing Rigid Insulation

Rigid insulation has many functions, some of which are not interchangeable. Take care when selecting the material if it will be installed near water. All foams are easily cut with a knife or saw. A table saw can also be used to cut rigid foam.

CAUTION

Wear a respirator to protect lungs and sinuses from the fine airborne particles released from insulation when it is cut with a power saw.

Rigid foams may be attached with fasteners or adhesives or applied by friction-fitting between the framing members. Fasteners include plastic-capped nails that are ring-shanked with large plastic washers. They hold the foam in place without pulling through the foam. Adhesives are used to glue the sheet to a surface. Some adhesives will dissolve the plastic foam, so be sure to use the correct adhesive as listed on the adhesive tube.

Polyurethane foam is sometimes used on roofs of cathedral ceilings (**Figure 54–17**). Various thicknesses may be used depending on the desired R-value (RSI). Roof sheathing should be installed on furring strips above the foam. Since the foam is an insulator, installing sheathing and shingles

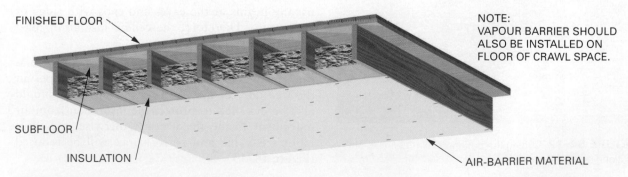

FIGURE 54–15 Air barrier material protects insulation between floor joists in a crawl space from air infiltration.

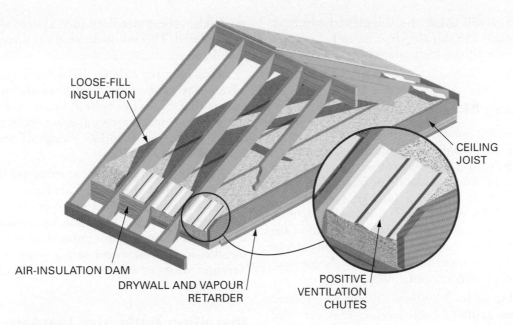

LOOSE-FILL
INSULATION

CEILING
JOIST

AIR-INSULATION DAM

DRYWALL AND VAPOUR
RETARDER

POSITIVE
VENTILATION
CHUTES

FIGURE 54–16 Loose-fill insulation is levelled to the desired depth.

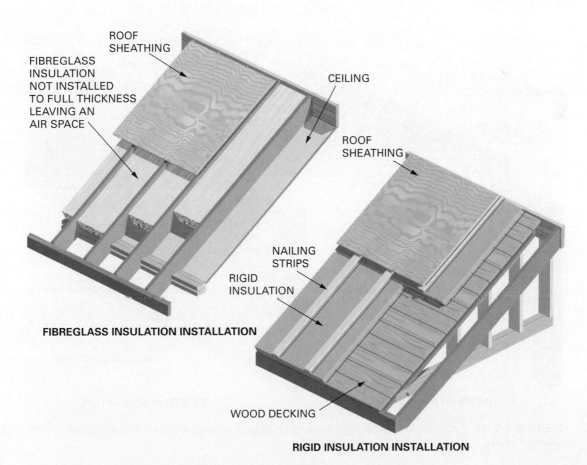

ROOF
SHEATHING

FIBREGLASS
INSULATION
NOT INSTALLED
TO FULL THICKNESS
LEAVING AN
AIR SPACE

CEILING

ROOF
SHEATHING

NAILING
STRIPS

RIGID
INSULATION

FIBREGLASS INSULATION INSTALLATION

WOOD DECKING

RIGID INSULATION INSTALLATION

FIGURE 54–17 Methods of installing rigid insulation under roof sheathing.

directly to it will cause the shingles to overheat. This will cause the shingle life span to be reduced. The air space allows for removal of moisture and excess heat.

Insulating Masonry Walls

Masonry walls can be insulated on the inside or outside of the building (**Figure 54–18**). Interior applications require the masonry wall to be furred. Furring can be wood or metal hat track. Fasten furring strips 16 or 24 inches (400 or 600 mm) on centre (OC). Dry walls may be insulated with fibreglass batts, expanded polystyrene, or polyurethane sheets. For damp walls, only extruded polystyrene boards should be used. To see a new basement insulation product, visit insofast.com.

Crawl space foundation walls may be insulated in the same fashion. Each method uses extruded polystyrene (**Figure 54–19**). The interior application does not need to be furred like basement walls. The band or rim joist area of the floor must be insulated. The entire ground area inside the building must be covered with a moisture barrier to keep ground moisture from entering the building. A layer of 6 mil (0.15 mm) polyethylene is typically used. Also, the entire floor joist system should not be insulated. This will help keep the space warmer and dryer.

Insulating the exterior has advantages and disadvantages over interior applications. An advantage is that the masonry wall becomes part of the thermal envelope. It serves as a heat battery, helping to reduce the high and low swings of temperature within the room.

The disadvantage is that a protective layer must be installed over the foam. This layer should be a masonry mastic trowelled on to protect the foam from physical abuse and the sun, which will degrade it over time. Also, exterior application is not recommended in regions where termites are a problem. Termites are more difficult to detect behind the insulation layer.

Installing Reflective Insulation

Reflective insulation usually is installed between framing members in the same manner as blanket insulation. It is attached to either the face or the side of the studs. However, an air space of at least ¾ inch (19 mm) must be maintained between its surface and the inside edge of the stud.

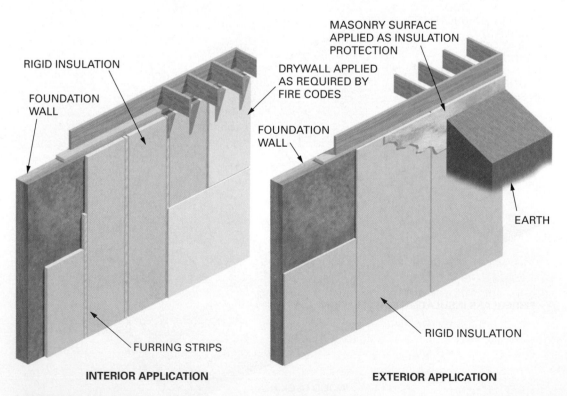

INTERIOR APPLICATION

RIGID INSULATION

FOUNDATION WALL

FURRING STRIPS

EXTERIOR APPLICATION

MASONRY SURFACE APPLIED AS INSULATION PROTECTION

DRYWALL APPLIED AS REQUIRED BY FIRE CODES

FOUNDATION WALL

RIGID INSULATION

EARTH

FIGURE 54–18 Extruded polystyrene rigid insulation may be applied to the exterior or interior of basement walls.

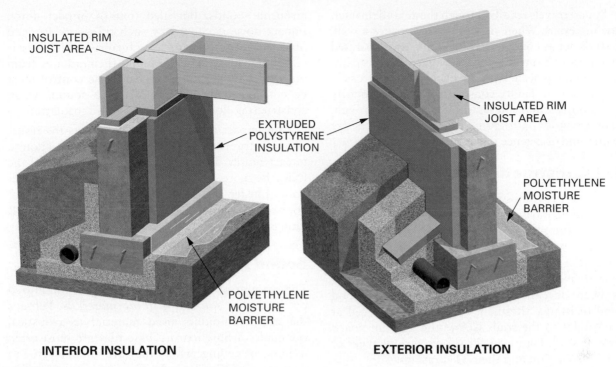

INSULATED RIM
JOIST AREA

EXTRUDED
POLYSTYRENE
INSULATION

POLYETHYLENE
MOISTURE
BARRIER

INTERIOR INSULATION

INSULATED RIM
JOIST AREA

POLYETHYLENE
MOISTURE
BARRIER

EXTERIOR INSULATION

FIGURE 54–19 Crawl space insulation methods.

ACOUSTICAL INSULATION

Acoustical or *sound* insulation resists the passage of noise through a building section. The reduction of sound transfer between rooms is important in offices, apartments, motels, and homes. Excessive noise is not only annoying but harmful. It not only causes fatigue and irritability, but also can damage the sensitive hearing nerves of the inner ear.

Sound insulation between active areas and quiet areas of the home is desirable. Sound insulation between the bedroom area and the living area is important. Bathrooms also should be insulated.

Sound Transmission

Noises create sound waves. These waves radiate outward from the source until they strike a wall, floor, or ceiling. These surfaces then begin to vibrate by the pressure of the sound waves in the air. Because the wall vibrates, it conducts sound to the other side in varying degrees, depending on the wall construction. There are three basic ways to reduce sound transmission: (1) interrupt the vibration path (e.g., staggered studs in a "party wall"); (2) provide a thick, solid barrier to abruptly stop the sound waves from vibrating (e.g., a concrete block wall); and (3) fill the space with

sound-absorbing materials (e.g., batt insulation). **Acoustical sealant** was first developed to create a perfect air seal under framed walls and around electrical outlets in order to constrict the transmission of sound waves.

Sound Transmission Class. The resistance of a building section, such as a wall, to the passage of sound is rated by its **sound transmission class (STC)**. The higher the number, the better the sound barrier. The approximate effectiveness of walls with varying STC numbers is shown in **Figure 54–20**.

25	Normal speech can be understood quite easily.
30	Loud speech can be understood fairly well.
35	Loud speech audible but not intelligible.
42	Loud speech audible as a murmur.
45	Must strain to hear loud speech.
48	Some loud speech barely audible.
50	Loud speech not audible.

This chart from the Acoustical and Insulating Materials Association illustrates the degree of noise control achieved with barriers having different STC numbers.

FIGURE 54–20 Approximate effectiveness of sound reduction in walls with varying STC ratings.

Sound travels readily through the air and through some materials. When airborne sound strikes a wall, the studs act as conductors unless they are separated in some way from the covering material. Electrical outlet boxes placed back to back in a wall easily transmit sound. Faulty construction, such as poorly fitted doors, often allows sound to pass through. Therefore, good, airtight construction practices are the first line of defence in controlling sound.

Wall Construction

A wall that provides sufficient resistance to airborne sound should have an STC rating of 45 or greater. At one time, the resistance usually was provided only by double walls, which resulted in increased costs. However, a system of using **sound-deadening insulating board** with a gypsum board outer covering has been developed. This system provides good sound resistance. Resilient steel channels placed at right angles to the studs isolate the gypsum board from the stud. **Figure 54–21** shows various types of wall construction and their STC ratings.

Floor and Ceiling Construction

Sound insulation between an upper floor and the ceiling below involves not only the resistance of airborne sounds, but also that of **impact noise**. Impact noise is caused by such things as dropped objects, footsteps, or moving furniture. The floor is vibrated by the impact. Sound is then radiated from both sides of the floor. Impact noise control must be considered as well as airborne sounds when constructing floor sections for sound insulation.

An **impact noise rating (INR)** shows the resistance of various types of floor-ceiling construction to impact noises. The higher the positive value of the INR, the more resistant the insulation is to impact noise transfer. **Figure 54–22** shows various types of floor-ceiling construction with their STC and INR ratings.

Sound Absorption

The amount of noise in a room can be minimized by the use of *sound-absorbing materials*. Perhaps the most commonly used material is **acoustical tile** made of fibreboard. These tiles are most often used in the ceiling, where they are not subjected to damage. The tiles are soft. The tile surface consists of small holes or fissures or a combination of both (**Figure 54–23**). These holes or fissures act as sound traps. The sound waves enter, bounce back and forth, and finally die out.

SOUND INSULATION OF SINGLE WALLS			SOUND INSULATION OF DOUBLE WALLS		
DETAIL	DESCRIPTION	STC RATING	DETAIL	DESCRIPTION	STC RATING
A 16" / 2" X 4"	1/2" GYPSUM WALLBOARD / 5/8" GYPSUM WALLBOARD	32 / 37	A 16" / 2" X 4"	1/2" GYPSUM WALLBOARD	38
B 16" / 2" X 4"	3/8" GYPSUM LATH (NAILED) PLUS 1/2" GYPSUM PLASTER WITH WHITECOAT FINISH (EACH SIDE)	39	B 16" / 2" X 4"	5/8" GYPSUM WALLBOARD (DOUBLE LAYER EACH SIDE)	45
C	8" CONCRETE BLOCK	45	C 16" / BETWEEN OR WOVEN / 2" X 4"	1/2" GYPSUM WALLBOARD 1 1/2" FIBROUS INSULATION	49
D 16" / 2" X 4"	1/2" SOUND-DEADENING BOARD (NAILED) 1/2" GYPSUM WALLBOARD (LAMINATED) (EACH SIDE)	46	D 16" / 2" X 4"	1/2" SOUND DEADENING BOARD (NAILED) 1/2" GYPSUM WALLBOARD (LAMINATED)	50
E 16" / 2" X 4"	RESILIENT CLIPS TO 3/8" GYPSUM BACKER BOARD 1/2" FIBREBOARD (LAMINATED) (EACH SIDE)	52			

FIGURE 54–21 STC ratings of various types of wall construction.

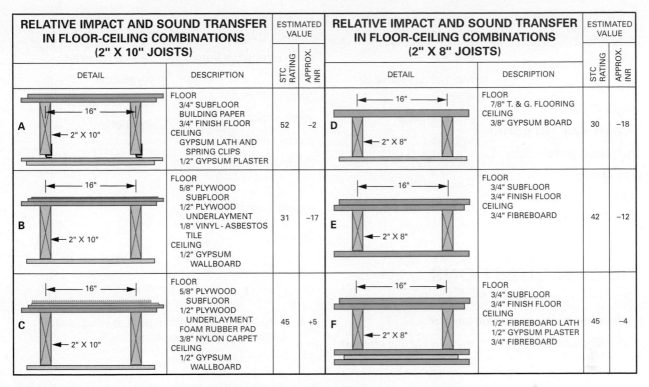

RELATIVE IMPACT AND SOUND TRANSFER IN FLOOR-CEILING COMBINATIONS (2" X 10" JOISTS)		ESTIMATED VALUE		RELATIVE IMPACT AND SOUND TRANSFER IN FLOOR-CEILING COMBINATIONS (2" X 8" JOISTS)		ESTIMATED VALUE	
DETAIL	DESCRIPTION	STC RATING	APPROX. INR	DETAIL	DESCRIPTION	STC RATING	APPROX. INR
A 16" 2" X 10"	FLOOR 3/4" SUBFLOOR BUILDING PAPER 3/4" FINISH FLOOR CEILING GYPSUM LATH AND SPRING CLIPS 1/2" GYPSUM PLASTER	52	−2	D 16" 2" X 8"	FLOOR 7/8" T. & G. FLOORING CEILING 3/8" GYPSUM BOARD	30	−18
B 16" 2" X 10"	FLOOR 5/8" PLYWOOD SUBFLOOR 1/2" PLYWOOD UNDERLAYMENT 1/8" VINYL - ASBESTOS TILE CEILING 1/2" GYPSUM WALLBOARD	31	−17	E 16" 2" X 8"	FLOOR 3/4" SUBFLOOR 3/4" FINISH FLOOR CEILING 3/4" FIBREBOARD	42	−12
C 16" 2" X 10"	FLOOR 5/8" PLYWOOD SUBFLOOR 1/2" PLYWOOD UNDERLAYMENT FOAM RUBBER PAD 3/8" NYLON CARPET CEILING 1/2" GYPSUM WALLBOARD	45	+5	F 16" 2" X 8"	FLOOR 3/4" SUBFLOOR 3/4" FINISH FLOOR CEILING 1/2" FIBREBOARD LATH 1/2" GYPSUM PLASTER 3/4" FIBREBOARD	45	−4

FIGURE 54–22 STC and INR for floor and ceiling constructions.

FIGURE 54–23 Sound-absorbing ceiling tiles.

KEY TERMS

acoustical sealant

acoustical tile

air leakage control (ALC)

capillary action

condensation

conduction

convection

degree-days rating

dew point

drying potential

eaves

ENERGY STAR® for New Homes initiative

expanded polystyrene (EPS)

extruded polystyrene (XPS)

flexible insulation

foamed-in-place

flue effect

house-as-a-system

ice dam

impact noise

impact noise rating (INR)

insulated glass

insulation

LEED® Canada for Homes

loose-fill insulation

low-emissivity coating

mechanical exhaust effect

polyurethane

radiation

rainscreen

R-value (RSI)

relative humidity

sound-deadening insulating board

sound transmission class (STC)

spray foams

stack effect

storm window

thermal envelope

vapour

vapour barrier

vapour diffusion retarder (VDR)

vapour retarder

ventilation

weatherstripping

wind effect

REVIEW QUESTIONS

Select the most appropriate answer.

1. Which structural building material has the greatest R-value?

   a. concrete
   b. steel
   c. stone
   d. wood

2. What is the definition of the insulating term *R-value*?

   a. the measure of the resistance of a material to the flow of heat
   b. the measure of the relative amount of the heat lost through a building section
   c. the measure of the conductivity of a material
   d. the measure of the total heat transfer through a building section

3. What is the boundary between conditioned and unconditioned air called?

   a. the thermal envelope
   b. the thermal resistance
   c. the thermal retarder
   d. the thermal dam

4. What material is used to protect insulation from "wind-washing" at the eaves?

   a. an air barrier
   b. an air dam
   c. an air retarder
   d. an air stopper

5. What is the term for the point at which water droplets form on a cooler surface?

   a. condensation point
   b. vapour point
   c. water point
   d. dew point

6. Which of the following does NOT reduce moisture migration into the insulated stud cavity?

   a. installing a vapour retarder on the inside of the wall
   b. placing sheathing tape on the seams of exterior sheathing
   c. using the airtight drywall approach
   d. applying acoustical sealant under the bottom.

7. Where should a vapour diffusion retarder be installed when building over a crawl space foundation?

   a. just below the subfloor
   b. just under the joists
   c. on top of the ground
   d. on the outside of the exterior foundation walls.

8. If air temperature suddenly increases, what happens to the relative humidity?

   a. It increases.
   b. It decreases.
   c. It remains the same.
   d. It depends on the heat source.

9. What happens if you tightly squeeze or compress flexible insulation into spaces?

    a. Its effectiveness is reduced.
    b. Its efficiency is increased by creating more air spaces.
    c. Nothing—it's necessary to hold the insulation in place.
    d. It helps prevent air leakage by sealing cracks.

10. When insulation is placed between roof framing members, how big an air space should there be between the insulation and the roof sheathing?

    a. at least ½ inch (12 mm)
    b. at least 1 inch (25 mm)
    c. at least 2 inches (50 mm)
    d. at least 3 inches (75 mm)

11. Which of the following, when properly installed, can reduce ice dams at the eaves?

    a. roof and soffit ventilation
    b. the choice of roofing material
    c. vapour retarder on the ceiling
    d. eave protection

12. What is the best method of venting an attic space?

    a. installing gable vents
    b. installing hip vents
    c. installing roof windows
    d. installing continuous ridge and soffit vents

Section 3

Exterior Finish

Tieg Martin

Title: Superintendent
Company: Ledcor Construction Limited, Vancouver

EDUCATION

After graduating from high school in West Vancouver in 1993, Tieg attended the University of British Columbia, where he interrupted his studies with manual work and travel overseas. While on campus, he led a hiking club on regular mountaineering expeditions. Tieg completed his double major in English literature and medieval history in 2002. In 2009, he attained his Red Seal Trade Certification in Carpentry and in 2012/2013, he completed coursework towards a B.Tech in Construction Management at the British Columbia Institute of Technology (BCIT).

HISTORY

After earning his B.A., Tieg began work on a master's degree in Islamic studies, wrote copy for a marketing firm, and served as a constituency assistant for a member of the B.C. Legislative Assembly. He found all of these undertakings fulfilling to some degree. Still, he resented the endless hours shackled to a computer and missed the outdoor activities he had pursued all his life. In 2004, he resigned his office jobs and found an apprenticeship with Ross Lauder, a residential home renovator based in North Vancouver.

ON THE JOB

Beginning his apprenticeship with a small company presented Tieg with almost limitless opportunity to learn. He was fortunate to have a mentor who allowed him to show initiative and assume responsibility. By the end of his second year of employment, Tieg was directing his own team in forming, framing, and finishing work and was coordinating the activities of other trades on-site. In 2005, he registered his own company to do small projects on weekends. After just over a year he was fully self-employed doing residential renovation projects of his own.

Tieg completed his on-the-job training as a lead carpenter, sub-contracting from Verne Glover Company Ltd., and attained his Red Seal Certificate of Apprenticeship in February 2009. He incorporated Kian Construction Ltd. in January 2009 and ran his own company for four years, specializing in residential renovations and architectural concrete installations.

After completing coursework in Construction Management, Tieg was hired by Ledcor Construction Limited as a project coordinator and promoted to superintendent in 2014.

BEST ASPECTS

Tieg loves doing work that is equal parts physical and mental. There is no limit to the knowledge and skills that a carpenter needs to do high-quality work at an efficient pace. Most of all, he has enjoyed working with long-time carpenters, with masters who take pride in their work and who are eager to teach. Tieg feels that without this valuable guidance he would never have enjoyed as much freedom and opportunity as he did while progressing in his apprenticeship. He also enjoys the excitement and camaraderie of teamwork.

Superintending for Ledcor is especially fulfilling in that it opens doors to larger, more complex projects using new materials and techniques.

CHALLENGES

The only aspect of carpentry that Tieg doesn't relish is the business management side (i.e., accounting). When the work is fulfilling and the crew is excited about the job, the last thing he wants to dwell on is the bookkeeping. Still, he believes his mentors when they tell him that a good tradesperson will not prosper without being a competent businessperson as well.

FUTURE

Tieg hopes to assume even more responsibility and to acquire experience with a broader range of construction materials and techniques in his current superintending role.

WORDS OF ADVICE

Know that you never stop learning and that honesty and dignity are your best assets.

Unit 17
Roofing

Chapter 55 Asphalt Shingles and Tile Roofing   **Chapter 57** Flashing

Chapter 56 Wood Shingles and Shakes

Materials used to cover a roof and make it tight are part of the exterior finish called *roofing*. Roofing adds beauty to the exterior and protects the interior. Before roofing is applied, the roof deck must be securely fastened. There must be no loose or protruding nails, and the deck must be clean of all debris. Properly applied roofing gives years of dependable service.

OBJECTIVES

After completing this unit, the student should be able to:

- define roofing terms.
- describe and apply roofing felt, organic or glass fibre-based asphalt shingles, and tile roofing.
- describe various grades and sizes of wood shingles and shakes and apply them.
- flash valleys, sidewalls, chimneys, and other roof obstructions.
- estimate needed roofing materials.

🪖 SAFETY REMINDER

A variety of roofing materials are available to protect a building from weather. Application of materials is usually straightforward and can be done with speed. Care must be taken to remember the dangers of falling.

Chapter 55

Asphalt Shingles and Tile Roofing

Asphalt shingles are the most commonly used roof covering for residential and light commercial construction. They are designed to provide protection from the weather for a period ranging from 20 to 30 years. They are available in many colours and styles.

ROOFING TERMS

An understanding of the terms most commonly used in connection with roofing is essential for efficient application of roofing material.

- A **square** is the amount of roofing required to cover 100 square feet (9.29 m²) of roof surface. There are usually three bundles of shingles per square or about 80 three-tab shingle strips (**Figure 55–1**). One square of shingles can weigh between 235 and 400 pounds (107 and 181 kg), depending on shingle quality.

- **Deck** is the wood roof surface to which roofing materials are applied.

- **Coverage** is the number of overlapping layers of roofing and the degree of weather protection offered by roofing material. Roofing may be called single, double, or triple coverage.

- A **shingle butt** is the bottom exposed edge of a shingle.

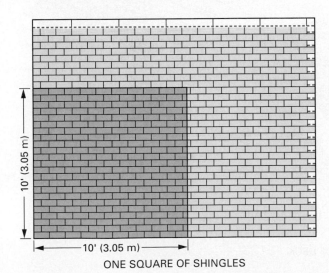

FIGURE 55–1 One square of shingles will cover 100 square feet (9.29 m²).

(labels: 10' (3.05 m) vertical and horizontal; ONE SQUARE OF SHINGLES)

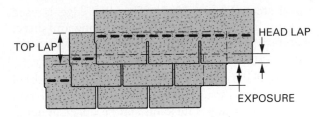

FIGURE 55–2 Asphalt strip exposure and lap.

(labels: TOP LAP, HEAD LAP, EXPOSURE)

- **Courses** are horizontal rows of shingles or roofing.

- **Exposure** is the distance between courses of roofing. It is the amount of roofing in each course exposed to the weather (**Figure 55–2**).

- The **top lap** is the height of the shingle or other roofing minus the exposure. In **roll roofing** this is also known as the **selvage**.

- The **head lap** is the distance from the bottom edge of an overlapping shingle to the top of a shingle two courses under, measured up the slope.

- **End lap** is the horizontal distance that the ends of roofing in the same course overlap each other.

- **Flashing** is strips of thin roofing material. It is usually made of lead, zinc, copper, vinyl, or aluminum. It may also be strips of roofing material used to make watertight joints on a roof. Metal flashing comes in rolls of various widths that are cut to the desired length.

- *Asphalt cements* and *coatings* are manufactured to various consistencies, depending on the purpose for which they are to be used. **Cements** are classified as *plastic*, *lap*, and *quick-setting*. They will not flow at summer temperatures. They are used as adhesives to bond asphalt roofing products and flashings. They are usually trowelled on the surface. **Coatings** are usually thin enough to be applied with a brush. They are used to resurface old roofing or metal that has become weathered.

- When two unlike metals come into contact with water, they form a **voltaic couple** that causes one of the metals to corrode. A simple way to prevent the disintegration caused by electrolysis is to secure metal roofing material with fasteners of the same metal.

PREPARING THE DECK

In preparation for installing roofing materials, the deck should be clean and clear of debris. This will prevent any damage to the roofing material and make a safer working surface. A metal **drip edge** is often installed along the roof edges. The metal drip edge is usually made of aluminum or galvanized steel (**Figure 55–3**). The drip edge is used to support the asphalt shingle overhang and direct water runoff away from the fascia and into the eavestrough.

Install the metal drip edge by using roofing nails of the same metal spaced 8 to 10 inches (203 to 254 mm) along its inner edge. End joints may be lapped by about 2 inches (50 mm). See IKO's site for video installation tips (*iko.com/blueprint-for-roofing-videos*).

Eave Protection

Eave protection is required for most roofs (with a slope of less than 8:12 or 1:1.5) over heated spaces in most of Canada (except in the very temperate zones, <3500 degree days [see table below]; recall that a degree day is the sum of the difference of the mean temperature and 18°C for all days that have a daily average temperature below 18°C).

Eave protection can be one of the following:

- #15 felt laid in two plies, lapped 19 inches (483 mm), cemented together
- type "S" (smooth) or type "M" (mineral) roll roofing with a minimum side lap and head lap of 4 inches (100 mm), cemented together
- glass-fibre or polyester-fibre coated base sheets
- self-sealing composite membranes consisting of modified bituminous coated material

The eave protection must extend up the slope of the roof to a point at least 12 inches (305 mm) horizontally in from the inside face of the outside wall. The minimum required width is 36 inches (914 mm), but in almost all cases, two rows of ice and water shield are needed. See *casma.ca/preventing-problems-from-ice-dams#.VM7pQ2jF98E for a diagram.*

DEGREE DAYS BELOW 18°C							
Vancouver BC	Kamloops BC	Toronto ON	Halifax NS	Ottawa ON	Regina SK	Winnipeg MN	Iqaluit NV
2926	3570	4065	4367	4602	5660	5777	10117

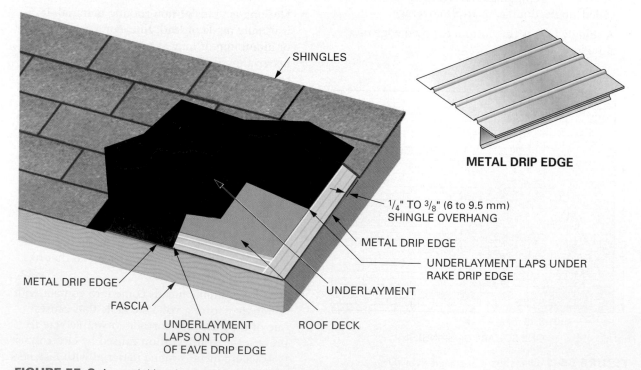

FIGURE 55–3 A metal drip edge may be used to support the shingle edge overhang.

An ice and water shield is a roofing membrane composed of two waterproofing materials bonded into one layer. It is composed of a rubberized asphalt adhesive backed by a layer of polyethylene. It comes in 3 × 75-foot (0.9 × 23 m) rolls. The rubberized asphalt surface is backed with a release paper to protect the sticky side. During installation, the release paper is peeled off, allowing the asphalt to bond to the roof deck.

This material is used in the trouble spots of a roof, such as along the eaves, in valleys, and in unique areas where leaks are more likely. Low-slope roofs and roofs exposed to severe blowing weather are at increased risk of leaks. An ice and water shield is often used during re-roofing applications of houses that experience ice damming at the eaves (**Figure 55–4**).

The ice and water shield is installed directly on the roof deck. The deck should be clean and clear of debris. There should be no voids in the deck. Vertical laps should be 6 inches (152 mm) and horizontal laps 4 inches (100 mm). Cut pieces to the desired size. After positioning the sheet, fold back enough membrane to peel off some release paper. Reposition the membrane and press the membrane to the deck. Pull remaining release paper off and press the membrane to the deck. This will ensure proper adherence to the roof deck.

Underlayment

The deck should next be covered with an asphalt shingle **underlayment**. The underlayment protects the sheathing from moisture until the roofing is applied. It also gives additional protection to the roof afterward.

Asphalt Felt

Asphalt felt consists of heavy felt paper saturated with asphalt or coal tar. It is usually made in various weights of pounds per square. Asphalt felt comes in 36-inch-wide (914 mm) rolls. The rolls are 72 or 144 feet (22 or 44 m) long, covering 2 or 4 squares. Usually the lightest-weight felt is used as an underlayment for asphalt shingles.

Apply a layer of asphalt felt underlayment over the deck starting at the bottom. Lay each course of felt over the lower course at least 2 inches (50 mm). Make any end laps at least 4 inches (100 mm). Lap the felt 6 inches (152 mm) from both sides over all hips and ridges.

Nail or staple through each lap and through the centre of each layer about 16 inches (406 mm) apart. Roofing nails driven through the centre of metal discs or specially designed, large-head felt fasteners hold the underlayment securely in strong winds until shingles are applied (**Figure 55–5**).

Plastic Felt

Plastic felt underlayments are made of two layers of polypropylene. Each roll covers about 10 squares. The advantages of plastic felt are that it is more tear-resistant and more slip-resistant, and it lays flatter than asphalt felt, which tends to buckle after it gets wet (**Figure 55–6**).

FIGURE 55–4 Application of ice and water shield.

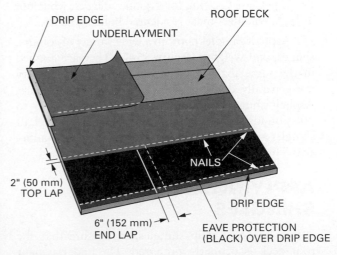

FIGURE 55–5 Application specifications for asphalt felt underlayment.

Courtesy of GAF Materials Corporation

FIGURE 55–6 Synthetic underlayments are installed under the shingle layer.

KINDS OF ASPHALT SHINGLES

Two types of asphalt shingles are manufactured. *Organic* shingles have a base made of heavy asphalt-saturated paper felt coated with additional asphalt. *Fibreglass* shingles have a base mat of glass fibres. The mat does not require the saturation process and requires only an asphalt coating. Both kinds of shingles are surfaced with selected mineral granules. The asphalt coating provides weatherproofing qualities. The granules protect the shingle from the sun, provide colour, and protect against fire.

Fibreglass-based asphalt shingles have an Underwriters Laboratories' Class A fire resistance rating. The Class A rating is the highest standard for resistance to fire. The Class C rating for organic shingles, while not as high, will meet most residential building codes.

Asphalt shingles come in a wide variety of colours, shapes, and weights (**Figure 55–7**). They are applied in the same manner. Shingle quality and longevity are generally determined by the weight per square. Asphalt shingles can weigh anywhere from 235 to 400 pounds (107 to 181 kg) per square. Most asphalt shingles are manufactured with factory-applied adhesive. This increases their resistance to the wind.

APPLYING ASPHALT SHINGLES

Before applying strip shingles, make sure that the roof deck is properly prepared. The underlayment and drip edge should be applied. Asphalt roofing products become soft in hot weather. Be careful not to damage them by digging in with heavily cleated shoes during application or by unnecessary walking on the surface after application. The slope of a roof should not be less than 4 inches per foot of run (1:3) when conventional methods of asphalt shingle application are used.

Asphalt Shingle Layout

On small roofs, strip shingles are applied by starting from either rake. On long buildings, a more accurate vertical alignment is ensured by starting at the centre and working both ways. Mark the centre of the roof at the eaves and at the ridge. Snap a chalk line between the marks. Snap a series of chalk lines 6 inches (152 mm) apart on each side of the centreline if the shingle tab cutouts are to break on the halves. Snap lines 4 inches (100 mm) apart if the cutouts are to break on the thirds. When applying the shingles, start the course with the end of the shingle to the vertical chalk line. Start succeeding courses in the same manner. Break the joints as necessary. Pyramid the shingles up in the centre. Work both ways toward the rakes (**Figure 55–8**).

If it is decided to start at the rakes and cutouts are to break on the halves, start the first course with a whole tab. The second course is started with a shingle from which 6 inches (152 mm) have been cut; the third course, with a strip from which the entire first tab is removed; the fourth, with one and one-half tabs removed, and so on (**Procedure 55–A**). These starting strips are precut for faster application. Waste from these strips is used on the opposite rake.

Product	Configuration	Per Square			
		Approx. Shipping Weight	Shingles	Bundles	Exposure
Wood appearance strip shingle more than one thickness per strip Laminated or job applied	Various edge, surface texture, & application treatments	285# to 390# (129–177 kg)	67 to 90	4 or 5	4″ to 6″ (100–152 mm)
Wood appearance strip shingle single thickness per strip	Various edge, surface texture, & application treatments	Various 250# to 350# (113–159 kg)	78 to 90	3 or 4	4″ to 5 1/8″ (100–130 mm)
Self-sealing strip shingle	Conventional 3 tab	205#–240# (93–109 kg)	78 or 80	3	5″ to 5 1/8″ (127–130 mm)
	2 or 4 tab	Various 215# to 325# (98–147 kg)	78 or 80	3 or 4	5″ to 5 1/8″ (127–130 mm)
Self-sealing strip shingle No cutout	Various edge and texture treatments	Various 215# to 290# (98–132 kg)	78 to 81	3 or 4	5″ (127 mm)
Individual lock-down Basic design	Several design variations	180# to 250# (82–113 kg)	72 to 120	3 or 4	

Courtesy of Asphalt Roofing Manufacturers Association

FIGURE 55–7 Asphalt shingles are available in a wide variety of sizes, shapes, and weights.

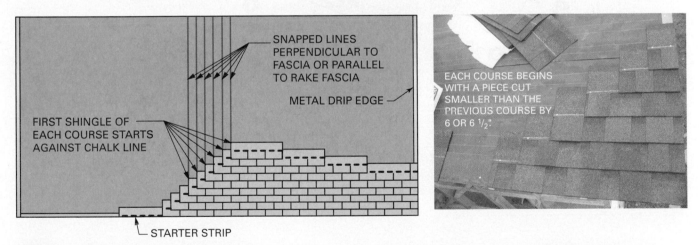

FIGURE 55–8 (A) On long roofs, start shingling from the centre working toward the rakes. (B) Rake starter pieces are installed in a stepped-back fashion.

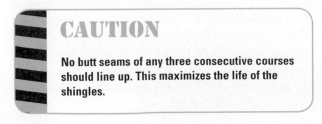

CAUTION

No butt seams of any three consecutive courses should line up. This maximizes the life of the shingles.

Cut the shingles by scoring them on the back side with a utility knife or use a hook blade. Use a square as a guide for the knife. Bend the shingle. It will break on the scored line.

The layouts may have to be adjusted so that tabs on opposite rakes will be of approximately equal

StepbyStep Procedures

Procedure 55–A Laying Out Rake Starters

STEP 1 BEGIN EAVE STARTER WITH A PIECE THAT HAS ABOUT 6" (150 mm) REMOVED. OVERHANG BOTH EDGES ABOUT ¼ OR ³⁄₈" (6 to 9.5 mm). INSTALL SEVERAL PIECES ALONG EAVE, MAINTAINING A STRAIGHT EDGE.

STEP 2 START THE FIRST FULL TAB SHINGLE FULL LENGTH. INSTALL SEVERAL MORE ALONG EAVE.

STEP 3 BEGIN THE SECOND COURSE WITH A FULL TAB SHINGLE THAT HAS 6 OR 6½" (150 or 165 mm) REMOVED FROM THE RAKE EDGE END OF THE SHINGLE. INSTALL SEVERAL FULL TAB SHINGLES OF THE SECOND COURSE.

STEP 4 BEGIN THE THIRD COURSE WITH A SHINGLE THAT HAS TWICE THE 6- OR 6½" (152 or 165 mm) INCREMENT. INSTALL SEVERAL FULL-LENGTH PIECES ON TOP OF FIRST COURSE. WATCH THAT EXPOSE IS UNIFORMLY SPACED.

STEP 5 FOURTH COURSE BEGINS WITH ONE-HALF A SHINGLE. REPEAT AS ABOVE, CUTTING MORE OFF EACH STARTER UNTIL LAST COURSE HAS A PIECE THAT IS 6 OR 6½" (152 or 165 mm) LONG.

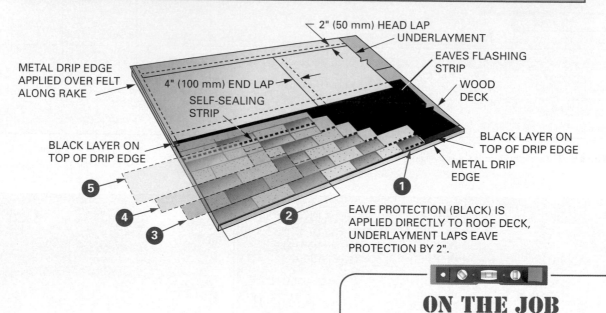

2" (50 mm) HEAD LAP
UNDERLAYMENT
EAVES FLASHING STRIP
WOOD DECK
METAL DRIP EDGE APPLIED OVER FELT ALONG RAKE
4" (100 mm) END LAP
SELF-SEALING STRIP
BLACK LAYER ON TOP OF DRIP EDGE
BLACK LAYER ON TOP OF DRIP EDGE
METAL DRIP EDGE
EAVE PROTECTION (BLACK) IS APPLIED DIRECTLY TO ROOF DECK, UNDERLAYMENT LAPS EAVE PROTECTION BY 2".

ON THE JOB

Cut off pieces from making the rake starter strips are used on the other end of the roof. This amounts to very little waste overall.

widths. No rake tab should be less than 3 inches (75 mm) in width.

Starter Course of Asphalt Shingles

The **starter course** backs up and fills in the spaces between tabs of the first regular course of shingles. Cut the exposed tabs off, or about 5 inches (127 mm) of exposure from a regular shingle. Apply these pieces so that the factory-applied sealing strip is near the eave edge. This will seal the first shingle course tabs. The first starter strip applied must be shorter in length by a minimum of 3 inches (75 mm). This step will keep the butt seams or end joints from aligning with the first course. Overhang the shingles past the drip edge ¼ to ³⁄₈ inch (6 to 9.5 mm).

Some roofers install the starter strips along the perimeter of the roof, keeping the sealing strips closest to the edge of the roof.

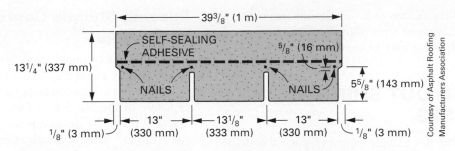

FIGURE 55–9 Recommended fastener locations for asphalt strip shingles.

Fastening Asphalt Shingles

Selecting suitable fasteners, using the recommended number, and putting them in the right places are important steps in the application of asphalt shingles. Lay the first regular course of shingles on top of the starter course. Keep their bottom edges flush with each other. Use a minimum of four fasteners in each strip shingle. Do not nail into or above the factory-applied adhesive (**Figure 55–9**).

The fastener length should be sufficient to penetrate the sheathing at least ¾ inch (19 mm), or through approved panel sheathing. Roofing nails should be 11- or 12-gauge galvanized steel or aluminum with barbed shanks. They should have ⅜- to ⁷⁄₁₆-inch (9.5 to 11mm) heads. Roofing nails may be driven by hand or with power nailers. In some locales, power-driven staples may be used in place of nails; however, their use is limited to shingles with factory-applied adhesive (**Figure 55–10**). Staples must be at least 16-gauge, and they should have a minimum crown width of ¹⁵⁄₁₆ inch (24 mm).

Align each shingle of the first course carefully. Fasten each shingle from the end nearest the shingle just laid. This prevents buckling. Drive fasteners straight so that the nail heads or staple crowns will not cut into the shingles. The entire crown of

FIGURE 55–10 Pneumatic staplers and nailers are often used to fasten asphalt shingles.

the staple or nail head should bear tightly against the shingle. It should not penetrate its surface (**Figure 55–11**). Continue applying shingles until the first course is complete.

Shingle Exposure

The maximum *exposure* of asphalt shingles to the weather depends on the type of shingle. Recommended maximum exposures range from 4 to 6 inches (100 to 152 mm). Most commonly used

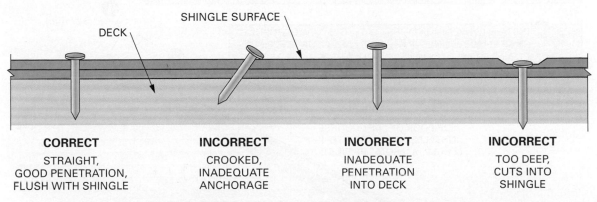

FASTENER SHOULD PENETRATE ¾" (19 mm) INTO ROOF DECK OR EXTEND ⅛" (3 mm) PAST.

FIGURE 55–11 It is important to fasten asphalt shingles correctly.

asphalt shingles have a maximum exposure of 5 inches (127 mm). Less than the maximum recommended exposure may be used, if desired.

LAYING OUT SHINGLE COURSES

When laying out shingle courses, space the desired exposure up each rake from the top edge of the first course of shingles. Snap lines across the roof for five or six courses. When snapping a long line, it may be necessary to thumb the line down against the roof at about centre. Then snap the lines on both sides of the thumb.

Lay succeeding courses so that their top edges are to the chalk line. Start each course so that the cutouts are staggered in the desired manner. Continue snapping lines and applying courses until a point 3 or 4 feet (1 or 1.2 m) below the ridge is reached. Some carpenters snap a line to straighten out the course after using the top of the cutout for the shingle exposure for a number of courses.

Spacing Shingle Courses to the Ridge

The last course of shingles below the *ridge cap* should be exposed by about the same amount as all the other shingle courses. This can be achieved by changing the exposure of the last 3 to 4 feet (1 to 1.2 m) of roof near the ridge. To determine the new exposure, first cut a full tab from a shingle. Centre it on the ridge. Then bend it over the ridge. Do this at both ends of the building. Mark the bottom edges on the roof.

Measure up 2 inches (50 mm). Snap a line between the marks. This line will be the top edge of the next to last course of shingles. It should be about 3½ inches (89 mm) down from the ridge. Divide the distance between this line and the top of the last course of shingles applied into spaces as close as possible to the exposure of previous courses. Do not exceed the maximum exposure (**Procedure 55–B**). Snap lines across the roof and shingle up to the ridge.

The line for the last course of shingles is snapped on the face of the course below. Lay out the exposure

---

StepbyStep Procedures

Procedure 55–B Evenly Spacing Upper Shingle Courses near the Ridge

STEP 1 MEASURE DOWN 3½" (89 mm) FROM THE TOP OF PLYWOOD SLOT IF ROOF IS TO HAVE A RIDGE VENT. MEASURE DOWN 5" (127 mm) FROM CENTRE OF RIDGE IF NO RIDGE VENT IS TO BE INSTALLED. DO THIS ON BOTH ENDS OF THE BUILDING.

STEP 2 MEASURE FROM MARKS IN STEP 1 TO THE TOP EDGE OF THE LAST COURSE OF SHINGLES INSTALLED 3 TO 4' (1 to 1.2 m) FROM RIDGE.

STEP 3 DIVIDE MEASURED DISTANCE BY NORMAL SHINGLE EXPOSURE. ROUND RESULT UP TO NEXT WHOLE NUMBER OF COURSES. DIVIDE THE MEASURED DISTANCE AGAIN NOW BY THIS ROUNDED NUMBER. THIS RESULT IS THE SLIGHTLY SMALLER EXPOSURE THAT WILL MAKE COURSES WORK OUT TO A FULL EXPOSURE AT THE RIDGE.

STEP 4 INSTALL REMAINING COURSE WITH THIS NEW EXPOSURE. SNAP SEVERAL LINES IF NEED BE. LAST COURSE IS EITHER CUT TO THE PLYWOOD IF A RIDGE VENT IS INSTALLED OR LAPPED OVER TO CLOSE THE RIDGE.

1 5" (127 mm)

2

TOP EDGE OF LAST COURSE INSTALLED AT LEAST 3 TO 4' (1 to 1.2 m) FROM THE RIDGE.

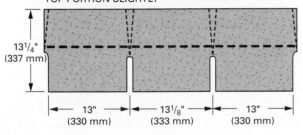

CUT ALONG DOTTED LINE, TAPERING
TOP PORTION SLIGHTLY

13¼"
(337 mm)

13"
(330 mm)   13⅛"
(333 mm)   13"
(330 mm)

FIGURE 55–12 Ridge and hip cap shingles are cut from strip shingles.

FIGURE 55–13 Applying ridge cap shingles to a vented ridge.

from the bottom edge of the course on both ends of the roof. Snap a line across the roof. Fasten the last course of shingles by placing their butts to the line. If the ridge is to be vented, cut the top edge of the shingle at the sheathing. If the ridge is not vented, bend the top shingle edges over the ridge. Fasten their top edges to the opposite slope.

Applying the Ridge Cap

Cut hip and ridge shingles from shingle strips to make approximately 13 × 13 inch (330 × 330 mm) squares. Cut shingles from the top of the cutout to the top edge on a slight taper. The top edge should be narrower than the bottom edge (**Figure 55–12**). Cutting the shingles in this manner keeps the top half of the shingle from protruding when it is bent over the ridge.

The ridge cap is applied after both sides of the roof have been shingled. At each end of the roof, centre a shingle on the ridge. Bend it over the ridge. Mark its bottom edge on the front slope or the one most visible. Snap a line between the marks.

Beginning at the bottom of a hip or at one end of the ridge, apply the shingles over the hip or ridge. Expose each 5 inches (127 mm). In cold weather, ridge cap shingles may have to be warmed in order to prevent cracking when bending them over the ridge. On the ridge, shingles are started from the end away from prevailing winds. The wind should blow over the shingle butts, not against them. Keep one edge, from the butt to the start of the tapered cut, to the

chalk line. Secure each shingle with one fastener on each side, 5½ inches (140 mm) from the butt and one inch (25 mm) up from each edge. Apply the cap across the ridge until 3 or 4 feet (1 or 1.2 m) from the end. Then space the cap to the end in the same manner as spacing the shingle course to the ridge. The last ridge shingle is cut to size. It is applied with one fastener on each side of the ridge. The two fasteners are covered with asphalt cement to prevent leakage.

Most roofs or attics will require ventilation. If they are above an insulated ceiling, the National Building Code of Canada (NBCC) has specific requirements for different types and slopes of roofs (see NBCC 9.19.1). In general, for asphalt shingle roofing, the ventilation requirement is 1/300 of the insulated attic space (i.e., one square of vent per 300 square feet of insulated ceiling area), with at least 25 percent at both the top of the roof and at the bottom of the eaves (**Figure 55–13**).

The NBCC also requires attic access if the area of the attic is greater than 108 ft² (10 m²) and the distance from the top of the ceiling joists to the underside of the rafters is greater than 23⅝ inches (600 mm) (NBCC 9.19.2). The most common size of the finished opening is 19⅝ × 27½ inches (500 × 700 mm).

CEMENT ROOF TILES

Another roofing option is cement roof tiles (**Figure 55–14**). They are designed for various architectural styles and will last 30 to 60 years. The lifetime range is determined by the amount of sun and wind-blown rain the roof receives.

Concrete and clay tiles are manufactured in a variety of styles and thicknesses. Consequently, their weight can range from 500 to 1200 pounds per square (24 to 60 kg/m²). Regular tiles range between 700 and 800 pounds per square (34 to 39 kg/m²). This increased weight affects the strength required

FIGURE 55–14 Cement tile roofing is applied to many homes in warm climates.

FIGURE 55–17 Areas of wind-blown rain have a second layer of hot-mopped felt.

from the roof system. Many truss manufacturers will require the roof to be loaded with the tiles before any finishes are applied (**Figure 55–15**). Only weeks later will the interior framing surfaces be shimmed and trimmed flush. This time allows the trusses to shift and adjust to the load.

Underlayment requirements vary with local codes. Most areas use at least one layer of 30# felt fastened with nails and metal caps (**Figure 55–16**).

Coastal areas prone to wind-blown rain typically use another layer comprised of 90# asphalt roll installed in hot, mopped in place bitumen (**Figure 55–17**).

The tile pieces range in sizes from 7 to 13 inches (178 to 330 mm) wide by 15 to 18 inches (381 to 472 mm) long. They typically have two pre-punched holes along the top edge that are used for fastening with screws. Tiles are installed to chalk lines with screws. At corners and hips, the tiles are cut with a circular saw with a diamond or composite blade (**Figure 55–18**).

FIGURE 55–15 Cement tile must be preloaded on roofs to allow the roof system to settle.

FIGURE 55–16 First step in tile roofing is felt applied with capped nails.

FIGURE 55–18 Tiles screwed in place.

Chapter 56

Wood Shingles and Shakes

Wood **shingles** and **shakes** are common roof coverings (**Figure 56–1**). Most shingles and shakes are produced from western red cedar. Cedar logs are first cut into desired lengths. They are then split into sections from which shingles and shakes are sawn or split. All shingles are sawn. Most shakes are split. Shingles, therefore, have a relatively smooth surface.

Most shakes have at least one highly textured, natural grain surface. Most shakes also have thicker butts than shingles (**Figure 56–2**).

Most wood shingles and shakes are produced by mills that are members of the Cedar Shake and Shingle Bureau. Their product label (**Figure 56–3**) is assurance that the products meet quality standards.

FIGURE 56–1 Wood shingles and shakes may be used as siding and roof covering.

FIGURE 56–2 All shingles are sawn from the log. Most shakes are split and have a rough surface.

See ICBO ES Evaluation Report Number 5109
for allowable values and/or conditions of use
concerning material presented in this document.
It is subject to re-examination,
revision and possible cancellation.

Courtesy of Fraser Cedar Products Ltd.

FIGURE 56–3 The product label ensures that quality requirements are met.

DESCRIPTION OF WOOD SHINGLES AND SHAKES

Wood shingles are available for use on roofs in four standard grades. Shakes are manufactured by different methods to produce four types. Both shingles and shakes may be treated to resist fire or premature decay in areas of high humidity (**Figure 56–4**). For more information, the Cedar Shake & Shingle Bureau publishes a Roof Manual with lengths in both metric and imperial (*www.cedarbureau.org/installation-and-maintenance/roof-manual/*).

Sizes and Coverage

Shingles come in lengths of 16, 18, and 24 inches (406, 457, and 610 mm). The butt thickness increases with the length. Shakes are available in lengths of 15, 18, and 24 inches (381, 457, and 610 mm). Their butt thicknesses are from $\frac{3}{8}$ to $\frac{3}{4}$ inch (9.5 to 19 mm). The 15-inch (381 mm) length is used for starter and finish courses. **Figure 56–5** shows the sizes and the amount of roof area that one square, laid at various exposures, will cover. See NBCC 9.26.9 for wood shingle application specifications and 9.26.10 for cedar shakes.

Western Red Cedar Shingles	Western Red Cedar Shakes
NO. 1 PREMIUM OFFICIAL LABEL COLOUR – BLUE The premium grade of shingles for roofs and sidewalls. These top-grade shingles are 100% heartwood, 100% clear, and 100% edge grain. Re-manufactured products: re-butted, re-jointed, and fancy butts.	PREMIUM GRADE & NO. 1 HANDSPLIT & RESAWN These shakes have split faces and sawn backs. Cedar logs are first cut into desired lengths. Blanks or boards of proper thickness are split and then run diagonally through a bandsaw to produce two tapered shakes from each blank.
NO. 2 OFFICIAL LABEL COLOUR – RED A good grade for many applications. Not less than 10" (254 mm) clear on 16" (406.4 mm) shingles, 11" (279.4 mm) clear on 18" (457.2 mm) shingles, and 16" (406.4 mm) clear on 24" (609.6 mm) shingles. Flat grain and limited sapwood are permitted in this grade.	PREMIUM GRADE & NO. 1 TAPERSAWN These shakes are sawn both sides. No. 2 and No. 3 are also available. No. 3 can be used as a starter for shake or shingle roofs and sidewalls. The No. 2 tapersawn can be applied to the roof at 7½" (190.5 mm) exposure and be cut into hip and ridge.
NO. 3 OFFICIAL LABEL COLOUR – BLACK A utility grade for economy applications and secondary buildings. Not less than 6" (152.4 mm) clear on 16" (406.4 mm) and 18" (457.2 mm) shingles, and 10" (254 mm) clear on 24" (609.6 mm) shingles.	NO. 1 TAPERSPLIT Produced largely by hand, using a sharp-bladed steel froe and a mallet. The natural shingle-like taper is achieved by reversing the block, end-for-end, with each split, which achieves the taper of the wood. This was the original shake that was handmade without any machinery.
DECORATIVE A utility decorative grade wood. Common uses: fences, bird houses, dividing walls. This is not a weather-resistant product. To accent interior walls, as a wall or door shim, as a filler to plumb walls, to help plumb for fancy butts. A product of 1000 uses.	NO. 1 STRAIGHTSPLIT Produced by machine or in the same manner as tapersplit shakes, except that by splitting from the same end of the block, the shakes acquire uniform thickness.

FIGURE 56–4 Description of grades and kinds of wood shingles and shakes.

Shingle Length Inches/ Millimetres	Shingle Grade Number	Minimum Shingle Width, [1] Inches/ Millimetres	Nominal Butt Thickness, Inches/ Millimetres	Edge Parallelism, Inches/ Millimetres
16/406.4	1	3/76.2	0.40/10.16	0.25/6.35
	2	3/76.2	0.40/10.16	0.375/9.525
	3	2.5/63.5	0.40/10.16	0.375/9.525
18/457.2	1	3/76.2	0.45/11.43	0.25/6.35
	2	3/76.2	0.45/11.43	0.375/9.525
	3	3/76.2	0.45/11.43	0.375/9.525
24/609.6	1	4/101.6	0.50/12.7	0.25/6.35
	2	3/76.2	0.50/12.7	0.375/9.525
	3	3/76.2	0.50/12.7	0.375/9.525

1. The maximum lineal inches per bundle comprised of shingles each less than 4 inches (101.6 mm) wide shall be the following:
   A. Shingle Grade No. 1–10%
   B. Shingle Grade No. 2–20%
   C. Shingle Grade No. 3–30%
Maximum shingle width for all lengths shall be 14 inches (355.6 mm).

FIGURE 56–5 Size and coverage of wood shingles.

Shingle Exposure									
Maximum Exposure Recommended for Roofs									
		Number 1 Label			Number 2 Label			Number 3 Label		
Shingle Length (inches)		16	18	24	16	18	24	16	18	24
Shingle Length (mm)		406.4	457.2	609.6	406.4	457.2	609.6	406.4	457.2	609.6
Pitch										
3"/12" to 4"/12" 76.2 mm/304.8 mm to 101.6 mm/ 304.8 mm	Exp (in.)	3.75	4.25	5.75	3.5	4	5.5	3	3.5	5
	Exp (mm)	95.25	107.95	146.05	88.9	101.6	139.7	76.2	88.9	127
4"/12" & steeper 101.6 mm/ 304.8 mm	Exp (in.)	5	5.5	7.5	4	4.5	6.5	3.5	4	5.5
	Exp (mm)	127	139.7	190.5	101.6	114.3	165.1	88.9	101.6	139.7

Shake Exposure			
Maximum Exposure Recommended for Roofs			
Shake Length (inches)		18	24
Shake Length (mm)		457.2	609.6
Pitch			
4"/12" & steeper	Exp (in.)	7.5	10(a)
101.6 mm/304.8 mm	Exp (mm)	190.5	254

(a) 24" × 3/8" (609.6 × 9.525 mm) handsplit shakes are limited to 5" (127 mm) maximum weather exposure, per UBC.

Formula for calculating material at reduced exposures:
Square footage divided by reduced coverage equals total material required

Example: You are estimating a roof that measures 3,200 square feet (297.40 m²), 32 squares. You have decided to put 16" (406.4 mm) shingles, Number 1 Label or Number 2 Label at 4" (101.6 mm) exposure. The coverage table tells you that a 4 bundle square at 4" (101.6 mm) exposure covers 80 square feet (7.43 m²). 3200 divided by 80 equals 40 squares of material.

FIGURE 56–6 Maximum exposures of wood shingles and shakes for various roof pitches.

Maximum Exposures

The area covered by one square of shingles or shakes depends on the amount exposed to the weather. The maximum amount of shingle exposure depends on the length and grade of the shingle or shake and the pitch of the roof. Shakes are not generally applied to roofs with slopes of less than 4 inches rise per foot (1:3). Shingles, with reduced exposures, may be used on slopes down to 3 inches rise per foot (1:4). **Figure 56–6** shows the maximum recommended roof exposure for wood shingles and shakes.

SHEATHING AND UNDERLAYMENT

Shingles and shakes may be applied over spaced or solid roof sheathing. **Spaced sheathing** or *skip sheathing* is usually 1 × 4 or 1 × 6 (19 × 89 or 19 × 140 mm) boards. **Solid sheathing** is usually APA-rated panels. Solid sheathing may be required in regions subject to frequent earthquakes or under treated shingles and shakes. It is also recommended for use with shakes in areas where wind-driven weather is common.

Spaced Sheathing

Solid wood sheathing is applied from the eaves up to a point that is plumb with a line 12 to 24 inches (305 to 610 mm) inside the wall line. An eaves flashing is installed, if required. Spaced sheathing may then be used above the solid sheathing to the ridge.

For shingles, either 1 × 4 or 1 × 6 (19 × 89 or 19 × 140 mm) spaced sheathing may be used. Space 4-inch (89 mm) boards the same amount as the shingles are exposed to the weather. In this method of application, each course of shingles is nailed to the centre of the board. If 6-inch (140 mm) boards are used, they are spaced two exposures. Two courses of shingles are nailed to the same board when courses are exposed up to, but not exceeding, 5½ inches (140 mm). For shingles with greater exposures, the sheathing is spaced a distance of one exposure (**Figure 56–7**).

In shake application, spaced sheathing is usually 1 × 6 (19 × 140 mm) boards spaced the same distance, on centre, as the shake exposure (**Figure 56–8**). The spacing should never be more

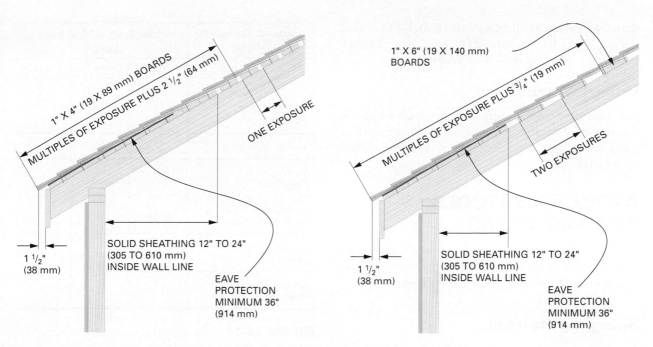

FIGURE 56-7 Application of wood shingles on spaced sheathing.

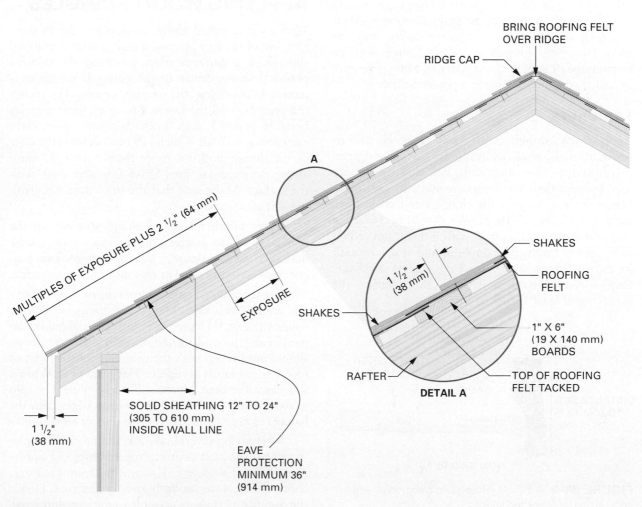

FIGURE 56-8 Method of applying shakes on spaced sheathing.

than 7½ inches (190 mm) for 18-inch (457 mm) shakes and 10 inches (254 mm) for 24-inch (610 mm) shakes installed on roofs.

Underlayment

No underlayment is required under wood shingles. A breather-type roofing felt may be used over solid or spaced sheathing. Underlayment is typically applied under shakes.

APPLICATION TOOLS AND FASTENERS

Shingles and shakes are usually applied with a shingling hatchet. Recommendations for the type and size of fasteners should be closely followed.

Shingling Hatchet

A shingling hatchet (**Figure 56–9**) should be light-weight. It should have both a sharp *blade* and a *heel*. A sliding gauge is sometimes used for fast and accurate checking of shingle exposure. The gauge permits several shingle courses to be laid at a time without snapping a chalk line. A power nailer may be used. However, shingles and shakes often need to be trimmed or split with the hatchet. More time may be lost than gained by using a power nailer.

Fasteners

Apply each shingle with only two corrosion-resistant nails, such as stainless steel, hot-dipped galvanized, or aluminum nails. Box nails usually are used because their smaller gauge minimizes splitting. Minimum nail lengths for shingles and shakes are shown in **Figure 56–10**. Staples should be 16-gauge aluminum or stainless steel with a minimum crown of $\frac{7}{16}$ inch (11 mm). The staple should be driven with its crown across the grain of the shingle or shake. Staples should be long enough to penetrate the sheathing at least ½ inch (12 mm).

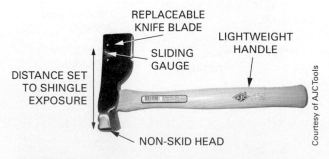

REPLACEABLE KNIFE BLADE

SLIDING GAUGE

LIGHTWEIGHT HANDLE

DISTANCE SET TO SHINGLE EXPOSURE

NON-SKID HEAD

Courtesy of AJC Tools

FIGURE 56–9 A shingling hatchet is commonly used to apply wood shingles and shakes.

Type of Shingle/ Shake	Nail Type & Minimum Length		
Shingles—New Roof	Type	in.	mm
15"/18" (381/457.2 mm)	3d box	$1\frac{1}{4}$	32
24" (609.6 mm)	4d box	$1\frac{1}{2}$	38
Shakes—New Roof	Type	in.	mm
18" (457.2 mm) Straight Split	5d box	$1\frac{3}{4}$	44
18"/24" (457.2/609.6 mm) Handsplit & Resawn	6d box	2	51
24" (609.6 mm) Tapersplit	5d box	$1\frac{3}{4}$	44
18"/24" (457.2/609.6 mm) Tapersawn	6d box	2	51

FIGURE 56–10 Recommended nail types and sizes for wood shingles.

APPLYING WOOD SHINGLES

With a string pulled along the bottom edge of shingles, install the first layer of a starter course of wood shingles. If a gutter is used, overhang the shingles plumb with the centre of the gutter. If no gutter is installed, overhang the starter course 1½ inches (38 mm) beyond the fascia. Space adjacent shingles between ¼ and ⅜ inch (6 and 9.5 mm) apart. Place each fastener about ¾ inch (19 mm) in from the edge of the shingle and not more than 1 inch (25 mm) above the exposure line. Drive fasteners flush with the surface. Make sure that the head does not crush the surface.

Apply another layer of shingles on top of the first layer of the starter course. The starter course may be tripled, if desired, for appearance. This procedure is recommended in regions of heavy snowfall.

Lay succeeding courses across the roof. Apply several courses at a time. Stagger the joints in adjacent courses at least 1½ inches (38 mm). There should be no joint in any three adjacent courses in alignment. Joints should not line up with the centreline of the heart of wood or any knots or defects. **Flat grain** or *slash grain* shingles wider than 8 inches (203 mm) should be split in two before fastening. Trim shingle edges with the hatchet to keep their butts in line (**Figure 56–11**).

After laying several courses, snap a chalk line to straighten the next course. Proceed shingling up the roof. When 3 or 4 feet (1 or 1.2 m) from the ridge, check the distance on both ends of the roof. Divide the distance as close as possible to the exposure used.

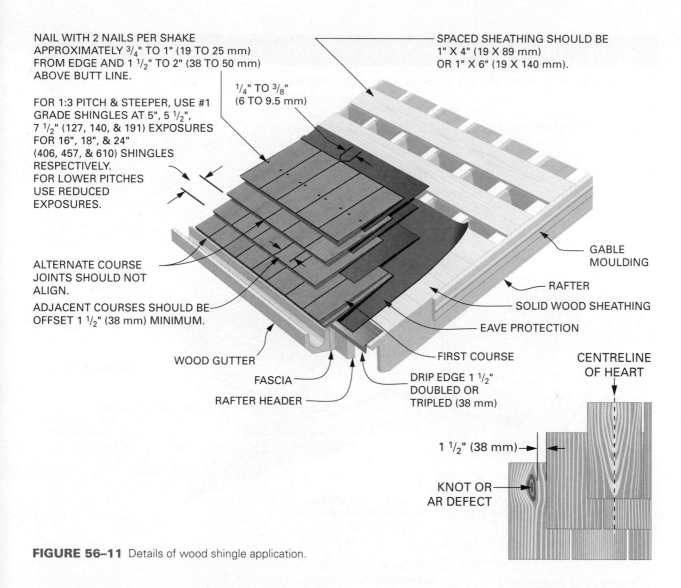

NAIL WITH 2 NAILS PER SHAKE APPROXIMATELY ³/₄" TO 1" (19 TO 25 mm) FROM EDGE AND 1 ¹/₂" TO 2" (38 TO 50 mm) ABOVE BUTT LINE.

FOR 1:3 PITCH & STEEPER, USE #1 GRADE SHINGLES AT 5", 5 ¹/₂", 7 ¹/₂" (127, 140, & 191) EXPOSURES FOR 16", 18", & 24" (406, 457, & 610) SHINGLES RESPECTIVELY. FOR LOWER PITCHES USE REDUCED EXPOSURES.

¹/₄" TO ³/₈" (6 TO 9.5 mm)

SPACED SHEATHING SHOULD BE 1" X 4" (19 X 89 mm) OR 1" X 6" (19 X 140 mm).

ALTERNATE COURSE JOINTS SHOULD NOT ALIGN.

ADJACENT COURSES SHOULD BE OFFSET 1 ¹/₂" (38 mm) MINIMUM.

GABLE MOULDING

RAFTER

SOLID WOOD SHEATHING

EAVE PROTECTION

WOOD GUTTER

FASCIA

RAFTER HEADER

FIRST COURSE

DRIP EDGE 1 ¹/₂" DOUBLED OR TRIPLED (38 mm)

CENTRELINE OF HEART

1 ¹/₂" (38 mm)

KNOT OR AR DEFECT

FIGURE 56–11 Details of wood shingle application.

A full course should show below the ridge cap. Shingle tips are cut flush with the ridge. Tips can be easily cut across the grain with a hatchet if cut on an angle to the side of the shingle.

On intersecting roofs, stop shingling the roof a few feet away from the valley. Select and cut a shingle at a proper taper. Apply it to the valley. Do not break joints in the valley. Do not lay shingles with their grain parallel to the valley centreline. Work back out. Fit a shingle to complete the course (**Procedure 56–A**).

Hips and Ridges

After the roof is shingled, 4- to 5-inch-wide (100 to 127 mm) hip and ridge caps are applied. Measure down on both sides of the hip or ridge at each end for a distance equal to the exposure. Snap a line between the marks.

Lay a shingle so that its bottom edge is to the line. Fasten with two nails. Use longer nails so that

they penetrate at least ¹/₂ inch (12 mm) into or through the sheathing. With the hatchet, trim the top edge flush with the opposite slope.

Lay another shingle on the other side with the butt even and its upper edge overlapping the first shingle laid. Trim its top edge at a bevel and flush with the side of the first shingle. Double this first set of shingles, alternating the joint. Apply succeeding layers of cap, with the same exposure as used on the roof. Alternate the overlap of each layer (**Figure 56–12**). Space the exposure when nearing the end so that all caps are about equally exposed.

APPLYING WOOD SHAKES

Shakes are applied in much the same manner as shingles (**Figure 56–13**). Mark the handle of the shingling hatchet at 7¹/₂ and 10 inches (191 and

StepbyStep Procedures

Procedure 56–A Applying Shingles or Shakes along an Open Valley

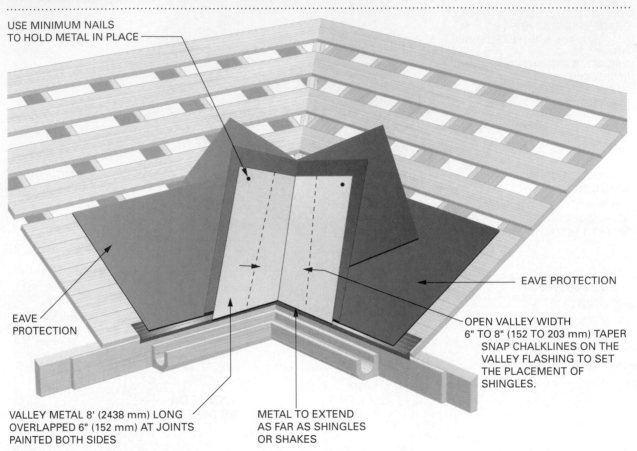

USE MINIMUM NAILS
TO HOLD METAL IN PLACE

EAVE PROTECTION

EAVE
PROTECTION

OPEN VALLEY WIDTH
6" TO 8" (152 TO 203 mm) TAPER
SNAP CHALKLINES ON THE
VALLEY FLASHING TO SET
THE PLACEMENT OF
SHINGLES.

VALLEY METAL 8' (2438 mm) LONG
OVERLAPPED 6" (152 mm) AT JOINTS
PAINTED BOTH SIDES

METAL TO EXTEND
AS FAR AS SHINGLES
OR SHAKES

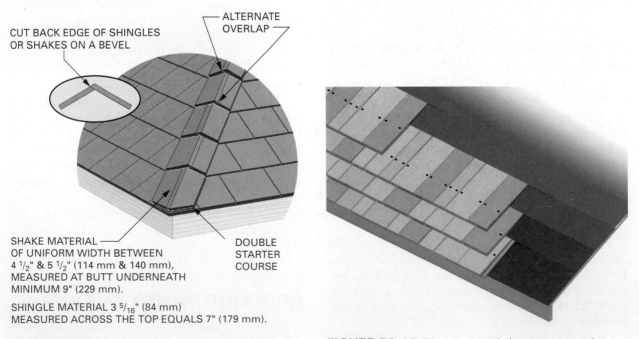

CUT BACK EDGE OF SHINGLES
OR SHAKES ON A BEVEL

ALTERNATE
OVERLAP

SHAKE MATERIAL
OF UNIFORM WIDTH BETWEEN
4 $\frac{1}{2}$" & 5 $\frac{1}{2}$" (114 mm & 140 mm),
MEASURED AT BUTT UNDERNEATH
MINIMUM 9" (229 mm).

SHINGLE MATERIAL 3 $\frac{5}{16}$" (84 mm)
MEASURED ACROSS THE TOP EQUALS 7" (179 mm).

DOUBLE
STARTER
COURSE

FIGURE 56–12 When applying hip and ridge shingles, alternate the overlap.

FIGURE 56–13 Three or four shake courses at a time are carried across the roof.

254 mm) from the top of the head. These are the exposures that are used most of the time when applying wood shakes.

Underlayment

A full width of felt underlayment is first applied under the starter course. Butts of the starter course should project 1½ inches (38 mm) beyond the fascia. Next, lay an 18-inch-wide (450 mm) strip of #30 roofing felt over the upper end of the starter course. Its bottom edge should be positioned at a distance equal to twice the shake exposure above the butt line of the first exposed course of shakes (**Figure 56–14**). For example, 24-inch (610 mm) shakes, laid with 10 inches (254 mm) of exposure, would have felt applied 20 inches (508 mm) above the butts of the shakes. The felt will cover the top 4 inches (100 mm) of the shakes. It will extend up to 14 inches (356 mm) on the sheathing. The top edge of the felt must rest on the spaced sheathing, if used.

Nail only the top edge of the felt. Fasten successive strips on their top edge only. Their bottom edges should be one shake exposure from the bottom of the previous strip. It is important to lay the felt straight. It serves as a guide for applying shakes. After the roof is felted, the tips of the shakes are tucked under the felt. The bottom should be exposed by the distance of twice the exposure.

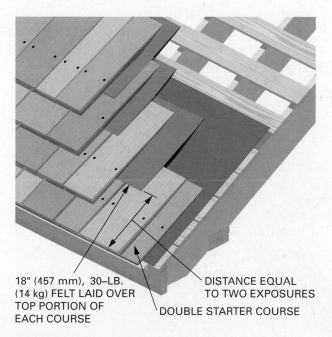

18" (457 mm), 30–LB. (14 kg) FELT LAID OVER TOP PORTION OF EACH COURSE

DISTANCE EQUAL TO TWO EXPOSURES

DOUBLE STARTER COURSE

FIGURE 56–14 An underlayment of felt is required when laying shakes.

Apply the second and successive courses with joints staggered and fasteners placed the same as for shingles. The spacing between shakes should be at least ⅜ inch (9.5 mm), but not more than ⅜ inch (16 mm) to allow wood to dry properly after rain. Lay straight-split shakes with their smooth end toward the ridge (**Figure 56–15**).

Maintaining Shake Exposure

There is a tendency to angle toward the eave. Therefore, check the exposure regularly with the hatchet handle. An easy way to be sure of correct exposure is to look through the joint between the edges of the shakes in the course below the one being nailed. The bottom edge of the felt will be visible. The butt of the shake being nailed is positioned directly above it (**Figure 56–16**).

Ridges and Hips

Adjust the exposure so that tips of shakes in the next-to-last course just come to the ridge. Use economical 15-inch (381 mm) starter-finish shakes for the last course. They save time by eliminating the need to trim shake tips at the ridge. Cap ridges and hips with shakes in the same manner as that used for wood shingles.

ESTIMATING ROOFING MATERIALS

Find the area of the roof in square feet. Divide the total by 100 to determine the number of squares needed. Add about 5 to 10 percent for waste. A simple roof with no dormers, valleys, or other obstructions requires less allowance for waste. A complicated roof requires more.

For wood shingles, add one square for every 240 linear feet of starter course. Add one square of shakes for 120 linear feet of starter course.

Add one extra square of shingles for every 100 linear feet of valleys and about two squares for shakes.

Add an extra bundle of shakes or shingles for every 16 feet (5 m) of hip and ridge to be covered.

Figure 2 pounds (0.9 kg) of nails per square at standard exposure.

Remember that a square of roofing will cover 100 square feet (9.3 m²) of roof surface only when applied at standard exposures. Allow proportionally more material when these exposures are reduced.

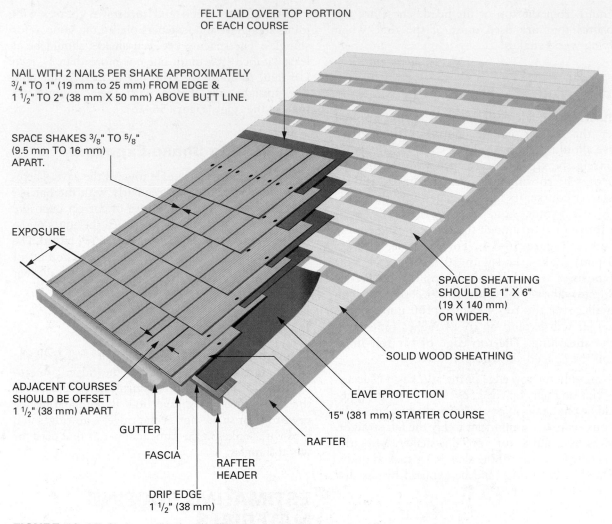

FELT LAID OVER TOP PORTION
OF EACH COURSE

NAIL WITH 2 NAILS PER SHAKE APPROXIMATELY
3/4" TO 1" (19 mm to 25 mm) FROM EDGE &
1 1/2" TO 2" (38 mm X 50 mm) ABOVE BUTT LINE.

SPACE SHAKES 3/8" TO 5/8"
(9.5 mm TO 16 mm)
APART.

EXPOSURE

SPACED SHEATHING
SHOULD BE 1" X 6"
(19 X 140 mm)
OR WIDER.

SOLID WOOD SHEATHING

ADJACENT COURSES
SHOULD BE OFFSET
1 1/2" (38 mm) APART

EAVE PROTECTION

15" (381 mm) STARTER COURSE

GUTTER

FASCIA

RAFTER

DRIP EDGE
1 1/2" (38 mm)

RAFTER
HEADER

FIGURE 56–15 Shake application details.

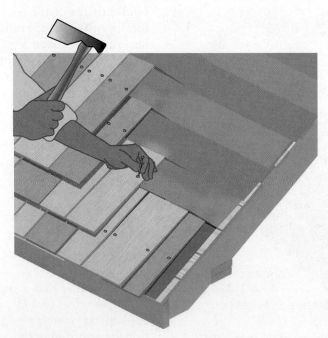

FIGURE 56–16 The tip of the shake is inserted between
the layers of underlayment.

Chapter 57
Flashing

Flashing is a material used in various locations that are susceptible to leaking. It prevents water from entering a building (**Figure 57–1**). The words *flash*, *flashed*, and *flashing* are also used as verbs to describe the installation of the material. Various kinds of flashing are applied at the eaves, valleys, chimneys, vents, and other roof projections. They prevent leakage at the intersections.

KINDS OF FLASHING

Flashing material may be sheet copper, zinc, aluminum, galvanized steel, vinyl, or mineral-surfaced asphalt roll roofing. Copper and zinc are high-quality flashing materials, but they are more expensive. Roll roofing is less expensive. Colours that match or contrast with the roof covering can be used. If properly applied, roll roofing used as a valley flashing will outlast the main roof covering. Sheet metal, especially copper, may last longer. However, it is good practice to replace all flashing when reroofing.

Eaves Flashing

Whenever there is a possibility of ice dams forming along the eaves and causing a backup of water, **ice and water shield flashing** is needed. Apply the flashing such that it overhangs the drip edge by ¼ to ⅜ inch (6 to 9.5 mm). The flashing should extend up the roof far enough to cover a point at least 12 inches (305 mm) inside the wall line of the building. If the overhang of the eaves requires that the flashing be wider than 36 inches (914 mm), the necessary horizontal lap joint is located on the portion of the roof that extends outside the wall line (**Figure 57–2**).

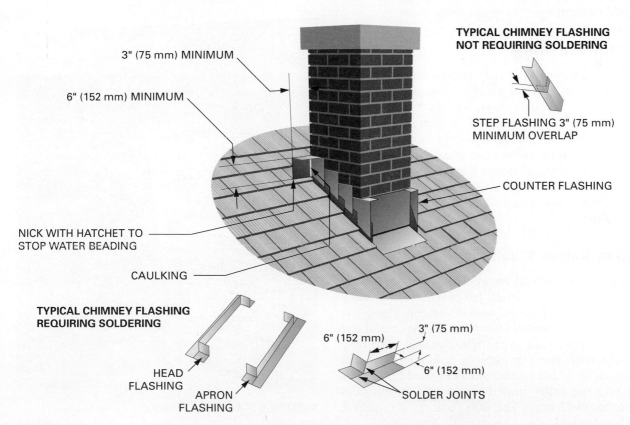

3" (75 mm) MINIMUM

6" (152 mm) MINIMUM

TYPICAL CHIMNEY FLASHING NOT REQUIRING SOLDERING

STEP FLASHING 3" (75 mm) MINIMUM OVERLAP

COUNTER FLASHING

NICK WITH HATCHET TO STOP WATER BEADING

CAULKING

TYPICAL CHIMNEY FLASHING REQUIRING SOLDERING

HEAD FLASHING

APRON FLASHING

6" (152 mm)

3" (75 mm)

6" (152 mm)

SOLDER JOINTS

FIGURE 57–1 Flashings are used to seal against leakage where the roofing butts against adjoining surfaces.

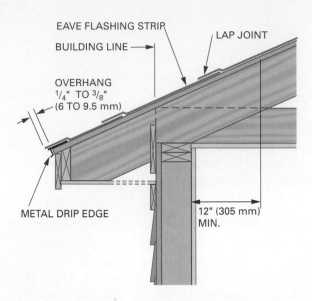

FIGURE 57–2 If there is danger of ice dams forming along the eaves, eave flashing is installed with seams away from the building line.

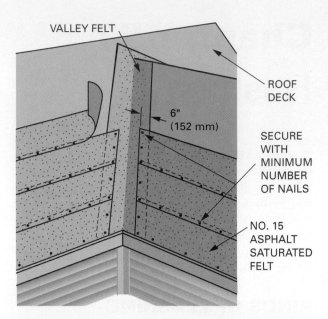

FIGURE 57–3 Felt underlayment, or preferably a peel-and-stick membrane, is applied in the valley before roof underlayment.

For a slope of at least 4 inches (1:3) rise per foot of run, install a single course of 36-inch (914 mm) wide flashing covering the underlayment and drip edge. On lower slopes, greater protection against water leakage is gained by applying ice and water shield flashing over the entire roof.

Valley Flashing

Roof valleys are especially vulnerable to leakage. This is because of the great volume of water that flows down through them. Valleys must be carefully flashed according to recommended procedures. Valleys are flashed in two ways: *open* or *closed*. **Open valleys** are constructed with no shingle or roofing material installed within several inches (75 mm) of the valley centre. **Closed valleys** have the roofing material covering the entire valley centreline.

Open Valley Flashing

Begin valley flashing by applying a 36-inch-wide (914 mm) strip of asphalt felt centred in the valley (**Figure 57–3**). Fold or crease the roll along the length and seat it well into the valley. Be careful not to cause any break in the felt. Fasten it with only enough nails along its edges to hold it in place. Let the courses of felt underlayment applied to the roof overlap the valley underlayment by not less than 6 inches (152 mm). The eave flashing, if required, is then applied.

Using Roll Roofing Flashing

Lay an 18-inch-wide (457 mm) layer of mineral-surfaced roll roofing centred in the valley (**Figure 57–4**). Its mineral-surfaced side should be down. Use only enough nails spaced 1 inch (25 mm)

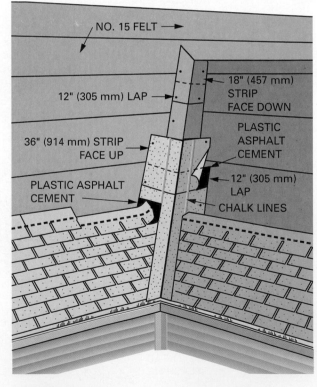

FIGURE 57–4 Method of applying roll roofing open valley flashing.

in from each edge to hold the strip smoothly in place. Press the roofing firmly in the centre of the valley when nailing the opposite edge. Next, lay a 36-inch-wide (914 mm) strip with its surfaced side up. Centre it in the valley. Fasten it in the same manner as the first strip, pressing firmly into the valley centre.

Snap a chalk line on each side of the valley. Place full shingles along the lines with the straight edge facing the centre of the valley. This will help to maintain a straight valley line and will also act as a guide for trimming the ends of the shingle courses. These lines will be 6 inches (152 mm) apart at the ridge. To allow for the flow of water and ice, these lines diverge at a rate of $\frac{1}{8}''$ per foot (1:100 mm) as they approach the eave. Thus, in a valley that is 16 feet (5 m) long, the lines will be 8 inches (200 mm) apart at the eaves.

The upper corner of each end asphalt shingle is clipped. This clip helps keep water from entering between the courses. Each roof shingle is cemented to the valley flashing with plastic asphalt cement.

Metal Flashing

Prepare the valley with ice and water shield in the same manner as described previously. Next, lay a strip of sheet metal flashing centred in the valley (**Figure 57–5**). The metal should extend at least 10 inches (154 mm) on each side of the valley centreline for slopes with a 6-inch (1:2) rise or less and

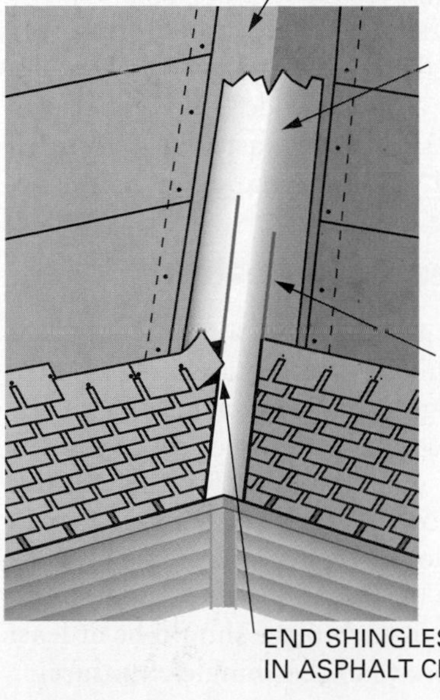

36" (914 mm) ICE AND WATER SHIELD CENTRED IN VALLEY AND OVERLAPPED BY THE ROOF UNDERLAYMENT

METAL FLASHING APPLIED 10" (254 mm) ON EACH SIDE OF CENTRE ON ROOFS WITH 6" 1:2 RISE OR LESS. APPLY 7" (178 mm) EACH SIDE OF CENTRE ON STEEPER ROOFS.

CHALK LINES TO GUIDE SHINGLES

END SHINGLES ARE BEDDED IN ASPHALT CEMENT

FIGURE 57–5 Method of applying metal open valley flashing.

7 inches (178 mm) on each side for a steeper slope. Carefully press and form it into the valley. Fasten the metal with nails of similar material spaced close to its outside edges. Use only enough fasteners to hold it smoothly in place.

Snap lines on each side of the valley, as described previously. Use them as guides for cutting the ends of the shingle courses. Trim the last shingle of each course to fit on the chalk line. Clip 1 inch (25 mm) from its upper corner at a 45-degree angle. To form a tight seal, cement the shingle to the metal flashing with a 3-inch (75 mm) width of asphalt plastic cement.

If a valley is formed by the intersection of a low-pitched roof and a much steeper one, a 1-inch-high (25 mm), crimped standing seam should be made in the centre of the metal flashing of an open valley. The seam will keep heavy rain water flowing down the steeper roof from overrunning the valley and possibly being forced under the shingles of the lower slope.

Closed Valley Flashing

A closed valley protects the valley flashing. It thus adds to the weather resistance at vulnerable points. Several methods are used to flash closed valleys.

The first step for any method is to apply the asphalt felt underlayment as previously described for open valleys. Then, centre a 36-inch (914 mm) width of smooth or mineral-surface roll roofing, 50-pound per square or heavier (type "S" or type "M"), in the valley over the underlayment. Form it smoothly in the valley. Secure it with only as many nails as necessary. Another method uses a strip of wide metal flashing on top of the felt underlayment.

Woven Valley Method

Valleys may be flashed by applying asphalt shingles on both sides of the valley and alternately weaving each course over and across the valley. This is called a **woven valley** (**Figure 57–6**).

Lay the first course of shingles along the edge of one roof up to and over the valley for a distance of at least 12 inches (305 mm). Lay the first course along the edge of the adjacent roof. Extend the shingles over the valley on top of the previously applied shingles (**Figure 57–7**).

Succeeding courses are then applied. Weave the valley shingles alternately, first on one roof and then on the other. When weaving the shingles, make sure they are pressed tightly into the valley. Also make sure that no nail is closer than 6 inches (152 mm) to the valley centreline. Fasten the end of the woven shingle with two nails. Most carpenters prefer to

FIGURE 57–6 Valleys are sometimes flashed by weaving shingles together.

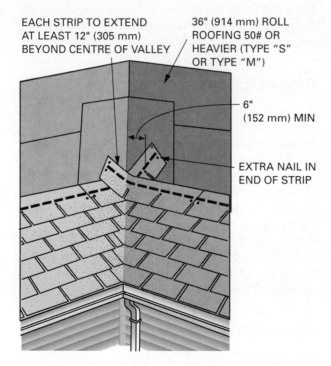

EACH STRIP TO EXTEND AT LEAST 12" (305 mm) BEYOND CENTRE OF VALLEY

36" (914 mm) ROLL ROOFING 50# OR HEAVIER (TYPE "S" OR TYPE "M")

6" (152 mm) MIN

EXTRA NAIL IN END OF STRIP

FIGURE 57–7 Details for applying a woven valley.

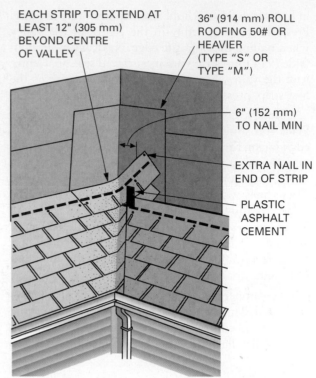

EACH STRIP TO EXTEND AT LEAST 12" (305 mm) BEYOND CENTRE OF VALLEY

36" (914 mm) ROLL ROOFING 50# OR HEAVIER (TYPE "S" OR TYPE "M")

6" (152 mm) TO NAIL MIN

EXTRA NAIL IN END OF STRIP

PLASTIC ASPHALT CEMENT

FIGURE 57–8 Details for applying a closed cut valley.

cover each roof area with shingles to a point approximately 3 feet (1 m) from the valley. They weave the valley shingles in place later.

No end joints should occur within 6 inches (152 mm) of the centre of the valley. Therefore, it may be necessary to occasionally cut a strip short that would otherwise end near the centre. Continue from this cut end with a full-length strip over the valley.

Closed Cut Valley Method

Apply the shingles to one roof area. Let the end shingle of every course overlap the valley by at least 12 inches (305 mm) (**Figure 57–8**). Make sure no

end joints occur within 6 inches (152 mm) of the centre of the valley. Place fasteners no closer than 6 inches (152 mm) from the centre of the valley. Form the end shingle of each course snugly in the valley. Secure its end with two fasteners.

Snap a chalk line 2 inches (50 mm) short of and parallel to the centre of the valley on top of the overlapping shingles. Apply shingles to the adjacent roof area. Cut the end shingle to the chalk line. Clip the upper corner of each shingle as described previously for open valleys. Bed the end of each shingle that lies in the valley in about a 3-inch-wide (75 mm) strip of asphalt cement. Make sure that no fasteners are located closer than 6 inches (152 mm) to the valley centreline.

Step Valley Flashing Method

Step flashings are individual metal pieces tucked between courses of shingles (see Figure 57–1). When applying step flashings in valleys, first estimate the number of shingle courses required to reach the ridge. Cut a piece of metal flashing for each course of shingles. Each piece should be at least 18 inches (457 mm) wide for slopes with a 6-inch rise (1:2) or greater and 24 inches (610 mm) wide for slopes with less pitch. The height of each piece should be at least 3 inches (75 mm) more than the shingle exposure.

Prepare the valley with underlayment as described previously. Snap a chalk line in the centre of the valley.

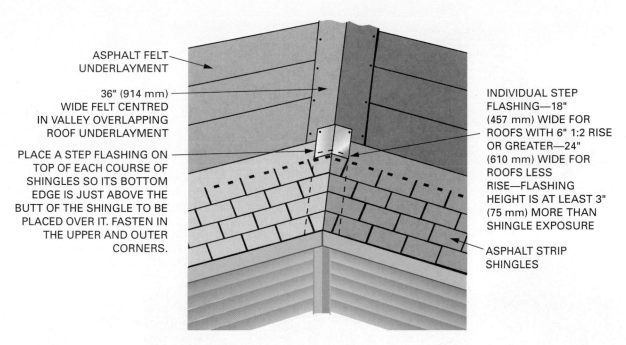

ASPHALT FELT
UNDERLAYMENT

36" (914 mm)
WIDE FELT CENTRED
IN VALLEY OVERLAPPING
ROOF UNDERLAYMENT

PLACE A STEP FLASHING ON
TOP OF EACH COURSE OF
SHINGLES SO ITS BOTTOM
EDGE IS JUST ABOVE THE
BUTT OF THE SHINGLE TO BE
PLACED OVER IT. FASTEN IN
THE UPPER AND OUTER
CORNERS.

INDIVIDUAL STEP
FLASHING—18"
(457 mm) WIDE FOR
ROOFS WITH 6" 1:2 RISE
OR GREATER—24"
(610 mm) WIDE FOR
ROOFS LESS
RISE—FLASHING
HEIGHT IS AT LEAST 3"
(75 mm) MORE THAN
SHINGLE EXPOSURE

ASPHALT STRIP
SHINGLES

FIGURE 57–9 Details for applying metal step flashings in a valley.

Apply the starter course on both roofs. Trim the end shingle of each course to the chalk line. Fit and form the first piece of flashing to the valley. Trim its bottom edge flush with the drip edge. Fasten it in the valley over the first layer of the starter course. Use fasteners of like material to prevent electrolysis. Fasten the upper corners of the flashing only.

Apply the first regular course of shingles to both roofs on each side of the valley. Trim the valley shingles so that their ends lay on the chalk line. Bed them in plastic asphalt cement. Do not drive nails through the metal flashing.

Apply the next piece of flashing in the valley over the first course of shingles. Keep its bottom edge about ½ inch (12 mm) above the butts of the next course of shingles. Apply the second course of shingles in the same manner as the first. Secure a flashing over the second course. Apply succeeding courses and flashings in this manner (**Figure 57–9**). Remember that flashing is placed over each course of shingles. Do not leave any flashings out. When the valley is completely flashed, no metal flashing surface is exposed. If the valley does not extend all the way to the ridge of the main roof, a *saddle* is applied over the ridge of the minor roof (**Figure 57–10**).

Flashing against a Wall

When roof shingles butt up against a vertical wall, the joint must be made watertight. The usual method

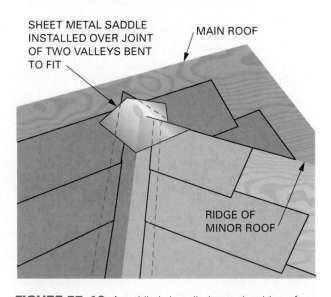

SHEET METAL SADDLE
INSTALLED OVER JOINT
OF TWO VALLEYS BENT
TO FIT

MAIN ROOF

RIDGE OF
MINOR ROOF

FIGURE 57–10 A saddle is installed over the ridge of a minor roof where it intersects with the main roof.

of making the joint tight is with the use of metal step flashings (**Figure 57–11**).

The flashings are purchased or cut about 8 inches (203 mm) in width. They are bent at right angles in the centre so that they will lay about 4 inches (100 mm) on the roof and extend about 4 inches (100 mm) up the sidewall. The length of the flashings is about 3 inches (75 mm) more than the exposure of the shingles. When used with shingles exposed 5 inches (127 mm) to the weather,

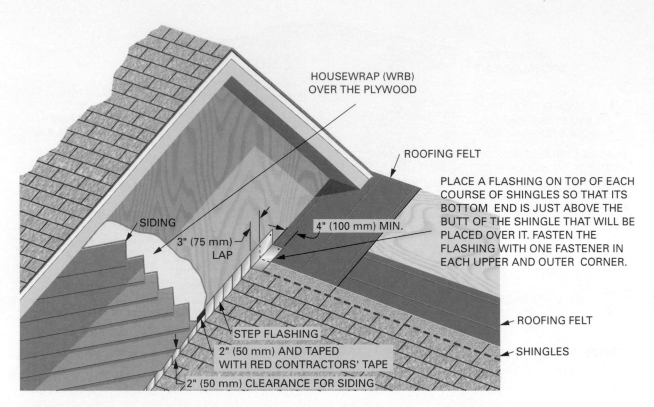

HOUSEWRAP (WRB) OVER THE PLYWOOD

ROOFING FELT

PLACE A FLASHING ON TOP OF EACH COURSE OF SHINGLES SO THAT ITS BOTTOM END IS JUST ABOVE THE BUTT OF THE SHINGLE THAT WILL BE PLACED OVER IT. FASTEN THE FLASHING WITH ONE FASTENER IN EACH UPPER AND OUTER CORNER.

SIDING

3" (75 mm) LAP

4" (100 mm) MIN.

ROOFING FELT

SHINGLES

STEP FLASHING

2" (50 mm) AND TAPED WITH RED CONTRACTORS' TAPE

2" (50 mm) CLEARANCE FOR SIDING

FIGURE 57–11 Using metal step flashing where a roof butts a wall.

they are made 8 inches (203 mm) in length. Cut and bend the necessary number of metal flashings.

The roofing is applied and flashed before the siding is applied to the vertical wall. First, apply an underlayment of asphalt felt to the roof deck. Turn the ends up on the vertical wall by about 3 to 4 inches (75 to 100 mm).

Apply the first layer of the starter course, working toward the vertical wall. Fasten a metal flashing on top of the starter course of shingles. Its bottom edge should be flush with the drip edge. Use one fastener in each top corner. Lay the first regular course with its end shingle over the flashing and against the sheathing of the sidewall. Do not drive any fasteners through the flashings. It is usually not necessary to bed the shingles to the flashings with asphalt cement. The step flashing holds down the end of the shingle below it.

Apply a flashing over the upper side of the first course and against the wall. Keep its bottom edge at a point that will be about ½ inch (12 mm) above the butt of the next course of shingles. Continue applying shingles and flashings in this manner until the ridge is reached. Some carpenters prefer to cut the shingles back if a tab cutout occurs over a flashing. This prevents metal from being exposed to view.

In the case where the roof ends but the wall continues on, it is important to install a **kick-out flashing**

or **water diverter** in order to deflect the water from the roof away from the sided wall (**Figure 57–12**). If the water gets behind the siding, significant wood rot will take place (**Figure 57–13**).

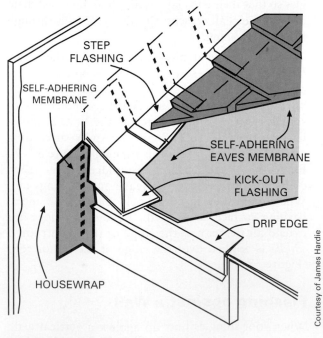

STEP FLASHING

SELF-ADHERING MEMBRANE

SELF-ADHERING EAVES MEMBRANE

KICK-OUT FLASHING

DRIP EDGE

HOUSEWRAP

Courtesy of James Hardie

FIGURE 57–12 Kickout flashing.

FIGURE 57-13 Preventing water damage by using a diverter.

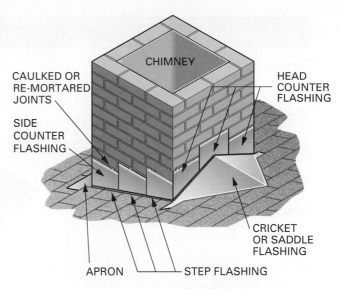

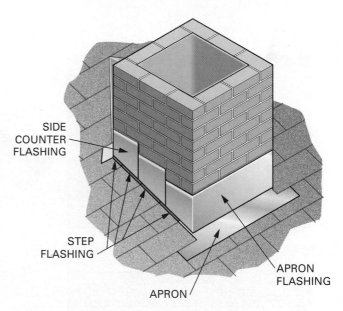

FIGURE 57-14 Chimneys are waterproofed with a series of layers of flashing.

Flashing a Chimney

Chimneys and other large protrusions through the roof, such as skylights, must be installed so that water and weather do not enter the building. Chimneys are a greater challenge to weather proofing because their height or length changes over the course of a year due to the masonry expanding and contracting from normal operations. A hot chimney (e.g., from a running furnace) is taller than a cool one. The change in distance for tall chimneys can be measured in inches. To accomplish waterproofing, a combination of flashing layers is used.

In many cases, especially on steep-sloped roof and wide chimneys, a cricket or saddle is built behind the chimney between the upper side of chimney and roof deck. The cricket prevents water and debris from accumulating behind the chimney (**Figure 57-14**). **Counter flashing** is installed by *masons* who build the chimney in conjunction with the step flashing that the carpenters apply. The top edge of the counter flashing is bent at a right angle at about ½ inch (12 mm) to be inserted into a mortar joint. This edge is then sealed in the joint with **mortar** or caulk. The lower edge of the counter flashing covers the step flashing that is installed on each course of shingles along the sides of the chimney. Flashing of a chimney may be installed while the roof is being shingled or after the roofing is complete. The mess of building a chimney is easier to clean up before shingles are installed. Bending metal flashing is best done on a brake, which is discussed in Chapter 66.

Underlayment is applied to the roof deck and tucked under any existing flashings. The shingle courses are brought up to the chimney. They are applied under the flashing on the lower side of the chimney. The top edges of the shingles are cut as necessary, until only a shingle exposure shows below (downhill from) the chimney. The **apron flashing** is applied on top of the lower course of shingles. Two cuts are made in it to allow a portion of the flashing to be bent vertically (**Procedure 57-A**). The lower edge of the apron flashing is pressed into a bed of asphalt or plastic cement on top of the shingles. The overall width should be about 12 inches (305 mm) and about 6 inches (152 mm) longer on both sides of the chimney. Where the

StepbyStep Procedures

Procedure 57–A Flashing a Chimney

STEP 1 APPLY UNDERLAYMENT TO THE ROOF DECK AND TUCK UNDER ANY EXISTING FLASHINGS. INSTALL SHINGLES UP TO CHIMNEY AND CUT THE TOP EDGES OF THE SHINGLES AROUND CHIMNEY.

STEP 2 PRECUT AND BEND APRON FLASHING, THEN INSTALL IT BY ADHERING WITH MASTIC TO THE SHINGLES. PLACE ONE NAIL INTO THE ROOF DECK THROUGH EACH OF THE UPPER, OUTERMOST CORNERS OF THE FLASHING.

STEP 3 INSTALL FIRST STEP FLASHING WITH A BEND THAT WRAPS TO THE FRONT SURFACE OF CHIMNEY. APPLY A SMALL AMOUNT OF MASTIC TO THE FOLDED EDGE OF STEP FLASHING TO ADHERE IT TO THE APRON AND NOT THE CHIMNEY.

STEP 4 INSTALL SHINGLES AND STEP FLASHINGS ALONG THE SIDE OF THE CHIMNEY. NAIL STEP FLASHING TO ROOF SHEATHING WITH ONE NAIL IN THE UPPER CORNER.

STEP 5 PRECUT AND BEND APRON, CUTTING IT LONG ENOUGH TO WRAP THE SIDES OF THE CHIMNEY ABOUT 4" (100 mm). ATTACH BY ADHERING THE VERTICAL SURFACES TO CHIMNEY BUT NOT THE STEP FLASHING.

STEP 6 INSTALL SIDE COUNTER FLASHING, BEGINNING WITH LOWER ONE, OVERLAPPING EACH PIECE ABOUT 3" (75 mm). POSTPONE LAST (UPPER) PIECE OF SIDE COUNTER FLASHING.

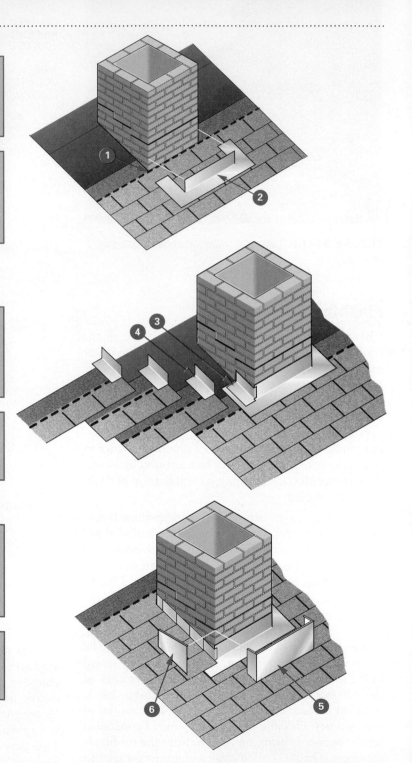

StepbyStep Procedures

Procedure 57–A (*Continued*)

STEP 7 INSTALL LAST SIDEWALL STEP FLASHING BY FITTING AROUND THE CORNER OF THE CHIMNEY.

STEP 8 INSTALL ONE STEP FLASHING ON THE BACKSIDE OF THE CHIMNEY WITH A PORTION OF IT BENT AROUND THE CORNER ABOUT 1" (25 mm). APPLY A LAYER OF MASTIC TO ADHERE THE STEP FLASHING TO THE PREVIOUS ONE.

STEP 9 CUT AND FIT CRICKET FROM PLYWOOD AND LUMBER AS NEEDED.

STEP 10 CUT AND FIT HEAD FLASHING TO COVER CRICKET WOOD FRAME. BEND IT UP THE BACKSIDE OF THE CHIMNEY ABOUT 4" (100 mm). ALSO ALLOW ABOUT 6" (150 mm) OR MORE TO RUN UP THE ROOF PAST THE VALLEY. ATTACH BY ADHERING FLASHING TO THE CRICKET WOOD FRAME AND ROOF DECK, BUT NOT TO THE CHIMNEY. LOWER END SHOULD BE ON THE SHINGLES.

STEP 11 ATTACH A SMALL FLASHING PIECE TO THE HEAD FLASHING VERTICAL SURFACE TO COVER THE V-SHAPED GAP. ADHERE TO FLASHING AND NOT CHIMNEY.

STEP 12 CUT, FIT, AND INSTALL LAST SIDE WALL COUNTER FLASHING TO WRAP CHIMNEY CORNER ABOUT 4" (100 mm).

STEP 13 INSTALL FIRST HEAD COUNTER TO COVER PREVIOUS FLASHING. ATTACH BY ADHERING VERTICAL SURFACE TO CHIMNEY WITH MASTIC.

STEP 14 INSTALL THE CENTRE-COUNTER FLASHING AT THE CRICKET RIDGE. MAKE SURE MASTIC ADHERES TO CHIMNEY ONLY.

STEP 15 INSTALL SHINGLES UP THE ROOF. INSTALL A SADDLE FLASHING ON TOP OF CRICKET AND UNDER NEAREST SHINGLE LAYER.

STEP 16 REPOINT THE MORTAR JOINTS.

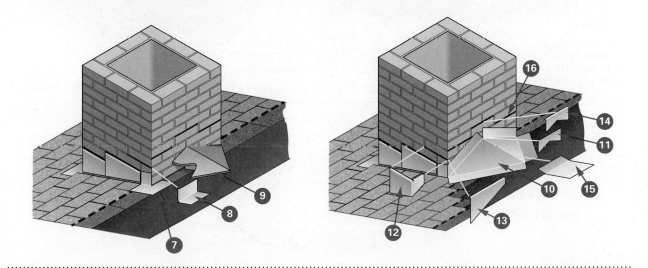

bend is made is not critical; the vertical bend or the surface touching the roof should be at least 4 inches (100 mm).

The first piece of step flashing is installed on top of the apron flashing with its end trimmed so that about 1 inch (25 mm) wraps around to the chimney's front surface. Step flashings are then installed by tucking them in between each shingle course and against the apron flashing or chimney as they progress up the roof. Step flashings are typically 8 × 8 inch (200 × 200 mm) pieces folded in half at a right angle. This is a similar process as in flashing against a wall. Nails are carefully placed into the roof sheathing along the flashing edge as far away from the chimney as possible.

The **apron** is counter flashing bent to fit in the mortar joint. It is also cut long enough to wrap around and up the sides of the chimney about 4 inches (100 mm) (**Figure 57–15**). **Side counter flashing**, also installed into mortar joints, overlaps the previous flashing by about 3 inches (75 mm). The mid and lower portions of the counter flashing

are bedded to the chimney with asphalt cement. It is very important not to attach the counter flashing to the step flashing. Remember, anything attached to the chimney will move up and down while anything attached to the roof will not move.

The last piece of sidewall step flashing is cut and fitted around the upper corners of the chimney. One more piece of step flashing is installed on the backside of the chimney and has a portion of it bent around to the side covering the last sidewall piece by about 1 inch (25 mm). Apply a layer of mastic to adhere the step flashings to themselves.

A cricket or saddle is typically built from plywood and lumber as needed for strength. Wide chimneys require more wood than smaller ones. The slope of the cricket is not critical; it needs only enough slope to adequately shed water. It looks better if it is sloped the same as the roof. The cricket subframe is installed on top of the last step flashing previously applied to the back side of the chimney. Large crickets may be covered with shingles creating

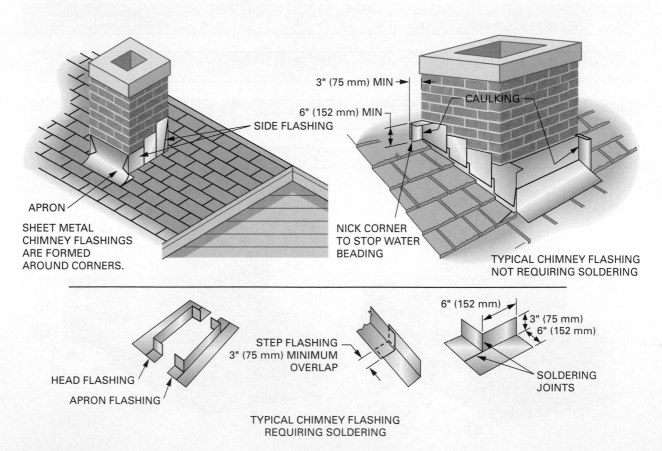

FIGURE 57–15 Chimney flashings wrap the entire chimney in overlapping pieces.

a small valley section. These courses of shingles are step flashed to the backside of the chimney. Smaller crickets are totally covered with flashing material. This flashing is called **head flashing**.

Head flashing covers the cricket and is bent up the backside of the chimney. It extends into the small valley and up the main roof far enough for the roof main shingles to cover it by at least 6 inches (152 mm). It also extends down to the valley past the cricket sub-frame by at least 6 inches (152 mm). It is cemented to the cricket sub-frame and roof deck, but not to the chimney. The lower edge of the cricket must be on top of shingles. The upper course of underlayment lies on top of the valley portion of the cricket.

Attach a flashing piece to cover the V-shaped gap in the vertical surfaces of the head flashing where the brick is exposed. It should be attached with mastic to the flashing only, not the chimney.

The head counter flashing is installed into the mortar joint and down to cover the cricket's vertical surfaces. It begins at the chimney corners with the last piece of side counter flashing that wraps the corner to the backside by about 4 inches (100 mm). The flashing process is completed when the centre counter flashing is installed at the cricket's ridge. Vertical surfaces are attached to the chimney with mastic.

Shingles are cemented to the head flashing and saddle flashing. A saddle flashing is installed over the upper-most portion of the cricket.

Working with metal flashing can be a little tricky to get right the first time; be patient. Gentle tapping with a hammer can help move and shape the metal. Chimneys may be flashed by methods and materials other than described previously, depending on the custom of certain geographical areas. The major thought to keep in mind when flashing is that water runs downhill. Every joint or seam should overlap toward the downhill side. Lapping waterproofing materials always begins at the lowest portion of the roof.

Flashing Vents

Flashings for round pipes, such as *stack vents* for plumbing systems and *roof ventilators*, usually come as *flashing collars* made for various roof pitches. They fit around the stack. They have a wide flange on the bottom that rests on the roof deck. The flashing is installed over the stack vent, with its flange on the

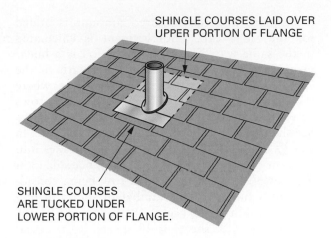

FIGURE 57–16 A vent stack flashing boot.

roof sheathing. It is fastened in place with one fastener in each upper corner.

Shingle up to the lower end of the stack vent flashing. Lift the lower part of the flange. Apply shingle courses under it. Cut the top edge of the shingles, where necessary, until the shingle exposure, or less, shows below the lower edge of the flashing. Apply asphalt cement under the lower end of the flashing. Press it into place on top of the shingle courses.

Apply shingles around the stack and over the flange. Do not drive nails through the flashing. Bed shingles to the flashing with asphalt cement, where necessary (**Figure 57–16**).

ESTIMATING ROOFING MATERIALS

Estimating the amount of material needed for roofing a building can start with the number of sheets of plywood that were required to deck the roof. One bundle of standard asphalt shingles cover 1 sheet (4 × 8′, or 1220 × 2440 mm) of plywood. So simply estimate 1 bundle per sheet plus 1 bundle for every 30 feet (9 m) of ridge.

Roof Area

The estimate starts with the roof area. The entire roof area is calculated in square feet (ft²) and then divided by 100 to get the number of squares. *Note that a hip roof has the same area as the equivalent sized gable roof.* The number of squares are then multiplied by the number of bundles of shingles per square and the required amount is added for the ridge cap. Underlayment is calculated in the same way. The eave protection is a linear measure that will depend on the length of the eaves and the eave projection. In cases where the eave projection is 2 feet (610 mm), two rows of eave protection are required along the eaves. See the table to the right for a sample estimate. In this example, the plywood application takes into account that there will be four rows per side (3 at 4′, 1 at 3′-3″) with the fourth row being considered as a full row. The number of sheets of plywood would then be $(16 \times 48' \times 2) \div 32$ ft², the area of a $4 \times 8'$ sheet, or 48 sheets. If the actual roof area is used $(48 \times 15.25' \times 2) \div 32$ ft², then the answer would be 45.75 or 46 sheets, which would not work well.

Gable Roof, 48′ Ridge, 15′-3″ Sloped Length, 2′ Eave Projection		
Item	**Calculations**	**Estimate**
Eave protection	48′ × 2 rows × 2 sides (÷ 65 ft/roll)	~200 ft or ~3 rolls
Eave protection coverage	48 × 5.5′ (2 rows) × 2	528 ft²
Roof area	48 × 15.25′ × 2 sides	1464 ft²
Number of squares	Roof area ÷ 100	14.64 squares
Underlayment	1464 ft² − 528 ft² = 936 ft²	936 ÷ 400 = 3 rolls
Asphalt shingles (3 bundles per square for 3-tab)	14.64 × 3 = 43.92 + 2 (for the ridge cap) = 45.92 = ~46	46 bundles
Wood shingles (7 bundles per square)	14.64 × 7 = 102.48 + 1 bundle per 240′ of eave, 96′ ÷ 240 = 0.4 + 1 bundle per 16′ of ridge, 48′ ÷ 16 = 3	102.48 + 0.4 + 3 = 105.88 = 106 bundles

KEY TERMS

apron

apron flashing

asphalt felt

asphalt shingles

cements

closed valleys

coatings

counter flashing

courses

coverage

cricket

deck

drip edge

end lap

exposure

flashing

flat grain

head flashing

head lap

ice and water shield flashing

kick-out flashing

mortar

open valleys

plastic felt

roll roofing

saddle

selvage

shakes

shingle butt

shingling hatchet

side counter flashing

sliding gauge

solid sheathing

spaced sheathing

square

starter course

step flashing

top lap

underlayment

voltaic couple

water diverter

wood shingles

woven valley

REVIEW QUESTIONS

Select the most appropriate answer.

1. How much area will a square of roofing cover?

   a. 1 square foot
   b. 100 square feet
   c. 150 square feet
   d. 200 square feet

2. How much area will one roll of #15 asphalt felt cover?

   a. 1 square
   b. 2 squares
   c. 3 squares
   d. 4 squares

3. When applying asphalt felt on a roof deck as under-layment, how much should each course lap over the lower course?

   a. at least 2 inches (50 mm)
   b. at least 4 inches (100 mm)
   c. at least 6 inches (152 mm)
   d. at least 12 inches (305 mm)

4. When laying out three tab shingles "on the thirds," how much shorter should the starter tabs be?

   a. 3 inches (75 mm) shorter
   b. 4 inches (100 mm) shorter
   c. 6 inches (152 mm) shorter
   d. 12 inches (305 mm) shorter

5. Where should nailing for asphalt shingles be located in relation to the self-sealing strip?

   a. above the self-sealing strip
   b. on the self-sealing strip
   c. below the self-sealing strip
   d. above or below the self-sealing strip

6. At what distance from the ridge should the horizontal shingle course be checked and adjusted for alignment?

   a. 1 to 2 feet (305 to 610 mm)
   b. 3 to 4 feet (914 to 1220 mm)
   c. 5 to 6 feet (1524 to 1829 mm)
   d. 7 to 8 feet (2134 to 2438 mm)

7. What is the term selvage used to describe?

   a. recycling building materials
   b. the portion of roofing material that is not to be exposed
   c. leftover unused roofing material
   d. the portion of roofing material that is exposed

8. What are most wood shingles and shakes made from?

   a. cypress
   b. redwood
   c. eastern white cedar
   d. western red cedar

9. What is the longest available length of wood shingles and shakes?

   a. 16 inches (406 mm)
   b. 18 inches (457 mm)
   c. 24 inches (610 mm)
   d. 28 inches (711 mm)

10. How much do wood shingles normally overhang the fascia?

    a. $3/_8$ inch (9.5 mm)
    b. 1 inch (25 mm)
    c. 1½ inches (38 mm)
    d. 2 inches (50 mm)

11. Why is it important to lay underlayment straight prior to applying shakes?

    a. so there are no wrinkles in the felt
    b. to obtain the proper lap
    c. to improve its looks after the roof is completed
    d. because the felt serves as an installation guide for shingles

12. What is the first step in installation of roll roof valley flashing?

    a. installing an 18-inch-wide (457 mm) flashing piece face up
    b. installing an 18-inch-wide (457 mm) flashing piece face down
    c. installing a full-width flashing piece face up
    d. installing a full-width underlayment

13. When nailing shingles, how far away from the valley centreline should nails be placed?

    a. at least 6 inches (152 mm)
    b. at least 8 inches (203 mm)
    c. at least 10 inches (254 mm)
    d. at least 12 inches (305 mm)

14. Step flashings used against a vertical wall are cut about 8 inches (203 mm) wide. They should be bent so that how much lies on the wall and how much lies on the roof?

    a. 3 inches (75 mm) lie on the wall and 5 inches (127 mm) lie on the roof
    b. 4 inches (100 mm) lie on the wall and 4 inches (100 mm) lie on the roof
    c. 5 inches (127 mm) lie on the wall and 3 inches (75 mm) lie on the roof
    d. 2 inches (50 mm) lie on the wall and 6 inches (152 mm) lie on the roof

15. What is the term for a built-up section between the roof and the upper side of a chimney?

    a. cricket
    b. dutchman
    c. furring
    d. counter flashing

Unit 18

Windows

Chapter 58   Window Terms and Types

Chapter 59   Window Installation

Windows are normally installed prior to the application of exterior *siding*. Care must be taken to provide easy-operating, weathertight, attractive units. Quality workmanship results in a more comfortable interior, saves energy by reducing fuel costs, minimizes maintenance, gives longer life to the units, and makes application of the exterior siding easier.

 SAFETY REMINDER

Windows are made in many styles, sizes, and shapes. They are often a focal point of large rooms. Follow manufacturer's recommendations and local code requirements carefully when installing them to avoid injury.

OBJECTIVES

After completing this unit, the student should be able to:

- describe the most popular styles of windows and name their parts.
- select and specify desired sizes and styles of windows from manufacturers' catalogues.
- install various types of windows in an approved manner.

Chapter 58

Window Terms and Types

Wood windows are one of many types of **millwork** (**Figure 58–1**). Millwork is a term used to describe products, such as windows, doors, and cabinets, fabricated in woodworking plants, that are used in the construction of a building. Windows are usually fully assembled and ready for installation when delivered to the construction site. Windows are also made of aluminum and steel. Windows made with exposed wood parts encased in vinyl are called **vinyl-clad** windows. The names given to various parts of a window are the same, in most cases, regardless of the window type.

PARTS OF A WINDOW

When shipped from the factory, the window is complete except for the interior trim. It is important that the installer know the names, location, and functions of the parts of a window in order to understand, or to give, instructions concerning them.

The Sash

The **sash** is a frame in a window that holds the glass. The type of window is generally determined by the way the sash operates. The sash may be installed in a fixed position, move vertically or horizontally, or swing outward or inward.

Sash Parts. Vertical edge members of the sash are called **stiles**. Top and bottom horizontal members are called **rails**. The pieces of glass in a sash are called **lights**. There may be more than one light of glass in a sash. Small strips of wood that divide the glass into smaller lights are called **muntins**. Muntins may divide the glass into rectangular, diamond, or other shapes (**Figure 58–2**).

Many windows come with false muntins called **grilles**. Grilles do not actually separate or support the glass. They are applied as an overlay to simulate small lights of glass. They are made of wood or plastic. They snap in and out of the sash for easy cleaning of the lights (**Figure 58–3**). They may also

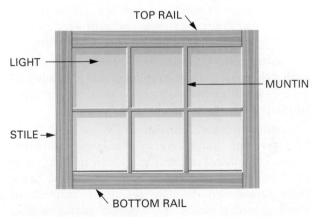

FIGURE 58–2 A window sash and its parts.

FIGURE 58–1 Windows of many types and sizes are fully assembled in millwork plants and ready for installation. *(Courtesy of Andersen Windows, Inc.)*

Courtesy of Andersen Windows, Inc.

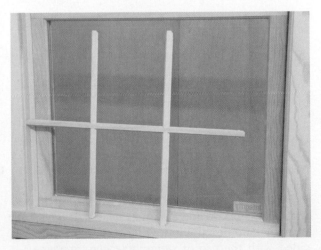

FIGURE 58–3 Removable grilles simulate true divided-light muntins.

be preinstalled between the layers of glass in double- or triple-glazed windows.

WINDOW GLASS

Several qualities and thicknesses of sheet glass are manufactured for **glazing** and other purposes. The installation of glass in a window sash is called glazing. **Single-strength (SS) glass** is about $3/32$ inch (2 mm) thick. It is used for small lights of glass. For larger lights, **double-strength (DS) glass** about $1/8$ inch (3 mm) thick may be used. *Heavy sheet glass* about $3/16$ inch (5 mm) thick is also manufactured. Many other kinds of glass are made for use in construction.

Safety Glass

Most residential windows are not glazed with safety glass, so if they break, they could fragment into large pieces and cause injury. Care must be taken to handle windows in a manner to prevent breaking the glass. Some codes, however, do require a type of **safety glass** in windows with low sill heights or located near doors. Skylights and roof windows are generally required to be glazed with safety glass.

Safety glass is constructed, treated, or combined with other materials to minimize the possibility of injuries resulting from contact with it. Several types of safety glass are manufactured.

Laminated glass consists of two or more layers of glass with inner layers of transparent plastic bonded together. **Tempered glass** is treated with heat or chemicals. When broken at any point, the entire piece immediately disintegrates into a multitude of small granular pieces. *Transparent plastic* is also used for safety glazing material.

Insulated Glass

To help prevent heat loss, and to avoid condensation of moisture on glass surfaces, **insulated glass**, or *thermal pane windows*, are used frequently in place of single-thickness glass.

Insulated glass consists of two or three (generally two) layers of glass separated by a sealed air space $3/16$ to 1 inch (5 to 25 mm) in thickness (**Figure 58–4**). Moisture is removed from the air between the layers before the edges are sealed. To raise the R-value of insulated glass, the space between the layers is filled with **argon gas**. Argon conducts heat at a lower rate than air. Additional window insulation may be provided with the use of *removable glass panels* or *a combination storm sash*.

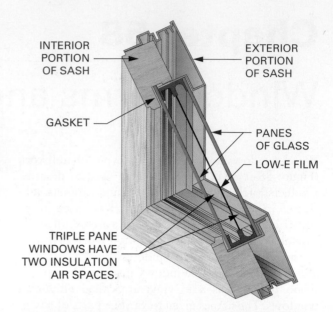

INTERIOR PORTION OF SASH
EXTERIOR PORTION OF SASH
GASKET
PANES OF GLASS
LOW-E FILM
TRIPLE PANE WINDOWS HAVE TWO INSULATION AIR SPACES.

FIGURE 58–4 Cross-sectional view of an energy-efficient window.

Some window companies have used krypton gas as the filler for the space between the panes of glass, with measured success.

Solar Control Glass. The R-value of windows may also be increased by using special *solar-control insulated glass*, called *high performance* or **low-emissivity glass (low-e)**. Low-e is used to designate a type of glazing that reflects heat back into the room in winter and blocks heat from entering in the summer (**Figure 58–5**). An invisible, thin, metallic coating is bonded to the air space side of the inner glass. This lets light through, but reflects heat.

THE WINDOW FRAME

The sash is hinged to, slides, or is fixed in a **window frame**. The frame sometimes comes with the exterior trim applied. It consists of several distinct parts (**Figure 58–6**).

The Sill

The bottom horizontal member of the window frame is called a **sill**. It is set or shaped at an angle to shed water. Its bottom side usually is grooved so that a weathertight joint can be made with the wall siding.

Jambs

The vertical sides of the window frame are called **side jambs**. The top horizontal member is called a **head jamb**.

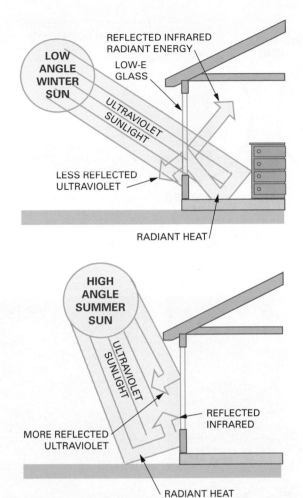

FIGURE 58-5 Low-e glass is used in windows to keep heat in during cold weather and out during hot weather.

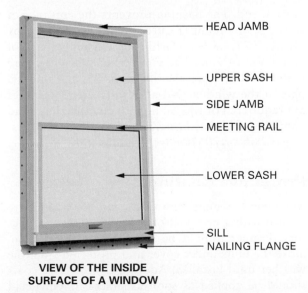

VIEW OF THE INSIDE SURFACE OF A WINDOW

FIGURE 58-6 A window frame consists of parts with specific terms.

Extension Jambs. The inside edge of the jamb should be flush with the finished interior wall surface when the window is installed. In some cases, windows can be ordered with jamb widths for standard wall thicknesses. In other cases, jambs are made narrow. **Extension jambs** (or window liners) are then provided with the window unit. The extensions are cut to width to accommodate various wall thicknesses. They are applied to the inside edge of the jambs of the window frame (**Figure 58–7**). The extension jambs are installed at a later stage of construction, when the interior trim is applied. They should be stored for safekeeping until needed.

Windows may also be purchased with extension jambs already installed. Care should always be taken to protect the jambs throughout the construction process.

Blind Stops

Blind stops are sometimes applied to certain types of window frames. They are strips of wood attached to the outside edges of the jambs. Their inside edges project about ½ inch (12 mm) inside the frame. They provide a weathertight joint between the

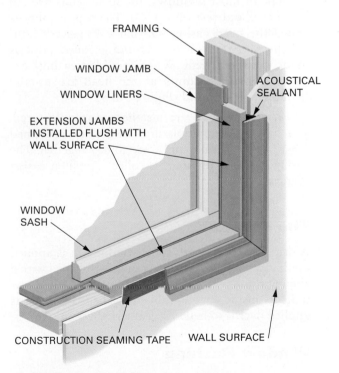

EXTENSION JAMBS MAY BE INSTALLED ON ALL FOUR SIDES AS SHOWN OR MAY EXCLUDE THE BOTTOM WHERE A STOOL IS INSTALLED.

FIGURE 58-7 To compensate for varying wall thicknesses, extension jambs are provided with some window units.

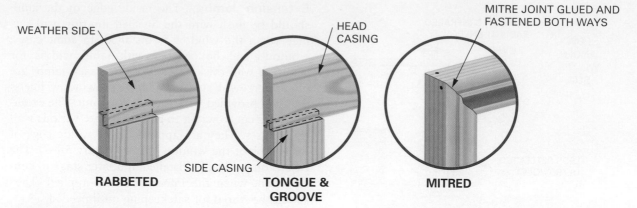

FIGURE 58–8 A weathertight joint is made between side and head casings.

outside casings and the frame. They also act as stops for screens and storm sash. They make the outer edge of the channel for the top sash of double-hung windows.

Casings or Brick Mould

Window units sometimes come with exterior casings applied. The side members are called **side casings**. In most windows, the lower ends are cut at a bevel and rest on the sill. The top member is called the **head casing**. On flat casings, a weathertight *rabbeted* or *tongue-and-grooved* joint is made between them. When moulded casings are used, the mitred joints at the head are usually bedded in compound (**Figure 58–8**).

When windows are installed or manufactured, side by side, in multiple units, a **mullion** is formed where the two side jambs are joined together. The casing covering the joint is called a mullion casing (**Figure 58–9**).

The Drip Cap

A **drip cap** comes with some windows. It is applied on the top edge of the head casing. It is shaped with a sloping top to carry rainwater out over the window unit. Its inside edge is rabbeted. The wall siding is applied over it to make a weathertight joint.

Window Flashing

In some cases, a **window flashing** is also provided. This is a piece of metal as long as the head casing, which is also called a *drip cap*. It is bent to fit over the head casing and against the exterior wall (**Figure 58–10**). To prevent the water that drips off of the drip cap, *end dams* are bent up

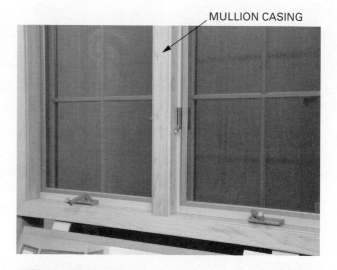

FIGURE 58–9 Window units that are joined create a mullion.

at each end. The flashing prevents the entrance of water at this point (**Figure 58–11**) as required by NBCC 9.27.3.8 (4). Flashings are usually made of aluminum or zinc. The vinyl flanges of vinyl-clad wood windows are usually formed as an integral part of the window. No additional head flashings are required. For tips on how to cut and shape end dams on-site, visit *http://www.joneakes.com/jons-fixit-database/2062#video_container_42.*

Protective Coatings

Most window units with wood exterior casings are primed with a first coat of paint applied at the factory. Priming should be done before installation. Store the units under cover and protected from the weather until installed. Additional protective coats should be applied as soon as practical. Vinyl-clad and **aluminum-clad** wood windows are designed to eliminate painting.

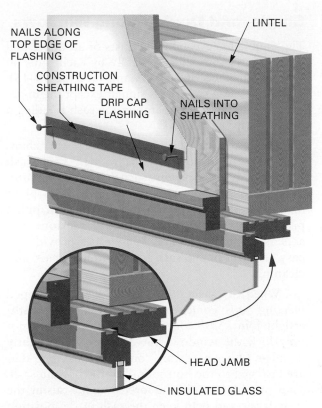

FIGURE 58–10 A window flashing covers the top edge of the head casing.

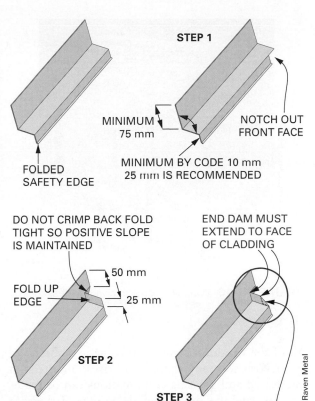

FIGURE 58–11 Window drip cap with end dams.

TYPES OF WINDOWS

Common types of windows are fixed, single- or double-hung, casement, sliding, awning, and hopper windows.

Fixed Windows

Fixed windows consist of a frame in which a sash is fitted in a fixed position. They are manufactured in many shapes (**Figure 58–12**).

Oval and *circular* windows are usually installed as individual units. *Elliptical*, *half rounds*, and *quarter rounds* are widely used in combination with other types. In addition, fixed windows are manufactured in other *geometric* shapes (*squares, rectangles, triangles, parallelograms, diamonds, trapezoids, pentagons, hexagons* and *octagons*). They may be assembled or combined with other types of windows in a great variety of shapes (**Figure 58–13**). In addition to factory-assembled units, lengths of the frame stock may be purchased for cutting and assembling odd shape or size units on the job.

Arch windows have a curved top or head that make them well suited to be joined in combination with a number of other types of windows or doors.

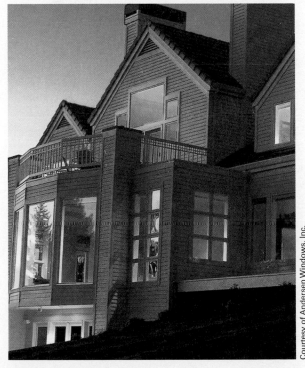

FIGURE 58–12 Fixed windows are often used in combinations.

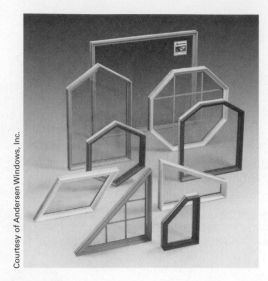

FIGURE 58–13 Windows come in a variety of shapes and sizes.

All of the windows mentioned come in a variety of sizes. With so many shapes and sizes, hundreds of interesting and pleasing combinations can be made. Arched windows may be made as part of the sash (**Figure 58–14**).

FIGURE 58–14 Arched windows can be part of a window sash.

Single- and Double-Hung Windows

The **double-hung window** consists of an upper and a lower sash that slide vertically by each other in separate channels of the side jambs (**Figure 58–15**). The **single-hung window** is similar except the upper sash is fixed. A strip separating the sash is called a **parting strip**.

In most units, the sash slides in metal channels that are installed in the frames. Each sash is provided with *springs*, *sash balances*, or *compression weather stripping* to hold it in place in any position. Compression weather stripping prevents air infiltration, provides tension, and acts as a counterbalance. Some types provide for easy removal of the sash for painting, repair, and cleaning.

When the sashes are closed, specially shaped *meeting rails* come together to make a weathertight joint. *Sash locks* located at this point not only lock the window, but draw the rails tightly together. Other hardware consists of *sash lifts* that are fastened to the bottom rail of the bottom sash. They provide an uplifting force to make raising the sash easier and help keep the sash in the position it is placed.

Most double-hung windows are also designed to be removed via a tilt-in action (**Figure 58–16**). This makes cleaning both inside and outside surfaces easy. Double-hung windows can be arranged in a number of ways. They can be installed side by side in multiple units or in combination with other types. **Bow window** units project out from the building, often creating more floor space (**Figure 58–17**). The look is of a smooth curve, usually with four to six equal sections. A **bay window** unit is similar to a bow except there are three sections at 135° to each other. The bay window is a truncated octagon.

Casement Windows

The **casement window** consists of a sash hinged at the side. It swings outward by means of a crank or lever. Most casements swing outward. The inswinging type is very difficult to make weathertight. An advantage of the casement type is that the entire sash can be opened for maximum ventilation. **Figure 58–18** shows the use of casement windows in a double window unit.

Sliding Windows

Sliding windows have sashes that slide horizontally in separate tracks located on the head jamb and

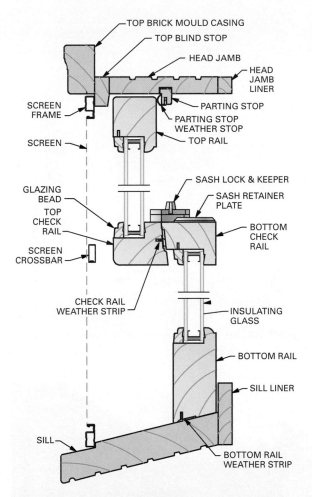

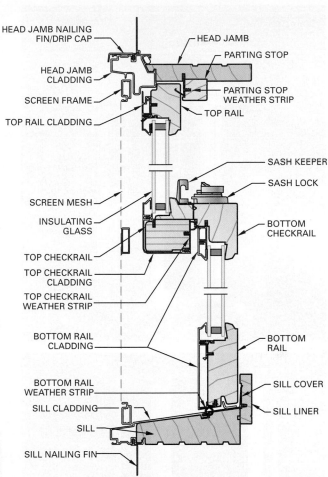

FIGURE 58–15 The double-hung window and its parts.

FIGURE 58–16 Double-hung windows may tilt in for easy cleaning.

sill (**Figure 58–19**). When a window-wall effect is desired, many units can be placed side by side. Most units come with all necessary hardware applied.

Awning and Hopper Windows

An **awning window** unit consists of a frame in which a sash hinged at the top swings outward by means of a crank or lever. A similar type, called the **hopper window**, is hinged at the bottom and swings inward.

Each sash is provided with an individual frame so that many combinations of widths and heights can be used. These windows are often used in combination with other types (**Figure 58–20**).

Skylight and Roof Windows

A **skylight** provides light only. **Roof windows** contain operating sashes to provide light and ventilation (**Figure 58–21**). One type of roof window comes with a tilting sash that allows access to the outside surface for cleaning. Special flashings are used when multiple skylights or roof windows are ganged together.

Courtesy of Loewen

Photo by M. Nauth

FIGURE 58–17 Bow and bay window units.

Courtesy of Loewen

FIGURE 58–18 Casement windows swing outward.

Courtesy of North Star Windows & Doors

FIGURE 58–19 The sashes in sliding windows move horizontally by each other.

Courtesy of Loewen

FIGURE 58–20 Awning windows are often used in stacks or in combination with other types of windows.

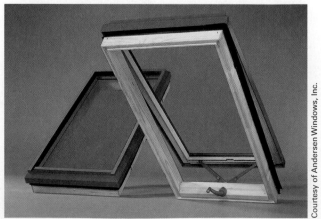

Courtesy of Andersen Windows, Inc.

FIGURE 58–21 Skylights and roof windows are made in a number of styles and sizes.

Chapter 59

Window Installation

There are numerous window manufacturers that produce hundreds of kinds, shapes, and sizes of windows. Because of the tremendous variety and design differences, follow the manufacturer's instructions closely to ensure a correct installation. Directions in this unit are basic to most window installations. They are intended as a guide, to be supplemented by procedures recommended by the manufacturer.

SELECTING AND ORDERING WINDOWS

The builder must study the plans to find the type and location of the windows to be installed. The floor plan shows the location of each unit. Each unit is usually identified by a number or a letter next to the window symbol.

Those responsible for designing and drawing plans for building or selecting windows must be aware of, and comply with, building codes that set certain standards in regard to windows. Most codes require minimum areas of natural light. Codes also require minimum ventilation by windows unless provided by other means. Some codes stipulate minimum window sizes in certain rooms for use as emergency egress (NBCC Table 9.32.2.1).

The National Building Code of Canada (NBCC) requires that the minimum unobstructed opening have an area of 3.8 square feet (0.35 m²), with no dimension less than 15 inches (381 mm). For example, a window with an opening of 20 × 28 inches (508 × 711 mm) would conform to the code. The sill height must be less than 5 feet (1.5 m) above the floor or above permanently fixed furniture.

Out-swinging windows, such as awning and casement windows, should not swing out over decks, patios, and similar areas unless they are high enough to permit persons to travel under them. When lower, the projecting sash could cause serious injury.

Window Schedule

The numbers or letters found in the floor plan identify the window in more detail in **window schedules**. This is, usually, part of a set of plans (**Figure 59–1**). This

Window Schedule				
Ident.	Quan.	Manufacturer	Size	Remarks
A	6	Andersen	TW28310	D.H. Tiltwash
B	1	Andersen	WDH2442	Woodwright D.H.
C	2	Andersen	3062	Narrowline D.H.
D	1	Andersen	CW24	Casement Single
E	1	Andersen	C34	Casement Triple
F	1	Andersen	C23	Casement Double

FIGURE 59–1 Typical window schedule found in a set of plans.

schedule normally includes window style, size, manufacturer's name, and unit number. Rough opening sizes may or may not be shown.

Manufacturers' Catalogues

Sometimes a window schedule is not included. Units are identified only by the manufacturer's name and number on the floor plan. To get more information, the builder must refer to the window manufacturer's catalogue.

The catalogue usually includes a complete description of the manufactured units and optional accessories, such as insect screens, glazing panels, and grilles. For a particular window style, the catalogue typically shows overall unit dimensions, rough opening widths and heights, and glass sizes of manufactured units. Large-scale, cross-section details of the window unit also usually are included so that the builder can more clearly understand its construction (**Figure 59–2**).

Order window units giving the type and identification letters and/or numbers found in the window schedule or manufacturer's catalogue. The size of all existing rough openings should be checked to make sure they correspond to the size given in the catalogue before windows are ordered.

INSTALLING WINDOWS

All rough window openings should be prepared to ensure weathertight window installations.

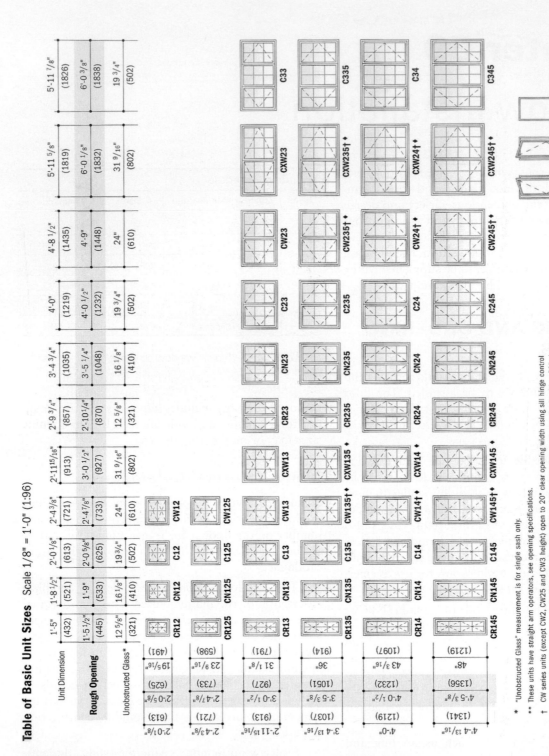

Table of Basic Unit Sizes Scale 1/8" = 1'-0" (1:96)

FIGURE 59–2 Typical page from a window manufacturer's catalogue.

Courtesy Andersen Windows, Inc.

Housewraps, Building Paper, and Other Vapour Permeable Membranes

Exterior walls are sometimes covered with a building paper, 15# asphalt felt, prior to the application of siding. This prevents the infiltration of air into the structure. Yet, it also allows the passage of water vapour to the outside. In place of building paper, exterior walls are often covered with a type of air infiltration barrier commonly called **housewrap**.

Housewrap is a very thin, tough plastic material. It is used to cover the sheathing on exterior walls for the same purpose as building paper (**Figure 59–3**). Housewraps are commonly known by the brand names of Typar and Tyvek. They are more resistant

Courtesy of TYPAR – a PGI brand

FIGURE 59–3 Housewrap is widely used as an air infiltration barrier on sidewalls.

than building paper to air leakage and are virtually tear-proof. Yet they are also breathable to allow water vapour to escape. Building paper comes in 36-inch-wide (914 mm) rolls. Housewrap rolls are 1.5, 3, 4.5, 5, 9, and 10 feet (0.45, 1, 1.4, 1.5, 2.7, and 3 m) wide.

Housewrap gets its name because it is completely wrapped around the building. It covers corners, window and door openings, plates, and sills. Housewrap is designed to survive prolonged periods of exposure to the weather. It can and usually is applied immediately after framing is completed, but before doors and windows have been installed. The wrap, then, serves also as a flashing for the sides of windows and doors.

Applying Housewrap. Begin at the corner, holding the roll vertically on the wall. Unroll it a short distance. Make sure the roll is plumb. Secure the sheet to the corner, leaving about a foot extending beyond the corner to overlap later. Continue to unroll. Make sure the sheet is straight, with no buckles. Fasten every 12 to 18 inches (305 to 457 mm) (**Figure 59–4**).

Unroll directly over window and door openings and around the entire perimeter of the building. Overlap all joints by at least 4″ (100 mm) (NBCC 9.27.3.3(2)). Secure them with a special housewrap tape. On horizontal joints, the upper layer should overlap the lower layer.

Make cuts in the housewrap in the shape of the capital letter "I" (**Figure 59–5**). Make a vertical slit in the centre of the window and then cut horizontally along the top and along the bottom of the rough opening. The side tabs are folded into the building and stapled to the jack studs. At the top of the opening, the paper is slit 8 inches (203 mm) upward at a 45-degree angle at each corner, and the flap that is created is folded up and stapled. A piece of bevelled siding is placed on the rough sill. Then a peel-and-stick

FIGURE 59–4 Housewrap is secured with plastic washers applied with handheld or pneumatic tackers.

membrane is applied to cover the siding and up each side by about 6 inches (152 mm), and then folded over onto the housewrap. Resilient sealant is applied to the edge of the top and sides of the opening, and then the window is installed and secured as per manufacturer's instructions. The peel-and-stick membrane is then applied to seal the window to the housewrap on the sides and to the wall sheathing on the top. For an alternative to the membrane sill pan, go to *https://www.youtube.com/watch?v=9XlZnVsbRT0*. (See also Figures 53–4, 53–5, 53–7, and 53–8.) A drip cap is installed, and the top flap of housewrap is folded down and taped to the drip cap. The space between the window and the rough sill is insulated and sealed from the inside, but it is allowed to drain freely on the outside. Doors are dealt with in a similar fashion (see Chapter 53).

CAUTION

Housewraps are slippery. They should not be used in any application where they will be walked on.

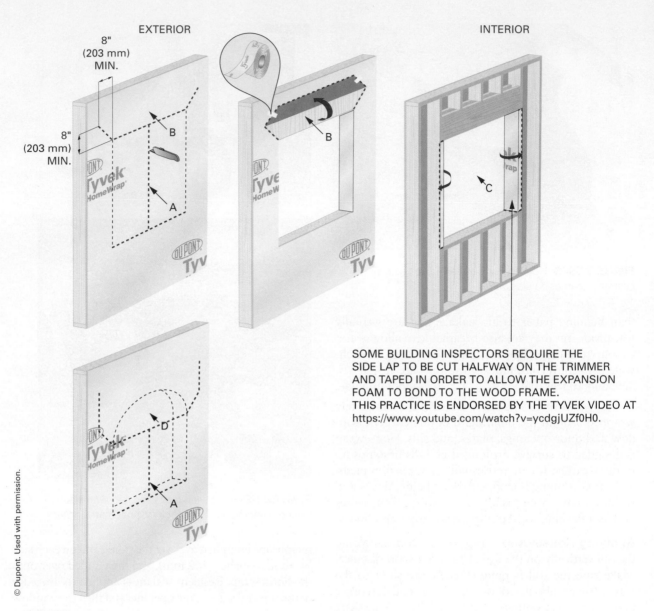

EXTERIOR

INTERIOR

8"
(203 mm)
MIN.

8"
(203 mm)
MIN.

SOME BUILDING INSPECTORS REQUIRE THE
SIDE LAP TO BE CUT HALFWAY ON THE TRIMMER
AND TAPED IN ORDER TO ALLOW THE EXPANSION
FOAM TO BOND TO THE WOOD FRAME.
THIS PRACTICE IS ENDORSED BY THE TYVEK VIDEO AT
https://www.youtube.com/watch?v=ycdgjUZf0H0.

© Dupont. Used with permission.

FIGURE 59–5 Installing weather barrier before window installation.

There are also peel-and-stick membrane air barriers that are fully water-resistant but at the same time are vapour-permeable. They are laminate polypropylene fabrics that fully adhere to the wall sheathing (**Figure 59–6**). As well, some commercially applied liquid membrane products have been modified to be vapour-permeable. Visit *wrmeadows.com*.

Fastening Windows

Remove all protection blocks from the window unit. Do not remove any diagonal braces applied at the factory. Close and lock the sash. If windows are stored inside, they can easily be moved through the openings and set in place. It is important to centre the unit in the opening on the rough sill with the window casing overlapping the wall sheathing.

CAUTION

Have sufficient help when setting large units. Handle them carefully to avoid damaging the unit or breaking the glass. Broken glass can cut through protective clothing and cause serious injury.

Place a level on the window sill. If not level, determine which side of the window is the highest. Check to see if the high side of the unit is at the desired height. If not, bring it up by shimming with a wood shim between the rough sill and the bottom end of the window's side jamb (**Figure 59–7**). Remove the window unit from the opening and apply sealant on the top

FIGURE 59–6 Peel-and-stick water-resistant barrier (WRB).

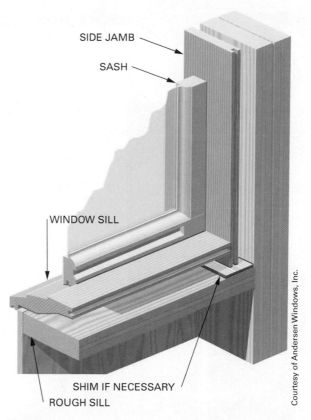

FIGURE 59–7 Use shims under the side jambs to level the window unit.

and sides to the backside of the casing or nailing flange. This will seal the unit to the housewrap. Replace unit and tack through the lower end of the casing into the sheathing on the high side. Shim the other side of the window in the same manner so that the window sill is level. Tack the lower end of the casing on that side.

On wide windows with long sills, shim at intermediate points so that the sill is perfectly straight and level, with no sag on wood-cased windows. Use either a long level or a shorter level in combination with a straightedge. Also, sight the sill by eye from end to end to make sure it is straight.

Plumb the ends of the side jambs with a level. Tack the top ends of the side casings. Straighten the side jambs between sill and head jamb. Tack through the side casings at intermediate points. Straighten and tack the head casing.

Check the joint between sash and jamb. Make sure the sash operates properly. If not, make necessary adjustments. Then fasten the window permanently in place. Use galvanized casing or common nails spaced about 16 inches (405 mm) apart. Keep nails about 2 inches (50 mm) back from the ends of the casings to avoid splitting. Nail length depends on the thickness of the casing. Nails should be long enough to penetrate the sheathing and into the framing members. Set the nails so that they can be puttied over later.

Vinyl-clad windows have a vinyl nailing flange. Large-head roofing nails are driven through the flange instead of through the casing (**Figure 59–8**).

Windows installed in masonry and brick veneer walls are usually attached to a treated wood buck. Adequate clearance should be left for applying resilient sealant around the entire perimeter between the window and masonry (**Figure 59–9**).

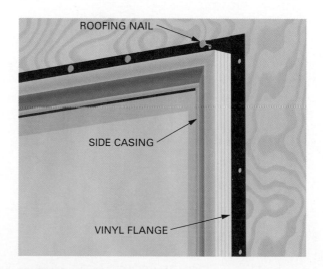

FIGURE 59–8 Roofing nails are used to fasten the flanges of vinyl-clad windows.

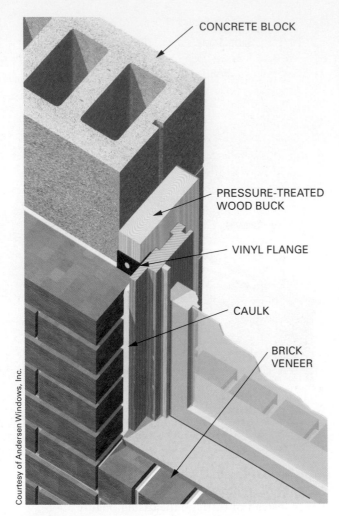

CONCRETE BLOCK

PRESSURE-TREATED WOOD BUCK

VINYL FLANGE

CAULK

BRICK VENEER

FIGURE 59–9 Windows are installed in masonry openings against wood bucks.

Flashing the Window

The window perimeter is sometimes flashed to the sheathing. The window flange is completely covered on the top and sides (**Figure 59–10**). Flashing material comes in rolls. It has a sticky side that is covered with a release paper. Begin with the bottom piece under the flange. Next, do the sides and finish with the top. This will keep the laps facing the correct direction in terms of any water runoff.

Additional head flashing may be applied over the window. This will help direct any water away from behind the siding (**Figure 59–11**). Its length should be equal to the overall width of the window. Do not let the ends project beyond the side casings. This will make the application of siding difficult. Place the flashing firmly on top of the head casing. Secure with fasteners along its top edge and into the wall sheathing.

METAL WINDOWS

Metal windows are available in the same styles as wood windows. The shape and sizes of the parts vary with the manufacturer and the intended use.

In frame construction, if metal windows are used, they may be set in a wood frame. The frame is then installed in the same manner as for wood windows. They may also be set in the opening with their flanges overlapping the siding or sheathing. Sealant is applied under the flanges. The unit is then screwed to the wall.

In masonry construction, wood **bucks** are fastened to the sides of the opening. Metal windows are installed against them. The flanges on the two sides

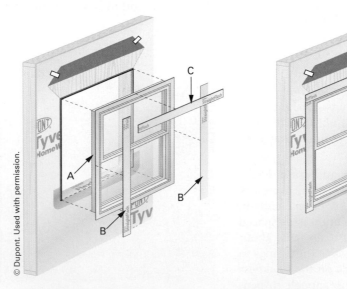

FIGURE 59–10 Applying sealant to the top and sides before window installation.

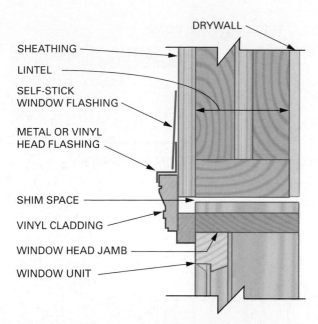

FIGURE 59–11 A drip cap is often installed as an extra layer of flashing over the top of the head casing.

Labels in figure:
- DRYWALL
- SHEATHING
- LINTEL
- SELF-STICK WINDOW FLASHING
- METAL OR VINYL HEAD FLASHING
- SHIM SPACE
- VINYL CLADDING
- WINDOW HEAD JAMB
- WINDOW UNIT

FIGURE 59–12 Windows in masonry walls are fastened to window bucks.

and top are bedded in resilient sealant (**Figure 59–12**). They are then screwed to the bucks.

Carefully follow the installation directions provided with the units, whether wood or metal. In areas prone to hurricanes, window installation must be approved by an inspector. This is done to ensure that every window installed is able to withstand the anticipated heavy wind loads.

RESILIENT SEALANT

Resilient sealant (or caulking) is a pliable material applied to fill gaps in building materials. It bonds to the surrounding material during cure and remains flexible after cure. This flexibility allows building materials to expand and contract with heat and moisture, maintaining the seal.

Sealants are made with a variety of materials, each with a different function. The materials include acrylic, butyl rubber, latex, polyurethane, and silicone. Many sealants are made with a mixture of these ingredients to blend the desirable characteristics (**Figure 59–13**). Read the manufacturer's recommendations on the tube to determine the best usage.

Generally speaking, acrylic and latex sealants are used when expected material movement is small. They perform best in interior applications. Silicone is used when resistance to mould and mildew are required, such as in kitchens and bathrooms. Silicone can be used for interior and exterior applications and has very good flexibility. Polyurethane is designed for exterior applications where severe material movement is expected. It can fill large gaps and has superior bonding and flexibility characteristics.

Resilient sealant performs best when it is installed with a backing material (**Figure 59–14**). Backing material allows the sealant to bond to the materials on opposite sides of the joint. When the building material moves, the sealant is compressed and stretched. If the sealant is allowed to bond on three surfaces, as the material moves, the bond begins to tear in the corners, and eventually the bond can be completely broken.

FIGURE 59–13 Resilient sealant is made of a variety of materials.

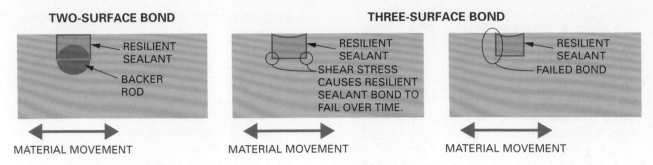

FIGURE 59–14 Large beads of resilient sealant should be installed with a backer rod.

DECONSTRUCT THIS

Carefully study **Figure 59–15** and think about what is wrong and/or what is right. Consider all possibilities. What construction practice or method is different in your area of the country?

FIGURE 59–15 This photo shows a window installed with flashing.

KEY TERMS

aluminum-clad	extension jambs	mullion	skylight
arch windows	fixed windows	muntins	sliding windows
argon gas	glazing	parting strip	stiles
awning window	grilles	rails	tempered glass
bay window	head casing	resilient sealant	vinyl-clad
blind stops	head jamb	roof windows	window flashing
bow window	hopper window	safety glass	window frame
bucks	housewrap	sash	window schedules
casement window	insulated glass	side casings	
double-hung window	laminated glass	side jambs	
double-strength (DS) glass	lights	sill	
	low-emissivity glass (low-e)	single-hung window	
drip cap		single-strength (SS) glass	
end dam	millwork		

REVIEW QUESTIONS

Select the most appropriate answer.

1. What is a frame holding a pane of glass called?

   a. light
   b. mullion
   c. sash
   d. stile

2. What is the term for small strips that divide the glass into smaller panes?

   a. mantels
   b. margins
   c. mullions
   d. muntins

3. When windows are installed in multiple units, what is formed by the joining of the side jambs?

   a. mantel
   b. margin
   c. mullion
   d. muntin

4. What type of window consists of an upper and a lower sash, both of which slide vertically?

   a. casement window
   b. double-hung window
   c. hopper window
   d. sliding window

5. What type of window has a sash hinged on one side and swings outward?

   a. an awning window
   b. a casement window
   c. a double-hung window
   d. a hopper window

6. What is the difference between a hopper and an awning window?

   a. The hopper window swings inward instead of outward.
   b. The hopper window swings outward instead of inward.
   c. The hopper window is hinged at the top rather than at the bottom.
   d. The hopper window is hinged on the side rather than on the bottom.

7. What does the term *fixed window* refer to?

   a. window with an unmovable sash
   b. repaired window
   c. window unit properly flashed
   d. All of the above.

8. Which of the following best describes single-hung windows?

   a. They have one sash.
   b. They have one sash that moves.
   c. They are installed with no other windows nearby.
   d. They slide from side to side.

9. What is the term for multiple window units fastened together to form a large curved window?

a. bay window
b. bow window
c. double casement window
d. fixed double awning window

10. What should be done to windows installed in areas prone to hurricanes?

a. They should be flashed on all sides of the opening.
b. They should be caulked into place.
c. They should be inspected by an inspector.
d. They should be approved by the regional municipality.

11. What is the best choice of caulk for interior bathroom applications?

a. latex
b. silicone
c. urethane
d. All of the above.

Unit 19

Exterior Doors

Exterior doors, like windows, are manufactured in millwork plants in a wide range of styles and sizes. Many entrance doors come prehung in frames, complete with exterior casings applied, and ready for installation. In other cases, the door is fitted and hinged to a door frame. Main entrance doors to dwelling units shall be provided with either a door viewer or transparent glazing in the door, or the door unit shall include a transparent sidelight [NBCC 9.7.2.1(2) a, b].

OBJECTIVES

After completing this unit, the student should be able to:
- describe the standard designs of exterior doors.
- name the parts of an exterior door frame.
- fit and hang a prehung exterior door.
- build and set a door frame.
- install locksets in doors.

Chapter 60

Exterior Door Styles

Exterior doors are often the central focus of a house. As one approaches it from the street, the architectural style of the house is revealed. Entrances to houses have many variations, yet the names of the parts and pieces are similar.

DOOR STYLES AND SIZES

Exterior flush and panel doors are available in many styles. There are many choices when designing entrances, and only a select few are considered here. To get a broad view of what is available, consult a manufacturer's literature.

Flush Doors

An exterior **flush door** has a smooth, flat surface of wood veneer or metal. It has a framed, *solid core* of staggered wood blocks or composition board. Wood *core blocks* are inserted in appropriate locations in composition cores. They serve as *backing* for door locks (**Figure 60–1**). Openings may be cut in flush doors either in the factory or on the job. Lights of various kinds and shapes are installed in them.

Moulding of various shapes may be applied in many designs to make the door more attractive.

Panel Doors

Panel doors are classified by one large manufacturer as *high-style, panel, sash, fire, insulated, French*, and *Dutch* doors. Sidelights, although not actually doors, constitute part of some entrances. They are fixed in the door frame on one or both sides of the door (**Figure 60–2**). A **transom** is similar to a sidelight. When used, transoms are installed above the door.

Panel Door Styles. High-style doors, as the name implies, are highly crafted designer doors. They may have a variety of cut-glass designs. A **panel door** is made with raised panels of various shapes. Sash doors have panels of tempered or insulated glass that allow the passage of light (**Figure 60–3**). Fire doors are used where required by codes. These doors prevent the spread of fire for a certain period of time. Insulated doors have thicker panels with low-E or argon-filled insulated glass. French doors may contain 1, 5, 10, 12, 15, or 18 lights of glass for the total

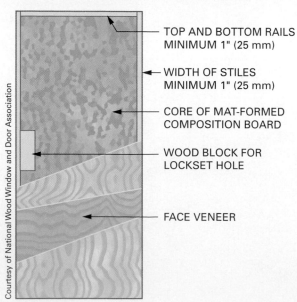

MAT-FORMED COMPOSITION BOARD CORE
7-PLY CONSTRUCTION ILLUSTRATED

Courtesy of National Wood Window and Door Association

TOP AND BOTTOM RAILS
MINIMUM 1" (25 mm)

WIDTH OF STILES
MINIMUM 1" (25 mm)

CORE OF MAT-FORMED
COMPOSITION BOARD

WOOD BLOCK FOR
LOCKSET HOLE

FACE VENEER

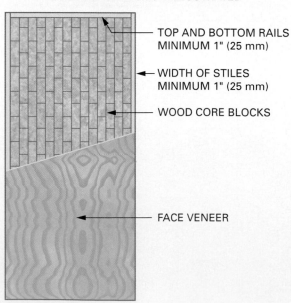

FRAMED BLOCK NON-GLUED CORE
5-PLY CONSTRUCTION ILLUSTRATED

TOP AND BOTTOM RAILS
MINIMUM 1" (25 mm)

WIDTH OF STILES
MINIMUM 1" (25 mm)

WOOD CORE BLOCKS

FACE VENEER

FIGURE 60–1 Composition or solid wood cores are possible in exterior flush doors.

Courtesy of Morgan Manufacturing

FIGURE 60–2 Sidelights are installed on one or both sides of the main entrance door.

width and length inside the frame. A Dutch door consists of top and bottom units, hinged independently of each other.

Parts of a Panel Door. A panel door consists of a frame that surrounds panels of solid wood and glass, or louvres. Some door parts are given the same terms as a window sash.

The outside vertical members are called *stiles*. Horizontal frame members are called *rails*. The *top rail* is generally the same width as the stiles. The *bottom rail* is the widest of all rails. A rail situated at lockset height, usually 38 inches (965 mm) from the finish floor to its centre, is called the **lock rail**. Almost all other rails are called *intermediate rails*. *Mullions* are vertical members between rails dividing panels in a door.

The moulded shape on the edges of stiles, rails, mullions, and bars, adjacent to panels, is called the **sticking**. The name is derived from the moulding machine, commonly called a sticker, used to shape the parts. Several shapes are used to *stick* frame members.

EXTERIOR PANEL DOORS   EXTERIOR SASH DOORS   SIDELIGHTS

Courtesy of Morgan Manufacturing

FIGURE 60–3 Several kinds of exterior doors are made in many designs.

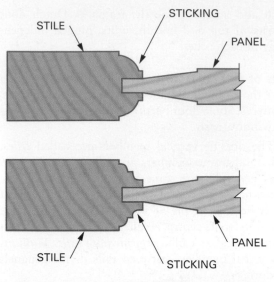

DOORS ARE STUCK WITH VARIOUS SHAPES.

FIGURE 60–4 The parts of an exterior panelled door.

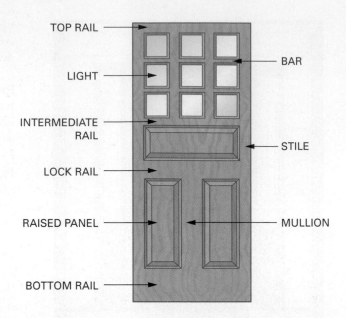

Bars are narrow horizontal or vertical rabbeted members. They extend the total length or width of a glass opening from rail to rail or from stile to stile. Door *muntins* are short members, similar to and extending from bars to a stile, rail, or another bar. Bars and muntins divide the overall length and width of the glass area into smaller lights.

Panels fit between and are usually thinner than the stiles, rails, and mullions. They may be raised on one side or on both sides for improved appearance (**Figure 60–4**).

PARTS OF AN EXTERIOR DOOR FRAME

Terms given to members of an exterior door frame are the same as similar members of a window frame. The bottom member is called a *sill* or **stool**. The vertical side members are called *side jambs*. The top horizontal part is a *head jamb*. The exterior door trim may consist of many parts to make an elaborate and eye-appealing entrance or a few parts for a simpler doorway. The **door casings**, if not too complex, are usually applied to the door frame before it is set (**Figure 60–5**). When more intricate trim is specified, it is usually applied after the frame is set (**Figure 60–6**).

Sills

In residential construction, door frames usually are designed and constructed for entrance doors that swing inward. Codes require that doors swing outward in buildings used by the general public. The shape of a wood door sill for an in-swinging door is different from that for an out-swinging door (**Figure 60–7**).

In addition to wood, extruded aluminum sills, also called *thresholds*, of many styles are manufactured for both in-swinging and out-swinging doors. They usually come with vinyl inserts to weatherstrip the bottom of the door. Some are adjustable for exact fitting between the sill and door (**Figure 60–8**).

Jambs

Side and head jambs are the same shape. Jambs may be square-edge pieces of stock to which door stops are later applied. Or, they may be **rabbeted jambs**, with single or double rabbets. On double-rabbeted jambs, one rabbet is used as a stop for the main door. The other is used as a stop for storm and screen doors (**Figure 60–9**). Several jamb widths are available for different wall thicknesses. For walls of odd thicknesses, jambs, except double-rabbeted ones, may be ripped to any desired width.

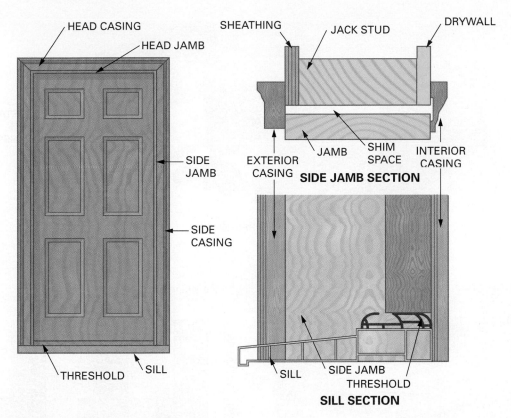

FIGURE 60–5 Parts of an exterior door frame.

FIGURE 60–6 Elaborate entrance trim is available.

Exterior Casings

Exterior casings may be plain square-edge stock. Mouldings are sometimes applied around the outside edges of the casings. This is done to improve the appearance of the entrance. Because the main entrance is such a distinctive feature of a building, the exterior casing may be enhanced on the job with fluted or otherwise shaped pieces and appropriate caps and mouldings applied (**Figure 60–10**). Flutes are narrow, closely spaced, concave grooves that run parallel to the edge of the trim. In addition, ornate main entrance trim may be purchased in knocked-down form. It is then assembled at the jobsite (**Figure 60–11**).

Weatherstripping

Exterior doors have weatherstripping to protect from air and weather infiltration. It is a soft pliable strip usually applied to the door jamb. When the door closes, it is compressed to seal the sides and top of

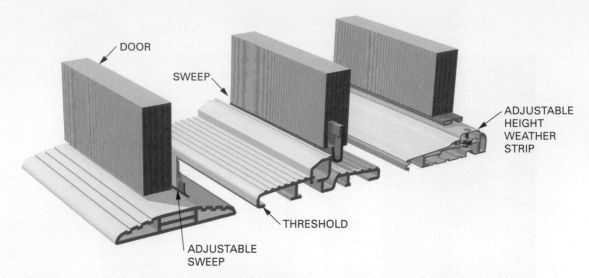

FIGURE 60–7 Wood sill shapes and styles vary according to the swing of the door.

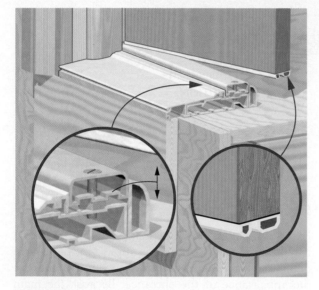

FIGURE 60–8 Some metal sills are adjustable for exact fitting at the bottom of the door.

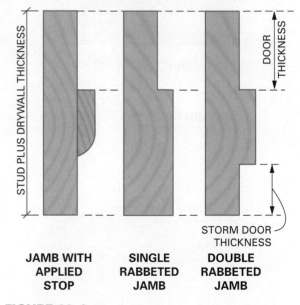

FIGURE 60–9 Door jamb cross-sections may be square edged, single rabbeted, or double rabbeted.

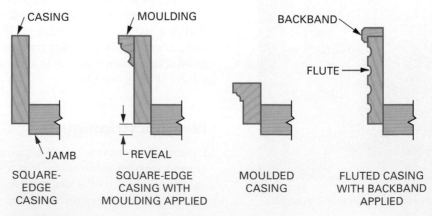

FIGURE 60–10 Exterior door casings may be enhanced by applying mouldings and by shaping.

FIGURE 60–11 A few samples of the many manufactured entrance door styles.

the door to the jamb (**Figure 60–12**). Weatherstripping acts with the threshold to keep weather and air from passing through while the door is closed.

EXTERIOR DOOR SIZES

Residential exterior entrance doors are typically 3'-0" wide × 6'-8" high (914 × 2032 mm). The National Building Code of Canada (NBCC) (9.5.11.1) states at least one entrance door must have a clear opening width of 32 inches (813 mm), as measured between the face of an open door and the doorstop on the opposite side. The height must be 78 inches (1981 mm), as measured between the threshold and the head jamb doorstop. The smallest door to accomplish this is a 3'-0" × 6'-8" (914 ×2032 mm) door. Entrance doors are normally manufactured in a thickness of 1¾ inches (44 mm).

Some styles, such as French, panel, and sash doors, are available in narrower widths, such as 1'-6", 2'-0", and 2'-4" (457, 610, and 711 mm), when they are installed as double doors. Sidelights are made in both 1⅜- and 1¾-inch (35 and 44 mm) thicknesses.

DETERMINING THE SWING OF THE DOOR

To order or make doors, the swing direction of the door must be noted. The set of prints for the building shows all doors and the direction of the door swing. The door schedule will list the doors, their styles, and their swings. The carpenter should be able to independently verify the swing of a door. It can be confusing because there are several ways of referring to swing. Exterior doors add more confusion because they have surfaces designed to be only the outside or the inside. Door swing is labelled either left-hand or right-hand.

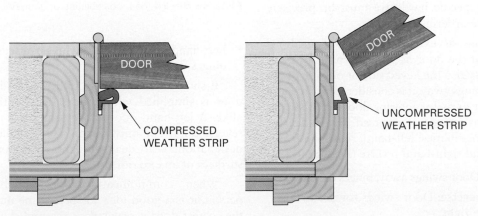

FIGURE 60–12 Plastic weatherstrips compress when the door closes to seal the opening.

LEFT-HAND SWING

RIGHT-HAND SWING

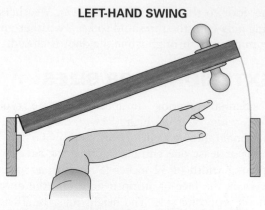
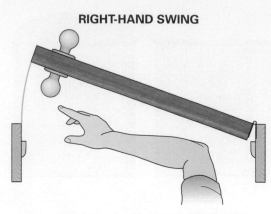

ELBOW NEAR HINGES, HAND NEAR KNOB

ELBOW NEAR HINGES, HAND NEAR KNOB

FIGURE 60–13 With forearm out, the elbow serves to identify the handing or swing of the door.

Interior Door Swing

A straightforward method for determining the swing or handing of interior doors is to consider your forearm when approaching the door as it swings away. The hand of the forearm that points your elbow to the hinge and your fingers to the knob gives the handing (**Figure 60–13**).

Some people consider the door with the viewer on the side of the door where the hinges are visible when the door is closed; this is the side of the door where the door opens toward you. The handing of the door is the side the knob is. In other words, the hand easiest to pull open the door with is the handing. This method is limiting when we consider exterior and commercial doors. We shall use the forearm method for exterior doors.

Exterior Door Swing

Exterior doors add a complication because they have an inside and outside surface. Door parts must be gasketed and caulked watertight on the outside. They may also swing inward or outward. Commercial doors that have pre-bored holes for crash bars and other side-specific hardware must be precisely described before ordering them.

The handing of exterior doors is considered from the side of the door as you approach to enter the room. This is also the keyed side or outside of the building. If it swings away, it is considered normal and may be thought of as for interior doors. If it swings towards you, it is considered reversed. This makes for four handing possibilities: left-hand, right-hand, left-hand reverse, and right-hand reverse (**Figure 60–14**):

- Right-hand: Door swings away, hinges on the right.
- Right-hand reverse: Door swings towards you, hinges on the right.
- Left-hand: Door swings away, hinges on the left.

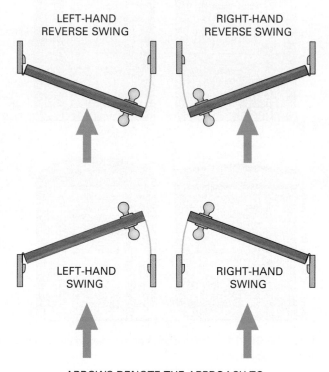

ARROWS DENOTE THE APPROACH TO THE ROOM THE DOORS ACCESS

FIGURE 60–14 Four possibilities of door swings.

- Left-hand reverse: Door swings towards you, hinges on the left.

It should be noted here that the handing of the door is simplified when doors have both surfaces alike. A left-hand door is the same as a right-hand reverse door. The only reason to make the different designations is to deal with the inside and outside surfaces of an exterior door.

When communicating doors with another person, it is a good idea to spend time making sure the correct door is ordered or assembled. Confusion here can cause lots of lost time and money.

OTHER EXTERIOR DOORS

Other exterior doors, such as swinging double doors, are fitted and hung in a similar manner to single doors. Allowance must be made, when fitting swinging double doors, for an **astragal** between them for weather-tightness (**Figure 60–15**). An astragal is a *moulding* that is rabbeted on both edges. It is designed to cover the joint between double doors. One edge has a square rabbet. The other has a bevelled rabbet to allow for the swing of one of the doors.

Patio Doors

Patio door units normally consist of two or three sliding or swinging glass doors completely assembled in the frame (**Figure 60–16**). Instructions for assembly are included with the unit. They should be followed carefully. Installation of patio door frames is similar to setting frames for swinging doors. After the frame is set, the doors are installed using special hardware supplied with the unit.

In sliding patio door units, one door is usually stationary. In two-door swinging units, either one or the other door is the swinging door. In three-door units, the centre door is usually the swinging door.

Garage Doors

Overhead garage doors come in many styles, kinds, and sizes. Two popular kinds used in residential construction are the *one-piece* and the *sectional* door. The rigid one-piece unit swings out at the bottom. It then slides up and overhead. The sectional type has hinged sections. These sections move upward on rollers and turn to a horizontal position overhead.

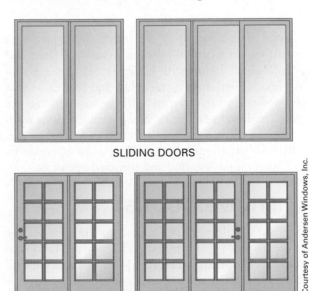

SLIDING DOORS

SWINGING DOORS

Courtesy of Andersen Windows, Inc.

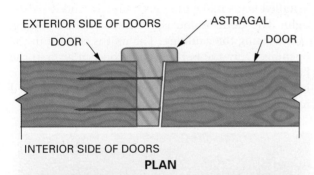

EXTERIOR SIDE OF DOORS
ASTRAGAL
DOOR
DOOR
INTERIOR SIDE OF DOORS
PLAN

Courtesy of Andersen Windows, Inc.

FIGURE 60–15 An astragal is required between double doors for weather-tightness.

Courtesy of Andersen Windows, Inc.

FIGURE 60–16 Two or three doors usually are used in sliding- or swinging-type patio door units.

A *rolling steel door*, used mostly in commercial construction, consists of narrow metal slats that roll up on a drum installed above the opening.

Special hardware, required for all types, is supplied with the door. Equipment is available for power operation of garage doors, including remote control. Also, supplied are the manufacturers' directions for installation. These should be followed carefully. There are differences in the door design and hardware of many manufacturers.

Chapter 61

Exterior Door Installation

Careful installation of a door frame and the hanging of a door are essential jobs for a carpenter. Fast and accurate installation is the hallmark of a craftsperson. A smooth operating door will last and protect against the weather for many years. Although most doors are purchased prehung, fitting and hanging a door from scratch is still an important part of the carpentry trade. There are situations in remodelling work where doors are hung on pre-existing and trimmed openings.

PREHUNG DOORS

The term **prehung door** is used to describe a door frame that is preassembled and ready to stand into the rough opening (**Figure 61–1**). It has side and head jambs fastened together with the sill. The door is bored or mortised for locksets and is attached to the frame via the hinges. The exterior casing and weatherstripping are also attached. The unit is ready for installation.

Installing a Prehung Door

Installation begins with unpacking the door unit. It comes with pads and blocks stapled to the frame, which are designed to protect the door frame during shipping. Remove all packing and be careful to remove all staples.

Verify that the opening is ready for the door unit. The width and height dimensions of the rough opening should be large enough for the unit to fit inside and also allow space to level and plumb the frame. The sheathing and drywall must not project into the opening, as it will interfere with fitting the unit. The floor must be checked for level. If it is not level, the sill must be shimmed and fully supported by the shim pieces.

Base flashing should be installed on top of the subfloor to protect the framing from water penetration. It should also slope to the outside slightly, which can be accomplished with bevel siding. This will encourage any water to exit to the outside. Housewrap must lap under the door side casing and over the head flashing. Sheathing surfaces must be sealed to the door in a similar fashion as for windows (see Chapter 59).

Applying caulking to the base flashing and to the back of the side and head casings before the unit is installed helps make the unit more air- and weather-tight. The door frame is then set in the opening, sandwiching the caulking. The unit is centred in the opening. Only the bottom is of concern at first; the head will be adjusted later. Tack the lower end of both side casings to the wall. The unit can be checked for level at the sill or head jamb.

FIGURE 61–1 Doors can be prehung to speed up installation.

The side jambs are then plumbed using a 6-foot (1.8 m) level. Read the level as close to perfect as possible. If a long level is not available, a combination of a straightedge and a carpenter's level will work (see Figure 40–25). A short carpenter's level is not suitable because any bows in the jambs are harder to realize and remove. If both jambs are bowed, small blocks of equal thickness may be placed at the top and bottom of the level. Once the unit is level and plumb, it may be tacked into the upper corners of the casing.

At this point, it is a good idea to check the operation of the door. Open the door and step to the inside. Close the door again and look at the gap between the door and the jamb. The gap should be the same all the way around the door. This will give the best operation of the door and weatherstrip. Once the operation is satisfactory, fasten the casing to the wall. Countersink casing fasteners as needed for finishing.

Slice the housewrap about ½ inch (12 mm) above the head casing. Insert an L-shaped piece of flashing material under the wrap and lap over the head casing (**Figure 61–2**). Make the flashing longer than the head casing to match the siding finish (J-moulding), and then bend and fold the ends up to form an end dam (see Figure 58–11). Apply flashing tape over the cut. Note that if flashing tape is applied to the sides of the door, it should lap under the head flashing tape.

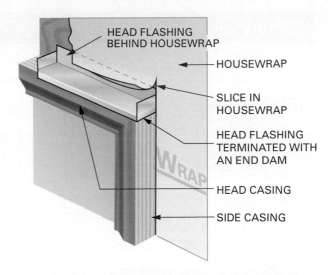

FIGURE 61–2 The housewrap laps over the drip cap (head flashing) and the drip cap over the head casing.

Jambs must now be shimmed to the trimmer studs. Shims are about 1½ × 12 inch (38 × 305 mm) long tapered wood pieces. They are installed in pairs with their tapers reversed. This makes the shim set thickness adjustable, yet always the same thickness throughout the set. Shim sets are installed from the inside (**Figure 61–3**).

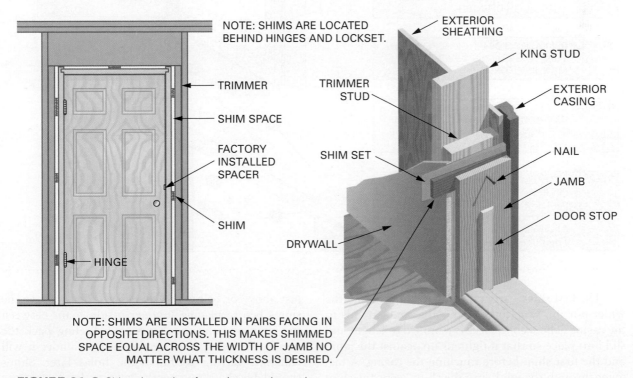

NOTE: SHIMS ARE INSTALLED IN PAIRS FACING IN OPPOSITE DIRECTIONS. THIS MAKES SHIMMED SPACE EQUAL ACROSS THE WIDTH OF JAMB NO MATTER WHAT THICKNESS IS DESIRED.

FIGURE 61–3 Shimming a door frame in a rough opening.

StepbyStep Procedures

PROCEDURE 61–A Installing a Prehung Exterior Door

STEP 1 REMOVE PACKING MATERIAL. CHECK OVER THE ROUGH OPENING FOR PROPER SIZE, DRYWALL AND SHEATHING CUT BACK, AND LEVELNESS OF FLOOR. LEVEL THE FLOOR IF NECESSARY WITH WIDE SHIMS.

STEP 2 INSTALL A BASE FLASHING. APPLY SEVERAL BEADS OF CHALK ON THE BASE FLASHING THAT WILL ADHERE TO THE SILL.

STEP 3 PLACE DOOR FRAME IN THE OPENING. CENTRE THE BOTTOM CASING IN THE OPENING AND LEVEL THE FRAME.

STEP 4 TACK THE BOTTOM ENDS OF EACH SIDE CASING TO THE WALL.

STEP 5 PLUMB EACH SIDE JAMB AND TACK THE TOP ENDS OF CASING TO THE WALL.

STEP 6 OPEN DOOR TO CHECK OPERATION. STRAIGHTEN JAMBS AS NECESSARY. FASTEN THE CASING TO THE WALL.

STEP 7 SHIM AND ADJUST THE JAMB TO THE ROUGH OPENING.

STEP 8 REPLACE A TOP HINGE SCREW WITH THE LONGER ONE PROVIDED BY MANUFACTURER.

STEP 9 CUT HOUSEWRAP (BLUE LINE). INSERT HEAD FLASHING ON TOP OF THE HEAD CASING AND UNDER THE WRAP.

STEP 10 INSTALL FLASHING TAPE ALONG PERIMETER OF THE DOOR STARTING AT THE BOTTOM AND WORKING TO THE TOP. OMIT THE BOTTOM TAPE TO ALLOW A PLACE FOR ANY WATER THAT LEAKS IN TO ESCAPE.

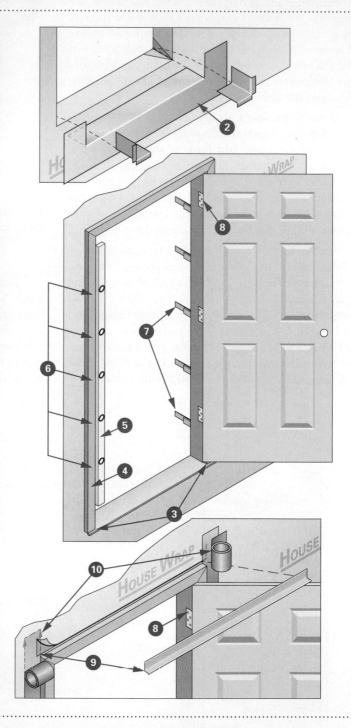

The first shim is trimmed as necessary so that the wider portion is as thick as possible and extends to the casing. The second piece is trimmed to length and slid into place so that it tightens up against the jamb and the first shim before touching the casing. A fastener through the jamb into the shim keeps the shim sets in place. Another way is to apply the fasteners just above or below the shims and then remove the shims before placing low-expansion foam between the jamb and the rough opening. Peeling back the weatherstripping to place the fastener under it will eliminate having to finish the holes later. Shims should be snug, not overly tight or loose, because they can create a bow in the jamb.

Wood jambs need at least six shim sets, three on each side. Other jambs, such as those made of plastic, should have more; five on each side and one on the head jamb shim set is recommended. It is also desirable to have a shim set behind each hinge and the lockset strike plate. These are *minimum* standards, increasing the number of shims makes the door more stable and secure.

Exterior doors come with a long screw attached in the packaging. This is to replace a top hinge screw. The long screw is driven through a shim set into the trimmer stud. It secures the top hinge and jamb back to the rough opening. This prevents the door hinge from sagging away, causing the door to rub the jamb as it closes (**Procedure 61–A**).

SETTING DOOR FRAMES IN MASONRY WALLS

In commercial construction, exterior wood or metal door frames are sometimes set in place before masonry walls are built. The frames must be set and firmly braced in a level and plumb position. The head jamb is checked for level. The bottom ends of the side jambs are secured in place. It may be necessary to shim one or the other side jamb in order to level the head jamb. The side jambs are then plumbed in a sideways direction. They are braced in position. Then the frame is plumbed and braced at a right angle to the wall (**Procedure 61–B**). It is important at this point to install a 2 × 4 (38 × 89) wood frame

StepbyStep Procedures

PROCEDURE 61–B Installing an Exterior Door Frame in a Masonry Wall

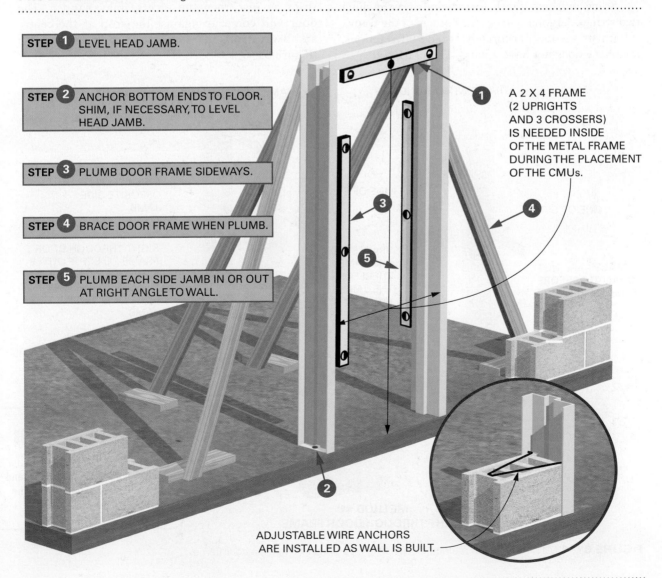

STEP 1 LEVEL HEAD JAMB.

STEP 2 ANCHOR BOTTOM ENDS TO FLOOR. SHIM, IF NECESSARY, TO LEVEL HEAD JAMB.

STEP 3 PLUMB DOOR FRAME SIDEWAYS.

STEP 4 BRACE DOOR FRAME WHEN PLUMB.

STEP 5 PLUMB EACH SIDE JAMB IN OR OUT AT RIGHT ANGLE TO WALL.

A 2 X 4 FRAME (2 UPRIGHTS AND 3 CROSSERS) IS NEEDED INSIDE OF THE METAL FRAME DURING THE PLACEMENT OF THE CMUs.

ADJUSTABLE WIRE ANCHORS ARE INSTALLED AS WALL IS BUILT.

(two uprights and three crossers) within the metal frame to help to withstand the pressure of the concrete masonry units (CMUs).

Finally, the frame is checked to see if it has a **wind**. The term wind is pronounced the same way that it is pronounced in the phrase "wind a clock." A wind is a twist in the door frame caused when the side jambs do not line up vertically with each other. No matter how carefully the side jambs of a door frame are plumbed, it is always best to check the frame to see if it has a wind.

One method of checking the door frame for a wind is to stand to one side. Look diagonally through the opening in the door frame to see if the outer edge of one side jamb lines up with the inner edge of the other side jamb. If they do not line up, the frame is in a wind. Make adjustments until they do line up. One way of making the adjustment is to plumb and brace one side at a right angle to the wall. Then sight, line up, and brace the other side jamb (**Figure 61–4**).

Some workers check for a wind by stretching two strings diagonally from the corners of the frame. If both strings meet accurately at their intersections, the frame does not have a wind (**Figure 61–5**).

MAKING A DOOR UNIT

The first step in making a door unit is to determine the side that will close against the stops on the door frame. If the design of the door permits that either side may be used toward the stops, sight along the door stiles from top to bottom to see if the door is bowed. Hardly any doors are perfectly straight. Most are bowed to some extent.

Determining Stop Side of Door

The door should be fitted to the opening so that the hollow side of the bow will be against the door stops. Hanging a door in this manner allows the top and bottom of the closed door to come up tight against stops. The centre comes up tight when the door is latched. Also, the door will not rattle.

If no attention is paid to which side the door stops against, then the reverse may happen. The door will come up against the stop at the centre and away from the stop at the bottom and top (**Figure 61–6**).

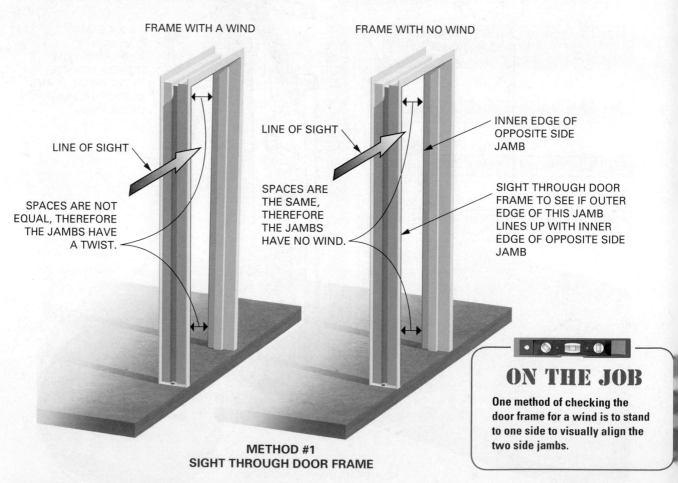

FRAME WITH A WIND

FRAME WITH NO WIND

LINE OF SIGHT

LINE OF SIGHT

SPACES ARE NOT EQUAL, THEREFORE THE JAMBS HAVE A TWIST.

SPACES ARE THE SAME, THEREFORE THE JAMBS HAVE NO WIND.

INNER EDGE OF OPPOSITE SIDE JAMB

SIGHT THROUGH DOOR FRAME TO SEE IF OUTER EDGE OF THIS JAMB LINES UP WITH INNER EDGE OF OPPOSITE SIDE JAMB

METHOD #1 SIGHT THROUGH DOOR FRAME

ON THE JOB

One method of checking the door frame for a wind is to stand to one side to visually align the two side jambs.

FIGURE 61–4 Visual inspection method for checking for a wind or twist in a door frame.

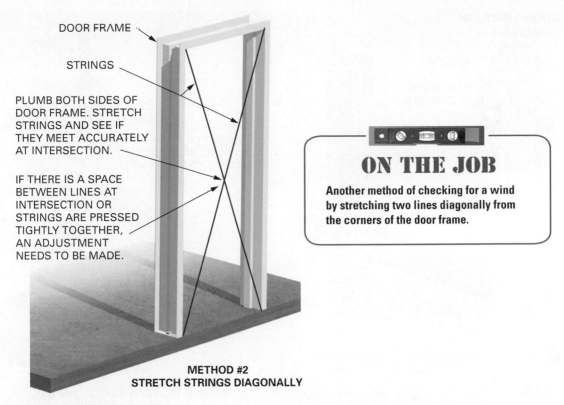

DOOR FRAME

STRINGS

PLUMB BOTH SIDES OF DOOR FRAME. STRETCH STRINGS AND SEE IF THEY MEET ACCURATELY AT INTERSECTION.

IF THERE IS A SPACE BETWEEN LINES AT INTERSECTION OR STRINGS ARE PRESSED TIGHTLY TOGETHER, AN ADJUSTMENT NEEDS TO BE MADE.

ON THE JOB

Another method of checking for a wind by stretching two lines diagonally from the corners of the door frame.

METHOD #2 STRETCH STRINGS DIAGONALLY

FIGURE 61–5 String method for checking for a wind or twist in a door frame.

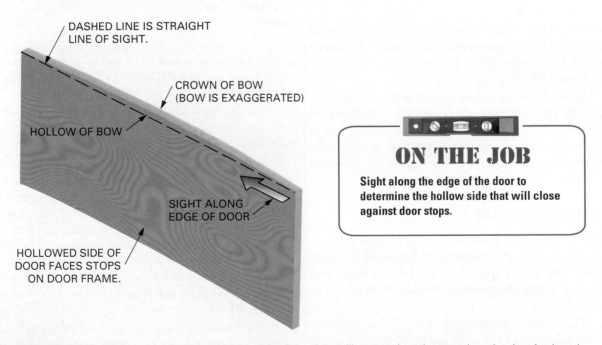

DASHED LINE IS STRAIGHT LINE OF SIGHT.

CROWN OF BOW (BOW IS EXAGGERATED)

HOLLOW OF BOW

SIGHT ALONG EDGE OF DOOR

HOLLOWED SIDE OF DOOR FACES STOPS ON DOOR FRAME.

ON THE JOB

Sight along the edge of the door to determine the hollow side that will close against door stops.

FIGURE 61–6 Method of determining which side of the door will rest against the top when the door is closed.

Determining Exposed Side of Sash Doors

It is important to hang exterior doors containing lights of glass with the proper side exposed to the weather. This prevents wind-driven rainwater from

seeping through joints. Manufacturers clearly indicate, with warning labels glued to the door, which side should face the exterior. Any door warranty is voided if the door is improperly hung. Do not hang exterior doors with the removable **glass bead** facing

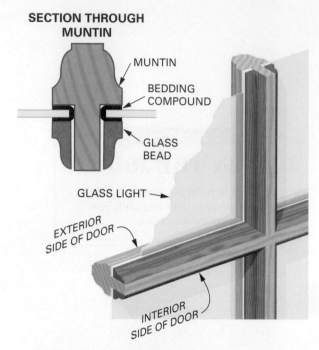

SECTION THROUGH MUNTIN

MUNTIN

BEDDING COMPOUND

GLASS BEAD

GLASS LIGHT

EXTERIOR SIDE OF DOOR

INTERIOR SIDE OF DOOR

GLASS BEAD IS THE MOULDING APPLIED AROUND LIGHTS OF GLASS.

FIGURE 61–7 Doors containing glass lights should face in the direction recommended by the manufacturer.

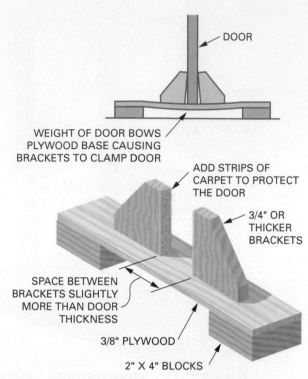

DOOR

WEIGHT OF DOOR BOWS PLYWOOD BASE CAUSING BRACKETS TO CLAMP DOOR

ADD STRIPS OF CARPET TO PROTECT THE DOOR

3/4" OR THICKER BRACKETS

SPACE BETWEEN BRACKETS SLIGHTLY MORE THAN DOOR THICKNESS

3/8" PLYWOOD

2" X 4" BLOCKS

A DOOR JACK CAN BE MADE ON THE JOB FROM SCRAP LUMBER.

FIGURE 61–8 Door jacks make working on a door edge easier.

outward. Glass bead is small moulding used to hold lights of glass in the opening. The bead can be identified by holes made when fasteners of the bead were set (**Figure 61–7**).

Some doors are manufactured with **compression glazing**. This virtually eliminates the possibility of any water seeping through the joints. When low-E insulating glass is used in a door, it is especially important to have the door facing in the direction indicated by the manufacturer.

Fitting the Door

Place the door on sawhorses, with its face side up. Measure carefully the width and height of the door frame. The frame should be level and plumb, but this may not be the case and should not be taken for granted.

The process of fitting a door into a frame is called jointing. The door must be carefully jointed. An even joint of approximately ³⁄₃₂ inch (2 mm) must be made between the door and the frame on all sides. A wider joint of approximately ⅛ inch (3 mm) must be made to allow for swelling of the door and frame in extremely damp weather.

Use a door jack to hold the door steady. A manufactured or job-made jack may be used (**Figure 61–8**).

Use a jointer plane, either hand or power. Plane the edges and ends of the door so that it fits snugly in the door frame.

Fit the top end against the head jamb. Then fit the bottom against the sill so that a proper joint is obtained. Careful jointing may require moving the door in and out of the frame several times.

Next, joint the hinge edge against the side jamb. Finally, plane the lock edge so that the desired joint is obtained on both sides. The lock edge must also be planed on a *bevel*. This is so the back edge does not strike the jamb when the door is opened. The bevel is determined by making the same joint between the back edge of the door and the side jamb when the door is slightly open, as when the door is closed (**Figure 61–9**).

Extreme care must be taken when fitting doors so as not to get them undersize. The door can always be jointed a little more. However, it cannot be made wider or longer (**Figure 61–10**). Do not cut more than ½ inch (12 mm) total from the width of a door. Cut no more than 2 inches (50 mm) from its height. Cut equal amounts from ends and edges when approaching maximum amounts. Check the fit frequently by placing the door in the opening, even

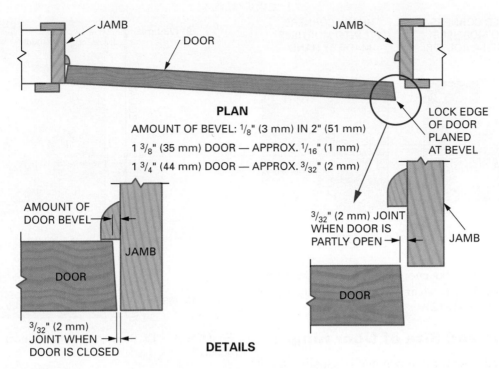

PLAN

AMOUNT OF BEVEL: $\frac{1}{8}$" (3 mm) IN 2" (51 mm)

1 $\frac{3}{8}$" (35 mm) DOOR — APPROX. $\frac{1}{16}$" (1 mm)

1 $\frac{3}{4}$" (44 mm) DOOR — APPROX. $\frac{3}{32}$" (2 mm)

DETAILS

FIGURE 61–9 The lock edge of a door must be planed at a bevel.

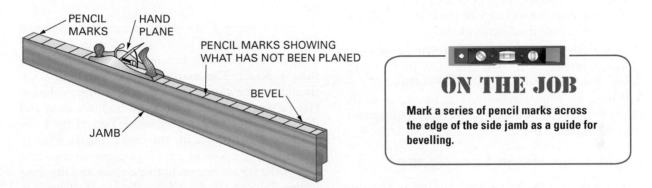

ON THE JOB

Mark a series of pencil marks across the edge of the side jamb as a guide for bevelling.

FIGURE 61–10 Technique for seeing the amount of material removed while planing.

if this takes a little extra time. Most entrance doors are very expensive. Care should be taken not to ruin one. Speed will come with practice. Handle the door carefully to avoid marring it or other finish. After the door is fitted, *ease* all sharp corners slightly with a block plane and with sandpaper. To ease sharp corners means to round them over slightly.

HANGING THE DOOR

On swinging doors, the loose-pin type **butt hinge** is ordinarily used. The pin is removed. Each *leaf* of the hinge is applied separately to the door and frame. The door is hung by placing it in the opening. The pins are inserted to rejoin the separated hinge leaves.

Extreme care must be taken so that the hinge leaves line up exactly on the door and frame. Three 4 × 4 (100 × 100 mm) hinges on 1¾-inch (44 mm) doors 7'-0" (2134 mm) or less in height are often used. Use four hinges on doors over 7'-0" (2134 mm) in height.

The hinge leaves are recessed flush with the door edge and only partway across. The recess for the hinge is called a **hinge gain**. Butt hinges come in two styles. They vary according to the method of cutting the gain (**Figure 61–11**). Hinge gains are only made partway across the edge of the door. This is so that the edge of the hinge is not exposed when the door is opened. The remaining distance, from the edge of the hinge to the side of the door, is called the **backset** of the hinge. Butt hinges must be wide enough so that

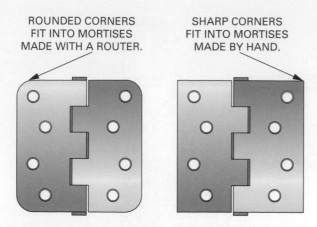

FIGURE 61–11 Two styles of butt hinges.

the pin is located far enough beyond the door face to allow the door to clear the door trim when fully opened (**Figure 61–12**).

Location and Size of Door Hinges

On panelled doors, the top hinge is usually placed with its upper end in line with the bottom edge of the top rail. The bottom hinge is placed with its lower end in line with the top edge of the bottom rail. The middle hinge is centred between them.

On **flush** doors, the usual placement of the hinge is approximately 7 to 9 inches (178 to 229 mm) down from the top and 11 to 13 inches (279 to 330 mm) up from the bottom, as measured to the centre of the hinge. A middle hinge is centred between the two (**Figure 61–13**).

Laying Out Hinge Locations

Place the door in the frame. Shim the top and bottom so the proper joint is obtained. Place shims between the lock edge of the door and side jamb of the frame.

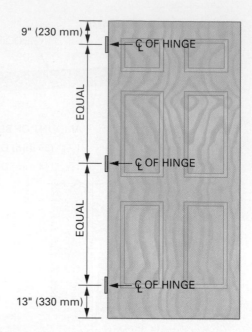

FIGURE 61–13 Recommended placement of hinges on doors.

The hinge edge should be tightly against the door jamb.

Use a sharp knife. Mark across the door and jamb at the desired location for one end of each hinge. A knife is used because it makes a finer line than a pencil. The marks on the door and jamb should not be any longer than the hinge thickness. Place a small X, with a pencil, on both the door and the jamb. This indicates on which side of the knife mark to cut the gain for the hinge (**Figure 61–14**). Care must be taken to cut hinge gains on the same side of the layout line on both the door and the door frame. Remove the door from the frame. Place it in the door jack with its hinge edge up in order to mark out and cut the hinge gains.

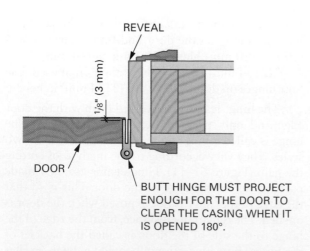

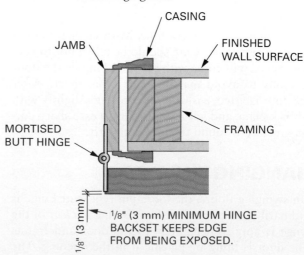

FIGURE 61–12 Hinges are set with the hinge barrel projecting out from the side of the door and jamb.

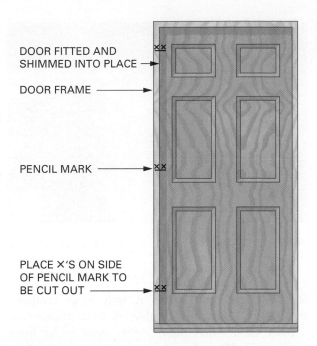

DOOR FITTED AND
SHIMMED INTO PLACE

DOOR FRAME

PENCIL MARK

PLACE ✕'S ON SIDE
OF PENCIL MARK TO
BE CUT OUT

FIGURE 61–14 Mark the location of hinges on the door and frame with a sharp knife.

Marking Out the Hinge Gain

The first step in marking out a hinge gain is to mark its ends. Place a hinge leaf on the door edge with its end on the knife mark previously made. With the *barrel* of the leaf against the side of the door, hold the leaf firmly. Score a line along one end with a sharp knife. Then tap the other end until the leaf just covers the line. Score a line along the other end (**Figure 61–15**). Score only partway across the door edge.

Cutting Hinge Gains

Use a sharp knife to deepen the scored backset line. It may be necessary to draw the knife along the line several times to score to the bottom of the gain. Take care if using a chisel for scoring. Using a chisel will easily split the edge of the door (**Figure 61–16**).

With the bevel of the chisel down, cut a small chip from each end of the gain. The chips will break off at the scored end marks (see **Procedure 12–B**, page 88). Holding the flat of the chisel on the shoulders of the gain, tap the chisel down to the bottom of the gain.

Make a series of small chisel cuts along the length of the gain. Brush off the chips. Then, with the flat of the chisel down, pare and smooth the excess down to the depth of the gain. Be careful not to slip and cut off the backset (**Figure 61–17**).

After the gain is made, the hinge leaf should press-fit into it. It should be flush with the door edge.

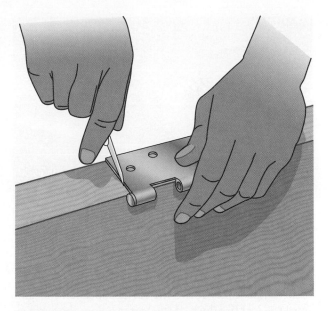

FIGURE 61–15 The ends of hinge gains may be laid out by scoring with a knife along a hinge leaf.

ON THE JOB

Use a knife to deepen the scored backset line when mortising for hinges.

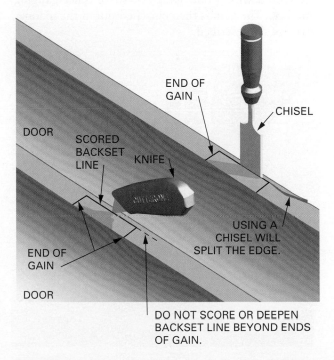

DOOR

SCORED
BACKSET
LINE

KNIFE

END OF
GAIN

CHISEL

USING A
CHISEL WILL
SPLIT THE EDGE.

END OF
GAIN

DOOR

DO NOT SCORE OR DEEPEN
BACKSET LINE BEYOND ENDS
OF GAIN.

FIGURE 61–16 A utility knife is useful in deepening the cut that is parallel to the grain.

FIGURE 61–17 Chiselling out the hinge gain.

If the hinge leaf is above the surface, deepen the gain until the leaf lies flush. If the leaf is below the surface, it may be shimmed flush with thin pieces of cardboard from the hinge carton.

Using Butt Hinge Markers. Instead of laying out hinge gains with a knife, butt hinge markers of several sizes are often used (see Figure 11–26, page 81). With markers, the location of the hinge leaves must still be marked on the door and jamb, as described. The width and length of the hinge gain are outlined by simply placing the marker in the proper location on the door edge and then tapping it with a hammer (**Figure 61–18**). However, the depth of the gain must be scored with a butt gauge or some other gauging method. The gain is then chiselled out in the manner previously described.

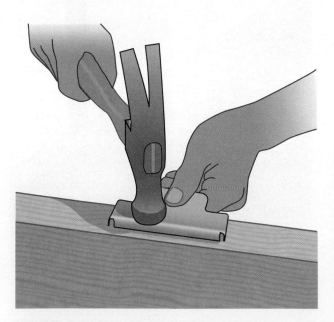

FIGURE 61–18 Sometimes, butt hinge markers are used to lay out hinge gains.

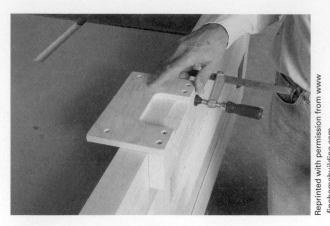

Reprinted with permission from www.finehomebuilding.com.

FIGURE 61–19 When using butt hinges, chisel the corners of the mortise square.

Routing a Butt Hinge. A butt hinge template fixture or jig and portable electric router are usually used when many doors need to be hung (**Figure 61–19**). Most hinge routing jigs contain three adjustable templates secured on a long rod. The templates are temporarily attached into position. The jig is used on both the door and frame as a guide to rout hinge gains. The templates are adjustable for different size hinges, hinge locations, and door thicknesses. An attachment positions the hinge template jig to provide the required joint between the top of the door and the frame. The gains are cut using a router with a special hinge mortising bit. The bit is set to cut the gain to the required depth. A template guide is attached to the base of the router. It rides against the jig template when routing the hinge gain (**Figure 61–20**).

Butt hinges with rounded corners are used in routed gains. Hinges with square corners require that the rounded corner of a routed gain be chiselled to a square corner.

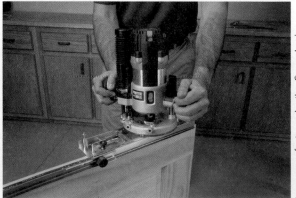

In memory of esteemed author Gaspar Lewis

FIGURE 61–20 A butt hinge template can be used on both door and frame when routing hinge gains.

Applying Hinges

Press the hinge leaf in the gain. Drill pilot holes for the screws. Centre the pilot holes carefully on the countersunk holes of the hinge leaf. A centring punch is often used for this purpose. Drilling off centre will cause the hinge to move from its position when the screw is driven. Fasten the hinge leaf with screws provided with the hinges. Cut gains and apply all hinge leafs on the door and frame in the same manner.

Shimming Hinges

Hang the door in the frame by inserting a pin in the barrels of the top hinge leaves first. Insert the other pins. Try the swing of the door. If the door binds against or is too close to the jamb on the hinged side, shim between the hinge leaves and the gain. Use a narrow strip of cardboard on the side of the screws nearest the pin. This will move the door toward the lock side of the door frame. If the door binds against or is too close to the jamb on the lock side, apply shims in the same manner. However, apply them on the opposite side of the hinge screws (**Figure 61–21**). Check the bevel on the lock edge of the door. Plane the lock edge to the proper bevel, if necessary. Ease all sharp, exposed corners.

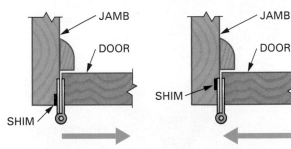

PLACING THE SHIM TOWARD THE OUTSIDE OF THE HINGE MOVES THE DOOR TOWARD THE LOCK EDGE.

PLACING THE SHIM TOWARD THE INSIDE OF THE HINGE MOVES THE DOOR AWAY FROM THE LOCK EDGE.

FIGURE 61–21 Shimming the hinge edges will move the door toward or away from the lock edge.

Chapter 62

Installing Exterior Door Hardware

After the door has been fitted and hung in the frame, the lockset and other door hardware are installed. A large variety of locks are available from several large manufacturers in numerous styles and qualities, providing a wide range of choices. Doors must be prepared to accept the lockset (**Figure 62–1**).

CYLINDRICAL LOCKSETS

Cylindrical locksets are often called *key in the knob locksets* (**Figure 62–2**). They are the most commonly used type in both residential and commercial construction. This is primarily because of the ease and speed of installing them. They may be obtained in several groups, from light-duty residential to extra-heavy-duty commercial and industrial applications.

In place of knobs, a **lever handle** is provided on locksets likely to be used by handicapped persons or in other situations where a lever is more suitable than a knob (**Figure 62–3**). **Deadbolt** locks are used for both primary and auxiliary locking of doors in residential and commercial buildings (**Figure 62–4**). They provide additional security. They also make an attractive design in combination with *grip-handle* locksets or latches (**Figure 62–5**). The deadbolt lock shall have a cylinder with no fewer than five pins and a bolt throw not less than 1 inch (25 mm) [NBCC 9.7.5.2(4) a, b]. Mortised locksets are more often

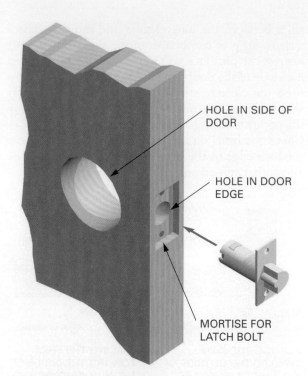

HOLE IN SIDE OF
DOOR

HOLE IN DOOR
EDGE

MORTISE FOR
LATCH BOLT

FIGURE 62–1 The method of preparing the door for
cylindrical locksets.

Courtesy of Schlage Lock Co.

FIGURE 62–2 Cylindrical locksets are the most commonly
used type in both residential and commercial construction.

Courtesy of Schlage Lock Co.

FIGURE 62–3 Locksets with lever handles are easier
to turn.

Courtesy of Schlage Lock Co.

FIGURE 62–4 Deadbolt locks are used primarily as
auxiliary locks for added security.

Courtesy of Schlage Lock Co.

FIGURE 62–5 A grip-handle lockset combines well with
a deadbolt lock.

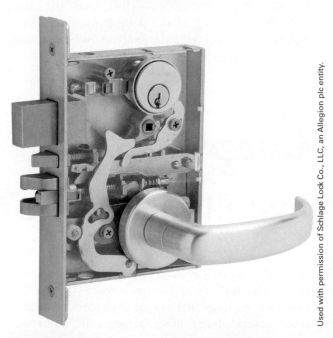

Used with permission of Schlage Lock Co., LLC, an Allegion plc entity.

FIGURE 62–6 Mortised locksets are designed for doors
with heavy traffic and use.

used in commercial doors than residential doors.
The name comes from the mortised hole needed in
the door to accept the lockset. Today, these doors
typically come precut with a mortised slot for the
lockset to slip into. Mortised locksets are designed
to be durable enough to withstand constant use
(**Figure 62–6**).

INSTALLING CYLINDRICAL LOCKSETS

To install a cylindrical lockset, first check the contents and read the manufacturer's directions carefully. There are so many kinds of locks manufactured that the mechanisms vary greatly. The directions included with the lockset must be followed carefully.

However, there are certain basic procedures. Open the door to a convenient position. Wedge the bottom to hold it in place (**Figure 62–7**). Measure up, from the floor, the recommended distance to the centreline of the lock. This is usually 36 to 40 inches (914 to 1016 mm). At this height, square a light line across the edge and stile of the door.

Marking and Boring Holes

Position the centre of the paper template supplied with the lock on the squared lines. Lay out the centres of the holes that need to be bored (**Figure 62–8**). It is important that the template be folded over the high corner of the bevelled door edge. The distance from the door edge to the centre of the hole through the side of the door is called the *backset* of the lock. Usual backsets are 2⅜ inches (60 mm) for residential and 2¾ inches (70 mm) for commercial. Make sure the

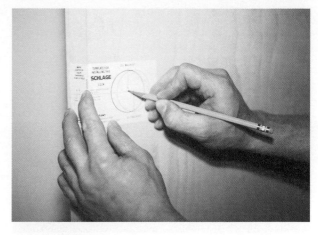

FIGURE 62–8 Using a template to lay out the centres of the holes for a lockset.

backset is marked correctly before boring the hole. One hole must be bored through the side and one into the edge of the door. The manufacturers' directions specify the hole sizes, where a 1-inch (25 mm) hole for bolts and 2⅛-inch (54 mm) hole for locksets are common.

The hole through the side of the door should be bored first. Stock for the centre of the boring bit is lost if the hole in the edge of the door is bored first. It can be bored with hand tools, using an expansion bit in a bit brace. However, it is a difficult job. If using hand tools, bore from one side until only the point of the bit comes through. Then bore from the other side to avoid splintering the door.

Using a Boring Jig. A boring jig is frequently used. It is clamped to the door to guide power-driven multispur bits. With a boring jig, holes can be bored completely through the door from one side. The clamping action of the jig prevents splintering (**Figure 62–9**).

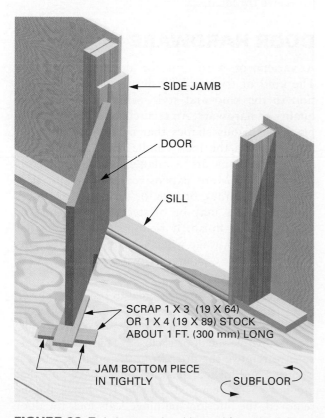

SIDE JAMB

DOOR

SILL

SCRAP 1 X 3 (19 X 64) OR 1 X 4 (19 X 89) STOCK ABOUT 1 FT. (300 mm) LONG

JAM BOTTOM PIECE IN TIGHTLY

SUBFLOOR

FIGURE 62–7 A door may be shimmed from the floor to hold it plumb during installation.

FIGURE 62–9 Boring jigs are frequently used to guide bits when boring holes for locksets.

FIGURE 62–10 Using a faceplate marker.

After the holes are bored, insert the latch bolt in the hole bored in the door edge. Hold it firmly and score around its faceplate with a sharp knife. Remove the latch unit. Deepen the vertical lines with the knife in the same manner as with hinges. Take great care when using a chisel along these lines. This may split out the edge of the door. Then chisel out the recess so that the faceplate of the latch lays flush with the door edge.

A **faceplate marker**, if available, may be used to lay out the mortise for the latch faceplate. A marker of the appropriate size is held in the bored latch hole and tapped with a hammer (**Figure 62–10**). Complete the installation of the lockset by following specific manufacturers' directions.

Installing the Striker Plate

The **striker plate** is installed on the door jamb so that when the door is closed it latches tightly with no play. If the plate is installed too far out, the door will not close tightly against the stop. It will then rattle. If the plate is installed too far in, the door will not latch.

To locate the striker plate in the correct position, place it over the latch in the door. Close the door snugly against the stops. Push the striker plate in against the latch. Draw a vertical line on the face of the plate flush with the outside face of the door (**Figure 62–11**).

Open the door. Place the striker plate on the jamb. The vertical line, previously drawn on it, should be in line with the edge of the jamb. Centre the plate on the latch. Hold it firmly while scoring a line around the plate with a sharp knife. Chisel

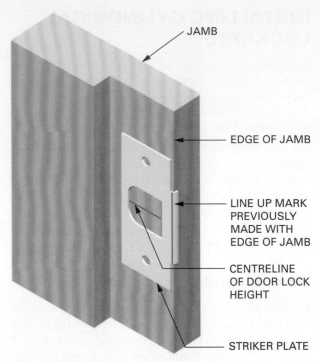

FIGURE 62–11 Installing the striker plate.

out the mortise so that the plate lies flush with the jamb. Screw the plate in place. Chisel out the centre to receive the latch.

DOOR HARDWARE

A variety of hardware for doors is available. The kind of trim and finish are factors, in addition to the kind and style, which determine the quality of hardware. An **escutcheon** is a decorative plate of various shapes that is installed between the door and the lock handle or knob. Locksets and escutcheons are available in various metals and finishes. More expensive locksets and trim are made of brass, bronze, or stainless steel. Less expensive ones may be of steel that is plated or coated with a finish. It is important to consider conditions and usage when selecting locksets. This is especially the case in areas with a humid climate or near salt water.

Security hinges are designed to prevent a closed door from being opened from the hinge side of the door. Most butt hinges have pins that may be removed while the door is closed. This makes it possible to remove the door from the opening when it is closed and locked. Security hinges have pins that are not easily removed. Should the hinge pin be removed, the interlocking hinge leafs lock the door

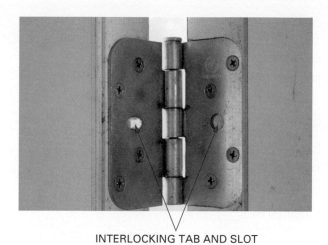

INTERLOCKING TAB AND SLOT

FIGURE 62–12 Security hinges are designed to keep hinge leaves from sliding apart if the pin is removed.

in the opening by resisting the leaves from sliding past each other (**Figure 62–12**).

Door closers are made to automatically close a door. They are attached to the door and the jamb. An adjustable spring mechanism allows variation to the speed and strength of the closer (**Figure 62–13**).

Door stops or keepers restrict the travel of the door. They may be installed to the door, floor, or the wall. They protect the door and wall from damage if door is opened too far (**Figure 62–14**).

Door viewers or peep holes allow view through the door from one side to the other [NBCC 9.7.2.1(2) a]. They offer a 190-degree limited view while blurring and obscuring the view from the other side. For this reason, it must be installed with the direction of the view in mind (**Figure 62–15**).

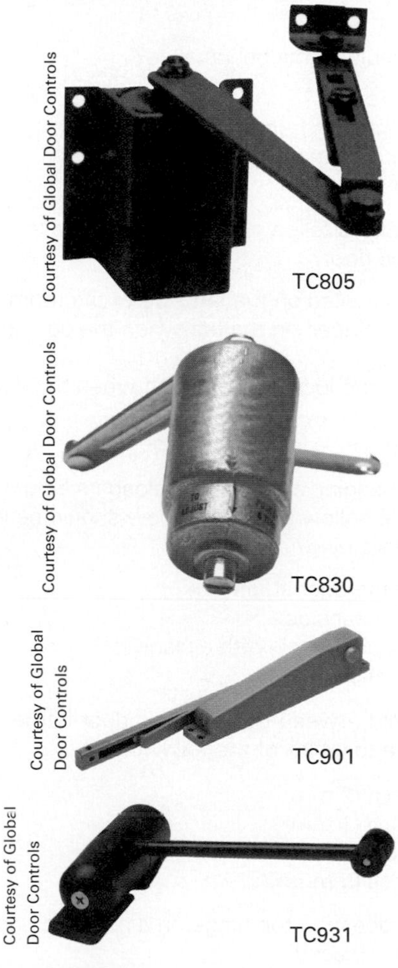

FIGURE 62–14 Stops are used to confine opening doors.

Courtesy of Global Door Controls

TC805

Courtesy of Global Door Controls

TC830

Courtesy of Global Door Controls

TC901

Courtesy of Global Door Controls

TC931

FIGURE 62–13 Door closers are designed to keep door closed after opening.

Courtesy of Schlage Lock Co.

FIGURE 62–15 Door viewers allow secure side occupant to view the other side without opening the door.

KEY TERMS

astragal	door casings	jointing	sticking
backset	door jack	lever handle	stool
boring jig	escutcheon	lock rail	striker plate
butt hinge	faceplate marker	lockset	transom
butt hinge template	flush	panel door	wind
compression glazing	flush door	patio door	
cylindrical locksets	glass bead	prehung door	
deadbolt	hinge gain	rabbeted jambs	

REVIEW QUESTIONS

Select the most appropriate answer.

1. What is the standard thickness of exterior wood doors in residential construction?

   a. 1⅜ inches (35 mm)
   b. 1½ inches (38 mm)
   c. 1¾ inches (44 mm)
   d. 2¼ inches (57 mm)

2. What is the typical minimum width of exterior entrance doors?

   a. 3'-0" (914 mm)
   b. 2'-8" (813 mm)
   c. 2'-6" (762 mm)
   d. 2'-4" (711 mm)

3. The height of exterior entrance doors in residential construction is generally not less than which of the following?

   a. 7'-0" (2134 mm)
   b. 6'-10" (2083 mm)
   c. 6'-8" (2032 mm)
   d. 6'-6" (1981 mm)

4. What is the term used to describe a door frame that is twisted?

   a. twist
   b. warp
   c. bowed
   d. wind

5. What is a narrow member dividing the glass in a door into smaller lights called?

   a. bar
   b. mullion
   c. rail
   d. all of the above

6. How should shims be installed?

   a. in pairs
   b. with nails through them
   c. behind hinges
   d. all of the above

7. Which of the following best describes a left-hand swinging door?

   a. It is installed on the left side of the room.
   b. It has hinges on the left when the door swings away.
   c. It has the lockset on the left when the door swings away.
   d. It requires a left hand to open it.

8. Before hanging a door, sight along its length for a bow. The hollow side of the bow should be which of the following?

   a. face the door stops
   b. face the inside
   c. be straightened with a plane
   d. all of the above

9. The joint between the door and door frame should be close to which of the following?

   a. 3/32 inch (2 mm)
   b. 3/64 inch (1 mm)
   c. ¼ inch (6 mm)
   d. 3/16 inch (5 mm)

10. Where does the top hinge on a panelled door have its centre placed?

    a. in line with the top edge of the bottom rail
    b. in line with the top edge of the intermediate rail
    c. 9 inches (229 mm) down from the top of the door
    d. 13 inches (330 mm) down from the top of the door

11. How far is the centre of the hole for a lockset in a residential door typically set back from the door edge?

    a. 2 inches (50 mm)
    b. 2⅛ inches (55 mm)
    c. 2⅜ inches (60 mm)
    d. 2¾ inches (70 mm)

12. What is the typical lockset hole diameter in residential doors?

    a. 2 inches (50 mm)
    b. 2⅛ inches (55 mm)
    c. 2⅜ inches (60 mm)
    d. 2¾ inches (70 mm)

13. What does the term *sidelights* refer to?

    a. outside lighting
    b. horizontal lights
    c. glass panels
    d. all of the above

14. What term refers to the part of a door frame located under a door?

    a. weatherstrip
    b. sole
    c. threshold
    d. escutcheon

15. What is the door hardware designed to prevent a closed door from being removed called?

    a. security hinge
    b. door keeper
    c. door closer
    d. mortised lockset

Unit 20

Siding and Cornice Construction

The exterior finish work is the major visible part of the architectural design of a building. Because the exterior is so prominent, it is important that all finish parts be installed straight and true with well-fitted joints.

The portion of the finish that covers the vertical area of a building is the siding. Siding does not include masonry covering, such as stucco or brick veneer. Siding is used extensively in both residential and commercial construction.

That area where the lower portion of the roof, or eaves, overhangs the walls is called the cornice. Variations in cornice design and detail can set the appearance of one building apart from another.

Protecting a building from weather is the major function of exterior finish. Use care to install siding with the nature of rain and wind-blown water in mind. This will maximize the life of the finish and the building.

OBJECTIVES

After completing this unit, the student should be able to:

- describe the shapes, sizes, and grades of various siding products.
- install corner boards and prepare sidewalls for siding.
- apply horizontal and vertical lumber siding.
- apply plywood and hardboard panel and lap siding.
- apply wood shingles and shakes to sidewalls.
- apply aluminum and vinyl siding.
- estimate required amounts of siding.
- describe various types of cornices and name their parts.
- install gutters and downspouts.

Chapter 63
Wood Siding Types and Sizes

Siding is manufactured from solid lumber, plywood, hardboard, aluminum, concrete, and vinyl. It comes in many different patterns. Prefinished types eliminate the need to refinish for many years, if at all. Siding may be applied horizontally, vertically, or in other directions to make many interesting designs (**Figure 63–1**).

WOOD SIDING

The natural beauty and durability of solid wood has long made it an ideal material for siding. The Western Red Cedar Lumber Association (WRCLA) and the California Redwood Association (CRA) are two major organizations whose member mills manufacture siding and other wood products. They have to meet standards supervised by their associations. Grade stamps of the WRCLA and CRA and other associations of lumber manufacturers are placed on siding produced by member mills. Grade stamps ensure the consumer of a quality product that complies with established standards (**Figure 63–2**). WRCLA member mills produce wood siding from species such as fir, larch, hemlock, pine, spruce, and cedar.

Wood Siding Grades

Grain. Some siding is available in *vertical grain*, *flat grain*, or *mixed grain*. In vertical grain siding, the

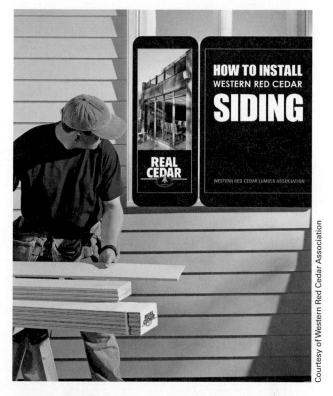

FIGURE 63–2 Western Red Cedar Lumber Association.

Courtesy of Western Red Cedar Association

annual growth rings, viewed in cross-section, must form an angle of 45 degrees or more with the surface of the piece. All other lumber is classified as flat grain (**Figure 63–3**). Vertical grain siding is the highest

FIGURE 63–1 Wood siding used in residential construction.

Courtesy of Fraser Wood Siding

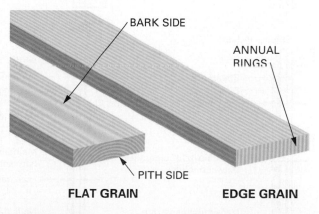

BARK SIDE

ANNUAL RINGS

PITH SIDE

FLAT GRAIN

EDGE GRAIN

FIGURE 63–3 In some species, siding is available in vertical, flat, or mixed grain.

quality. It warps less, takes and holds finishes better, has fewer defects, and is easier to work.

Surface Texture. Sidings are manufactured with *smooth*, *rough*, or *resawn* surfaces. Saw-textured surface finishes are obtained by resawing in the mill. They generally hold finishes longer than smooth surfaces.

WWPA Grades. Grades published by the Western Wood Products Association (WWPA) for siding products are shown in **Figure 63–4**. Siding graded as *premium* has fewer defects such as knots and pitch pockets. The highest premium grade is produced from clear, all-heart lumber. *Knotty* grade siding is divided into #1, #2, and #3 *common*. The grade depends on the type and number of knots and other defects.

Siding Patterns and Sizes

The names, descriptions, and sizes of siding patterns are shown in **Figure 63–5**. Some patterns can be used only for either horizontal or vertical applications. Others can be used for both. *Drop* and *tongue-and-grooved* sidings are manufactured in a variety of patterns other than shown. *Bevel* siding, more

General Categories (Note that there are additional grades for bevel pattern)		Grades		
		Western Species		Cedar
		Selects	Finish	Western & Canadian
All Patterns	Premium Grades	C Select	Superior	Clear Heart A Grade
		D Select	Prime	
				B Grade
Additional Grades for Bevel Patterns	Premium		Superior Bevel	Clear VG Heart A Bevel B Bevel Rustic C Bevel
	Knotty		Prime Bevel	Select Knotty Quality Knotty
All Patterns	Knotty Grades	Commons	Alternate Boards	
		#2 Common	Select Merch.	Select Knotty Quality Knotty
		#3 Common	Construction	
			Standard	

Courtesy of Western Wood Products Association

FIGURE 63–4 WWPA grade rules for siding products.

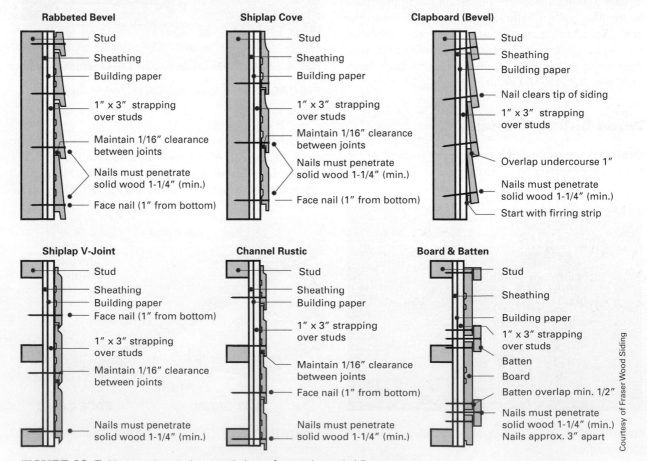

FIGURE 63–5 Names, descriptions, and sizes of natural wood siding patterns.

Courtesy of Fraser Wood Siding

Courtesy of Maibec Siding

FIGURE 63–6 Bevel siding is commonly known as clapboards.

commonly known as *clapboards*, is a widely used kind (**Figure 63–6**).

PANEL AND LAP SIDING

Most *panel* and *lap* siding is manufactured from plywood, hardboard, and fibre-cement boards. They come in a variety of sizes, patterns, and surface textures. Plywood siding manufactured by American Plywood Association (APA) member mills is known as APA303 siding. It is produced in a variety of surface textures and patterns (**Figure 63–7**).

Hardboard siding is made of high-density fibreboard with a hard, tempered surface. It typically comes primed or prefinished in a variety of colours. It also comes in a variety of surface styles (**Figure 63–8**).

Fibre-cement siding is made of a fibre-reinforced cementitious material. It also comes in a variety of finishes and has excellent decay and termite resistance properties. Special considerations must be made when cutting this type of siding. Cutting of fibre-cement boards produces silica dust, which is known to cause cancer. All cutting should be done with a special dust-reducing circular saw blade or a set of shears. Cutting should be done outside and downwind of any other workers.

Panel Siding

Panel siding comes in 4-foot (1.2 m) widths and lengths of 8 and 10 feet (2.4 to 3 m), and sometimes 9 feet (2.7 m). It is usually applied vertically, but

may also be applied horizontally. Most panel siding is shaped with shiplapped edges for weathertight joints.

Lap Siding

Lap siding is applied horizontally and is manufactured in styles with rough-sawn, weathered wood grain, or other embossed surface textures so that they look like wood. Some surfaces are grooved or beaded with square or bevelled edges. They come in widths from 6 to 12 inches (152 to 305 mm), and lengths of 12 or 16 feet (3.6 or 5 m).

Lap siding should overlap by at least 1¼ inches (32 mm). It may be face-nailed or blind-nailed (**Figure 63–9**). Check local codes for feasibility. Blind nails are placed 1 inch (25 mm) down from the top. Face nails are placed 1 inch (25 mm) up from the bottom. Nails on hardboard siding should penetrate only one layer of siding. This will allow the material to expand and contract with moisture. Fibre-cement siding may have two nails per board in vertical alignment.

Nails are driven just snug or flush with the surface. If nails are driven too deep, they lose holding power. These should be caulked closed and another nail driven nearby.

Butt seams may be made with either moderate contact or with a gap. The gap is later caulked shut.

The National Building Code of Canada (NBCC) requires that exterior siding provide first and second planes of protection—the first plane to deflect rain-driven moisture and the second plane to drain away any moisture that gets by the first layer (**Figure 63–10**). This is generally accomplished with the application of a base flashing, housewrap taped to the base flashing, taped and sealed connections at openings, ¾-inch (19-mm) strapping over the studs, and then the application of the siding. At the bottom of the strapping, a screen is installed to keep the insects out (Figure 64–2).

ESTIMATING SIDING

First, calculate the wall area by multiplying its length by its height. To calculate the area of a gable end, multiply the length by the height. Then divide the result by two. Add the square foot areas together with the areas of other parts that will be sided, such as dormers, bays, and porches.

Because windows and doors will not be covered, their total surface areas must be deducted. Multiply the width by the height (large openings only). Subtract the results from the total area to be covered by siding (**Figure 63–11**).

Rough Sawn

Manufactured with a slight rough-sawn texture running across panel. Available without grooves, or with grooves of various styles in lap or panel form. Generally available in 11/32-inch, 3/8-inch, 15/32-inch, 1/2-inch, 19/32-inch and 5/8-inch thicknesses. Rough sawn also available in Texture 1-11, reverse board-and-batten, channel groove and V-groove. Available in Douglas-fir, cedar, southern pine and other species.

APA Texture 1-11

APA 303 Siding panel with shiplapped edges and parallel grooves 1/4-inch deep, 3/8-inch wide; grooves 4-inch or 8-inch o.c. are standard. Other spacing sometimes available are 2-inch, 6-inch, and 12-inch o.c., check local availability. T1-11 is available in 19/32-inch and 5/8-inch thicknesses. Also available with scratch-sanded, unsanded, overlaid, rough sawn, brushed and other surfaces. Available in Douglas-fir, cedar, southern pine and other species.

Kerfed Rough Sawn

Rough-sawn surface with narrow grooves, providing a distinctive effect. Long edges shiplapped for continuous pattern. Grooves are typically 4-inch o.c. Also available with grooves in multiples of 2-inch o.c. Generally available in 11/32-inch, 3/8-inch, 15/32-inch, 1/2-inch, 19/32-inch and 5/8-inch thicknesses. Depth of kerf grooves varies with panel thickness.

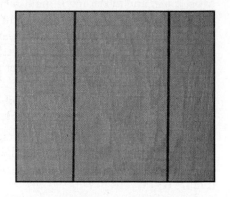

Reverse Board-and-Batten

Deep, wide grooves cut into brushed, rough sawn, scratch-sanded or other textured surfaces. Grooves about 1/4-inch deep, 1-inch to 1-1/2-inch wide, spaced 8-inch or 12-inch o.c. with panel thickness of 19/32-inch and 5/8-inch. Provides deep, sharp shadow lines. Long edges shiplapped for continuous pattern. Available in cedar, Douglas-fir, southern pine and other species.

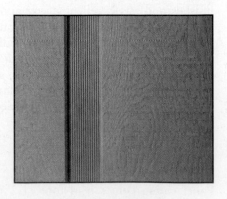

APA – The Engineered Wood Association

FIGURE 63–7 APA303 plywood panel siding is produced in a wide variety of sizes, surface textures, and patterns.

APA Lap Siding

Rough-sawn, smooth, overlaid or embossed surfaces, with square or beveled edges. Available in 3/8-inch, 7/16-inch and 19/32-inch thicknesses.

Brushed

Brushed or relief-grain textures accent the natural grain pattern to create striking surfaces. Generally available in 11/32-inch, 3/8-inch, 15/32-inch, 1/2-inch, 19/32-inch and 5/8-inch thicknesses. Available in Douglas-fir, cedar and other species.

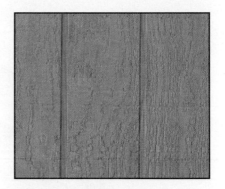

Channel Groove

Shallow grooves about 1/16-inch deep, 3/8-inch wide, cut into faces of panels, 4-inch or 8-inch o.c. Other groove spacings available. Shiplapped for continuous patterns. Generally available in surface patterns and textures similar to Texture 1-11 and in 11/32-inch, 3/8-inch, 15/32-inch and 1/2-inch thicknesses. Available in Douglas-fir, cedar, southern pine and other species.

Overlaid Siding

Overlaid Siding, such as Medium Density Overlay (MDO) and overlaid oriented strand board (OSB), is available without grooving; with V-grooves* (spaced 6-inch or 8-inch o.c. usually standard); or in T1-11 or reverse board-and-batten type grooving, as illustrated at right. Overlaid panel siding available in 11/32-inch, 3/8-inch, 7/16-inch, 15/32-inch, 1/2-inch, 19/32-inch and 5/8-inch thicknesses; also in lap siding. Overlaid siding is available factory-primed. Siding is overlaid on one side and available with texture-embossed or smooth surface.

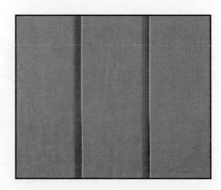

*Not available as OSB siding

FIGURE 63–7 *(Continued)*

FIGURE 63–8 Hardboard lap siding. *(Courtesy of 2009 Louisiana-Pacific Corporation. All rights reserved. LP and CanExel are trademarks of Louisiana-Pacific Corporation Co.)*

The amount of siding to order depends on the kind of siding used. When calculating, use the *factor* for either linear feet or board feet, according to the way it is sold (**Figure 63–12**). Multiply square feet by factor to convert to lineal feet or board feet.

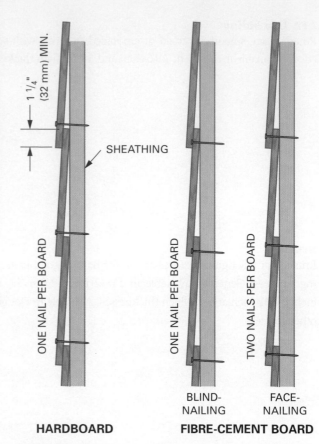

FIGURE 63–9 Recommended nailing for hardboard and fibre-cement board lap siding.

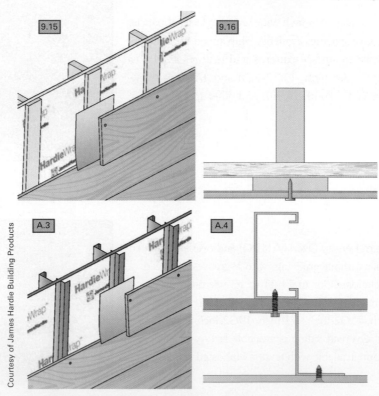

FIGURE 63–10 Rainscreen application of fibre-cement board lap siding.

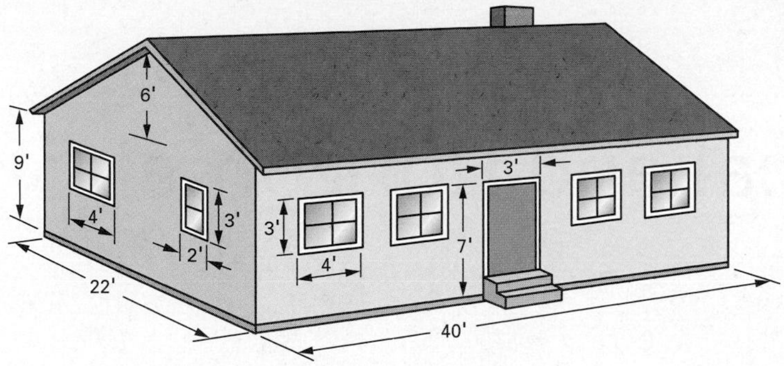

HOUSE AREA:

BUILDING PERIMETER 2 X (40' + 22') = 124 FT.
WALL AREA 124' X 9' = 1116 SQ. FT.
LEFT GABLE $\frac{22' \text{ X } 6'}{2}$ = 66 SQ. FT.
RIGHT GABLE $\frac{22' \text{ X } 6'}{2}$ = 66 SQ. FT.

TOTAL HOUSE AREA = 1248 SQ. FT.

OPENING AREA:
10 WINDOWS 3' X 4' = 120 SQ. FT.
2 WINDOWS 2' X 4' = 16 SQ. FT.
2 DOORS 3' X 7' = 42 SQ. FT.
TOTAL OPENING AREA = 178 SQ. FT.

TOTAL SIDING AREA = 1248 – 178 = 1070 SQ. FT.

HOUSE AREA:

BUILDING PERIMETER 2 X (7 + 12 m) = 38 m^2
WALL AREA 38 m^2 X 2.75 m = 104.5 m^2
GABLE ENDS 2 X 1/2 (7 X 1.8 m) = 12.6 m^2

TOTAL HOUSE AREA = 155.1 m^2

OPENING AREA:
10 WINDOWS 1.2 m X 1 m = 12 m^2
2 DOORS 1 m X 2.1 m = 4.2 m^2

TOTAL OPENING AREA = 16.2 m^2

TOTAL SIDING AREA = 155.1 m^2 – 16.2 m^2 = 138.9 m^2

FIGURE 63–11 Estimating the area to be covered by siding.

Pattern	Nominal Width	Width: Dressed	Width: Exposed Face	Factor for Converting SF to Lineal Feet	Factor for Converting SF to Board Feet
Bevel & Bungalow	4	3½	2½	4.8	1.60
	6	5½	4½	2.67	1.33
	8	7¼	6¼	1.92	1.28
	10	9¼	8¼	1.45	1.21
Dolly Varden	4	3½	3	4.0	1.33
	6	5½	5	2.4	1.2
	8	7¼	6¾	1.78	1.19
	10	9¼	8¾	1.37	1.14
	12	11¼	10¾	1.12	1.12
Drop T&G & Channel Rustic	4	3⅜	3⅛	3.84	1.28
	6	5⅜	5⅛	2.34	1.17
	8	7⅛	6⅞	1.75	1.16
	10	9⅛	8⅞	1.35	1.13
Log Cabin	6	$5\frac{7}{16}$	$4\frac{15}{16}$	2.43	2.43
	8	7⅛	6⅝	1.81	2.42
	10	9⅛	8⅝	1.39	2.32
Boards	2	1½	The exposed face width will vary depending on size selected and on how the boards-and-battens or boards-on-boards are applied.		
	4	3½			
	6	5½			
	8	7¼			
	10	9¼			

Courtesy of Western Wood Products

Coverage Estimator

FIGURE 63–12 Estimating information for converting area to dimensions required for material purchase.

Chapter 64

Applying Vertical and Horizontal Wood Siding

The method of siding application varies with the type. This chapter describes the application procedures for the most commonly used kinds of solid and engineered wood siding.

PREPARATION FOR SIDING APPLICATION

To maximize the siding finish longevity, drying potential must be built into the siding application. This can be achieved by applying a breathable **water-resistant barrier (WRB)** to the sheathing and furring the siding off the sheathing, thereby allowing air to circulate behind the siding (**Figure 64–1**). As air circulates behind the siding, it mixes with soffit air and eventually leaves at the ridge.

This also serves as the second plane of protection (rainscreen) that allows any moisture that penetrates the siding to drain down the WRB and away from the house. The rainscreen can be made into a **pressurized rainscreen** by closing the furring at the top with a top course of furring. The siding is then sealed to the top and to the corner pieces of furring (**Figure 64–2**).

Furring may be 1 × 3, 1 × 4, or 4-inch (19 × 64, 19 × 89, or 100 mm) strips of plywood nailed to the sheathing over each stud. Screen is used to protect the airspace from insects. Furring should be installed before windows are installed. This maintains normal exterior finish details.

A 6- to 8-inch (152 to 203 mm) wide screen is smoothly stapled to the sheathing about 4 inches (100 mm) above the bottom edge of the siding. The extra screen is folded up and over after the furring is applied. Caulking or adhesive applied to the screen before the siding is applied seals the screen to the backside of the siding.

The order of installation will be as follows:

- Seal the junction at the header wrap and the foundation wall.
- Install a drip base flashing 1 inch (25 mm) below the top of the foundation.
- Lap the housewrap over the drip edge and seal ½ inch (12 mm) above the ledge.
- Staple a 6-inch (152 mm) strip of insect screen just above the top of the drip edge.
- Lay an 1½-inch-wide (38 mm) piece of strapping flat on the drip edge's ledge.
- Install wall furring (strapping) over the screen butted to the horizontal piece.
- Install short pieces of strapping in between the furring for a starter nailing base.
- Fold up the insect screen and staple to the furring and short pieces.
- Install the starter strip in line with the bottoms of the furring strips.
- Remove horizontal piece and commence siding installation.

Next, it must be determined how the siding will be ended or treated at the foundation, eaves, and corners. The installation of various kinds of exterior wall trim may first be required.

Foundation Trim

In most cases, no additional trim is applied at the foundation. The siding is started so that it extends slightly below the top of the foundation. However, a **water table** may be installed for appearance. It sheds water a little farther away from the foundation. The water table usually consists of a board and a drip cap installed around the perimeter. Its bottom edge is slightly below the top of the foundation. The siding is started on top of the water table.

Eaves Treatment

At the eaves, the siding may end against the bottom edge of the frieze. The width of the frieze may vary. It is necessary to know its width when laying out horizontal siding courses. The siding may also terminate against the soffit, if no frieze is used. The joint between is then covered by a cornice moulding. The size of the moulding must be known to plan exposures on courses of horizontal siding.

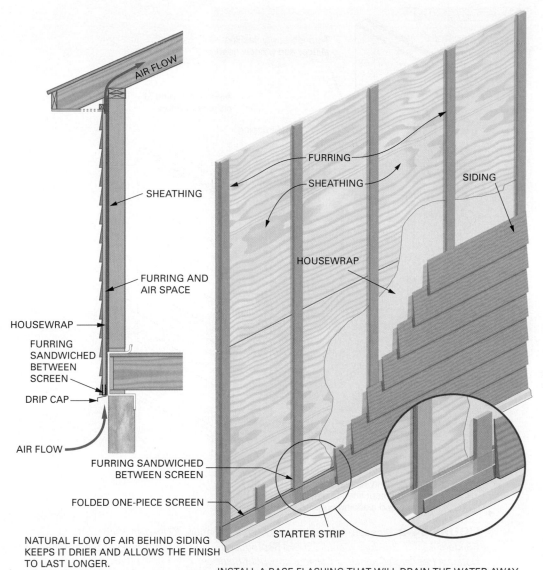

AIR FLOW

SHEATHING

FURRING AND
AIR SPACE

HOUSEWRAP

FURRING
SANDWICHED
BETWEEN
SCREEN

DRIP CAP

AIR FLOW

FURRING SANDWICHED
BETWEEN SCREEN

FOLDED ONE-PIECE SCREEN

NATURAL FLOW OF AIR BEHIND SIDING
KEEPS IT DRIER AND ALLOWS THE FINISH
TO LAST LONGER.

FURRING

SHEATHING

SIDING

HOUSEWRAP

STARTER STRIP

INSTALL A BASE FLASHING THAT WILL DRAIN THE WATER AWAY
FROM THE WALL. THE HOUSEWRAP IS TAPED (RED TAPE) TO THE
TOP OF THE BASE FLASHING, THE WOOD FURRING IS INSTALLED,
AND THEN THE INSECT SCREEN IS SLIPPED INTO PLACE.

A

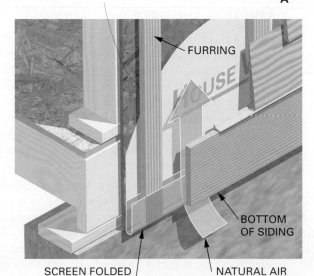

FURRING

SCREEN FOLDED
IN U-SHAPE

NATURAL AIR
CIRCULATION
BEHIND SIDING

BOTTOM
OF SIDING

B

FIGURE 64–1 (A) Siding lasts longer
when it is furred off the sheathing.
(B) Air flow behind siding begins though
the screen.

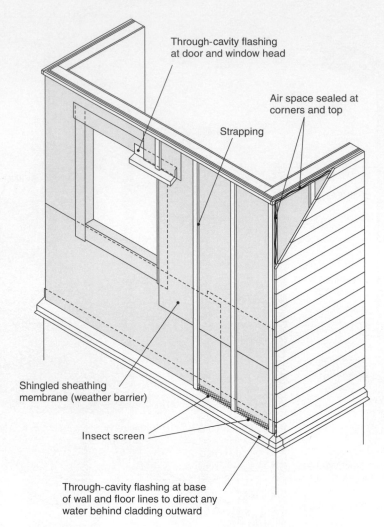

Through-cavity flashing
at door and window head

Air space sealed at
corners and top

Strapping

Shingled sheathing
membrane (weather barrier)

Insect screen

Through-cavity flashing at base
of wall and floor lines to direct any
water behind cladding outward

FIGURE 64–2 Pressure-equalized rainscreen installation. *(Canadian Home Builders Assoc, Builders Manual, 2008. Reproduced with the permission of the National Research Council of Canada, copyright holder.)*

Rake Trim

At the rakes, the siding may be applied under a furred-out rake fascia. When the rake overhangs the sidewall, the siding may be fitted against the rake frieze. When fitted against the rake soffit, the joint is covered with a moulding (**Figure 64–3**).

Gable End Treatment

Sometimes a different kind of siding is used on the gable ends than on the sidewalls below the plate. The joint between the two types must be weathertight. One method of making the joint is to use a drip cap and flashing between the two types of material. Furring strips may also be used on the gable end framing in place of extending the gable plate and studs. The nailing pattern for various types of wood siding are shown in Figure 64–12.

Treating Corners

One method of treating corners is with the use of **corner boards**. Horizontal siding may be mitred around exterior corners, or metal corners may be used on each course of siding. In interior corners, siding courses may butt against a square corner board or against each other (**Figure 64–4**). The thickness of corner boards depends on the type of siding used. The corner boards should be thick enough so that the siding does not project beyond the face of the corner board.

The width of the corner boards depends on the effect desired. However, one of the two pieces, making up an outside corner, should be narrower than the other by the thickness of the stock. Then, after the wider piece is fastened to the narrower one, the same width is exposed on both sides of the corner. The joint between the two pieces should be on the side of the building that is least viewed.

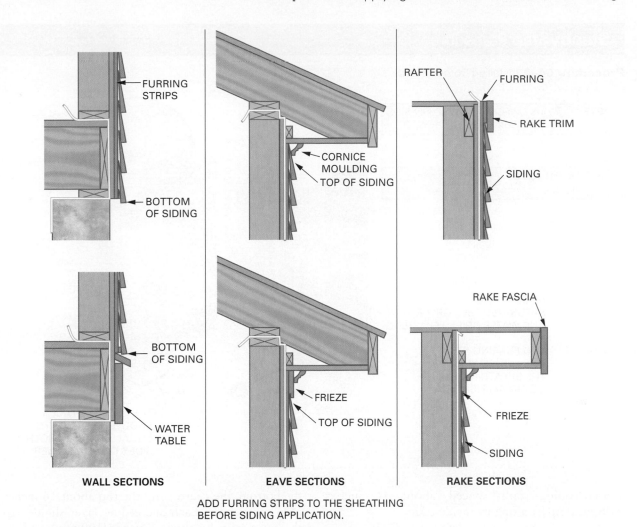

ADD FURRING STRIPS TO THE SHEATHING
BEFORE SIDING APPLICATION.

FIGURE 64–3 Methods of ending the siding at the foundation, eaves, and rakes.

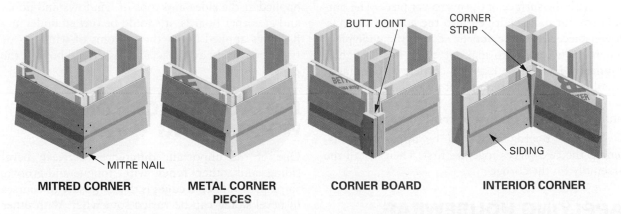

ADD FURRING STRIPS AND HOUSEWRAP TO THE SHEATHING BEFORE SIDING APPLICATION.

FIGURE 64–4 Methods for returning and ending courses of horizontal siding at corners.

INSTALLING CORNER BOARDS

Before installing corner boards, flash both sides of the corner. Install a strip of #15 felt vertically on each side. One edge should extend beyond the edge of the corner board at least 2 inches (50 mm). Tuck the top ends under any previously applied felt.

With a sharp plane, slightly back-bevel one edge of the narrower of the two pieces that make up the outside corner board. This ensures a tight fit between the two boards. Cut, fit, and fasten the narrow piece. Start at one end and work toward the bottom. Keep the bevelled edge flush with the corner. Fasten with galvanized or other

Procedure 64–A Installing Corner Boards

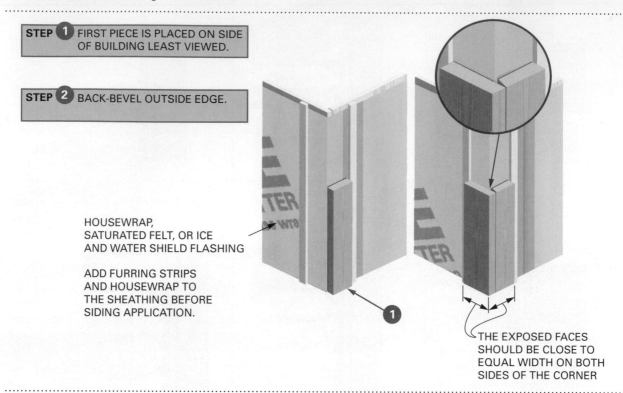

STEP **1** FIRST PIECE IS PLACED ON SIDE OF BUILDING LEAST VIEWED.

STEP **2** BACK-BEVEL OUTSIDE EDGE.

HOUSEWRAP, SATURATED FELT, OR ICE AND WATER SHIELD FLASHING

ADD FURRING STRIPS AND HOUSEWRAP TO THE SHEATHING BEFORE SIDING APPLICATION.

THE EXPOSED FACES SHOULD BE CLOSE TO EQUAL WIDTH ON BOTH SIDES OF THE CORNER

non-corroding nails spaced about 16 inches (406 mm) apart along its inside edge.

Cut, fit, and fasten the wider piece to the corner in a similar manner. Make sure its outside edge is flush with the surface of the narrower piece. The outside row of nails is driven into the edge of the narrower piece. Plane the outside edge of the wide piece wherever necessary to make it come flush. Slightly round over all sharp exposed corners by planing a small chamfer and sanding. Set all nails so that they can be filled over later. Make sure a tight joint is obtained between the two pieces (**Procedure 64–A**).

Corner boards may also be applied by fastening the two pieces together first. Then install the assembly on the corner.

APPLYING HOUSEWRAP

A description of the kinds and purposes of housewrap was given in Chapter 59. A *breathable* type material should be applied to sidewalls. It is typically applied to the whole building all at once. This serves as an immediate and temporary weather barrier as well as a backup water shed if a future leak develops in the siding.

Apply the paper horizontally. Start at the bottom of the wall. Make sure the paper lies flat with no wrinkles. Fasten it in position. If nailing, use large-head roofing nails. Fasten in rows near

the bottom, the centre, and the top about 16 inches (406 mm) apart. Each succeeding layer should lap the lower layer by about 4 inches (100 mm).

The sheathing paper should lap over any flashing applied at the sides and tops of windows and doors and at corner boards. It should be tucked under any flashings applied under the bottoms of windows or frieze boards. In any case, all laps should be over the paper below.

INSTALLING HORIZONTAL WOOD SIDING

One of the important differences between bevel siding and other types with tongue-and-groove, shiplap, or rabbeted edges is that exposure of courses of bevel siding can be varied somewhat. With other types, the amount exposed to the weather is constant with every course and cannot vary.

The ability to vary the exposure is a decided advantage. It is desirable from the standpoint of appearance, weathertightness, and ease of application to have a full course of horizontal siding exposed above and below windows and over the tops of doors (**Figure 64–5**). The exposure of the siding may vary gradually up to a total of ½ inch (12 mm) over the entire wall, but the width of each exposure should not vary more than ¼ inch (6 mm) from its neighbour.

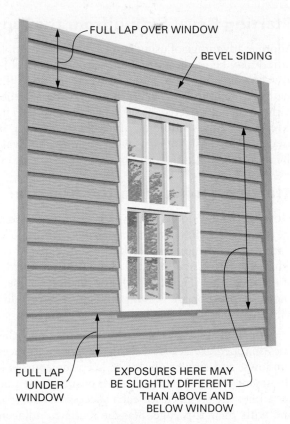

FULL LAP OVER WINDOW

BEVEL SIDING

FULL LAP UNDER WINDOW

EXPOSURES HERE MAY BE SLIGHTLY DIFFERENT THAN ABOVE AND BELOW WINDOW

FIGURE 64–5 Bevel siding exposure may be varied from top to bottom.

Determining Siding Exposure

To determine the siding exposure so that it is about equal both above and below the window sill, divide the overall height of the window frame by the amount of exposure. For example, consider a window that is 52 inches in height with the coverage required above and below the window being 12½ inches and 40½ inches, respectively (**Figure 64–6**). For the metric example, consider the window height to be 1320 mm and the wall coverage above and below the window to be 318 mm and 1032 mm, respectively. The exposure of each section may be adjusted to allow full laps above and below the window.

First, the number of courses for each section must be determined. This is done by dividing the coverage distance by the maximum exposure of the siding, and then rounding that number up to the next whole number of courses.

Next, the number of courses is divided into the coverage distances to find the exposure for that area. Note that because windows vary in height around the house, this process does not always work out neatly. Sometimes all that can be done is to adjust the last exposure to the largest that it can be.

Layout lines are then transferred to a **storey pole**. The storey pole is used to lay out the courses all around the building (**Figure 64–7**).

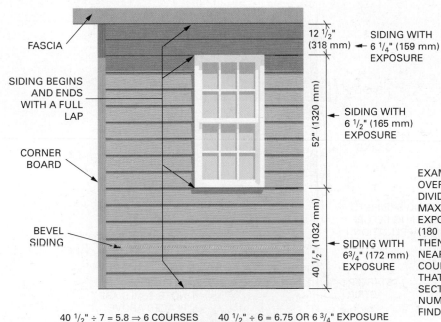

FASCIA

SIDING BEGINS AND ENDS WITH A FULL LAP

CORNER BOARD

BEVEL SIDING

12 ½" (318 mm) — SIDING WITH 6 ¼" (159 mm) EXPOSURE

52" (1320 mm)

SIDING WITH 6 ½" (165 mm) EXPOSURE

40 ½" (1032 mm) — SIDING WITH 6¾" (172 mm) EXPOSURE

EXAMPLE: CONSIDER THE OVERALL DIMENSIONS. DIVIDE THE HEIGHTS BY THE MAXIMUM ALLOWABLE EXPOSURE, 7 INCHES (180 mm) IN THIS EXAMPLE. THEN ROUND UP TO THE NEAREST NUMBER OF COURSES THAT WILL COVER THAT SECTION. DIVIDE THE SECTION HEIGHT BY THE NUMBER OF COURSES TO FIND THE EXPOSURE.

40 ½" ÷ 7 = 5.8 ⇒ 6 COURSES   40 ½" ÷ 6 = 6.75 OR 6 ¾" EXPOSURE

52" ÷ 7 = 7.4 ⇒ 8 COURSES   52" ÷ 8 = 6.5 OR 6 ½" EXPOSURE

12 ½" ÷ 7 = 1.8 ⇒ 2 COURSES   12 ½" ÷ 2 = 6.25 OR 6 ¼" EXPOSURE

1032 mm ÷ 180 = 5.73 ⇒ 6 COURSES   1032 ÷ 6 = 172 mm EXPOSURE

1320 mm ÷ 180 = 7.33 ⇒ 8 COURSES   1320 ÷ 8 = 165 mm EXPOSURE

318 mm ÷ 200 = 1.59 ⇒ 2 COURSES   318 ÷ 2 = 159 mm EXPOSURE

FIGURE 64–6 Example of determining siding exposures around a window.

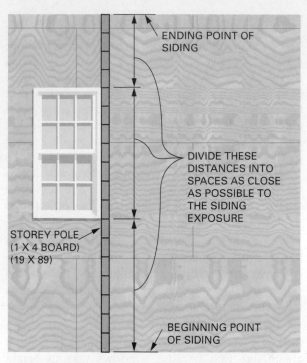

FIGURE 64–7 Laying out a storey pole for courses of horizontal siding.

Instead of calculating the number and height of siding courses mathematically, *dividers* may be used to space off the distances.

Starting Bevel Siding from the Top

Another advantage of using bevel siding is that application may be made starting at the top and working toward the bottom if more convenient. With this method, a number of chalk lines may be snapped without being covered by a previous course. This saves time. Any scaffolding already erected may be used and then dismantled as work progresses toward the bottom.

Starting Siding from the Bottom

Most horizontal siding, however, is usually started at the bottom. For bevel siding, a *furring strip*, called a **starter strip**, of the same thickness and width of the siding **headlap** is first fastened along the bottom edge of the sheathing (**Figure 64–8**). For the first course, a line is snapped on the wall at a height that is in line with the top edge of the first course of siding.

For siding with constant exposure, the only other lines snapped are across the tops of wide entrances, windows, and similar objects to keep the courses in alignment. For lap siding, with exposures that may vary, lines are snapped for each successive course in line with their top edge. Stagger joints in adjacent courses as far apart as possible. A small piece of felt paper often is used behind the butt seams to ensure the weathertightness of the siding.

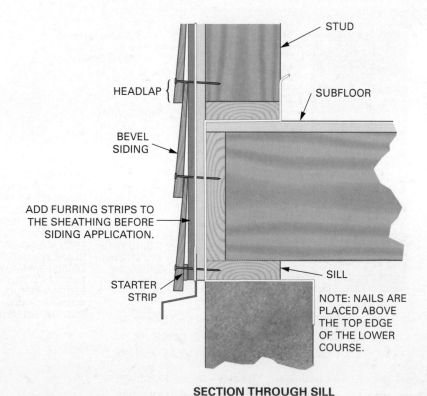

SECTION THROUGH SILL

FIGURE 64–8 For bevel siding, a strip of wood the same thickness and width of the headlap is used as a starter strip.

Fitting Siding

When applying a course of siding, start from one end and work toward the other end. With this procedure, only the last piece will have to be fitted. Tight-fitting butt joints must be made between pieces. If an end joint must be fitted, use a block plane to trim the end as needed. When a piece has to be fitted between other pieces, measure carefully. Cut it slightly long. Place one end in position. Bow the piece outward, position the other end, and snap into place (**Figure 64–9**).

A **preacher** is often used for accurate layout of siding where it butts against corner boards, casings, and similar trim. The siding is allowed to overlap the trim. The preacher is held against the trim. A line is then marked on the siding along the face of the preacher (**Figure 64–10**).

When fitting siding under windows, make sure the siding fits snugly in the groove on the underside of the window sill for weathertightness (**Figure 64–11**).

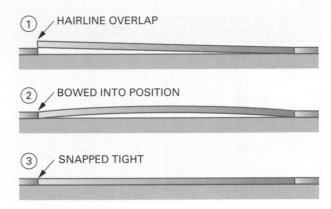

FIGURE 64–9 Method of cutting horizontal siding to fit snugly.

A groove called a *quirk* is cut in the underside of the window sill to prevent water runoff from curling back under the sill and damaging the siding.

Fastening Siding

Siding is fastened to each bearing stud, or about every 16 inches (406 mm). On bevel siding, fasten through the butt edge just above the top edge of the course below. Do not fasten through the lap. This prevents splitting of the siding that might be caused by slight swelling or shrinking due to moisture changes. Care must be taken to fasten as low as possible to avoid splitting the siding in the centre. The location and number of fasteners recommended for siding are shown in **Figure 64–12**.

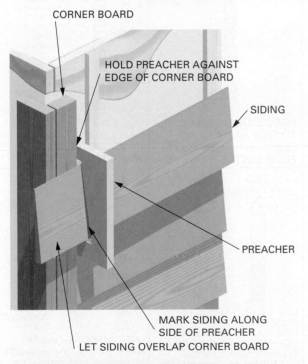

ON THE JOB

Use a preacher for accurate markings of bevel siding to fit against corner boards and casings.

FIGURE 64–10 A preacher may be used for accurately marking a siding piece for length.

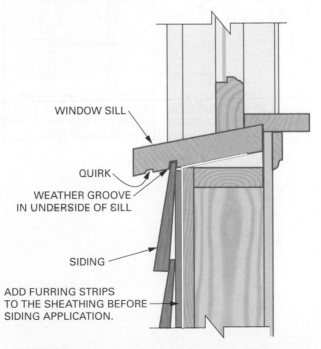

FIGURE 64–11 Fit siding into weather groove on the underside of the window sill.

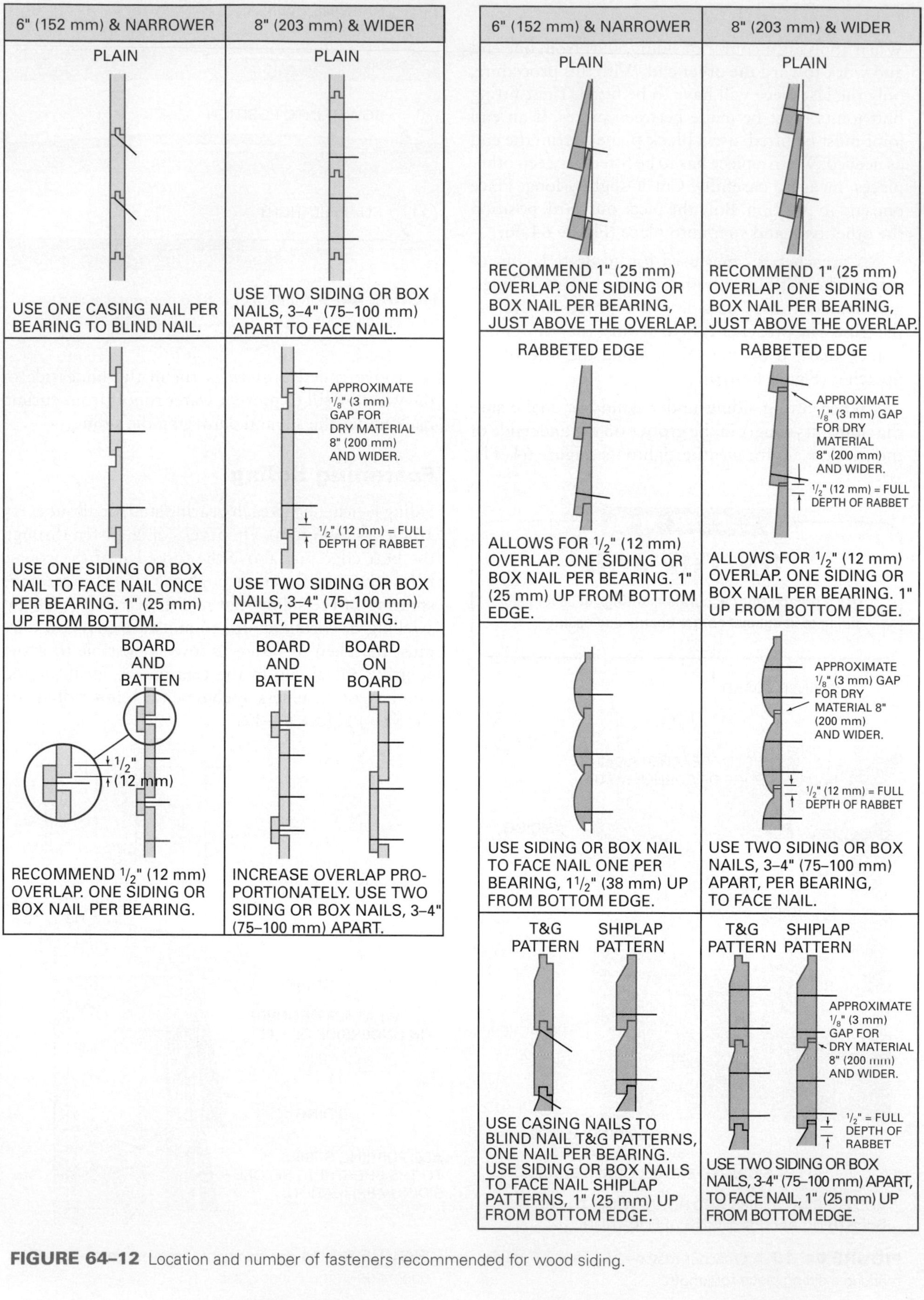

6" (152 mm) & NARROWER	8" (203 mm) & WIDER	6" (152 mm) & NARROWER	8" (203 mm) & WIDER
PLAIN	PLAIN	PLAIN	PLAIN
USE ONE CASING NAIL PER BEARING TO BLIND NAIL.	USE TWO SIDING OR BOX NAILS, 3–4" (75–100 mm) APART TO FACE NAIL.	RECOMMEND 1" (25 mm) OVERLAP. ONE SIDING OR BOX NAIL PER BEARING, JUST ABOVE THE OVERLAP.	RECOMMEND 1" (25 mm) OVERLAP. ONE SIDING OR BOX NAIL PER BEARING, JUST ABOVE THE OVERLAP.
	APPROXIMATE ¹/₈" (3 mm) GAP FOR DRY MATERIAL 8" (200 mm) AND WIDER. ¹/₂" (12 mm) = FULL DEPTH OF RABBET	RABBETED EDGE	RABBETED EDGE — APPROXIMATE ¹/₈" (3 mm) GAP FOR DRY MATERIAL 8" (200 mm) AND WIDER. ¹/₂" (12 mm) = FULL DEPTH OF RABBET
USE ONE SIDING OR BOX NAIL TO FACE NAIL ONCE PER BEARING. 1" (25 mm) UP FROM BOTTOM.	USE TWO SIDING OR BOX NAILS, 3–4" (75–100 mm) APART, PER BEARING.	ALLOWS FOR ¹/₂" (12 mm) OVERLAP. ONE SIDING OR BOX NAIL PER BEARING. 1" (25 mm) UP FROM BOTTOM EDGE.	ALLOWS FOR ¹/₂" (12 mm) OVERLAP. ONE SIDING OR BOX NAIL PER BEARING. 1" UP FROM BOTTOM EDGE.
BOARD AND BATTEN ¹/₂" (12 mm)	BOARD AND BATTEN / BOARD ON BOARD		APPROXIMATE ¹/₈" (3 mm) GAP FOR DRY MATERIAL 8" (200 mm) AND WIDER. ¹/₂" (12 mm) = FULL DEPTH OF RABBET
RECOMMEND ¹/₂" (12 mm) OVERLAP. ONE SIDING OR BOX NAIL PER BEARING.	INCREASE OVERLAP PROPORTIONATELY. USE TWO SIDING OR BOX NAILS, 3–4" (75–100 mm) APART.	USE SIDING OR BOX NAIL TO FACE NAIL ONE PER BEARING, 1¹/₂" (38 mm) UP FROM BOTTOM EDGE.	USE TWO SIDING OR BOX NAILS, 3–4" (75–100 mm) APART, PER BEARING, TO FACE NAIL.
		T&G PATTERN / SHIPLAP PATTERN	T&G PATTERN / SHIPLAP PATTERN — APPROXIMATE ¹/₈" (3 mm) GAP FOR DRY MATERIAL 8" (200 mm) AND WIDER. ¹/₂" = FULL DEPTH OF RABBET
		USE CASING NAILS TO BLIND NAIL T&G PATTERNS, ONE NAIL PER BEARING. USE SIDING OR BOX NAILS TO FACE NAIL SHIPLAP PATTERNS, 1" (25 mm) UP FROM BOTTOM EDGE.	USE TWO SIDING OR BOX NAILS, 3-4" (75–100 mm) APART, TO FACE NAIL, 1" (25 mm) UP FROM BOTTOM EDGE.

Courtesy of Western Wood Products Association

FIGURE 64–12 Location and number of fasteners recommended for wood siding.

VERTICAL APPLICATION OF WOOD SIDING

Bevel sidings are designed for horizontal applications only. **Board on board** and **board and batten** are applied only vertically. Almost all other patterns may be applied in either direction.

INSTALLING VERTICAL TONGUE-AND-GROOVE SIDING

Corner boards usually are not used when wood siding is applied vertically. The siding boards are fitted around the corner (**Figure 64–13**). Rip the grooved edge from the starting piece. Slightly back-bevel the ripped edge. Place it vertically on the wall with the bevelled edge flush with the corner, similar to making a corner board.

The tongue edge should be plumb, the bottom end should be about 1 inch (25 mm) below the sheathing (**Figure 64–14**). The top end should butt or be tucked under any trim above. **Face-nail** the edge nearest the corner. **Blind-nail** into the tongue edge. Nails should be placed from 16 to 24 inches (406 to 610 mm) apart. Blocking must be provided between studs if siding is applied directly to the frame.

Fasten a temporary piece on the other end of the wall, projecting below the sheathing by the same amount. Stretch a line to keep the bottom ends of other pieces in a straight line. Apply succeeding pieces by tocnailing into the tongue edge of each piece.

Make sure the edges between boards come up tight. If they do not come up tight by nailing alone,

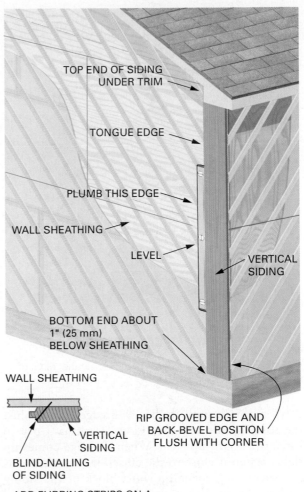

FIGURE 64–14 Starting the application of vertical board siding.

FIGURE 64–13 Vertical tongue-and-groove siding needs little accessory trim, such as corner boards.

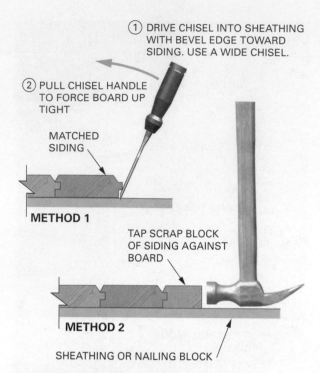

FIGURE 64-15 Techniques for tightening the joints of boards during installation.

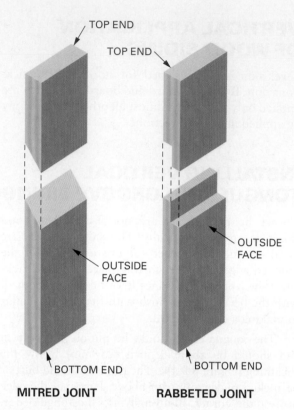

FIGURE 64-16 Use mitred or rabbeted end joints between lengths of vertical board siding.

drive a chisel, with its bevelled edge side against the tongue, into the sheathing. Use it as a pry to force the board up tight. When the edge comes up tight, fasten it close to the chisel. If this method is not successful, toenail a short block of the siding with its grooved edge into the tongue of the board (**Figure 64-15**). Drive the nail home until it forces the board up tight. Drive nails into the siding on both sides of the scrap block. Remove the scrap block.

Continue applying pieces in the same manner. Make sure to keep the bottom ends in a straight line. Avoid making horizontal joints between lengths. If joints are necessary, use a mitred or rabbeted joint for weathertightness (**Figure 64-16**).

Fitting around Doors and Windows

Vertical siding is fitted tightly around window and door casings with different methods than those used for horizontal siding.

Approaching a Wall Opening. When approaching a door or window, cut and fit the piece just before the one to be fitted against the casing. Then remove it. Set it aside for the time being.

Cut, fit, and tack the piece to be fitted against the casing in place of the next to last piece. Level from the top of the window casing and the bottom of the sill. Mark the piece.

To lay out the piece so it will fit snugly against the side casing, first cut a scrap block of the siding material, about 6 inches (152 mm) long (**Procedure 64-B**). Remove the tongue from one edge. Be careful to remove all of the tongue, but no more. Hold the block so that its grooved edge is against the side casing and the other edge is on top of the siding to be fitted. Mark the piece by holding a pencil against the outer edge of the block while moving the block along the length of the side casing.

Cut the piece, following the layout lines carefully. When laying out to fit against the bottom of the sill, make allowance to rabbet the siding to fit in the weather groove on the bottom side of the window sill. Place and fasten the pieces in position.

Continue to apply the siding with short lengths across the top and bottom of the window. Each length under a window must be rabbeted at the top end to fit in the weather groove at the sill.

Leaving a Wall Opening. A full length must also be fitted to the casing on the other side of the window. To mark the piece, first tack a short length of scrap siding above and below the window and against the last pieces of siding installed. Tack the length of siding to be fitted against these blocks. Level from the top and bottom of the window. Mark the piece for the horizontal cuts.

StepbyStep Procedures

Procedure 64–B Method for Fitting Vertical Board Siding When Approaching a Window Casing

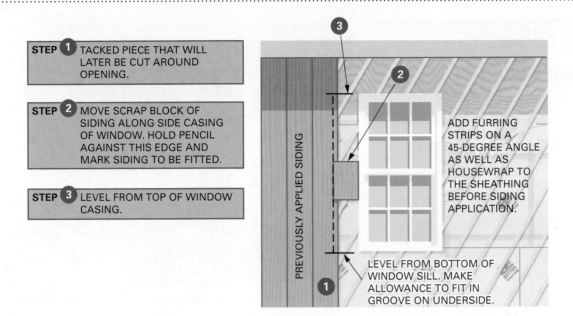

STEP 1 TACKED PIECE THAT WILL LATER BE CUT AROUND OPENING.

STEP 2 MOVE SCRAP BLOCK OF SIDING ALONG SIDE CASING OF WINDOW. HOLD PENCIL AGAINST THIS EDGE AND MARK SIDING TO BE FITTED.

STEP 3 LEVEL FROM TOP OF WINDOW CASING.

PREVIOUSLY APPLIED SIDING

ADD FURRING STRIPS ON A 45-DEGREE ANGLE AS WELL AS HOUSEWRAP TO THE SHEATHING BEFORE SIDING APPLICATION.

LEVEL FROM BOTTOM OF WINDOW SILL. MAKE ALLOWANCE TO FIT IN GROOVE ON UNDERSIDE.

To lay out the piece for the vertical cut that fits against the side casing, use the same block with the tongue removed that was used previously. Hold the grooved edge against the side casing. With a pencil against the other edge, ride the block along the side casing while marking the piece to be fitted (**Procedure 64–C**).

Remove the piece and the scrap blocks from the wall. Carefully cut the piece to the layout lines. Then fasten in position. Continue applying the rest of the siding until you are almost to the other end of the wall.

Method of Ending Vertical Siding

The last piece of vertical siding should be close to the same width of previously installed pieces. If siding is installed in random widths, plan the application. The width of the last piece should be equal, at least, to the width of the narrowest piece. It is not good practice to allow vertical siding to end with a narrow sliver.

Stop several feet from the end. Space off, and determine the width of the last piece. If it will not be a satisfactory width, install narrower or wider pieces for the remainder, as required. It may be necessary to rip available siding to narrower widths and reshape the grooves (**Figure 64–17**).

When the corner is reached, the board is ripped to width along its tongue edge. It is slightly bev-elled for the first piece on the next wall to butt against. When the last corner is reached, the board is ripped in a similar manner. However, it is smoothed to a square edge to fit flush with the surface of the first piece installed. All exposed sharp corners should be eased or slightly rounded.

INSTALLING PANEL SIDING

Plywood, hardboard, and other panel siding is usually installed vertically. It can be installed horizontally, if desired. Lap siding panels are ordinarily applied horizontally.

Installing Vertical Panel Siding

Start a vertical panel so that it is plumb, with one edge squared and flush with the starting corner. The inner edge should fall on the centre of a stud. Fasten panels, of ½-inch (12 mm) thickness or less, with 2-inch (50 mm) siding nails. Use 2½-inch (64 mm) siding nails for thicker panels. Fasten panel edges about every 6 inches (152 mm) and about every 12 inches (305 mm) along intermediate studs.

Apply successive sheets. Leave a ⅛-inch (3 mm) space between panels. Panels must be installed with their bottom ends in a straight line. There should be a minimum of 8 inches (203 mm) above the finished grade line.

StepbyStep Procedures

Procedure 64–C Method of Fitting Vertical Board Siding When Leaving a Window Casing

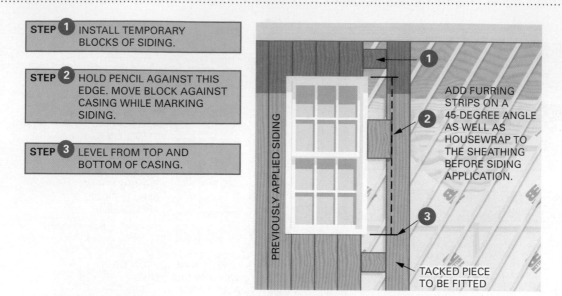

STEP 1 INSTALL TEMPORARY BLOCKS OF SIDING.

STEP 2 HOLD PENCIL AGAINST THIS EDGE. MOVE BLOCK AGAINST CASING WHILE MARKING SIDING.

STEP 3 LEVEL FROM TOP AND BOTTOM OF CASING.

PREVIOUSLY APPLIED SIDING

ADD FURRING STRIPS ON A 45-DEGREE ANGLE AS WELL AS HOUSEWRAP TO THE SHEATHING BEFORE SIDING APPLICATION.

TACKED PIECE TO BE FITTED

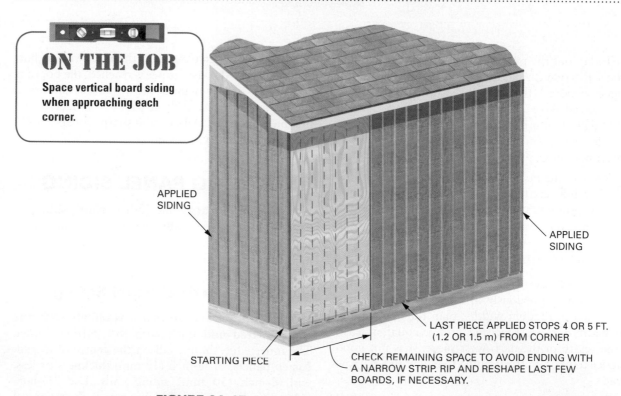

ON THE JOB

Space vertical board siding when approaching each corner.

APPLIED SIDING

APPLIED SIDING

STARTING PIECE

LAST PIECE APPLIED STOPS 4 OR 5 FT. (1.2 OR 1.5 m) FROM CORNER

CHECK REMAINING SPACE TO AVOID ENDING WITH A NARROW STRIP. RIP AND RESHAPE LAST FEW BOARDS, IF NECESSARY.

FIGURE 64–17 Joints in vertical siding may be adjusted slightly to ensure that the corner piece is nearly as wide as the others.

Installing Horizontal Panel Siding

Mark the height of the first course of horizontal panel siding on both ends of the wall. Snap a chalk line between marks. Fasten a full-length panel with its top edge to the line, its inner end on the centre of a stud, and its outer end flush with the corner. Fasten in the same way as for vertical panels.

Apply the remaining sheets in the first course in like manner. Trim the end of the last sheet flush with the corner. Start the next course so that joints will line up with those in the course below.

Both vertical and horizontal panels may be applied to sheathing or directly to studs if backing is provided for all joints (**Figure 64–18**).

Carefully fit and caulk around doors and windows. It is important that horizontal butt joints be either offset and lapped, rabbeted, or flashed (**Figure 64–19**). Vertical joints are either shiplapped or covered with **battens**.

Applying Lap Siding Panels

Panels of lap siding are applied in much the same manner as wood bevel siding with some exceptions. First, install a strip, of the same thickness and width of the siding headlap, along the bottom of the wall. Determine the height of the top edge of the first course. Snap a chalk line across the wall. Apply the first course with its top edge to the snapped line.

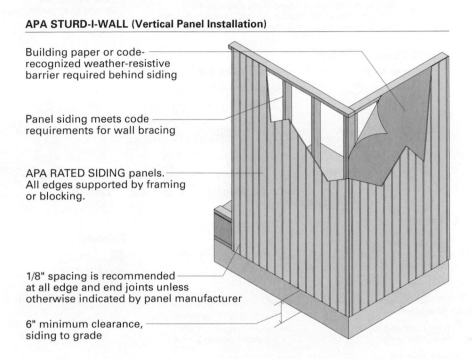

APA STURD-I-WALL (Vertical Panel Installation)

Building paper or code-recognized weather-resistive barrier required behind siding

Panel siding meets code requirements for wall bracing

APA RATED SIDING panels. All edges supported by framing or blocking.

1/8" spacing is recommended at all edge and end joints unless otherwise indicated by panel manufacturer

6" minimum clearance, siding to grade

APA STURD-I-WALL (Horizontal Panel Siding Installation)

Building paper or other code-recognized weather-resistive barrier

Battens at 4' or 8' o.c. to conceal butt joints at panel ends. Nails through battens must penetrate studs at least 1".

APA RATED SIDING panels (nailing as required for vertical installation)

1/8" spacing is recommended at all edge and end joints unless otherwise indicated by panel manufacturer

6" minimum clearance, siding to grade

Panel siding meets code requirements for wall bracing

Seal panel edges

2x4 blocking at horizontal joints

APA—The Engineered Wood Association

FIGURE 64–18 Panel siding may be applied vertically or horizontally to sheathing or directly to studs.

TYPICAL PANEL SIDING JOINT DETAILS
(Note: Water-resistive barrier [building paper or house wrap] is required behind siding.)

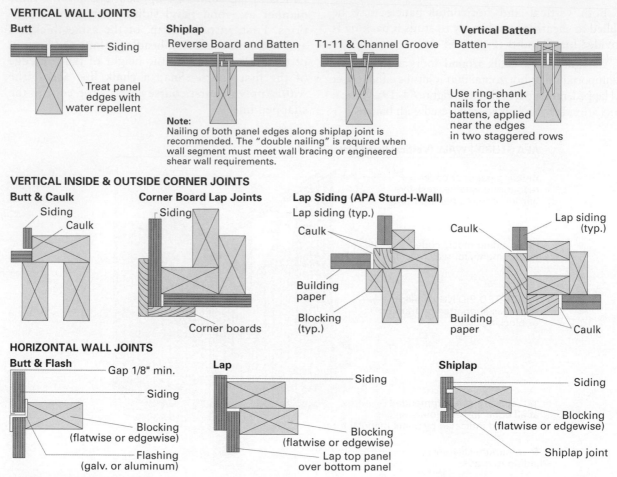

VERTICAL WALL JOINTS

Butt
— Siding
Treat panel edges with water repellent

Shiplap
Reverse Board and Batten
Note:
Nailing of both panel edges along shiplap joint is recommended. The "double nailing" is required when wall segment must meet wall bracing or engineered shear wall requirements.

T1-11 & Channel Groove

Vertical Batten
Batten
Use ring-shank nails for the battens, applied near the edges in two staggered rows

VERTICAL INSIDE & OUTSIDE CORNER JOINTS

Butt & Caulk
Siding
Caulk

Corner Board Lap Joints
Siding
Corner boards

Lap Siding (APA Sturd-I-Wall)
Lap siding (typ.)
Caulk
Building paper
Blocking (typ.)

Caulk
Lap siding (typ.)
Building paper
Caulk

HORIZONTAL WALL JOINTS

Butt & Flash
Gap 1/8" min.
Siding
Blocking (flatwise or edgewise)
Flashing (galv. or aluminum)

Lap
Siding
Blocking (flatwise or edgewise)
Lap top panel over bottom panel

Shiplap
Siding
Blocking (flatwise or edgewise)
Shiplap joint

HORIZONTAL BELTLINE JOINTS
(For multistory buildings, when conventional lumber floor joists and rim boards are used, make provisions at horizontal joints for shrinkage of framing, especially when applying siding direct to studs.)

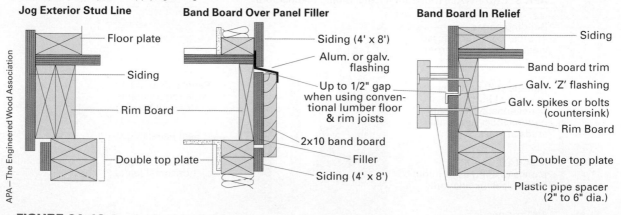

Jog Exterior Stud Line
Floor plate
Siding
Rim Board
Double top plate

APA—The Engineered Wood Association

Band Board Over Panel Filler
Siding (4' x 8')
Alum. or galv. flashing
Up to 1/2" gap when using conventional lumber floor & rim joists
2x10 band board
Filler
Siding (4' x 8')

Band Board In Relief
Siding
Band board trim
Galv. 'Z' flashing
Galv. spikes or bolts (countersink)
Rim Board
Double top plate
Plastic pipe spacer (2" to 6" dia.)

FIGURE 64–19 Panel siding joint details.

When applied over sheathing that is thick enough to securely hold fasteners, space nails 8 inches (203 mm) apart in a line about ¾ inch (19 mm) above the bottom edge of the siding. When applied directly to framing, fasten at each stud location. A 1/8-inch (3 mm) caulked joint between the ends of siding and trim is recommended. Joints between ends of siding may also be flashed with a narrow strip of #15 felt centred behind the joint and backed with a wood shingle wedge (**Figure 64–20**).

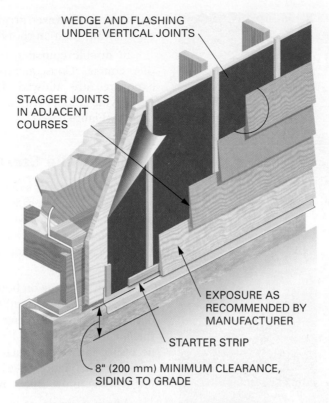

WEDGE AND FLASHING
UNDER VERTICAL JOINTS

STAGGER JOINTS
IN ADJACENT
COURSES

EXPOSURE AS
RECOMMENDED BY
MANUFACTURER

STARTER STRIP

8" (200 mm) MINIMUM CLEARANCE,
SIDING TO GRADE

ADD FURRING STRIPS AND HOUSEWRAP TO
THE SHEATHING BEFORE SIDING APPLICATION.

FIGURE 64–20 Lap siding application details.

Chapter 65

Wood Shingle and Shake Siding

Wood shingles and shakes may be used for siding, as well as roofing (**Figure 65–1**). Those previously described in Chapter 56 for roofing may also be applied to sidewalls.

SIDEWALL SHINGLES AND SHAKES

Some kinds of shingles and shakes are designed for sidewall use only (**Figure 65–2**). Rebutted and rejointed ones are machine-trimmed with parallel edges and square butts for sidewall application. Rebutted and rejointed machine-grooved, sidewall shakes have striated faces.

Special fancy butt shingles are available in a variety of designs. They provide interesting patterns, in combination with square butts or other types of siding (**Figure 65–3**).

APPLYING WOOD SHINGLES AND SHAKES

Wood shingles and shakes may be applied to sidewalls in either single-layer or double-layer courses. In **single coursing**, shingles are applied to walls with a single layer in each course, in a way similar to roof

FIGURE 65–1 Wood shingles and shakes may be used as siding.

Courtesy of BC Shake and Shingle Association

application. However, greater exposures are allowed on sidewalls than on roofs (**Figure 65–4**).

In double coursing, two layers are applied in one course. Consequently, even greater weather exposures are allowed. Double coursing is used when wide courses with deep, bold shadow lines are desired (**Figure 65–5**).

Applying the Starter Course

The starter course of sidewall shingles and shakes is applied in much the same way as the starter course on roofs. A double layer is used for single-course applications. A triple layer is used for triple coursing. Less expensive undercourse shingles are used for underlayers.

Fasten a shingle on both ends of the wall with its butt about 1 inch (25 mm) below the top of the foundation. Stretch a line between them at the butts. Sight the line for straightness. Fasten additional shingles at necessary intervals. Attach the line to their butts to straighten it (**Figure 65–6**). Even a tightly stretched line will sag in the centre over a long distance.

Apply a single course of shingles so that the butts are as close to the chalk line as possible without touching it. Remove the line. Apply another course on top of the first course. Offset the joints in the outer layer at least 1½ inches (38 mm)

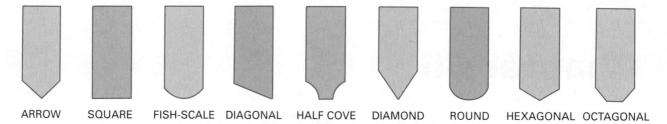

ARROW   SQUARE   FISH-SCALE   DIAGONAL   HALF COVE   DIAMOND   ROUND   HEXAGONAL   OCTAGONAL

FANCY BUTT RED CEDAR SHINGLES. NINE OF THE MOST POPULAR DESIGNS ARE SHOWN. FANCY BUTT SHINGLES CAN BE CUSTOM PRODUCED TO INDIVIDUAL ORDERS.

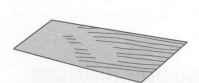

REBUTTED AND REJOINTED. MACHINE TRIMMED FOR PARALLEL EDGES WITH BUTTS SAWN AT RIGHT ANGLES. FOR SIDEWALL APPLICATION WHERE TIGHTLY FITTING JOINTS ARE DESIRED.

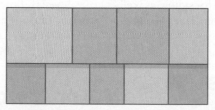

PANELS. WESTERN RED CEDAR SHINGLES ARE AVAILABLE IN 4- AND 8-FOOT (1.2 AND 2.4 m) PANELIZED FORM.

MACHINE GROOVED. MACHINE-GROOVED SHAKES ARE MANUFACTURED FROM SHINGLES AND HAVE STRIATED FACES AND PARALLEL EDGES. USED DOUBLE-COURSED ON EXTERIOR SIDEWALLS.

FIGURE 65–2 Some wood shingles and shakes are made for sidewall applications only.

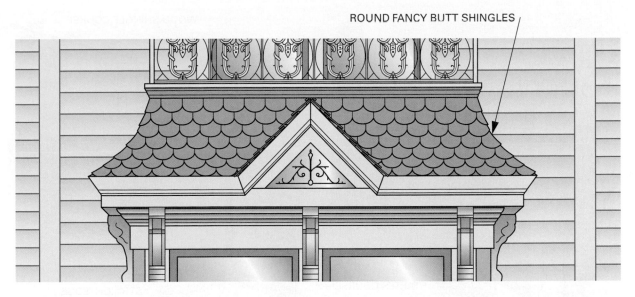

ROUND FANCY BUTT SHINGLES

FIGURE 65–3 Fancy butt shingles are still used to accent sidewalls with distinctive designs.

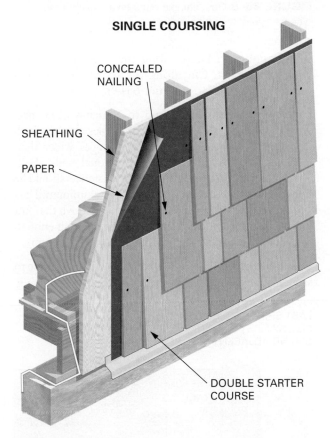

SINGLE COURSING

CONCEALED NAILING

SHEATHING

PAPER

DOUBLE STARTER COURSE

FIGURE 65–4 Single-coursed shingle wall application is similar to roof application, with greater weather exposures allowed.

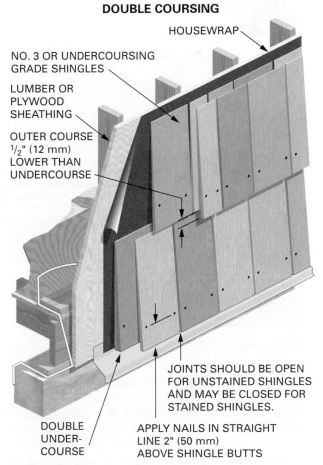

DOUBLE COURSING

HOUSEWRAP

NO. 3 OR UNDERCOURSING GRADE SHINGLES

LUMBER OR PLYWOOD SHEATHING

OUTER COURSE $1/2$" (12 mm) LOWER THAN UNDERCOURSE

JOINTS SHOULD BE OPEN FOR UNSTAINED SHINGLES AND MAY BE CLOSED FOR STAINED SHINGLES.

DOUBLE UNDER-COURSE

APPLY NAILS IN STRAIGHT LINE 2" (50 mm) ABOVE SHINGLE BUTTS

FIGURE 65–5 Double-coursed shingles, with two layers in each row, permit even greater weather exposures.

from those in the bottom layer. Untreated shingles should be spaced $1/8$ to $1/4$ inch (3 to 6 mm) apart to allow for swelling and to prevent buckling. Shingles can be applied close together if factory primed or if treated soon after application (**Figure 65–7**).

Single Coursing

A storey pole may be used to lay out shingle courses in the same manner as with horizontal wood siding.

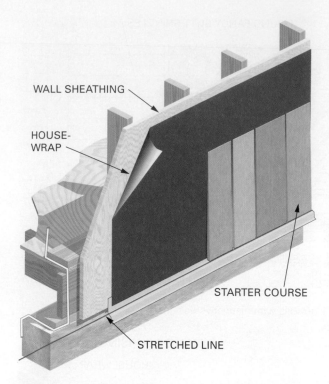

WALL SHEATHING

HOUSE-WRAP

STARTER COURSE

STRETCHED LINE

FIGURE 65–6 Stretch a straight line as a guide for the butts of the starter course.

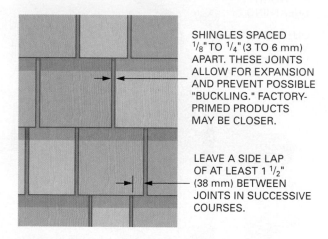

SHINGLES SPACED ⅛" TO ¼" (3 TO 6 mm) APART. THESE JOINTS ALLOW FOR EXPANSION AND PREVENT POSSIBLE "BUCKLING." FACTORY-PRIMED PRODUCTS MAY BE CLOSER.

LEAVE A SIDE LAP OF AT LEAST 1½" (38 mm) BETWEEN JOINTS IN SUCCESSIVE COURSES.

FIGURE 65–7 Stagger joints between shingle courses.

Snap a chalk line across the wall, at the shingle butt line, to apply the first course. Using only as many finish nails as necessary, tack 1 × 3 (19 × 64 mm) straightedges to the wall with their top edges to the line. Lay individual shingles with their butts on the straightedge (**Figure 65–8**). Use a shingling hatchet to trim and fit the edges, if necessary. Butt ends are not trimmed. If rebutted and rejointed shingles are used, no trimming should be necessary.

At times it may be necessary to fit a shingle between others in the same course. Tack the next-to-last shingle in place with one nail. Slip the last

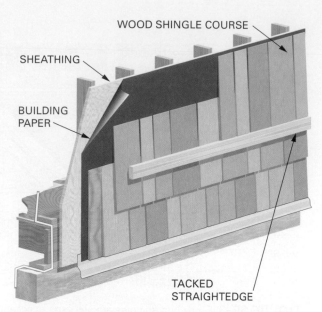

WOOD SHINGLE COURSE

SHEATHING

BUILDING PAPER

TACKED STRAIGHTEDGE

FIGURE 65–8 Rest shingle butts on a straightedge when single coursing on sidewalls.

shingle under it. Score along the overlapping edge with the hatchet. Cut along the scored line. Fasten both shingles in place (**Figure 65–9**).

Fasteners. Each shingle, up to 8 inches (203 mm) wide, is fastened with two nails or staples about ¾ inch (19 mm) in from each edge. On shingles wider than 8 inches (203 mm), drive two additional nails about 1 inch (25 mm) apart near the centre. Fasteners should be hot-dipped galvanized, stainless steel, or aluminum. They should be driven about 1 inch (25 mm) above the butt line of the next course. Fasteners must be long enough to penetrate the sheathing by at least ¾ inch (19 mm).

Staggered and Ribbon Coursing. An alternative to straight-line courses are staggered and ribbon coursing.

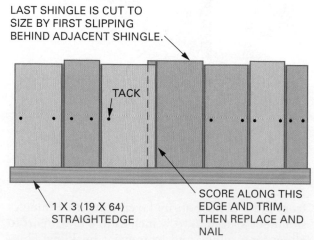

LAST SHINGLE IS CUT TO SIZE BY FIRST SLIPPING BEHIND ADJACENT SHINGLE.

TACK

1 × 3 (19 × 64) STRAIGHTEDGE

SCORE ALONG THIS EDGE AND TRIM, THEN REPLACE AND NAIL

FIGURE 65–9 Technique for cutting last shingle to the proper size.

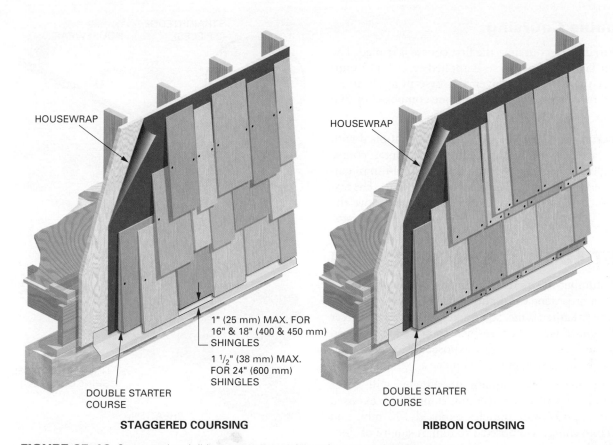

HOUSEWRAP

1" (25 mm) MAX. FOR
16" & 18" (400 & 450 mm)
SHINGLES

1 ½" (38 mm) MAX.
FOR 24" (600 mm)
SHINGLES

DOUBLE STARTER
COURSE

STAGGERED COURSING

HOUSEWRAP

DOUBLE STARTER
COURSE

RIBBON COURSING

FIGURE 65–10 Staggered and ribbon courses are alternatives to straight-line courses.

In staggered coursing, the butt lines of the shingles are alternately offset below, but not above, the horizontal line. Maximum offsets are 1 inch for 16- and 18-inch shingles and 1½ inches for 24-inch shingles. In metric, maximum offsets are 25 mm for 406 and 457 mm shingles and 38 mm for 610 mm shingles.

In ribbon coursing, both layers are applied in straight lines. The outer course is raised about 1 inch (25 mm) above the inner course (**Figure 65–10**).

Corners

Shingles may be butted to corner boards like any horizontal wood siding. On outside corners, they may also be applied by alternately overlapping each course in the same manner as in applying a wood shingle ridge. Inside corners may be woven by alternating the corner shingle first on one wall and then the other (**Figure 65–11**).

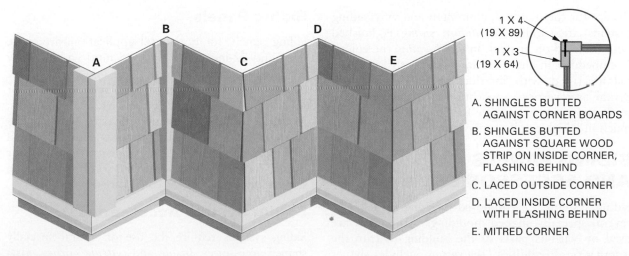

A B C D E

1 X 4
(19 X 89)

1 X 3
(19 X 64)

A. SHINGLES BUTTED
AGAINST CORNER BOARDS

B. SHINGLES BUTTED
AGAINST SQUARE WOOD
STRIP ON INSIDE CORNER,
FLASHING BEHIND

C. LACED OUTSIDE CORNER

D. LACED INSIDE CORNER
WITH FLASHING BEHIND

E. MITRED CORNER

FIGURE 65–11 Corner details for wood shingle siding.

Double Coursing

When double coursing, the first course is tripled. The outer layer of the course is applied ½ inch (12 mm) lower than the inner layer. For ease in application, use a rabbeted straightedge or one composed of two pieces with offset edges (**Figure 65–12**).

Fastening. Each inner layer shingle is applied with one fastener at the top centre. Each outer course shingle is face-nailed with two 1¾-inch (44 mm) galvanized box or special 14-gauge shingle nails. The fasteners are driven about 2 inches (50 mm) above the butts, and about ¾ inch (19 mm) in from each edge.

ESTIMATING SHINGLE SIDING

The number of squares of shingles needed to cover a certain area depends on how much they are exposed to the weather. One square of shingles will cover 100 square feet when 16-inch shingles are exposed 5 inches, 18-inch shingles exposed 5½ inches, and 24-inch shingles exposed 7½ inches.

In metric, one square of shingles will cover 9.29 m² of wall when 406 mm shingles are exposed 127 mm, 457 mm shingles exposed 140 mm, and 610 mm shingles exposed 190 mm. A square of shingles will cover more area with greater exposures and less area with smaller exposures.

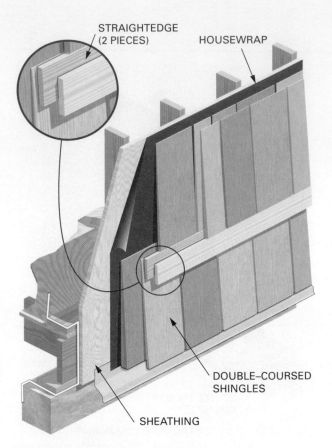

FIGURE 65–12 Use a straightedge made of two pieces with offset edges for double-coursed application.

Chapter 66

Aluminum and Vinyl Siding

Except for the material, aluminum and vinyl siding systems are similar. *Aluminum siding* is finished with a baked-on enamel. In *vinyl siding*, the colour is embedded in the material itself. Both kinds are manufactured with interlocking edges, for horizontal and vertical applications. Descriptions and instructions are given here for vinyl siding systems, much of which can be applied to aluminum systems.

SIDING PANELS AND ACCESSORIES

Siding systems are composed of siding panels and several specially shaped mouldings. Mouldings are used on various parts of the building to trim the installation. In addition, the system includes shapes for use on soffits.

Siding Panels

Siding panels, for horizontal application, are made in 8- and 12-inch (203 and 305 mm) widths. They come in configurations to simulate one, two, or three courses of bevel or drop siding. Panels designed for vertical application come in 12-inch (305 mm) widths. They are shaped to resemble boards. They can be used in combination with horizontal siding. Vertical siding panels with solid surfaces may also be used for soffits. For ventilation, perforated soffit panels of the same configuration are used (**Figure 66–1**).

Siding System Accessories

Siding systems require the use of several specially shaped accessories. *Inside* and *outside corner posts* are used to provide a weather-resistant joint to corners.

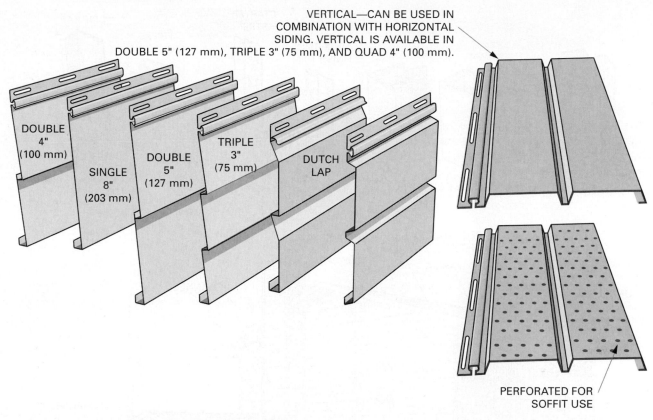

VERTICAL—CAN BE USED IN COMBINATION WITH HORIZONTAL SIDING. VERTICAL IS AVAILABLE IN DOUBLE 5" (127 mm), TRIPLE 3" (75 mm), AND QUAD 4" (100 mm).

DOUBLE 4" (100 mm)

SINGLE 8" (203 mm)

DOUBLE 5" (127 mm)

TRIPLE 3" (75 mm)

DUTCH LAP

PERFORATED FOR SOFFIT USE

FIGURE 66–1 Commonly used configurations of horizontal and vertical vinyl siding. *(Courtesy of Vinyl Siding Institute)*

Corner posts are available with channels of appropriate widths to accommodate various configurations of siding.

Some other accessories include **J-channels**, *starter strips*, and **undersill trim**, also known as *finish trim*. J-channels are made with several opening sizes. They are used in a number of places such as around doors and windows, at transition of materials, against soffits, and in many other places (**Figure 66–2**). The majority of vinyl siding panels and accessories are manufactured in 12′-6″ (3.81 m) lengths. Trim accessories and moulding are typically 12 feet (3.66 m) long.

APPLYING HORIZONTAL SIDING

The siding may expand and contract as much as ¼ inch (6 mm) over a 12′-6″ (3.81 m) length with changes in temperature. For this reason, it is important to centre fasteners in the slots. Do not drive them too tightly. There should be about 1/32 inch (1 mm) between the head of the fastener, when driven, and the siding (**Figure 66–3**). After the panel is fastened, it should be easily moved from side to side in the nail slots. Space fasteners 16 inches (406 mm) apart for horizontal siding and 6 to 12 inches (152 to 305 mm) apart for accessories unless otherwise specified by the manufacturer.

Applying the Starter Strip

Snap a level line to the height of the starter strip all around the bottom of the building. Fasten the strips to the wall with their top edges to the chalk line. Leave a ¼-inch (6 mm) space between them and other accessories to allow for expansion (**Figure 66–4**). Make sure the starter strip is applied as straight as possible. It controls the straightness of entire installation.

Installing Corner Posts

Corner posts are installed in corners ¼ inch (6 mm) below the starting strip and ¼ inch (6 mm) from the top. Attach the posts by fastening in the top of the upper slot on each side. The posts will hang on these fasteners. The rest of the fasteners should be centred on the slots. Make sure the posts are straight and true from top to bottom (**Figure 66–5**).

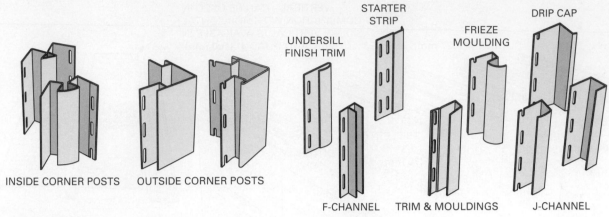

INSIDE CORNER POSTS   OUTSIDE CORNER POSTS

UNDERSILL FINISH TRIM

STARTER STRIP

FRIEZE MOULDING

DRIP CAP

F-CHANNEL   TRIM & MOULDINGS   J-CHANNEL

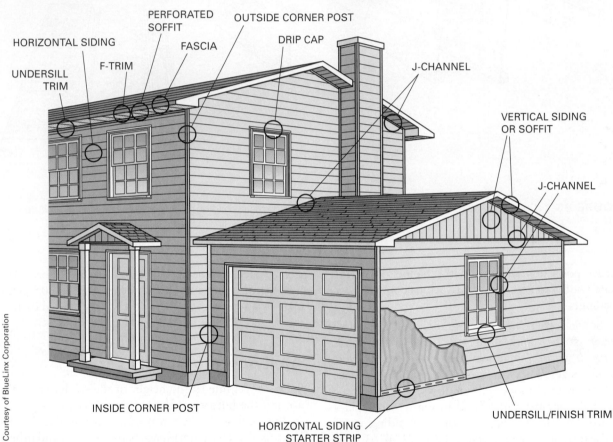

UNDERSILL TRIM

HORIZONTAL SIDING

F-TRIM

PERFORATED SOFFIT

FASCIA

OUTSIDE CORNER POST

DRIP CAP

J-CHANNEL

VERTICAL SIDING OR SOFFIT

J-CHANNEL

UNDERSILL/FINISH TRIM

INSIDE CORNER POST

HORIZONTAL SIDING STARTER STRIP

Courtesy of BlueLinx Corporation

FIGURE 66–2 Various accessories are used to trim a vinyl siding installation.

Installing J-Channel

Install J-channel pieces across the top and along the sides of window and door casings. They may also be installed under windows or doors with the undersill trim nailed inside of the channel. To mitre the corners, cut all pieces to extend, on both ends, beyond the casings and sills a distance equal to the width of the channel face. On both ends of the side J-channels, cut a ¾-inch (19 mm) notch out of the bottom of the J-channel. Fasten in place.

On both ends of the top and bottom channels, make ¾-inch (19 mm) cuts. Bend down the tabs and mitre the faces. Install them so that the mitred faces are in front of the faces of the side channels (**Figure 66–6**).

Installing Siding Panels

Snap the bottom of the first panel into the starter strip. Fasten it to the wall. Start from a back corner, leaving a ¼-inch (6 mm) space in the corner

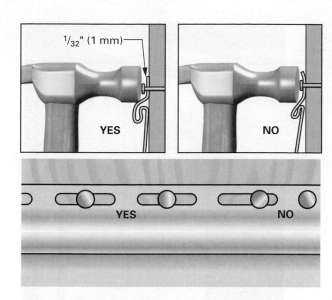

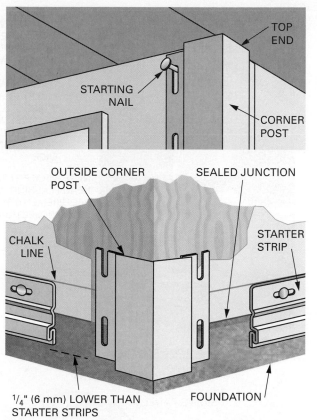

FIGURE 66–3 Fasten siding so as to allow for expansion and contraction.

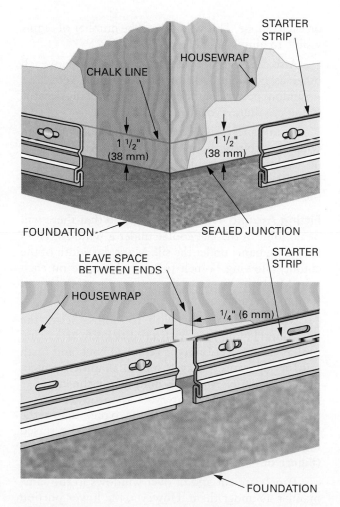

FIGURE 66–4 Installation of the starter strip.

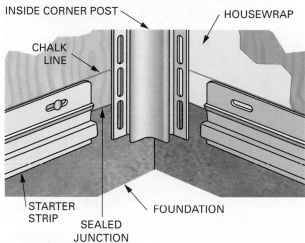

FIGURE 66–5 Inside and outside corner posts are installed in a similar way.

post channel. Work toward the front with other panels. Overlap each panel about 1 inch (25 mm) (**Figure 66–7**).

The seams of overlapped panels are more visible from one direction and less so from the other. Lap the panels so that they are visible from the direction least travelled. This will put the best side of the siding toward the most often viewed side. If either direction is equally travelled, alternate the direction of the lap

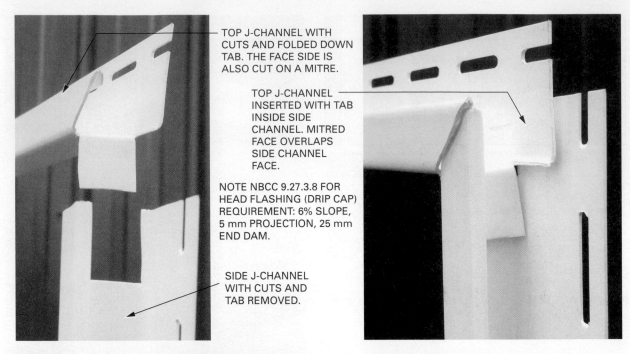

TOP J-CHANNEL WITH CUTS AND FOLDED DOWN TAB. THE FACE SIDE IS ALSO CUT ON A MITRE.

TOP J-CHANNEL INSERTED WITH TAB INSIDE SIDE CHANNEL. MITRED FACE OVERLAPS SIDE CHANNEL FACE.

NOTE NBCC 9.27.3.8 FOR HEAD FLASHING (DRIP CAP) REQUIREMENT: 6% SLOPE, 5 mm PROJECTION, 25 mm END DAM.

SIDE J-CHANNEL WITH CUTS AND TAB REMOVED.

FIGURE 66–6 Cutting J-channel to fit around windows and doors so that water does not get behind siding.

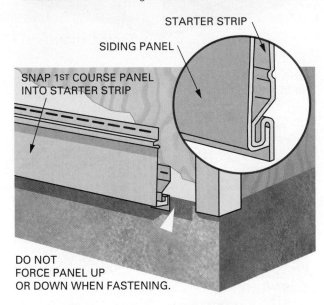

STARTER STRIP

SIDING PANEL

SNAP 1ST COURSE PANEL INTO STARTER STRIP

DO NOT FORCE PANEL UP OR DOWN WHEN FASTENING.

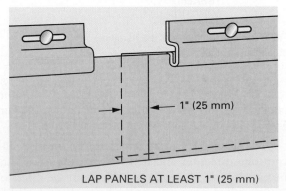

1" (25 mm)

LAP PANELS AT LEAST 1" (25 mm)

FIGURE 66–7 Installing the first course of horizontal siding panels.

on each course. This will reduce the number of seams visible to almost a half.

Also, keep vertical seams as far away from each other as possible. Do not let them align vertically. A random pattern in the seams will be less noticeable to the eye than vertically aligned seams.

Install successive courses by interlocking them with the course below. Use tin snips, hacksaw, utility knife, or circular saw. Reverse the blade if a circular saw is used, for smooth cutting through the vinyl.

Fitting Around Windows. Plan so that there will be no joint in the last course under a window. Hold the siding panel under the sill. Mark the width of the cutout, allowing ¼-inch (6 mm) clearance on each side. Mark the height of the cutout, allowing ¼-inch (6 mm) clearance below the sill.

Make vertical cuts with tin snips. Score the horizontal layout lines with a utility knife or scoring tool. Bend the section to be removed back and forth until it separates. Using a special snaplock punch, punch the panel ¼ inch (6 mm) below the cut edge at 6-inch (150 mm) intervals to produce *raised lugs* facing outward. Install the panel under the window and up in the undersill trim. The raised lugs cause the panel to snap snugly into the trim (**Figure 66–8**).

Panels are cut and fit over windows in the same manner as under them. However, the lower portion is cut instead of the top. Install the panel by placing

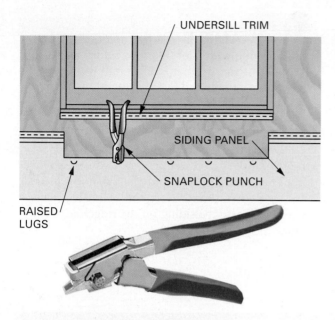

FIGURE 66–8 Method of fitting a panel under a window.

CUT EDGE OF PANEL FITS INTO
J-CHANNEL OVER TOP OF WINDOW.

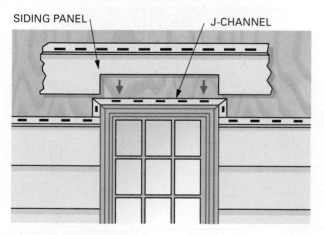

FIGURE 66–9 Fitting a panel over a window.

it into the J-channel that runs across the top of the window (**Figure 66–9**).

Installing the Last Course under the Soffit

The last course of siding panels under the soffit is installed in a manner similar to that for fitting under a window. An undersill trim is applied on the wall, up against the soffit. Panels in the last course are cut to width. Lugs are punched along the cut edges. The panels are then snapped firmly into place into the undersill trim (**Procedure 66–A**).

Gable End Installation

The rakes of a gable end are first trimmed with J-channels. The panel ends are inserted into the channel with a ¼-inch (6 mm) expansion gap. Make a pattern for cutting gable end panels at an angle where they intersect with the rake. Use two scrap pieces of siding to make the pattern (**Figure 66–10**). Interlock one piece with an installed siding panel below. Hold the other piece on top of it and against the rake. Mark along the bottom edge of the slanted piece on the face of the level piece.

APPLYING VERTICAL SIDING

The installation of vertical siding is similar to that for horizontal siding with a few exceptions. The method of fastening is the same. However, space fasteners about 12 inches (305 mm) apart for vertical siding panels. The starter strip is different. It may be ½-inch (12 mm) J-channel or drip cap flush with and fitted into the corner posts (**Figure 66–11**). Around windows and doors, under soffits, against rakes, and other locations, ½-inch (12 mm) J-channel is used. One of the major differences is that a vertical layout should be planned so that the same and widest possible piece is exposed at both ends of the wall.

Installing the First Panel

To install the first panel, start by determining the widest possible width of the first and last panel. This is done by measuring, between the corner posts, the width of the face to be sided. Divide this number by the exposure of one panel. Take the decimal remainder and add one to it. Divide this number by two. This will be the size of the first and last panel (**Figure 66–12**).

Install the first vertical panel plumb on the starter strip with one edge into the corner post. Allow ¼ inch (6 mm) at top and bottom. Place the first nails in the uppermost end of the top nail slots to hold it in position.

The edge of the panel may need to be cut in order for equal widths to be exposed on both ends of the wall. If the panel is cut on the flat surface, place a piece of undersill trim backed by furring into the channel of the corner post. Punch lugs along the cut edge of the panel at 6-inch (152 mm) intervals. Snap the panel into the undersill trim. Edges of vertical panels cut to fit in J-channels around windows and doors are treated in the same way (**Figure 66–13**).

EXAMPLE What are the starting and finishing widths for a wall section that measures 18′-9″ for siding that is 12″ wide?

Convert this measurement to a decimal by first dividing the inches portion by 12 and then adding it to the feet to get 18.75′.

Divide this by the siding exposure, in feet: 18.75 ÷ 1 foot = 18.75 pieces.

Subtract the decimal portion along with one full piece giving 1.75 pieces. Next 1.75 ÷ 2 = 0.875, multiplied by 12 gives 10½″.

This is the size of the starting and finishing piece. Thus there are 17 full-width pieces and two 10½″ wide pieces.

EXAMPLE What are the starting and finishing widths of 305 mm siding that is to be applied to a wall that is 5715 mm wide?

Wall width divided by siding width → 5715 ÷ 305 = 18.74 pieces.

Subtract 1 full piece plus the decimal → 18 − 1.74 = 17 full pieces.

Divide the 1.74 by 2 and multiply by 305 → 1.74/2 × 305 = 261 mm.

The starting piece and the finishing piece will be 261 mm and there will 17 full pieces. This will centre the siding on the ridgeline of the roof.

Step**by**Step Procedures

Procedure 66–A Fitting the Last Course of Horizontal Siding under the Soffit

STEP 1 MEASURE FOR LAST COURSE OF SIDING.

STEP 2 CUT PIECE AND CREATE RAISED LUGS WITH SNAPLOCK PUNCH.

STEP 3 SNAP CUT EDGE INTO UNDERSILL TRIM AND BOTTOM EDGE INTO COURSE BELOW.

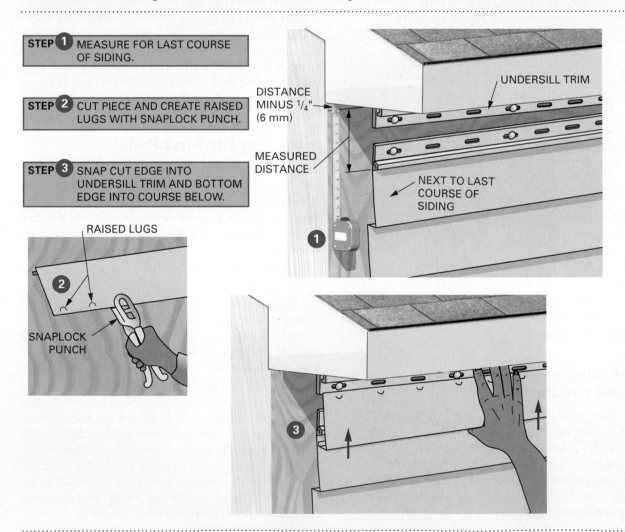

DISTANCE MINUS ¼″ (6 mm)

MEASURED DISTANCE

UNDERSILL TRIM

NEXT TO LAST COURSE OF SIDING

RAISED LUGS

SNAPLOCK PUNCH

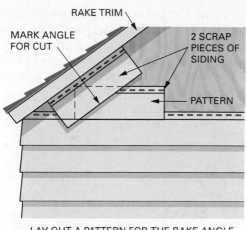

LAY OUT A PATTERN FOR THE RAKE ANGLE
ON A SCRAP PIECE OF SIDING

USE THE PATTERN TO MAKE THE
RAKE ANGLE ON SIDING PANELS

FIGURE 66–10 Fitting horizontal siding panels to the rakes.

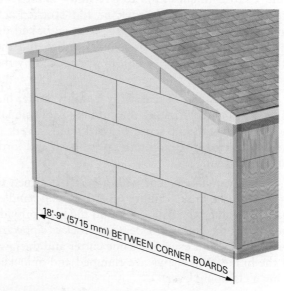

FIGURE 66–12 Example for finding first and last panel width of vertical siding.

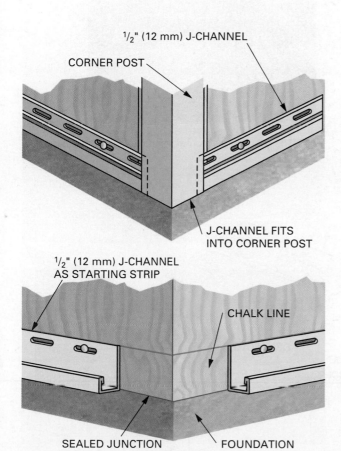

FIGURE 66–11 The starter strip shape and its intersection with corner posts is different for vertical application of vinyl siding compared to horizontal.

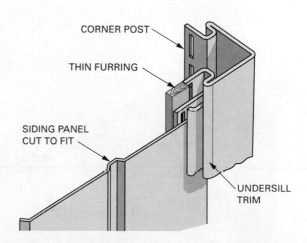

FIGURE 66–13 Undersill trim and furring are required when vertical siding is cut to fit into corner posts and J-channels.

ALUMINUM ACCESSORIES

Fascia is completed by covering the wood subfascia with an L-shaped piece of siding. This piece may be made of either metal or vinyl. The top edge is installed by slipping it under the metal drip edge. The bottom is held in place with nails (**Figure 66–14**). Nails are aluminum or stainless steel painted to the same colour as the trim.

Aluminum trim pieces are often fabricated on the job. This is done by using a tool, called a brake, to bend sheet metal (**Figure 66–15**). Light-duty brakes are designed for aluminum only, whereas others will bend heavy sheet metal.

Aluminum stock is sold in 50-foot (15.2 m) or 100-foot (30 m) rolls of various widths ranging from 12 to 24 inches (305 to 610 mm). These rolls are referred to as coil stock or flat stock. Each side of the sheet is coloured with a baked-on enamel finish. One side is usually white and the other one of a variety of colours produced by the manufacturer.

Coil stock can be cut with a utility knife, but using a straightedge makes the cut look professional. Stock is cut with one score of the knife. The cut is then completed by bending the piece back and forth through the cut.

Stock can also be cut with a cutter designed to work with the brake (**Figure 66–16**). A coil handler makes unrolling the coil easier and the jaws provide the rail for the cutter to ride. Coil stock is ripped to the desired width using the same cutter and the jaws of the brake. Brake jaws are clamped and unlocked using a clamping lever.

Once the pieces are cut to width, they are then bent to the desired configuration. Care must be taken to visualize the piece as it is bent. Mistakes are easy to make. For example, the piece may have the correct shape but the wrong colour facing outward.

Bending the stock begins with careful positioning of the piece in the jaws. The same amount

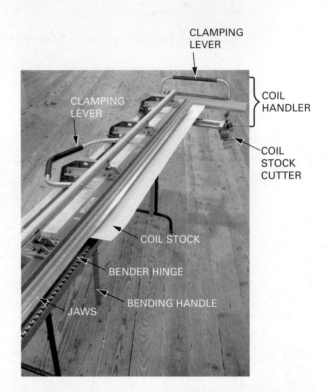

FIGURE 66–15 An aluminum brake is used to shape coil stock.

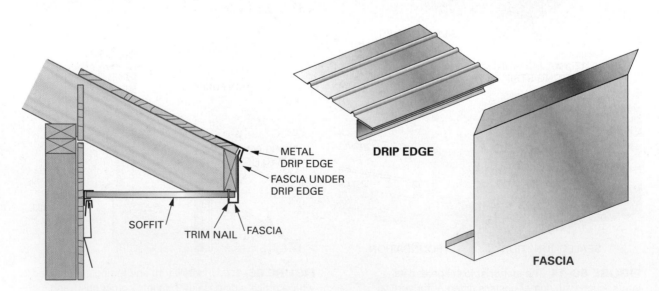

FIGURE 66–14 Metal fascia material is fitted under metal drip edge.

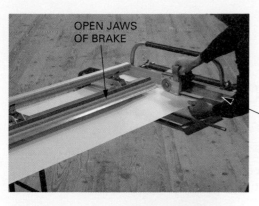

NOTE: CUTTER MUST BE KEPT
TIGHT TO GUIDE RAIL AS CUT IS MADE.

FIGURE 66–16 A special cutter can be used to cut stock to length and width.

FIGURE 66–17 Bending metal is done by clamping jaws and lifting handles.

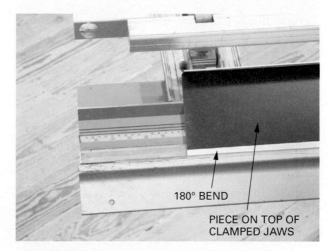

FIGURE 66–18 Full 180-degree bends are completed on top of the jaws.

of stock should be revealed from the jaws on both ends. This will ensure that the piece is not tapered. Locking the jaws tight secures the piece during the bend. The bend is made by raising the handles of the brake (**Figure 66–17**). The bend may be any angle from 0 to 180 degrees. Care must be taken to bend the stock to the desired angle. Making repetitive stops at the same angle takes some practice.

Making an edge return uses a full 180-degree bend, which takes two steps. First, align the piece to as small a bend as possible, usually about ⅝ inch (16 mm). The piece should be flush with the brake bender. Then lock and bend as far as the brake will allow. This is about 150 degrees. The jaws are then unlocked, the piece is removed, and the jaws relocked. The piece is then placed on top of the jaws (**Figure 66–18**). The final bend to 180 degrees is made while the piece is placed on top of the jaws.

ESTIMATING ALUMINUM AND VINYL SIDING

Aluminum and vinyl siding panels are sold by the square. Determine the wall area to be covered. Add 10 percent of the area for waste. Divide by 100. This gives you the number of squares needed.

Become familiar with accessories and how they are used. Measure the total linear feet (metres) required for each item.

Chapter 67

Cornices and Eavestroughs

Cornice terms from earlier times remain in use. However, cornice design has changed considerably. In earlier times, cornices were very elaborate and required much time to build. Now, cornice design, in most cases, is much more simplified. Only occasionally is a building designed with an ornate cornice similar to those built in years gone by.

PARTS OF A CORNICE

Several finish parts are used to build the cornice (**Figure 67–1**). In some cases, additional framing members are required.

Subfascia

The **subfascia** is sometimes called the false fascia or rough fascia. It is a horizontal framing member fastened to the rafter tails. It provides an even, solid, and continuous surface for the attachment of other cornice members. When used, the subfascia is generally a nominal 1- or 2-inch-thick (19 or 38 mm)

piece. Its width depends on the slope of the roof, the tail cut of the rafters, or the type of cornice construction.

Soffit

The finished member on the underside of the cornice is called a **plancier** and is often referred to as a soffit (**Figure 67–2**). Soffit material may include solid lumber, plywood, strand board, fibreboard, or corrugated aluminum and vinyl panels. Soffits may be perforated or constructed with screen openings to allow for ventilation of the rafter cavities. Soffits may be fastened to the bottom edge of the rafter tails to the slope of the roof. The soffit is an ideal location for the placement of attic ventilation.

Fascia

The fascia is fastened to the subfascia or to the ends of the rafter tails. It may be a piece of lumber grooved

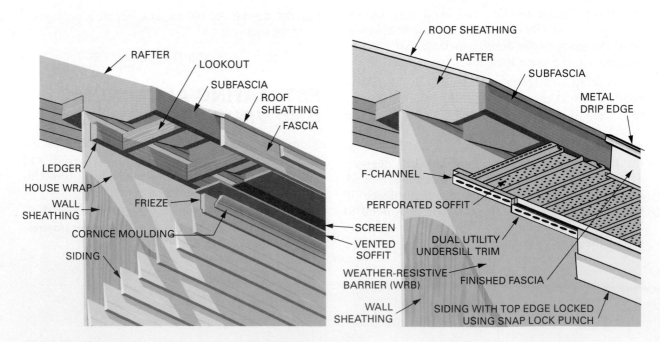

FIGURE 67–1 Cornices may be constructed of different materials.

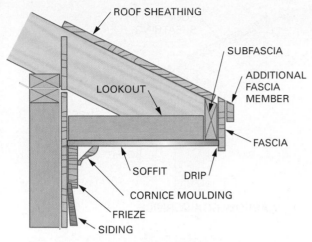

WOOD CORNICE

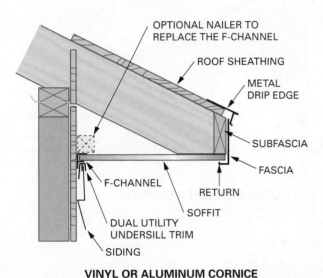

VINYL OR ALUMINUM CORNICE

FIGURE 67–2 The soffit is the bottom finish member of the cornice.

to receive the soffit. It also may be made from bent aluminum and vinyl material used to wrap the subfascia. Fascia provides a surface for the attachment of a gutter. The fascia may be built up from one or more members to enhance the beauty of the cornice.

The bottom edge of the fascia usually extends below the soffit by ¼ to ⅜ inch (6 to 9.5 mm). The portion of the fascia that extends below the soffit is called the **drip**. The drip is necessary to prevent rainwater from being swept back against the walls of the building. In addition, a drip makes the cornice more attractive.

Frieze

The **frieze** is fastened to the sidewall with its top edge against the soffit. Its bottom edge is sometimes

rabbeted to receive the sidewall finish. In other cases, the frieze may be furred away from the sidewall to allow the siding to extend above and behind its bottom edge. However, the frieze is not always used. The sidewall finish may be allowed to come up to the soffit. The joint between the siding and the soffit is then covered by a moulding.

Cornice Moulding (Bed Moulding)

The **cornice moulding (bed moulding)** is used to cover the joint between the frieze and the soffit. If the frieze is not used, the cornice moulding covers the joint between the siding and the soffit.

Lookouts

Lookouts are framing members, usually 2 × 4 (38 × 89 mm) stock, that are used to provide a fastening surface for the soffit. They run horizontally from the end of the rafter to the wall, adding extra strength to larger overhangs. Lookouts may be installed at every rafter or spaced 48 inches (1220 m) on centre (OC), depending on the material being used for the soffit (Figure 67–2).

CORNICE DESIGN

Cornices are generally classified into three main types: box, snub, and open (**Figure 67–3**).

The Box Cornice

The **box cornice** is probably most common. It gives a finished appearance to this section of the exterior. Because of its overhang, it helps protect the sidewalls from the weather. It also provides shade for windows.

Box cornices may be designated as narrow or wide. They may be constructed with level or sloping soffits. A *narrow box cornice* is one in which the cuts on the rafters serve as nailing surfaces for the cornice members. A *wide box cornice* may be constructed with a level or sloping soffit. A wide, level soffit requires the installation of lookouts.

The Snub Cornice

The **snub cornice** is not as attractive as some of the other designs, nor does it give as much protection to the sidewalls of a building because of its small overhang. There is no rafter projection beyond the wall. The snub cornice is chosen primarily to cut down the cost of material and labour. A snub cornice is frequently used on the rakes of a gable end in

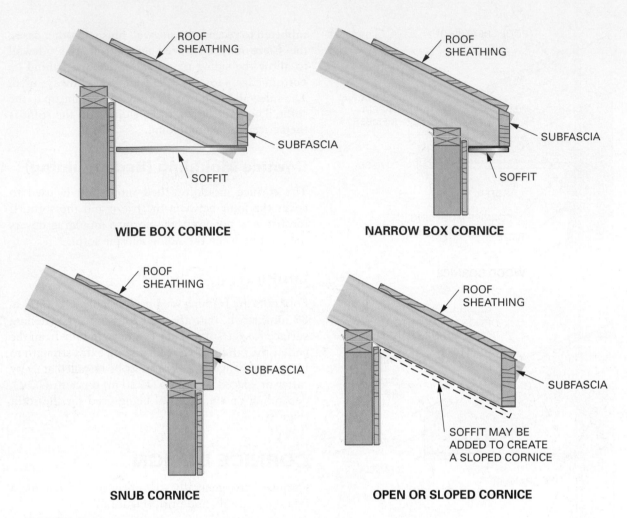

ROOF SHEATHING

SUBFASCIA

SOFFIT

WIDE BOX CORNICE

ROOF SHEATHING

SUBFASCIA

SOFFIT

NARROW BOX CORNICE

ROOF SHEATHING

SUBFASCIA

SNUB CORNICE

ROOF SHEATHING

SUBFASCIA

SOFFIT MAY BE ADDED TO CREATE A SLOPED CORNICE

OPEN OR SLOPED CORNICE

FIGURE 67–3 The cornice may be constructed in various styles.

combination with a boxed cornice on the sides of a building.

The Open Cornice

The **open cornice** has no soffit. It is used when it is desirable to expose the rafter tails. It is often used when the rafters are large, laminated or solid beams with a wide overhang that exposes the roof decking on the bottom side. Open cornices give a contemporary or rustic design look to post-and-beam framing. They provide protection to side-walls at low cost. This cornice might also be used for conventionally framed buildings for reasons of design and to reduce costs.

By adding a soffit, a *sloped cornice* is created. The soffit is installed directly to the underside of the rafter tails. This is sometimes done to simplify the cornice detail when there is also an overhang over the gable end of the building.

Rake Cornices

The main cornice is constructed on the rafter tails where they meet the walls of a building. On buildings with hip or mansard roofs, the main cornice, regardless of the type, extends around the entire building.

On buildings with gable roofs, a boxed main cornice with a sloping soffit, attached to the bottom edge of the rafter tails, may be returned up the rakes to the ridge. The cornice that runs up the rakes is called a **rake cornice** (**Figure 67–4**).

Cornice Returns. A main cornice with a horizontal soffit attached to level lookouts may, at times, be terminated at each end wall against a snub rake cornice (**Figure 67–5**). At other times, a **cornice return** must be constructed to change the level box cornice to the angle of the roof.

A main cornice with a level soffit may also be returned upon itself. That is, the main cornice is

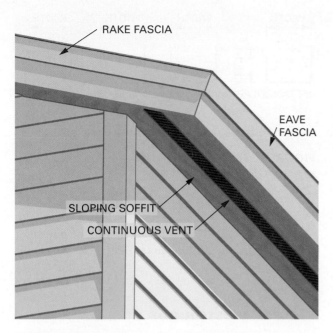

FIGURE 67–4 A boxed cornice with a sloping soffit may be returned up the rakes of a gable roof.

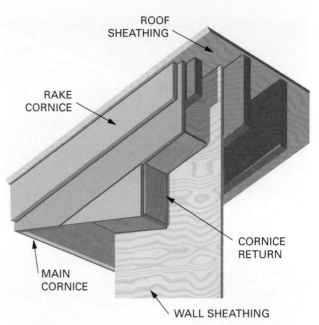

FIGURE 67–6 A boxed main cornice with a level soffit may be changed to the angle of the roof with a cornice return.

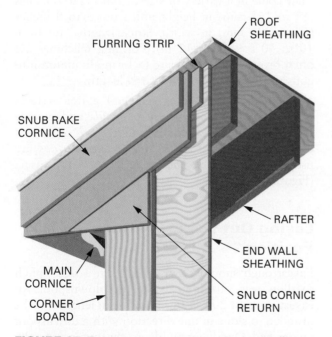

FIGURE 67–5 A boxed main cornice with a level soffit may be terminated at the gable ends against the rake cornice.

mitred at each end. It is turned 90 degrees toward and beyond the corner as much as it overhangs on the side of the building. This short section on each end of a gable roof is called a cornice return. The cornice return provides a stop for the rake cornice. It adds to the design of the building at this point (**Figure 67–6**). However, cornice returns of this type are rarely built today. A large amount of labour is

involved in their construction. The main cornice is returned on the rakes of the gable end in a much more simple fashion, as described.

PRACTICAL TIPS FOR INSTALLING CORNICES

Install cornice members in a straight and true line with well-fitting and tight joints. Do not dismiss the use of hand tools for cutting, fitting, and fastening exterior trim.

All fasteners should be non-corrosive. Stainless steel fasteners offer the best protection against corrosion. Hot-dipped galvanized fasteners provide better protection than plated ones. Fasteners used in wood should be well-set. They should be puttied to conceal the fastener. This also prevents the heads from corroding. Setting and concealing fasteners in exterior trim is a mark of quality.

Paint or otherwise seal and protect all exterior trim as soon as possible after installation. Properly installed and protected wood exterior trim will last indefinitely.

EAVESTROUGHS

An **eavestrough** is a shallow trough or conduit set below the edge of the roof along the fascia. It catches and carries off rainwater from the roof. A **downspout**, also called a *conductor*, is a rectangular or round pipe.

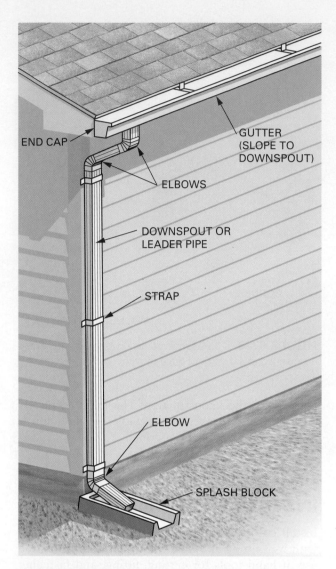

FIGURE 67–7 Eavestroughs and downspouts form an important system for conducting water away from the building.

It carries water from the eavestrough downward and away from the foundation (**Figure 67–7**).

Eavestroughs, or gutters, may be made of wood, galvanized iron, aluminum, copper, or vinyl. Copper gutters require no finishing. Vinyl and aluminum gutters are prefinished and ready to install. Wood and galvanized metal gutters need several protective coats of finish after installation.

The size of an eavestrough is determined by the area of the roof for which it handles the water runoff. Under ordinary conditions, 1 square inch (645 mm²) of gutter cross-section is required for every 100 square feet (9.3 m²) of roof area. For instance, a 4 × 5 inch (100 × 125 mm) gutter has a cross-section area of 20 square inches

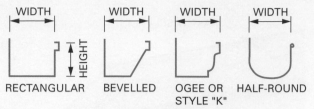

FIGURE 67–8 Metal eavestroughs are available in several shapes.

(12 500 mm²). It is, therefore, capable of handling the runoff from 2000 square feet (187.5 m²) of roof surface.

Metal Eavestroughs

Metal eavestroughs are made in rectangular, bevelled, ogee, or semicircular shapes (**Figure 67–8**). They come in a variety of sizes, from 2½ to 6 inches (63 to 152 mm) in height and from 3 to 8 inches (75 to 203 mm) in width. Stock lengths run from 10 to 40 feet (3 to 12 m). Forming machines are often brought to the jobsite to form aluminum into gutters of practically any desired length.

Besides straight lengths, metal gutter systems have components comprised of *inside* and *outside corners*, *joint connectors*, *outlets*, *end caps*, and others. *Metal brackets* or *spikes and ferrules* are used to support the gutter sections on the fascia (**Figure 67–9**).

Laying Out the Eavestrough Position

Eavestrough should be installed with a slight pitch to allow water to drain toward the downspout. An eavestrough of 20 feet (6 m) or less in length may be installed to slope in one direction with a downspout on one end. On longer buildings, the gutter is usually crowned in the centre. This allows water to drain to both ends.

On both ends of the fascia, mark the location of the top edge of the eavestrough. The top outside edge of the eavestrough should be in relation to a straight line projected from the top surface of the roof. The height of the eavestrough depends on the pitch of the roof (**Figure 67–10**).

Stretch a chalk line between the two marks. Move the centre of the chalk line up enough to give the eavestrough the proper pitch. Thumb the line down.

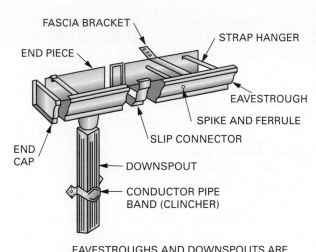

EAVESTROUGHS AND DOWNSPOUTS ARE
MADE UP OF MANY SEPARATE PARTS.

PIECE NEEDED	DESCRIPTION
	EAVESTROUGH COMES IN VARIOUS LENGTHS
	SLIP JOINT CONNECTOR: USED TO CONNECT JOINTS OF EAVESTROUGH
	END CAPS—WITH OUTLET: USED AT ENDS OF EAVESTROUGH RUNS
	END PIECE—WITH OUTLET USED WHERE DOWNSPOUT CONNECTS
	OUTSIDE MITRE: USED FOR OUTSIDE TURN IN EAVESTROUGH
	INSIDE MITRE: USED FOR INSIDE TURN IN EAVESTROUGH

PIECE NEEDED	DESCRIPTION	PIECE NEEDED	DESCRIPTION
	FASCIA BRACKET USED TO HOLD EAVESTROUGH TO FASCIA ON WALL		ELBOW—STYLE B FOR DIVERTING DOWNSPOUT TO LEFT OR RIGHT
	STRAP HANGER CONNECTS TO EAVE OF ROOF TO HOLD EAVESTROUGH		CONNECTOR PIPE BAND OR CLINCHER USED TO HOLD DOWNSPOUT TO LEFT OR RIGHT
	STRAINER CAP SLIPS OVER OUTLET IN END PIECE AS A STRAINER		SHOE USED TO LEAD WATER TO SPLASHER BLOCK
	DOWNSPOUT COMES IN 10' (3 m) LENGTHS		MASTIC USED TO SEAL ALUMINUM EAVESTROUGH AT JOINTS
	ELBOW—STYLE A FOR DIVERTING DOWNSPOUT IN OR OUT FROM WALL		SPIKE AND FERRULE USED TO HOLD EAVESTROUGH TO FASCIA OF ROOF

FIGURE 67–9 Components of a metal eavestrough system.

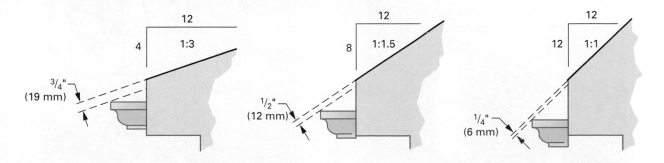

FIGURE 67–10 The height of the eavestrough on the fascia is in relation to the slope of the roof.

Snap it on both sides of the centre. For a slope in one direction only, snap a straight line lower on one end than the other to obtain the proper pitch. It is important to install eavestrough on the chalk line. This avoids any dips that may prevent complete draining of the gutter.

INSTALLING METAL AND VINYL EAVESTROUGHS

Fasten the eavestrough brackets to the chalk line on the fascia with screws. All screws should be made of stainless steel or other material that is corrosion resistant. Aluminum brackets may be spaced up to 30 inches (762 mm) OC. Steel brackets may be spaced up to 48 inches (1.2 m) OC. Install the gutter sections in the brackets. Use slip-joint connectors to join the sections. Apply the recommended gutter sealant to connectors before joining.

Locate the outlet tubes as required, keeping in mind that the downspout should be positioned plumb with the building corner and square with the building. Join with a connector. Add the end cap. Use either inside or outside corners where eavestroughs make a turn.

Vinyl eavestroughs and components are installed in a manner similar to metal ones.

INSTALLING DOWNSPOUTS

Metal or vinyl downspouts are fastened to the wall of the building in specified locations with aluminum

straps. Downspouts should be fastened at the top and bottom and every 6 feet (1.8 m) in between for long lengths.

The connection between the downspout and an eavestrough is made with 45-degree elbows and short straight lengths of downspout (**Figure 67–11**). The connection will depend on the offset of the eavestrough from the downspout. Because water runs downhill, care should be taken when putting the downspout pieces together. The downspout components are assembled where the upper piece is inserted into the lower one (**Figure 67–12**). This makes the joint lap such that the water cannot escape until it reaches the bottom-most piece. An elbow, called a *shoe*, should be used with a splash block at the bottom of the downspout. This leads water away from the foundation. An alternative method is to connect the downspout with underground piping that

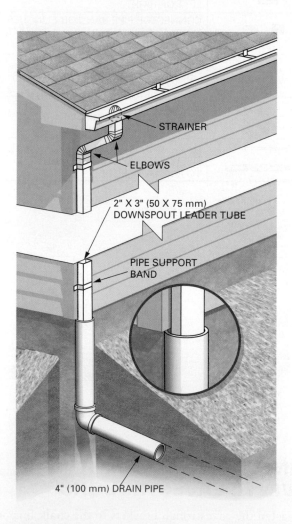

FIGURE 67–11 Typical downspout leader tubes fastened in place with support bands.

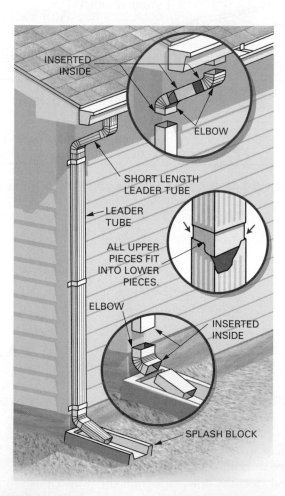

FIGURE 67–12 Upper eavestrough components are inserted into lower ones to ensure that the downspout does not leak.

carries the water away from the foundation. The piping can be connected to storm drains and dry-wells or piped to the surface elsewhere. Storm water as found in eavestrough downspouts should never be connected to footing or foundation drains. Strainer caps should be placed over eavestrough outlets if water is conducted by this alternative method. Leaves and other debris that fall into eavestrough flow into the drainage system and can cause clogging problems. Gutter protection accessories will prevent debris from getting into the eavestrough (**Figure 67-13**). More information can be found at *https://www.youtube.com/user/alurexinc/videos*.

Photo by M. Nauth

FIGURE 67–13 Gutter Protection keeps leaves and snow out of the eavestrough.

KEY TERMS

battens	downspout	lookouts	staggered coursing
bed moulding	drip	open cornice	starter strip
blind-nail	eavestrough	panel siding	storey pole
board and batten	face-nail	plancier	striated
board on board	fancy butt	preacher	subfascia
box cornice	fibre-cement siding	pressurized rainscreen	undercourse
brake	flat grain	rake	undersill trim
coil stock	frieze	rake cornice	vertical grain
corner boards	hardboard siding	ribbon coursing	water-resistant barrier (WRB)
cornice moulding	headlap	single coursing	
cornice return	J-channels	snaplock punch	water table
double coursing	lap siding	snub cornice	

REVIEW QUESTIONS

Select the most appropriate answer.

1. How is bevel siding applied?

   a. horizontally
   b. vertically
   c. horizontally or vertically
   d. horizontally, vertically, or diagonally

2. Which of the following statements about fibre-cement siding products is true?

   a. They are made to look like concrete blocks.
   b. They are installed similar to wood siding.
   c. They are only installed vertically.
   d. They are only installed horizontally.

3. What is one particular advantage of bevel siding over other types of lap siding?

   a. It comes in a variety of prefinished colours.
   b. It has a constant weather exposure.
   c. The weather exposure can be varied slightly.
   d. Application can be made in any direction.

4. When applying horizontal siding, which of the following should be done?

   a. Maintain exactly the same exposure with every course.
   b. Apply full courses above and below windows.
   c. Work from the top down.
   d. Use a water table.

5. In order to lay out a storey pole for courses of horizontal siding, which of the following must be known?

   a. the width of windows and doors
   b. the kind and size of finish at the eaves and foundation
   c. the location of windows, doors, and other openings
   d. the length of the wall to which siding is to be applied

6. Which of the following statements about wood shingle siding is true?

   a. The exposure may be varied.
   b. Butt seams should line up vertically.
   c. Each piece should be fastened with four nails.
   d. Each piece should be fastened with three nails.

7. When installing aluminum or vinyl siding, how should nails be driven?

   a. tightly against the flange
   b. just up to the flange
   c. into the wall sheathing only
   d. into the wall studs only

8. How is vinyl siding installed?

   a. with vertical butt seams randomly placed
   b. with butt seam overlapping away from view
   c. with a loose fit to trim pieces
   d. all of the above

9. To allow for expansion when installing solid vinyl starter strips, how large a space should be left between the ends?

   a. at least $\frac{1}{8}$ inch (3 mm)
   b. at least $\frac{1}{4}$ inch (6 mm)
   c. at least $\frac{3}{8}$ inch (9.5 mm)
   d. at least $\frac{1}{2}$ inch (12 mm)

10. What is the term for the exterior trim that extends up the slope of the roof on a gable end?

    a. box finish
    b. rake finish
    c. return finish
    d. snub finish

11. What is the term for a member of the cornice fastened in the vertical position to the rafter tails?

    a. drip
    b. fascia
    c. soffit
    d. frieze

12. What is a soffit?

    a. the part of a cornice that is attached horizontally under the rafter tails
    b. the part of a cornice that is vertical and attached to the ends of rafters
    c. the part of a cornice that serves as a drip cap
    d. the part of a cornice that separates the frieze from the lookout

13. How should eavestroughs be installed?

    a. level with the fascia
    b. with downspouts in the centre
    c. with downspout leader tubes connected to the foundation drains
    d. with the idea that water runs downhill

Unit 21
Decks, Porches, and Fences

| Chapter 68 | Deck and Porch Construction | Chapter 69 | Fence Design and Erection |

Among the final steps in finishing the exterior is the building of porches, decks, fences, and other accessory structures. Plans may not always show specific construction details. Therefore, it is important to know some of the techniques used to build these structures.

Accents to a building may come in the form of decks and porches. They serve as reminders to the overall workmanship of the house. Take care to install material in a neat and professional manner.

OBJECTIVES

After completing this unit, the student should be able to:

- describe the construction of and kinds of materials used in decks and porches.
- lay out and construct footings for decks and porches.
- install supporting posts, beams, and joists.
- apply decking in the recommended manner and install flashing, for an exposed deck, against a wall.
- construct deck stairs and railings.
- describe several basic fence styles.
- design and build a straight and sturdy fence.

Chapter 68

Deck and Porch Construction

Wood porches and decks are built to provide outdoor living areas for various reasons, in both residential and commercial construction. The construction of both is similar. However, a porch is covered by a roof. Its walls may be enclosed with wire mesh screens for protection against insects. With screen and storm window combinations, glass replaces screens to keep the porch comfortably warm in the cold months.

DECK MATERIALS

Decking materials must be chosen for strength and durability, as well as appearance and resistance to decay. Redwood, cedar, and pressure-treated pine are often used as decking boards. Other manufactured decking materials available include EON7, Timber Tech®, and Trex®. These decking products can be made from a mixture of plastic and sawdust or from just vinyl. They are cut, fit, and fastened in the same manner as wood and can have the added benefit of being made mostly from recycled material. The carpenter must follow each manufacturer's specifications for proper installation (**Figure 68–1**).

If not specified, the kind, grade, and sizes of material must be selected before building a deck. Also, the size and kind of fasteners, connectors, anchors, and other hardware must be determined.

Lumber

All lumber used to build decks should be **pressure-treated** with preservatives or be from a decay-resistant species, such as redwood or cedar. Remember, it is the heartwood of these species that is resistant to decay, not the sapwood. Either *all-heart* or pressure-treated

lumber should be used wherever there is a potential for decay. This is essential for posts that are close to the ground and other parts subject to constant moisture. (A description of pressure-treated lumber and its uses can be found in Chapter 32.)

Lumber Grades. For pressure-treated (PT) pine and western cedar, #2 grade is structurally adequate for most applications. Appearance can be a deciding factor when choosing a grade (**Figure 68–2**). If a better appearance is desired, higher grades should be considered.

Lumber Sizes. Specific sizes of supporting posts, beams, and joists depend on the spacing and height of supporting posts and the spacing of beams and joists. In addition, the sizes of structural members

FIGURE 68–2 A deck constructed with pressure-treated lumber.

Photo by M. Nauth. Courtesy of Shayne Mitchell and Julie Mahoney

FIGURE 68–1 Wood-vinyl composite deck.

Courtesy of The Deck Store, Inc.

depend on the type of wood used and the weight imposed on the members. Too many factors prohibit generalization about sizes of structural members. Check with local building officials or with a professional to determine the sizes of structural members for specific deck construction. Determining sizes with incomplete information may result in failure of undersized members. Unnecessary expense is incurred with the use of oversized members.

Fasteners

All nails, fasteners, and hardware should be stainless steel or hot-dipped galvanized (HDG). Electroplated galvanizing is not acceptable because the coating is too thin (less than one-tenth the coating thickness and bond to substrate steel than HDG). Screws can be coloured-matched with a ceramic coating. In addition to corroding and eventual failure, poor-quality fasteners will react with substances in decay-resistant woods and cause unsightly stains.

BUILDING A DECK

Most wood decks consist of posts, set on footings, supporting a platform of *beams* and *joists* covered with deck boards. Posts, rails, balusters, and other special parts make up the railing. Other parts, such as shading devices, privacy screens, benches, and planters, lend finishing touches to the area.

Installing a Ledger

If the deck is to be built against a building, a ledger, usually the same size as the joists, is bolted to the wall for the entire length of the deck. The ledger should be held off the building using spacer washers in order to prevent the transfer of moisture and to allow the ledger to dry in the event of water infiltration (**Figure 68–3**). Holes are pre-drilled into the rim joist of the house and filled with resilient sealant before the lag screws are inserted and tightened. The ledger acts as a beam to support joists that run at right angles to the wall. It is installed to a level line. Its top edge is located to provide a comfortable step down from the building after the decking is applied, about 4 inches (100 mm). Some carpenters prefer to run the joists parallel to the wall so that the deck boards can run perpendicular to the building. In this way, they can slope the boards slightly away from the wall to allow for water runoff. The ledger height may be used as a benchmark for establishing the elevations of supporting posts and beams.

After the decking is applied, a flashing is installed under the siding and on top of the deck board. Caulking is applied between the deck and the flashing. The flashing is then fastened, close to and along its top

Photo by M. Nauth

FIGURE 68–3 The ledger is spaced off the wall with hockey pucks and anchored with HDG ½-inch (12-mm) lags at 24 inches (610 mm) on centre (OC).

flange, with nails spaced closely together. The outside edge of the flashing should extend beyond the ledger (**Figure 68–4**).

Concrete Pier and Footing Layout and Construction

Concrete piers for the supporting posts must be accurately located (**Figure 68–5**). To determine their location, erect batter boards and stretch lines in a manner previously described in Chapter 28. All piers must have footing pads, which will require digging a hole that will accommodate the footing. There are several pre-manufactured plastic footing forms, as well as fibre forms for the piers (**Figure 68–6**). The holes are dug to the required depth, and the forms are located, plumbed, aligned, backfilled, and then filled with concrete.

Pier and Footing Size and Style

When placed in stable soil, the diameter of the concrete pier is usually twice the width of the post it is supporting. The footing depth reaches undisturbed soil, at least 12 inches (305 mm) below grade. In cold climates, the footing should extend below the frost line.

Several methods are commonly used. One method is to partially fill the footing hole with concrete, and then set a cylindrical fibre-form into the wet concrete. After the forms are aligned and plumbed, the concrete is poured to the desired height. The lines are strung

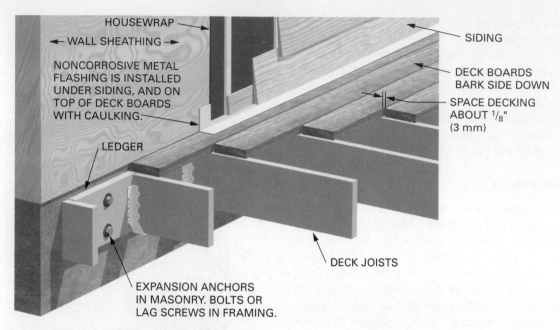

HOUSEWRAP

◄── WALL SHEATHING ──►

NONCORROSIVE METAL
FLASHING IS INSTALLED
UNDER SIDING, AND ON
TOP OF DECK BOARDS
WITH CAULKING.

LEDGER

EXPANSION ANCHORS
IN MASONRY. BOLTS OR
LAG SCREWS IN FRAMING.

SIDING

DECK BOARDS
BARK SIDE DOWN

SPACE DECKING
ABOUT 1/8"
(3 mm)

DECK JOISTS

FIGURE 68–4 A ledger is made weathertight with flashing.

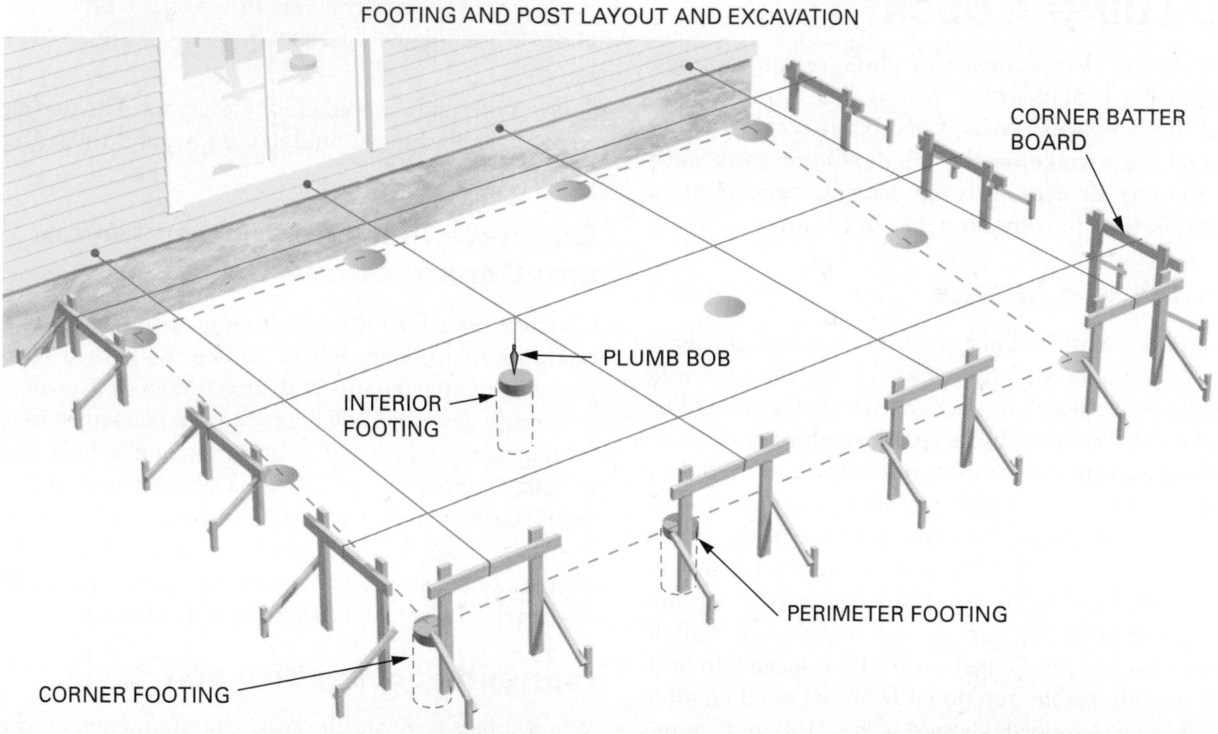

FOOTING AND POST LAYOUT AND EXCAVATION

CORNER BATTER
BOARD

PLUMB BOB

INTERIOR
FOOTING

PERIMETER FOOTING

CORNER FOOTING

FIGURE 68–5 Pier and footing layout and excavation.

again and the anchors are set, aligned to the layout lines. Set post anchors while the concrete is still wet (**Figure 68–7**).

Another method is to attach plastic "feet" to the base of the fibre-forms; place them into the excavated holes; align, plumb, and brace the forms; backfill around them; and then pour the concrete. The post anchors can also be set after the concrete has cured (**Figure 68–8**).

Recently, a B.C.-based company, Fab-Form, has created a tube out of a high-strength fabric that is cut to length (with a knife or scissors) and becomes the form for the concrete pier. It is called FAST-TUBE™ (**Figure 68–9**).

FIGURE 68–6 Excavating for the plastic footing form.

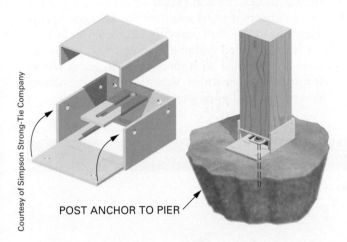

POST ANCHOR TO PIER

Courtesy of Simpson Strong-Tie Company

FIGURE 68–7 Post anchors may be fastened to concrete using an expansion bolt.

Erecting Supporting Posts

All supporting posts are set on piers. They are then plumbed and braced. Cut posts, for each footing, a little longer than their final length. Tack the bottom of each post to the anchor. Brace them in a plumb position in both directions (**Procedure 68–A**).

When all posts are plumbed and braced, the tops must be cut level to the proper height. From the height of the deck, deduct the deck thickness and the depth of the beam. Mark on a corner post. Mark the other posts by levelling from the first post marked. Mark each post completely around using a square. Cut the tops with a portable circular saw (**Figure 68–10**).

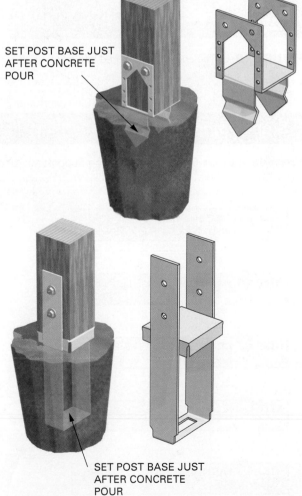

SET POST BASE JUST AFTER CONCRETE POUR

SET POST BASE JUST AFTER CONCRETE POUR

Courtesy of Simpson Strong-Tie Company

FIGURE 68–8 Post anchors may be set into concrete after it is placed.

Courtesy of Fab-Form Industries Ltd.

FIGURE 68–9 The concrete is poured into a FAST-TUBE.

Installing Beams

Install the beams on the posts using post-and-beam metal connectors. The deck should slope slightly, about ⅛ inch per foot (1:100), away from the building. The size of the connector will depend on the size of the posts and beams. Install beams with the crowned edge up. Any splice joints should fall over the centre of the post.

Installing Joists

Joists may be placed over the top or between the beams. When joists are hung between the beams, the overall depth of the deck is reduced. This provides more clearance between the frame and the ground. For decking run at right angles, joists may be evenly spaced close to 20 inches (508 mm) OC. Joists should be spaced 16 inches (406 mm) OC for diagonal decking.

Lay out and install the joists in the same manner as described in earlier chapters. Use appropriate hangers if joists are installed between beams (**Figure 68–11**). When joists are installed over beams, use recommended framing anchors. Make sure all joists are installed with their crowned edges facing upward, and remember that the framing anchors have to be hot-dipped galvanized (HDG).

Step**by**Step Procedures

Procedure 68–A Erecting and Bracing Supporting Posts

STEP 1 FASTEN POST TO ANCHOR WITH NAILS OR BOLTS.

STEP 2 NAIL BRACE TOPS TO POSTS.

STEP 3 DRIVE STAKE AT THE END OF EACH BRACE.

STEP 4 PLUMB POSTS EACH WAY.

STEP 5 FASTEN BRACES TO STAKES.

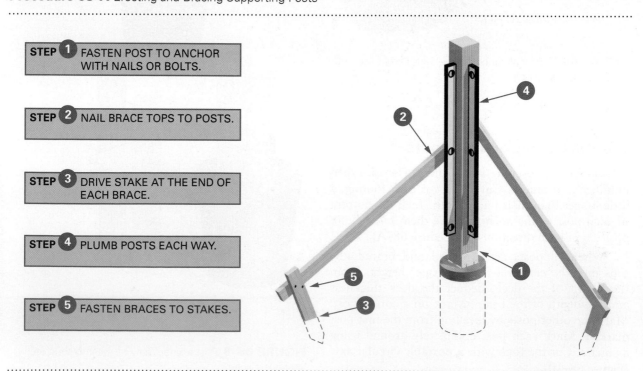

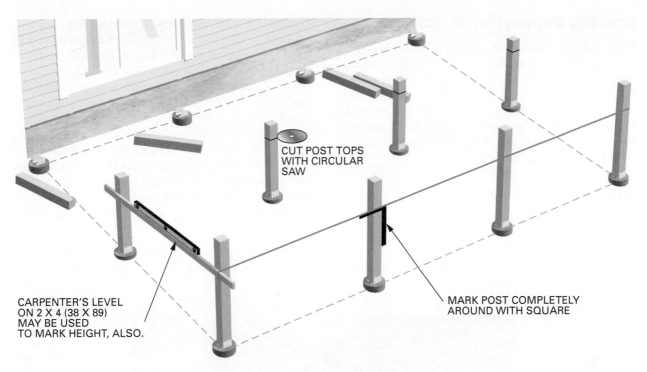

CUT POST TOPS
WITH CIRCULAR
SAW

CARPENTER'S LEVEL
ON 2 X 4 (38 X 89)
MAY BE USED
TO MARK HEIGHT, ALSO.

MARK POST COMPLETELY
AROUND WITH SQUARE

FIGURE 68–10 The post height is determined and the tops cut level with each other.

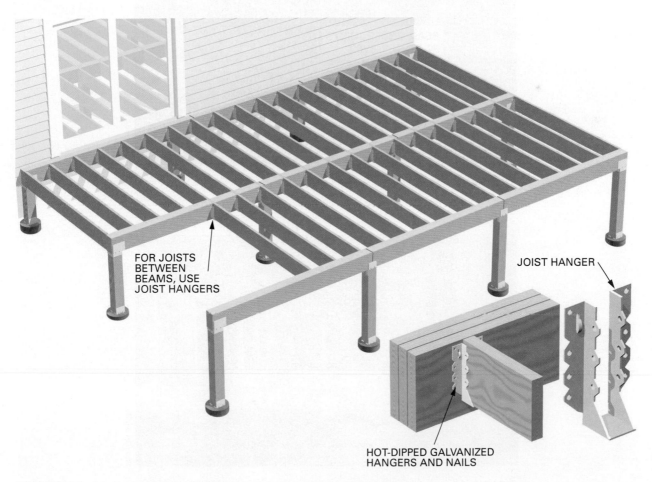

FOR JOISTS
BETWEEN
BEAMS, USE
JOIST HANGERS

JOIST HANGER

HOT-DIPPED GALVANIZED
HANGERS AND NAILS

FIGURE 68–11 Deck joists are installed between beams with joist hangers.

Bracing Supporting Posts

If the deck is 2 feet (610 mm) or more above grade, the supporting posts should be permanently braced in a manner similar to that shown in **Figure 68–12**. Use minimum 2 × 4 (38 × 89 mm) braces for heights up to 4 feet (1.2 m). If auxiliary posts are extended above the deck to support the guardrail, then they must be braced with full depth blocking and secured with HDG carriage bolts.

Applying Deck Boards

Specially shaped radius edge decking is available in both pressure-treated and natural decay-resistant lumber (**Figure 68–13**). It is usually used to provide the surface and walking area of the deck.

Dimension lumber that is nominally 2 × 4 (38 × 89 mm) or 2 × 6 (38 × 140 mm) is also widely used. Plain-sawn lumber has a natural tendency to shrink more in the sapwood than in the heartwood. That is, the annular rings tend to flatten out (see Figure 2–13, page 13). Because of this tendency, and also for the sake of appearance, it is best to install deck boards with the rings facing up and the bark side down. When installed this way, the boards will shed water if they cup. When hot-dipped galvanized finish nails are angle-nailed into the deck boards and used in conjunction with construction adhesive, there will be very little movement.

It has been common practice to install the deck boards with the bark side up and to use fasteners

FIGURE 68–12 Beam under cantilevered joists.

FIGURE 68–13 Special radius edge decking (5⁄4 premium decking).

and adhesives to "pin" the outer edges down. This keeps the denser side of the deck board in contact with the joist. This helps reduce decay because moisture accumulates at these connection points. The Southern Pine Council emphasizes the need for proper fastening and deck maintenance to ensure longevity of the structure. The recommended fastener is 3¼-inch (83 mm) hot-dipped galvanized or stainless-steel ringed or spiral nails or deck screws.

Boards are usually laid parallel with the long dimension of the deck. Because deck boards usually do not come longer than 16 feet (4.9 m), it may be desirable to lay the boards parallel to the short dimension, if their length will span it, to eliminate end joints in the decking. Framing the substructure so that deck boards slope and run at right angles to the house facilitates water runoff and snow removal. This requires that the beams be placed perpendicular to the house wall. The deck can then be unattached to the house, or the first joist can be fastened to the house and thereby bind the substructure. Deck joists are usually placed on top of the beams. Boards may also be laid in a variety of patterns including diagonal, herringbone, and parquet. Make sure the supporting structure has been designed and built to accommodate the design (**Figure 68–14**).

Much care should be taken with the application of deck boards. Snap a straight line as a guide to apply the starting row. Start at the outside edge if the deck is built against a building. A ripped and narrower ending row of decking is not as noticeable against the building as it is on the outside edge of the deck (**Figure 68–15**).

Straighten boards as they are installed. Maintain about a ⅛-inch (3 mm) space between dry boards. If the decking boards are wet, as with most pressure-treated boards, they will shrink as they dry. Nailing them tight together is the preferred method, as the space will appear when the lumber reaches equilibrium moisture content.

If the deck boards do not span the entire width of the deck, cut the boards on a 15-degree angle so that their ends are centred over joists. Make tight-fitting end joints. Stagger them between adjacent rows. Pre-drill holes for fasteners to prevent splitting the ends. Let the end of each row overhang the deck. Use two screws or nails in each joist. Drive nails at an angle. Set the heads below the surface. This will keep the nails from working loose and the heads from staining the surface. Hot-dipped galvanized finish nails can be used in conjunction with exterior construction adhesive, and when the heads are set, they are hidden from view.

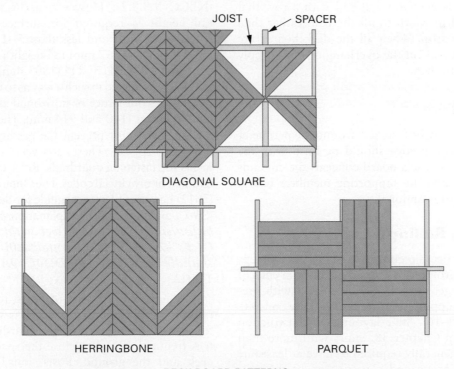

DECK BOARD PATTERNS

FIGURE 68–14 Deck boards may be installed in various arrangements. (From SFPA Booklet #301-20M-12/90, *Answers to Consumer Questions about CCA Pressure-Treated Southern Pine, p. 2. Courtesy of Southern Pine Council*)

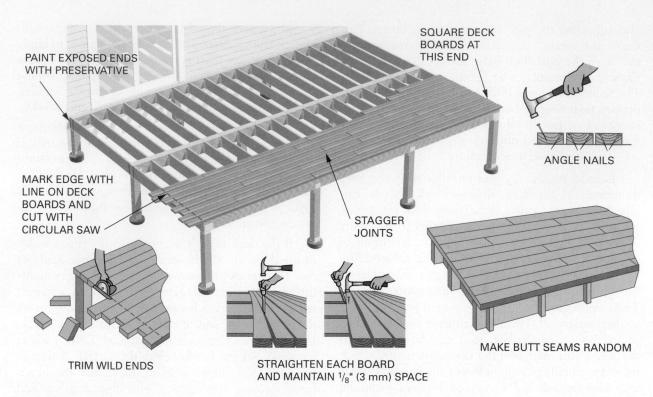

FIGURE 68–15 Techniques for installing deck boards.

When approaching the end, it is better to increase or decrease the spacing of the last six or seven rows of decking. This way you will end with a row that is nearly equal in width to all the rest, rather than with a narrow strip. When all the deck boards are laid, snap lines and cut the overhanging ends. Apply a preservative to them.

Applying Trim

A fascia board may be fastened around the perimeter of the deck. Its top edge should be flush with the deck surface. The fascia board conceals the cut ends of the decking and the supporting members below. The fascia board is optional.

Stairs and Railings

Stairs. Most decks require at least one or two steps leading to the ground. To protect the bottom ends of the stair carriage, they should be treated with preservative and supported by an above-grade concrete pad (**Figure 68–16**). Stair layout and construction are described in Chapter 38. Stairs with more than two risers are generally required to have at least one handrail. The design and construction of the stair handrail should conform to that of the deck railing.

Rails. There are numerous designs for deck railings. The National Building Code of Canada (NBCC)

does not require a guardrail if the surface of the deck is less than 2 feet (610 mm) off the ground (NBCC 9.8.8.2). However, a 36-inch (914 mm) rail height is required for surfaces greater than 2 feet (610 mm) and less than 5'-11" (1803 mm), and a 42-inch (1067 mm) rail height is required when the surface exceeds 5'-11" (1803 mm). The guardrail must be constructed in such a way as to inhibit climbing, which rules out lattice or horizontal rails between 5½ and 36 inches (140 and 914 mm). The space between the balusters must prevent the passage of a 3⅞-inch (98 mm) sphere. There are very specific requirements for fastening guardrails. Refer to the applicable national or provincial codes. The Ontario building code (2012) has a supplemental guide for deck construction (SB-7), and a thorough explanation can be found at *https://siouxlookout.civicweb.net/document/2063/ Decks%20-%20Supplimental%20Information.pdf ?handle=CED5291A4FDD4C19B1B2BDB620F 157C2.*

Railings may consist of posts, top and bottom rails, and balusters. **Lattice work** is sometimes used to close off the area above the guardrail for privacy. It is frequently used to close the space between the deck and the ground. Posts, rails, balusters, and other deck parts are manufactured in several shapes especially for use on decks (**Figure 68–17**).

Stanchions or posts are sometimes notched on their bottom ends to fit over the edge of the deck.

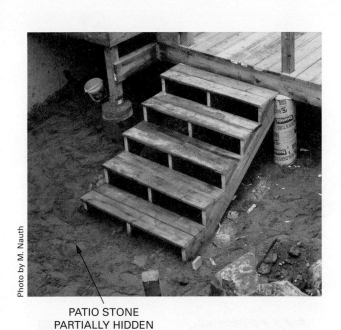

Photo by M. Nauth

PATIO STONE
PARTIALLY HIDDEN

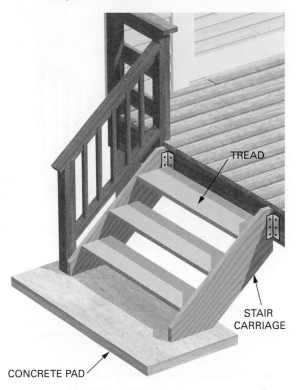

TREAD

STAIR
CARRIAGE

CONCRETE PAD

FIGURE 68–16 Stairs for decks are usually constructed with a simple basement or utility design.

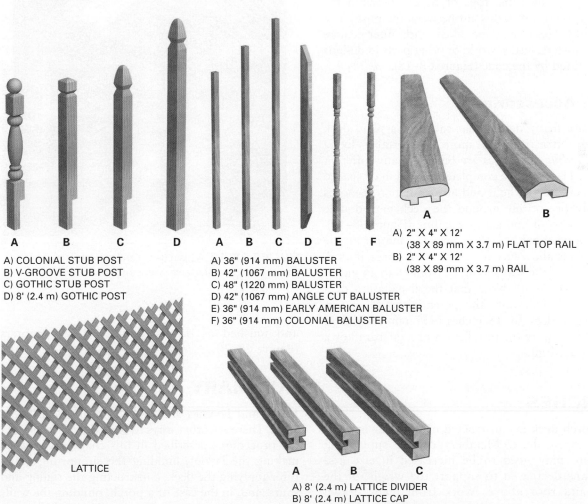

A
B
C
D

A) COLONIAL STUB POST
B) V-GROOVE STUB POST
C) GOTHIC STUB POST
D) 8' (2.4 m) GOTHIC POST

A
B
C
D
E
F

A) 36" (914 mm) BALUSTER
B) 42" (1067 mm) BALUSTER
C) 48" (1220 mm) BALUSTER
D) 42" (1067 mm) ANGLE CUT BALUSTER
E) 36" (914 mm) EARLY AMERICAN BALUSTER
F) 36" (914 mm) COLONIAL BALUSTER

A
B

A) 2" X 4" X 12'
(38 X 89 mm X 3.7 m) FLAT TOP RAIL
B) 2" X 4" X 12'
(38 X 89 mm X 3.7 m) RAIL

LATTICE

A
B
C

A) 8' (2.4 m) LATTICE DIVIDER
B) 8' (2.4 m) LATTICE CAP
C) 8' (2.4 m) HEAVY-DUTY LATTICE CAP

FIGURE 68–17 Railing parts are manufactured in many shapes.

SQUARE EDGE POST AND BALUSTERS

TURNED POST AND BALUSTERS

FIGURE 68–18 Deck railings are constructed with various designs.

They are usually spaced about 4 feet (1.2 m) apart. They are fastened with lag screws or bolts. The top rail may go over the tops or be cut between the posts. The bottom rail is cut between the posts. The remaining space may be filled with intermediate rails, balusters, lattice work, or other parts in designs as permitted by the code (**Figure 68–18**).

Deck Accessories

There are many details that can turn a plain deck into an attractive and more comfortable living area. *Shading structures* are built in many different designs. They may be completely closed in or spaced to provide filtered light and air circulation. Benches partially or entirely around the deck may double as a place to sit and act as a railing (**Figure 68–19**). Bench seats should be 18 inches (457 mm) from the deck. Make allowance for cushion thickness, if used. The depth of the seat should be from 18 to 24 inches (457 to 610 mm). Note that the installation of a fixed bench will raise the potential standing surface of the deck by 18 inches (457 mm), which in turn may require the installation of a guardrail or a higher guardrail.

PORCHES

The porch deck is constructed in a similar manner to an open deck. Members of the supporting structure may need to be increased in size and the spans decreased to support the weight of the walls and roof above. Work from plans drawn by professionals or check with building officials before starting. Porch walls and roofs are framed

FIGURE 68–19 A plain deck can be made more attractive and comfortable with benches and shading devices.

and finished as described in previous chapters (**Figure 68–20**).

SUMMARY

Decks and porches are designed in many different ways. There are other ways to construct them besides the procedures described in this chapter. However, making the layout, building the supporting structure, applying the deck, constructing the railing and stairs, and, in the case of a porch, building the walls and roof are basic steps that can be applied to the construction of practically any deck or porch.

FIGURE 68–20 A porch is composed of a deck enclosed by walls and a roof.

Chapter 69

Fence Design and Erection

A superior fence combines utility and beauty (**Figure 69–1**). It may define space, create privacy, provide shade or shelter, screen areas from view, or form required barriers around swimming pools and other areas for the protection of small children.

The design of a fence is often the responsibility of the builder. With creativity and imagination, he or she can construct an object of beauty and elegance that also fulfills its function. The objective of this chapter is to create an awareness of the importance

Courtesy of California Redwood Association

FIGURE 69–1 A fence can serve its intended purpose and also enhance the surroundings.

of design by showing several styles of fences and the methods used to erect them.

FENCE MATERIAL

Fences are not supporting structures. Lower, knotty grades of lumber may be used to build them. This is not the case if appearance is a factor and you want to show the natural grain of the wood. Many kinds of softwood may be used as long as they are shielded from the weather by protective coatings of paint, stains, or preservatives. However, for wood posts that are set into the ground and other parts that may be subjected to constant moisture, pressure-treated or all-heart, decay-resistant lumber must be used.

The same type of fasteners that are used on exposed decks should be used to build fences. Inferior hardware and fasteners will corrode and cause unsightly stains when in contact with moisture.

FENCE DESIGN

Fences consist of posts, rails, and fenceboards. Fences may be constructed in almost limitless designs. Zoning regulations sometimes restrict their height or placement. Often the site will affect the design. For example, the fence may have to be stepped like a staircase on a steep slope (**Figure 69–2**). Fences on property lines can be designed to look attractive from either side. Fence designs may block wind and sunlight. Fenceboards may be spaced in many attractive patterns. Most fences are constructed in several basic styles, each of which can be designed in numerous ways. Provisions should be made in the fence design to drain water from any area where it may otherwise be trapped.

FIGURE 69–2 Sometimes the site, such as a steep slope, may affect the design of a fence.

Picket Fence

The picket fence is commonly used on boundary lines or as barriers for pets and small children. Usually not more than 4 feet (1.2 m) high, the pickets are spaced to provide plenty of air and also to conserve material. The tops of the pickets may be shaped in various styles. Or they may be cut square, with ends exposed or capped with a moulding. The pickets may be applied with their tops in a straight line or in curves between posts (**Figure 69–3**). When pickets are applied with their edges tightly together, the assembly is called a stockade fence. Stockade fences are usually higher and are used when privacy is desired.

Board-on-Board Fence

The board-on-board fence is similar to a picket fence. However, the boards are alternated from side to side so that the fence looks the same from both sides (**Figure 69–4**). The boards may vary in height and spacing according to the degree of privacy and protection from wind and sun desired. The tops or edges of the boards may be shaped in many different designs.

Lattice Fence

This fence gets its name from narrow strips of wood called *lattice*. Strips are spaced by their own width. Two layers are applied at right angles to each other, either diagonally or in a horizontal and vertical fashion, to form a lattice work panel. Panels of various sizes can be prefabricated and installed between posts and rails.

FIGURE 69–3 The picket fence is constructed in many styles.

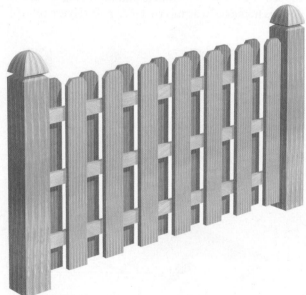

FIGURE 69–4 A board-on-board fence is similar to a picket fence; however, it looks the same on both sides.

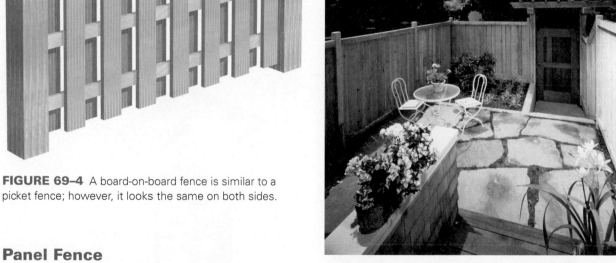

FIGURE 69–5 The panel fence provides shade and privacy, and it also restricts air movement.

Panel Fence

The panel fence creates a solid barrier with boards or panels fitted between top and bottom rails (**Figure 69–5**). Fenceboards may be installed diagonally or in other appealing designs. Alternating the panel design provides variety and adds to the visual appeal of the fence. A small space should be left between panel boards to allow for swelling in periods of high humidity. In many cases, two or more basic styles may be combined to enhance the design. **Figure 69–6** shows lattice panels combined with board panels.

Louvred Fence

The louvred fence is a panel fence with vertical boards set at an angle (**Figure 69–7**). The fence permits the flow of air through it and yet provides privacy. This fence is usually used around patios and pools.

FIGURE 69–6 Solid panels are combined with lattice work panels for an attractive design.

FIGURE 69–7 The louvred fence provides privacy and lets air through.

Post-and-Rail Fence

The post-and-rail fence is a basic and inexpensive style normally used for long boundaries (**Figure 69–8**). Designs include two or more square-edge or round rails, of various thicknesses, widths, and diameters, cut between the posts or fastened to their edges. Most post-and-rail fence designs have large openings. They are not intended as barriers.

FENCE POST DESIGN

Fence posts are usually wood or iron. Wood posts are usually 4 × 4 (89 × 89 mm). Larger sizes may be used, depending on the design. The post tops may be shaped in various ways to enhance the design of the fence (**Figure 69–9**). To conserve material and reduce expenses, a 4 × 4 (89 × 89 mm) post may be made to appear larger by applying furring and then boxing it in with 1-inch (19 mm) boards. The top may then be capped with shaped members and moulding in various designs to make an attractive fence post (**Figure 69–10**).

FIGURE 69–9 Fence post tops may be shaped in various ways to enhance the design of the fence.

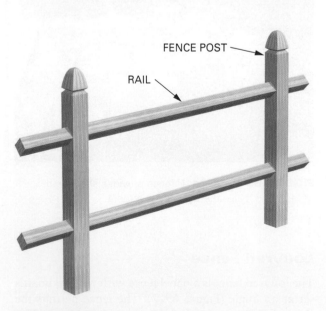

FIGURE 69–8 The post-and-rail fence is an inexpensive design for long boundary lines.

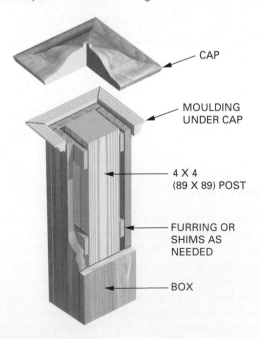

CAP

MOULDING UNDER CAP

4 X 4 (89 X 89) POST

FURRING OR SHIMS AS NEEDED

BOX

FIGURE 69–10 Wood posts may be boxed and capped to give a more solid look.

Iron posts may be pipe or solid rod ranging in size from 1 to 1¼ inches (25 to 32 mm) in diameter for fences 3 to 4 feet (914 to 1220 mm) tall. Larger diameters should be used for higher fences. Iron fence posts may be boxed in to simulate large wood posts. The tops are then usually capped with various shaped members similar to boxed wood posts (**Figure 69–11**). Iron posts should be galvanized or otherwise coated to resist corrosion.

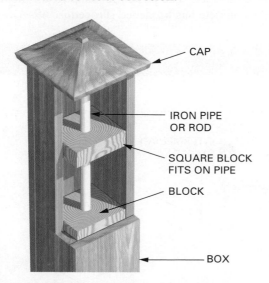

FIGURE 69–11 An iron fence post may be boxed and capped.

BUILDING A FENCE

The first step in building a fence is to set the fence posts. Locate and stretch a line between the end posts. If it is not possible to stretch a line because of steep sloping land, set up a transit-level to lay out a straight line (see Chapter 28). If the fence is to be built on a property line, make sure that the exact locations of the boundary markers are known.

Setting Posts

Posts are generally placed about 8 feet (2.4 m) apart, to their centrelines. Mark the post locations with stakes along the fence line. Dig holes about 10 inches (254 mm) in diameter with a post-hole digger. The depth of the hole depends on the height of the fence. Higher fences require that posts be set deeper (**Figure 69–12**). The bottom of the hole should be filled with gravel or stone. This provides drainage and helps eliminate moisture to extend the life of the post.

Posts may be set in the earth. For the strongest fences, however, set the posts in concrete. Set the end posts first. Use a level to ensure that the posts are plumb in both directions. Brace them securely. The height of the post is determined by measuring from the ground, especially when the ground slopes.

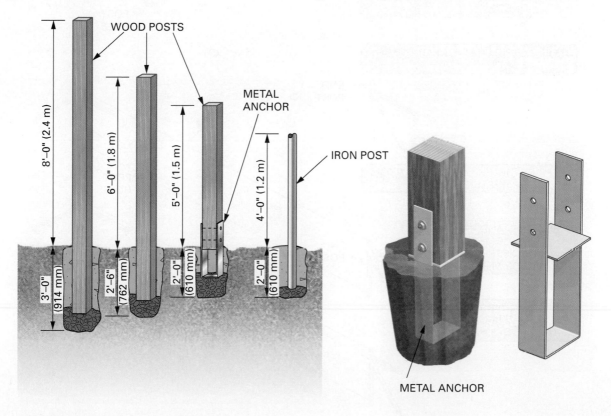

FIGURE 69–12 Design requirements for fence posts.

If top fence rails run across the tops of the posts, then the posts are left long. The tops are cut later in a straight or contoured line, as necessary. Make sure the face edges of the posts are aligned with the length of the fence. After the posts are braced, stretch lines between the end posts at the top and at the bottom.

Set and brace intermediate posts. The face edges of the posts are kept to the stretched line at the top and bottom. They are plumbed in the other direction with a level. Lag screws or spikes partially driven

into the bottom of the post strengthen the set of fence posts in concrete.

Place concrete around the posts. The concrete may be placed in the hole dry, as the moisture from the ground will provide enough water to hydrate the cement. Tamp it well into the holes. Form the top so that the surface pitches down from the posts. Instead of setting the posts in concrete, metal anchors may be embedded. The posts are installed on them after the concrete has hardened (**Procedure 69–A**).

Step**by**Step Procedures

Procedure 69–A Setting Fence Posts

STEP 1 LOCATE AND STRETCH LINE BETWEEN END MARKERS.

STEP 2 DRIVE INTERMEDIATE STAKES TO LOCATE POSTS.

STEP 3 SET AND BRACE END POSTS PLUMB.

STEP 4 EXCAVATE FOR INTERMEDIATE POSTS.

STEP 5 STRETCH LINE NEAR TOP OF POSTS.

STEP 6 STRETCH LINE NEAR BOTTOM OF POSTS.

STEP 7 SET AND BRACE INTERMEDIATE POSTS TO STRETCHED LINES TOP AND BOTTOM.

STEP 8 PLUMB EDGES OF INTERMEDIATE POSTS AND BRACE.

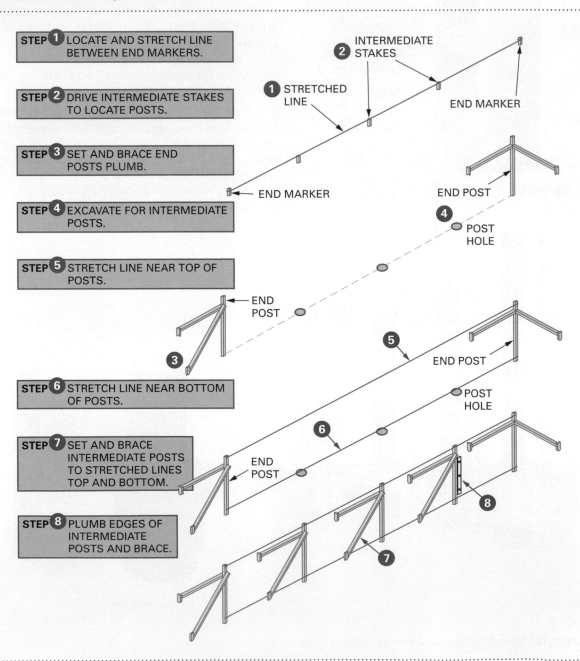

Installing Fence Rails

Usually two or three horizontal rails are used on most fences. However, the height or the design may require more. Rails may run across the face or be cut in between the fence posts. If rails are cut between posts, they are installed to a snapped chalk line across the edges of the posts. They are secured by toenailing or with metal framing connectors. The bottom rail should be kept at least 6 inches (152 mm) above the ground.

Very often, the faces of wood posts are not in the same line as the rails. Therefore, rail ends must be cut with other than square ends to fit between posts.

Procedure 69–B shows a method of laying out rail ends to fit between twisted posts.

When iron fence posts are used, holes are bored in the rails. The rails are then installed over the previously set iron posts. Splices are made on the ends to continue the rail in a straight, unbroken line. Special metal pipe grips are made to fasten fence rails to iron posts (**Figure 69–13**). If iron posts are boxed with wood, then the rails are installed in the same manner as for wood posts.

On some rustic-style post-and-rail fences, the rails are dowelled into the posts. In this case, post and rails must be installed together, one section at a time.

StepbyStep Procedures

Procedure 69–B Cutting Railing to Meet a Twisted Post

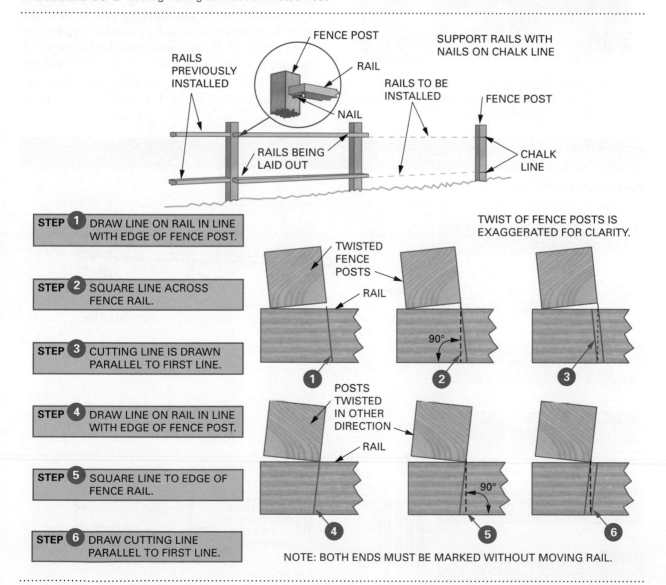

STEP 1 DRAW LINE ON RAIL IN LINE WITH EDGE OF FENCE POST.

STEP 2 SQUARE LINE ACROSS FENCE RAIL.

STEP 3 CUTTING LINE IS DRAWN PARALLEL TO FIRST LINE.

STEP 4 DRAW LINE ON RAIL IN LINE WITH EDGE OF FENCE POST.

STEP 5 SQUARE LINE TO EDGE OF FENCE RAIL.

STEP 6 DRAW CUTTING LINE PARALLEL TO FIRST LINE.

NOTE: BOTH ENDS MUST BE MARKED WITHOUT MOVING RAIL.

METAL GRIP TIES ARE MANUFACTURED TO
ATTACH FENCE RAILS TO PIPE FENCE POSTS.

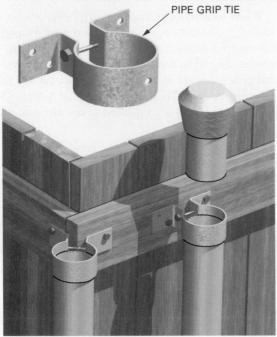

FIGURE 69–13 Fence rails may be attached to iron
fence posts. Metal grip ties are manufactured to attach
fence rails to pipe fence posts.

Applying Spaced Fenceboards and Pickets

Fasten pickets in plumb positions with their tops to
the correct height at the starting and ending points.
Stretch a line tightly between the tops of the two
pickets. If the fence is long, temporarily install intermediate pickets to support the line from sagging.
Sight the line by eye to see if it is straight. If not,
make adjustments and add more support pickets
if necessary. Use a picket for a spacer, and fasten
pickets to the rails. If the spacing is different, rip a
piece of lumber for use as a spacer.

Cut only the bottom end of the pickets when trimming their height. The bottom of pickets should not
touch the ground. Place a 2-inch (38 mm) block on the
ground. Turn the picket upside down with its top end
on the block. Mark it at the chalk line. Fasten the picket
with its top end to the stretched line (**Procedure 69–C**).

Continue cutting and fastening pickets using the
spacer. Keep their tops to the line. Check the pickets for
plumb frequently. If not plumb, bring back into plumb
gradually with the installation of three or four pickets.

Stop 3 or 4 feet (914 or 1220 mm) from the
end. Check to see if the spacing will come out even.

StepbyStep Procedures

Procedure 69–C Installing Spaced Pickets

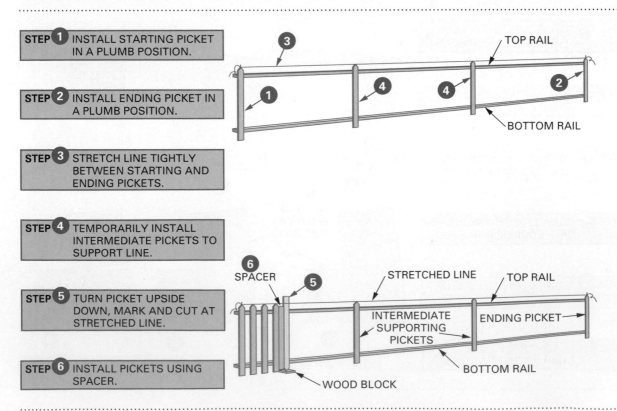

STEP 1 INSTALL STARTING PICKET IN A PLUMB POSITION.

STEP 2 INSTALL ENDING PICKET IN A PLUMB POSITION.

STEP 3 STRETCH LINE TIGHTLY BETWEEN STARTING AND ENDING PICKETS.

STEP 4 TEMPORARILY INSTALL INTERMEDIATE PICKETS TO SUPPORT LINE.

STEP 5 TURN PICKET UPSIDE DOWN, MARK AND CUT AT STRETCHED LINE.

STEP 6 INSTALL PICKETS USING SPACER.

Usually the spacing has to be either increased or decreased slightly. Set the dividers for the width of a picket plus a space, increased or decreased slightly, whichever is appropriate. Space off the remaining distance. Adjust the dividers until the spacing comes

out even. Any slight difference in a few spaces is not noticeable. This is much better than ending up with one narrow, conspicuous space (**Figure 69–14**).

Installing Pickets in Concave Curves

In some cases, the pickets or other fenceboards are installed with their tops in concave curved lines between fence posts. If the fenceboards have shaped tops that cannot be cut, erect the fence using the following procedure.

Install a picket on each end at the high point of the curve. In the centre, temporarily install a picket with its top to the low point of the curve. Fasten a flexible strip of wood in a curve to the top of the three pickets. Start from the centre. Work both ways to install the remainder of the pickets with their tops to the curved strip. Space the pickets to each end (**Figure 69–15**). Other fenceboards, such as board-on-board and louvres, are installed in a similar manner.

If fenceboard tops are to be cut in the shape of the curve, tack them in place with their tops above the curve. Bend the flexible strip to the curve. Mark all the fenceboard tops. Remove the fenceboards, if necessary. Cut the tops and replace them.

Usually, the most difficult part of fence building is the construction of the gate. The gate itself has to be

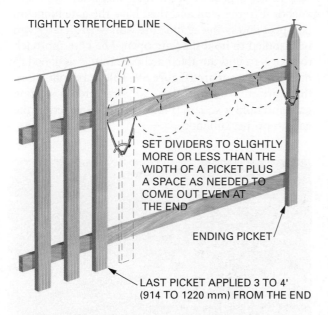

TIGHTLY STRETCHED LINE

SET DIVIDERS TO SLIGHTLY MORE OR LESS THAN THE WIDTH OF A PICKET PLUS A SPACE AS NEEDED TO COME OUT EVEN AT THE END

ENDING PICKET

LAST PICKET APPLIED 3 TO 4' (914 TO 1220 mm) FROM THE END

FIGURE 69–14 Technique of dividing the remaining distance into equal spaces when ending a picket fence installation.

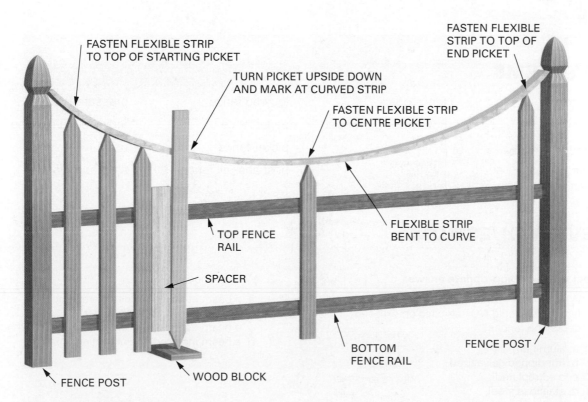

FASTEN FLEXIBLE STRIP TO TOP OF STARTING PICKET

TURN PICKET UPSIDE DOWN AND MARK AT CURVED STRIP

FASTEN FLEXIBLE STRIP TO TOP OF END PICKET

FASTEN FLEXIBLE STRIP TO CENTRE PICKET

FLEXIBLE STRIP BENT TO CURVE

TOP FENCE RAIL

SPACER

BOTTOM FENCE RAIL

FENCE POST

FENCE POST

WOOD BLOCK

FENCE POST

FIGURE 69–15 A method for installing spaced fenceboards with their tops in a concave curve.

Photo by M. Nauth

Photo by M. Nauth

FIGURE 69–16 Substantive gate hardware.

rigid and square, and it has to be installed between two plumb posts or between a plumb post and a foundation wall. The post on which the gate hinges has to be solidly fixed into the ground, and this usually requires a concrete pier below the frost line. It usually works well to set a post close to the foundation wall and anchor it to the wall and then have the gate hinge on that post and swing against the wall where a latch can be installed to hold the gate open. The best approach to take is to find durable hardware that is designed to hold the gate square (**Figure 69–16**). In retrofit cases, there are turn-buckles that can straighten a sagging gate and hold it square (**Figure 69–17**) and return the gate to proper function.

Courtesy of *Today's Homeowner*

FIGURE 69–17 Retrofit gate hardware.

KEY TERMS

balusters

board-on-board fence

deck boards

electroplated galvanizing

fence posts

hot-dipped galvanized (HDG)

lattice work

ledger

louvred fence

panel fence

picket fence

post-and-rail fence

pressure-treated

radius edge decking

rails

stockade fence

REVIEW QUESTIONS

Select the most appropriate answer.

1. Which of the following types of hardware and fasteners should NOT be used on exposed decks and fences?

   a. aluminum
   b. hot-dipped galvanized
   c. electroplated
   d. stainless steel

2. What is a ledger?

   a. a beam attached to the side of a building
   b. a beam supported by a girder
   c. a beam used to support girders
   d. a beam installed on supporting posts

3. How deep must a footing for supporting fence posts extend into the ground?

   a. at least 12 inches (305 mm) below grade
   b. below the frost line
   c. to stable soil
   d. at least 24 inches (610 mm) below grade

4. How must deck joists be installed?

   a. between beams
   b. over beams
   c. crowned edge up
   d. using adhesive or glue

5. When is a railing required on deck stairs?

   a. on stairs with more than two risers
   b. on stairs with more than four risers
   c. on stairs with over 30 inches (762 mm) total rise
   d. on stairs with over 3 feet (914 mm) total rise

6. When is a guardrail required on decks?

   a. when they are more than three stair risers above the ground
   b. when they are 24 inches (610 mm) or more above the ground
   c. when they are 4 feet (1.2 m) or more above the ground
   d. depends on local building codes

7. What is the usual height of a bench seat without a cushion?

   a. 14 inches (356 mm)
   b. 16 inches (406 mm)
   c. 18 inches (457 mm)
   d. 20 inches (508 mm)

8. What is a high fence with picket edges applied tightly together called?

   a. board-on-board fence
   b. panel fence
   c. post-and-rail fence
   d. stockade fence

9. For all types of fences, how high above the ground should the bottom rail be installed?

   a. at least the thickness of a 2 × 4 (38 × 89 mm) block
   b. at least the width of a 2 × 4 (38 × 89 mm) block
   c. at least 6 inches (152 mm)
   d. at least 8 inches (203 mm)

10. Which of the following should be done when applying fence pickets to rails?

   a. plumb each one before fastening
   b. plumb them frequently
   c. cut the top ends to the line
   d. fasten bottom ends flush with bottom rail

Section 4

Interior Finish

Allyson Woodside

Title: Lead Hand
Company: Amsted Construction, Stittsville, Ontario

EDUCATION

Allyson was born and raised in New Brunswick and completed high school in Fredericton. She finished her postsecondary education at St. Thomas University and graduated with a BA in English. Only later in life did she fully appreciate that her undergrad education was part of a much bigger picture.

HISTORY

After graduation, Allyson worked as a cook for ten years. Her ability to assess, plan, and organize led to repeated requests for her to take on a leadership role. Later she wrote the provincial exam and received her papers as a chef.

In her search for a niche in life, she moved to Ottawa, where she continued working as a chef. After taking a night course in cabinetmaking, she enrolled in the Heritage Carpentry program at Algonquin College, Perth Campus. In 2002, after graduating at the top of her class, she sent out 32 résumés to prospective employers. She was hired full-time by Amsted Construction, where she had had a summer job as a labourer. She has never looked back since.

ON THE JOB

On a typical day at Amsted, Allyson may find herself digging holes, doing demolition, framing, roofing, installing doors and windows, or installing crown moulding. The company does mainly renovations, so every job, every day, is new and exciting.

BEST ASPECTS

The greatest challenge of renovation is to make new work blend into the existing structure so that it does *not* look like an add-on. That is also the work's most rewarding aspect. "The thinking, the analyzing, the planning, the teamwork, and the standing back and looking at a job well done," says Allyson. "That's what keeps me coming back for more." Allyson credits her mentor Steve Mulley for his great patience: He is constantly training her in all areas of carpentry.

IMPORTANCE OF EDUCATION

Allyson completed her second level of in-school apprenticeship training in the winter of 2007. She finished with the highest mark in residential construction theory out of a class of 40 apprentices. "Learning through 'book' reading, classroom instruction, and shop practice has provided me with a solid base and has given me the confidence and the resources to tackle jobs that I haven't done before."

THE FUTURE

Allyson has successfully completed the Inter-Provincial Red Seal Exam for General Carpenter. She continues to work for Amsted, training apprentices in the same way she was trained—on the job. She tried her hand at teaching a pre-trades shop class at the college level and estimating to second-year carpenter apprentices.

WORDS OF ADVICE

"A person entering the trade must be physically fit (or must get fit), be willing to wholeheartedly do *any* kind of work, and must be willing to work on a team. The three Rs of *reading, writing,* and *arithmetic* are as important to the carpenter as they are to any other vocation. A good place to start is to work in the summers as a labourer on a construction site."

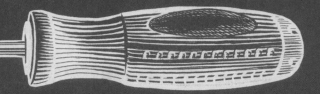

Unit 22

Drywall Construction

The term *drywall construction* generally means the application of gypsum board. Drywall is used extensively as interior finish and is produced from gypsum, which is mined from the earth.

Installing the interior finish covers up other work performed. Check that all the work—framing, insulating, and mechanicals—is truly complete before proceeding.

OBJECTIVES

After completing this unit, the student should be able to:

- describe various kinds, sizes, and uses of gypsum panels.
- describe the kinds and sizes of nails, screws, and adhesives used to attach gypsum panels.
- make single-ply and multi-ply gypsum board applications to interior walls and ceilings.
- conceal gypsum board fasteners and corner beads.
- reinforce and conceal joints with tape and compound.

Chapter 70

Gypsum Board

Gypsum board is sometimes called *wallboard*, *plasterboard*, **drywall**, or *Sheetrock*®. It is used extensively in construction (**Figure 70–1**). The term Sheetrock is a brand name for gypsum panels made by the U.S. Gypsum Company. However, the brand name is in such popular use that it has become a generic name for gypsum panels. Gypsum board makes a strong, high-quality, fire-resistant wall and ceiling covering. It is readily available; is easy to apply, decorate, or repair; and is relatively inexpensive.

GYPSUM BOARD

Many types of gypsum board are available for a variety of applications. The board or panel is composed of a gypsum core encased in a strong, smooth-finish paper on the **face** side and a natural-finish paper on the back side. The face paper is folded around the long edges. This reinforces and protects the core. The long edges are usually tapered. This allows the joints to be concealed with compound without any noticeable crown joint (**Figure 70–2**). A crowned joint is a buildup of the compound above the surface.

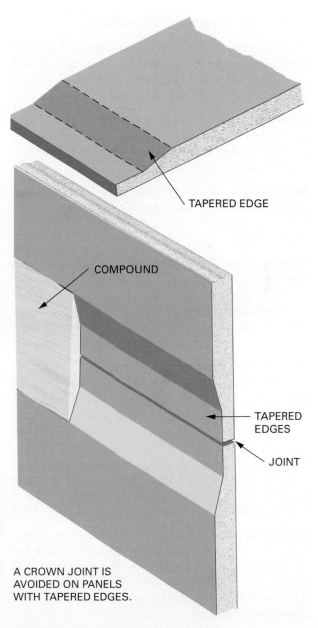

COMPOUND

TAPERED EDGE

TAPERED EDGES

JOINT

A CROWN JOINT IS AVOIDED ON PANELS WITH TAPERED EDGES.

FIGURE 70–2 The long edges of gypsum panels usually are tapered for effective joint concealment.

Courtesy Gypsum Association, gypsum.org

FIGURE 70–1 The application of gypsum board to interior walls and ceilings is called drywall construction.

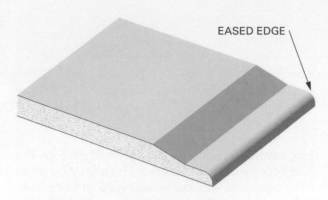

EASED EDGE

FIGURE 70–3 An eased edge panel has a rounded corner that produces a stronger concealed joint.

Types of Gypsum Panels

Most gypsum panels can be purchased, if desired, with an aluminum foil backing. The backing functions as a vapour retarder. It helps prevent the passage of interior water vapour into wall and ceiling spaces.

Regular. Regular gypsum panels are most commonly used for single-layer or multilayer application. They are applied to interior walls and ceilings in new construction and remodelling.

Eased Edge. An eased edge gypsum board has a special tapered, rounded edge. This produces a much stronger concealed joint than a tapered, square edge (**Figure 70–3**).

Type X. Type X gypsum board is typically known as firecode board. It has greater resistance to fire because of special additives in the core. Type X gypsum board is manufactured in several degrees of resistance to fire. Type X looks the same as regular gypsum board; however, it is labelled Type X on the edge or on the back.

Water-Resistant. Water-resistant or *moisture-resistant* (MR) gypsum board consists of a special moisture-resistant core and paper cover that is chemically treated to repel moisture. It is used frequently as a base for application of wall tile in tub, shower, and other areas subjected to considerable moisture. It is easily recognized by its distinctive green face. It is frequently called *green board* by workers in the field. Water-resistant panels are available with a Type X core for increased fire resistance.

Ceiling Drywall. Ceiling drywall is now manufactured in ½-inch (12.5 mm) thickness and it has the same properties as ⅝-inch (16 mm) board. CD board, as it is called, has glass fibres in the gypsum that give it the structural integrity to resist sag due to gravity and moisture.

Special Purpose. Backing board is designed to be used as a base layer in multilayer systems. It is available

with regular or Type X cores. *Coreboard* is available in 1-inch (25 mm) thicknesses. It is used for various applications, including the core of solid gypsum partitions. It comes in 24-inch (610 mm) widths with a variety of edge shapes. *Predecorated* panels have coated, printed, or overlay surfaces that require no further treatment. *Liner board* has a special fire-resistant core encased in a moisture-repellent paper. It is used to cover shaft walls, stairwells, chaseways, and similar areas.

Veneer Plaster Base. Veneer plaster bases are commonly called *blue board*. They are large, 4-foot-wide (1.2 m) gypsum board panels faced with a specially treated blue paper. This paper is designed to receive applications of veneer plaster. Conventional plaster is applied about ⅜ inch (9 mm) thick and takes considerable time to dry. In contrast, specially formulated veneer plaster is applied in one coat of about ¹⁄₁₆ inch (1.5 mm), or two coats totalling about ⅛ inch (3 mm). It takes only about 48 hours to dry.

Gypsum lath is used as a base to receive conventional plaster. Other gypsum panels, such as soffit board and sheathing, are manufactured for exterior use.

Sizes of Gypsum Panels

Widths and Lengths. Coreboards and liner boards come in 2-foot (610 mm) widths and from 8 to 12 feet (2.4 to 3.6 mm) long. Other gypsum panels are manufactured 4 feet or 48 inches (1220 mm) or 54 inches (1372 mm) wide and in lengths of 8, 9, 10, 12, 14, or 16 feet (2.4, 2.7, 3, 3.6, 4.2, and 5 m). Gypsum board is made in a number of thicknesses. Not all lengths are available in every thickness. The longest sheets are available in a new lightweight drywall product (up to 16 feet [4.9 m] in length).

Thicknesses.

- A ¼-inch (6 mm) thickness is used as a base layer in multilayer applications. It is also used to cover existing walls and ceilings in remodelling work. It can be applied in several layers for forming curved surfaces with short radii.

- A ⅜-inch (9.5 mm) thickness is usually applied as a face layer in repair and remodelling work over existing surfaces. It is also used in multilayer applications in new construction.

- Both ½ inch and ⅝ inch (12.5 and 16 mm) are commonly used thicknesses of gypsum panels for single-layer wall and ceiling application in residential and commercial construction. The ⅝-inch-thick (16 mm) panel is more rigid and has greater resistance to impacts and fire than does the ½-inch (12.5 mm) panel.

- Coreboards and liner boards come in thicknesses of ¾ and 1 inch (19 and 25 mm).

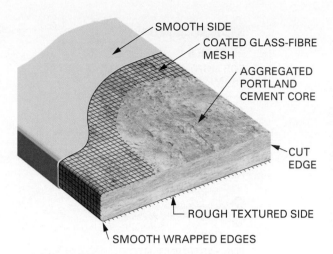

FIGURE 70–4 Composition of cement board.

CEMENT BOARD

Like gypsum board, **cement board** and *wonder board* are panel products. However, they have a core of Portland cement reinforced with a glass-fibre mesh embedded in both sides (**Figure 70–4**). The core resists water penetration and will not deteriorate when wet. It is designed for use in areas that may be subjected to high-moisture conditions. It is used extensively in bathtub, shower, kitchen, and laundry areas as a base for ceramic tile. In fact, some building codes require its use in these areas.

Panels are manufactured in sizes designed for easy installation in tub and shower areas with a minimum of cutting. Standard cement board panels come in a thickness of ½ inch (12.5 mm), in widths of 32 or 36 inches (0.8 to 0.9 m), and in 5-foot (1.5 m) lengths. Custom panels are available in a thickness of $^5/_8$ inch (16 mm), widths of 32 or 48 inches (0.8 or 1.2 m), and lengths from 32 to 96 inches (0.8 m to 2.4 m).

Cement board is also manufactured in a $^5/_{16}$-inch (8 mm) thickness. It is used as an underlayment for floors and countertops. Exterior cement board is used primarily as a base for various finishes on building exteriors.

DRYWALL FASTENERS

Specially designed nails and screws are used to fasten drywall panels. Ordinary nails or screws are not recommended. The heads of common nails are too small in relation to the shank. They are likely to cut the paper surface when driven. Staples may be used to fasten only the base layer in multilayer applications. They must penetrate at least $^5/_8$ inch (16 mm) into supports. Using the correct fastener is extremely important for proper performance of the application. Fasteners with corrosion-resistant coatings must be used when

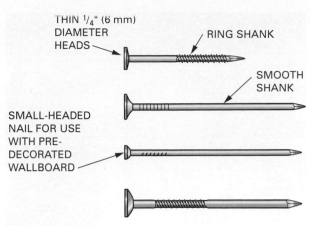

FIGURE 70–5 Special nails are required to fasten gypsum board.

applying water-resistant gypsum board or cement board. Care should be taken to drive the fasteners straight and at right angles to the wallboard to prevent the fastener head from breaking the face paper.

Nails

Gypsum board nails should have flat or concave heads that taper to thin edges at the rim. Nails should have relatively small-diameter shanks with heads at least ¼ inch (6 mm), but no more than $^5/_{16}$ inch (8 mm) in diameter. For greater holding power, nails with annular ring shanks are used (**Figure 70–5**).

Smooth shank nails should penetrate at least $^7/_8$ inch (22 mm) into framing members. Only ¾-inch (19 mm) penetration is required when ring shank nails are used. Greater nail penetrations are required for fire-rated applications.

Nails should be driven with a drywall hammer that has a convex face. This hammer is designed to compress the gypsum panel face to form a dimple of not more than $^1/_{16}$ inch (1.5 mm) when the nail is driven home (**Figure 70–6**). The dimple is made

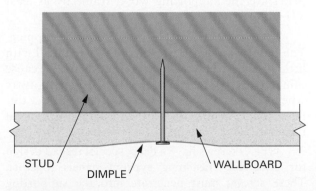

FIGURE 70–6 Fasteners are set with a dimple in the board for easier concealing with compound.

FIGURE 70–7 Drywall screws are driven with a screwgun to the desired depth.

Courtesy Gypsum Association, gypsum.org

so that the nail head can later be covered with compound.

Screws

Special drywall screws are used to fasten gypsum panels to steel or wood framing or to other panels. They are made with Phillips heads designed to be driven with a drywall screwgun (**Figure 70–7**). A proper setting of the nosepiece on the power screwdriver ensures correct countersinking of the screwhead. When driven correctly, the specially contoured bugle head makes a uniform depression in the panel surface without breaking the paper.

Different kinds of drywall screws are used for fastening into wood, metal, and gypsum panels. *Type W* screws are used for wood, *Type S* and *Type S-12* for metal framing, and *Type G* for fastening to gypsum backing boards (**Figure 70–8**).

Type W screws have sharp points. They have specially designed threads for easy penetration and excellent holding power. The screw should penetrate the supporting wood frame by at least $5/8$ inch (16 mm).

Type S screws are self-drilling and self-tapping. The point is designed to penetrate sheet metal with little pressure. This is an important feature because thin metal studs have a tendency to bend away when driving screws. Type S-12 screws have a different drill point designed for heavier gauge metal framing.

Type G screws have a deep, special thread design for effectively fastening gypsum panels together. These screws must penetrate into the supporting board by at least ½ inch (12.5 mm). If the supporting

TYPE S
FOR LIGHT-GAUGE METAL FRAMING

TYPE S-12
FOR 20-GAUGE OR HEAVIER METAL FRAMING

TYPE G
FOR FASTENING INTO BASE LAYERS OF GYPSUM BOARD

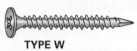

TYPE W
FOR WOOD FRAMING

FIGURE 70–8 Several types of screws are used to fasten gypsum panels. Selection of the proper type is important.

board is not thick enough, longer fasteners should be used. Make sure there is sufficient penetration into framing members.

Adhesives

Drywall adhesives are used to bond single layers directly to supports or to laminate gypsum board to base layers. Adhesives used to apply gypsum board are classified as stud adhesives and laminating adhesives.

For bonding gypsum board directly to supports, special drywall stud adhesive or approved *construction adhesive* is used. Supplemental fasteners must be used with stud adhesives. Stud adhesives are available in large cartridges. They are applied to framing members with hand or powered adhesive guns (**Figure 70–9**).

CAUTION

Some types of drywall adhesives may contain a flammable solvent. Do not use these types of adhesives near an open flame or in poorly ventilated areas.

FIGURE 70–9 Applying drywall adhesive to studs.

For laminating gypsum boards to each other, joint compound adhesives and **contact adhesives** are used. Joint compound adhesives are applied over the entire board with a suitable spreader prior to lamination. Boards laminated with joint compound adhesive require supplemental fasteners.

When contact adhesives are used, no supplemental fasteners are necessary. However, the board cannot be moved after contact has been made. The adhesive is applied to both surfaces by brush, roller, or spray gun. It is allowed to dry before laminating. A modified contact adhesive is also used. It permits an open time of up to 30 minutes, during which the board can be repositioned, if necessary.

Chapter 71

Single-Layer and Multilayer Drywall Application

Single-layer gypsum board applications are widely used in light commercial and residential construction. They adequately meet building code requirements. Multilayer applications are more often used in commercial construction. They have increased resistance to fire and sound transmission. Both systems provide a smooth, unbroken, quality surface if recommended application procedures are followed.

SINGLE-LAYER APPLICATION

Drywall should be not be delivered to the jobsite until shortly before installation begins. Boards stored on the job for long periods are subject to damage. The boards must be stored under cover and stacked flat on supports. Supports should be at least 4 inches (100 mm) wide and placed fairly close together (**Figure 71–1**). Leaning boards against framing for long periods may cause the boards to warp. This makes application more difficult. To avoid damaging the edges, carry the boards. Do not drag them. Then, set the boards down gently. Be careful not to drop them.

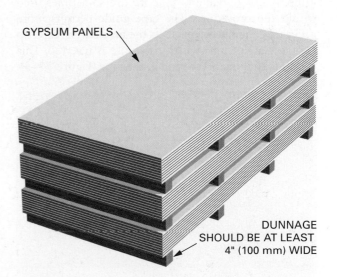

GYPSUM PANELS

DUNNAGE
SHOULD BE AT LEAST
4" (100 mm) WIDE

FIGURE 71–1 Correct method of stacking gypsum board.

Cutting and Fitting Gypsum Board

Take measurements accurately. Cut the board by first scoring the face side through the paper to the core. Use a utility knife. Guide it with a *drywall T-square*, if cutting a square end (**Figure 71–2**). The board is then

The Gypsum Construction Handbook, Sixth Edition. Reproduced with permission of John Wiley & Sons, Inc.

FIGURE 71–2 Using a drywall T-square as a guide when scoring across the width of a board.

broken along the scored face. The back paper is scored along the fold. The sheet is then broken by snapping the board in the reverse direction (**Procedure 71–A**).

To make cuts parallel to the long edges, the board is often gauged with a tape and scored with a utility knife. A *tape guide* and *tape tip* are sometimes used to aid the procedure. The tape guide permits more accurate gauging and protects the fingers. The tape tip contains a slot into which the knife is inserted. This prevents slipping off the end of the tape (**Figure 71–3**).

FIGURE 71–3 Technique for cutting gypsum board parallel to the edge.

When making parallel cuts close to long edges, it is usually necessary to score both sides of the board to obtain a clean break.

Smooth ragged edges with a drywall rasp, coarse sanding block, or knife. A job-made drywall rasp can be made by fastening a piece of metal lath to a wood block. Cut panels should fit easily into place without being forced. Forcing the panel may cause it to break.

Aligning Framing Members

Before applying the gypsum board, check the framing members for alignment. Stud edges should not be out of alignment more than $1/8$ inch (3 mm) with adjacent studs. A wood stud that is out of alignment can be straightened by the procedure shown in **Figure 71–4**. For a load-bearing wall, add a straight 4-foot (1.2 m) piece of stud material to stiffen the stud. Ceiling joists

ON THE JOB

A crowned stud can be straightened.

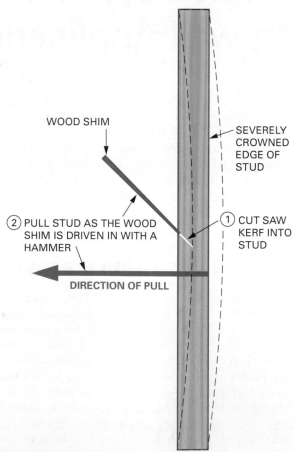

WOOD SHIM

SEVERELY CROWNED EDGE OF STUD

② PULL STUD AS THE WOOD SHIM IS DRIVEN IN WITH A HAMMER

① CUT SAW KERF INTO STUD

DIRECTION OF PULL

FIGURE 71–4 Technique for straightening a severely crowned stud.

StepbyStep Procedures

Procedure 71–A Scoring and Breaking Drywall

STEP 1 BREAK BY LIFTING SHEET AND STEPPING BACK WITH THE CUT.

STEP 2 SCORE BACK SIDE OF SHEET, LEAVING TOP AND BOTTOM PAPER TO ACT AS A HINGE.

STEP 3 RAISE SHEET ONTO TOE, THEN SNAP SHEET BACK BY SWINGING IT TOWARD YOU.

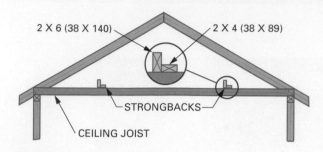

FIGURE 71-5 A strongback is sometimes used to align ceiling joists or the bottom chord of roof trusses.

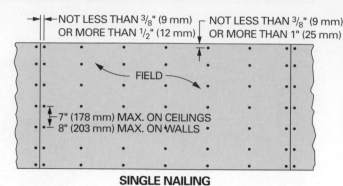

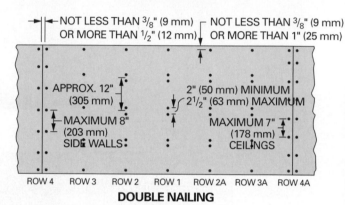

FIGURE 71-6 Spacing of single-nailed or double-nailed panels. Greater fastener spacing is used when panels are screwed.

are sometimes brought into alignment with the installation of a strongback across the tops of the joists at about the centre of the span (**Figure 71-5**).

Fastening Gypsum Panels

Drywall is fastened to framing members with nails or screws. Hand pressure should be applied on the panel next to the fastener being driven. This ensures that the panel is in tight contact with the framing member. The use of adhesives reduces the number of nails or screws required. A single or double method of nailing may be used. Check provincial building code for minimum fastening requirements.

Single Nailing Method. With this method, nails are spaced a maximum of 7 inches (178 mm) on centre (OC) on ceilings and 8 inches (203 mm) OC on walls into frame members. Nails should be first driven in the centre of the board and then outward toward the edges. Perimeter fasteners should be at least $^3/_8$ inch (9.5 mm), but not more than 1 inch (25 mm) from the edge.

Double Nailing Method. In double nailing, the perimeter fasteners are spaced as for single nailing. In the field of the panel, space a first set of nails 12 inches (305 mm) OC. Space a second set 2 to 2½ inches (50 to 63 mm) from the first set. The first nail driven is reseated after driving the second nail of each set. This assures solid contact with framing members (**Figure 71-6**).

Screw Attachment. Screws are spaced 12 inches (305 mm) OC on ceilings and 16 inches (406 mm) OC on walls when framing members are spaced 16 inches (406 mm) OC. If framing members are spaced 24 inches (610 mm) OC, then screws are spaced a maximum of 12 inches (305 mm) OC on both walls and ceilings.

Using Adhesives. Apply a straight bead about ¼ inch (6 mm) in diameter to the centreline of the stud edge. On studs where panels are joined, two parallel beads of adhesive are applied, one on each side of the centreline. Zigzag beads should be avoided, to prevent the adhesive from squeezing out at the joint.

On wall applications, supplemental fasteners are used around the perimeter. Space them about 16 inches (406 mm) apart. On ceilings, in addition to perimeter fastening, the field is fastened at about 24-inch (610 mm) intervals (**Figure 71-7**).

Gypsum panels may be prebowed. This reduces the number of supplemental fasteners required. Prebow the panels by one of the methods shown in **Figure 71-8**. Make sure the finish side of the panel faces in the correct direction. Allow them to remain overnight or until the boards have a 2-inch (50 mm) permanent bow. Apply adhesive to the studs. Fasten the panel at top and bottom plates. The bow keeps the centre of the board in tight contact with the adhesive until bonded.

Ceiling Application

Gypsum panels are applied first to ceilings and then to the walls. Panels may be applied parallel, or at right angles, to joists or furring. If applied parallel, edges and ends must bear completely on framing. If applied at right angles, the edges are fastened where they cross over each framing member. Ends must be fastened completely to joists or furring strips.

Carefully measure and cut the first board to width and length. Cut edges should be against the wall. Lay out lines on the panel face indicating the location of the framing in order to place fasteners accurately.

Gypsum board panels are heavy. At least two or more people are needed for ceiling application unless a drywall panel lifter is available. Lift the panel overhead and place it in position (**Figure 71–9**). Install two **deadmen** under the panel to hold it in position.

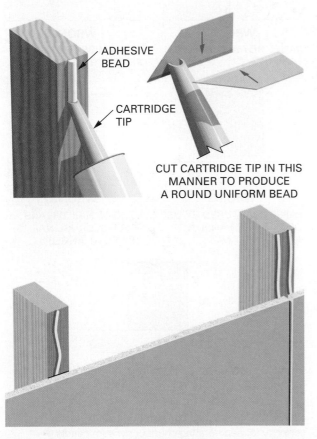

FIGURE 71–7 Two beads of adhesive are applied under joints in the board.

FIGURE 71–9 Drywall lifters are often used to raise drywall panels to the ceiling while the worker positions and fastens them in place.

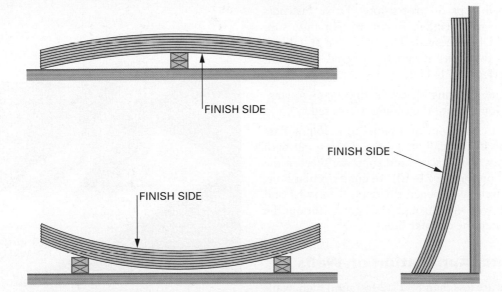

FIGURE 71–8 Prebowing keeps the board in tight contact with the adhesive until bonded and reduces the number of fasteners required.

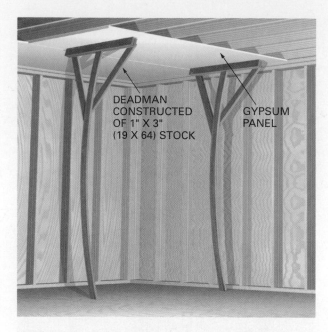

FIGURE 71–10 A deadman may be used to hold a board in place while fastening.

Deadmen are supports made in the form of a "T." They are easily made on the job using 1 × 3 (19 × 64 mm) lumber with short braces from the vertical member to the horizontal member. The leg of the support is made about ¼ inch (6 mm) longer than the floor-to-ceiling height. The deadmen are wedged between the floor and the panel. They hold the panel in position while it is being fastened (**Figure 71–10**). Using deadmen is much easier than trying to hold the sheet in position and fasten it at the same time.

Fasten the sheet in one of the recommended manners. Hold the board firmly against framing to avoid nail pops or protrusions. Drive fasteners straight into the member. Fasteners that miss supports should be removed. The nail hole should be dimpled so that later it can be covered with joint compound (**Figure 71–11**).

Continue applying sheets in this manner, staggering end joints, until the ceiling is covered.

To cut a corner out of a panel to accommodate a protrusion in the wall, make the shortest cut with a drywall saw. Then, score and snap the sheet in the other direction (**Figure 71–12**). To cut a circular hole, mark the circle with pencil dividers, then twist and push the drywall saw through the board. Cut out the hole, following the circular line.

Horizontal Application on Walls

When walls are less than 8′-1″ (2464 mm) high, wallboard is usually installed horizontally, at right angles to the studs. If possible, use a board of sufficient

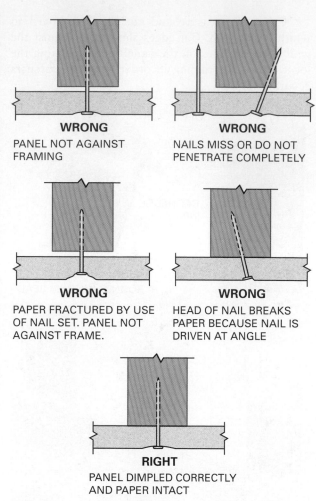

WRONG
PANEL NOT AGAINST FRAMING

WRONG
NAILS MISS OR DO NOT PENETRATE COMPLETELY

WRONG
PAPER FRACTURED BY USE OF NAIL SET. PANEL NOT AGAINST FRAME.

WRONG
HEAD OF NAIL BREAKS PAPER BECAUSE NAIL IS DRIVEN AT ANGLE

RIGHT
PANEL DIMPLED CORRECTLY AND PAPER INTACT

FIGURE 71–11 It is important to drive fasteners correctly for secure attachment of gypsum panels.

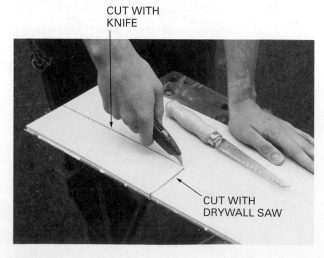

CUT WITH KNIFE

CUT WITH DRYWALL SAW

FIGURE 71–12 Use a knife and drywall saw to cut the corner out of a gypsum panel.

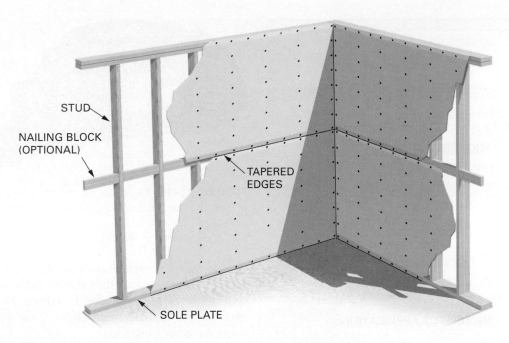

STUD

NAILING BLOCK
(OPTIONAL)

TAPERED
EDGES

SOLE PLATE

FIGURE 71–13 Horizontal application of gypsum panels to walls.

length to go from corner to corner. Otherwise, use as long a board as possible to minimize end joints, because they are difficult to conceal. Stagger end joints or centre them over and below window and door openings if possible. That way, not so much of the joint is visible. End joints should not fall on the same stud as those on the opposite side of the partition.

The top panel is installed first. Cut the board to length to fit easily into place without forcing it. Stand the board on edge against the wall. Start fasteners along the top edge opposite each stud. Raise the sheet so that the top edge is firmly against the ceiling and drive nails. Fasten the rest of the sheet in the recommended manner (**Figure 71–13**).

Measure and cut the bottom panel to width and length. Cut the width about ¼ inch (6 mm) narrower than the distance measured. Lay the panel against the wall. Raise it with a drywall foot lifter against the bottom edge of the previously installed top panel. A **drywall foot lifter** is a tool especially designed for this purpose (**Figure 71–14**). However, one can be made on the job by tapering the end of a short piece of 1 × 3 or 1 × 4 (19 × 64 or 19 × 89 mm) lumber. Fasten the sheet as recommended. Install all others in a similar manner. Stagger any necessary end joints. Locate them as far from the centre of the wall as possible so that they will be less conspicuous. Avoid placing end joints over the ends of window and door lintels. This will reduce the potential for wallboard cracks when the building settles.

FIGURE 71–14 A drywall foot lifter is used to lift gypsum panels in the bottom course up against the top panels.

Where end joints occur on steel studs, attach the end of the first panel to the open or unsupported edge of the stud. This holds the stud flange in a rigid position for the attachment of the end of the adjoining panel. Making end joints in the opposite manner usually causes the stud edge to deflect. This results in an uneven surface at the joint (**Figure 71–15**).

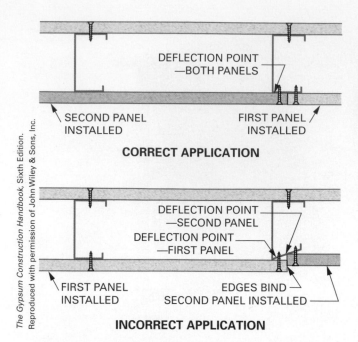

The Gypsum Construction Handbook, Sixth Edition.
Reproduced with permission of John Wiley & Sons, Inc.

FIGURE 71–15 Sequence of making end joints when attaching gypsum panels to steel studs.

Courtesy of Robert Bosch Tool Corporation

FIGURE 71–16 A portable electric drywall cutout tool often is used to make cutouts for outlet boxes and similar objects.

Making Cutouts in Wall Panels. There are several ways of making cutouts in wall panels for electrical outlet boxes, ducts, and similar objects. Care must be taken not to make the cutout much larger than the outlet. Most cover plates do not cover by much. If cut too large, much extra time has to be taken to patch up around the outlet, replace the panel, or install oversize outlet cover plates.

Plumb the sides of the outlet box down to the floor, or up to the previously installed top panel, whichever is more convenient. The panel is placed in position. Lines for the sides of the box are plumbed on it from the marks on the floor or on the panel. The top and bottom of the box are laid out by measuring down from the bottom edge of the top panel. With a saw or utility knife, cut the outline of the box. Take care not to damage the vapour retarder by pulling the lower end of the sheet away from the wall as you cut.

A fast, easy, and accurate way of making cutouts is with the use of a portable electric **drywall cutout tool (Figure 71–16)**. The approximate location of the centre of the outlet box is determined and marked on the panel. The panel is then installed over the box. Using the cutout tool, a hole is plunged through the panel in the approximate centre of the outlet box. Care must be taken not to make contact with wiring. The tool is not recommended for use around live wires. The tool is moved in any direction until the bit hits a side of the box. It is then withdrawn slightly

to ride over the edge to the outside of the box. The tool is then moved so that the bit rides around the outside of the box to make the cutout. Usually cutouts are made for outlets after all the panels in a room have been installed.

To make cutouts around door openings, either mark and cut out the panel before it is applied, or make the cutout after it is applied. To make the cutout after the panel is applied, use a saw to cut in one direction. Then score it flush with the opening on the back side in the other direction. Bend and score it on the face to make the cutout. Another method uses the drywall cutout tool around framing.

Vertical Application on Walls

Vertical application of gypsum panels on walls, with long edges parallel to studs, is more practical if the ceiling height is more than 8′-1″ (2464 mm) or the wall is 4′-0″ (1220 mm) wide or less. Note that vertical application requires more lineal feet of drywall seams, and finishing the seams is more physically demanding than with horizontal application.

To install vertical panels, cut the first board, in the corner, to length and width. Its length should be about ¼ inch (6 mm) shorter than the height from floor to ceiling. It should be cut to width so that the edge away from the corner falls on the centre of a stud. All cut edges must be in the corners. None should butt edges of adjacent panels.

With a foot lifter, raise the sheet so that it is snug against the ceiling. The tapered edge should be plumb and centred on the stud. Fasten it in the specified manner. Continue applying sheets around the room with tapered, uncut edges against each other. There should be no horizontal joints between floor and ceiling (**Figure 71–17**). Make any necessary cutouts as previously described.

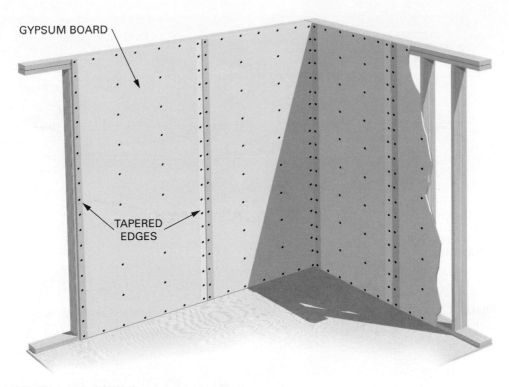

FIGURE 71–17 Applying gypsum panels vertically.

Truss Uplift and Floating Corners

In the colder regions of Canada, where the ceiling insulation keeps the bottom chord of trusses warm and the exposed top chord is cold and dry, there is differential shrinkage in the truss members, causing the bottom chord to arch upward. This is called *truss uplift* and it leads to nail-popping and seam-tearing. The strategy of *floating drywall corners* helps hide the visible aspects of the arching bottom chord. The chord will still arch, but the drywall defects will not appear.

Generally, the bottom chords of the trusses are strapped or furred with 1 × 3s or 1 × 4s (19 × 64s or 19 × 89s) on the underside, and the drywall is attached to the strapping or furring strips. When the interior partitions run parallel to the strapping, the ceiling drywall is not fastened to the piece of strapping that is beside the interior partition. When the partition is perpendicular to the strapping, the ceiling drywall is fastened 12 inches (305 mm) away from the wall for ½-inch (12 mm) drywall and 16 inches (406 mm) away from the wall for ⅝-inch (16 mm) drywall. (Along the exterior walls, the ceiling drywall is fastened as close to the wall as possible.)

The upper row of wall drywall is fitted tightly to the ceiling drywall and fastened 8 inches (203 mm) down from the ceiling into the partition studs. The lower row of wall drywall is then lifted snug to the bottom edge of the upper row and fastened as per the manufacturer's specifications.

The ceiling partition joint is "mudded" and taped with paper tape or fibre tape, as per usual practice. Care must be taken not to score the paper tape. When truss uplift does take place, the bottom chords of the trusses move but the drywall joints remain intact. The movement of the trusses is not apparent to the homeowner.

At interior wall corners, the underlying wallboard is not fastened. The overlapping board is fitted snugly against the underlying board. This brings it in firm contact with the face of the stud. The overlapping panel is nailed or screwed into the interior corner stud (**Figure 71–18**).

APPLYING GYPSUM PANELS TO CURVED SURFACES

Gypsum panels may be applied to curved surfaces. However, closer spacing of the frame members may be required to prevent flat areas from occurring on the face of the panel. If the paper and core of gypsum panels are moistened, they may be bent to curves with shorter radii than when dry. After the boards are thoroughly moistened, they should be stacked on

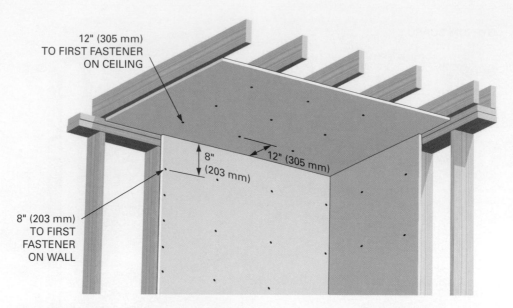

12" (305 mm)
TO FIRST FASTENER
ON CEILING

8"
(203 mm)

12" (305 mm)

8" (203 mm)
TO FIRST
FASTENER
ON WALL

FIGURE 71-18 Floating angle method of applying drywall has no fasteners at the corners of sheets.

a flat surface. They should be allowed to stand for at least one hour before bending. Moistened boards must be handled very carefully. They will regain their original hardness after drying. Wallboard marketed as *bendable* does not need to be wet before it is shaped. The minimum bending radii for dry and wet gypsum panels are shown in **Figure 71–19**.

To apply panels to a convex surface, fasten one end to the framing. Gradually work to the other end by gently pushing and fastening the panel progressively to each framing member. When applying panels to a concave curve, fasten a stop at one end. Carefully push on the other end to force the centre of the panel against the framing. Work from the end against the stop. Fasten the panel successively to each framing member.

Gypsum board may be applied to the curved inner surfaces of arched openings. If the dry board cannot be bent to the desired curve, it may be moistened or parallel knife scores made about 1 inch (25 mm) apart across its width.

Board Thickness in Inches	Dry	Wet
¼ (6 mm)	5 ft. (1.5 m)	2 to 2½ ft. (610 to 762 mm)
⅜ (9 mm)	7½ ft. (2.3 m)	3 to 3½ ft. (914 to 1067 mm)
½ (12 mm)	20 ft. (6 m)	4 to 4½ ft. (1.2 to 1.4 m)

FIGURE 71–19 Minimum bending radii of gypsum panels.

DRYWALL APPLICATION TO BATH AND SHOWER AREAS

Water-resistant gypsum board and cement board panels are used in bath and shower areas as bases for the application of ceramic tile. (Some areas permit only cement board; check your local codes.) Framing should be 16 inches (406 mm) OC. Steel framing should be at least 20-gauge thickness.

Apply panels horizontally with the bottom edge not less than ¼ inch (6 mm) above the lip of the shower pan or tub. The bottom edges of gypsum panels should be uncut and paper-covered.

Check the alignment of the framing. If necessary, apply furring strips to bring the face of the board flush with the lip of the tub or shower pan (**Figure 71–20**).

Provide blocking between studs about 1 inch (25 mm) above the top of the tub or shower pan. Install additional blocking between studs behind the horizontal joint of the panels above the tub or shower pan.

Cement board panels are cut using a masonry blade in a circular saw. Care must be taken to reduce exposure to dust. Before attaching panels, apply thinned ceramic-tile adhesive to all cut edges around holes and other locations.

Attach panels with corrosion-resistant screws or nails spaced not more than 8 inches (203 mm) apart. When ceramic tile more than ³/₈ inch (9 mm) thick is to be applied, the nail or screw spacing should not exceed 4 inches (100 mm) OC.

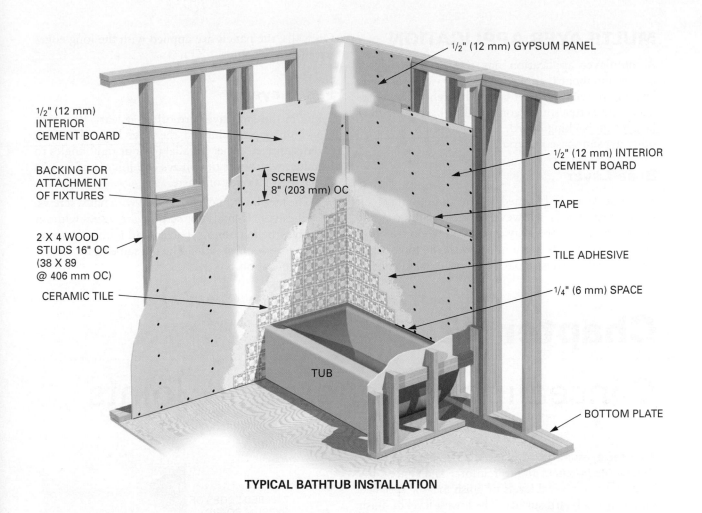

½" (12 mm) GYPSUM PANEL

½" (12 mm) INTERIOR CEMENT BOARD

BACKING FOR ATTACHMENT OF FIXTURES

2 X 4 WOOD STUDS 16" OC (38 X 89 @ 406 mm OC)

CERAMIC TILE

SCREWS 8" (203 mm) OC

½" (12 mm) INTERIOR CEMENT BOARD

TAPE

TILE ADHESIVE

¼" (6 mm) SPACE

TUB

BOTTOM PLATE

TYPICAL BATHTUB INSTALLATION

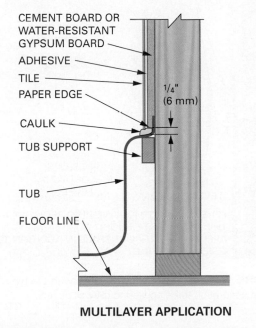

CEMENT BOARD OR WATER-RESISTANT GYPSUM BOARD

ADHESIVE

TILE

PAPER EDGE

CAULK

TUB SUPPORT

TUB

FLOOR LINE

¼" (6 mm)

MULTILAYER APPLICATION

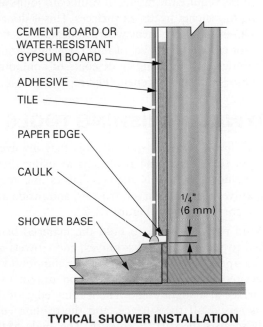

CEMENT BOARD OR WATER-RESISTANT GYPSUM BOARD

ADHESIVE

TILE

PAPER EDGE

CAULK

SHOWER BASE

¼" (6 mm)

TYPICAL SHOWER INSTALLATION

FIGURE 71–20 Installation details around bathtubs and showers.

MULTILAYER APPLICATION

A multilayer application has one or more layers of gypsum board applied over a base layer. This layering provides greater strength, higher fire resistance, and better sound control. The base layer may be gypsum backing board, regular gypsum board, or other gypsum base material.

Base Layer

The base layer is fastened in the same manner as single-layer panels. However, double nailing is not necessary and staples may be used in wood framing. On ceilings, panels are applied with the long edges either at right angles or parallel to framing members.

On walls, the panels are applied with the long edges parallel to the studs.

Face Layer

Joints in the face layer are offset at least 10 inches (254 mm) from joints in the base layer. The face layer is applied either parallel to or at right angles to framing, whichever minimizes end joints and results in the least amount of waste.

The face layer may be attached with nails, screws, or adhesives. If nails or screws are used without adhesive, the maximum spacing and minimum penetration into framing should be the same as for single-layer application.

Chapter 72

Concealing Fasteners and Joints

After the gypsum board is installed, it is necessary to conceal the fasteners and to reinforce and conceal the joints. One of several levels of finish may be specified for a gypsum board surface. The lowest level of finish may simply require the taping of wallboard joints and *spotting* of fastener heads on surfaces. This is done in warehouses and other areas where appearance is normally not critical. The level of finish depends, among other things, on the number of coats of compound applied to joints and fasteners (**Figure 72–1**).

DRYWALL FINISHING TOOLS

Drywall finish requires special tools that involve a little practice to become proficient at using them. They include hand tools such as hawks and mudpans, knives, flat and curved trowels, and inside and outside corner trowels (**Figure 72–2**).

Mud pans are used to hold the material being applied to the wall. Hawks, knives, and trowels are used to apply, smooth, and shape the compound to the desired effect. They are held and drawn over the surface at a slight angle. The leading edge of the tool is raised off the surface, and the trailing edge lightly scrapes the surface smooth. The angle of the tool changes the amount of compound left behind. As the angle is reduced, more material is allowed to flow past the trowel. A shaper angle scrapes the surface more, pulling material away. To use these tools

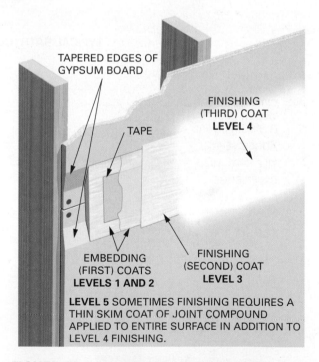

TAPERED EDGES OF GYPSUM BOARD

TAPE

FINISHING (THIRD) COAT **LEVEL 4**

EMBEDDING (FIRST) COATS **LEVELS 1 AND 2**

FINISHING (SECOND) COAT **LEVEL 3**

LEVEL 5 SOMETIMES FINISHING REQUIRES A THIN SKIM COAT OF JOINT COMPOUND APPLIED TO ENTIRE SURFACE IN ADDITION TO LEVEL 4 FINISHING.

FIGURE 72–1 The level of finish varies with the type of final decoration to be applied to drywall panels.

effectively, the angle held is constant and adjusted as needed during the entire pass of the tool. Automatic or mechanical tools are designed to speed up the finishing process. The banjo is designed to apply joint

FIGURE 72–2 Hand tools used to finish drywall.

compound and tape in one pass. The bazooka® provides a similar junction, but allows for a longer reach to the ceiling (**Figure 72–3**). It is filled with compound with a specially designed hand pump.

The nail spotter applies compound over the fasteners, smoothing the surface of extra material. Flat finishers apply a wide, thin layer of compound over flat seams (**Figure 72–4**). These tools are used on the third layer (second coat). They have a slight crown built into them to apply a layer slightly raised in the middle. When it dries, it is nearly flat. This makes a fourth layer often unnecessary.

BAZOOKA®

Courtesy of AMES Taping Tools

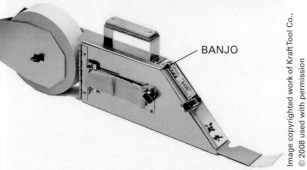

BANJO

Image copyrighted work of Kraft Tool Co., © 2008 used with permission

FIGURE 72–3 Bazooka® and banjo are automatic drywall tools designed to apply tape and compound in one pass.

FLAT FINISHER

Courtesy of AMES Taping Tools

NAIL SPOTTER

Courtesy of AMES Taping Tools

FIGURE 72–4 Automatic drywall tools designed to apply joint compound and smooth the surface in one pass.

DESCRIPTION OF MATERIALS

Fasteners are concealed with *joint compound*. Joints are reinforced with *joint tape* and covered with joint compound. Exterior corners are reinforced with *corner bead*. Other kinds of drywall trim may be used around doors, windows, and other openings (**Figure 72–5**).

Joint Compounds

Drying-type joint compounds for joint finishing and fastener spotting are made in both a dry powder form and a ready-mixed form in three general types. Drying-type compounds provide smooth application and ample working time. A *taping compound* is used to embed and adhere tape to the board over the joint. A *topping compound* is used for second and third coats over taped joints. An *all-purpose compound* is used for both bedding the tape and finishing the joint. All-purpose compounds do not possess the strength or workability of two-step taping and topping compound systems.

Setting-type joint compounds are used when a faster setting time than that of drying types is desired. Drying-type compounds harden through the loss of water by evaporation. They usually cannot be recoated until the next day. Setting-type compounds harden through a chemical reaction when water is added to the dry powder. Therefore, they come only in a dry powder form and not ready-mixed. They are formulated in several different setting times. The fastest setting type will harden in as little as 20 to 30 minutes. The slowest type takes four to six hours to set up. Setting-type joint compounds permit finishing of drywall interiors in a single day.

Joint Reinforcing Tape

Joint reinforcing tape is used to cover, strengthen, and provide crack resistance to drywall joints. One type is made of *high-strength fibre paper*. It is designed for use with joint compounds on gypsum panels. It is creased along its centre to simplify folding for application in corners (**Figure 72–6**).

Another type is made of *glass-fibre mesh*. It is designed to reinforce joints on veneer plaster gypsum panels. It is not recommended for use with conventional compounds for general drywall joint finishing. It may be used with special high-strength setting compounds. Glass-fibre mesh tape is available with a plain back or with an adhesive backing for quick application (**Figure 72–7**). Joint tape is normally available 2 and 2½ inches (50 and 63 mm) wide in 300-foot (90 m) rolls.

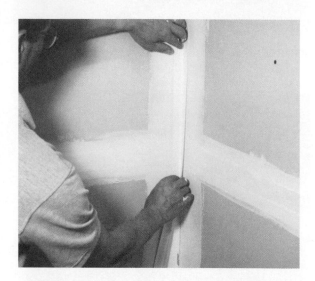

FIGURE 72–6 Applying joint tape to an interior corner.

FIGURE 72–7 An adhesive-backed glass-fibre mesh tape is quickly applied to drywall joints.

Courtesy of Saint-Gobain ADFORS

FIGURE 72–5 Materials typically used to finish drywall.

Corner Bead and Other Drywall Trim

Corner beads are applied to protect exterior corners of drywall construction from damage by impact. One type with solid metal flanges is widely used. Another type has flanges of expanded metal with a fine mesh. This provides excellent keying of the compound (**Figure 72–8**).

Corner bead is fastened through the drywall panel into the framing with nails or staples. Instead of using fasteners, a *clinching tool* is sometimes used. It crimps the solid flanges and locks the bead to the corner (**Figure 72–9**). Corner bead can also be applied using contact cement. After dry fitting, contact cement is applied to the drywall corner and to the corner bead. The surfaces are allowed to dry, and then the metal bead is positioned exactly in place. There is no room for error because the two surfaces will bond on contact.

Metal corner tape is applied by embedding it in joint compound. It is used for corner protection on arches, windows with no trim, and other locations (**Figure 72–10**).

A variety of *metal trim* is used to provide protection and finished edges to drywall panels. Metal trim is used at windows, doors, inside corners, and intersections. Such trim is fastened through their flanges into the framing (**Figure 72–11**).

Control joints are metal strips with flanges on both sides of a ¼-inch (6-mm) V-shaped slot. Control joints are placed in large drywall areas. They relieve stresses induced by expansion or contraction. They are used from floor to ceiling in long partitions and from wall to wall in large ceiling areas (**Figure 72–12**). The flanges are concealed with compound in a manner similar to corner beads and other trim.

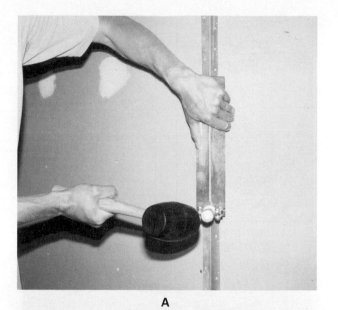

A

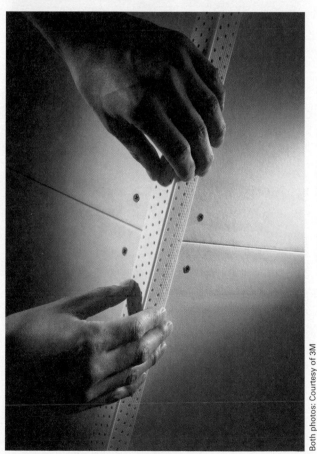

B

Both photos: Courtesy of 3M

FIGURE 72–9 Corner bead applied with (A) a clinching tool and (B) a spray-on adhesive.

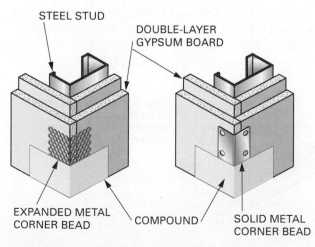

STEEL STUD

DOUBLE-LAYER GYPSUM BOARD

EXPANDED METAL CORNER BEAD

COMPOUND

SOLID METAL CORNER BEAD

FIGURE 72–8 Corner beads are used to finish and protect exterior corners of drywall panels.

A wide assortment of rigid vinyl drywall accessories is available (**Figure 72–13**), including the metal trim previously discussed. These accessories are designed for easy installation and workability to reduce installation

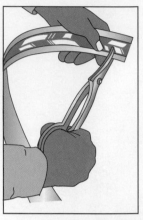

CUT TAPE
WITH SNIPS

EMBED IN JOINT
COMPOUND

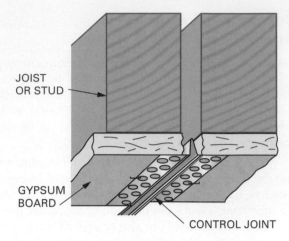

JOIST
OR STUD

GYPSUM
BOARD

CONTROL JOINT

FIGURE 72–12 Control joints are used in large wall and ceiling areas subject to movement by expansion and contraction.

The Gypsum Construction Handbook, Sixth Edition. Reproduced with permission of John Wiley & Sons, Inc.

FIGURE 72–10 Flexible metal corner tape is applied to exterior corners by embedding in compound. (Gypsum Construction Handbook, *U.S. Gypsum Corporation*)

time. Most have edges to guide the drywall knife, which allow for an even application of joint compound. Some have edges that are later torn away when the painting is done. This allows the finish to be applied more quickly and at the same time more uniformly. Vinyl accessories make it possible to create smooth joints easily, whether they are curved or straight.

Some companies now make a paper-faced metal or vinyl corner bead. This product comes in a variety of profiles and has a variety of accessories that allow for easy transitioning from corner bead to baseboard or from corner bead to corner bead (**Figure 72–14**).

APPLYING JOINT COMPOUND AND TAPE

In cold weather, care should be taken to maintain the interior temperature at a minimum of 10°C (50°F) for 24 hours before and during application of joint compound, and for at least four days after application has been completed.

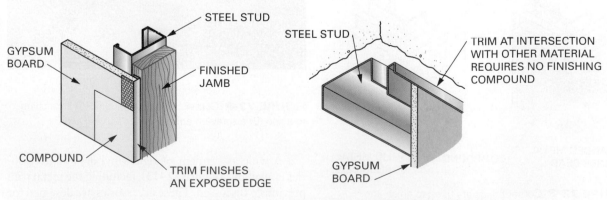

STEEL STUD

GYPSUM
BOARD

FINISHED
JAMB

COMPOUND

TRIM FINISHES
AN EXPOSED EDGE

STEEL STUD

TRIM AT INTERSECTION
WITH OTHER MATERIAL
REQUIRES NO FINISHING
COMPOUND

GYPSUM
BOARD

FIGURE 72–11 Various types of metal trim are used to provide finished edges to gypsum panels.

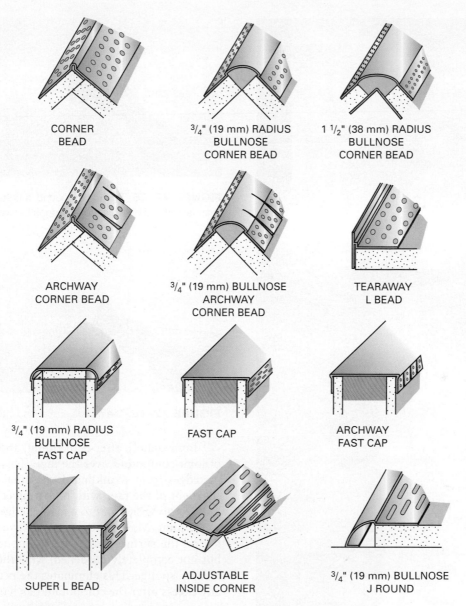

CORNER
BEAD

3/4" (19 mm) RADIUS
BULLNOSE
CORNER BEAD

1 1/2" (38 mm) RADIUS
BULLNOSE
CORNER BEAD

ARCHWAY
CORNER BEAD

3/4" (19 mm) BULLNOSE
ARCHWAY
CORNER BEAD

TEARAWAY
L BEAD

3/4" (19 mm) RADIUS
BULLNOSE
FAST CAP

FAST CAP

ARCHWAY
FAST CAP

SUPER L BEAD

ADJUSTABLE
INSIDE CORNER

3/4" (19 mm) BULLNOSE
J ROUND

FIGURE 72–13 Many rigid vinyl drywall accessories are available.

Care should also be taken to use clean tools. Avoid contamination of the compound by foreign material, such as sawdust, hardened plaster, or different types of compounds.

Pre-filling Joints

Before applying compound to drywall panels, check the surface for fasteners that have not been sufficiently recessed. Also look for other conditions that may affect the finishing. Pre-fill any joints between panels of ¼ inch (6 mm) or more and all V-groove joints between eased-edged panels with compound. A 24-hour drying period can be eliminated with the use of setting compounds for pre-filling operations. Normally the flat, tapered seams are finished first before the corners.

Embedding Joint Tape

Fill the recess formed by the tapered edges of the sheets with the specified type of joint compound. Use a joint knife (**Figure 72–15**). Centre the tape on the joint. Lightly press it into the compound. Draw the knife along the joint with sufficient pressure to *embed* the tape and remove excess compound (**Figure 72–16**).

There should be enough compound under the tape for a proper bond, but not over 1/32 inch (1 mm) under the edges. Make sure there are no air bubbles under the tape. The tape edges should be well adhered to the compound. If not satisfactory, lift the portion. Add compound and embed the tape again. A *taping tool* sometimes is used. It applies the compound and embeds the tape at the same time (**Figure 72–17**).

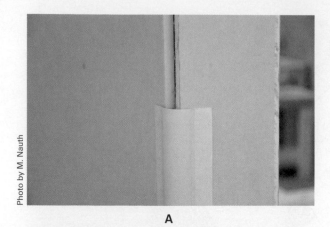

A

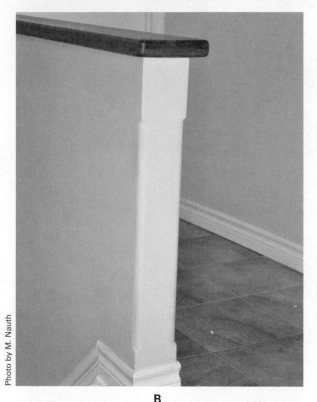

B

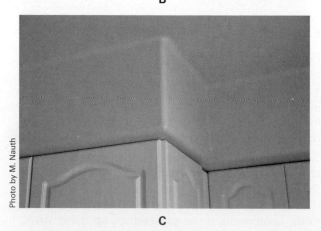

C

FIGURE 72–14. (A) Wallboard is held back at the corners for bull-nosed corner bead; (B) top and base accessories; (C) exposed corner and coved ceiling.

FIGURE 72–15 Taping compound is first applied to the channel formed by the tapered edges between panels.

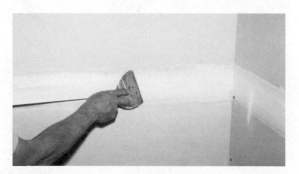

FIGURE 72–16 The tape is embedded into the compound.

Immediately after embedding, apply a thin coat of joint compound over the tape. This helps prevent the edges from wrinkling. It also makes easier concealment of the tape with following coats. Draw the knife to bring the coat to a feather edge on both sides of the joint. Make sure there is no excess compound left on the surface. After the compound has set up, but not completely hardened, wipe the surface with a damp sponge. This eliminates the need for sanding any excess after the compound has hardened.

Spotting Fasteners

Fasteners should be *spotted* immediately before or after embedding joint tape. Spotting is the application of compound to conceal fastener heads. Apply enough pressure on the taping knife to fill only the depression. Level the compound with the panel surface. Spotting is repeated each time additional coats of compound are applied to joints.

Applying Compound to Corner Beads and Other Trim

The first coat of compound is applied to corner beads and other metal trim when first coats are given to joints and fasteners. The nose of the bead or trim serves as a guide for applying the compound. The compound is applied about 6 inches (152 mm) wide from the nose of the bead to a feather edge on the

A

FIGURE 72–18 Applying a coat of compound to a drywall joint.

any excess, if necessary, to avoid interfering with the next coat of compound.

The second coat is sometimes called a **fill coat** (**Figure 72–18**). It is feathered out about 2 inches (50 mm) beyond the edges of all first coats, approximately 7 to 10 inches (178 to 254 mm) wide. Care must always be taken to remove all excess compound so that it does not harden on the surface. Professional drywall finishers rarely have to sand any excess in preparation for following coats. Remember, a damp sponge rubbed over the joint after the compound starts to set up will remove any small particles of excess. It will also help bring edges to a feather edge.

If the level of finish requires it, apply a third and **finishing coat** of compound over all fill coats. The edges of the finishing coat should be feathered out about 2 inches (50 mm) beyond the edges of the second coat. Some drywall finishers apply a skim coat of compound over the entire wall surface (**Figure 72–19**). This provides a more uniform surface where reflected light shows variations between the panels and compound.

The Gypsum Construction Handbook, Sixth Edition. Reproduced with permission of John Wiley & Sons, Inc.

Courtesy Gypsum Association, gypsum.org

B

FIGURE 72–17 (A) A taping tool applies tape and compound at the same time. (B) The corner is then smoothed and finished before compound sets.

wall. Each subsequent finishing coat is applied about 2 inches (50 mm) wider than the previous one.

Fill and Finishing Coats

Allow the first coat to dry thoroughly. This may take 24 hours or more depending on temperature and humidity, unless a setting-type compound has been used. It is common to use setting-type compounds for first coats and slower setting types for finishing coats. Feel the entire surface to find out whether any excess compound has hardened on the surface. Sand

FIGURE 72–19 Some walls are finished with a skim coat of compound.

Courtesy Gypsum Association, gypsum.org

Butt Joints

Butt joints occur when the length of the wall exceeds the length of the longest sheet of drywall available. Because the ends of the drywall sheets are *not* tapered, butt joints would therefore be raised when the seam is taped and finished. There are several techniques for dealing with this problem, but the **back-blocking** application developed by the Canadian Gypsum Corporation has proven to work well.

The sheets of drywall are allowed to meet in the space between the studs or joists. About six strips of drywall *blocking* are cut about 8 inches (203 mm) wide and slightly shorter than the space between the studs or joists. Strips of drywall are fastened to the sides of the studs to act as stops for the drywall blocks so that the blocks are held slightly back of the face of the studs or joists. For a 2 × 4 (38 × 89 mm) wall, these strips would be about 2¾ inches (70 mm) and are held flush to the back of the studs. Spread joint compound (or a quick-setting compound) in ½-inch-high (12.5 mm) by ³/₈-inch-wide (9.5 mm) beads at 1½ inches (38 mm) on centre on the face of the blocks. Install the first drywall sheet so that it ends within the stud or joist space. Place the blocks (with the compound already applied) in the space along the full length of the end and edge of the sheet.

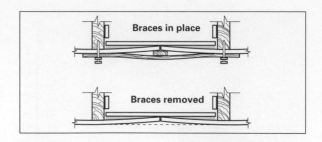

FIGURE 72–20 Back-blocking butt joints in drywall installation. *(The Gypsum Construction Handbook, Sixth Edition. Reproduced with permission of John Wiley & Sons, Inc.)*

Immediately install the second sheet into position so that the butt ends meet loosely. Next, place a 1 × 3 (19 × 63 mm) strap along the full length of the butt joint and hold it in place with three other 1 × 3 (19 × 63 mm) straps that are securely screwed to each adjacent stud or joist. The straps are removed after the compound has dried and a *taper* is formed that will allow for an invisible joint (**Figure 72–20**).

Another method also butts the end joints over the stud space and uses a 9-inch-wide (229 mm) strip of plywood with cardboard strips to create raised edges (see *http://www.finehomebuilding.com/how-to/tips/invisible-drywall-butt-joints.aspx*) (**Figure 72–21**), or a manufactured product called

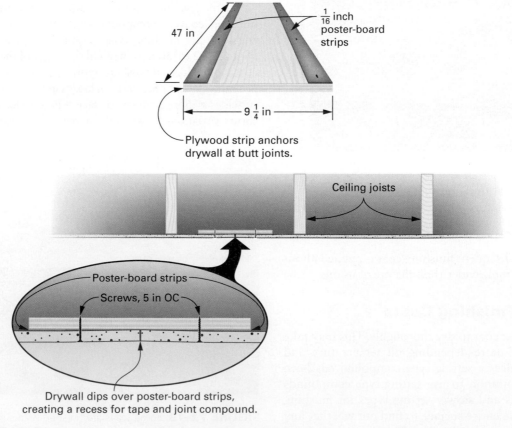

FIGURE 72–21 Drywall butt joints. *(Reprinted with permission from www.finehomebuilding.com.)*

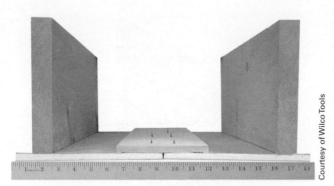

FIGURE 72–22 Drywall butt joints by Rocksplicer.

a Rocksplicer (*http://rocksplicer.com/rock_splicer .html*) to produce the same effect of a depressed butt joint that will receive tape and mud without producing a "bump" on the surface (see *http://www .thisisdrywall.com/?p=362*) (**Figure 72–22**).

Interior Corners

Interior corners are finished in a similar way. However, the tape is folded in the centre to fit in the corner. After the tape is embedded, drywall finishers usually apply a setting compound to one side only of each interior corner. By the time they have finished all interior corners in a room, the compound has set enough to finish the other side of the corners.

Sanding

The final step in finishing is sanding. All surfaces should be smooth and clean before paint is applied. Sanding may be done by hand using a pole sander or, on larger jobs, with a portable electric sander (**Figure 72–23**). Sandpaper with a grit of 100, 120, or 150 is appropriate to erase any lumps, bumps, or extra material.

Care should be taken not to raise a nap on the drywall paper with too much sanding. A damp sponge can also be used to smooth the surface. It does not remove material as fast as sandpaper or raise a nap on the drywall paper. It also produces virtually no dust.

Textures and Coatings

Textures are sometimes added to a wall or ceiling surface to add character and charm. They are applied to a mostly finished wall surface; sanding is not usually necessary. The texture itself is joint compound applied in various thicknesses and methods to create a desired effect (**Figure 72–24**).

FIGURE 72–23 A pole and electric sander for finishing drywall.

FIGURE 72–24 Texture may be added to the wall surface.

REMEDIES FOR DRYWALL PROBLEMS

In drywall finishing, often a situation arises that must be repaired. The surface may not yet acceptable for paint because the drywall finish requires an extra process.

Bubbles in Tape

Joint tape sometimes will bubble away from the drywall surface as it dries. This happens because the tape has not bonded to the drywall, which is usually since there is no joint compound between it and the drywall. Small bubbles may be sliced and opened with a utility knife, then more compound added under the flap. It is all then closed up and smoothed with a taping knife. Large long bubbles are best cut and removed. Scrape out dried compound enough to make a new compound and tape application.

Fastener Pops

Fastener or nail pops occur when fastener heads work themselves out from beneath the finished surface. This happens for several reasons, most often because the fastener was not installed correctly. Pops occur when the fastener does not penetrate solidly into the framing or when the fastener is driven too far, breaking through the surface paper. It also happens when there is a space between the drywall and framing. Either the drywall was not tight in the first place or the framing material shrank because it was not previously dry.

Repair of fastener pops begins by removing the old fastener. Drive screws nearby making sure they hit solid framing while pressing the drywall tight to the framing. Spot finish the heads as before, making sure several coats are applied to compensate for normal compound shrinkage.

Patching Drywall

Damage often happens to drywall such that the surface is broken and holes are formed. Small holes about 2 inches (50 mm) or less may be repaired by applying joint compound. A small piece of tape embedded over the hole prevents the patch from cracking.

Medium-sized holes that occur between framing are patched with a new piece of drywall (**Figure 72–25**). The hole is cut larger to make a square hole. Then a new piece of drywall is cut to fit and finished. The new piece may be secured by first screwing pieces of plywood inside the hole to the existing drywall. The new

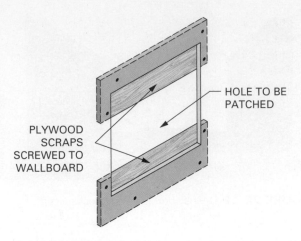

FIGURE 72–25 Patch techniques for medium-sized holes in drywall using plywood supports.

piece is then secured to the portions of plywood that span over the hole.

Another method uses a larger repair piece to create bonding flaps. The piece is cut about 3 inches (75 mm) larger than the opening (**Figure 72–26**). Dimensions of the hole size are scored on the backside of the piece. Breaking the edges and removing the gypsum while leaving the face paper intact creates the bonding flaps. Joint compound is applied to both the flaps and the hole perimeter. Pressing and smoothing the piece into place with a knife beds the flaps to the wall.

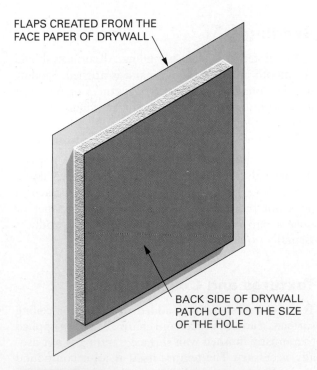

FIGURE 72–26 Patch techniques for medium-sized holes in drywall using bonding flaps.

Large holes are repaired by cutting the hole large enough to expose the framing. Portions of the hole perimeter have one-half the thickness of the framing member available to secure the new piece.

ESTIMATING DRYWALL

To estimate the amount of drywall material needed, determine the area of the walls and ceilings to be covered. To find the ceiling area, multiply the length of the room by its width. To find the wall area, multiply the perimeter of the room by the height. Subtract all large wall openings. Combine all areas to find the total number of square feet of drywall required. Add about 5 percent of the total for waste. The number of drywall panels can then be determined by dividing the total area to be covered by the area of one panel.

It is good practice to list the number of panels and the lengths of the panels by room so that when the entire load arrives on-site, the sheets can be distributed to each room location.

About 1000 screws are needed for 1000 square feet of drywall applied to framing at 16 inches OC (for 93 m² at 406 mm OC), and 850 screws are needed for 24-inch (610-mm) OC framing. About 5 pounds (2.5 kg) of nails are required to fasten each 1000 square feet (93 m²) of drywall.

Approximately 1½ five-gallon (35 L) pails of joint compound and 370 feet (115 m) of joint tape will be needed to finish each 1000 square feet of drywall.

KEY TERMS

back-blocking	contact adhesives	drywall foot lifter	joint compounds
backing board	control joints	eased edge	joint reinforcing tape
banjo	corner beads	face	metal corner tape
bazooka®	deadmen	fill coat	Type X
butt joints	drywall	finishing coat	
cement board	drywall cutout tool	gypsum lath	

REVIEW QUESTIONS

Select the most appropriate answer.

1. What is the standard width of gypsum board?

   a. 36 inches (914 mm)
   b. 48 inches (1220 mm)
   c. 54 inches (1372 mm)
   d. 60 inches (1524 mm)

2. What are the standard lengths of gypsum board?

   a. 8, 10, and 12 feet (2.4, 3.0, and 3.6 m)
   b. 8, 10, 12, and 14 feet (2.4, 3.0, 3.6, and 4.3 m)
   c. 8, 9, 10, 12, and 14 feet (2.4, 2.7, 3.0, 3.6, and 4.3 m)
   d. 8, 9, 10, 12, 14, and 16 feet (2.4, 2.7, 3.0, 3.6, 4.3, and 4.9 m)

3. When fastening drywall, what is the minimum penetration for ring-shanked nails into the framing member?

   a. ½ inch (12 mm)
   b. ¾ inch (19 mm)
   c. ⅞ inch (22 mm)
   d. 1 inch (25 mm)

4. When is gypsum board usually installed vertically on walls?

   a. when 2 × 3 studs are used
   b. when 2 × 4 studs are used
   c. when 2 × 6 studs are used
   d. when steel studs are used

5. Which of the following is sometimes used to align ceiling joists?

   a. deadman
   b. dutchman
   c. strongback
   d. straightedge

6. In the single nailing method, what is the maximum spacing for nails?

   a. 8 inches (203 mm) OC on walls; 7 inches (178 mm) OC on ceilings
   b. 10 inches (254 mm) OC on walls; 8 inches (203 mm) OC on ceilings
   c. 12 inches (305 mm) OC on walls; 10 inches (254 mm) OC on ceilings
   d. 12 inches (305 mm) OC on walls and ceilings

7. What is the spacing for screws?

   a. 12 inches (305 mm) OC on walls; 10 inches (254 mm) OC on ceilings
   b. 12 inches (305 mm) OC on walls and ceilings
   c. 16 inches (406 mm) OC on walls and ceilings
   d. 16 inches (406 mm) OC on walls; 12 inches (305 mm) OC on ceilings

8. What is the least offset required for joints in the face layer of a multilayer application from joints in the base layer?

   a. at least 6 inches (152 mm)
   b. at least 8 inches (203 mm)
   c. at least 10 inches (254 mm)
   d. at least 12 inches (305 mm)

9. What is the minimum distance between the paper-covered edge of water-resistant gypsum board and the lip of the tub or shower pan?

   a. not less than ¼ inch (6 mm) above the lip
   b. not less than $3/8$ inch (9.5 mm) above the lip
   c. not less than ½ inch (12 mm) above the lip
   d. not less than ¾ inch (19 mm) above the lip

10. When ceramic tile more than $3/8$-inch (9.5 mm) thick is to be applied over water-resistant gypsum board, what is the maximum spacing between screws or nails used to fasten the board?

    a. not more than 4 inches (100 mm) OC
    b. not more than 6 inches (152 mm) OC
    c. not more than 8 inches (203 mm) OC
    d. not more than 10 inches (254 mm) OC

Unit 23

Wall Panelling and Ceramic Tile

Chapter 73 Wall Panelling: Types and Application

Chapter 74 Ceramic Tile

Plans and specifications often call for the installation of *wall panelling* in certain rooms of both residential and commercial construction. *Ceramic tile* is widely used in washrooms, baths, showers, kitchens, and similar areas.

Interior finishes serve as accents to design and protection of the structure from water. Long-term use and function relies on proper workmanship.

OBJECTIVES

After completing this unit, the student should be able to:

- describe and apply several kinds of sheet wall panelling.
- describe and apply various patterns of solid lumber wall panelling.
- describe and install ceramic wall tile to bathroom walls.
- estimate quantities of wall panelling and ceramic wall tile.

Chapter 73

Wall Panelling: Types and Application

Two basic kinds of wall panelling are sheets of various prefinished material and solid *wood boards*. Many compositions, colours, textures, and patterns are available in sheet form. Solid wood boards of many species and shapes are used for both rustic and elegant interiors (**Figure 73–1**).

DESCRIPTION OF SHEET PANELLING

Sheets of prefinished plywood, **hardboard**, **particleboard**, **plastic laminate**, and other materials are used to panel walls.

Photographer: Mitch Lenet. Copyright © Aanischaaukamikw Cree Cultural Institute

FIGURE 73–1 Solid wood board ceiling panels provide warmth and beauty to interiors of buildings.

Plywood

Prefinished plywood is probably the most widely used sheet panelling. A tremendous variety is available in both **hardwood** and **softwood**. The more expensive types have a face veneer of real wood. The less expensive kinds of plywood panelling are prefinished with a printed wood grain or other design on a thin vinyl covering. Care must be taken not to scratch or scrape the surface when handling these types. Unfinished plywood panels are also available.

Some sheets are scored lengthwise at random intervals to imitate solid wood panelling. There is always a score 16, 24, and 32 inches (406, 610, and 813 mm) from the edge. This facilitates fastening of the sheets and in case the sheet has to be ripped lengthwise to fit stud spacing.

Most commonly used panel thicknesses are $3/16$ and ¼ inch (5 and 6 mm). Sheets are normally 4 feet wide and 7 to 10 feet long (1.2 m wide and 2.1 to 3 m long). An 8-foot (2.4 m) length is most commonly used. Panels may be shaped with square, bevelled, or shiplapped edges (**Figure 73–2**). Matching moulding is available to cover edges, corners, and joints. Thin ring-shanked nails, called *colour pins*, are available in colours to match panels. They are used when exposed fastening is necessary.

Hardboard

Hardboard is available in many man-made surface colours, textures, and designs. Some of these designs simulate stone, brick, stucco, leather, weathered or smooth wood, and other materials. Unfinished hardboard is also used, which has a smooth, dark brown surface suitable for painting and other decorating. Unfinished hardboard may be solid or perforated in a number of designs.

Tileboard is a hardboard panel with a baked-on plastic finish. It is embossed to simulate ceramic wall tile. The sheets come in a variety of solid colours, marble, floral, and other patterns. Tileboard is designed for use in bathrooms, kitchens, and similar areas.

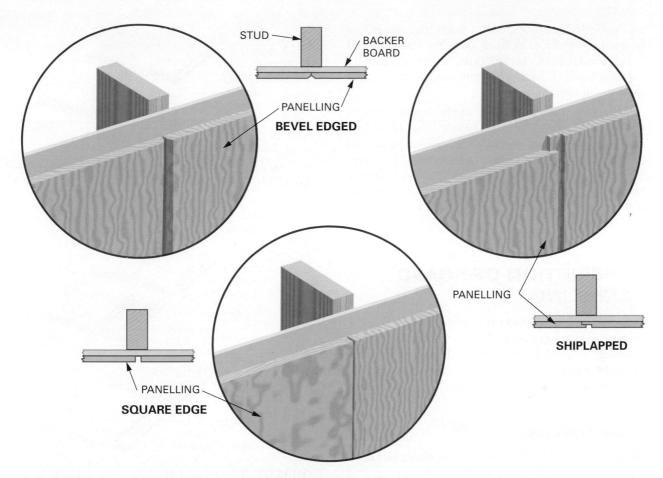

FIGURE 73–2 Sheet panelling comes with various edge shapes.

Hardboard panelling comes in widths of 4 feet (1.2 m) and in lengths of from 8 to 12 feet (2.4 to 3.6 m). Commonly used thicknesses are from ⅛ to ¼ inch (3 to 6 mm). Colour-coordinated moulding and trim are available for use with hardboard panelling.

Particleboard and MDF

Panels of particleboard and MDF (medium-density fibreboard) come with wood grain or other designs applied to the surface, similar to plywood and hardboard. Sheets are usually ¼ inch (6 mm) thick, 4 feet (1.2 m) wide, and 8 feet (2.4 m) long. Prefinished particleboard and MDF panels are used when an inexpensive wall covering is desired. Because the sheets are brittle and break easily, care must be taken when handling them. They must be applied only on a solid wall backing, usually on drywall for fire-protection reasons.

Unfinished particleboard is not usually used as an interior wall finish. One exception, made from aromatic cedar chips, is used to cover walls in closets to repel moths.

Plastic Laminates

Plastic laminates are widely used for surfacing kitchen cabinets and post-formed countertops. They are also used to cover walls or parts of walls in kitchens, bathrooms, washrooms, and similar areas where a durable, easy-to-clean surface is desired. Laminates can be scorched by an open flame. However, they resist mild heat, alcohol, acids, and stains. They clean easily with a mild detergent.

Laminates are manufactured in many colours and designs, including wood grain patterns. Surfaces are available in gloss, satin, and textured finishes, among others.

Laminates are ordinarily used in two thicknesses. Vertical-type laminate is relatively thin (about ¹⁄₃₂ inch [1 mm]). It is used for vertical surfaces, such as walls and cabinet sides. Vertical-type laminate is available only in widths of 4 or 8 feet (1.2 or 2.4 m). Regular or standard laminate is about ¹⁄₁₆ inch (1.5 mm) thick. It comes in widths of 24, 36, 48, and 60 inches (0.6, 0.9, 1.2, and 1.5 m) and in lengths of 5, 6, 8, 10, and 12 feet (1.5, 1.8, 2.4, 3.0, and 3.6 m). It is generally used on horizontal surfaces, such as countertops. It can be used on walls, if desired, or if the size required is not available in vertical type. Sheets are usually manufactured 1 inch (25 mm) wider and longer than the nominal size.

Laminates are difficult to apply to wall surfaces because they are so thin and brittle. Also, because a *contact cement* is used, the sheet cannot be moved once it makes contact with the surface. Thus, prefabricated panels, with sheets of laminate already bonded to a backer, are normally used to panel walls. See Chapter 88 for installation techniques for laminates.

DESCRIPTION OF BOARD PANELLING

Board panelling is used on interior walls when the warmth and beauty of solid wood is desired. Wood panelling is available in softwoods and hardwoods of many species. Each has its own distinctive appearance, unique grain, and knot pattern.

Wood Species

Woods may be described as light, medium, and dark toned. Light tones include birch, maple, spruce, and white pine. Some medium tones are cherry, cypress, hemlock, oak, ponderosa pine, and fir. Among the darker-toned woods are cedar, mahogany, redwood, and walnut. For special effects, knotty pine, wormy chestnut, pecky cypress, and white-pocketed Douglas fir board panelling may be used.

Surface Textures and Patterns

Wood panelling is available in many shapes. It is either planed for smooth finishing or rough-sawn for a rustic, informal effect. Square-edge boards may be joined edge to edge, spaced on a dark background, or applied in *board-and-batten* or *board-on-board* patterns. *Tongue-and-grooved* or *shiplapped* panelling comes in patterns, a few of which are illustrated in **Figure 73–3**.

Sizes

Most wood panelling comes in a ¾-inch (19 mm) thickness and in nominal widths of 4, 6, 8, 10, and 12 inches (100, 152, 203, 254, and 305 mm). A few patterns are manufactured in a ⁹⁄₁₆-inch (14 mm) thickness. Aromatic cedar panelling is used in clothes closets. It runs

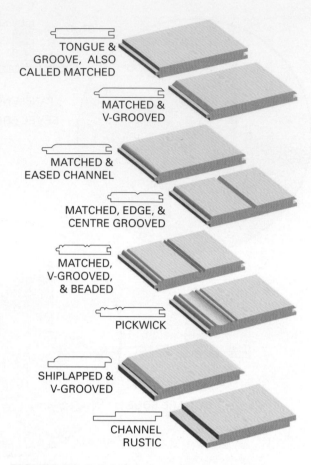

FIGURE 73–3 Solid wood panelling is available in a number of patterns.

from ⅜ to ⅝ inch (9.5 to 16 mm) thick, depending on the mill. It is usually *edge-* and *end-matched* (tongue-and-grooved) for application to a backing surface.

Moisture Content

To avoid shrinkage, wood panelling, like all interior finish, should be dried to a *moisture equilibrium* content. That is, its moisture content should be about the same as the area in which it is to be used. Interior finish applied with an excessive moisture content will eventually shrink, causing open joints, warping, loose fasteners, and many other problems.

OTHER PANELLING

The types of panelling described in this chapter are those that are most commonly used. To become acquainted with other types and methods of application, a study of manufacturers' catalogues dealing with sheet and board panelling, found in *Sweet's Architectural File*, is suggested.

FIGURE 73–4 Wood panelling can be applied on walls and ceilings.

Sheet panelling, such as prefinished plywood and hardboard, is usually applied to walls with long edges vertical. *Board panelling* may be installed vertically, horizontally, diagonally, or in many interesting patterns (**Figure 73–4**).

INSTALLATION OF SHEET PANELLING

A backer board layer, at least ³⁄₈ inch (9.5 mm) thick, should be installed on walls prior to the application of sheet panelling. The backing makes a stronger and more fire-resistant wall, helps block sound transmission, tends to bring studs into alignment, and provides a rigid finished surface for application of panelling (**Figure 73–5**).

Furring strips must be applied to masonry walls before panelling is installed. The strips are usually applied by driving hardened nails. Instead of using furring strips, a freestanding wood wall close to the masonry wall can be built, if enough space is available.

Preparation for Application

Mark the location of each stud in the wall on the floor and ceiling. Panelling edges must fall on stud centres, even if applied with adhesive over a backer board, in case supplemental nailing of the edges is necessary. Panels are usually fastened with a combination of colour pins and adhesive.

Apply narrow strips of colour on the wall from floor to ceiling where joints between panelling will occur. Use paint or magic marker coloured the same as the joint of the panel. If joints between sheets open slightly because of shrinkage during extended dry periods or heating seasons, it

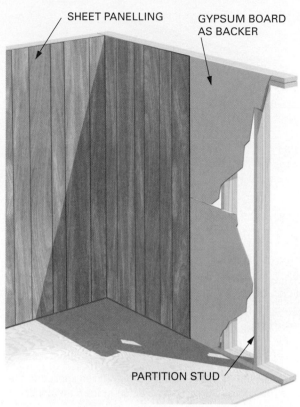

SHEET PANELLING   GYPSUM BOARD AS BACKER

PARTITION STUD

FIGURE 73–5 Apply sheet panelling over a gypsum wallboard base.

is not so noticeable with a similar colour behind the joint.

Sometimes panelling does not extend to the ceiling but covers only the lower portion of the wall. This partial panelling is called **wainscotting**. It usually extends to about 3 feet (914 mm) above the floor (**Figure 73–6**). If the wall is to be wainscotted, snap a horizontal line across the wall to indicate its height.

Stand panels on their long edge against the wall for at least 48 hours before installation. This allows them to adjust to room temperature and humidity. Otherwise, sheets may buckle after installation, especially if they were very dry when installed.

Just prior to application, stand panels on end, side by side, around the room. Arrange them by matching the grain and colour to obtain the most pleasing appearance.

Starting the Application

Start in a corner and continue installing consecutive sheets around the room. Select the starting corner, remembering that it will also be the ending

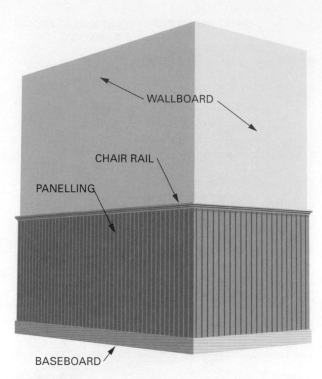

FIGURE 73–6 Wainscotting is a wall finish applied to the lower portion of the wall that is different from the upper portion.

point. This corner should be the least visible, such as behind an often-open door. Cut the first sheet to a length about ¼ inch (6 mm) less than the wall height. Place the sheet in the corner. Plumb

the outside edge and tack it in the plumb position. Set the distance between the points of the dividers the same as the amount the sheet overlaps the centre of the stud (**Figure 73–7**). Scribe this amount on the edge of the sheet butting the corner (**Figure 73–8**).

Remove the sheet from the wall. Place it on sawhorses on which a sheet of plywood (or two 8-foot [2.4 m] lengths of 2 × 4s [38 × 89s]) has been placed for support. Cut to the scribed lines with a sharp, fine-toothed, hand crosscut saw. Handsaws may be used on the face side because they cut on the downstroke. This action will keep the splintering of the cut on the backside of the panel. Power saws should be used to cut from the backside.

Replace the sheet with the cut edge fitting snugly in the corner. The joint at the ceiling need not be fitted if a moulding is to be used. If a tight fit between the panel and ceiling is desired, set the dividers and scribe a small amount at the ceiling line. Remove the sheet again. Cut to the scribed line. Replace the sheet, and raise it snugly against the ceiling. The space at the bottom will be covered later by a baseboard.

Fastening

If only nails are used, fasten about 6 inches (152 mm) apart along edges and about 12 inches (305 mm)

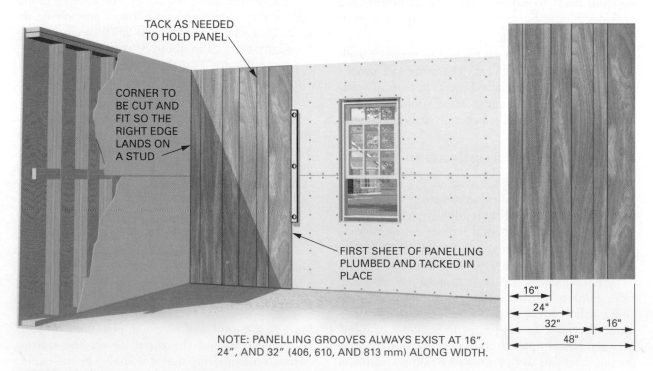

NOTE: PANELLING GROOVES ALWAYS EXIST AT 16", 24", AND 32" (406, 610, AND 813 mm) ALONG WIDTH.

FIGURE 73–7 The first sheet must be set plumb in the corner.

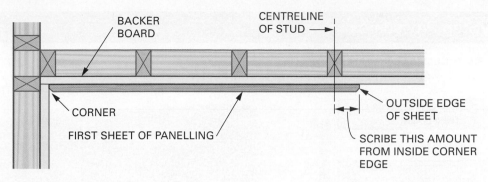

FIGURE 73–8 Set scribers equal to the largest space and scribe the edge of the first sheet to the corner.

apart on intermediate studs for ¼-inch (6 mm) thick panelling. Nails may be spaced farther apart on thicker panelling. Drive nails at a slight angle for better holding power (**Figure 73–9**).

If adhesives are used, apply a ⅛-inch (3 mm) continuous bead where panel edges and ends make contact. Apply beads 3 inches (75 mm) long and about 6 inches (152 mm) apart on all intermediate studs. Put the panel in place. Tack it at the top. Be sure the panel is properly in position. Press on the panel surface to make contact with the adhesive. Use firm, uniform pressure to spread the adhesive beads evenly between the wall and the panel. Then, grasp the panel and slowly pull the bottom of the sheet a few inches (50 mm) away from the wall (**Figure 73–10**). Press the sheet back into position after about two minutes. After about 20 minutes, recheck the panel. Apply pressure to ensure thorough adhesion and to smooth the panel surface. Apply successive sheets in the same manner. Do not force panels into position. Panels should touch very lightly at joints.

Wall Outlets

To lay out for wall outlets, plumb and mark both sides of the outlet to the floor. If the opening is close to the ceiling, plumb upward and mark lightly on the ceiling. Level the top and bottom of the outlet on the wall beyond the edge of the sheet to be installed. Or level on the adjacent sheet, if closer. Cut, fit, and tack the sheet in position. Level and plumb marks from the wall and floor onto the sheet for the location of the opening (**Procedure 73–A**).

Another method is to rub a cake of carpenter's chalk on the edges of the outlet box. Fit and tack the sheet in position. Tap on the sheet directly over the outlet to transfer the chalked edges to the back of the sheet. Remove the sheet. Cut the opening for the outlet. Openings for wall outlets, such as electrical boxes must be cut fairly close to the location and size. The cover plate may not cover if the cutout

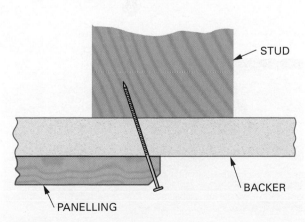

FIGURE 73–9 Drive nails or colour pins at a slight angle for better holding power.

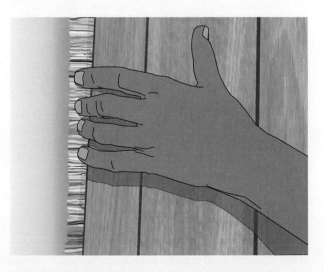

FIGURE 73–10 Pull the sheet a short distance away from the wall to allow the adhesive to dry slightly.

StepbyStep Procedures

Procedure 73–A Cutting Outlet Holes in Panelling

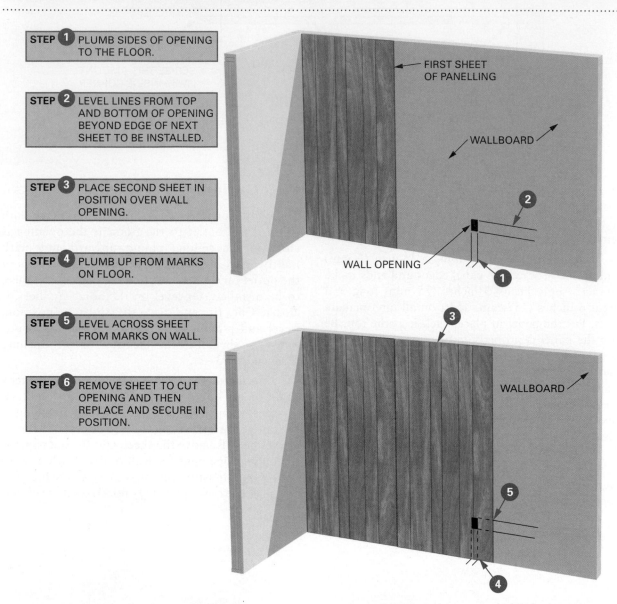

STEP 1 PLUMB SIDES OF OPENING TO THE FLOOR.

STEP 2 LEVEL LINES FROM TOP AND BOTTOM OF OPENING BEYOND EDGE OF NEXT SHEET TO BE INSTALLED.

STEP 3 PLACE SECOND SHEET IN POSITION OVER WALL OPENING.

STEP 4 PLUMB UP FROM MARKS ON FLOOR.

STEP 5 LEVEL ACROSS SHEET FROM MARKS ON WALL.

STEP 6 REMOVE SHEET TO CUT OPENING AND THEN REPLACE AND SECURE IN POSITION.

FIRST SHEET OF PANELLING

WALLBOARD

WALL OPENING

WALLBOARD

is not made accurately. This could require replacement of the sheet. A jigsaw may be used to cut these openings. When using the sabre saw, cut from the back of the panel to avoid splintering the face (**Figure 73–11**).

Ending the Application

The final sheet in the wall need not fit snugly in the corner if the adjacent wall is to be panelled or if interior corner moulding is to be used. Take measurements at the top, centre, and bottom. Cut the sheet to width, and install.

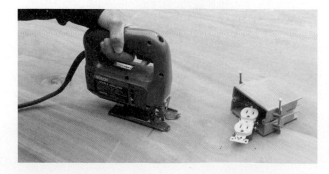

FIGURE 73–11 A jigsaw may be used to cut an opening for an electrical outlet box.

If the last sheet butts against a finished wall and no corner moulding is used, the sheet must be cut to fit snugly in the corner. To mark the sheet accurately, first measure the remaining space at the top, bottom, and about the centre. Rip the panel about ½ inch (12 mm) wider than the greatest distance. Place the sheet with the cut edge in the corner and the other edge overlapping the last sheet installed. Tack the sheet in position. The amount of overlap should be exactly the same from top to bottom. Set the dividers for the amount of overlap. Scribe this amount on the edge in the corner (**Procedure 73–B**). Instead of dividers, it is sometimes more exact to use a small block of wood for scribing. The width is cut the same as the amount of overlap. Care must be taken to keep from turning the dividers while scribing along a surface (**Figure 73–12**).

StepbyStep Procedures

Procedure 73–B Scribing the Last Piece of Panelling

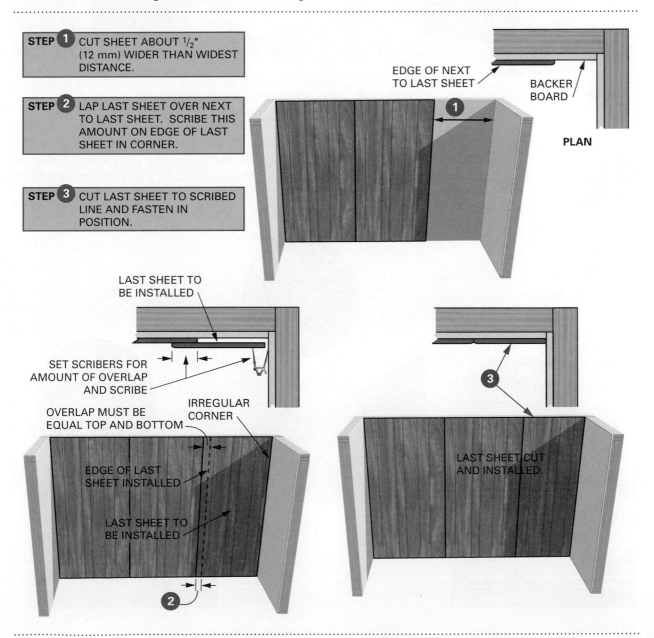

STEP 1 CUT SHEET ABOUT ½" (12 mm) WIDER THAN WIDEST DISTANCE.

STEP 2 LAP LAST SHEET OVER NEXT TO LAST SHEET. SCRIBE THIS AMOUNT ON EDGE OF LAST SHEET IN CORNER.

STEP 3 CUT LAST SHEET TO SCRIBED LINE AND FASTEN IN POSITION.

EDGE OF NEXT TO LAST SHEET

BACKER BOARD

PLAN

LAST SHEET TO BE INSTALLED

SET SCRIBERS FOR AMOUNT OF OVERLAP AND SCRIBE

OVERLAP MUST BE EQUAL TOP AND BOTTOM

IRREGULAR CORNER

EDGE OF LAST SHEET INSTALLED

LAST SHEET TO BE INSTALLED

LAST SHEET CUT AND INSTALLED.

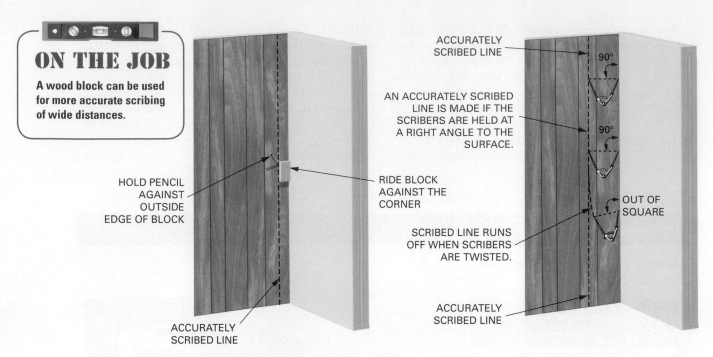

ON THE JOB

A wood block can be used for more accurate scribing of wide distances.

HOLD PENCIL AGAINST OUTSIDE EDGE OF BLOCK

RIDE BLOCK AGAINST THE CORNER

ACCURATELY SCRIBED LINE

ACCURATELY SCRIBED LINE

AN ACCURATELY SCRIBED LINE IS MADE IF THE SCRIBERS ARE HELD AT A RIGHT ANGLE TO THE SURFACE.

90°

90°

OUT OF SQUARE

SCRIBED LINE RUNS OFF WHEN SCRIBERS ARE TWISTED.

ACCURATELY SCRIBED LINE

FIGURE 73–12 Accurate scribing requires that the marked line be made perpendicular to the corner.

Cut to the scribed line. If the line is followed carefully, the sheet should fit snugly between the last sheet installed and the corner, regardless of any irregularities. On exterior corners, a quarter-round moulding is sometimes installed against the edges of the sheets. Or the joint may be covered with a wood, metal, or vinyl corner moulding (**Figure 73–13**).

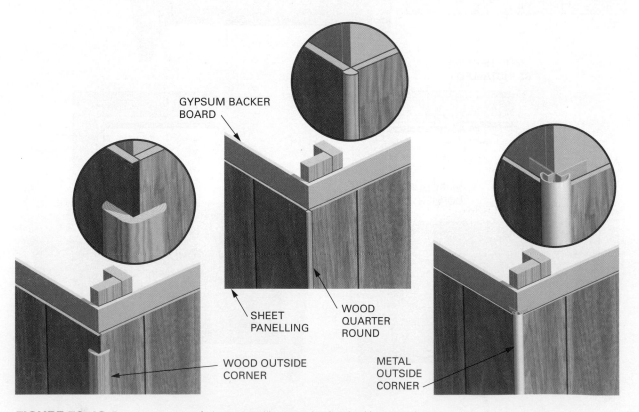

GYPSUM BACKER BOARD

SHEET PANELLING

WOOD QUARTER ROUND

WOOD OUTSIDE CORNER

METAL OUTSIDE CORNER

FIGURE 73–13 Exterior corners of sheet panelling may be finished in several ways.

INSTALLING SOLID WOOD BOARD PANELLING

Horizontal board panelling may be fastened to studs in new and existing walls (**Figure 73–14**). For vertical application of board panelling in a frame wall, blocking must be provided between studs (**Figure 73–15**). On existing and masonry walls, horizontal furring strips must be installed. Blocking or furring must be provided in appropriate locations for diagonal or pattern applications of board panelling.

Allow the boards to adjust to room temperature and humidity by standing them against the walls around the room. At the same time, put them in the order of application. Match them for grain and colour. If tongue-and-grooved boards are to be eventually stained or painted, apply the same finish to the tongues so that an unfinished surface is not exposed if the boards shrink after installation.

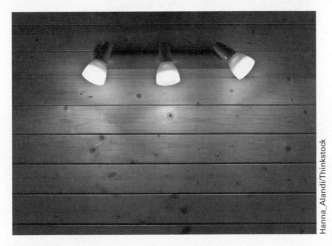

FIGURE 73–14 No wall blocking is required for horizontal application of board panelling.

Hanna_Alandi/Thinkstock

Starting the Application

Select a straight board to start with. Cut it to length, about ¼ inch (6 mm) less than the height of the wall. If tongue-and-grooved stock is used, tack it in a plumb position with the grooved edge in the corner. If a tight fit is desired, adjust the dividers to scribe an amount a little more than the depth of the groove. Rip to the scribed line. Face-nail along the cut edge into the corner with finish nails about 16 inches (406 mm) apart. Blind-nail the other edge through the tongue.

Continuing the Application

Apply succeeding boards by blind-nailing into the tongue only (**Figure 73–16**). Make sure the joints

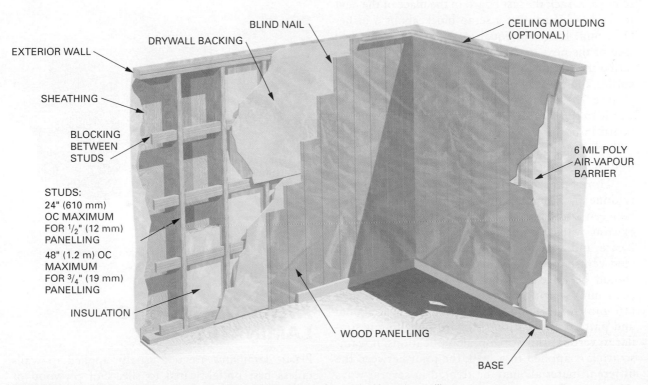

FIGURE 73–15 Blocking must be provided between studs for vertical board panelling.

Labels in figure:
EXTERIOR WALL
SHEATHING
BLOCKING BETWEEN STUDS
STUDS: 24" (610 mm) OC MAXIMUM FOR 1/2" (12 mm) PANELLING / 48" (1.2 m) OC MAXIMUM FOR 3/4" (19 mm) PANELLING
INSULATION
DRYWALL BACKING
BLIND NAIL
WOOD PANELLING
CEILING MOULDING (OPTIONAL)
6 MIL POLY AIR-VAPOUR BARRIER
BASE

FIGURE 73–16 Tongue-and-grooved panelling is blind-nailed.

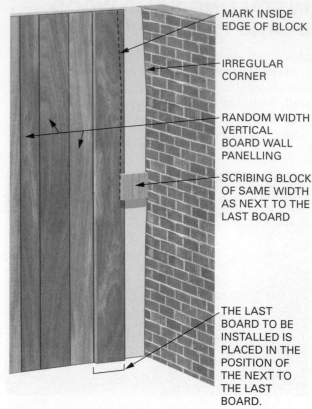

MARK INSIDE EDGE OF BLOCK

IRREGULAR CORNER

RANDOM WIDTH VERTICAL BOARD WALL PANELLING

SCRIBING BLOCK OF SAME WIDTH AS NEXT TO THE LAST BOARD

THE LAST BOARD TO BE INSTALLED IS PLACED IN THE POSITION OF THE NEXT TO THE LAST BOARD.

FIGURE 73–17 Laying out the last board to fit against a finished corner.

between boards come up tightly. See Figure 64–15 on page 730 for methods of bringing edge joints of matched boards up tightly if they are slightly crooked. Severely warped boards should not be used. As installation progresses, check the panelling for plumb. If out of plumb, gradually bring back by driving one end of several boards a little tighter than the other end. Cut out openings in the same manner as described for sheet panelling.

Applying the Last Board

If the last board in the installation must fit snugly in the corner without a moulding, the layout should be planned so that the last board will be as wide as possible. If boards are a uniform width, the width of the starting board must be planned to avoid ending with a narrow strip. If random widths are used, they can be arranged when nearing the end.

Cut and fit the next to the last board. Then remove it. Tack the last board in the place of the next to the last board. Cut a scrap block about 6 inches (152 mm) long and equal in width to the finished face of the next to the last board. Use this block to scribe the last board by running one edge along the corner and holding a pencil against the other edge (**Figure 73–17**). Remove the board from the wall. Cut it to the scribed line. Fasten the next to the last board in position. Fasten the last board in position with the cut edge in the corner. Face-nail the edge nearest the corner.

Horizontal application of wood panelling is done in a similar manner. However, blocking between studs on open walls or furring strips on existing walls are not necessary. On existing walls, locate and snap lines to indicate the position of stud centrelines. The thickness of wood panelling should be at least ⅜ inch (9.5 mm) for 16-inch (406 mm) spacing of frame members and ⅝ inch (16 mm) for 24-inch (610 mm) spacing. Diagonal and pattern application of board panelling is similar to vertical and horizontal applications. If wainscotting is applied to a wall, the joint between the different materials may be treated in several ways (**Figure 73–18**).

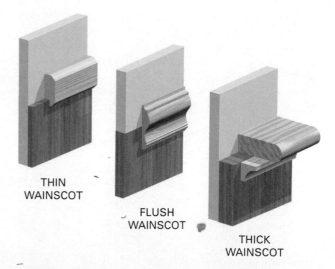

THIN WAINSCOT

FLUSH WAINSCOT

THICK WAINSCOT

FIGURE 73–18 Methods of finishing the joint at the top of wainscotting.

APPLYING PLASTIC LAMINATES

Plastic laminates are not usually applied to walls unless first prefabricated to sheets of plywood or similar material. They are then installed on walls in the same manner as sheet panelling. Special matching moulding is used between sheets and on interior and exterior corners. (The application of plastic laminates is described in greater detail in Unit 29 on kitchen cabinets and countertops.)

ESTIMATING PANELLING

Sheet Panelling

To determine the number of sheets of panelling needed, measure the perimeter of the room. Divide the perimeter by the width of the panels to be used. Deduct from this number any large openings such as doors, windows, or fireplaces. Deduct ⅔ of a panel for a door and ½ for a window or fireplace. Round off any remainder to the next highest number.

Board Panelling

Determine the square area to be covered by multiplying the perimeter by the height of each room. Deduct the area of any large openings. An additional percentage of the total area to be covered is needed because of the difference in the nominal size of lumber and its actual size.

Multiply the area to be covered by the area factor shown in **Figure 73–19**. Add 5 percent for waste in cutting. For example, the total area to be covered is 850 square feet, and 1 × 8 tongue-and-groove board panelling is to be used. Multiply 850 by 1.21, the sum of the coverage factor of 1.16 found in the table and 0.05 for waste in cutting. Round the answer of 1028.50 to 1029 for the number of board feet of panelling needed. To reduce waste in cutting, order suitable lengths.

Nominal Size	Width		Area Factor*
	Dress	Face	
Shiplap			
1 × 6	5½	5⅛	1.17
1 × 8	7¼	6⅞	1.16
1 × 10	9¼	8⅞	1.13
1 × 12	11¼	10⅞	1.10
Tongue-and-Groove			
1 × 4	3⅜	3⅛	1.28
1 × 6	5⅜	5⅛	1.17
1 × 8	7⅛	6⅞	1.16
1 × 10	9⅛	8⅞	1.13
1 × 12	11⅛	10⅞	1.10
S4S			
1 × 4	3½	3½	1.14
1 × 6	5½	5½	1.09
1 × 8	7¼	7¼	1.10
1 × 10	9¼	9¼	1.08
1 × 12	11¼	11¼	1.07
Panelling and Siding Patterns			
1 × 6	5⁷⁄₁₆	5¹⁄₁₆	1.19
1 × 8	7⅛	6¾	1.19
1 × 10	9⅛	8¾	1.14
1 × 12	11⅛	10¾	1.12

*Number multiplied by square feet to convert square feet to board feet.

FIGURE 73–19 Factors used to estimate amounts of board panelling.

Chapter 74

Ceramic Tile

Ceramic tile is used to cover floors and walls in washrooms, baths, showers, and other high-moisture areas that need to be cleaned easily and frequently (**Figure 74–1**). On large jobs, the tile is usually applied by specialists. On smaller jobs, it is sometimes more expedient for the general carpenter to install tile. Ceramic tile is usually set in place using thin set, a mortar-type adhesive, but is sometimes set using an organic adhesive. Thin set is used when tile is likely to get soaked with water, such as when used in showers. Organic adhesive is used when tile will not get soaked, such as on a kitchen counter backsplash.

DESCRIPTION OF CERAMIC TILE

Ceramic tiles are usually rectangular or square, but many geometric shapes, such as hexagons and octagons, are manufactured. Many solid colours,

FIGURE 74–1 Ceramic tile is used extensively in high-moisture areas. An acrylic base is used for the floor and aluminum trim finishes the edges.

Photo by M. Nauth

© chandlerphoto/Thinkstock

FIGURE 74–2 Special pieces are used to trim a ceramic tile installation.

patterns, designs, and sizes give a wide choice to achieve the desired wall effect.

The most commonly used tiles are nominal 4- and 6-inch (100 and 152 mm) squares, in ¼-inch (6 mm) thickness. Many other sizes are also available including 1- through 12-inch (25 through 305 mm) square tiles. Special pieces such as *base*, *caps*, *inside corners*, and *outside corners* are used to trim the installation (**Figure 74–2**).

For many tile installations today, there are special aluminum finishing trim pieces that are set as part of the tile installation job. These trim pieces can be mitred for both inside and outside corners of wall installations as well as for tile "baseboards." This type of trim allows the installer to use up many small pieces and to maintain the flow from floor to base to wall (see Figure 74–1).

WALL PREPARATION FOR TILE APPLICATION

Cement board is the recommended backing for ceramic tile when tile will become wet daily. Otherwise, water-resistant gypsum board may be used. Installation instructions for these products are given

in Chapters 70 and 71. Many new architectural drawings for commercial application are requiring glass-fibre reinforced tile backer board (see *http://www.buildgp.com/densshield-tilebacker-board#* and *http://www.certainteed.com/products/gypsum/board/glass-mat/340924* for product specifications and installation requirements).

Minimum Backer and Tile Area in Baths and Showers

Tiles should overlap the lip and be applied down to the shower floor or top edge of the bathtub. On tubs without a showerhead, they should be installed to extend to a minimum of 6 inches (152 mm) above the rim. Around bathtubs with no showerheads, tiles should extend a minimum of 15¾″ (400 mm) (**Figure 74–3**). Around bathtubs with showerheads, tiles should extend a minimum of 3′ 11″ (1.2 m) above the rim or 6 inches (1.8 mm) above the showerhead, whichever is higher.

In shower stalls, tiles should be a minimum of 5′ 11″ (1.8 m) above the shower floor or 6 inches (152 mm) above the showerhead, whichever is higher. A 4-inch (100 mm) minimum extension of

Photo by M. Nauth

FIGURE 74–3 Ceramic tile used as a backsplash around a bathtub.

© Bryngelzon/Thinkstock

FIGURE 74–4 Border tiles should be greater than ½ a tile width and equal on both sides of the wall.

the full height is recommended beyond the outside face of the tub or shower.

Calculating Border Tiles

Before beginning the application of ceramic wall tile, the width of the *border tiles* must be calculated. Border tiles are those that fit to the corners and edges of the tiled area. The installation has a professional appearance if the border tiles are the same width and are greater than one-half of a tile width (**Figure 74–4**).

Measure the width of the wall from corner to corner. Divide it by the width of a tile. Measure the tile accurately. Sometimes, the actual size of a tile is different than its nominal size. Add the width of a tile to the remainder. Divide by 2 to find the width of border tiles.

> **EXAMPLE** A wall section measures 8′-4″, or 100 inches from wall to wall. If 4¼-inch, actual size, tiles are used, dividing 100 by 4¼ equals 23 full tiles with 0.53 of a tile remainder. Add 1 to the decimal remainder and multiply it by the width of a tile, or 1.53 × 4.25 = 6.5 inches. Divide this by 2 to get 3¼ inches. This will give 22 full tiles and two 3¼-inch border tiles across the tiled area.

> **METRIC EXAMPLE** A wall section measures 2540 mm from wall to wall. If the actual size of the tile is 108 mm, dividing 2540 by 108 equals 23.52, which is 23 full tiles plus 0.52 of a tile. Adding 1 to 0.52 gives 1.52, dividing by 2 gives 0.76 of a tile border or (0.76 × 108 mm) an 82 mm border. This will give 22 full tiles and an 82 mm border on either side.

Tile Layout

A layout line needs to be placed as a guide for the individual tiles. Measure out from the corner and mark a point that will be the edge of a full tile. If the section to be tiled is a wall, plumb a line from top to bottom. If the section is a floor, measure and mark another point from the other corner. Snap a line through these points.

For wall applications, check the level of the floor. If the floor is level, tiles may be applied by placing the bottom edge of the first row on the floor using plain tile or base tile. If the floor is not level, the bottom edge of the tiles must be cut to fit the floor while keeping the top edges level. Place a level on the floor and find the low point of the tile installation. From this point, measure up and mark on the wall the height of a tile. Draw a level line on the wall through the mark. Tack a straightedge on the wall with its top edge to the line. Tiles are then laid to the straightedge. When tiling is completed above, the straightedge is removed. The tiles in the bottom row are then cut and fitted to the floor.

For floor applications of tile, the subfloor has to be rigid enough to withstand flexing under load and thereby cracking the grout lines or even the tiles. This can be done in several ways: a double layer of subfloor panels (min. thickness of 1¼″ or 31 mm), cement tile backer board, or the application of uncoupling membranes such as Schluter-DITRA (**Figure 74–5**). These membranes are available in various thicknesses that allow a perfect level match with adjoining floors and that allow the installer to place the tile on top of single-layer subfloor on 24-inch (610 mm) centres. Electric heat pads can be placed under the tiles to create a warm heated floor. (For more information, see *www.schluter.com.*)

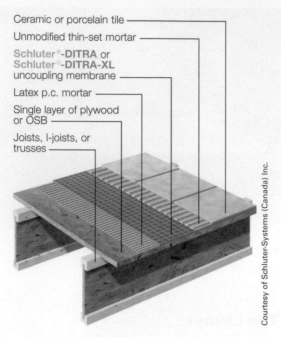

Ceramic or porcelain tile
Unmodified thin-set mortar
Schluter®-**DITRA** or Schluter®-**DITRA-XL** uncoupling membrane
Latex p.c. mortar
Single layer of plywood or OSB
Joists, I-joists, or trusses

Courtesy of Schluter-Systems (Canada) Inc.

FIGURE 74–5 Application of an uncoupling membrane (Schluter-DITRA).

Before laying the tile, check that the line is parallel to the opposite wall. Adjust the line to split any differences. This line is best placed where the tile will be most often viewed from a distance. This will ensure that the tiles are as straight as possible where they are most visible.

TILE APPLICATION

When tiling walls such as a shower, it usually is best to install tile on the back wall first. This way, the joint in the corner is the least visible and the most watertight. Apply the recommended adhesive to the wall or floor with a trowel. Use a flat trowel with grooved edges that allows the recommended amount of adhesive to remain on the surface (**Figure 74–6**). Too heavy a coat results in adhesive being squeezed out between the joints of applied tile, causing a mess. Too little adhesive results in failure of tiles to adhere to the wall. Follow the manufacturer's directions for the type of trowel to use and the amount of adhesive to be spread at any one time. Be careful not to spread adhesive beyond the area to be covered.

Apply the first tile to the guide line. Press the tile firmly into the adhesive. Apply other tiles in the same manner. Start from the centre guide line and pyramid upward and outward. As tiles are applied, slight adjustments may need to be made to keep them lined up. Tile spacers may be used to help keep tile properly spaced (**Figure 74–7**). Tile spacers are

FIGURE 74–6 Use a trowel that is properly grooved to allow the correct amount of adhesive to remain on the surface.

FIGURE 74–7 Rubber tile spacers may be used to maintain a uniform grout spacing.

rubber pieces with the same dimensions as the joints between tiles. Keep fingers clean and adhesive off the face of the tiles. Clean tiles with a damp cloth.

Cutting Border Tiles

After all *field tiles* are applied, it is necessary to cut and apply border tiles. Field tiles are whole tiles that are applied in the centre of the wall. Ceramic tile may be cut in any of several ways. Tile saws can be used to cut all types of tile (**Figure 74–8**). They are operated in a manner similar to that of a power mitre box except that the material is eased into the blade on a sliding tray. Water is pumped from the reservoir below to the blade, keeping it cool during the cutting process. Thin ceramic tile is often cut using a hand cutter (**Figure 74–9**). First the tool scores the tile. The

FIGURE 74–8 Tile cutters are often used for cutting ceramic tile.

FIGURE 74–10 Nibblers are used to cut curves.

FIGURE 74–9 Hand cutter for thin ceramic tile.

FIGURE 74–11 Special trim pieces are often used to add accents to a tiled surface.

back edge of the cutter is then pressed against the tile to break it. This tool cannot make small or narrow cuts. A nibbler is used to make small or irregular cuts (**Figure 74–10**). Nibblers chip small pieces off in sometimes random directions. Care should be taken to cut only small pieces at a time. This will allow a successful cut to be made gradually.

To finish the edges and ends of the installation, *caps* are sometimes used. Caps may be 2 × 6 or 4 × 4 (50 × 152 or 100 × 100 mm) pieces with one rounded finished edge. Special trim pieces are used to finish interior and exterior corners (**Figure 74–11**).

Grouting Tile Joints

After all tile has been applied and the glue has set, the joints are filled with **tile grout**. Grout comes in a

powder form. It is mixed with water to form a paste of the desired consistency. It is spread over the face of the tile with a *rubber trowel* (or float) to fill the joints. The grout is worked into the joints. Then, the surface is wiped as clean as possible with the trowel.

Unsanded grout is used for joints between 1/16″ to 1/8″ (1.5–3 mm) and sanded grout is used for larger joints, 1/8″ to 5/8″ (3–16 mm). Epoxy grouts (2 part) are used in high traffic areas such as commercial kitchens and meat plants where waterproofing and steam cleaning are essential. Because they are mould and mildew resistant they are also used for countertops.

After floor tile grout has set up slightly (but not hardened), the excess is removed. Wipe the tile across the grout lines at a 45-degree angle with a damp sponge. Wipe once, then turn the sponge over and wipe another area. Rinse and repeat. The key is to keep the sponge clean for one wiping at a time. Let it set up more and repeat. Finish the grouting by buffing with a dry clean rag.

After the grout has set and cured for several days, silicone grout sealer may be applied. This product seals tiny pores in the grout, making it more resistant to staining. Sealer is liberally brushed on the grout, allowing it to soak in. The excess is wiped clean before it dries.

Other Types of Tile

Besides ceramic, there are tiles that are made of porcelain, slate, glass, marble, and metal. Large tiles require an absolutely flat substrate. For installation procedures contact the product supplier.

ESTIMATING TILE

First, determine the square foot area to be covered for the amount of tile to order. To find the number of tiles needed, multiply the area covered by the number of tiles in one square foot.

EXAMPLE 4×4 tiles are being used to cover 120 square feet of surface. Each 4×4 tile = 16 square inches. Then one square foot or 144 square inches is divided by 16, equalling 9 tiles needed to cover 1 square foot. Thus, if 120 square feet are to be covered, then 120 times 9 equals 1080 tiles.

METRIC EXAMPLE 100×100 mm tiles are being used to cover 11.4 m^2 of surface. The area of 1 tile in metres squared is 0.1 m $\times$ 0.1 m or 0.01 m^2. To find the number of tiles, simply divide the surface area by the area of each tile to get 1140 tiles ($11.4 \div 0.01 = 1140$).

The number of straight pieces of cap is found by determining the total linear feet to be covered and dividing by the length of the cap. The number of interior and exterior corners is determined by counting from a layout of the installation.

KEY TERMS

board panelling	hardwood	sheet panelling	tile grout
ceramic tile	particleboard	softwood	wainscotting
hardboard	plastic laminate	thin set	

REVIEW QUESTIONS

Select the most appropriate answer.

1. On prefinished plywood panelling that is scored to simulate boards, how far from the edge are some scores always placed?

   a. 12 and 16 inches (305 and 406 mm)
   b. 16 and 20 inches (406 and 508 mm)
   c. 16 and 24 inches (406 and 610 mm)
   d. 16 and 32 inches (406 and 813 mm)

2. How is the wall finish called wainscotting applied?

   a. diagonally
   b. partway up the wall
   c. as a coating on prefinished wall panels
   d. around tubs and showers

3. For most parts of Canada, about what moisture content should wood used for interior finish be dried to?

   a. 8 percent
   b. 12 percent
   c. 15 percent
   d. 20 percent

4. What is the thickest plastic laminate described in this unit?

   a. vertical type
   b. regular type
   c. backer type
   d. all-purpose type

5. If ¾-inch (19 mm) thick wood panelling is to be applied vertically over open studs, at what intervals must wood blocking be provided between studs for nailing?

   a. 16 inches (406 mm)
   b. 24 inches (610 mm)
   c. 32 inches (813 mm)
   d. 48 inches (1220 mm)

6. How should board panelling be applied vertically to an existing wall?

   a. nail panelling to existing studs
   b. apply horizontal furring strips
   c. remove the wall covering and install blocking between studs
   d. use adhesives

7. What type of panel backing is best for use behind ceramic tile in showers?

   a. waterproof drywall
   b. water-resistant gypsum board
   c. waterproof cement board
   d. exterior-grade plywood

8. What tool is used to cut a partial circle in the side of a ceramic tile?

   a. hole saw
   b. hand cutter
   c. ceramic tile saw
   d. nibbler

9. What is the minimum that bathroom tile should extend over the tops of showerheads?

   a. 4 inches (100 mm)
   b. 6 inches (152 mm)
   c. 8 inches (203 mm)
   d. 12 inches (305 mm)

10. How many 4 × 4 ceramic tiles are needed to cover 150 square feet of wall area?

    a. 1674
    b. 1500
    c. 1350
    d. 1152

11. How many 100 × 100 mm tiles are needed to cover 13.6 m² of wall?

    a. 1420
    b. 1360
    c. 1300
    d. 1200

Unit 24
Ceiling Finish

Chapter 75 Suspended Ceilings   **Chapter 76** Ceiling Tile

Inexpensive and highly attractive ceilings can be created by installing suspended ceilings or ceiling tiles. They can be installed in new construction beneath exposed joists or when remodelling, below existing ceilings.

Suspended or tile ceiling finish is installed in sections and pieces. The layout of the pattern and border tiles is important for the ceiling to look professionally installed.

Chapter 75

Suspended Ceilings

Suspended ceilings are widely used in commercial and residential construction as a ceiling finish. They also provide space for recessed lighting, ducts, pipes, and other necessary conduits (**Figure 75–1**). Besides improving the appearance of a room, a suspended ceiling conserves energy by increasing the insulating value of the ceiling. It also aids in controlling sound transmission. In remodelling work, a suspended system can be easily installed beneath an existing ceiling. In basements, where overhead pipes and ducts may make other types of ceiling application difficult, a suspended type is easily installed. In addition, removable panels make pipes, ducts, and wiring accessible.

Courtesy of Armstrong World Industries, Inc.

FIGURE 75–1 Installing panels in a suspended ceiling grid.

SUSPENDED CEILING COMPONENTS

A suspended ceiling system consists of panels that are laid into a metal grid. The grid consists of **main runners**, **cross tees**, and **wall angles**. It is constructed in a 2-foot × 4-foot (610 × 1220 mm) rectangular or 2-foot × 2-foot (610 × 610 mm) square pattern (**Figure 75–2**). Grid members come pre-finished in white, black, brass, chrome, and wood grain patterns, among others.

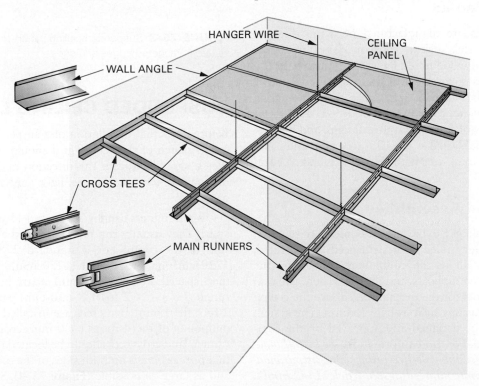

HANGER WIRE

CEILING PANEL

WALL ANGLE

CROSS TEES

MAIN RUNNERS

FIGURE 75–2 A suspended ceiling consists of grid members and ceiling panels.

Wall Angles

Wall angles (also called wall mould) are L-shaped pieces that are fastened to the wall to support the ends of main runners and cross tees. They come in 10- and 12-foot (3 and 3.6 m) lengths. They provide a continuous finished edge around the perimeter of the ceiling where it meets the wall.

Main Runners

Main runners or tees are shaped in the form of an upside-down T, giving the system the common use name of "T-bar ceiling." They come in 12-foot (3.6 m) lengths. End splices make it possible to join lengths of main runners together. Slots are punched in the side of the runner at 12-inch (305 mm) intervals to receive cross tees. Along the top edge, punched holes are spaced at intervals for suspending main runners with *hanger wire*. Main runners extend from wall to wall. They are the primary support of a ceiling's weight.

Cross Tees

Cross tees come in 2- and 4-foot (610 and 1220 mm) lengths. A slot, of similar shape and size as those in main runners, is centred on the 4-foot (1220 mm) cross tees for use when turning a 2 × 4 (610 × 1220 mm) grid into a 2 × 2 (610 × 610 mm) grid. They come with connecting tabs on each end. These tabs are inserted and locked into main runners and other cross tees.

Ceiling Panels

Ceiling panels are manufactured of many different kinds of material, such as gypsum, glass fibres, mineral fibres, and wood fibres. Panel selection is based on considerations such as fire resistance, sound control, thermal insulation, light reflectance, moisture resistance, maintenance, appearance, and cost. Panels are given a variety of surface textures, designs, and finishes. They are available in 2 × 2 (610 × 610 mm) and 2 × 4 (610 × 1220 mm) sizes with square or rabbeted edges (**Figure 75–3**).

Fire-Rated Assemblies

Suspended ceilings can be made to conform to required fire ratings, but the entire system needs to be rated, which means all of the components, including the hanger wire, the supports, and the tiles. There will also be requirements for the mechanicals above the ceiling and for protrusions such as light fixtures. The system will have to be designed and approved by the engineers involved. For more information, see *http://www.ceilume.com/ceiling-tile-fire-ratings.cfm*, *http://www.armstrong.com/commceilingsna/article21342.html#*, and *http://www.certainteed.com/products/ceilings/by-material/suspension-systems/344264*.

Courtesy of Armstrong Commercial Ceilings and Walls

FIGURE 75–3 Suspended ceiling panels may have square or rabbeted edges and ends.

SUSPENDED CEILING LAYOUT

Before the actual installation of a suspended ceiling, a scaled sketch of the ceiling grid should be made. The sketch should indicate the direction and location of the main runners, cross tees, light panels, and border panels.

Main runners usually are spaced 4 feet (1220 mm) apart. They usually run parallel with the long dimension of the room. For a standard 2 × 4 (610 × 1220 mm) pattern, 4-foot (1220 mm) cross tees are then spaced 2 feet (610 mm) apart between main runners. If a 2 × 2 (610 × 610 mm) pattern is used, 2-foot (610 mm) cross tees are installed between the midpoints of the 4-foot (1220 mm) cross tees. Main runners and cross tees should be located in such a way that *border panels* on both sides of the room are equal and as large as possible (**Figure 75–4**). Sketching the ceiling layout also helps when estimating materials, especially for specialized ceiling layouts (**Figure 75–5**).

FIGURE 75–4 A typical layout for a suspended ceiling grid.

FIGURE 75–5 TechZone™ ceiling system.

Courtesy of Armstrong Commercial Ceilings and Walls

Sketching the Layout

Sketch a grid plan by first drawing the overall size of the ceiling to a convenient scale. Use special care in measuring around irregular walls.

Locating Main Runners. To locate main runners, change the width of room to inches and divide by 48. Add 48 inches to any remainder. Divide the sum by 2 to find the distance from the wall to the first main runner. This distance is also the length of border panels.

For example, if the width of the room is 15'-8", changing to inches equals 188. Dividing 188 by 48 equals 3, with a remainder of 44 inches. Adding 48 to 44 equals 92 inches. Dividing 92 by 2 equals 46 inches, the distance from the wall to the first main runner.

Draw a main runner the calculated distance from, and parallel to, the long dimension of the

ceiling. Draw the rest of the main runners parallel to the first, and at 4-foot (1220 mm) intervals. The distance between the last main runner and the wall should be the same as the distance between the first main runner and the opposite wall (**Figure 75–6**).

Locating Cross Tees. To locate 4-foot cross tees between main runners, first change the long dimension of the ceiling to inches. Divide by 24. Add 24 to the remainder. Divide the sum by 2 to find the width of the border panels on the other walls.

For example, if the long dimension of the room is 27'-10", changing it to inches equals 334. Dividing 334 by 24 equals 13, with a remainder of 22 inches. Adding 24 to 22 equals 46 inches. Dividing 46 by 2 equals 23 inches, the distance from the wall to the first row of cross tees.

In metric, if the length of the room is 8484 mm, divide by 610 mm to get 13.9082. Add 1 to the decimal (1.9082) and divide by 2 to get 0.954098. Multiply this number by 610 to get a distance of 582 mm to set the first 1220 cross tee.

Draw the first row of cross tees the calculated distance from, and parallel to, the short wall. Draw the remaining rows of cross tees parallel to each other at 2-foot (610 mm) intervals. The distance from the last row of cross tees to the wall should be the same as the distance from the first row of cross tees to the opposite wall (**Figure 75–7**). There are other acceptable ways to lay out this ceiling; it mostly depends on room usage and in-track lighting (**Figure 75–8**).

CONSTRUCTING THE CEILING GRID

The ceiling grid is constructed by first installing *wall angles*, then installing *suspended ceiling lags* and *hanger wires*, suspending the *main runners*, inserting full-length *cross tees*, and finally, cutting and inserting *border* cross tees.

A suspended ceiling must be installed with at least 3 inches (75 mm) for clearance below the lowest air duct, pipe, or beam. This clearance provides enough room to insert ceiling panels in the grid. If recessed lighting is to be used, allow a minimum of 6 inches (152 mm) for clearance. The height of the ceiling may be located by measuring up from the floor. If the floor is rough or out of level, the ceiling line may be located with various levelling devices previously described. A combination of a hand level and straightedge, or a water level, builders' level, transit-level, or laser level, can be used (**Figure 75–9**). Snap chalk lines on all walls around the room to the height of the top edge of the wall angle.

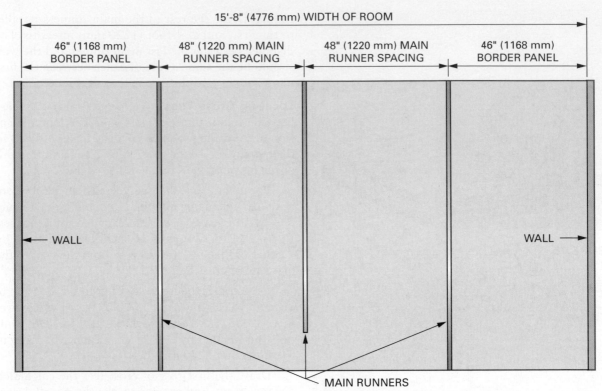

1. CHANGE ROOM WIDTH DIMENSION TO INCHES:
   15 X 12 + 8 = 188"
2. DIVIDE 188 BY 48 = 3.91667 TILES
3. ADD 1 TO REMAINDER = 1.91667 TILES
4. DIVIDE THIS NUMBER BY 2: 1.91667 ÷ 2 = 0.95833
5. MULTIPLY FRACTION OF A TILE BY 48:
   0.95833 X 48 = 46" BORDER TILE LENGTH

1. ROOM WIDTH DIMENSION IS 4776 mm.
2. DIVIDE 4776 BY 1220 = 3.914754 TILES
3. ADD 1 TO THE DECIMAL = 1.914754
4. DIVIDE 1.914754 BY 2 = 0.957377
5. MULTIPLY THE ANSWER BY 1220 = 1168 mm

FIGURE 75–6 Method of determining the location of main runners.

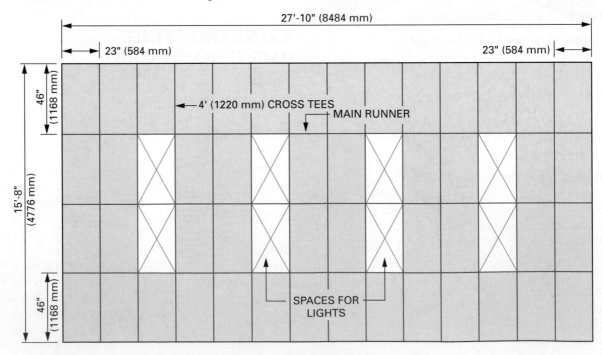

FIGURE 75–7 Completed sketch of a suspended ceiling layout.

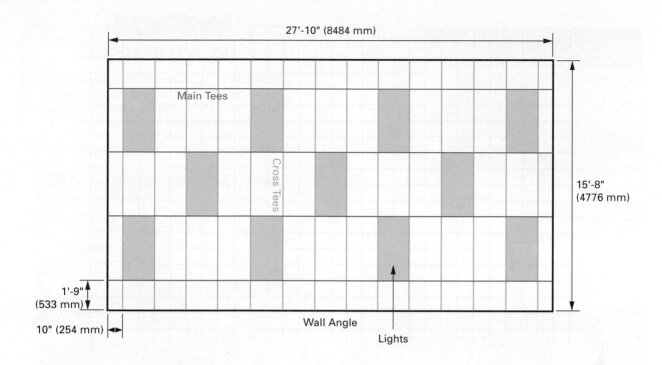

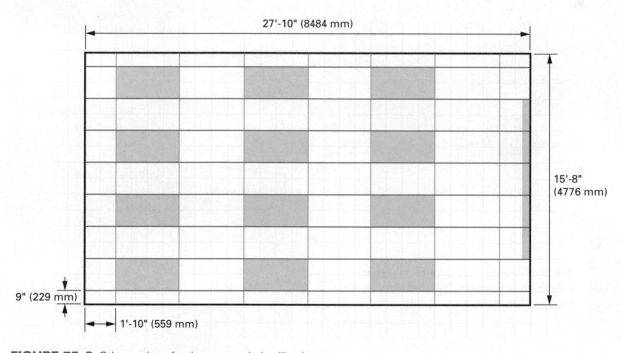

FIGURE 75–8 Other options for the suspended ceiling layout.

The visible line laser level eliminates the need for chalk lines and string lines. Most lasers are self-levelling, and they are attached to the wall angle with a bracket. The bracket can be reversed and adjusted so that the visible line hits the hanger wires at the spot that they need to be bent. The wires are bent at 90 degrees and the main "T" runners are hung on them. Alternatively, the laser can be hung below the wall angle, a sensor is held at each fastening point, and the track is raised or lowered as required.

Fasten wall angles around the room with their top edge lined up with the chalk line. It may be easier to fasten the wall angle by pre-punching holes with a centre punch or spike. Fasten into framing wherever possible, not more than 24 inches (610 mm) apart (**Figure 75–10**).

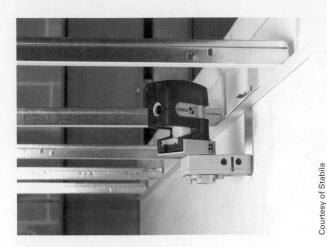

FIGURE 75–9 A laser level may be used to install a suspended ceiling's main runners.

Courtesy of Stabila

FIGURE 75–10 Installing wall angles.

CAUTION

Use care in handling the cut ends of wall angle and other grid members. Cut metal ends are very sharp and can cause serious injury.

To fasten wall angles to concrete walls, short masonry nails sometimes are used. However, they are difficult to hold and drive. Use a small strip of cardboard to hold the nail while driving it with the hammer (**Figure 75–11**). Lead or plastic inserts and screws may also be used to fasten the wall angles. Their use does require more time. If available, power nailers can be used for efficient fastening of wall angles to masonry walls.

Make mitre joints on outside corners. Make butt (or lapped) joints in interior corners and between straight lengths of wall angle (**Figure 75–12**). Use a

ON THE JOB

To fasten wall angles to concrete walls, short masonry nails sometimes are used. However, they are difficult to hold and drive. Use a small strip of cardboard to hold the nail while driving it with the hammer.

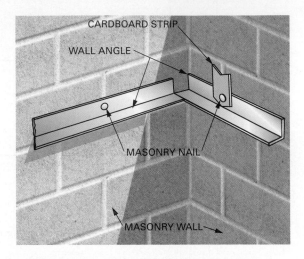

CARDBOARD STRIP

WALL ANGLE

MASONRY NAIL

MASONRY WALL

FIGURE 75–11 Technique for driving short nails.

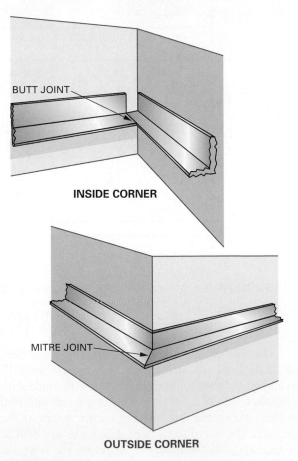

BUTT JOINT

INSIDE CORNER

MITRE JOINT

OUTSIDE CORNER

FIGURE 75–12 Methods of joining wall angle at corners.

combination square to lay out and draw the square and angled lines. Cut carefully along the lines with snips.

Installing Hanger Lags

From the ceiling sketch, determine the position of the first main runner. Stretch a line at this location across the room from the top edges of the wall angle. Stretch the line tightly on nails inserted between the wall and wall angle (**Figure 75–13**). The line serves as a guide for installing *hanger lags* or *screw eyes* and *hanger wires* from which main runners are suspended.

Install hanger lags not over 4 feet (1.2 m) apart and directly over the stretched line. Hanger lags should be of the type commonly used for suspended ceilings. They must be long enough to penetrate wood joists a minimum of 1 inch (25 mm) to provide strong support. *Eye pins* are driven into concrete with a *powder-actuated* fastening tool (see Figure 16–18, page 127). Hanger wires may also be attached directly around the lower chord of bar joists or trusses.

Installing Hanger Wire

Cut a number of hanger wires using wire cutters. The wires should be about 12 inches (305 mm) longer than the distance between the overhead construction and the stretched line. For residential work, 16-gauge wire is usually used. For commercial work, 12-gauge and heavier wire is used.

Attach the hanger wires to the hanger lags. Insert about 6 inches (152 mm) of the wire through the screw eye. Securely wrap the wire around itself three

ON THE JOB

Stretch lines for main runners on nails inserted between the wall and wall angle.

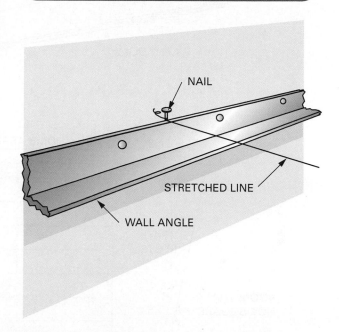

FIGURE 75–13 Technique for stretching a line between wall angles.

times. Pull on each wire to remove any kinks. Then make a 90-degree bend where it crosses the stretched line (**Figure 75–14**). Stretch lines, install hanger lags, and attach and bend hanger wires in the same

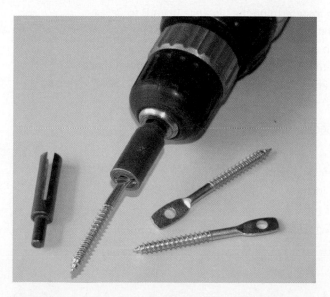

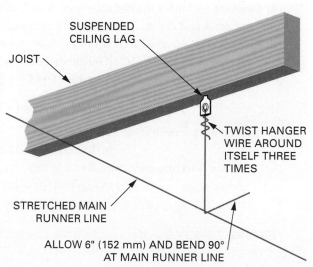

FIGURE 75–14 Suspended ceiling lags are used to support hanger wire. Wire may be pre-bent to accept main runners.

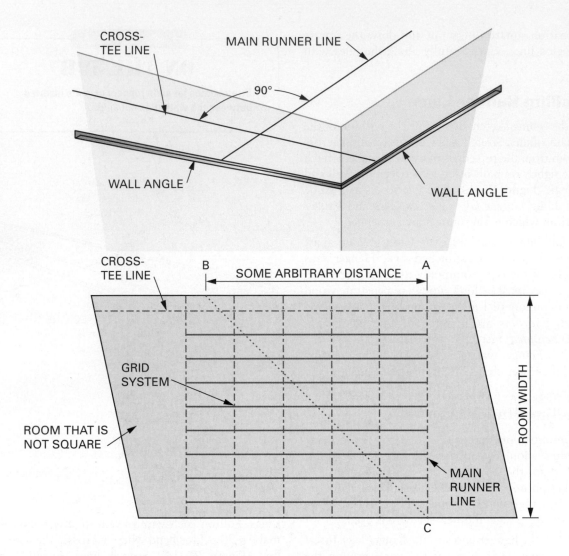

1) Starting at point A, measure the room width, keeping your tape as square as possible with the wall.
2) From A, measure some distance to B. The actual distance does not matter; it merely needs to be large enough to make a big triangle.
3) Use these numbers in the Pythagorean theorem to determine the distance from B to C.
4) Measure and mark the distance from B to C and mark C. C is now square with point A.
5) Connect A and C with a string and measure each successive row of main runner from it.
6) **EXAMPLE:** If AC, the room width, is 18'-9", then measure AB to be, say, 16'-0". Convert these dimensions to inches. 16'-0" becomes 192" and 18'-9" becomes 225" ($18 \times 12 + 9 = 225$"). Put these dimensions in the Pythagorean theorem. $C^2 = A^2 + B^2$:

$$C = \sqrt{192^2 + 225^2} = \sqrt{36864 + 50625} = \sqrt{87489} = 295.7853952"$$

To convert the decimal to sixteenths:

$$0.7853952 \times 16 = 12.566 \text{ sixteenths} \Rightarrow {}^{13}/_{16}"$$

Thus, the measurement from B to C is 295 $^{13}/_{16}$ inches.

In metric units, let AC = 5715 mm and AB = 4877 mm; then:

$$BC = \sqrt{5715^2 + 4877^2} = \sqrt{32661225 + 23785129} = \sqrt{56446354} = 7513 \text{ mm}$$

FIGURE 75–15 Method of stretching two perpendicular lines using the Pythagorean theorem.

manner at each main runner location. Leave the last line stretched tightly in position.

CAUTION

Wear eye protection when installing bent wire hangers. During installation they are often at the same elevation as your eyes.

Installing Main Runners

The ends of the main runners rest on the wall angles. They must be cut so that a cross-tee slot in the web of the runner lines up with the first row of cross tees. A cross-tee line must be stretched, at wall angle height, across the short dimension of the room to line up the slots in the main runners. The line must run exactly at right angles to the main runner line and at a distance from the wall equal to the width of the border panels. If the walls are at right angles to each other, the location of the cross-tee line can be determined by measuring out from both ends of the wall.

When the walls are not at right angles, the Pythagorean theorem (see Figure 28–6, page 285) is used to square the grid system (**Figure 75–15**). After the main runner line is installed, measure out from the short wall, along the stretched main runner line, a distance equal to the width of the border panel. Mark the line. Stretch the cross-tee line through this mark and at right angles to the main runner line.

At each main runner location, measure from the short wall to the stretched cross-tee line. Transfer the measurement to the main runner. Measure from the first cross-tee slot beyond the measurement, so as to cut as little as possible from the end of the main runner (**Procedure 75–A**). Cut the main runners about ⅛ inch (3 mm) less to allow for the thickness of the wall angle. Backcut the web slightly for easier installation at the wall. Measure and cut main runners individually. Do not use the first one as a pattern to cut the rest.

Hang the main runners by resting the cut end on the wall angle and inserting suspension wires in the appropriate holes in the top of the main runner. Bring the runners up level and bend the wires (**Figure 75–16**). Twist the wires with at least three turns to hold the main runners securely.

More than one length of main runner may be needed to reach the opposite wall. Connect lengths of main runners together by inserting tabs into matching ends. Make sure end joints come up tight. The length of the last section is measured from the end of the last one installed to the opposite wall, allowing about ⅛ inch (3 mm) less for the thickness of the wall angle.

Installing Cross Tees

Cross tees are installed by inserting the tabs on the ends into the slots in the main runners. These fit into position easily, although the method of attaching varies from one manufacturer to another.

Install all full-length cross tees between main runners first. Lay in a few full-size ceiling panels. This stabilizes the grid while installing cross tees for border panels. Cut and install cross tees along the border. Insert the connecting tab of one end in the main runner and rest the cut end on the wall angle (**Figure 75–17**). If the walls are not straight or square, it is necessary to cut cross tees for border tiles individually. For 2 × 2 (610 × 610 mm) panels, install 2-foot (610 mm) cross tees at the midpoints of the 4-foot (1220 mm) cross tees. After the grid is complete, sight sections by eye. Straighten where necessary by making adjustments to border cross tees or hanger wires.

Installing Ceiling Panels

Ceiling panels are placed in position by tilting them slightly, lifting them above the grid, and letting them fall into place (**Figure 75–18**). When handling panels, be careful to avoid marring the finished surface. Install all border panels first. Then install full-sized field panels. Measure each border panel individually, if necessary. Cut them slightly smaller than measured so that they can drop into place easily. Cut the panels with a sharp utility knife using a straightedge as a guide. A scrap piece of cross-tee material can be used as a straightedge. Always cut with the finished side of the panel up.

Cutting Ceiling Panels around Columns

When a column is near the centre of a ceiling panel, cut the panel at the midpoint of the column. Cut semicircles from the cut edge to the size required for the panel pieces to fit snugly around the column. After the two pieces are rejoined around the column, glue scrap pieces of panel material to the back of the installed panel.

If the column is close to the edge or end of a panel, cut the panel from the nearest edge or end to

Procedure 75–A Cutting Main Runners So Cross-Tee Slots Align

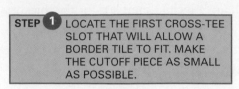

STEP ❶ LOCATE THE FIRST CROSS-TEE SLOT THAT WILL ALLOW A BORDER TILE TO FIT. MAKE THE CUTOFF PIECE AS SMALL AS POSSIBLE.

STEP ❷ MEASURE BACK WIDTH OF BORDER PANEL.

STEP ❸ CUT MAIN RUNNER HERE.

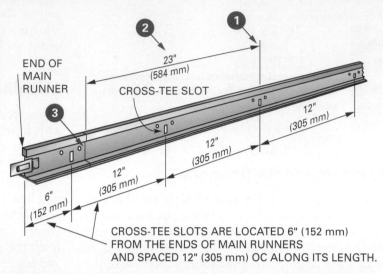

END OF MAIN RUNNER

23" (584 mm)

CROSS-TEE SLOT

12" (305 mm)

12" (305 mm)

12" (305 mm)

6" (152 mm)

CROSS-TEE SLOTS ARE LOCATED 6" (152 mm) FROM THE ENDS OF MAIN RUNNERS AND SPACED 12" (305 mm) OC ALONG ITS LENGTH.

Courtesy of Trimble Navigation Limited

FIGURE 75–16 Method of adjusting the main runner level.

FIGURE 75–18 Installing ceiling panels.

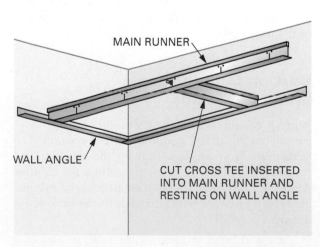

MAIN RUNNER

WALL ANGLE

CUT CROSS TEE INSERTED INTO MAIN RUNNER AND RESTING ON WALL ANGLE

FIGURE 75–17 Inserting a cross tee in the main runner.

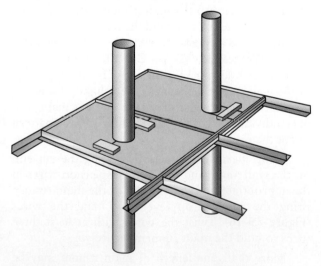

FIGURE 75–19 Fitting ceiling panels around columns.

fit around the column. The small piece is also fitted around the column and joined to the panel by gluing scrap pieces to its back side (**Figure 75–19**).

LINEAR METAL CEILING SYSTEM

The linear metal ceiling system is used in commercial work to provide an architecturally attractive ceiling. It can be formed into flat or curved ceilings (**Figure 75–20**). Main tees are installed in a similar manner as for the drop ceiling. The cross tees are placed 4 feet (1.2 m) on centre or closer as needed to keep the main tees secure.

The ceiling panels are 3¼ or 7¼ inches wide (80 or 184 mm) and 12 feet long (3.6 m) metal pans. They are shaped to snap into place leaving a ¾-inch (19 mm) space between adjacent panels. They attach to the main tee that has special clips to hold each panel (**Figure 75–21**).

There are other propriety systems, such as SNAPCLIC, which are suspended ceiling systems that are using materials other than metal and softboard (see *snapclipsystem.com*).

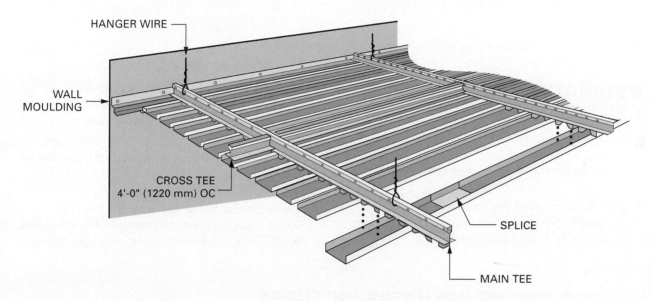

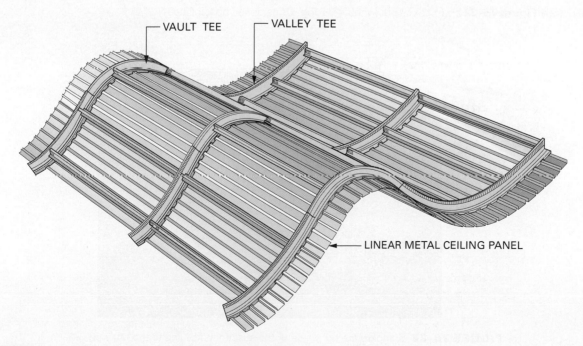

FIGURE 75–20 Linear metal ceilings may be flat or follow curved layouts.

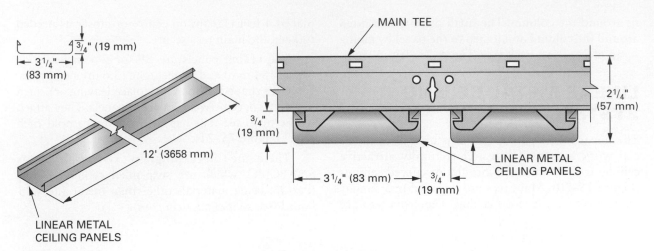

FIGURE 75–21 Linear panels are attached to the main tees by clips in the main tee.

ESTIMATING SUSPENDED CEILING MATERIALS

- Divide the perimeter of the room (in feet or metres) by 10 or 12 feet (3 or 3.6 m) in order to determine the number of pieces of wall angle required.

- Find the number of main runners needed from the sketch. No more than two pieces can be cut from one 12-foot (3660 mm) length.

- Count the number of 2- and 4-foot (610 and 1220 mm) cross tees from the sketch. Border cross tees must be cut from full-length tees.

- From the sketch, count the number of hanger wires and screw eyes needed. Multiply the number needed by the length of each hanger wire to find the total linear feet of hanger wire needed.

- Count the number of ceiling panels from the sketch. Each border panel requires a full-size ceiling panel. There will be one less for each light fixture.

WHAT'S WRONG WITH THIS PICTURE?

Carefully study **Figure 75–22** and think about what is wrong. Consider all possibilities.

FIGURE 75–22 Compare the right side of the suspended ceiling to the left side. The border tiles should be the same size and wider than a half tile. Layout centring a full tile instead of a main runner would look better. Also, the light is off centre.

FIGURE 75–23 This ceiling has been laid out with equal sized border tiles. The borders are also wider than one-half a tile width. This also centres the light in the room.

Chapter 76

Ceiling Tile

Ceiling tile is usually stapled to furring strips that are fastened to exposed joists. They may also be cemented to existing ceilings, provided the ceilings are solid and level. If the existing ceiling is not sound, furring strips should be installed and fastened through the ceiling into the joists above (**Figure 76–1**).

DESCRIPTION OF CEILING TILES

Most ceiling tiles are made of wood fibre or mineral fibre. Wood fibre tiles are lowest in cost and are adequate for many installations. Mineral fibre tiles are used when a more fire-resistant type is required.

FIGURE 76–1 Installing ceiling tile on wood furring strips.

Courtesy of Armstrong World Industries, Inc.

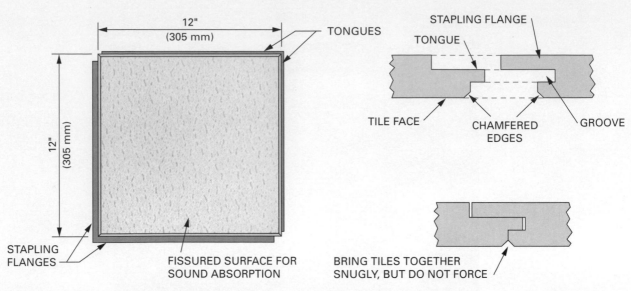

FIGURE 76–2 A typical ceiling tile has tongue-and-grooved edges with stapling flanges.

Manufacture

Wood fibres are pressed into large sheets that are $7/16$ to ¾ inch (11 to 19 mm) thick. Mineral fibre tiles are made of rock that is heated to a molten state. The fibres are then sprayed into a sheet form. The surfaces of some sheets are **fissured** or **perforated** for sound absorption. The surfaces of other sheets are embossed with different designs or left smooth. Then they are given a factory finish and cut into individual tiles. Most tiles are cut with **chamfered**, *tongue-and-grooved* edges with two adjacent *stapling flanges* for concealed fastening (**Figure 76–2**).

Sizes

The most popular sizes of ceiling tile are a 12-inch (3050 mm) square and a 12- × 24-inch (305 × 610 mm) rectangle in thicknesses of ½ inch (12 mm). Tiles are also manufactured in squares of 16 inches (406 mm) and in rectangles of 16 × 32 inches (406 × 813 mm).

CEILING TILE LAYOUT

Before installation begins, it is necessary to calculate the size of *border tiles* that run along the walls. It is desirable for border tiles to be as wide as possible and of equal widths on opposite walls.

Calculating Border Tile Sizes

To find the width of the border tiles along the long walls of a room, first determine the dimension of the short wall. In most cases, the measurement will be a number of full feet, plus a few inches. Not counting the foot measurement, add 12 more inches to the remaining inch measurement. Divide the sum by 2 to find the width of border tiles for the long wall. The following example is applicable for a distance between the long walls of 10′-6″.

EXAMPLE

Room width	10′-6″
Add to inches	12″
Width of border tiles	18″ ÷ 2 = 9″

The width of border tiles along the short walls is calculated in the same manner. The following example applies when the distance between the short walls is 19′-8″.

EXAMPLE

Room length	19′-8″
Add to inches	12″
Width of border tiles	20″ ÷ 2 = 10″

METRIC EXAMPLE For a room width of 3200 mm, first divide by 305 to get 10.4918, then add 1 to the decimal to get 1.4918. Divide by 2 to get 0.7459, and then multiply by 305 to get 227.5 mm for a border tile.

For a room length of 5994 mm, first divide by 305 to get 19.65246, then add 1 to the decimal to get 1.65246. Divide by 2 to get 0.82623, and then multiply by 305 to get 252 mm for a border tile.

PREPARATION FOR CEILING TILE APPLICATION

Unless an adhesive application to an existing ceiling is to be made, **furring strips** must be provided on which to fasten ceiling tiles. Furring strips are usually fastened directly to exposed joists. They are sometimes applied to an existing ceiling and fastened into the concealed joists above.

Locating Concealed Joists

If the joists are hidden by an existing ceiling, tap on the ceiling with a hammer. To find a concealed joist, drive a nail into the spot where a dull thud is heard. Locate other joists by the same method or by measuring from the first location. Usually, ceiling joists are spaced 16 inches (406 mm) on centre (OC) and run parallel to the short dimension of the room. When all joists are located, snap lines on the existing ceiling directly below and in line with the concealed joists.

LAYING OUT AND APPLYING FURRING STRIPS

For fastening 12-inch (305 mm) tiles, furring strips must be installed 12 inches (305 mm) OC. From the corner, measure out the width of the border tiles. This measurement is the centre of the first furring strip away from the wall. To mark the edge, measure from the centre, in either direction, half the width of the furring strip. Mark an *X* on the side of the mark toward the centre of the furring strip. From the edge of the first furring strip, measure and mark, every 12 inches (305 mm), across the room. Place *X*s on the same side of the mark as the first one (**Figure 76–3**).

Lay out the other end of the room in the same manner. Snap lines between the marks for the location of the furring strips. The strips are fastened by keeping one edge to the chalk line with the strip on the side of the line indicated by the *X*. Fasteners must penetrate at least 1 inch (25 mm) into the joist. Starting and ending furring strips are also installed against both walls.

Squaring the Room

First, snap a chalk line on a furring strip as a guide for installing border tiles against the long wall. The line is snapped parallel to, and the width of the border tiles, away from the wall.

A second chalk line must be snapped to guide the application of the short wall border tiles. The line

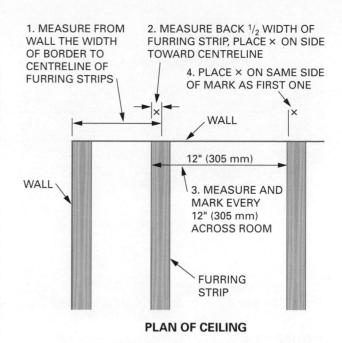

1. MEASURE FROM WALL THE WIDTH OF BORDER TO CENTRELINE OF FURRING STRIPS

2. MEASURE BACK ½ WIDTH OF FURRING STRIP, PLACE × ON SIDE TOWARD CENTRELINE

4. PLACE × ON SAME SIDE OF MARK AS FIRST ONE

WALL

WALL

12" (305 mm)

3. MEASURE AND MARK EVERY 12" (305 mm) ACROSS ROOM

FURRING STRIP

PLAN OF CEILING

FIGURE 76–3 Furring strip layout for ceiling tile.

must be snapped at exactly 90 degrees to the first chalk line. Otherwise tiles will not line up properly. From the short wall, measure in along the first chalk line the width of the short wall border tiles. From this point, use the Pythagorean theorem and snap another line at a right angle to the first line. This method of squaring lines has been previously described in Figure 75–15 (page 840) and Figure 28–6 (page 285).

INSTALLING CEILING TILE

Ceiling tiles should be allowed to adjust to normal interior conditions for 24 hours before installation. Some carpenters sprinkle talcum powder or corn starch on their hands to keep them dry. This prevents fingerprints and smudges on the finished ceiling. Cut tiles face up with a sharp utility knife guided by a straightedge. All cut edges should be against a wall.

Starting the Installation

To start the installation, cut a tile to fit in the corner. The outside edges of the tile should line up exactly with both chalk lines. Because this tile fits in the corner, it must be cut to the size of border tiles on both long and short sides of the room. For example, if the border tiles on the long wall are 9 inches (229 mm) and those on the short wall are 10 inches (254 mm), the corner tile should be cut twice to make it 9 × 10 inches (229 × 254 mm). Staple the tile in position. Be careful to line up the outside edges with both chalk lines (**Figure 76–4**).

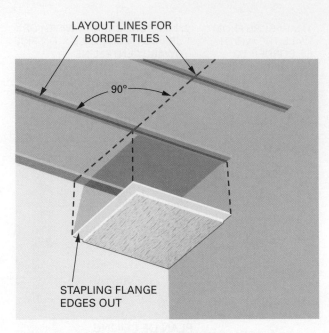

FIGURE 76–4 Install the first tile in the corner with its outside edges to the layout lines.

Completing the Installation

After the corner tile is in place, work across the ceiling. Install two or three border tiles at a time. Then fill in with full-sized field tiles (**Figure 76–5**). Tiles are applied so that they are snug to each other, but not jammed tightly. Fasten each tile with

four ½- or ⁹⁄₁₆-inch (12 or 13 mm) staples. Place two in each flange, using a hand stapler. Use six staples in 12 × 24-inch (305 × 610 mm) tiles. Continue applying tiles in this manner until the last row is reached.

When the last row is reached, measure and cut each border tile individually. Cut the tiles slightly less than measured for easy installation. Do not force the tiles in place. Face-nail the last tile in the corner near the wall where the nailhead will be covered by the wall moulding. After all tiles are in place, the ceiling is finished by applying moulding between the wall and ceiling. (The application of wall moulding is discussed in a later unit.)

Adhesive Application of Ceiling Tile

Ceiling tile is sometimes cemented to existing plaster and drywall ceilings. These ceilings must be completely dry, solid, level, and free of dust and dirt. If the existing ceiling is in poor condition or has loose paint, adhesive application is not recommended.

Special tile adhesive is used to cement tiles to ceilings. Four daubs of cement are used on 12 × 12-inch (305 × 305 mm) tile. Six daubs are used on 12 × 24-inch (305 × 610 mm) tile. Before applying the daubs, prime each spot by using the putty knife to force a thin layer into the back surface of the tile. Apply the daubs, about the size of a walnut. Press the tile into position. Keep the adhesive away from the edges to allow for spreading when the tile is pressed into position. No staples are required to hold the tile in place.

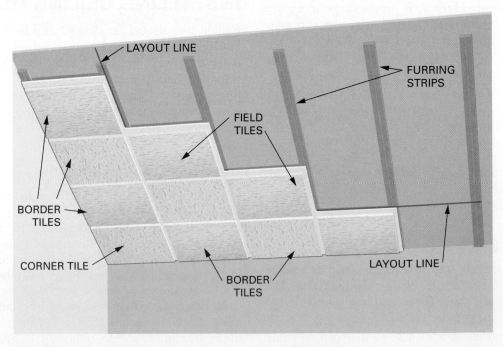

FIGURE 76–5 Install a few border tiles, and then fill in with full-sized field tiles.

ESTIMATING MATERIALS FOR TILE INSTALLATION

To estimate ceiling tile, measure the width and length of the room to the next whole foot measurement. Multiply these figures together to find the area of the ceiling in square feet. Divide the ceiling area by the number of square feet contained in one of the ceiling tiles being used to find the number of ceiling tiles needed.

To estimate furring strips, measure to the next whole foot the length of the room in the direction that the furring strips are to run. Multiply this by the number of rows of furring strips to find the total number of linear feet of furring strip stock needed. To find the number of rows, divide the width of the room by the furring strip spacing and add one.

KEY TERMS

chamfered

cross tees

fissured

furring strips

main runners

perforated

wall angles

REVIEW QUESTIONS

Select the most appropriate answer.

1. What are the most common sizes of suspended ceiling panels?

   a. 8″ × 12″ and 12″ × 12″
   b. 12″ × 12″ and 16″ × 16″
   c. 12″ × 12″ and 12″ × 24″
   d. 24″ × 24″ and 24″ × 48″

2. What is the dimension of a border panel for a 12′-6″-wide room, when 24 × 24-inch suspended ceiling panels are used?

   a. 3 inches
   b. 6 inches
   c. 15 inches
   d. 16 inches

3. What is the diagonal measurement of a 16′ × 24′ rectangle?

   a. 28⅞″
   b. 346⅛″
   c. 384″
   d. 832″

4. How many usable pieces may be cut from one main runner?

   a. 2
   b. 3
   c. 4
   d. more than 4

5. In a suspended ceiling, what is hanger wire used to suspend?

   a. cross tees
   b. main runners
   c. wall angle
   d. furring strips

6. What is the first step in installing a suspended ceiling?

   a. Square the room.
   b. Make a sketch of the planned ceiling.
   c. Calculate border tiles.
   d. Install the wall angle.

7. Approximately how many 2 × 2 suspended ceiling tiles are required for a room that measures 17′-9″ × 23′-9″?

   a. 88
   b. 108
   c. 130
   d. 422

8. What are the most common sizes of stapled-place ceiling tile?

   a. 8″ × 12″ and 12″ × 12″
   b. 12″ × 12″ and 16″ × 16″
   c. 12″ × 12″ and 12″ × 24″
   d. 14″ × 24″ and 24″ × 48″

9. How far into the joists must nails used to fasten furring to ceiling joists penetrate?

   a. at least ¾ inch (19 mm)
   b. at least 1 inch (25 mm)
   c. at least 1¼ inches (32 mm)
   d. at least 1½ inches (38 mm)

10. How many staples are required to fasten 12 × 12-inch (305 × 305 mm) ceiling tiles?

    a. two staples
    b. three staples
    c. four staples
    d. six staples

11. What is the dimension of a border panel for a 4532 mm wide room, when 610 × 610 mm suspended ceiling panels are used?

   a. 262 mm
   b. 436 mm
   c. 524 mm
   d. 567 mm

12. What is the diagonal measure of a 3452 × 5678 mm rectangle?

   a. 6645 mm
   b. 13 512 mm
   c. 7698 mm
   d. 15 396 mm

13. How many 610 × 610 mm suspended ceiling tiles are required for a room that measures 3452 × 5678 mm?

   a. 50
   b. 60
   c. 70
   d. 80

Unit 25

Interior Doors and Door Frames

| **Chapter 77** Description of Interior Doors | **Chapter 78** Installation of Interior Doors and Door Frames |

Interior doors used in residential and light commercial buildings are less dense than exterior doors. They are not ordinarily subjected to as much use and are not exposed to the weather. In commercial buildings, such as hospitals and schools, heavier and larger interior doors are specified that meet special conditions.

Doors must be installed level and plumb to operate properly. Otherwise they may not remain in the position last placed, moving on their own.

OBJECTIVES

After completing this unit, the student should be able to:

- describe the sizes and kinds of interior doors.
- make and set interior door frames.
- hang an interior swinging door.
- install locksets on interior swinging doors.
- set a prehung door and frame.
- install bypass, bifold, pocket, and folding doors.

Chapter 77

Description of Interior Doors

Interior doors are classified by style as *flush*, *panel*, *French*, *louvre*, and *café* doors. Interior flush doors have a smooth surface, are usually less expensive, and are widely used when a plain appearance is desired. Some of the other styles have special uses. "Dutch" doors are created when a full-size door, usually solid core, is cut horizontally in the middle and each half gets its own hinges and latches. They can operate separately or together when connected with a barrel bolt. Doors are also classified by the way they operate, such as *swinging*, *sliding*, or *folding*.

INTERIOR DOOR SIZES AND STYLES

For residential and light commercial use, most interior doors are manufactured in 1⅜-inch (35 mm) thickness. Some, like café and bifold doors, are also made in 1¼- and 1⅛-inch (32 and 28 mm) thicknesses. Most doors are manufactured in 6'-8" (2032 mm) heights. Some types may be obtained in heights of 6'-0" and 6'-6" (1829 and 1981 mm). Door widths range from 1'-0" to 3'-0" (305 to 914 mm) in increments of 2 inches (50 mm). However, not all sizes are available in every style.

Flush Doors

A flush door is made with a *solid* or *hollow* core. Solid core doors are generally used as entrance or fire-rated doors. (They have previously been described in Unit 19.) Hollow-core doors are commonly used in the interior except when fire resistance or sound transmission is critical.

A hollow-core door consists of a light perimeter frame. This frame encloses a mesh of thin wood or composition material supporting the faces of the door. Solid wood blocks are appropriately placed in the core for the installation of locksets. The frame and mesh are covered with a thin plywood called a *skin*. *Lauan* plywood is used extensively for flush door skins. Flush doors are also available with veneer faces of *birch*, *gum*, *oak*, and *mahogany*, among others (**Figure 77–1**). When flush doors are to be painted, an overlay plywood or tempered hardboard may be used for the skin.

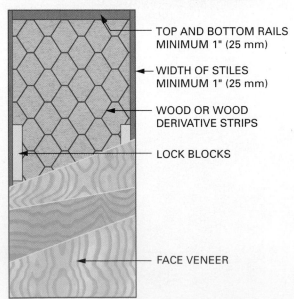

**MESH OR CELLULAR CORE
7 PLY CONSTRUCTION ILLUSTRATED**

TOP AND BOTTOM RAILS MINIMUM 1" (25 mm)

WIDTH OF STILES MINIMUM 1" (25 mm)

WOOD OR WOOD DERIVATIVE STRIPS

LOCK BLOCKS

FACE VENEER

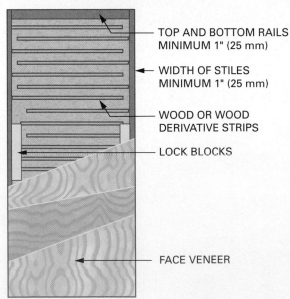

**LADDER CORE
7 PLY CONSTRUCTION ILLUSTRATED**

TOP AND BOTTOM RAILS MINIMUM 1" (25 mm)

WIDTH OF STILES MINIMUM 1" (25 mm)

WOOD OR WOOD DERIVATIVE STRIPS

LOCK BLOCKS

FACE VENEER

FIGURE 77–1 The construction of hollow-core flush doors.

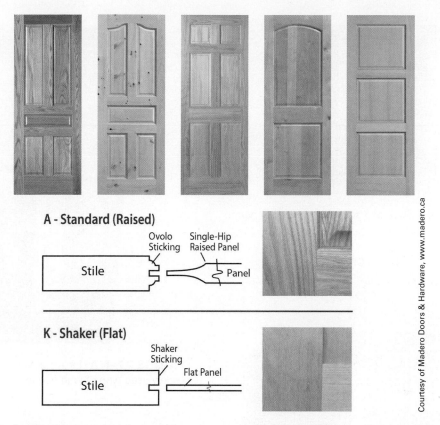

FIGURE 77–2 Styles of commonly used interior panel doors.

Panel Doors

An interior panel door consists of a frame with usually one to eight wood panels in various designs (**Figure 77–2**). They are similar in style to some exterior panel doors. The panels can sometimes be replaced with glass inserts, though large pieces of glass should be safety glass. (The construction of panel doors has been previously described in Unit 19.)

French Doors

French doors may contain from one to fifteen lights of glass. They are made in a 1¾-inch (44 mm)

thickness for exterior doors and 1⅜-inch (35 mm) thickness for interior doors (**Figure 77–3**).

Louvre Doors

Louvre doors are made with spaced horizontal slats called louvres used in place of panels. The louvres are installed at an angle to obstruct vision but permit the flow of air through the door. Louvred doors are widely used on clothes closets (**Figure 77–4**).

Café Doors

Café doors (also called saloon doors) are short panel or louvre doors. They are hung in pairs that swing

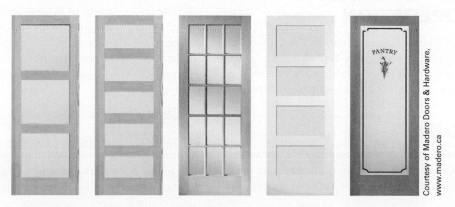

FIGURE 77–3 French doors are used in the interior as well as for entrances.

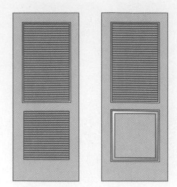

FIGURE 77–4 Louvre doors obstruct vision but permit the circulation of air.

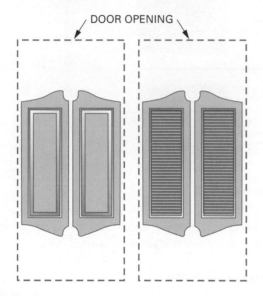

FIGURE 77–5 Café doors usually are used between kitchens and dining areas.

in both directions. They are used to partially screen an area, yet allow easy and safe passage through the opening. The tops and bottoms of the doors are usually shaped in a pleasing design (**Figure 77–5**).

METHODS OF INTERIOR DOOR OPERATION

Doors are also identified by their method of operation. Doors either swing on hinges or slide on tracks. The choice of door operation depends on such factors as convenience, cost, safety, and space.

Swinging Doors

Swinging doors are hinged on one edge. They swing out of the opening. When closed, they cover the total opening. Swinging doors that swing in one direction

FIGURE 77–6 A single-acting swinging door is the most widely used type of interior door.

are called *single-acting doors* (**Figure 77–6**). With special hinges, they can swing in both directions. They are then called *double-acting doors*. Swinging doors are the most commonly used type of door. They have the disadvantage of requiring space for the swing.

Bypass Doors

Bypass doors are commonly used on wide clothes-closet openings. A double track is mounted on the head jamb of the door frame. Rollers that ride in the track are attached to the doors so that they slide by each other. A floor guide keeps the doors in alignment at the bottom (**Figure 77–7**). Usually two doors are used in a single opening. Three or more doors may be used, depending on the situation.

The disadvantage of bypass doors is that although they do not project out into the room, access to the complete width of the opening is not possible. They are easy to install, but are not practical in openings less than 6 feet (1.8 m) wide.

Pocket Doors

The pocket door is opened by sliding it sideways into the interior of the partition. When opened, only the lock edge of the door is visible (**Figure 77–8**). Pocket doors may be installed as a single unit, sliding in one direction,

FIGURE 77–7 Bypass doors are used on wide closet openings.

Courtesy of L.E. Johnson Products, Inc.

or as a double unit sliding in opposite directions. When opened, the total width of the opening is obtained, and the door does not project out into the room. Pocket doors are used when these advantages are desired.

FIGURE 77–8 The pocket door slides into the interior of the partition.

FIGURE 77–9 A pocket door frame comes preassembled from the factory. It is installed when the interior partitions are framed.

The installation of pocket doors requires more time and material than other methods of door operation. A special pocket door frame unit and track must be installed during the rough framing stage (**Figure 77–9**). The rough opening in the partition must be large enough for the door opening and the pocket.

Bifold Doors

Bifold doors are made in flush, panel, louvre, or combination panel and louvre styles. They are made in narrower widths than other doors. This allows them to operate in a folding fashion on closet and similar-type openings (**Figure 77–10**).

Bifold doors consist of pairs of doors hinged at their edges. The doors on the jamb side swing on pivots installed at the top and bottom. Other doors fold up against the jamb door as it is swung open. The end door has a guide pin installed at the top. The pin rides in a track to guide the set when opening or closing (**Figure 77–11**). On very wide openings the guide pin is replaced by a combination guide and support to keep the doors from sagging.

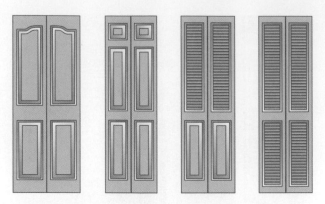

FIGURE 77–10 Bifold doors are manufactured in many styles.

FIGURE 77–11 Bifold doors provide access to almost the total width of the opening.

Bifold doors may be installed in double sets, opening and closing from each side of the opening. They have the advantage of providing access to almost the total width of the opening, yet they do not project out much into the room.

Chapter 78

Installation of Interior Doors and Door Frames

Many interior doors come **prehung** in their frames for easier and faster installation on the job. However, it is often necessary to build and set the door frame, hang the door, and install the locksets.

INTERIOR DOOR FRAMES

Special rabbeted jamb stock or nominal 1-inch square-edge lumber is used to make interior door frames. If square-edge lumber is used, a separate *stop*, if needed, is applied to the inside faces of the door frame.

Checking Rough Openings

The first step in making an interior door frame is to measure the door opening to make sure it is the correct width and height. The rough opening width for single-acting swinging doors should be the width of the door plus twice the thickness of the side jamb, plus ½ inch (12 mm) each side for shimming between the door frame and the opening. For example, if the thickness of the side jamb beyond the door is ¾ inch (19 mm), the rough opening width is 2½ inches (63 mm) more than the door width.

The rough opening height should be the height of the door, plus the thickness of the head jamb, plus ½ inch (12 mm) for clearance at the top, plus the thickness of the finished floor, plus a desired clearance under the door. An allowance of ¼ to ½ inch (6 to 12 mm) is usually made for clearance between the finished floor and the door.

For example, if the head jamb and finished floor thickness are both ¾ inch (19 mm), the rough opening height should be 2½ inches (63 mm) over the door height, if ½ inch (12 mm) is allowed for clearance under the door (**Figure 78–1**).

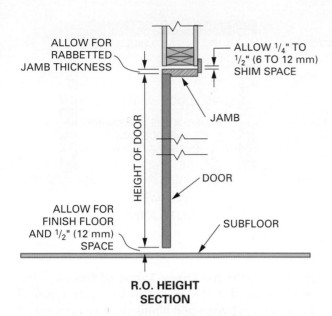

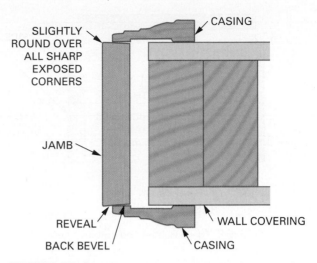

FIGURE 78–2 Back bevel jamb edges slightly to permit casings to fit snugly against them.

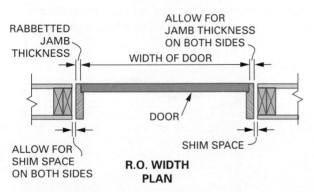

FIGURE 78–1 The size of rough openings for doors needs to take into account space for jambs and shimming.

The rough opening size for other than single-acting swinging doors, such as bypass and bifold doors, should be checked against the manufacturer's directions. The sizes of the doors and allowances for hardware may differ with the manufacturer.

Making an Interior Door Frame

Interior door frames are constructed like exterior door frames except they have no sill. Interior door frames usually are installed after the interior wall covering has been applied. Measure the total thickness of the wall, including the wall covering, to find the jamb width.

Cutting Jambs to Width

If necessary, rip the door jamb stock so that its width is the same as the wall thickness. If rabbeted jamb stock is used, cut the edge opposite the rabbet. Plane and smooth both edges to a slight *back bevel*. The back bevel permits the door casings, when later applied, to fit tightly against the edges of the door

frame in case there are irregularities in the wall (**Figure 78–2**). *Ease* all sharp exposed corners.

Cutting Jambs to Length

On interior door frames, the *head jamb* is usually cut to fit between, and is dadoed into, the *side jambs*. The side jambs run the total height of the rough opening. Cut both side jambs to a length equal to the height of the opening.

Head Jambs of Door Frames for Hinged Doors. If rabbeted jambs are used, the length of the head jamb is the width of the door plus $3/16$ inch (4 mm). The extra $3/16$ inch (4 mm) is for joints of $3/32$ inch (2 mm) on each side, between the edges of the door and the side jambs. If square-edge lumber is used, its length is the same as a rabbeted head jamb. However, ½ inch (12 mm) is added for dadoing ¼ inch (6 mm) deep into each side jamb (**Figure 78–3**). Cut the head jamb to length with both ends square.

Side Jambs of Door Frames for Hinged Doors. Measure up from the bottom ends. Square lines across the side jambs to mark the location of the bottom side of the head jamb. This dimension is the sum of:

- the thickness of the finish floor, if the door frame rests on the subfloor,
- an allowance of ½ inch (12 mm) minimum between the door and the finish floor,
- the height of the door
- $3/32$ inch (2 mm) for a joint between the door and the head jamb, and
- on rabbeted jambs, subtract ½ inch (12 mm) for the depth of the rabbet.

Hold a scrap piece of jamb stock to the squared lines. Mark its other side to lay out the width of the

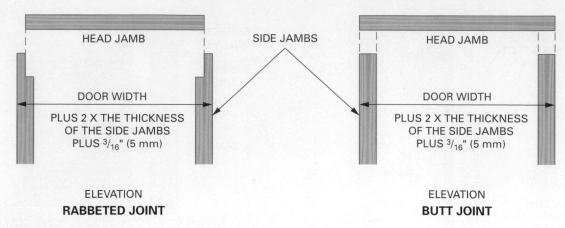

FIGURE 78–3 Rabbet and butt joint head jamb connections.

dado. Mark the depth of the dadoes on both edges of the side jambs. Cut the dadoes to receive the head jamb. On rabbeted jambs, dado depth is to the face of the rabbet. A dado depth of ¼ inch (6 mm) is sufficient on plain jambs (**Figure 78–4**).

Jamb Lengths for Other Types of Doors. The length of head and side jambs for other types of doors, such as bypass and bifold, must be determined from instructions provided by the manufacturer of the hardware and the door. Door hardware and door

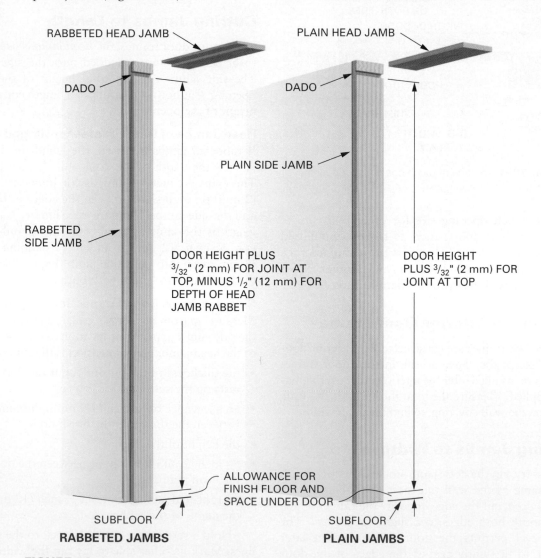

FIGURE 78–4 Laying out plain and rabbeted side jambs of door frames for swinging doors.

sizes differ with the manufacturer. This affects the length of the door jambs.

Plain, square-edge lumber jambs are used to make door frames for doors other than single-acting swinging doors. The rest of the procedure, such as checking rough openings, cutting jambs to width, assembling, and setting, is the same for all door frames.

ASSEMBLING THE DOOR FRAME

Fasten the side jambs to the head jamb, keeping the edges flush. If there is play in the dado, first wedge the head jamb with a chisel so that the face side comes up tight against the dado shoulders before fastening.

Cut a narrow strip of wood. Tack it to the side jambs a few inches up from the bottom so that the frame width is the same at the bottom as it is at the top. These strips are commonly called spreaders (**Figure 78–5**).

SETTING THE DOOR FRAME

Door frames must be set so that the jambs are straight, level, and plumb (**Procedure 78–A**). They are usually set before the finish floor is laid. If a rabbeted frame is used, determine the swing of the door so that the rabbet is facing toward the correct side.

Cut any *horns* from the top ends of side jambs. Place the frame in the opening. The horns are cut off in case the side jambs need to be shimmed to level the head jamb.

Install shims directly opposite the ends of the head jamb between the opening and the side jambs. Shim an equal amount on both sides so that the frame is close to being centred at the top of the opening. Drive shims up snugly but not tightly.

Levelling the Head Jamb

Keep the edges of the frame flush with the wall. With the bottom ends of both side jambs resting on the subfloor, check whether the head jamb is level. Level the head jamb, if necessary, by placing shims between the bottom end of the appropriate side jamb and the subfloor. When the head jamb is level, tack the frame in place on both sides, close to the top. Drive fasteners through side jambs and shims into the studs. Fasteners, either finish nails or screws, are placed in the same part of the jamb as the door stop in order to keep them out of sight. It is *a better practice* to level across the door opening first and then adjust the length of one side jamb in order to level the head jamb. This way both side jambs rest solidly on the subfloor.

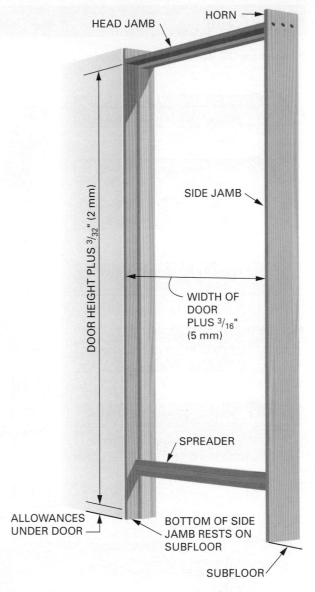

FIGURE 78–5 An assembled interior rabbeted door frame.

If the door frame rests on a finish floor, then a tight joint must be made between the bottom ends of the side jambs and the finish floor. If the floor is level, side jambs will fit the floor if their lengths are exactly the same and their bottom ends have been cut square. Once the frame is set, the head jamb should be level when the ends of the side jamb are resting on the floor.

If the floor is not quite level, or if the side jambs are of unequal length, level the head jamb by shimming under the bottom end of the side jamb on the low side. Set the dividers for the amount the jamb has been shimmed. Scribe that amount on the bottom of the opposite side jamb. Remove the frame from the opening. Cut to the scribed line. Replace the frame in the opening. The head jamb should be level. The

StepbyStep Procedures

Procedure 78–A Installing a Door Jamb

STEP 1 CUT HORNS FROM TOP OF SIDE JAMBS IF NECESSARY.

STEP 2 SET FRAME IN OPENING. SHIM ON BOTH SIDES OPPOSITE HEAD JAMB. LEVEL HEAD JAMB AND FASTEN AT TOP THROUGH SIDE JAMB SHIMS.

STEP 3 PLUMB SIDE JAMBS SHIM AND TACK AT BOTTOM.

STEP 4 STRAIGHTEN SIDE JAMBS, INSTALL INTERMEDIATE SHIMS AND TACK IN PLACE.

STEP 5 SIGHT THROUGH THE FRAME TO CHECK FOR A WIND. DRIVE AND SET ALL NAILS WHEN EDGES LINE UP.

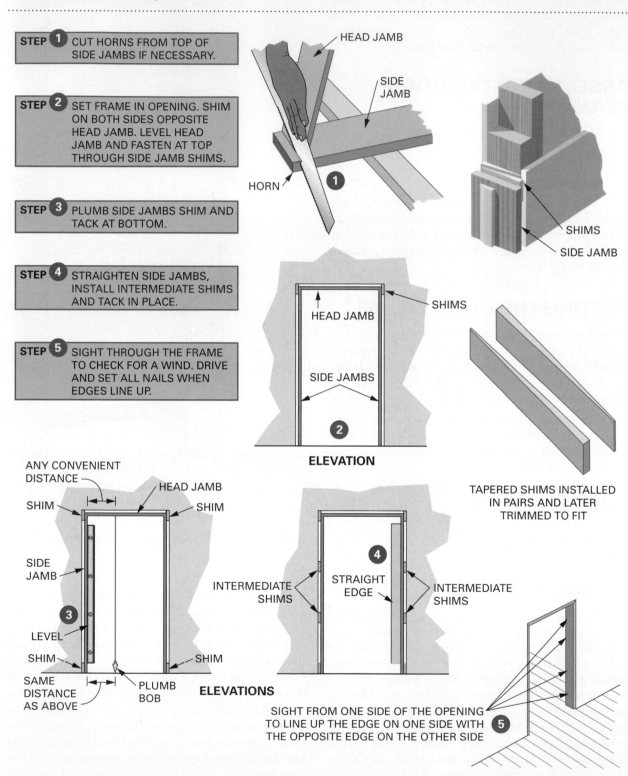

HEAD JAMB

SIDE JAMB

HORN

1

SHIMS

SIDE JAMB

HEAD JAMB

SHIMS

SIDE JAMBS

2

ELEVATION

TAPERED SHIMS INSTALLED IN PAIRS AND LATER TRIMMED TO FIT

ANY CONVENIENT DISTANCE

SHIM

HEAD JAMB

SHIM

SIDE JAMB

3

LEVEL

SHIM

SHIM

SAME DISTANCE AS ABOVE

PLUMB BOB

INTERMEDIATE SHIMS

STRAIGHT EDGE

4

INTERMEDIATE SHIMS

ELEVATIONS

SIGHT FROM ONE SIDE OF THE OPENING TO LINE UP THE EDGE ON ONE SIDE WITH THE OPPOSITE EDGE ON THE OTHER SIDE

5

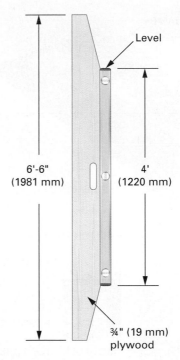

FIGURE 78–7 A ¾-inch (19 mm) straightedge is used with a 4-foot (1220 mm) level to plumb hinge-side door jambs.

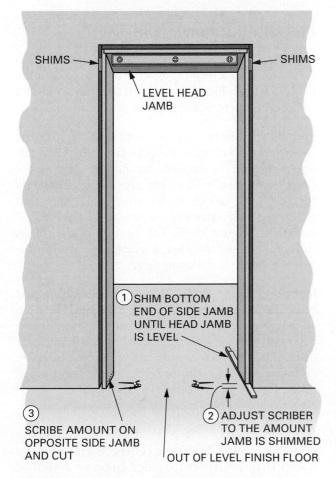

FIGURE 78–6 Technique for cutting side jambs to make the head jamb level.

bottom ends of both side jambs should fit snugly against the finish floor (**Figure 78–6**).

Plumbing the Side Jambs

Several ways of plumbing door frames have been previously described. An accurate and fast method is with the use of a 6-foot (1.8 m) level. An alternative method is to use a 4-foot level (1220 mm) and a ¾-inch (19 mm) plywood straightedge that is about 6½ feet (1981 mm long) (**Figure 78–7**). When the hinge-side jamb is plumb, shim and tack its bottom end in place. Only this side needs to be plumbed. Locate the bottom end of the other side jamb by

measuring across. The door frame width at the bottom should be the same as on the top. Shim and tack the bottom end of the other side jamb in place.

Straightening Jambs

Use a 6-foot (1.8 m) straightedge against the side jambs. Many carpenters hang the door at this point and use the door as the straightedge that guides the shimming of the jambs. Straighten them by shimming at intermediate points. Besides other points, shims should be placed opposite hinge and lockset locations. Fasten the jambs by nailing through the shims. Head jambs on wide door frames are straightened, shimmed, and fastened in a similar way.

Sighting the Door Frame for a Wind

Before any nails are set, sight the door frame to see if it has a *wind*. The frame must be sighted by eye to make sure that side jambs line up vertically with each other and that the frame is not twisted. This is important when installing rabbeted jambs. The method of checking for a wind in door frames has been previously described (see Unit 19).

If the frame has a wind, move the top or bottom ends of the side jambs slightly until they line up with each other. Fasten top and bottom ends of the side jambs securely. Set the nails.

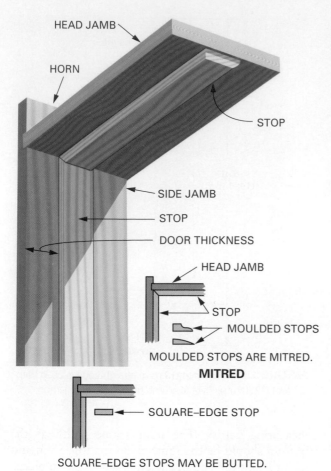

HEAD JAMB

HORN

STOP

SIDE JAMB

STOP

DOOR THICKNESS

HEAD JAMB

STOP

MOULDED STOPS

MOULDED STOPS ARE MITRED.
MITRED

SQUARE–EDGE STOP

SQUARE–EDGE STOPS MAY BE BUTTED.
BUTTED

FIGURE 78–8 A back mitre joint is used for moulded stops, and a butt joint is usually used on square-edge stops.

Applying Door Stops

At this time, *door stops* may be applied to plain jambs. The stops are not permanently fastened, in case they have to be adjusted when locksets are installed. A **back mitre** joint is usually made between moulded side and header stops. This joint can also be coped. A butt joint is made between square-edge stops (**Figure 78–8**).

HANGING INTERIOR DOORS

The method of fitting and hanging single-acting, hinged, interior doors is similar to that for exterior doors.

Double-Acting Doors

Double-acting doors are installed with either special pivoting hardware installed on the floor and the head jamb or with spring-loaded double-acting hinges. Both types return the door to a closed position after being opened. When opened wide, the doors can be held in the open position. A different type of light-duty, double-acting hardware is used on café doors. To install double-acting door hardware, follow the manufacturer's directions.

Installing Bypass Doors

Bypass doors are installed so that they overlap each other by about 1 inch (25 mm) when closed. Cut the track to length. Install it on the head jamb according to the manufacturer's directions.

Installing Rollers. Install pairs of *roller hangers* on each door. The roller hangers may be offset a different amount for the door on the outside than the door on the inside. They are also offset differently for doors of various thicknesses. Make sure that rollers with the same and correct offset are used on each door (**Figure 78–9**). The location of the rollers from the edge of the door is usually specified in the manufacturer's instruction sheet.

Installing Door Pulls. Mark the location and bore holes for door pulls. Flush pulls must be used so that bypassing is not obstructed (**Figure 78–10**). The proper size hole is bored partway into the door. The pull is tapped into place with a hammer and wood block. The press fit holds the pull in place. Rectangular flush pulls, also used on bypass doors, are held in place with small recessed screws.

Hanging Doors. Hang the doors by holding the bottom outward. Insert the rollers in the overhead track. Then gently let the door come to a vertical position. Install the inside door first, then the outside door (**Figure 78–11**).

Fitting Doors. Test the door operation and the fit against side jambs. Door edges must fit against side jambs evenly from top to bottom. If the top or bottom portion of the edge strikes the side jamb first, it may cause the door to jump from the track. The door rollers have adjustments for raising and lowering. Adjust one or the other to make the door edges fit against side jambs.

Installing Floor Guides. A floor guide is included with bypass door hardware to keep the doors in alignment. The guide is centred on the lap to steady the doors at the bottom. Mark the location of the guide. Remove the doors. Install the inside section of the guide. Replace the inside door.

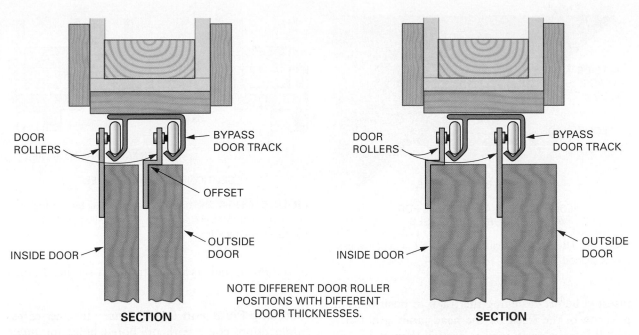

FIGURE 78–9 Bypass door rollers are offset different distances for use on doors of various thicknesses.

FIGURE 78–10 Bypass doors must have flush pulls.

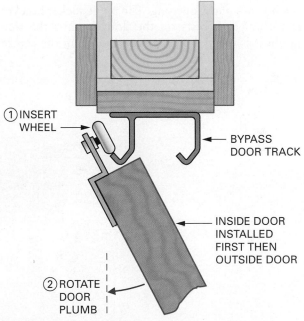

FIGURE 78–11 Bypass doors are hung on the overhead track by holding the bottom of the door outward.

Replace the outside door. Install the rest of the guide (**Figure 78–12**).

Installing Bifold Doors

Before installing bifold doors, make sure the opening size is as specified by the hardware or door manufacturer. Usually bifold doors come hinged together in pairs. The hardware consists of the track, pivot

sockets, pivot pins and guides, door aligners, door pulls, and necessary fasteners (**Figure 78–13**).

Installing the Track. Cut the track to length, if necessary. Fasten it to the head jamb with screws provided in the kit. The track contains adjustable sockets for the door *pivot pins*. Make sure these are

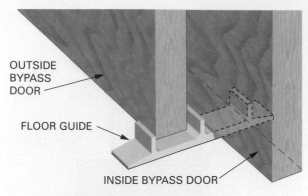

FLOOR GUIDE IS ADJUSTABLE FOR
VARIOUS DOOR THICKNESSES.

FIGURE 78–12 A floor guide is installed to keep bypass doors aligned.

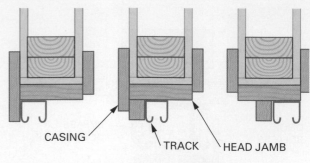

SECTIONS THROUGH LINTEL

FIGURE 78–14 The bifold door track may be located in any position on the head jamb in several ways. Trim conceals the track from view.

inserted before fastening the track in position. The position of the track on the head jamb is not critical. It may be positioned as desired (**Figure 78–14**).

Installing Bottom Pivot Sockets. Locate the bottom pivot sockets. Fasten one on each side, at the bottom of the opening. The pivot socket bracket is L-shaped. It rests on the floor against the side jamb. It is centred on a plumb line from the centre

of the pivot sockets in the track on the head jamb above.

Installing Pivot and Guide Pins. In most cases, bifold doors come with pre-bored holes for *pivot* and *guide pins*. If not, it is necessary to bore them. Follow the manufacturer's directions as to size and location. Install pivot pins at the top and bottom ends of the door in the pre-bored holes closest to the jamb. Sometimes the top pivot pin is spring-loaded. It can then be depressed for easier installation of the

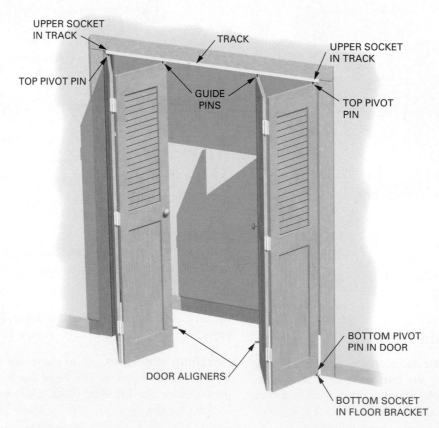

FIGURE 78–13 Installation of the bifold door requires several kinds of special hardware.

door. The bottom pivot pin is threaded and can be adjusted for height. The guide pin rides in the track. It is installed in the hole provided at the top end of the door farthest away from the jamb.

Hanging the Doors. After all the necessary hardware has been applied, the doors are ready for installation. Loosen the set screw in the top pivot socket. Slide it along the track toward the centre of the opening about one foot (305 mm) away from the side jamb. Place the door in position by inserting the bottom pivot pin in the bottom pivot socket. Tilt the doors to an upright position. At the same time insert the top pivot pin in the top socket, and the guide pin in the track, while sliding the socket toward the jamb.

Adjusting the Doors. Adjust top and bottom pivot sockets in or out so that the desired joint is obtained between the door and the jamb. Lock top and bottom pivot sockets in position. Adjust the bottom pivot pin to raise or lower the doors, if necessary.

If more than one set of bifold doors is to be installed in an opening, install the other set on the opposite side in the same manner. Install knobs in the manner and location recommended by the manufacturer.

Where sets of bifold doors meet at the middle of an opening, door aligners are installed, near the bottom, on the inside of each of the meeting doors. The door aligners keep the faces of the centre doors lined up when closed (**Figure 78–15**).

Installing Pocket Doors

The pocket door frame, complete with track, is installed when interior partitions are framed. The pocket consists of two ladder-like frames between which the door slides. A steel channel is fastened to the floor. The channel keeps the pocket opening spread the proper distance apart.

The frame, which is usually preassembled at the factory, is made of ¾-inch (19 mm) stock. The pocket is covered by the interior wall finish. Care must be taken when covering the pocket frame not to use fasteners that are so long that they penetrate the frame. If fasteners penetrate through the pocket door frame, they will probably scratch the side of the door as it is operated or stop its complete entrance into the pocket.

Installing Door Hardware. Attach rollers to the top of the door in the location specified by the manufacturer. Install pulls on the door. On pocket doors, an edge pull is necessary, in addition to recessed pulls on the sides of the door. A special pocket door pull contains edge and side pulls. It is mortised in the edge of the door. In most cases, all

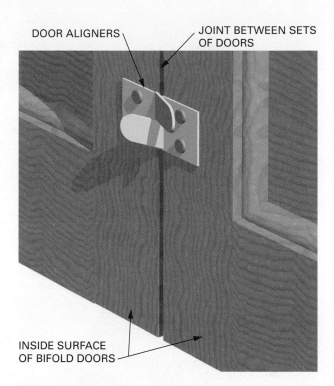

DOOR ALIGNERS

JOINT BETWEEN SETS OF DOORS

INSIDE SURFACE OF BIFOLD DOORS

FIGURE 78–15 Door aligners are used near the bottom where sets of bifold doors meet.

the necessary hardware is supplied when the pocket door frame is purchased.

Hanging the Door. Engage the rollers in the track by holding the bottom of the door outward in a way similar to that used with bypass doors. Test the operation of the door to make sure it slides easily and butts against the side jamb evenly. Make adjustments to the rollers, if necessary. Stops are later applied to the jambs on both sides of the door. The stops serve as guides for the door. When the door is closed, the stops prevent it from being pushed out of the opening (**Figure 78–16**).

INSTALLING A PREHUNG DOOR

A prehung single-acting, hinged door unit consists of a door with hinge gains, a door jamb set with hinge gains, and door stop. Some units are made with a predetermined hinge side, and others can be universally installed. Small cardboard shims are stapled to the lock edge and top end of the door to maintain proper clearance between the door and frame.

Prehung units are available in several jamb widths to accommodate various wall thicknesses. Some prehung units have split jambs that are adjustable for varying wall thicknesses (**Figure 78–17**).

A prehung door unit can be set in a matter of minutes. Many prehung units come without the casing

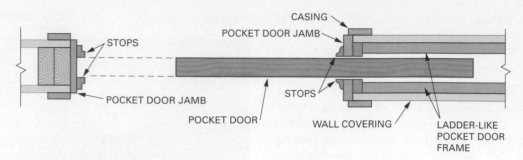

FIGURE 78–16 Plan view of a pocket door.

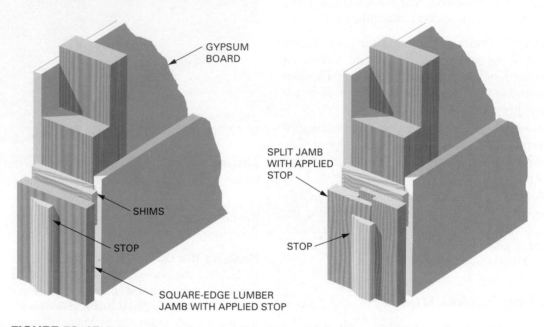

FIGURE 78–17 Prehung door units come with solid or split jambs.

attached, but if it is attached, remove the casings carefully from one side of the solid jamb units. Centre the unit in the opening so that the door will swing in the desired direction. Be sure the door is closed and spacer shims are in place between the jamb and door. Plumb the hinge side of the door unit. Tack it to the wall through the casing.

Open the door and move to the other side. Install shims between the side jambs and the rough opening at intermediate points, keeping side jambs straight. Nail through the side jambs and shims. Remove spacers. Check the operation of the door. Make any necessary adjustments. Replace the previously removed casings. Drive and set all nails (**Procedure 78–B**).

Prehung door units with split jambs are set in a similar manner. However, there is no need to remove the casings. One section is installed as described earlier. The remaining section is inserted into the one already in place.

INSTALLING LOCKSETS

Locksets are installed on interior doors in the same manner as for exterior doors, as described in Chapter 62. Although their installation is basically the same, some locks are used exclusively on interior doors.

The **privacy lockset** is often used on bathroom and bedroom doors. It is locked by pushing or turning a button on the room side. On most privacy locksets, a turn of the knob on the room side unlocks the door. On the opposite side, the door can be unlocked by a pin or key inserted into a hole in the knob. The unlocking device should be kept close by, in a prominent location, in case the door needs to be opened quickly in an emergency.

The **passage lockset** has knobs on both sides that are turned to unlatch the door. This lockset is used when it is not desirable to lock the door.

StepbyStep Procedures

Procedure 78–B Installing a Prehung Door

STEP ❶ REMOVE CASINGS FROM ONE SIDE OF UNIT.

STEP ❷ PLACE UNIT IN OPENING AND PLUMB SIDE CASING.

STEP ❸ FASTEN THROUGH CASINGS INTO WALL.

STEP ❹ MOVE TO OTHER SIDE OF DOOR.

STEP ❺ INSTALL SHIMS BETWEEN JAMB AND WALL.

STEP ❻ FASTEN THROUGH JAMB AND SHIMS. CHECK THAT JAMBS ARE PLUMB AND STRAIGHT.

STEP ❼ REPLACE CASINGS THAT WERE PREVIOUSLY REMOVED.

STEP ❽ FASTEN THROUGH CASINGS INTO WALL.

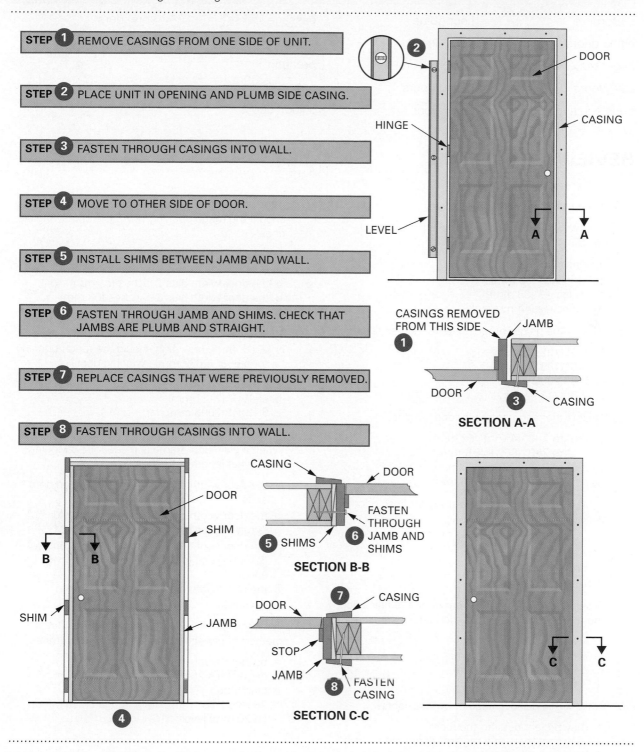

KEY TERMS

back mitre

bifold doors

bypass doors

café doors

double-acting doors

floor guide

flush door

French doors

louvre doors

panel door

passage lockset

pocket door

prehung

privacy lockset

saloon doors

spreaders

swinging doors

REVIEW QUESTIONS

Select the most appropriate answer.

1. What thickness are most manufactured interior doors?

   a. 1 inch (25 mm)
   b. 1⅜ inches (35 mm)
   c. 1½ inches (38 mm)
   d. 1¾ inches (44 mm)

2. What is the height of most interior doors?

   a. 6'-0" (1829 mm)
   b. 6'-6" (1981 mm)
   c. 6'-8" (2032 mm)
   d. 7'-0" (2134 mm)

3. What is the range of interior door widths?

   a. 1'-6" to 2'-6" (457 to 762 mm)
   b. 2'-2" to 2'-8" (660 to 813 mm)
   c. 2'-6" to 2'-8" (762 to 813 mm)
   d. 1'-0" to 3'-0" (305 to 914 mm)

4. What material is used extensively for flush door skins?

   a. fir plywood
   b. lauan plywood
   c. metal
   d. plastic laminate

5. What is the usual distance between the finish floor and the bottom of swinging doors for clearance?

   a. ¼ to ½ inch (6 to 12 mm)
   b. ½ to ¾ inch (12 to 19 mm)
   c. ½ to 1 inch (12 to 25 mm)
   d. ¾ to 1 inch (19 to 25 mm)

6. What is one disadvantage of bypass doors?

   a. They project out into the room.
   b. They cost more and require more time to install.
   c. They are difficult to operate.
   d. They do not provide total access to the opening.

7. If the jamb stock is ¾ inch (19 mm) thick, what size should the rough opening width be for a swinging door?

   a. the door width plus ¾ inch (19 mm)
   b. the door width plus 1½ inches (38 mm)
   c. the door width plus 2 inches (51 mm)
   d. the door width plus 2½ inches (63 mm)

8. If the jamb stock and the finished floor are both ¾ inch (19 mm) thick and the space under the door is ½" (12 mm), how high should the rough opening be for a 6'-8" (2032 mm) swinging door?

   a. 7'-0" (2134 mm)
   b. 6'-11½" (2121 mm)
   c. 6'-10½" (2096 mm)
   d. 6'-9½" (2070 mm)

9. When a plain door frame is made for a swinging door, what length is the head jamb?

   a. the door width plus the rabbet width on both ends
   b. the door width plus the rabbet width plus ³⁄₃₂" (2 mm) space on both ends
   c. the door width plus the rabbet width plus ³⁄₃₂" (2 mm) space plus ½" (12 mm) shim space both ends
   d. the door width plus twice the side jamb thickness

10. What tool can be used for an accurate and fast method of plumbing side jambs of a door frame?

   a. builder's level
   b. carpenter's 26-inch (660 mm) hand level
   c. plumb bob
   d. 78-inch (1981 mm) straightedge and a 4-foot (1220 mm) level

Unit 26

Interior Trim

Chapter 79	Description and Application of Moulding	Chapter 80	Application of Door Casings, Base, and Window Trim

Interior trim, also called *interior finish*, involves the application of moulding around windows and doors; at the intersection of walls, floor, and ceilings; and to other inside surfaces. Mouldings are strips of material, shaped in numerous patterns, for use in a specific location. Wood is used to make most mouldings, but some are made of plastic or metal.

Interior trim is among the final materials installed and requires the installer to take care not to mar the finish. Blemishes and dings in the finish may be visible for the life of the building.

OBJECTIVES

After completing this unit, the student should be able to:

- identify standard mouldings and describe their use.
- apply ceiling and wall moulding.
- apply interior door casings, baseboard, base cap, and base shoe.
- install window trim, including stools, aprons, jamb extensions, casings, and stop beads.
- install closet shelves and closet poles.
- install mantels.

Chapter 79
Description and Application of Moulding

Mouldings are available in many standard types. Each type is manufactured in several sizes and patterns. Standard patterns are usually made only from softwood. When other kinds of wood, or special patterns, are desired, mills make custom mouldings to order. All mouldings must be applied with tight-fitting joints to present a suitable appearance.

STANDARD MOULDING PATTERNS

Standard mouldings are designated as bed, crown, cove, full round, half round, quarter round, base, base shoe, base cap, casing, chair rail, back band, apron, stool, stop, and others (**Figure 79–1**).

Moulding usually comes in lengths of 8, 10, 12, 14, and 16 feet (2.4, 3.0, 3.6, 4.2, and 4.8 m). Some mouldings are available in odd lengths. Door casings, in particular, are available in lengths of 7 feet (2.1 m) to reduce waste.

Finger-jointed lengths are made of short pieces joined together. These are used only when a paint finish is to be applied. The joints show through a stained or natural finish.

MOULDING SHAPE AND USE

Some mouldings are classified by the way they are shaped. Others are designated by location. For example, *beds, crowns,* and *coves* are terms related

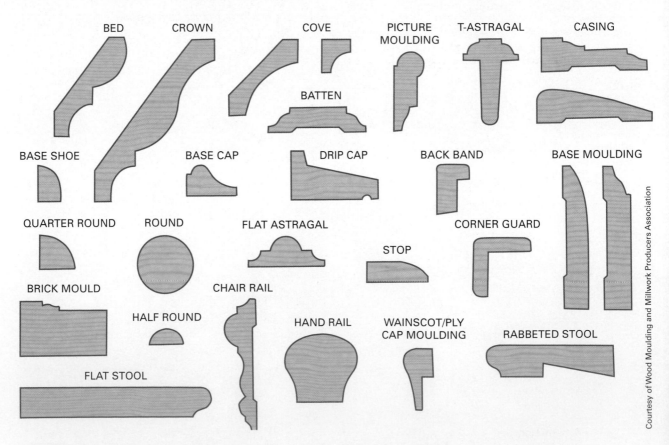

FIGURE 79–1 Standard moulding patterns.

Courtesy of Wood Moulding and Millwork Producers Association

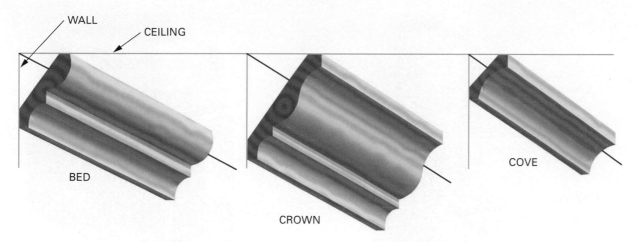

FIGURE 79–2 Bed, crown, and cove mouldings are often used at the intersections of walls and ceilings.

to shape. Although they may be placed in other locations, they are usually used at the intersections of walls and ceilings (**Figure 79–2**). Also classified by their shape are **full-round, half-round,** and **quarter-round.** They are used in many locations. Full rounds are used for such things as closet poles. Half rounds may be used to conceal joints between panels or to trim shelf edges. Quarter rounds may be used to trim outside corners of wall panelling and for many other purposes (**Figure 79–3**).

Designated by location, **base moulding, base shoe moulding,** and **base cap moulding** are applied at the bottom of walls where they meet the floor. When square-edge base is used, a base cap is usually applied to its top edge. Base shoe is normally used to

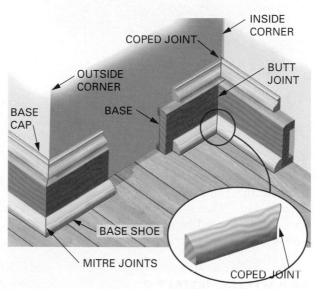

FIGURE 79–4 Base, base shoe, and base cap mouldings are used to trim the bottom of the wall.

conceal the joint between the bottom of the base and the finish floor (**Figure 79–4**).

A **Casing** is used to trim around windows, doors, and other openings. They cover the space between the frame and the wall. **Back bands** are applied to the outside edges of casings for a more decorative appearance.

A **apron,** a **stool,** and a **stop** are all parts of window trim. Stops are also applied to door frames. On the same window, aprons should have the same moulded shape as casings. Aprons butt up against the bottom of the stool, and they run from the outside edge of the left casing to the outside edge of the right casing (**Figure 79–5**).

Corner guards are also called *outside corners.* They are used to finish the outside corners of interior

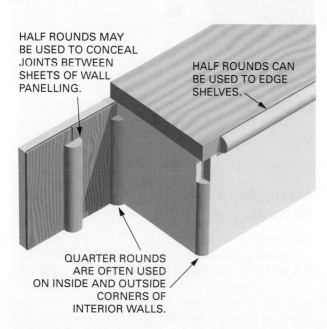

FIGURE 79–3 Half-round and quarter-round mouldings are used for many purposes.

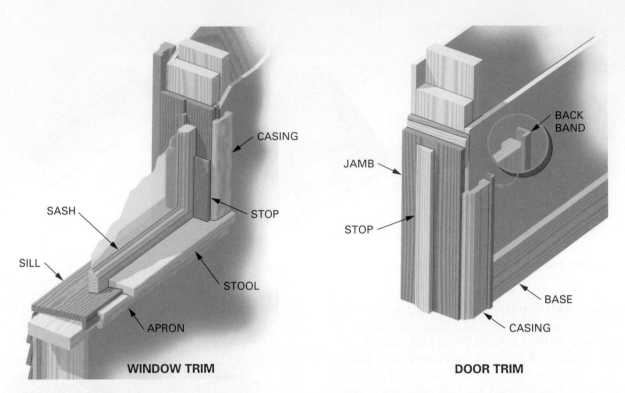

FIGURE 79–5 Casing, back bands, and stops are used for window and door trim. Stools and aprons are part of window trim.

wall finish. **Caps** and **chair rail** trim the top edge of wainscotting. (These mouldings have been previously described in Chapter 73.) Others, such as *astragals*, *battens*, *panel*, and *picture* mouldings, are used for various purposes.

MAKING JOINTS ON MOULDING

Because wood will shrink and expand with the varying humidity levels of the heating and cooling seasons, end joints between lengths of all moulding need to be **scarf joints**, that is cut at a lap angle of 45 degrees, or as is often done, at 22½ degrees, to reduce the visible cut line. These angles are positive stops on all motorized mitre saws, both left and right. After the moulding has been fastened, joints between lengths should be sanded flush, except on prefinished mouldings.

Joints on exterior corners are **mitred**. (Mitred joints are defined in Chapter 12.) Joints on interior corners are usually **coped**, especially on large mouldings. A coped joint is made by fitting the piece on one wall with a square end into the corner. The end of the moulding on the other wall is cut to fit against the shaped face of the moulding on the first wall (**Figure 79–6**).

FIGURE 79–6 A coped joint is made by fitting the end of one piece of moulding against the shaped face of the other piece.

Methods of Mitring Using Mitre Boxes

Mouldings of all types may be mitred by using either a hand or a power **mitre box**. A handsaw mitre box is a tool that cuts a piece of material at an angle. The most common angle is 45 degrees. A job-built version of a mitre box used to guide a handsaw is made from wood scraps. Another style of box is metal with a backsaw attached and is easily adjustable to cut different angles (**Figure 79–7**).

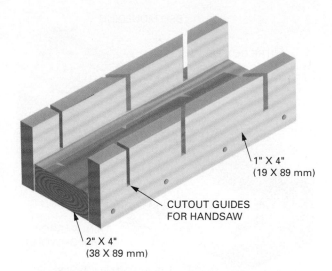

1" X 4"
(19 X 89 mm)

CUTOUT GUIDES
FOR HANDSAW

2" X 4"
(38 X 89 mm)

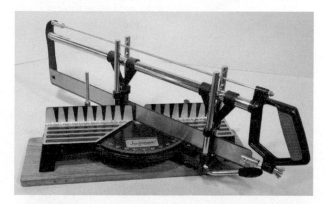

FIGURE 79–7 Mitre boxes for a handsaw.

The most popular way to cut mitres and other end cuts on trim is with a power mitre box (**Figure 79–8**). With this tool, a carpenter is able to cut virtually any angle with ease, whether it is a simple or a compound mitre (one with two angles). Fine adjustments

FIGURE 79–8 A power mitre box makes easy work of cutting moulding.

to a piece of trim, $\pm\frac{1}{64}$ inch (0.4 mm), can be made with great speed and accuracy. The power mitre box is discussed in more depth in Chapter 18.

Positioning Moulding in the Mitre Box

It is important to visualize the cut ahead of time. Sometimes it is best to hold the back of the moulding against the fence and sometimes against the base. Placing moulding in the correct position in the mitre saw is essential for accurate mitring. Cut all mouldings with their face sides or edges up or toward the operator so that the wood splinters out the back side, not the face side.

Flat mitres are cut by holding the moulding with its face side up and its thicker edge against the side of the mitre box. Some mouldings, such as base, base cap and shoe, and chair rail, are held right side up. The moulding bottom edge should be against the bottom of the mitre box and the moulding back against the side of the mitre box (**Figure 79–9**).

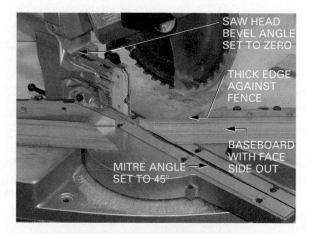

SAW HEAD BEVEL ANGLE SET TO ZERO

THICK EDGE AGAINST FENCE

BASEBOARD WITH FACE SIDE OUT

MITRE ANGLE SET TO 45°

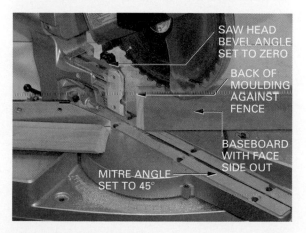

SAW HEAD BEVEL ANGLE SET TO ZERO

BACK OF MOULDING AGAINST FENCE

BASEBOARD WITH FACE SIDE OUT

MITRE ANGLE SET TO 45°

FIGURE 79–9 The position of moulding in a mitre box depends on how the saw is set up and where the moulding is to be installed.

Cutting Bed, Crown, and Cove Mouldings

Bed moulding, crown moulding, and cove moulding may be done using two methods. Each method will produce accurate, clean-fitted joints.

Upside-Down Method. The first method positions the moulding upside down in the mitre saw from how it will be installed. The portion of the moulding's back side that touches the wall (W) will rest against the fence. The portion of the back side that touches the ceiling (C) will rest on the base of the saw (**Figure 79–10**). Care should be taken to ensure that the back sides of the moulding are resting flat and square against the surfaces of the mitre saw. This position is convenient because the bottom edge of this type of moulding is the edge that is usually marked for cutting to length. With the bottom edge up, the mark on which to start the cut can be easily seen. This method is a bit awkward because the piece must be held at the precise angle before cutting. In addition to just holding the piece by hand, it can be clamped into position and/or stop blocks can be fastened to an auxiliary base to keep it from sliding down.

A thin, narrow strip of wood fastened to the bottom of the wood mitre box and against the moulding helps prevent the moulding from moving when being mitred. The strip also ensures that subsequent pieces of the same type of moulding will be positioned at the same angle. Therefore, they will be mitred the same as the first piece (**Figure 79–11**).

Flat-Cut Method. The second method of cutting ceiling mouldings, especially large crown mouldings, places the back side of the moulding flat against the base. The mitre saw is set to a specific compound angle in order to make the cut (**Figure 79–12**). These angles are different for different angles of mouldings, but the two most common are 45 degrees for bed

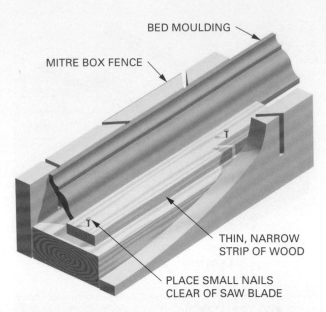

FIGURE 79–11 Technique for holding a wide piece of moulding in the proper position for cutting.

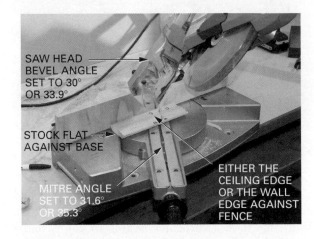

FIGURE 79–12 Large ceiling mouldings are held flat to mitre saw base and cut at a compound angle.

moulding and 38 degrees for crown. Most of the professional-grade mitre saws on the market will have detents in the scales to make the angles easy to find.

There are three main concepts to cutting moulding on the flat in a compound mitre saw. First, determine the angle at which the moulding is designed to rest against the wall. Second, determine which way you want the mitre saw bevel angle to be tilted, left or right. Third, all cuts are made with the saw at the same bevel angle while changing only the base mitre angles.

Tilting the saw head left or right to make a bevel cut is entirely user preference. Right-handed users may prefer the head tilted to the right so that the entire blade is visible, whereas left-handed users may prefer the tilt to the left. Angle settings for the saw depend on angle of the manufactured moulding (**Figure 79–13**). You can determine the angle of the crown by eye

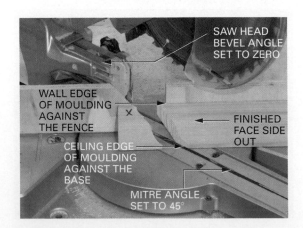

FIGURE 79–10 To mitre ceiling mouldings, they must be positioned upside down in the mitre saw. It may be helpful to think of the fence as the wall and the base as the ceiling.

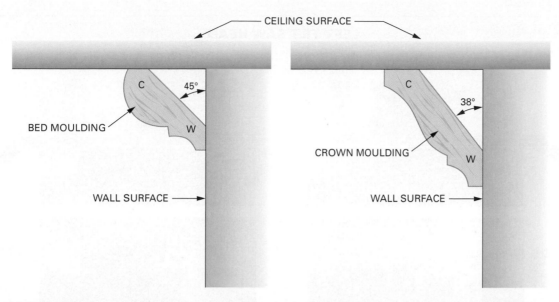

FIGURE 79–13 The profile of bed mouldings and crown mouldings are typically 38 degrees and 45 degrees.

or with a speed square. A symmetrical angle means 45 degrees. The angles for setting the compound mitre saw are shown in **Figure 79–14**. Begin setup by tilting the saw head in the desired direction and at the correct bevel angle. Then choose the mitre angle direction.

Setting the base mitre angle direction determines which side of the mitre is being cut. There are essentially

Moulding Design Angle	Base Mitre Angle (left or right)	Saw Head Bevel Angle (left or right)
38°	31.6°	33.9°
45°	35.3°	30°

FIGURE 79–14 The angles to set a compound mitre saw to cut crown moulding on the flat.

four types of cuts. Two cuts come from either side of the blade of the left-swing mitre, and two more from the right-swing mitre. Labelling each cut #1, #2, #3, and #4 makes referring to them easier (**Figure 79–15**).

The last step in understanding the compound mitre flat-cut method is to note which edge of the moulding touches the fence. Either the wall (W) or ceiling (C) side of the moulding's cross-section touches the fence in each setup (**Figure 79–16**). It is important to cut small test pieces to make sure that the corners are at 90 degrees, and if not to make adjustments for a perfect fit. There are angle finders commercially available.

Procedure 79–A shows the steps to cut the moulding shown in Figure 79–15.

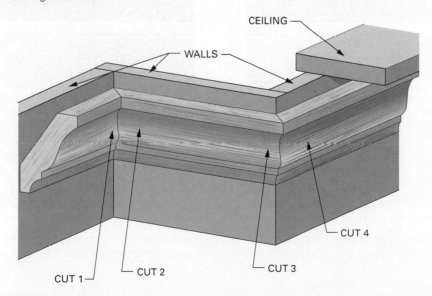

FIGURE 79–15 There are four types of cuts for mitres—two for inside and two for outside mitres. The inside mitre is generally replaced by a square cut #1 and a coped joint #2.

LEFT TILT SAW HEAD

RIGHT TILT SAW HEAD

FIGURE 79–16 To cut large ceiling mouldings on the flat, first choose a saw tilt direction, then make all cuts by changing the mitre angle direction and which side of the moulding touches the saw fence, ceiling (C) or wall (W).

StepbyStep Procedures

Procedure 79–A Cutting Crown Moulding on the Flat. The procedure assumes a right saw head bevel angle and a 38-degree moulding profile.

STEP 1 LIGHTLY MARK MOULDING WITH A W AND C. PLACE MOULDING ON SAW BASE WITH W AGAINST THE FENCE. FIRST CUT WILL HAVE #1 ON THE RIGHT AND #4 ON THE LEFT.

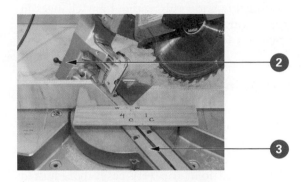

STEP 2 FROM FIGURE 79–14 SET SAW HEAD BEVEL ANGLE TO 33.9°.

STEP 3 SET BASE MITRE ANGLE TO 31.6° TO THE RIGHT. CUT PIECE AS NEEDED.

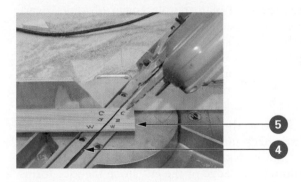

STEP 4 ROTATE MITRE ANGLE TO OPPOSITE SIDE AND SET TO THE SAME ANGLE AS BEFORE. DO NOT READJUST THE SAW HEAD BEVEL ANGLE.

STEP 5 MEASURE AND LIGHTLY MARK NEW PIECE TO CUT #2 AND #3. PLACE NEW MOULDING PIECE ON BASE SO CEILING EDGE OF MOULDING TOUCHES THE FENCE.

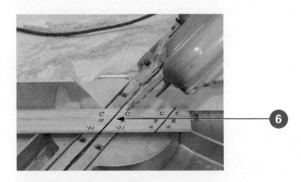

STEP 6 MEASURE THE LENGTH AND CUT AT THE SAME ANGLE SETTINGS. THIS SIMPLY MEANS SLIDE THE STOCK TO THE RIGHT FOR A NEW CUT.

STEP 7 ASSEMBLE THE PIECES.

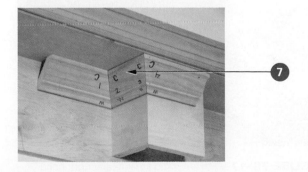

Mitring with a Table Saw

Mitres may also be made by using the table saw or the radial arm saw. The use of mitring jigs is helpful when making flat mitres on window and door casings. The jigs allow both right- and left-hand mitres to be cut quickly and easily without any changes in the setup (**Figure 79–17**). (The construction and use of mitring jigs for saws are more fully described in Unit 7.)

Coped Joints

Coped joints not only improve initial appearance of an inside joint but maintain a tight-looking fit when the wood shrinks. After the drywall compound is placed, inside corners are rarely an exact right angle. The coped joint can very easily be adjusted to compensate for this. Before mouldings are cut, it is important to plan the placement of joints. Scarf joints and coped joints are to be cut so that they are the least visible from the point of entry to the room. When you face a wall, the inside corner of the piece of moulding for that wall is cut square to the corner and the other piece that meets it is coped (**Figure 79–18**).

To cope the end of a piece of moulding, first make a **back mitre** on the end. A back mitre starts from the end and is cut back on the face of the moulding. The edge of the cut along the face forms the profile of the cope. Rub the side of a pencil point lightly along the profile to outline it more clearly.

Use a coping saw with a fine-tooth blade. Cut along the outlined profile with a slight undercut. Cut with the handle of the coping saw above the work and the teeth of the blade pointing away from

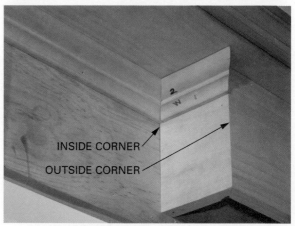

INSIDE CORNER

OUTSIDE CORNER

FIGURE 79–18 Coped joints form inside angles that look like mitres.

the handle (**Figure 79–19**). Hold the moulding so that it is over the end of a sawhorse. The side of the moulding that will butt the wall should be lying flat on the top of the sawhorse. This way the cut is done with the coping saw plumb. Cut carefully along the outlined profile, keeping the saw blade plumb. A slight under cut of the blade helps ensure that the joint will fit nicely (Figure 79–19). It may be necessary to touch up the cut with a wood file or sharp utility knife or sandpaper. Coping large crown mouldings can also be done with jigs and a jigsaw (**Figure 79–20**).

APPLYING MOULDING

To apply chair rail, caps, or some other type of moulding located on the wall, string lines should be drawn level and small marks made at intervals on the wall or chalk lines could be snapped if plain white chalk is used. This ensures that moulding is applied in a straight line. No lines need to be snapped for base mouldings or for small-sized mouldings applied at the intersection of walls and ceiling.

FIGURE 79–17 With a mitring jig, left- and right-hand mitres can be made quickly and easily without changing the setup (*guard has been removed for clarity*).

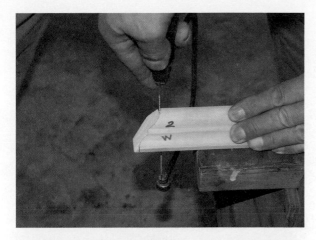

FIGURE 79–19 To cut a coped joint, first cut an inside or "back" mitre, and then place the moulding (with the wall side down) on a sawhorse and cut along the mitre profile.

FIGURE 79–20 A crown moulding coping fixture for a jigsaw. *(Courtesy of Lee Valley Tools Ltd.; www.leevalley .com. © Copyright Lee Valley Tools Ltd.)*

For large-sized ceiling mouldings, such as beds, crowns, and coves, again the same procedure can be followed with a string line, a level, or a chalk line. This ensures straight application of the moulding

and easier joining of the pieces without leaving visible markings on the surface. Without a straight line to guide application, the moulding may be forced at different angles along its length. This results in a noticeably crooked bottom edge and difficulty making tight-fitting mitres and coped joints.

Hold a short scrap piece of the large-sized moulding at the proper angle at the wall and ceiling intersection (**Figure 79–21**). Lightly mark the wall along the bottom edge of the moulding. Measure the distance from the ceiling down to the mark. Measure and mark this same distance down from the ceiling on each end of each wall to which the moulding is to be applied. Snap lines between the marks. Apply the moulding so that its bottom edge is to the (white) chalk line.

Apply the moulding to the first wall with square ends in both corners. If more than one piece is required to go from corner to corner, install the first piece with one end square and the other back-mitred at 22½ degrees. Scarf joints between lengths are desired because they make a less visible joint and they remain so even when the wood shrinks.

On some mouldings, such as quarter rounds and small cove mouldings, the straight, back surfaces should, but may not always, be of equal width. One of the back surfaces of these mouldings should be marked with a pencil to ensure positioning them in the mitre box the same way each time. Mitring the moulding with the same side down each time helps make fitting more accurate, faster, and easier (**Figure 79–22**).

If a small-sized moulding is used, fasten it with finish nails in the centre. Use nails of sufficient length to penetrate into solid wood at least 1 inch (25 mm). If large-sized moulding is used, fastening is required along both edges (**Figure 79–23**).

Press the moulding in against the wall or intersection with one hand while driving the nail almost home. Then set the nail below the surface. Nail at about 16-inch (406 mm) intervals and in other locations as necessary to bring the moulding tight against the surface. End nails should be placed 2 to 3 inches (50 to 75 mm) from the end to keep from splitting the moulding. If it is likely that the moulding may split, blunt the pointed end of the nail. Or, drill a hole slightly smaller than the nail diameter.

Install the last piece on the first wall by first squaring one end. Place the square end in the corner. Let the other end overlap the first piece. Mark and cut it at the overlap. This method is more accurate than measuring and then transferring the measurement to the piece. Mark all pieces of interior trim for length in this manner whenever possible. Cut

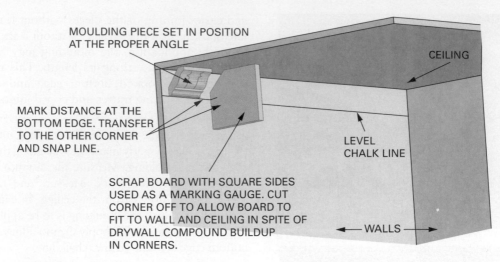

MOULDING PIECE SET IN POSITION AT THE PROPER ANGLE

CEILING

MARK DISTANCE AT THE BOTTOM EDGE. TRANSFER TO THE OTHER CORNER AND SNAP LINE.

LEVEL CHALK LINE

SCRAP BOARD WITH SQUARE SIDES USED AS A MARKING GAUGE. CUT CORNER OFF TO ALLOW BOARD TO FIT TO WALL AND CEILING IN SPITE OF DRYWALL COMPOUND BUILDUP IN CORNERS.

WALLS

FIGURE 79–21 Hold a scrap piece of moulding against the ceiling and wall to determine the distance from the ceiling to its bottom edge.

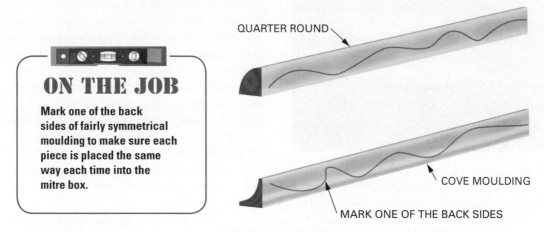

ON THE JOB

Mark one of the back sides of fairly symmetrical moulding to make sure each piece is placed the same way each time into the mitre box.

QUARTER ROUND

COVE MOULDING

MARK ONE OF THE BACK SIDES

FIGURE 79–22 Technique for reducing the confusion of working with moulding that has a fairly symmetrical cross-section.

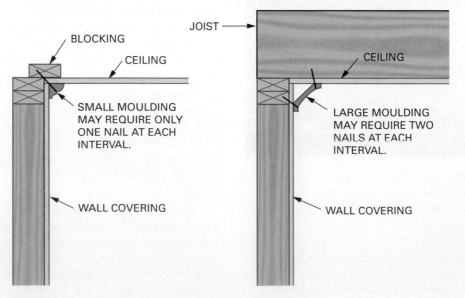

BLOCKING

CEILING

SMALL MOULDING MAY REQUIRE ONLY ONE NAIL AT EACH INTERVAL.

WALL COVERING

JOIST

CEILING

LARGE MOULDING MAY REQUIRE TWO NAILS AT EACH INTERVAL.

WALL COVERING

FIGURE 79–23 Methods of fastening moulding.

and fasten the last piece with its square end into the corner.

Cope the starting end of the piece on each succeeding wall against the face of the last piece installed on the previous wall. Work around the room in one direction, either clockwise or counterclockwise. The end of the last piece installed must be coped to fit against the face of the previous piece and then cut to meet the door casing or the outside mitre as warranted. The primary rule is that the pieces of moulding are placed so that the joints are least visible.

Most carpenters today are using air-powered trim nailers or brad nailers. This allows for rapid placement of nails as well as eliminates the need to either pre-drill or set the nails below the surface. Care must be taken to place the nails so that the head of the "T" is parallel to the wood grain to avoid crushing. This also makes removal of the nails easier when they are pulled through from the back of the moulding. In cases where steel studs are used or where there is minimal backing, trim nails can be applied close together and angled toward each other in order to "stitch" the moulding to the drywall. In the ICI sector, in large institutions and stores, heavier baseboards are fastened to the steel studs with trim head screws.

Chapter 80

Application of Door Casings, Base, and Window Trim

In addition to wall and ceiling moulding, the application of door casings, base, base cap, base shoe, and window trim is a major part of interior finish work. Care must be taken to avoid marring the work and to make neat, tight-fitting joints.

DOOR CASINGS

Door casings are mouldings applied around the door opening. They trim and cover the space between the door frame and the wall. Casings must be applied before any base mouldings because the base butts against the edge of door casing (**Figure 80–1**). Door casings extend to the floor.

Design of Door Casings

Mouldings or *S4S* stock may be used for door casings. S4S is the abbreviation for **surfaced four sides**. It is used to describe smooth, square-edge lumber. When moulded casings are used, the joint at the head must be mitred unless butted against **plinth blocks**. Plinth blocks are small decorative blocks. They are thicker and wider than the door casing. They are used as part of the door trim at the base and at the head (**Figure 80–2**).

When using S4S lumber, the joint may be mitred or butted. If a butt joint is used, the head casing

overlaps the side casing. The appearance of S4S casings and some moulded casings may be enhanced with the application of back bands (**Figure 80–3**).

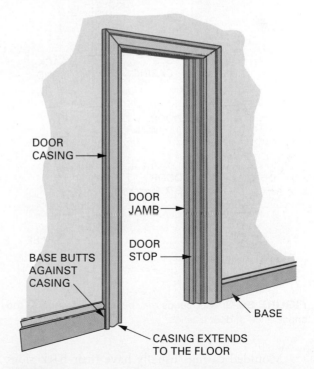

DOOR CASING

DOOR JAMB

DOOR STOP

BASE BUTTS AGAINST CASING

BASE

CASING EXTENDS TO THE FLOOR

FIGURE 80–1 Door casings are applied before the base is installed.

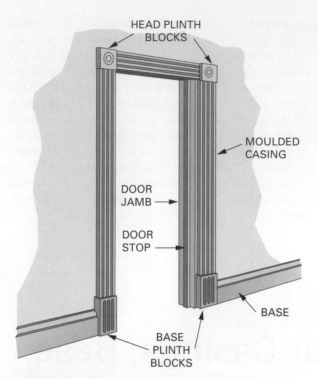

FIGURE 80–2 Moulded casings are mitred at the head unless plinth blocks are used.

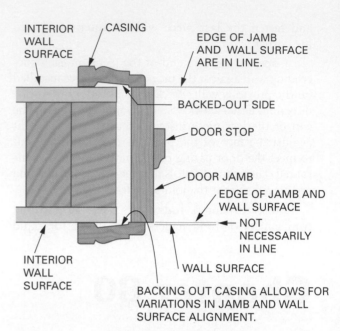

PLAN VIEW OF SIDE JAMB

FIGURE 80–4 Backing out door casings allows for a tight fit on wall and jamb.

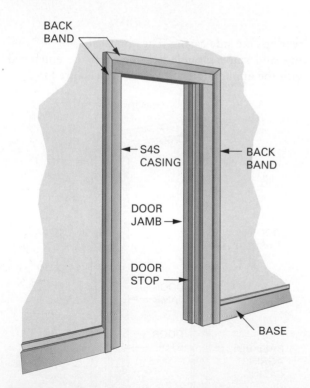

FIGURE 80–3 Back bands may be applied to improve the appearance of door casings.

Moulded casings usually have their back sides backed out. In cases where the jamb edges and the wall surfaces may not be exactly flush with each

other, the backed-out surfaces allow the casing to come up tight on both wall and jamb (**Figure 80–4**). If S4S casings are used, they must be backed out on the job. (A method of backing out S4S lumber using the table saw and a dado blade is illustrated in Figure 19–15.)

Applying Door Casings

Door casings are set back from the inside face of the door frame a distance of about $3/16$ inch (5 mm). This allows room for the door hinges and the striker plate of the door lock. This setback is called a reveal (**Figure 80–5**). The reveal also improves the appearance of the door trim.

Set the blade of the combination square so that it extends $5/16$ inch (8 mm) beyond the body of the square. Gauge lines at intervals along the side and head jamb edges by riding the square against the inside face of the jamb. Let the lines intersect where side and head jambs meet. Mark lightly with a sharp pencil or mark with a utility knife. The knife leaves no pencil lines to erase later.

The following procedure applies to moulded door casings mitred at the head. If several door openings are to be cased, cut the necessary number of casings to rough lengths with a mitre cut on one end of each piece. Rough lengths are a few inches longer than actually needed. For each

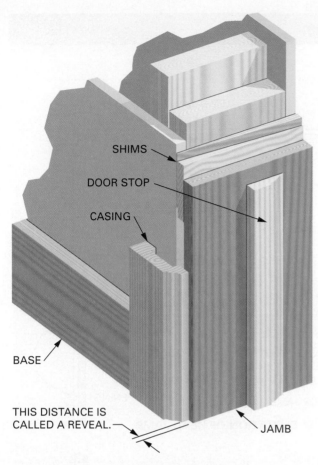

SHIMS

DOOR STOP

CASING

BASE

THIS DISTANCE IS
CALLED A REVEAL.

JAMB

FIGURE 80–5 The setback of the door casing on the jamb is called a reveal.

interior door opening, four side casings and two head casings are required. Cut side casings in pairs with right- and left-hand mitres for use on both sides of the opening.

Applying the Head Casing

Mitre one end of the head casing. Hold it against the head jamb of the door frame so that the mitre is on the intersection of the gauged lines. Mark the length of the head casing at the intersection of the gauged lines on the opposite side of the door frame. Mitre the casing to length at the mark.

Fasten the head casing in position. Its inside edge should be to the gauged lines on the head jamb. The mitred ends should be in line with the gauged lines on the side jambs. Use finish nails along the inside edge of the casing into the head jamb. If the casing edge is thin, use 1¼- or 1½-inch (32 or 38 mm) finish nails spaced about 12 inches (305 mm) apart. Keep the edge of the casing to the gauged lines on the jamb. Straighten the casing as necessary as nailing progresses. Drive nails at the proper angle to keep

them from coming through the face or back side of the jamb. Pneumatic finish nailers speed up the job of fastening interior trim.

Fasten the top edge of the casing into the framing. The outside edge is thicker, so longer nails are used, usually 2- or 2½-inch (51 or 63 mm) finish nails. They are spaced farther apart, about 16 inches (406 mm) on centre (OC) (**Procedure 80–A**). Do not drive end nails at this time. It may be necessary to move the ends slightly to fit the mitred joint between head and side casings.

Applying the Side Casings

Mark one of the previously mitred side casings by turning it upside down with the point of the mitre touching the floor. If the finish floor has not been laid, hold the point of the mitre on a scrap block of wood that is equal in thickness to the finish floor. Mark the side casing in line with the top edge of the head casing (**Procedure 80–B**). Make a square cut on the casing at the mark.

Place the side casing in position. Try the fit at the mitred joint. If the joint needs fitting, trim the mitred end of the side casing by planing thin shavings with a sharp block plane. The joint may also be fitted by shimming the casing away from the side of the mitre saw and making a thin corrective cut. Shim either near or far from the saw blade as needed to hold the casing at the desired angle (**Figure 80–6**). Most mitre saws now cut angles steeper than 45 degrees, usually to 47 degrees. When fitted, apply a little glue to the joint. Nail the side casing in the same manner as the head casing.

Avoid sanding the joint to bring the casing faces flush. It is difficult to keep from sanding across the grain on one or the other of the pieces. Cross-grain scratches will be very noticeable, especially if the trim is to have a stained finish. Bring the faces flush, if necessary, by shimming between the back of the casing and the wall. Usually, only very thin shims are needed. Any small space between the casing and the wall can be filled later with latex caulking. Also, the back side of the thicker piece may be planed or chiselled thinner. Most carpenters prefer to do these rather than try to sand the joint.

Drive a 1¼-inch (32 mm) finish nail into the edge of the casing and through the mitred joint. Drive end nails. Then set all fasteners. Keep nails 2 or 3 inches (50 or 75 mm) from the end to avoid splitting the casing. If there is danger of splitting, blunt the pointed end or drill a hole slightly smaller in diameter than the nail.

StepbyStep Procedures

Procedure 80–A Cutting a Head Casing to Fit

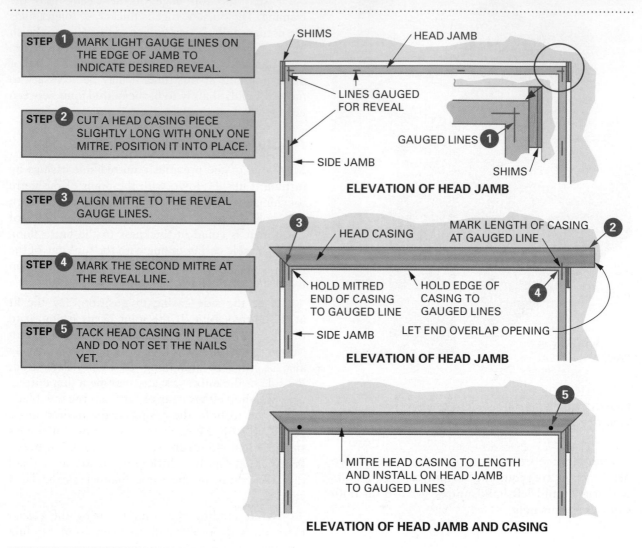

STEP **1** MARK LIGHT GAUGE LINES ON THE EDGE OF JAMB TO INDICATE DESIRED REVEAL.

STEP **2** CUT A HEAD CASING PIECE SLIGHTLY LONG WITH ONLY ONE MITRE. POSITION IT INTO PLACE.

STEP **3** ALIGN MITRE TO THE REVEAL GAUGE LINES.

STEP **4** MARK THE SECOND MITRE AT THE REVEAL LINE.

STEP **5** TACK HEAD CASING IN PLACE AND DO NOT SET THE NAILS YET.

SHIMS   HEAD JAMB

LINES GAUGED FOR REVEAL

GAUGED LINES **1**

SIDE JAMB

SHIMS

ELEVATION OF HEAD JAMB

3   HEAD CASING   MARK LENGTH OF CASING AT GAUGED LINE   **2**

HOLD MITRED END OF CASING TO GAUGED LINE   HOLD EDGE OF CASING TO GAUGED LINES

4

SIDE JAMB   LET END OVERLAP OPENING

ELEVATION OF HEAD JAMB

5

MITRE HEAD CASING TO LENGTH AND INSTALL ON HEAD JAMB TO GAUGED LINES

ELEVATION OF HEAD JAMB AND CASING

APPLYING BASE MOULDINGS

Moulded or S4S stock may be used for base. If S4S base is used, it should be backed out. A base cap should be applied to its top edge. The base cap conforms easily to the wall surface, resulting in a tight fit against the wall. The base trim should be thinner than the door casings against which it butts. This makes a more attractive appearance.

The base is applied in a manner similar to wall and ceiling moulding. However, copes are laid out for joints in interior corners. Instead of back-mitring to outline the cope, it is usually more accurate to lay out the cope by **scribing** (**Figure 80–7**).

When placed against the wall, the face of the base may not always be square with the floor. Therefore, if the base is tilted slightly, back-mitring to obtain the outline of the cope will result in a poor fit against it.

Apply the base to the first wall with square ends in each corner. Drive and set two finishing nails, of sufficient length, at each stud location. Nailing blocks previously installed during framing provide solid wood for fastening the ends of the base in interior corners.

Cut the base to go on the next wall about an inch longer than required. Lay the base against the

StepbyStep Procedures

Procedure 80–B Cutting Side Casings to Fit

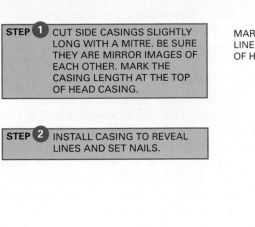

STEP 1 CUT SIDE CASINGS SLIGHTLY LONG WITH A MITRE. BE SURE THEY ARE MIRROR IMAGES OF EACH OTHER. MARK THE CASING LENGTH AT THE TOP OF HEAD CASING.

STEP 2 INSTALL CASING TO REVEAL LINES AND SET NAILS.

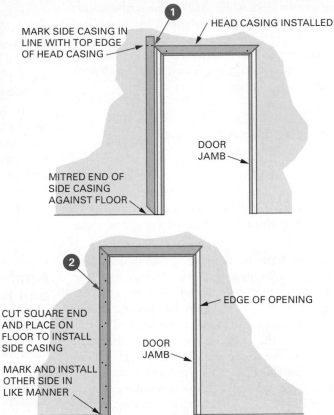

MARK SIDE CASING IN LINE WITH TOP EDGE OF HEAD CASING

HEAD CASING INSTALLED

DOOR JAMB

MITRED END OF SIDE CASING AGAINST FLOOR

CUT SQUARE END AND PLACE ON FLOOR TO INSTALL SIDE CASING

MARK AND INSTALL OTHER SIDE IN LIKE MANNER

EDGE OF OPENING

DOOR JAMB

ELEVATIONS OF DOOR OPENING

ON THE JOB

Slight adjustments may be made in mitre box cuts by shimming moulding against fence.

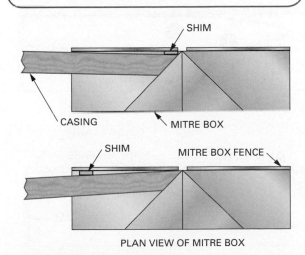

SHIM

CASING

MITRE BOX

SHIM

MITRE BOX FENCE

PLAN VIEW OF MITRE BOX

FIGURE 80–6 Technique for making small adjustments to the angle of a mitre.

wall by bending it so that the end to be scribed lies flat against the wall and against the first base. Set the dividers to scribe about ½ inch (12 mm). Lay out the cope by riding the dividers along the face of the base on the first wall.

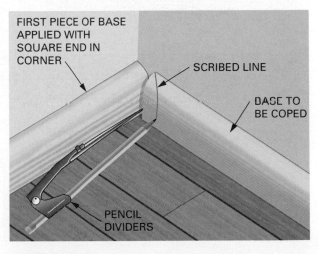

FIRST PIECE OF BASE APPLIED WITH SQUARE END IN CORNER

SCRIBED LINE

BASE TO BE COPED

PENCIL DIVIDERS

FIGURE 80–7 Laying out a coped joint on base moulding by scribing.

Hold the dividers while scribing so that a line between the two points is parallel to the floor. Twisting the dividers while making the scribe results in an inaccurate layout. Cut the end to the scribed line with a slight undercut. Bend the base back in position and try the fit. If scribed and cut accurately, no adjustments should be necessary. Its overall length must now be determined.

Cutting the Base to Length

The length of a baseboard that fits between two walls may be determined by measuring from corner to corner. Then transfer the measurement to the baseboard. Another method of determining its length eliminates using the rule. It may be faster and more accurate.

With the base in the last position described above, place marks, near the centre, on the top edge of the base and the wall so that they line up with each other. Place the other end in the opposite corner. Press the base against the wall at the mark. The difference between the mark on the wall and the mark on the base is the amount to scribe off the end in the corner. Set the dividers to this distance. Scribe the end. Cut to the scribed line (**Procedure 80–C**). If a tighter fit is desired, set the dividers slightly less than the distance between the marks. This method of fitting long lengths between walls may be applied to

other kinds of trim. However, this works especially well with the base.

Place one end in the corner, and bow out the centre. Place the other end in the opposite corner, and press the centre against the wall. Fasten in place. Continue in this manner around the room in a previously planned order. Make regular mitre joints on outside corners.

If both ends of a single piece are to have regular mitres for outside corners, it is imperative that it be fastened in the same position as it was marked. Tack the rough length in position with one finish nail in the centre. Mark both ends. Remove and cut the mitres. Installing the piece by first fastening into the original nail hole ensures that the piece is fastened in the same position as marked (**Procedure 80–D**).

Applying the Base Cap and Base Shoe

The base cap is applied in the same manner as most wall or ceiling moulding. Cope interior corners and mitre exterior corners. The base shoe is also applied in a similar manner as other moulding. It is nailed into the baseboard in order to allow the hardwood flooring to expand and contract without restriction (**Figure 80–8**).

Because the base shoe is a small-sized moulding and has solid backing, both interior and exterior corners are

---

StepbyStep Procedures

Procedure 80–C Cutting Moulding to Fit between Walls

| STEP **1** | BEND BASE AGAINST WALL WITH ONE END IN CORNER. |

| STEP **2** | MARK WALL AND TOP EDGE OF BASE AT SOME POINT NEAR THE CENTRE. |

| STEP **3** | PLACE OTHER END OF BASE IN OPPOSITE CORNER. |

| STEP **4** | SCRIBE THE DIFFERENCE BETWEEN MARKS OFF THE END IN THE CORNER. |

StepbyStep Procedures

Procedure 80–D Cutting Outside Mitres

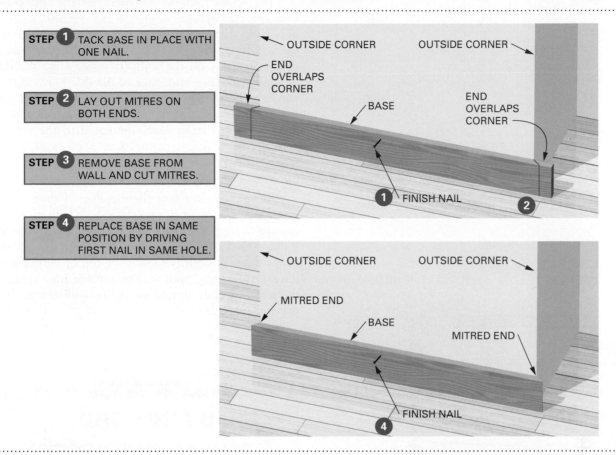

STEP 1 TACK BASE IN PLACE WITH ONE NAIL.

STEP 2 LAY OUT MITRES ON BOTH ENDS.

STEP 3 REMOVE BASE FROM WALL AND CUT MITRES.

STEP 4 REPLACE BASE IN SAME POSITION BY DRIVING FIRST NAIL IN SAME HOLE.

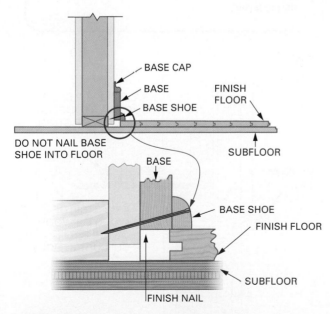

FIGURE 80–8 The base shoe is fastened into the baseboard, not into the floor. This allows the hardwood floor to expand and contract without hindrance.

mitred. When the base shoe must be stopped at a door opening or other location, with nothing to butt against, its exposed end is generally back-mitred and sanded smooth (**Figure 80–9**). A better finish would be to face-mitre the base shoe and return it to the finish floor. Alternatively, the cut end can be shaped and sanded to match the profile of the shoe. No base shoe is required if carpeting is to be used as a finish flooring, as the carpet is stretched on to the "tackless" strip and tucked under the baseboard.

INSTALLING WINDOW TRIM

Interior window trim, in order of installation, consists of the *stool*—also called *stool cap*—*apron*, *jamb extensions*, *casings*, and *stops* or *stop bead* (**Figure 80–10**). Although the kind and amount of trim may differ, depending on the style of window, the application is basically the same. The procedure described in this chapter applies to most double-hung windows.

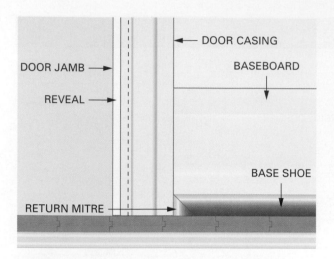

FIGURE 80–9 The end of base shoe moulding is usually return-mitred, glued, and taped with masking tape.

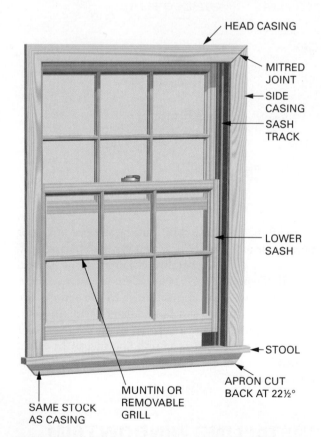

FIGURE 80–10 Component parts of window trim.

Installing the Stool

The bottom side of the stool is rabbeted at an angle to fit on the sill of the window frame so that its top side will be level. Its final position has the outside edge against the sash. Both ends are notched around

the side jambs of the window frame. Each end projects beyond the casings by an amount equal to the casing thickness.

The stool length is equal to the distance between the outside edges of the vertical casing plus twice the casing thickness. On both sides of the window, just above the sill, hold a scrap piece of casing stock on the wall. Its inside edge should be flush with the inside face of the side jamb of the window frame. Draw a light line on the wall along the outside edge of the casing stock. Lay out a distance outward from each line equal to the thickness of the window casing. Cut a piece of stool stock to length equal to the distance between the outermost marks.

Raise the lower sash slightly. Place a short, thin strip of wood under it, on each side, which projects inward to support the stool while it is being laid out (**Figure 80–11**). Place the stool on the strips. Raise or lower the sash slightly so that the top of the stool is level. Position the stool with its outside edge against the wall. Its ends should be in line with the marks previously made on the wall.

ON THE JOB

Wood shingles or trim wood strips placed under the window sash support the stool in a level position during layout.

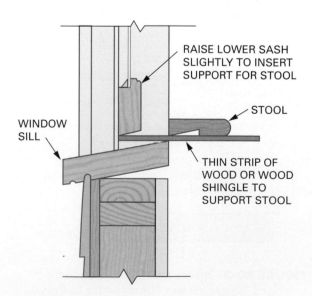

FIGURE 80–11 Technique to hold a stool for easy layout.

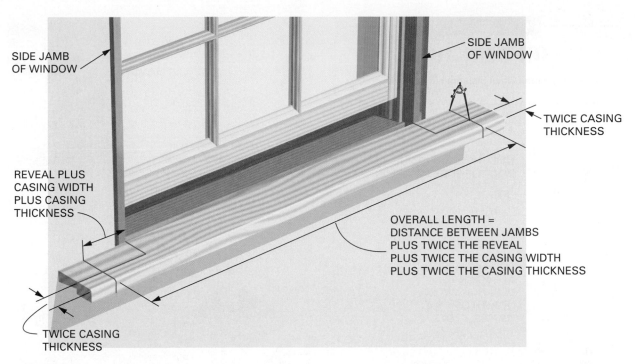

FIGURE 80–12 Method of laying out a stool.

Square lines, across the face of the stool, even with the inside face of each side jamb of the window frame. Set the pencil dividers or scribers so that, on both sides, an amount equal to twice the casing thickness will be left on the stool. Scribe the stool by riding the dividers along the wall on both sides and along the bottom rail of the window sash (**Figure 80–12**).

Cut to the lines, using a handsaw. Smooth the sawn edge that will be nearest to the sash. Shape and smooth both ends of the stool the same as the inside edge. Apply a small amount of caulking compound to the bottom of the stool along its outside edge. Fasten the stool in position by driving finish nails along its outside edge into the sill. Set the nails.

Applying the Apron

The *apron* covers the joint between the sill and the wall. It is applied with its ends in line with the outside edges of the window casing. Cut a length of apron stock equal to the distance between the outer edges of the window casings.

Each end of the apron is then *returned upon itself*. This means that the ends are shaped the same as its face. To return an end upon itself, hold a scrap piece on the apron. Draw its profile flush with the end. Cut to the line with a coping saw. Sand the cut end smooth (**Figure 80–13**). Return the other end upon itself in the same manner.

FIGURE 80–13 Returning the end of an apron upon itself.

Some carpenters face-mitre the apron (at 45 degrees) and return it to itself so that it sits flat to the wall finish. The small piece of return is usually glued and held in place with masking tape until the glue dries.

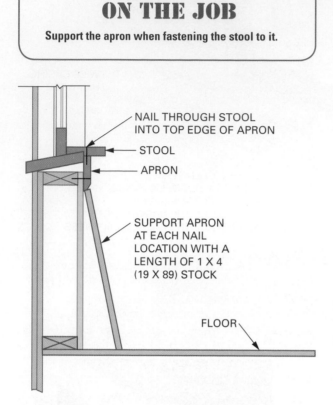

FIGURE 80–14 Technique for holding an apron in place for nailing.

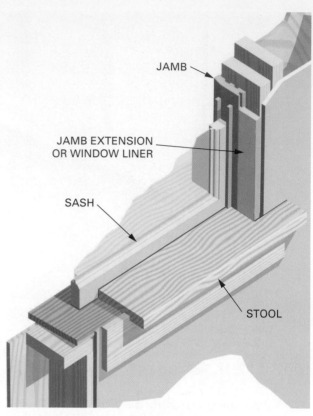

FIGURE 80–15 Jamb extensions are used to widen the window jamb.

Place the apron in position with its upper edge against the bottom of the stool. Be careful not to force the stool upward. Keep the top side of the stool level by holding a square between it and the edge of the side jamb. Fasten the apron along its bottom edge into the wall. Then drive nails through the stool into the top edge of the apron. When nailing through the stool, wedge a short length of 1 × 4 (19 × 89 mm) stock between the apron and the floor at each nail location. This supports the apron while nails are being driven. Failure to support the apron results in an open joint between it and the stool. Take care not to damage the bottom edge of the apron with the supporting piece (**Figure 80–14**).

Installing Jamb Extensions

Windows are often installed with jambs that are narrower than the wall thickness. Strips must be fastened to these narrow jambs to bring the inside edges flush with inside wall surface. These strips are called **jamb extensions**.

Some manufacturers provide jamb extensions with the window unit. However, they are not always applied when the window is installed, but when the window is trimmed. Therefore, when windows are set, these pieces should be carefully stored and then retrieved when it is time to apply the trim. They are usually precut to length and need only to be cut to width.

Measure the distance from the inside edge of the jamb to the finished wall. Rip the jamb extensions to this width with a slight back-bevel on the inside edge. Cut the pieces to length, if necessary, and apply them to the head and side jambs. Drive finish nails through the edges into the edge of the jambs (**Figure 80–15**).

Applying the Casings

Window *casings* usually are installed with a reveal similar to that of door casings. They also may be installed flush with the inside face. In either case, the bottom ends of the side casings rest on the stool. The window casing pattern is usually the same as the door casings. Window casings are applied in the same manner as door casings.

Cut the number of window casings needed to a rough length with a mitre on one end. Cut side casings with left- and right-hand mitres. Install the header casing first and then the side casings. Find the length of side casings by turning them upside down with the point of the mitre on the stool in the same manner as door casings. Fasten casings with their inside edges flush with the inside face of the jamb. Make neat, tight-fitting joints at the stool and at the head.

INSTALLING CLOSET TRIM

A simple clothes closet is normally furnished with a shelf and a rod for hanging clothes. Usually a piece of 1 × 4 (19 × 89 mm) stock is installed around the walls of the closet to support the shelf and the rod. These pieces are called **cleats**. The shelf is installed on top of a cleat. The closet pole is installed in the centre of it. Shelves are not fastened to the cleat. Rods are installed for easy removal in case the closet walls need refinishing.

Shelves are usually 1 × 12 (19 × 286 mm) or melamine boards. Rods may be ¾-inch (19 mm) steel pipe, $^{15}/_{16}$-inch (24 mm) full, round wood poles, or chrome-plated rods manufactured for this purpose. On long spans, the rod may be supported in its centre by special metal closet pole supports. On each end, the closet pole is supported by plastic or metal closet pole *sockets*. In place of sockets, holes and notches are made in the cleat to support the ends of the closet pole.

For ordinary clothes closets, the height from the floor to the top edge of the cleat is 66 inches (1676 mm) (**Figure 80–16**). Measure up from the

floor this distance. Draw a level line on the back wall and two end walls of the closet. Ease the bottom outside corner and install the cleat so that its top edges are to the line. The cleat is installed in the same manner as baseboard. Butt the interior corners. Fasten with two finish nails at each stud.

Install the closet pole sockets on the end cleats. The centre of the socket should be at least 12 inches (300 mm) from the back wall and centred on the width of the cleat. Fasten the sockets through the predrilled holes with the screws provided.

Installing Closet Shelves

For a professional job, fit the ends and back edge of the shelf to the wall. Cut the shelf about ½ inch (12 mm) longer than the distance between end walls.

Place the shelf in position by laying one end on the cleat and tilting the other end up and resting against the wall. Scribe about ¼ inch (6 mm) off the end resting on the cleat. Remove the shelf. Cut to the scribed line. Measure the distance between corners along the back wall. Transfer this measurement to the shelf, measuring from the scribed cut along the back edge of the shelf.

Place the shelf in position, tilted in the opposite direction. Set the dividers to scribe the distance from the wall to the mark on the shelf. Scribe and cut the other end of the shelf. Place the shelf into position, resting it on the cleats. Scribe the back edge to the wall to take off as little as possible. Cut to the scribed line. Ease the corners on the front edge of the shelf with a hand plane. Sand and place the shelf in position (**Procedure 80–E**).

Installing the Closet Pole

Measure the distance between pole sockets. Cut the pole to length. Install the pole on the sockets. One socket is closed. The opposite socket has an open top. Place one end of the pole in the closed socket. Then rest the other end on the opposite socket.

Linen Closets

Linen closets usually consist of a series of shelves spaced 12 to 16 inches (305 to 406 mm) apart. Cleats used to support shelves are 1 × 2 (19 × 38 mm) stock, chamfered on the bottom outside corner. A *chamfer* is a bevel on the edge of a board that extends only partway through the thickness of the stock.

Lay out level lines for the top edges of each set of cleats. Install the cleats and shelves in the same manner as described for clothes closets.

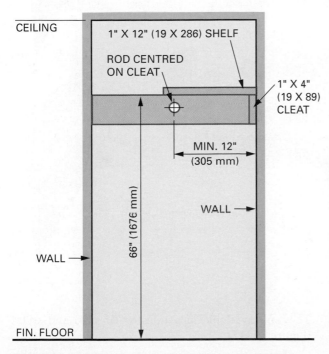

FIGURE 80–16 Specifications for an ordinary clothes closet.

StepbyStep Procedures

Procedure 80–E Fitting a Closet Shelf

STEP 1 CUT SHELF ABOUT ¹/₂" (12 mm) LONGER THAN WIDTH OF CLOSET.

STEP 2 TILT SHELF IN POSITION WITH ONE END ON CLEAT. SCRIBE ABOUT ¹/₄" (6 mm) ON THIS END. REMOVE SHELF AND CUT TO SCRIBED LINE.

STEP 3 FROM SCRIBED END, LAY OUT LENGTH OF SHELF ON BACK EDGE.

STEP 4 REPLACE SHELF IN TILTED POSITION WITH OPPOSITE END ON CLEAT.

STEP 5 SET DIVIDERS FOR DISTANCE FROM WALL TO MARK INDICATING SHELF LENGTH AT BACK EDGE. SCRIBE ALONG END OF SHELF.

STEP 6 REMOVE SHELF AND CUT END TO SCRIBED LINE. REPLACE SHELF WITH BOTH ENDS ON CLEAT.

STEP 7 SCRIBE BACK EDGE TO BACK WALL. SCRIBE ONLY ENOUGH TO FIT SHELF. REMOVE SHELF AND CUT TO SCRIBED LINE. EASE CORNERS ON FRONT EDGE. REPLACE SHELF.

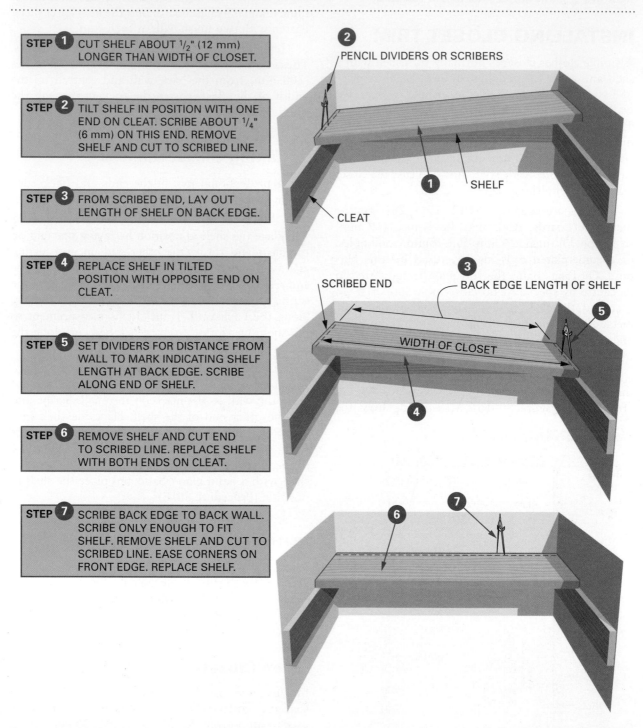

PENCIL DIVIDERS OR SCRIBERS

SHELF

CLEAT

SCRIBED END

BACK EDGE LENGTH OF SHELF

WIDTH OF CLOSET

MANTELS

Mantels are used to decorate fireplaces and to cover the joint between the fireplace and the wall. Most mantels come preassembled from the factory. They are available in a number of sizes and styles (**Figure 80–17**).

Study the manufacturer's directions carefully. Place the mantel against the wall. Centre it on the

FIGURE 80–17 Mantels may come preassembled in a number of styles and sizes.

fireplace. Scribe it to the floor or wall as necessary. Carefully fasten the mantel in place and set all nails.

CONCLUSION

All pieces of interior trim should be sanded smooth after they have been cut and fitted, and before they are fastened. The sanding of interior finish provides a smooth base for the application of stains, paints, and clear coatings. Always sand with the grain, never across the grain.

All sharp, exposed corners of trim should be rounded over slightly. Use a block plane to make a slight chamfer. Then round over with sandpaper.

If the trim is to be stained, make sure every trace of glue is removed. If excess glue is allowed to dry, it seals the surface and does not allow the stain to penetrate resulting in a blotchy finish.

Be careful not to make hammer marks in the finish. Occasionally rubbing the face of the hammer with sandpaper to clean it helps prevent it from glancing off the head of a nail.

Make sure any pencil lines left along the edge of a cut are removed before fastening the pieces. Pencil marks in interior corners are difficult to remove after the pieces are fastened in position. Pencil marks show through a stained or clear finish and make the joint appear open. When marking interior trim make light, fine pencil marks.

Note: Layout lines in the illustrations are purposely made dark and heavy only for the sake of clarity.

Make sure all joints are tight fitting. Measure, mark, and cut carefully. Do not leave a poor fit. Do it over, if necessary!

KEY TERMS

apron	casing	finger-jointed	reveal
back bands	chair rail	full-round	scarf joints
back mitre	cleat	half-round	scribing
base moulding	coped	jamb extensions	standard mouldings
base cap moulding	corner guards	mitre box	stool
base shoe moulding	cove moulding	mitred	stop
bed moulding	crown moulding	plinth blocks	surfaced four sides
caps	door casings	quarter-round	

REVIEW QUESTIONS

Select the most appropriate answer.

1. What are bed, crown, and cove mouldings frequently used as?

   a. window trim
   b. ceiling moulding
   c. part of the base
   d. door casings

2. What are back bands applied to?

   a. wainscotting
   b. exterior corners
   c. casings
   d. interior corners

3. What is a *stool* part of?

a. soffit
b. door trim
c. base
d. window trim

4. What is usually done to the joint between mouldings in interior corners?

a. It is coped.
b. It is mitred.
c. It is butted.
d. It is bisected.

5. What is the setback of door casings from the face of the jamb often referred to as?

a. gain
b. backset
c. reveal
d. quirk

6. How do you find the length of door side casings?

a. by measuring the distance from floor to the header casing
b. by marking the length on a scrap strip and transferring it to the side casing
c. by inverting the side casing with the point of the mitre cut on the floor
d. by holding the side casing with the right end up and marking the mitre

7. What should be done if the joint between a head and side casing is not a tight fit?

a. Plane the mitred surfaces.
b. Fill the gap with glue.
c. Sand the casing face.
d. Nail the casing tighter.

8. How can the cope on baseboard be laid out more accurately?

a. by back-mitring
b. by returning it
c. by using a combination square
d. by scribing

9. What is the base shoe fastened to?

a. the baseboard only
b. both the base and the floor
c. the floor only
d. directly to the wall

10. When the end of a moulding has no material to butt against, what should be done to the end?

a. It should be back-mitred.
b. It should be mitred.
c. It should be returned onto itself.
d. It should be coped.

Unit 27

Stair Finish

The staircase is usually the most outstanding feature of a building's interior. It is a showplace for architectural appeal and carpentry skills. All stair finish work must be done in a first-class manner. Joints between stair finish members must be accurate and tight-fitting. Balustrades are installed with perfectly fitting joints.

 SAFETY REMINDER

Staircases are used by many different sizes of people. They must be built to safety standards. They are also assembled from many small pieces. Each piece should be examined for defects, looking for strength as well as appearance characteristics. Each must be installed securely enough to provide support and prevent accidents.

OBJECTIVES

After completing this unit, the student should be able to:

- name various stair finish parts and describe their location and function.
- lay out, dado, and assemble a housed-stringer staircase.
- apply finish to open and closed staircases.
- lay out treads for winding steps.
- install a post-to-post balustrade, without fittings, from floor to balcony on the open end of a staircase.
- install an over-the-post balustrade, with fittings, on an open staircase that runs from a starting step to an intermediate landing and then to a balcony.

Chapter 81

Laying Out Open and Closed Staircases

Many kinds of stair finish parts are manufactured in a wide variety of wood species, such as oak, beech, cherry, poplar, pine, and hemlock. It is important to identify each of the parts, know their location, and understand their function when learning to apply stair finish.

TYPES OF STAIRCASES

The stair finish may be separated in two parts: the **stair body** and the **balustrade**. Important components of the stair body finish are *treads*, *risers*, and *finish stringers*. The stair body may be constructed as an *open* or *closed staircase*. In an open staircase, the ends of the treads are exposed to view. In a closed staircase, they butt against the wall. Staircases may be open or closed on one or both sides.

FIGURE 81–2 An open-one-side staircase with an over-the-post balustrade.

Major parts of the balustrade include *hand-rails*, *newel posts*, and *balusters*. Balustrades are constructed in either a **post-to-post** or **over-the-post** method. In the post-to-post method, the handrail is fitted between the newel posts (**Figure 81–1**). In the over-the-post method, the handrail runs continuously from top to bottom. It requires special curved sections, called *fittings*, where the handrail changes height or direction (**Figure 81–2**).

STAIR BODY PARTS

Many kinds of stair parts are required to finish the stair body.

Risers

Risers are vertical members that enclose the space between treads. They are manufactured in a thickness

FIGURE 81–1 A closed staircase with a post-to-post balustrade on a knee wall.

of ¾ inch (19 mm) and in widths of 7½ and 8 inches (191 and 203 mm).

Treads

Treads are horizontal finish members on which the feet are placed when ascending or descending stairs. High-quality treads are made from oak. Others are made from poplar or hard pine. They normally come in ¾- and 1¹/₃₂-inch (19 and 26 mm) thicknesses, in 10½- and 11½-inch (267 and 292 mm) widths, and in lengths from 36 up to 72 inches (914 to 1829 mm).

Nosings. The outside edge of the tread beyond the riser has a half-round shape. It is called the nosing. The National Building Code of Canada (NBCC) limits the projection of the tread nosing to 1 inch (25 mm) for residential stairs.

A *return nosing* is a separate piece mitred to the open end of a tread to conceal its end grain. Return nosings are available in the same thickness as treads and in 1¼-inch (32 mm) widths. Treads are available with the return nosing already applied to one end (**Figure 81–3**).

Landing Treads. Landing treads are used at the edge of landings and balconies. They match the tread thickness at the nosing. However, they are rabbeted to match the finish floor thickness on the landing. They come in 3½- and 5½-inch (89 and 140 mm) widths. The wider landing tread is used when newel posts are more than 3½ inches (89 mm) wide (**Figure 81–4**).

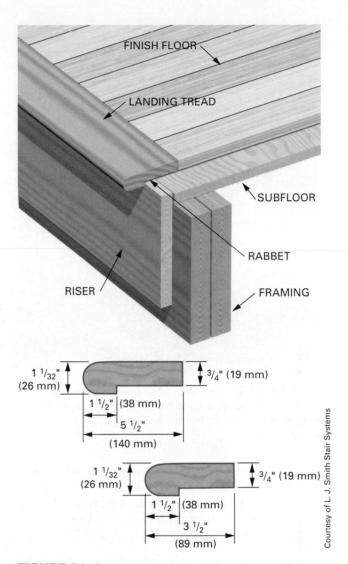

FIGURE 81–4 Landing treads are rabbeted to match the thickness of the finish floor.

Courtesy of L. J. Smith Stair Systems

Tread Moulding. The tread moulding is a small cove moulding used to finish the joint under stair and landing treads. The moulding should be the same kind of wood as the treads. Its usual size is ⅝ × ¹³/₁₆ inch (16 × 21 mm).

Finish Stringers

Finish stringers are sometimes called skirt boards. They are members of the stair body used to trim the intersection of risers and treads with the wall. When placing the stairs against a framed wall, a 2 × 4 (38 × 89 mm) is nailed between the carriage and the wall. This allows the drywall and skirt boards to fit neatly between the studs and the treads. They are called *closed finish stringers* when they are located above treads that butt the wall.

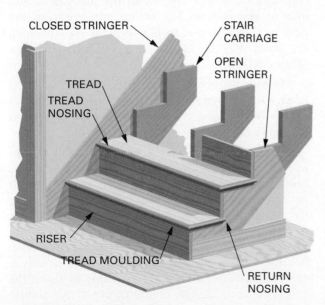

FIGURE 81–3 Treads and risers are fastened with glue and screws to supporting stair carriages.

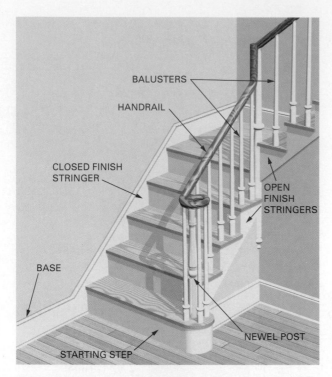

FIGURE 81–5 Open and closed finish stringers are finished trim pieces used to cover the intersections of the treads and risers with the wall and stair body.

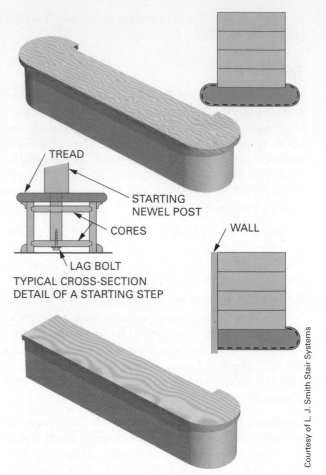

Courtesy of L. J. Smith Stair Systems

FIGURE 81–6 Starting steps are available in a number of styles and sizes.

They are termed *open finish stringers* when they are placed on the open side of a stairway below the treads (**Figure 81–5**). Finish stringer lineal stock is available in a ¾-inch (19 mm) thickness and in widths of 9¼ and 11¼ inches (235 and 286 mm).

Starting Steps

A **starting step** or *bull-nose step* is the first tread-and-riser unit, sometimes used at the bottom of a stairway. The starting step is used when the staircase is open on one or both sides and the handrail curves outward at the bottom. They are available in a number of styles with *bull-nosed* ends, preassembled and ready for installation (**Figure 81–6**).

BALUSTRADE MEMBERS

Finish members of the balustrade are available in many designs that are combined to complement each other. Various types of fittings are sometimes joined to straight lengths of handrail when turns in direction are required.

Newel Posts

A **newel post** is anchored securely to the staircase to support the handrail. In post-to-post balustrades,

the newel posts have flat, square surfaces near the top, against which the handrails are fitted, and also at the bottom for fitting and securing the post to the staircase. In between the flat surfaces, the posts may be *turned* in a variety of designs (**Figure 81–7**).

In over-the-post systems, a round pin at the top of each newel post fits into the underside of handrail fittings. The posts are tapered toward the top end in a number of turned designs (**Figure 81–8**).

Three types of newel posts are used in a post-to-post balustrade. *Starting newels* are used at the bottom of a staircase. They are fitted against the first or second riser. If fitted against the second riser, the flat, square surface at the bottom must be longer. At the top of the staircase, *second-floor newels* are used. *Intermediate landing newels* are also available. Because part of the bottom end of these newels is exposed, turned *buttons* are available to finish the end. The same design (a decorative finish similar to the ones at the top of the newel) is used in the same staircase for each of the three types of posts.

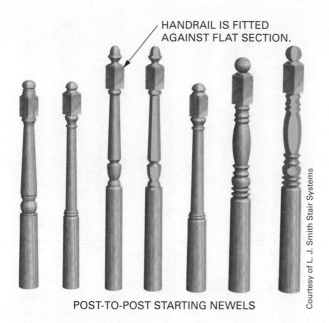

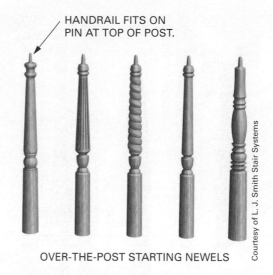

OVER-THE-POST STARTING NEWELS

FIGURE 81–8 Newels in over-the-post balustrades are made with a pin at the top.

POST-TO-POST STARTING NEWELS

FIGURE 81–7 Newels in post-to-post balustrades must have flat surfaces where the handrails attach.

They differ only in their overall length and in the length of the flat surfaces (**Figure 81–9**).

Four types of newel posts are used in an over-the-post balustrade. There are three types of *starting newels* depending on the type of handrail fitting used.

If a *volute* or *turnout* is used, a newel post with a dowel at the bottom is installed on top of a required starting step. The fourth type is a longer newel for landings, where a gooseneck handrail fitting is used (**Figure 81–10**).

When the balustrade ends against a wall, a *half newel* is sometimes fastened to the wall. The handrail is then butted to it. In place of a half newel, the

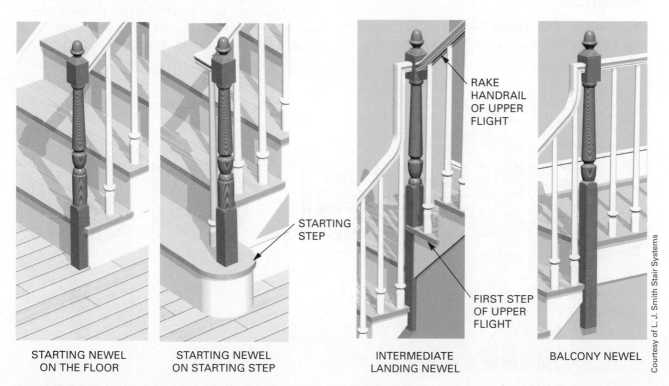

FIGURE 81–9 Three types of newel posts are used in a post-to-post balustrade.

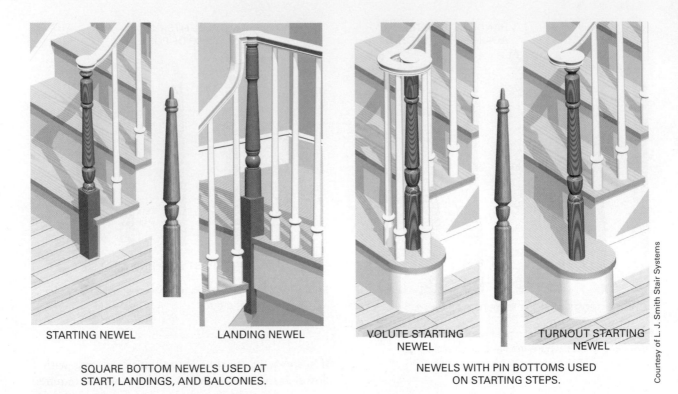

STARTING NEWEL   LANDING NEWEL

VOLUTE STARTING
NEWEL

TURNOUT STARTING
NEWEL

SQUARE BOTTOM NEWELS USED AT
START, LANDINGS, AND BALCONIES.

NEWELS WITH PIN BOTTOMS USED
ON STARTING STEPS.

Courtesy of L. J. Smith Stair Systems

FIGURE 81–10 Newels for over-the-post balustrades either have pinned or square bottoms.

handrail may butt against an oval or round *rosette* (**Figure 81–11**).

Handrails

The handrail is the sloping finish member grasped by the hand of the person ascending or descending the stairs. It is installed horizontally when it runs along the edge of a balcony. Handrail heights are 34 to 38 inches (864 to 965 mm) [NBCC 9.8.7.4(2)] vertically above the nosing edge of the tread. There should be a continuous 2-inch (50 mm) finger clearance between the rail and the wall. Several styles of handrails come in lineal lengths that are cut to fit on the job. Some handrails

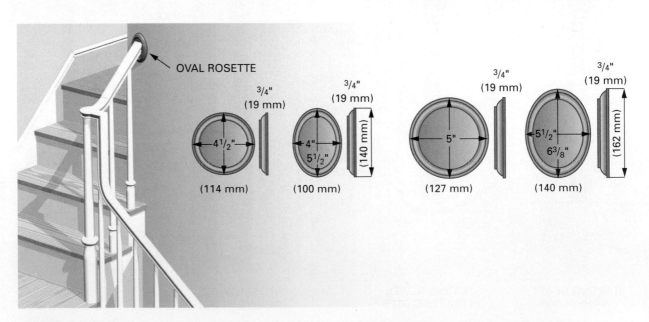

OVAL ROSETTE

3/4"
(19 mm)

3/4"
(19 mm)

4 1/2"
(114 mm)

4"
5 1/2"
(100 mm)

140 mm

3/4"
(19 mm)

5"
(127 mm)

3/4"
(19 mm)

5 1/2"
6 3/8"
(140 mm)

162 mm

Courtesy of L. J. Smith Stair Systems

FIGURE 81–11 Rosettes are fastened to the wall to provide a decorative attaching surface for the handrail.

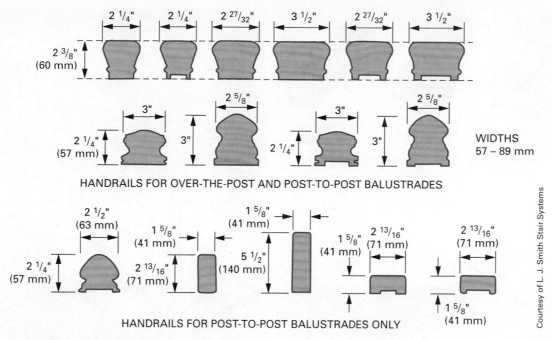

HANDRAILS FOR OVER-THE-POST AND POST-TO-POST BALUSTRADES

WIDTHS
57 – 89 mm

HANDRAILS FOR POST-TO-POST BALUSTRADES ONLY

Courtesy of L. J. Smith Stair Systems

FIGURE 81–12 Straight lengths of handrail are manufactured in many styles.

are *ploughed* with a wide groove on the bottom side to hold square top balusters in place (**Figure 81–12**).

On closed staircases, a balustrade may be installed on top of a **knee wall** or buttress. In relation to stairs, a knee wall is a short wall that projects a short distance above and on the same rake as the stair body. A **shoe rail** or buttress cap, which is ploughed on the top side, is usually applied to the top of the knee wall on which the bottom ends of balusters are fastened (**Figure 81–13**). Narrow strips, called **fillets**, are used between balusters to fill the ploughed groove on handrails and shoe rails.

Handrail Fittings

Short sections of specially curved handrail are called **fittings**. They are used at various locations, joined to straight sections, to change the direction of the handrail. They are classified as *starting*, *gooseneck*, and *miscellaneous* fittings.

Starting Fittings. To start an over-the-post handrail, starting fittings called **volutes**, **turnouts**, or *starting easings* may be used. In a post-to-post system, a straight length of handrail may be used at the bottom. To start with a soft, graceful curve, an *upeasing* is used (**Figure 81–14**).

Gooseneck Fittings. In over-the-post systems, in which the handrail is continuous, fittings called **goosenecks** are required at intermediate landings and at the top. This is because of changes in the handrail height or direction. In post-to-post systems, goosenecks are not required (**Figure 81–15**).

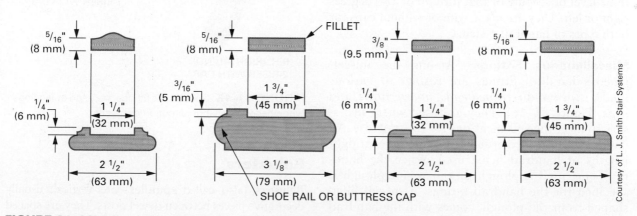

FILLET

SHOE RAIL OR BUTTRESS CAP

Courtesy of L. J. Smith Stair Systems

FIGURE 81–13 A shoe rail is often used at the bottom of a balustrade that is constructed on a knee wall.

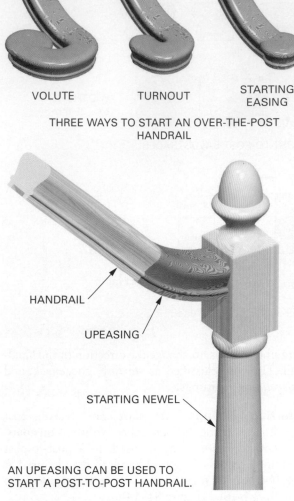

VOLUTE TURNOUT STARTING EASING

THREE WAYS TO START AN OVER-THE-POST HANDRAIL

HANDRAIL

UPEASING

STARTING NEWEL

AN UPEASING CAN BE USED TO START A POST-TO-POST HANDRAIL.

Courtesy of L. J. Smith Stair Systems

FIGURE 81–14 A handrail fitting is used to start an over-the-post handrail. An upeasing is sometimes used to start a post-to-post handrail.

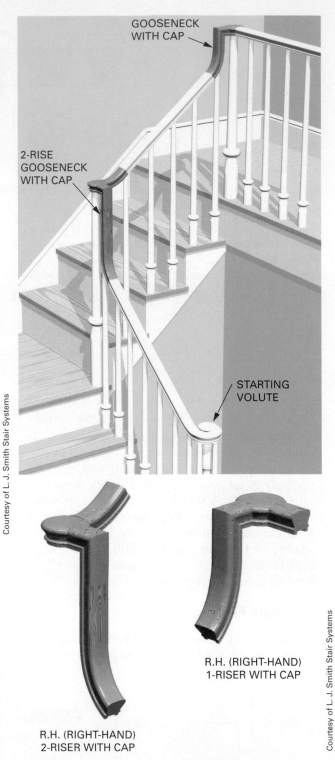

GOOSENECK WITH CAP

2-RISE GOOSENECK WITH CAP

STARTING VOLUTE

R.H. (RIGHT-HAND) 2-RISER WITH CAP

R.H. (RIGHT-HAND) 1-RISER WITH CAP

Courtesy of L. J. Smith Stair Systems

FIGURE 81–15 Gooseneck fittings are used at landings when handrails change direction or height.

Goosenecks are available for handrails that continue level or sloping or that turn 90 or 180 degrees right or left. They are made with or without caps for both types of handrail systems.

Miscellaneous Fittings. Among the miscellaneous handrail fittings are *easings* of various kinds, *coped* and *returned ends*, *quarterturns*, and *caps* (**Figure 81–16**). They are used where necessary to meet the specifications for the staircase. All handrail fittings are ordered to match the straight lengths of handrail being used. When the handrail being used is ploughed, a matching plough is specified for the handrail fittings. A special fillet, shaped to fit the plough, comes with the handrail fitting.

Balusters

Balusters (also called **spindles**) are vertical, usually decorative pieces between newel posts. They are spaced close together and support the handrail. On a knee wall,

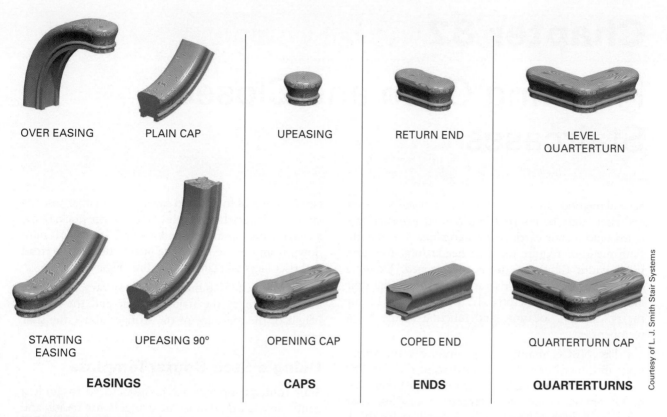

OVER EASING   PLAIN CAP   UPEASING   RETURN END   LEVEL QUARTERTURN

STARTING EASING   UPEASING 90°   OPENING CAP   COPED END   QUARTERTURN CAP

EASINGS   **CAPS**   **ENDS**   **QUARTERTURNS**

Courtesy of L. J. Smith Stair Systems

FIGURE 81–16 Many miscellaneous fittings are available to meet the needs of any type of handrail installation.

they run from the handrail to the shoe rail. On an open staircase, they run from the handrail to the treads (see Figure 81–2).

Balusters are manufactured in many styles. They should be selected to complement the newel posts being used (**Figure 81–17**). For example, in a post-to-post staircase, balusters with square tops are usually used. In an over-the-post staircase,

balusters that are tapered at the top complement the newel posts.

Most balusters are made in lengths of 31, 34, 36, 39, and 42 inches (787, 864, 914, 991, and 1067 mm) for use in any part of the stairway. Several lengths of the same style of baluster are needed for each tread of the staircase because of the rake of the handrail.

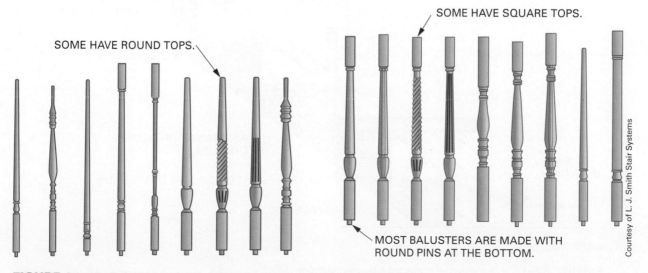

SOME HAVE SQUARE TOPS.

SOME HAVE ROUND TOPS.

MOST BALUSTERS ARE MADE WITH ROUND PINS AT THE BOTTOM.

Courtesy of L. J. Smith Stair Systems

FIGURE 81–17 Balusters are made in designs that match newel post design.

Chapter 82

Finishing Open and Closed Staircases

Several methods are used to finish a set of stairs. Treads and risers may be inserted into housed stringers, fastened onto a stair carriage, or a combination of both. Both methods require the riser height (unit rise) and tread width (unit run) to be determined. Example calculations of the unit rise and unit run of the stair may be found in Chapter 37. Review Unit 12 for the description, location, and function of the finish members.

The NBCC maintains requirements for variations in dimensions (9.8.4.4). Adjacent riser heights must be within $\frac{3}{16}$ inch (5 mm) of each other. The maximum variation from largest to smallest riser must be $\frac{3}{8}$ inch (10 mm). Check local codes. With care, all riser dimensions of a set of stairs can be built with only $\frac{1}{16}$-inch (1.5 mm) variations.

MAKING A HOUSED STRINGER

Housed stringers can be laid out using a stair router template or a pitch board and job-made router template (see Chapter 38 for dadoed carriages). In either case the layout begins on the face side of the stringer stock. Draw a setback line parallel to and about 2 inches (50 mm) down from the top edge. The intersection of the tread and riser faces will land on this line (**Figure 82–1**).

The 2-inch (50 mm) distance may vary, depending on the width of the stringer stock and the desired height of the top edge of the stringer above the stair treads.

Using a Stair Router Template

Stair router templates are manufactured to guide a router in making dadoes in stringers for treads and risers. The router must be equipped with a straight bit and a template guide of the correct size. Stair templates are adjustable for different rises and runs. They are easily clamped to the stock for routing the dadoes and then moved.

The template is shaped so that the dadoes will be the exact tread width at the nosing and wider toward the back side of the stair. The template has non-parallel sides so that the finished dadoes will be tapered.

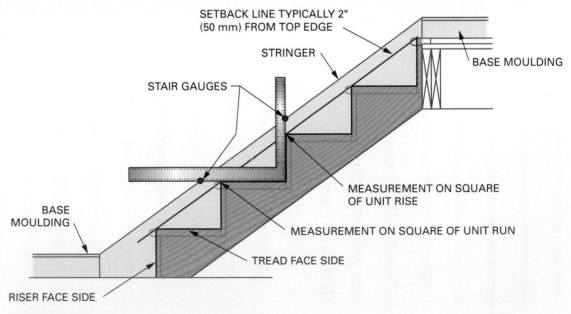

FIGURE 82–1 Layout considerations for a housed stringer.

The treads and risers are then wedged tightly against the face side shoulders of the dadoes (**Figure 82–2**).

Full layout of all treads and risers is not necessary while using the template. An alignment gauge is designed to position the template along the setback line according to the unit length of the stair (**Figure 82–3**). Using the Pythagorean theorem, calculate the unit length. Lightly mark squared lines on the setback line spaced out the unit length distance.

Using a framing square, mark the unit rise and unit run on the board. Mark them such that a rise–run intersection lands on the setback line. Then, using these lines, mark the thickness of a tread and riser pair. Loosen template shoulder clamp bolts and position the square edges of the template to fit parallel to the tread-riser layout. Retighten shoulder clamps.

Move and clamp the template to the stringer with the alignment gauge on a unit length line. Rout out the stringer about ¼ inch (6 mm) deep. Place the router on the template where it does not touch the stock material. Start router and ease it into the stringer. Press the

router guide firmly against the template on all four sides. This will ensure the dado is completely removed.

Let the router come to a complete stop before removing it from the template. This will reduce the danger of personal injury and damage to the template and bit. Loosen the template clamp and move the template to the next unit length line. Make sure the template is resting entirely on the face of the stock. Rout the dado and repeat for the remaining treads and risers.

Using a Pitch Board

A **pitch board** may be used to lay out a housed stringer. This process requires that each tread and riser pair be laid out. A pitch board is a piece of stock, usually ¾ inch (19 mm) thick. It is cut to the rise and run of the stairs. A strip of wood is fastened to the rake edge of the pitch board. This is used to hold the pitch board against the edge while laying out the stringer (**Figure 82–4**). Care should be taken when making the pitch board because many layouts will be made from it.

Using the pitch board, lay out the risers and treads for each step of the staircase. These lines show the location of the face side of each riser and tread and are the outside edges of the housing (**Figure 82–5**).

After the stringer has been laid out, make a template to guide the router by cutting out the shape of the dadoes from a piece of thin plywood or hardboard. The cut is made slightly larger than the dadoes to allow for the router guide to follow the template. Take care to make cuts smooth and clean as the router guide will transfer onto the stringer every deviation in the template.

Completing the Stringer

Cut and fit the top and bottom ends of the stringer to the floor and the top end to the landing. Equalize the

> **CAUTION**
>
> Take great care when removing the router from the template. The bit will damage the template if it is touched and the operator is at risk from flying metal fragments.

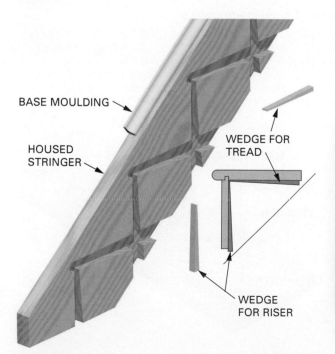

BASE MOULDING

HOUSED STRINGER

WEDGE FOR TREAD

WEDGE FOR RISER

FIGURE 82–2 A housed stringer is dadoed to accept treads, risers, and wedges.

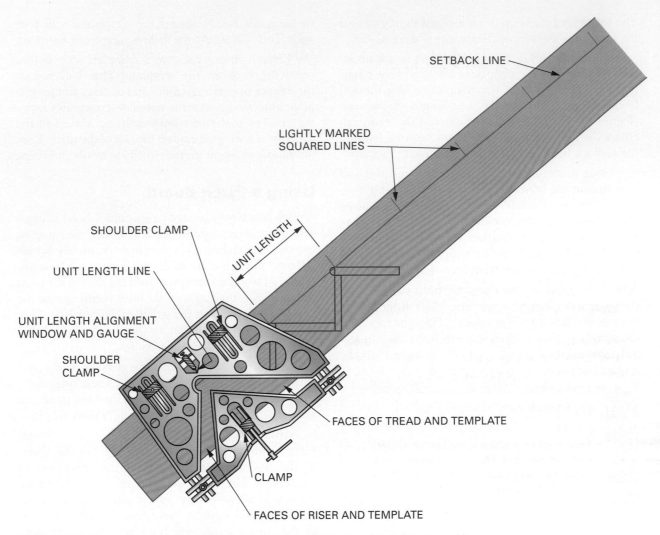

SETBACK LINE

LIGHTLY MARKED
SQUARED LINES

SHOULDER CLAMP

UNIT LENGTH LINE

UNIT LENGTH

UNIT LENGTH ALIGNMENT
WINDOW AND GAUGE

SHOULDER
CLAMP

FACES OF TREAD AND TEMPLATE

CLAMP

FACES OF RISER AND TEMPLATE

FIGURE 82–3 A stair router template is clamped to the stock at each unit length marking.

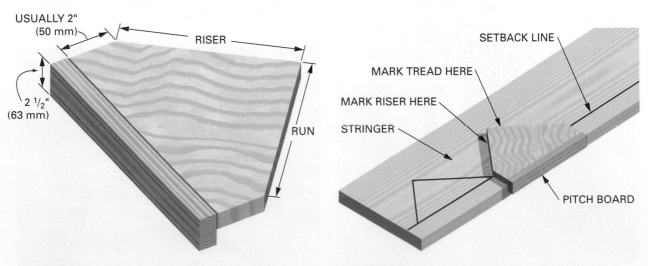

USUALLY 2"
(50 mm)

RISER

2 1/2"
(63 mm)

RUN

FIGURE 82–4 A pitch board can be used for laying out a housed stringer.

SETBACK LINE

MARK TREAD HERE

MARK RISER HERE

STRINGER

PITCH BOARD

FIGURE 82–5 Use the pitch board to step off each tread and riser on a housed stringer.

bottom riser to account for the finished floor thickness. Make end cuts that will properly join with the baseboard. This joint should be made in a professional manner to provide a continuous line of finish from one floor to the next. Because the stringer is usually S4S stock and the base often is not, the stringer should be cut to allow the base to end against it.

LAYING OUT AN OPEN STRINGER

The layout of an open (or *mitred*) stringer is similar to that of a housed stringer. However, riser and tread layout lines intersect at the top edge of the stringer, instead of against a line in from the edge. The riser layout line is the outside face of the riser. This layout line is mitred to fit the mitred end of the riser. The tread layout is to the face side of the tread. The risers and treads are marked lightly with a sharp pencil (**Figure 82–6**).

To lay out the *mitre cut* for the risers, measure in at right angles from the riser layout line a distance equal to the thickness of the riser stock. Draw another plumb line at this point. Square both lines across the top edge of the stringer stock. Draw a diagonal line on the top edge to mark the mitre angle (**Figure 82–7**).

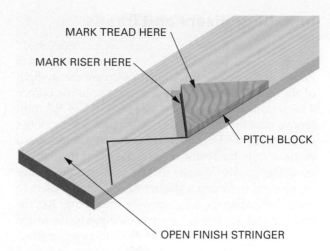

FIGURE 82–6 Laying out the open finish stringer of a housed staircase.

To mark the tread cut on the stringer, measure down from the tread layout line a distance equal to the thickness of the tread stock. Draw a level line at this point for the tread cut. The tread cut is square through the thickness of the stringer. Fit the bottom end to the floor. Fit the top end against the landing. Make the mitred plumb cuts for the risers and the square level cuts for the treads.

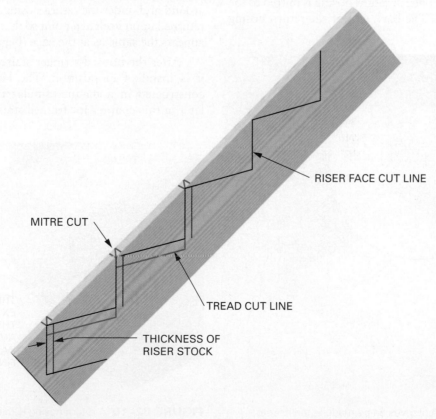

FIGURE 82–7 Laying out the mitre angle on an open finish stringer.

Installing Risers and Treads

Cut the required number of risers to a rough length. Determine the face side of each piece. Sometimes a rabbet and a groove are made to increase the strength of the tread and riser joint forming the inside corner. This is done by cutting a ⅜ × ⅜-inch (9.5 × 9.5 mm) groove near the bottom edge on the face side of all but the starting riser. The groove is located so that its top is a distance from the bottom edge equal to the tread thickness. The groove is made for the rabbeted inner edge of the tread to fit into it (**Figure 82–8**). Rip the risers to width by cutting the edge opposite the groove. Rip the treads to width. Rabbet their back edges to fit in the riser grooves. Cut the risers and tread to exact lengths.

On a closed staircase, the risers are installed with wedges, glue, and screws between housed stringers. On the open side of a staircase, where the riser and open stringer meet, a mitred joint is made so that no end grain is exposed (**Figure 82–9**). The treads are then installed with wedges, glue, and screws on the closed side and with screws through screw blocks on the open side. Screw blocks reinforce interior corners on the underside at appropriate locations and intervals.

Applying Return Nosings and Tread Moulding

If the staircase is open, a return nosing is mitred to the end of the tread. The back end of the return nosing

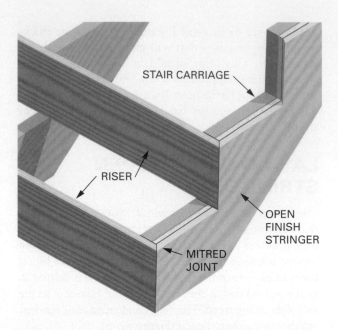

FIGURE 82–9 A mitred joint is made between the risers and open finish stringer so that no end grain is exposed.

projects past the riser the same amount as the tread overhangs the riser. The end is returned upon itself.

The tread moulding is then applied under the overhang of the tread. If the staircase is closed on both sides, the moulding is cut to fit between finish stringers. On the open end of a staircase, the moulding is mitred around and under the return nosing. It is stopped and returned upon itself at a point so that the end assembly appears the same as at the edge (**Figure 82–10**).

After the housed-stringer staircase is assembled, it is installed in position. The balustrade is then constructed in a manner similar to that described later in this chapter for framed staircases.

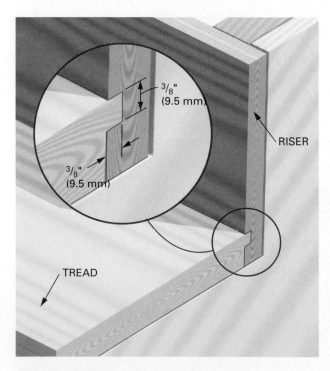

FIGURE 82–8 A groove can be made in the riser to receive the rabbeted inner edge of the tread.

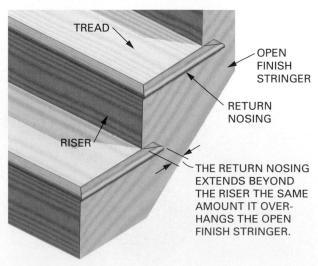

FIGURE 82–10 A return nosing is mitred to the open end of a tread.

FINISHING THE BODY OF A CLOSED STAIRCASE SUPPORTED BY STAIR FRAMING

This section describes the installation of finish to a closed staircase in which the supporting stair carriages have already been installed between two walls that extend from floor to ceiling.

Applying Risers

The first trim members applied to the stair carriages are the risers. Rip the riser stock to the proper width. Cut grooves as described previously for housed-stringer construction. Cut the risers to a length, with square ends, about ¼ inch (6 mm) less than the distance between walls.

Fasten the risers in position with three 2½-inch (63 mm) finish nails into each stair carriage. Start at the top and work down. Remove the temporary treads installed previously as work progresses downward (**Figure 82–11**).

> **CAUTION**
>
> Put up positive barriers at the top and bottom of the stairs so that the stairs cannot be used while the finish is applied. A serious accident can happen if a person who does not realize that the temporary treads have been removed uses the stairs.

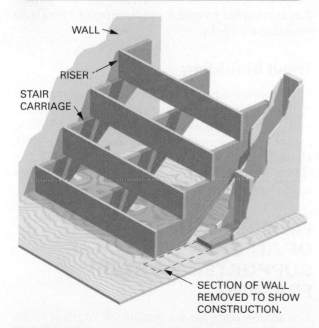

FIGURE 82–11 Risers are typically the first finish members applied to the stair carriage in a closed staircase.

Laying Out and Installing the Closed Finish Stringer

After the risers have been installed, the *closed finish stringer* is cut around the previously installed risers. Usually 1 × 10 (19 × 235 mm) lumber is used. When installed, its top edge will be about 3 inches (75 mm) above the tread nosing. A 1 × 12 (19 × 286 mm) may be used if a wider finish stringer is desired.

Tack a length of stringer stock to the wall. Its bottom edge should rest on the top edges of the previously installed risers. Its bottom end should rest on the floor. The top end should extend about 6 inches (150 mm) beyond the landing.

Lay out plumb lines, from the face of each riser, across the face of the finish stringer. If the riser itself is out of plumb, then plumb upward from that part of the riser that projects farthest outward. Then lay out level lines on the stringer, from each tread cut of the stair carriage and also from the floor of the landing above (**Figure 82–12**).

Remove the stringer from the wall. Cut to the layout lines. Follow the plumb lines carefully. Plumb cuts will butt against the face of the risers, so a careful cut needs to be made. Not as much care needs to be taken with level cuts because treads will later butt against and cover them. A circular saw may be used

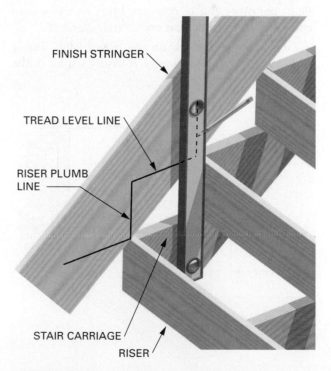

FIGURE 82–12 The closed finish stringer is laid out using a level to extend plumb and level lines from the stair carriage.

to make most of the cut and then a handsaw is used to finish the cut.

After the cutouts are made in the finish stringer, tack it back in position. Fit it to the floor. Then lay out top and bottom ends to join the base that will later be installed on the walls. Remove the stringer. Make the end cuts. Sand the board, and place it back in position. Fasten the stringer securely to the wall with finishing nails. Do not nail too low to avoid splitting the lower end of the stringer. Install the finish stringer on the other wall in the same manner.

Drive shims, at each step, between the back side of the risers and the stair carriage. The shims force the risers tightly against the plumb cut of the finish stringer. Shim at intermediate stair carriages to straighten the risers, from end to end, between walls.

Installing Treads

Treads are cut on both ends to fit snugly between the finish stringers. The nosed edge of the tread projects beyond the face of the riser by 1 inch (25 mm) (**Figure 82–13**).

Along the top edge of the riser, measure carefully the distance between finish stringers. Transfer the measurement and square lines across the tread. Cut in from the nosed edge. Square through the thickness for a short distance. Then undercut slightly. Smooth the cut ends with a block plane. Rub one end with wax. Place the other end in position. Press on the waxed end until the tread lies flat on the stair carriages.

Place a short block on the nosed edge. Tap it until the inner rabbeted edge is firmly seated in the groove of the riser.

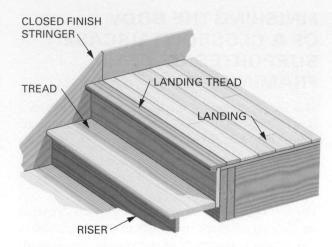

FIGURE 82–14 A rabbeted landing tread is used at the top of the stairway.

If it is possible to work from the underside, the tread may be fastened by the use of screw blocks at each stair carriage and at intermediate locations.

If it is not possible to work from the underside, the treads must be face-nailed. Fasten each tread in place with three 2½-inch (63 mm) finish nails into each stair carriage. It may be necessary to drill holes in hardwood treads to prevent splitting the tread or bending the nail. A little wax applied to the nail makes driving easier and helps keep the nail from bending.

Start from the bottom and work up, installing the treads in a similar manner. At the top of the stairs, install a landing tread. If 1¹/₃₂-inch-thick (26 mm) treads are used on the staircase, use a landing tread that is rabbeted to match the thickness of the finish floor (**Figure 82–14**).

Tread Moulding

The tread moulding is installed under the overhang of the tread and against the riser. Cut the moulding to the same length as the treads, using a mitre box. Predrill holes. Fasten the moulding in place with 1¼-inch (32 mm) finish nails spaced about 12 inches (305 mm) apart. Nails are driven at about a 45-degree angle through the centre of the moulding.

FINISHING THE BODY OF AN OPEN STAIRCASE SUPPORTED BY STAIR FRAMING

This section describes the installation of finish to the stair body of a staircase, supported by stair carriages, which is closed on one side and open on the other side.

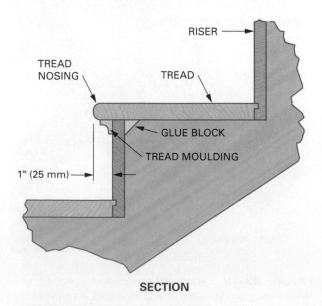

SECTION

FIGURE 82–13 Tread and riser details.

Installing the Finish Stringers

The *open finish stringer* must be installed before the *risers* and the *closed finish stringer*. To lay out the open finish stringer, cut a length of finish stringer stock. Fit it to the floor and against the landing. Its top edge should be flush with the top edge of the stair carriage. Tack it in this position to keep it from moving while it is being laid out.

First, lay out level lines on the face of the stringer in line with the tread cut on the stair carriage. Next, plumb lines must be laid out on the face of the finish stringer for making mitre joints with risers.

Using a Preacher to Lay Out Plumb Lines. Use a **preacher** to lay out the plumb lines on the open finish stringer. A preacher is made from a piece of nominal 4-inch (89 mm) stock about 12 inches (305 mm) long. Its thickness must be the same as the riser stock. The preacher is notched in the centre. It should be wide enough to fit over the finish stringer. It should be long enough to allow the preacher to rest on the tread cut of the stair carriage when held against the rise cut.

Place the preacher over the stringer and against the rise cut of the stair carriage. Plumb the preacher with a hand level. Lay out the plumb cut on the stringer by marking along the side of the preacher that faces the bottom of the staircase (**Figure 82–15**).

Mark the top edge of the stringer along the side of the preacher that faces the top of the staircase. Draw a diagonal line across the top edge of the stringer for the mitre cut. Lay out all plumb lines on the stringer in this manner.

Remove the stringer. Cut to the layout lines. Make mitre cuts along the plumb lines. Cut square through the thickness along the level lines. Sand the piece. Fasten it in position. To ensure the piece will be in the same position as it was when laid out, fasten it first in the same holes where the piece was originally tacked.

Installing Risers

Cut risers to length by making a square cut on the end that goes against the wall. Make mitres on the other end to fit the mitred plumb cuts of the open finish stringer. Sand all pieces before installation. Apply a small amount of glue to the mitres. Fasten them in position to each stair carriage. Drive finish nails both ways through the mitre to hold the joint tight (**Figure 82–16**). Wipe off any excess glue. Set all nails. Lay out and install the closed finish stringer in the same manner as described previously.

Installing the Treads

Rip the treads to width. Rabbet the back edges. Make allowance for the rabbet when ripping treads

ON THE JOB

Use a preacher to lay out plumb cuts on an open finish stringer.

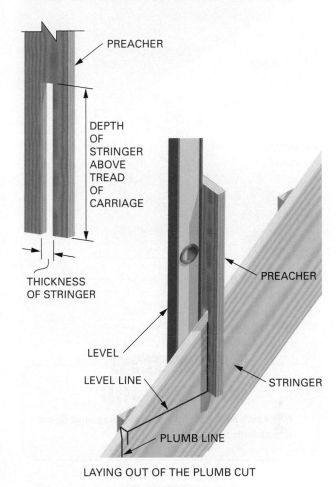

LAYING OUT OF THE PLUMB CUT

FIGURE 82–15 Technique for easily marking a stringer on both faces.

to width. Cut one end to fit against the closed finish stringer. Make a cut on the other end to receive the return nosing. This is a combination square and mitre cut. The square cut is made flush with the outside face of the open finish stringer. The mitre starts a distance equal to the width of the return nosing beyond the square cut (**Figure 82–17**).

Applying the Return Nosings

The return nosings are applied to the open ends of the treads. Nosings are returned to hide the dovetail pin cut where the balusters are inserted into the treads and to give a finished look to the stairs with no visible end grain. Mitre one end of the return nosing to fit against the mitre on the tread. Return the end on

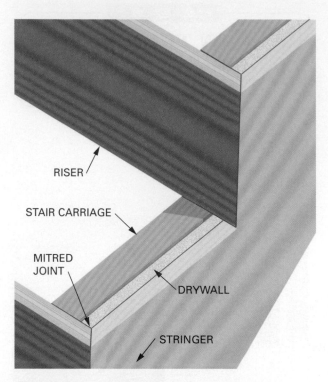

FIGURE 82–16 Open finish stringers are mitred to receive mitred risers.

ON THE JOB

Make sure nails used to fasten return nosings do not line up with balusters.

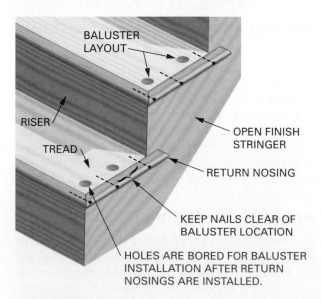

FIGURE 82–17 Alignment of trim nails should be positioned to avoid future baluster holes.

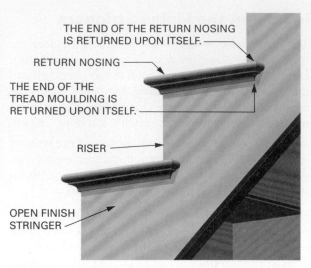

FIGURE 82–18 The back ends of the return nosing and moulding are returned upon themselves.

itself. The end of the return nosing extends beyond the face of the riser, the same amount as its width.

Predrill pilot holes in the return nosing for nails. Locate the holes so that they are not in line with any balusters that will later be installed on the treads. Holes must be bored in the treads to receive the balusters. Any nails in line with the holes will damage the boring tool (Figure 82–17).

Apply glue to the joint. Fasten the return nosing to the end of the tread with three 2½-inch (63 mm) finishing nails. Set all nails. Sand the joint flush. Apply all other return nosings in the same manner. Treads may be purchased with the return nosing applied in the factory. If used, the closed end of the tread is cut so that the nosed end overhangs the finish stringer by the proper amount.

Applying the Tread Moulding

The tread moulding is applied in the same manner as for closed staircases. However, it is mitred on the open end and returned back onto the open stringer. The back end of the return moulding is cut and returned upon itself at a point so that the end assembly shows the same as at the edge (**Figure 82–18**). Predrill pilot holes in the moulding. Fasten it in place. Moulding on starting and landing treads should be tacked only in case it needs to be removed for fitting after newel posts have been installed.

INSTALLING TREADS ON WINDERS

Treads on winding steps are especially difficult to fit because of the angles on both ends. A method used by many carpenters involves the use of a pattern and a scribing block. Cut a thin piece of plywood so that it fits in the tread space within ½ inch (12 mm) on the ends and back edge. The outside edge should be straight and in line with the nosed edge of the tread when installed.

Tack the plywood pattern in position. Use a ¾-inch (19 mm) block, rabbeted on one side by the thickness of the pattern, to scribe the ends and back edge. Scribe by riding the block against stringers, riser, and post while marking its inside edge on the pattern. Remove the pattern. Tack it on the tread stock. Place the block with its rabbeted side down and inside edge to the scribed lines on the pattern. Mark the

tread stock on the outside edge of the block (**Figure 82–19**).

PROTECTING THE FINISHED STAIRS

Protect the risers and treads by applying a width of building paper to them. Unroll a length down the stairway. Hold the paper in position by tacking thin strips of wood to the risers. This assumes the tack hole may be filled later before the finish is applied. Sometimes when appearance is paramount, the finished set of stairs is installed last after all tradespeople have completed their work.

ON THE JOB

Use a scribing block and pattern to lay out winding treads.

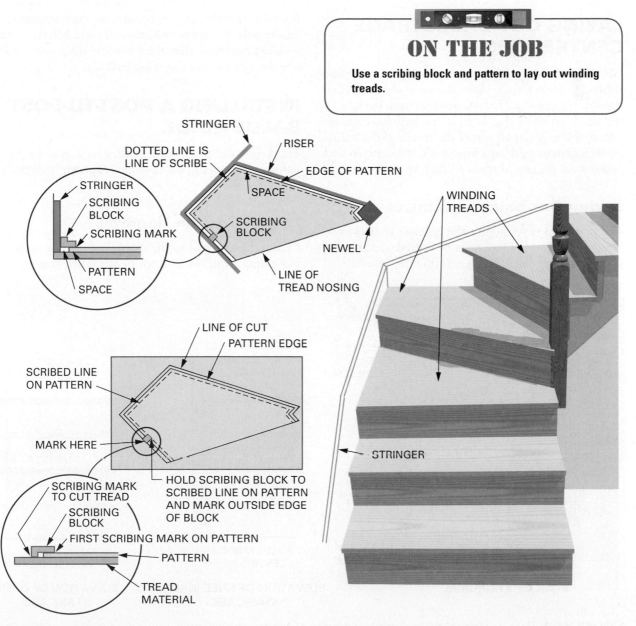

FIGURE 82–19 Technique for scribing winding treads to the stringers, risers, and posts.

Chapter 83

Installing Balustrades

Balustrades are the most visible and complex component of a staircase (see Introduction, Figure I–8). Installing them is one of the most intricate kinds of interior finish work. This chapter describes installation of post-to-post and over-the-post balustrades. Mastering the techniques described in this chapter will enable the student to install balustrades for practically any situation.

LAYING OUT BALUSTRADE CENTRELINES

For the installation of any balustrade, its centreline is first laid out. On an open staircase, the centreline should be located a distance inward, from the face of the finish stringer, that is equal to half the baluster width. It is laid out on top of the treads. If the balustrade is constructed on a knee wall, it is centred and laid out on the top of the wall (**Figure 83–1**).

Laying Out Baluster Centres

The next step is to lay out the baluster centres. The NBCC requires that balusters be spaced so as to prevent the passage of a $3\frac{7}{8}$-inch (100 mm) sphere.

On open treads, the face of the front baluster is placed flush with the face of the riser below. If two balusters can be used on each tread, the middle baluster is centred between two consecutive front balusters. For a set of stairs with a 10-inch (254 mm) run and a $1\frac{1}{4}$-inch (32 mm) square baluster, the space between the base of the balusters is $[10'' - 2(1\frac{1}{4}'')] \div 2$, or $3\frac{3}{4}''$; $[254 \text{ mm} - 2(32 \text{ mm})] \div 2$, or 95 mm. Check the space at the narrowest part of the spindle, because the maximum space can be only $3\frac{7}{8}$ inches (100 mm). If three balusters are required per tread, then the on-centre (OC) spacing is one-third of the unit run (**Figure 83–2**).

INSTALLING A POST-TO-POST BALUSTRADE

The following procedure applies to a post-to-post balustrade running, without interruption, from floor to floor.

Laying Out the Handrail

Clamp the handrail to the tread nosings. Use a short bar clamp from the bottom of the finish stringer to

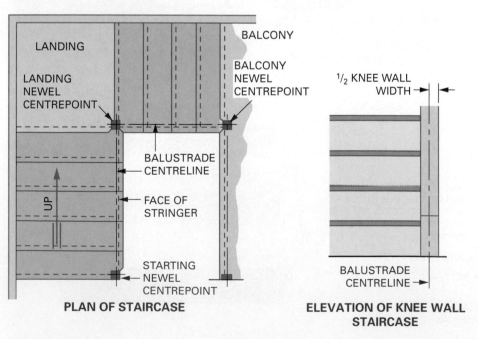
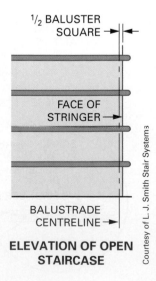

FIGURE 83–1 The centreline of the balustrade is laid out on a knee wall or open treads.

PLAN OF STAIRCASE

ELEVATION OF KNEE WALL STAIRCASE

ELEVATION OF OPEN STAIRCASE

Courtesy of L. J. Smith Stair Systems

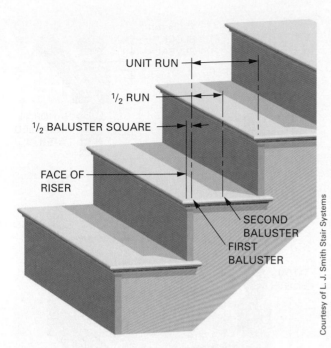

FIGURE 83–2 Layout of baluster centres on open treads.

the top of the handrail. Clamp opposite a nosing to avoid bowing the handrail. Use only enough pressure to keep the handrail from moving. Protect the edges of the handrail and finish stringer with blocks to avoid marring the pieces. Use a framing square to mark the handrail where it will fit between starting and balcony newel posts (**Figure 83–3**).

While the handrail is clamped in this position, use a framing square at the landing nosing to measure the vertical thickness of the rake handrail. Also, at the bottom, measure the height from the first tread to the top of the handrail where it butts the newel post. Record and save the measurements for later use (**Figure 83–4**).

Determining the Height of the Starting Newel

The NBCC states that the handrail height has to be between 34 and 38 inches (864 and 965 mm) [NBCC 9.8.7.4(2)] above the tread at the leading edge line (**Figure 83–5**). All stairways with three or more risers must have a handrail from first level to second level and a handrail on both sides if the width of the stairs exceeds 43 inches (1092 mm). For stairs built beside a wall, the handrail must be fastened every 4 feet (1.2 m) with at least two screws that penetrate at least 1¼ inches (32 mm) into solid framing members. The shape, size, and

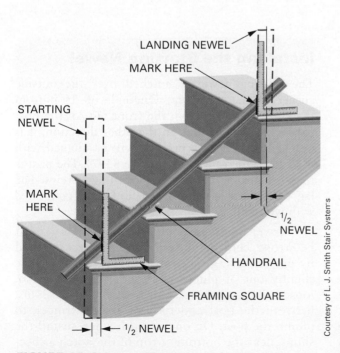

FIGURE 83–3 The handrail is laid out to fit between starting and balcony newel posts.

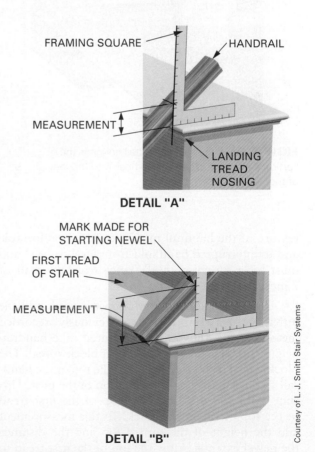

FIGURE 83–4 Determine the two measurements shown and record for future use.

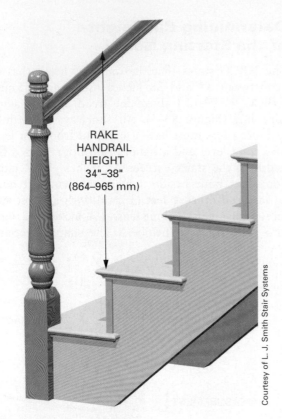

FIGURE 83–5 The rake handrail height is the vertical distance from the tread nosing to the top of the handrail.

ON THE JOB

Always check local building codes to determine handrail height requirements.

RAKE HANDRAIL HEIGHT 34"–38" (864–965 mm)

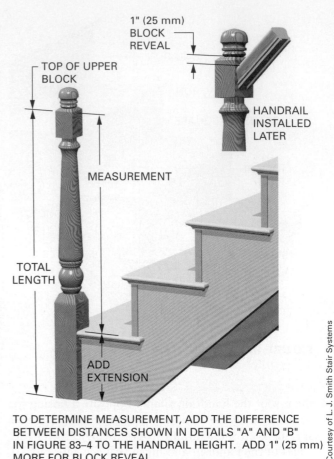

1" (25 mm) BLOCK REVEAL

TOP OF UPPER BLOCK

MEASUREMENT

TOTAL LENGTH

ADD EXTENSION

HANDRAIL INSTALLED LATER

TO DETERMINE MEASUREMENT, ADD THE DIFFERENCE BETWEEN DISTANCES SHOWN IN DETAILS "A" AND "B" IN FIGURE 83–4 TO THE HANDRAIL HEIGHT. ADD 1" (25 mm) MORE FOR BLOCK REVEAL.

FIGURE 83–6 Determining the height of the starting newel.

texture of the handrail must provide a comfortable and uninterrupted handhold from level to level and must have a minimum clearance from the wall of 2 inches (50 mm).

If a turned starting newel post is used, add the difference between the two previously recorded measurements above to the required rake handrail height. Add 1 inch (25 mm) for a block reveal. The block reveal is the distance from the top of the handrail to the top of the square section of the post. This sum is the distance from the top of the first tread to the top of the upper block. To this measurement add the height of the turned top and the distance the newel extends below the top of the first tread to the floor (**Figure 83–6**). Cut the starting newel to its total length.

Installing the Starting Newel

The starting newel is notched over the outside corner of the first step. One-half of its bottom thickness is left on from the front face of the post to the face of the riser. In the other direction, it is notched so that its centreline will be aligned with the balustrade centreline (**Figure 83–7**). The post is then fastened to the first step with lag screws. The lag screws are counterbored and later concealed with wood plugs.

Newel posts must be set plumb. They must be strong enough to resist lateral force applied by persons using the staircase. The post may be slightly out of plumb after it is fastened. If so, loosen the lag screws slightly. Install thin shims, between the post and riser or finish stringer, to plumb the post. On one or both sides, install the shims, near the bottom or top of the notch as necessary, to plumb the post. When plumb, retighten the lag screws.

Installing the Balcony Newel

The NBCC requires a guardrail for floors or landings that are more than 600 mm (2') above the adjacent surface. For landings between 600 and 1800 mm

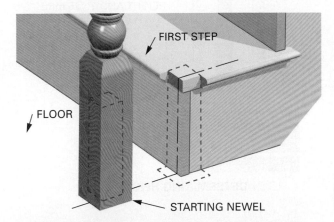

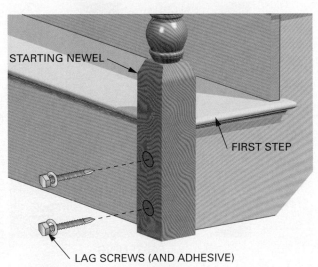

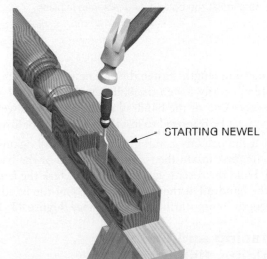

FIGURE 83–7 The starting newel is notched to fit over the first step.

(2' and 5'-11"), the required height of guardrail is 900 mm (35¼"). When the difference in floor height exceeds 1800 mm (5'-11"), as is the case with a second-floor balcony, the guardrail height is required to be 1070 mm (42").

The height of the balcony newel is determined by finding the sum of the required balcony handrail height, a block reveal of 1 inch (25 mm), the height of the turned top, and the distance the newel extends below the balcony floor.

Trim the balcony newel to the calculated height. Notch and fit it over the top riser with its centrelines aligned with both the rake and balcony handrail centrelines. Plumb it in both directions. Fasten it in place with counterbored lag bolts (**Figure 83–8**).

Preparing Treads and Handrail for Baluster Installation

Bore holes in the treads at the centre of each baluster. The diameter of the hole should be equal to the diameter of the pin at the bottom end of the baluster. The depth of the hole should be slightly more than the length of the pin (**Figure 83–9**).

Cut the handrail to fit between starting and balcony newels. Lay it back on the tread nosings. The handrail can be cut with a handsaw, a radial arm saw, or a compound mitre saw with the blade tilted. If using a power saw, make a practice cut to be sure the setup is correct before cutting the handrail. Transfer the baluster centrelines from the treads to the handrail (**Figure 83–10**).

Turn the handrail upside down and end for end. Set it back on the tread nosings with the starting newel end facing up the stairs. Bore holes at baluster centres at least ¾ inch (19 mm) deep, if balusters with round tops are to be used (**Figure 83–11**).

Installing Handrail and Balusters

Prepare the posts for fastening the handrail by counterboring and drilling shank holes for lag bolts through the posts. Place the handrail at the correct height between newel posts. Drill pilot holes. Temporarily fasten the handrail to the posts (**Figure 83–12**).

Cut the balusters to length. Allow ¾ inch (19 mm) for insertion in the hole in the bottom of the handrail. The handrail may have to be removed for baluster installation and then fastened permanently. The bottom pin is inserted in the holes in the treads. The top of the baluster is inserted in the holes in the handrail bottom.

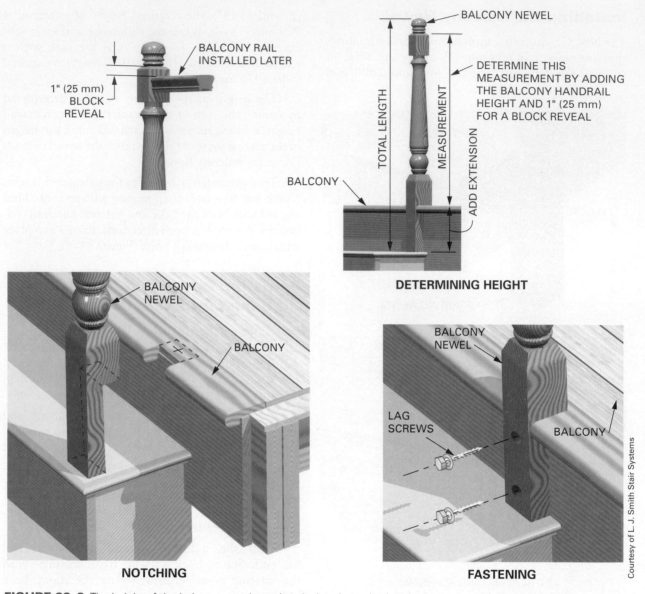

FIGURE 83–8 The height of the balcony newel post is calculated, notched at the bottom, and fastened in place.

If *square-top balusters* are used, they are trimmed to length at the rake angle. They are inserted into a *ploughed handrail*. The balusters are then fastened to the handrail with finish nails and glue (**Figure 83–13**). Care must be taken to keep the handrail in a straight line from top to bottom when fastening square-top balusters. Care must also be taken to keep each baluster in a plumb line. Install fillets in the plough of the handrail, between the balusters.

Installing the Balcony Balustrade

Cut a half newel to the same height as the balcony newel. Temporarily place it against the wall. Mark the length of the balcony handrail (**Figure 83–14**). Cut the handrail to length. Fasten the half newel to one end of it. Temporarily fasten the half newel to the wall and the other end of the handrail to the landing newel, if they must be removed to install the balcony balusters.

If the balcony handrail ends at the wall against a rosette, first fasten the rosette to the end of the handrail. Hold the rosette against the wall. Mark the length of the handrail at the landing newel. Cut the handrail to length. Temporarily fasten it in place (**Figure 83–15**).

Spacing and Installing Balcony Balusters

The balcony balusters are spaced by adding the thickness of one baluster to the distance between

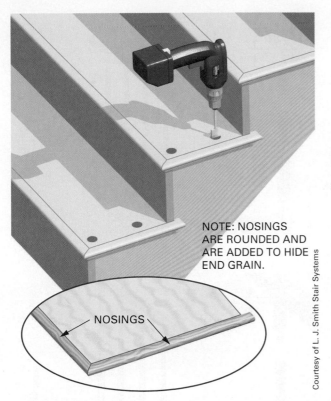

NOTE: NOSINGS ARE ROUNDED AND ARE ADDED TO HIDE END GRAIN.

NOSINGS

Courtesy of L. J. Smith Stair Systems

FIGURE 83–9 Holes are bored in the top of the treads at each baluster centre point.

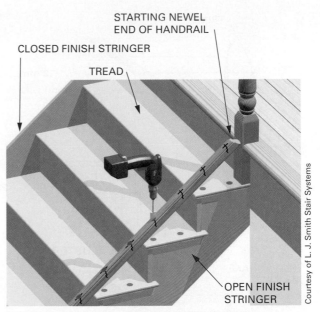

STARTING NEWEL END OF HANDRAIL

CLOSED FINISH STRINGER

TREAD

OPEN FINISH STRINGER

Courtesy of L. J. Smith Stair Systems

FIGURE 83–11 Rotate and invert rake handrail to bore holes in the bottom to receive round-top balusters.

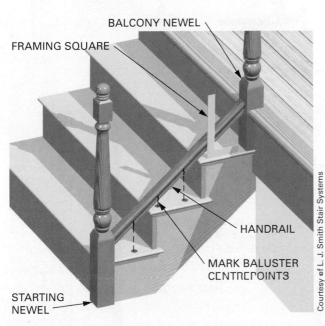

BALCONY NEWEL

FRAMING SQUARE

HANDRAIL

MARK BALUSTER CENTREPOINTS

STARTING NEWEL

Courtesy of L. J. Smith Stair Systems

FIGURE 83–10 The handrail is fitted between newel posts and then the baluster centres are transferred to it.

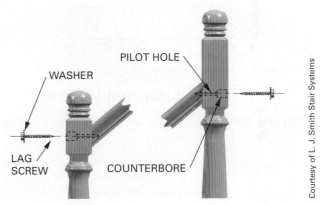

PILOT HOLE

WASHER

LAG SCREW

COUNTERBORE

Courtesy of L. J. Smith Stair Systems

FIGURE 83–12 The handrail is fastened to newel posts with lag screws. The use of nails when constructing balustrades is discouraged.

to be narrow enough to prevent the passage of a sphere that is $3^{15}/_{16}''$ (100 mm) in diameter [NBCC 9.8.8.5 (1)]. The balcony balusters are then installed in a manner similar to the rake balusters.

INSTALLING AN OVER-THE-POST BALUSTRADE

The following procedures apply to an over-the-post balustrade running from floor to floor with an intermediate landing (see Figure 81–2). An over-the-post

the balcony newel and the half newel. The overall distance is then divided into spaces that equal, as close as possible, the spacing of the rake balusters (**Figure 83–16**). The space between the balusters has

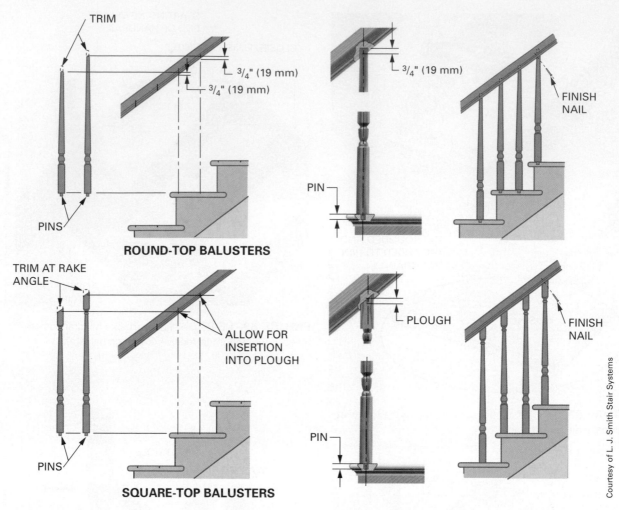

FIGURE 83–13 Balusters are cut to length and installed between handrail and treads.

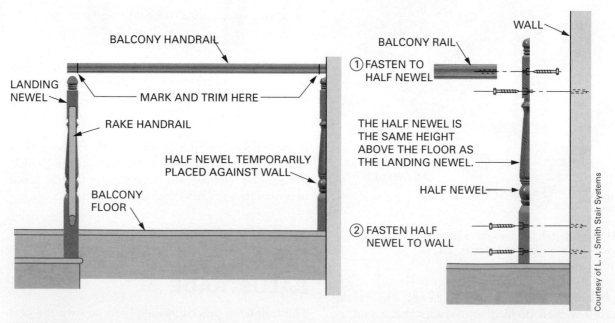

FIGURE 83–14 The balcony rail is fitted between the landing newel and a half newel placed against the wall.

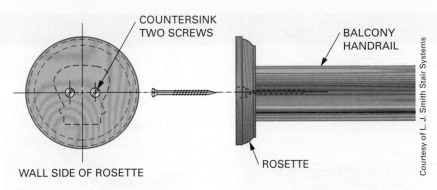

COUNTERSINK TWO SCREWS

BALCONY HANDRAIL

WALL SIDE OF ROSETTE

ROSETTE

Courtesy of L. J. Smith Stair Systems

FIGURE 83–15 A rosette is sometimes used to end the balcony handrail instead of a half newel.

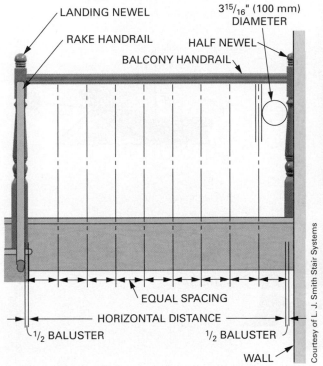

LANDING NEWEL

RAKE HANDRAIL

BALCONY HANDRAIL

HALF NEWEL

3¹⁵/₁₆" (100 mm) DIAMETER

EQUAL SPACING

HORIZONTAL DISTANCE

½ BALUSTER

½ BALUSTER

WALL

Courtesy of L. J. Smith Stair Systems

FIGURE 83–16 Balcony balusters are installed as close as possible to the same spacing as the rake balusters.

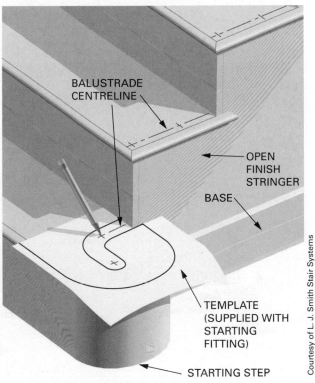

BALUSTRADE CENTRELINE

OPEN FINISH STRINGER

BASE

TEMPLATE (SUPPLIED WITH STARTING FITTING)

STARTING STEP

Courtesy of L. J. Smith Stair Systems

FIGURE 83–17 Baluster and newel post centres are laid out on a starting step with a template.

balustrade is more complicated. Handrail fittings are required to be joined to straight sections to construct a continuous handrail from start to end. The procedure consists of constructing the entire handrail first, and then setting newel posts and installing balusters.

Constructing the Handrail

The first step is to lay out balustrade and baluster centrelines on the stair treads, as described previously. If a starting step is used, lay out the baluster and starting newel centres using a template provided with the starting fitting (**Figure 83–17**).

Laying Out the Starting Fitting

Make a pitch block by cutting a piece of wood in the shape of a right triangle whose sides are equal in length to the rise and tread run of the stairs. Hold the cap of the starting fitting and the run side of the pitch block on a flat surface. The rake edge of the pitch block should be against the bottom side of the fitting. Mark the fitting at the tangent point, where its curved surface touches the pitch block. A straight line, tangent to a curved line, touches the curved line at one point only.

StepbyStep Procedures

Procedure 83–A Cutting the Starting Fitting

STEP 1 MAKE A PITCH BLOCK USING UNIT RISE AND UNIT RUN OF STAIRS.

STEP 2 MARK TANGENT POINT WHERE PITCH BLOCK TOUCHES RAIL.

STEP 3 ROTATE PITCH BLOCK AND MARK ANGLE ALONG RAKE EDGE OF PITCH BLOCK.

STEP 4 TRIM CONNECTING EASEMENT USING PITCH BLOCK AS A GUIDE. BE SURE TO SECURE RAIL BEFORE CUTTING.

EASEMENT
RAKE
RISE
RUN
PITCH BLOCK

RUN
RAKE
RISE

PITCH BLOCK
RAKE
RISE
RUN
MITRE SAW

Courtesy of L. J. Smith Stair Systems

Then turn the pitch block so that its rise side is on the flat surface. Mark the fitting along the rake edge of the pitch block. Place the fitting in a power mitre box, supported by the pitch block. Cut it to the layout line (**Procedure 83–A**).

CAUTION

When cutting handrail fittings supported by a pitch block in a power mitre box, clamp the pitch block and fitting securely to make sure they will not move when cut. If either one moves during cutting, a serious injury could result. Even if no injury occurs, the fitting would most likely be ruined.

Joining the Starting Fitting to the Handrail

The starting fitting is joined to a straight section of handrail by means of a special handrail bolt. The bolt has threads on one end designed for fastening into wood. The other end is threaded for a nut. Holes must be drilled in the end of both the fitting and the handrail in a manner that ensures their alignment when joined.

To mark the hole locations, make a template by cutting about a ⅛-inch (3 mm) piece of handrail. Drill a 1/16-inch (1.5 mm) hole centred on the template width and 5/16 inch (8 mm) from the bottom side. Mark one side rail and the other side fitting. This will ensure that the template is facing in the right direction when making the layout. If the hole in the template is drilled slightly off centre and the template turned when making the layout, the handrail and fitting will not be in alignment when joined. Use the template to mark the location of the rail bolt on the end of the fitting and the handrail.

Drill all holes to the depth and diameter shown in **Figure 83–18**. Double-nut the rail bolt and turn it into the fitting. Remove the nuts and place the handrail on the bolt. Install a washer and nut, and tighten to join the sections. Clamp the assembly to the tread nosings. The newel centre points on the fitting and the starting tread should be in a plumb line (**Figure 83–19**). Note the measurement in Detail "C" of Figure 83–19.

Installing the Landing Fitting

The second flight of stairs turns at the landing, so a right-hand, two-riser gooseneck fitting is used. It is laid out and joined to the bottom end of the handrail of the second flight in a similar manner as the starting fitting (**Procedure 83–B**). The assembled fitting and rail are then clamped to the nosings of the treads in the second flight. Position the rail so that the *newel cap* of the fitting is in a plumb line with the baluster centrelines of both flights (**Figure 83–20**).

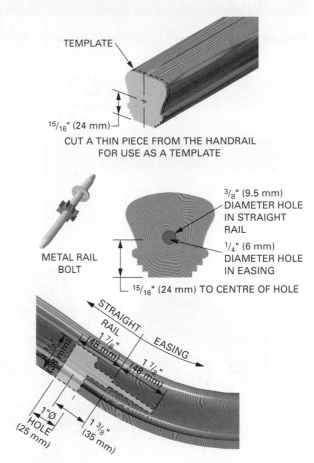

CUT A THIN PIECE FROM THE HANDRAIL FOR USE AS A TEMPLATE

TEMPLATE

$^{15}/_{16}$" (24 mm)

METAL RAIL BOLT

$^3/_8$" (9.5 mm) DIAMETER HOLE IN STRAIGHT RAIL

$^1/_4$" (6 mm) DIAMETER HOLE IN EASING

$^{15}/_{16}$" (24 mm) TO CENTRE OF HOLE

STRAIGHT RAIL $1^7/_8$" (48 mm)   EASING $1^7/_8$" (48 mm)

$1^1/_2$" (38 mm)

1"Ø HOLE (25 mm) $1^3/_8$" (35 mm)

FIGURE 83–18 A template is made to mark the ends of handrails and fittings for joining with rail bolts. The ends are drilled to specific depths and diameters.

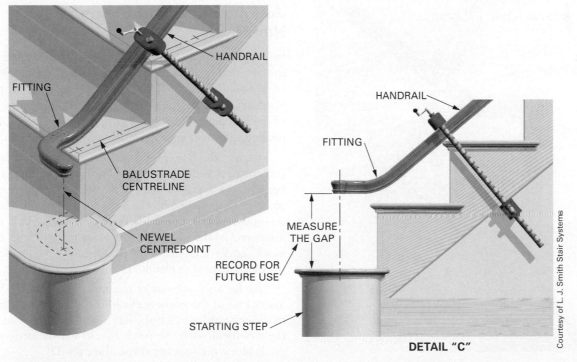

HANDRAIL

FITTING

BALUSTRADE CENTRELINE

NEWEL CENTREPOINT

STARTING STEP

HANDRAIL

FITTING

MEASURE THE GAP

RECORD FOR FUTURE USE

Courtesy of L. J. Smith Stair Systems

DETAIL "C"

FIGURE 83–19 The starting fitting and handrail assembly are clamped to the tread nosings of the first flight of stairs.

Step**by**Step Procedures

Procedure 83–B Cutting the Landing Fitting to the Second-Flight Handrail

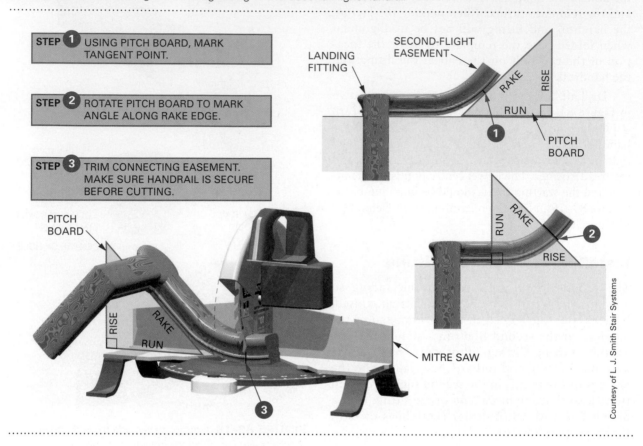

STEP **1** USING PITCH BOARD, MARK TANGENT POINT.

STEP **2** ROTATE PITCH BOARD TO MARK ANGLE ALONG RAKE EDGE.

STEP **3** TRIM CONNECTING EASEMENT. MAKE SURE HANDRAIL IS SECURE BEFORE CUTTING.

Courtesy of L. J. Smith Stair Systems

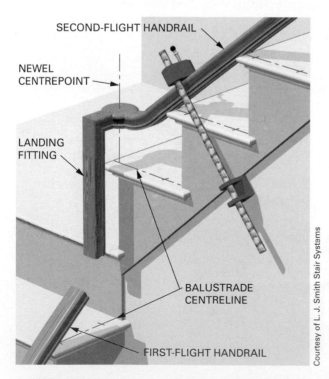

Courtesy of L. J. Smith Stair Systems

FIGURE 83–20 The second-flight handrail is clamped to the tread nosings after the landing fitting is joined to it.

Joining First- and Second-Flight Handrails

In preparation for laying out the easement used to join the first- and second-flight handrails, tack a piece of plywood about 5 inches (127 mm) wide to the bottom side of the gooseneck fitting and the handrail of the first flight. These pieces are used to rest the connecting easement against when laying out the joint.

Clamp the handrails back on the treads. Make sure the newel cap centres are plumb with starting and landing newel post centres. Rest the connecting easement on the plywood blocks. Level its upper end. Mark the gooseneck in line with the upper end. Mark the lower end of the easement at a point tangent with the block under the handrail (**Procedure 83–C**). Note the measurement in Detail "D" of Procedure 83–C.

Make a square cut at the mark laid out on the lower end of the gooseneck. Join the gooseneck and easement with a rail bolt. Lay out the lower end of the easement with the pitch block. Cut it using the pitch block for support (**Procedure 83–D**).

Place the assembled landing fitting and handrail back on the stair nosings. Mark the handrail of the

StepbyStep Procedures

Procedure 83–C Marking the Landing Fitting

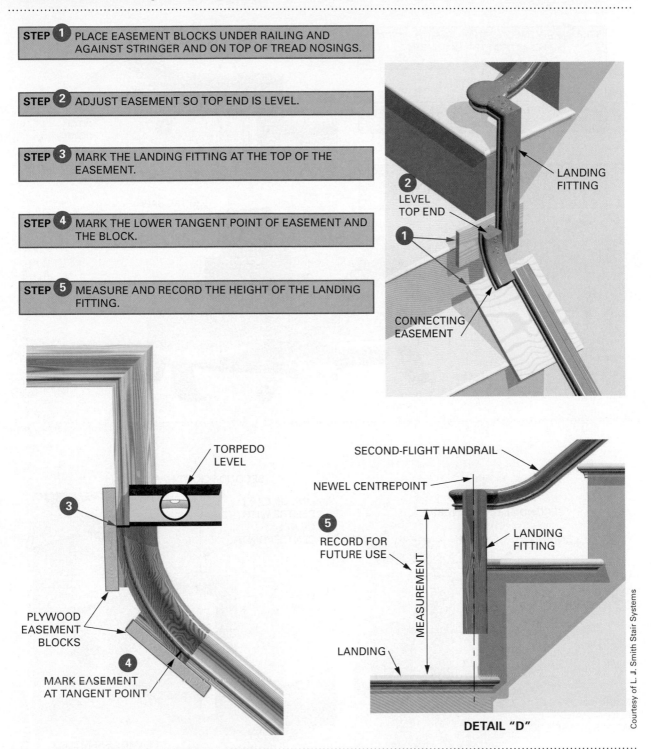

STEP 1 PLACE EASEMENT BLOCKS UNDER RAILING AND AGAINST STRINGER AND ON TOP OF TREAD NOSINGS.

STEP 2 ADJUST EASEMENT SO TOP END IS LEVEL.

STEP 3 MARK THE LANDING FITTING AT THE TOP OF THE EASEMENT.

STEP 4 MARK THE LOWER TANGENT POINT OF EASEMENT AND THE BLOCK.

STEP 5 MEASURE AND RECORD THE HEIGHT OF THE LANDING FITTING.

LANDING FITTING

2 LEVEL TOP END

1

CONNECTING EASEMENT

TORPEDO LEVEL

3

PLYWOOD EASEMENT BLOCKS

4

MARK EASEMENT AT TANGENT POINT

SECOND-FLIGHT HANDRAIL

NEWEL CENTREPOINT

5 RECORD FOR FUTURE USE

MEASUREMENT

LANDING

LANDING FITTING

DETAIL "D"

Courtesy of L. J. Smith Stair Systems

lower flight where it meets the end of the fitting. Cut the handrail square at the mark. Join it to the landing fitting with a rail bolt. Clamp the entire handrail assembly to the tread nosings with newel cap centres plumb with newel post centrelines (**Figure 83–21**).

Installing the Balcony Gooseneck Fitting

A one-riser balcony gooseneck fitting is used when balcony rails are 36 inches (914 mm) high. Two-riser

StepbyStep Procedures

Procedure 83–D Connecting the Easement and Landing Fitting

STEP 1 SQUARE CUT THE LANDING FITTING AND ATTACH THE EASEMENT FITTING.

STEP 2 MARK THE ANGLE ALONG THE RAKE AT THE TANGENT POINT.

STEP 3 TRIM CONNECTING EASEMENT. MAKE SURE THE RAIL IS SECURE BEFORE CUTTING.

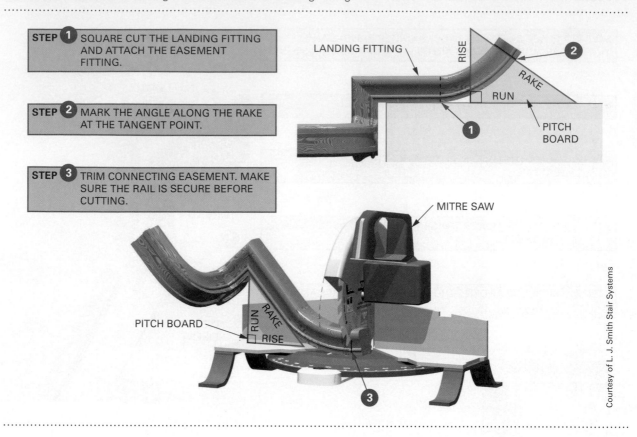

LANDING FITTING

RISE
RAKE
RUN
PITCH BOARD

MITRE SAW

PITCH BOARD
RUN
RAKE
RISE

Courtesy of L. J. Smith Stair Systems

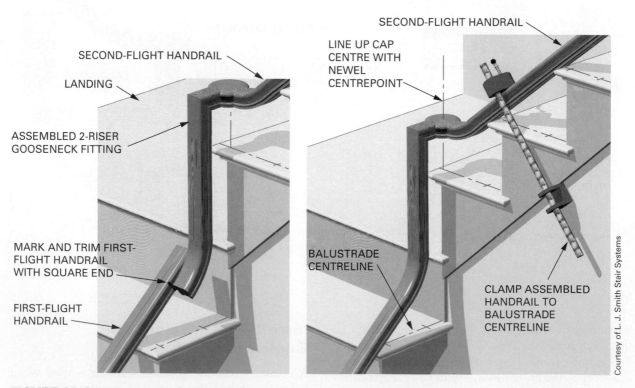

SECOND-FLIGHT HANDRAIL

LANDING

ASSEMBLED 2-RISER GOOSENECK FITTING

MARK AND TRIM FIRST-FLIGHT HANDRAIL WITH SQUARE END

FIRST-FLIGHT HANDRAIL

LINE UP CAP CENTRE WITH NEWEL CENTREPOINT

SECOND-FLIGHT HANDRAIL

BALUSTRADE CENTRELINE

CLAMP ASSEMBLED HANDRAIL TO BALUSTRADE CENTRELINE

Courtesy of L. J. Smith Stair Systems

FIGURE 83–21 Lower and upper handrail assemblies are joined. They are then clamped to the tread nosings.

fittings are used for rails 42 inches (1067 mm) high. In this case, a one-riser fitting is used at the balcony. Laying out and fitting a two-riser fitting at the landing has been previously described.

Hold the fitting so that the centre of its cap is directly above the balcony newel post centreline. Hold it against the handrail of the upper flight. Mark it and the handrail at the point of tangent (**Figure 83–22**). Lay out and make the cut on the gooseneck with the use of a pitch block (**Procedure 83–E**). Cut the handrail square at the mark. Join the gooseneck fitting and the handrail with a rail bolt.

Clamp the entire rail assembly back on the nosing of the treads in line with the balustrade centrelines and the three newel post centres. Use a framing square to transfer the baluster centres from the treads to the side of the handrails. Remove the handrail assembly out of the way of newel post installation.

INSTALLING NEWEL POSTS

In this staircase, the starting newel is installed on a starting step. The height of the rake handrail is calculated, from the height of the starting newel, to make sure the handrail will conform to the height required by the building code.

From the height of the starting newel to be used, subtract the previously recorded distance between the starting fitting and the starting tread as shown in Detail "C" of Figure 83–19. Then add the vertical thickness of the rake handrail shown in Figure 83–4, Detail "A." The result is the rake handrail height (**Procedure 83–F**). If the height does not conform to the building code, the starting newel post height must be changed.

Installing the Starting Newel Post

Before installing the starting step, measure the diameter of the dowel at the bottom of the starting newel post. At the centrepoint of the newel post, bore a hole for the dowel through the tread and floor. Install the post. Wedge it under the floor. Wedges are driven in a through-mortise cut in the dowelled end of the post.

An alternative method is used when there is no access under the floor. Bore holes only through the tread and upper riser block of the starting step. Cut the dowel to fit against the lower riser block so that the newel post rests snugly on the tread. Fasten the end of the dowel with a lag screw through the lower riser block (**Figure 83–23**). Fasten the assembled starting newel and starting step in position.

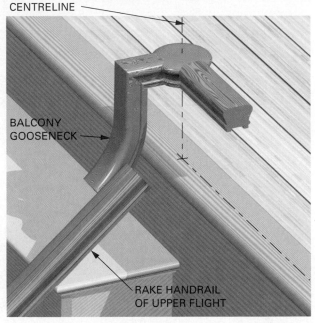

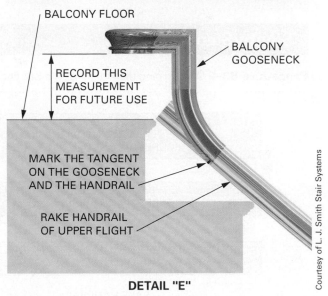

DETAIL "E"

FIGURE 83–22 Method used to mark the balcony gooseneck fitting and handrail of the upper flight.

Courtesy of L. J. Smith Stair Systems

Installing the Landing Newel Post

The height of the landing newel above the landing is found by subtracting the previously recorded distance between the starting fitting and the starting tread (Figure 83–19, Detail "C") from the height of the starting newel. Then add the distance between the landing fitting and the landing as previously recorded and shown in Procedure 83–C, Detail "D." To this length, add the distance that the landing newel extends below the landing.

Step**by**Step Procedures

Procedure 83–E Cutting the Balcony Gooseneck Fitting

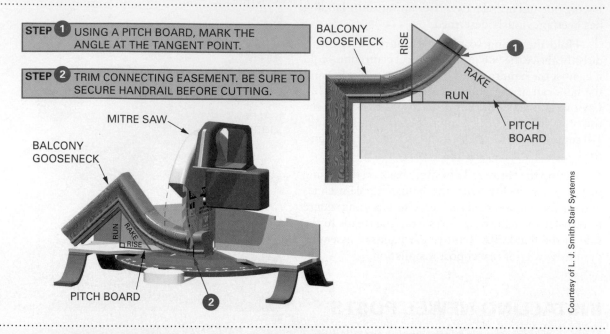

STEP ❶ USING A PITCH BOARD, MARK THE ANGLE AT THE TANGENT POINT.

STEP ❷ TRIM CONNECTING EASEMENT. BE SURE TO SECURE HANDRAIL BEFORE CUTTING.

BALCONY GOOSENECK

RISE

RAKE

RUN

PITCH BOARD

MITRE SAW

BALCONY GOOSENECK

RUN

RAKE

RISE

PITCH BOARD

Courtesy of L. J. Smith Stair Systems

Step**by**Step Procedures

Procedure 83–F Determining the Height of the Starting Newel Post

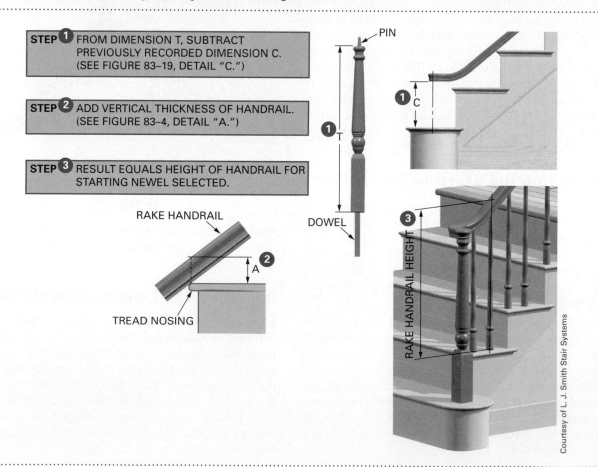

STEP ❶ FROM DIMENSION T, SUBTRACT PREVIOUSLY RECORDED DIMENSION C. (SEE FIGURE 83–19, DETAIL "C.")

STEP ❷ ADD VERTICAL THICKNESS OF HANDRAIL. (SEE FIGURE 83–4, DETAIL "A.")

STEP ❸ RESULT EQUALS HEIGHT OF HANDRAIL FOR STARTING NEWEL SELECTED.

RAKE HANDRAIL

A ❷

TREAD NOSING

PIN

T

❶

DOWEL

C ❶

❸

RAKE HANDRAIL HEIGHT

Courtesy of L. J. Smith Stair Systems

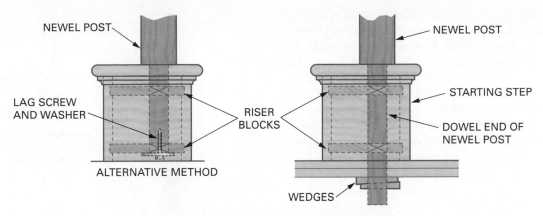

SECTIONS THROUGH STARTING STEP

FIGURE 83–23 Methods of installing the starting newel on a starting step.

Notch the landing newel to fit over the landing and the first step of the upper flight. Fasten the post in position with lag bolts in counterbored holes (**Procedure 83–G**).

Installing the Balcony Newel Post

The height of the balcony handrail must be calculated before the height of the balcony newel can be

---

Step**by**Step Procedures

Procedure 83–G Determining the Height of the Landing Newel

STEP 1 FROM THE STARTING NEWEL HEIGHT, DETERMINED FROM PROCEDURE 83–F, SUBTRACT PREVIOUSLY RECORDED MEASUREMENT. (SEE FIGURE 83–19, DETAIL "C.")

STEP 2 ADD PREVIOUSLY RECORDED MEASUREMENT. (SEE PROCEDURE 83–C, DETAIL "D.")

STEP 3 CUT AND FIT LANDING NEWEL BY NOTCHING THE LANDING, TREAD, AND NEWEL AS NEEDED.

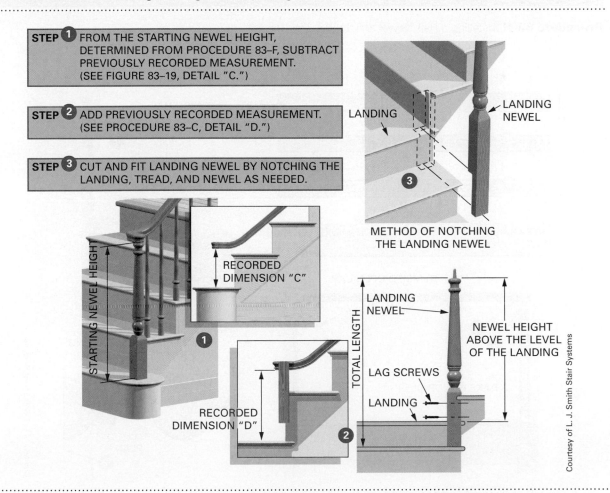

METHOD OF NOTCHING THE LANDING NEWEL

determined. The height of the balcony handrail is found by subtracting the previously recorded distance between the starting fitting and the starting tread (Figure 83–19, Detail "C") from the height of the starting newel. Then add the previously recorded distance between the balcony gooseneck fitting and the landing (Figure 83–22, Detail "E"). Then add the thickness of the balcony handrail. The balcony handrail height must conform to the building code. If not, substitute a two-riser gooseneck fitting instead of a one-riser fitting.

The height of the balcony newel above the balcony floor is found by subtracting the handrail thickness from the handrail height. To this length, add the distance the post extends below the floor. Notch and install the post over the balcony riser in line with balustrade centrelines (**Figure 83–24**).

Installing Balusters

Bore holes for balusters in the tread and bottom edge of the handrail. No holes are bored in the handrail

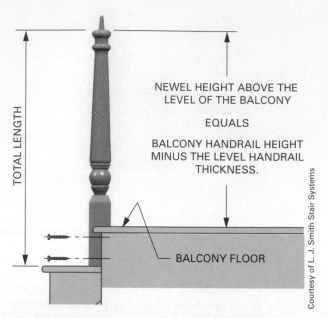

FanNEWEL HEIGHT ABOVE THE
LEVEL OF THE BALCONY

EQUALS

BALCONY HANDRAIL HEIGHT
MINUS THE LEVEL HANDRAIL
THICKNESS.

TOTAL LENGTH

BALCONY FLOOR

Courtesy of L. J. Smith Stair Systems

FIGURE 83–24 The height of the balcony newel is determined. The post is then fastened in place.

StepbyStep Procedures

Procedure 83–H Installing a Half Newel and the Balcony Handrail

STEP ❶ INSTALL HALF NEWEL TO WALL.

STEP ❷ TRIM CAP TO CENTRELINE.

STEP ❸ MARK AND TRIM RAIL TO LENGTH.

STEP ❹ JOIN RAIL TO GOOSENECK AND HALF NEWEL.

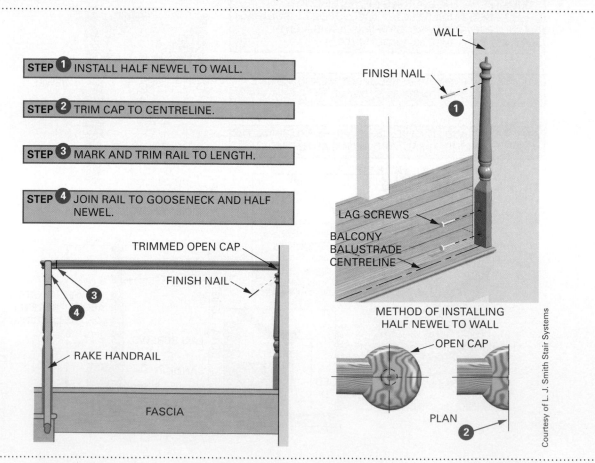

TRIMMED OPEN CAP

FINISH NAIL

RAKE HANDRAIL

FASCIA

WALL

FINISH NAIL

LAG SCREWS

BALCONY
BALUSTRADE
CENTRELINE

METHOD OF INSTALLING
HALF NEWEL TO WALL

OPEN CAP

PLAN

Courtesy of L. J. Smith Stair Systems

if square-top balusters are used. Install the handrail on the posts. Cut the balusters to length. Install them in the manner described previously for post-to-post balustrades.

Installing the Balcony Balustrade

Cut a half newel to the same height as the balcony newel extends above the floor. Install it against the wall on the balustrade centreline. Cut an opening cap so that it fits on top of the half newel. Join the cap to the end of the balcony handrail with a rail bolt. Place the cap on the half newel. Mark the length of the handrail at the balcony newel. Cut the handrail and join it, on one end, to the balcony newel post and, on the other end to the half newel (**Procedure 83–H**).

A rosette may be used against the wall instead of the half newel and opening cap. The procedure for installing a rosette has been previously described (see Figure 83–15).

Installing Balcony Balusters

Balcony balusters are laid out and installed in the same manner as described previously for post-to-post balustrades. Balconies with a span of 10 feet (3 m) or more should have intermediate balcony newels installed every 5 or 6 feet (1.5 or 1.8 m).

KEY TERMS

balusters	handrail	pitch board	stair body
balustrade	handrail bolt	post-to-post	starting step
easement	knee wall	preacher	tread moulding
fillets	landing treads	risers	treads
finish stringers	newel post	shoe rail	turnouts
fittings	nosing	skirt boards	volutes
goosenecks	over-the-post	spindles	

REVIEW QUESTIONS

Select the most appropriate answer.

1. What is the rounded outside edge of a tread that extends beyond the riser called?

   a. housing
   b. turnout
   c. coving
   d. nosing

2. What are finish boards between the staircase and the wall finish called?

   a. returns
   b. balusters
   c. skirt boards
   d. casings

3. What are treads rabbeted on their back edge to fit into?

   a. risers
   b. housed stringers
   c. return nosings
   d. newel posts

4. What is done to an open stringer?

   a. It is housed to receive risers.
   b. It is mitred to receive risers.
   c. It is housed to receive treads.
   d. It is mitred to receive treads.

5. What is a *volute* part of?

   a. a tread
   b. a baluster
   c. a newel post
   d. a handrail

6. What is the entire rail assembly on the open side of a stairway called?

   a. baluster
   b. balustrade
   c. guardrail
   d. finish stringer assembly

7. In a framed staircase, what are the treads and risers supported by?

   a. stair carriages
   b. housed stringers
   c. each other
   d. blocking

8. What is one of the first things to do when trimming a staircase?

   a. Check the rough framing for rise and run.
   b. Block the staircase so that no one can use it.
   c. Straighten the stair carriages.
   d. Install all the risers.

9. According to the NBCC, what is the maximum distance tread nosing can project beyond the riser face?

   a. a maximum of ¾ inch (19 mm)
   b. a maximum of 1 inch (25 mm)
   c. a maximum of 1¼ inches (32 mm)
   d. a maximum of 1½ inches (38 mm)

10. When newel posts are notched around the stairs, what are their centrelines lined up with?

    a. the centreline of the stair carriage
    b. the centreline of the balustrade
    c. the outside face of the open stringer
    d. the outside face of the stair carriage

Unit 28

Finish Floors

Many times a layer of underlayment is installed over the subfloor as part of the finished floor. A number of materials are then used for finish floors. Each may be applied by a flooring specialist, such as carpet by carpet installers. But in some areas, carpenters are asked to install various flooring materials.

Resilient flooring sheets and tiles are often used in kitchens and bathrooms. Wood flooring comes as strips of solid wood or a wood-plastic combination. Solid wood flooring is a long-time favourite because of its durability, beauty, and warmth.

OBJECTIVES

After completing this unit, the student should be able to:

- describe the kinds, sizes, and grades of hardwood finish flooring.
- apply strip, plank, and parquet finish flooring.
- estimate quantities of wood finish flooring required for various installations.
- apply underlayment and resilient tile flooring.
- estimate required amounts of underlayment and resilient tile for various installations.

SAFETY REMINDER

Installing flooring requires the installer to bend or work on knees for long periods of time. Take care to protect your body from injury due to repetitive movements and contact with hard surfaces.

Chapter 84

Description of Wood Finish Floors

Most hardwood finish flooring is made from white or red oak. Beech, birch, hard maple, and pecan finish flooring are also manufactured. For less expensive finish floors, some softwoods such as Douglas fir, hemlock, tamarack, and southern yellow pine are used.

KINDS OF HARDWOOD FLOORING

The four basic types of solid wood finish flooring are *strip*, *plank*, *parquet strip*, and *parquet block*. *Engineered wood flooring* is a relatively new type that is gaining in popularity. *Laminated parquet blocks* are also manufactured.

Solid Wood Strip Flooring

Solid wood **strip flooring** is probably the most widely used type. Most strips are tongue-and-grooved on edges and ends to fit precisely together.

Unfinished strip flooring is milled with square, sharp corners at the intersections of the face and edges. After the floor is laid, any unevenness in the faces of adjoining pieces is removed by sanding the surface so that strips are flush with each other.

Prefinished strips are sanded, finished, and waxed at the factory. They cannot be sanded after installation. A **chamfer** is machined between the face side and edges of the flooring prior to prefinishing. When installed, these chamfered edges form small V-grooves between adjoining pieces. This obscures any unevenness in the surface.

The most popular size of hardwood strip flooring is ¾ inch (19 mm) thick with a face width of 2¼ inches (57 mm). The face width is the width of the exposed surface between adjoining strips. It does not include the tongue. Other thicknesses and widths are manufactured (**Figure 84–1**).

Engineered Wood Flooring

Engineered wood flooring, sometimes referred to as **laminated strip flooring**, is a five-ply prefinished wood assembly. Each board is ⁹⁄₁₆ inch thick, 7½ inches wide, and 7 feet, 11½ inches long (14 × 190 × 2426 mm). The board consists of a bottom veneer, a three-ply cross-laminated core, and a face layer. The face layer consists of three rows of hardwood strips joined snugly edge to edge (**Figure 84–2**).

The uniqueness of this flooring is in the exact milling of edge and end tongue and grooves. This precision allows the boards to be joined with no noticeable unevenness of the prefinished surface. This

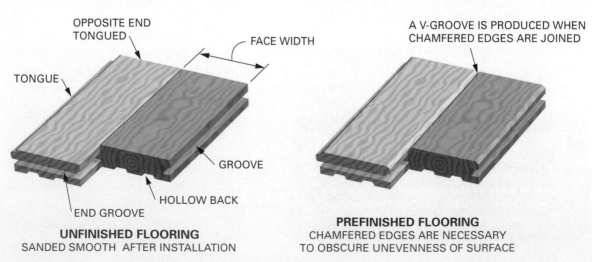

FIGURE 84–1 Hardwood strip flooring is edge and end matched. The edges of prefinished flooring are chamfered.

FIGURE 84-2 Engineered hardwood flooring.

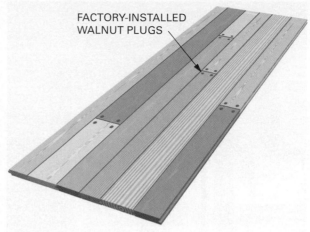

FACTORY-INSTALLED
WALNUT PLUGS

FIGURE 84-3 Plank flooring usually is applied in rows of alternating widths. Plugs of contrasting colour may be added to simulate screw fastening.

eliminates the need to chamfer the edges. Without the chamfers, a smooth continuous surface without V-grooves results when the floor is laid.

In the construction of engineered wood floors, the wood plies are stacked on top of each other but in the opposite directions. This creates a wood floor that is dimensionally stable and less affected by moisture than a ¾-inch (19 mm) solid wood floor. This means you can install these floors over concrete slabs in basements, as well as anywhere else in the home.

Plank Flooring

Solid wood **plank flooring** is similar to strip flooring. However, it comes in various mixed combinations ranging from 3 to 8 inches (75 to 203 mm) in width. For instance, plank flooring may be laid with alternating widths of 3 and 4 inches (75 and 100 mm); 3, 4, and 6 inches (75, 100, and 152 mm); and 3, 5, and 7 inches (75, 127, and 178 mm), or in any random-width combination.

Like strips, planks are available unfinished or prefinished. The edges of some prefinished planks have deeper chamfers to accentuate the plank widths. The surface of some prefinished plank flooring may have plugs of contrasting colour already installed to simulate screw fastening. One or more plugs, depending on the width of the plank, are used across the face at each end (**Figure 84-3**).

Unfinished plank flooring comes with either square or chamfered edges and with or without plugs. The planks may be bored for plugs on the job, if desired.

Bamboo is a grass that can be harvested for flooring every 6 years, whereas hardwood trees can take 40 years to grow in order to harvest for flooring (**Figure 84-4**). Bamboo is two times as hard as oak and it expands due to moisture at one-half the rate as most hardwoods (e.g., go to *youtube.com* and search for "DassoSWB Hardness"). It is very dense and strong and when a scratch-resistant coating is added, it is hard to beat for durability. There are many colour variations available and it can be installed in the same fashion as hardwood flooring (**Figure 84-5**).

Laminate flooring is manufactured in a variety of colours and patterns (**Figure 84-6**). Most of the brands have a snap-and-click tongue-and-groove joint, though there are still some that require gluing between the joints. This type of wood-simulated flooring is classified as a floating floor because it is not fastened to the subfloor. It "floats" on a pad that is specified by each manufacturer (**Figure 84-7**).

FIGURE 84-4 Bamboo, a renewable, sustainable resource.

Courtesy of DuroDesign, Inc.

FIGURE 84–5 Bamboo flooring.

Courtesy of Torlys

FIGURE 84–6 Laminate flooring.

Parquet Strips

Parquet strip flooring has short strips that are laid to form various mosaic designs. The original parquet floors were laid by using short strips. Some, at the present time, are laid in the same manner. This type is manufactured in precise, short lengths, which are multiples of its width. For instance, 2¼-inch (57 mm) parquet strips come in lengths of 9, 11¼, 13½, and 15¾ inches (229, 286, 343, and 400 mm). Each piece is tongue-and-grooved on the edges and ends. Herringbone, basket weave, and other interesting patterns can be made using parquet strips (**Figure 84–8**).

Parquet Blocks

Parquet block flooring consists of square or rectangular blocks, sometimes installed in combination with strips, to form mosaic designs. The three basic types are the *unit*, *laminated*, and *slat* block (**Figure 84–9**).

Unit Blocks. The highest-quality parquet block is made with ¾-inch-thick (19 mm), tongue-and-groove, solid hardwood, usually oak. The widely

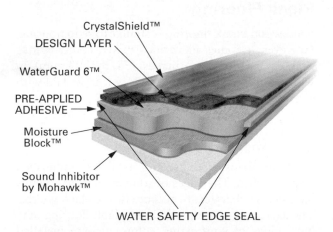

FIGURE 84–7 Laminate flooring details.

used 9 × 9 (229 × 229 mm) **unit block** is made with six strips, 1½ inches (38 mm) wide, or with four strips 2¼ inches (57 mm) wide. Unit blocks are laid with the direction of the strips at right angles to adjacent blocks (**Figure 84–10**). Unit blocks are made in other sizes and used in combination with parquet strips. Several patterns have gained popularity.

FRENCH HERRINGBONE
PATTERN

BASKET WEAVE
PATTERN

HERRINGBONE
PATTERN

STONE
PATTERN

FIGURE 84–8 Parquet strips are made in lengths that are multiples of its width.

Monticello is the name of a parquet originally designed by Thomas Jefferson, the third president of the United States. The pattern consists of a 6 × 6 (152 × 152 mm) centre unit block surrounded by 2¼-inch-wide (57 mm) pointed *pickets*. Each centre unit block is made of four 1½-inch-wide (38 mm) strips (**Figure 84–11**). Each block comes with three pickets joined to it at the factory.

FIGURE 84–10 Unit blocks are widely used in an alternating pattern.

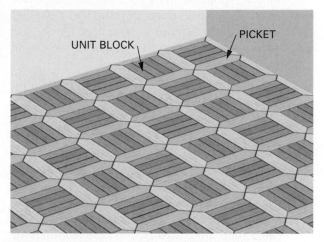

UNIT BLOCK

PICKET

FIGURE 84–11 The *Monticello* pattern is a famous parquet that uses square centre blocks and picket-shaped strips.

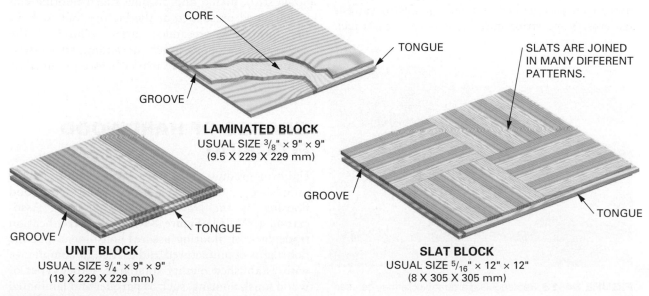

CORE

TONGUE

GROOVE

LAMINATED BLOCK
USUAL SIZE 3/8" × 9" × 9"
(9.5 X 229 X 229 mm)

SLATS ARE JOINED
IN MANY DIFFERENT
PATTERNS.

GROOVE

TONGUE

UNIT BLOCK
USUAL SIZE 3/4" × 9" × 9"
(19 X 229 X 229 mm)

GROOVE

TONGUE

SLAT BLOCK
USUAL SIZE 5/16" × 12" × 12"
(8 X 305 X 305 mm)

FIGURE 84–9 Three basic types of parquet block.

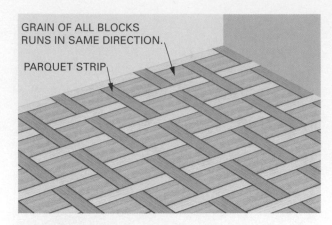

FIGURE 84-12 A popular parquet called the *Marie Antoinette* uses centre blocks and parquet strips.

Another popular parquet, called the *Marie Antoinette*, is copied from part of the Versailles Palace in France. Square centre unit blocks are enclosed by strips applied in a basket weave design (**Figure 84-12**).

Rectangular parquet blocks are often used in a *herringbone* pattern. One commonly used block is 4½ × 9 (114 × 229 mm) and made of three strips each 1½ inches (38 mm) wide and 9 inches (229 mm) long (**Figure 84-13**).

Laminated Blocks. Laminated blocks are generally made of three-ply laminated oak in a ³⁄₈-inch (9.5 mm) thickness. Most blocks come in 8 × 8 or 9 × 9 (203 × 203 or 229 × 229 mm) sizes. A 2 × 12 (50 × 305 mm) laminated strip is manufactured for use in herringbone and similar patterns.

Slat Blocks. Slat blocks are also called *finger blocks*. They are made by joining many short, narrow strips together in small squares of various patterns. Some strips may be as narrow as ⁵⁄₈ inch

FIGURE 84-13 Rectangular parquet blocks may be used to make herringbone and similar patterns.

17 ½" × 17 ½" (445 × 445 mm)
EACH PIECE IN THIS BLOCK
MEASURES ⁵⁄₈" × 4 ³⁄₈" (16 × 111 mm)

12" × 12"
(305 × 305 mm)

12" × 12"
(305 × 305 mm)

12" × 12"
(305 × 305 mm)

FIGURE 84-14 Slat blocks are made of short, narrow strips joined in various patterns.

(16 mm) and as short as 2 inches (50 mm) or less. Several squares are assembled to make the block (**Figure 84-14**). The squares are held together with a mesh backing or with a paper on the face side that is removed after the block is laid.

Presanded Flooring

Some strip, plank, and parquet finish flooring can be obtained presanded at the factory, but without the finish. The presanded surface eliminates the necessity for all but a touch-up sanding after installation to remove surface marks before the finish is applied.

GRADES OF HARDWOOD FLOORING

Uniform grading rules have been established for unfinished and factory-finished solid hardwood flooring by the National Wood Flooring Association (NWFA) (**Figure 84-15**). The certification trademark on flooring assures consumers that the flooring is manufactured and graded in compliance with established quality standards. Other types of wood finish flooring, such as parquet and laminated flooring, have no official grade rules.

Unfinished Oak Flooring (Red & White Separated)	Unfinished Hard Maple, Beech & Birch	Unfinished Pecan Flooring	Prefinished Oak Flooring
Clear Plain or CLear Quartered* Best appearance, mostly heartwood. Best grade, most uniform colour, limited small character marks. Bundles 1¼ ft and up. Average length 3¾ ft.	**First Grade** Best appearance. Natural colour variation, limited character marks. Bundles 1¼ and up. Average length 3 ft.	**First Grade** Excellent appearance. Natural colour variation, limited character marks, unlimited sapwood. Bundles 1¼ ft & up. Average length 3 ft.	**Prime Prefinished Oak (Special Order)** Excellent appearance. Natural colour variation, limited character marks, unlimited sapwood. Bundles 1¼ ft & up. Average length 3¼ ft.
Select Plain or Select Quartered Excellent appearance. Limited character marks, unlimited sound sapwood. Bundles 1¼ ft and up. Average length 3¼ ft. **Select & Better** A combination of Clear and Select grades.	**Second Grade** colour variations in appearance. Varying sound wood characteristics of species. Bundles 1¼ ft and up. Average length 2¾ ft.	**Second Grade** colour variation in appearance. Varying sound wood characteristics of species. Bundles 1¼ ft and up. Average length 2½ ft.	**Standard Prefinished Oak** Variegated appearance. Varying sound wood characteristics of species. Bundles 1¼ ft & up. Average length 2½ ft. **Standard & Better Grade** Bundles 1¼ ft & up. Average length 2½ ft.
No. 1 Common Variegated appearance. Light and dark colours; knots, flag worm holes and other character marks allowed to provide a variegated appearance after imperfections are filled and finished. Bundles 1¼ ft and up. Average length 2½ ft.	**Second & Better Grade** A combination of First and Second grades. Lengths equivalent to Second Grade. Average length 2¾ ft **Third Grade** Rustic appearance. All wood characteristics of species. Serviceable economical floor after filling. Bundles 1¼ ft and up. Average length 2¼ ft.	**Second & Better Grade** A combination of First and Second grades. Bundles 1¼ ft and up. Average length 2½ ft.	
No. 2 Common Rustic appearance. All wood characteristics of species. A serviceable economical floor after knot holes, worm holes, checks and other imperfections are filled and finished. Bundles 1¼ ft and up. Average length 2¼ ft. Red and White may be mixed.	**Third & Better Grade** A combination of First, Second, and Third grades. Bundles 1¼ ft and up. Average length 2¼ ft.	**Third Grade** Rustic appearance. All wood characteristics of species. A serviceable, economical floor after filling. Bundles 1¼ ft. and up. Average length 2 ft. **Third & Better Grade** A combination of First, Second, and Third grades. Bundles 1¼ ft and up.	**Tavern Prefinished Oak** Rustic appearance. All wood characteristics of species. A serviceable, economical floor. Bundles 1¼ ft & up. Average length 2 ft. **Tavern & Better Prefinished Oak** All wood characteristics of species. Bundles 1¼ ft & up. Average length 2 ft.
1¼ Shorts Pieces 9 to 18 inches. Bundles average nominal 1¼ ft. **No. 1 Common & Better Shorts** A combination grade of Clear, Select, and No. 1 Common. **No. 2 Common Shorts** Same as No. 2 Common.			

SELECT AND BETTER / SECOND AND BETTER / THIRD AND BETTER / STANDARD AND BETTER / TAVERN AND BETTER

NOFMA grading standards provided by NWFA

FIGURE 84–15 Guide to hardwood flooring grades.

Unfinished Flooring

Oak flooring is available quarter-sawn and plain-sawn. The grades for unfinished oak flooring, in declining order, are clear, select, No. 1 common, No. 2 common, and 1¼-foot (380 mm) shorts. Quarter-sawn flooring is available in clear and select grades only.

Birch, beech, and hard maple flooring are graded in declining order as first grade, second grade, third grade, and special grade. Grades of pecan flooring

are first grade, first grade red, first grade white, second grade, second grade red, and third grade. Red grades contain all heartwood. White grades are all bright sapwood.

In addition to appearance, grades are based on length. For instance, bundles of 1¼-foot (380 mm) shorts contain pieces from 9 to 18 inches (229 to 457 mm) long. The average length of clear bundles is 3¾ feet (1150 mm). The flooring comes in bundles in lengths of 1¼ feet (380 mm) and up. Pieces in each bundle are not of equal lengths. A bundle may include pieces from 6 inches (152 mm) under to 6 inches (152 mm) over the nominal length of the bundle. No pieces shorter than 9 inches (229 mm) are allowed.

Prefinished Flooring

Grades of prefinished flooring are determined after it has been sanded and finished. In declining order, they are prime, standard and better, standard, and tavern. Prefinished beech and pecan are furnished only in a combination grade called tavern or better.

ESTIMATING HARDWOOD FLOORING

To estimate the amount of hardwood flooring material needed, first determine the area to be covered. Add to this a percentage of the area depending on the width of the flooring to be used. The percentages include an additional 5 percent for end matching and normal waste (this gives the number of board feet needed):

55 percent for flooring 1½ inches (38 mm) wide

42.5 percent for flooring 2 inches (50 mm) wide

38.33 percent for flooring 2¼ inches (57 mm) wide

For example, the area of a room 16 by 24 feet is 384 square feet. If 2¼-inch flooring is to be used, multiply 384 by 0.3833 to get 147.18. Round this off to 147 square feet. Add 147 to 384 to get 531, which is the number of board feet of flooring required.

Metric example: The area of a room 5 by 7.3 m is 36.5 m². Add 5 percent for normal waste. Unfinished hardwood tongue-and-groove flooring is available dressed to size, and the retailer has the conversion charts to estimate square footage coverage. Prefinished hardwood flooring comes in boxes with the square footage coverage on the box. It is therefore an easy estimate.

Chapter 85

Laying Wood Finish Floor

Lumber used in the manufacture of hardwood flooring has been air dried, kiln dried, cooled, and then accurately machined to exacting standards. It is a fine product that should receive proper care during handling and installation.

HANDLING AND STORAGE OF WOOD FINISH FLOOR

Maintain moisture content of the flooring by observing recommended procedures. Flooring should not be delivered to the jobsite until the building has been closed in. Outside windows and doors should be in place. Cement work, plastering, and other materials must be thoroughly dry. In warm seasons, the building should be well ventilated. During cold months, the building should be heated, not exceeding 22°C (72°F), for at least five days before delivery and until flooring is installed and finished.

Do not unload flooring in the rain. Stack the bundles in small lots in the rooms where the flooring is to be laid. Leave adequate space around the bundles for good air circulation. Let the flooring become acclimated to the atmosphere of the building for four or five days or more before installation.

Concrete Slab Preparation

Wood finish floors can be installed on an on-grade or above-grade concrete slab. New slabs should be at least 90 days old. Flooring should not be installed when tests indicate excessive moisture in the slab. Floors should not be installed on below-grade slabs unless adequate subfloor mechanical ventilation can be accomplished. There are manufactured tiles on the market that are designed as a subfloor for below-grade concrete slabs. These tongue-and-groove tiles are 2 × 2 feet (610 × 610 mm) OSB panels with either an air-gap membrane (*Dricore.com*) or extruded polystyrene (XPS) (*green-barricade.org*) adhered to the underside. Each manufacturer defends its choice of substrate on its website.

Testing for Moisture

A test can be made by laying a smooth rubber mat on the slab. Put weight on it to prevent moisture from escaping. Allow the mat to remain in place for

24 hours. If moisture shows when the mat is removed, the slab is too wet. Another method of testing is by taping and sealing the edges of about a 1 × 1 foot (305 × 305 mm) square of 6-mil (0.15 mm) clear polyethylene film to the slab. If moisture is present, it can be easily seen on the inside of the film after 24 hours.

If no moisture is present, prepare the slab. Grind off any high spots. Fill low spots with levelling compound. The slab must be free of grease, oil, or dust.

Applying a Moisture Barrier

A moisture barrier must be installed over all concrete slabs. This ensures a trouble-free finish floor installation. Spread a skim coat of mastic with a straight trowel over the entire slab. Allow to dry at least two to three hours. Then cover the slab with polyethylene film. Lap the edges of the film 4 to 6 inches (100 to 152 mm) and extending up all walls enough to be covered by the baseboard, when installed.

When the film is in place, walk it in. Step on the film, over every square inch of the floor, to make sure it is completely adhered to the cement. Small bubbles of trapped air may appear. The film may be punctured, without concern, to let the air escape.

Applying Plywood Subfloor

A plywood subfloor may be installed over the moisture barrier on which to fasten the finish floor. In cold climates, it would be advisable to separate the subfloor from the concrete with an approved below-grade board insulation to reduce condensation. If there is a history of water seepage into the basement, it would be advisable to raise the subfloor off of the concrete, either with sleepers or an air-gap membrane, and to provide directional drainage to the floor drain. Exterior-grade sheathing plywood of at least ¾-inch (19 mm) thickness is used. The plywood is laid with staggered joints. Leave a ¾-inch (19 mm) space at walls and ¼- to ½-inch (6 to 12 mm) space between panel edges and ends. Fasten the plywood to the concrete with at least nine nails per panel (**Figure 85–1**).

Instead of driving fasteners, the plywood may be cemented to the moisture barrier. Cut the plywood in 4 × 4 foot (1.2 × 1.2 m) squares. Use a portable electric circular saw. Make scores ⅜ inch (9 mm) deep on the back of each panel to form a 12 × 12 inch (305 × 305 mm) grid. Lay the panel in asphalt cement spread with a ¼-inch (6 mm) notched trowel.

Applying Sleepers

Finish flooring may also be fastened to **sleepers** installed on the slab. Sleepers are short lengths of

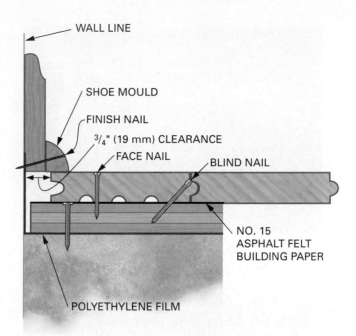

FIGURE 85–1 Installation details of a plywood subfloor over a concrete slab.

lumber cemented to the slab. They must be pressure-treated and dried to a suitable moisture content. Usually, 2 × 4 (38 × 89 mm) lumber, from 18 to 48 inches (457 to 1220 mm) long, is used.

Sleepers are laid on their side and cemented to the slab with mastic. They are staggered, with end laps of at least 4 inches (100 mm), in rows 12 inches (305 mm) on centre (OC) and at right angles to the direction of the finish floor. If the slab was not installed with a vapour retarder, a polyethylene vapour barrier is then placed over the sleepers. The edges are lapped over the rows (**Figure 85–2**). With end-matched flooring, end joints need not meet over the sleepers.

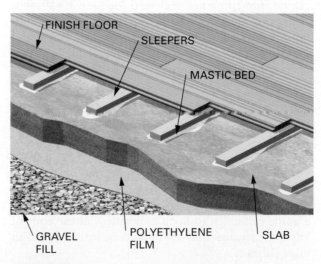

FIGURE 85–2 Sleepers are cemented to a concrete slab to provide fastening for strip or plank finish flooring.

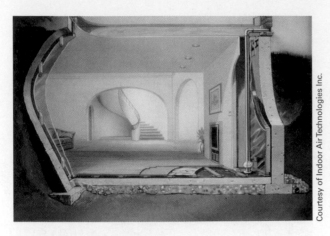

FIGURE 85–3 Sealed, drained, and depressurized low "E" cavity floor.

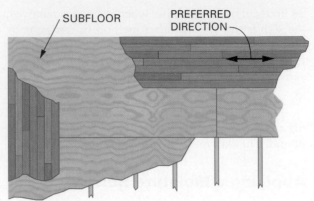

FINISH FLOOR MAY RUN IN EITHER DIRECTION, BUT PERPENDICULAR TO JOISTS IS PREFERRED.

PANEL SUBFLOOR

FIGURE 85–4 Several factors determine the direction in which strip flooring is laid.

It is not advisable to place a wood or wood-product floor over a concrete slab unless a system is devised that will circulate and vacate the damp air out from under the floor. Moisture and darkness will promote mould growth, and this will adversely affect the health of the occupants. Consider the design engineering that has generated the Canadian system—the enclosed conditioned housing (ECHO) basement (**Figure 85–3**). See *http://www.indoorair.ca/iat/pdf/echobasement_photo_np.pdf* for more about that system.

SUBFLOORS ON JOISTS

Exterior plywood or boards are recommended for use as subfloors on joists when wood finish floors are installed on them. If plywood is used, a full ½-inch (12 mm) thickness is required. Thicknesses of ⅝ inch (16 mm) or more are preferred for ¾-inch (19 mm) strip finish flooring. Use ¾-inch-thick (19 mm) subfloor for ½-inch (12 mm) strip flooring. The Wood Flooring Manufacturers Association (NOFMA) recognizes OSB as a suitable subfloor for finish wood flooring.

LAYING WOOD FINISH FLOOR

In new construction, the base or door casings are not usually applied yet for easier application of the finish floor. In remodelling, the base and base shoe must be removed. Use a scrap piece of finish flooring as a guide on which to lay a handsaw. Cut the ends of any door casings that are extending below the finish floor surface.

Before laying any type of wood finish flooring, nail any loose areas. Sweep the subfloor clean.

Scraping may be necessary to remove all plaster, taping compound, or other materials.

Laying Strip Flooring

Strip flooring laid in the direction of the longest dimension of the room gives the best appearance. The flooring may be laid in either direction on a plywood subfloor (**Figure 85–4**).

When the subfloor is clean, cover it with building paper. Lap it 4 inches (100 mm) at the seams, and at right angles to the direction of the finish floor. The paper helps keep out dust, prevents squeaks in dry seasons, and retards moisture from below that could cause warping of the floor.

Snap chalk lines on the paper showing the centreline of floor joists so that flooring can be nailed into them. For better holding power, fasten flooring through the subfloor and into the floor joists whenever possible. On ½-inch (12 mm) plywood subfloors, flooring fasteners must penetrate into the joists.

Starting Strip. The location and straight alignment of the first course is important. Place a strip of flooring on each end of the room, ¾ inch (19 mm) from the starter wall with the groove side toward the wall. Mark along the edge of the flooring tongue. Snap a chalk line between the two points. The gap between the flooring and the wall is needed for expansion. It will eventually be covered by the base.

Hold the strip with its tongue edge to the chalk line. Face-nail it with 2½-inch (63 mm) finish nails,

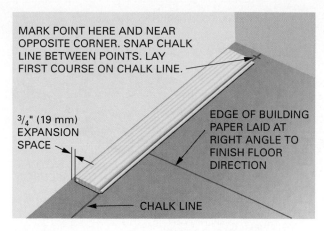

MARK POINT HERE AND NEAR OPPOSITE CORNER. SNAP CHALK LINE BETWEEN POINTS. LAY FIRST COURSE ON CHALK LINE.

³/₄" (19 mm) EXPANSION SPACE

EDGE OF BUILDING PAPER LAID AT RIGHT ANGLE TO FINISH FLOOR DIRECTION

CHALK LINE

FIGURE 85–5 A chalk line is snapped on the floor for alignment of the starting row of strip flooring.

alternating from one edge to the other, not more than 8 inches (203 mm) apart. Work from left to right with the grooved end of the first piece toward the wall. Left is determined by having the back of the person laying the floor to the wall where the starting strip is laid. Make sure end joints between strips are driven up tight (**Figure 85–5**).

When necessary to cut a strip to fit to the right wall, use a strip long enough so that the cut-off piece is 8 inches (203 mm) or longer. Start the next course on the left wall with this piece.

Blind-Nailing

Flooring is blind-nailed by driving nails at about a 45-degree angle through the flooring. Start the nail in the corner at the top edge of the tongue. Usually 2¼-inch (57 mm) hardened cut or spiral screw nails are used. Recommendations for fastening are shown in **Figure 85–6**.

For the first two or three courses of flooring, a hammer must be used to drive the fasteners. For floor laying, a heavier than usual hammer, from 20 to 28 ounces (567 to 794 g), is generally used for extra driving power. Care must be taken not to let the hammer glance off the nail. This may damage the edge of the flooring. Care also must be taken that, on the final blows, the hammer head does not hit the top corner of the flooring. To prevent this, raise the hammer handle slightly on the final blow so that the hammer head hits the nail head and the tongue but not the corner of the flooring (**Figure 85–7**).

After the nail is driven home, its head must be set slightly. This allows adjoining strips to come up tightly against each other. Floor layers use the head of the next nail to be driven to set the nail just driven.

Strip T & G		Blind-Nail Shading along the Length of Strips. Minimum Two Nails per Piece Near the Ends. (1″–3″) (25–75 mm)
Size Flooring	**Size Nail to Be Used**	
¾ × 1½″, 2¼″, & 3¼″ (19 × 38/57/83 mm)	2″ (50 mm) serrated-edge barbed fastener, 2¼″ or 2½″ (57 or 63 mm) screw or cut nail, 2″ (50 mm) 15-gauge staples with ¼″ (6 mm) crown.	In addition-10–12″ (254–305 mm) apart-8–10″ (203–254 mm) preferred
	On slab with ¾″ (19 mm) plywood subfloor use 1½″ (38 mm) barbed fastener, ½″ (12 mm) plywood subfloor with joists a maximum 16″ (406 mm) OC, fasten into each joist with additional fastening between, or 8″ (203 mm) apart.	
½ × 1½″ & 2″ (12 × 38/50 mm)	1½″ (38 mm) serrated-edge barbed fastener, 1½″ (38 mm) screw, cut steel, or wire casing nail.	10″ (254 mm) apart ½″ (12 mm) flooring must be installed over a minimum ⅝″ (16 mm) thick plywood subfloor.
⅜ × 1¼″ & 2″ (9.5 × 32/50 mm)	1¼″ (32 mm) serrated-edge barbed fastener, 1½″ (38 mm) bright wire casing nail	8″ (203 mm) apart
Square-Edge Flooring		
⁵⁄₁₆ × 1½″ & 2″ (8 × 38/50 mm)	1″ (25 mm) 15-gauge fully barbed flooring brad.	2 nails every 7″ (178 mm)
⁵⁄₁₆ × 1⅓″ (8 × 34 mm)	1″ (25 mm) 15-gauge fully barbed flooring brad.	1 nail every 5″ (127 mm) on alternate sides of strip
Plank ¾ × 3″ to 8″ (19 × 75 to 203 mm)	2″ (50 mm) serrated-edge barbed fastener, 2¼″ (57 mm) or 2½″ (63 mm) screw, or cut nail, use 1½″ (38 mm) length with ¾″ (19 mm) plywood subfloor on slab.	8″ (203 mm) apart
FOLLOW MANUFACTURER'S INSTRUCTIONS FOR INSTALLING PLANK FLOORING		

Courtesy of The Wood Flooring Manufacturers Association

FIGURE 85–6 Nailing guide for strip and plank finish flooring.

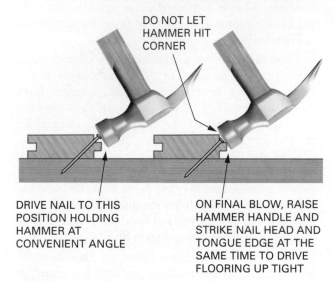

DO NOT LET HAMMER HIT CORNER

DRIVE NAIL TO THIS POSITION HOLDING HAMMER AT CONVENIENT ANGLE

ON FINAL BLOW, RAISE HAMMER HANDLE AND STRIKE NAIL HEAD AND TONGUE EDGE AT THE SAME TIME TO DRIVE FLOORING UP TIGHT

FIGURE 85–7 Technique for driving a blind nail.

When fastening flooring, the floor layer holds a hammer in one hand and a number of nails in the other. While driving one nail, the floor layer fingers the next nail to be driven into position to be used as a set. When the nail being driven is most of the way, the fingered nail is laid on edge with its head on the nail to be set. With one sharp blow, the nail is set (**Figure 85–8**). The setting nail is then the next nail to be driven. In this manner, the floor layer maintains a smooth, continuous motion when fastening flooring.

Note: A nail set should not be used to set hardened flooring nails. If used, the tip of the nail set will be flattened, thus rendering the nail set useless. Do not lay the nail set flat along the tongue, on top of the nail head, and then set the nail by hitting the side of the nail set with a hammer. Not only is this method slower, but it invariably breaks the nail set, possibly causing an injury.

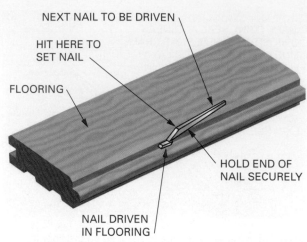

NEXT NAIL TO BE DRIVEN

HIT HERE TO SET NAIL

FLOORING

HOLD END OF NAIL SECURELY

NAIL DRIVEN IN FLOORING

FIGURE 85–8 Method to set nails driven by hand.

Racking the Floor

After the second course of flooring is fastened, lay out seven or eight loose rows of flooring, end to end. Lay out in a staggered pattern. End joints should be at least 6 inches (152 mm) apart. Find or cut pieces to fit within ½ inch (12 mm) of the end wall. Distribute long and short pieces evenly for the best appearance. Avoid clusters of short strips. Laying out loose flooring in this manner is called **racking the floor**. Racking is done to save time and material (**Figure 85–9**).

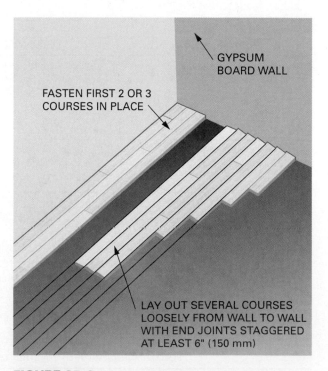

GYPSUM BOARD WALL

FASTEN FIRST 2 OR 3 COURSES IN PLACE

LAY OUT SEVERAL COURSES LOOSELY FROM WALL TO WALL WITH END JOINTS STAGGERED AT LEAST 6" (150 mm)

FIGURE 85–9 Racking the floor places the strips in position for efficient installation.

Using the Power Nailer

At least two courses of flooring must be laid by hand to provide clearance from the wall before a power nailer can be used. The power nailer holds strips of special barbed fasteners. The fasteners are driven and set through the tongue of the flooring at the proper angle. Although it is called a power nailer, a heavy hammer is swung by the operator against a plunger to drive the fastener (**Figure 85–10**).

The hammer is double ended. One end is rubber and the other end is steel. The flooring strip is placed in position. The steel end of the hammer is used to drive the edges and ends of the strips up tight. The rubber end is used against the plunger of the power nailer to drive the fasteners. Slide the power nailer along the tongue edge. Drive fasteners about 8 to 10 inches (203 to 254 mm) apart or as needed to bring the strip up tight against previously laid strips.

Note: When using the power nailer, one heavy blow is used to drive the fastener. Some nailer models have a ratcheting drive shaft that allows for multiple blows to set the nail. In either case, after the nail is set, the shaft returns and another fastener drops into place ready to be driven. Make sure the wood strip is fit fairly tight before nailing.

Whether laying floor with a power nailer or driving nails with a hammer, the floor layer stands with heels on strips already fastened, and toes on the loose strip to be fastened. With weight applied to the joint, easier alignment of the tongue and groove is

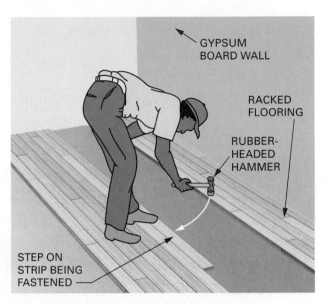

ON THE JOB

Step on the strip being fastened to align tongue-and-groove edges before drilling pieces together.

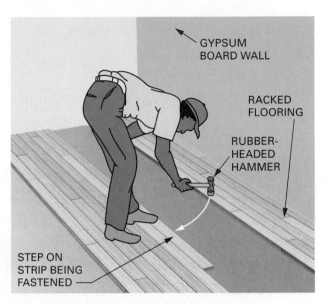

- GYPSUM BOARD WALL
- RACKED FLOORING
- RUBBER-HEADED HAMMER
- STEP ON STRIP BEING FASTENED

FIGURE 85–11 Technique of aligning boards together tightly before nailing.

possible (**Figure 85–11**). The weight of the worker also prevents the loose strip from bouncing when it is driven to make the edge joint tight. Avoid using a power nailer, pneumatic nailer, and hammer-driven fasteners on the same strip of flooring. Each method of fastening places the strips together with varying degrees of tightness. This variation, compounded over multiple strips, will cause waves in the straightness of the flooring.

Ending the Flooring

Continue across the room. Rack seven or eight courses as work progresses. The last three or four courses from the opposite wall must be nailed by hand. This is because of limited room to place the power nailer and swing the hammer. The next-to-the-last row can be blind-nailed if care is taken. However, the flooring must be brought up tightly by prying between the flooring and the wall. Use a bar to pry the pieces tight at each nail location (**Figure 85–12**).

The last course is installed in a similar manner. However, it must be face-nailed. It may need to be ripped to the proper width. If it appears that the installation will end with an undesirable, difficult to

FIGURE 85–10 A power nailer is widely used to fasten strip flooring.

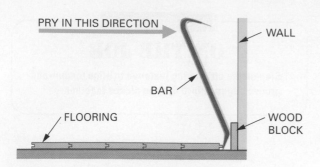

FIGURE 85–12 The last two courses of strip flooring may be brought tight with a pry bar.

apply, narrow strip, lay wider strips in the last row (**Procedure 85–A**).

Framing around Obstructions

A much more professional and finished look is given to a strip flooring installation if **hearths** and other floor obstructions are framed. Use flooring, with

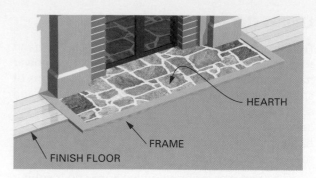

FIGURE 85–13 Frame around floor obstructions, such as hearths, with strips that are mitred at the corners.

mitred joints at the corners, as framework around the obstructions (**Figure 85–13**).

Changing Direction of Flooring

Sometimes it is necessary to change the direction of flooring when it extends from a room into another

---

Step**by**Step Procedures

Procedure 85–A Installing the Last Strip of Flooring

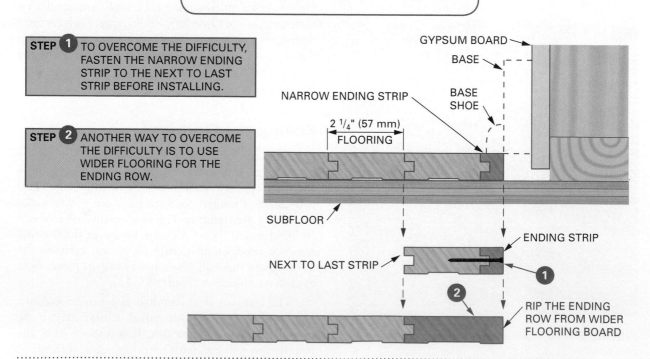

ON THE JOB

Use one of these methods to overcome the difficulty of ending with a narrow strip.

STEP ❶ TO OVERCOME THE DIFFICULTY, FASTEN THE NARROW ENDING STRIP TO THE NEXT TO LAST STRIP BEFORE INSTALLING.

STEP ❷ ANOTHER WAY TO OVERCOME THE DIFFICULTY IS TO USE WIDER FLOORING FOR THE ENDING ROW.

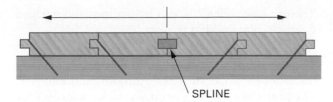

FIGURE 85–14 The direction of strip flooring can be changed by the use of a spline.

room, hallway, or closet. To do this, face-nail the extended piece to a chalk line. Change directions by joining groove edge to groove edge and inserting a *spline*, ordinarily supplied with the flooring (**Figure 85–14**). For best appearance, avoid bunching short or long strips. Open extra bundles, if necessary, to get the right selection of lengths. The completed floor installation is now ready to sand, stain (if required), seal, and finish (**Figure 85–15**).

Bamboo flooring can be installed in the same nail-down fashion as hardwood flooring. It can also be glued down and installed as a floating floor (**Figure 85–16**).

FIGURE 85–15 Power sanding and preparing the hardwood floor for its finish.

FIGURE 85–16 Bamboo flooring.

Laying Laminate Strip Flooring

Before laying laminate strip flooring, a ⅛-inch (3 mm) foam underlayment, or other material supplied and approved by the manufacturer, is applied to the subfloor or slab (**Figure 85–17**). The flooring is "floating" and is not fastened or cemented to the floor. Today, most laminate flooring is manufactured with a snap-and-click tongue-and-groove profile that eliminates the need for glue (**Figure 85–18**).

FIGURE 85–17 Types of underlayment for laminate flooring.

Courtesy of Armstrong-Swiftlock

FIGURE 85–18 Tongue-and-groove snap-and-click laminate flooring.

The product comes in varying thicknesses and degrees of hardness. It is best to cut it with a sabre saw (jigsaw) equipped with a metal-cutting blade.

The first row is laid in a straight line with end joints clicked together. The tongue edge is placed toward the wall, using blocks to create about a ½-inch (12 mm) expansion space. Subsequent courses are also laid in a straight line across the room and then the entire row is tipped so that the tongue sits in the groove of the previous row and then it is lowered and snapped into place. Usually the last course must be cut to width. To lay out the last course to fit, lay a complete row of boards with the groove toward the wall, directly on top of the already installed next to last course. Cut a short piece of flooring for use as a scribing block. Hold the groove edge against the wall. Move the block along the wall while holding a pencil against the other edge to lay out the flooring. The width of the tongue of the scribing block provides the necessary expansion space (**Figure 85–19**).

FIGURE 85–19 Technique for scribing the last strip on flooring using a scrap piece of flooring.

The last row, when cut, can be snapped and wedged tightly in place with a pry bar.

It is recommended by the manufacturers that glue should be used for the installation of laminate floors in wet areas such as bathrooms and kitchens. The procedure will be essentially the same except for extra care for the gluing operation. For an example of an installation guide, go to *beaulieucanada.com/ EN/Residential.awp* and click on Download Center for a list of installation documents for products such as laminate, hardwood, and more. For an installation video, go to *youtube.com* and search for "DassoSWB Hardness."

Installing Plank Flooring

Plank flooring is installed like strip flooring. Alternate the courses by widths. Start with the narrowest pieces. Then use increasingly wider courses, and repeat the pattern. Stagger the joints in adjacent courses. Use lengths so that they present the best appearance.

Manufacturers' instructions for fastening the flooring vary and should be followed. Generally, the flooring is blind-nailed through the tongue of the plank and at intervals along the plank in a manner similar to strip flooring.

INSTALLING PARQUET FLOORING

Procedures for the application of parquet flooring vary with the style and the manufacturer. Detailed installation directions are usually provided with the flooring. Generally, both parquet blocks and strips are laid in *mastic*. Use the recommended type. Apply with a notched trowel. The depth and spacing of the notches are important to leave the correct amount of mastic on the floor. Parquet may be installed either square with the walls or diagonally.

Square Pattern

When laying unit blocks in a square pattern, two layout lines are snapped, at right angles to each other, and parallel to the walls. Blocks are laid with their edges to the lines. Lines are usually laid out so that rows of blocks are either centred on the floor or half the width of a block off centre. This depends on which layout produces border blocks of equal and maximum widths against opposite walls.

To determine the location of the layout lines, measure the distance to the centre of the room's width. Divide this distance by the width of a block. If the remainder is half or more, snap the layout line in the centre. If the remainder is less than half, snap the layout line off centre by half the width of the block.

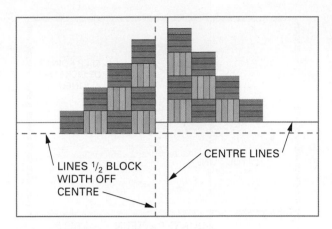

FIGURE 85–20 Parquet blocks are laid to the centre line or off the centre lines. This depends on which produces the best size of border blocks.

Find the location of the other layout line in the same way. It is possible that one of the layout lines will be centred, while the other must be snapped off centre.

Other factors may determine the location of layout lines, such as ending with full blocks under a door, or where they meet another type of floor. Regardless of the location, two lines, at right angles to each other, must be snapped.

Place one unit at the intersection of the lines. Position the grooved edges exactly on the lines. Lay the next units ahead and to one side of the first one and along the lines. Install blocks in a pyramid. Work from the centre outward toward the walls in all directions. Make adjustments as installation progresses to prevent misalignment (**Figure 85–20**).

Diagonal Pattern

To lay unit blocks in a diagonal pattern, measure an equal distance from one corner of the room along both walls. Snap a starting line between the two marks. At the centre of the starting line, snap another line at right angles to it. The location of both lines may need to be changed in order to end with border blocks of equal and largest possible size against opposite walls. The diagonal pattern is then laid in a manner similar to the square pattern (**Procedure 85–B**).

Special Patterns

Many parquet patterns can be laid out with square and diagonal layout lines. The *herringbone* pattern requires three layout lines. One will be the 90-degree line used for a square pattern. The other line crosses the intersection at a 45-degree angle. Align the first block or strip with its edge on the diagonal line. The corner of the piece should be lined up at the intersection. Continue the pattern in rows of three units wide, aligning the units with the layout lines (**Figure 85–21**).

The *Monticello* pattern can be laid square or diagonally in the same way as unit blocks. For best results, lay the parquet in a pyramid pattern. Alternate the grain of the centre blocks. Keep *picket* points in precise alignment (**Figure 85–22**).

The *Marie Antoinette* pattern may also be laid square or diagonally. It is started by laying a *band* with its grooved edge and end aligned with the

---

StepbyStep Procedures

Procedure 85–B Laying Out a Diagonal Pattern

STEP 1 LAY OUT EQUAL DISTANCES FROM A SQUARE CORNER.

STEP 2 SNAP A LAYOUT LINE BETWEEN THE TWO POINTS.

STEP 3 SNAP A LAYOUT LINE AT RIGHT ANGLES TO THE FIRST LAYOUT LINE.

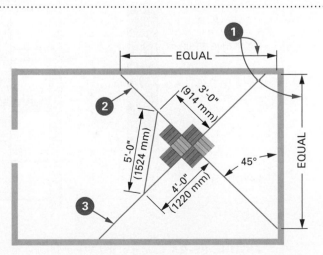

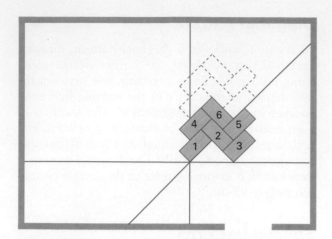

FIGURE 85–21 The herringbone pattern requires 90- and 45-degree layout lines.

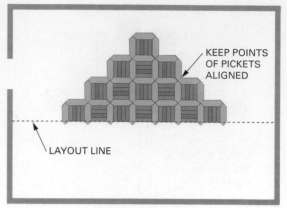

SQUARE PATTERN

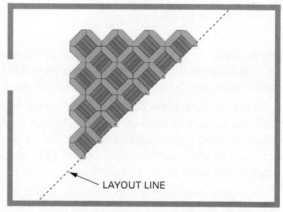

DIAGONAL PATTERN

layout lines with the tongued edge to the right. Continue laying the pattern by placing centre blocks and bands in a sequence so that bands appear woven (**Figure 85–23**). Tongues of all members face toward the right or ahead. In this pattern, the grain of centre blocks runs in the same direction.

Many other parquet patterns are manufactured. With the use of parquet blocks and strips, the design possibilities are almost endless. Competence in laying the popular patterns described in this chapter will enable students to apply the principles for professional installations of many different parquet floors.

FIGURE 85–22 Layout of the *Monticello* parquet pattern.

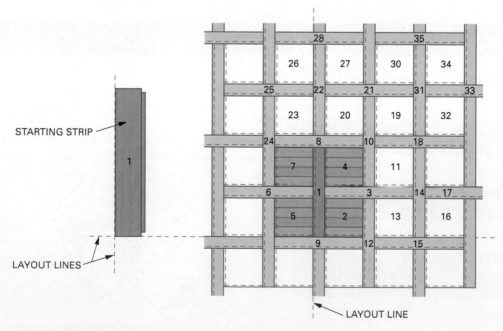

FIGURE 85–23 Layout of the *Marie Antoinette* parquet pattern.

Chapter 86

Underlayment and Resilient Tile

Resilient flooring is widely used in residential and commercial buildings. It is a thin, flexible material that comes in sheet or tile form. It is applied on a smooth concrete slab or an underlayment.

UNDERLAYMENT

Plywood, strandboard, hardboard, or particleboard may be used for **underlayment**. It is installed on top of the subfloor. This provides a base for the application of resilient sheet or tile flooring. The number of underlayment panels required is found by dividing the area to be covered by the area of the panel.

There are also underlayment materials that reduce sound transmission (see *http://www .acousticalsolutions.com/33~enkasonic-floor -underlayment*).

Underlayment thickness may range from ¼ to ¾ inch (6 to 19 mm), depending on the material and job requirements. In many cases, where a finish wood floor meets a tile floor, the underlayment thickness is determined by the difference in the thickness of the two types of finish floor. Both floor surfaces should come flush with each other.

Installing Underlayment

Sweep the subfloor as clean as possible. Stagger the joints between the subfloor and the underlayment. If installation is started in the same corner as the subfloor, the thickness of the walls will be enough to offset the joints. Leave about ¹⁄₃₂ inch (1 mm) between underlayment panels and a ¼-inch (6 mm) gap around the room perimeter and around doors to allow for expansion. For an installation guide, see *http://www .armstrong.com/content2/resam/files/18882.pdf*.

If the underlayment is to go over a board subfloor, install the underlayment with its face grain across the boards. In all other cases, a stiffer and stronger floor is obtained if the face grain is across the floor joists (**Figure 86–1**).

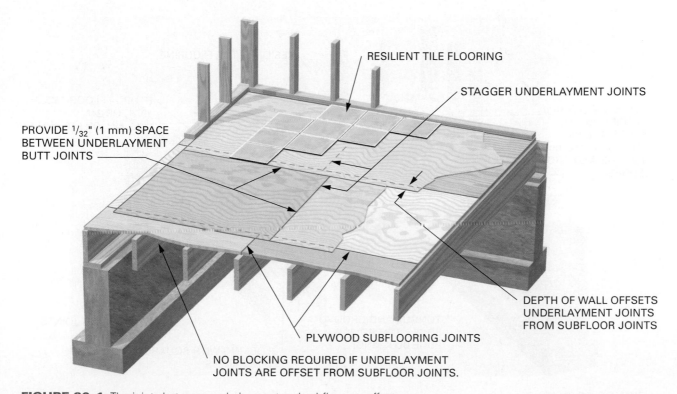

RESILIENT TILE FLOORING

STAGGER UNDERLAYMENT JOINTS

PROVIDE ¹⁄₃₂" (1 mm) SPACE BETWEEN UNDERLAYMENT BUTT JOINTS

DEPTH OF WALL OFFSETS UNDERLAYMENT JOINTS FROM SUBFLOOR JOINTS

PLYWOOD SUBFLOORING JOINTS

NO BLOCKING REQUIRED IF UNDERLAYMENT JOINTS ARE OFFSET FROM SUBFLOOR JOINTS.

FIGURE 86–1 The joints between underlayment and subfloor are offset.

Fasten the first row in place with staples or nails. Underlayment requires more nails than subfloor to provide a squeak-free and stiff floor. Nail spacing and size are shown in **Figure 86–2** for various thicknesses of plywood underlayment. Similar nail spacing and size are appropriate for other types of underlayment.

Install remaining rows of underlayment with end joints staggered from the previous courses. The last row of panels is ripped to width to fit the remaining space.

If APA Sturd-I-Floor is used, it can double as a subfloor and underlayment. If square-edged panels are used, blocking must be provided under the joints to support the edges. If tongue-and-groove panels are used, blocking is not required (**Figure 86–3**).

RESILIENT TILE FLOORING

Most resilient floor tiles are made of vinyl. Many different colours, textures, and patterns are available. Tiles come in 12 × 12 inch (305 × 305 mm) squares. They are applied to the floor in a manner similar to that used for applying parquet blocks. The most common tile thicknesses are $\frac{1}{16}$ and $\frac{1}{8}$ inch (1.5 and 3.2 mm). Thicker tiles are used in commercial applications subjected to considerable use.

Plywood Thickness (mm)	Fastener Size (approx.) and Type	Fastener Spacing (mm)	
		Panel Edges	Panel Interior
6 mm	32 ring-shank nails	75	150 each way
	18-gauge staples	75	150 each way
9.5/12 mm	32 ring-shank nails	150	150 each way
	16-gauge staples	75	150 each way
16/19 mm	38 ring-shank nails	150	150 each way
	16-gauge staples	75	150 each way
6 mm	32 ring-shank nails	75	150 each way
	18-gauge staples	75	150 each way

Plywood Thickness (inch)	Fastener Size (approx.) and Type	Fastener Spacing (inches)	
		Panel Edges	Panel Interior
¼	1¼" ring-shank nails	3	6 each way
	18-gauge staples	3	6 each way
⅜ or ½	1¼" ring-shank nails	6	8 each way
	16-gauge staples	3	6 each way
⅝ or ¾	1½" ring-shank nails	6	8 each way
	16-gauge staples	3	6 each way
¼	1¼" ring-shank nails	3	6 each way
	18-gauge staples	3	6 each way

FIGURE 86–2 Nailing specifications for plywood underlayment.

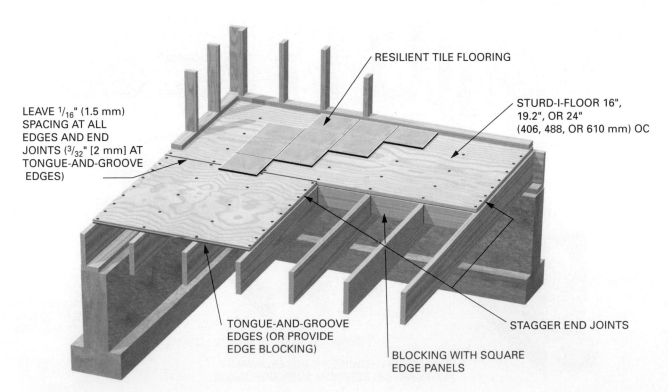

LEAVE $\frac{1}{16}$" (1.5 mm) SPACING AT ALL EDGES AND END JOINTS ($\frac{3}{32}$" [2 mm] AT TONGUE-AND-GROOVE EDGES)

RESILIENT TILE FLOORING

STURD-I-FLOOR 16", 19.2", OR 24" (406, 488, OR 610 mm) OC

TONGUE-AND-GROOVE EDGES (OR PROVIDE EDGE BLOCKING)

BLOCKING WITH SQUARE EDGE PANELS

STAGGER END JOINTS

FIGURE 86–3 APA Sturd-I-Floor requires no underlayment for the installation of resilient floors.

Long strips of the same material are called feature strips. They are available from the manufacturer. The strips vary in width from ¼ inch to 2 inches (6 to 50 mm). They are used between tiles to create unique floor patterns.

INSTALLING RESILIENT TILE

Before installing resilient tile, make sure underlayment fasteners are not projecting above the surface. Fill any open areas, such as splits, with a floor levelling compound (**Figure 86–4**). On underlayment, use a portable disc sander to bring the joints flush. Skim the entire surface with the sander to make sure the surface is smooth. Sweep the floor clean. For an example of an installation guide, go to *beaulieucanada.com/EN/ Residential.awp* and click on Download Center for a list of installation documents for products such as LVT, laminate, hardwood, and more.

Check to see if the spaces between the door stops and casing are sufficient to allow tile to slide under. They may be trimmed using a handsaw on a scrap piece of tile as a gauge (**Figure 86–5**).

FIGURE 86–4 Uneven floor surfaces are patched with a patching compound before tile is installed.

FIGURE 86–5 The bottom end of moulding may be trimmed to allow the floor tile to slide under.

CAUTION

If the application involves the removal of existing resilient floor covering, be aware that it, and the adhesive used, may contain asbestos. The presence of asbestos in the material is not easily determined. If there is any doubt, always assume that the existing flooring and adhesive do contain asbestos. Practices for removal of existing flooring or any other building material containing asbestos should comply with standards set by federal and provincial health and safety regulations.

If the application is over an existing resilient floor covering, do not sand the existing surface unless absolutely sure it does not contain asbestos. Inhalation of asbestos dust can cause serious harm.

Layout

Snap layout lines across the floor at right angles to each other. The lines may be centred if border tiles are a half width or more. If border tiles are less than a half width, layout lines may be snapped half the width of a tile off centre. The important thing is to have two perpendicular lines that serve to guide the joints in the tile.

Applying Adhesive

Adhesive is applied to the entire floor area (**Figure 86–6**). When the adhesive becomes transparent and appears to be dry, it is ready for tile to be installed. It will remain tacky and ready for application for up to 72 hours.

FIGURE 86–6 Floor adhesive is often trowelled onto the entire floor area before tile application begins.

FIGURE 86–7 Tile is applied from the layout lines toward the walls.

FIGURE 86–8 A tile cutter is often used to install border tile.

A notched trowel is used to spread the adhesive. It must be properly sized according to the manufacturer's recommendation. If the notches are too deep, more adhesive than necessary will be applied. This will result in the adhesive squeezing up through the joints onto the face of the tile and will probably require removal of the application.

Laying Tiles

Start by applying a tile to the intersection of the layout lines with the two adjacent edges on the lines. Lay tiles with edges tight. Work toward the walls (**Figure 86–7**). Watch the grain pattern. It may be desired to alternate the run of the patterns or to lay the patterns in one direction. Some tiles are stamped with an arrow on the back. They are placed so that the arrows on all tiles point in the same direction.

Lay tiles in place instead of sliding them into position. Sliding the tile pushes the adhesive through the joint. With most types of adhesives, it may be difficult or impossible to slide the tile.

Applying Border Tiles

Border tiles are often cut using a tile cutter (**Figure 86–8**). It cuts fast and the cut edges are clean and straight. Each piece is measured, marked, and cut by rotating the handle.

Border tiles may also be cut by scoring with a sharp utility knife and bending. To lay out and score a border tile to fit snugly, first place the tile to be cut directly on top of the last tile installed. Make sure all edges are in line. Place a full tile with its edge against the wall and on top of the one to be fitted.

Score the border tile along the outside edge of the top tile (**Figure 86–9**). Bend and break the tile along the scored line.

If the base has not been installed yet, the border tiles are fit roughly into place. Then when the base is installed later, it covers the cut edge. If the base has already been installed, the tile must be fit closely. Scored cuts may need to be smoothed with a file or sandpaper to improve their look. After the floor is completely tiled, it is rolled with a heavy steel roller to remove air bubbles and securely adhere the tiles to the underlay.

For tiles that require curved cuts, a propane torch or a heat gun can be used on the back side of the tile to warm it. This makes the knife cut easily and the resulting cut is smooth (**Figure 86–10**). Use

FULL TILE

BORDER TILE

FIGURE 86–9 Fitting a border tile by placing it under a guide piece of tile while scoring.

FIGURE 86–10 A propane torch is used to soften the tile for cutting. Use care not to overheat and burn the tile.

FIGURE 86–12 Vinyl base may be bent to fit around a corner.

FIGURE 86–11 Fitting a tile around a corner involves two cuts.

FIGURE 86–13 Cork flooring.

care not to overheat and burn the tile. This is also done when fitting a tile around a corner where two or more cuts are required (**Figure 86–11**).

Applying a Vinyl Cove Base

Many times a vinyl cove base is used to trim a tile floor. A special vinyl base cement is applied to its back. The base is pressed into place (**Figure 86–12**).

Cork Flooring

Cork flooring is fast becoming the flooring of choice in high-end homes across Canada. Its first selling feature is that it is natural, renewable, and biodegradable (**Figure 86–13**). The cork tree is unique in that its bark can be stripped several times without damaging the tree, and the trees can live up to 200 years (**Figure 86–14**).

FIGURE 86–14 The outer bark of the "cork" oak tree.

There are many benefits to a cork floor. It is soft, quiet, warm, and resilient, which means it is comfortable to stand on, and if dishes are dropped, the floor absorbs the impact. It is impermeable and naturally hygienic, and with a UV-resistant finish, it is durable and easy to clean. It is used in hospitals for this reason.

Installation procedures are straightforward and similar to tiles. Layout marks are made on the floor with equal borders that are larger than one-half of a tile, the adhesive is spread, and the tiles are set in place (**Figure 86–15**). Start ½ inch (12 mm) away from the wall (**Figure 86–16**) and use a block to securely fit the tongue-and-groove joints (**Figure 86–17**). The final row is pressed into place with a flat prybar (**Figure 86–18**).

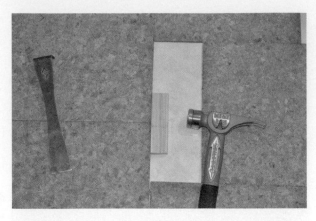

FIGURE 86–17 Tapping a block to snug up successive rows of cork flooring.

Courtesy of DAP Products Inc.

FIGURE 86–15 Adhesive for cork flooring.

FIGURE 86–18 The final row of the cork flooring is pressed into place using a flat prybar.

ESTIMATING RESILIENT TILE

To estimate the amount of tile flooring needed, find the area of the room in square feet. To do this, measure the length and width of the room to the next whole foot. Multiply these figures to find the area. For 12 × 12 inch tiles, the result is the number of pieces needed. Divide the number of pieces by 45, or whatever the number of pieces that the box contains, to determine the number of boxes. Rounding this number up to the next whole box usually takes care of any waste factor.

In metric units, multiply the length of the room by its width (in metres) and divide by 0.093 to get the number of tiles (305 × 305 mm). Divide this number by 45 or by the number of tiles in a box and round up to the next whole number to get the number of boxes to order.

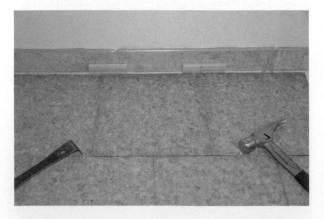

FIGURE 86–16 Start ½" (12 mm) away from the baseboard.

SHEET GOODS

Sheet goods is a general term that refers to roll flooring that comes in 6- and 12-foot widths (1.8 to 3.6 m). There are several different types, each with its own backing, adhesive, and adhesive spreader. The installation techniques are similar. There are **vinyl sheet goods**, **fibreglass sheet goods** (**Figure 86–19**), and **linoleum**.

Linoleum is considered a "green" product and is recognized by the LEED® certification group when applied with solvent-free, low-VOC (volatile organic compounds) adhesives. It is made with a jute backing and is comprised of linseed oil, rosin, wood flour, and powdered cork. It is renewable, bio-degradable, and burns with high heat and low emissions.

Installation involves cutting the roll to the correct width and length with the predominant lines parallel to the wall and laying it into its place. Half of the roll is then rolled up, the adhesive is spread, and then the rolled-portion is unfurled and pressed into the adhesive. This procedure is repeated for the second half of the roll and then the entire floor is rolled with a heavy steel roller, starting at the centre of the room and working to the walls (**Figure 86–20**). The adhesive is allowed to dry for a specified time before traffic is allowed on the floor area. For an example of an installation guide, go to *beaulieucanada.com/EN/Residential.awp* and click on Download Center for a list of installation documents for products such as sheet vinyl, laminate, and more.

FIGURE 86–19 Fibreglass-based sheet goods.

Courtesy of Armstrong World Industries, Inc.

Courtesy of Crain Cutter Co., Inc

FIGURE 86–20 A steel floor roller.

KEY TERMS

bamboo	hearths	plank flooring	strip flooring
blind-nailed	laminate flooring	prefinished strips	underlayment
chamfer	laminated blocks	racking the floor	unfinished strip flooring
cork flooring	laminated strip flooring	resilient flooring	unit block
engineered wood flooring	linoleum	sheet goods	vinyl sheet goods
fibreglass sheet goods	parquet block flooring	slat blocks	
	parquet strip flooring	sleepers	

REVIEW QUESTIONS

Select the most appropriate answer.

1. What is the maximum temperature at which hardwood flooring should be stored in a heated building?

   a. 22°C (72°F)
   b. 25.5°C (78°F)
   c. 29.5°C (85°F)
   d. 32°C (90°F)

2. What is most hardwood finish flooring made from?

   a. Douglas fir
   b. hemlock
   c. southern pine
   d. oak

3. Bundles of strip flooring may contain pieces over and under the nominal length of the bundle. How much longer or shorter are they allowed to be?

   a. 4 inches (100 mm)
   b. 6 inches (152 mm)
   c. 8 inches (203 mm)
   d. 9 inches (229 mm)

4. What are the shortest pieces allowed in bundles of hardwood strip flooring?

   a. 4 inches (100 mm)
   b. 6 inches (152 mm)
   c. 8 inches (203 mm)
   d. 9 inches (229 mm)

5. Why are the edges of prefinished strip flooring chamfered?

   a. to prevent splitting
   b. to apply the finish
   c. to simulate cracks between adjoining pieces
   d. to obscure any unevenness in the floor surface

6. What is the best grade of unfinished oak strip flooring?

   a. prime
   b. clear
   c. select
   d. quarter-sawn

7. When estimating the amount of 2¼-inch (57 mm) face hardwood flooring required, what percentage of the area to be covered should be added to the calculation?

   a. 42.5 percent
   b. 38.33 percent
   c. 33.33 percent
   d. 29 percent

8. When it is necessary to cut the last strip in a course of flooring, the waste is used to start the next course. What length should it be?

   a. at least 8 inches (203 mm) long
   b. at least 10 inches (254 mm) long
   c. at least 12 inches (305 mm) long
   d. at least 16 inches (206 mm) long

9. For floor laying, what weight of hammer is generally used?

   a. 13 to 16 ounces (370 to 450 g)
   b. 16 to 20 ounces (450 to 570 g)
   c. 20 to 28 ounces (570 to 800 g)
   d. 25 to 30 ounces (710 to 850 g)

10. What should be done when changing direction of strip flooring?

    a. Face-nail both strips.
    b. Turn the extended strip around.
    c. Blind-nail both strips.
    d. Use a spline.

WHAT'S WRONG WITH THIS PICTURE?

Carefully study **Figure 86–21** and think about what is wrong. Consider all possibilities.

FIGURE 86–21 This figure shows 2¼-inch (57 mm) red oak strip flooring that has many joints too close together. The butt joints should be staggered by at least 6 inches (152 mm) between rows. Having a row between the joints improves the appearance of the floor and reduces the potential for floor squeaks.

FIGURE 86–22 This floor has been laid out with fewer joints in one location. These joints are offset to have a row between them.

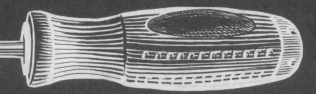

Unit 29

Cabinets and Countertops

| Chapter 87 | Description and Installation of Manufactured Cabinets | Chapter 88 | Countertop and Cabinet Construction |

Cabinets and countertops usually are purchased in preassembled units and may be installed by a carpenter. Manufactured cabinets are often used because of the great variety and shorter installation time than job-built cabinets. Cabinets can be custom-made to meet the specifications of almost any job, but they are usually made in a cabinet shop. Countertops, cabinet doors, and drawers may be customized in a wide variety of styles and sizes.

SAFETY REMINDER

Kitchen cabinets are installed in large units that are cumbersome and often heavy to move. Watch that other wall and floor finishes are not marred when positioning the cabinets. Also, remember to lift with your legs and not your back.

OBJECTIVES

After completing this unit, the student should be able to:

- state the sizes and describe the construction of typical base and wall kitchen cabinet units.
- plan, order, and install manufactured kitchen cabinets.
- construct and install a laminate countertop.
- identify cabinet doors and drawers according to the type of construction and method of installation.
- identify overlay, lipped, and flush cabinet doors and proper drawer construction.
- apply cabinet hinges, pulls, and door catches.

Chapter 87

Description and Installation of Manufactured Cabinets

Manufactured kitchen and bath cabinets come in a wide variety of styles, materials, and finishes. The carpenter must be familiar with the various kinds, sizes, uses, and construction of the cabinets to know how to plan, order, and install them.

DESCRIPTION OF MANUFACTURED CABINETS

For commercial buildings, many kinds of specialty cabinets are manufactured. They are designed for specific uses in offices, hospitals, laboratories, schools, libraries, and other buildings. Most cabinets used in residential construction are manufactured for the kitchen or bathroom. All cabinets, whether for commercial or residential use, consist of a case that is fitted with shelves, doors, and/or drawers. Cabinets are manufactured and installed in essentially the same way. Designs vary considerably with the manufacturer, but sizes are close to the same.

Kinds and Sizes

One method of cabinet construction utilizes a **face frame** (**Figure 87–1**). This frame provides openings for doors and drawers. Another method, called *European*

FIGURE 87–1 Face-frame kitchen cabinets.

FIGURE 87–2 Frameless kitchen cabinets.

or **frameless**, eliminates the face frame (**Figure 87–2**). Face-framed cabinets usually give a traditional look. The frameless system is a simple box with full-sized doors that cover the face edge of the box. Because of ease of construction, many cabinet shops use the same box construction and attach a face frame to it to give the cabinets the traditional look (**Figure 87–3**). A thorough treatment of the layout and construction of cabinets using this 1¼-inch (32 mm) system, European hinges, and melamine-coated particle board is covered in the book *Building Frameless Kitchen Cabinets* by Danny Proulx (ISBN 0-09731869-0-9). This is the most popular system used in Canada.

The two basic kinds of kitchen cabinets are the **wall unit** and the **base unit**. The surface of the countertop is usually about 36 inches (914 mm) from the floor. Wall units are installed about 18 inches (457 mm) above the countertop. This distance is enough to accommodate such articles as coffeemakers, toasters, blenders, and mixers, yet it keeps the top shelf within reach, not over 6 feet (1.8 m) from the floor. The usual overall height of a kitchen cabinet installation is 7 feet (2.1 m) (**Figure 87–4**).

Wall Cabinets. Standard wall cabinets are 12 inches (305 mm) deep. They normally come in heights of 42, 30, 24, 18, 15, and 12 inches (1067, 762, 610, 457,

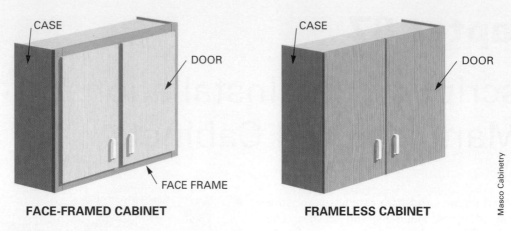

FACE-FRAMED CABINET

FRAMELESS CABINET

Masco Cabinetry

FIGURE 87–3 Two basic methods of cabinet construction.

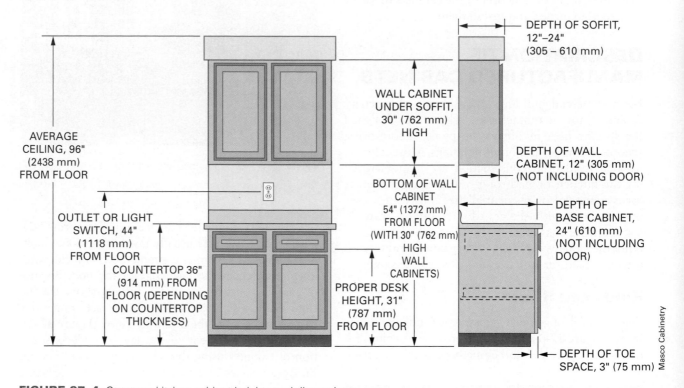

Masco Cabinetry

FIGURE 87–4 Common kitchen cabinet heights and dimensions.

381, and 305 mm). The standard height is 30 inches (762 mm). Shorter cabinets are used above sinks, refrigerators, and ranges. The 42-inch (1067 mm) cabinets are for use in kitchens without soffits where more storage space is desired. A standard-height wall unit usually contains two adjustable shelves.

Usual wall cabinet widths range from 9 to 48 inches (229 to 1220 mm) in 3-inch (75 mm) increments. They come with single or double doors depending on their width. Single-door cabinets can be hung so that doors can swing in either direction.

Wall corner cabinets make access into corners easier. **Double-faced cabinets** have doors on both sides for use above island and peninsular bases. Some wall cabinets are made 24 inches (610 mm) deep for installation above refrigerators. A microwave oven case, with a 30-inch-wide (762 mm) shelf, is available (**Figure 87–5**).

Base Cabinets. Most base cabinets are manufactured 34½ inches (876 mm) high and 24 inches (610 mm) deep. By adding the usual countertop thickness of 1½ inches (38 mm), the work surface is at the standard height of 36 inches (914 mm) from the floor. Base cabinets come in widths to match wall cabinets. Single-door cabinets are manufactured in widths from 9 to 24 inches (229 to 610 mm). Double-door cabinets come in widths

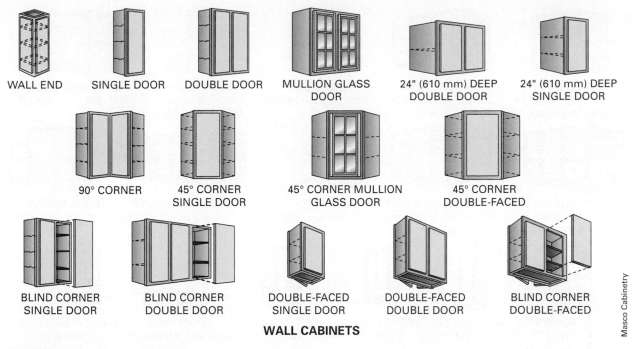

WALL END SINGLE DOOR DOUBLE DOOR MULLION GLASS DOOR 24" (610 mm) DEEP DOUBLE DOOR 24" (610 mm) DEEP SINGLE DOOR

90° CORNER 45° CORNER SINGLE DOOR 45° CORNER MULLION GLASS DOOR 45° CORNER DOUBLE-FACED

BLIND CORNER SINGLE DOOR BLIND CORNER DOUBLE DOOR DOUBLE-FACED SINGLE DOOR DOUBLE-FACED DOUBLE DOOR BLIND CORNER DOUBLE-FACED

WALL CABINETS

Masco Cabinetry

FIGURE 87–5 Kinds and sizes of manufactured wall cabinets.

from 27 to 48 inches (686 to 1220 mm). A recess called a **toe space** is provided at the bottom of the cabinet.

The standard base cabinet contains one drawer, one door, and an adjustable shelf. Some base units have no drawers; others contain all drawers.

Double-faced cabinets provide access from both sides. Corner units, with round revolving shelves, make corner storage easily accessible (**Figure 87–6**).

Tall Cabinets. Tall cabinets are usually manufactured 24 inches (610 mm) deep, the same depth as base cabinets. Some utility cabinets are 12 inches

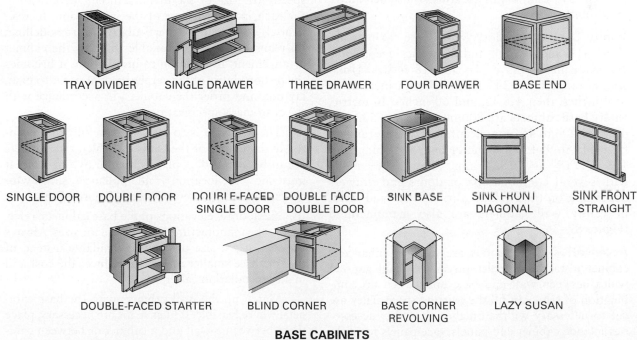

TRAY DIVIDER SINGLE DRAWER THREE DRAWER FOUR DRAWER BASE END

SINGLE DOOR DOUBLE DOOR DOUBLE-FACED SINGLE DOOR DOUBLE FACED DOUBLE DOOR SINK BASE SINK FRONT DIAGONAL SINK FRONT STRAIGHT

DOUBLE-FACED STARTER BLIND CORNER BASE CORNER REVOLVING LAZY SUSAN

BASE CABINETS

Masco Cabinetry

FIGURE 87–6 Most base cabinets are manufactured to match wall units.

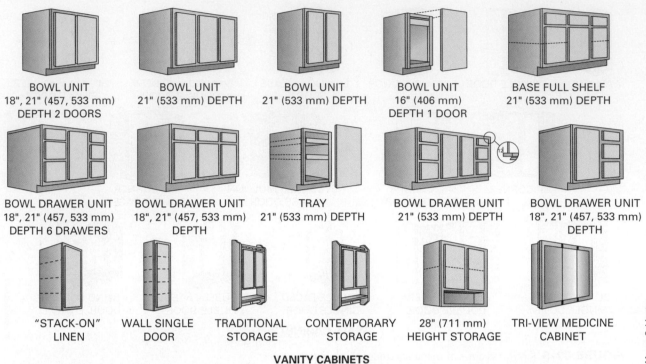

BOWL UNIT
18", 21" (457, 533 mm)
DEPTH 2 DOORS

BOWL UNIT
21" (533 mm) DEPTH

BOWL UNIT
21" (533 mm) DEPTH

BOWL UNIT
16" (406 mm)
DEPTH 1 DOOR

BASE FULL SHELF
21" (533 mm) DEPTH

BOWL DRAWER UNIT
18", 21" (457, 533 mm)
DEPTH 6 DRAWERS

BOWL DRAWER UNIT
18", 21" (457, 533 mm)
DEPTH

TRAY
21" (533 mm) DEPTH

BOWL DRAWER UNIT
21" (533 mm) DEPTH

BOWL DRAWER UNIT
18", 21" (457, 533 mm)
DEPTH

"STACK-ON"
LINEN

WALL SINGLE
DOOR

TRADITIONAL
STORAGE

CONTEMPORARY
STORAGE

28" (711 mm)
HEIGHT STORAGE

TRI-VIEW MEDICINE
CABINET

VANITY CABINETS

Masco Cabinetry

FIGURE 87–7 Vanity cabinets are made similar to kitchen cabinets, but differ in size.

(305 mm) deep. They are made 66 inches (1676 mm) high and in widths of 27, 30, and 33 inches (686, 762, and 838 mm) for use as oven cabinets. Single-door utility cabinets are made 18 and 24 inches (457 and 610 mm) wide. Double-door pantry cabinets are made 36 inches (914 mm) wide. Wall cabinets with a 24-inch (610 mm) depth are usually installed above tall cabinets.

Vanity Cabinets. Most vanity base cabinets are made 31½ inches high and 21 inches deep. Some are made in depths of 16 and 18 inches. Usual widths range from 24 to 36 inches in increments of 3 inches, then 42, 48, and 60 inches. In metric, vanity base cabinets are 800 mm high and 533 mm deep (or 406 to 457 mm deep), with widths ranging from 610 to 914 mm in increments of 75 mm, then 1067, 1219, and 1524 mm. They are available with several combinations of doors and drawers depending on their width. Various sizes and styles of vanity wall cabinets are also manufactured (**Figure 87–7**).

Accessories. Accessories are used to enhance a cabinet installation. Filler pieces fill small gaps in width between wall and base units when no combination of sizes can fill the existing space. They are cut to necessary widths on the job. Other accessories include cabinet end panels, face panels for dishwashers and refrigerators, open shelves for cabinet ends, and spice racks.

LAYING OUT MANUFACTURED KITCHEN CABINETS

The blueprints for a building contain plans, elevations, and details that show the cabinet layout. Architects may draw the layout, but they may not specify the size or the manufacturer's identification for each individual unit of the installation. In residential construction, particularly in remodelling, no plans are usually available to show the cabinet arrangement. In addition to installation, it becomes the responsibility of the carpentry contractor to plan, lay out, and order the cabinets, in accordance with the customer's specifications.

The first step is to measure carefully and accurately the length of the walls on which the cabinets are to be installed. A plan is then drawn to scale. It must show the location of all appliances, sinks, windows, and other necessary items (**Figure 87–8**).

Next, draw elevations of the base cabinets, referring to the manufacturer's catalogue for sizes. Always use the largest size of cabinets available instead of two or three smaller ones. This reduces the cost and makes installation easier.

Match up the wall cabinets with the base cabinets, where feasible. If filler strips are necessary, place them between a wall and a cabinet or between cabinets in the corner. Identify each unit on the elevations with the manufacturer's identification (**Figure 87–9**).

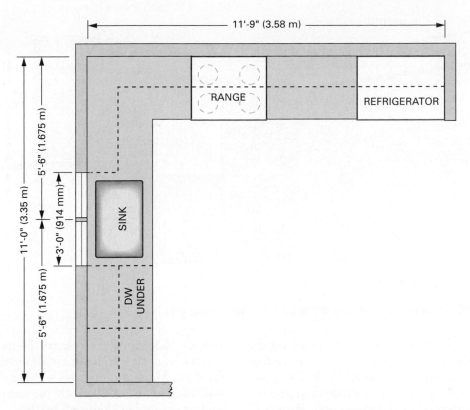

FIGURE 87–8 Typical plan of a kitchen cabinet layout showing location of walls, windows, and appliances.

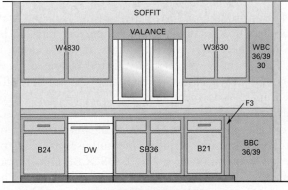

SINK WALL ELEVATION

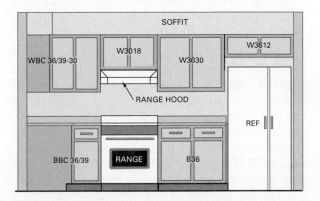

FIGURE 87–9 Elevations of the installation are sometimes drawn and the cabinets identified.

Make a list of the units in the layout. Order from the distributor.

Computer Layouts

Computer programs, such as AutoCAD and 2020 Design, are available to help in laying out manufactured kitchen cabinets. When the required information is fed into the computer, a number of different layouts can be quickly made. When an acceptable layout is made, it can be printed with each of the cabinets in the layout identified and priced. Most large kitchen cabinet distributors will supply computerized layouts on request. The larger companies will send the CAD drawings directly to a CNC (computer numerical control machine, which will cut the various components of the cabinet boxes.

INSTALLING MANUFACTURED CABINETS

Cabinets must be installed level and plumb even though floors are not always level and walls not always plumb. Level lines are first drawn on the wall for base and wall cabinets. To level base cabinets that sit on an unlevel floor, either shim the cabinets from the high point of the floor or scribe and fit the

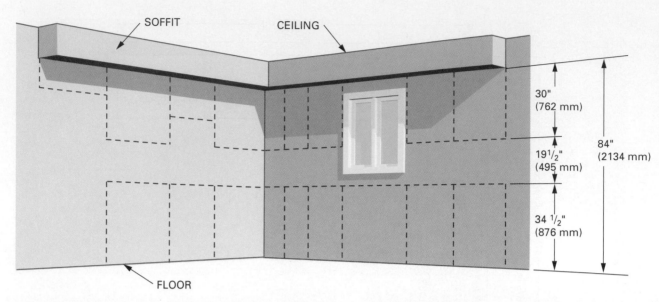

FIGURE 87–10 The wall is laid out with outlines for the cabinet locations.

cabinets to the floor from the lowest point on the floor. Shimming the base cabinets leaves a space that must be later covered by a moulding. Scribing and fitting the cabinets to the floor eliminate the need for a moulding. The method used depends on various conditions of the job. If shimming base cabinets, lay out the level lines on the wall from the highest point on the floor where cabinets are to be installed. If fitting cabinets to the floor, measure up from the low point.

Laying Out the Wall

Measure 34½ inches (876 mm) up the wall. Draw a level line to indicate the tops of the base cabinets. Use the most accurate method of levelling available (described in Chapter 27). Another level line must be made on the wall 54 inches (1372 mm) from the floor. The bottoms of the wall units are installed to this line. It is more accurate to measure 19½ inches (495 mm) up from the first level line and snap lines parallel to it than to level another line.

The next step is to mark the stud locations in a framed wall. (Cabinet mounting screws will be driven into the studs.) An electronic stud finder works well to locate framing members. The other method is to lightly tap on and across a short distance of the wall with a hammer. Drive a finish nail in at the point where a solid sound is heard. Drive the nail where holes are later covered by a cabinet. If a stud is found, mark the location with a pencil. If no stud is found, try a little over to one side or the other.

Measure at 16-inch (406 mm) intervals in both directions from the first stud to locate other studs. Drive a finish nail to test for solid wood. Mark each stud location. If studs are not found at

16-inch (406 mm) centres, try 24-inch (610 mm) centres. At each stud location, draw plumb lines on the wall. Mark the outlines of all cabinets on the wall to visualize and check the cabinet locations against the layout (**Figure 87–10**).

Installing Wall Units

Many installers prefer to install the soffit and wall cabinets first so that the work does not have to be done leaning over the base units. If this is done, it is important to place the wall and ceiling drywall first and any necessary fire blocking before the soffit is framed. The soffit may be framed using any of several methods (**Figure 87–11**). One method uses drywall to cover 2 × 2 (38 × 38 mm) framing or steel stud framing. Another method uses panelling or wood strips to cover a 2 × 2 (38 × 38 mm) frame. In either case, the ceiling and wall drywall should be installed completely to the corner and taped beforehand. The drywall corners on the soffit can be eased with bull-nosed corners and coved to the ceiling for a softer look (**Figure 87–12**).

A **cabinet lift** may be used to hold the cabinets in position for fastening to the wall. If a lift is not available, the doors and shelves may be removed to make the cabinet lighter and easier to clamp together. Frameless cabinets use **European hinges** on the doors, and they can be removed with the simple depression of a latch (see Figure 88–27, page 981). If possible, screw a 1 × 3 (19 × 64 mm) strip of lumber so that its top edge is on the level line for the bottom of the wall cabinets. This is used to support the wall units while they are being fastened. If it is not possible to screw to the wall,

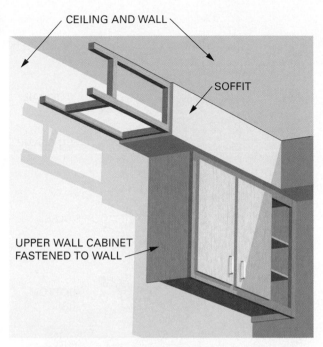

SOFFIT BUILT PRIOR TO CABINET INSTALLATION

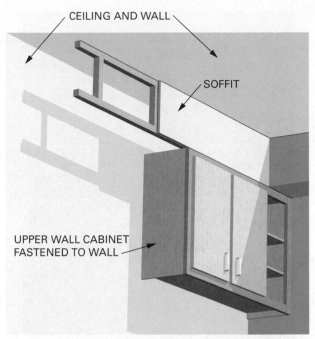

SOFFIT BUILT AFTER CABINET INSTALLATION

FIGURE 87–11 Two methods of finishing a soffit.

Courtesy of M. Nauth

FIGURE 87–12 Soffits with bull-nosed corners and coved tops.

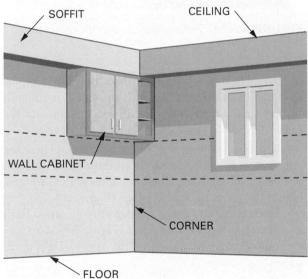

FIGURE 87–13 Installation of wall cabinets is started in the corner.

build a stand on which to support the unit near the line of installation.

Start the installation of wall cabinets in a corner **Figures 87–13** and **87–14**. Check for square in the corner and decide how you are going to make adjustments, by shimming and by scribing. Many corner units cut across the back with a 6-inch (152 mm) vertical strip at 45 degrees, leaving the top and bottom square. These are scribed to fit the corner to accommodate the build-up of drywall compound. On the wall, measure from the line representing the outside of the cabinet to the stud centres. Transfer the measurements to the cabinets. Drill shank holes for mounting screws through mounting rails usually

installed at the top and bottom of the cabinet. Place the cabinet on the supporting strip or stand so that its bottom is on the level layout line. Wood and steel framing typically do not need predrilled screw holes. Concrete screws used in masonry walls do need to have predrilled holes. This is normally done with the cabinet held in place. Make sure holes are sufficiently deep to prevent the screw from bottoming out in the hole. Fasten the cabinet in place with mounting screws of sufficient length to hold the cabinet

Courtesy of M. Nauth

FIGURE 87–14 Start installation with the corner unit first.

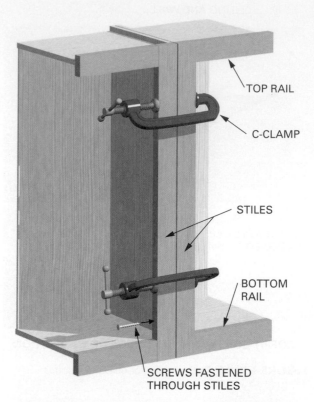

TOP RAIL

C-CLAMP

STILES

BOTTOM RAIL

SCREWS FASTENED THROUGH STILES

FIGURE 87–15 The stiles of adjoining cabinets are joined together with screws.

securely. Do not fully tighten the screws. It is important at this point to observe if the back of the unit lies flat against the wall, and if not, set shims at the fastening points.

The next cabinet is installed in the same manner. Align the adjoining stiles so that their faces are flush with each other. Clamp them together with C-clamps. Screw the stiles tightly together (**Figure 87–15**). For frameless cabinets, the face edges are clamped flush and they are fastened together (**Figure 87–16**). Continue this procedure around the room. Tighten all mounting screws.

If a filler needs to be used, it is better to add it next to a blind corner cabinet or at the end of a run. It may be necessary to scribe the filler to the wall (**Procedure 87–A**).

Installing Base Cabinets

Start the installation of base cabinets in a corner. Shim the bottom until the cabinet top is on the layout line. Then level and shim the cabinet from back to front.

If cabinets are to be fitted to the floor, shim until their tops are level across width and depth. This will bring the tops above the layout line that was measured from the low point of the floor. Adjust the pencil

Photo by M. Nauth

FIGURE 87–16 Fastening frameless "boxes" keeping face edges flush.

dividers so that the distance between the points is equal to the amount the top of the unit is above the layout line. Scribe this amount on the bottom end of the cabinets by running the dividers along the floor (**Figure 87–17**).

Cut both ends and toeboard to the scribed lines. There is no need to cut the cabinet backs because they do not, ordinarily, extend to the floor.

StepbyStep Procedures

Procedure 87-A Scribing a Filler Piece Using a Block

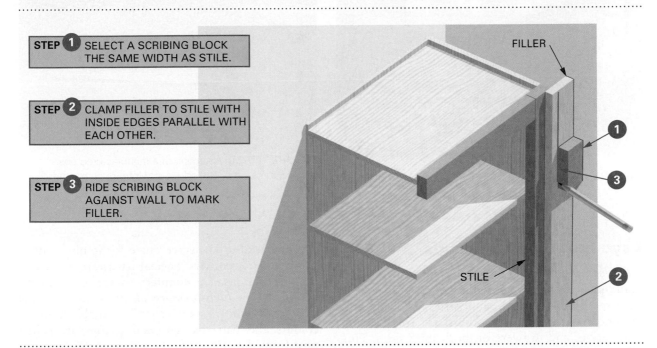

STEP 1 SELECT A SCRIBING BLOCK THE SAME WIDTH AS STILE.

STEP 2 CLAMP FILLER TO STILE WITH INSIDE EDGES PARALLEL WITH EACH OTHER.

STEP 3 RIDE SCRIBING BLOCK AGAINST WALL TO MARK FILLER.

FILLER

STILE

Place the cabinet in position. The top ends should be on the layout line. Fasten it loosely to the wall.

The remaining base cabinets are installed in the same manner. Align and clamp the stiles of adjoining cabinets or the face edges of frameless cabinets. Fasten them together. Finally, fasten all units securely to the wall, shimming where necessary (**Figure 87–18**).

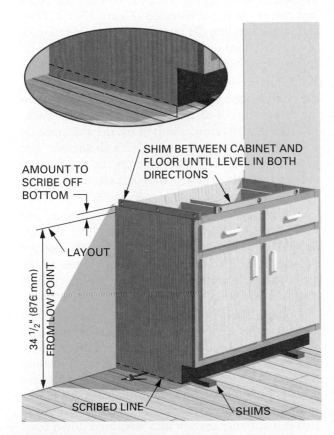

SHIM BETWEEN CABINET AND FLOOR UNTIL LEVEL IN BOTH DIRECTIONS

AMOUNT TO SCRIBE OFF BOTTOM

LAYOUT

34 1/2" (876 mm) FROM LOW POINT

SCRIBED LINE

SHIMS

FIGURE 87–17 Method of scribing base cabinets to the floor.

Courtesy of Amerock Corporation

FIGURE 87–18 Base cabinets are secured with screws to wall studs.

FIGURE 87–19 Some islands are freestanding cabinets.

FIGURE 87–20 A section of a manufactured post-formed countertop. The edges and interior corner are rounded.

Layout of Cabinet Islands

Islands are cabinet units not attached to a wall. They stand alone in the room and may be accessed from all sides. Some are backed up and attached to short walls; others are freestanding with doors and drawers on both long sides. This allows for a wider countertop as a possibility (**Figure 87–19**).

The walking space around the island should be at least 36 inches (914 mm). This allows a minimum of space to access all parts of the island and other cabinets. Increasing this space to 42 or 48 inches (1067 or 1220 mm) provides more space for multiple occupants of the kitchen.

Islands with a sink must be plumbed for water and drains. Typically, the plumbing rough-in is done before the installation of cabinets. The carpenter must then position the island cabinets over the pipes in such a way that the plumber may later perform the hookup.

The finished floor is typically installed after the cabinets are installed. This allows the floor to be replaced without affecting the cabinets. Tile floors are usually installed butting to the island toe board after the cabinets have been adequately anchored to the floor.

INSTALLING MANUFACTURED COUNTERTOPS

Countertops are manufactured in various standard lengths. They can be cut to fit any installation against walls. They are also available with one end precut at a 45-degree angle for joining with a similar one at corners. Special hardware is used to join the sections. The countertops are covered with a thin, tough *high-pressure plastic laminate*. This is generally known as **mica**. It is available in many colours and patterns. The countertops are called **post-formed** countertops. This term comes from the method of forming the mica to the rounded edges and corners of the countertop (**Figure 87–20**). Post-forming is bending the mica with heat to a radius of ¾ inch (19 mm) or less. This can be done only with special equipment.

After the base units are fastened in position, the pre-finished countertop is fastened on top of the base units and against the wall. The backsplash can be scribed, limited by the thickness of its scribing strip, to an irregular wall surface. Use pencil dividers to scribe a line on the top edge of the backsplash. Then plane or belt sand to the scribed line.

Other types of countertops, such as granite and seamless acrylic-based solid surfaces (like the one on the island cabinet in Figure 87–19), do not use a backsplash but instead apply ceramic tile to the wall space between the upper and lower cabinets (**Figure 87–21**).

Fasten the countertop to the base cabinets with screws up through triangular blocks usually installed in the top corners of base units. Use a stop on the drill bit. This prevents drilling through the countertop. Use screws of sufficient length, but not so long that they penetrate the countertop. Some synthetic countertops are set in a bed of silicone sealant and weighted down until the sealant is dry.

Courtesy of M. Nauth

FIGURE 87–21 Ceramic tile backsplash.

Exposed cut ends of post-formed countertops are covered by specially shaped pieces of plastic laminate.

Sink cutouts are made by carefully outlining the cutout and cutting with a jigsaw. The cutout pattern usually comes with the sink. Use a fine-tooth blade (or a blade that cuts on the down stroke) to prevent chipping out the face of the mica beyond the sink. Some tape applied to the metal base of the saw will prevent scratching of the countertop when making the cutout. Some jigsaws come with a plastic shoe for this purpose.

Chapter 88

Countertop and Cabinet Construction

Cabinets are usually purchased from cabinet shops and installed by carpenters on the job. Occasionally they are made by the carpenter who installs them. Nevertheless, all carpenters should understand how cabinets are constructed.

FACE-FRAME CABINET CONSTRUCTION

Cabinets are constructed by pre-cutting, machining, and shaping pieces from dimensions on a set of prints. The process begins with the sides, bottom, and back. They are then joined and assembled as a unit (**Figure 88–1**). The height of kitchen base cabinet is normally 34½ inches (876 mm). This allows for the countertop to be 1½ inches (38 mm) thick, making the overall height 36 inches (914 mm). Wall units may be 30 or 36 inches (762 or 914 mm) high depending on customer's desire. The depth of these cabinets is found on the set of prints.

The sides and bottom pieces are usually made from ¾-inch (19 mm) plywood. The quality of material for the more visible side pieces should be the highest of the material being used because of their visual prominence. Pieces are joined together in a

dado joint (**Figure 88–2**). This joint offers superior strength to the cabinet. The bottom is fastened to the sides with pocket screws from the bottom side or with corner glue blocks. The concern is to hide the fasteners from normal view and still have a strong joint.

The back may be made from thin, inexpensive plywood because little stress is placed on it and it is difficult to see after the cabinet is in use. The back is fitted to the sides and bottom in a rabbet joint. This allows the joint to be invisible from the sides of the finished cabinet.

Mounting rails are installed for the countertop to be attached later as a separate unit. The top is fastened from underneath through the mounting rails. These are nothing more than 2-inch-wide (50 mm) strips of wood or plywood fastened to the cabinet's upper perimeter. These strips need not be made from high-quality material as they will be hidden from view.

The face frame is a visible and attractive part of a cabinet. It is often assembled as a unit, and then attached to the sides, bottom, and mounting rail. The face frame is pieced together according to an age-old pattern, where the styles are full length and top and bottom rails butt to them. The mullions and drawer rails are joined later (**Figure 88–3**).

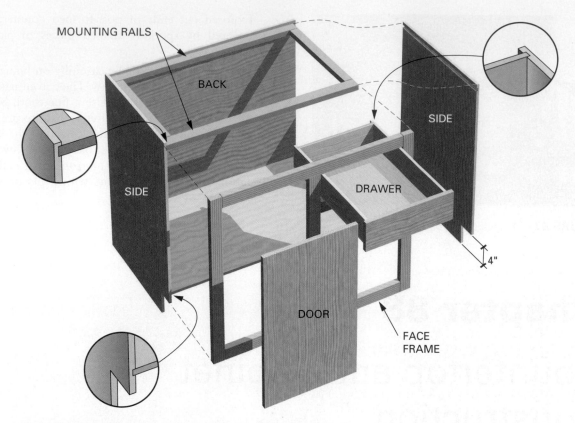

FIGURE 88–1 Cabinets are assembled from pre-machined parts and pieces.

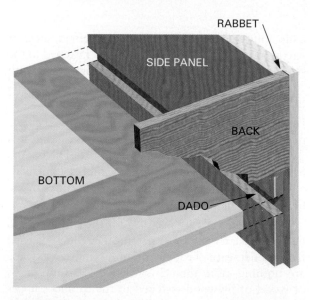

FIGURE 88–2 Sides are rabbeted and dadoed to accept bottom and back panels.

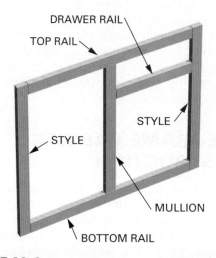

FIGURE 88–3 Typical arrangement of face-framed pieces.

The joints used to make the connections in the face-framed pieces usually vary according to the carpenter's tools, expertise, and preference. In any case, the joints are glued. Joints appear to be butted, but they are often joined with a mortise and tenon, biscuits, or pocket-screwed connections (**Figure 88–4**). This gives more strength to the joint by making a pining effect and increasing the surface area of the glued wood. It is important when making these joints to keep the face-piece surfaces aligned and flat. They are sanded after assembly to make the surfaces smooth. Frames are then attached to the mounting rails using glue and pocket screws from the back side.

Mounting rails are secured in a dado joint with an angled screw into the cabinet side panels. Rails are butted and attached to the back panel and

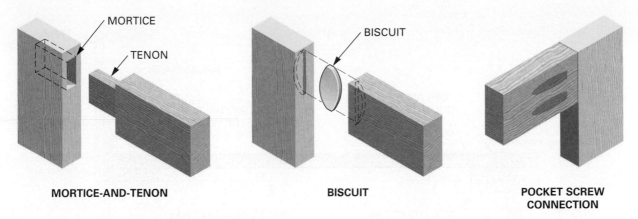

MORTICE-AND-TENON

BISCUIT

POCKET SCREW CONNECTION

FIGURE 88–4 Three styles of joining face-framed pieces.

face-frame top rail with pocket screws. This makes a secure mounting-rail connection that will resist pullout when the countertop is later attached.

COUNTERTOP CONSTRUCTION

A laminated countertop is made from two layers of ⅝- or ¾-inch (15 or 19 mm) panels, plywood, particleboard, or medium density fibreboard (MDF). The top layer is the full width and length of the countertop. The bottom layer is only a few inches wide and along the perimeter. This gives the illusion that the countertop is 1½ inches (38 mm) thick. Splices in the top pieces are reinforced with bottom layer pieces and glued and screwed. The overall width of the countertop should be about 25 inches (635 mm) to allow for an overhang of the front surface of the cabinet.

Fitting the Countertop

After the top panel has been cut to size and the bottom layer attached, it should be checked to see if it fits to the wall. If the wall has an irregular surface, it may need to be scribed on the top to the wall. Place the top panel assembly on the base cabinets, against the wall. Its outside edge should overhang the face frame the same amount along the entire length. Open the pencil dividers or scribers to the amount of overhang. Scribe the back edge of the countertop to the wall. Cut the countertop to the scribed line. Place it back on top of the base cabinets. The ends should be flush with the ends of the base cabinets. The front edge should be flush with the face of the face frame (**Figure 88–5**). Install a 1 × 2 (19 × 38) on the front edge and at the ends, if an end overhang is desired. Keep the top edge flush with the top side of the countertop.

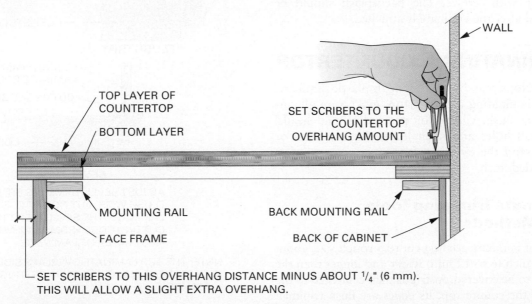

WALL

TOP LAYER OF COUNTERTOP

BOTTOM LAYER

SET SCRIBERS TO THE COUNTERTOP OVERHANG AMOUNT

MOUNTING RAIL

BACK MOUNTING RAIL

FACE FRAME

BACK OF CABINET

SET SCRIBERS TO THIS OVERHANG DISTANCE MINUS ABOUT ¼" (6 mm). THIS WILL ALLOW A SLIGHT EXTRA OVERHANG.

FIGURE 88–5 Scribing the countertop to fit the wall with its outside edge flush with the face of the cabinet.

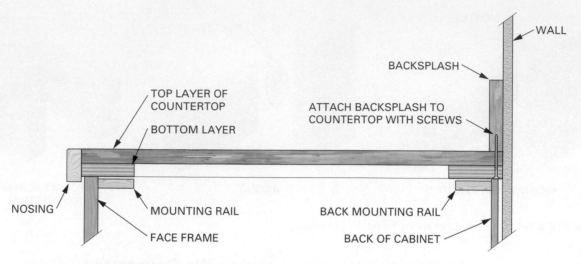

NOSING MAY BE A FINISHED STRIP OF
CABINET WOOD OR OTHER MATERIAL
TO BE COVERED WITH LAMINATE.

FIGURE 88–6 Technique for fastening the backsplash to the countertop.

Applying the Backsplash

If a backsplash is used, rip a 4-inch-wide (100 mm) length of ¾-inch (19 mm) stock that is the same length as the countertop. Use lumber for the backsplash, if lengths over 8 feet (2.4 m) are required, to eliminate joints. Fasten the backsplash on top of and flush with the back edge of the countertop by driving screws up through the countertop and into the bottom edge of the backsplash (**Figure 88–6**). In the corners, fasten the ends of the backsplash together with screws. The backsplash should be installed after the laminate is attached.

LAMINATING A COUNTERTOP

Countertops may be covered with plastic laminate. Before laminating a countertop, make sure all surfaces are flush. Check for protruding nail heads. Fill in all holes and open joints. Lightly hand or power sand the entire surface, making sure joints are sanded flush.

Laminate Trimming Tools and Methods

Pieces of laminate are first cut to a rough size, about ¼ to ½ inch (6 to 12 mm) wider and longer than the surface to be covered. A strip is then cemented to the edge of the countertop. Its edges are flush trimmed even with the top and bottom surfaces. Laminate is then cemented to the top surface, overhanging the

edge strip. The overhang is then bevel trimmed even with the laminated edge. A **laminate trimmer** or a small router fitted with laminate trimming bits is used for rough cutting and flush and bevel trimming of the laminate (**Figure 88–7**).

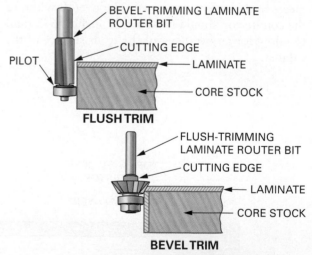

ADJUST BEVEL-TRIMMING BIT TO CUT
FLUSH WITH, BUT NOT INTO, EDGE
LAMINATE. THE BEVEL KEEPS THE
CUTTING EDGE FROM GRAZING THE
FIRST LAYER OF LAMINATE.

NOTE: THE TOP LAMINATE OVERLAPS SIDE TO
HELP PREVENT MOISTURE FROM GETTING UNDER
LAMINATE.

FIGURE 88–7 The laminate trimmer is used with flush and bevel bits to trim overhanging edges of laminate.

Cutting Laminate to Rough Sizes

Sheets of laminate are large, thin, and flexible. This makes them difficult to cut on a table saw. One method of cutting laminates to rough sizes is by clamping a straightedge to the sheet. Cut it by guiding a laminate trimmer with a flush-trimming bit along the straightedge (**Figure 88–8**). It is easier to run the trimmer across the sheet than to run the sheet across the table saw. Also, the router bit leaves a smooth, clean-cut edge. Use a solid carbide trimming bit, which is smaller in diameter than one with ball bear-

ings. It makes a narrower cut, is easier to control, and creates less waste. With this method, cut all the pieces of laminate needed to a rough width and length. Cut the narrow edge strips from the sheet first.

Another method for cutting laminate to rough size is to use a carbide-tipped hand cutter (**Figure 88–9**). It scores the laminate sufficiently in three passes to be able to break the piece away. Make the cut on the face side and then bend the piece up to create the cleanest break. When used with a straightedge, the cutter is fast and effective.

Using Contact Cement

Contact cement is used for bonding plastic laminates and other thin, flexible material to surfaces. A coat of cement is applied to the back side of the laminate and to the countertop surface. The cement must be dry before the laminate is bonded to the core. The bond is made on contact without the need to use clamps. A contact cement bond may fail for several reasons:

- Not enough cement is applied. If the material is porous, like the edge of particleboard or plywood, a second coat is required after the first coat dries. When enough cement has been applied, a glossy film appears over the entire surface when dry.

- Too little time is allowed for the cement to dry. Both surfaces must be dry before contact is made. To test for dryness, lightly press your finger on the surface. Although it may feel sticky, the cement is dry if no cement remains on the finger.

- The cement is allowed to dry too long. If contact cement dries too long (more than about two hours, depending on the humidity), it will not bond properly. To correct this condition, merely apply another coat of cement and let it dry.

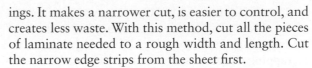

ON THE JOB

Clamp the laminate to a straightedge. Cut rough sizes with a laminate trimmer.

FIGURE 88–8 Technique for using a laminate trimmer to cut laminate to size.

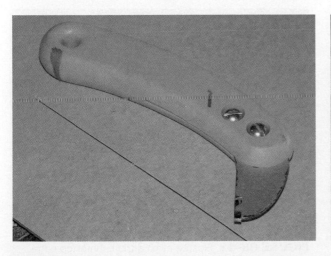

FIGURE 88–9 A hand cutter is often used to cut laminate to a rough size.

• The surface is not rolled out or tapped after the bond is made. Pressure must be applied to the entire surface using a 3-inch (75 mm) **J-roller** or by tapping with a hammer on a small block of wood.

> # CAUTION
>
> **Some contact cements are flammable. Apply only in a well-ventilated area around no open flame. Avoid inhaling the fumes.**

Laminating the Countertop Edges

Remove the backsplash from the countertop. Apply coats of cement to the countertop edges and the back of the edge laminate with a narrow brush or small paint roller. After the cement is dry, apply the laminate to the front edge of the countertop (**Figure 88–10**). Position it so that the bottom edge, top edge, and ends overhang. A permanent bond is made when the two surfaces make contact. A mistake in positioning means removing the bonded piece—a time-consuming, frustrating, and difficult job. Roll out or tap the surface.

Apply the laminate to the ends in the same manner as to the front edge piece. Make sure that the square ends butt up firmly against the back side of the overhanging ends of the front edge piece to make a tight joint.

Trimming Laminated Edges

The overhanging ends of the edge laminate must be trimmed before the top and bottom edges. If the laminate has been applied to the ends,

a bevel-trimming bit must be used to trim the overhanging ends.

Using a Bevel-Trimming Bit. When using a bevel-trimming bit, the router base is gradually adjusted to expose the bit so that the laminate is trimmed flush with the first piece but not cutting into it. The bevel of the cutting edge allows the laminate to be trimmed without cutting into the adjacent piece (see Figure 88–7). A flush-trimming bit cannot be used when the pilot rides against another piece of laminate because the cutting edge may damage it.

Ball-bearing trimming bits have *live pilots*. Solid carbide bits have *dead pilots* that turn with the bit. When using a trimming bit with a dead pilot, the laminate must be lubricated where the pilot will ride. Rub a short piece of white candle or some solid shortening on the laminate to prevent marring the laminate with the bit.

Using the bevel-trimming bit, trim the overhanging ends of the edge laminate. Then, using the flush-trimming bit, trim off the bottom and top edges of both front and end edge pieces (**Figure 88–11**). To save the time required to change and adjust trimming bits, some installers use two laminate trimmers, one with a flush bit and the other with a bevel bit.

Use a cabinet scraper or a file to smooth the top edge flush with the surface. Scrape or file *flat* on the countertop core so that a sharp square edge is made. This ensures a tight joint with the countertop laminate. File *toward* the core to prevent chipping the laminate. Smooth the bottom edge. Ease the sharp outside corner with the file and a sanding block.

FIGURE 88–10 Applying laminate to the edge of the countertop.

FIGURE 88–11 Flush trimming the countertop edge laminate.

Laminating the Countertop Surface

Apply contact bond cement to the countertop and the back side of the laminate. Let dry. To position large pieces of countertop laminate, first place thin strips of wood or metal venetian blind slats about a foot apart on the surface. Lay the laminate to be bonded on the strips or slats. Then position the laminate correctly (**Figure 88–12**).

Make contact on one end. Gradually remove the slats one by one until all are removed. The laminate should then be positioned correctly with no costly errors. Roll out the laminate (**Figure 88–13**). Trim the overhanging back edge with a flush-trimming bit. Trim the ends and front edge with a bevel-trimming bit (**Figure 88–14**). Use a flat file to smooth the trimmed edge. Slightly ease the sharp corner.

FIGURE 88–14 The outside edge of the countertop laminate is bevel trimmed flush to the edge laminate.

FIGURE 88–12 Position the laminate on the countertop using strips before allowing cemented surfaces to contact each other.

LAMINATING A COUNTERTOP WITH TWO OR MORE PIECES

When the countertop is laminated with two or more lengths, tight joints must be made between them. Tight joints can be made by clamping the two pieces of laminate in a straight line on some strips of ¾-inch (19 mm) stock. Butt the ends together or leave a space less than ¼ inch (6 mm) between them. Note that particleboard sheets are available in widths up to 5 feet (1524 mm) and in lengths up to 12 feet (3658 mm). Also note that 4 × 8 sheets will measure 49 × 97 inches (1245 × 2464 mm).

Using one of the strips as a guide, run the laminate trimmer, with a flush-trimming bit installed, through the joint. Keep the pilot of the bit against the straightedge. Cut the ends of both pieces at the same time to ensure making a tight joint (**Figure 88–15**). Bond the sheets as previously described. Apply seam-filling compound, especially made for laminates, to make a practically invisible joint. Wipe off excess compound with the recommended solvent.

Laminating Backsplashes

Backsplashes are laminated in the same manner as countertops. Laminate the backsplash. Then reattach it to the countertop with the same screws. Use a little silicone sealant between the backsplash and countertop. This prevents any water from seeping through the joint (**Figure 88–16**).

Laminating Rounded Corners

If the edge of a countertop has a rounded corner, the laminate can be bent. Strips of laminate can be

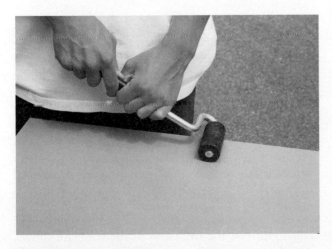

FIGURE 88–13 Rolling out the laminate with a J-roller is required to ensure a proper bond.

ON THE JOB

Tight joints are required between the ends of two lengths of laminate.

A GUIDE STRIP FOR ROUTER MUST BE INSTALLED ON OPERATOR'S RIGHT WHEN PULLING ROUTER THROUGH CUT.

BOTH PIECES ARE HELD SECURELY AND CUT AT THE SAME TIME.

SUPPORTING STRIPS

FIGURE 88-15 Making a tight laminate butt seam.

FIGURE 88-16 Apply the laminate to the backsplash and then fasten it to the laminated countertop.

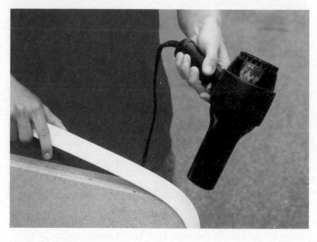

FIGURE 88-17 A heat gun makes laminate bend easily.

cold bent to a minimum radius of about 6 inches (152 mm). Heating the laminate to 190°C (325°F) uniformly over the entire bend will facilitate bending to a minimum radius of about 2½ inches (63 mm). Heat the laminate carefully with a *heat gun*. Bend it until the desired radius is obtained (**Figure 88–17**). Experimentation may be necessary until success in bending is achieved.

CAUTION

Keep fingers away from the heated area of the laminate. Remember that the laminate retains heat for some time.

There are many types of solid-surface countertops on the market—for example, granite and concrete. Products such as Corian and Gibraltar are composite materials. Ceramic tile is also used on countertops (**Figure 88–18**).

Another attractive alternative is to use a wood edge for the face of a laminate countertop. The wood strip, ¾ × 1½ inches (19 × 38 mm), is fastened to the face edge of the particleboard countertop using a biscuit joiner or dowels. It is then sanded or scraped flush with the surface and then the plastic laminate is applied. The laminate is trimmed flush, and then an ogee or a round-over finish is routed into the wood edge until just the thickness of the laminate shows. This has to be done in small increments to avoid splintering the wood. Oak and maple are popular

FIGURE 88–18 Ceramic countertop with maple bullnose.

FIGURE 88–19 Laminate countertop with oak edge.

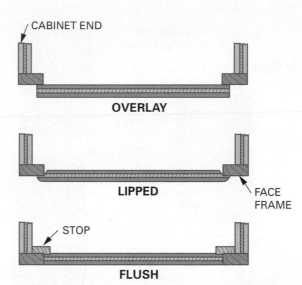

FIGURE 88–20 Plan views of types of cabinet doors.

face frame. The overlay door is easy to install. It does not require fitting in the opening, and the face frame of the cabinet acts as a stop for the door. *European-style* cabinets omit the face frame. Doors completely overlay the front edges of the cabinet (**Figure 88–21**).

Lipped Doors

The lipped door has rabbeted edges that overlap the opening by about $3/8$ inch (9.5 mm) on all sides.

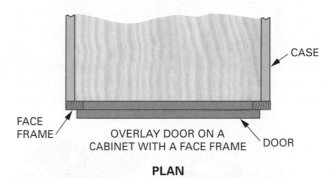

choices for the wood edge (**Figure 88–19**). Laminate is now being manufactured with a light-coloured substrate so that lighter finishes can be used without an exposed black line.

KINDS OF DOORS

Cabinet doors are classified by their construction and also by the method of installation. Sliding doors are occasionally installed, but most cabinets are fitted with hinged doors that swing.

Hinged cabinet doors are classified as overlay, lipped, and flush, based on the method of installation (**Figure 88–20**). The overlay method of hanging cabinet doors is the most widely used.

Overlay Doors

The **overlay door** laps over the opening, usually $3/8$ inch (9.5 mm) on all sides. However, it may overlay any amount. In many cases, it may cover the entire

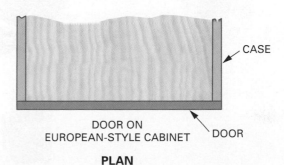

FIGURE 88–21 Overlay doors lap the face frame by varying amounts.

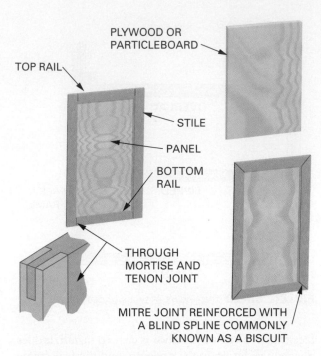

FIGURE 88–22 Panelled doors of simple design can be made on the job.

Usually the ends and edges are rounded over to give a more pleasing appearance. Lipped doors and drawers are easy to install. No fitting is required and the rabbeted edges stop against the face frame of the cabinet. However, a little more time is required to shape the rabbeted edges.

Flush Type

The flush-type door fits into and flush with the face of the opening. They are a little more difficult to hang because they must be fitted in the opening. A fine joint,

about the thickness of a dime, must be made between the opening and the door. Stops must be provided in the cabinet against which to close the door.

Door Construction

Panelled doors have an exterior framework of solid wood with panels of solid wood, plywood, hardboard, plastic, glass, or other panel material. Many complicated designs are manufactured by millworkers with specialized equipment. With the equipment available, carpenters can make panelled doors of simple design only (**Figure 88–22**). Both solid doors and panelled doors may be hinged in overlay, lipped, or flush fashion.

TYPES OF HINGES

Several types of cabinet hinges are *surface*, *offset*, *overlay*, *pivot*, and *butt*. For each type there are many styles and finishes (**Figure 88–23**). Some types are *self-closing* hinges that hold the door closed and eliminate the need for door catches.

Surface Hinges

Surface hinges are applied to the exterior surface of the door and frame. The back side of the hinge leaves may lie in a straight line for flush doors. One leaf may be offset for lipped doors (**Figure 88–24**). The surface type is used when it is desired to expose the hardware, as in the case of wrought iron and other decorative hinges.

Offset Hinges

Offset hinges are used on lipped doors. They are called *offset surface* hinges when both leaves are

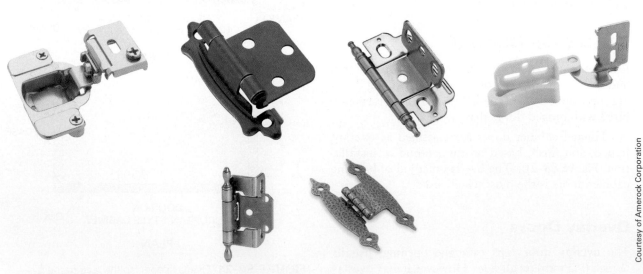

FIGURE 88–23 Cabinet door hinges come in many styles and finishes.

Courtesy of Amerock Corporation

NEL

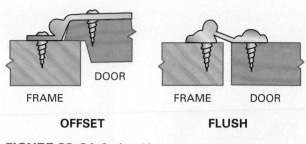

FIGURE 88–24 Surface hinges.

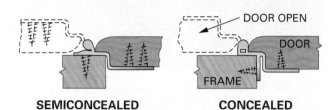

FIGURE 88–25 Offset hinges.

FIGURE 88–27 A 170° European hinge.

fastened to outside surfaces. The *semiconcealed offset* hinge has one leaf bent to a $\frac{3}{8}$-inch (9.5 mm) offset that is screwed to the back of the door. The other leaf screws to the exterior surface of the face frame. A *concealed offset* type is designed in which only the pin is exposed when the door is closed (**Figure 88–25**).

Overlay Hinges

Overlay hinges are the most common type and are available in *semiconcealed* and *concealed* types. With semiconcealed types, the amount of overlay is variable. Certain concealed overlay hinges are made for a

specific amount of overlay, such as $\frac{1}{4}$, $\frac{5}{16}$, $\frac{3}{8}$, and $\frac{1}{2}$ inch (6, 8, 9.5, and 12 mm). Some overlay hinges, with one leaf bent at a 30-degree angle, are used on doors with reverse bevelled edges (**Figure 88–26**). *European hinges* are completely concealed (**Figure 88–27**). The doors are usually drilled to receive the hinge with the hinge manufacturer's proprietary equipment that also bores the holes to receive the adjustable shelves. The carpenter can use the specified flat-bottomed drill bit (called a Forstner bit) in a drill press to accomplish the same.

Pivot Hinges

Pivot hinges are usually used on overlay doors. They are fastened to the top and bottom of the door and to the inside of the case. They are frequently used when there is no face frame and the door completely covers the face of the case (**Figure 88–28**).

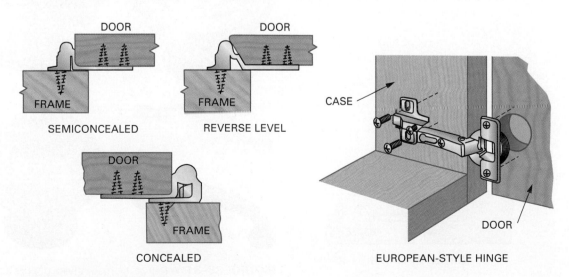

FIGURE 88–26 Overlay hinges.

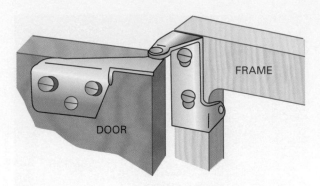

FIGURE 88–28 Pivot hinges for an overlay door.

Butt Hinges

Butt hinges are used on flush doors. **Butt hinge** for a cabinet door is a smaller version of those used on entrance doors. The leaves of the hinge are set into **gains** in the edges of the frame and the door, in the same manner as for entrance doors. Butt hinges are used on flush doors when it is desired to conceal most of the hardware. They are not often used on cabinets because they take more time to install than other types (**Figure 88–29**).

Hanging flush cabinet doors with butt hinges is done in the same manner as hanging entrance doors. Drill pilot holes for all screws so that they are centred on the holes in the hinge leaf. Drilling the holes off centre throws the hinge to one side when the screws are driven. This usually causes the door to be out of alignment when hung. Many carpenters use a self-centring tool, called a *VIX bit*, when drilling pilot holes for screw fastening of cabinet door hinges of all types (**Figure 88–30**). The tool centres a twist drill on the hinge leaf screw hole. It also stops at a set depth to prevent drilling through the door or face frame.

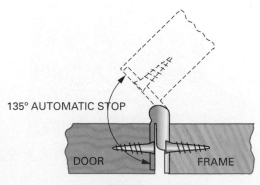

135° AUTOMATIC STOP

DOOR   FRAME

FOR FLUSH DOORS

FIGURE 88–29 Butt hinges.

FIGURE 88–30 The VIX bit is a self-centring drill stop used for drilling holes for cabinet hinges.

INSTALLING PULLS AND KNOBS

Cabinet pulls or knobs are used on cabinet doors and drawers. They come in many styles and designs. They are made of metal, plastic, wood, porcelain, or other material (**Figure 88–31**).

Pulls and knobs are installed by drilling holes through the door. Then fasten them with machine screws from the inside. When two screws are used to fasten a pull, the holes are

Courtesy of Amerock Corporation

FIGURE 88–31 A few of the many styles of pulls and knobs used on cabinet doors and drawers.

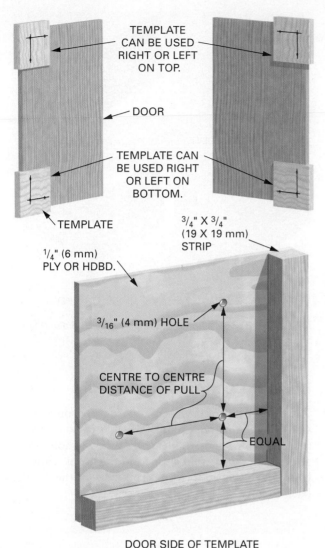

TEMPLATE CAN BE USED RIGHT OR LEFT ON TOP.

DOOR

TEMPLATE CAN BE USED RIGHT OR LEFT ON BOTTOM.

TEMPLATE

¹/₄" (6 mm) PLY OR HDBD.

³/₄" X ³/₄" (19 X 19 mm) STRIP

³/₁₆" (4 mm) HOLE

CENTRE TO CENTRE DISTANCE OF PULL

EQUAL

DOOR SIDE OF TEMPLATE

FIGURE 88–32 Technique for making a jig (template) to speed installation of door pulls.

FIGURE 88–33 Several types of catches are available for use on cabinet doors.

Courtesy of Amerock Corporation

DOOR CATCHES

Almost all kitchen and bathroom cabinets have self-closing hinges today. These require the installation of bumper pads made of felt or plastic. Older cabinets will have a variety of **catches**, such as the magnetic, friction, elbow, and bullet types (**Figure 88–33**).

DRAWER CONSTRUCTION

Drawers are classified as overlay, lipped, and flush in the same way as doors. In a cabinet installation, the drawer type should match the door type.

Drawer fronts are generally made from the same material as the cabinet doors. Drawer sides and backs are generally ½ inch (12 mm) thick. They may be made of solid lumber, plywood, or particleboard.

Medium-density fiberboard with a printed wood grain is also manufactured for use as drawer sides and backs. The drawer bottom is usually made of ¼-inch (6 mm) plywood or hardboard. Small drawers may have ⅛-inch (3 mm) hardboard bottoms.

Drawer Joints

Typical joints between the front and sides of drawers are the dovetail, lock, and rabbet joints. The dovetail joint is used in higher-quality drawer construction. It takes a longer time to make, but is the strongest. Dovetail drawer joints may be made using a router and a dovetail template (**Figure 88–34**). The lock joint is simpler. It can be easily made using a table saw. The rabbet joint is the easiest to make. However,

drilled slightly oversized in case they are a little off centre. This allows the pulls to be fastened easily without cross-threading the screws. Usually ³/₁₆-inch (4 mm) diameter holes are drilled for ⅛-inch (3 mm) machine screws. To drill holes quickly and accurately, make a template from scrap wood that fits over the door. The template can be made so that holes can be drilled for doors that swing in either direction (**Figure 88–32**).

ON THE JOB

Use a template to drill holes for cabinet pulls.

FIGURE 88–34 Dovetail joints can be made with a router and a dovetail template.

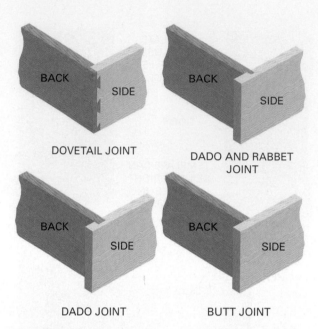

DOVETAIL JOINT   DADO AND RABBET JOINT

DADO JOINT   BUTT JOINT

FIGURE 88–36 Various joints between the drawer back and side.

it must be strengthened with fasteners in addition to glue (**Figure 88–35**).

Joints normally used between the sides and back are the dovetail, dado and rabbet, dado, and butt joints. With the exception of the dovetail joint, the drawer back is usually set in at least ½ inch (12 mm) from the back ends of the sides to provide added strength. This helps prevent the drawer back from being pulled off if the contents get stuck while opening the drawer (**Figure 88–36**).

Today in Canada, most drawers are "boxes" made of ⅝-inch (15 mm) melamine, and a pre-manufactured drawer face that matches the cabinet doors is then secured to the front of the box.

Drawer Bottom Joints

The drawer bottom is fitted into a groove on all four sides of the drawer (**Figure 88–37**). In some cases, the drawer back is made narrower, the four sides assembled, the bottom slipped in the groove, and its back edge fastened to the bottom edge of the drawer back (**Figure 88–38**).

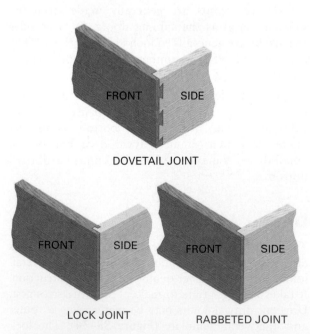

DOVETAIL JOINT

LOCK JOINT   RABBETED JOINT

FIGURE 88–35 Various joints between the drawer front and side.

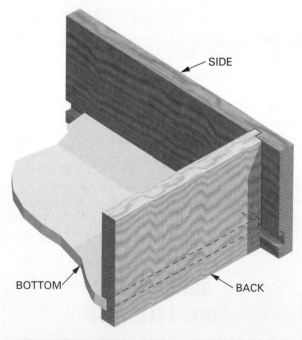

FIGURE 88–37 A drawer bottom may be fitted into a groove at the drawer back.

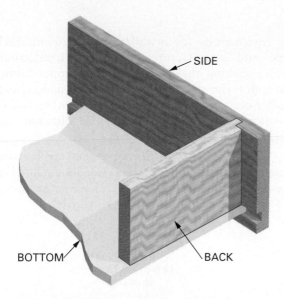

FIGURE 88–38 A drawer bottom may be fastened to the bottom edge of the drawer back.

DRAWER GUIDES

There are many ways of guiding drawers. Drawer guides are designed to make the drawer easy to pull out and return. They are often pre-manufactured metal guides. Sometimes the carpenter will build them of hardwood.

Wood Guides

The type of drawer guide selected affects the size of the drawer. The drawer must be supported level and guided sideways. It must also be kept from tilting down when opened. Probably the simplest wood guide is the centre strip. It is installed in the bottom centre of the opening from front to back (**Figure 88–39**). The strip projects above the bottom of the opening about ¼ inch (6 mm). The bottom edge of the drawer back

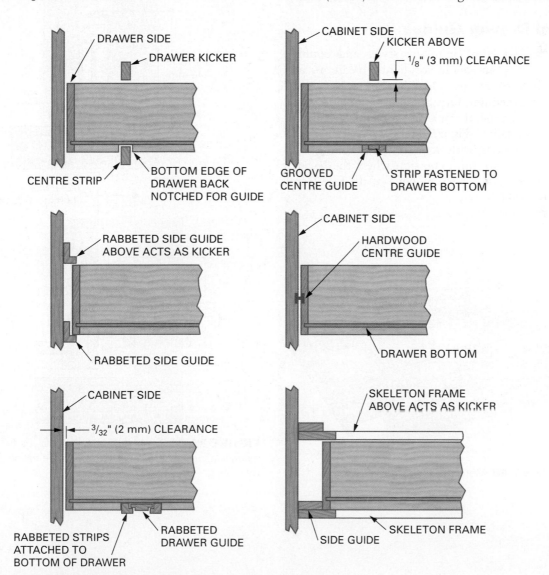

FIGURE 88–39 Wood drawer guides are installed in several ways.

is notched to ride in the guide. A kicker is installed. It is centred above the drawer to keep it from tilting downward when opened.

Another type of wood guide is the grooved centre strip. The strip is placed in the centre of the opening from front to back. A matching strip is fastened to the drawer bottom. In addition to guiding the drawer, this system keeps it from tilting when opened, eliminating the need for drawer kickers.

Another type of wood guide is a rabbeted strip. Strips are used on each side of the drawer opening. The drawer sides fit into and slide along the rabbeted pieces. Sometimes these guides are made up of two pieces instead of rabbeting one piece. A kicker above the drawer is necessary with this type of guide.

Metal Drawer Guides

Metal guides come in many styles and configurations. They are made for light-, medium-, and heavy-duty situations. Most kitchen cabinets will need to be medium duty. Some have a nylon wheel that rides in a metal track mounted on each side of the drawer guide (**Figure 88–40**). Others are self-contained guides with steel bearings that ride in tracks. These are full extension where the back of the drawer rides out to the edge of the face frame (**Figure 88–41**).

Instructions for installation differ with each type and manufacturer. Read the instructions before making the drawer so that proper allowances for the drawer guide can be made. Most guides need only ½ inch (12 mm) on either side of the drawer box and face frame, plus allowance for guide tolerance.

The main types of drawer guides are bottom-mounted (**Figure 88–42**) and side-mounted (**Figure 88–43**). These types are three-quarter extension (75 percent) and full extension, respectively.

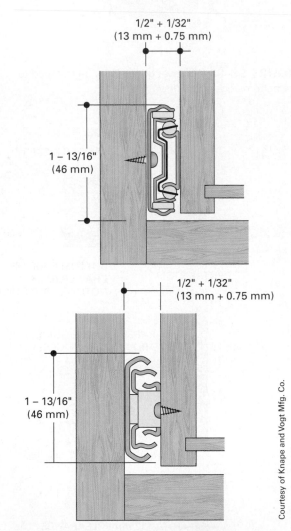

1/2" + 1/32"
(13 mm + 0.75 mm)

1 – 13/16"
(46 mm)

1/2" + 1/32"
(13 mm + 0.75 mm)

1 – 13/16"
(46 mm)

Courtesy of Knape and Vogt Mfg. Co.

FIGURE 88–41 Metal drawer guides may be full extension, usually accomplished with three telescoping members.

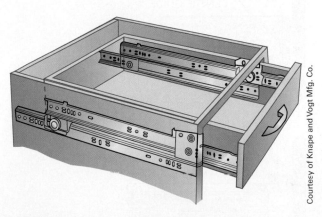

Courtesy of Knape and Vogt Mfg. Co.

FIGURE 88–40 Metal drawer guides.

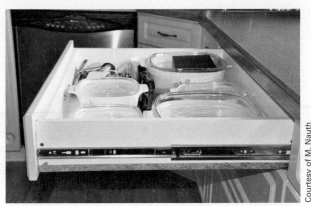

Courtesy of M. Nauth

FIGURE 88–43 Side-mounted drawer guides (full extension).

FIGURE 88–42 Bottom-mounted drawer guides (75 percent extension).

KEY TERMS

accessories	face frame	mica	rough size
base unit	filler	offset hinges	surface hinges
butt hinge	flush-type door	overlay doors	toe space
cabinet lift	frameless	overlay hinges	wall unit
catches	gains	panelled doors	
contact cement	J-roller	pilot	
double-faced cabinets	laminate trimmer	pivot hinges	
European hinges	lipped door	post-formed	

REVIEW QUESTIONS

Select the most appropriate answer.

1. What is the usual vertical distance between the base unit and a wall unit?

   a. 12 inches (305 mm)
   b. 15 inches (381 mm)
   c. 18 inches (457 mm)
   d. 24 inches (610 mm)

2. What is the usual distance from the floor to the surface of a countertop?

   a. 30 inches (762 mm)
   b. 32 inches (813 mm)
   c. 36 inches (914 mm)
   d. 42 inches (1067 mm)

3. To accommodate sinks and provide adequate working space, what is the width of a countertop in a kitchen?

   a. 25 inches (635 mm)
   b. 28 inches (711 mm)
   c. 30 inches (762 mm)
   d. 32 inches (813 mm)

4. What is the standard height of a wall cabinet?

   a. 24 inches (610 mm)
   b. 30 inches (762 mm)
   c. 32 inches (813 mm)
   d. 36 inches (914 mm)

5. What is the height of most manufactured base kitchen cabinets?

   a. 30¾ inches (781 mm)
   b. 32½ inches (826 mm)
   c. 34½ inches (876 mm)
   d. 35¼ inches (895 mm)

6. What style of drawer front or door has its edges and ends rabbeted to fit over the opening?

   a. overlay
   b. lipped
   c. flush
   d. rabbeted

7. What type of doors are offset hinges used on?

   a. panelled doors
   b. lipped doors
   c. flush doors
   d. overlay doors

8. What type of doors are butt hinges used on?

   a. flush doors
   b. overlay doors
   c. lipped doors
   d. solid doors

9. What type of joint is used in the box construction of frameless cabinets?

   a. dado joint
   b. box joint
   c. dovetail joint
   d. butt joint

10. What percentage of full opening do bottom-mounted drawer slides allow?

    a. 50 percent
    b. 60 percent
    c. 75 percent
    d. 100 percent

Appendix

IMPERIAL–METRIC CONVERSION FACTORS

Framing Terms

| | Imperial | | Systeme Internationale – Metric | |
	Nominal	Actual	Standard (mm)	Actual (mm)
Lumber	$2'' \times 2''$	$1\frac{1}{2}'' \times 1\frac{1}{2}''$	40×40	38×38
	$2'' \times 4''$	$1\frac{1}{2}'' \times 3\frac{1}{2}''$	40×90	38×89
	$2'' \times 6''$	$1\frac{1}{2}'' \times 5\frac{1}{2}''$	40×140	38×140
	$2'' \times 8''$	$1\frac{1}{2}'' \times 7\frac{1}{4}''$	40×185	38×184
	$2'' \times 10''$	$1\frac{1}{2}'' \times 9\frac{1}{4}''$	40×235	38×235
	$2'' \times 12''$	$1\frac{1}{2}'' \times 11\frac{1}{4}''$	40×285	38×286
Panels	$2' \times 8'$	$24'' \times 96''$	600×2400	610×2440
	$4' \times 8'$	$48'' \times 96''$	1200×2400	1220×2440
Spacings	12" o.c.		300 o.c.	305 o.c.
	16" o.c.		400 o.c.	406 o.c.
	19.2" o.c.		480 o.c.	488 o.c.
	24" o.c.		600 o.c.	610 o.c.

Units of Measure

Metric		Imperial
°C	$\times\ 1.8 + 32 =$	°F
kg	$\times\ 2.205 =$	lb
kPa	$\times\ 0.1450 =$	lbf/in² (psi)
kPa	$\times\ 20.88 =$	lbf/ft²
L	$\times\ 0.2200 =$	gal (imp.)
L/s	$\times\ 13.20 =$	gal/min (gmp)
lx	$\times\ 0.09290 =$	ft-candle
m	$\times\ 3.281 =$	ft
m²	$\times\ 10.76 =$	ft²
m³	$\times\ 35.32 =$	ft³
mm	$\times\ 0.03937 =$	in.
m³/h	$\times\ 0.5886 =$	ft³/min (cfm)
m/s	$\times\ 196.8 =$	ft/min
MJ	$\times\ 947.8 =$	Btu
N	$\times\ 0.2248 =$	lbf
watts	$\times\ 3.412 =$	Btu/h
ng/(pa.s.m²)	$\times\ 0.0174 =$	perms
Pa	$\times\ 0.004014 =$	in. of water

Index

Note: Page numbers preceded by *f* or *t* refer to figures and tables.